PROCEEDINGS OF THE TWENTIETH LUNAR AND PLANETARY SCIENCE CONFERENCE

PROCEEDINGS OF THE TWENTIETH LUNAR AND PLANETARY SCIENCE CONFERENCE

Virgil L. Sharpton and Graham Ryder, editors

LUNAR AND PLANETARY INSTITUTE

Published by the Lunar and Planetary Institute,
3303 NASA Road 1, Houston, TX 77058

ISBN 0-942862-04-X
ISSN 0270-9511

Logo and cover design by Donna Jalufka

In dedication
to the scientists and engineers of Apollo
who initiated twenty years of scientific discovery and
enabled a new vision of the Moon

Preface

The Apollo Program heralded an unprecedented age of exploration of the solar system. In the 20 years since the return of the first lunar samples, we have witnessed 10 additional lunar expeditions (5 manned), 15 missions to Venus (including soft landers, atmospheric probes, and orbital radars), 8 missions to Mars (including the amazingly successful Viking landers), and numerous flyby missions such as the spectacular Voyager encounters with the giant outer planets and their satellites. Of all the planets of our solar system, only Pluto remains unvisited. Nurtured by this tremendous harvest of planetary data (and the ever-present hunger for more) a vast, international multidisciplinary community of planetary scientists has emerged from the nucleus of pre-Apollo lunar enthusiasts.

Nowhere is the record of planetary exploration and the evolution of planetary sciences more completely expressed than in the 20-year history of this *Proceedings* series. Since its inception 20 years ago, this series has published nearly 3000 papers documenting the stepwise scientific exploration of the solar system and the ensuing changes in the planetary community. In its efforts to serve this ever-changing community the *Proceedings* has throughout its history undergone several changes in title, format, and publishers. This year marks the sixth such change.

There was serious discussion this year about making this volume the last in the series. The *Proceedings* is clearly an attractive vehicle for publication of studies relevant to the themes of the annual Lunar and Planetary Science Conference; each year more planetary papers are published in this single volume than in any other publication. For instance, this year there are 50 papers spanning a broad range of topics from sample analyses, to studies of Mars, to analysis of comets and interplanetary dust, to studies of Venus. Unfortunately, institutional and individual sales of the *18th Proceedings* and *19th Proceedings* volumes are down. Why is this?

Some point to the growth of planetary science, which has naturally led to subdivision into groups along conventional disciplinary lines: geochemistry, petrology, geophysics, geology, remote sensing, astronomy, and astrophysics. This in turn has prompted the major scientific journals in these fields to accept, if not specifically cater to, planetary studies. Thus it has been argued that it is a tribute to the health of the planetary community that it has surpassed its need for a *Proceedings*. The dispersal of planetary papers through numerous journals forces buyers to pick and choose; the *Proceedings* is simply too expensive in this age of limited resources and wide selection.

Others counter that the *Proceedings* continues to play a critical role as a vehicle for many kinds of papers that still are not considered germane for nonplanetary journals. To many ardent supporters, the *Proceedings* represents the identity of the planetary community; the more the community diversifies, the more critical the need for that identity to be preserved through a single publication.

The Lunar and Planetary Institute recognizes that the *Proceedings* performs a unique service in its annual compilation of planetary studies from around the world and is committed to its continued publication. We have elected to publish the *Proceedings of the 20th Lunar and Planetary Science Conference* ourselves in an effort to substantially lower the purchase price while at the same time continuing to bring the community the quality volume it deserves and expects. We have elected to retain the format of the last two *Proceedings* volumes in order to facilitate recognition of these volumes as a continuing series. On this retrospective occasion of the 20th anniversary of the first lunar landing, it is perhaps most fitting to look to the next 20 years and to the new scientific challenges that lay ahead for future planetary scientists: to prospects of lunar bases and manned expeditions to Mars, to mapping the surface of Venus at unprecedented resolution, to probing the atmospheres of Jupiter. This *Proceedings* series is a tradition born of the Apollo program and hopefully one that future generations can share.

Getting this book out in less than a year is a grueling task. The Editors gratefully acknowledge the dedication and expertise of the Board of Associate Editors in overseeing the review process and the tireless efforts and talents of the staff of the LPI Publications Services Department. In particular, we are indebted to Renee Dotson, whose dedication and guidance make timely production of this volume possible.

Virgil L. Sharpton
Graham Ryder

TABLE OF CONTENTS

Petrology and Geochemistry of the Moon

Geology of the Moon

Shock and Terrestrial Cratering

Geology of Mars

Petrology and Geochemistry of the Moon

Proceedings of the 20th Lunar and Planetary Science Conference, pp. 3-12
Lunar and Planetary Institute, Houston, 1990

Buoyancy-driven Melt Segregation in the Earth's Moon, I. Numerical Results

J. W. Delano

Department of Geological Sciences, State University of New York, Albany, NY 12222

The densities of lunar mare magmas have been estimated at liquidus temperatures for pressures from 0 to 47 kbar (0-4.7 GPa; center of the Moon) using a third-order Birch-Murnaghan equation and compositionally dependent parameters from Lange and Carmichael (1987). Results on primary magmatic compositions represented by pristine volcanic glasses suggest that the density contrast between very-high-Ti melts and their liquidus olivines may approach zero at pressures of about 25 kbar (2.5 GPa). Since this is the pressure regime of the mantle source regions for these magmas, a compositional limit of eruptability for mare liquids may exist that is similar to the highest Ti melt yet observed among the lunar samples. Although the Moon may have generated magmas having >16.4 wt.% TiO_2, those melts would probably not have reached the lunar surface due to their high densities, and may have even sunk deeper into the Moon's interior as negatively buoyant diapirs. This process may have been important for assimilative interactions in the lunar mantle. The phenomenon of melt/solid density crossover may therefore occur not only in large terrestrial-type objects (e.g., Earth) but also in small objects (i.e., Moon) where, despite low pressures, the range of melt compositions is extreme.

INTRODUCTION

In a landmark publication, *Stolper et al.* (1981) proposed that olivine may float in basic magmas at high pressures (~80 kbar; ~8.0 GPa) in the Earth's upper mantle. Other investigators have provided important additional constraints on the likelihood of this process occurring in the Earth (e.g., *Agee and Walker,* 1988; *Lange and Carmichael,* 1987; *Herzberg,* 1984, 1987; *Ohtani,* 1984; *Rigden et al.,* 1984, 1988, 1989). The geochemical and petrological implications of this phenomenon are of profound geologic importance (e.g., *Nisbet and Walker,* 1982; *Agee and Walker,* 1988).

The possibility for basic melts to become denser than the residual crystals, with which they are in chemical equilibrium in their source regions, arises from the fact that silicate melts are about an order of magnitude more compressible than mantle crystals (e.g., *Stolper et al.,* 1981). This densification of silicate melts with increasing pressure appears to be gradual and continuous due to a progressive change in the Si-O-Si bond angle (*Stolper and Ahrens,* 1987; *Itie et al.,* 1989; *Williams and Jeanloz,* 1988). With increasing pressure, the Si atoms in adjacent corner-shared tetrahedra may go from being tetrahedrally coordinated at low pressures to being octahedrally coordinated at high pressures. This transition may be complete in most basic melts at pressures in the vicinity of 250 kbar (25 GPa), whereupon the melt stiffens (*Stolper and Ahrens,* 1987; *Rigden et al.,* 1988; *Williams and Jeanloz,* 1988).

Since the maximum pressure in the Moon is only about 47 kbar (4.7 GPa), *Rigden et al.* (1984) concluded that melt/solid density crossovers would not have occurred in a primordial lunar magma ocean having the mafic composition estimated by *Ringwood* (1979). Although that view is certainly correct, it does not address the issue of whether other mantle-derived melts (e.g., mare magmas) in the Moon might exhibit this behavior. This paper presents the results of numerical calculations performed on mare compositions represented by picritic lunar volcanic glasses (*Delano,* 1986) and crystalline mare basalts. Experiments are presently underway to test the veracity of the views derived from these calculations (J. W. Delano and E. B. Watson, unpublished data, 1989).

NUMERICAL PROCEDURES

Liquidus Temperatures

The liquidus temperature of each picritic glass and mare basalt composition was estimated in a fashion analogous to that used by *Stolper and Walker* (1980) in their work on terrestrial

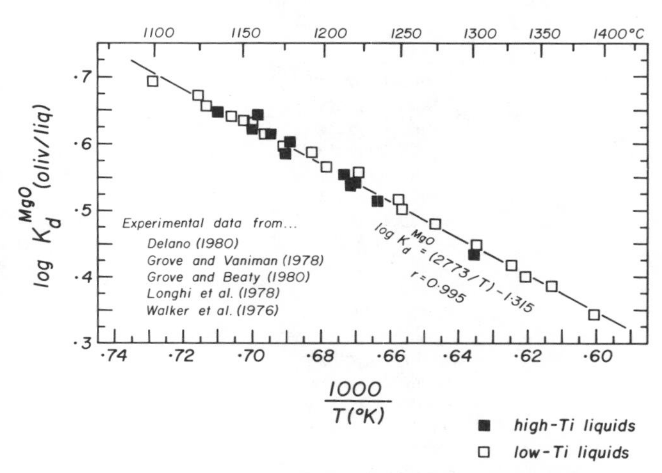

Fig. 1. An Arrhenius plot showing the linear relationship of reciprocal temperature (°K) and the ratio of *mole fractions* MgO between olivine and melt. Note that this relationship appears to be independent of the Ti-abundance in the melt. This relationship was used in this study to estimate the liquidus temperatures (at zero pressure) of specific lunar mare compositions for which no experimental data are presently available.

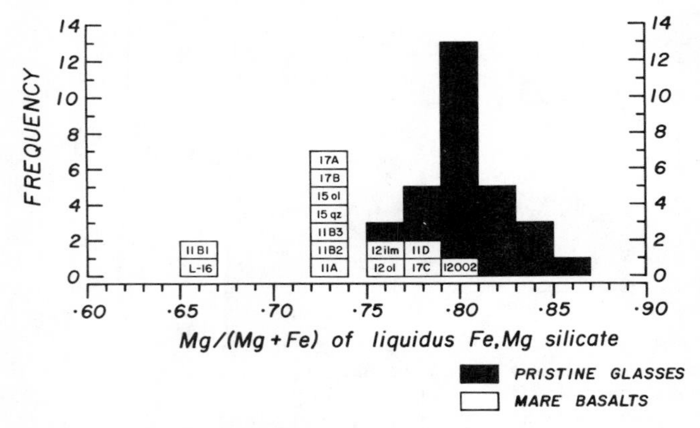

Fig. 2. The compositions of olivines (mole fraction of forsterite) estimated to occur on the liquidi of pristine lunar glasses and crystalline mare basalts are displayed on this histogram. The pristine glasses often have liquidus olivines that are more magnesian than those in the crystalline mare basalts. This is consistent with the pristine glasses being representatives of primary magmas whereas the mare basalts are commonly differentiated during emplacement (e.g., *Longhi,* 1987).

basalts. Melting experiments on lunar mare compositions, in which microprobe data on coexisting olivines and liquids were published (e.g., *Delano,* 1980; *Grove and Vaniman,* 1978; *Grove and Beaty,* 1980; *Longhi et al.,* 1978; *Walker et al.,* 1976), were used to define a systematic relation between 1000/T(K) and log K_d^{MgO} (olivine/liquid) involving the ratio of *mole fractions.* These experimental data (Fig. 1) form a well-constrained equation (1) having a correlation coefficient of 0.995 on 30 pairs of olivine + liquid

$$\log K_d^{MgO} = (2773/T) - 1.315 \quad (1)$$

In order to use this equation to estimate the liquidus temperature of any lunar mare composition that has not been directly investigated experimentally, it is necessary to derive the composition of the liquidus olivine. The olivine composition can be estimated by knowing the value for the distribution coefficient (K_D) as defined below

$$K_D = \left(\frac{X_{Fe}}{X_{Mg}}\right)_{olivine} \cdot \left(\frac{X_{Mg}}{X_{Fe}}\right)_{melt} \quad (2)$$

Experiments on lunar compositions (*Delano,* 1980; *Green et al.,* 1975; *Longhi et al.,* 1978) have demonstrated that the K_D is sensitive to the abundance of TiO_2 in the melt. This compositional dependence of K_D may be caused by the presence of $FeTi_2O_5$ complexes in high-Ti melts, that places limits on the Fe available for partitioning into olivine (*Jones,* 1988).

The equations used in the present study were: (3) for mare basalts (data from *Grove and Vaniman,* 1978; *Grove and Beaty,* 1980; *Longhi et al.,* 1978; *Walker et al.,* 1976)

TABLE 1. Summary of important physical and chemical parameters at 0 kbar for lunar picritic magmas represented by pristine glasses (*Delano,* 1986).

	Liquidus Temperature (°C)	Liquidus Olivine (mol.% Fo)	Isothermal Bulk Modulus (K_T^o) at Liquidus Temperature (Kilobars)	Viscosity at Liquidus Temperature (Poise)	Density of Melt at Liquidus Temperature (g/cc)
Apollo 15 green C	1409	85.6	216.5	6.4	2.817
Apollo 15 green A	1405	82.5	218.4	4.7	2.875
Apollo 16 green	1409	80.6	218.7	4.6	2.914
Apollo 15 green B	1402	82.9	218.5	5.6	2.864
Apollo 15 green D	1414	82.4	217.5	4.9	2.886
Apollo 15 green E	1425	83.3	215.9	4.6	2.879
Apollo 14 green B	1439	83.9	214.4	3.8	2.882
Apollo 14 VLT	1375	82.5	220.3	9.6	2.850
Apollo 11 green	1410	81.0	218.5	4.6	2.912
Apollo 17 VLT	1375	80.5	219.3	10.2	2.869
Apollo 17 green	1448	84.0	211.9	3.1	2.891
Apollo 14 green A	1406	79.6	216.6	4.3	2.930
Apollo 15 yellow	1342	77.6	217.1	8.6	2.931
Apollo 14 yellow	1377	77.6	213.7	4.1	2.991
Apollo 17 yellow	1320	77.3	214.6	6.8	2.976
Apollo 17 orange	1362	80.5	209.6	4.6	2.993
74220	1369	80.3	209.2	4.1	3.015
Apollo 15 orange	1372	79.8	209.5	3.8	3.031
Apollo 17 orange	1298	76.2	212.3	6.8	3.015
Apollo 11 orange	1360	79.4	209.6	4.1	3.044
Apollo 14 orange	1353	81.4	204.3	4.3	3.033
Apollo 15 red	1303	79.3	206.1	5.8	3.053
Apollo 14 red	1325	79.7	202.1	4.3	3.084
Apollo 14 black	1331	79.9	202.4	3.7	3.121
Apollo 12 red	1331	80.1	201.0	—	3.122

TABLE 2. Summary of important physical and chemical parameters at 0 kbar for mare basalts.

	Liquidus Temperature (°C)	Liquidus Olivine (mol.% Fo)	Isothermal Bulk Modulus (K°_{T}) at Liquidus Temperature (Kilobars)	Density of Melt at Liquidus Temperature (g/cc)
12002	1373	78.6	216.0	2.919
15499	1255	72.5	218.7	2.855
15016	1301	72.4	219.6	2.925
12008	1320	75.7	216.6	2.939
12009	1299	74.7	218.1	2.899
70255 (A)	1179	72.2	209.3	2.976
70215 (b)	1190	72.7	208.6	3.005
74245 (C)	1234	77.5	210.7	2.978
Apollo 11 B1	1123	65.3	206.6	2.959
Apollo 11 B2	1183	71.7	208.8	2.991
Apollo 11 B3	1197	73.1	211.1	2.966
Apollo 11 D	1239	77.3	217.7	2.871
Apollo 11 high-K	1179	72.4	204.2	2.974

$$K_D = 0.348 - 0.0075 \text{ (mol.\% TiO}_2\text{ in melt)} \quad (3)$$

$$r = -0.989$$

(4) for picritic glasses (data from *Delano,* 1980)

$$K_D = 0.333 - 0.0071 \text{ (mol.\% TiO}_2\text{ in melt)} \quad (4)$$

$$r = -0.995$$

Jones (1988) derived the following relation

$$K_D = 0.320 + 0.120\, X_{Mg} + 0.106\, X_{Fe} - 0.863\, X_{Ti} - 0.00007\, P \quad (5)$$

where P = pressure in kilobars and X = cation fraction in liquid.

Equation (5) from *Jones* (1988) yields olivine compositions that agree to within ±1 mol.% forsterite with those derived using equations (3) and (4). In the present study, it has been assumed that the K_D is not sensitive to pressures in the range relevant to the Moon ($\lesssim$47 kbar), in agreement with *Jones* (1988). However, since *Delano* (1980) reported the possibility of a pressure dependence among high-Ti melt compositions, the K_D is being reexamined experimentally (J. W. Delano and E. B. Watson, unpublished data, 1989).

By using equations (3)-(5) to estimate the value for the K_D applicable to each mare composition, the compositions of liquidus olivines were derived (Fig. 2). The *mole fractions* of MgO in this olivine and in the melt were used to define a value for the log K_d^{MgO}, which led to an estimate for the liquidus temperature (Table 1) using equation (1). The liquidus temperatures at zero pressure are thought to be accurate to within ±20°C. Since few of the lunar compositions in Tables 1 and 2 have been experimentally investigated, the values for dT/dP of their liquidi are not known. Based upon high-pressure melting experiments (e.g., *Delano,* 1980; *Green et al.,* 1975; *Walker et al.,* 1976), the slope of the liquidus for mare compositions in the pressure range from 0 to 47 kbar was assumed to lie between 7°C/kbar and 10°C/kbar. All densities discussed later in this paper were estimated within this large range of liquidus temperatures.

Zero-Pressure Melt Density

Lange and Carmichael (1987) published the molar volumes of the major rock-forming oxides as a function of temperature. When the gram formula weight (gfw) of each oxide is divided

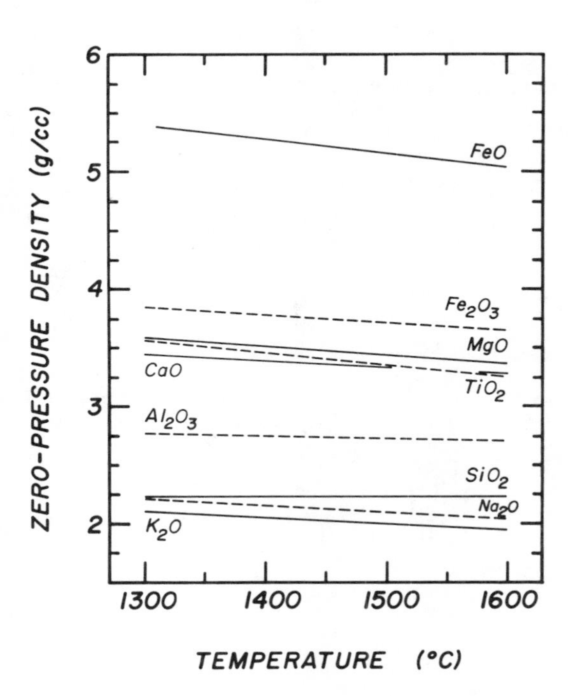

Fig. 3. The major rock-forming oxides have zero-pressure densities ranging from about 5 g/cc for FeO to about 2 g/cc for K_2O at magmatic temperatures. These data are derived from *Lange and Carmichael* (1987).

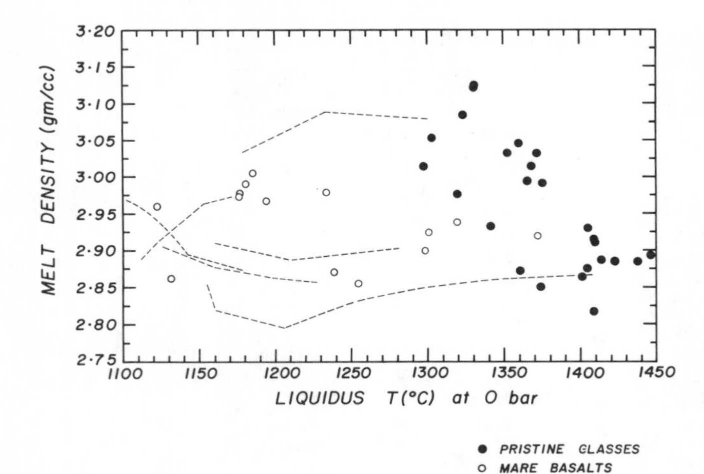

Fig. 4. The melt densities at liquidus temperatures have been estimated for pristine glasses (solid circles) and mare basalts. The dashed lines show the density of melts generated experimentally by zero-pressure fractionation. The experiments used to construct the dashed lines are from *Delano* (1980), *Grove and Beaty* (1980), *Grove and Bence* (1977), *Grove and Vaniman* (1978), and *Walker et al.* (1977). The melt densities were estimated using the parameters of *Lange and Carmichael* (1987).

by its volume, the density of each component is found (Fig. 3). The molar volume of each mare composition was computed at its appropriate liquidus temperature. The gfw of each mare sample was then divided by this molar volume to estimate the melt density at zero pressure and liquidus temperature (Fig. 4).

Zero-Pressure Olivine Density

The thermodynamic parameters for olivine from *Sumino* (1979) and *Suzuki et al.* (1981, 1983) were used to estimate the density of liquidus olivine in each mare composition at its specific liquidus temperature. These parameters have been extrapolated from the Fo_{85} to Fo_{95} explored by those authors to the compositions (Fig. 2) applicable to these mare compositions.

High-Pressure Density of Melt and Olivine

The approach followed in this portion of the numerical exercise was similar to that used by *Stolper et al.* (1981) involving iterative solutions of a third-order Birch-Murnaghan equation

$$P = 3/2 \cdot K_T^\circ \cdot \left[\left(\frac{\rho}{\rho_o}\right)^{7/3} - \left(\frac{\rho}{\rho_o}\right)^{5/3} \right] \cdot \left[1 - 3/4 \cdot (4 - K_T') \cdot \left[\left(\frac{\rho}{\rho_o}\right)^{2/3} - 1 \right] \right] \quad (6)$$

where P = pressure in bars; K_T° = isothermal bulk modulus (bars) of the liquid at 1-bar pressure and temperature, T; ρ_o = density of the liquid at 1-bar pressure and temperature, T; ρ = density of the liquid at high pressure, P and temperature, T; K_T' = pressure derivative of the isothermal bulk modulus.

Lange and Carmichael (1987) developed a set of compositional parameters for estimating the isothermal bulk modulus (K_T°) of silicate liquids at liquidus temperatures. The values of K_T° for the mare melts were generally in the vicinity of 220 kbar (Tables 1 and 2), which are similar to values experimentally measured in terrestrial basalts (e.g., *Rigden et al.,* 1984, 1988). At present, no experimental determinations of K_T° have been made for lunar compositions.

The pressure derivative of the bulk modulus has been measured in terrestrial basalt compositions and found to have a value of ~4-5 (*Rigden et al.,* 1984, 1988). Although Rigden et al. measured the pressure derivative of the adiabatic bulk modulus (K_S'), rather than that of the isothermal bulk modulus (K_T'), the difference is small. For the present study of lunar compositions, the melt densities were estimated over a range of assumed K_T' from 5 to 7. This was done for two reasons. First, since no experimental measurements have been made on lunar compositions, a substantial range for K_T' was assumed that might bracket the actual value. Second, since lunar compositions generally have lower abundances of SiO_2 and Al_2O_3 than terrestrial basalts, the rate at which the density of a lunar melt increases with increasing pressure ($\delta\rho/\delta P$) may be lower than that of a terrestrial basalt, and hence have higher K_T', if the concepts of *Stolper and Ahrens* (1987) are correct. To accommodate this possibility, values for K_T' of up to 7 were used in all calculations on all calculations on lunar compositions.

The third-order Birch-Murnaghan equation (equation (6)) was solved iteratively for ρ using (1) K_T' from 5 to 7, and (2) dT/dP of liquidus temperature from 7°C/kbar to 10°C/kbar (7-10°C/0.1 GPa). This range of conditions is substantial and has been explored in the hope of bracketing the true, but as yet unknown, densities of these lunar melts.

In using the parameters recommended by *Lange and Carmichael* (1987), the author observed that $\delta K_T/\delta T$ for the picritic lunar glass compositions was typically in the vicinity of -120 bars/°K. This large negative value had the effect of causing a rapid increase of density with pressure (i.e., densification) of the picritic lunar glass compositions, which the author felt might be an artifact resulting from the large uncertainty associated with the temperature dependence of the sound speed coefficient for MgO (*Rivers and Carmichael,* 1987, their Table 9b, p. 9261). A value for $\delta KT/\delta T$ = -25 bars/°K was assumed. The difference in calculated density that results from using $\delta K_T/\delta T$ = -120 bars/°K vs. -25 bars/°K is modest, however (i.e., -0.1 g/cc at 50 kbar, 5.0 GPa).

The density of liquidus olivine coexisting with each lunar mare composition at elevated pressures and at liquidus temperatures was also estimated using equation (6) and the extrapolated thermodynamic parameters from *Sumino* (1979) and *Suzuki et al.* (1981, 1983). The range of calculated densities for each olivine shown in Fig. 7 reflect the fact that the liquidus temperatures of the magmas were assumed to increase at between 7°C/kbar to 10°C/kbar (7-10°C/0.1 GPa). This caused the uncertainty in liquidus temperature to increase with increasing pressure (±15°C at 10 kbar; ±38°C at 25 kbar) beyond the 1-bar uncertainty of ±20°C already mentioned. Furthermore, a slightly increased uncertainty in the olivine density occurs with increasing pressure (i.e., ±0.01 g/

cc at 25 kbar, 2.5 GPa). Diagrams showing the pressure dependence of melt densities at liquidus temperature (Figs. 7a-c) have been constructed to illustrate the effects of these stated uncertainties.

RESULTS

Low Pressure

The compositional range in lunar mare magmas evaluated in this study is extreme. For example, the abundance of TiO_2 among the picritic lunar glasses ranges by a factor of 60 from 0.26 wt.% to 16.4 wt.% (*Delano,* 1986). This results in a considerable spectrum of calculated melt densities at zero-pressure liquidus temperatures from 2.82 g/cc for the lowest-Ti melt to 3.12 g/cc for the highest-Ti melt (Table 1). Figure 4 summarizes the estimated density and zero-pressure liquidus temperature for each of the mare basalts and pristine glasses evaluated in this study. Note that the pristine glasses consistently have higher liquidus temperatures and a larger range of densities than the crystalline mare basalts.

Figure 5a shows the difference in densities between olivine and melt at zero-pressure liquidus temperatures among the 25 varieties of picritic magmas represented by the pristine glasses. Although liquidus olivine is always denser than the coexisting melt at zero pressure, note that the density difference ($\Delta\rho$) decreases from ~0.4 g/cc for low-Ti melts to ~0.2 g/cc for high-Ti melts. In contrast, the magmas represented by the crystalline mare basalts (Fig. 5b) do not display this relationship, but rather maintain a large $\Delta\rho$ (~0.5 g/cc) that is independent of Ti abundance in the melt.

The viscosity of the picritic magmas represented by the pristine glasses at their respective liquidus temperatures have been estimated (Fig. 6) using the compositionally dependent parameters of *Bottinga and Weill* (1972). These results are consistent with experiments (e.g., *Murase and McBirney,* 1970; *Uhlmann et al.,* 1974) showing that mare magmas have

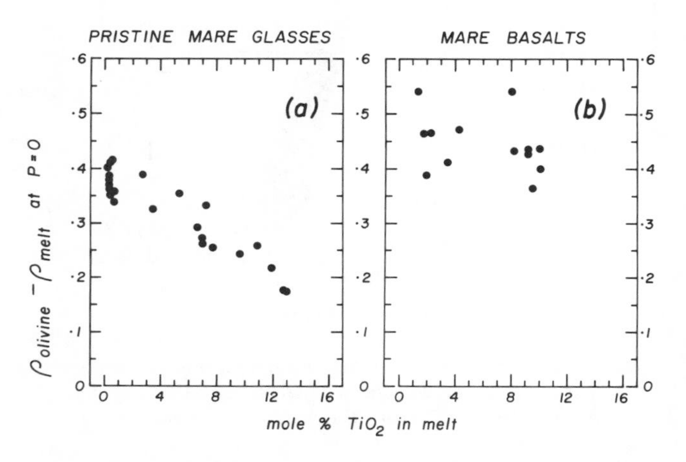

Fig. 5. The density contrast (at zero pressure) at liquidus temperatures between equilibrium olivine and melt are shown plotted against the mol.% TiO_2 in the melt. Note that the density contrast diminishes with increasing TiO_2 in the melt among the pristine glasses, but not among the mare basalt compositions.

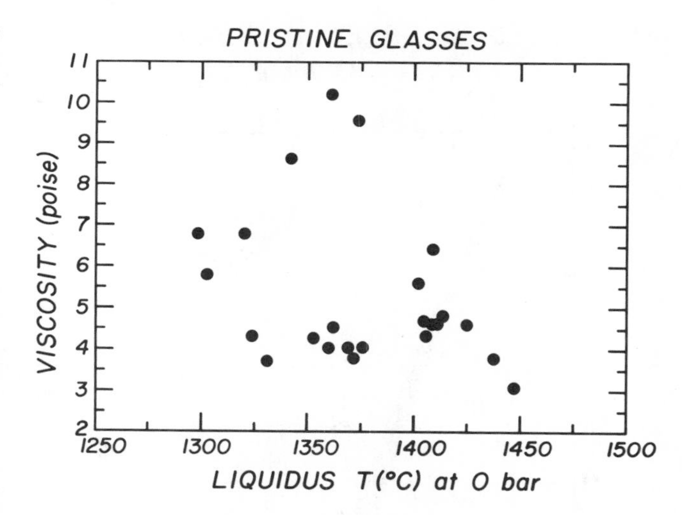

Fig. 6. The viscosities of mare magmas represented by the pristine glasses (*Delano,* 1986) have been estimated at liquidus temperatures using the method of *Bottinga and Weill* (1972).

viscosities of $\lesssim$10 p at their liquidus temperatures. There appears to be a slight tendency for the viscosity of the melt to decrease with increasing TiO_2 abundance among the pristine lunar glasses. In addition, these viscosities may decrease further with increasing pressure (*Waff,* 1975).

As indicated in Fig. 4, the pristine glasses generally exhibit higher liquidus temperatures and a wider range of melt densities than the crystalline mare basalts. Fractional crystallization of six mare compositions have been experimentally investigated (*Delano,* 1980; *Grove and Vaniman,* 1978; *Grove and Bence,* 1977; *Grove and Beaty,* 1980; *Walker et al.,* 1977). Figure 4 shows the calculated densities of fractionated melts generated during differentiation of these mare compositions. Note that the fractionated liquids and mare basalts have densities in the vicinity of 2.95 ± 0.05 g/cc (Fig. 4), which is also the density of the lunar highlands crust (e.g., *Haines and Metzger,* 1980; *Hood,* 1986; *Solomon,* 1975).

High Pressure

Figure 7 shows the estimated densities of three picritic melts and their equilibrium olivine and orthopyroxenes at liquidus temperatures in the pressure range from 0 to 47 kbar (0-4.7 GPa). These three melts (Apollo 15 green A; Apollo 17 orange glass 74220; Apollo 14 black glass) were selected as being illustrative of the effect that TiO_2 has on the $\Delta\rho$ between melt and equilibrium olivine with increasing pressure. For example, the low-Ti picritic melt (Fig. 7a) is seen to have a density significantly less than its liquidus olivine in the pressure interval of ≤25 kbar. However, with increasing TiO_2 in the melt (Fig. 7b), the $\Delta\rho$ at 20-25 kbar decreases to nearly zero (Fig. 7c). At pressures greater than 20-25 kbar, the liquidus phase of these picritic melts is assumed to change from olivine to low-Ca pyroxene based on experimental data from similar systems (*Chen et al.,* 1982; *Green et al.,* 1975; *Grove and Lindsley,* 1978; *Kesson,* 1975; *Chen and Lindsley,* 1983;

Delano, 1980). This change has been illustrated in Figs. 7a-c by showing a transitional region at 20-25 kbar (2.0-2.5 GPa) between olivine and orthopyroxene. Since garnet may replace orthopyroxene as the liquidus phase at $\geq$ 30 kbar (e.g., *Chen et al.,* 1982), the extension of orthopyroxene to 47 kbar on these diagrams is not strictly correct. However, since all investigators agree that primary mare magmas (e.g., pristine glasses; *Longhi,* 1987) were derived from depths of about 400 to 500 km (*Chen et al.,* 1982; *Chen and Lindsley,* 1983; *Delano,* 1980; *Green et al.,* 1975; *Kesson,* 1975; *Marvin and Walker,* 1978; *Stolper et al.,* 1981) or *less* (e.g., *Binder,* 1980), the portion of Figs. 7a-c most meaningful in assessing the relative densities of melt and equilibrium solid (i.e., olivine) is at pressures of $\leq$25kbar.

DISCUSSION

Melt/Solid Density Crossovers

Figure 8 contains four cartoons that may prove helpful for qualitatively understanding the effects that may accompany the densification of mafic melts with increasing pressure. In Fig. 8a,

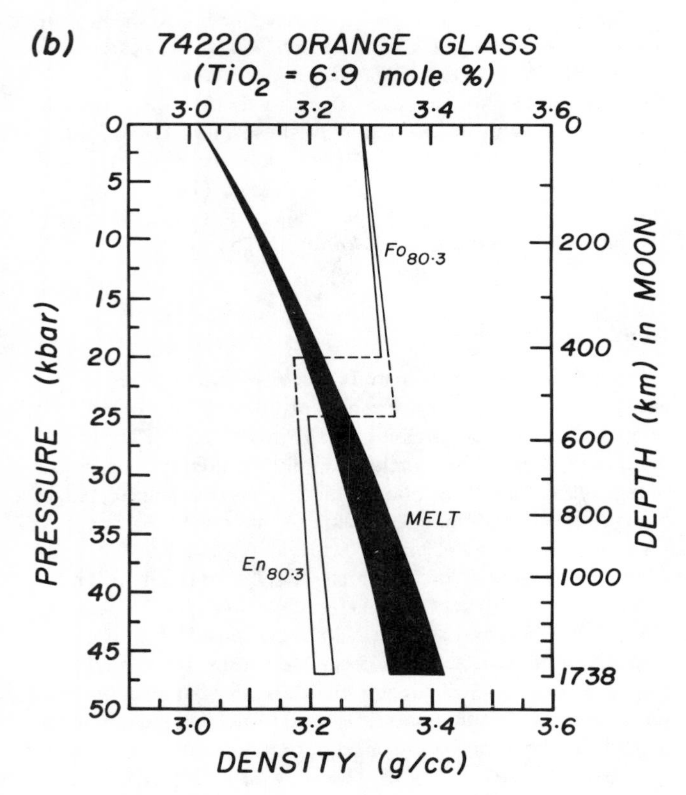

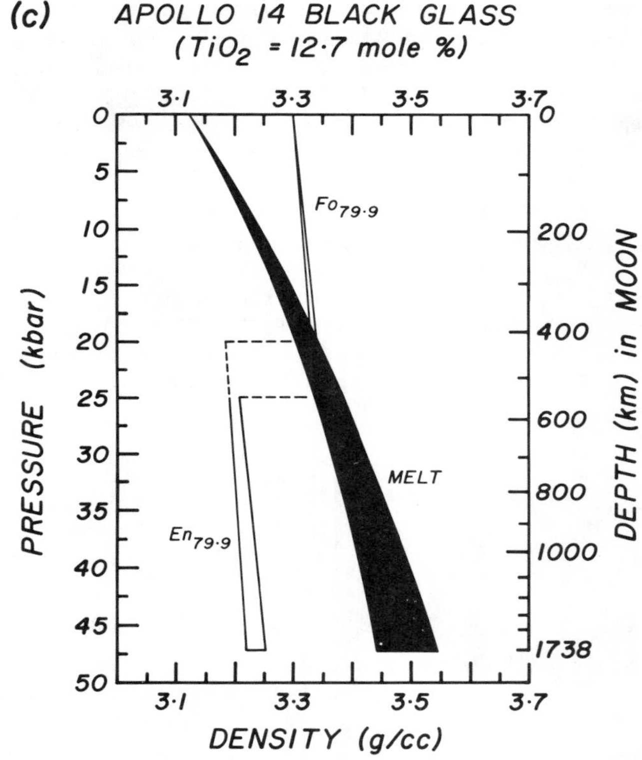

Fig. 7. The densities of mare magmas and equilibrium olivine (+ orthopyroxene) at liquidus temperatures in the pressure interval from 0 to 47 kbar. The liquidus temperature was assumed to increase at 7° to 10°C per kbar. The pressure derivative of the isothermal bulk modulus (K_T') was assumed to fall within the range of 5 to 7 using a third-order Birch-Murnaghan equation (see equation (6)). The results for three pristine glass compositions are shown: **(a)** Apollo 15 green A; **(b)** Apollo 17 orange glass 74220; and **(c)** Apollo 14 black glass.

the melt's density is less than that of either the residuum or the surrounding mantle. Under such circumstances, the melt will tend to segregate from the residuum by ascending buoyantly. A more quantitative description of this phenomenon is provided in Fig. 9, where Darcy's Law (modified after *Walker et al.,* 1978) has been used to estimate the velocity of melt migration as a function of the density difference between residuum and liquid for some lunar conditions [e.g., acceleration of gravity = 160 cm/sec^2; viscosity of melt = 5 p; effective radius of crystals in residuum = 0.2 cm, based on dunite clast in mare basalt 74275 (*Meyer and Wilshire,* 1974; *Delano and Lindsley,* 1982); fraction of partial melting from 0.05 to 0.30].

Darcy's Law (as modified after *Walker et al.,* 1978)

$$V = \frac{g\Delta\rho R^2 f^2}{73.5\eta}$$

where g = acceleration of gravity; $\Delta\rho$ = density contrast between crystal and melt; R = effective radius (cm) of crystals in source region; f = fraction of melt present in source region; and η = liquid viscosity. Note in Fig. 9 that for a given density contrast between residuum and melt, the velocity of melt segregation by buoyancy forces increases with increasing fraction of partial melting (f). As the density difference approaches zero (Fig. 8b), the velocity of melt segregation also approaches zero (Fig. 9). When the melt/solid density crossover occurs (Fig. 8c), the velocity changes sign and the melt sinks with respect to the residuum. With increasing pressure, the melt may also become denser than the surrounding asthenosphere (Fig. 8d) and generate a descending melt diapir.

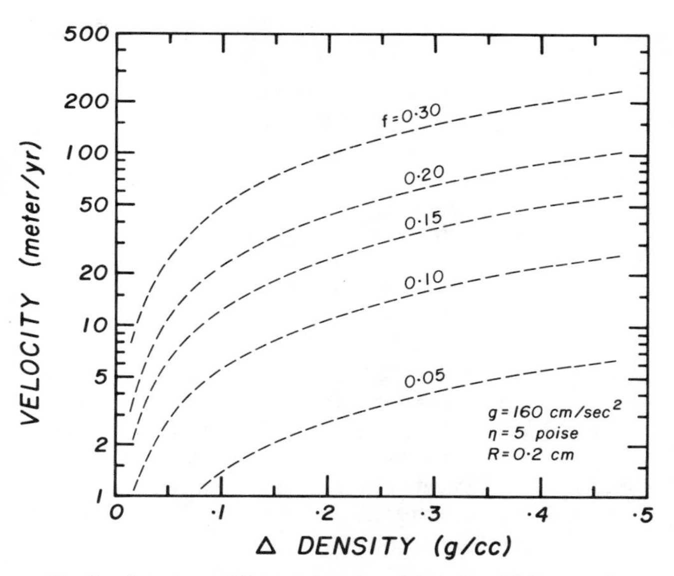

Fig. 9. Solutions of Darcy's Law (modified after *Walker et al.,* 1978) for a variety of lunar conditions: acceleration of gravity = 160 cm/sec^2; viscosity of melt = 5 p, effective radius of crystals in residuum = 0.2 cm; fraction of melt in source region from 0.05 to 0.30.

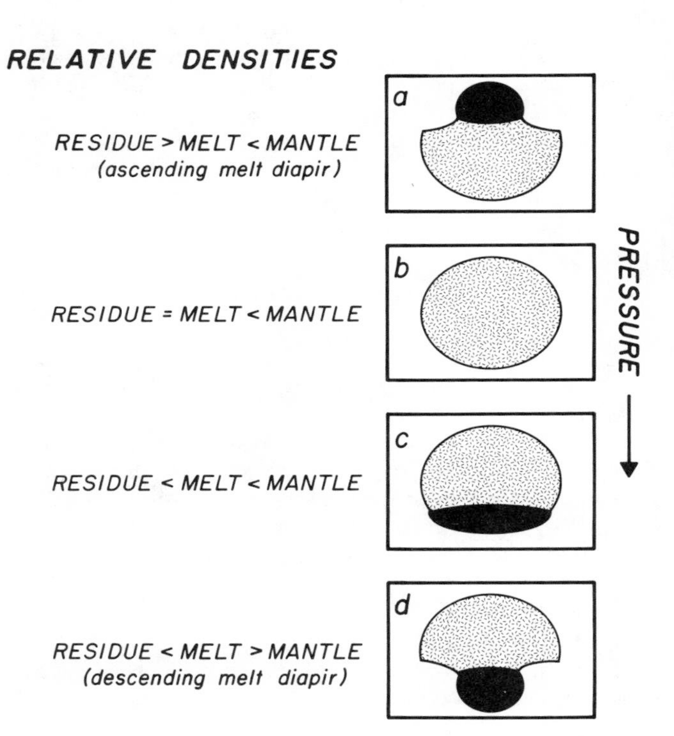

Fig. 8. A series of four cartoons illustrating the qualitative effects of melt densification with increasing pressure.

Figures 7a-c illustrate that the density contrast between melt and residual olivine in source regions at pressures of 20 to 25 kbar (e.g., *Delano,* 1980) decreases toward zero with increasing abundance of TiO_2 in the melt. If these numerical estimates are approximately correct, it suggests that the very-high-Ti magma represented by the Apollo 14 black pristine glass may have been a "marginal escapee" from its source region as the density contrast approached zero (Fig. 8b). In light of this result, the fact that this is the highest Ti melt yet observed (*Delano,* 1986; *Marvin and Walker,* 1978) to have been erupted onto the lunar surface may be significant. For example, if the Moon had generated melts with higher abundances of TiO_2 than the Apollo 14 black pristine glass, it seems unlikely that such melts would have been erupted onto the lunar surface. Instead, such ultrahigh TiO_2 melts (>16.4 wt.% TiO_2) may have segregated from their residuum by sinking rather than rising (Figs. 8c,d). This implies that for magmas generated by partial melting at pressures of 20 to 25 kbar (2.0-2.5 GPa) in the Moon there may be a compositional limit whereby magmas cannot be erupted onto the lunar surface. The Apollo 14 black glass may be close to that compositional limit of eruptability.

Attenuating Zone: A Speculation

If the Moon generated mafic liquids having densities greater than the limit discussed previously, then those magmas would have sunk with respect to the residua in their source regions and may under some stringent circumstances have sunk to considerably greater depths in the Moon. If this process of negatively buoyant diapirism occurred (Fig. 8d), it would represent a novel means for redistributing elements inside the Moon. Titanium and other incompatible elements (e.g., K, U, Th) might have been cycled back into the deep interior after the magma ocean event at 4.4 Ga by the sinking of not only dense cumulate blocks (e.g., *Ringwood and Kesson,* 1976; *Herbert,* 1980) but also dense, ultrahigh TiO_2 (>16.4 wt.% TiO_2) melts. Although it has been argued that the former process is likely to have occurred based on experimental evidence that high-Ti mare magmas came from deep (~500 km) source regions (e.g., *Green et al.,* 1975; *Delano,* 1980; *Marvin and Walker,* 1978), is there any evidence in support of the latter process? A speculative response to this question is that the seismically defined attenuating zone (e.g., *Goins et al.,* 1978, 1979; *Nakamura et al.,* 1973, 1982) may be that evidence. The attenuating zone (Fig. 10) has been interpreted as being a region of partial melting (e.g., *Nakamura et al.,* 1973), although the temperatures appear modest (e.g., *Toksoz et al.,* 1978). Perhaps this region of the Moon is a partial melt zone where the high-density melt has descended from shallower regions. This melt would not only have an ultrahigh Ti composition with a relatively low liquidus temperature but would also occupy <10% (by volume) of the attenuating zone as an intergranular fluid. This upper limit of 10% (by volume) is based on (1) an estimated abundance of 0.15-0.20 wt.% TiO_2 in the bulk Moon (e.g., *Jones and Delano,* 1989; *Rasmussen and Warren,* 1985) and (2) the assumption that all of this titanium had been sequestered into the attenuating zone in the form of dense melts having 20 wt.% TiO_2. Since (2) is clearly false, the 10% (by volume) of the attenuating zone is an upper limit.

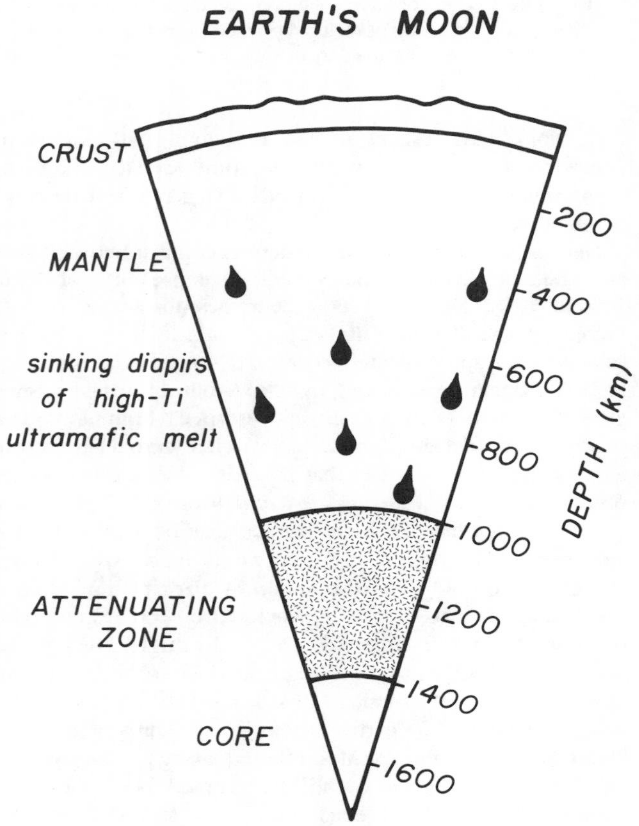

Fig. 10. A speculative cartoon illustrating the idea that negatively buoyant diapirs consisting of ultrahigh Ti melts may have sunk into the deep interior to form a zone of partial melt. The seismically defined attenuating zone (e.g., *Nakamura et al.,* 1973) may be a manifestation of this process. Since these high-Ti melts could occupy <10% (by volume) of the attenuating zone (refer to text), the sizes of the negatively buoyant diapirs shown in this figure are greatly exaggerated.

Planetary Implications

The reasons that melt/solid density-crossovers may occur at modest pressures in the Moon seems to be due to several properties of these high-Ti lunar melt compositions: (1) high abundances of Fe and Ti, which are relatively dense components compared to SiO_2 and Al_2O_3, as shown in Fig. 3; (2) the low oxidation state of lunar magmas, such that nearly all of the total iron occurs as dense FeO, rather than as the lower density Fe_2O_3; (3) small abundances of low-density components, such as K_2O and Na_2O; (4) the low abundance of volatiles in lunar magmas compared to terrestrial systems; and (5) the relatively magnesian (i.e., low density) nature of the liquidus olivines (Fo_{78-80}) in equilibrium with these high-Ti melts.

If our experiments (J. W. Delano and E. B. Watson, unpublished data, 1989) ultimately demonstrate that melt/solid density crossovers can occur in a body as small as the Moon, then this process may be a geochemically important mechanism for terrestrial-type objects in general. In assessing the possible role of density crossovers in other celestial objects, pressure is only one of the parameters that should be addressed. The compositional spectrum of primary melts produced by partial fusion of planetary mantles is also important.

CONCLUSIONS

If the melt densities estimated in this study are approximately correct, melt/solid density crossovers may occur for the very-high-Ti melt compositions at pressures of 20 to 25 kbar. Since this pressure interval coincides with the presumed location of the mare source regions (e.g., *Delano,* 1980; *Green et al.,* 1975; *Chen et al.,* 1982), there may be a compositional limit of eruptability among mare magmas approaching ~16 wt.% TiO_2. Although the lunar mantle may have produced melts with higher Ti, and hence higher density, those hypothetical melts would not have risen to the lunar surface and might have even sunk into the deep Moon to generate the presently observed attenuating zone located at depths >1000 km.

Acknowledgments. This work benefited from reviews by G. Miller (California Institute of Technology), J. Jones (NASA Johnson Space Center), and especially F. Richter (University of Chicago). Communications with P. Hess (Brown University), D. Walker (Lamont-Doherty Geological Observatory), and A. E. Ringwood (Australian National University) are also gratefully acknowledged. This work was supported by NASA grant NAG 9-78.

REFERENCES

Agee C. B. and Walker D. (1988) Static compression and olivine flotation in ultrabasic silicate liquids. *J. Geophys. Res., 93,* 3437-3449.

Binder A. B. (1980) On the mare basalt magma source region. *Proc. Lunar Planet. Sci. Conf. 11th,* pp. 1-22.

Bottinga Y. and Weill D. (1972) The viscosity of magmatic silicate liquids; a model for calculation. *Am. J. Sci., 272,* 438-476.

Chen H.-K. and Lindsley D. H. (1983) Apollo 14 very low titanium glasses: Melting experiments in iron-platinum alloy capsules. *Proc. Lunar Planet. Sci. Conf. 14th,* in *J. Geophys., 88,* B335-B342.

Chen H.-K., Delano J. W., and Lindsley D. H. (1982) Chemistry and phase relations of VLT volcanic glasses from Apollo 14 and Apollo 17. *Proc. Lunar Planet. Sci. Conf. 13th,* in *J. Geophys. Res., 87,* A171-A181.

Delano J. W. (1980) Chemistry and liquidus phase relations of Apollo 15 red glass: Implications for the deep lunar interior. *Proc. Lunar Planet. Sci. Conf. 11th,* pp. 251-288.

Delano J. W. (1986) Pristine lunar glasses: Criteria, data, and implications. *Proc. Lunar Planet. Sci. Conf. 16th,* in *J. Geophys. Res., 91,* D201-D213.

Delano J. W. and Lindsley D. H. (1982) Chromium, nickel, and titanium abundances in 74275 olivines: More evidence for a high-pressure origin of high-titanium mare basalts (abstract). In *Lunar and Planetary Science XIII,* pp. 160-161. Lunar and Planetary Institute, Houston.

Goins N. R., Toksoz M. N., and Dainty A. M. (1978) Seismic structure of the lunar mantle: an overview. *Proc. Lunar Planet. Sci. Conf. 9th,* pp. 3575-3588.

Goins N. R., Toksoz M. N., and Dainty A. M. (1979) The lunar interior: A summary report. *Proc. Lunar Planet. Sci. Conf. 10th,* pp. 2421-2439.

Green D. H., Ringwood A. E., Hibberson W. O., and Ware N. G. (1975) Experimental petrology of Apollo 17 mare basalts. *Proc. Lunar Sci. Conf. 6th,* pp. 871-893.

Grove T. L. and Beaty D. W. (1980) Classification, experimental petrology and possible volcanic histories of the Apollo 11 high-K basalts. *Proc. Lunar Planet. Sci. Conf. 11th,* pp. 149-177.

Grove T. L. and Bence A. E. (1977) Experimental study of pyroxene-liquid interaction in quartz-normative basalt 15597. *Proc. Lunar Sci. Conf. 8th,* pp. 1549-1579.

Grove T. L. and Lindsley D. H. (1978) Compositional variation and origin of lunar ultramafic green glasses (abstract). In *Lunar and Planetary Science IX,* pp. 430-432. Lunar and Planetary Institute, Houston.

Grove T. L. and Vaniman D. T. (1978) Experimental petrology of very low Ti (VLT) basalts. In *Mare Crisium: The View from Luna 24* (R. B. Merrill and J. J. Papike, eds.), pp. 445-471. Pergamon, New York.

Haines E. L. and Metzger A. E. (1980) Lunar highland crustal models based on iron concentrations: Isostasy and center-of-mass displacement. *Proc. Lunar Planet. Sci. Conf. 11th,* pp. 689-718.

Herbert F. (1980) Time-dependent lunar density models. *Proc. Lunar Planet. Sci. Conf. 11th,* pp. 2015-2030.

Herzberg C. T. (1984) Chemical stratification in the silicate earth. *Earth Planet. Sci. Lett., 67,* 249-260.

Hood L. L. (1986) Geophysical constraints on the lunar interior. In *Origin of the Moon* (W. K. Hartmann, R. J. Phillips, and G. J. Taylor, eds.), pp. 361-410. Lunar and Planetary Institute, Houston.

Itie J. P., Polian A., Calas G., Petiau J., Fontaine A., and Tolentino H. (1989) Pressure-induced coordination changes in crystalline and vitreous GeO_2. *Phys. Rev. Lett., 63,* 398-401.

Jones J. H. (1988) Partitioning of Mg and Fe between olivine and liquids of lunar compositions: The role of composition, pressure and Ti speciation (abstract). In *Lunar and Planetary Science XIX,* pp. 561-562. Lunar and Planetary Institute, Houston.

Jones J. H. and Delano J. W. (1989) A three-component model for the bulk composition of the Moon. *Geochim. Cosmochim. Acta, 53,* 513-527.

Kesson S. E. (1975) Mare basalts: Melting experiments and petrogenetic interpretations. *Proc. Lunar Sci. Conf. 6th,* pp. 921-944.

Lange R. A. and Carmichael I. S. E. (1987) Densities of Na_2O-K_2O-CaO-MgO-FeO-Fe_2O_3-Al_2O_3-TiO_2-SiO_2 liquids: New measurements and derived partial molar properties. *Geochim. Cosmochim. Acta, 51,* 2931-2946.

Longhi J. (1987) On the connection between mare basalts and picritic volcanic glasses. *Proc. Lunar Planet. Sci. Conf. 17th,* in *J. Geophys. Res., 92,* E349-E360.

Longhi J., Walker D., and Hays J. F. (1978) The distribution of Fe and Mg between olivine and lunar basaltic liquids. *Geochim. Cosmochim. Acta, 42,* 1545-1558.

Marvin U. B. and Walker D. (1978) Implications of a titanium-rich glass clod at Oceanus Procellarum. *Am. Mineral. 63,* 924-929.

Meyer C. E. and Wilshire H. G. (1974) "Dunite" inclusion in lunar basalt 74275 (abstract). In *Lunar Science V,* pp. 503-505. The Lunar Science Institute, Houston.

Murase T. and McBirney A. R. (1970) Viscosity of lunar lavas. *Science, 167,* 1491-1493.

Nakamura Y., Lammlein D., Latham G., Ewing M., Dorman J., Press F., and Toksoz N. (1973) New seismic data on the state of the deep lunar interior. *Science, 181,* 49-51.

Nakamura Y., Latham G. V., and Dorman H. J. (1982) Apollo lunar seismic experiment—Final summary. *Proc. Lunar Planet. Sci. Conf. 13th,* in *J. Geophys. Res., 87,* A117-A123.

Nisbet E. G. and Walker D. (1982) Komatiites and the structure of the Archaen mantle. *Earth Planet. Sci. Lett., 60,* 105-113.

Ohtani E. (1984) Generation of komatiite magma and gravitational differentiation in the deep upper mantle. *Earth Planet. Sci. Lett., 67,* 261-272.

Rasmussen K. L. and Warren P. H. (1985) Megaregolith thickness, heat flow, and the bulk composition of the Moon. *Nature, 313,* 121-124.

Rigden S. M., Ahrens T. J., and Stolper E. M. (1984) Density of liquid silicates at high pressures. *Science, 226,* 1071-1074.

Rigden S. M., Ahrens T. J., and Stolper E. M. (1988) Shock compression of molten silicate: Results for a model basaltic composition. *J. Geophys. Res., 93,* 367-382.

Rigden S. M., Ahrens T. J., and Stolper E. M. (1989) High-pressure equation of state of molten anorthite and diopside. *J. Geophys. Res., 94,* 9508-9522.

Ringwood A. E. (1979) *Origin of the Earth and Moon.* Springer-Verlag, New York. 295 pp.

Ringwood A. E. and Kesson S. E. (1976) A dynamic model for mare basalt petrogenesis. *Proc. Lunar Sci. Conf. 7th,* pp. 1697-1722.

Rivers M. L. and Carmichael I. S. E. (1987) Ultrasonic studies of silicate melts. *J. Geophys. Res., 92,* 9247-9270.

Solomon S. C. (1975) Mare volcanism and lunar crustal structure. *Proc. Lunar. Sci. Conf. 6th,* pp. 1021-1042.

Stolper E. M. and Ahrens T. J. (1987) On the nature of pressure-induced coordination changes in silicate melts and glasses. *Geophys. Res. Lett., 14,* 1231-1233.

Stolper E. and Walker D. (1980) Melt density and the average composition of basalt. *Contrib. Mineral. Petrol., 74,* 7-12.

Stolper E., Walker D., Hager B. H., and Hays J. F. (1981) Melt segregation from partially molten source regions: The importance of melt density and source region size. *J. Geophys. Res., 86,* 6261-6271.

Sumino Y. (1979) The elastic constants of Mn_2SiO_4, Fe_2SiO_4, and Co_2SiO_4 and the elastic properties of olivine group minerals at high temperature. *J. Phys. Earth, 27,* 209-238.

Suzuki I., Seka K., Takei H., and Sumino Y. (1981) Thermal expansion of fayalite, Fe_2SiO_4. *Phys. Chem. Mineral., 7,* 60-63.

Suzuki I., Anderson O. L., and Sumino Y. (1983) Elastic properties of a single-crystal forsterite Mg_2SiO_4, up to 1200 K. *Phys. Chem. Mineral., 10,* 38-46.

Toksoz M. N., Hsui A. T., and Johnston D. H. (1978) Thermal evolutions of the terrestrial planets. *Moon and Planets, 18,* 281-320.

Uhlmann D. R., Klein L., Kritchevsky G., and Hopper R. W. (1974) The formation of lunar glasses. *Proc. Lunar Sci. Conf. 5th,* pp. 2317-2331.

Waff H. S. (1975) Pressure-induced coordination changes in magmatic liquids. *Geophys. Res. Lett., 2,* 193-196.

Walker D., Kirkpatrick R. J., Longhi J., and Hays J. F. (1976) Crystallization history of lunar picritic basalt sample 12002: Phase-equilibria and cooling-rate studies. *Geol. Soc. Am. Bull., 87,* 646-656.

Walker D., Longhi J., Lasaga A. C., Stolper E. M., Grove T. L., and Hays J. F. (1977) Slowly cooled microgabbros 15555 and 15065. *Proc. Lunar Sci. Conf. 8th,* pp. 1521-1547.

Walker D., Stolper E. M., and Hays J. F. (1978) A numerical treatment of melt/solid segregation: Size of the eucrite parent body and stability of the terrestrial low-velocity zone. *J. Geophys. Res., 83,* 6005-6013.

Williams Q. and Jeanloz R. (1988) Spectroscopic evidence for pressure-induced coordination changes in silicate glasses and melts. *Science, 239,* 902-905.

Proceedings of the 20th Lunar and Planetary Science Conference, pp. 13-24
Lunar and Planetary Institute, Houston, 1990

Silicate Liquid Immiscibility in Isothermal Crystallization Experiments

J. Longhi

Lamont-Doherty Geological Observatory of Columbia University, Palisades, NY 10964

Isothermal crystallization experiments involving various mixtures and configurations of synthetic two-liquid pairs in iron capsules sealed in evacuated silica glass tubes show incompatibility under some conditions, compositional convergence under other run conditions, and further unmixing under still others. Results show that natural silicate liquid immiscibility may coexist stably with plagioclase, contrary to what is observed in simple systems. Comparison of results of isothermal and controlled-cooling-rate experiments on similar compositions suggests that kinetic barriers may have prevented the crystallization of silica minerals in the controlled-cooling-rate experiments. Multiple regression of the observed two-liquid separations against various chemical parameters indicates that P_2O_5 concentration, *Mg'*, and the plagioclase and olivine contents of the mafic liquid have significant effects on unmixing. Calculation of two-liquid separation in model melts calculated to be residual to ferroan anorthosites and presumed to be parental to *ur*KREEP indicates that silicate liquid immiscibility very probably developed during the latest stages of early lunar differentiation. However, there appears to be little evidence of two-liquid fractionation in the K_2O/P_2O_5 ratio of KREEP that lies within the range of values of low-Ti mare basalts. Immiscibility may have played a role in the formation of some lunar granites, but because these rocks have mafic minerals with intermediate *Mg'*, they could not have formed as residua of the lunar "magma ocean," which was hyperferroan. Lunar granites most probably crystallized as the plutonic residua of KREEP and high-Mg magmas.

INTRODUCTION

There are long-standing proposals that silicate liquid immiscibility (SLI) played a role in the petrogenesis of lunar granitic rocks (*Hess et al.,* 1975; *Taylor et al.,* 1979). *Neal and Taylor* (1989a) repeated this proposal recently with the additional suggestion (*Neal and Taylor,* 1989b) that the liquid precursor of these granites was *ur*KREEP (*Warren and Wasson,* 1979). Given the apparent importance of lunar granites and primitive KREEP as assimilants in mare basalts (*Shervais et al.,* 1985) and in high-Mg magmas (*Longhi,* 1981; *Warren,* 1988), it seems prudent to learn as much as possible about the development and effects of SLI in lunar magmas. There already is a considerable body of work describing trace element partitioning between immiscible liquid pairs in simple (*Watson,* 1976) and natural compositional systems (*Ryerson and Hess,* 1978), and these studies are in general accord. However, the existing experimental data on major elements suggests a possible disparity in phase equilibria between the simple laboratory systems, which petrologists consult as models of the complex natural system, and the natural system itself. Furthermore, the experimental major element data on natural lunar immiscible liquids are somewhat erratic and, consequently, the bounds of the two-liquid field are difficult to parameterize (see below). These considerations prompted this study.

Figure 1 illustrates two of the relatively simple phase diagrams used in discussions of SLI and depicts some of the two-liquid pairs developed in controlled cooling rate experiments on natural and synthetic lunar compositions (*Rutherford et al.,* 1974; *Hess et al.,* 1975, 1978). The liquidus boundaries in Fig. 1a are taken from a portion of the Fe_2SiO_4 (Fa) - $KAlSi_2O_6$ (Lc) - SiO_2 (Sil) system (*Roedder,* 1951; *Visser and Koster van Groos,* 1979a). Figure 1b shows that the effect of adding $CaAl_2Si_2O_8$ (An) to the Ol (Fa) - KS join in Fig. 1a (*Irvine,* 1976) is to suppress the two-liquid field. This is a mildly disturbing circumstance given the ubiquity of plagioclase in late stage magmas. *Visser and Koster van Groos* (1979b) have shown that adding less than 1 wt.% P_2O_5 to the Fa-Lc-Sil join expands the two-liquid field sufficiently for fayalite and K-feldspar to coexist with an immiscible liquid pair. Other studies on the Fa-Lc-Sil system show that increasing pressure (*Watson and Naslund,* 1977) and increasing oxygen fugacity (*Naslund,* 1976) also expand the two-liquid field, but not to the point of reaching the K-feldspar field. These data suggest that natural levels of P_2O_5 and other high field-strength elements will expand the two-liquid field in differentiating magmas, but leave unanswered the question of whether SLI will coexist with plagioclase. The only published experimental data in which plagioclase coexists with immiscible silicate liquids are those produced in controlled cooling rate experiments. Figure 1a depicts the two-liquid pairs coexisting with plagioclase in runs on lunar rock powders and on synthetic lunar compositions that have not been doped with trace elements.

There is considerable variation in both the length and orientation of the two-liquid tielines that does not correlate with temperature or any simple compositional parameter, such as *Mg'* [molar MgO/(FeO+MgO)] or P_2O_5 concentration. Some of the scatter is no doubt due to difficulties in analyzing the characteristically small globules that develop in these experiments. However, there is an additional concern: Despite the fact that most of these compositions project into the silica field, no silica phase was reported in the run tables of

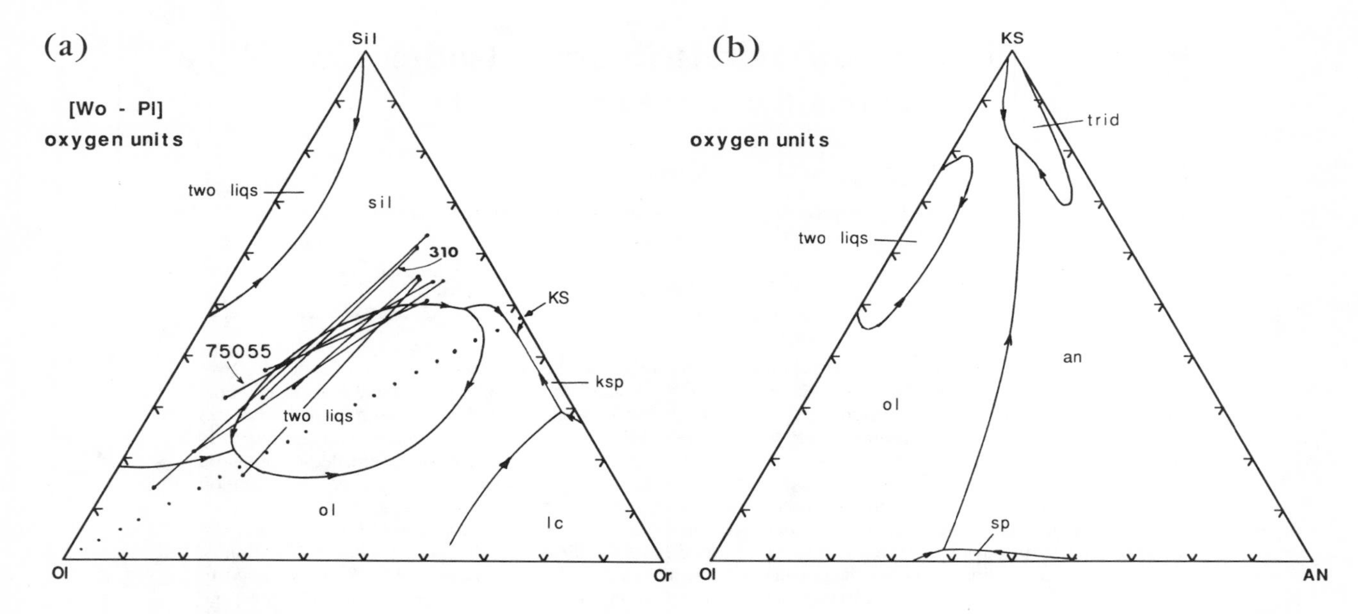

Fig. 1. **(a)** Projection of liquidus boundaries and compositions onto the $(Mg, Fe, Mn)_2SiO_4$(01) - $KA1Si_3O_8$(Or) - SiO_2(Sil) plane in oxygen units. Solid curves are liquidus boundaries in the system Fe_2SiO_4 (Fa) - $KA1Si_2O_6$(Lc) - Sil after *Roedder* (1951). KS is the $Or_{56}Q_{44}$ composition from *Irvine* (1976). Dots are the Fa-KS join. Tielines join two-liquid pairs reported by *Rutherford et al.* (1974) and *Hess et al.* (1975, 1978). These compositions are projected from $CaSiO_3$(Wo) and $NaA1Si_3O_8$(Ab) + $CaA1_2Si_2O_8$(An) = P1 components, as well as TiO_2 and P_2O_5. Projection equations are listed below. **(b)** Liquidus boundaries in the system Fa-An-KS after *Irvine* (1976). Projection equations for **(a)** (oxygen units): Ol[Wo-P1] = 2(FeO + MgO + MnO)/sum; Or[Wo-P1] = $16K_2O$/sum; sum = $2SiO_2$ + FeO + MgO + MnO - 2CaO - $2A1_2O_3$ + $6K_2O$ - $10Na_2O$, where oxides are in mole percent.

Rutherford et al. (1974) and *Hess et al.* (1975). The absence of a silica phase after prolonged fractional crystallization raises the possibility of metastability in these experiments. After all, silica minerals are present in the mesostases of the mare basalts (75055 - *Dymek et al.,* 1975; 12038 - *Keil et al.,* 1971) that were investigated, and although silica minerals were not reported in 14310 (e.g., *Ridley et al.,* 1972), they are commonly present in KREEP basalts (*Powell et al.,* 1973) and other KREEP-rich impact melts such as 68415 (*Walker et al.,* 1973), so their absence in 14310 is most probably due to quenching of the incompletely crystallized residuum. More importantly, metastability might also account for the coexistence of plagioclase and immiscible liquids. For example, *Rutherford et al.* (1974) noted a tendency for immiscibility to develop marginal to plagioclase crystals. Figure 1b suggests that if olivine failed to nucleate marginal to a growing feldspar, metastable crystallization of plagioclase might drive the liquid composition into the metastable extension of the two-liquid field. These considerations led me to question whether both the controlled cooling rate and natural SLI were metastable (*Longhi,* 1989a) and to initiate some simple experiments to resolve the issue.

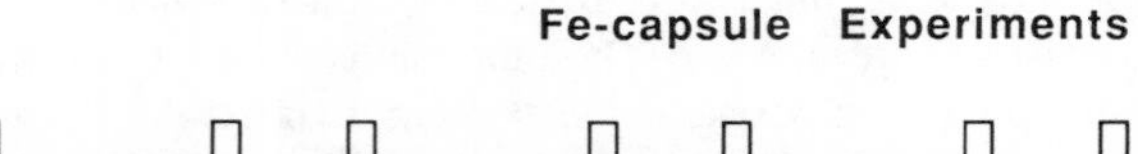

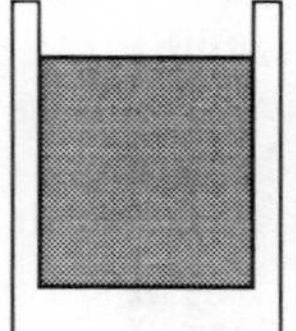
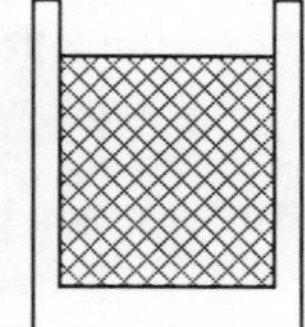
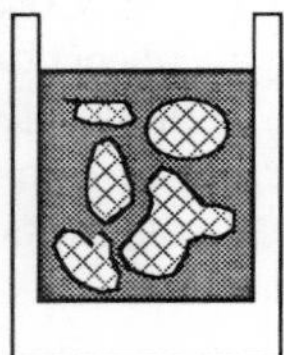
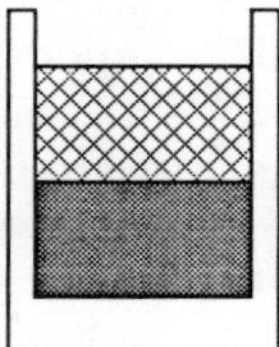
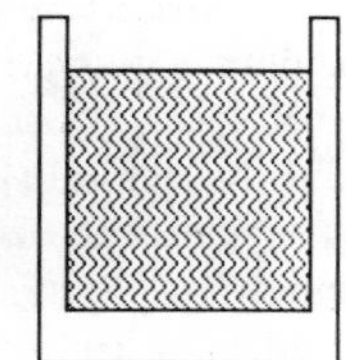

Fig. 2. Configuration of experimental charges.

EXPERIMENTAL METHODS

Four glasses were synthesized from reagent grade oxides and carbonates with the compositions of two of the sets of coexisting liquids reported by *Hess et al.* (1975): a KREEP basalt derivative (14310-102,6) and a mare basalt derivative (75055-41,14). These pairs are referred to in the run table as HS1, HF1, HS2, and HF2 respectively where HS signifies high-Si, HF signifies high-Fe, 1 refers to 14310, and 2 refers to 75055. Following subsolidus decarbonation, the oxide mixtures were fused several times in a Pt crucible in air with intervening grinding in an agate mortar and then fused again in the Pt crucible in a H_2-CO_2 stream with fO2 controlled at approximately 1 log unit below the Fe/FeO buffer. The reported liquidus temperatures of the HS1 and HS2 compositions are 1050°C and 990°C, respectively, but repeated fusions at temperatures as high as 1350°C failed to dissolve all of the quartz starting material. Minor fayalite and ilmenite also crystallized in the high-Si compositions. The relic quartz grains were easily distinguished in the subsequent runs by their size (50-200 μm), shape (anhedral, subequant), and orange fluorescence under the microprobe beam. Silica minerals that grew from the fused glasses during the runs were much smaller (5-30 μm), euhedral (equant—cristobalite, elongated—tridymite), and fluoresced blue. Quartz did dissolve completely during the fusions of the high-Fe compositions resulting in crystal-free, homogeneous glasses; however, comparison of the nominal and analyzed glass compositions shows obvious Fe-loss to the Pt crucibles during the syntheses: 7% of the Fe present in HF1 and 17% in HF2. The CaO concentrations of HS1 and HF1 are low due to errors in weighing.

The experiments were run in high purity Fe-capsules sealed in evacuated silica tubes with the various sample configurations illustrated in Fig. 2. All charges were run at constant temperature except for those in the fused micromixture configuration. These were held initially at 1200°C for 18 hours, quenched in air, and then put back in the furnace at the run temperautre. The charges were quenched in water, mounted in epoxy, polished, examined in reflected light, and analyzed with a Cameca CAMEBAX/MICRO electron microprobe. Analytical conditions were 15 kv acceleration voltage and 25 namp beam current for 20 sec for all elements except Na and K which were analyzed first at 5 namp. High resolution backscattered electron (BSE) imaging photographs were taken with a Cambridge Stereoscan 250 Mk2 electron microscope.

RESULTS

The two sets of composition, 14310 and 75055 derivatives (Table 1), provided an opportunity to compare the compositional effects of KREEP [intermediate *Mg'* (0.3), high P_2O_5] with mare basalt [low *Mg'* (0.03), moderate P_2O_5]. In this first series of experiments the starting materials were synthesized according to the analyses of *Hess et al.* (1975) and no attempt was made to include additional trace elements that were present in the rock samples and probably had concentrations in the derivatives measurable with the microprobe. Increasing concentrations of high field-strength elements should increase the immiscible separation (*Ryerson and Hess,* 1978), so any two-liquid separations observed in the present study should be minima. In the case of 75055, which has low concentrations of Zr and REE, the effects of these high field-strength elements should not be pronounced. However, a rough estimate of the concentrations of these elements prior to unmixing in the 14310 residual liquids reported by *Hess et al.* (1975) indicates ~0.5 wt.% ZrO_2 and ~0.25 wt.% ΣREE. Therefore, noticeable differences in the extent of unmixing between that observed in the present study and that produced from the natural composition run at equilibrium are possible.

14310

Liquids in the first set of runs with the HS1-HF1 (1049°C) series of compositions gained iron. This meant that the starting material had not equilibrated completely with the H_2-CO_2

TABLE 1. Composition of starting materials (wt.%).

	HS1		HF1		HS2		HF2	
No. of Pts.	*	4	*	4	*	3	*	3
SiO_2	69.48	68.0 (0.9)	45.14	47.1 (0.7)	69.52	63.7 (1.4)	46.0	48.1 (0.32)
TiO_2	2.64	2.37 (0.09)	6.84	7.56 (0.05)	0.97	1.07 (0.11)	2.67	2.83 (0.02)
Al_2O_3	11.24	13.0 (0.5)	9.14	9.99 (0.17)	12.25	14.8 (0.6)	8.57	9.19 (0.09)
Cr_2O_3	0.02	0.01 (0.01)	0.08	0.10 (0.01)	0.02	0.01 (0.01)	0.03	0.01 (0.01)
FeO	5.11	4.61 (0.33)	16.34	15.06 (0.35)	8.65	6.63 (0.22)	27.60	23.0 (0.4)
MgO	1.59	2.32 (0.13)	3.75	4.88 (0.08)	0.12	0.29 (0.04)	0.28	0.60 (0.02)
MnO	0.08	0.08 (0.01)	0.22	0.20 (0.02)	0.34	0.40 (0.03)	0.92	0.75 (0.03)
CaO	4.55	2.93 (0.17)	10.76	8.53 (0.18)	4.16	5.77 (0.63)	10.42	11.29 (0.09)
K_2O	3.54	3.51 (0.12)	1.28	1.17 (0.07)	4.55	5.20 (0.34)	0.90	0.87 (0.06)
Na_2O	0.87	0.63 (0.06)	0.47	0.54 (0.04)	0.49	0.63 (0.04)	0.25	0.31 (0.01)
P_2O_5	0.92	0.62 (0.06)	4.23	3.99 (0.40)	0.19	0.49 (0.13)	2.36	2.70 (0.42)
Mg'	0.357	0.473	0.290	0.366	0.024	0.072	0.018	0.044
Other Phases		qtz, ilm		—		qtz, ol		—

*Compositions as reported by *Hess et al.* (1975) in Table 3.
1 = 14310, 102-6; 2 = 75055, 41-14.
( ) = standard deviation.

atmosphere during the synthesis and so the reduction process was repeated at a higher temperature. Subsequent runs in the iron capsules lost a small amount (1-3%) of the iron present via reduction; however, the iron gain in this first series of runs provided an opportunity to examine the effects of varying *Mg'* on the 14310 derivatives. In the first series of runs SLI was not observed in the simple high-Fe or high-Si configurations, but sharp two-liquid interfaces (Figs. 3a,b) were observed in the run with the random macromixture configuration (HS1/HF1-1). This run produced large (>>100 mm) enclaves of relatively crystal-free high-Fe and high-Si glass separated by crystal-rich (plag, sil, pyx, ilm) reaction zones in which the glass was high-Fe. Analyses show that the glasses in the crystal-free zones are relatively homogenous (filled squares, Fig. 4a) whereas glasses in and near the reaction zones vary in composition (open squares, Fig. 4a); the compositional gaps (tielines) across the two high-Si and high-Fe liquid interfaces are similar and sharp. The compositional separation in run HS1/HF1-1 is >11 wt.% SiO_2 (Table 2), but is obviously much less than the 24 wt.% reported by *Hess et al.* (1975) as shown by the tieline joining filled circles in Fig. 4a. Also more extensive and opposite in sense is the gap in *Mg'* between the high-Si and high-Fe liquids produced in the two studies; *Hess et al.* (1975) reported *Mg'* values of 0.36 and 0.29 for the high-Si and high-Fe liquids, respectively, whereas the values of *Mg'* at the interface shown in Fig. 3b are 0.25 and 0.27.

Following the second reductions of the HS1 and HF1 glasses, the first series of runs (1053°C) produced no observable SLI. There was no evidence of unmixing in HS1-2, while both layered macromixture, HS1/HF1-2 (Fig. 3c), and the mechanical micromixture, HS1/HF1-3, configurations produced regions of high-Si and high-Fe glass completely separated by crystalline zones. Figure 4b illustrates liquid compositions from these runs. On the basis of the high P_2O_5 concentrations (>6 wt.%) in the high-Fe glasses (Table 2), it seems possible that SLI developed initially but that some equilibration process, perhaps Fe-loss, rendered the two liquids incompatible. Another series of runs at 1041°C, consisting of a fused mixture of the two endmember glasses (HS1/HF1-5) and a mechanical micromixture (HS1/HF1-6) produced new evidence of SLI. The fused mixture produced two groups of glass compositions whose averages differ by less than 2.5 wt.% SiO_2 (Table 2; solid squares in Fig. 4c). The mechanical micromixture produced a heavily crystalline run product that had a wide range of liquid compositions (open squares in Fig. 4c) and locally two-liquid interfaces. Unlike run HS1/HF1-1, the regions with the

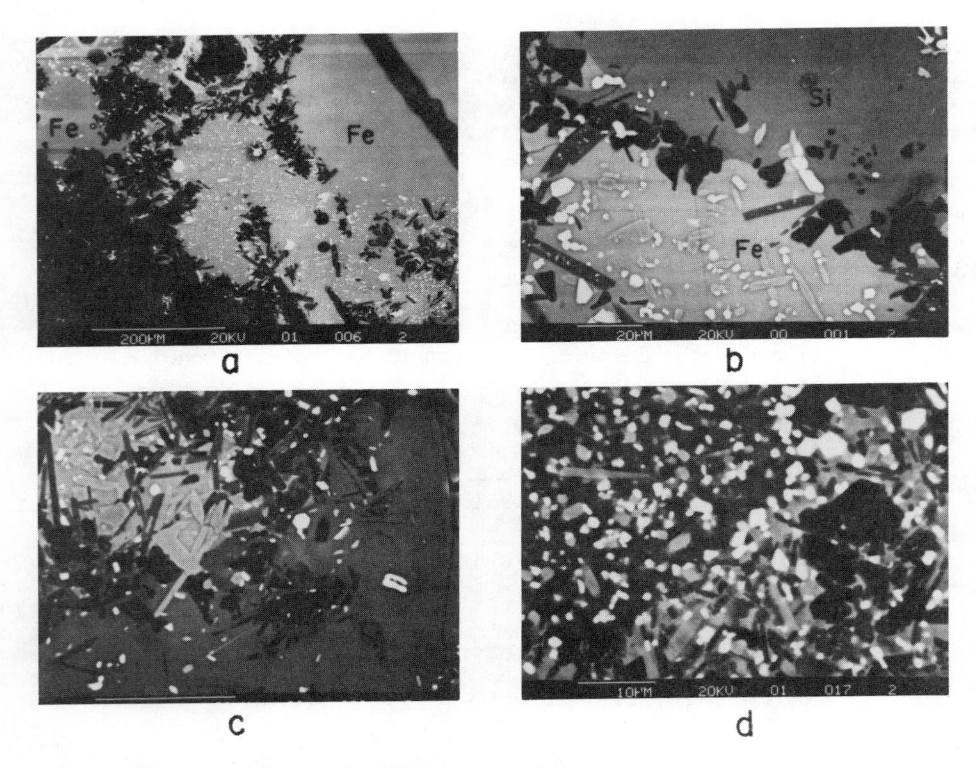

Fig. 3. Backscattered electron photomicrographs of HS1-HF1 (14310) series run products. **(a)** HS1/HF1-1 (1049°): high-Si glass is dark gray; high-Fe glass is light gray; **(b)** Close-up two-liquid contact in **(a)**: brightest phase is ilmenite, low-Ca pyroxene (1pyx) is slightly lighter than high-Fe glass; plagioclase is medium gray, and silica is black; **(c)** HS1/HF1-2 (1053°): no two-liquid contact is evident; **(d)** HF1-2 (1023°): contact between high-Si (dark matrix) and high-Fe (light matrix) regions.

narrowest compositional separations do not have readily observable two-liquid contacts; instead the observable two-liquid contacts correspond to intermediate compositional separations. Although two-liquid contacts may be present at the narrowest compositional separations and simply are not visible in BSE imaging because of the small atomic number contrast, it is nonetheless clear from Fig. 4c that two-liquid contacts may occur over a range of compositions. Most importantly though, the convergence of liquid compositions from opposite directions suggests that the equilibrium separation lies between the two closest HS1/HF1-6 liquids and the HS1/HF1-5 liquids (solid squares). This two-liquid separation is smaller than that observed in the 1049°C run (Fig. 4a). However, the absence of plagioclase in HS1/HF1-5

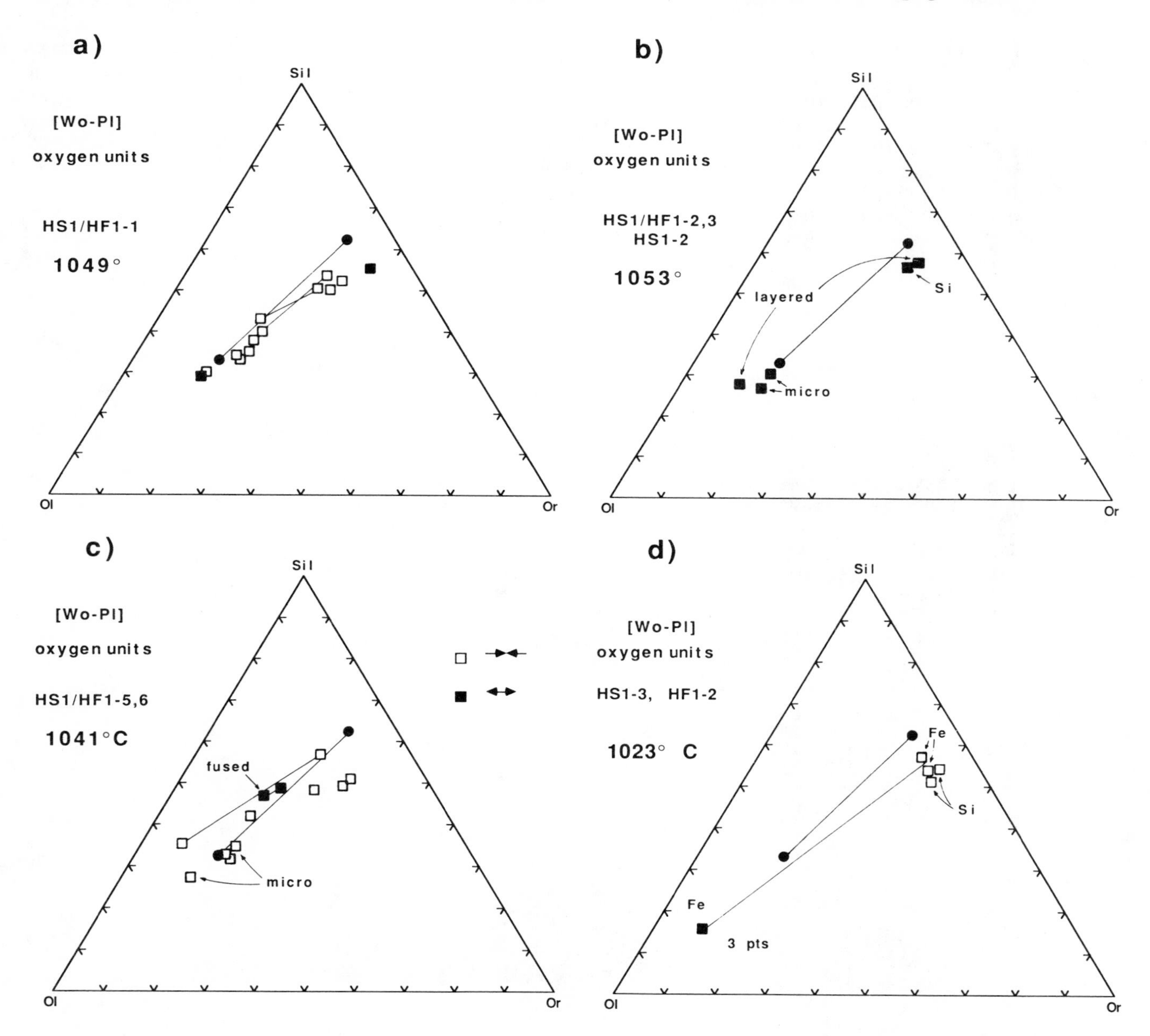

Fig. 4. Projected compositions of experimental liquids of HS1-HF1 series (squares). Large circles are 14310, 102-6 compositions from *Hess et al.* (1975). Filled symbols are average analyses; open symbols are individual analyses. **(a)** HS1/HF1-1 (1049°), random macromixture configuration: filled squares are averages of crystal-free regions; **(b)** HS1/HF1-2 (1053°), layered macromixture (filled squares); HS1/HF1-3 (1053°), mechanical micromixture (filled squares respresent average analyses on different chips; **(c)** HS1/HF1-4 (1041°), fused micromixture (filled squares); HS1/HF1-5 (1041°), mechanical micromixutre (open squares); **(d)** HF1-2 (1023°), single composition (squares); HS1-3 (1023°), single composition (squares).

TABLE 2. Run conditions and phase compositions (wt.%) for HS1-HF1 series.

Run Configuration	HS1/HF1-2 Layered Macro-Mixture			HS1/HF1-3 Mechanical Micro-Mixture	HS1/HF1-1 Random Macro-Mixture		HF1-2 Single Composition		HS1-3 Single Composition	HS1/HF1-5 Fused Micro-Mixture		HS1/HF1-6 Mechanical Micro-Mixture			
T(°C)	1053			1053	1049		1023		1023	1041		1041			
Phase	liq	liq	lpyx(Fe)	liq	liq	liq	liq	liq	liq	liq	liq	liq a	liq b	liq c	liq d
No. of Pts.	2	3	3	1	1	3	2	1	1	4	4	1	1	1	1
SiO_2	68.67 (0.13)	42.91 (0.26)	52.66 (0.45)	43.70 (0.92)	62.01	50.94	38.55 (0.57)	67.78	68.80	55.88 (0.80)	53.54 (0.57)	58.45	50.51	63.80	44.48
TiO_2	1.69 (0.08)	4.50 (0.05)	0.80 (0.13)	4.02 (0.02)	1.92	2.92	3.66 (0.06)	1.57	1.16	2.66 (0.06)	3.02 (0.19)	2.21	3.14	1.72	3.80
Al_2O_3	11.95 (0.11)	9.94 (0.06)	0.93 (0.19)	10.75 (0.07)	11.73	11.57	8.51 (0.10)	11.25	13.01	12.10 (0.15)	11.99 (0.14)	11.48	10.34	10.36	9.96
Cr_2O_3	0.00 (0.00)	0.06 (0.01)	0.14 (0.06)	0.07 (0.03)	0.02	0.04	0.03 (0.00)	0.03	0.02	0.01 (0.01)	0.03 (0.01)	0.02	0.02	0.02	0.08
FeO	5.20 (0.17)	18.19 (0.16)	24.47 (0.13)	18.52 (0.23)	9.83	15.95	27.46 (0.72)	7.33	5.12	13.85 (0.67)	15.00 (0.50)	12.36	17.39	9.72	19.93
MgO	1.28 (0.08)	4.81 (0.02)	20.10 (0.15)	4.39 (0.16)	1.80	3.31	2.85 (0.09)	0.60	0.93	2.11 (0.10)	2.34 (0.09)	1.71	2.62	2.04	3.67
MnO	0.06 (0.01)	0.23 (0.02)	0.32 (0.02)	0.23 (0.02)	0.11	1.17	0.21 (0.03)	0.06	0.07	0.14 (0.01)	0.14 (0.01)	0.13	0.25	0.13	0.24
CaO	3.18 (0.19)	10.08 (0.08)	1.79 (0.12)	9.02 (0.08)	4.44	6.65	9.41 (0.07)	3.10	1.32	7.02 (0.09)	7.65 (0.34)	5.44	7.39	3.84	9.18
K_2O	4.21 (0.07)	0.85 (0.03)	-	1.24 (0.05)	3.22	2.09	0.69 (0.01)	4.66	5.20	2.03 (0.09)	1.71 (0.13)	2.96	1.65	2.98	0.60
Na_2O	0.75 (0.01)	0.38 (0.04)	0.04 (0.08)	0.44 (0.03)	0.56	0.42	0.36 (0.01)	0.81	0.47	0.66 (0.04)	0.53 (0.10)	0.80	0.47	0.64	0.36
P_2O_5	0.69 (0.12)	6.34 (0.11)	-	6.07 (0.02)	2.01	4.10	6.56 (0.11)	0.88	0.59	1.04 (0.08)	1.41 (0.04)	1.66	2.80	1.09	5.23
Sum	97.69	98.30	101.25	98.46	97.65	98.16	98.30	98.06	96.69	97.48	97.39	97.33	96.57	96.34	97.52
Other Phases	sil,pl,ilm			sil,pl, lpyx,ilm	sil,pl,ilm,lpyx		ol,pl,aug,ilm, sil,wht		ol,pl,aug, ilm,sil	lpyx,sil,ilm		lpyx,sil,ilm,pl			
Time (hrs.)	136			136	114		116		116	160		160			
Comments	no two-liquid contact; cf. Fig. 3c			no two-liquid contacts	sharp two-liquid contact with crystals; cf. Figs. 3a,b		cf. Fig. 3d		no two-liquid separation obvious			a,b = two closest compositions; c,d = two-liquid contact			

( )=standard deviation.

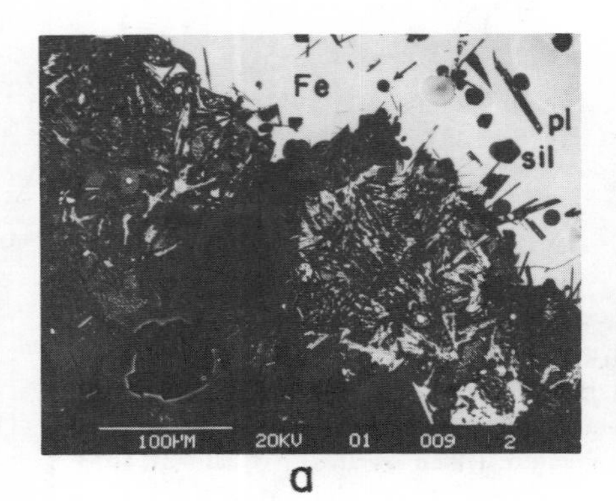

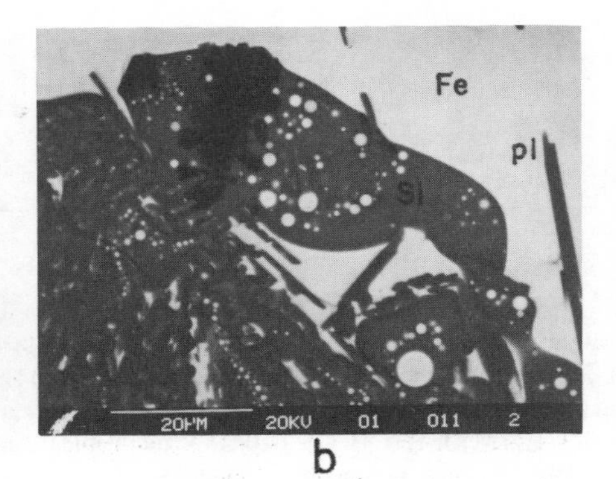

Fig. 5. Backscattered electron photomicrographs of HS2/HF2 (75055) series run products. **(a)** HS2/HF2-1 (990°): arrows point to high-Si droplets (medium gray) in a high-Fe glass (light gray) near artifical two-liquid contact; **(b)** Also HS2/HF2-1: high-Fe glass droplets in matrix of high-Si glass near artificial two-liquid contact.

TABLE 3. Run conditions and phase compositions (wt. %) for HS2-HF2 series.

Run Configuration	HS2/HF2-1 Layered Macro-mixture				HS2/HF2-3 Fused* Macro-mixture		HS2/HF2-4 Mechanical Micro-mixture	
T(°C)	990				975		975	
Phase	liq	pl(Si)	liq	pl(Fe)	liq	liq	liq	liq
No. of Pts	8	1	6	6	1	1	8	4
SiO_2	70.05 (0.48)	50.93	43.60 (0.44)	48.98 (1.82)	64.90	53.09	70.82 (0.75)	43.20 (0.56)
TiO_2	0.77 (0.02)	0.21	3.13 (0.06)	0.12 (0.03)	1.00	1.92	0.72 (0.03)	3.09 (0.03)
Al_2O_3	10.64 (0.06)	28.52	7.00 (0.06)	30.65 (0.67)	11.04	10.45	10.62 (0.11)	6.83 (0.12)
Cr_2O_3	0.00 (0.00)	0.00	0.02 (0.02)	0.00 (0.00)	0.00	0.01	0.01 (0.02)	0.01 (0.02)
FeO	6.37 (0.12)	2.16	27.25 (0.36)	2.00 (0.20)	9.89	19.20	5.96 (0.32)	27.80 (0.48)
MgO	0.11 (0.01)	0.06	0.58 (0.02)	0.07 (0.01)	0.20	0.44	0.04 (0.01)	0.22 (0.02)
MnO	0.18 (0.01)	0.06	0.87 (0.04)	0.06 (0.01)	0.30	0.57	0.18 (0.03)	0.92 (0.04)
CaO	2.92 (0.09)	15.13	11.55 (0.14)	16.37 (0.02)	4.95	8.65	2.79 (0.09)	11.80 (0.20)
K_2O	5.93 (0.13)	3.20	1.00 (0.04)	0.65 (0.63)	4.55	2.48	6.22 (0.17)	1.04 (0.10)
Na_2O	0.61 (0.06)	0.65	0.27 (0.05)	1.21 (0.22)	0.57	0.36	0.59 (0.07)	0.26 (0.06)
P_2O_5	0.22 (0.04)	—	3.19 (0.12)	—	0.45	1.16	0.19 (0.03)	2.92 (0.09)
Sum	99.52	100.92	98.46	100.11	97.85	98.16	98.15	98.40
Other Phases	sil,ol,ilm,aug				sil,pl,ilm,ol		ol,pl,aug,ilm,sil	
Time (hrs.)	140				160		160	
Comments	matrix + spherules; cf. Figs. 5a,b		matrix only	no secondary fluorescence correction	sharp two-liquid contact			

*Fused at 1250°C for 12 hrs, quenched, and run at 975°C.
( ) = standard deviation.

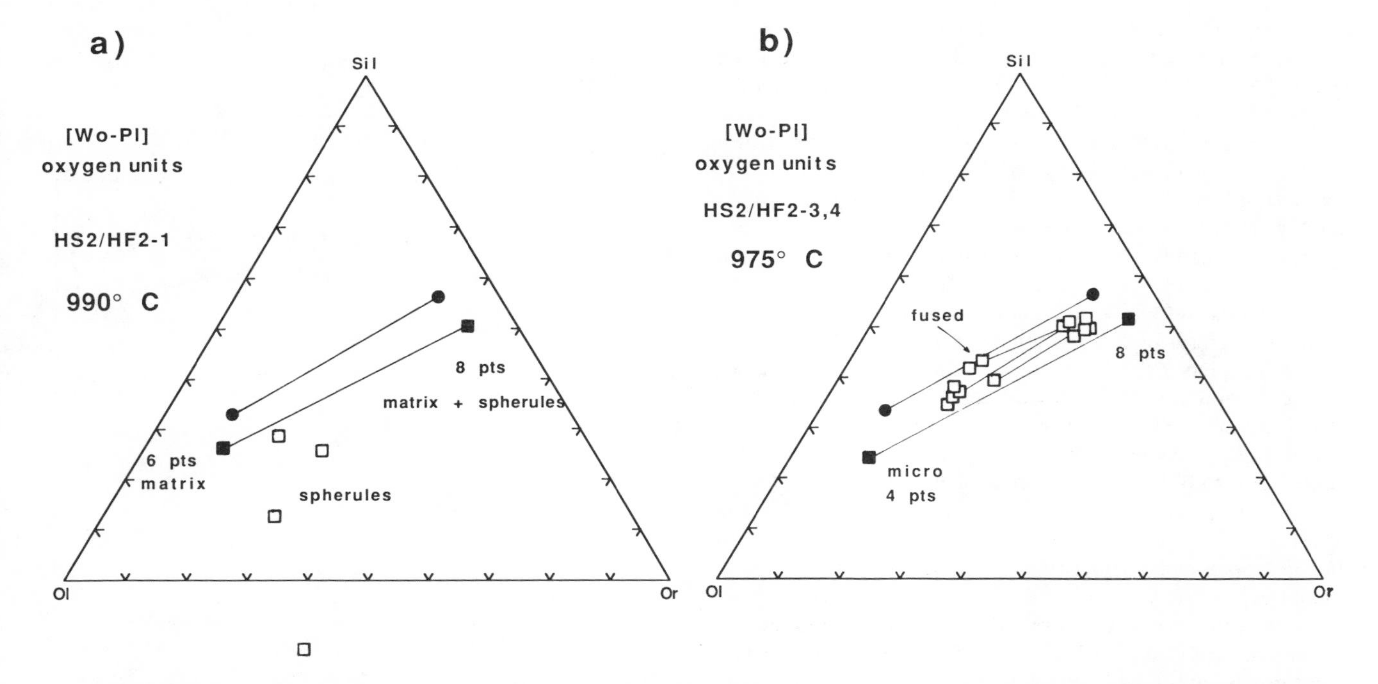

Fig. 6. Projected compositions of experimental liquids of HS2-HF2 series (squares). Large circles are 75055, compositions from *Hess et al.* (1975). Filling/open criteria as in Fig. 4. **(a)** HS2/HF2-1 (990°), layered macromixture; **(b)** HS2/HF2-3 (975°), fused micromixture (open squares); HS2/HF2-4 (975°), mechanical micromixture (filled squares).

complicates the interpretation of equilibrium. Apparently, plagioclase failed to nucleate from the fused mixture of glasses and the run achieved a metastable state. Examination of the liquid compositions in other projections that include the plagioclase component indicates that there is no more than 5% to 10% excess plagioclase component dissolved in the HS1/HF1-5 liquids. Given the data of *Irvine* (1976) illustrated in Fig. 1b that shows the plagioclase component to surpress immiscibility, HS1/HF1-5 may be regarded as no better than a bracketing run.

SLI also occurred in the lowest temperature (1023°C) run of the HS1-HF1 series, HF1-2. No evidence of SLI was observed in its companion run, HS1-3, although the high-Si liquids in this run are generally similar to those generated by SLI in HF1-2 (Fig. 4d). Figure 3d illustrates a heavily crystalline two-liquid region of HF1-2. Figure 4d shows that the high-Si liquids in this run vary perceptibly in composition, whereas the high-Fe liquids are relatively homogeneous. Table 2 shows that the high-Fe liquids also have the most extreme compositions produced in this study with ~38.5 wt.% SiO_2 and 6.6% P_2O_5 (whitlockite is present). The empirical curve (P_2O_5 vs. SiO_2) for whitlockite saturation drawn by *Dickinson and Hess* (1982) predicts ~9% P_2O_5 when extrapolated to this SiO_2 concentration. This discrepancy suggests that additional compositional parameters are probably necessary to represent the whitlockite saturation surface adequately, much as *Dickinson and Hess* (1982) warned.

75055

Two series of experiments on the 75055 derivatives produced some evidence of SLI, although the results differ somewhat from those of the 14310 derivatives. BSE imaging revealed SLI in the 990°C run to be restricted to the vicinity of the artificial two-liquid contact produced in a layered macromixture configuration (Figs. 5a,b). Immiscibility developed in both the high-Si and high-Fe layers. The high-Si liquid droplets in the high-Fe layer were relatively few in number and large enough (10-20 μm) to probe reliably, whereas the high-Fe droplets were much more numerous and smaller (<10 μm). The high-Si liquid droplets were relatively homogeneous and identical in composition to the liquid in the high-Si layer. Erratic concentrations of K_2O and P_2O_5 in the high-Fe droplets (Table 3) suggest the presence of either submicroscopic crystals or second generation high-Si droplets. Second generation high-Fe droplets are present locally in larger high-Si liquid droplets. The two-liquid tieline (Fig. 6a) is similar in length and orientation to that reported by *Hess et al.* (1975), although the tieline joining the data from the present study is offset from the Hess et al. tieline away from the Sil component. The presence of numerous crystals of second generation silica mineral (cristobalite) in this (Fig. 5a) and other runs demonstrates that the offset of the two tielines cannot be due to a deficiency of SiO_2 in the Fe-capsule experiments caused by incomplete dissolution of quartz in the starting materials.

At 975°C a mechanical micromixture run, HS2/HF2-4, produced SLI with relatively restricted compositional ranges for the high-Si and high-Fe liquids in a well-crystallized matrix (Fig. 6b, solid squares). Its companion run, HS2/HF2-3, a fused micromixture produced easily visible regions of SLI (50-100 μm) with sharp contacts, but more variable in composition and having a much smaller compositional separation (open squares). As with the other fused micromixture run, HS1/HF1-5, plagioclase was not observed. Inasmuch as there is little compositional convergence between these two runs, it is difficult to judge which run represents the closer approach to equilibrium. Given that the separation in HS2/HF2-4 (solid squares) is virtually identical to that in the 990°C run (Fig. 6a), that the liquids in HS2/HF2-4 are relatively homogeneous while those in HS2/HF2-3 are not, and that plagioclase failed to nucleate from the fused mixture, I believe that HS2/HF2-4 achieved the closer approach to equilibrium.

Solid Solution

In theory, a straightforward test of equilibrium in immiscible liquid pairs is to compare the compositions of common phases. In practice, this test has not proven feasible: Phases with potentially wide ranges of composition, such as plagioclase and pyroxene, are generally too small in the high-Si liquids to probe reliably, whereas the phases that are generally large enough to be probed, such as silica and ilmenite, have restricted ranges of composition. Figure 7 illustrates part of this dilemma in run HS2/HF2-1. In Fig. 7, wt.% Na_2O represents plagioclase composition, whereas wt.% FeO is a measure of contamination of the analysis either with glass or pyroxene. Larger plagioclase crystals in the high-Fe layer have a very restricted range of FeO contents near 2%, whereas those in the high-Si layer have a wide range of FeO contents. Extrapolation of the plagioclase trend developed in the high-Si zone to 1.5 wt.% FeO (to allow

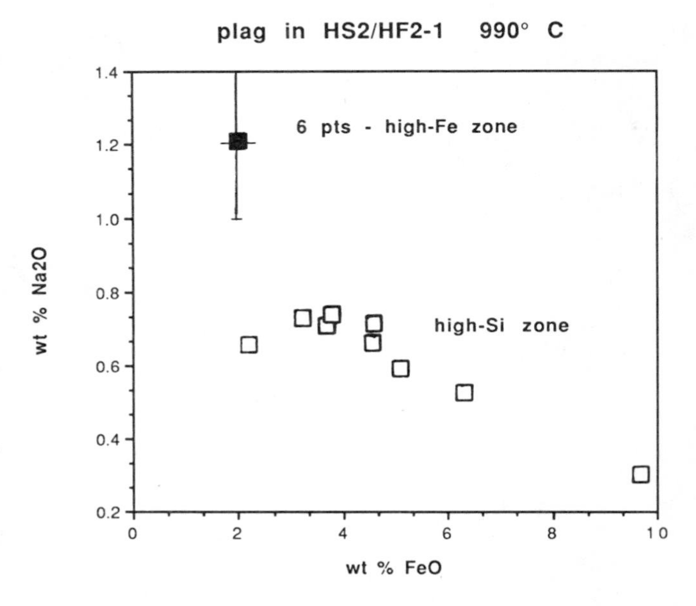

Fig. 7. Comparison of plagioclase compositions from high-Fe and high-Si portions of run HS2/HF2-1 (990°).

for secondary fluorescence of Fe in plagioclase; *Longhi et al.*, 1976) indicates that ~0.8 wt.% is the average concentration of Na_2O in plagioclase in the high-Si zone. This value is distinctly less than the average of 1.2% Na_2O observed in plagioclase from the high-Fe zone. However, not only do none of the analyzed feldspars lie within the immiscible droplets, but all of the analyzed crystals lie more than 100 μm from the artificial two-liquid interface where SLI developed. It is thus possible that one (or both) of these groups of plagioclase crystal grew before the immiscibility developed and has not reequilibrated completely. Given the facts that stable two-liquid interfaces develop by both unmixing and convergence and that the same phases are present on both sides of the interfaces, incomplete crystal-liquid equilibrium seems rather more likely than complete disequilibrium between the liquids.

DISCUSSION

The compositions employed in this study are not exact duplicates of natural lunar magmas, in that they lack certain high field-strength elements such as Zr and REE that are likely to enhance silicate liquid immiscibility (*Ryerson and Hess,* 1978). Nonetheless the experiments described above demonstrate that SLI may develop stably in derivatives of mare basalt and KREEP-rich liquids in the presence of plagioclase. Thus the controlled-cooling-rate experiments of *Rutherford et al.* (1974) and *Hess et al.* (1975) were substantially correct analogs of the development of SLI during natural fractional crystallization. In detail, however, these experiments do appear to have produced liquids metastable with regard to the crystallization of silica minerals. No mention of silica minerals is made in their run tables, whereas euhedral silica minerals are ubiquitous in the isothermal runs of the present study. Although it is possible that the silica minerals were present in the runs of *Rutherford et al.* (1974) and *Hess et al.* (1975) but were simply not observed, the compositions of liquids produced in this study are lower in Sil component than their closest analogs in the *Hess et al.* (1975) study (Figs. 4 and 6), despite similar assemblages of other minerals (plag, ilm, pyx, ol). This difference is consistent with the crystallization of silica minerals in the isothermal experiments and their failure to crystallize in the controlled-cooling-rate experiments. Given that silica minerals are nearly ubiquitous constituents of the mesostases of mare and KREEP basalts, the absence of silica minerals in the controlled-cooling-rate experiments cannot simply be explained by the presence of minor amounts of Zr and REE somehow destablizing silica. Rather the controlled-cooling-rate experiments appear to be prone to metastability, at least with respect to silica crystallization. Careful reexamination of the run products of the controlled-cooling-rate experiments will resolve the issue.

Experiments from this study also show that run times on the order of a week are not sufficient to homogenize high-Si/high-Fe heterogeneities on the scale of tens of microns (the

TABLE 4. Summary of two-liquid data.

Run	T(°C)	P_2O_5* (wt.%)	*Mg'**	NOR*	QPl* (×100)	Ol [Wo-Pl]* (×100)	ΔOl (obsv.)	ΔOl (calc.)
HS1/HF1-2	1053	6.34	0.320	0.160	33.0	60.5	<0	2.3
HS1/HF1-1	1049	4.80	0.270	0.312	31.5	38.2	19.7	29.3
HS1/HF1-5,6	1041	2.10†	0.215†	0.215†	31.7†	37.6†	12.0†	6.2
HF1-2	1023	6.56	0.156	0.152	30.2	74.6	63.9	52.5
HS2/HF2-1	990	3.19	0.037	0.254	21.9	61.1	53.2	55.8
HS2/HF2-4	975	2.92	0.014	0.269	21.3	60.4	53.3	60.1
MKM8-3[a]	1025	0.0	0.035	0.0	28.6	67.2	<0	-0.9
QMD-9[b]	1050	0.28	0.529	0.518	33.8	6.0	<0	-152
FAN(ilm)[c]	1005	0.40	0.015	0.080	30.2	67.2	-	12.1
FAN(wht)[d]	895	3.98	<0.005	0.390	34.6	34.4	-	112
15386(aug)[e]	1025	2.36	0.228	0.325	29.6	44.5	-	-2.3
15386(wht)[f]	931	3.83	0.106	0.390	34.4	33.0	-	75.9
MN(ilm)[g]	989	2.43	0.176	0.345	29.2	39.7	-	22.0

$$\Delta Ol = 15.7P_2O_5 - 339Mg' - 1.18Ol[Wo\text{-}Pl] + 90.4 \quad (R^2 = 0.927)$$
$$(3.0) \quad (63) \quad (0.46) \quad (27.1)$$

*Data from high-Fe liquid.
†Average values.
( ) = standard error.
a = included in regression (*Longhi and Pan,* 1988).
b = not in regression (*Rutherford et al.,* 1976).
c = calculated residual liquid to ferroan anorthosities at 94% crystallization (*Longhi,* 1989b).
d = calulated residual liquid to ferroan anorthosities at 99% crystallization (*Longhi,* 1989b).
e = calculated residual liquid to 15386 at 72% crystallization (*Longhi,* 1989b).
f = calculated residual liquid to 15386 at 84% crystallization (*Longhi,* 1989b).
g = calculated residual liquid to Mg-norite magma at 89% crystallization (*Longhi,* 1989b).
NOR = $2K_2O/(Al_2O_3 + Na_2O + K_2O)$ - mole fractions.
QPl = $8(Al_2O_3 + Na_2O - K_2O)/(2SiO_2 + MgO + FeO + MnO + CaO + 3Al_2O_3 + Na_2O + K_2O)$.

size of particles in the runs with mechanical mixtures of high-Si and high-Fe glasses). Thus, once produced, immiscible liquid pairs should prove difficult to react away on the timescale of normal laboratory experiments at low temperature (<1050°C) or during the cooling of thick flows and shallow intrusions. Also, the failure of plagioclase to nucleate from fused mixtures of high-Si and high-Fe glasses suggests that SLI may develop before plagioclase nucleates in hybrid liquids produced by assimilation of granitic material.

Having established that SLI is not a metastable feature in natural systems, the next step is to quantify the compositional bounds of the natural two liquid field. Table 4 summarizes some of the important chemical parameters of immiscible liquids from this study (all of the parameters pertain to the high-Fe liquid). I have used these data in an attempt to parameterize the multisaturated (plag + sil + ilm + ol/lpyx) portion of the two-liquid field. From the fact that the only ternary system that has a central field of immiscibility is the Fa-Lc-Sil system (*Roedder,* 1951), it is reasonable to expect that Fe and K components are likely to be important variables controlling the development of SLI. Also, the work of *Ryerson and Hess* (1978, 1980) and *Visser and Koster van Groos* (1979b) has demonstrated that increasing P_2O_5 enhances immiscibility, whereas the phase diagram (Fig. 1b) produced by *Irvine* (1976) demonstrates that increasing plagioclase component surpresses immiscibility. These components are represented as Ol[Wo-Pl] (the olivine coordinate in Figs. 4 and 6), *Mg',* NOR (the orthoclase fraction of the normative feldspar), wt.% P_2O_5, and QPl (the Pl coordinate of the Ol-Pl-Or-Wo-Qtz system) respectively. I have taken the difference in the projected Ol[Wo-Pl] coordinates of the immiscible liquid pairs in Fig. 4 and 6 as a convenient measure of unmixing. In order to provide some limits on the extent of immiscibility, I have included data from two runs in which SLI was not observed. One is from this study (HS1/HF1-2); the other, MKM8-3 (*Longhi and Pan,* 1988), is silica-saturated (+plag+ol) and has very low *Mg',* but no P_2O_5 or K_2O. Table 4 lists the results of one of the multiple regression calculations carried out with ΔOl as the dependent variable. The variables *Mg',* P_2O_5, and Ol[Wo-Pl] account for most of the variation, as shown. Surprisingly, NOR does not have a statistically significant control on ΔOl when introduced at any stage of the regression. On the other hand, QPl is marginally significant but when included its coefficent has a positive sign, erroneously suggesting that increasing plagioclase component enhances immiscibility. Consequently, I have rejected QPl as a viable variable for the present dataset. Clearly, the statistical insignificance of K indicates that the database employed in the regression is incomplete. I attempted to incorporate data from simpler systems (e.g., *Irvine,* 1976; *Visser and Koster van Groos,* 1979b); however, these data produced regressions with much higher overall uncertainties and poor fits to the data from this study, so I have not presented them. Also excluded from the data base are most runs that have not unmixed because the regression requires negative values of ΔOl for such runs and, therefore, they cannot be included in the regression unless they are fortuitously on the verge of unmixing (ΔOl ~ 0). For example, the regression equation predicts a ΔOl of -91.4 for an unmixed liquid saturated with pyroxene, plagioclase, and K-feldspar (*Rutherford et al.,* 1976). Inclusion of this run in the regression with ΔOl = 0 worsens the fit considerably (R^2 = 0.8). On the other hand run MKM8-3 is apparently close to the two-liquid field, so I left it in the regression. Despite the drawbacks of a limited compositional database, the regression equation nonetheless represents a first step toward quantifying the two-liquid field in lunar liquids and some of its predictions are instructive.

First, compare the predicted ΔOl for the runs HS1/HF-1 (29.3), which gained iron, and HS1/HF-2 (2.3), which was run at a similar temperature and did not develop a stable two-liquid interface. The calculation overestimates the observed separations, 19.7 and ≤0 respectively, but preserves the sense of the difference between the ΔOl values; also, given that ΔOl in HS1/HF1-2 is negative, but probably close to zero, the calculation probably comes within 50% of predicting the magnitude of the difference. Analysis of the contributions of the various terms in the regression equation shows that an important reason for a stable two-liquid interface in HS1/HF-1, but not in HS1/HF-2 despite its higher P_2O_5 concentration, is the difference in *Mg'* caused by the iron gain in HS1/HF-1. Similarly, the redevelopment of a moderate two-liquid separation in HS1/HF1-5 and -6 and very extensive separation in HF1-2 correlates well with increasing P_2O_5 and decreasing *Mg'.* Higher concentrations of P_2O_5 in KREEP basalts are an important reason why KREEP compositions develop SLI at higher temperatures than mare basalts (e.g., *Hess et al.,* 1975, 1978). However, alkalis play an indirect role as well. It is well known that increasing alkalis shift olivine/plagioclase and pyroxene/plagioclase cotectics away from the Ol component. Thus the higher alkali contents of the KREEP compositions contribute to lower Ol[Wo-Pl] coordinates of cotectic, residual liquids that work in concert with higher P_2O_5 to stabilize SLI at higher temperatures than in mare basalts.

Second, when applied to late-stage differentiates of the lunar "magma ocean" (or at least the magmas that crystallized the ferroan anorthosites), the regression equation predicts stable SLI as suggested by *Neal and Taylor* (1989a). Table 4 lists compositional data for two calculated liquids residual to the anorthosite parent magma (*Longhi,* 1989b). The one with higher *Mg'* represents the first appearance of ilmenite (>94 vol.% crystallization) and the other represents the residual liquid at the first appearance of whitlockite (>99% crystallization). The former liquid is marginally within the two-liquid field, whereas the latter is almost certainly within the two-liquid field. That SLI occurred during the late stages of earliest lunar differentiation now seems likely. Whether SLI played an important petrogenetic role still is problematical. *Hess* (1989) warned that unless there is efficient separation of the two liquids, continued crystallization would consume the high-Fe liquid and eventually produce a granitic liquid with complementary mafic cumulates. This liquid and mass of crystals could be related by fractional crystallization without necessarily appealing to SLI. Such a process may explain why typical KREEP, which developed its incompatible element ratios during extreme fractional crystallization (*Warren and Wasson,* 1979; *Warren,* 1988), has a K_2O/P_2O_5 ratio (1.2) that is well within the range (0.4-1.6) of low-Ti mare basalts (*Cuttita et al.,* 1971, 1973), whose source regions developed as cumulates from less extremely fractionated liquids (e.g., *Taylor,* 1982). According to the calculations of fractional crystallization, any immiscible granitic liquids that did develop during the formation of *ur*KREEP would have mafic minerals with *Mg'* <0.02 (cf. Fig. 1, *Longhi,* 1989b) and thus would have been quite different from most Apollo14 granites, such as 14303,204 (*Warren et al.,* 1983) which has *Mg'* in augite ~0.5.

Third, the regression equation predicts SLI for differentiates of KREEP basalt magmas such as 15386 (in agreement with the experiments of *Hess et al.,* 1989) and differentiates of hypothetical Mg-norite parent magmas (*Longhi,* 1989b). These calculated liquids would crystallize mafic minerals with values of *Mg'* similar to those in typical Apollo 14 granites. Whether or not SLI played a role in the petrogenesis of particular granites is beyond the scope of this study, but it is clear that development of SLI would have been likely in KREEP basalt and Mg-norite plutons.

CONCLUSIONS

Isothermal crystallization experiments show silicate liquid immiscibility to be a stable phenomenon in residual lunar liquids saturated with plagioclase. However, unmixing and equilibration are sluggish at temperatures in the range of 975°C to 1050°C, so if a metastable two-liquid pair should develop, it could easily persist despite an annealing time of a week. Natural immiscibility is likely to produce even larger compositional separations than those observed in this study because of the presence in natural liquids, particularly KREEP-rich varieties, of a few tenths of a percent of high field-strength elements, such as Zr and REE, that enhance unmixing. The present results indicate that controlled-cooling-rate experiments (*Rutherford et al.,* 1974; *Hess at al.,* 1975, 1978) were substantially correct analogs of the natural process of liquid immiscibility. However, the ubiquitous presence of silica minerals in the isothermal experiments and their absence in the controlled-cooling-rate experiments suggest that the controlled-cooling-rate experiments were metastable with respect to silica crystallization.

An attempt to parameterize the two-liquid field by regressing the compositional separation against various chemical parameters shows that P_2O_5 concentration, *Mg',* and olivine and plagioclase contents of the mafic liquid are significant variables, whereas K content apparently is not. Increasing P_2O_5 increases two-liquid separation, whereas increasing *Mg'* and olivine and plagioclase contents tend to diminish it. A combination of high P_2O_5 and low Ol content in residual KREEP liquids helps to explain why immiscibility develops at higher temperatures in KREEP compositions than in mare basalts (*Hess et al.,* 1975). Application of the regression equation to model liquids (*Longhi,* 1989b) residual to ferroan anorthosites and possibly parental to *ur*KREEP leads to the conclusion that SLI very probably developed during the course of extreme fractional crystallization following the crystallization of ferroan anorthosites, much as *Neal and Taylor* (1989b) suggested. However, because the K_2O/P_2O_5 ratio in KREEP is within the range of values in low-Ti mare basalts, whose source regions developed through less extreme frationation, it seems likely that SLI did not have much of an effect on the composition of *ur*KREEP. Although SLI may have played a more important role in the petrogensis of some lunar "granites" (*Taylor et al.,* 1979; *Warren et al.,* 1983; *Neal and Taylor,* 1989b), these rocks have mafic minerals with intermediate values of *Mg'*, which precludes them from having formed as derivatives of the extreme differentiation process that produced *ur*KREEP because mafic minerals crystallized from *ur*KREEP would have had extremely low *Mg'* values. It is rather more likely that the granitic rocks are the plutonic residua of KREEP-basalt and high-Mg magmas.

Acknowledgments. Thoughtful and informative reviews by P. C. Hess and C. R. Neal helped to improve this paper considerably. This research was supported by NASA grant NAG-9-93. Lamont-Doherty Geological Observatory Contribution #4540.

REFERENCES

Cuttitta F., Rose H. J. Jr., Annell C. S., Carron M. K., Christian R. P., Dwornik E. J., Greenland L. P., Helz A. W., and Ligon D. T. Jr. (1971) Elemental composition of some Apollo 12 lunar rocks and soils. *Proc. Lunar Sci. Conf. 2nd,* pp. 1217-1129.

Cuttitta F., Rose H. J. Jr., Annell C. S., Carron M. K., Christian R. P., Ligon D. T. Jr., Dwornik E. J., Wright T. L., and Greenland L. P. (1973) Chemistry of twenty-one igneous rocks and soils returned by the Apollo 15 mission. *Proc. Lunar Sci. Conf. 4th,* pp. 1081-1096.

Dickinson J. E. and Hess P. C. (1982) Whitlockite saturation in lunar basalts (abstract). In *Lunar and Planetary Science XIII,* pp. 172-173. Lunar and Planetary Institute, Houston.

Dymek R. F., Albee A. L., and Chodos A. A. (1975) Comparative mineralogy and petrology of Apollo 17 mare basalts: samples 70215, 71055, 74255, and 75055. *Proc. Lunar Sci. Conf. 6th,* pp. 49-77.

Hess P. C. (1989) Highly evolved liquids from the fractionation of mare and nonmare basalts. In *Workshop on Moon in Transition: Apollo 14, KREEP, and Evolved Pristine Rocks* (G. J. Taylor and P. H. Warren, eds.), pp. 46-52. Lunar and Planetary Institute, Houston.

Hess P. C., Rutherford M. J., Guillemette R. N., Ryerson F. J., and Tuchfeld H. A. (1975) Residual products of fractional crystallization in lunar magmas: An experimental study. *Proc. Lunar Sci. Conf. 6th,* pp. 895-910.

Hess P.C., Rutherford M.J., and Campbell H.W. (1978) Ilmenite crystallization in non-mare basalt: Genesis of KREEP and high-Ti basalt. *Proc. Lunar Planet. Sci. Conf. 9th,* pp. 705-724.

Hess P. C. Horzempa P., and Rutherford M. J. (1989) Fractionation of Apollo 15 KREEP basalts (abstract). In *Lunar and Planetary Science XX,* pp. 408-409. Lunar and Planetary Institute, Houston.

Irvine T. N. (1976) Metastable liquid immiscibility and MgO-FeO-SiO_2 fractionation patterns in the system Mg_2SiO_4-Fe_2SiO_4-$CaA1_2Si_2O_8$-$KA_1Si_3O_8$-SiO_2. *Carnegie Inst. Wash. Yearb., 75,* pp. 597-611.

Keil K., Prinz M., and Bunch T. E. (1971) Mineralogy, petrology, and chemistry of some Apollo 12 samples. *Proc. Lunar Sci. Conf. 2nd,* pp. 319-341.

Longhi J. (1981) Preliminary modeling of high pressure partial melting: implications for early lunar differentiation *Proc. Lunar Planet. Sci. 12B,* pp. 1001-1018.

Longhi J. (1989a) Is natural silicate liquid immiscibility metastable? (abstract). In *Lunar and Planetary Science XX,* pp. 586-587. Lunar and Planetary Institute, Houston.

Longhi J. (1989b) Fractionation trends of evolved lunar magmas (abstract). In *Lunar and Planetary Science XX,* pp. 584-585. Lunar and Planetary Institute, Houston.

Longhi J. and Pan V. (1988) A reconnaissance study of phase boundaries in low-alkali basaltic liquids. *J. Petrol., 29,* 115-148.

Longhi J., Walker D., and Hays J. F. (1976) Fe and Mg in plagioclase. *Proc. Lunar Sci. Conf. 7th,* pp. 1281-1300.

Nashlund H. R. (1976) Liquid immiscibility in the system $KA_1Si_3O_8$-$NaA_1Si_3O_8$-FeO-Fe_2O_3-SiO_2 and its application to natural magmas. *Carnegie Inst. Wash. Yearb., 75,* pp. 592-597.

Neal C. R. and Taylor L. A. (1989a) The nature of barium partitioning between immiscible melts: A comparison of experimental and natural systems with reference to lunar granite petrogenesis. *Proc. Lunar Planet. Sci. Conf. 19th,* pp. 209-218.

Neal C. R. and Taylor L. A. (1989b) Metasomatic products of the lunar magma ocean: The role of KREEP dissemination. *Geochim. Cosmochim. Acta, 53,* 529-541.

Powell B. N., Aitken F. K., and Weiblen P. W. (1973) Classification, distribution, and origin of lithic fragments from the Hadley-Apennine region. *Proc. Lunar Sci. Conf. 4th,* pp. 445-460.

Ridley W. I., Brett R., Williams R. J., Takeda H., and Brown R. W. (1972) Petrology of Fra Mauro basalt 14310. *Proc. Lunar Sci. Conf. 3rd,* pp. 159-170.

Roedder E. W. (1951) Low temperature liquid immiscibility in the system K_2O-FeO-$A1_2O_3$-SiO_2. *Am. Mineral., 36,* 282-286.

Rutherford M. J., Hess P. C., and Daniel G. H. (1974) Experimental liquid line of descent and liquid immiscibility for basalt 70017. *Proc. Lunar Sci. Conf. 5th,* pp. 569-583.

Rutherford M. J., Hess P. C., Ryerson F. J., Campbell H. W., and Dick P. A. (1976) The chemistry, origin and petrogenetic implications of lunar granite and monzodiorite. *Proc. Lunar Sci. Conf. 7th,* pp. 1723-1740.

Ryerson F. J. and Hess P. C. (1978) Implications of liquid-liquid distribution coefficients to mineral-liquid partitioning. *Geochim. Cosmochim. Acta, 42,* pp. 921-932.

Ryerson F. J. and Hess P. C. (1980) The role of P_2O_5 in silicate melts. *Geochim. Cosmochim. Acta, 44,* pp. 611-624.

Shervais J. W., Taylor L. A., Laul J. C., Shih C.-Y., and Nyquist L. E. (1985) Very high potassium (VHK) basalt: complications in mare basalt petrogenesis. *Proc. Lunar Planet. Sci. Conf. 16th,* in *J. Geophys. Res., 90,* D3-D18.

Taylor G. J., Warner R. D., Keil K., Ma M.-S., and Schmitt R. A. (1979) Silicate liquid immiscibility, evolved lunar rocks and the formation of KREEP. *Proceedings of the Conference on the Lunar Highlands Crust* (J. J. Papike and R. B. Merrill, eds.), pp. 339-352. Pergamon, New York.

Taylor S. R. (1982) *Planetary Science: A Lunar Perspective.* Lunar and Planetary Institute, Houston. 481 pp.

Visser W. and Koster van Groos A. F. (1979a) Phase relations in the system K_2O-FeO-Al_2O_3-SiO_2 at 1 atmosphere with special emphasis on low temperature liquid immiscibility. *Am. J. Sci., 279,* 70-91.

Visser W. and Koster van Groos A. F. (1979b) Effects of P_2O_5 and TiO_2 on liquid-liquid equilibria in the system K_2O-FeO-Al_2O_3-SiO_2. *Am. J. Sci., 279,* 970-988.

Walker D., Longhi J., Grove T. L., Stolper E. M., and Hays J. F. (1973) Experimental petrology and origin of rocks from the Descartes Highlands. *Proc. Lunar Sci. Conf. 4th,* pp. 1013-1032.

Warren P. W. (1988) KREEP: Major-element diversity, trace-element uniformity (almost). In *Workshop on Moon in Transition: Apollo 14, KREEP, and Evolved Lunar Rocks* (G. J. Taylor and P. H. Warren, eds.), pp. 106-110. Lunar and Planetary Institute, Houston.

Warren P. W. and Wasson J. T. (1979) The origin of KREEP. *Rev. Geophys. Space Phys., 17,* 73-88.

Warren P. W., Taylor G. J., Keil K., Shirley D. N., and Wasson J. T. (1983) Petrology and chemistry of two "large" granite clasts from the Moon. *Earth Planet. Sci. Lett., 64,* 175-185.

Watson E. B. (1976) Two-liquid partition coefficients: experimental data and geochemical implications. *Contrib. Mineral. Petrol., 56,* 119-134.

Watson E. B. and Naslund H. R. (1977) The effect of pressure on liquid immiscibility in the system K_2O-FeO-Al_2O_3-SiO_2-CO_2. *Carnegie Inst. Wash. Yearb., 76,* pp. 410-414.

Proceedings of the 20th Lunar and Planetary Science Conference, pp. 25-30
Lunar and Planetary Institute, Houston, 1990

Europium Anomalies in Mare Basalts as a Consequence of Mafic Cumulate Fractionation from an Initial Lunar Magma

J. G. Brophy and A. Basu

Department of Geology, Indiana University, Bloomington, IN 47405

A hypothesis is tested that partial melting of mafic mineral cumulates derived from the early crystallization of a lunar magma ocean or a magma pocket, large or small, can yield liquids with negative europium anomalies (Eu/Eu^*) similar to those observed in lunar mare basalts. The magnitudes of Eu/Eu^* in different solids and liquids as they separate from an initial lunar magma body have been calculated using current experimental data on solid/liquid partition coefficients at low oxygen fugacities. A negative Eu anomaly develops in a pyroxene-rich cumulate from an initial lunar magma, which is retained in partial melts derived from this cumulate. The magnitudes of calculated anomalies agree with observed Eu/Eu^* of lunar mare basalts if the cumulate consists of nearly 100% Ca-poor pyroxene. This process requires the initial magma to be extremely enriched in Eu (and by inference in all REE) unless the mare basalts were produced from about <5% partial melting of the Ca-poor pyroxene cumulates, which in turn must also have accounted for about 90% or more of the initial magma. If Ca-rich pyroxenes were to represent the cumulates of the initial magma, more than 50% partial melting is required to fit observed Eu/Eu^* values of common mare basalts and realistic estimates of REE abundances of the initial magma. Additional mechanisms, such as plagioclase flotation in a global magma ocean, may be necessary to explain Eu/Eu^* and Eu abundance in common lunar rocks.

INTRODUCTION

The complementary patterns of rare earth element (REE) distribution and europium anomalies (Eu/Eu^*, where Eu = combined total of Eu^{+3} and Eu^{+2} present in a sample, and, Eu^* = estimated amount of Eu in the sample had it been all Eu^{+3}, i.e., amount of Eu in a sample with no Eu anomaly) in the old anorthositic crustal material and in the younger mare basalts in the Moon are well established. A simple model of plagioclase flotation and mafic cumulate settling may explain the observed patterns if there were to be an initial magma ocean and if the mafic cumulates were to be the source regions of the mare basalts (*Wood et al.,* 1970; *Wood,* 1975; *Taylor,* 1982). Conversely, the REE patterns and Eu/Eu^* could be used as evidence for this model. However, the Eu^{+2}/Eu^{+3} ratio in a magma is dependent on, and is enhanced, by low oxygen fugacity (*Towell et al.,* 1965); and lunar mare basalt compositions indicate very low oxygen fugacities (*Sato,* 1976; *Sato et al.,* 1973). Further, the crystal structures of pyroxenes are such that they exclude Eu^{+2} relative to Eu^{+3} and any pyroxene-rich cumulate would acquire negative Eu/Eu^* (*Papike et al.,* 1988). Therefore, *Shearer and Papike* (1989) have proposed that a "combined crystal chemical -$^fO_2$ effect . . . rather than prior removal of plagioclase" may have produced the negative Eu/Eu^* in the source regions of mare basalts.

The purpose of this work is to quantitatively calculate and examine both Eu/Eu^* and total Eu abundances in partial melts of mafic cumulates that could have settled from an initial lunar magma body. For the sake of convenience and convention we call this the lunar magma ocean (MO). We have made some model calculations to trace the fate of Eu/Eu^* and total Eu in different solids and liquids as they separate from an initial magma ocean and subsequently undergo partial melting to give rise to younger basaltic magmas in the moon. We have then compared our model results with observed values of both Eu/Eu^* and total Eu in common mare basalts and pristine glasses to evaluate the recent proposal of *Shearer and Papike* (1989) and *Shearer et al.* (1989).

MODEL CALCULATIONS

For modeling purposes we have assumed that perfect fractional crystallization of olivine, orthopyroxene, and/or clinopyroxene (e.g., *Hodges and Kushiro,* 1974; *Drake and Consolmagno,* 1976; *Jones and Delano,* 1989) from a global magma ocean formed cumulates that subsequently underwent batch (equilibrium) partial melting.

$(Eu/Eu^*)_{sol}$ values and Eu abundances in the solid cumulate have been calculated from

$$(Eu/Eu^*)_{sol} = (Eu/Eu^*)_{MO}\,(D_{Eu}/D_{Eu}{}^{+3})\,(1\text{-}F)^{(D_{Eu}\text{-}D_{Eu}{}^{+3})} \quad (1)$$

or

$$(Eu/Eu^*)_{sol} = (Eu/Eu^*)_{MO}\,(D_{Eu}/D_{Eu}{}^{+3})\,(1\text{-}F)^{\text{-}\Delta X_{Eu}{}^{+2}}$$

and

$$Eu_{sol} = Eu_{MO}D_{Eu}\,(1\text{-}F)^{(D_{Eu}\text{-}1)} \quad (2)$$

where $(Eu/Eu^*)_{MO}$ = initial ratio value in the MO; Eu_{MO} = initial Eu abundance in the MO; $D_{Eu} = (X_{Eu}{}^{+2})(D_{Eu}{}^{+2}) + (1\text{-}X_{Eu}{}^{+2})(D_{Eu}{}^{+3})$; $X_{Eu}{}^{+2}$ = fraction of total Eu present as Eu^{+2} in the liquid; $D_{Eu}{}^{+2}$ = bulk solid/liquid partition coefficient for Eu^{+2}; $D_{Eu}{}^{+3}$ = bulk solid/liquid partition coefficient for Eu^{+3}; $\Delta = D_{Eu}{}^{+3} - D_{Eu}{}^{+2}$; and F = fraction of solids crystallized. Values of $D_{Eu}{}^{+2}$ and $D_{Eu}{}^{+3}$ used in the modeling are listed in Table 1

TABLE 1. Partition coefficients used for the model calculations.

Mineral	D_{Eu}^{+3}	D_{Eu}^{+2}
Olivine	0.001	0.001
Orthopyroxene	0.029	0.011
Clinopyroxene	0.600	0.330

and justified in Appendix A. $(Eu/Eu^*)_{liq}$ values and Eu abundances in cumulate partial melts have been calculated from

$$(Eu/Eu^*)_{liq} = (Eu/Eu^*)_{sol} \{F_l + D_{Eu}^{+3}(1-F_l)\}/\{F_l + D_{Eu}(1-F_l)\} \quad (3)$$

and

$$Eu_{liq} = Eu_{sol}/(F_l + D_{Eu}(1 - F_l) \quad (4)$$

where $(Eu/Eu^*)_{sol}$ is from equation (1), Eu_{sol} is from equation (2), and, F_l = the fraction of melt formed. Unless otherwise specified, we have assumed that $(Eu/Eu^*)_{MO} = 1.0$, and that $X_{Eu}^{+2} = 0.7$ (*Schreiber,* 1977) in both the initial magma ocean and all initial partial melts of cumulates.

RESULTS

Variation in cumulate $(Eu/Eu^*)_{sol}$ with increasing degrees of MO crystallization for the three end-member cases of monomineralic olivine, orthopyroxene, and clinopyroxene crystallization are summarized in Fig. 1. The magnitude of $(Eu/Eu^*)_{sol}$ in the cumulate lies somewhere between the upper and lower bounds set by the three end-member cases depending upon the actual mineral proportions crystallizing from the magma body. A 100% olivine cumulate exhibits no europium anomaly. A 100% orthopyroxene cumulate exhibits a negative anomaly with a magnitude $[(Eu/Eu^*)_{sol} = 0.57\text{-}0.59]$ that is largely independent of the degree of magma ocean crystallization. A 100% clinopyroxene cumulate displays a moderately negative anomaly $[(Eu/Eu^*)_{sol} <0.7]$ at small amounts of crystallization (<10%). The anomaly dissappears $[(Eu/Eu^*)_{sol} = 1.0]$ at 87% crystallization and becomes positive $[(Eu/Eu^*) >1]$ for larger degrees of crystallization. For example, at 98% crystallization, $(Eu/Eu^*)_{sol}$ equals 1.4.

Figure 2 shows the variation of $(Eu/Eu^*)_{liq}$ in the subsequent partial melt as a function of the degree of cumulate melting. As in Fig. 1, results are shown only for the three end-member cases of a monomineralic olivine, orthopyroxene, and clinopyroxene cumulate. Partial melting results are a function of the initial value of $(Eu/Eu^*)_{sol}$ in the cumulate, which itself is a function of the degree of MO crystallization. Because any estimate of this parameter is model dependent, we have chosen to show results for the three cases of 10%, 50% and 90% MO crystallization.

As in Fig. 1, melts obtained by batch partial melting of different proportions of cumulate olivine, orthopyroxene and clinopyroxene have $(Eu/Eu^*)_{liq}$ values somewhere between the three end-member extremes. For an initial cumulate that is 100% olivine, $(Eu/Eu^*)_{liq} = 1.0$ for the partial melt regardless of the degree of MO crystallization or degree of cumulate melting. Partial melts of a 100% orthopyroxene cumulate display $(Eu/Eu^*)_{liq}$ values that decrease sharply from 0.77 to 0.60 in the first 20% melting and then drop further, at a lower rate, to 0.58 at 98% melting. $(Eu/Eu^*)_{liq}$ in partial melts of a 100% clinopyroxene cumulate is strongly dependent on the degree of initial MO crystallization. Cumulates that represent small (<10%) to intermediate (50%) degrees of crystallization yield $(Eu/Eu^*)_{liq} = 1.0\text{-}1.2$ in the early stages of melting (<5-20%). $(Eu/Eu^*)_{liq}$ decreases to 0.7-0.8 at 98% melting. Clinopyroxene cumulates formed by greater than 87% MO crystallization yield partial melts with $(Eu/Eu^*)_{liq} >1.0$ at all degrees of melting.

Though a general consensus in the lunar science community is that approximately 70% of the Eu in the lunar mare basalts exists as Eu^{+2} (e.g., *Philpotts,* 1970; *Drake,* 1975; *Schreiber,* 1977; *Taylor,* 1982), we have also considered the effect of variable X_{Eu}^{+2} in both the initial MO and subsequent partial melts. X_{Eu}^{+2} in either of the two liquids is controlled by the prevailing oxygen fugacity conditions, with lower oxygen fugacities favoring an increase in X_{Eu}^{+2}. An increase in X_{Eu}^{+2} should lead to smaller bulk Eu partition coefficients for both orthopyroxene and clinopyroxene; this would translate into more pronounced negative Eu anomalies in both the cumulates and their partial melts. It may be noted that this is the basis for the proposal of *Shearer and Papike* (1989) and *Shearer et al.* (1989). Figure 3 shows the calculated effect of varying X_{Eu}^{+2} in both the MO and subsequent partial melts from 0.6 to 1.0. For simplicity, only the results for melting of a 10% MO cumulate are shown. As predicted, an increase in the proportion of Eu^{+2} increases the magnitude of the negative Eu anomaly in partial melts of both clinopyroxene and orthopyroxene cumulates. Smaller amounts of MO crystallization yield nearly identical results while larger amounts of MO crystallization predict smaller negative anomalies (Fig. 3).

We have also calculated the abundances of total Eu in partial melts normalized to an initial, but unspecified abundance of

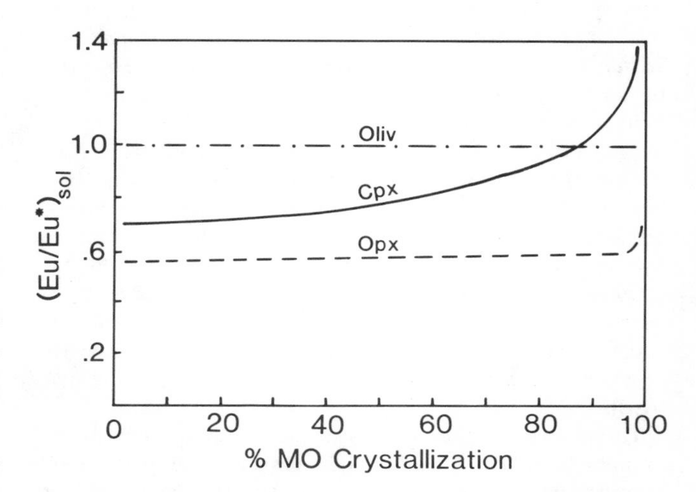

Fig. 1. Calculated values of Eu/Eu* in hypothetical monomineralic olivine, orthopyroxene, and clinopyroxene cumulates formed by fractional crystallization of a magma ocean with initial Eu/Eu* = 1.0 and X_{Eu}^{+2} [i.e., $Eu^{+2}/(Eu^{+2} + Eu^{+3})$] = 0.7.

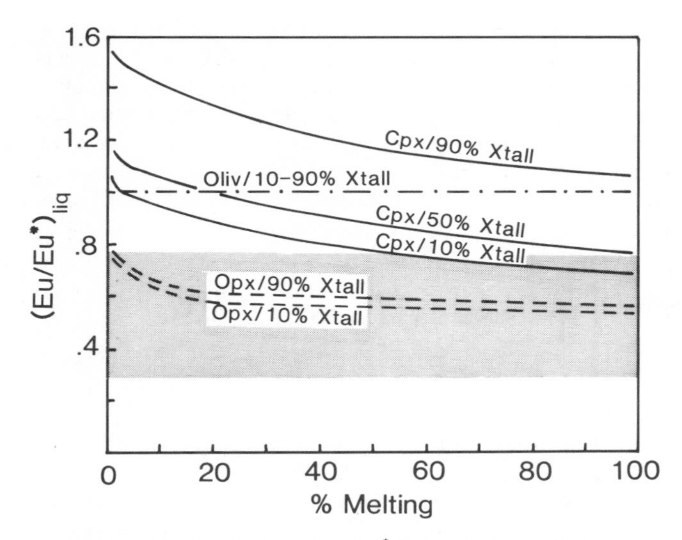

Fig. 2. Calculated values of Eu/Eu* in liquids formed by batch (equilibrium) partial melting of hypothetical monomineralic olivine, orthopyroxene, and clinopyroxene cumulates. $X_{Eu}{}^{+2}$ in the *initial* liquids is assumed to be 0.7. Results are shown for cumulates that represent 10%, 50%, and 90% crystallization of the magma ocean. The shaded region represents the typical ranges of Eu/Eu* observed in common low-titanium, high-titanium and high-alumina mare basalts, and in orange and green glasses (*Taylor,* 1982; *Hughes et al.,* 1989).

Eu in MO. We express this normalized Eu abundance as Eu/Eu_{MO}. Figure 4 shows predicted MO-normalized Eu abundances in partial melts generated under the same conditions as in Fig. 2. A 100% olivine cumulate yields partial melts with very small normalized Eu abundances ($Eu/Eu_{MO} << 1.0$) at modest degrees of partial melting (fraction of liquid > 0.01), regardless of the degree of MO crystallization. For most conditions a 100% orthopyroxene cumulate yields partial melts with similarly small normalized Eu abundances. Larger abundances ($Eu/Eu_{MO} > 1.0$) are predicted for small degrees of melting (e.g., <10%) of a cumulate formed by large amounts of MO crystallization (e.g., >90%). A 100% clinopyroxene cumulate yields partial melts with normalized Eu abundances that depend largely on the initial amount of MO crystallization. Cumulates that represent small to intermediate (<50%) amounts of crystallization yield partial melts with (Eu/Eu_{MO}) of >1.0 during the early stages of melting (<30%), which then diminish to <0.78 at 98% melting. Partial melts of cumulates that represent greater than 87% MO crystallization always predict $Eu/Eu_{MO} > 1.0$.

DISCUSSION

In addition to predicted values of $(Eu/Eu^*)_{liq}$ in cumulate partial melts, Fig. 2 also shows the typical ranges of Eu/Eu* (0.27 to 0.77) measured in common low-Ti, high-Ti and high-Al mare basalts, and in orange and some green glasses (*Basaltic Volcanism Study Project,* 1981; *Taylor,* 1982; *Hughes et al.,* 1989). A comparison of predicted and observed values indicates that partial melting of a 100% olivine cumulate does not, under any circumstances, reproduce observed values of Eu/Eu*. For most conditions, partial melting of a 100% clinopyroxene cumulate does not satisfy the observed values. However, high degrees of melting (>50%) of a clinopyroxene cumulate that represents small (<10%) to intermediate degrees of MO crystallization (<50%) does yield Eu/Eu* values (0.70-0.88) that overlap the upper range of observed values. In contrast, partial melting of a 100% orthopyroxene cumulate predicts Eu/Eu* values (0.57-0.78) that fall within the range of observed values regardless of the degree of MO crystallization or cumulate partial melting. Although we do not consider this to be very realistic, the predicted values in general agree with the conclusions of *Shearer and Papike* (1989) and *Shearer et al.* (1989).

For the *Shearer and Papike* (1989) hypothesis to be fully acceptable, the appropriate conditions of MO crystallization, cumulate mineralogy, and degree of cumulate partial melting that satisfy observed values of Eu/Eu* must also satisfy the observed abundances of total Eu in common mare basalts. A typical range of Eu abundances in both mare basalts and orange glasses is 0.30 to 2.29 ppm (*Taylor,* 1982; *Hughes et al.,* 1989), which translates to chondrite-normalized abundances of 4.35 to 33.19 respectively. Any prediction of absolute Eu abundances is necessarily set by the assumed abundance of Eu present in the initial magma ocean. Although the commonly assumed or modeled composition of an initial lunar magma ocean is <10× chondrite (e.g., *Wood,* 1972; *Jones and Delano,* 1989), we have preferred not to accept any one value as starting point. Instead, we have chosen the opposite path of predicting the initial magma ocean abundances necessary to explain the observed range of Eu abundances in mare basalts that might have been generated by the partial melting of mafic cumulates from a lunar MO. Therefore, the results illustrated in Fig. 4 have been combined with the range of observed Eu abundances to calculate the required ranges of initial chondrite-normalized Eu abundances in the initial MO. Again, results are

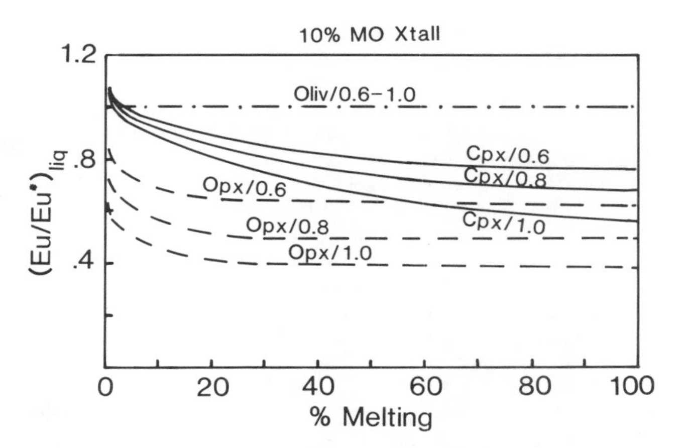

Fig. 3. Calculated effect of variable liquid $X_{Eu}{}^{+2}$ (0.6, 0.8, and 1.0) on the results shown in Fig. 2. For clarity, results are shown only for melting of cumulates that represent 10% crystallization of the magma ocean. Cumulates that represent smaller degrees of initial crystallization yield almost identical results while cumulates that represent greater degrees of initial crystallization predict larger values of magmatic Eu/Eu*.

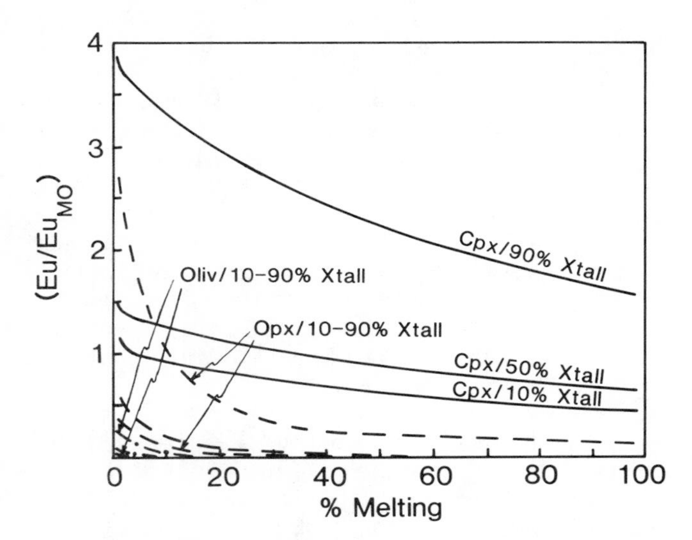

Fig. 4. Calculated abundances of total Eu (Eu^{+2} + Eu^{+3}) in liquids formed by partial melting of hypothetical monomineralic olivine, orthopyroxene, and clinopyroxene cumulates. All abundances are normalized to an unspecified total Eu abundance in the initial magma ocean. $X_{Eu}{}^{+2}$ in the liquids is assumed to be 0.7. Results are shown for 10%, 50%, and 90% crystallization of the initial magma ocean.

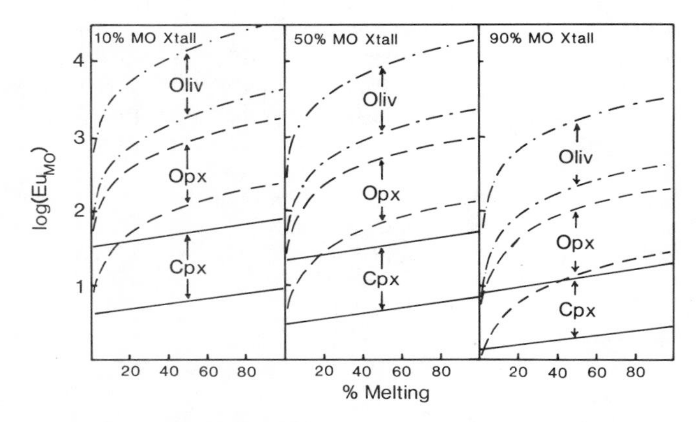

Fig. 5. Calculated range of *required* total Eu abundances in the initial magma ocean necessary to satisfy a typical range of absolute Eu abundances observed in mare basalts (*Taylor,* 1982). Required abundances are shown as a function of the degree of partial melting of hypothetical monomineralic olivine, orthopyroxene, and clinopyroxene cumulates that represent 10%, 50%, and 90% crystallization of the initial magma ocean. Note the log scale of the ordinate and the required high abundance of Eu enrichment in the MO.

shown only for the end-member cases of monomineralic olivine, orthopyroxene, and clinopyroxene cumulates (Fig. 5; note change of scale on the ordinate).

For cumulates that represent up to 50% MO crystallization, we find (1) a 100% olivine cumulate requires Eu/Eu$_{MO}$ of 50×-7,000× chondrite depending on the degree of cumulate melting; (2) a 100% orthopyroxene cumulate requires Eu/Eu$_{MO}$ values of 7×-400× chondrite at <10% melting and 100×-1000× at 90% melting; and, (3) a 100% clinopyroxene cumulate requires Eu/Eu$_{MO}$ values of 5×-50× at <10% melting, and 8×-70× at 90% melting. For a cumulate that represents 90% MO crystallization, we find (1) a 100% olivine cumulate requires Eu/Eu$_{MO}$ values of 10×-500× at <10% melting and 600×-5000× at 90% melting; (2) a 100% orthopyroxene cumulate requires Eu/Eu$_{MO}$ values of 1×-40× at <10% melting and 40×-350- at 90% melting; and, (3) a 100% clinopyroxene cumulate requires Eu/Eu$_{MO}$ values of 2×-10× at <10% melting and 4×-30× at 90% melting.

SUMMARY AND CONCLUSIONS

The results of our model calculations show that indeed a negative europium anomaly will develop in a pyroxene-rich cumulate from an initial lunar magma with $(Eu/Eu^*)_{MO}$ = 1.0, and that a partial melt derived from this cumulate will retain a negative europium anomaly as suggested by *Shearer and Papike* (1989). Our results suggest that an orthopyroxene-rich cumulate can satisfy observed Eu/Eu* values for any degree of initial magma ocean crystallization and any degree of cumulate partial melting. Small (<10%) to intermediate (<50%) degrees of melting of such a cumulate that represents intermediate (~50%) to large (~90%) degrees of MO crystallization respectively, predicts a required MO Eu abundances of around 10× chondrites. A clinopyroxene-rich cumulate can satisfy the upper range of observed Eu/Eu* values provided that the initial degree of magma ocean crystallization is much less than 90% and the degree of cumulate melting is extremely large (>50%). This scenario also predicts MO abundances around 10X chondrites. Thus, our results indicate that the only conditions of mafic cumulate formation and partial melting that are consistent with both observed Eu/Eu* and total Eu abundances are (1) small degrees of melting of an orthopyroxene-rich cumulate that represents intermediate (50%) to large (90%) degrees of MO crystallization respectively and (2) very large degrees of melting of a clinopyroxene-rich cumulate that represents small to moderate degrees of MO crystallization. In our opinion, these conditions are not petrogenetically very realistic, although it is possible that the second condition involving clinopyroxene might be fulfilled if clinopyroxene were a minor phase in the material undergoing extensive partial melting. Additional mechanisms, such as plagioclase flotation in an initial magma ocean, or a magma body, may be necessary to quantitatively explain the observed Eu/Eu* and Eu/Eu$_{MO}$ in lunar mare basalts and volcanic glasses.

APPENDIX A. Eu^{+3}, Eu^{+2} Partition Coefficients

The most important factor controlling the evolution of Eu/Eu* during mafic cumulate formation from an initial magma and subsequent partial melting of the cumulates is the *relative difference* between the Eu^{+3} and Eu^{+2} solid/liquid partition coefficients (D) for olivine, orthopyroxene, and clinopyroxene. Because Eu^{+3} and Eu^{+2} concentrations may not be directly measured, $D_{Eu}{}^{+3}$ is estimated by interpolating between D(Sm) and D(Gd) or D(Tb), and $D_{Eu}{}^{+2}$ is assumed to be equal to D(Sr). Thus, in choosing appropriate $D_{Eu}{}^{+3}$ and $D_{Eu}{}^{+2}$ values for our work we have tried to rely on investigations that simultaneously determined D values for all rare earth elements and Sr. To minimize possible effects of contamination in natural samples, consideration has been further restricted to experimental studies. D_{Eu}^{+3} and D_{Eu}^{+2} pairs of olivine, orthopyroxene, and clinopyroxene, derived from those datasets meeting the stated criteria are listed in Table A1. Also listed for each pair are the absolute difference between D_{Eu}^{+2} and D_{Eu}^{+3}, and the 100×Mg/(Mg+Fe) atomic ratio (mg′) of the host mineral.

The olivine data cover a narrow range of mineral mg′ that falls below that likely to be in equilibrium with a magma ocean mg′ of around 77 a recently suggested by *Jones and Delano* (1989). Within this narrow compositin range, experimental D_{Eu}^{+3} values vary from around 0.01 to as high as 0.23. The recent study of *McKay* (1986) suggests that a value of around 0.001 is more appropriate. This uncertainty aside, the most important feature of the olivine data is that the difference between D_{Eu}^{+3} and D_{E}^{+2} is always small, but may be either positive or negative. This indicates that the olivine structure does not discriminate between Eu^{+3} and Eu^{+2}. For the present study, a value of 0.001 has been assumed for both and D_{Eu}^{+3} and D_{Eu}^{+2}.

TABLE A1. Experimental Eu^{+3}, Eu^{+2} partition coefficients.

Mineral	mg′	D_{Eu}^{+3}	D_{Eu}^{+2}	Δ*	Ref	Comments
Oliv	82	0.227	0.232	-0.005	1	Alkali Ol Basalt 10 kb/ol+gl
Oliv	82	0.108	0.099	0.009	1	Alkali Ol Basalt 10 kb/ol+gl
Oliv	85	0.126	0.134	-0.008	1	Alkali Ol Basalt 10 kb/ol+gl
Oliv	86	0.016	0.010	0.006	2	Ol-Pl-Q Synth. 1 atm/pl+ol+cpx+gl
Oliv	88	0.012	0.003	0.009	3	Synth. Lunar Basalt 1 atm/ol+pl+gl
Opx	88	0.028	0.018	0.010	4	Synth. Lunar Basalt 1 atm/ol+opx+gl
Opx	92	0.015	0.009	0.006	4	Synth. Lunar Basalt 1 atm/ol+opx+pl+gl
Opx	—	0.029	0.011	0.018	5	Synth. Lunar Basalt 1 atm/pig+gl
Cpx	81	0.693	0.301	0.392	6	Hi-Alumina Basalt 30 kb/cpx+gl
Cpx	78	0.739	0.351	0.388	6	Alkali Ol Basalt 20 kb/cpx+gl
Cpx	77	0.662	0.292	0.370	6	Alkali Ol Basalt 20 kb/cpx+ol+gl
Cpx	74	0.654	0.239	0.415	6	Alkali Ol Basalt 20 kb/cpx+ol+gl
Cpx	62	0.615	0.308	0.307	6	Hi-Alumina Basalt 30 kb/cpx+gl
Cpx	72	0.591	0.360	0.231	6	Hi-Alumina Basalt 30 kb/cpx+gl
Cpx	86	0.517	0.296	0.221	6	Alkali Ol Basalt 30 kb/cpx+gl
Cpx	79	0.714	0.348	0.366	6	Alkali Ol Basalt 30 kb/cpx+gl
Cpx	82	0.517	0.260	0.257	1	Picrite 10 kb/cpx+gl
Cpx	84	0.594	0.432	0.162	7	Thol Basalt 20 kb/cpx+gl

*$\Delta = D_{Eu}^{+3} - D_{Eu}^{+2}$.
References: (1) *Shimizu et al.,* 1982; (2) *McKay and Weill,* 1976; (3) *McKay and Weill,* 1977; (4) *Weill and McKay,* 1975; (5) *McKay et al.,* 1989; (6) *Shimizu,* 1980; (7) *Tanaka and Nishizawa,* 1975.

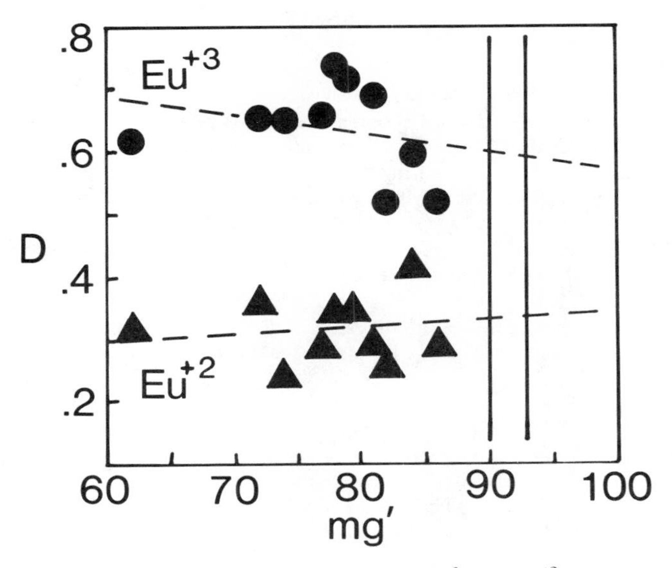

Fig. A1. Experimental clinopyroxene D_{Eu}^{+3} and D_{Eu}^{+2} values as a function of host mineral mg′. The dashed lines represent separate linear regressions of the two datasets. The vertical solid lines indicate the range of mg′ predicted to be in equilibrium with the suggested magma ocean compositions (mg′ = 77) of *Jones and Delano* (1989) assuming a (Fe/Mg) K_D range of 0.25 to 0.40 (*Grove and Baker,* 1984).

The clinopyroxene data span a wide range of mg′s. D_{Eu}^{+3} and D_{Eu}^{+2} values are much higher than for olivine. In Fig. A1 all clinopyroxene data are plotted against mineral mg′. Separate regression lines for D_{Eu}^{+3} and D_{Eu}^{+2} suggest a slight compositional control. A predicted mineral mg′ of 90-93 in equilibrium with the *Jones and Delano* (1989) magma ocean composition yields the adopted D_{Eu}^{+3} and D_{Eu}^{+2} values of 0.60 and 0.33 respectively.

Although two datasets for orthopyroxene (*Weill and McKay,* 1975) fully satisfy the stated requirements, *McKay* (1986) has since demonstrated that these data are unreliable. A more recent study (*McKay et al.,* 1989) reports Sm, Gd, and total Eu (Eu^{+2} + Eu^{+3}) D values for low-Ca pyroxene (pigeonite) crystallized from a synthetic lunar basalt composition at low oxygen fugacities (IW). Based on this latest study, we have adopted D_{Eu}^{+2} and D_{Eu}^{+3} values of 0.011 and 0.029 respectively.

Acknowledgments. We were motivated by the penetrating questions asked by Dr. J. J. Papike at the Workshop on the Moon in Transition (held at the Lunar and Planetary Institute in November 1988). We are grateful to G. McKay for sharing unpublished experimental data. Discussions with L. Haskin, R. Korotev, and D. Lindstrom were very helpful. Detailed reviews by G. McKay and R. Korotev, and the editorial assistance of J. Jones (with impeccable modeling insights), improved an earlier version of the paper. B. Hill and K. Schulte assisted in manuscript preparation. This research was supported in part by NASA grant NSG-9077.

REFERENCES

Basaltic Volcanism Study Project (1981) *Basaltic Volcanism on the Terrestrial Planets.* Pergamon. New York. 1286 pp.

Drake M. J. (1975) The oxidation state of europium as an indicator of oxygen fugacity. *Geochim. Cosmochim. Acta, 39,* 55-64.

Drake M. J. and Consolmagno G. J. (1976) Critical review of models for the evolution of high-Ti mare basalts. *Proc. Lunar Sci. Conf. 7th,* pp. 1633-1658.

Grove T. L. and Baker M. B. (1984) Phase equilibria controls on the tholeiitic versus calc-alkaline differentiation trends. *J. Geophys. Res., 89,* 3253-3274.

Haskin L. A., Frey F. A., Schmitt R. A., and Smith R. H. (1966) Meteoritic, solar, and terrestrial rare-earth distribution. *Phys. Chem. Earth, 7,* 167-321.

Hodges F. M. and Kushiro I. (1974) Apollo 17 petrology and experimental determination of differentiation sequences in model moon compositions. *Proc. Lunar Sci. Conf. 5th,* pp. 505-520.

Hughes S. S., Delano J. W., and Schmitt R. A. (1989) Petrogenetic modeling of 74220 high-Ti orange volcanic glasses and the Apollo 11 and 17 high-Ti mare basalts. *Proc. Lunar Planet. Sci. Conf. 19th,* pp. 175-188.

Jones J. H. and Delano J. W. (1989) A three-component model for the bulk composition of the Moon. *Geochim. Cosmochim. Acta, 53,* 513-527.

McKay G. A. (1986) Crystal/liquid partitioning of REE in basaltic systems: Extreme fractionation of REE in olivine. *Geochim. Cosmochim. Acta, 50,* 69-79.

McKay G. A. and Weill D. F. (1976) The petrogenesis of KREEP. *Proc. Lunar Sci. Conf. 7th,* pp. 2427-2447.

McKay G. A. and Weill D. F. (1977) KREEP petrogenesis revisited. *Proc. Lunar Sci. Conf. 8th,* pp. 2339-2355.

McKay G. A., Wagstaff J., and Le L. (1989) The source of the mare basalt europium anomaly: REE distribution coefficients for pigeonite. In *Papers Presented to the Workshop on Lunar Volcanic Glasses.* Lunar and Planetary Institute, Houston, in press.

Papike J. J., Shearer C. K., Simon S. B., and Shimizu N. (1988) Lunar pyroxenes: crystal chemical rationalization of REE zoning, pattern shapes, and abundances—an ion microprobe investigation (abstract). In *Lunar and Planetary Science XIX,* pp. 901-902. Lunar and Planetary Institute, Houston.

Philpotts J. A. (1970) Redox estimation from a calculation of Eu^{+2} and Eu^{+3} concentrations in natural phases. *Earth Planet. Sci. Lett., 9,* 257-268.

Sato M. (1976) Oxygen fugacity and other thermochemical parameters of Apollo 17 high-Ti basalts and their implications on the reduction mechanism. *Proc. Lunar Sci. Conf. 7th,* pp. 1323-1344.

Sato M., Hickling N. L., and McLane J. E. (1973) Oxygen fugacity values of Apollo 12, 14, and 15 lunar samples and reduced state of lunar magmas. *Proc. Lunar Sci. Conf. 4th,* pp. 1061-1079.

Schreiber H. D. (1977) Redox states of Ti, Zr, Hf, Cr, and Eu in basaltic magmas: An experimental study. *Proc. Lunar Sci. Conf. 8th,* pp. 1785-1807.

Shearer C. K. and Papike J. J. (1989) Is plagioclase removal responsible for the negative Eu anomaly in the source regions of mare basalts? (abstract). In *Lunar and Planetary Science XX,* pp. 994-995. Lunar and Planetary Institute, Houston.

Shearer C. K., Papike J. J., Simon S. B. and Shimizu N. (1989) An ion microprobe study of the intra-crystalline behaviour of REE and selected trace elements in pyroxene from mare basalts with different cooling and crystallization histories. *Geochim. Cosmochim. Acta, 53,* 1041-1054.

Shimizu H. (1980) Experimental study on rare-earth element partitioning in minerals formed at 20 and 30 kb for basaltic systems. *Geochem. J., 14,* 185-202.

Shimizu H., Sangen K., and Masuda A. (1982) Experimental study on rare-earth element partitioning in olivine and clinopyroxene formed at 10 and 20 kb for basaltic systems. *Geochem. J., 16,* 107-117.

Tanaka T. and Nishizawa O. (1975) Partitioning of REE, Ba and Sr between crystal and liquid phases for a natural silicate system at 20 Kb pressure. *Geochem. J., 9,* 161-166.

Taylor S. R. (1982) *Planetary Science: A Lunar Perspective.* Lunar and Planetary Institute, Houston. 481 pp.

Towell D. G., Winchester J. W., and Spirn R. V. (1965) Rare-earth distribution in some rocks and associated minerals of the batholith of southern California. *J. Geophys. Res., 70,* 3485-3496.

Weill D. F. and McKay G. A. (1975) The partitioning of Mg, Fe, Sr, Ce, Sm, En, and Yb in lunar igneous systems and a possible origin of KREEP by equilibrium partial melting. *Proc. Lunar Sci. Conf. 6th,* pp. 1143-1158.

Wood J. A. (1972) Thermal history and early magmatism in the Moon. *Icarus, 16,* 229-240.

Wood J. A. (1975) Lunar petrogenesis in a well-stirred magma ocean. *Proc. Lunar Sci. Conf. 6th,* pp. 1087-1102.

Wood J. A., Marvin U. B., Powell B. N. and Dickey J. S. Jr. (1970) Lunar anorthosites. *Science, 167,* 602-604.

Proceedings of the 20th Lunar and Planetary Science Conference, pp. 31-59
Lunar and Planetary Institute, Houston, 1990

Pristine Moon Rocks: An Alkali Anorthosite with Coarse Augite Exsolution from Plagioclase, a Magnesian Harzburgite, and Other Oddities

P. H. Warren, E. A. Jerde, and G. W. Kallemeyn

Institute of Geophysics and Planetary Physics, University of California, Los Angeles, CA 90024

We report new chemical analyses and petrographic descriptions for 18 lunar rock samples, most of which were previously unstudied. Several of the samples studied are extraordinary. Two anorthosites from Apollo 12 are similar in composition to most other anorthosites from the west-central nearside region (i.e., alkali, whereas most anorthosites from the eastern nearside are ferroan). However, the texture of alkali anorthosite 12033,425 features a long (330 μm) and narrow (~17 μm) crystal of augite surrounded by a single crystal of plagioclase, with the long dimension of the augite exactly parallel to the long axes of the Carlsbad-albite twins of the plagioclase. This texture clearly suggests that the augite formed by exsolution out of the plagioclase. Apollo 12 rocklet 12037,174 is a lightly brecciated norite (average pyroxene composition $En_{64.9}Wo_{4.5}$) that may represent a cumulate from a transitional mare-nonmare type magma. Another Apollo 12 rocklet, 12033,503, is a magnesian harzburgite, with subequal amounts of enstatite and olivine, traces of Cr-Fe spinel, and FeNi metal, but no plagioclase. The FeNi metal is extremely Ni-rich (53 wt.%), and the enstatite and olivine are both remarkably Ca-poor. The bulk composition is remarkably Ir-rich (19 ng/g) for a pristine rock. The texture of this lithology is also unusual. Coarse, largely intact olivines, anhedral with gently curving outlines, are set in a groundmass of monomineralic enstatite. The enstatite has been granulated to a strongly bimodal grain size, with large, brick-like grains, typically about 140 × 70 μm, but in optical continuity with one another up to 3 mm apart, set in a groundmass of much finer grains, typically about 3 μm across. The origin of this bizarre texture is enigmatic, but it may have formed as a result of sudden decomposition of protoenstatite, triggered by shock metamorphism. Other samples investigated include an Apollo 15 ferroan anorthosite, two Apollo 15 norites, several uncommonly Al-rich and REE-poor impact melt breccias from Apollo 14, and several granulitic polymict breccias. One of the granulitic breccias, 12032,277, is uncommonly low in incompatible elements, for an Apollo 12 lithology, and thus supports the hypothesis that lunar granulitic breccias formed mainly in an era when KREEP was not yet abundant in the lunar crust.

INTRODUCTION

Most of the upper crust of the Moon has been profoundly altered by brecciation and melting caused by meteoritic impacts. Pervasive, fine-scale mixing has obscured the original igneous compositions and textures of most nonmare rock samples. However, a minor proportion of the nonmare material accessible at the surface has survived in unmixed or "pristine" form. Studies of pristine rocks have revealed great diversity in their compositions, including some systematic groupings that appear to reflect a complex, multistage genesis for the lunar crust (*Warren,* 1985). The set of roughly 100 compositionally pristine rocks analyzed to date represents a poor sampling of such a diverse population, however, and as additional pristine fragments are found among the Apollo breccia clasts and coarse soil particles, unique and surprising lithologies continue to be found.

Previous studies of pristine rocks (e.g., *Warren et al.,* 1983a,b; *Shervais and Taylor,* 1986) have revealed compositional variations that appear to be related to geographic (i.e, Apollo landing site) provenance of the samples. Regional variations in the gross composition of the crust are also apparent from orbital-spectrometric (e.g., *Davis and Spudis,* 1985) and Earth-based (*Pieters,* 1989) remote-sensing results. Investigation of these variations, as well as the general diversity of the pristine crust, can be advanced by finding additional pristine rocks from the sites that have thus far yielded few pristine nonmare samples. Relatively large numbers of pristine nonmare rocks are already known from Apollo 15, Apollo 17, and especially the "pure highlands" mission, Apollo 16. We have concentrated on samples from the primarily mare Apollo 12 site, and from Apollo 14. Even though Apollo 14 went to a primarily highlands site, the rocks sampled are dominated by a single compositional variety of impact melt breccia, and include few pristine lithologies, and none larger than ~1.8 g (*Warren et al.,* 1983b). The only previous study of a large number of nonmare samples from Apollo 12 was the collaborative work of *Simon and Papike* (1985) and *Laul* (1986).

ANALYTICAL PROCEDURES

All samples were first analyzed by INAA, using a procedure modified from that of *Kallemeyn et al.* (1989). Many of the samples studied are small, virtually irreplaceable rocklets taken from breccia clasts or coarse sieve fractions of soils. We have endeavored to gather maximum scientific return from each of these small rocklets. Depending upon availability of sample mass, one of three procedures was used after INAA: (1) For the largest samples, the INAA sample was powdered (in most

TABLE 1. Concentrations of 39 elements in Apollo lunar rocklets (r) and breccia clasts (c).

	Mass mg	Na mg/g	Mg mg/g	Al mg/g	Si mg/g	K μg/g	Ca mg/g	Sc μg/g	Ti mg/g	V μg/g	Cr μg/g	Mn μg/g	Fe mg/g	Co μg/g
10010,138(r)	63	3.21	71§	90	nd	3400	76‡	13.8	3.2‡	63	1860	880	55	18.0
12003,210(r)	48	11.7	<50	170‡	nd	2060	110§	6.4	<13	<61	320	350	24	3.6
12032,277(r)	81	2.70	49	128	203	250	101	11.6	9‡	<100	740	740	61	36
12033,501(r)	62	11.6	1.8	174	221	1540	121	0.36	0.33§	<55	23.1	44	2.34	0.25¶
12033,503a(r)	43	0.181	250	2.4¶	nd	67‡	1.5¶	3.2	<3.8	48	2550	850	63	54
12033,503b(r)	9.5	0.232	260	2.4¶	nd	131‡	<14	3.3	<3.8	42	2160	860	66	74
12033,503wm	52	0.19	252	2.4¶	nd	79‡	1.5	3.2	<3.8	47	2480	850	64	58
12037,174a(r)	46	2.09	105§	69	nd	560	60‡	24.1	<6	104§	3360	1830	113	45‡
12037,174b(r)	16	5.2	<90	145§	nd	3370	115	9.4	<20	44§	1070	650	38	12.2
12037,174wm	61	2.89	nd	89§	nd	1290	74	20.3	<10	88§	2770	1520	94	37
12042,278(r)	41	6.8	60	75	232	6800	70	24.5	15.7	<100	1380	1120	89	35
12044,134(r)	39	1.90	46§	62	nd	490	80‡	54	23‡	260	5000	2310	166	52‡
12071,10(c)	32	2.41	65	132	213	163§	102	7.8	1.3‡	<66	710	600	39	17.0
12071,10(ic)	163	2.35	57	140	207	100§	107	6.5	1.2	nd	570	480	33.6	13.9
14076,7(c)	30	4.38	32¶	128	nd	2140	103	12.7	5.3§	<34	880	580	48	20.1
14077,3(r)	211	5.10	48	111	224	4110	88	18.6	7.8	nd	1100	870	63	28.5
14315,28(c)	17	5.1	86§	131	nd	2150	89‡	11.7	<18	<80	1400	680	52	17.0
15015,179(c)	11	4.60	73¶	87	nd	2850	62§	22.5	13§	75‡	1820	1100	85	27
15295,41(c)	106	3.08	1.8	183	211	85	139	5.3	<0.5	nd	24.2	60	2.7	0.31
15360,11(c)	99	2.76	51	131	223	510	95	8.2	0.7‡	nd	1820	610	34	13.3
15361,1(r)	134	2.40	94	96	233	280	68	9.4	1.09‡	93	2300	620	42	23.1
15364,15(r)	195	2.20	101	120	204	162	86	2.35	0.5‡	nd	730	440	38	19.2
67215;14(r)	239	2.21	nd	nd	nd	157	98	11.6	nd	nd	690	630	45	12.4
Rel. uncert.†		2	4	4	3	7	3	3	8	9	3	3	4	4

TABLE 1. (continued).

	Ni μg/g	Zn μg/g	Ga μg/g	Ge ng/g	Br μg/g	Rb μg/g	Sr μg/g	Zr μg/g	Cs ng/g	Ba μg/g	La μg/g	Ce μg/g	Nd μg/g	Sm μg/g	Eu μg/g
10010,138(r)	<37	nd	3.6‡	nd	nd	20‡	120	300	910	253	19.0	45	29	7.5	1.15
12003,210(r)	<64	nd	11.3	nd	nd	<15	480	250‡	<720	460	20.6	47.8	26	6.1	6.6
12032,277(r)	160§	nd	5.0	47	nd	<14	195§	177‡	<660	47§	3.9	10.4	5.9‡	2.00	1.09
12033,501(r)	0.36	<6	10.6	3.3	nd	<5	430	50§	49§	360	11.6	23.3	11.0	2.41	6.2
12033,503a(r)	194	nd	0.36§	nd	0.57	nd	nd	nd	nd	<35	1.19	3.6‡	<4	0.49	0.061
12033,503b(r)	131§	nd	<1.1	nd	2.0	nd	<500	<700	nd	<200	1.69	4.2§	nd	0.62	0.075‡
12033,503wm	183	nd	0.36§	nd	0.8	nd	<500	<700	nd	<35	1.28	3.7‡	<4	0.51	0.064
12037,174a(r)	<108	<15	3.1‡	nd	nd	<23	102§	84§	170§	80§	13.6	37	21	6.2	0.68
12037,174b(r)	<136	20§	4.6‡	nd	14‡	nd	104‡	350‡	590	370	31.4	79	45	12.3	1.75
12037,174wm	<120	<17	3.5‡	nd	14§	nd	103§	153§	280§	155§	18.2	48	27	7.8	0.96
12042,278(r)	320	nd	5.4§	nd	nd	24	320§	1410	990‡	1040	118	310	200	48	3.1
12044,134(r)	<160	<21	3.8	nd	nd	<23	204‡	<350	nd	38¶	6.5	19.6	12.4	4.4	0.98
12071,10(c)	37	nd	2.7‡	1190	nd	nd	164	<90	<290	27§	1.61	4.9	2.2§	0.69	0.89
12071,10(ic)	28‡	nd	2.8‡	nd	nd	<7	139	<18	<190	20‡	1.50	5.1	1.9‡	0.58	0.82
14076,7(c)	234	33‡	3.1‡	nd	nd	6.8§	149§	470	250§	350	31.0	80	47	12.5	1.47
14077,3(r)	274	nd	3.9	nd	nd	11.9	207	780	510	670	60	144	88	23.9	2.30
14315,28(c)	<110	19.9	5.9	nd	nd	<15	179§	360	520‡	490	21.0	51	30‡	8.4	2.22
15015,179(c)	179	12‡	4.6	nd	nd	12§	190‡	650	530	490	56	160	81	22.5	2.07
15295,41(c)	0.17‡	2.0‡	3.5	1.9	nd	nd	191	45	54	8.9§	0.23	0.56	0.35¶	0.072	0.94
15360,11(c)	1.25‡	2.0	3.9	4.3	nd	nd	124	<65	<134	62	3.23	8.1	2.8‡	0.90	1.19
15361,1(r)	28	<5.6	3.2§	9.9	nd	<5	107‡	42§	140‡	41§	2.96	8.0	3.5	1.15	0.78
15364,15(r)	17	1.8§	2.9	6.2	nd	nd	130	44‡	45§	43	2.01	5.7	2.9	0.73	0.87
7215;14(r)	32*	nd	1.9‡	37*	nd	<5	140‡	<93	nd	24§	0.91	2.14	1.73‡	0.47	0.76
Rel. uncert.†	8	6	8	7	10	10	9	9	10	6	4	7	8	4	5

TABLE 1. (continued).

	Tb µg/g	Dy µg/g	Ho µg/g	Yb µg/g	Lu µg/g	Hf µg/g	Ta µg/g	Re pg/g	Os ng/g	Ir ng/g	Au ng/g	Th µg/g	U µg/g	*mg**
10010,138(r)	1.60	11.5	nd	7.3	1.02	6.5	0.90	nd	nd	<4.9	<1.4	4.8	1.27	0.75
12003,210(r)	1.1§	7.5†	1.54	3.5	0.53	2.89	0.31	nd	nd	<6.7	<4	1.49	0.33§	nd
12032,277(r)	0.49	3.8‡	0.68	1.82	0.283	1.68	0.28	149	2.01	1.57	0.45	0.93	0.21¶	0.648
12033,501(r)	0.31	1.56‡	nd	0.62	0.065‡	0.85	0.065§	<13	<0.03	<0.03	0.075	0.162§	<0.08	0.639
12033,503a(r)	<0.19	nd	nd	0.38	0.054‡	0.27§	<0.16	nd	nd	<6.3	1.9	0.18§	<0.38	0.901
12033,503b(r)	nd	nd	nd	0.41‡	0.065§	0.55¶	<0.9	nd	nd	74	12.4	<0.7	<1.4	nd
12033,503wm	<0.19	nd	nd	0.39	0.056‡	0.32§	<0.16	nd	nd	19	3.8	0.18§	<0.38	0.910
12037,174a(r)	1.54	9.5	nd	7.1	1.00	2.95	0.35‡	nd	nd	<8.3	<5	1.87	0.44	0.68
12037,174b(r)	2.26	14.7	3.3§	7.0	1.00	8.9	1.39	nd	nd	<11	<7	4.7	1.00	nd
12037,174wm	1.73	10.8	nd	7.1	1.00	4.5	0.62‡	nd	nd	<9	<6	2.6	0.58	0.68
12042,278(r)	10.3	70	nd	35	5.0	38	4.4	1080	9.6	11.5	5.3	18.4	nd	0.608
12044,134(r)	1.04	6.7	nd	4.0	0.54	3.2	0.46	nd	nd	<12	<5	0.77	0.18¶	0.389
12071,10(c)	0.145‡	1.10§	nd	0.63	0.095	0.45	0.07¶	<15	0.097‡	0.080‡	0.038‡	0.16§	<0.24	0.793
12071,10(ic)	0.129	0.83‡	nd	0.53	0.076	0.39	0.051§	nd	nd	<2	<1.2	0.132‡	<0.20	0.796
14076,7(c)	2.87	16.3	3.5‡	9.9	1.37	9.9	1.10	nd	nd	11.3	8§	6.2	1.6	0.60
14077,3(r)	4.9	33.7	nd	17.9	2.54	19.3	2.19	nd	nd	9.5	4.3‡	10.4	2.7	0.636
14315,28(c)	1.89	15.0	2.9	10.5	1.59	8.0	0.81	nd	nd	<9	<7	3.7	0.97‡	0.79
15015,179(c)	5.4	36	nd	17.1	2.42	15.8	1.91	nd	nd	5.0	<18	9.2	2.3	0.66
15295,41(c)	0.014§	nd	nd	0.058‡	0.008‡	1.03	<0.028	2.2§	0.030‡	0.050	0.094	0.026§	nd	0.605
15360,11(c)	0.213	1.5§	0.30‡	1.13	0.160	0.43	<0.10	<16	0.053‡	0.042§	0.065	<0.11	nd	0.775
15361,1(r)	0.25	1.9‡	0.40	1.27	0.19	0.80	0.13§	1.8‡	0.014‡	0.008	0.25	0.43	0.11§	0.837
15364,15(r)	0.173	nd	0.20	0.67	0.099	0.96	0.049	20	0.44	0.28	0.068	0.36	0.086‡	0.859
67215;14(r)	0.111‡	nd	0.17§	0.60	0.090	0.30	0.028§	64*	1.21*	1.21*	0.23	<0.09	nd	nd
Rel. uncert.†	5	5	5	5	4	5	5	10	9	10	8	5	6	

*RNAA of a separate 238-mg chip of 67215,14 gave results of 7.8 ng/g Ge, 0.38 ng/g Ir, 13.3 µg/g Ni, 1.2 ng/g Os, and 71 pg/g Re.
†Normal uncertainty limits (% relative) for 70% confidence.
‡Uncertainty limits 1.1-2.0 × greater than normal.
§Uncertainty limits 2.1-4.0 × greater than normal.
¶Uncertainty limits 4.1-6.0 × greater than normal.
Abbreviations: wm = weighted mean, ic = impure clast, *mg** = molar Mg/(Mg+Fe), nd = not determined.

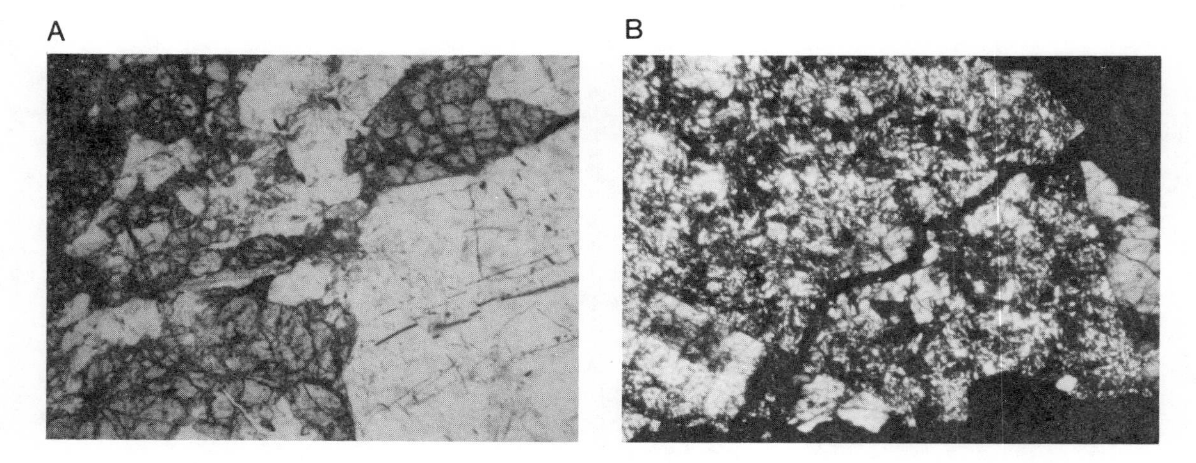

Fig 1. Photomicrographs of lithologies that are only "possibly" pristine. **(a)** Relatively unbrecciated area of KREEPy diabase 10010,19, dominated by plagioclase (light) and pyroxene (dark), with one small grain of troilite (black). View is 0.63 × 0.48 mm; nicols uncrossed. **(b)** Clast 14315,28, showing generally micropoikilitic texture (typical of impact melts), but also several essentially intact coarse grains of plagioclase, at right and lower left edges of this cross-nicols, 1.6 × 1.2 mm, view.

TABLE 2. Compositions of minerals determined by electron probe microanalysis of thin sections.

Sample		Pyroxene										Olivine					Plagioclase				
		Low-Ca					Medium- or High-Ca					Fo = molar Mg/(Mg+Fe)					An = molar Ca/(Ca+Na+K)				
		N	$\bar{x}$	SD	min.	max.	N	$\bar{x}$	SD	min.	max.	N	$\bar{x}$	SD	min.	max.	N	$\bar{x}$	SD	min.	max.
10010;19	Wo	14	3.70	0.35	3.24	4.49	1	32.24	—	—	—	—	—	—	—	—	13	92.19	0.78	91.13	93.55
	*mg**		71.57	2.84	67.46	77.12		72.10	—	—	—										
12003;179	(no px)	—	—	—	—	—	—	—	—	—	—	—	—	—	—	—	14	82.12	1.37	78.40	83.60
12032;277	Wo	8	2.23	0.27	1.77	2.71	4	43.61	1.30	42.65	45.43	4	62.53	0.61	62.11	63.44	7	93.72	0.42	93.13	94.36
	*mg**		69.20	0.96	68.25	71.27		79.34	2.71	76.81	82.01										
12033;425	Wo	—	—	—	—	—	16	43.73	0.50	43.00	44.56	—	—	—	—	—	15	82.79	0.87	81.50	84.52
	*mg**	—	—	—	—	—		74.48	0.51	73.65	75.29										
12033;503	Wo	13	0.40	0.16	0.20	0.68	—	—	—	—	—	16	89.49	0.18	89.23	89.88	—	—	—	—	—
	*mg**		90.60	0.24	90.22	90.95		—	—	—	—										
12037;174	Wo	19	4.48	1.78	3.31	10.94	—	—	—	—	—	—	—	—	—	—	14	91.72	1.33	89.45	93.40
	*mg**		67.95	3.48	60.67	73.23		—	—	—	—										
12042;243	Wo	4	3.49	0.79	2.82	4.51	—	—	—	—	—	2	78.85	0.01	78.84	78.85	3	89.27	2.89	87.49	92.61
	*mg**		75.65	11.18	65.97	88.30		—	—	—	—										

TABLE 2. (continued).

Sample		Pyroxene										Olivine					Plagioclase				
		Low-Ca					Medium- or High-Ca					Fo = molar Mg/(Mg+Fe)					An = molar Ca/(Ca+Na+K)				
		N	$\bar{x}$	SD	min.	max.	N	$\bar{x}$	SD	min.	max.	N	$\bar{x}$	SD	min.	max.	N	$\bar{x}$	SD	min.	max.
14315;28	Wo	11	3.57	0.26	3.10	4.06	—	—	—	—	—	—	—	—	—	—	18	91.67	2.30	86.33	94.95
	*mg**		80.14	0.57	79.30	80.92		—	—	—	—										
15015;179	Wo	4	3.85	0.63	2.98	4.46	5	17.25	8.96	9.78	28.78	—	—	—	—	—	13	89.14	4.13	82.29	95.24
	*mg**		70.69	0.81	70.18	71.90		70.56	1.12	69.13	71.65										
15295;41	Wo	3	4.04	0.31	3.74	4.36	5	40.52	2.73	37.37	43.12	—	—	—	—	—	5	95.80	0.67	95.29	96.97
	*mg**		43.77	0.51	43.25	44.27		58.88	2.51	55.66	62.03										
15360;11	Wo	14	3.15	0.21	2.80	3.66	—	—	—	—	—	—	—	—	—	—	15	93.34	0.64	91.77	94.14
	*mg**		78.46	0.24	78.13	79.04		—	—	—	—										
15361;4	Wo	19	2.34	0.18	1.85	2.63	2	11.79	3.97	8.99	14.60	—	—	—	—	—	20	94.03	0.43	93.40	94.64
	*mg**		83.76	0.84	82.22	85.63		81.27	0.44	80.96	81.58										
15364;14	Wo	8	2.92	0.95	1.78	4.38	9	38.89	2.88	33.55	43.16	23	71.68	0.32	70.89	72.19	40	94.08	1.08	91.60	96.19
	*mg**		75.84	0.35	75.28	76.35		80.31	2.19	75.12	82.86										

Abbreviations: $\bar{x}$ = average, SD = standard deviation, *mg** = molar Mg/(Mg+Fe).

cases before the INAA), and split into two aliquots, one 5-10 × larger than the other. The larger aliquot was used for RNAA, following our standard procedure (*Warren et al.,* 1986) to determine siderophile elements Au, Ge, Ir, Ni, Os, and Re. The smaller aliquot was used for determination of major elements and Ti with an electron probe, after conversion of the aliquot into a fused bead (we call this technique electron probe fused bead analysis, or EPFBA). (2) For intermediate-sized samples, the sample from INAA was broken into two subequal chunks, one of which was powdered and then split into RNAA and EPFBA aliquots, as in procedure (1); the other chunk was converted into a thin section for petrographic studies. (3) For the smallest samples, the entire INAA sample was used for conversion into a thin section for petrographic studies.

All samples investigated with procedure (3) can be recognized from the lack of data for Si (determined exclusively by EPFBA), Ge, Os, and Re (determined exclusively by RNAA) among our results (Table 1). Most of the larger samples were studied by procedure (1). The only procedure (2) samples were 12032,277, 12033,501, and 15360,11. In general, for the procedure (2) and (3) samples, our INAA sample (or a portion of it) was converted into the only thin section studied from each rocklet. However, in a few cases (12032,277, 14315,28, 15015,179, and 15360,11) a second thin section was produced using separate chunks (not studied by INAA), in which the lithology of interest was "contaminated" with adhering host-breccia matrix or, in case of 12032,277, glass veins.

SAMPLE DESCRIPTIONS

Diabase (?) or Impact Melt Breccia 10010,19

This rocklet came from the Apollo 11 "contingency" regolith-scoop sample. We studied a 60-mg chip (10010,138) that represented more than half of 10010,19 (originally 110 mg), described by *Kramer and Twedell* (1977) as a "small anorthosite breccia chip."

The thin section is far less than standard thickness, but it shows a cataclastic texture (Fig. 1), and a mode with feldspar (~55%), pyroxene (~45%), 1-2% of a silica phase, and traces of troilite, ilmenite, rutile, Fe-Cr-spinel, and a phosphate mineral. The pyroxene is dominantly orthopyroxene, but roughly 1/10 is augite. The feldspar is dominantly plagioclase, but roughly 1/10 is either K-feldspar or "ternary" Ca-Na-K-feldspar. Original grain sizes have been obscured by cataclasis. The coarsest essentially intact plagioclase is 2.3 × 1.1 mm, although its boundaries on two sides are edges of the thin section. The coarsest remaining pyroxene is 1.2 mm across, but bounded largely by the edge of the section, and seemingly in optical continuity with another exposure of pyroxene that would make the total breadth of the exposed crystal 2.1 mm. Elsewhere, cataclasis followed by a modicum of thermal annealing, which produced a mildly granulitic texture, has reduced the grains to scattered fragments, mostly within a factor or two of 0.2 mm across. The largest troilite is a grain 0.12 mm across, the largest ilmenite is a 120 × 12 μm lath, and the largest rutile is 50 μm across. The largest, most nearly intact feldspars tend to be subhedral, blocky-prismatic

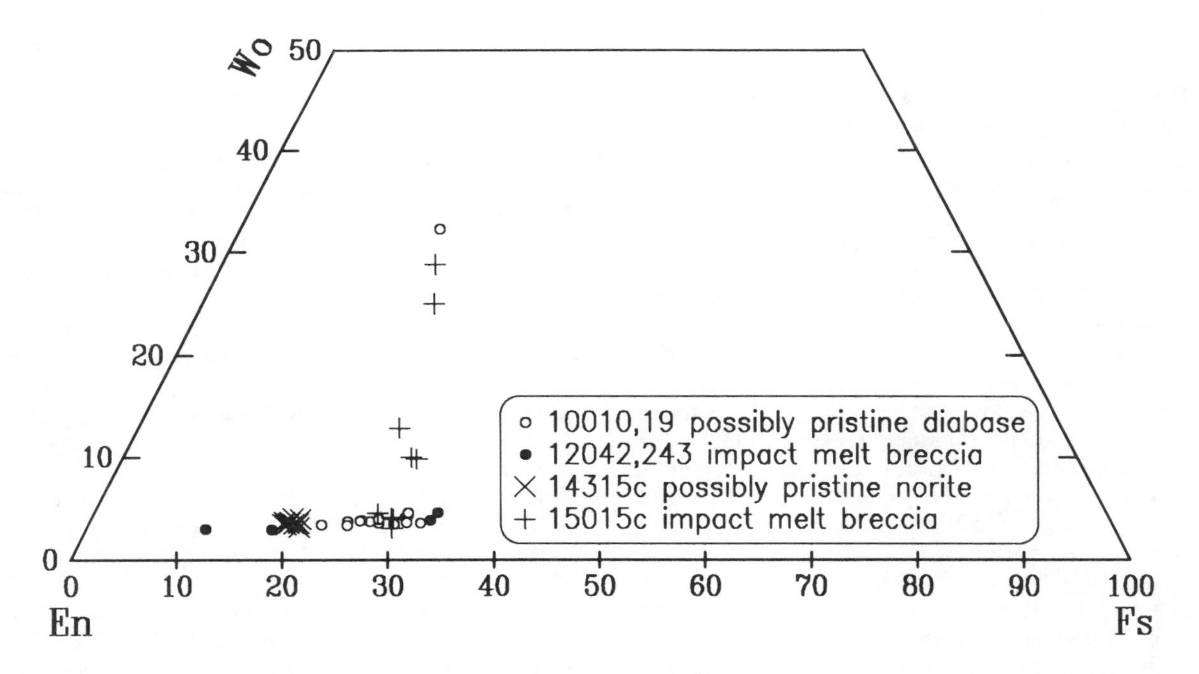

Fig. 2. Pyroxene molar En-Fs-Wo proportions for two impact melt breccias and two possibly pristine lithologies that otherwise might be considered impact melt breccias.

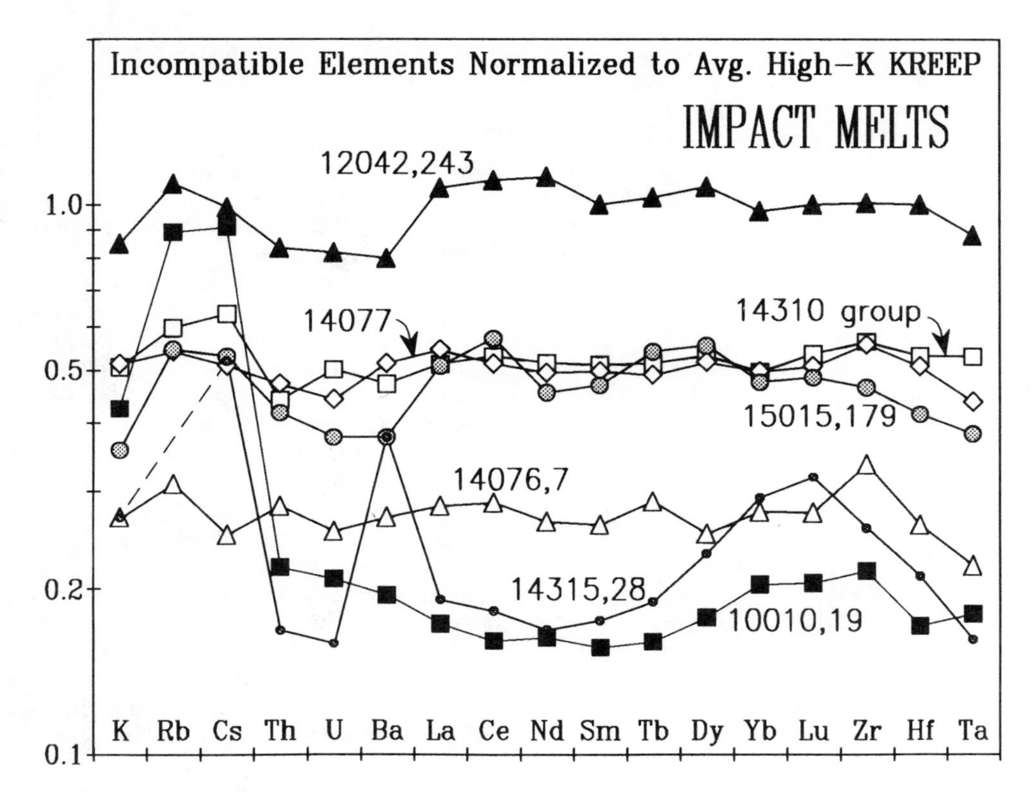

Fig. 3. Concentrations of incompatible elements in impact melt lithologies normalized to average high-K KREEP (*Warren,* 1989). Included for comparison is an average of data (mainly from *McKay et al.,* 1978) for the five 14310-group impact melts described by *McKay et al.* (1978). Note: although conservatively included among the impact melts here, both 10010,19 and 14315,28 (especially 10010,19) are possibly pristine.

plagioclase grains, whereas the largest, most nearly intact pyroxenes are mainly anhedral grains that apparently formed interstitially to the plagioclase.

Mineral-composition data are shown in Table 2 and Fig. 2. The absence of metal and the low Ni content of 17 ± 8 μg/g suggest, but hardly prove, that the lithology is pristine. Few polymict breccias or impact melt rocks have Ni contents <<100 μg/g (*Warren,* 1990b), but some do. The thin section shows no apparent foreign clasts. A wide range of minerals was found, with wide-ranging solid-solution compositions, implying that the lithology is not plutonic. A mare affinity seems unlikely, based on the high Al, low Fe, and low Ti concentrations (Table 1). Rutile is common in lunar rocks of KREEP affinity. However, the bulk-rock incompatible element concentrations are moderate, and their pattern is only vaguely KREEP-like (Fig. 3). No mineral grains with manifestly exotic compositions (suggestive of xenocrystic origins) were detected. We infer that the 10010,19 lithology is most likely a pristine rock, probably a partial cumulate from either a hypabyssal intrusion or the deepest portion of a thick lava flow. However, at least until a similar but more obviously pristine sample is discovered, 10010,19 should be excluded from compilations of rocks for which certitude of pristinity is essential.

Poorly Sampled Alkali Anorthosite 12003,179

This 100-mg rocklet was separated by *Marvin* (1978) from a soil sample collected just a few meters from the Apollo 12 landing point. We studied a 50-mg chip (12003,210). *Marvin* (1978) noted that roughly half of the rocklet's surface is a "dark gray" material. The thin section produced from our sample after INAA shows that the dark material is a layer of impact-melt glass, and that this glassy material constituted ~3 vol.% of the INAA sample. The glassy lithology is found exclusively at, or within a few tens of micrometers of, the margins of the thin section. Apparently it formed as a partial coating, ~0.3 mm thick, over the rocklet. The rest of the section is virtually all a single 4.4×3.3 mm crystal of plagioclase, with a tiny (0.60×0.2 mm) fragment of another plagioclase in one corner, and possibly an even smaller fragment of a third plagioclase in the opposite corner. No mafic silicates are discernible.

Our only results for "noble" siderophile elements (Table 1) are upper limits. Considering 12003,179 as a whole (i.e., including the ~3% glassy material), the rocklet is obviously not pristine. However, the interior beneath the thin glass coating is extremely coarse grained, and apparently monomict

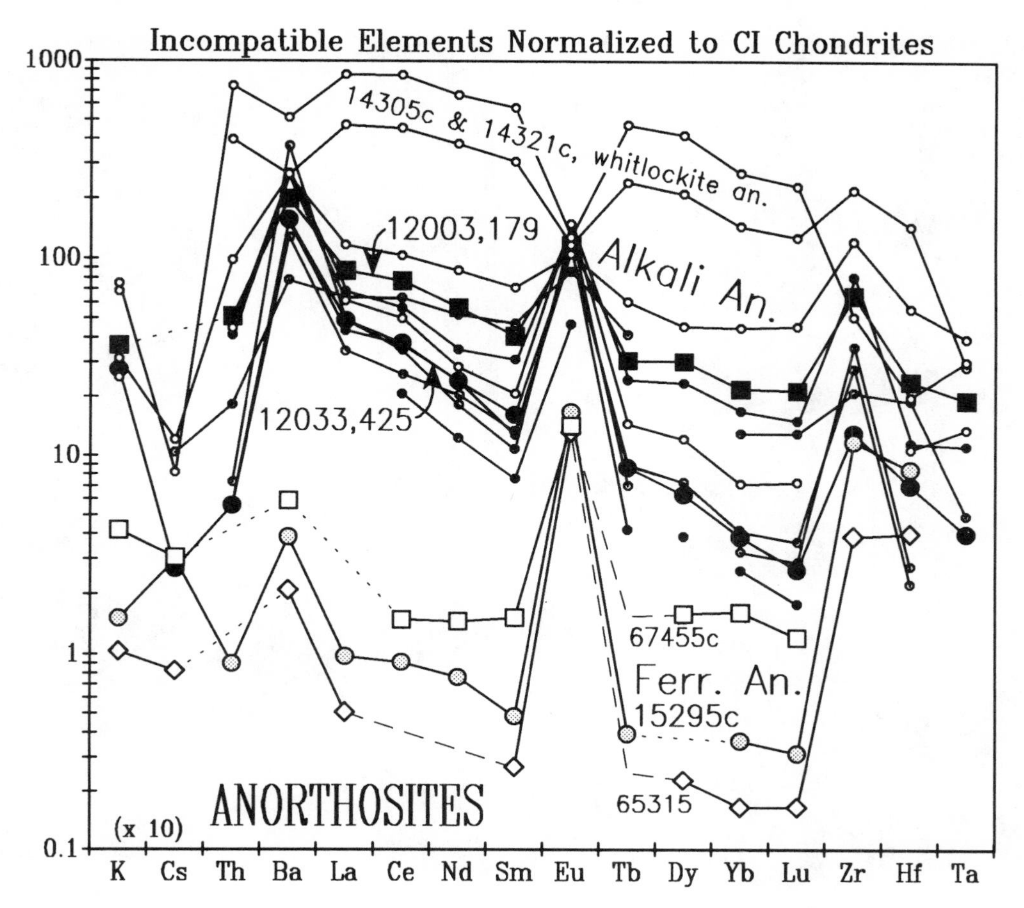

Fig. 4. Concentrations of incompatible elements in alkali anorthosites 12003,179 and 12033,425 (large black symbols) and the 15295,41 ferroan anorthosite, normalized to CI chondrites. Shown for comparison are two ferroan anorthosites (*Wänke et al.,* 1974; *Lindstrom et al.,* 1977) that, for the elements shown, are representative of virtually the entire range for definitely pristine ferroan anorthosites, and a compilation of literature data for alkali anorthosites (small circular symbols). The alkali anorthosites, in order of increasing Sm concentration, are 12033,97,7, 14160,106, 12033,532, 14304c, 12033,425, 14305c2, 12073c, 12003,179, 15405c, 14047c, 14321c, and 14305c1. Data are from *Laul* (1986), *Warren et al.* (1987), *Lindstrom et al.* (1988), and earlier papers cited therein.

(pristine). Our bulk- and mineral-composition results (Tables 1 and 2, Fig. 4) indicate that the pristine interior lithology is an alkali anorthosite. As such, for trace-lithophile elements its composition was probably not significantly affected by the ~3% of contamination. However, our sample size was obviously too small, considering the coarse granularity, to be representative for much besides the plagioclase portion of the parent lithology.

Granulitic Polymict Breccia 12032,277

This 200-mg rocklet was separated by *Marvin* (1978) from a soil sample collected at the north rim of Bench crater. *Marvin* (1978) noted that the rocklet is crossed by numerous thin (<0.1 mm) veinlets of black, glassy material. For our studies, we broke the rocklet into seven smaller fragments, of which only the three most glass-poor fragments, totaling 81 mg, were used for INAA. The glass content of the INAA sample was estimated to be ~2-3 vol.%. The remaining four fragments were returned to Houston, where they were used to produce a thin section that we studied in addition to a thin section produced from a major portion (58 mg) of the INAA sample. The remainder of the INAA sample was powdered, and a 6 mg aliquot was used to produce a fused bead for EPFBA, while another aliquot (15 mg) was consumed for RNAA.

This lithology is a fine-grained, feldspathic granulitic breccia (Fig. 5). Its mode (excluding the several percent of dark-glassy vein material) shows roughly 75% plagioclase, 8% low-Ca pyroxene, 7% high-Ca pyroxene, 7% olivine, and 2% opaques, mainly ilmenite, but also a large trace of FeNi-metal and a small trace of troilite. The grain size is uniformly close to 0.1 mm, although there are many scattered domains roughly 1 mm

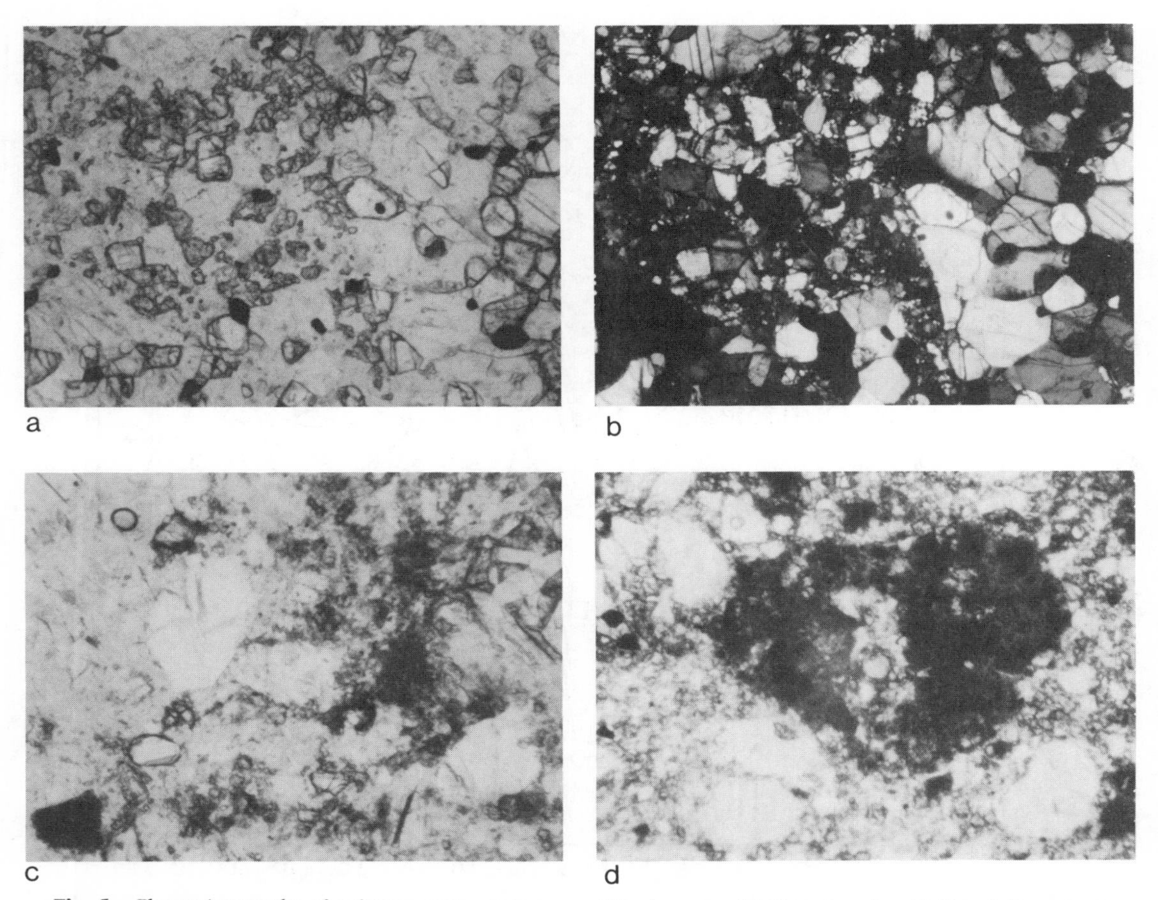

Fig. 5. Photomicrographs of polymict, more or less granulitic, breccias. **(a)** Typical region (0.63 × 0.48 mm) of 12032,277, with nicols uncrossed. **(b)** Same view, through crossed nicols. This texture is typical of granulitic polymict breccias. **(c)** This 0.98 ×0.75 mm view of 15364,14 (nicols uncrossed) shows several features that suggest the breccia is polymict. In lower left and slightly right of center can be seen two aphanitic clasts. In upper right corner can be seen part of a diabasic-textured clast [cf. the "euhedral plagioclase" clast described by *Steele et al.* (1977)]. **(d)** For comparison, a 0.98 × 0.75 mm view of 67215,18 (nicols uncrossed), showing a large aphanitic clast similar to (albeit bigger than) those in 15364.

across that have grains several times finer or coarser than the generally prevalent size. It seems likely that most of these domains are lithic clasts whose boundaries have been obscured by thermal metamorphism. Most of the larger grains are plagioclase, and the largest is 1.0 mm across. Mineral-composition data are shown in Table 2 and Fig. 6. The thermal metamorphism implied by the granulitic texture presumably tended to homogenize the compositions of all minerals in this rock.

Conceivably, the lithology is chemically pristine, its fine-grained granulitic texture notwithstanding. The composition of the metal is unusually Ni- and Co-rich for a polymict rock; among 10 analyses, Ni ranges from 16.9 to 20.5 wt.%, and Co ranges from 1.48 to 1.94 wt.%. However, similar compositions were found for metals in feldspathic granulitic breccia 79215 by *Bickel et al.* (1976), and 79215 is clearly polymict (*Warner et al.*, 1977; *McGee et al.*, 1978). The siderophile element concentrations (Table 1) seem high for a pristine rock, even allowing for contamination by 1-2% of glassy material. (The sample used for RNAA appeared to be a relatively glass-free subsample of the INAA sample; the Ni content of the RNAA sample was 100 μg/g, whereas the INAA sample, with ~2-3% glassy material, contained 220 μg/g of Ni.) Most pristine rocks have "noble" siderophile elements such as Ir, Os, and Re at concentrations $<3 \times 10^{-4}$ times CI-chondrites (*Warren and Wasson,* 1977). In order to have Ir, Os, and Re be $<3 \times 10^{-4}$ times CI-chondrites in the glass-free portion of 12032,277, and assuming, for the sake of argument, that fully 3% of the RNAA sample was glassy material, the glassy material would have to contain these elements at concentrations of 48, 62, and 4.6 ng/g, respectively. Virtually all bulk-rock analyses of lunar polymict materials have Ir, Os, and Re at substantially lower concentrations than these (*Warren,* 1990b). Hence, even without the glassy veins, the 12032,277 lithology is probably not pristine.

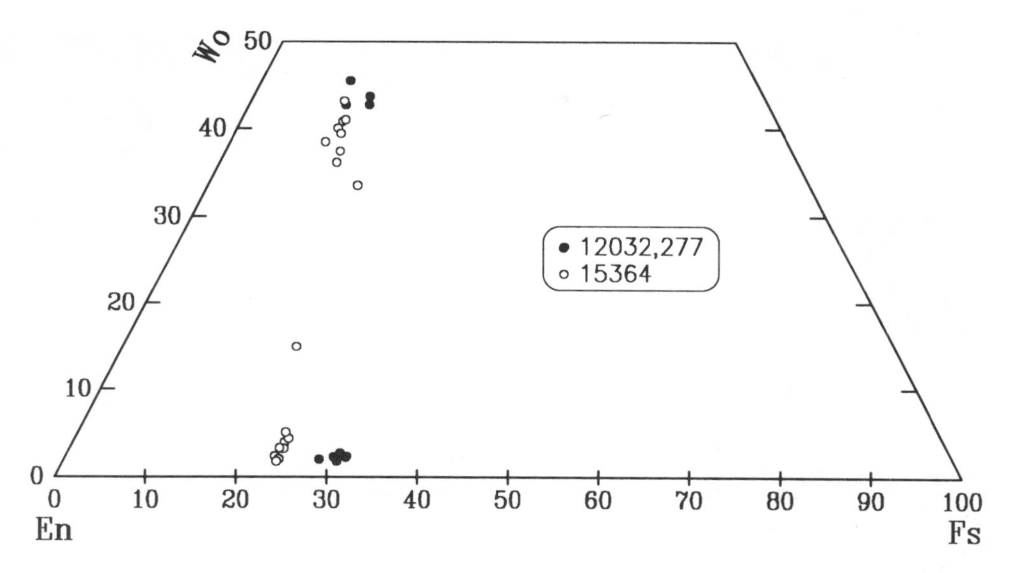

Fig. 6. Pyroxene molar En-Fs-Wo proportions for two granulitic breccias.

Alkali Anorthosite with Augite Exsolution from Plagioclase: 12033,425

This 130-mg rocklet was separated by *Marvin* (1978) from a trench soil sample collected at the NW rim of Head crater. We studied a 62-mg chip (12033,501). After INAA, most of the sample was used to make thin section 12033,575, but 21 mg of it was powdered, and aliquots of the powder were used for RNAA (15 mg) and to make a fused bead for EPFBA (6 mg).

This lithology is a monomict, cataclastic anorthosite (Fig. 7). Most of the thin section has been brecciated to grain sizes <<1 mm, but one relatively unbrecciated zone shows a plagioclase crystal fragment 1.8 mm across (bounded by the edge of the thin section), and it seems likely that the original plagioclase crystals were generally about this size. The mode shows 99% plagioclase, 1% augite, and a trace of ilmenite.

Most of the augite and ilmenite occurs as inclusions within plagioclase. The augite with the longest single dimension is a

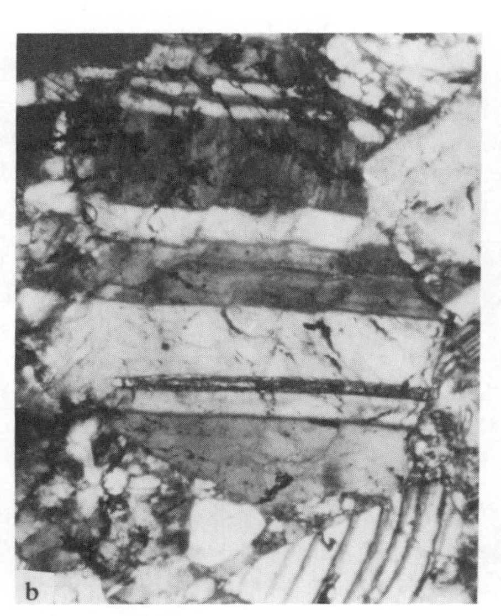

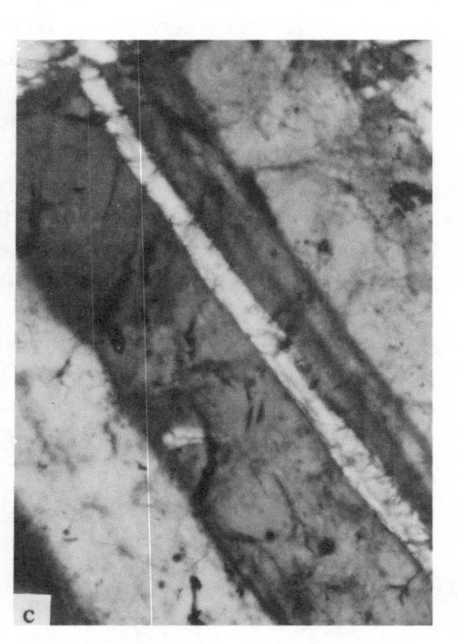

Fig. 7. Photomicrographs of the 12033,425 alkali anorthosite. **(a)** General view of the cataclastic texture. View is 2.0 × 1.5 mm; nicols partly crossed. **(b)** Lamella of augite within plagioclase. The augite is seen in the lower-central area, as a dark, high-relief lamella extending almost fully across the large plagioclase grain in a direction parallel with its Carlsbad-albite twinning (i.e., nearly horizontal, in this 0.63 × 0.48 mm view). **(c)** Close-up view (0.31 × 0.24 mm) of the same augite lamella, seen as white region extending from upper left to lower right.

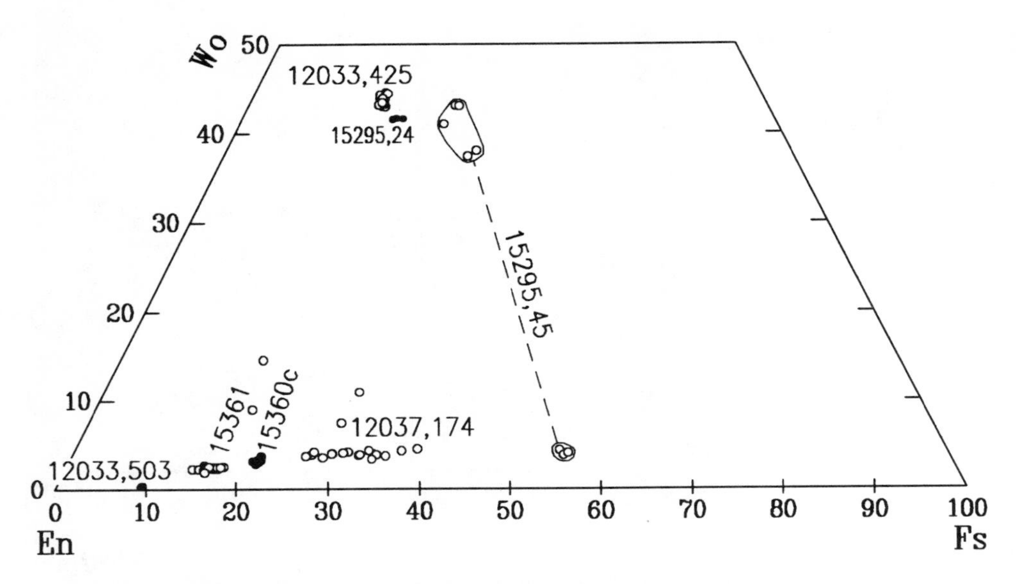

Fig. 8. Pyroxene molar En-Fs-Wo proportions for six clearly pristine lithologies. Also shown for comparison are data for a different (?) clast of 15295 ferroan anorthosite (*Warren and Wasson,* 1978).

lamellar grain that appears to have exsolved from plagioclase (Figs. 7b,c). This augite lamella is 330 μm long but its maximum width is only 17 μm. Its long dimension is virtually parallel to the twinning in the plagioclase crystal (or crystal fragment) that surrounds it, and one end of the lamella is coincident with the boundary of that plagioclase crystal. In composition, the augite lamella is indistinguishable from more equant augites elsewhere in the section (Table 2, Fig. 8). The ilmenite has a moderate MgO concentration, 4.2 wt.%. Several other augites within single crystals of plagioclase have similar, albeit far less clear-cut, lamellar shapes; e.g., a combination of two elongate, optically-continuous augite outcrops, with combined length 132 μm and maximum width 22 μm, found within a plagioclase crystal with little visible twinning. The largest augite, in terms of area exposed, is a roughly teardrop-shaped grain, whose long dimension, 240 μm, is oriented nearly perpendicular to the twinning of the large (1.8 mm) plagioclase that completely surrounds it. The largest augite not included within plagioclase is a blocky, 75 × 45 μm grain, which appears to contain submicron-scale exsolution lamellae. This grain is found in one of the most severely cataclasized regions of the thin section. Possibly it, too, originally formed within a plagioclase crystal.

Our RNAA results for siderophile elements (Table 1) confirm that this lithology is pristine. Based on the uniformly low An ratio of its plagioclase (Table 2), it is clearly an alkali anorthosite. Note that our small sample was probably far more representative for this relatively fine-grained (thoroughly cataclasized) alkali anorthosite than was our even smaller sample of the similar but extremely coarse-grained 12003,179 lithology.

Harzburgite 12033,503

We studied a 52-mg sample (12033,503), representing roughly half the original mass of the smallest of four rocklets collectively designated 12033,428 (*Marvin,* 1978). (The total mass of 12033,428 was originally 900 mg. However, based on Marvin's description, as well as our own binocular-microscopic observations, it seems unlikely that any of the other three 13033,428 rocklets is similar to the 12033,503 rocklet.) Marvin noted that the fragment from which 12033,503 was taken is coarse-grained, with "a waxy luster," and that it appeared to consist of about 55% plagioclase and 45% "yellow-brown olivine or pyroxene." Binocular-microscopic identification of extremely magnesian (and thus extremely pale) lunar olivine is difficult, however, and it now appears that the phase identified by Marvin as plagioclase is actually olivine, while the other phase is low-Ca pyroxene. Thus, the lithology is a harzburgite. Our allocation came in the form of two fragments, which we elected to study separately, because the lithology appeared coarse-grained and we wanted to assess the impact of its heterogeneity on our results. The two thin sections are similar. Both are dominantly olivine and low-Ca pyroxene. The larger thin section (corresponding to INAA sample 12033,503a) also contains a large trace of Cr-Fe-spinel and a small trace of FeNi-metal.

The lithology has been partially comminuted in a highly unusual fashion. Olivine crystals tend to be largely intact, with gently curving boundaries (Fig. 9a), and up to 1.4 mm across. Probably the original olivine grains were typically 1-2 mm. The Cr-Fe-spinel occurs in grains up to 0.3 μm across. The pyroxene tends to be fragmented into a bizarre, mottled

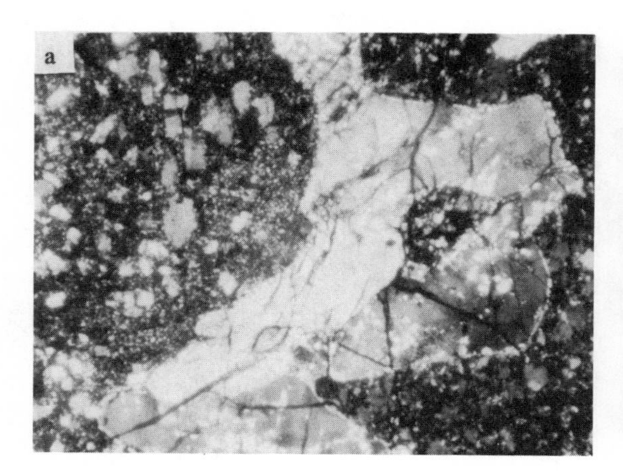

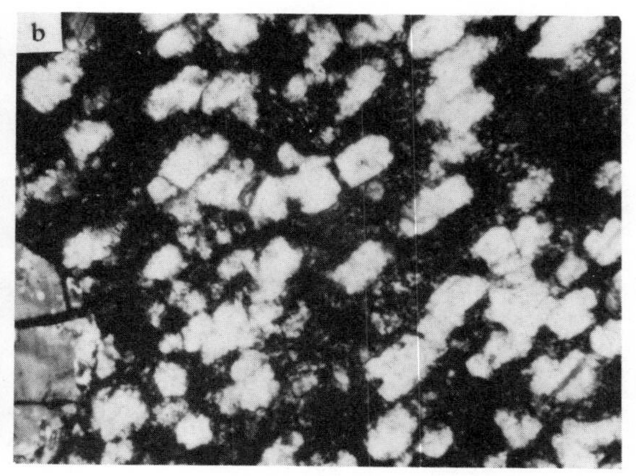

Fig. 9. Photomicrographs of the 12033,503 harzburgite. **(a)** A cluster of five olivine crystals (contiguous light-colored zones in center and right), showing extremely anhedral shapes with gently curving boundaries, set amidst a groundmass of granulated enstatite (dark grey except for scattered light "bricks"). Note in right-center area where a pair of olivine crystals together surround (in this two-dimensional view) a 0.2-mm patch of enstatite. View is 1.3 × 1.0 mm; nicols partly crossed. **(b)** A texture reminiscent of a brick wall is seen within the enstatite regions when the larger, optically continuous "brick" grains are rotated to 45° from extinction. Except for part of an olivine crystal in the lower left corner, this entire 0.63 × 0.48 mm (nicols crossed) view is enstatite.

texture. About half of it occurs as an extremely fine-grained groundmass, composed of crystallites typically about 3 μm across. A few areas of the groundmass are mixtures of olivine and pyroxene, but generally it appears to be exclusively pyroxene. The remainder of the pyroxene occurs as much larger unbroken domains, typically about 140 × 70 μm, scattered fairly evenly throughout the groundmass, and in optical continuity with one another over distances of up to 3.0 mm (limited by the edge of the thin section). Thus, the overall appearance of the pyroxene resembles a wall of massive bricks, of uneven size, set within a relatively generous supply of fine-grained mortar (Fig. 9b). Within each zone of optical continuity, the long dimensions of the "bricks" tend to be aligned parallel to one another. Assuming that the zones of optical continuity (or at least large regions within the zones of monomineralic pyroxene) correspond to originally coarse grains, then the original texture was that of a poikilitic cumulate, with ovoid olivine chadacrysts surrounded by pyroxene oikocrysts (the overall shapes of which cannot be discerned, mainly due to their coarseness in comparison to the small sizes of the two thin sections).

The compositions of the two forms of pyroxene are indistinguishable. Like the olivine (Table 2), the pyroxene is extremely magnesian. It is also remarkably Ca-poor, averaging $Wo_{0.40}$ (range among 13 analyses $Wo_{0.20-0.68}$). Similar mineral compositions have been reported for a 4 × 3 mm olivine pyroxenite clast from Apollo 14 breccia 14305 (*Shervais et al.,* 1984).

For most elements, our INAA results (Table 1) show only mild disparities between the two samples, but the results for siderophile elements Ir and Au are grossly disparate. The 12033,503 lithology is obviously monomict (it completely lacks plagioclase), and despite the high Ir and Au concentrations, we classify it as pristine. Laboratory contamination is only remotely conceivable as a cause of the high Ir and Au results for 12033,503b. Nothing was unusual about the procedures used to prepare this sample for INAA, and the level of contamination would never have been even remotely precedented among analyses from our laboratory. This sample contains FeNi-metal that is extremely Ni-rich. We obtained five analyses, only two of which yielded sums ≥ 94.1 wt.%, due mainly to the fine grain size (comparable to the electron beam diameter) of the metals probed. If we normalize the sums to 100 wt.%, the average composition is, in wt.%, Fe = 44.63, Co = 2.25, and Ni = 53.12; if only the two best-sum analyses are used, the average becomes Fe = 44.90, Co = 2.25, and Ni = 52.84. Another ultramafic lunar rock, the 72415/72417 dunite, also contains extremely Ni-rich FeNi-metal (average Ni content roughly 30 wt.% (*Dymek et al.,* 1975; *Ryder et al.,* 1980)), and has been analyzed repeatedly for bulk-rock Ir and Au concentrations by *Morgan and Wandless* (1988). Based on 10 analyses, these authors found Ir = 0.62 ± 1.14 ng/g and Au = 2.33 ± 1.66 ng/g (1-σ standard deviations). These 72417 concentrations are higher than the poorly constrained "averages" for other pristine rocks (*Warren,* 1990b) by more than an order of magnitude for Ir, and by several orders of magnitude for Au. Based on the weighted mean of our two analyses of 12033,503 (considering the tiny mass of 12033,503b, it would be inappropriate to use it alone for comparison with results for much larger samples), the Au concentration of 12033,503 is less than 2 of the 10 results for 72417. The Ir concentration of 12033,503 remains

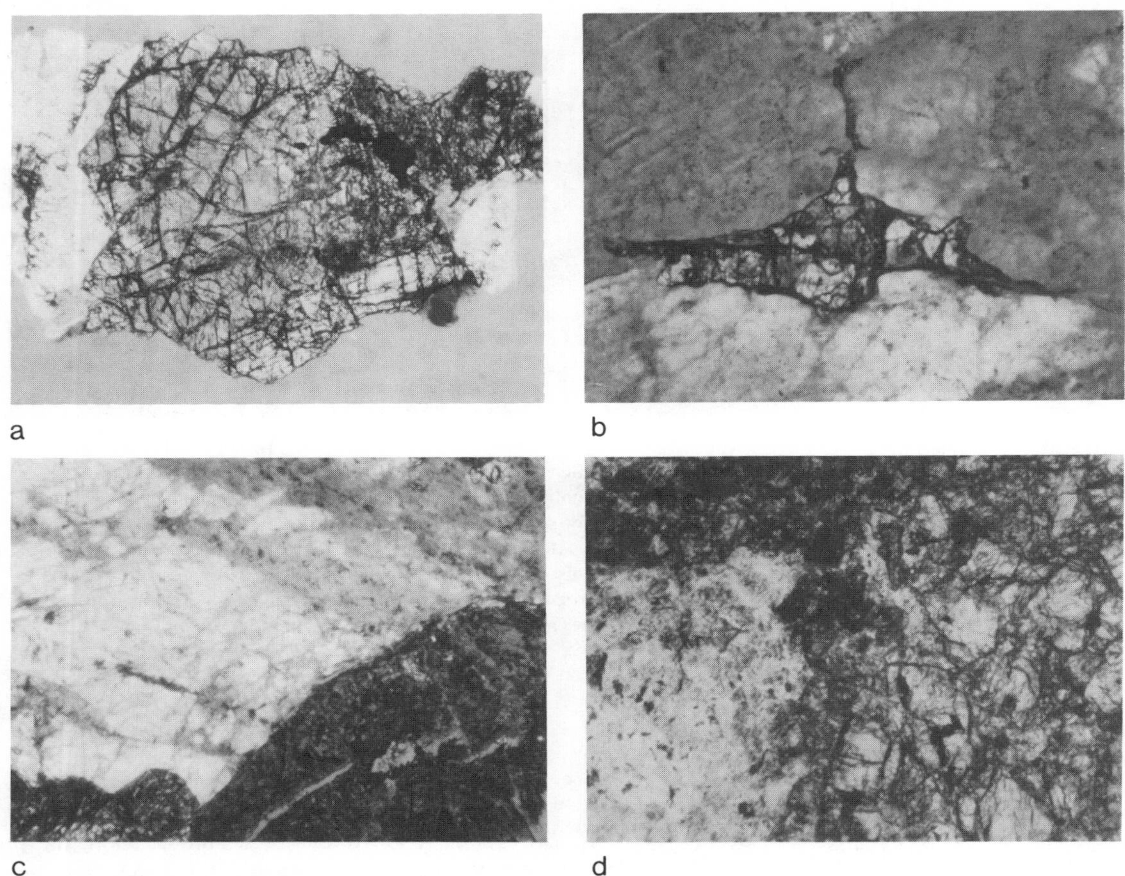

Fig. 10. Photomicrographs of three pristine norites and one ferroan anorthosite. **(a)** Largest contiguous outcrop of 12037,174 among two thin sections produced after it partially disaggregated. Parts of two pyroxenes, three plagioclases, and one ilmenite are visible in this 2.0 × 1.5 mm view (nicols partly crossed). **(b)** Relict cumulate texture in 15295,41: Portions of three coarse, cumulus plagioclase crystals, and (in center) a small interstitial, extremely anhedral augite. View is 0.49 × 0.38 mm (partly crossed nicols). **(c)** Relict cumulate texture in 15360,11: Portions of two large cumulus pyroxene crystals (dark greys) and three large, at least partly cumulus, plagioclase crystals (light greys) are visible. View is 1.3 × 1.0 mm (partly crossed nicols). **(d)** Cataclastic texture of 15361,4. Except for zone of monomineralic plagioclase (light grey, lower left), this view is essentially pure low-Ca pyroxene. View is 1.3 × 1.0 mm (uncrossed nicols).

unprecedented among pristine lunar rocks, but considering how much easier it is to contaminate with Au than with Ir, we find this comparison with the comparably ultramafic 72417 reassuring.

Mare-Nonmare Transitional (?) Norite 12037,174

We studied two complete fragments (12037,174a, 46 mg, and 12037,174b, 16 mg), selected from 12037,166, a set of miscellaneous rocklets from the NW rim of Bench crater. *Marvin* (1978) described the two 12037,174 rocklets together as "friable, coarse-grained green and white anorthositic" material. Our own binocular-microscopic observations of the rocklets just prior to INAA indicated that they were uncommonly coarse-grained (grains typically 2-3 mm, up to 4 mm); and their general appearance led us to infer a 99% probability that the two were originally part of a single former rock. However, the tiny ,174b rocklet appeared to comprise only parts of two to four crystals, and had a lower proportion of pyroxene ("green") than the ,174a rocklet. The combined mode appeared to be roughly 50% pyroxene, 50% plagioclase, with only a trace of minute opaques.

Unfortunately, the ,174a sample broke into several pieces during INAA, and consequently the two thin sections produced from it consist of 10 separate areas, none more than 2.2 mm across. Worse still, no thin section was produced from the ,174b rocklet, as all of its tiny mass was lost during polishing. Nonetheless, the available thin sections suffice to indicate that the lithology is a mildly-brecciated, coarse-grained pristine rock, probably a cumulate (Fig. 10). The thin section mode shows 70% pyroxene, which is entirely low-Ca and at least partially orthopyroxene, 30% plagioclase, 1% ilmenite, and a trace of troilite. Note, however, that the high bulk Al concentration of the smaller ,174b fragment (Table 1) implies that it was dominantly plagioclase. The pyroxenes tend to occur as large, blocky masses, ophitic or oikocrystic relative

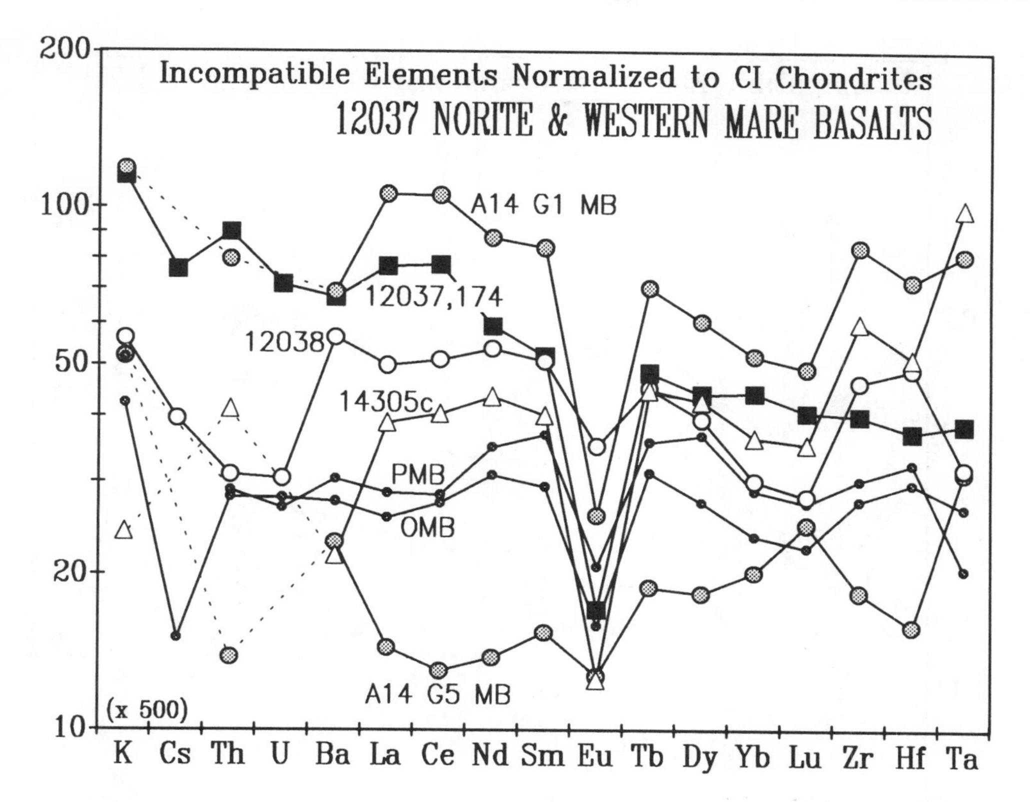

Fig. 11. Concentrations of incompatible elements in 12037,174 normalized to CI chondrites. Shown for comparison is a selection of mare lithologies from the west-central nearside (i.e., Apollo 12 and 14) region: Apollo 14 G1 MB and Apollo 14 G5 MB are the average compositions of Apollo 14 aluminous mare basalt groups 1 and 5 of *Dickinson et al.* (1985); 12038 is an aluminous mare basalt from Apollo 12 (*S. R. Taylor,* 1982); PMB and OMB are *Taylor's* (1982) compositions for Apollo 12 pigeonite and olivine mare basalts, respectively; and 14305c is the mare cumulate clast analyzed by *Taylor et al.* (1983).

to the plagioclase. The pyroxenes are observed to be up to 2.0 mm across, and considering the fragmentary nature of the thin section, we infer that most of them were originally at least this coarse. Most of the plagioclase occurs as subhedral laths, typically about 0.2 mm across, with longer dimensions unmeasurable due to the fragmentary nature of the thin sections. However, the largest plagioclase fragment (1.3 × 0.5 mm) is relatively equant (albeit its shorter dimension is entirely limited by the edge of the section). The ilmenite occurs as anhedral masses, the largest of which is 35 μm across.

This lithology is manifestly pristine, but we are not certain whether it is of mare or nonmare affinity. The Wo ratio of the pyroxene is mostly <4.0 mol.%, and the mean based on 19 analyses is $En_{64.9}Wo_{4.5}$ (Fig. 8, Table 2). Mare basalts generally contain pyroxene with average Wo ratio >>10 mol.% (*Papike et al.,* 1976). Even a cumulate mare lithology such as the ancient (4.23 Ga) olivine gabbronorite clast from 14305 (*Taylor et al.,* 1983) contains no pyroxene with Wo <7 mol.%, and has an average Wo ratio of roughly 20 mol.% [cf. Apollo 15 mare cumulates, such as 15385, 15387, and the 15459 picrite clast (*Ryder,* 1985)]. Of course, conceivably our tiny sample happens to be an unrepresentative piece of a parent rock with abundant augite elsewhere. The plagioclase in mare cumulates is generally far more heterogeneous in composition than observed for 12037,174 (Table 2); e.g., in the 14305 clast, *Taylor et al.* (1983) found a range of An_{82-93}. The FeO content of the 12037,174 plagioclase (0.14 ± 0.03 wt.%) would also be unprecedentedly low for a mare lithology. The ilmenite content of 12037,174 would be low, but far from uniquely so, for a mare lithology.

The narrow widths of most of the plagioclase grains tend to suggest that the lithology is an orthocumulate (a cumulate with a high "trapped liquid" proportion), and therefore most likely not plutonic, but rather hypabyssal, i.e., more mare-like than most pristine nonmare rocks. The pattern of incompatible element concentrations (Fig. 11) is comparable to those of Apollo 14 mare basalts from "Group 3" of *Dickinson et al.* (1985). It is also comparable to those of several Apollo 17 plutonic norites, especially 72255c (Fig. 12), but plutonic norites do not have such pronounced negative Eu anomalies as 12037,174. The Eu concentration of 12037,174 is only 0.55 × that of 72255c. A 5 × 3 mm gabbro clast from Apollo 17 breccia 76255 (*Warner et al.,* 1976) appears roughly

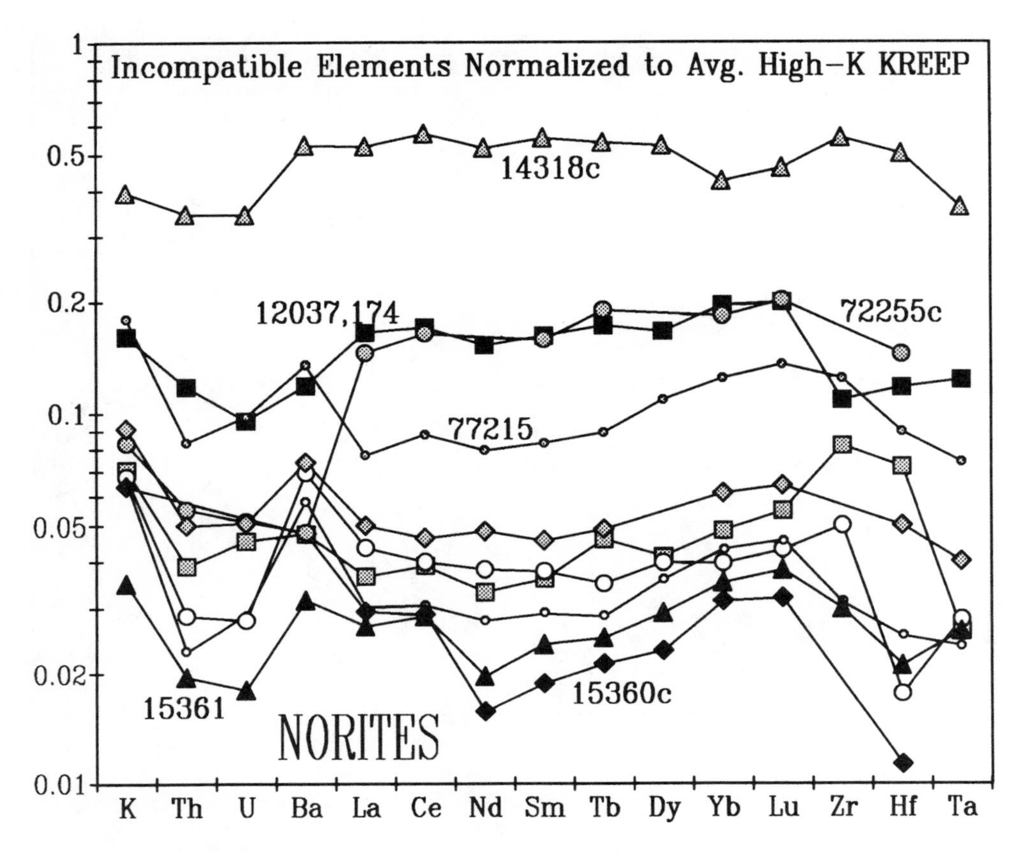

Fig. 12. Concentrations of incompatible elements in three pristine norites (black symbols), normalized to average high-K KREEP (*Warren,* 1989). Also shown for comparison is a selection of pristine norites from the literature. In order of increasing Sm concentration, samples plotted are: 15360c, 15361, 78235, 15445c, 15455c, 77035c, 77215, 72255c, 12037,174, and 14318c. Data are mainly from compilations of *Ryder and Norman* (1979), *Ryder* (1985), and *Warren* (1990a).

similar in texture and major-element composition, especially Mg/(Mg+Fe) (hereafter abbreviated as mg^*), to 12037,174 (the 76255 clast has never been analyzed for trace elements). Differences include the nature of the pyroxene, which in 76255 is orthopyroxene and augite exsolved from primary pigeonite and the far greater compositional heterogeneity of the 76255 plagioclase (*Warner et al.,* 1976). Considering that the 76255 gabbro has traditionally been classified as nonmare (*Warner et al.,* 1976; *James and Flohr,* 1983; *Warren,* 1985), we are inclined to regard 12037,174 as nonmare. (Strictly speaking, the term "mare" only applies to the basaltic lavas of the lunar "seas," which might well have intrusive counterparts that are identical in bulk composition, are yet are not of "mare" origin. The 12037,174 lithology does not appear to be extrusive.) However, like the 14305 clast discussed by *Taylor et al.* (1983), lithologies such as 12037,174 and the 76255 gabbro tend to blur the boundary between relatively primitive mare and relatively evolved nonmare materials. We defer discussion of 12037,174 as a nonmare, or mare-nonmare transitional, lithology to a later section.

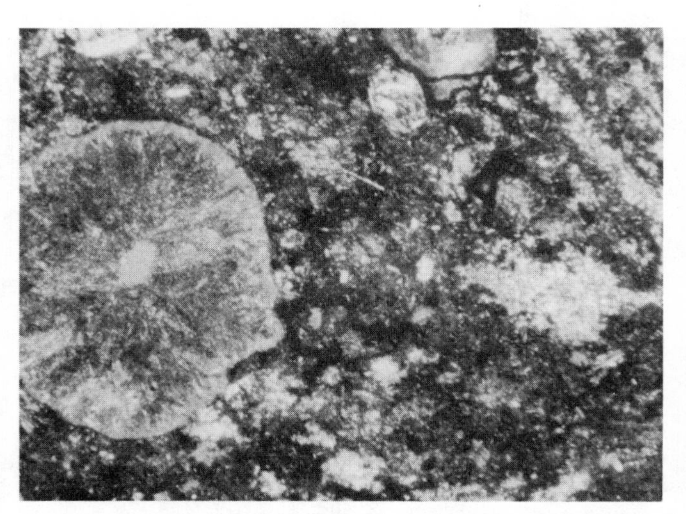

Fig. 13. Photomicrograph of 14076,7, showing micropoikilitic matrix and large chondrule-like clast at left. View is 1.3 × 1.0 mm (partly crossed nicols).

Noritic Impact Melt Breccia 12042,243

Marvin (1978) classified this 260-mg rocklet from near Surveyor crater as a fine-grained norite. We studied a 41-mg fragment (12042,278) by INAA. After INAA, the sample was powdered, and aliquots were used for RNAA (31 mg) and EPFBA (10 mg). A separate 15-mg fragment was used to produce a thin section. The lithology is a fine-grained, poikilitic impact melt breccia; for examples of textural analogs, cf. 60616 and 65015 (*Ryder and Norman,* 1980). Siderophile elements are at high concentrations (Table 1), and the composition of the abundant FeNi-metal in the thin section is in the "meteoritic range" typical of polymict lunar samples, clustering around 7 wt.% Ni, 0.6 wt.% Co. The bulk composition of the sample is distinctly KREEPy (Fig. 3). The depletion of Th, U, and Ba relative to the REE is a noteworthy example of significant (albeit slight) diversity in the trace-element patterns of KREEPy materials (cf. *Warren,* 1989).

Mare Basalt 12044,134

The sample we studied was a complete 39-mg rocklet from the south rim of Surveyor crater, the most leucocratic of 16 "igneous-looking" and fine-grained fragments grouped by *Marvin* (1978) as 12044,117. The thin section and our analysis (Table 1) indicate that it is an ordinary Apollo 12 mare basalt.

Mostly-shock-melted but Compositionally Near-pristine Troctolite from 12071,5

The petrography and the lunar-petrologic implications of complex 1.38-g rocklet 12071,5 are described by P. H. Warren and E. A. Jerde (unpublished data, 1989). For completeness (i.e., for the convenience of future pristine rock data compilers), we include our bulk-compositional results for two samples (one matrix-free, the other virtually matrix-free) of a large troctolitic clast in Table 1. This clast appears to represent a troctolitic cumulate that has been severely shocked, to the extent that most of its crystals have been transformed into brownish glass. However, the clast seems to have remained compositionally at least nearly pristine, in the sense that there appears to have been little segregation of melt apart from crystals, and little or no mixing with external ("contamination") material.

High-Al, Low-K KREEP Polymict Breccia (Clast) 14076,7

Apollo 14 rocklet 14076 is a 2.0-g sample from the Station G trench. It consists mainly of two distinct regolith breccia lithologies: a coarse-grained noritic regolith breccia that is compositionally similar to the local soils, and a fine-grained anorthositic regolith breccia that is compositionally unlike any other Apollo 14 regolith sample (*Vaniman,* 1990; E. A. Jerde, R. V. Morris, and P. H. Warren, unpublished data, 1989). Each of these lithologies formed roughly half of the original rocklet. However, a major portion of the noritic regolith breccia was a relatively large (5 × 3 mm) clast of a fine-grained, dark grey lithology. Our sample 14076,7 represents roughly half of this clast. The thin section produced from 14076,7 after INAA shows a micropoikilitic impact melt breccia, mainly composed of oikocrysts typically ~0.5 mm across, embaying chadacrysts typically ~10 μm across (Fig. 13). However, roughly 10% of the clast consists of lithic and mineral clasts, typically ~100 μm across. The lithic clasts are themselves impact melt breccias, ranging in texture from devitrified-glassy to micropoikilitic to subophitic (basaltic). Many of them appear rounded, and at least superficially resemble the "chondrules" described from Apollo 14 regolith breccias (e.g., *Kurat et al.,* 1972). The largest lithic clast by far is an almost perfectly round micropoikilitic "chondrule" ~0.6 mm across (Fig. 13).

Classification of this lithology is problematical. The presence of spherules in a breccia is ordinarily taken as evidence for a regolith derivation (*Stöffler et al.,* 1980; *McKay et al.,* 1986). It seems likely that the 14076,7 lithology formed as a mixture of impact melt with an uncommonly spherule-rich regolith (or regolith breccia), in proportions such that most of the larger clasts in the regolith survived essentially intact, even though the melt, its crystallization products, and the finer components of the regolith merged into a fairly uniform micropoikilitic groundmass. Thus, the lithology is a clast-rich impact melt breccia, but apparently with a former regolith (or regolith breccia) as a major, and perhaps even dominant, component.

Despite the small size of our sample, this lithology is so fine-grained that our analysis (Table 1) should be fairly representative. The Al content (128 mg/g) is remarkably high by Apollo 14 polymict standards. The only large Apollo 14 rocks of comparable Al content are the "white" fragmental breccia 14063, which has 116 mg/g (*Helmke et al.,* 1972; *Laul et al.,* 1972; *Rose et al.,* 1972; *Taylor et al.,* 1972), the unique regolith breccia 14315, which has 115 mg/g (*Rose et al.,* 1972; *Jerde et al.,* 1987), and a group of distinctive high-Al, medium-K KREEPy impact melt breccias (14073, 14078, 14310, 14152,5,102, and 14276) that apparently were derived from a single impact melt, and have 107-115 mg/g (*McKay et al.,* 1978). The 14076,7 impact melt also appears distinctly poorer in incompatible elements than the 14310-group impact melt; e.g., Sm = 12.5 μg/g in 14076,7, vs. 20-28 μg/g in the five high-Al, medium-K KREEPy impact melt breccias of *McKay et al.* (1978).

High-Al, Medium-K KREEP (14310-like) Impact Melt Breccia 14077

This 2.77-g rocklet was collected from the same Station G trench as 14076, and according to *Carlson and Walton* (1978) it was "described during PET as being composed of 99% feldspar." Our INAA results (Table 1) indicate that it is uncommonly Al-rich by Apollo 14 polymict standards, but its composition corresponds precisely to the 14310 group of high-Al, medium-K KREEPy impact melt breccias reviewed by *McKay et al.* (1978). Its texture is also similar to that of 14310 (albeit 14077 grain sizes are slightly finer). The strong compositional similarity between 14077 and 14310 includes trace siderophile elements such as Au, Ga, Ni, and Ir (*Baedecker et al.,* 1972; *Morgan et al.,* 1972; *Taylor et al.,*

1972). Our results support the suggestion of *McKay et al.* (1978) that the entire 14310 group (14073, 14078, 14310, 14152,5,102, 14276, and now also 14077) was derived from a single impact melt. Note that all these rocks were collected at Station G, except 14276, which was collected ~280 m west of Station G. Thus, considering that the entire Apollo 14 traverse spanned ~1630 m, it appears that the parental impact melt sheet was localized to within a few hundred meters of Station G. If so, then the relatively Al-rich composition of this melt sheet (compared to the local regolith and most polymict rocks from Apollo 14) in turn suggests that the upper crust is significantly more anorthositic than the uniformly mafic Apollo 14 regolith, within only a few hundred meters of the site.

High-Al Impact Melt Breccia (Clast) 14315,28

The clast we studied outcropped as a 6 × 4 mm white area on 14315,3. We were allocated a 37-mg sample that included essentially the entire clast, plus some adhering matrix material. By mechanically cleaning the matrix off the clast, we obtained a 17-mg sample, estimated to be 99.9% matrix-free, for INAA and petrographic study.

The thin section shows a mode dominated by plagioclase and lesser low-Ca pyroxene, with little else except a trace of troilite. The texture is unusual. It is clearly a product of impact melting, as most of the section is roughly micropoikilitic, with prismatic chadacrysts typically about 25 μm across surrounded by oikocrysts typically about 0.3 mm across. The volume ratio of chadacryts to oikocrysts is uncommonly high, and in some chadacryst-rich areas the texture is almost microdiabasic. Roughly 10% of the section consists of scattered mineral clasts that are relatively coarse-grained (Fig. 1), with low-Ca pyroxene up to 0.35 mm, and plagioclase up to 1.1 mm (limited by edge of the section). However, the pyroxene clasts grade imperceptibly into oikocrysts. The few (3-4) plagioclase clasts tend to be far larger than any of the oikocrysts, and for the most part they have sharp boundaries with the micropoikilitic groundmass. The largest plagioclase has a sharp boundary on most sides, but on one side its boundary is indistinct and appears to grade from pure plagioclase into a micropoikilitic (groundmass-like) mixture of plagioclase and pyroxene. Thus, the overall impression gleaned from the thin section is that while the lithology is definitely a clast-rich impact melt, it conceivably represents a pristine anorthositic norite that has endured shock-melting of its grains in situ, but avoided mixing with any external material. The low-Ca pyroxene is uniform in composition (Fig. 2), but the plagioclase is compositionally diverse. The coarse plagioclase mineral fragments are Na-poor (average of 6 analyses: $An_{93.29}$) compared to the groundmass plagioclase. Plagioclase An ratio also tends to be more compositionally diverse than pyroxene mg^* among Apollo 15 impact melt breccias (*Laul et al.,* 1988).

Unfortunately, our INAA results for siderophile elements (Table 1) are only upper limits that, although consistent with an unmixed (pristine) composition, do not prove much except that the lithology is probably not, based on Ni, a regolith breccia. Thus, we must assume that the lithology is thoroughly polymict. For an Apollo 14 impact melt, the clast is unusually Al-rich and incompatible-element-poor. However, if the impact melt had a monomict protolith, as the petrographic evidence weakly suggests, it might be expected to deviate markedly in composition from the more common polymict varieties of impact melt, in which original diversity has been "smoothed out" by mixing. If any one pristine lithology dominates 14315,28, it would seem to be a relatively magnesian and mafic variety of alkali anorthosite, based on its high Na/Ca, Eu/Al, etc. (Tables 1 and 2).

Low-K KREEP Impact Melt Breccia (Clast) 15015,179

The sample we studied was a 5 × 4 mm clast found on 15015,4. After cleaning off matrix material, we were left with only a 10.5-mg chip of pure clast for our studies. The lithology is a poikilitic impact melt, compositionally (Table 1) much like numerous others from Apollo 15, especially the "Group B" samples of *Ryder and Spudis* (1987). As with most Apollo 15 impact melt breccias (*Laul et al.,* 1988), far more diversity is shown by the plagioclase An ratio than by the pyroxene mg^* ratio (Table 2, Fig. 2).

Ferroan Anorthosite (Clast) 15295,41

Large regolith breccia 15295 was broken into several pieces during processing in 1971 (*Ryder,* 1985). A small (~0.25-g) clast of ferroan anorthosite found on 3.8-g fragment 15295,5 was studied by *Warren and Wasson* (1978). The clast we studied was found as a 20 × 8 mm outcrop on 15295,0. As noted by *Ryder* (1985), the *Warren and Wasson* (1978) clast may have been part of this larger clast, before 15295 was split apart. However, according to notes in the 15295 "Data Pack" (stored at NASA Johnson Space Center), the 15295,5 fragment could not "be fitted to" the 15295,0 fragment.

Our bulk compositional data (Table 1) indicate that the 15295,41 clast is similar to the 15295,5 clast. Both are nearly monomineralic plagioclase, with <1% pyroxene, which is dominantly high-Ca. Both are cataclastic but monomict. However, the 15295,41 clast is not as extremely cataclastic as the 15295,5 clast. In the latter, the coarsest surviving plagioclase fragment is ~1.1 mm across. In the thin section corresponding to 15295,41, surviving plagioclase is up to 4 × 2 mm; original grains were probably at least this coarse. This section also shows vestiges of an original cumulate texture, with subhedral cumulus plagioclase and anhedral interstitial pyroxene (Fig. 10). The largest pyroxene is an elongate grain, ~0.90 × 0.05 mm, sandwiched between two much larger plagioclase crystals. The pyroxene of the 15295,41 clast has significantly lower mg^* than the pyroxene of the 15295,5 clast (Fig. 8); the range between them is greater than the typical range for individual pristine ferroan anorthosites, although the range for 60665 (*Dowty et al.,* 1974) is comparable. Thus, it seems likely (though far from definite) that we are dealing with two separate, albeit similar, pristine clasts.

Mg-suite Norite (Clast) 15360,11

Regolith breccia 15360, a 9.3-g rock collected at the rim of Spur crater, contains a 1-cm clast (*Ryder,* 1985) that was hitherto unstudied. Our 100-mg allocation included a substantial rind of matrix material. We mechanically separated a 76-mg sample of pure clast material (15360,11) to be the main focus of our studies, and used the remaining material for conversion into a mainly-matrix thin section. After INAA, the 15360,11 sample was divided into a 24-mg sample for RNAA, a 14-mg sample for EPFBA, and a 36-mg sample for conversion into a pure-clast thin section (15360,15).

The thin sections show a brecciated but monomict norite, with an original coarse-grained, igneous texture plainly discernible (Fig. 10). Estimation of original grain size is hampered by the small size of the thin sections. Section 15360,15 is only 5 mm^2, and comprises parts of only 3-4 plagioclase crystals and 2-3 orthopyroxene crystals. However, plagioclase is up to 2.9 mm across, and pyroxene is up to 1.9 mm (in both cases limited by edges of the thin section). Grains observed during binocular-microscopic inspection of the sample prior to INAA appeared to be up to 5 mm. The estimated mode is 65% plagioclase and 35% pyroxene. The equant shapes and gently-curving shapes of the grain boundaries suggest that the lithology originated as a plagioclase-orthopyroxene cumulate. Both plagioclase and pyroxene are compositionally uniform (Table 2, Fig. 8), and they indicate that the lithology is of Mg-suite affinity. Our results for siderophile elements (Table 1) confirm that this norite is pristine.

Mg-suite Norite 15361

Like the 15360 clast, rocklet 15361 (0.9 g) was collected at the rim of Spur crater, and hitherto unstudied. Their modes, textures, and bulk compositions are also similar. Thin section 15361,4 shows a texture more cataclastic than the 15360 norite (Fig. 10), but monomineralic crush zones are up to 1.8 mm across for pyroxene and 1.5 mm for plagioclase (in both cases limited by edges of the thin section). Much of the plagioclase has been maskelynitized. After cataclasis, the lithology was mildly annealed (it was remarkably difficult to crush in preparation for INAA). Nonetheless, it appears monomict. The mode is roughly 60% pyroxene and 40% plagioclase. Both plagioclase and pyroxene are compositionally uniform (Table 2, Fig. 8), and indicate an Mg-suite affinity. In fact, the orthopyroxene is remarkably magnesian, for an olivine-free lithology. Our results for siderophile elements (Table 1) confirm that this norite is pristine.

Granulitic-fragmental-poikilitic Polymict Breccia 15364

As reviewed by *Ryder* (1985), 15364 is a little-studied 1.5-g rock collected at the rim of Spur crater. The only previous subdivision of the rock was a single 0.2-g chip used to produce all existing thin sections—including the one we studied, 15364,14. Aside from *Ryder* (1985), the only previous studies were by the Chicago mineralogy group (*Steele et al.,* 1977; *Hansen et al.,* 1979; *Smith et al.,* 1980). *Smith and Steele* (1980) classified 15364 as "plutonic," and *Ryder* (1985) classified it as a "monomict (?)" fine-grained fragmental breccia. The classification remains problematical, but our results suggest that the lithology is most likely polymict.

Previous workers described the fine grain size of the rock's groundmass (*Ryder,* 1985, Fig. 2a), and some of the diversity of its clast textures. Recognizable clasts make up ~20% of the rock. Several clasts of medium-grained (typically 0.2-0.4 mm), igneous-textured (roughly diabasic) material (e.g., Fig. 2b of *Ryder,* 1985; Fig. 5c of *Steele et al.,* 1977) seem out of place, beside scattered grains of coarse, equant plagioclase, the largest of which in 15364,14 is 1.6 mm across (albeit limited by the edge of the thin section). Adding to the general appearance of incongruity are fine-grained (~0.1 mm) granulitic clasts (e.g., Fig. 2c of *Ryder,* 1985), fine-grained poikilitic clasts (chadacrysts typically ~0.1 mm across) (e.g., Fig. 2d of *Ryder,* 1985), and several small clasts in 15364,14 that are largely aphanitic (Fig. 5). This range of clast textures is very similar to that seen in 67215, a quasipristine but nonetheless polymict (*McGee,* 1988) breccia (see below). Fine-grained granulitic domains are also found in pristine breccias such as ferroan anorthosite 62236 (*Nord and Wandless,* 1983), but the mingling of diabasic, finely-poikilitic, and especially the aphanitic clasts suggest that 15364 is a mixture of granulated cumulate materials and impact melt products, i.e., that it is polymict.

We estimate the thin section mode to be roughly 80% plagioclase, 10% olivine, and 10% pyroxene. However, our bulk-compositional results imply a norm that, translated into vol.%, has significantly lower plagioclase: ~66 vol.%. The mafic silicates, at least within the thin sections (all from a single "butt"), are rather uniform in *mg** ratio [Fig. 6, Table 2; cf. *Steele et al.* (1977) and *Smith et al.* (1980)]. However, the *mg** ratio found for the bulk composition (Table 1) is far higher than that of any mafic silicate found in the thin section. We have no reason to doubt the accuracy of the bulk-compositional data. Agreement between INAA and EPFBA results, where they overlap, is fine: INAA gave Fe = 39 mg/g, EPFBA gave 38 mg/g; INAA gave Ca = 86 mg/g, EPFBA gave 86 mg/g. Apparently, the bulk-analysis sample contained either Mg-spinel (which would be incongruous in a rock with such a high pyroxene/olivine ratio), or mafic silicates far more magnesian than any in the thin sections. Such heterogeneity tends to support the inference that the lithology is polymict. Plagioclase compositions are also diverse (Table 2). Note, however, that although in their text *Steele et al.* (1977) report a composition of An_{97} for "euhedral" plagioclase, the composition shown in their Table 3 is actually $An_{94.7}$.

The siderophile element concentrations (Table 1) are extremely low, for a polymict breccia. Relative to the "cut-off" concentrations (3×10^{-4} times CI chondrites) recommended by *Warren and Wasson* (1977) for distinguishing pristine rocks from polymict breccias, 15364 has Au = 1.6, Ir = 2.0, Ni = 5.3, Os = 3.0, and Re = 1.8. Also, Ge, which is siderophile but also relatively volatile and "labile," is at 0.63× the "cut-off." For further comparison, concentrations of the "noble" siderophile elements Ir, Os, and Re average 3.0× higher in

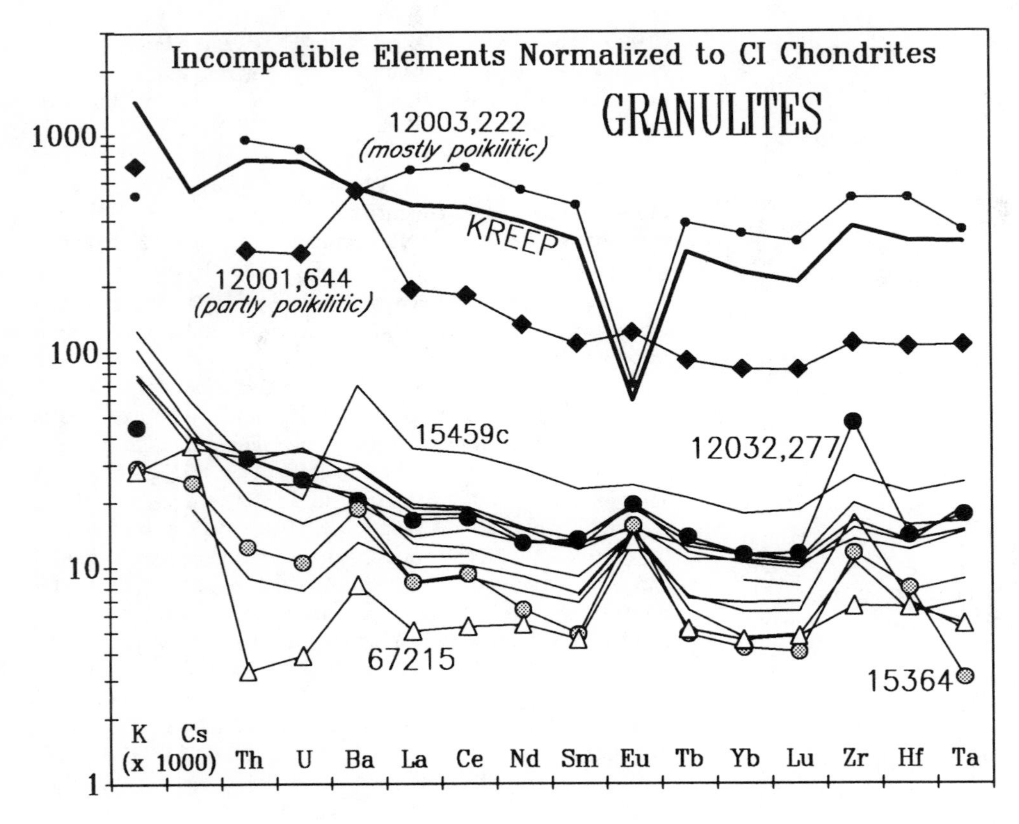

Fig. 14. Concentrations of incompatible elements in granulitic breccias, normalized to CI chondrites. The 67215 composition is an average of our data (Table 1) and data of *Lindstrom and Lindstrom* (1986). The 12003,222 and 12001,644 "poikilitic/granulitic rocks" are from *Laul* (1986), with petrographic classifications from *Simon and Papike* (1985). Other granulites plotted include 15418, 67415, 67955, 76230/5, 77017, 78155, and 79215, all from compilation of *Lindstrom and Lindstrom* (1986), and two 15459 clasts from *Lindstrom et al.* (1988).

67215 (Table 1) than in 15364. On balance, 15364 appears more likely polymict than pristine. However, it probably represents a mixture of a relatively limited range of precursor rock types.

Anorthositic Norite Granulitic Breccia 67215

Although the bulk composition of the upper lunar crust corresponds to anorthositic norite (*S. R. Taylor,* 1982), anorthositic norites are not as abundant among pristine rocks as more extreme compositional types such as norite, anorthosite, troctolite, and noritic/troctolitic anorthosite (*Warren and Wasson,* 1977). The 276-g granulitic breccia 67215 has been widely discussed as a pristine or near-pristine anorthositic norite ever since *G. J. Taylor* (1982) suggested that it appears monomict; see, for example, *Warren et al.* (1983a), or *Lindstrom and Lindstrom* (1986). Detailed petrography by *McGee* (1988) indicates that the breccia is polymict (cf. *Stöffler et al.,* 1985), but it probably represents a mixture of relatively few precursor rock types, dominated by ferroan anorthosite and noritic anorthosite. *Lindstrom and Lindstrom* (1986) had earlier concluded that the breccia is contaminated by meteoritic Ir. We have applied RNAA to determine several additional highly-siderophile elements: Au, Ge, Os, and Re, while supplementing the *Lindstrom and Lindstrom* (1986) data for Ir, Ni, and numerous other elements (Table 1). Except for the volatile-labile Ge, the four additional highly-siderophile elements are found to be at roughly the same chondrite-normalized concentrations as Ir. Thus, these data tend to confirm that 67215 is polymict.

DISCUSSION

Granulitic-fragmental Polymict Breccias (12032,277, 15364, and 67215)

Warner et al. (1977) noted that polymict breccias with granulitic textures tend to be essentially free of KREEP—unlike impact melt breccias, which tend to have high incompatible element concentrations in KREEP-like patterns. *Warner et al.* (1977) also noted, based on ages for only a few granulitic breccias (all from Apollo 17), a tendency for them to be older (4.1-4.3 Ga) than impact melt breccias (3.85-4.1 Ga). Six granulitic breccias from Apollo 16 Station 11 (North Ray crater) give ^{40}Ar-^{39}Ar ages restricted to 3.8-4.0 Ga (*Stöffler et al.,* 1985). However, several pristine ferroan anorthosites dated in the same study appear equally "young." *Warner et al.* (1977) suggested that the absence of KREEP in granulitic breccias reflected an absence of KREEP in the pre-4.1 Ga crust. *Warner et al.* (1977) postulated that the older lunar impacts tended to generate granulites, instead of melt breccias, because the fragment-laden impact melts tended to cool more slowly in the early lunar crust than in the younger lunar crust. Warner et al. suggested that the ambient crustal temperature was higher in the early Moon, due to "a higher impact flux and a higher heat flow." *Bickel and Warner* (1978) attributed the origin of granulitic breccias to "thermal metamorphism in early breccia sheets." Granulitic breccias generally preserve recognizable (albeit barely so) lithic clasts, so a crucial assumption in this model is that most impact melts (even in the early, hotter Moon) tend to be fragment-laden.

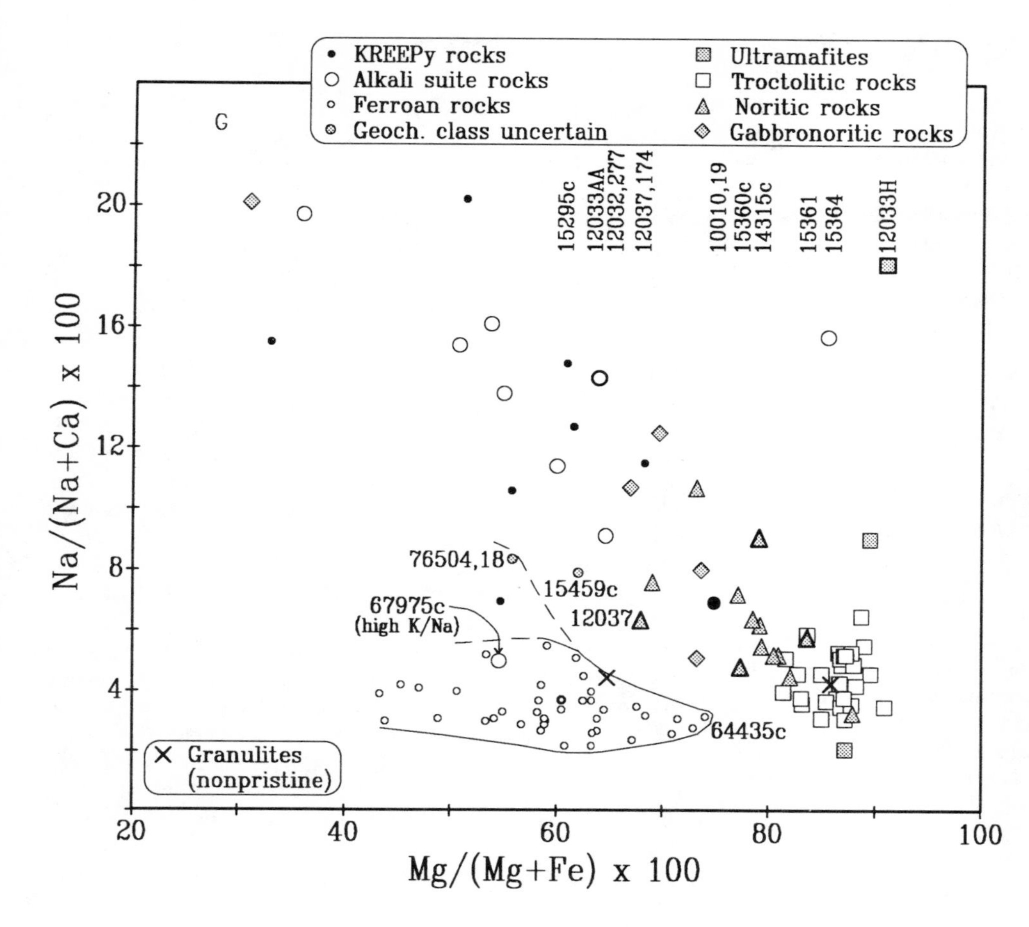

Fig. 15. Bulk-rock molar Mg/(Mg+Fe) (*mg**) ratio vs. molar Na/(Na+Ca) for pristine nonmare rocks. Samples from this study are shown with heavy lines around symbols, and identified by sample numbers in upper right. Note that dubiously pristine samples 10010,19 and 14315c are included here for completeness, albeit they are not recommended for future compilations of pristine rocks. Other data souces include *James et al.* (1987) for 67975c; *Warren et al.* (1986) for 76504,18; *Lindstrom et al.* (1988) for 15459c; and *James et al.* (1989) for 64435c.

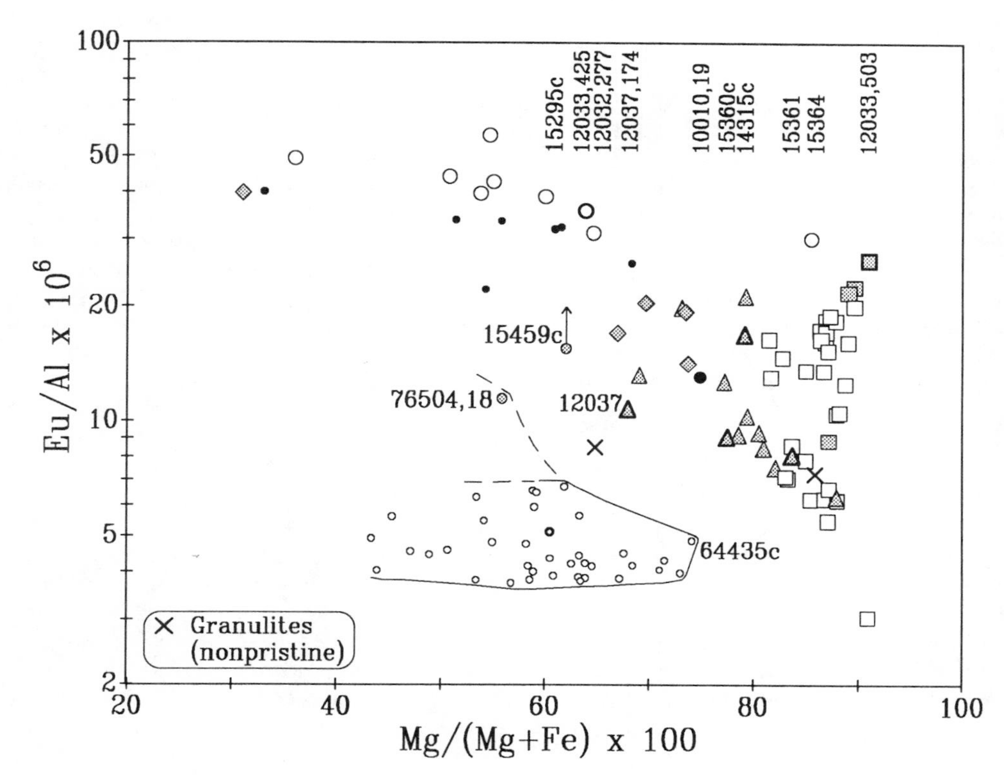

Fig. 16. Bulk-rock *mg** ratio vs. Eu/Al for pristine nonmare rocks. Symbols and data sources as for Fig. 15. The arrow drawn upward from 15459c indicates the Eu/Al ratio implied by the more credible of two discrepant Al data that were averaged for this sample (more credible because the higher Al datum tends to imply a sum for the analysis of >>100%).

Bickel and Warner (1978) studied 50 granulitic breccias by thin-section petrography, and noted that their silicate solid solution compositions tend to cluster in between the two main groups of pristine rocks (Mg-rich rocks and ferroan anorthosites). *Lindstrom and Lindstrom* (1986) reviewed the chemistry and petrography of eight large granulitic breccias and suggested that they exhibit a significant compositional bimodality, linked to *mg** ratio. Thus, *Lindstrom and Lindstrom* (1986) distinguished ferroan (15418, 67215, 77017, and 78155) from magnesian (67415, 67955, 79215, and 76230/5) granulites. However, *Bickel and Warner* (1978) found little or no indication of bimodality in *mg** among their 50 granulite samples. If the three roughly granulitic breccias we have studied are classified following *Lindstrom and Lindstrom* (1986), 12032,277 and (by definition) 67215 are ferroan, while 15364 is magnesian.

Few granulitic breccias have been studied from either Apollo 12 or Apollo 14, which are located in a far more KREEP-rich region of the Moon than the eastern-hemisphere sources (Apollo 16 and 17) of most previously-studied granulites. Two "poikilitic/granulitic" Apollo 12 rocklets analyzed by *Laul* (1986) have vastly higher incompatible element contents than analyzed granulitic breccias from the eastern sites (Fig. 14). However, the petrographic descriptions of these two rocklets (*Simon and Papike,* 1985) indicate that neither is very granulitic; both (and especially 12003,222, the one with the highest REE content) are probably best classified as impact melt breccias. In contrast, 12032,277, which is classically granulitic (Fig. 5), is no more incompatible-element-rich than several of the eastern granulites (Fig. 14). The correlation between texture and composition even extends to the one outlier among the eastern granulites, 15459c, which appears more like a relatively coarse-grained poikilitic impact melt breccia than a classically granulitic breccia (*Lindstrom et al.,* 1988).

Thus, our data for 12032,277 (and 15364 and 67215) appear to support the suggestion of *Warner et al.* (1977) that granulitic breccias generally formed in environments with relatively little KREEP. More age data for granulites are needed to test the further suggestion of *Warner et al.* (1977) that this composition-texture correlation reflects an absence of KREEP in the pre-4.1 Ga crust. It might simply reflect a tendency for granulites to form deeper in the crust, coupled with a tendency for KREEP content to increase towards the surface. The difference between these two models may have profound implications regarding lunar and planetary cratering history. *Ryder* (1989) has suggested that the scarcity of impact melts older than 3.9 Ga implies that the rate of cratering was "negligible" for 500 Ma before a cratering cataclysm occurred at ~3.85 Ga. However, if the *Warner et al.* (1977) model of granulitic breccia formation is correct, then the scarcity of ancient impact melt breccias may simply reflect the higher ambient temperature of the pre-3.9 Ga crust. *Warren et al.* (1989a) show that insulation by a megaregolith may have caused ambient temperatures to remain within a few hundred degrees of the solidus within all but the outermost few kilometers of the crust for several hundred Ma after the formative "magma ocean" episode.

Pristine Anorthosites (12003,179, 12033,425, and 15295,41)

Exsolution of augite from plagioclase. Geochemically, 12003,179 and 12033,425 are fairly typical alkali anorthosites, and 15295,41 is a typical ferroan anorthosite (Figs. 4, 15, and 16). However, the texture of 12033,425 is extraordinary. The only other clear instances of pyroxene exsolution out of feldspar among lunar rocks have been reported for the ancient 76535 pristine-troctolitic granulite (*Nord,* 1976), and the ferroan anorthosite 62236 (*Nord and Wandless,* 1983). Also, *Lally et al.* (1972) obtained electron micrographs of roundish, submicron-sized pyroxene grains concentrated along twin boundaries in plagioclase of the "genesis rock" 15415 ferroan-anorthositic granulite. The scale of exsolution in 76535 is comparably small; straight rows of mainly-pyroxene exsolution products are up to ~500 μm long, but the rows of inclusions are strictly discontinuous, the particles that form the rows are never more than ~2 μm wide (*Nord,* 1976), and they consist of many different crystals, comprising three different varieties of pyroxene, FeNi metal, and another, unidentified phase. The exsolution in 62236 is more nearly comparable to that in 12033,425. *Nord and Wandless* (1983) report augite lamellae in 62236 up to 220×5 μm. The 330×17 μm augite lamella in 12033,425 is a single, optically-continuous crystal. The absence of compositional disparity between the augite exsolution lamella and the augite elsewhere in the thin section (Table 2, Fig. 8) suggests that all, or nearly all, of the pyroxene in the rocklet originated by exsolution. As discussed above, most of the augite in the section occurs as elongate inclusions within plagioclase, albeit generally not as straight and narrow grains oriented parallel to plagioclase twin lamellae.

The reaction that presumably drives the exsolution is $Ca(Mg,Fe)Si_3O_8 \rightarrow Ca(Mg,Fe)Si_2O_6 + SiO_2$ (*Smith and Steele,* 1974). The component on the left contains tetrahedral Mg and Fe instead of Al. However, we detected no SiO_2 in thin section 12033,575. *Smith and Steele* (1974) noted that exsolution of amphibole is fairly common in plagioclase from anorthosites in terrestrial layered intrusions exposed to mild thermal metamorphism. Based mainly on this analogy, these authors suggested that many lunar ferroan anorthosites are completely lacking in primary (not formed by exsolution) mafic phases. They further suggested that calculated *mg** ratios of the parent melts are far higher if the pyroxenes are strictly exsolution products, with *mg** determined indirectly by plagioclase/melt equilibria, instead of by pyroxene/melt equilibria. *Smith and Steele* (1974) argued that the resultant *mg** ratios inferred for the parent melts of 15415 and other ferroan anorthosites are 0.7-0.8, which they viewed as more "reasonable for derivation of a feldspar-rich crust during early differentiation of the Moon" than the ratios (~0.3-0.4) implied by assuming that the pyroxenes are primary. At first glance, the discovery of 12033,425 might appear to support these speculations. However, many ferroan anorthosites contain mafic silicates far too abundant and coarse-grained to represent exsolution products (e.g., *James et al.,* 1989; *Warren,* 1990c). Moreover, a study of partitioning of Mg and Fe into plagioclase under appropriately reducing conditions (*Longhi et al.,* 1976)

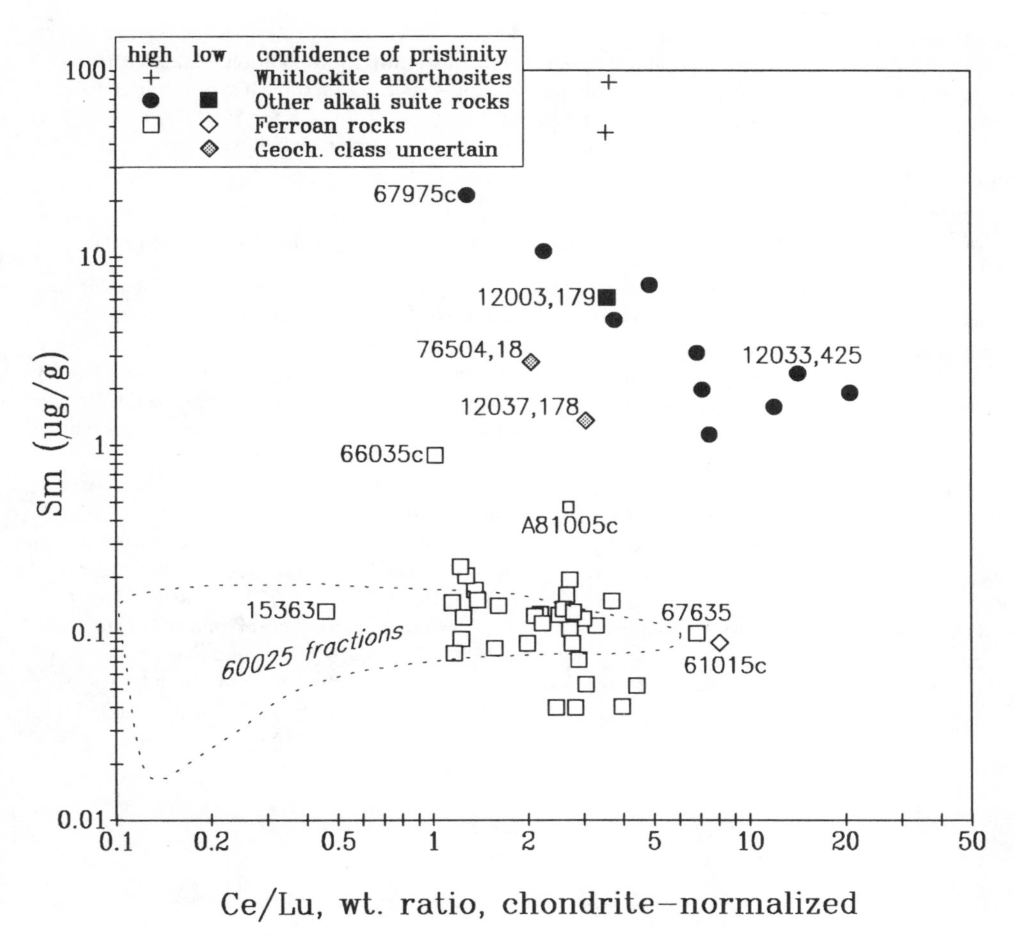

Fig. 17. Ce/Lu vs. Sm for pristine anorthosites (data sources as for Fig. 4), alkali suite lithology 67975c (*James et al.,* 1987), and geochemically anomalous, possibly pristine anorthosite 12037,178 (*Laul,* 1986). Note inverse correlation among whitlockite-poor alkali suite rocks.

indicates that for any given plagioclase *mg**, the equilibrium melt *mg** is far lower than assumed by *Smith and Steele* (1974); e.g., attributing the pyroxenes of 15415 to plagioclase exsolution implies a parent melt *mg** of 0.41, not 0.7-0.8.

Extensive exsolution of augite from plagioclase probably requires an unusually slow cooling history. *Nord* (1976) argued that the 76535 troctolite probably cooled at a rate of less than 1°/Ma, averaged over 600 Ma, and *Nord and Wandless* (1983) suggested that 62236 probably cooled at a similarly slow rate. A range of evidence reviewed by *Ryder* (1985) suggests a similar history of prolonged metamorphism, presumably deep within the lunar crust, for 15415. By analogy with these other three lunar rocks with pyroxene exsolution out of plagioclase, 12033,425 is probably also a slowly-cooled derivative of a relatively deep pluton. Its original texture is obscured by cataclasis, but vestiges of 120° triple junctions (Figs. 7a,b) hint at a former coarsely-granulitic texture comparable to that of 15415. This inferred deep origin for 12033,425 contrasts with the shallow origins inferred by *Warren et al.* (1983a) for several other alkali anorthosites, based on limited exsolution in their pigeonites. It now appears that alkali anorthosite probably formed in a variety of crustal intrusions, spanning a wide range in depth.

Origin of alkali anorthosite. Enough alkali-suite rocks have now been analyzed for a clear trend to emerge relating the slopes of their REE patterns to their overall REE abundance levels. Taking Ce as a frequently determined light REE, and Lu as a heavy REE, the Ce/Lu ratio shows a strong inverse correlation with the concentration of the middle REE, e.g., Sm (Figs. 4 and 17). This trend is consistent with a model (*Warren et al.,* 1983a) in which the most REE-poor alkali anorthosites are plagioclase adcumulates, and the more REE-rich alkali-suite lithologies formed from similar magmas, but with relatively high contents of "trapped liquid" (TL). The average chondrite-normalized Ce/Lu ratio, $(Ce/Lu)_{cn}$, of the five most REE-poor alkali anorthosites is ~12.3, and the average Sm content of the same samples is 1.8 μg/g. We can estimate the REE pattern of the parent melt by assuming that these samples formed as pure cumulus (and "adcumulus") plagioclase, and applying the plagioclase/melt distribution coefficients of *McKay* (1982) (D_{Ce} = 0.048, D_{Sm} = 0.029 and D_{Lu} = 0.008; value for Lu derived by slight extrapolation from Yb). The result is an implied parent melt with $(Ce/Lu)_{cn}$ = 2.05 and 62 μg/g Sm. This composition closely resembles the mean of five analyses (slightly contaminated with a melt-rock lithology) for the 67975c alkali gabbronorite (*James et al.,* 1987): (Ce/

$Lu)_{cn}$ = 2.05 and Sm = 45.2 μg/g. It also resembles average high-K KREEP: $(Ce/Lu)_{cn}$ = 2.23 and Sm = 48 μg/g (*Warren,* 1989); and the 15405c quartz monzodiorite "superKREEP" composition: $(Ce/Lu)_{cn}$ = 2.64 and Sm = 87.5 μg/g (*Ryder,* 1985).

If the most mafic and REE-rich of the whitlockite-poor alkali anorthosites, 14047c, is modeled as a mixture of pure cumulus plagioclase (no cumulus mafic material) plus TL with $(Ce/Lu)_{cn}$ = 2.05 and 62 μg/g of Sm, then its Sm content of 10.7 μg/g implies a TL content of 15 wt.%; and the implied $(Ce/Lu)_{cn}$ is ~2.50, in good agreement with the measured value of 2.25 (*Warren et al.,* 1983a). If the even more Sm-rich 67975c alkali gabbronorite is modeled in similar fashion, its TL content is implied to be 72 wt.%. However, 67975c contains a large trace of whitlockite, which complicates any interpretation of its REE composition. For 14047c, the TL component is implied to contain ~200 mg/g Fe, which is more Fe-rich than 67975c or any pristine KREEP sample (Fig. 18). Thus, it seems likely that 14047c contains several percent cumulus pigeonite, albeit its nonplagioclase component is probably dominated by a KREEP-like TL.

The case for a link between the alkali suite of intrusive (dominantly cumulate) rocks and KREEP, suggested earlier by numerous studies (e.g., *Warren et al.,* 1983a,b; *Shervais et al.,* 1984; *James et al.,* 1987) now seems stronger than ever. However, the detailed mechanism of the linkage is still unclear, and will likely remain so until more isotopic data are produced for alkali suite samples. As discussed by *Warren et al.* (1983a), the sole existing isotopic datum, a $^{87}Sr/^{86}Sr$ ratio reported by *Hubbard et al.* (1971), implies an unbelievably low initial $^{87}Sr/^{86}Sr$ of ≤0.6982.

Pristine Norites (12037,174, 15360,11, and 15361)

The 15360c and 15361 norites are noteworthy for their low contents of incompatible trace elements, by lunar norite standards (Fig. 12). Almost equally low REE contents were found for several arguably pristine Apollo 15 norites by *Lindstrom et al.* (1989), but the samples analyzed were so small (4 and 15 mg) that they could scarcely be representative for plutonic rocks.

Aside from ferroan anorthosite 15295,41, most of the pristine and possibly pristine samples of this study are clearly affiliated with the Mg-suite, or else the alkali suite (which arguably is best considered as a subset of the Mg-suite) (Figs. 15 and 16). However, the 12037,174 norite, and to a lesser extent the 15360c norite, have combinations of mg^* and plagiophile ratios that lead to a slight diminution of the "gap" between the ferroan rocks and other pristine nonmare rocks.

The "gap" among lithologies with mg^* <0.7 has also been

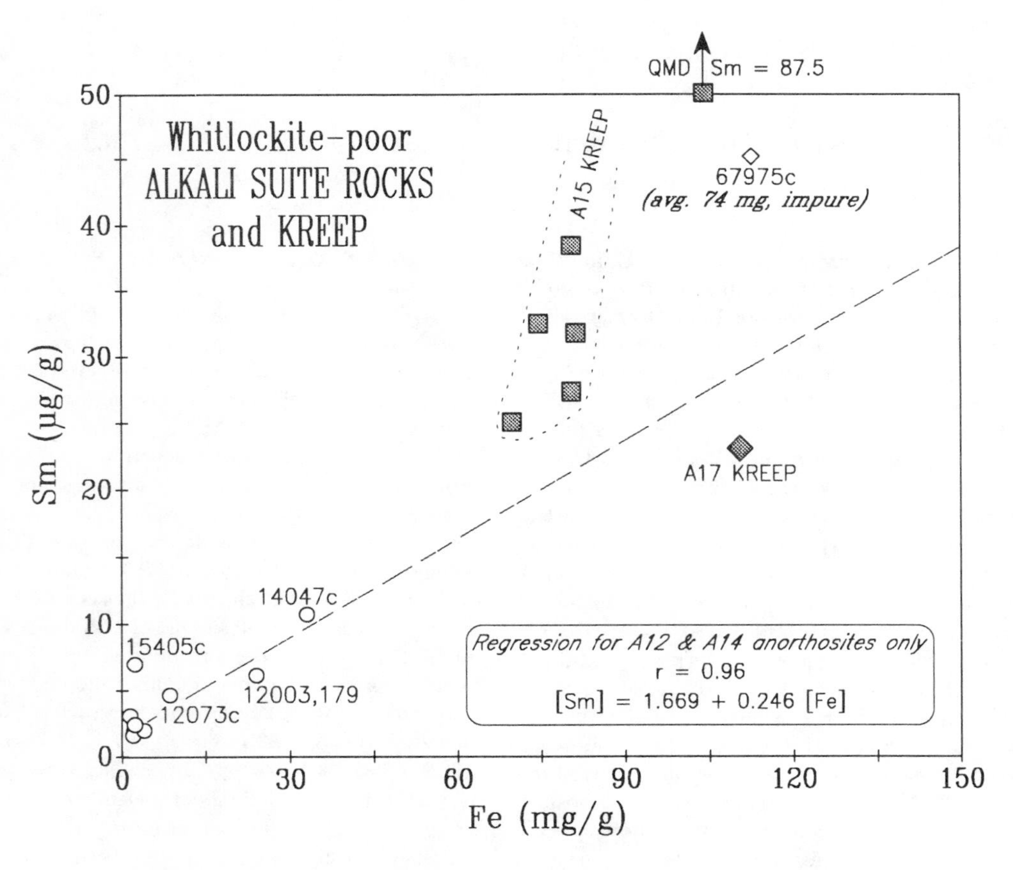

Fig. 18. Fe vs. Sm for pristine alkali suite and KREEP rocks, showing rough correlation.

partially filled in recent years by the probably (but none too assuredly) pristine anorthosite 76504,18 (*Warren et al.,* 1986), and the probably pristine "ferroan norite" 15459c (*Lindstrom et al.,* 1988). Actually, this clast is far from truly ferroan, according to the traditional lunar meaning of that term (which implies an ultralow, ferroan anorthosite-like Na/(Na+Ca) ratio). *Lindstrom et al.* (1988, 1989) also claim that several other "ferroan norite" clasts are "plutonic" (i.e., chemically pristine). However, they provide little evidence to support these claims. The grain sizes they describe hardly warrant the designation "plutonic." They mention the presence of metal in several instances, but give no hint as to its composition (a key test for meteoritic contamination). Even 15459,292, which is the clast included in our Figs. 15 and 16, does not look far different in texture from "poik" clast 15459,309 (Fig. 1J and 1L of *Lindstrom et al.,* 1988). All of these lithologies *may* well be pristine, but at this stage, it seems liberal to include even 15459,292 in the category of very probably pristine rocks.

Analogous problems arise with several "plutonic highland" lithologies described from Apollo 12 rocklets by *Simon and Papike* (1985) and analyzed by *Laul* (1986). At least one, gabbro 12003,250, appears at be of mare affinity. Another "plutonic" norite, 12001,647, has orthopyroxene *mg** ranging from 0.62 to 0.81, and plagioclase An ranging from 87 to 95—hardly the sort of mineralogy expected for a truly plutonic lithology. Perhaps the most interesting sample studied by *Simon and Papike* (1985) and *Laul* (1986) is 12037,178, a tiny anorthosite that has a texture that does indeed look igneous (albeit the plagioclase is far too heterogeneous in composition for a truly plutonic lithology), and if pristine would also help to fill the extreme low-*mg** end of the "gap" between the ferroan anorthosites and other pristine nonmare rocks.

Realistically, we should not expect the "gap" to hold at its low-*mg** end. Very-low-Ti mare basalts are known to appear geochemically ferroan if plotted on diagrams such as Figs. 15 and 16 (*Warren et al.,* 1986). Even if 12037,174 is not (as we tentatively infer above) a mare-nonmare transitional type of lithology, it stands to reason that such lithologies probably exist. Mare magmatism is known to have begun, at least in the west-central nearside (Apollo 14) region, by 4.23 Ga (*Taylor et al.,* 1983). Several nonmare noritic and gabbronoritic cumulates have Sm-Nd ages of between 4.18 and 4.23 Ga (the youngest is gabbronorite 67667); excluding the oldest of six Sm-Nd ages for pristine Mg-rich rocks, the average of the remaining five is 4.25 Ga (*Carlson and Lugmair,* 1988). Thus, mare and nonmare magmatism probably overlapped to a large extent. Even if we assume that the source regions of mare basalts were totally separated from the source regions for all other lunar magmas, magma mixing must have occasionally

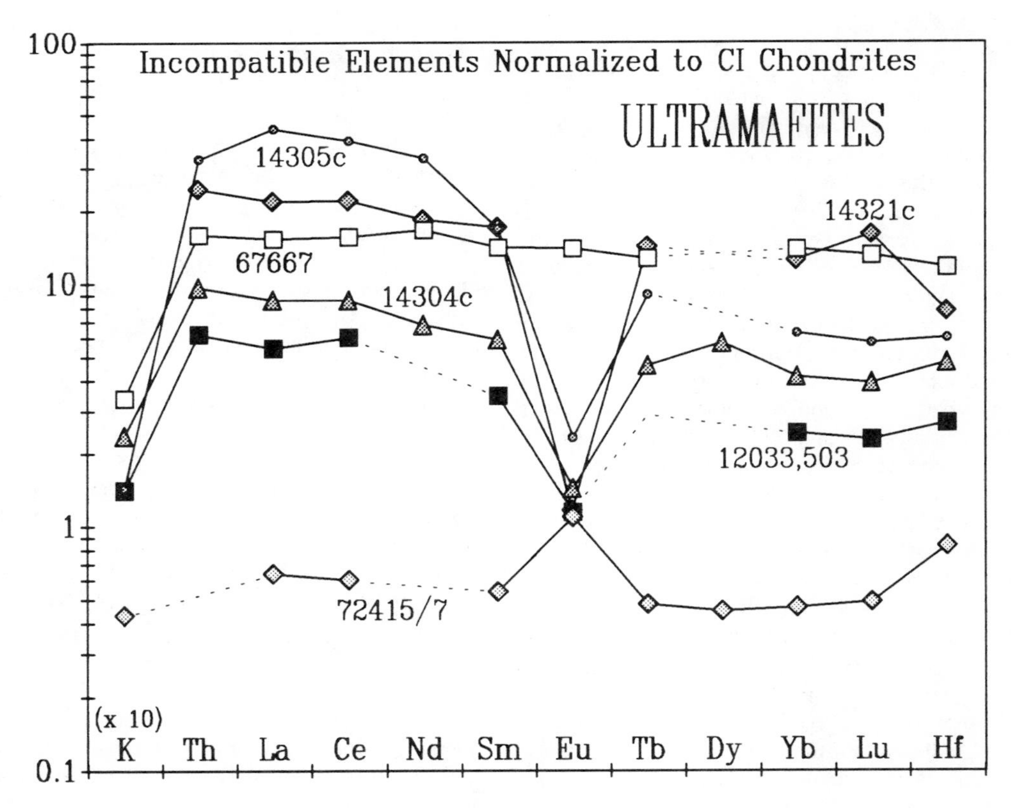

Fig. 19. Concentrations of incompatible elements in 12033,503 and other ultramafic lunar samples, normalized to CI chondrites. Data sources are *Warren* (1990a) for 72415/7, *Warren et al.* (1987) for 14304c, *Warren and Wasson* (1979) for 67667, *Lindstrom et al.* (1984) for 14321c, and *Shervais et al.* (1984) for 14305c.

engendered intermediate lithologies. In the cases of parent melts with affinities toward VLT mare basalts, such transitional lithologies can be expected to fill the "gap" between ferroan and other nonmare rocks. However, if the "gap" truly represents a fundamental bimodality in the geochemistry of nonmare rocks, as has generally been assumed, then even mare-nonmare transitional lithologies should only "fill" the "gap" near its low-*mg** end, because all known mare basalts have far lower mg than the highest-*mg** members of the ferroan suite of pristine nonmare rocks.

Harzburgite 12033,503: Comparison with Other Lunar Ultramafites

Aside from the well-known 72415/7 dunite, and the relatively feldspathic and low-*mg** lherzolite (gabbronorite) 67667, the only previously-described lunar ultramafites have all been clasts so tiny and texturally ambiguous, they might conceivably all represent crushed equivalents of just one or two former crystals from former lithologies that, at a cm scale, would look more like common Al-rich troctolites. The 14304 "dunite" clast of *Goodrich et al.* (1986) and *Warren et al.* (1987) has experienced as complex history of shock and annealing, and might (conceivably) have originally formed by granulation of a single, extremely coarse olivine. The 14305 "olivine pyroxenite" clast of *Shervais et al.* (1984) consists of a single grain of enstatite (98 vol.% of the sample), with a single inclusion of olivine (2 vol.%), and a single, tiny inclusion of a metal/phosphide grain. Before the 14321 "dunite" clast of *Lindstrom et al.* (1984) could be made into a thin section, it disaggregated into 80 tiny fragments of pure olivine (or a mixture of olivine with minor plagioclase; the evidence is ambiguous); again, this might have originally formed by granulation of a single, extremely coarse olivine.

Despite the strange partial granulation of its pyroxene, the 12033,503 harzburgite preserves clear vestiges of a poikilitic igneous cumulate texture (Fig. 9a), comprising many separate olivine chadacrysts, and probably more than one pyroxene oikocryst. However, this harzburgite from Apollo 12 resembles the three analyzed ultramafites from nearby Apollo 14, in that they all show remarkably high concentrations of incompatible trace elements, considering their ultramafic compositions (Fig. 19). These compositions imply remarkably "evolved" incompatible element contents for their otherwise "primitive" (high-*mg**) parent melts. In the case of 12033,503, we can set a firm upper limit on its content of TL by noting that the bulk-rock Al concentration (0.45 wt.% Al_2O_3) is lower by at least a factor of 10, and most likely by a factor of 20, than that of any plausible lunar-nonmare melt. Thus, 10 wt.%, and more likely 5 wt.%, is the upper limit for TL in this sample. It follows that the parent melt had to have light REE $\geq 60\times$ CI chondrites (and probably $\geq 120\times$).

Analogous calculations applied to the 14305c olivine pyroxenite and the 14321c dunite imply parent melts with light REE contents at least $\geq 170\times$ CI chondrites and $\geq 180\times$ CI chondrites, respectively. In other words, relative to average high-K KREEP (*Warren,* 1989), the 12033 harzburgite, 14305c, and 14321c parent melts are implied to have light REE contents of $\geq 0.13\times$, $\geq 0.37\times$, and $\geq 0.39\times$, respectively. The actual REE contents were probably higher than these lower limits, especially for the pyroxene-rich 12033,503 and 14305c, because at least some of the Al in their bulk-rock compositions probably reflects Al partitioned into pyroxene, for which D is far higher for Al (roughly 0.1) than for light REE (roughly 0.01). The paradox that the inferred parent melts were extremely magnesian and yet uncommonly REE-rich seems best explained by assuming that a magnesian melt rising from the deep lunar interior was mixed with a small amount of extremely REE-rich magma, as the intrusive melt stopped within, or passed through, the lower crust. As discussed by *Warren* (1988), Mg-rich magmas may have commonly mixed with, or assimilated, urKREEP in this fashion. Such mixing may have been especially common in the KREEP-rich west-central nearside (Apollo 12/14) region. In the case of 14305c, the light REE are also greatly enriched relatively to the heavy REE, and *Shervais et al.* (1984) suggest that the parent melt may have formed by extremely low-degree partial melting of a garnet-bearing source. However, 14305c remains unique in this respect, among the analyzed ultramafites.

Another similarity between 12033,503 and the 14305c olivine pyroxenite is that both have extremely low-Ca pyroxene and low-Ca olivine. Shervais et al. found 0.06 wt.% CaO in 14305c olivine; we find 0.039 ± 0.019 in 12033,503 olivine. Low olivine CaO contents might reflect origins in relatively deep, slow-cooling environments, or they may simply be consequences of the extremely magnesian compositions of the olivines (*Ryder,* 1984).

The origin of the "brick wall" granulation texture of the 12033,503 pyroxene is enigmatic. A. E. Rubin (personal communication, 1989) suggests that the texture resulted from recrystallization in the aftermath of a severe shock that at one time converted most of the pyroxene oikocryst material into diaplectic glass. H. Takeda (personal communication, 1989) suggests that the partial granulation may be a result of decomposition of the high-temperature protoenstatite polymorph of pyroxene into the lower-temperature orthoenstatite polymorph. Possibly shock and decomposition both played roles. Shear pressure is known to greatly influence the stabilities of the various enstatite polymorphs (*Smyth,* 1974; *Buseck et al.,* 1980).

As discussed above, the high Ir concentration of 12033,503 is probably not attributable to contamination or otherwise spurious measurement. Unpalatable though it may seem to those of us who would like to use siderophile elements as gages of chemical pristinity of all lunar rock samples, the siderophile element contents of 12033,503, and the roughly comparable concentrations in the 72415/7 dunite (*Morgan and Wandless,* 1988), indicate that the diagnostic utility of siderophile elements does not extend to magnesian ultramafites (at least, not if they contain extremely high-Ni FeNi metals). Other implications of these new data will not be discussed in detail here. Suffice it to say that our new data for 12033,503, the new data for 72415/7 (*Morgan and Wandless,* 1988), and the relatively high Os and Ir contents recently found for even a ferroan anorthosite (*Warren et al.,* 1987) (the metal of which is extremely Ni-poor), all necessitate reevaluation of the average siderophile element contents of the lunar crust. However, other than a few ultramafic zones (volumetrically minor, at least in the upper crust), most of the pristine lunar crust still appears to be characterized by extremely low siderophile element contents. Except for magnesian ultramafites, it still seems best to assume that rocks with contents of "noble" siderophile elements such as Ir, Os, and Re at levels $>3 \times 10^{-4}$ times CI chondrites are not pristine, unless very strong petrographic evidence says otherwise.

The Nonmare Crust of the Apollo 12 Region

Previous work has indicated a similarity between the nonmare rocks from the Apollo 12 site and those from the nearby Apollo 14 site (*Warren et al.,* 1983a; *Simon and Papike,* 1985; *Laul,* 1986). This similarity is also apparent from the Apollo 12 samples described above. The four pristine nonmare samples comprise two alkali anorthosites, one magnesian ultramafite, and one norite. Although rare among rocks from most other sites, alkali anorthosite and magnesian ultramafite are relatively common from Apollo 14. However, granulitic polymict breccias similar to 12032,277 seem rare among Apollo 14 lithologies. The relatively high ratio of high-Ca to low-Ca pyroxene in 12032,277 (essentially 1:1) would be even more surprising among Apollo 14 rocks. However, Earth-based reflectance spectra (*Pieters,* 1989) indicate that the crust is generally more gabbroic in the western nearside than it is in the eastern nearside, where Apollo 15, 17, and especially 16 obtained the bulk of our highlands rock samples.

"Excess" Eu and the Representativeness of the Known Pristine Rocks

It has recently been suggested that the known types of pristine rocks are grossly unrepresentative of the actual pristine crust, and in particular that ferroan anorthosite must be far less common than the known pristine rocks would seem to indicate (*Korotev and Haskin,* 1988). This suggestion is based on a supposed "excess" of Eu in polymict rocks, relative to the Eu derived by theoretical mixing of ferroan anorthosite plagioclase with what was alleged to be representative plagioclase for other pristine rocks. According to *Korotev and Haskin* (1988), these mass-balance calculations require a "missing" Eu-rich component, which is so important that the "need for this component limits the proportion of ferroan anorthosite in the average upper crust to be less than 40% by weight." We would be the last to suggest that the existing dataset for pristine rocks should be assumed perfectly representative of the full diversity of the lunar crust. However, we disagree with the stated conclusions of *Korotev and Haskin* (1988) regarding "excess" Eu and its implications for the representativeness of the known pristine rocks, in part for semantic reasons, and in part for scientific reasons.

Korotev and Haskin (1988) chose to construe the term "ferroan anorthosite" to mean anorthosite in the strictest possible sense (>90 vol.% plagioclase). However, semiofficial lunar nomenclature includes noritic/troctolitic rocks with as little as 77.5 vol.% plagioclase as varieties of anorthosite (*Stöffler et al.,* 1980). More importantly, the most active investigators of pristine rocks have traditionally included noritic and troctolitic anorthosites (77.5-90 vol.% plagioclase) under the general heading of ferroan anorthosite (e.g., *Warren and Wasson,* 1979; *James,* 1980; *Warren et al.,* 1983b). Since the term "*ferroan* anorthosite" has seldom been used except *in connection with pristine rocks,* it seems ill-advised, and potentially misleading, for a paper dedicated to evaluating the representativeness of known pristine rocks to use the term "ferroan anorthosite" in a far narrower sense than generally used by the most active investigators of pristine rocks.

Underlying the Eu mass-balance calculations described by *Korotev and Haskin* (1988) are various pivotal assumptions regarding the average composition of the crust, and the average composition of the non-ferroan-anorthosite components of the crust. For example, *Korotev and Haskin* (1988) constrained the average Eu content of the crust on the basis of plots of Th vs. Sm/Eu and Eu vs. Sm. However, this approach is sensitive to the assumed crustal average for Th. As noted by *Warren and Kallemeyn* (1988) and *Eugster* (1989), data from lunar meteorites reinforce an indication already evident in the Apollo γ-ray results (*Metzger et al.,* 1977) that the distribution of Th in the lunar crust is regionally heterogeneous. All three of the analyzed lunar meteorites contain well under 0.4 μg/g Th, whereas the data of *Metzger et al.* (1977) indicate that a large region of the west-central nearside highlands contains 2.5 ± 0.2 μg/g. Thus, the same Th vs. Eu (technically Sm vs. Eu) relationship used by *Korotev and Haskin* (1988) implies that Eu must also show great regional heterogeneity. It follows that one should hardly expect to achieve mass balance by comparing the Eu contents of a small number of pristine rocks from a single locale vs. the average Eu content of the entire crust. Yet, apparently because they preferred to calculate mass balance for plagioclase instead of taking a more direct approach dealing in rocks, *Korotev and Haskin* (1988) chose to regard (p. 1807) only two samples as representatives of the Mg-rich pristine rocks: 76535, with a plagioclase Eu concentration of 1.26 μg/g, and 78235, with a nearly identical plagioclase Eu concentration, ~1.6 μg/g. Note that these rocks are both from Apollo 17, and thus might be products of a single small layered intrusion.

We prefer to evaluate the crustal mass balance implications of Eu vis-a-vis pristine rocks by a more direct approach. As long as we are concerned with gross-crustal mass balance, we want to model the most thoroughly polymict (mixed) samples available: highlands soils. Our dataset includes the three discrete lunar meteorites that have so far been analyzed (ALHA81005, Y791197, and Y82192/86032), averages for soils from the Apollo 14 and 16 sites, an average of data for the most Al-rich Apollo 17 soil, 73141, and an average for Luna 20 soil. The Eu contents of these regolith samples of diverse lunar provenance are shown plotted vs. Al in Fig. 20, which also includes data for pristine nonmare rocks. This diagram is largely self-explanatory. Except for the most KREEP-rich soils (Apollo 14, Apollo 16, and Apollo 17), the lunar highland regolith shows Eu concentrations within the range of ferroan anorthosites (however defined), and if anything, embarrassingly *low* for mass balance involving mixtures of all known pristine nonmare rocks.

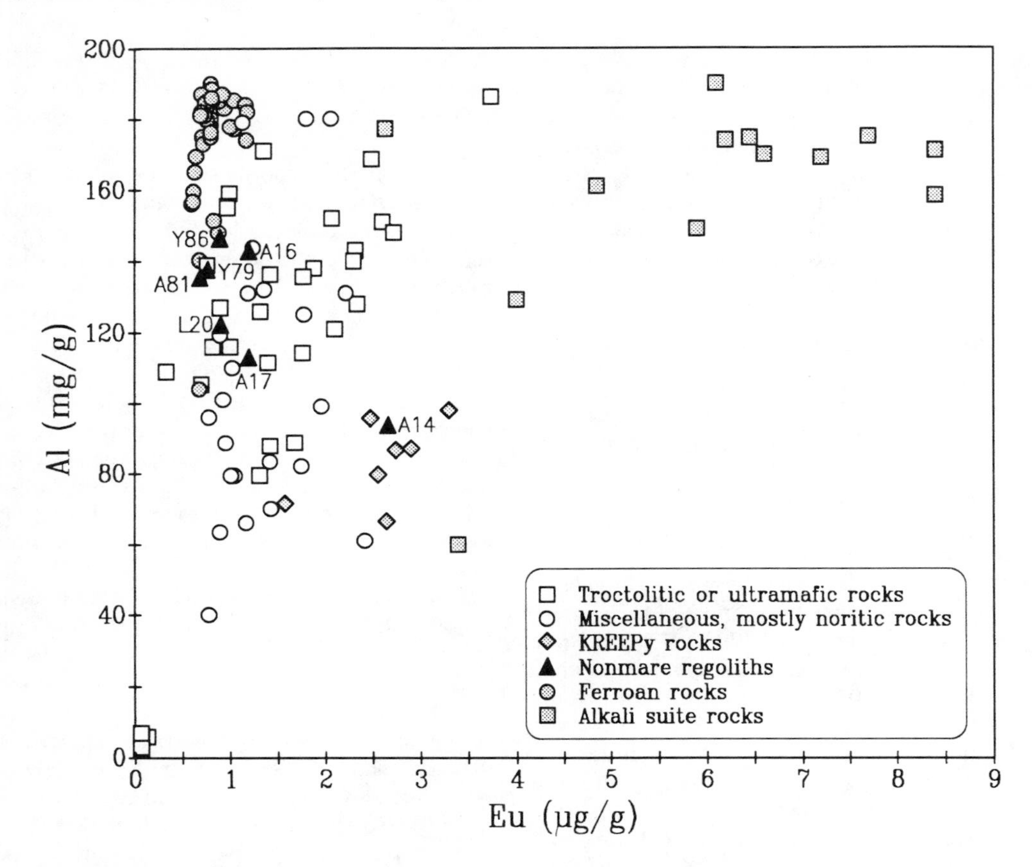

Fig. 20. Eu vs. Al for pristine nonmare rocks and highlands regolith samples. Data for lunar regolith samples are averages from sources cited by *Warren et al.* (1989b).

A basically valid argument regarding the representativeness of the known pristine rocks was outlined by *Korotev and Haskin* (1988) in their subsidiary discussion of mass balance for Fe and Mg (viewed in connection with Al). The argument is based on the low *mg** ratios of the polymict samples (epitomized by highlands regoliths), in comparison to possible mixtures of pristine rocks. This argument can be traced back to *Korotev et al.* (1980), and numerous other papers by the same authors (e.g., *Korotev,* 1983; *Lindstrom and Lindstrom,* 1986). However, the mass-balance models used to support the argument have been based on an unrealistically mafic-poor composition for ferroan anorthosite (35.5 wt.% Al_2O_3, i.e., 100% plagioclase), which considerably exacerbates the mass balance problem for Fe and Mg. A detailed assessment of the average mode of all known relatively large (>10 g) pristine ferroan anorthosites (in the full, broad sense) indicates that they contain 8.1 ± 8.6 wt.% mafic silicates (*Warren,* 1990c). Note the huge 1-σ uncertainty. Any model sensitive to this parameter should allow for an enormous possible range, say 4 to 20 wt.%. Mass balance models of the type discussed by *Korotev and Haskin* (1988) work well if the actual ferroan component of the crust has a mafic silicate content roughly 1-σ higher than the mean (*Warren,* 1990c). Models for crustal genesis by flotation of plagioclase along with rafted mafic silicates over a ferroan (i.e., FeO-rich, and thus, dense) magma ocean also predict mafic silicate contents of roughly 19 wt.% (*Warren,* 1990c). Thus, these two approaches agree in suggesting that the mean modal mafic silicate content of ferroan anorthosite is higher than the imprecisely constrained arithmetic mean of the few available large samples. However, this statistically insignificant disparity can hardly be viewed as a significant contradiction of the hypothesis that ferroan anorthosite, in the broad (and customary) sense of rocks with "ferroan" geochemistry and $\geq$ 77.5 vol.% (~73 wt.%) plagioclase (*Stöffler et al.,* 1980) represents a major component of the lunar crust.

In large measure, our disagreement with *Korotev and Haskin* (1988) is semantic regarding "ferroan anorthosite." We completely agree that the crust must contain a major proportion of material with a mafic-silicate/plagioclase ratio vastly higher than the sort of ferroan anorthosite used as a mixing component by *Korotev and Haskin* (1988). In fact, it was in our "foray" series (*Warren and Wasson,* 1978, 1979; *Warren et al.,* 1983b) that numerous pristine rocks containing $\geq$ 10 vol.% mafic silicates (including several large rocks: 60025, 60135, 62236, and 62237) were first classified as ferroan anorthosites. *Korotev and Haskin* (1988) suggest that a major proportion of the lunar crust must be either ferroan norite or ferroan anorthositic norite (i.e., that it has <77.5 vol.% plagioclase), whereas our preferred interpretation is that the precise average composition of ferroan anorthosite corresponds to a ferroan noritic/troctolitic anorthosite (or possibly a marginal anorthosite, *sensu stricto*). In terms of petrogenetic models, this difference is not merely semantic; it is potentially crucial. Density modeling (*Warren and Wasson,* 1979; *Warren,* 1990c) indicates that a large mass of ferroan noritic/troctolitic anorthosite might form by plagioclase flotation (e.g., over a primordial magma ocean), whereas a large mass of ferroan norite or anorthositic norite would tend to sink.

Acknowledgments. This study benefited from very helpful discussions with A. E. Rubin and H. Takeda, and from reviews by R. L. Korotev and C. R. Neal. The research was supported by NASA grant NAG 9-87.

REFERENCES

Baedecker P. A., Chou C.-L., and Wasson J. T. (1972) The extralunar component in lunar soils and breccias. *Proc. Lunar Sci. Conf. 3rd,* pp. 1343-1359.

Bickel C. E. and Warner J. L. (1978) Survey of lunar plutonic and granulitic lithic fragments. *Proc. Lunar Planet. Sci. Conf. 9th,* pp. 629-652.

Bickel C. E., Warner J. L., and Phinney W. C. (1976) Petrology of 79215: Brecciation of a lunar cumulate. *Proc. Lunar Sci. Conf. 7th,* pp. 1793-1819.

Buseck P. R., Nord G. L. Jr., and Veblen D. R. (1980) Subsolidus phenomena in pyroxenes. In *Reviews in Mineralogy, Volume 7, Pyroxenes* (C. T. Prewitt, ed.), pp. 117-211. Mineral. Soc. Amer., Washington, D.C.

Carlson R. W. and Lugmair G. W. (1988) The age of ferroan anorthosite 60025: Oldest crust on a young Moon? *Earth Planet. Sci. Lett., 90,* 119-130.

Carlson I. C. and Walton W. J. A. (1978) *Apollo 14 Rock Samples.* JSC Publ. No. 14240, NASA Johnson Space Center, Houston. 413 pp.

Davis P. A. and Spudis P. D. (1985) Petrologic province maps of the lunar highlands derived from orbital geochemical data. *Proc. Lunar Planet. Sci. Conf. 16th,* in *J. Geophys. Res., 90,* D61-D74.

Dickinson T., Taylor G. J., Keil K., Schmitt R. A., Hughes S. S., and Smith M. R. (1985) Apollo 14 aluminous mare basalts and their possible relationship to KREEP. *Proc. Lunar Planet. Sci. Conf. 15th,* in *J. Geophys. Res., 90,* C365-C374.

Dowty E., Prinz M., and Keil K. (1974) Ferroan anorthosite: A widespread and distinctive lunar rock type. *Earth Planet. Sci. Lett., 24,* 15-25.

Dymek R. F., Albee A. L., and Chodos A. A. (1975) Comparative petrology of lunar cumulate rocks of possible primary origin: Dunite 72415, troctolite 76535, norite 78235, and anorthosite 62237. *Proc. Lunar Sci. Conf. 6th,* pp. 301-341.

Eugster O. (1989) History of meteorites from the Moon collected in Antarctica. *Science, 245,* 1197-1202.

Goodrich, C. A., Taylor G. J., Keil K., Kallemeyn G. W., and Warren P. H. (1986) Alkali norite, troctolites, and VHK mare basalts from breccia 14304. *Proc. Lunar Planet. Sci. Conf. 16th,* in *J. Geophys. Res., 91,* D305-D318.

Hansen E. C., Steele I. M., and Smith J. V. (1979). Lunar highland rocks: Element partitioning among minerals I: Electron microprobe analyses of Na, Mg, K and Fe in plagioclase; *mg* partitioning with orthopyroxene. *Proc. Lunar Planet. Sci. Conf. 10th,* pp. 627-638.

Helmke P. A., Haskin L. A., Korotev R. L., and Ziege K. (1972) Rare earths and other trace elements in Apollo 14 samples. *Proc. Lunar Sci. Conf. 6th,* pp. 1275-1292.

Hubbard, N. J., Gast P. W., Meyer C., Nyquist L. E., Shih C.-Y., and Weismann H. (1971) Chemical composition of lunar anorthosites and their parent liquids. *Earth Planet. Sci. Lett., 13,* 71-75.

James O. B. (1980) Rocks of the early lunar crust. *Proc. Lunar Planet. Sci. Conf. 11th,* pp. 365-393.

James O. B. and Flohr M. K. (1983) Subdivision of the Mg-suite noritic rocks into Mg-gabbronorites and Mg-norites. *Proc. Lunar Planet. Sci. Conf. 13th,* in *J. Geophys. Res., 88,* A603-A614.

James O. B., Lindstrom M. M., and Flohr M. K. (1987) Petrology and geochemistry of alkali gabbronorites from lunar breccia 67975. *Proc. Lunar Planet. Sci. Conf. 17th,* in *J. Geophys. Res., 92,* E314-E330.

James O. B., Lindstrom M. M., and Flohr M. K. (1989) Ferroan anorthosite from lunar breccia 64435: Implications for the origin and history of lunar ferroan anorthosites. *Proc. Lunar Planet. Sci. Conf. 19th,* pp. 219-243.

Jerde E. A., Warren P. H., Morris R. V., Heiken G. H., and Vaniman D. T. (1987) A potpourri of regolith breccias: "New" samples from the Apollo 14, 16 and 17 landing sites. *Proc. Lunar Planet. Sci. Conf. 17th,* in *J. Geophys. Res., 92,* E526-E536.

Kallemeyn G. W., Rubin A. E., Wang D., and Wasson J. T. (1989) Ordinary chondrites: bulk compositions, classification, lithophile-element fractionations, and composition-petrographic type relationships. *Geochim. Cosmochim. Acta,* in press.

Korotev R.L. (1983) Compositional relationships of the pristine nonmare rocks to the highlands soils and breccias (abstract). In *Workshop on Pristine Highlands Rocks and the Early History of the Moon* (J. Longhi and G. Ryder, eds.), pp. 52-55. LPI Tech. Rpt. 83-02, Lunar and Planetary Institute, Houston.

Korotev R. L. and Haskin L. A. (1988) Europium mass balance in polymict samples and implications for plutonic rocks of the lunar crust. *Geochim. Cosmochim. Acta, 52,* 1795-1813.

Korotev R. L., Lindstrom M. M., and Haskin L. A. (1980) A synthesis of lunar highlands compositional data. *Proc. Lunar Planet. Sci. Conf. 11th,* pp. 395-429.

Kramer F. E. and Twedell D. B. (1977) *Apollo 14 Coarse Fines (4-10 mm): Sample Location and Classification.* JSC Publ. No. 12992, NASA Johnson Space Center, Houston. 91 pp.

Kurat G., Keil K., Prinz M., and Nehru C. E. (1972) Chondrules of lunar origin. *Proc. Lunar Sci. Conf. 3rd,* pp. 707-721.

Lally J. S., Fisher R. M., Christie J. M., Griggs D. T., Heuer A. H., Nord G. L. Jr., and Radcliffe S. V. (1972) Electron petrography of Apollo 14 and 15 rocks. *Proc. Lunar Sci. Conf. 3rd,* pp. 401-422.

Laul J. C. (1986) Chemistry of the Apollo 12 highland component. *Proc. Lunar Planet. Sci. Conf. 16th,* in *J. Geophys. Res., 91,* D251-D261.

Laul J. C., Wakita H., Showalter D. L., Boynton W. V., and Schmitt R. A. (1972) Bulk, rare earth, and other trace elements in Apollo 14 and 15 and Luna 16 samples. *Proc. Lunar Sci. Conf. 3rd,* pp. 1181-1200.

Laul J. C., Simon S. B., and Papike J. J. (1988) Chemistry and petrology of the Apennine Front, Apollo 15, Part II: Impact melt rocks. *Proc. Lunar Planet. Sci. Conf. 18th,* pp. 203-217.

Lindstrom M. M. and Lindstrom D. J. (1986) Lunar granulites and their precursor anorthositic norites of the early lunar crust. *Proc. Lunar Planet. Sci. Conf. 16th,* in *J. Geophys. Res., 91,* D263-D276.

Lindstrom M. M., Nava D. F., Lindstrom D. J., Winzer S. R., Lum R. K. L., Schuhmann P. J., Schuhmann S., and Philpotts J. A. (1977) Geochemical studies of the White Breccia Boulders at North Ray Crater, Descartes region of the lunar highlands. *Proc. Lunar Sci. Conf. 8th,* pp. 2137-2151.

Lindstrom M. M., Knapp S. A., Shervais J. W., and Taylor L. A. (1984) Magnesian anorthosites and associated troctolites and dunite in Apollo 14 breccias. *Proc. Lunar Planet. Sci. Conf. 15th,* in *J. Geophys. Res., 89,* C41-C49.

Lindstrom M. M., Marvin U. B., Vetter S. K., and Shervais J. W. (1988) Apennine Front revisted: Diversity of Apollo 15 highland rock types. *Proc. Lunar Planet. Sci. Conf. 18th,* pp. 169-185.

Lindstrom M. M., Marvin U. B., and Mittlefehldt D. W. (1989) Apollo 15 Mg- and Fe-norites: A redefinition of the Mg-suite differentiation trend. *Proc. Lunar Planet. Sci. Conf. 19th,* pp. 245-254.

Longhi J., Walker D., and Hays J. F. (1976) Fe and Mg in plagioclase. *Proc. Lunar Sci. Conf. 7th,* pp. 1281-1300.

Marvin U. B. (1978) *Apollo 12 Coarse Fines (2-10): Sample Locations, Description, and Inventory,* JSC Publ. No. 14434, NASA Johnson Space Center, Houston. 143 pp.

McGee J. J. (1988) Petrology of brecciated ferroan noritic anorthosite 67215. *Proc. Lunar Planet. Sci. Conf. 18th,* pp. 21-31.

McGee J. J., Bence A. E., Eichhorn G., and Schaeffer O. (1978) Feldspathic granulite 79215: Limitations in T-fO_2 conditions and time of metamorphism. *Proc. Lunar Planet. Sci. Conf. 9th,* pp. 743-772.

McKay D. S., Bogard D. D., Morris R. V., Korotev R. L., Johnson P., and Wentworth S. J. (1986) Apollo 16 regolith breccias: Characterization and evidence for early formation in the mega-regolith. *Proc. Lunar Planet. Sci. Conf. 16th,* in *J. Geophys. Res., 91,* D277-D303.

McKay G.A. (1982) Partitioning of REE between olivine, plagioclase, and synthetic basaltic melts: implications for the origin of lunar anorthosites (abstract). In *Lunar and Planetary Science XIII,* pp. 493-494. Lunar and Planetary Institute, Houston.

McKay G. A., Weismann H., Nyquist L. E., Wooden J. L., and Bansal B. M. (1978) Petrology, chemistry, and chronology of 14078: Chemical constraints on the origin of KREEP. *Proc. Lunar Planet. Sci. Conf. 9th,* pp. 661-687.

Metzger A. E., Haines E. L., Parker R. E., and Radocinski R. G. (1977) Thorium concentrations in the lunar surface, I, Regional values and crustal content. *Proc. Lunar Sci. Conf. 8th,* pp. 949-999.

Morgan J. W. and Wandless G. A. (1988) Lunar dunite 72415-72417: siderophile and volatile trace elements (abstract). In *Lunar and Planetary Science XIX,* pp. 804-805. Lunar and Planetary Institute, Houston.

Morgan J. W., Laul J. C., Krähenbühl U., Ganapathy R., and Anders E. (1972) Major impacts on the Moon: Characterization from trace elements in Apollo 12 and 14 samples. *Proc. Lunar Sci. Conf. 3rd,* pp. 1377-1395.

Nord G. L. Jr. (1976) 76535: Thermal history deduced from pyroxene precipitation in anorthite. *Proc. Lunar Sci. Conf. 7th,* pp. 1875-1888.

Nord G. L. Jr. and M.-V. Wandless (1983) Petrology and comparative thermal and mechanical histories of clasts in breccia 62236. *Proc. Lunar Planet. Sci. Conf. 13th,* in *J. Geophys. Res., 88,* A645-A657.

Papike J. J., Hodges F. N., Bence A. E., Cameron M., and Rhodes J. M. (1976) Mare basalts: crystal chemistry, mineralogy, and petrology. *Rev. Geophys. Space Phys., 14,* 475-540.

Pieters C. M. (1989) Compositional stratigraphy of the lunar highland crust (abstract). In *Lunar and Planetary Science XX,* pp. 848-849. Lunar and Planetary Institute, Houston.

Rose H. J. Jr., Cuttitta F., Annell C. S., Carron M. K., Christian R. P., Dwornik E. J., Greenland L. P., and Ligon D. T. Jr. (1972) Compositional data for twenty-one Fra Mauro lunar materials. *Proc. Lunar Sci. Conf. 3rd,* pp. 1215-1229.

Ryder G. (1984) Most olivine in the lunar highlands is of shallow origin (abstract). In *Lunar and Planetary Science XV,* pp. 707-708. Lunar and Planetary Institute, Houston.

Ryder G. (1985) *Catalog of Apollo 15 Rocks.* JSC Publ. 20787, NASA Johnson Space Center Curatorial Branch, Houston. 1296 pp.

Ryder G. (1989) Lunar samples, lunar accretion, and the early bombardment of the Moon. *Eos Trans. AGU,* in press.

Ryder G. and Norman M. (1979) *Catalog of Pristine Non-mare Materials, Part 1: Non-anorthosites (Revised).* JSC Publ. 14565, NASA Johnson Space Center Curatorial Facility, Houston. 147 pp.

Ryder G. and Norman M. D. (1980) *Catalog of Apollo 16 Rocks.* JSC Publ. 16904, NASA Johnson Space Center, Curatorial Branch, Houston. 1144 pp.

Ryder G. and Spudis P. D. (1987) Chemical composition and origin of Apollo 15 impact melts. *Proc. Lunar Planet. Sci. Conf. 17th,* in *J. Geophys. Res., 92,* E432-E446.

Ryder G., Norman M. D., and Score R. A. (1980) The distinction of pristine from meteorite-contaminated highlands rocks using metal compositions. *Proc. Lunar Planet. Sci. Conf. 11th,* pp. 471-479.

Shervais J. W. and Taylor L. A. (1986) Petrologic constraints on the origin of the Moon. In *Origin of the Moon* (W. K. Hartmann, R. J. Phillips, and G. J. Taylor, eds.), pp. 173-201. Lunar and Planetary Institute, Houston.

Shervais J. W., Taylor L. A., Laul J. C., and Smith M. R. (1984) Pristine highland clasts in consortium breccia 14305: petrology and geochemistry. *Proc. Lunar Planet. Sci. Conf. 15th,* in *J. Geophys. Res., 89,* C25-C40.

Simon S. B. and Papike J. J. (1985) Petrology of the Apollo 12 highland component. *Proc. Lunar Planet. Sci. Conf. 16th,* in *J. Geophys. Res., 89,* D47-D60.

Smith J. V. and Steele I. M. (1974) Intergrowths in lunar and terrestrial anorthosites with implications for lunar differentiation. *Am. Mineral., 59,* 673-680.

Smith J. V. and Steele I. M. (1980) Lunar highland rocks: Element partitioning among minerals II: Electron microprobe analyses of Al, P, Ca, Ti, Cr, Mn and Fe in olivine. *Proc. Lunar Planet. Sci. Conf. 11th,* pp. 555-569.

Smyth J. R. (1974) Experimental study of the polymorphism of enstatite. *Am. Mineral., 59,* 345-352.

Steele I. M., Irving A. J., and Smith J. V. (1977) Apollo 15 breccia rake samples—mineralogy of lithic and mineral clasts. *Proc. Lunar Sci. Conf. 8th,* pp. 1925-1941.

Stöffler D., Knöll H.-D., Marvin U. B., Simonds C. H., and Warren P. H. (1980) Recommended classification and nomenclature of lunar highland rocks—a committee report. In *Proceedings of the Conference on the Lunar Highlands Crust* (R. B. Merrill and J. J. Papike, eds.), pp. 51-70. Pergamon, New York.

Stöffler D., Bischoff A., Borchardt R., Burghele A., Deutsch A., Jessberger E. K., Ostertag R., Palme H., Spettel B., Reimold W. U., Wacker K., and Wänke H. (1985) Composition and evolution of the lunar crust in the Descartes highlands, Apollo 16. *Proc. Lunar Planet. Sci. Conf. 15th,* in *J. Geophys. Res., 90,* C449-C506.

Taylor G. J. (1982) A possibly pristine, ferroan anorthosite (abstract). In *Lunar and Planetary Science XIII,* p. 798. Lunar and Planetary Institute, Houston.

Taylor L. A. Shervais J. W., Hunter R. H., Shih C.-Y., Nyquist L., Bansal B., Wooden J., and Laul J. C. (1983) Pre-4.2 AE mare basalt volcanism in the lunar highlands. *Earth Planet. Sci. Lett., 66,* 33-47.

Taylor S. R. (1982) *Planetary Science: A Lunar Perspective.* Lunar and Planetary Institute, Houston. 481 pp.

Taylor S. R., Kaye M., Muir P., Nance W., Rudowski R., and Ware N. (1972) Composition of the lunar uplands: Chemistry of Apollo 14 samples from Fra Mauro. *Proc. Lunar Sci. Conf. 3rd,* pp. 1231-1249.

Vaniman D. T. (1990) Glass variants and multiple HASP trends in Apollo 14 regolith breccias. Proc. *Lunar Planet. Sci. Conf. 20th,* this volume.

Wänke H., Palme H., Baddenhausen H., Dreibus G., Jagoutz E., Kruse H., Spettel B., Teschke F., and Thacker R. (1974) Chemistry of

Apollo 16 and 17 samples: Bulk composition, late stage accumulation and early differentiation of the Moon. *Proc. Lunar Sci. Conf. 5th,* pp. 1307-1335.

Warner J. L., Simonds C. H., and Phinney W. C. (1976) Apollo 17, Station 6 boulder sample 76255: Absolute petrology of breccia matrix and igneous clasts. *Proc. Lunar Sci. Conf. 7th,* pp. 2233-2250.

Warner J. L., Phinney W. C., Bickel C. E., and Simonds C. H. (1977) Feldspathic granulitic impactites and pre-final bombardment lunar evolution. *Proc. Lunar Sci. Conf. 8th,* pp. 2051-2066.

Warren P. H. (1985) The magma ocean concept and lunar evolution. *Annu. Rev. Earth Planet. Sci., 13,* 201-240.

Warren P. H. (1988) The origin of pristine KREEP: Effects of mixing between urKREEP and the magmas parental to the Mg-rich cumulates. *Proc. Lunar Planet. Sci. Conf. 18th,* pp. 233-241.

Warren P. H. (1989) KREEP: major-element diversity, trace-element uniformity (almost) (abstract). In *Workshop on Moon in Transition: Apollo 14, KREEP, and Evolved Lunar Rocks* (G. J. Taylor and P. H. Warren, eds.), pp. 149-153. LPI Tech. Rpt. 89-03, Lunar and Planetary Institute, Houston.

Warren P. H. (1990a) Highland igneous rocks and monomict breccias. In *Lunar Sourcebook* (G. H. Heiken, D. T. Vaniman, and B. M. French, eds.), Chapter 6. Cambridge Univ., New York, in press.

Warren P. H. (1990b) Siderophile elements. In *Lunar Sourcebook* (G. H. Heiken, D. T. Vaniman, and B. M. French, eds.), Chapter 8. Cambridge Univ., New York, in press.

Warren P. H. (1990c) Lunar anorthosites and the magma ocean hypothesis: Importance of FeO enrichment in the parent magma. *Am. Mineral.,* in press.

Warren P. H. and Kallemeyn G. W. (1988) Lunar meteorites: constraints on lunar composition and evolution (abstract). In *Lunar and Planetary Science XIX,* pp. 1236-1237. Lunar and Planetary Institute, Houston.

Warren P. H. and Wasson J. T. (1977) Pristine nonmare rocks and the nature of the lunar crust. *Proc. Lunar Sci. Conf. 8th,* pp. 2215-2235.

Warren P. H. and Wasson J. T. (1978) Compositional-petrographic investigation of pristine nonmare rocks. *Proc. Lunar Planet. Sci. Conf. 9th,* pp. 185-217.

Warren P. H. and J. T. Wasson (1979) The compositional-petrographic search for pristine nonmare rocks: Third foray. *Proc. Lunar Planet. Sci. Conf. 10th,* pp. 583-610.

Warren P. H., Taylor G. J., Keil K., Kallemeyn G. W., Rosener P. S., and Wasson J. T. (1983a) Sixth foray for pristine nonmare rocks and an assessment of the diversity of lunar anorthosites. *Proc. Lunar Planet. Sci. Conf. 13th,* in *J. Geophys. Res., 88,* A615-A630.

Warren P. H., Taylor G. J., Keil K., Kallemeyn G. W., Shirley D. N., and Wasson J. T. (1983b) Seventh foray: Whitlockite-rich lithologies, a diopside-bearing troctolitic anorthosite, ferroan anorthosites, and KREEP. *Proc. Lunar Planet. Sci. Conf. 14th,* in *J. Geophys. Res., 88,* B151-B164.

Warren P. H., Shirley D. N., and Kallemeyn G. W. (1986) A potpourri of pristine Moon rocks, including a VHK mare basalt and a unique, augite-rich Apollo 17 anorthosite. *Proc. Lunar Planet. Sci. Conf. 16th,* in *J. Geophys. Res., 91,* D319-D330.

Warren P. H., Jerde E. A., and Kallemeyn G. W. (1987) Pristine Moon rocks: A "large" felsite and a metal-rich ferroan anorthosite. *Proc. Lunar Planet. Sci. Conf. 17th,* in *J. Geophys. Res., 92,* E303-E313.

Warren P. H., Haack H., and Rasmussen K. (1989a) Effects of megaregolith insulation on Sm-Nd cooling ages of deep-crustal cumulates from the Moon and large asteroids (abstract). In *Lunar and Planetary Science XX,* pp. 1179-1180. Lunar and Planetary Institute, Houston.

Warren P. H., Jerde E. A., and Kallemeyn G. W. (1989b) Lunar meteorites: siderophile element contents, and implications for the composition and origin of the Moon. *Earth Planet. Sci. Lett., 91,* 245-260.

Proceedings of the 20th Lunar and Planetary Science Conference, pp. 61-75
Lunar and Planetary Institute, Houston, 1990

Highly Evolved and Ultramafic Lithologies from Apollo 14 Soils

R. W. Morris, G. J. Taylor, H. E. Newsom, and K. Keil

Department of Geology and Institute of Meteoritics, University of New Mexico, Albuquerque, NM 87131

S. R. Garcia

Research Reactor Group, INC-5, Mail Stop G776, Los Alamos National Laboratory, Los Alamos, NM 87545

The Apollo 14 sample collection contains a wide variety of lithologies, ranging from highly evolved KREEP and granite to ultramafic dunite. Our studies of 2-4 mm fragments from several Apollo 14 soil samples discovered a unique collection of rock fragments that span almost the whole range of magma evolution, from ultramafic cumulates to extreme differentiates apparently produced by silicate liquid immiscibility. Three of the fragments, from soil sample 14001,28, are dimict granitic breccias, composed of granitic clasts intruded by a ferropyroxenitic glass. These samples may represent an immiscible melt pair that has been remixed by impact. REE abundances in these breccias are extremely high, with La up to 830× chondrites, and show evidence of fractionation of apatite and zircon from a KREEPy parent magma. Two of the samples from 14161,212 are ultramafic rocks, including a dunite and a peridotite/impact melt breccia. The dunite has very low REE abundances (≈0.5× chondrites in approximately chondritic relative abundances) and appears to be a cumulate from an Mg-suite magma with very low incompatible element abundances, implying that 14161,212.4 was unaffected by the assimilation of urKREEP seen in the other Apollo 14 dunites. The peridotite is a unique rock fragment composed of forsteritic olivine and enstatitic pyroxene with anomalously low Al_2O_3, CaO, TiO_2, and Cr_2O_3 abundances (0.20, 0.30, 0.05, and 0.34 wt.% respectively).

INTRODUCTION

Most of the rocks returned from the lunar highlands are complex breccias composed of a variety of older rocks. These polymict breccias have been produced by impact into the lunar crust, leading to pervasive, fine-scale mixing of originally igneous rocks. A number of small clasts and rock fragments separated from highlands breccias and soils have textures and chemistries that are considered "pristine" [produced by endogenous igneous activity and not affected by impact-induced mixing (*Warren and Wasson,* 1979)]. Studies of these pristine rocks have shown that the highlands are composed of a complex suite that almost certainly is a poor sample of the wide range of rock types actually present in the lunar highlands. Analysis of new clasts and rock fragments from highlands breccias and soils continues to uncover unique rocks that can lead to insights into the origin and early evolution of the lunar crust.

A fascinating range of lithologies has been described in the Apollo 14 sample collection, including granites, ultramafic rocks, a variety of anorthosites, troctolites, an assortment of high-Alumina mare basalts, KREEPy impact melts, and KREEP basalts. Important among these are the highly evolved granites that, along with the KREEP component, probably formed during the late stages of the initial lunar differentiation, in the putative magma ocean (*Warren and Wasson,* 1979; *Neal and Taylor,* 1989a), and therefore contain important information about the Moon's primary differentiation and about the early igneous processes operating on it. To increase our understanding of how these interesting lithologies formed and to expand our knowledge about lunar differentiation, we extracted eighteen fragments that appeared to have igneous textures from the 2-4-mm size fractions of several Apollo 14 soil samples. We examined each fragment by instrumental neutron activation analysis and studied them petrographically. The suite contains a wide variety of rock types: granitic breccias, ultramafic rocks, mare basalts, and impact melt breccias.

MATERIALS AND METHODS

Eighteen 2- to 4-mm particles were selected from sieve samples of Apollo 14 soils 14161, 14167, 14257, and 14001. Particles studied included four from sample 14001,28; two from 14161,203; five from 14161,212; one from 14161,229; two from 14257,80; and four from 14167,69. One sample (14161,212.1) was determined to be a breccia by preliminary microscopic examination and was cut into two parts: a black crust (14161,212.1b) and a light-colored clast (14161,212.1a). In this paper, we will describe five of the samples that were unique highly evolved or ultramafic rocks. Samples described here include three granitic breccias (14001,28.2, 14001,28.3, and 14001,28.4), a dunite (14161,212.4), and a peridotite/impact melt breccia (14161,212.1). The mare basalts and impact melt rocks will be described in a later paper.

Major and trace element abundances in all samples were determined by nondestructive, sequential instrumental neutron activation analysis (Table 1). Irradiation and counting was conducted in two stages. First, samples and standards were irradiated for 5 minutes and counted for 20 minutes at Los Alamos National Laboratory by S. Garcia. The short count procedure is described in *Minor et al.* (1981). After two to four weeks samples and standards were repackaged with flux monitors and re-irradiated for four hours at a neutron flux of

TABLE 1. Whole rock compositions of Apollo 14 rock fragments analyzed by INAA.

Rock Type	Granitic Breccia			Impact Melt	Peridotite	Dunite
Sample #	001,28.3	001,28.4	001,28.2	161,212.11	161,212.12	161,212.4
Weight (mg)	15.3	15.0	34.2	22.8	34.7	16.2
Major Elements (wt.%)						
TiO_2	1.4	1.0	1.8	0.32	<0.70	<1.2
Al_2O_3	8.8	11.3	9.6	1.7	0.57	0.87
FeO	12.2	11.0	9.1	9.5	8.0	10.7
MnO	0.13	0.13	0.11	0.12	0.11	0.14
MgO	<0.83	<0.99	<1.2	30.0	33.7	38.1
CaO	5.5	5.7	3.9	0.85	0.22	0.28
Na_2O	1.4	1.9	1.7	0.070	0.0045	0.0078
K_2O	2.1	2.5	3.4	<1.3	<0.20	<1.7
Cr_2O_3	0.022	0.015	0.014	0.37	0.33	0.46
Trace Elements (ppm)						
Sc	20	20	15	5.6	3.7	2.8
V	<7	<7	<8	47	49	49
Co	3.8	3.9	2.7	43	47	120
Zn	46	31	—	—	17	—
Se	10	9.2	15	0.49	0.29	<0.28
Rb	87	89	110	<2.8	<6.4	8.6
Sr	170	190	170	<32	<20	<31
Sb	—	0.40	0.23	0.12	0.074	0.11
Cs	2.0	2.2	2.3	0.064	0.027	0.051
Ba	2300	3100	3600	39	<40	<320
La	200	130	110	12	0.10	0.091
Ce	460	270	250	30	0.38	0.40
Sm	66	40	36	5.0	0.087	0.055
Eu	3.3	3.0	2.8	0.27	0.021	0.026
Tb	15	9.3	8.1	0.99	0.021	<0.024
Yb	64	43	37	3.6	0.074	0.10
Lu	7.7	5.9	5.0	0.47	0.0093	0.016
Hf	120	43	41	2.9	0.054	<.080
Ta	7.4	6.3	10	0.40	0.011	0.027
Ir (ppb)	<5.7	<5.1	<1.6	2.1	<1.0	<1.9
Th	42	28	35	2.0	<0.34	0.39
U	13	8.2	9.1	0.55	<0.10	0.14
Mg#	<6.4	<8.3	<11	76.0	80.7	78.1

9.7×10^{12} neutrons per cm^2 per second in the Omega West research reactor at Los Alamos. Samples and standards were returned to the counting laboratory at the University of New Mexico after four to seven days and were counted at least three times on a coaxial Ge gamma ray detector, including long counts of up to four days starting several weeks after irradiation. Since our counting laboratory is fairly new, great care was taken to ensure the quality of the standardization. Five standards were used, with NBS SRM 1633A Coal Fly Ash being the primary standard. Secondary standards used mainly for Ca, Cr, K, Na, Sc, Fe, and Ir include USGS BHVO-1, USGS SCO-1, USGS SGR-1 (*Gladney et al.,* 1984), and the Allende meteorite reference sample (*Jarosewich et al.,* 1987). Spectral data were reduced using the latest version of the SELEX programs of *Kruse and Spettel* (1982). These programs allow interactive fitting of difficult background regions and small peaks. Estimated uncertainties due to counting statistics are ± 1% to 5% for Fe, Na, Sc, Cr, Mn, Co, Cs, La, Ce, Sm, Eu, Tb, Lu, Hf, Ta, and Th; ± 6% to 10% for Ca, Ti, Al, Se, Rb, Ba, Yb, and U; ± 11% to 20% for K, Zn, and Sr. The detection limit for a given element is reported as the abundance that would give a 50% one-sigma uncertainty due to counting statistics. The extremely low abundances of trace elements in the dunite and peridotite generally caused uncertainties about double those seen in the other samples. Estimated uncertainties for the dunite and peridotite analyses are ± 1% to 5% for Fe, Sc, Cr, Co, and Sm; ± 6% to 10% for Mg, Al, Na, Mn, V, La, and Th; ± 11% to 20% for Sb, Ba, Eu, Tb, Lu, and Hf; ± 21% to 30% for Ca, Ti, Zn, Ta, Yb, and Ir; and ± 31% to 50% for Cs, Se, Rb, Ce, and U.

Samples were then made into polished thin sections and characterized using a petrographic microscope equipped with an automatic point-counting stage and using back-scattered electron imaging. Mineral compositions were determined using an automated JEOL 733 Superprobe electron microprobe. Data were corrected for differential matrix effects by the method of *Bence and Albee* (1968). Bulk abundances of major and minor elements in some of the samples were determined by broad beam microprobe analysis using a 50-μm beam. These results must be considered only approximate due to the difficulties inherent in correcting data gathered from multiphase targets using mineral standards. Special long counts

Fig. 1. Photomicrographs of granitic breccia 14001,28.3: **(a)** Entire section, uncrossed nicols. Dark material is the brown, high-Fe glass. Light colored granitic clasts are composed of colorless glass, quartz/feldspar granophyres, and isolated grains of quartz, feldspar, and maskelynite. Field of view is about 3.0 mm; **(b)** Veins of high-Fe glass intruding granitic clasts, uncrossed nicols. The chemical heterogeneity and mixing of the high-Fe and high-Si glasses is reflected in the color gradation from dark brown to colorless. Field of view is about 0.75 mm; Backscattered electron images of granitic breccia 14001,28.4; **(c)** Granophyre clast surrounded by high-Fe glass (white). Dark gray lamellae are a silica mineral, probably quartz. Light gray material consists of both feldspar lamellae and colorless feldspathic glass. Scale bar is 100 μm; **(d)** High magnification image of quartz/feldspar granophyre. Scale bar is 100 μm.

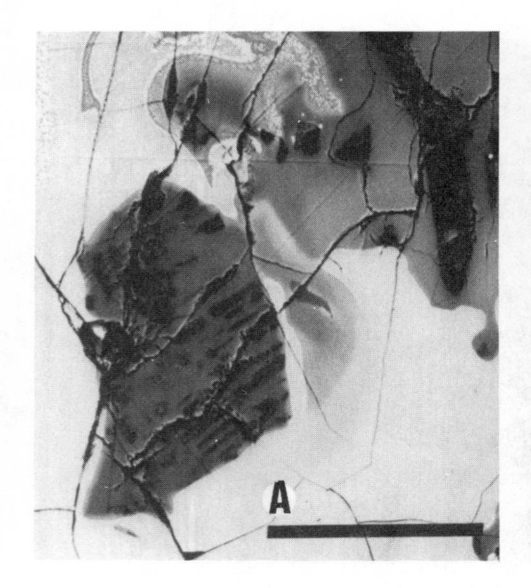

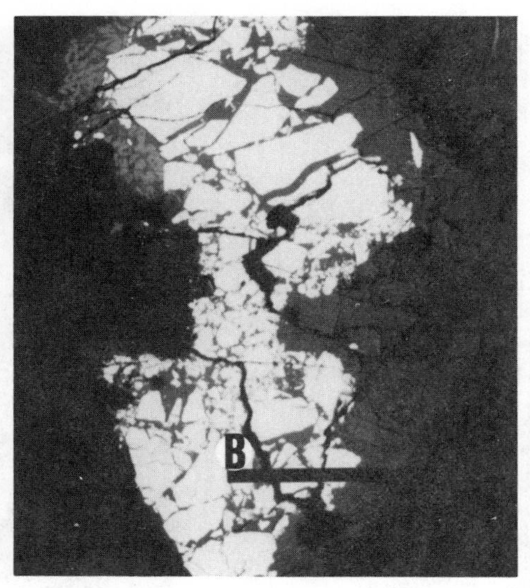

Fig. 2. Backscattered electron images of granitic breccia 14001,28.3: **(a)** High magnification image of a partially digested granophyre clast surrounded by high-Fe (white) and high-Si (light gray) glass. The mixing of these glasses is visible along their mutual contact; **(b)** High magnification image of the zircon-rich clast in 14001,28.3. Numerous grains of zircon (white) and phosphate (light gray) are surrounded and intruded by the colorless glass (dark gray). The jigsaw puzzle-like outlines of the zircons are plainly visible here. Scale bars are 100 μm.

(200 sec per REE) at high beam currents ($\approx$100 na) were used to measure the abundances of several of the REE in the brown glass in sample 14001,28.3. The beam size was increased to 10 μm in these analyses to minimize volatilization and spot to spot heterogeneity. The REE microprobe standards are a series of four doped glasses (*Drake and Weill,* 1972). Uncertainties for the REE analyses were quite variable, depending on the amount of each REE present in the analyzed spot. Approximate uncertainties are as follows: La $\pm$ 9% to 47% (La_2O_3 = 0.051 to 0.0093 wt.%); Ce $\pm$ 2% to 3% (Ce_2O_3 = 0.23 to 0.16 wt.%); Dy $\pm$ 6% to 11% (Dy_2O_3 = 0.14 to 0.081 wt.%); Er $\pm$ 8% to 25% (Er_2O_3 = 0.11 to 0.038 wt.%); Yb $\pm$ 18% to 75% (Yb_2O_3 = 0.053 to 0.013 wt.%). Detection limits for the REE were about 0.009 oxide wt.%, or about 38 to 44 ppm of each REE.

GRANITIC BRECCIAS

Petrography

Three of the samples, all from soil 14001,25, are granitic breccias, with unique trace element signatures. These are dimict breccias composed of clasts of shocked and shock melted granite/feldspathic breccia set in a brown ferropyroxenitic glass (Fig. 1). The brown glass makes up between 10% and 40% by area of the thin sections of the breccias and occurs as large masses and ropy coatings draped over the granitic clasts and as veins between and intruding clasts. The centers of the larger masses of glass appear to have cooled slowly enough to form crystallites smaller than 1 μm that have the same composition as the surrounding glass. This brown glass is very heterogeneous in color and chemistry, and is characterized by high, but variable abundances of FeO ($\approx$25%), TiO_2 ($\approx$3%), P_2O_5 ($\approx$2%), and MgO ($\approx$2%).

Despite the severe brecciation, patches of the original texture can be seen in the granitic clasts. Between 10% and 30% of each breccia is made up of shocked, granophyric intergrowths of ternary feldspar ($An_{24}Ab_{25}Or_{51}$) and a silica mineral, probably quartz (Fig. 1). These granophyres are identical in texture to granophyres from known pristine granites (e.g., *Ryder,* 1976) and indicate that at least part of each breccia is pristine. Intergrowth grains are up to 1.0 mm in size, although most of the grains are less than 0.4 mm. Individual silica and feldspar lamellae in the granophyres range from $\approx$10 to 40 μm in width. The outlines of these intergrowth grains tend to be indistinct and are often in contact with a colorless glass of similar composition (Fig. 2), suggesting that the colorless glass was probably derived by shock melting of the granite. An average of 12 microprobe analyses of these granophyres appears in Table 2 but, as noted before, these broad beam results are only approximate. Most of the granitic clasts are complex breccias composed of a matrix of colorless,

TABLE 2. Microprobe analyses of components of granitic breccias—brown glass in felsite 12033 shown for comparison.

	Average* Granite	Average† Granophyre	Colorless‡ Glass	Brown§ Glass	Brown¶ Glass
SiO_2	73.9	76.5	65.5	48.3	47.7
Al_2O_3	14.3	12.0	15.0	6.8	5.3
FeO	1.8	0.99	6.1	26.9	36.1
CaO	1.6	1.4	2.3	7.1	2.4
MgO	0.03	0.07	0.44	2.2	1.6
K_2O	4.6	5.1	4.4	1.3	2.5
Na_2O	2.0	2.1	2.0	0.73	1.1
P_2O_5	—	0.15	0.31	1.8	—
TiO_2	1.2	0.60	0.74	3.0	2.4
MnO	—	—	0.06	0.32	0.48
Cr_2O_3	—	—	—	0.03	—
BaO	—	—	0.49	—	—
ZrO_2	—	—	0.08	0.29	—
Total	99.4	98.8	99.4%	98.8	99.5

*Average of 30 analyses, 50-μm beam.
†Average of 12 analyses, 50-μm beam.
‡Average of 30 analyses, 10-μm beam.
§Average of 36 analyses, 10-μm beam.
¶*Warren et al.* (1987).
**Total on individual analyses are between 98% and 102%.

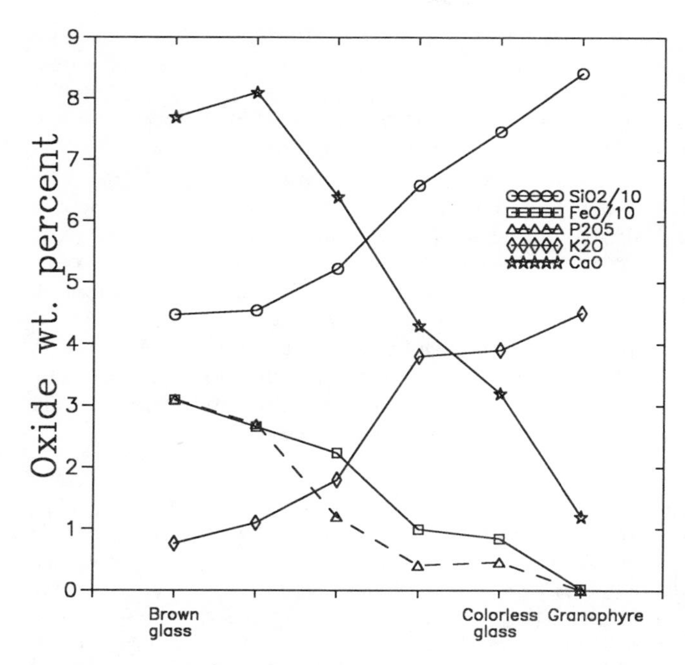

Fig. 3. Microprobe traverse from the center of the high-Fe glass to a granophyre, showing the mixing of glasses seen in thin section 14001,28.3 (Fig. 2). Each step represents approximately 100 μm.

feldspathic glass enclosing a variety of materials. Partially digested quartz/feldspar intergrowths, rounded grains of shocked quartz (<0.4 mm in diameter), subangular to subrounded grains (<0.3 mm) of shocked feldspar ($An_{72}Ab_{26}Or_2$ to $An_{90}Ab_9Or_1$), and grains (<0.3 mm) of isotropic maskelynitized feldspar of similar composition are all included in the colorless glass matrix. This glass is extremely variable in composition and the glass is enriched in Ca, Al, and Na when adjacent to feldspars. Near the granophyres, this colorless glass is rich in Si, K, Na, and Al and almost devoid of Fe, Ti, P, Mg, and Ca, but the glass gets progressively richer in the latter group of elements closer to the brown glass. This indicates that there was some mixing between the brown and colorless glasses, presumably during impact-induced emplacement of the brown glass into the granite. This mixing also affected the composition of the brown glass. In the centers of the large masses of glass, the brown glass is very rich in Fe, Ti, P, and Mg and almost devoid of K, Na, and Al. Adjacent to the granitic clasts, the brown glass contains much higher concentrations of K, Na, and Al. A profile of microprobe analyses from the brown glass to a granophyre demonstrates the effects of this mixing (Fig. 3). This mixing of glasses was also described by *Warren et al.* (1987) in felsite 12033. However, compared to our breccias, their most "primary" (*Neal et al.,* 1989) brown glass is still contaminated by high Si, Al, and K felsite material (Table 2).

Angular grains of zircon, up to 40 × 60 μm, are also found in the colorless glass, making up about 0.5% by area of sample 14001,28.3. They are concentrated mainly in a 0.6 mm × 0.3 mm lens-shaped clast, making up about 80% of that clast (Fig 2). In many places it appears that the zircon fragments could be fit together like a puzzle, implying that the clast may have been a few large zircon grains that were broken up and intruded by the colorless glass. A few small, isolated zircons are dispersed throughout the rest of sample 14001,28.3. Zircons were not observed in any of the other sections. Traces of a phosphate mineral were seen in the zircon-rich clast (Fig. 2), but were not observed elsewhere in the section or in any of the other granitic breccias. We were unable to determine the composition of the phosphate mineral since its grain size (≈2 μm) was approximately the same size as the electron beam.

Whole Rock Compositions

Neutron activation analyses of the granitic breccias are given in Table 1. K_2O and Na_2O abundances are high in these samples, but not as high as in most granites, while CaO is fairly low. FeO is moderately abundant while MgO is below the detection limit of ≈1.2 wt.%, producing the extremely low Mg# [MgO/(MgO + FeO)] of these highly evolved samples. Cr and Co abundances are also extremely low, while incompatible elements such as Th, U, Rb, Ba, and REE are highly enriched. Ir is below the detection limit of ≈2-5 ppb in these breccias, suggesting that they might be pristine, although they have brecciated textures.

The unique nature of these granitic breccias is most obvious in their incompatible trace element abundances, particularly the REE. Chondrite-normalized (Orgueil abundances, *Anders*

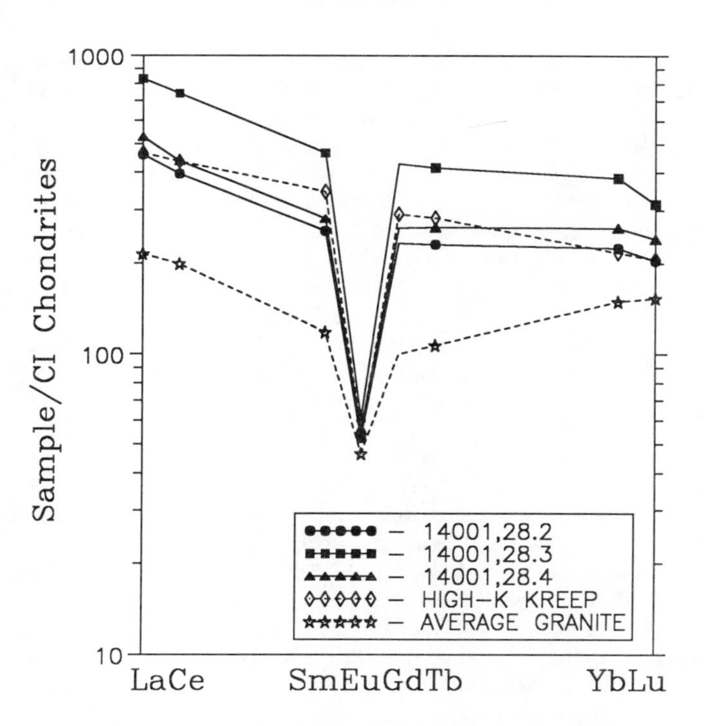

Fig. 4. Chondrite-normalized (*Anders and Grevesse,* 1989) REE abundances in the granitic breccias. High-K KREEP (*Warren and Wasson,* 1979) and "Average Granite" (*Neal and Taylor,* 1989a) are shown for comparison.

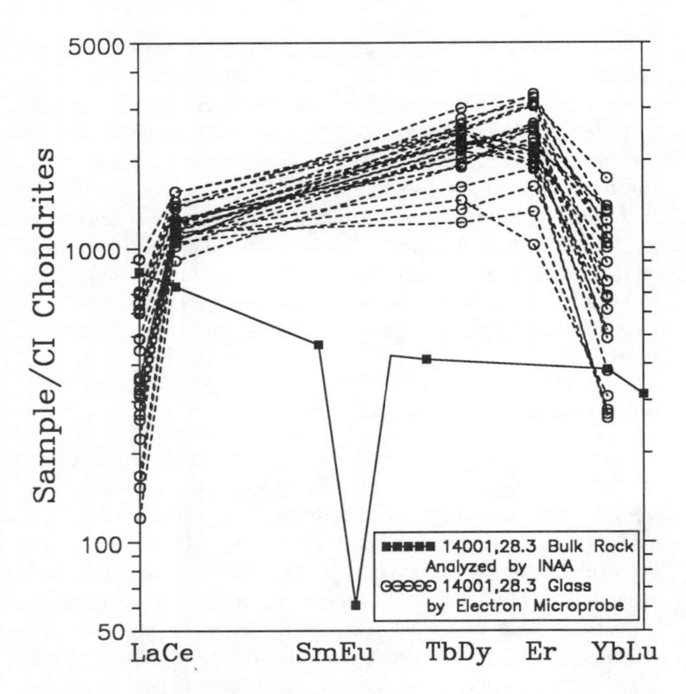

Fig. 5. Chondrite normalized REE abundances in the brown glass in 14001,28.3 measured by electron microprobe. Analytical uncertainties are 8-50% for La, 2-3% for Ce, 5-11% for Dy, 8-25% for Er, and 30-75% for Yb. In spite of the large errors, it appears that the pattern is bow-shaped and concave-down, similar to apatite (*Henderson,* 1982).

and Grevesse, 1989) REE abundances of the three breccias are shown in Fig. 4, compared with high-K KREEP (*Warren and Wasson,* 1979) and "average granite" (*Neal and Taylor,* 1989b). REE patterns in the breccias are quite distinct from any previously measured lunar rock. The REE pattern is not like KREEP, and is very different from the V-shaped pattern of known lunar granites and felsites. Light REE in the granitic breccias are strongly enriched over the middle REE, with an average chondrite-normalized La/Sm ratio of 1.81, which is much higher than the ratio of 1.35 seen in KREEP. Heavy REE occur in almost chondritic relative abundances (though at much higher absolute abundances), with an average normalized Tb/Yb ratio of 1.04, much lower than the KREEP ratio of 1.37. This pattern is, however, similar to the "urKREEP" pattern recently proposed by *Neal and Taylor* (1989a), including the small depletion of Lu relative to Tb and Yb. This similarity in REE pattern has led *Neal and Taylor* (1989b) to propose that these granitic breccias may represent the first unadulterated samples of urKREEP among lunar samples.

Absolute abundances of REE in the breccias are also quite distinct from KREEP and average granite. Sample 14001,28.3 is extremely enriched in REE, with La abundance ≈830× chondritic abundances, much higher than high-K KREEP (*Warren and Wasson,* 1979). Only two previously measured rocks have higher REE abundances than 14001,28.3: the "white clast" 14313,34 (*Haskin et al.* 1973), which may simply be an unrepresentative phosphate-rich portion of an alkali anorthosite (*Warren et al.* 1983), and the quartz monzodiorite 15405 (*Ryder and Norman,* 1979). Whole rock REE abundance appears to correlate with the amount of brown glass present in thin section. 14001,28.3 has the highest REE abundance and the most brown glass (≈40%). 14001,28.4 and 14001,28.2 have progressively lower REE abundances and ≈15% and ≈10% glass, respectively. The extreme REE abundance in 14001,28.3 made it possible to measure several of the more abundant REE (La, Ce, Dy, Er, and Yb) by electron microprobe, using long counts at high beam currents. Chondrite-normalized REE plots of a range of analyses of brown glass with La above detection limit and FeO >25 wt.% (arbitrary cut-off chosen because of the mixing) are shown in Fig. 5. Although the analytical errors are fairly large, it can be seen that this REE pattern is bow-shaped and concave down, roughly similar to a pattern dominated by apatite.

Discussion

These granitic breccias are an important addition to the analyzed collection of rocks from Apollo 14, since they are among the most evolved lithologies in the sample collection. The first question that must be answered is "are these granitic breccias pristine?" The rocks are breccias containing shock-melted glasses and cannot be texturally pristine, as a whole. However, the low siderophile element abundances (Ir <2 ppb) and the presence of texturally pristine clasts suggest that they might be chemically pristine. Chemical pristinity is also supported by the REE pattern and abundances that are very different from KREEP. The extreme compositions of the glasses also support chemical pristinity, since they do not resemble any of the commonly observed components of lunar breccias (FAN, MG, KREEP). Thus, we consider these breccias to be chemically pristine, but texturally pristine only in part.

The petrography of these breccias suggests that they might be simple mechanical mixtures of two components: a KREEPy

TABLE 3. Microprobe analyses of immiscible melts.

	Brown Glass	Colorless Glass	High Fe*	High Si*	High Fe†	High Si†	KREEP‡
SiO_2	48.3	65.5	42.2	76.3	43.1	66.2	47.9
Al_2O_3	6.8	15.0	5.4	11.5	10.3	12.6	16.6
FeO	26.9	6.1	32.9	3.0	20.8	6.8	10.6
CaO	7.1	2.3	10.7	1.6	10.8	5.8	9.5
MgO	2.2	0.44	0.79	0.07	3.8	1.3	10.6
K_2O	1.3	4.4	0.41	6.7	1.6	3.6	0.83
Na_2O	0.73	2.0	0.16	0.14	0.5	1.1	0.86
P_2O_5	1.8	0.31	0.57	0.15	1.9	0.5	0.78
TiO_2	3.0	0.74	4.2	0.68	6.7	1.4	1.7
MnO	0.32	0.06	0.32	0.05	—	—	0.14
Total	98.8	97.4	97.7	100.	99.5	99.3	99.5
K_2O/P_2O_5	0.72	14.2	0.7	44.7	0.84	7.2	1.1
Al_2O_3/P_2O_5	3.8	48.4	9.5	76.7	5.4	25.2	21.3
K_2O/MgO	0.59	10.0	0.52	95.7	0.42	2.8	0.078
SiO_2/FeO	1.8	10.7	1.3	25.4	2.1	9.7	4.5

**Roedder and Weiblen* (1972).
†*Hess et al.* (1989).
‡*Warren and Wasson* (1979).
Brown glass = Average of 36 analyses, 10-μm beam.
Colorless glass = Average of 30 analyses, 10-μm beam.

impact melt (the brown glass) and granite. However, this is unlikely for a number of reasons. First, the breccias contain extremely high abundances of REE, much higher than KREEP and granite. A least-squares mixing calculation indicates that sample 14001,28.3 could be a mixture of 65% of a hypothetical "supergranite" (La ≈ 590× chondrites) with 35% of a hypothetical "superKREEP" (La ≈ 1280× chondrites), both of which contain far higher REE abundances than observed KREEPy or granitic lithologies. Second, microprobe analysis of the brown glass (Fig. 5) shows that its REE pattern is actually concave-down and bow-shaped, not the KREEPy pattern assumed for the mixing calculation. Finally, none of our modeled mixtures could simultaneously reproduce both the LREE and HREE patterns in the breccias.

The occurrence of granites in the dominantly basaltic lunar environment, without intermediate rocks, suggests that silicate liquid immiscibility may play a role in their genesis (e.g., *Hess et al.,* 1975; *Rutherford et al.,* 1976; *Roedder,* 1978; *Neal and Taylor,* 1987, 1989a). However, other authors have stressed the need for apatite or whitlockite fractionation to produce the V-shaped REE profile of lunar granite (*Dickinson and Hess,* 1983; *Salpas et al.,* 1985). Although severely brecciated, our granitic breccias contain evidence for the occurrence of both of these processes in granite genesis.

The dichotomy between the high-Si, high K granite clasts and the high-Fe glass is strongly suggestive of silicate liquid immiscibility. Experimental work (*Hess et al.,* 1975; *Rutherford et al.,* 1976; *Roedder,* 1978) and observations of natural rocks (*Roedder and Weiblen,* 1970, 1971, 1972; *Crawford and Hollister,* 1977) have shown that extensive (90% to 98%) Fenner trend fractional crystallization of a basaltic melt can lead to silicate liquid immiscibility, producing one melt rich in FeO, P_2O_5, TiO_2, REE, MgO, and one melt rich in SiO_2, K_2O, Al_2O_3, and Na_2O. This is essentially the distribution of elements seen in these breccias, although this distribution has been somewhat modified by later mixing of the brown and colorless glasses. As described above, the granitic component (SiO_2, K_2O, Al_2O_3, and Na_2O) is enriched in the brown glass nearest the granitic clasts and colorless glass, while the mafic component (FeO, P_2O_5, TiO_2, REE, MgO, and CaO) is enriched in the colorless glass adjacent to brown glass. This implies that there must have been some reaction between the intruded brown glass and the granite. The areas of the brown glass most distant from the granite will be least affected by this mixing and will be taken to represent the primary composition of the brown glass in the following discussion. This same pattern of elemental variation was seen in the brown glass veins in felsite 12033 (*Warren et al.* 1987), although none of the glass in the felsite contains as much MgO, TiO_2, and CaO and as little K_2O and Na_2O as the centers of some of the larger masses of glass in 14001,28.3 (Table 2). The quartz/feldspar intergrowths, although texturally pristine, lack the small amounts of accessory minerals seen in other granites. The colorless glass is probably more representative of bulk granite due to the mixing of melted components. For these reasons, we will represent the bulk granite by an average of 30 analyses of the colorless glass. In addition, the composition of the colorless glass is much less extreme that the granophyre clasts and provides a more conservative representation of the composition of the granite. Microprobe analyses of the brown and colorless glasses are compared to immiscible melt pairs analyzed by *Roedder and Weiblen* (1971) and *Hess et al.* (1989) in Table 3. Also shown for comparison is the high-K KREEP composition of *Warren and Wasson* (1979). Examination of liquid/liquid distribution coefficients compiled by *Neal and Taylor* (1989a), shown in Table 4, indicates that P_2O_5 is most strongly partitioned into the high-Fe melt and K_2O is most strongly partitioned into the high-Si melt. Since K_2O and P_2O_5 tend to be conserved during fractional crystallization, the K_2O/P_2O_5 ratio should be most affected by liquid immiscibility. A calculation of this ratio shows that the brown glass and average granite have K_2O/P_2O_5 ratios similar to those in well-described immiscible melt pairs (*Roedder and Weiblen,* 1971; *Hess et al.,* 1989) and quite different from KREEP (Table 3). Other oxide ratios also are similar to known immiscible melts and very different from KREEP.

TABLE 4. Liquid/liquid distribution coefficients from immiscible silicate melt pairs.

Major Element Oxide	Distribution coefficient $D_{Basic/Acid}$*	$D_{Basic/Acid}$ calculated from 14001,28.3 (brown/colorless glass)
SiO_2	0.61	0.74
TiO_2	4.2	4.1
Al_2O_3	0.60	0.45
FeO	6.7	4.4
MgO	4.7	5.0
CaO	4.6	3.1
Na_2O	0.44	0.37
K_2O	0.24	0.30
P_2O_5	18.5	5.8

*Compiled by *Neal and Taylor* (1989a).

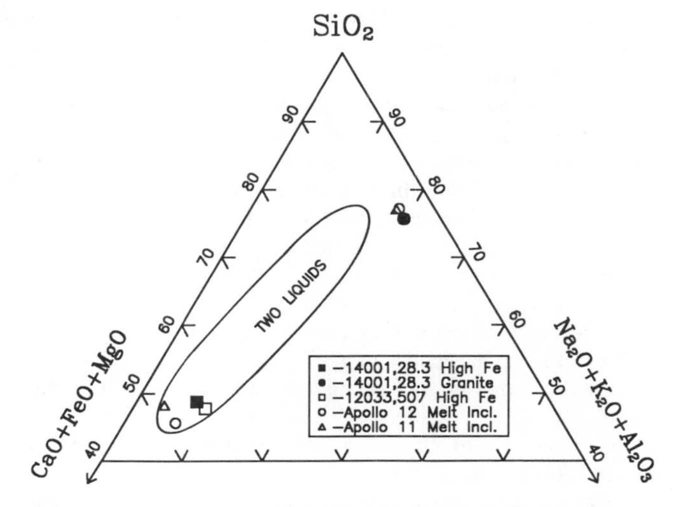

Fig. 6. The high-Fe (brown) and high-Si (colorless) glasses in 14001,28.3 plot near opposite ends of a miscibility gap on the pseudoternary diagram of *Weiblen and Roedder* (1973). Also plotted are immiscible melt pairs from Apollo 11 and Apollo 12 mare basalts (*Roedder and Weiblen,* 1971, 1972), and the high-Fe glass from felsite 12033 (*Warren et al.,* 1987).

These primary compositions plot near Apollo 11 and Apollo 12 immiscible melt inclusions at opposite ends of the immiscibility gap proposed by *Weiblen and Roedder* (1973), implying that both the granite and the high-Fe glass are products of silicate liquid immiscibility (Fig. 6). The fact that the immiscible melt compositions do not plot exactly on the boundaries of the miscibility gap, although not surprising, should be explained. The diagram was drawn for the pure K_2O-FeO-Al_2O_3-SiO_2 system and it is probably not exactly applicable to the more complex chemical system in these breccias. In addition, mixing may be drawing the high-Fe melt compositions off the curve.

The similarity between the REE pattern in the granitic breccias and the pattern inferred for the pre-immiscibility melt parental to lunar granite led *Neal and Taylor* (1989b) to propose that our granitic breccias may represent the first unadulterated samples of urKREEP among lunar samples. A comparison of major, minor and trace element chemistries of the breccias and the inferred composition of urKREEP (*Neal and Taylor* 1989a) shows a number of important differences and some interesting similarities. Most of the elements that partition into the high-Fe melt (TiO_2, FeO, REE, Th, and U) occur in our average granitic breccia at about 50% of the abundance inferred for urKREEP. This may indicate that much of the high-Fe melt was not injected back into the granite, leading to an underrepresentation of elements that partition into the high-Fe melt. This is supported by the enrichments relative to urKREEP of SiO_2, Al_2O_3, Cs, Ba, Rb, and Hf in the granitic breccias. However, a number of elements do not follow this pattern. Na_2O is extremely enriched in the breccias ($\approx 3\times$ urKREEP), while K_2O is slightly depleted ($\approx 0.9\times$ urKREEP), probably indicating that the feldspars in our breccias are much more sodic than those in "average granite." Since most of the granites contain K-feldspars (up to Or_{96}) while our granophyres have no feldspars over Or_{60}, this seems a reasonable conclusion. MnO in our breccias is strongly enriched relative to urKREEP ($\approx 2\times$ urKREEP) while Cr_2O_3 is strongly depleted ($\approx 0.05\times$ urKREEP). The reason for these differences is uncertain, but probably reflects differences in granite parent magma composition and evolutionary history. For example, chromite fractionation may have occurred in the parent magma of the three granitic breccias but did not significantly affect the composition of the parent magmas of the other lunar granites. Rb is enriched slightly above Ba and the rest of the felsic elements while the K_2O abundance is strongly depleted. This variability also indicates that many of the differences between urKREEP and our breccias are probably due to differences in granite chemistry, coupled with differences in the amount of high-Fe melt intruded into the granite.

One of the main difficulties with an origin of granite by silicate liquid immiscibility is the absence of any rocks identified to have crystallized from the granites' high-Fe melt partner. In response to this problem, *Neal and Taylor* (1989a) proposed a "K-frac/REEP-frac" model. In this model, the low viscosity, high-Fe melt (REEP-frac) fractionates fayalite and ilmenite (*Neal et al.,* 1989) and percolates upwards into the lunar crust. This P- and Ca-rich melt matasomatizes various highlands rocks (e.g., *Lindstrom et al.,* 1984), producing REE-rich rocks with apatite dominated (bow-shaped) REE patterns. The concave-down bow-shaped pattern in the high-Fe (REEP-frac) glass in 14001,28.3 supports this model.

One of the most distinctive features of lunar granites and felsites is their V-shaped REE pattern (Fig. 4). The origin of this pattern has been a matter of some debate, but its similarity to an inverted apatite pattern has led some authors (e.g., *Dickinson and Hess,* 1983; *Salpas et al.,* 1985) to suggest that the REE pattern in granite is controlled by apatite fractionation from a crystallizing magma. *Ryder* (1976), in his study of breccia 15405, proposed that a KREEP basalt magma could be a parent liquid for granite. However, the REE patterns and abundances in granite and KREEP basalt make this unlikely. The main problem with this hypothesis is that the KREEP pattern is depleted in HREE relative to the MREE, ($Gd/Lu \approx 11$), making it difficult to produce the granitic HREE enrichment by any amount of apatite fractionation (*Dickinson and Hess,* 1983). Also, the REE abundances in granite are less than those in KREEP basalts. *Salpas et al.* (1985) showed that lowering the REE abundances by apatite fractionation would require the removal of an assemblage of at least 10% apatite

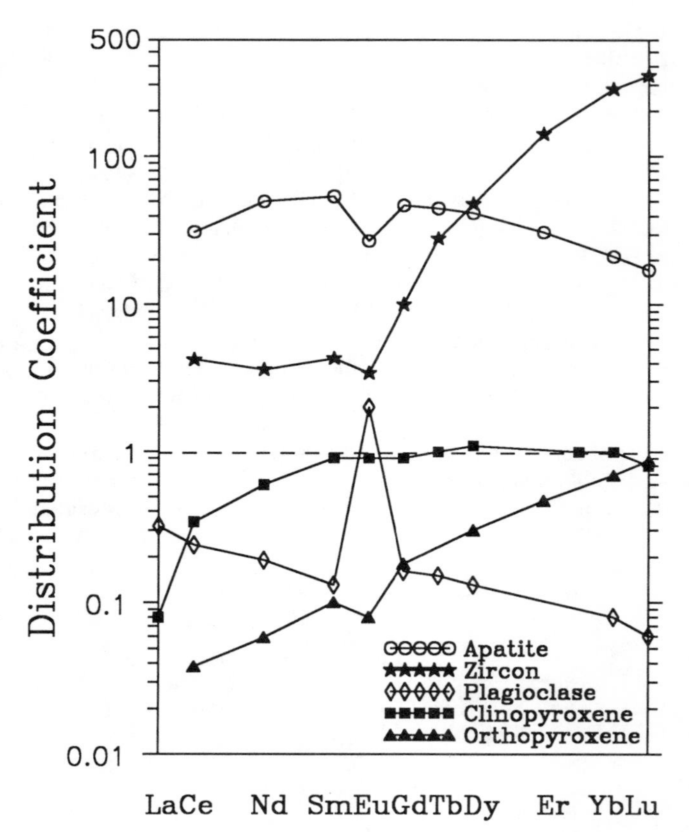

Fig. 7. Examples of mineral/melt distribution coefficients for some of the phases in basaltic and rhyolitic rocks. Data are taken from *Henderson* (1982). The REE are incompatible in most of the major phases in basaltic rocks, but are strongly compatible in some accessory phases.

for 80-90% fractionation. This amount of apatite seems rather large, considering that the normative mineralogy of KREEP basalt 15386 contains only 1.5% apatite.

Fractionation of apatite from a mare basalt magma is much more effective in producing a V-shaped REE pattern since mare basalts can have an essentially flat REE pattern (*Dickinson and Hess,* 1983). However, granite is associated primarily with KREEPy highland rocks at Apollo 14 and is not associated with mare basalts at the mare landing sites. In addition, a calculation (using the revised partition coefficients of *McKay et al.,* 1986) of Rayleigh fractionation from a high-Al mare basalt composition (*Dickinson et al.,* 1985) indicated that raising the La abundance from ≈20 to ≈400× chondrites would require about 97% fractionation of the olivine + chromite + orthopyroxene + plagioclase + ilmenite assemblage that *Neal et al.* (1988) proposed would fractionate from a high-Al mare basalt. However, even this high degree of fractionation only raises the K_2O content to ≈1 wt.%, much lower than the >3 wt.% K_2O seen in lunar granites.

Since crystal fractionation has been proposed for the origin of the V-shaped granite REE pattern, we tried to model the REE in our granitic breccias by crystal fractionation. The breccias have a unique REE pattern characterized by a LREE pattern steeper than KREEP and an essentially flat HREE pattern, all at extreme REE abundances (Fig. 4). Since KREEP has elevated REE abundances and is a widespread component at the Apollo 14 site (*Warren and Wasson,* 1979; *Shervais and Taylor,* 1986), we tried to model the REE pattern in the granitic breccias by fractional crystallization from a magma with high-K KREEP elemental abundances. Partition coefficients (Fig. 7) of some of the phases seen in basalts and granites show that while most of the major phases exclude REE, minor phases can strongly partition the REE and may greatly alter the REE pattern by fractionation. The presence of zircon and phosphate in our breccias and apatite in many of the previously described granites suggested that fractionation of these two minor phases might produce the unusual REE pattern in 14001,28.3 from a KREEPy parent. The coexistence of these two phases in the zircon-rich clast in 14001,28.3 supports the idea that they could have fractionated together. Modeling of this fractionation showed that 1% fractionation of an assemblage composed of 98.5% apatite and 1.5% zircon from high-K KREEP could reproduce the REE pattern seen in the granitic breccias, but at much lower absolute abundance (La ≈ 280× CI) than observed in the breccias (La ≈ 830× CI). If the fractionation involved only apatite and zircon, the REE pattern in 14001,28.3 would require a parent "superKREEP" with La ≈ 1020× chondrites. The depletion of phosphate in the zircon-rich clast, relative to our calculated assemblage, may be due to preferential melting of the phosphate leaving a clast with abundant refractory zircon, rather than the apatite-dominated assemblage calculated. This is supported by the unusually P_2O_5-rich chemistry of the colorless glass in the region of the clast.

Of course, these minor phases would not fractionate alone. Using the phase diagram of *Longhi and Pan* (1988) for Mg# of 0.05, we estimate that an assemblage of 40% fayalite, 35% plagioclase, and 25% silica would fractionate from the magma, in addition to the small amounts of apatite and zircon. If we use KREEP basalt 15386 (*Vaniman and Papike,* 1980) as the parent, 66% fractionation of the above assemblage can reproduce the REE pattern in 14001,28.3. However, significant fractionation of any Fe-bearing phase, like fayalite or ilmenite, can be ruled out since 14001,28.3 contains more FeO (12.2%) than KREEP basalt 15386 (10.6%) or high-K KREEP (10.6%), indicating that the KREEPy parent lithology either did not fractionate large amounts of any Fe-bearing phase or started with REE abundances much higher than high-K KREEP. In addition, the granitic breccias contain much more silica (≈68 wt.%) than high-K KREEP (47.9 wt.%) and KREEP basalt (50.8 wt.%), indicating that significant fractionation of silica can also be ruled out. This also implies that the granitic breccia parent magma contained REE abundances much higher than high-K KREEP or any KREEPy lithology.

ULTRAMAFIC ROCKS

Two of the samples, both from soil 14161,212, were determined to be ultramafic rocks with extremely low REE abundances. One of the samples, 14161,212.1 is a breccia, which was cut into two parts (14161,212.1a and 14161,212.1b) prior to INAA to separate what appeared to be a KREEPy impact melt crust from the possible pristine clast. After chemical analysis, 14161,212.1a was made into a thin section. Unfortunately, 14161,212.1b was lost during thin sectioning due to its friable nature. Chemical and petrographic analysis shows that 14161,212.4 is a dunite, 14161,212.1a is a peridotite, and 14161,212.1b is an ultramafic impact melt rock. Each of these rocks will be discussed separately in the following sections.

DUNITE

Dunite 14161,212.4 is an important addition to the three previously described lunar dunites: 72417 (*Dymek et al.,* 1975; *Laul and Schmitt,* 1975); 14321,1141 (*Lindstrom et al.,* 1984); and 14304,121 (*Goodrich et al.,* 1986; *Warren et al.,* 1987). Surprisingly, three of the four dunites come from the KREEP-rich Apollo 14 site. The only large sample, however, is 72417, which comes from the Apollo 17 site.

Petrography

Sample 14161,212.4 is composed almost exclusively of olivine with a felty, highly shocked texture that appears to have been strongly cataclastized. Cr-spinel is also present in small amounts (≈3% by area). Although the grain boundaries are irregular and indistinct and the section has been extensively plucked, there appear to be about 6 to 8 grains of olivine and 2 to 3 grains of spinel present in the thin section. Extinction in the olivine grains is patchy and diffuse and the high birefringence expected of olivine is absent, possibly due to shock and a thinner than usual section (Fig. 8). The olivine composition is very uniform and magnesian, at Fo 85 ± 0.8 (average of 15 analyses, Table 5). The Cr-spinel appears to be a solid solution, primarily between chromite and Fe-spinel (Table 5). Bulk analysis by INAA suggests that a small amount

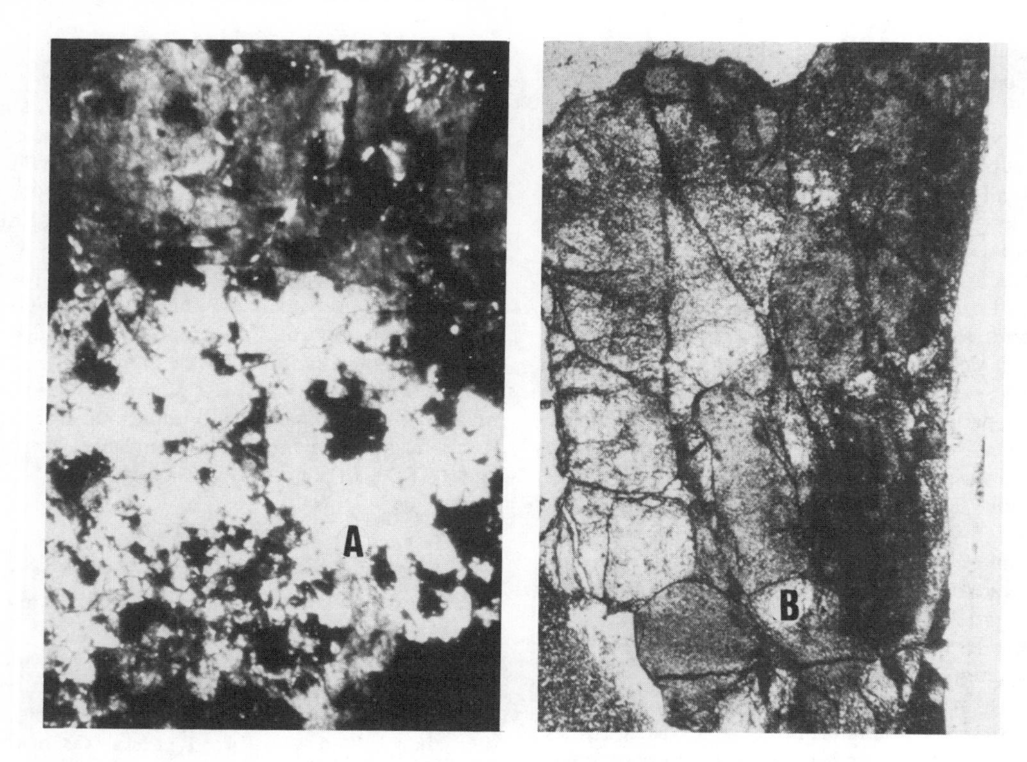

Fig. 8. Photomicrographs of ultramafic rock fragments: **(a)** Dunite 14161,212.4, crossed nicols. Two olivine grains with irregular, patchy extinction are visible here. Field of view is about 0.75 mm; **(b)** Peridotite 14161,212.1a, uncrossed nicols. A large grain of olivine takes up most of the top of the photo. The darker part of the sample is felty textured enstatite. Long dimension of field of view is about 1.5 mm.

of plagioclase (<2%) may be present in all Apollo 14 dunites but no plagioclase has been observed in thin section. The mineralogy of dunite 72417 is quite different from the three Apollo 14 dunites. Rock sample 72417 contains a much higher abundance of plagioclase (≈4%), in addition to small amounts of pyroxene (≈3%), Cr-spinel, and Fe-Ni metal (*Dymek et al.,* 1975). These minor components are not observed in the Apollo 14 dunites, except for the Cr-spinel in 14161,212.4, although this may simply represent poor sampling in the small samples from Apollo 14.

Whole Rock Composition

In contrast to the mineralogy, the bulk chemistry of 14161,212.4 (Table 1) is quite distinct from the other two Apollo 14 dunites, and very similar to some of the plagioclase-poor splits of 72417 (*Laul and Schmitt,* 1975). Elemental abundances measured in dunite 14161,212.4 are within the range of the values seen in the very heterogeneous dunite 72417, except for a few elements. FeO, MgO, and Sc are less abundant in 14161,212.4 and the Mg# is slightly lower, implying that our dunite may have crystallized from a somewhat more evolved magma than 72417. The slightly less Fo-rich olivine is consistent with a more evolved nature. The MnO abundance in 14161,212.4 is also slightly higher than in

TABLE 5. Mineral compositions in ultramafic rocks.

	Peridotite 14161,212.1a		Dunite 14161,212.4	
	Pyroxene*	Olivine†	Olivine‡	Spinel§
SiO_2	57.2	40.3	40.8	0.26
MgO	34.4	47.3	47.8	8.5
FeO	6.4	10.7	10.0	20.9
CaO	0.30	0.03	0.08	0.13
Na_2O	0.01	—	—	—
TiO_2	0.05	—	—	0.52
Al_2O_3	0.20	0.01	0.08	14.0
Cr_2O_3	0.34	0.12	0.04	54.4
MnO	0.11	0.10	0.12	0.54
NiO	—	0.03	0.02	0.12
Total	99.0	98.6	98.9	99.4
En	83.7	—	—	—
Fs	15.6	—	—	—
Wo	0.7	—	—	—
Fo	—	81.6	82.7	—
Fa	—	18.4	17.3	—

*Average of 23 analyses.
†Average of 14 analyses.
‡Average of 22 analyses.
§Average of 4 analyses.

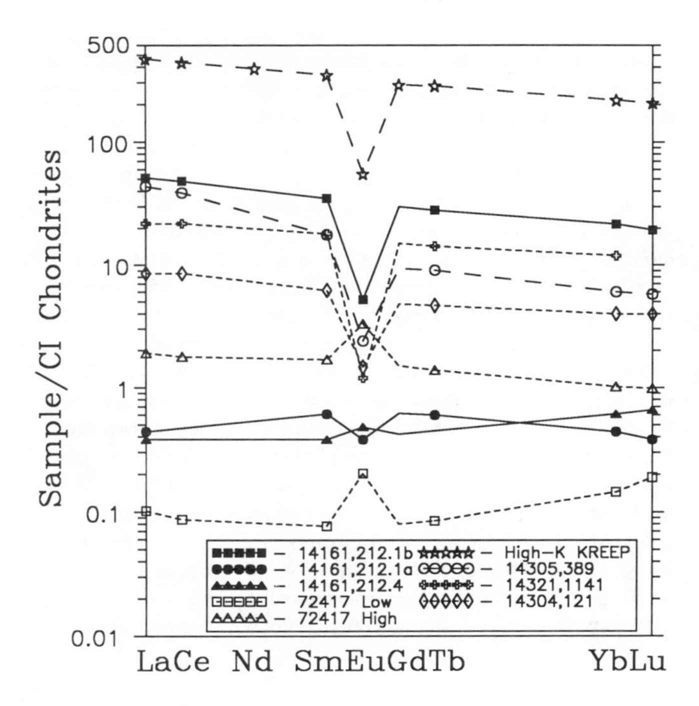

Fig. 9. Chondrite normalized REE abundances in the ultramafic rocks mentioned in the text. Samples measured in this work are shown with filled symbols and solid lines. Other samples are shown with open symbols and dashed lines. Sources of data include: 72417 - *Laul and Schmitt* (1975); 14321,1141 - *Lindstrom et al.* (1984); 14304,121 - *Warren et al.* (1987); 14305,389 - *Shervais et al.* (1984).

the other dunites. The Co abundance in 14161,212.4 is extremely high, over twice the average amount seen in 72417 and almost twice the amounts in 14321,1141 and 14304,121. This seems to indicate that the magma from which 14161,212.4 crystallized had a very high Co abundance, although the presence of unobserved Co-rich metal grains cannot be ruled out. How a Co-rich magma of this type could be produced is unknown.

The REE abundance and pattern in 14161,212.4 (Fig. 9) is similar to the more olivine-rich samples measured by *Laul and Schmitt* (1975), except that 14161,212.4 has no Eu anomaly (or a very small positive anomaly), rather than the large positive anomaly seen in splits of 72417. *Laul and Schmitt* (1975) showed that the REE pattern in 72417 was controlled primarily by the amount of plagioclase in the sample. Thus, the HREE enriched pattern and lack of an Eu anomaly in 14161,212.4 indicate that plagioclase is absent from the bulk rock, as in thin section. The lower abundances of Al_2O_3, Na_2O, and CaO in 14161,212.4 also support this conclusion. The other two Apollo 14 dunites, 14321,1141, and 14304,121 have much different REE patterns and abundances. These two dunites are characterized by elevated REE abundances, with La ≈20 to ≈50 times the amount seen in 14161,212.4. The REE pattern is also quite different, being strongly enriched in LREE, and having a large negative Eu anomaly (Fig. 9). Other lithophile elements, e.g., Ta, Hf, and Th, are also much lower in 14161,212.4 than in the other Apollo 14 dunites (*Lindstrom et al.,* 1984; *Goodrich et al.,* 1986).

Discussion

The question of pristinity is one of the first questions that must be answered with lunar rock samples. Textural evidence for pristinity is ambiguous in 14161,212.4 due to the severe shock and thin section plucking. However, its chemical similarity to the texturally pristine 72417 implies that the dunite can not be reproduced by mixing of known end-members, except for pure dunite. Thus, 14161,212.4 is probably a chemically, if not texturally, pristine rock. The low abundance of siderophile elements (Ir $<$ 1.9 ppb), although not low enough to be conclusive, is consistent with this conclusion. In addition, the REE pattern appears unaffected by the mixing of KREEP seen in most impact melts. 14161,212.4 is very magnesian (Mg# ≈ 78) and very calcic [Ca/(Ca+Na) ≈ 0.97]. The dunite plots near the magnesian end of the Mg-rich trend on an (Mg#) vs. [Ca/(Ca+Na) × 100] diagram, very near dunite 72417, implying that 14161,212.4 may be a member of the Mg-suite of pristine lunar rocks. The Mg-suite contains a variety of lithologies that are thought to represent parts of layered mafic intrusions emplaced up to 30 km deep in the FAN crust (*Shervais and Taylor,* 1986). In this model, dunites represent olivine-rich layers of these plutons.

Goodrich et al. (1986) noted that the olivine grain size and compositions in 14304,121 were similar to those in a troctolite from the same breccia, implying that the "dunite" may actually be an unrepresentative, plagioclase-poor portion of a troctolite. 14321,114 has similar olivine compositions and REE abundances and may also be an unrepresentative portion of a troctolite. The large sample size (≈4 cm) and large olivine grain size (up to 10 mm) of 72417 effectively rule out the possibility that it is an unrepresentative portion of another rock, and it must be an actual cumulate dunite. Since the REE abundances in 14161,212.4 are far lower than any known troctolite, and its composition is so similar to some of the plagioclase-poor splits of 72417, it seems likely that 14161,212.4 is also a cumulate dunite and not a piece of a troctolite.

The main difference between 14161,212.4 and the other Apollo 14 dunites is that 14321,1141 and 14304,121 both have much higher abundances of incompatible elements (Fig. 9). It is thought that the elevated REE abundances, negative Eu anomaly, and LREE enrichment in the other Apollo 14 dunites were produced by the assimilation of urKREEP (*Warren et al.,* 1987), a process that has been called upon to explain the chemistry of the aluminous mare basalts (*Dickinson et al.,* 1985; *Neal et al.,* 1988), and VHK basalts (*Shervais et al.,* 1985) at the same site. According to this assimilation model, dunite 14161,212.4 may represent a dunite crystallized from an Mg-suite magma that did not assimilate urKREEP during ascent and crystallization. Assuming that the magma that crystallized dunite 14161,212.4 is parental to the other dunites, a least-squares mixing calculation shows that the REE abundances in 14321,1141 and 14304,121 could be reproduced by assimilating ≈2% and ≈5% high-K KREEP respectively, or somewhat smaller amounts of urKREEP.

PERIDOTITE

Petrology

Sample 14161,212.1a is composed almost exclusively of low-Ca pyroxene (≈65%) and olivine (≈35%). Small amounts of plagioclase (<1%) may be present in the bulk rock, given its Ca, Al, and Na abundances, but are not observed in thin section. This composition makes 14161,212.1a the first lunar peridotite according to the classification scheme of *Prinz and Keil* (1977) and *Stöffler et al.* (1980), although it should be noted that the lack of clinopyroxene in this sample may make the term harzburgite a more appropriate name (*McBirney,* 1984). The only other previously described "peridotite" is 67667 (*Steele and Smith,* 1973; *Warren and Wasson,* 1979), but this sample contains about 20% plagioclase and is more properly called a gabbronorite or "feldspathic lherzolite" (*Ryder and Norman,* 1979).

The texture of 14161,212.1a resembles a severely shocked cumulate. It is composed of rounded to subrounded grains of shocked olivine, up to 0.2 by 0.1 mm in size, poikilitically enclosed by a severely shocked matrix of pyroxene with a "felty" texture. The pyroxene is so severely shocked that individual grains are not visible and extinction and birefringence are absent (Fig 8).

Both minerals are very homogenous and very magnesian. Olivines are forsteritic (Fo 85 ± 0.8) while the pyroxenes are strongly enstatitic (En 86), with almost no (0.8%) wollastonite component. Both minerals contain very low abundances of minor elements, with less than 1.2 wt.% total $TiO_2 + Al_2O_3 + K_2O + MnO + Cr_2O_3$. Average microprobe analyses of each mineral are shown in Table 5. Comparing these analyses to the large data base of minor element abundances obtained by *Bersch et al.* (1989) shows that the olivines plot at the low end, but within the range of olivines in pristine rocks. The pyroxenes, however, have much lower minor element abundances than any analyzed by *Bersch et al.* (1989), and plot far below the field of pyroxenes from known pristine rocks. The only other rock with pyroxenes of similar minor element contents is the olivine orthopyroxenite 14305,389 (*Shervais et al.,* 1984). This small sample is composed of ≈90% enstatite (En 90.9, Wo 0.25), ≈9% forsterite (Fo 89.6), ≈1% plagioclase (An 88), and trace amounts of shreibersite and tetrataenite. The pyroxenes in this sample are, like those in 14161,212.1a, anomalously low in the same minor elements (<1 wt.% total), implying that these unusual rocks may be related.

Whole Rock Chemistry

The chemistry of the peridotite is strongly ultramafic, with about 34 wt.% MgO and 8 wt.% FeO, giving an Mg# of 81. The minor element content of the peridotite is extremely low, consistent with the low minor element contents of each mineral as described above, indicating that the bulk rock did not contain trace minerals not observed in thin section.

Incompatible element abundances are very low in peridotite 14161,212.1a, with REE abundances about 0.5 times chondrites, in approximately chondritic relative abundances (La/Lu ≈ 12), as shown in Fig. 9. Abundances of other incompatible elements, e.g., Ta, Hf, Th, are also extremely low (Table 1). The REE pattern in the peridotite appears to be slightly bow shaped, with a small negative Eu anomaly, possibly indicating the presence of a trace of apatite. However, the 1-σ uncertainties on the analyses are far too large (±5% to 40%) to be certain of the shape of the REE pattern. If there is apatite present, it may support the suggestion of *Lindstrom et al.* (1984) that the elevated REE abundances in dunite 14304,121 were produced by apatite metasomatism. However, we must stress that the evidence for metasomatism in the peridotite is highly speculative. Incompatible element abundances are one of the main differences between the peridotite and the olivine orthopyroxenite. The orthopyroxenite has rather high REE abundances, with La ≈ 44× chondrites (Fig. 9). Its REE pattern is highly fractionated, with La/Lu ≈ 74 (*Shervais et al.,* 1984). Abundances of the other incompatible elements in the orthopyroxenite are also much higher than in the peridotite.

Discussion

It seems likely that 14161,212.1a is a pristine rock. Its poikilitic cumulate texture and large olivine grain size show that it is probably pristine. The low abundance of siderophile elements (Ir < 1.0 ppb) and the lack of a KREEPy REE pattern are consistent with this conclusion. On an (Mg#) vs. [Ca/(Ca+Na)] diagram, 14161,212.1a plots near the magnesian end of the Mg-rich trend, indicating that the peridotite is a member of the Mg-suite of rocks. The relatively large grain size, low-Ca content of the olivine, and the low pyroxene equilibration temperatures (*Shervais et al.,* 1984) all support an origin as a slowly cooled, deep crustal cumulate. However, an origin of the peridotite as a partial-melt residue should also be considered.

Given the small size of peridotite 14161,212.1a and the olivine orthopyroxenite, it is possible that they are nonrepresentative portions of some other rock. However, the very low minor element contents in the pyroxenes of these samples and the poikilitic texture of the peridotite effectively rule out the possibility that they could be unrepresentative portions of any known lunar troctolite or gabbro. The peridotite major and trace element composition is very similar to the dunites, 14161,212.4 and 72417. All have ultramafic compositions, very low abundances of incompatible trace elements, and forsteritic olivines. This similarity in composition, and the small size of this sample make it possible that the peridotite is simply an unrepresentative portion of a dunite like 72417. However, this is unlikely since the pyroxenes in the previously described dunites have a significantly lower Mg# and much higher abundances of minor elements like Ca, Al, and Ti. In addition, the pyroxenes in dunite 72417 are sparse and occur as simplectites and as tiny grains interstitial to the plagioclase (*Dymek et al.,* 1975), not as large masses poikilitically enclosing olivine as in the peridotite. Since these unique samples cannot be parts of any previously described lunar lithology, they must represent unique lithologies that crystallized from unusual magmas with very low minor element contents.

Shervais et al. (1984) proposed that the parent magma for the olivine orthopyroxenite was probably derived by partial melting of deep magma ocean cumulates in equilibrium with residual garnet. The peridotite has a similar composition and may also have come from partial melting of deep magma ocean cumulates. However, in contrast to the peridotite, the high trace element contents of the orthopyroxenite and its inferred parent magma are difficult to reconcile with an origin from a deep cumulate. The peridotite does not have high trace element abundances, making its parent magma more likely to be a partial melt of deep cumulates. The peridotite has a small, but definite negative Eu anomaly, indicating that its parent magma could not be derived by melting primitive Moon.

Another possibility, given the ultramafic chemistry and extremely low contents of minor and trace elements in 14161,212.1a, is that the peridotite may be a partial-melt residue. It is generally accepted that mare basalt magmas were produced by partial melting of magma ocean cumulates in the lunar mantle. *Delano* (1980) showed that pyroclastic glasses were produced by partial melting from about 400-500 km depth in the mantle, leaving residues of olivine + orthopyroxene, just the assemblage seen in the peridotite. In addition, a partial-melt residue will tend to have low abundances of incompatible elements. Thus, the peridotite may be a mantle rock that was excavated by impact, presumably the Imbrium event. *Spudis et al.* (1988) have calculated that a small percentage (≈0.5%) of mantle rocks may have been excavated by the Imbrium basin impact and some of this hypothetical mantle ejecta could exist at the Apollo 14 site. The severely shocked nature of the sample supports the idea that it was excavated from depth. However, experimental work by *Delano* (1980) and *Chen et al.* (1982) has shown that residual pyroxenes from pyroclastic glass partial melting have over 3.0 wt.% Al_2O_3, more than an order of magnitude greater than in the pyroxenes in the peridotite. Thus, the very low Al_2O_3 content of the pyroxenes argues against an origin as a mantle partial-melt residue, although it is possible that the peridotite is a residue produced by multiple episodes of partial melting. *Walker et al.* (1973) proposed that KREEP basalts were produced by partial melting of "average highlands" crust, leaving olivine/pyroxene residues, indicating that a crustal, rather than mantle, origin is possible for 14161,212.1a. However, the extremely low Al_2O_3 abundances in the pyroxenes makes this even more unlikely, given the feldspathic nature of average lunar crustal material.

ULTRAMAFIC IMPACT MELT

Since 14161,212.1b was lost during thin sectioning, we must rely on the INAA data alone for our interpretation of this sample; however, it is probably an ultramafic impact melt with elevated siderophile abundances and a KREEPy REE pattern.

Whole Rock Chemistry

14161,212.1b is a strongly ultramafic rock, with ≈30 wt.% MgO and ≈9.5 wt.% FeO. Minor elements are scarce, making up less than 4% of the bulk rock. Al_2O_3, CaO, and Na_2O occur in low concentrations, but much higher than in the coexisting pristine peridotite 14161,212.1a. MgO, FeO, Cr_2O_3, and MnO are all less abundant in the impact melt than in the peridotite. Despite these differences, the Mg# of the impact melt is still fairly close to that of the peridotite, supporting the idea that the impact melt and the pristine peridotite clast are closely related. Compatible trace elements (e.g., Sc, Co, and V) are depleted in 14161,212.1b relative to the peridotite while all incompatible trace elements are strongly enriched in the impact melt. The REE abundance in 14161,212.1b is very high for an ultramafic rock, with La ≈ 52× chondrites. The REE profile is a fairly typical KREEPy pattern, with strong LREE enrichment (La/Lu ≈ 24) and a large negative Eu anomaly (Fig. 9).

Discussion

The KREEPy REE pattern of 14161,212.1b and its occurrence as a breccia crust imply that it might be an impact melt. The presence of measurable siderophile element contamination (2 ppb Ir) supports this conclusion. Its major element chemistry is very similar to the peridotite (14161,212.1a) and the dunite (14161,212.4), suggesting that it is probably an impact-produced mixture of these rocks and KREEP. A least-squares mixing calculation shows that the REE pattern and abundance in 14161,212.1b could be reproduced by mixing the peridotite and/or dunite with about 10% high-K KREEP (*Warren and Wasson,* 1979) or approximately 15% KREEP basalt 15386 (*Vaniman and Papike,* 1980).

However, these mixing proportions do not exactly reproduce the measured major and trace element abundances in 14161,212.1b. Our mixing calculations show that the best fits for most elements are obtained by mixing KREEP with the peridotite alone or with a mixture of peridotite and dunite. Mixtures involving the KREEP basalt generally resulted in poor fits. A few of the elements show large differences between the calculated and observed abundances. Good fits to the MgO abundance in 14161,212.1b are only produced by mixing the peridotite alone with KREEP or KREEP basalt, suggesting that the target rock was probably mostly peridotite. However, the FeO abundance can only be fit by mixing KREEP or KREEP basalt with the 14161,212.1a/14161,212.4 mixture, suggesting that pure peridotite could not be the parent material. This fact along with similarity in major and trace element chemistries suggest that the peridotite and dunite may be related. CaO and Al_2O_3 were much higher in all our calculated mixtures than in 14161,212.1b, suggesting that plagioclase may have been preferentially removed from the observed impact melt. *Simonds et al.* (1978) noted that mafic mineral clasts were underrepresented in impact melts and ascribed this to differential digestion of clasts by the superheated impact melt. Thus, the Ca and Al deficiency in 14161,212.1b relative to our calculated mixtures may be due to the lack of feldspar clasts in the preserved portion of the impact melt. The Co abundance in all mixtures involving 14161,212.4 was much higher than in 14161,212.1b, indicating that the dunite has an anomalously high Co abundance, as was shown above. The Co abundance in the impact melt crust was, however, well fit using the peridotite. This anomalous Co abundance in the dunite casts doubt on the existence of a close relationship between the dunite and peridotite.

CONCLUSIONS

In a suite of 18 2-4-mm rock fragments separated from several Apollo 14 soil samples, we have discovered a number of unique fragments, including three highly evolved granitic breccias and two ultramafic rocks. INAA and petrographic analysis of these rocks has led to a number of interesting conclusions, listed below.

1. The granitic breccias are dimict breccias composed of an immiscible melt pair (brown glass = high-Fe, granite = high-Si) that have been remixed by impact. They are the first samples that have a texturally pristine granite juxtaposed with its high-Fe melt partner.

2. The unique REE pattern of the breccias, characterized by strongly enriched LREE and flat HREE, can be modeled by pre-immiscibility fractional crystallization of apatite and zircon from a magma with a KREEP-like pattern. However, the absolute abundances of REE in these breccias are so high that their parent magma must have been strongly enriched in REE relative to KREEP.

3. These two lines of evidence support the hypothesis (*Neal and Taylor,* 1987) that lunar granite was formed by extreme fractional crystallization combined with subordinate silicate liquid immiscibility.

4. The chemistry of the brown glass (high P_2O_5, FeO, TiO_2, and CaO) and its concave-down, bow-shaped REE pattern support the "K-frac/REEP-frac" hypothesis of *Neal and Taylor* (1989b) that assumes that the high-Fe immiscible melt would produce apatite/whitlockite + plagioclase assemblages after fractionation of fayalite and ilmenite. However, the occurrence of both the high-Fe and high-Si components in these tiny breccias does not support the extensive separation of melts envisioned in their model.

5. The ultramafic rocks analyzed include the fourth described lunar dunite and the first true peridotite, in addition to an impact melt mixture of these lithologies.

6. The dunite is an olivine + Cr-spinel cumulate that differs from the other Apollo 14 dunites in that it crystallized from a magma that did not assimilate crustal material.

7. The peridotite is a unique rock fragment, with pyroxenes containing much lower minor element contents (<1.1 wt.% Al_2O_3 + CaO + TiO_2 + Na_2O + Cr_2O_3 + MnO) than pyroxenes in any pristine rock, except for the olivine orthopyroxenite described by *Shervais et al.* (1984).

8. The origin of this peridotite is uncertain, but it may be a cumulate from an unusual Mg-suite magma of unknown origin, or may be a residue of a partial melting event, either in the mantle or the crust. The former seems more likely, since experimental partial melt residues contain about an order of magnitude more Al_2O_3 in their pyroxenes. However, an origin as a residue from multiple episodes of partial melting cannot be ruled out.

9. 14161,212.1b is an impact melt rock produced by mixing ultramafic target rocks similar to the peridotite and a KREEP lithology similar to the high-K KREEP composition.

Acknowledgments. We would like to thank T. Servilla, G. Conrad, and M. Jercinovic for technical assistance. The manuscript was greatly improved by careful reviews by C. Neal, T. Dickinson, and D. Lindstrom. The research was supported by NASA grant NAG 9-30.

REFERENCES

Anders E. and Grevesse N. (1989) Abundances of the elements: Meteoritic and solar. *Geochim. Cosmochim. Acta, 53,* 197-214.

Bence A. E. and Albee A. L. (1968) Empirical correction factors for the electron microanalysis of silicates and oxides. *J. Geol., 76,* 382-403.

Bersch M. G., Taylor G. J., and Keil K. (1989) Ferroan anorthosites from an evolving magma ocean (abstract). In *Lunar and Planetary Science XX,* pp. 67-68. Lunar and Planetary Institute, Houston.

Chen H. -K., Delano J. W., and Lindsley D. H. (1982) Chemistry and phase relations of VLT volcanic glasses from Apollo 14 and Apollo 17. *Proc. Lunar Planet. Sci. Conf. 13th,* in *J. Geophys. Res., 87,* A171-A181.

Crawford M. L. and Hollister L. S. (1977) Evolution of KREEP: Further petrologic evidence. *Proc. Lunar Sci. Conf. 8th,* pp. 2403-2417.

Delano J. W. (1980) Chemistry and liquidus phase relations of Apollo 15 red glass: Implications for the deep lunar interior. *Proc. Lunar Planet. Sci. Conf. 11th,* pp. 251-288.

Dickinson J. E. and Hess P. C. (1983) Role of whitlockite and apatite in lunar felsite (abstract). In *Lunar and Planetary Science XIV,* pp. 158-159. Lunar and Planetary Institute, Houston.

Dickinson T., Taylor G. J., Keil K., Schmitt R. A., Hughes S. S., and Smith M. R. (1985) Apollo 14 aluminous mare basalts and their possible relationship to KREEP. *Proc. Lunar Planet. Sci. Conf. 15th,* in *J. Geophys. Res., 90,* C365-C374.

Drake M. J. and Weill D. F (1972) New rare-earth element standards for microprobe analysis. *Chem. Geol., 20,* 179-181.

Dymek R. F., Albee A. L., and Chodos A. A. (1975) Comparative petrology of lunar cumulate rocks of possible primary origin: Dunite 72415, troctolite 76535, norite 78235, and anorthosite 62237. *Proc. Lunar Sci. Conf. 6th,* pp. 301-341.

Gladney E. S., Burns C. E., Perrin D. R., Roelandts I., and Gills T. E. (1984) 1982 compilation of elemental concentration data for NBS biological, geological, and environmental standard reference materials. *National Bureau of Standards Special Publication 260-88.* 231 pp.

Goodrich C. A., Taylor G. J., Keil K., Kallemeyn G. W., and Warren P. H. (1986) Alkali norite, troctolites, and VHK mare basalts from breccia 14304. *Proc. Lunar Planet. Sci. Conf. 16th,* in *J. Geophys. Res., 91,* D305-318.

Haskin L. A., Helmke P. A., Blanchard D. P., Jacobs J. W., and Telander K. (1973) Major and trace element abundances in samples from the lunar highlands. *Proc. Lunar Sci. Conf. 4th,* pp. 1274-1296.

Henderson P. (1982) *Inorganic Chemistry.* Pergamon, Oxford. 353 pp.

Hess P. C., Rutherford M. J., Guillemetted R. N., Ryerson F. J., and Tuchfeld H. A. (1975) Residual products of fractional crystallization of lunar magmas: An experimental study. *Proc. Lunar Sci. Conf. 6th,* pp. 895-909.

Hess P. C., Horzempa P., and Rutherford M. J. (1989) Fractionation of Apollo 15 KREEP basalts (abstract). In *Lunar and Planetary Science XX,* pp. 408-409. Lunar and Planetary Institute, Houston.

Jarosewich E., Clarke R. S. Jr., and Barrows J. N. (1987) The Allende meteorite reference sample. *Smithson. Contrib. Earth Sci., 27.*

Kruse H. and Spettel B. (1982) A combined set of automatic and interactive programs for instrumental neutron activation analysis. *J. Radioanal. Chem., 70,* 427-434.

Laul J. C. and Schmitt R. A. (1975) Dunite 72417: A chemical study and interpretation. *Proc. Lunar Sci. Conf. 6th,* pp. 1231-1254.

Lindstrom M. M., Knapp S. A., Shervais J. W., and Taylor L. A. (1984) Magnesium anorthosites and associated troctolites and dunite in Apollo 14 breccias. *Proc. Lunar Planet Sci. Conf. 15th,* in *J. Geophys. Res., 89,* C41-C49.

Longhi J. and Pan V. (1988) A reconnaissance study of phase boundaries in low-alkali basaltic liquids. *J. Petrology, 26,* 115-147.

McBirney A. R. (ed.) (1984) *Igneous Petrology.* Freeman, Cooper and Company, San Francisco. 504 pp.

McKay G. A., Wagstaff J., and Yang S.-R. (1986) Zr, Hf, and REE partition coefficients for ilmenite and other minerals in high-Ti lunar mare basalts: An experimental study. *Proc. Lunar Planet. Sci. Conf. 16th,* in *J. Geophys. Res., 91,* D229-D237.

Minor M. M., Hensley W. K., Denton M. M., and Garcia S. R. (1981) An automated analysis system. *Radioanal. Chem., 70,* 459-472.

Neal C. R. and Taylor L. A. (1987) The petrogenesis of lunar granite from a basaltic magma: Extreme fractional crystallization with subordinate silicate liquid immiscibility (abstract). *Meteoritics, 22,* 470-471.

Neal C. R. and Taylor L. A. (1989a) Metasomatic products of the lunar magma ocean: The role of KREEP dissemination. *Geochim. Cosmochim. Acta, 53,* 529-541.

Neal C. R. and Taylor L. A. (1989b) Definition of a pristine, unadulterated urKREEP composition using the "K-frac/REEP-frac" hypothesis (abstract). In *Lunar and Planetary Science XX,* pp. 772-773. Lunar and Planetary Institute, Houston.

Neal C. R., Taylor L. A., and Lindstrom M. M. (1988) Apollo 14 mare basalt petrogenesis: Assimilation of KREEP-like components by a fractionating magma. *Proc. Lunar Planet. Sci. Conf. 18th,* pp. 139-153.

Neal C. R., Taylor L. A., and Patchen A. D. (1989) The "K-frac/REEP-frac" hypothesis: Evidence for both KREEP components in 12033 felsite with post-sli fractionation of the REEP-frac (abstract). In *Lunar and Planetary Science XX,* pp. 778-779. Lunar and Planetary Institute, Houston.

Prinz M. and Keil K. (1977) Mineralogy, petrology, and chemistry of ANT-suite rocks from the lunar highlands. *Phys. Chem. Earth, 10,* 215-237.

Roedder E. (1978) Silicate liquid immiscibility in magmas and in the system K_2O-FeO-Al_2O_3-SiO_2: An example of serendipity. *Geochim. Cosmochim. Acta, 42,* 1597-1617.

Roedder E. and Weiblen P. W. (1970) Lunar petrology of silicate melt inclusions, Apollo 11 rocks. *Proc. Apollo 11 Lunar Sci. Conf.,* pp. 801-837.

Roedder E. and Weiblen P. W. (1971) Petrology of silicate melt inclusions, Apollo 11 and Apollo 12 and terrestrial equivalents. *Proc. Lunar Sci. Conf. 2nd,* pp. 507-528.

Roedder E. and Weiblen P. W. (1972) Petrographic features and petrologic significance of melt inclusions in Apollo 14 and 15 rocks. *Proc. Lunar Sci. Conf. 3rd,* pp. 251-279.

Rutherford M. J., Hess P. C., Ryerson F. J., Campbell H. W., and Dick P. A. (1976) The chemistry, origin, and petrogenetic implications of lunar granite and monzonite. *Proc. Lunar Sci. Conf. 7th,* pp. 1723-1740.

Ryder G. (1976) Lunar sample 15405: Remnant of a KREEP basalt-granite differentiated pluton. *Earth Planet. Sci. Lett., 29,* 255-268.

Ryder G. and Norman M. (1979) *Catalog of Pristine Non-mare Materials, Part 1: Non-anorthosites Revised.* JSC Publ. No. 14565, NASA Johnson Space Center, Houston.

Salpas P. A., Shervais J. W., Kinapo S. A., and Taylor L. A. (1985) Petrogenesis of lunar granite: The result of apatite fractionation (abstract). In *Lunar and Planetary Science XVI,* pp. 726-727. Lunar and Planetary Institute, Houston.

Shervais J. W. and Taylor L. A. (1986) Petrologic constraints on the origin of the moon. In *Origin of the Moon* (W. K. Hartmann, R. J. Phillips, and G. J. Taylor, eds.), pp. 173-201. Lunar and Planetary Institute, Houston.

Shervais J. W., Taylor L. A., Laul J. C., and Smith M. R. (1984) Pristine highland clasts in consortium breccia 14305: Petrology and geochemistry. *Proc. Lunar Planet. Sci. Conf. 15th,* in *J. Geophys. Res., 89,* C25-C40.

Shervais J. W., Taylor L. A., Laul J. C., Shih C.-Y., and Nyquist L. E. (1985) Very high potassium (VHK) basalt: Complications in mare basalt petrogenesis. *Proc. Lunar Planet. Sci. Conf. 16th,* in *J. Geophys. Res., 90,* D3-D18.

Simonds C. H., Floran R. J., McGee P. E., Phinney W. C., and Warner J. L. (1978) Petrogenesis of melt rocks, Manicouagan impact structure, Quebec. *J. Geophys. Res., 83,* 2773-2788.

Spudis P. D., Hawke B. R., and Lucy P. G. (1988) Materials and formation of the Imbrium basin. *Proc. Lunar Planet. Sci. Conf. 18th,* pp. 115-168.

Steele I. M. and Smith J. V. (1973) Mineralogy and petrology of some Apollo 16 rocks and fines: General petrologic model of Moon. *Proc. Lunar Sci. Conf. 6th,* pp. 451-467.

Stöffler D., Knöll H.-D., Marvin U. B., Simonds C. H., and Warren P. H. (1980) Recommended classification and nomenclature of lunar highland rocks-A committee report. *Proceedings of the Conference on the Lunar Highlands Crust* (J. J. Papike and R. B. Merrill, eds.), pp. 51-70. Pergamon, New York.

Vaniman P. T. and Papike J. J. (1980) Lunar highland melt rocks: Chemistry, petrology, and silicate mineral chemistry. In *Proceedings on the Conference on the Lunar Highlands Crust* (J. J. Papike and R. B. Merrill, eds.), pp. 271-337. Pergamon, New York.

Walker D., Grove T. L., Longhi J., Stolper E. M., and Hays J. F. (1973) Origin of lunar feldspathic rocks. *Earth Planet. Sci. Lett., 20,* 325-336.

Warren P. H. and Wasson J. T. (1979) The origin of KREEP. *Rev. Geophys. Space Phys., 17,* 73-88.

Warren P. H., Taylor G. J., Keil K., Shirley D. N., and Wasson J. T. (1983) Seventh foray: Whitlockite-rich lithologies, a diopside-bearing troctolitic anorthosite, ferroan anorthosites and KREEP. *Proc. Lunar Planet. Sci. Conf. 13th,* in *J. Geophys. Res., 88,* B151-B164.

Warren P. H., Jerde E. A., and Kallemeyn G. W. (1987) Pristine moon rocks: A "large" felsite and a metal-rich ferroan anorthosite. *Proc. Lunar Planet. Sci. Conf. 17th,* in *J. Geophys. Res., 92,* E303-E313.

Weiblen P. W. and Roedder E. (1973) Petrology of melt inclusions in Apollo samples 15598 and 62295, and of clasts in 67915 and several lunar soils. *Proc. Lunar Sci. Conf. 4th,* pp. 681-703.

Proceedings of the 20th Lunar and Planetary Science Conference, pp. 77-90
Lunar and Planetary Institute, Houston, 1990

Apollo 15 KREEP-poor Impact Melts

M. M. Lindstrom

Mail Code SN2, NASA Johnson Space Center, Houston, TX 77058

U. B. Marvin and B. B. Holmberg

Harvard-Smithsonian Center for Astrophysics, 60 Garden Street, Cambridge, MA 02138

D. W. Mittlefehldt

Lockheed Engineering and Science Company, 2400 NASA Road 1, Houston, TX 77058

The Apollo 15 site at the Apennine Front includes ejecta from the Imbrium and Serenitatis impacts. Such major-basin impacts excavate deep into the crust and are the sources of low-K Fra Mauro (LKFM) impact melts common at all highland sites. Here we describe a new type of Apollo 15 impact melt rock, the KREEP-poor impact melts. We present analyses of 18 fragments of this melt and identify 9 more in the literature. These KREEP-poor impact melts form a relatively tight cluster in bulk and mineral composition and generally have poikilitic texture. This melt has gabbroic bulk composition, with a moderate mg′ of 71. It has very low incompatible-element concentrations and a slope in its heavy REE pattern that is distinct from that of KREEP. This KREEP-poor impact melt also has the highest Ti contents of LKFM impact melts from all highland sites. This melt is fairly common, constituting about one third of 75 analyzed Apollo 15 impact melt samples. KREEP-poor impact melts may be basin-related melts which are mixtures of various Mg-suite noritic or gabbroic lower crustal rocks. It contains the highest proportion of an unsampled Ti-rich component that has previously been postulated for LKFM impact melts and is the best source of clues to that lithology. KREEP-poor impact melts probably are part of the pre-Serenitatis section at the Apennine Front and represent an early impact into the lower crust.

INTRODUCTION

Characterization of the lunar highlands at the Apennine Front was one of the major goals of the Apollo 15 mission. The site lies on the rim of the Imbrium basin and within the rings of the Serenitatis basin. The returned samples were expected to include Imbrium and Serenitatis ejecta and pre-Serenitatis materials (*Swann et al.,* 1972; *Spudis and Ryder,* 1985). Because major basin-forming impacts such as Imbrium and Serenitatis bring to the surface rocks of deep-crustal origin, it was hoped that the site would yield many samples from the deep crust. Two such deep-crustal rocks are the black and white breccias 15445 and 15455, which are impact melt rocks containing clasts of plutonic norites and troctolites (*Ryder and Bower,* 1977).

Unfortunately, however, such deep-seated rocks are rare among the large samples collected at the base of the Apennine Front. Instead, most of the large rocks from highland stations are regolith breccias containing abundant clasts of younger volcanic rocks, KREEP basalt, mare basalt, and green glass. Thus, in order to obtain a comprehensive view of the nature of the highlands at the Apennine Front it is necessary to evaluate the highlands components of the regolith breccias and soils by studying small lithic fragments that are found as clasts in breccias and coarse particles in soils. Several recent studies of Apollo 15 lithic fragments have contributed to our understanding of the Apennine Front highlands. *Lindstrom et al.* (1988) and *Ryder et al.* (1988) evaluated the variety of highland rock types at the site, while *Simon et al.* (1988) and *Lindstrom et al.* (1989a) focused on highland igneous rocks. *Ryder and Spudis* (1987) and *Laul et al.* (1988) discussed Apollo 15 impact melt rocks, which are the topic of this study.

Impact melt rocks are produced by mixing and melting of a variety of rocks in the target area. Their composition represents an average composition of part of the target. The compositions of mafic melt rocks produced by basin-forming impacts are thought to represent an average of deep-crustal materials (*Ryder and Wood,* 1977; *McCormick et al.,* 1989). Thus, study of such rocks may provide insight into the nature of the deep crust. *Ryder and Spudis* (1987) presented the first survey of Apollo 15 impact melt rocks. They showed that most Apollo 15 melts are mafic rocks having 15-20% Al_2O_3, but that they show remarkable diversity in trace element composition. Based on REE concentrations and Ti-Sc-Sm variations, these authors divided the melts into five groups, designated A through E. One (group A) has the composition of Apollo 15 KREEP basalt and is interpreted as resulting from post-Imbrium impacts onto KREEP basalt flows of the Apennine Bench formation. Three groups (B-D) have low-K Fra Mauro (LKFM) noritic compositions. Their REE patterns are KREEP-dominated but REE concentrations are variable and lower than in KREEP. Ryder and Spudis interpreted the group D melts (15445, 15455) as Imbrium impact melt and concluded that the Apollo 15 impact melts are unlike melts from other highland sites and are of distinct provenance. Their fifth group (E) consists of a single sample that is a KREEP-free feldspathic rock with very low REE concentrations. Later studies of additional Apollo 15 impact melts (*Laul et al.,* 1988; *Lindstrom et al.,* 1988; *Ryder et al.,* 1988) suggested that groups B-D form a continuum in REE concentration and found a few melt rocks with REE concentrations between groups D and E.

The impact melt samples described here were studied as part of consortium investigations of regolith breccias 15295 and 15459. Both breccias were sawn, exposing new surfaces. The surfaces were mapped and all clasts of appropriate size were extracted for study. Clasts were split for petrographic and chemical analyses. Samples were divided into groups based on hand specimen and preliminary thin section examination. Poikilitic breccias were by far the most abundant group, followed by granular and fine-grained breccias, plutonic anorthosites and norites, and volcanic KREEP and mare basalts. Further petrographic examination showed that essentially all the poikilitic, granular, and fine-grained breccias are impact melts. All samples were analyzed by INAA and the data combined with petrographic observations to classify the clasts. The compositions of plutonic and volcanic rocks corresponded with their textural classifications. However, for the impact melts, textural and compositional classifications are not closely related. Impact melt compositions span nearly the entire range of group A-E impact melts of *Ryder and Spudis* (1987), and include a few unusual melt compositions. Approximately half of the 25 impact melts we studied are generally consistent with groups A-D melts described by *Ryder and Spudis* (1987). Our petrographic and major element studies of these clasts are not complete and they will be reported later. The other half of the melts are intermediate in composition between groups D and E. Although the group might be simply designated as group E melts, they span such a wide range in composition that they are unlikely to represent a single impact melt. We prefer instead the general name KREEP-poor impact melts (KPIM). This paper describes the suite of KREEP-poor impact melts and identifies a subgroup of them that might represent a single impact. This group has not been described previously [except in our recent abstract that identified them as recrystallized gabbroic breccias (*Lindstrom et al.,* 1989b)]. We suggest that this nearly KREEP-free mafic melt rock may be an important pre-Serenitatis component of the Apennine Front.

MAJOR AND TRACE ELEMENT COMPOSITION

Splits of 13 small clasts of KREEP-poor impact melts ranging in weight from 6 to 77 mg (most under 20 mg) were analyzed for major and trace elements at JSC. All 13 samples were analyzed by neutron activation using the INAA procedures of *Lindstrom et al.* (1989a). Major element concentrations were determined in 12 of the samples by electron microprobe analysis of fused glass beads. Because such small samples may not be representative of even relatively fine-grained melt, we undertook a more thorough investigation of the only large example, a 1 × 3 cm white clast in breccia 15459. We had previously analyzed three ~100 mg splits of this clast [15459,231-1 and 2, 320 (*Lindstrom et al.,* 1988)] that showed minor compositional variability. In order to acquire a representative analysis we obtained an additional 2 g of the clast, which was ground in a diamonite mortar and homogenized. We split two ~100 mg samples for INAA and two ~10 mg samples for fused bead analysis. An additional split has been taken for $^{39}Ar/^{40}Ar$ analysis by Don Bogard (NASA/JSC). The remaining homogenized powder is available for other studies.

Major and trace element analyses of all these clasts of KREEP-poor impact melts are given in Table 1, together with our previous analyses of the large clast from 15459. The five analyses of the large clast (15459,414-a and b ,231-1 and 2 and ,320) can be used to evaluate heterogeneity in ~100 mg splits of a single specimen. The two homogenized splits (,414-a and b) have compositions that are identical within analytical uncertainties and are assumed to represent the bulk sample. The other two splits were taken from different parts of the clast and analyzed at different times in L. A. Haskin's laboratory at Washington University. Split ,320 was a clean 99-mg sample of bulk clast material. It is very similar in composition to the homogenized samples. Split ,231, which originally weighed 344 mg, was cleaned of adhering breccia matrix and split for analysis of two melt samples. The two analyzed samples of ,231 are very similar and slightly more mafic and about 10% richer in REE and other incompatible elements than the homogenized samples. Thus, there may be some real heterogeneity in this large clast on a scale of 100-300 mg. Nonetheless, the five analyses of the large clast resemble each other more than they resemble analyses of other KREEP-poor impact melt clasts.

Nine additional samples of KREEP-poor impact melts have been reported previously; one in *Ryder and Spudis* (1987), five in *Laul et al.* (1988), and three in *Ryder et al.* (1988). Examination of the entire dataset for these impact melt rocks shows that they exhibit a fairly restricted variation in bulk composition. They are relatively mafic with 17-22% Al_2O_3, 11-14% CaO, 6-10% FeO and mg′ [molar Mg/(Mg+Fe)] 67-77. Because major element analyses are incomplete for two of the compositonally extreme samples (*Ryder et al.,* 1988), the actual ranges for these parameters may be greater than reported here. Minor and trace element contents are more variable than major element contents. Variations of a factor of three are observed in K_2O (0.05-0.16%), TiO_2 (0.86-2.4%), Sc (9-30 ppm), Sm (1.8-6.1 ppm), and other incompatible elements.

Ryder and Spudis (1987) used REE patterns and Ti-Sc-Sm variations as the basis for their compositional groupings of Apollo 15 impact melts. Figure 1 is a REE plot for Apollo 15 impact melts giving the ranges for groups A-D and a number of individual KREEP-poor impact melt samples. Literature data are plotted in Fig. 1a, which shows the range of typical KREEP-poor impact melt samples and two atypical samples, 15414,35 (*Ryder and Spudis,* 1987) and 15243,62 (*Ryder et al.,* 1988). Our new data are plotted in Fig. 1b, which shows the range of typical KREEP-poor impact melts, the mean of these samples (discussed later), and the composition of the large clast from 15459. The REE patterns of the two unusual samples in Fig. 1a are clearly distinct from all the other KREEP-poor impact melt samples. The remainder of the 27 KREEP-poor impact melt samples have generally similar REE patterns that vary slightly in slope, but significantly in concentration level, and Eu anomaly. Europium anomalies range from slightly negative in samples having the highest REE concentrations to positive in samples having the lowest REE concentrations. The mean composition has no Eu anomaly at all. Light REE (LREE) patterns for most KREEP-poor impact melt samples are negative and approximately parallel, while heavy REE (HREE) patterns are essentially flat, with slopes varying from slightly negative to slightly positive. Compared to the KREEP-dominated REE patterns of group A-D melts, the patterns of the KREEP-poor impact melt samples have similar LREE slopes, but distinctly flatter HREE slopes. The low REE concentrations and deviations from the typical KREEP REE patterns limit the amount of KREEP component in these breccias to a very small fraction, probably not more than 5%.

Figure 2 is a plot of Ti/Sm vs. Sc/Sm that *Ryder and Spudis* (1987) used to further characterize the Apollo 15 impact melts. Our data and all data for Apollo 15 impact melts published since 1986 are plotted (samples that have not been analyzed for Ti are plotted on the Sc/Sm axis). This plot separates groups B-D better than the REE alone. Except for the group B-C melts, the Sc/Sm ratios are themselves a fairly good discriminant for the groups. KREEP-poor impact melts form a roughly linear array between the group D melts and Ryder and Spudis' group E sample, with most samples clustering near Ti/Sm of 2500 and Sc/Sm of 4. Several KREEP-poor impact melt samples deviate from this cluster. The deviant samples include the two that have unusual REE patterns: 15414,35 (*Ryder and Spudis,* 1987), which is also more feldspathic and more ferroan; and 15243,62 (*Ryder et al.,* 1988), which has extremely high Sc. Another highly unusual sample is 15459,386, which has very low Ti and REE and high mg′. Three samples [15295,90 (this study), 15223,45, and 15204,102 (both *Laul et al.,* 1988)] plot along the trend in Fig. 2, but below the typical KREEP-poor impact melt cluster and parallel to group D impact melts. These samples have very low Ti and Sc and high mg′. We consider them to be a distinct subgroup of KREEP-poor impact melts. Sample 15243,65 (which has not been analyzed for Ti and Mg) may also be a part of this group and sample 15459,386 may be a member of this group that is unusually depleted in incompatible elements. The impact melt sample having the highest REE concentrations (15295,94) is also much more ferroan (mg′ 67 vs. 71) and distinct from the typical samples.

The remaining 19 samples are similar in all respects except for their variation in REE concentrations. Sample 15295,64, which plots above the cluster in Fig. 2, has the lowest REE concentrations, but is otherwise similar to typical KREEP-poor impact melts. Most of the samples cluster fairly tightly in REE. Ten of the 19 samples have REE concentrations within 10%

TABLE 1. Major and trace element composition of KREEP-poor impact melts.

Sample	15295	15295	15295	15295	15295	15295	15295	15295	15459	15459
INAA	,64	,68	,70	,82	,90	,94	,104	,110	,301	,358
TS	,121	,69	,71	,81	,89	,95	,105	,130	,301	,357
WT (mg)	20.0	33.0	69.1	19.7	20.7	10.5	9.8	77.3	8.9	7.7
SiO_2 (%)	45.4	45.5	45.8	45.1	46.6	45.2	45.3	45.7		45.7
TiO_2	1.78	1.74	1.74	1.89	0.64	1.81	1.78	1.73		2.15
Al_2O_3	20.7	20.4	21.1	19.1	22.2	20.4	19.5	20.3		18.6
FeO	7.53	7.51	7.66	8.25	5.58	7.75	7.80	7.53		8.40
MgO	10.4	10.7	10.9	11.3	10.8	8.98	10.8	10.2		11.3
CaO	12.0	12.2	12.2	11.7	12.9	12.6	12.1	12.4		11.6
Na_2O	0.55	0.62	0.71	0.62	0.46	0.63	0.60	0.62		0.60
K_2O	0.05	0.11	0.13	0.14	0.14	0.16	0.12	0.14		0.11
P_2O_5	0.04	0.08	0.08	0.08	0.04	0.09	0.10	0.08		0.06
mg′	71	72	72	71	78	67	71	71		71
Na_2O	0.583	0.667	0.693	0.645	0.463	0.730	0.629	0.639	0.691	0.646
CaO	12.3	13.1	12.9	12.6	13.0	13.4	12.1	12.6	11.7	12.1
FeO	8.03	7.70	8.06	8.45	5.62	7.60	7.87	7.63	7.90	8.75
Sc (ppm)	18.7	17.5	17.4	17.8	10.6	17.8	17.6	17.0	16.1	20.4
Cr	1300	1175	1100	1130	1220	1160	1180	1180	1150	1290
Co	22.0	20.3	20.6	21.4	15.1	19.2	20.3	19.2	22.3	21.1
Ni	79	58	78	70	138	100	61	57	80	90
Sr	190	200	230	240	180	240	185	195	190	210
Ba	94	140	135	160	150	170	160	130	125	145
La	5.27	9.76	9.60	10.35	11.1	12.67	12.08	9.03	9.04	8.36
Ce	14.1	26.0	25.2	25.8	28.6	32.7	31.5	23.8	24.5	22.1
Nd	8	12	15	20	17	18	15	13	16	14
Sm	2.78	4.88	4.77	5.11	5.01	5.87	5.80	4.42	4.38	4.49
Eu	1.50	1.73	1.74	1.71	1.15	1.74	1.61	1.63	1.67	1.65
Tb	0.65	1.01	1.04	1.12	0.98	1.18	1.21	0.98	0.87	0.99
Yb	2.78	4.12	3.95	4.33	4.04	4.78	4.89	4.18	3.30	3.90
Lu	0.429	0.607	0.601	0.644	0.54	0.702	0.718	0.580	0.528	0.596
Hf	2.78	3.86	3.66	4.10	4.00	4.47	4.57	3.78	3.05	3.80
Ta	0.43	0.62	0.60	0.66	0.5	0.67	0.65	0.61	0.49	0.63
Th	0.79	1.97	1.79	1.98	1.99	2.15	2.12	1.75	1.66	1.39
U	0.22	0.42	0.48	0.47	0.55	0.66	0.62	0.45	0.40	0.37
Ir (ppb)	2.5	2	1.5	2.5	6.6	<3	<6	1.5		3.L.

TABLE 1. (continued).

Sample	15459	15459	15459	15459	15459	15459	15459	15459	mean	std dev
INAA	,382	,386	,394	,414a	,414b	,231-1	,231-2	,320		
TS	,400	,387	,393	,125	,125	,125	,125	,125		
WT (mg)	6.6	6.3	7.2	117.3	130.4	118.0	99.7	98.8		
SiO_2	44.7	47.9	46.1	44.7	44.6				45.3	0.5
TiO_2	2.09	0.19	1.70	1.70	1.57		2.14		1.77	0.26
Al_2O_3	17.5	22.4	19.2	20.9	21.1		20.1		19.9	1.0
FeO	8.95	5.65	8.30	7.94	7.77		8.07		7.82	0.51
MgO	11.9	10.4	11.2	10.8	10.4		11.6		10.7	0.8
CaO	11.4	12.4	12.0	12.3	12.4		11.7		12.04	0.34
Na_2O	0.57	0.36	0.60	0.61	0.61		0.69		0.62	0.04
K_2O	0.09	0.062	0.13	0.10	0.11		<0.3		0.10	0.04
P_2O_5	0.08	0.04	0.07	0.07	0.06				0.07	0.01
mg′	70	77	71	71	71		72		70.9	1.1
Na_2O	0.661	0.386	0.639	0.678	0.688	0.661	0.693	0.679	0.66	0.03
CaO	12.0	12.5	12.0	13.0	13.2	11.7	11.7	12.3	12.37	0.53
FeO	9.42	5.82	8.28	7.41	7.48	8.22	8.07	7.58	7.94	0.59
Sc	21.5	9.0	19.5	15.7	16.0	17.1	17.4	15.4	17.4	1.7
Cr	1360	2040	1190	980	1000	1100	1090	1000	1121	112
Co	27.4	16.2	21.1	19.0	19.6	20.0	20.5	19.5	20.3	2.2
Ni	110	62	90	81	70	66	82	76	75	14
Sr	200	160	180	205	195	220	200	190	201	19
Ba	140	92	140	115	120	120	130	120	134	18
La	9.48	3.72	9.55	7.47	7.69	8.10	8.49	7.49	9.0	1.7
Ce	25.0	9.6	24.2	19.1	19.7	21.0	22.0	20.6	23.5	4.3
Nd	17	<10	13	10	12	13	14	10	13.9	3.0
Sm	5.15	1.60	4.85	3.81	3.91	4.02	4.35	3.74	4.49	0.74
Eu	1.60	1.11	1.57	1.57	1.61	1.62	1.64	1.58	1.64	0.09
Tb	1.10	0.32	1.02	0.90	0.90	0.94	1.00	0.85	0.98	0.14
Yb	3.86	1.48	4.39	3.39	3.45	3.60	3.87	3.4	3.91	0.54
Lu	0.595	0.205	0.647	0.489	0.498	0.538	0.578	0.489	0.58	0.08
Hf	3.88	0.64	3.52	2.94	3.04	3.25	3.55	2.78	3.53	0.56
Ta	0.64	0.11	0.48	0.53	0.55	0.57	0.61	0.53	0.58	0.07
Th	1.61	0.35	1.54	1.32	1.32	1.38	1.46	1.60	1.63	0.35
U	0.38	0.19	0.55	0.33	0.41	0.50	0.44	0.35	0.44	0.10
Ir (ppb)	5			2.5	2.2			1.5	1.7	1.3

of the mean, and only 3 have REE concentrations more than 25% from the mean. Unrepresentative sampling may be the cause of the extreme REE variations because all of our extreme samples are less than 20 mg in weight. We have not attempted to evaluate heterogeneity for such small samples; however, we observed only about 10% variation in 100-mg samples of the large clast. We also note that the composition of the large clast does not represent the mean of the compositions of the smaller clasts. All analyzed splits of the large clast have REE concentrations lower than those of the mean, with the homogenized samples having REE lower than all but two of the typical KREEP-poor impact melt samples.

In conclusion, using REE patterns, Ti-Sc-Sm variations, and bulk composition, we have identified a group of 19 of the 27 KREEP-poor impact melt samples that we consider to be a typical subgroup of KREEP-poor impact melt rocks. The mean composition and standard deviations for the remaining 19 samples are given in Table 1. Although the individual sample compositions exhibit real variations in REE, at least on the scale sampled, they cluster quite tightly for most other elements and we feel that the typical KREEP-poor impact melt samples represent a distinct KREEP-poor mafic melt rock unlike any other impact melt from Apollo 15 or other highland sites. A second subgroup consisting of 3-5 samples is distinct from typical KREEP-poor impact melts and may represent a second such melt type.

PETROGRAPHY AND MINERAL COMPOSITION

We have performed detailed petrologic and mineral compositional analyses of 12 KREEP-poor impact melt clasts using the JEOL 733 microprobe at Harvard-SAO. Mineral assemblages and compositions are summarized in Table 2. The samples are all polymict microbreccias of broadly gabbroic composition with virtually identical mineral assemblages, although they vary in grain sizes, textures, and modes. They contain about 52% to 65% plagioclase, and most of them show a range of feldspar compositions from about An_{80} to An_{96} (Table 2). Plagioclases typically have very low K_2O contents (0.0-0.15 wt.%). Plagioclase compositions show a patchy distribution. Individual grains are homogeneous. No evidence of zoning or of exsolution lamellae has been detected either by microprobe analyses or X-ray maps of sodium distributions. However, both calcic and sodic grains occur both as angular fragments and as small matrix grains. Rare grains of feldspar richer in alkalis (An_{64-76}) were found in four of the samples.

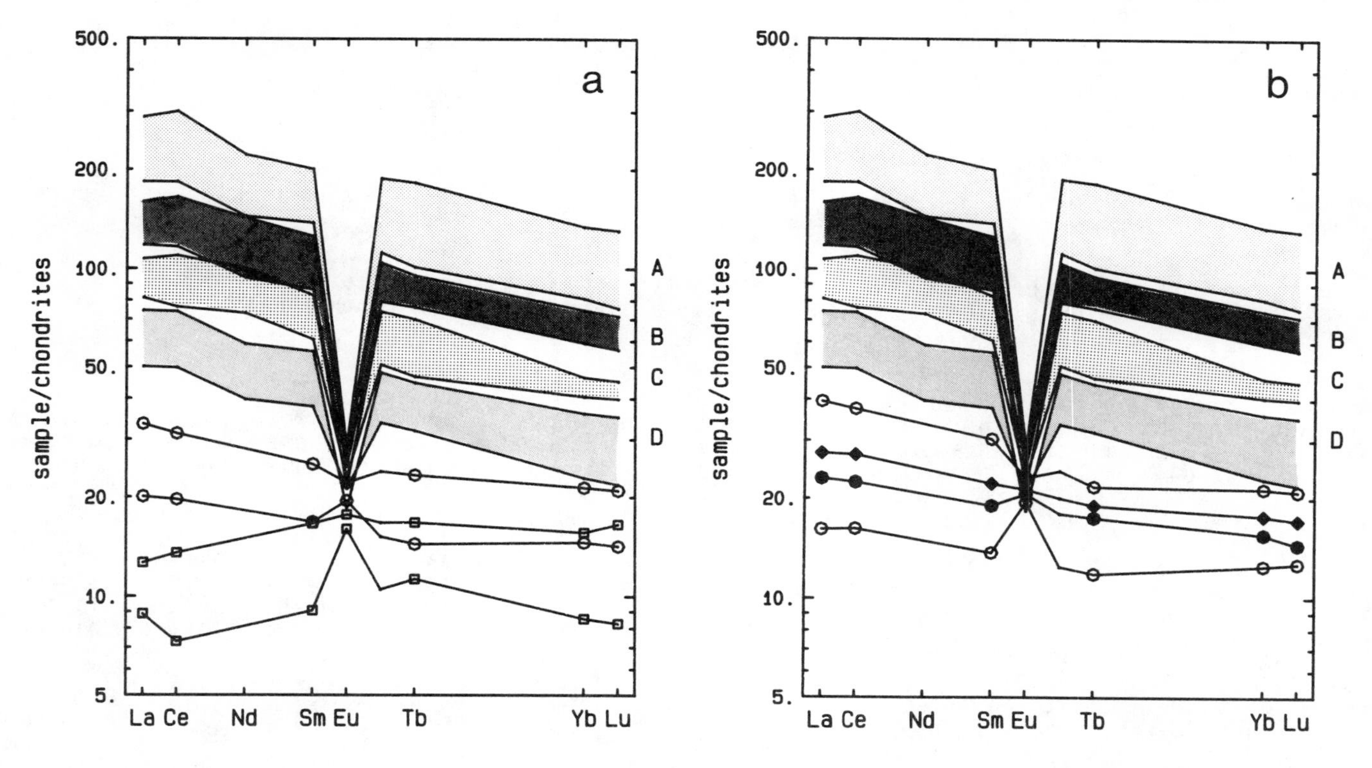

Fig. 1. Chondrite-normalized REE plots for Apollo 15 impact melts. Groups A-E are based on the definitions of *Ryder and Spudis* (1987). Shaded areas are ranges for groups A-D based on literature data and our unpublished analyses. Individual samples are plotted for KREEP-poor impact melts. **(a)** Literature data. Squares are unique samples 15414,35 (*Ryder and Spudis,* 1987) and 15243,62 (*Ryder et al.,* 1988). Open circles are the KREEP-poor impact melts with highest and lowest REE concentrations. **(b)** This study. Open circles are again the KREEP-poor impact melts with highest and lowest REE concentrations. Filled diamonds represent the mean of typical KREEP-poor impact melts. Filled circles represent the composition of the homogenized large KREEP-poor impact melt clast (15459,414).

Mafic mineral compositions are illustrated in Fig. 3. Unzoned olivine grains occur in all of the samples, where they range in modal abundance from about 2% in 15295,130 to 16% in 15459,301. Their average composition is $Fo_{71.5}$, with a range of Fo_{69-74}. Olivines in all samples contain <0.25 wt.% CaO, values consistent with those reported in annealed ANT suite breccias (*Smith and Steele,* 1976). Pyroxene typically constitutes 25-40% of the samples. Most of the samples contain clinobronzite clustering around En_{70-78}, and all but one (15295,121) also contain augite ($En_{40-50}Fs_{10-14}Wo_{40-46}$). Low-Ca pyroxene predominates in all samples. Two clast fragments in thin section 15459,400 have more ferroan orthopyroxenes, which are comparable to pyroxenes in two relatively ferroan Mg-norite clasts (15459,298 and ,375) we studied previously (*Lindstrom et al.,* 1989a). Sparse grains (1-2%) of Mg-ilmenite occur in all of the samples. Other accessory minerals, including minute grains of troilite, baddeleyite (ZrO_2) and chromian spinel, are very rare.

Most of the samples are micropoikilitic in texture with small, euhedral to subhedral laths and tablets of plagioclase embedded within optically continuous low-Ca pyroxenes. In several samples (e.g., 15459,125, and 15295,121 and ,130) the largest pyroxene domains are optically continuous for 0.7 to 1 mm. A few samples, including 15459,301, are somewhat

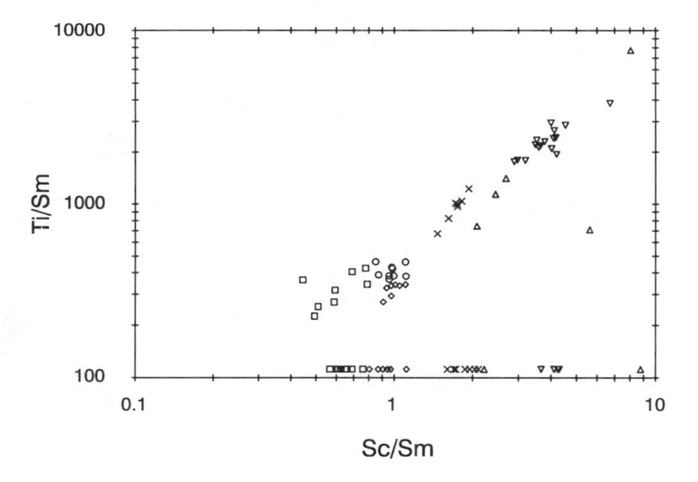

Fig. 2. Ti/Sm vs. Sc/Sm for Apollo 15 impact melts. Symbols represent groups A-E as defined by *Ryder and Spudis* (1987). Samples for which no Ti analysis is available are plotted along the Sc/Sm axis. Squares are group A; circles are group B; diamonds are group C; Xs are group D; triangles are KREEP-poor impact melts. Triangles pointing up are atypical KREEP-poor impact melts, while those pointing down are typical KREEP-poor impact melts.

TABLE 2. Modes and mineral compositions of KREEP-poor impact melts.

Mineralogy	15459,125	15295,121	15295,130	15459,393	15459,301	15295,81	15459,400A	15459,400B	15459,400C	15459,400D
Mode %										
Plagioclase	65	60	55	53	54	54	52		60	
Pyroxene	27	36	42	40	28	39	7		39	
Olivine	7	3	2	5	16	5	40		0	
Opaques	1	1	1	2	2	2	1		1	
Plagioclase (An-Ab-Or)										
Calcic	95-4-0.3	96-3-0.3	94-5-0.3	96-3-0.3	96-3-0.3	96-3-0.2	92-7-0.4	95-4-0.2	97-2-0.2	91-8-0.8
Sodic	83-16-1	87-12-0.6	80-18-4	83-15-2	80-18-2	83-16-1	88-11-0.6	88-11-0.6		81-15-4
Alkalic			64-23-13	74-19-7		68-20-12	76-19-5			
Low-Ca Pyroxene										
En-Fs-Wo	73-24-3	74-24-2	74-24-2	75-23-2	76-22-2	73-25-2	73-25-2	70-26-4	61-37-2	65-31-4
Al_2O_3 wt.%	1.17-1.30	0.71-2.21	0.81-1.40	0.54-1.71	1.28-1.72	0.64-1.45	0.60-0.94	1.05-1.24	0.44-0.53	0.80-1.20
TiO_2	0.88-1.12	0.47-1.01	0.25-1.68	0.48-0.93	0.15-0.73	0.25-0.73	0.25-0.78	0.66-0.72	0.19-0.25	0.57-0.88
Cr_2O_3	0.34-0.44	0.24-0.58	0.22-0.47	0.00-0.43	0.30-0.42	0.15-0.39	0.09-0.28	0.00-0.13	0.00-0.23	0.00-0.24
High-Ca Pyroxene	minor	none	trace	subordinate	subordinate	subordinate		subequal		
En-Fs-Wo	47-12-41		49-11-40	46-10-44	49-10-41	45-10-45	47-13-40	47-12-41	40-14-46	42-13-45
Al_2O_3 wt.%	2.25-2.62		1.46	0.58-2.80	0.92-2.39	1.62	1.19-2.20	2.28-2.43	0.66-1.07	1.38-2.48
TiO_2	2.12-2.35		0.96	0.47-1.85	0.59-2.31	0.98	1.67-2.13	1.88-2.06	0.29-0.51	1.13-2.18
Cr_2O_3	0.53-0.62		0.36	0.20-0.51	0.26-0.45	0.18	0.35-0.47	0.45-0.52	0.00-0.26	0.19-0.37
Olivine										
Fo	71-72	70-71	70-74	71-73	73-74	71-73	69-70	68-70		68-70
CaO wt.%	0.04-0.12	0.07-0.08	0.04-0.09	0.02-0.16	0.06-0.16	0.04-0.23	0.05-0.16	0.12-0.24		0.06-0.14
Ilmenite										
MgO wt.%	4.3-5.8	5.4-5.8	4.6-5.0	4.7-5.7	4.6-5.3	4.7-5.2	3.88-3.97		2.50	
Chromite										
MgO wt.%			3.52							
FeO			29.10							
TiO_2			1.15							
Cr_2O_3			46.04							
Al_2O_3			14.22							

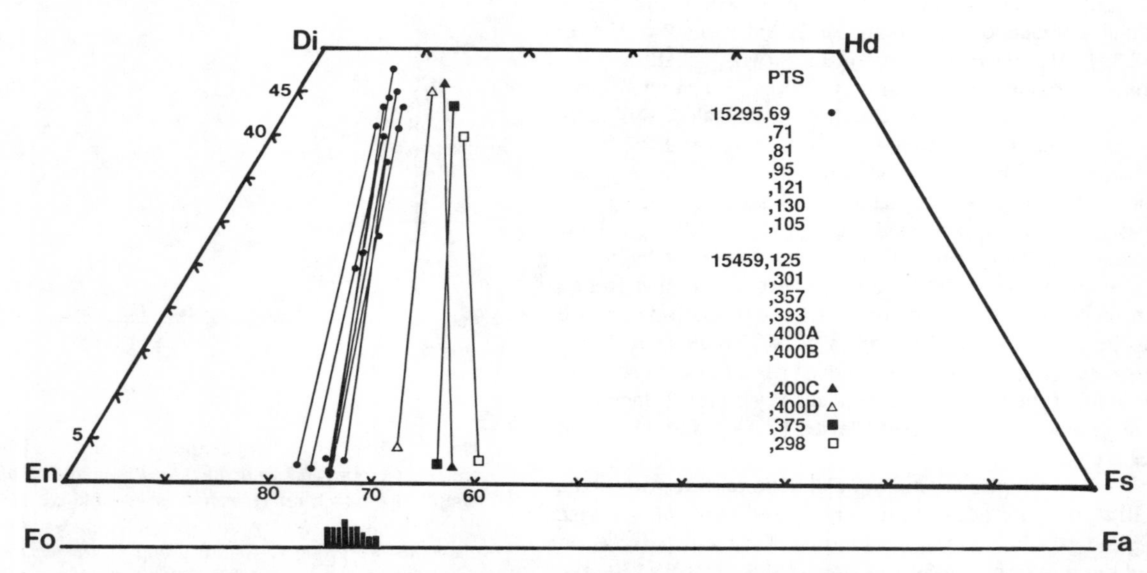

Fig. 3. Pyroxene and olivine compositions in KREEP-poor impact melt clasts. Maximum and minimum Wo contents in pyroxenes are indicated by solid dots for the 13 clasts listed by number, and by other symbols for 4 others.

richer than the others in mafic minerals. These more mafic clasts display a mosaicized texture that obscures any well-developed poikilitic areas that are visible only at high magnification in backscattered electron images. Some of the pyroxenes retain a patchy optical continuity for distances of up to 225 μm. Angular fragments of plagioclase, which frequently are larger than the ambient grain size, occur in all of the samples. As a group these KREEP-poor impact melts show a marked diversity of microtextures. Most of them have poikilitic areas and a few also show patches of either igneous or granulitic textures. Their similarity in bulk composition is in marked contrast to their diversity in texture. Seven KREEP-poor impact melt samples, exhibiting both typical and atypical textures, are described below and photomicrographs are shown in Fig. 4.

15459,125

This thin section sampled the large clast in 15459, which we homogenized and used to test heterogeneity. It is described in the Apollo 15 Catalog (*Ryder,* 1985) as a poikilitic impact melt. It is the most plagioclase-rich of the KREEP-poor impact melts and has the coarsest poikilitic texture, with optically continuous grains of orthopyroxene, up to 1 mm long, enclosing euhedral-subhedral crystals of plagioclase and rounded grains of olivine (Fig. 4a,b). Plagioclase also occurs in angular clasts up to 600 μm across. Ilmenite and sparse grains of troilite occur as accessory minerals. Plagioclase composition ranges from An_{95} to An_{83}, with the more Ca-rich members ($>An_{90}$) making up about two-thirds of the total. Strongly zoned plagioclase clasts with cores of An_{92} and rims of An_{69} were reported by *Ryder* (1985), but we observed nothing of this description. Unzoned olivines occur in small, individual grains up to 30 μm across. The predominant pyroxene is clinobronzite which contains thin lamellae of augite. Despite the evidence for relatively slow cooling and equilibration in the well-separated pyroxene pairs and the lack of zoned minerals, the range of matrix plagioclase compositions shows that this sample is not an equilibrated rock.

15295,121

This clast lacks augite, but is otherwise similar to the rest of the KREEP-poor impact melts. Plagioclases show the narrowest range observed in KREEP-poor impact melt samples. Ilmenite is the only accessory mineral we have found. Euhedral to subhedral plagioclase and olivine grains are poikilitically enclosed in pyroxene grains 350 μm long. The resulting texture (Fig. 4c) is strongly suggestive of crystallization from a melt.

15295,130

This clast has a mode and texture typical of KREEP-poor impact melts. Clinobronzites form optically continuous grains, up to 750 μm long, enclosing euhedral grains of plagioclase and olivine (Fig. 4d). Augite is rare. In addition to ilmenite, opaque minerals include two minute grains of Mg-Al chromite. Mafic mineral compositions are typical of KREEP-poor impact melts. Plagioclase compositional range (An_{94-80}) is larger than most KREEP-poor impact melts and a single grain has the most alkalic composition measured in any KREEP-poor impact melts.

15459,393

This clast is a typical KREEP-poor impact melt. For the most part, it has an orderly micropoikilitic texture with optically continuous pyroxenes up to 320 μm long. In addition to plagioclase clasts, 15295,393 contains a large, angular olivine grain that is broken at one end and measures 240×320 μm (Fig. 4e). The composition of this olivine (Fo_{73}) is similar to that of the small olivine crystals poikilitically enclosed in pyroxenes (Table 2). Ilmenite in dendritic grains is the main accessory mineral, but sparse grains of troilite and three or four 5 μm grains of ZrO_2 also are present, the latter embedded in ilmenite (Fig. 4f). The presence of the large olivine grain prompted us to analyze olivine clasts in the adjacent breccia matrix. We found them to be markedly more magnesian (Fo_{88-89}) and clearly distinguishable from the olivines in the melt rock. The matrix orthopyroxenes ($En_{61}Fs_{36}Wo_3$) also differ from those in the melt rock, being much more ferroan.

15459,301

This clast is markedly richer in olivine than all of the other samples except fragments 15459,400A-B. Although the sample contains 54% modal plagioclase, it has abundant plagioclase clasts, and it is estimated that the melt matrix contains 60% pyroxene and olivine. Olivine occurs in small (10-40 μm) grains in pyroxenes and also in larger grains (up to 120 μm) that appear to be clasts. Clinobronzite and augite are both present. In the microscope, small, twinned laths of feldspar almost overwhelm the pyroxenes, although some of the latter show optical continuity for up to 400 μm. Backscattered electron images provide clearer views of the texture, with small, feldspar laths enclosed in pyroxene (Fig. 4g). This texture depicts, beyond any doubt, the crystallization of feldspars from a melt that later formed pyroxene, yet lying only a few micrometers away is a large metamorphic plagioclase enclosing a necklace of tiny olivine grains.

15295,81

This clast consists of two fragments that have a mode typical of KREEP-poor impact melts, but have a distinct texture. The texture is disorderly but two or three patches of optically continuous pyroxene grains, up to 450 μm long, are visible. These pyroxenes poikilitically enclose abundant $<$80-μm laths of plagioclase (Fig. 4h). A 270-μm zone of crushed plagioclase occurs within the clast, suggesting that 15295,81 contains both a poikilitic component and an incorporated cataclastic anorthosite fragment. Much of the feldspar, occurring outside the poikilitic areas, has been optically randomized by shock pressures. The plagioclase shows a compositional range of An_{83-96} plus one small patch of a more alkali-rich material. Pyroxenes, with low-Ca compositions predominating, cluster with those of other KREEP-poor impact melt samples (Table 2 and Fig. 3). Sparse olivines occur in the pyroxenes and

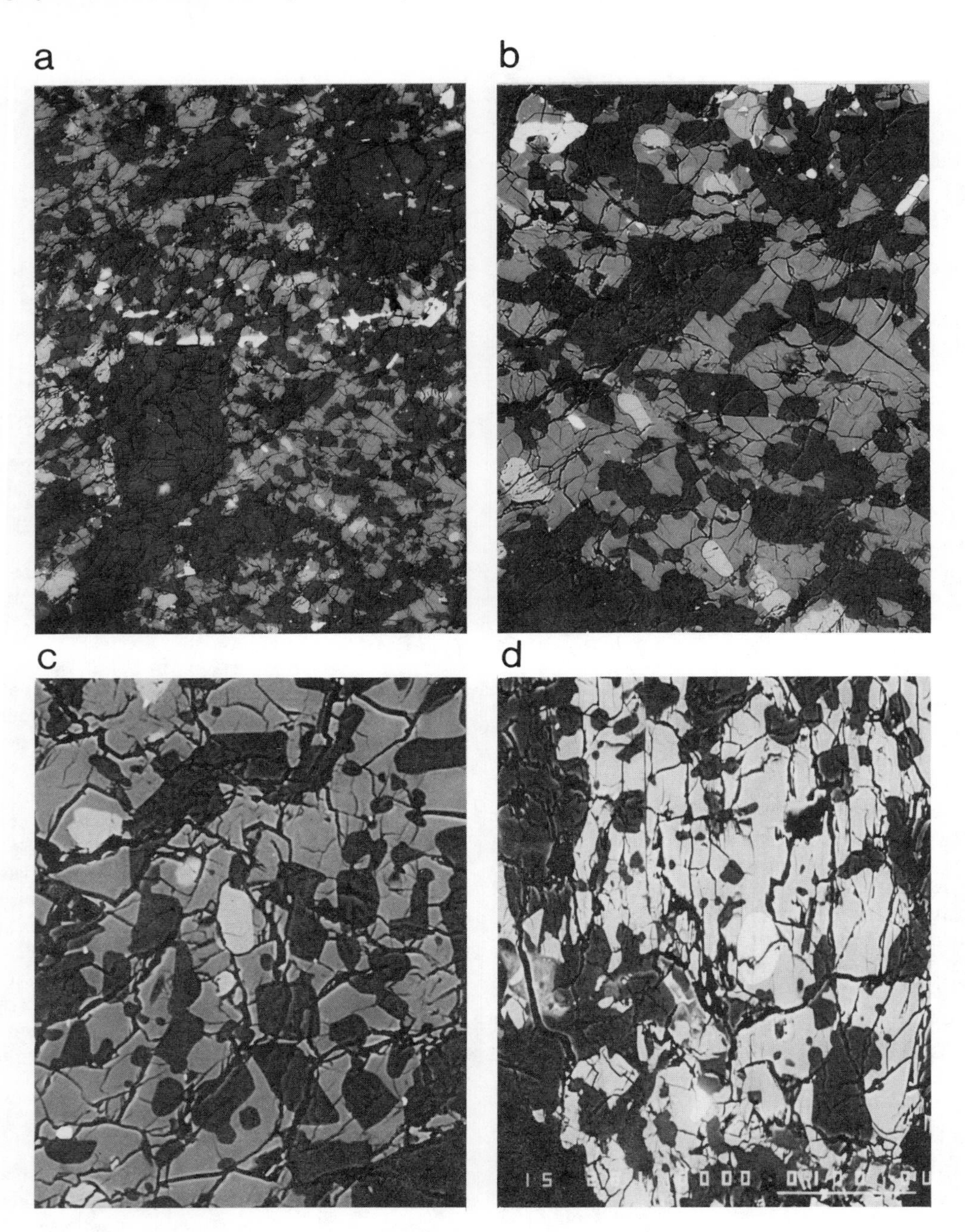

Fig. 4. Backscattered electron images of KREEP-poor impact melt clasts. Plagioclases are dark gray, pyroxenes are medium gray, olivines are light gray, and ilmenites and troilites are very light gray to white. **(a)** and **(b)** Two views of 15459,125 showing a large, optically continuous pyroxene (lower right) poikilitically enclosing euhedral-subhedral grains of plagioclase and olivines. Two large clasts of plagioclase are visible in **(a)** (left center and upper right). Ilmenite occurs mainly in scattered, irregular grains. The line of ilmenites crossing the center of **(a)** may represent one or two large dendritic grains that were linked in three dimensions. [Width of field, left to right, is 800 μm in **(a)**, 300 μm in **(b)**.] **(c)** 15295,121 showing euhedral character of plagioclase crystals and olivines poikilitically enclosed within clinobronzite. (Width of field is 110 μm.) **(d)** 15295,130 clinobronzite enclosing plagioclase and olivine grains.

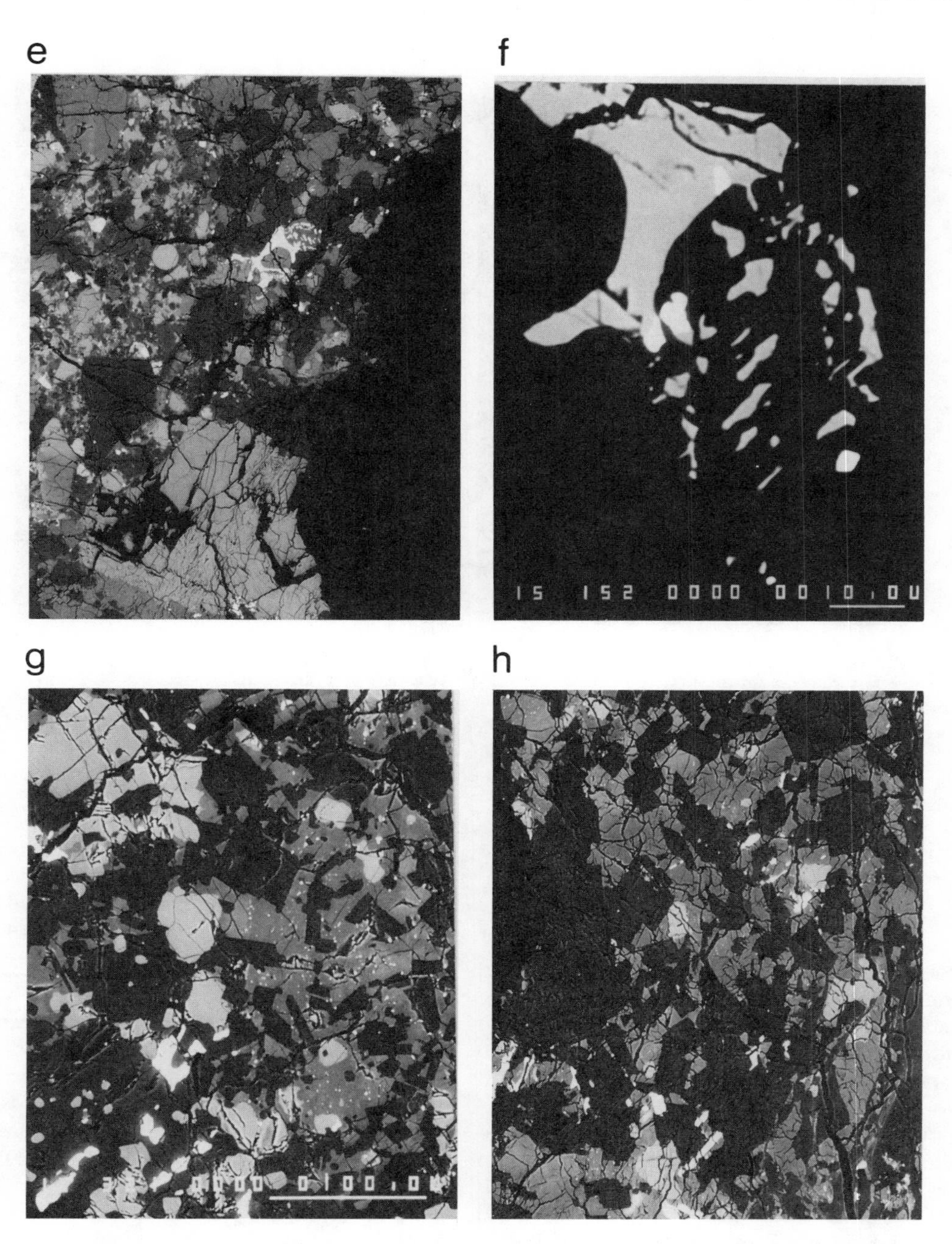

Fig. 4. **(e)** and **(f)** A general and a detailed view of 15459,393. In **(e)** an area with well-developed poikilitic texture is interrupted by the large, angular olivine clast at lower right. At center is a dendritic ilmenite attached to a graphic intergrowth of olivine with ilmenite crystallites. The breccia matrix, which includes a conspicuous glass spherule, occupies the upper left; **(f)** is a close-up of the olivine-ilmenite intergrowth. This image shows three minute, white grains of baddeleyite (ZrO_2) embedded in the ilmenite (gray). The contrast is too high, however, for distinguishing between the olivine that lies to the right of the large forked ilmenite grain and encloses the small ilmenite crystallites, and orthopyroxene that lies to the left of the forked grain. Both the olivine (Fo_{76}) and the orthopyroxene ($En_{77}Fs_{22}Wo_{0.2}$) are slightly richer in MgO than their counterparts in the rest of the clast (Table 2). The white grain at lower right is FeS. **(g)** 15459,301, which displays a persuasively igneous looking array of plagioclase laths poikilitically enclosed in clinobronzite. Olivine occurs in small grains within the feldspar and the pyroxene, and in larger grains, which are clasts (upper left). Conspicuous white grains are ilmenite. **(h)** 15295,81, a clast in which large and small euhedral-subhedral plagioclase grains are enclosed within optically continuous clinobronzite. Outside this view, the clast contains an angular remnant of cataclastic anorthosite. (Width of field is 300 μm).

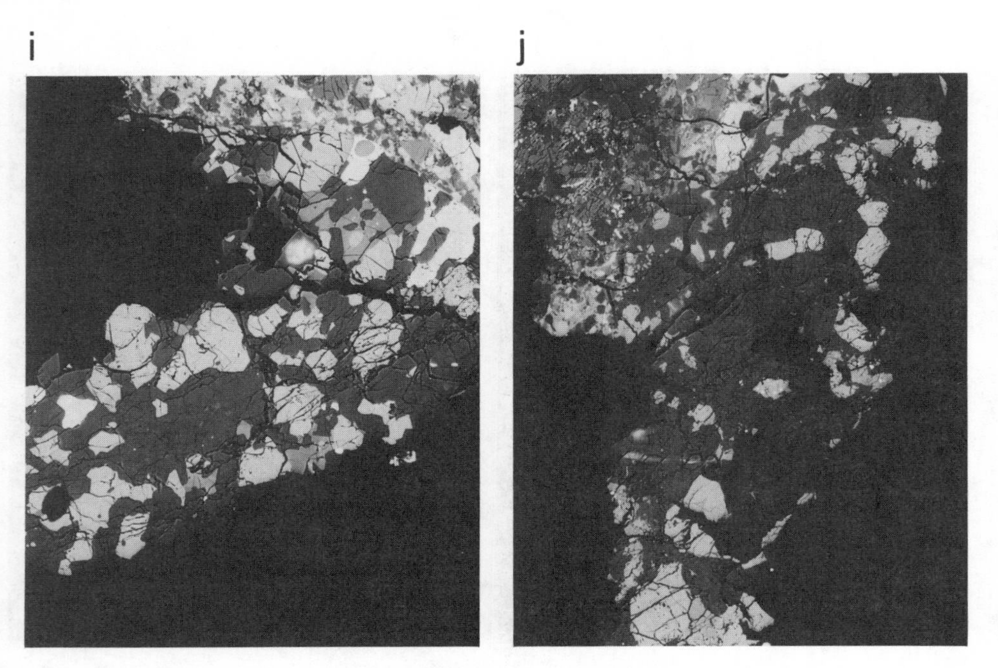

Fig. 4. **(i)** and **(j)** Two of the four clast fragments from 15459,400, with attached breccia matrix at top. **(i)** Fragment A, which consists mainly of plagioclase and olivine, in conspicuous light gray grains. The few pyroxenes occur as small, irregular, medium gray grains. Bright grains are ilmenite plus a trace of FeS. **(j)** Fragment C with about 60% plagioclase, which has a granulitic texture, and no olivine at all. The light gray grains are clinohypersthene and augite. (Width of fields is 500 μm.)

scattered through the rest of the sample. Some of the mafic minerals have irregular margins suggestive of reaction zones.

15459,400

This thin section has four small clast fragments (A, B, C, and D) with differing mineral assemblages and textures. Only one of the fragments has a small patch with poikilitic texture, and none has both modes and mineral compositions similar to typical KREEP-poor impact melts. Fragments A (Fig. 4i) and B are extremely olivine-rich but have mineral compositions similar to typical KREEP-poor impact melts (Fig. 3). The two fragments differ only in their pyroxene proportions. In fragment A the small pyroxene grains are clinobronzite with thin lamellae of augite, while in fragment B clinobronzite and augite are subequal. Fragment A also contains a single Fe-rich pyroxene grain similar to those in fragment C. Sparse opaques are ilmenite and troilite. Fragment C (Fig. 4j) has no olivine, and no bytownite. The anorthite has a well-developed granulitic texture that contrasts with that of all other samples of this study. The low-Ca pyroxene is markedly richer in FeO than those of other members of the KREEP-poor impact melt group and similar to those of norites 15459,298 and ,375, (*Lindstrom et al.,* 1989a). Pyroxenes from the latter two clasts are plotted in Fig. 3 for comparison with those of fragment C. Fragment D is intermediate in mineral composition between fragments A and B and fragment C. Some of the differences, such as the overabundance of olivine in fragments A-B or lack of olivine in fragment C, may be attributable to the small sample sizes. However, the differences in pyroxene compositions are real. We are uncertain of the relationship between fragments, but the composition of the analyzed sample is similar to that of typical KREEP-poor impact melts.

ORIGIN OF POIKILITIC TEXTURE

Most of these breccias have poikilitic textures and exhibit some degree of annealing or recrystallization. All four typical KREEP-poor impact melt samples from other studies (*Laul et al.,* 1988; *Ryder et al.,* 1988) are described as poikilitic impact melts. The origin of poikilitic textures in lunar breccias has been a matter of debate. *Albee et al.* (1973) and *Bence et al.* (1973) argued for a metamorphic origin for Apollo 16 and 17 poikilitic breccias, while *Simonds et al.* (1973, 1974) argued that their textures resulted from crystallization from impact melts. Eventually the impact melt origin was accepted for the LKFM breccias so common at the Apollo 15, 16 and 17 sites.

A separate group of highland breccias that frequently has poikilitic texture was identified as feldspathic granulitic impactites by *Warner et al.* (1977) and granulitic breccias by *Stöffler et al.* (1980). These metamorphic breccias have more equilibrated mineral compositions and differ from poikilitic impact melts in being generally more feldspathic and distinctly lower in incompatible element concentrations (*Warner et al.,* 1977; *Lindstrom and Lindstrom,* 1986). We evaluated the possibility that the KREEP-poor poikilitic breccias described here might be more mafic metamorphic breccias similar in origin to the typical feldspathic granulitic breccias. *Ryder and Spudis* (1987) discuss petrographic criteria for distinguishing between impact melts and granulitic breccias. The presence of euhedral crystals, rounded clasts, interstitial glass, vesicles, and mineral compositional variations are indications of impact melt origin.

The textures of KREEP-poor impact melt samples range from those with clear evidence of crystallization from a melt to those with considerable recrystallization. *Ridley* (1977) attributes the poikilitic texture in the largest and coarsest-grained example (15459,125) to metamorphism because calculated equilibration temperatures of 1000°-1100°C are well below the liquidus temperature. However, *Ryder* (1985) described the same clast as an impact melt closely resembling melt-sheet rocks from the Apollo 17 site. Despite the textural variability in KREEP-poor impact melt samples, all exhibit the same mineral assemblages and bulk and mineral compositions. None of the samples have interstitial glass or vesicles. Mineral compositions are homogeneous for mafic minerals, yet plagioclase is clearly unequilibrated. Not only the larger clasts, but smaller matrix grains have a wide range of plagioclase compositions in each sample. Based largely on the more recrystallized samples, we previously identified these rocks as recrystallized gabbroic breccias (*Lindstrom et al.,* 1989b) that we interpreted to be metamorphic breccias. Based on the clear evidence in some clasts for crystallization from a melt and the unequilibrated plagioclase compositions in all samples, we now interpret the clasts to be impact melts, some of which were later annealed.

ABUNDANCE OF KREEP-POOR IMPACT MELTS

The abundance of KREEP-poor impact melts at the Apennine Front can be estimated by combining petrographic surveys of the Apollo 15 regolith with recent compositional/petrographic studies of Apollo 15 impact melts. Early petrographic studies of the Apollo 15 regolith (*Cameron et al.,* 1973; *Phinney et al.,* 1972; *Powell et al.,* 1973) and recent surveys of Apollo 15 regolith breccias (*Simon et al.,* 1986; *McKay et al.,* 1989) show that impact melt rocks and granulitic breccias are the most common highland lithic fragments other than KREEP basalts. These petrographic surveys give little indication of fragment compositions, but recent compositional/petrographic studies of individual Apollo 15 breccia clasts and coarse fines (*Ryder and Spudis,* 1987; *Ryder et al.,* 1988; *Laul et al.,* 1988; *Lindstrom et al.,* 1988, this work, and unpublished data) can be used to estimate proportions of various compositional groups. Among 75 fragments of Apollo 15 impact melt rocks, 23 are KREEP-poor samples. The KREEP-poor impact melts described here appear to make up approximately one third of the impact melt population and be a relatively common component at the Apennine Front.

LKFM AND HIGHLAND IMPACT MELTS

Low-K Fra Mauro (LKFM) basalt is an enigmatic and ubiquitous lunar highland component. It was first defined by the Apollo Soil Survey (*Reid et al.,* 1972) as a group of Apollo 15 glasses that have major element composition similar to Apollo 14 KREEP but have much lower K_2O (and presumably REE and P) concentrations. It has been observed as a glass composition in all highland sites and identified as a common impact melt composition at the Apollo 15, 16, and 17 sites, but LKFM has never been identified as an igneous rock (*Reid et al.,* 1977). Its occurrence as the dominant impact melt rocks at the Apollo 15 and 17 sites near the rims of major basins led *Ryder and Wood* (1977) and *Spudis* (1984) to propose that LKFM melts are produced mostly by basin-forming impacts. LKFM composition is more mafic than the feldspathic upper crustal composition observed in most of the highlands, leading those authors to suggest that LKFM melts represent lower crustal composition.

There is no single dominant LKFM composition at the Apollo 15 site where the Apennine Front has approximately LKFM bulk composition. Instead there are a number of separate melt rocks each with generally similar LKFM compositions. Table 3 gives the major element composition and Sc and Sm concentrations of LKFM glass, Apollo 15 KREEP basalt, and averages for each of the major Apollo 15 impact melt groups. The impact melt groups include KREEP-poor impact melts described here, three LKFM impact melts (B-C-D), and one KREEP basalt melt (A). All of the compositions are generally similar in being mafic rocks with 15-20% Al_2O_3, but they exhibit significant variations in mg′, TiO_2 and Na_2O, and major variations in K_2O and Sm. Variations in K_2O and Sm are assumed to represent variations in the proportions of a KREEP component. KREEP-poor impact melts have the lowest K_2O and Sm concentrations of Apollo 15 impact melts, and the K_2O content closest to that of LKFM glass. (Sm has not been measured in LKFM glass.) In fact, the 0.2-0.4% K_2O commonly assumed to be typical of LKFM melts (*Ryder and Spudis,* 1987) is distinctly higher than the range of about 0.05-0.2% K_2O in LKFM glasses as originally defined by *Reid et al.* (1972). We do not suggest, however, a return to the original definition of LKFM or that KREEP-poor impact melts be considered to be LKFM rocks, because KREEP-poor impact melt composition

TABLE 3. LKFM and APOLLO 15 impact melts.

Ref.	LKFM Glass 1	KPIM 2	GpD 2	GpC 2	GpB 2	GpA 2	KREEP Basalt 3
SiO_2	46.6	45.3	46.2	46.8	48.5	49.8	50.8
TiO_2	1.25	1.77	1.59	0.90	1.20	1.88	2.13
Al_2O_3	18.8	19.9	18.0	19.1	15.7	16.1	15.4
FeO	9.67	7.82	9.01	7.8	9.6	10.4	10.1
MgO	11.0	10.7	14.7	11.6	13.8	8.5	9.0
CaO	11.6	12.0	10.4	11.3	11.0	10.3	9.2
Na_2O	0.37	0.62	0.55	0.57	0.58	0.79	0.80
K_2O	0.12	0.10	0.17	0.25	0.34	0.79	0.58
mg′	67	71	74	72	73	63	65
Sc	—	17.4	16.8	15.0	17.2	20.8	20.5
Sm	—	4.5	9.8	13.6	20.4	35.8	32.0

References: 1. *Reid et al.* (1972). 2. KPIM mean from Table 1. Means of other impact melt groups based on data of *Ryder and Spudis* (1987), *Laul et al.* (1988), *Lindstrom et al.* (1988), *Ryder et al.* (1988). 3. KREEP basalt mean of 15382, 15386 (*Ryder,* 1985).

does not match LKFM composition in detail any better than do those of other Apollo 15 impact melts. It is richer in TiO_2, Na_2O, and more magnesian than LKFM. LKFM has come to mean a noritic impact melt with low to moderate K_2O and a KREEP-dominated REE pattern at moderately low REE concentrations. In this sense KREEP-poor impact melt is not a LKFM melt because its REE pattern is distinct from that of KREEP in its flat slope in HREE. KREEP-poor impact melt is the low-KREEP member of a series of Apollo 15 mafic impact melts.

Variations in other elements do not correlate so well with a single component. For example, mg′ is distinctly higher in KREEP-poor impact melts and all three LKFM melts than in either LKFM glass or KREEP basalt. The impact melts have a higher proportion of a magnesian mafic component. TiO_2 concentrations are quite variable. TiO_2 is high in KREEP-poor impact melts and KREEP basalt, and low in LKFM impact melts having intermediate amounts of KREEP component. This is illustrated in Fig. 5, which plots Ti and Sm for compositions in Table 3 plus typical impact melts from other highland sites. Overall there is no strong correlation in the data, but six of the impact melts do form a linear trend showing a negative correlation from KREEP-poor impact melts to Apollo 14-16(1) melts. If KREEP were the major source of Ti in LKFM impact melts, as has previously been proposed (*Ryder,* 1979), there should be a positive correlation between Ti and Sm for LKFM melts. There appear to be at least two high Ti components

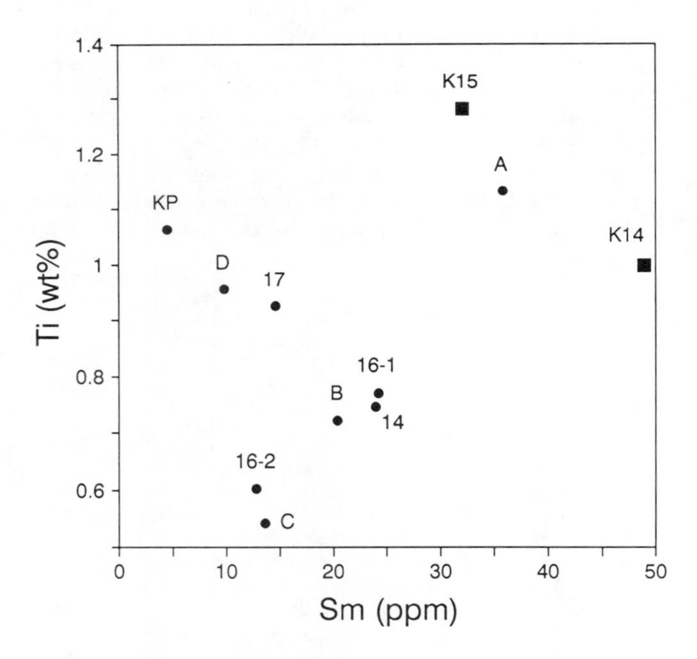

Fig. 5. Ti vs. Sm in highland impact melts. Circles are impact melts, squares are KREEP. Data for Apollo 15 impact melts are given in Table 3 and identified by letters A-D (*Ryder and Spudis,* 1987) and KP for KREEP-poor impact melts. Impact melts from other sites are identified by site number and group number for Apollo 16. Data are for samples cited in *Korotev* (1987b). Apollo 14 KREEP data are from *Warren and Wasson* (1979).

in lunar highlands rocks. KREEP-poor impact melts have the highest Ti contents and lowest Sm concentration (and hence KREEP component) of highlands impact melts and provide the best evidence for the high-Ti low-KREEP component.

Korotev (1987a) discussed compositional trends in Apollo 15 soils and evaluated the wide variety of mixing components used to model soil composition. He showed that the composition of the most mare-basalt-free soil, which he called the Apennine Front soil component (AFSC), is a suitable mafic component for modeling soil compositions. AFSC has a bulk composition similar to LKFM with REE concentrations between groups C and D LKFM impact melts (Table 3). He also concluded that mixing models using only local igneous rocks are geologically unacceptable because they require 20% mare basalt and 30% KREEP basalt in Apennine Front soils. Models using group D LKFM impact melts as a mafic component require another more ferroan mafic component to match mg′ and Ti and Sc concentrations. Using KREEP-poor impact melts as the mafic component in these calculations may resolve some of these problems because it is both higher in Ti and lower in mg′.

Using the compositions of polymict breccias and soils such as AFSC, LKFM, or KREEP-poor impact melts as mixing components tells us something about near-surface mixing processes. In order to use highland breccias to learn about crustal igneous rocks we need to evaluate their igneous components. Previous attempts to model the igneous components of lunar highlands breccias, including LKFM impact melts, concluded that they consist of an anorthositic component, a KREEP component, a KREEP-poor mafic component, and a meteorite component. Attempts to model the compositions in detail have been generally unsatisfactory because of the difficulty in identifying the KREEP-poor mafic component. *Ryder* (1979) discussed the components of highland breccias of various compositions. He rejected the postulated mafic components of *Taylor and Bence* (1975) and *Wänke et al.* (1977) and the empirically-derived component of *Wasson et al.* (1977), using instead actual highland mafic rocks. He concluded that plutonic norites are the dominant mafic component, with dunites or troctolites representing a minor component. He noted a major problem in matching the Ti and Sc contents of highland breccias using either hypothetical or Mg-suite mafic components. He rejected the possibility that mare basalt is the major source of these transition metals, and concluded that KREEP in varying forms is the major source of Ti and Sc in highland breccias. Although we agree with most of Ryder's observations and conclusions, we disagree with this last conclusion. KREEP-poor impact melts have the highest Ti contents of highland impact melts, yet they also have the smallest KREEP component. A non-KREEPy mafic component is the major source of Ti in KREEP-poor impact melts.

Meaningful quantitative modeling of the igneous components of KREEP-poor impact melts is impossible because no appropriate high-Ti highland igneous rock has been identified. However, we can set some limits on various components. The mafic bulk composition and low incompatible element concentrations of KREEP-poor impact melts suggest that it is strongly dominated by its mafic highland component. The amounts of all other components are small. A very small amount of a meteorite component is indicated by the siderophile element concentrations. We cannot exclude a small amount of mare basalt, but this is limited to only a few percent by the moderately magnesian (mg′ 71) bulk composition, and the identification of a high-Ti highland component would eliminate the need for mare basalts in highland impact melts. The amount of KREEP component is limited to only a few percent by the low incompatible-element concentrations and non-KREEP REE patterns. The presence of large calcic plagioclase clasts in the melt indicates that there is a minor anorthositic component, but it too is limited to a small proportion by the mafic bulk composition. The calcic plagioclase clasts may be relicts of anorthositic rocks added as the mafic melt cooled. This is suggested by *McCormick et al.* (1989), who conclude from a study of clasts in terrestrial impact melts that LKFM melts may incorporate most of their clasts in the late stages of the cratering process, perhaps during emplacement.

The mafic component of KREEP-poor impact melts is largely noritic or gabbroic. Using observed highland plutonic rocks we suggest that KREEP-poor impact melts are dominated by a Mg-norite component, with small amounts of Mg-troctolites and gabbronorites providing sources of olivine and augite not available in Mg-norites. The Ti-rich igneous source remains unknown. Alternately, KREEP-poor impact melts may consist largely of an unsampled gabbroic lithology in which all minerals are present in a single lithology. The mode of that lithology might be similar to that of KREEP-poor impact melts that consist of approximately 56% plagioclase, 36% pyroxene, 6% olivine, and 2% ilmenite, but the actual mode of this unsampled gabbroic lithology would probably be more mafic than KREEP-poor impact melts because anorthositic clasts are present in the bulk samples used to derive this mode. *McCormick et al.* (1989) also suggest that an unsampled rock type(s) that is the dominant component of LKFM melts exists in the lower crust. They describe that unsampled rock type as having relatively high mg′ (>70), high concentrations of transition metals (Ti > about 1.5%) and variable KREEP contents. KREEP-poor impact melt is the highland melt rock with the highest proportion of this unsampled mafic rock type and provides the best source of clues to the nature of the Ti-rich highland component.

Evaluation of the provenance of KREEP-poor impact melts is model dependent. Their mafic bulk composition and absence (or near-absence) of KREEP and mare basalt components suggest that they are not mixtures of surficial components. They are very similar to LKFM impact melts and may therefore represent the melt from a basin-sized impact into the lower crust. Their low incompatible element concentrations distinguish them from the LKFM Apollo 15 and 17 impact melts thought to represent Imbrium and Serenitatis melts (*Ryder and Wood,* 1977; *Ryder and Spudis,* 1987). In the Apennine Front cross section of *Spudis* (1980) we suggest that KREEP-poor impact melts represent a mafic melt from an unidentified impact that is now part of the pre-Serenitatis section. The acquisition of an age on the large KREEP-poor impact melts clast may help to evaluate their provenance.

Acknowledgments. C. Galindo and K. Willis of the JSC-LESC curatorial staff carefully extracted the breccia clasts. This work was supported by NASA's Planetary Materials and Geochemistry Program through RTOP 152-13-40-21 to M.M.L. and grant NAG 9-29 to U.B.M. O. James and R. Korotev provided thoughtful reviews of the manuscript.

REFERENCES

Albee A. L., Gancarz A. J., and Chodos A. A. (1973) Metamorphism of Apollo 16 and 17 and Luna 20 metaclastic rocks at about 3.95 AE: Samples 61156, 64423,14-2, 65015, 67483,15-2, 76055, 22006, and 22007. *Proc. Lunar Sci. Conf. 4th,* pp. 569-595.

Bence A. E., Papike J. J., Sueno S., and Delano J. W. (1973) Pyroxene poikiloblastic rocks from the lunar highlands. *Proc. Lunar Sci. Conf. 4th,* pp. 597-611.

Cameron K. L., Delano J. W., Bence A. E., and Papike J. J. (1973) Petrology of the 2-4 mm soil fraction from the Hadley-Apennine Region of the Moon. *Earth Planet. Sci. Lett., 19,* 9-21.

Korotev R. L. (1987a) Mixing levels, the Apennine Front soil component, and compositional trends in the Apollo 15 soils. *Proc. Lunar Planet. Sci. Conf. 17th,* in *J. Geophys. Res., 92,* E411-E431.

Korotev R. L. (1987b) The meteorite component of Apollo 16 noritic impact breccias. *Proc. Lunar Planet. Sci. Conf. 17th,* in *J. Geophys. Res., 92,* E491-E512.

Laul J. C., Simon S. B., and Papike J. J. (1988) Chemistry and petrology of the Apennine Front, Apollo 15, Part II: Impact melt rocks. *Proc. Lunar Planet. Sci. Conf. 18th,* pp. 203-217.

Lindstrom M. M. and Lindstrom D. J. (1986) Lunar granulites and their precursor anorthositic norites of the early lunar crust. *Proc. Lunar Planet. Sci. Conf. 16th,* in *J. Geophys. Res., 91,* D263-D276.

Lindstrom M. M., Marvin U. B., Vetter S. K., and Shervais J. W. (1988) Apennine front revisited: Diversity of Apollo 15 highland rock types. *Proc. Lunar Planet. Sci. Conf. 18th,* pp. 169-185.

Lindstrom M. M., Marvin U. B., and Mittlefehldt D. W. (1989a) Apollo 15 Mg- and Fe-rich norites: A redefinition of the Mg-suite differentiation trend. *Proc. Lunar Planet. Sci. Conf. 19th,* pp. 245-254.

Lindstrom M. M., Marvin U. B., Holmberg B. B., and Mittlefehldt D. W. (1989b) Geochemistry and petrology of recrystallized gabbroic breccias from the Apollo 15 site (abstract). In *Lunar and Planetary Science XX,* pp. 576-577. Lunar and Planetary Institute, Houston.

McCormick K. A., Taylor G. J., Keil K., Spudis P. D., Grieve R. A. F., and Ryder G. (1989) Sources of clasts in terrestrial impact melts: Clues to the origin of LKFM. *Proc. Lunar Planet. Sci. Conf. 19th,* pp. 691-696.

McKay D. S., Bogard D. D., Morris R. V., Korotev R. L., Wentworth S. J., and Johnson P. (1989) Apollo 15 regolith breccias; Window to a KREEP regolith. *Proc. Lunar Planet. Sci. Conf. 19th,* pp. 19-41.

Phinney W. C., Warner J. L., Simonds C. H., and Lofgren G. E. (1972) Classification and distribution of rock types at Spur Crater. In *The Apollo 15 Lunar Samples* (J. W. Chamberlain and C. Watkins, eds.), pp. 149-153. The Lunar Science Institute, Houston.

Powell B., Aitken F., and Weiblen P. (1973) Classification, distribution, and origin of lithic fragments from the Hadley-Apennine region. *Proc. Lunar Sci. Conf. 4th,* pp. 445-460.

Reid A. M., Warner J., Ridley W. I., and Brown R. W. (1972) Major element composition of glasses in three Apollo 15 soils. *Meteoritics, 7,* 395-415.

Reid A. M., Duncan A. R., and Richardson S. H. (1977) In search of LKFM. *Proc. Lunar Sci. Conf. 8th,* pp. 2321-2338.

Ridley W. I. (1977) Some petrologic aspects of Imbrium stratigraphy. *Philos. Trans. R. Soc. Lond., A285,* 105-114.

Ryder G. (1979) The chemical components of highlands breccias. *Proc. Lunar Planet. Sci. Conf. 10th,* pp. 561-581.

Ryder G. (1985) *Catalog of Apollo 15 rocks, Parts 1, 2, and 3.* Curatorial Branch Publication 72, JSC 20787. 1296 pp.

Ryder G. and Bower J. F. (1977) Petrology of the Apollo 15 black-and-white rocks 15445 and 15455-fragments of the Imbrium melt sheet. *Proc. Lunar Sci. Conf. 8th,* pp. 1895-1923.

Ryder G. and Spudis P. (1987) Chemical composition and origin of Apollo 15 impact melts. *Proc. Lunar Planet. Sci. Conf. 17th,* in *J. Geophys. Res., 92,* E432-E446.

Ryder G. and Wood J. A. (1977) Serenitatis and Imbrium impact melts: Implications for large-scale layering in the lunar crust. *Proc. Lunar Sci. Conf. 8th,* pp. 655-668.

Ryder G., Lindstrom M., and Willis K. (1988) The reliability of macroscopic identifications of lunar coarse-fines particles and the petrogenesis of 2-4 mm particles in Apennine Front sample 15243. *Proc. Lunar Planet. Sci. Conf. 18th,* pp. 219-232.

Simon S. B., Papike J. J., Gosselin D. C., and Laul J. C. (1986) Petrology, chemistry, and origin of Apollo 15 regolith breccias. *Geochim. Cosmochim. Acta, 50,* 2675-1591.

Simon S. B., Papike J. J., and Laul J. C. (1988) Chemistry and petrology of the Apennine Front, Apollo 15, Part II: KREEP basalts and plutonic rocks. *Proc. Lunar Planet. Sci. Conf. 18th,* pp. 187-201.

Simonds C. H., Warner J. L., and Phinney W. C. (1973) Petrology of Apollo 16 poikilitic rocks. *Proc. Lunar Sci. Conf. 4th,* pp. 613-632.

Simonds C. H., Phinney W. C., and Warner J. L. (1974) Petrography and classification of Apollo 17 non-mare rocks with emphasis on samples from the Station 6 boulder. *Proc. Lunar Sci. Conf. 5th,* pp. 337-353.

Smith J. V. and Steele I. M. (1976) Lunar mineralogy: A heavenly detective story. *Am. Mineral., 61,* 1059-1116.

Spudis P. D. (1980) Petrology of the Apennine Front, Apollo 15: Implications for the geology of the Imbrium impact basin. In *Papers Presented to the Conference on Multi-ring Basins,* pp. 83-85. Lunar and Planetary Institute, Houston.

Spudis P. D. (1984) Apollo 16 site geology and impact melts: Implications for the geologic history of the lunar highlands. *Proc. Lunar Planet. Sci. Conf. 15th,* in *J. Geophys. Res., 89,* C95-C107.

Spudis P. D. and Ryder G. (1985) Geology and petrology of the Apollo 15 landing site: past, present and future understanding. *Eos Trans. AGU, 66,* 721-726.

Stöffler D., Knoll H.-D., Marvin U. B., Simonds C. H., and Warren P. H. (1980) Recommended classification and nomenclature of lunar highland rocks—a committee report. In *Proceedings of the Conference on the Lunar Highlands Crust* (J. J. Papike and R. B. Merrill, eds.), pp. 51-70. Pergamon, New York.

Swann G. A. et al. (1972) Preliminary geologic investigation of the Apollo 15 landing site. In *Apollo 15 Preliminary Science Report,* pp. 5-1 to 5-112. NASA SP-289.

Taylor S. R. and Bence A. E. (1975) Evolution of the lunar highland crust. *Proc. Lunar Sci. Conf. 6th,* pp. 1121-1141.

Wänke H., Baddenhausen H., Blum K., Cendales M., Dreibus G., Hofmeister H., Kruse H., Jagoutz E., Palme H., Spettel B., Thacker R., and Vilesek E. (1977) On the chemistry of lunar samples and achondrites. Primary matter in the lunar highlands: A re-evaluation. *Proc. Lunar Sci. Conf. 8th,* pp. 2191-2213.

Warner J. L., Phinney W. C., Bickel C. E., and Simonds C. H. (1977) Feldspathic granulitic impactites and pre-final bombardment lunar evolution. *Proc. Lunar Sci. Conf. 8th,* pp. 2051-2066.

Warren P. H. and Wasson J. T. (1979) The origin of KREEP. *Rev. Geophys. Space Phys., 17,* 73-88.

Wasson J. T., Warren P. H., Kallemeyn G. W., McEwing C. E., Mittlefehldt D. W., and Boynton W. V. (1977) SCCRV, a major component of the highlands rocks. *Proc. Lunar Sci. Conf. 8th,* pp. 2237-2252.

Proceedings of the 20th Lunar and Planetary Science Conference, pp. 91-100
Lunar and Planetary Institute, Houston, 1990

Mineralogical Comparison of the Y86032-type Lunar Meteorites to Feldspathic Fragmental Breccia 67016

H. Takeda[1], M. Miyamoto[2], H. Mori[1], S. J. Wentworth[3], and D. S. McKay[4]

We investigated lunar meteorites Y82193 and Y86032 and Apollo 16 feldspathic fragmental breccia 67016 by mineralogical techniques to gain a better understanding of the origin and evolution of the lunar meteorites and the feldspathic fragmental breccias. The mineralogical characteristics and thermal histories of 67016 and the Y86032-type lunar meteorites (which include Y82192, Y82193, and Y86032) are very similar. They are feldspathic fragmental breccias with granulitic clasts, clast-laden vitric breccia clasts, devitrified glasses, and comminuted mineral fragments. These samples, especially 67016, contain little evidence of a regolith component. On a transmission electron microscopy scale, the fine-grained matrix of 67016,304 is composed of monomineralic fragments of plagioclase, olivine, and pyroxene with glassy interstitial materials. The high porosity of the matrix indicates that 67016 has not been heated enough to drive off noble gases, so the low solar wind component suggests that 67016 precursor material was never exposed at the lunar surface. Because the Y86032-type meteorites contain a small regolith component, indicated by the presence of regolith-derived spherules, these meteorites may be intermediate between feldspathic fragmental breccias, which lack a regolith component, and true regolith breccias. Some pyroxene fragments in Y86032 have exsolution and inversion textures typical of plutonic rocks. Cooling rates for granulitic clasts in 67016 and Y82193 were estimated from chemical zoning of olivines. Olivine Mg/Fe ratios are constant, but Ca zoning with Ca enrichment towards the rims has been preserved. Cooling rates from 1000°C, estimated from Ca zoning profiles for 50-μm-sized grains, are between 0.3°C/day and 3° C/day. The parental materials of the granulitic clasts may have been initially quickly cooled, regardless of their subsequent complex thermal histories.

INTRODUCTION

All known lunar meteorites have been identified as regolith breccias of highlands origin (e.g., *Ryder and Ostertag,* 1983; *Ostertag et al.,* 1986; *Takeda et al.,* 1988a). The meteorites have many compositional and petrological similarities to highlands regolith breccias collected at the lunar surface, particularly to the Apollo 16 regolith breccias (*Bischoff et al.,* 1987).

We have previously compared lithic clasts and mineral fragments in lunar meteorites Y791197, Y82192, Y82193, and ALH81005 with those in Apollo 16 regolith breccia 60019 (*Takeda et al.,* 1988a). The most noteworthy differences in terms of lithic clast types are higher abundances of granulitic clasts in the lunar meteorites and a higher abundance of poikilitic clasts in 60019. In addition, Y791197 and ALH81005 contain VLT (very low Ti)-type mare basalt (*Ryder and Ostertag,* 1983; *Treiman and Drake,* 1983; *Lindstrom et al.,* 1986; *Takeda et al.,* 1986). The mineralogical characteristics of highlands regolith breccias have been detailed in our study of 60016 (*Takeda et al.,* 1979). The compositions of pyroxene fragments in 60016 are distributed over a wide range, covering almost all known pyroxenes in nonmare pristine rocks (*Warren et al.,* 1983; *Ryder and Norman,* 1978a,b). Pyroxenes in the lunar meteorites show similar features (*Takeda et al.,* 1988a).

Three of the lunar meteorites (Y82192, Y82193, and Y86032) are very closely related. Large (648 g) meteorite Y86032 was recovered from the area where Y82192 and Y82193 were found (*Yanai and Kojima,* 1987). From the cosmic-ray exposure age and trapped noble gas ratios, *Eugster et al.* (1988) concluded that Y86032 is paired with Y82192/3. Previous workers (*Bischoff et al.,* 1987; *Eugster et al.,* 1988; *Takeda et al.,* 1987a, 1988b) have noted that these Y86032-type meteorites have some affinities to feldspathic fragmental breccias. Many Apollo 16 feldspathic fragmental breccias are similar to the Apollo 16 regolith breccias with respect to petrology and composition, except that the feldspathic fragmental breccias do not contain an identifiable regolith component. The Y86032-type meteorites contain very little evidence of a regolith component, such as agglutinates and spheres (*Bischoff et al.,* 1987; *Takeda et al.,* 1987a). In addition, their abundances of trapped solar wind gases are extremely low (*Eugster and Niedermann,* 1988; *Eugster et al.,* 1988). Trapped gas abundances are even lower than those in the ancient (~4 G.y.) Apollo 16 regolith breccias, which are very immature (*McKay et al.,* 1986). Therefore, the Y86032-type lunar meteorites might be fragmental breccias never exposed to solar wind or they might be immature regolith breccias that may have lost part of their former contents of noble gases due to shock-metamorphism.

During the course of our cooperative comparative studies, it was noted that Y86032-type meteorites bear more petrographic resemblance to feldspathic fragmental breccia 67016 than to regolith breccia 60019 (D. S. McKay, personal communication, 1987; *Takeda and Miyamoto,* 1988). Feldspathic fragmental breccia 67016 was studied previously by *Norman* (1981) and *Lindstrom and Salpas* (1983) but these studies did not compare 67016 with lunar meteorites.

[1]Mineralogical Institute, Faculty of Science, University of Tokyo, Hongo, Tokyo 113, Japan
[2]College of Arts and Sciences, University of Tokyo, Komaba, Meguro-ku, Tokyo 153, Japan
[3]Lockheed, 2400 NASA Road 1, Houston, TX 77058
[4]Mail Code SN14, NASA Johnson Space Center, Houston, TX 77058

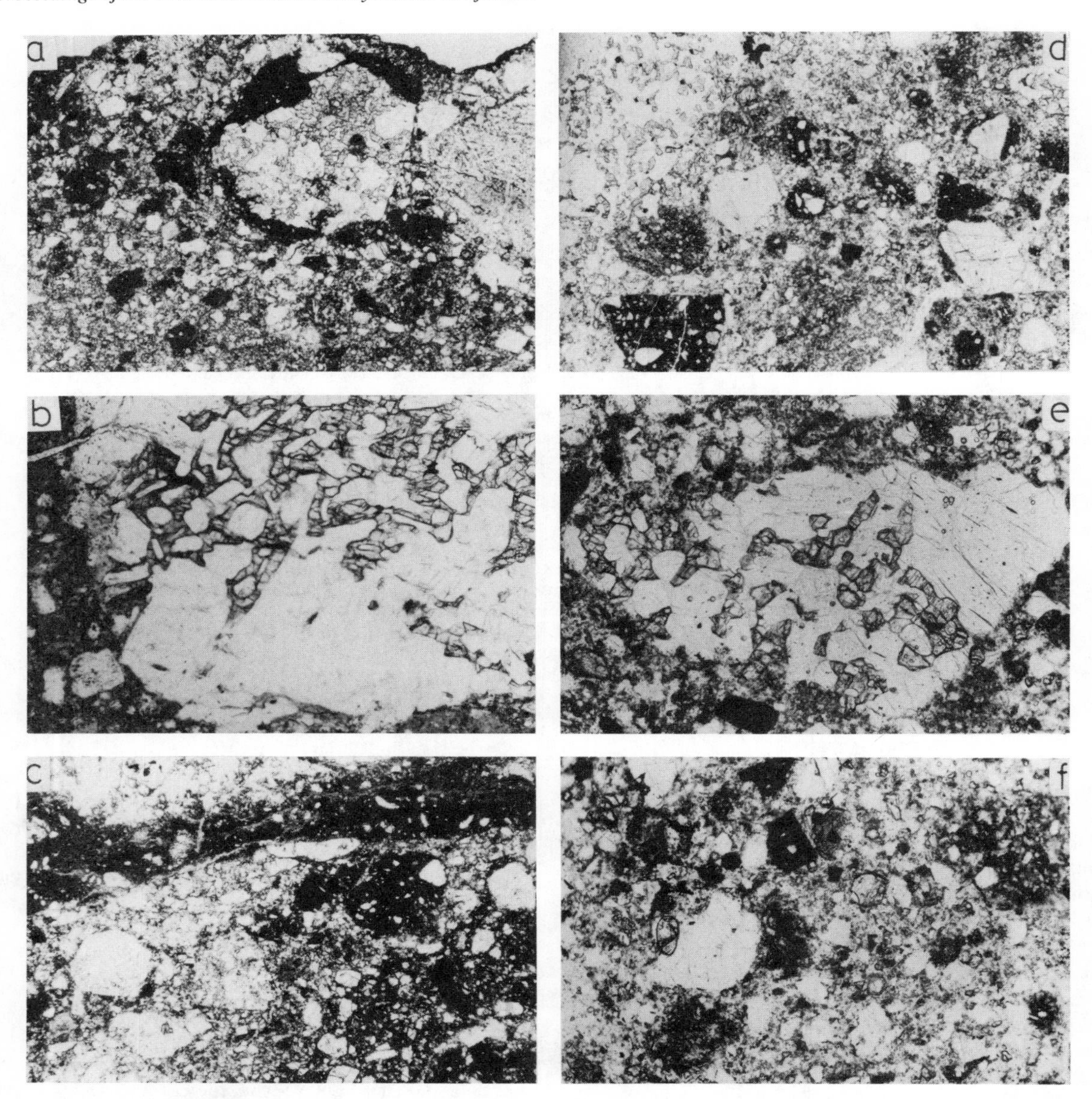

Fig. 1. Photomicrographs of clasts and matrices in feldspathic fragmental breccia 67016,111 and lunar meteorites Y82193 and Y86032; unpolarized light. **(a)** General view of Y82193,91-1. Two clasts are TR1 and GR2. Width = 3.3 mm. **(b)** Granulitic clast GR1 in Y82193,91-1. Width = 1.3 mm. **(c)** A portion of PTS Y86032,51-3. Width = 3.3 mm. **(d)** General view of 67016,111. Width = 3.3 mm. **(e)** Granulitic clast in 67016,111. **(f)** A matrix portion of PTS 67016,304. Width =1.3 mm.

We have investigated Apollo 16 feldspathic fragmental breccia 67016 and lunar meteorites Y82193 and Y86032 by mineralogical techniques, which include electron microprobe analysis and analytical transmission electron microscopy (TEM), in order to compare the feldspathic fragmental breccias to lunar meteorite counterparts and to gain a better understanding of the origin and evolution of lunar meteorites. This information will be of importance in understanding the unsampled lunar highlands and the evolution of the lunar crust including the farside.

SAMPLES AND EXPERIMENTAL TECHNIQUES

We studied thin section 67016,111, two chips of 67016 (,302 and ,304), and thin section Y82193,91-1, supplied by the National Institute of Polar Research (NIPR). Another thin section, Y86032,51-3, has been studied as a part of the consortium study (*Takeda et al.,* 1988b). The polished thin sections were examined by electron probe microanalyzer (EPMA) and photomicroscope. Small rock matrix fragments 67016,304 and ,302 and Y86032,110 and ,111 were studied by analytical TEM. Chemical analyses were made with a JEOL 8600 Super Probe at the Geological Institute of the University of Tokyo. Phases unmixed by exsolution in pyroxenes were examined by measuring the chemical compositions at 10-50-μm intervals.

We also measured Mg-Fe and CaO chemical zoning profiles of olivines in granulitic clasts in 67016 and Yamato 82193 with a JEOL electron microprobe (JCXA-733) at Ocean Research Institute, University of Tokyo. We measured zoning profiles selected by backscattered electron images. The acceleration voltage was 15 kV and beam current was 30 nA on a Faraday cage. Counting times at peak wavelengths for Si, Mg, Fe, and Ca were 10, 10, 10, and 60 sec, respectively. We counted the background intensity of each element at both sides of the peak wavelength for the same counting time as that of the peak intensity for every point on a profile.

The glass bulk compositons of the breccia matrices and matrices of glassy clasts of Y82193,91-3 and Y86032,51-3 were obtained by broad beam (5 μm) microprobe analyses (average of 5 to 10 spots per grain) to avoid vaporization of volatiles. These studies were performed as a part of the consortium study (*Takeda et al.,* 1988b).

Glass data were obtained at JSC from one polished thin section of Y82193 (section 91-1) and several thin sections of 67016. Techniques included optical petrography and scanning electron microscopy (with a JEOL 35 CF SEM). Major element compositions of glass clasts were determined with a Cameca Camebax electron microprobe, using standard operating and data reduction procedures as described in *Wentworth and McKay* (1988b). Glass data for comparison are from the Apollo 16 regolith breccias and soils (*Wentworth and McKay,* 1988b).

We also investigated glassy matrix in small chips of 67016,304, Y86032,110, and Y86032,111 with a Hitachi H-600 analytical TEM equipped with a Kevex 8000 system, which is capable of analyzing the chemical composition, texture, and atomic arrangements of regions as small as 800 Å. The method is the same as that for ALHA81005 (*Takeda et al.,* 1986). The chip mounted in resin was sliced into two sections. One slice was used for the EPMA study and another was polished to a thickness of about 10 μm. The sample was glued to a 3-mm molybdenum TEM grid for support and then thinned in a GATAN ion-thinning machine until perforation occurred. Examination of microtextures of the sample was carried out by the analytical TEM.

RESULTS

Y82193

Thin section Y82193,91-1 consists of granulitic clasts, clast-laden vitric (devitrified) breccia clasts, and fragments of plagioclase, pyroxene, olivine, and devitrified glass, set in a matrix of comminuted mineral and rock fragments with some glassy materials between the fragments (Fig. 1a). The matrix of the whole breccia is compact but not as compact as that of regolith breccia 60019. Shock features, such as undulatory extinction (present in all clasts) and fracturing, are common in Y82193,91-1 (*Bischoff et al.,* 1987; *Takeda et al.,* 1987a). This texture is that of a typical feldspathic fragmental breccia, similar to Y82192 (*Takeda et al.,* 1987a) and the Apollo 16 regolith breccias (*McKay et al.,* 1986). The texture is different from that of the previously studied thin section Y82193,91-4 (*Takeda et al.,* 1987a), however, which is enriched in glassy matrix and contains fewer granulitic clasts. Thin section Y82193,91-1 contains no obvious agglutinates, but it does contain rare spheres and possible sphere fragments; therefore, this sample is just barely a regolith breccia according to standard petrographic criteria (*Stöffler et al.,* 1979), which require the presence of agglutinates or spheres.

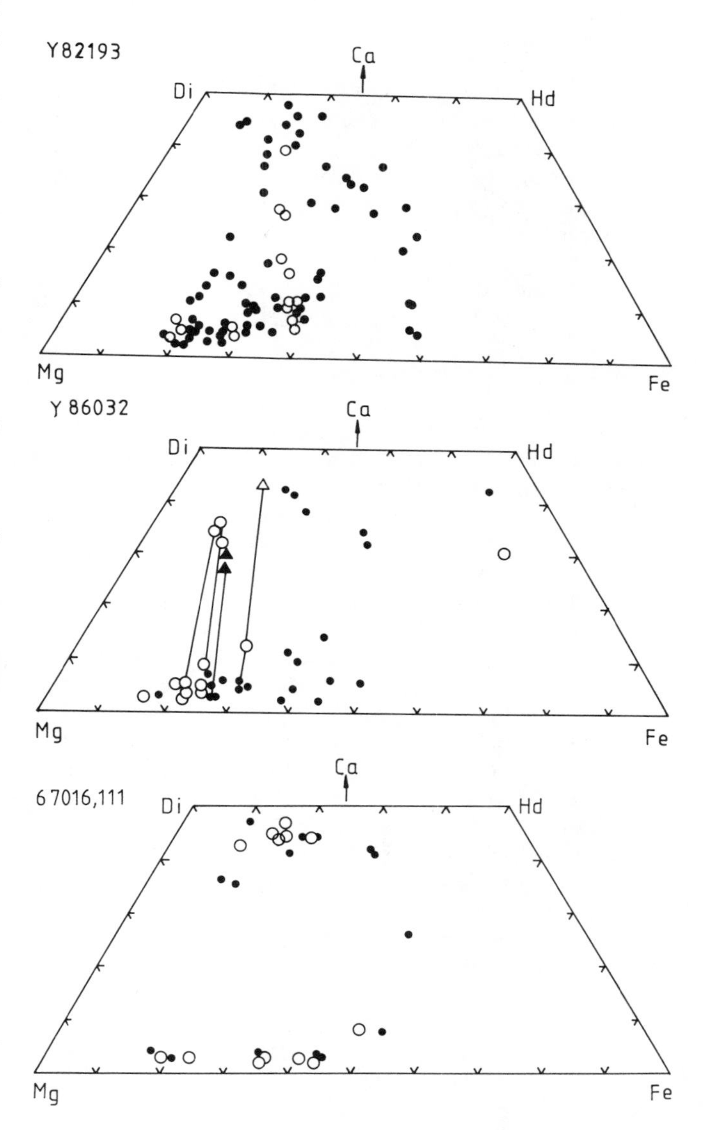

Fig. 2. Pyroxene quadrilaterals for 67016,111, Y82193,91-1, and Y86032. Open circles = clasts; solid circles = fragments in matrix; triangles = augite exsolution. Tie line connects coexisting or exsolved pairs.

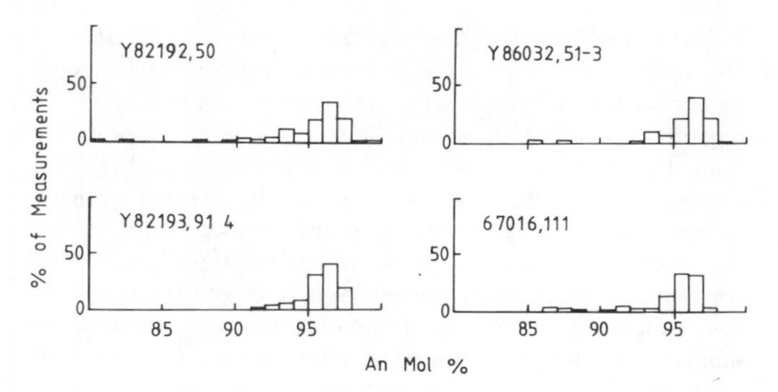

Fig. 3. Histograms of plagioclase An mol% distributions in 67016,111 and lunar meteorites Y82192,50, Y82193,91-4, and Y86032,51-3.

The largest clast (AN1), 1.9 × 1.3 mm in size, contains numerous small rounded crystals of mafic silicates, less than 0.02 mm in diameter, aligned along lath-shaped plagioclase crystals of An_{94} (Fig. 1a). The mafic silicates are mostly olivine (Fa_{20}) and minor pyroxene. A small clast of this kind contains more mafic particles.

Two other granulitic clasts (GR1 and GR2) are more coarse-grained crystalline breccias. GR1 is 1.8 × 1.2 mm in size. Distribution of mafic silicates is not uniform and a portion of the clast is all plagioclase (Fig. 1b). The texture is subpoikilitic and is not igneous, and the matrix is not granoblastic. The mafic silicates are mostly olivines (Fa_{34}) and minor pyroxenes ($Ca_5Mg_{67}Fe_{28}$). Plagioclase compositions range from An_{94} to An_{97}. The olivine is uniform in Mg/Fe but Ca zoning is still preserved. GR2 is another coarse-grained granulitic clast, 1.36 × 1.21 mm in size; it is glass-coated (Fig. 1a). Large poikilitic pyroxenes and olivines are present and plagioclase is more rounded and equant, and less granulitic. The amounts of olivine, pyroxene, and plagioclase are nearly equal. The olivine composition is Fa_{26}, pyroxene is $Ca_{4.7}Mg_{72}Fe_{23.3}$, and plagioclase is An_{96}. The texture rather resembles that of spinel troctolite but without spinel. The shock texture is not well annealed.

The compositions of pyroxenes and plagioclase in Y82193 are summarized in Figs. 2 and 3, which also show summaries of pyroxene and plagioclase compositions for other lunar meteorites and feldspathic fragmental breccia 67016. Selected pyroxene, olivine, and plagioclase compositions are given in Table 1. Compositions of coexisting plagioclase, olivine, and low-Ca pyroxene in lithic clasts are given in Fig. 4.

Clast-laden vitric clasts are smaller than the granulitic clasts, up to 0.63 × 0.38 mm in size. They are shock melted, compact, nonvesicular, and devitrified. A few fractures are present and a few of the clasts have incipient flow features.

All of the glass clasts in Y82193,91-1 are at least partly recrystallized and some are clast-bearing; no homogeneous glasses were found in this sample. One clast has a bead-like morphology. We used a compositional classification (*Wentworth and McKay,* 1988b), which was modified from the classification of *Reid et al.* (1972). All 31 glass clasts analyzed in Y82193,91-1 have highlands compositions. Average compositions for the glass types are given in Table 2. Most of

TABLE 1a. Chemical compositions (wt.%) of pyroxenes in Y82193 and 67016.

Sample No.	Y82193		67016			
Clasts	GR1	GR2	GR1	GR2		GB1
Phases	Opx	Opx	Opx	Opx	Aug	Aug
SiO_2	52.9	53.9	54.3	52.0	51.9	50.4
TiO_2	0.91	0.71	0.88	0.21	0.37	2.46
Al_2O_3	0.97	0.93	1.19	0.44	0.63	2.93
FeO	16.71	14.52	11.97	26.7	10.82	6.30
MnO	0.38	0.33	0.27	0.32	0.15	0.13
MgO	23.2	27.8	29.5	19.42	13.78	15.78
CaO	2.94	1.42	1.70	0.86	21.8	20.7
Na_2O	0.03	0.02	0.00	0.02	0.04	0.17
Cr_2O_3	0.43	0.30	0.47	0.12	0.14	0.83
Total	98.47	99.93	100.28	100.09	99.63	99.70
Ca*	6.1	2.7	3.3	1.8	44.1	43.5
Mg	66.9	75.2	78.7	55.4	38.8	46.2
Fe	27.0	22.1	18.0	42.8	17.1	10.3

*Atomic percent.

TABLE 1b. Chemical compositions (wt.%) of olivines in Y82193 and 67016.

Samples	Y82193		67016			
Clasts	GR1	GR2	GR1	GR2	AN1	GB1
SiO_2	37.0	37.1	37.8	36.6	36.1	36.3
TiO_2	0.04	0.06	0.03	0.03	0.05	0.04
Al_2O_3	0.01	0.05	0.00	0.00	0.00	0.03
FeO	29.8	23.9	25.4	32.5	33.4	31.8
MnO	0.32	0.35	0.02	0.41	0.37	0.26
MgO	32.5	37.7	35.9	29.9	29.2	31.2
CaO	0.17	0.07	0.11	0.06	0.09	0.09
Cr_2O_3	0.02	0.03	0.00	0.00	0.01	0.03
Total	99.86	99.26	99.58	99.50	99.22	99.75
Fo*	66.0	73.8	71.6	62.1	60.9	63.6
Fa	34.0	26.2	28.4	37.9	39.1	36.4

*Atomic percent.

TABLE 1c. Chemical compositions (wt.%) of plagioclases in Y82193 and 67016.

Samples	Y82193		67016		
Clasts	GR1	GR2	GR1	GR2	GB1
SiO_2	44.1	44.1	44.6	43.7	44.6
TiO_2	0.00	0.02	0.01	0.04	0.05
Al_2O_3	35.7	35.3	35.6	36.0	35.2
FeO	0.06	0.09	0.04	0.15	0.14
MnO	0.05	0.05	0.00	0.00	0.00
MgO	0.07	0.08	0.11	0.00	0.05
CaO	19.63	19.47	19.20	19.80	18.90
Na_2O	0.38	0.39	0.70	0.39	0.68
K_2O	0.00	0.01	0.06	0.00	0.07
Total	99.99	99.51	100.32	100.08	99.69
An*	96.6	96.5	93.5	96.6	93.6
Ab	3.4	3.5	6.1	3.4	6.0
Or	0.0	0.0	0.4	0.0	0.4

*Atomic percent.

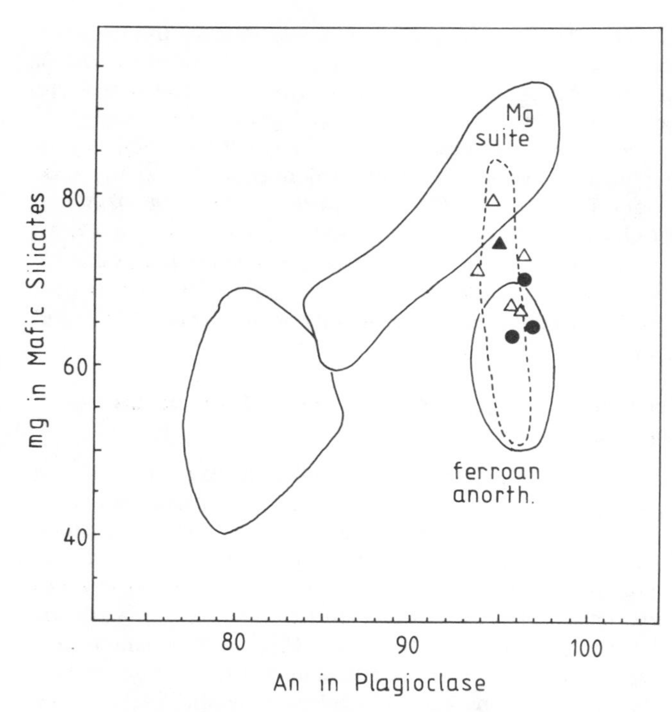

Fig. 4. Compositions of coexisting plagioclase, olivine, and low-Ca pyroxene in lithic clasts in Y82193 (open symbols) and 67016 (solid symbols). Mg-suite and ferroan anorthosite trends from *Lindstrom and Lindstrom* (1986). Triangles = olivine; circles = pyroxene.

the glasses are of highland basalt (i.e., feldspathic) composition, with $Al_2O_3 \geq 23$ wt.% and $K_2O < 0.25$ wt.%. A few glasses have low-K Fra Mauro (LKFM) compositions, which include glasses with $Al_2O_3 < 23$ wt.% and $K_2O < 0.25$ wt.%. One glass is of KREEP composition ($Al_2O_3 < 23$ wt.% and $K_2O \geq 0.25$ wt.%).

Y86032

A preliminary report on Y86032 has been published as a part of the consortium study (*Takeda et al.,* 1988a). The polished thin section of the representative portion of Y86032,51-3 has petrographic characteristics very much like those of Y82192/3. Granulitic clasts and clast-laden vitric (devitrified) breccia clasts are dominant (Fig. 1c). The mineral fragments consist of plagioclase, pyroxene, and olivine. The distribution of pyroxene compositions in the pyroxene quadrilateral is similar to those of Y82192/3 and 67016,111 (Fig. 2). One large fragment of inverted pigeonite in Y86032 contains blebby augites and another of orthopyroxene shows exsolution lamellae of augite on (100). They are from plutonic lunar crustal rocks. Plagioclase compositions (Fig. 3) are also much like those in Y82192/3 and 67016. Y86032 is characterized by brown clast-laden glassy veins, which penetrate into breccia matrices (Fig. 1c).

TEM observation of Y86032,110 and ,111 showed that the fine-grained matrix material of both samples is composed of micrometer-sized angular fragments of plagioclase and very

TABLE 2. Average compositions of glass clasts in Y82193,91-1.

	Highland Basalt		Low-K Fra Mauro		KREEP
SiO_2	44.2	(0.93)	44.8	(2.54)	47.2
TiO_2	0.18	(0.08)	0.22	(0.13)	1.52
Al_2O_3	27.8	(3.13)	19.4	(1.71)	18.7
Cr_2O_3	0.12	(0.06)	0.23	(0.11)	0.28
FeO	5.25	(2.02)	11.0	(1.88)	6.94
MnO	0.07	(0.04)	0.15	(0.02)	0.15
MgO	5.62	(2.70)	10.8	(2.86)	10.4
CaO	16.0	(1.55)	12.2	(0.89)	13.5
Na_2O	0.34	(0.16)	0.46	(0.21)	0.81
K_2O	0.02	(0.01)	0.03	(0.02)	0.32
P_2O_5	0.03	(0.01)	0.05	(0.03)	0.71
Total	99.6		99.3		100.5
CaO/Al_2O_3	0.58		0.63		0.72
Mg' (atomic)	0.65		0.63		0.73
No. of Grains	25		5		1

Standard deviations are given in parentheses.

fine-grained (submicrometer-sized) interstitial material. The very fine-grained material is mainly composed of plagioclase and minor mafic silicate minerals with a recrystallized texture. No glassy material was observed in the matrix.

67016

The 67016,111 polished thin section consists of impact melt breccia clasts with plagioclase fragments set in a dark devitrified glassy matrix, granulitic clasts with different mafic silicate/plagioclase ratios, fragments of shocked plagioclase, and mafic silicates (Fig. 1d). The texture is similar to those of other feldspathic fragmental breccias and of Y82192/3 and Y86032. The matrix of the entire breccia is not as glassy as that of regolith breccia 60019 (*Takeda et al.,* 1988a).

Granulitic clast GR1 is similar to granulitic clasts in Y82193 (Fig. 1e). The mafic silicates are mostly orthopyroxene ($Ca_{2.5}Mg_{78}Fe_{19.5}$). Plagioclase compositions range from An_{92} to An_{96} with a mean of $An_{93.5}$ (Table 1). The largest granulitic clast (GR2) includes frequently connected chains of subrounded mafic silicate grains up to 0.12 mm in diameter in plagioclase with a composition of An_{96}. Orthopyroxene ($Ca_2Mg_{55}Fe_{43}$) and augite ($Ca_{44}Mg_{38}Fe_{18}$) are present with a modal ratio of 5:3. Mafic silicates in other clasts (e.g., AN1) are less abundant and individual crystals are more rounded and isolated than those in GR2. The grain sizes of mafic silicates in one clast are only about 0.01 mm in diameter and are densely distributed in plagioclase. These features are similar to those observed in Y82193 (Fig. 1a). One clast (GB1), 0.63 × 0.42 mm in size, is gabbroic and consists of equigranular olivine (Fa_{36}), augite ($Ca_{43}Mg_{46}Fe_{10}$), and plagioclase (An_{90}-An_{94}). Fragments of such rocks are not rare. One noritic clast contains pyroxene with exsolution texture. Some anorthositic clasts (AN1) include several rounded olivine and pyroxene grains (Table 1).

The compositional distributions of the pyroxene fragments in matrices and of the pyroxene crystals in the above clasts in 67016 (Fig. 2) are similar to those of the Y86032 group, 60016, and 60019 (*Takeda et al.,* 1988a). Neither 67016 nor

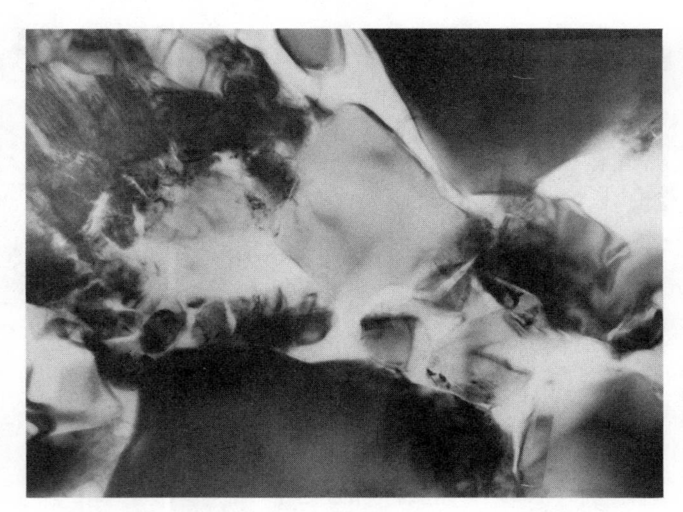

Fig. 5. Electron photomicrograph of matrix of 67016,304. White areas are voids. Width = 3 μm.

the Y86032 group contains extremely Mg-poor pyroxenes like those found in the VLT clasts in ALH81005 and Y791197 (*Treiman and Drake,* 1983; *Lindstrom et al.,* 1986; *Takeda et al.,* 1986). The presence of gabbroic clasts is particularly interesting because gabbro units in the lunar highlands have been inferred from remote-sensing data (*Lucey and Hawke,* 1987).

Plagioclase in 67016 is also similar to plagioclase in the 86032 group, as shown by histograms of An contents (Fig. 3). For crystalline lithic clasts in 67016, plagioclase has a narrow range in composition, from An_{92} to An_{97}. Mg numbers [$100 \times Mg/(Mg + Fe)$] for mafic silicates in 67016 crystalline clasts have a wide range, from 81 (pyroxene in GR1) to 56 (pyroxene in GR2). The plagioclase and pyroxene compositions indicate that clasts range from a magnesium-rich part of the Mg suite to an iron-rich part of the ferroan anorthosite trend, with some compositions falling between the suites. This trend is also similar to that for lunar granulites studied by *Lindstrom and Lindstrom* (1986), and to the lunar meteorites (*Lindstrom et al.,* 1986; *Takeda et al.,* 1987a). Granulitic clasts found in Y82193,91-1 and Y86032,51 are also similar.

The matrix sample of 67016,304 contains subangular fragments of plagioclase, pyroxene, and olivine set in much finer transparent plagioclase and fine light brown glassy materials (Fig. 1f). Modal abundances, obtained by a point analysis, are plagioclase 61%, pyroxene 11%, olivine 5%, and glass and mixtures 23%. The amount of glass is not as high as that in ALH81005, Y791197, or 60019. Some parts of the matrix have a finely recrystallized texture. TEM observation showed that the fine-grained matrix of 67016,304 is composed of mineral fragments of plagioclase, olivine, and pyroxene with glassy interstitial material. The matrix has high porosity and a less sintered, clastic appearance (Fig. 5).

Glass fragments are extremely rare in 67016. Only 23 fragments were found in 8 thin sections. Similar to glasses in Y82193, none of the 67016 glasses are optically homogeneous; all of them are crystalline (mostly aphanitic) and some contain mineral clasts. Spherules, which would indicate the presence of a regolith component, have not been identified in 67016. Average compositions of the glass types found in 67016 are given in Table 3. All the analyzed glasses have highlands compositions and most (18 out of 23) fall into the category of highland basalt. This composition is similar to the bulk composition of 67016 (*Ryder and Norman,* 1980). A few glasses (a total of four) have LKFM compositions. No KREEP glasses were found but one glass is enriched in silica (Table 3); this glass is probably related to the granitic clasts in 67016, previously noted (but not analyzed) by *Nord et al.* (1975) and *Norman* (1981).

Estimation of Cooling Rates from Chemical Zoning of Olivines

Chemical zoning of olivines in granulitic clasts in 67016 and Y82193 has been studied to estimate cooling rates and burial depths of some clasts in their parental materials. The method is the same as that used for the chondrite study (*Miyamoto et al.,* 1986; *Miyamoto,* 1987). As shown in Fig. 6, Mg/Fe ratios are constant, but Ca zoning with Ca enrichment towards the rim has been detected; that is, the Mg-Fe zoning is homogenized but the Ca zoning is not homogenized during cooling. These observations can be explained by the fact that the diffusion coefficient of Ca in olivine is smaller than that of Fe in olivine (*Buening and Buseck,* 1973; *Morioka,* 1981), and give some constraints on cooling histories of olivine.

There are coexisting orthopyroxene and augite in the clast, and the lowest temperature of equilibration estimated from their chemistries (*Lindsley and Anderson,* 1983) is a little below 700°C. The crystallization of primary orthopyroxene with mg number = 56 implies that the initial crystallization temperature is below 1000°C. The orthopyroxene and augite crystals do not show significant chemical zoning within experimental error. The diffusion coefficients and nature of chemical zoning in these rocks are not known well enough to determine cooling rates.

TABLE 3. Average compositions of glass clasts in 67016 thin sections.

	Highland Basalt		Low-K Fra Mauro		High Silica
SiO_2	44.8	(1.40)	45.9	(2.51)	80.4
TiO_2	0.32	(0.31)	0.82	(0.68)	0.57
Al_2O_3	29.5	(3.01)	21.2	(1.49)	11.5
Cr_2O_3	0.06	(0.02)	0.18	(0.04)	0.01
FeO	4.30	(2.03)	7.86	(2.59)	0.50
MnO	0.06	(0.03)	0.10	(0.02)	0.01
MgO	4.06	(2.09)	9.09	(2.87)	0.16
CaO	16.4	(1.65)	14.3	(2.45)	5.55
Na_2O	0.61	(0.14)	0.60	(0.19)	0.33
K_2O	0.07	(0.03)	0.10	(0.05)	0.60
P_2O_5	0.06	(0.10)	0.15	(0.13)	0.04
Total	100.3		100.3		99.7
CaO/Al_2O_3	0.56		0.68		0.48
Mg' (atomic)	0.62		0.67		0.36
No. of Grains	18		4		1

Standard deviations are given in parentheses.

The cooling rate to homogenize the Mg-Fe zoning in olivines in the clasts is slower than about 3°C/day and the cooling rate to preserve the Ca zoning is faster than 0.5°C/day. We have calculated burial depths in a sheet needed to homogenize the Mg-Fe zoning but to preserve the Ca zoning. We used two values of thermal diffusivity of the sheet. One is 0.004 cm^2/sec for rock-like material (*Horai and Winkler,* 1974); the other is 0.00001 cm^2/sec for regolith-like material (*Cremers and Hsia,* 1974). The maximum (initial) temperature of the sheet is assumed to be 1000°C and the ambient temperature is 200°K. Under the constraints that the Mg-Fe zoning is homogenized and the Ca zoning is not homogenized, we obtained a burial depth of about 3 m for rock-like material and about 20 cm for regolith-like material. Our calculations suggest that these olivines were buried in a shallow zone of a cooling unit. This result is consistent with the hypothesis that they may be an annealed product of an impact melt rock that was part of a hot ejecta blanket.

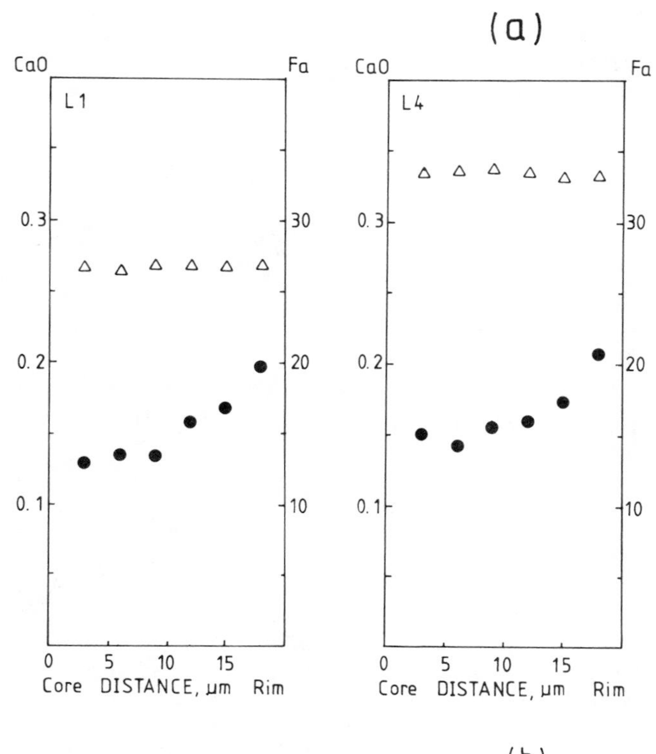

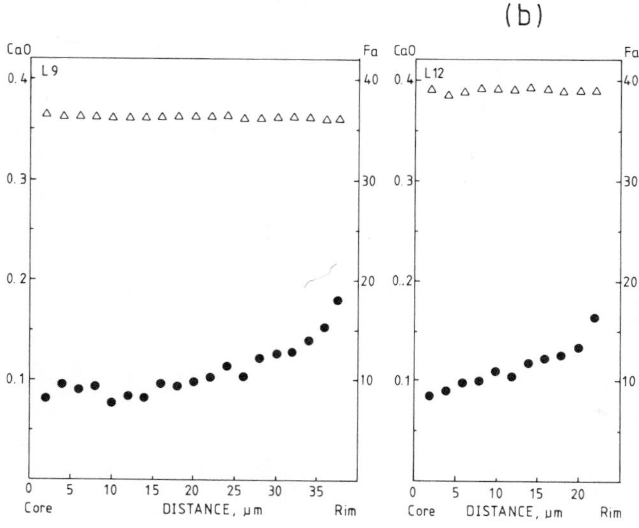

Fig. 6. Olivine chemical zoning profiles from core to rim of molecular percent Fa [Fe × 100/(Mg+Fe)] (triangles, scale on right) and weight percent CaO (solid circles, scale on left) for **(a)** Y82193 granulitic clasts; **(b)** 67016 clasts.

The thermal history of the clasts would be more complex than that assumed by our calculations. The clasts have experienced at least two thermal events, cooling of the original rock and cooling of the impact sheet into which they were placed. However, Ca zoning of olivines in the clasts survived the complex thermal episodes. We can say, therefore, that the clasts have not experienced any slow cooling events to homogenize the Ca zoning of olivines; for example, the clasts could not have been deeply buried in a hot ejecta blanket.

DISCUSSION

The relationships among feldspathic fragmental breccias, regolith breccias, and present-day lunar soils are not well known. Comparisons of clast compositions and populations are an important part of understanding the relationships of the lunar meteorites to the Apollo feldspathic fragmental and regolith breccias. In addition, the presence of a lunar meteorite-like breccia (67016) at the Apollo 16 site is useful in interpreting the impact sites of the Y86032-type lunar meteorites. Some differences between different polished thin sections of Y82193 and those between Y82192 and Y82193 can be explained by studying the heterogeneity of clasts and melt rocks within the new large lunar meteorite, Y86032 (*Yanai and Kojima,* 1987).

Among the Apollo breccias, the pyroxene mineralogy and chemistry of granulitic clasts in 67016 are the most like clasts in the Y86032-type lunar meteorites. Poikilitic clasts, dominant in 60019, are rare in 67016. Although the lunar meteorites and the lunar granulites generally are proposed to have originated from the farside of the Moon (*Treiman and Drake,* 1983; *Lindstrom et al.,* 1986; *Lindstrom and Lindstrom,* 1986), the preservation of such breccias on the nearside will be useful in interpreting the heterogeneity and origin of lunar meteorites.

Assuming that the premise of an old formation history for lunar granulites (*Warner et al.,* 1977; *Lindstrom and Lindstrom,* 1986) is correct, and because of the absence of poikilitic clasts dominant in Apollo 16 regolith breccias such as 60019, the lunar meteorites and 67016 might have formed before the basin-forming event that produced the poikilitic clasts in 60019 (*Takeda et al.,* 1988a). The preservation of Ca zoning in the granulitic olivine implies that an original rock or an impact deposit was cooled rapidly from high temperature, with little modification by later events. This process may correspond to crystallization of impact melt rocks after a basin-forming event. If this hypothesis is accepted, the granulitic texture, homogenization of Fe/Mg zoning, and the final low temperature deduced from the orthopyroxene-augite geothermometer, may be combined into a scenario in which prolonged annealing took place before these rocks became clasts in the present breccias. This possible scenario is in line with some basin-formation events, which probably occurred before the basin-forming event related to the poikilitic breccias such as those in 60019.

The differences between the microtextures of 67016,304 and Y86032 matrices suggest that the mechanism to form these breccias is different. The presence of recrystallized interstitial material, which gives hardness to the Y86032-type samples, indicates a higher temperature formation of the breccia matrix in comparison with 67016, which still contains amorphous interstitial material. The fine-grained (submicrometer-sized) amorphous material in 67016 indicates that it has not undergone intense heating, which would have induced devitrification. This suggests that the lack of solar wind gases in 67016 cannot be explained by a heating event intense enough to drive solar wind gases out of regolith or regolith breccia material and that the gases were never present. Therefore, it is unlikely that the 67016 materials were ever exposed at the lunar surface under conditions that permitted the acquisition of solar wind gases.

The compositions of glass clasts in feldspathic fragmental breccia 67016 and meteorite Y82193 are typical of lunar highlands glasses. The glass populations in 67016 and Y82193 are similar to each other in several ways. Both samples have glasses that are predominantly of highland basalt (feldspathic) composition, and neither sample contains homogeneous glass or mare glass. These similarities indicate that the samples underwent similar histories in the highlands regolith or megaregolith. There are also some distinct differences between the 67016 and Y82193 glasses, however. The biggest difference is that Y82193 contains a trace of spherules derived from the regolith, whereas definite regolith-derived glasses have not been identified in 67016. In addition, Y82193 contains more glass clasts than 67016. The small regolith component in Y82193 indicates that the Y86032-type meteorites may be transitional between the feldspathic fragmental breccias and true regolith breccias.

The ancient Apollo 16 regolith breccias, which are otherwise very similar to 67016, contain more abundant glasses with a wider compositional range (*Wentworth and McKay,* 1988b). Because of the similarities, it is possible that the ancient regolith breccias were derived from feldspathic fragmental breccia-type material, which subsequently underwent regolith exposure or was mixed with a regolith component. If the feldspathic fragmental breccias are precursors of the ancient regolith breccia suite at the Apollo 16 site, then 67016 may represent very old megaregolith. The possibly transitional Y86032-type meteorites may fill in the sampling gap and help us determine whether such a relationship exists between the feldspathic fragmental and regolith breccias.

In many respects, the lunar meteorites are consistently unlike the Apollo and Luna highlands samples, which are from a comparatively small central region of the nearside. Compared with the other lunar meteorites, Y82192/3 and Y86032 have low Ni contents; in addition, the Y86032-type meteorites have low contents of siderophile elements in general, and Au in particular, compared to Apollo and Luna highland soils (*Warren et al.,* 1989; *Koeberl et al.,* 1989). The low K, Th, U, Au, etc., contents of the lunar meteorites imply that Th, U, and other refractory lithophile elements, and siderophile elements such as Au, may be low on the farside (*Warren et al.*, 1989). Since KREEP-type rocks are abundant around Mare Imbrium, located near the middle of the nearside, the lunar meteorites with low KREEP components probably were derived from lunar cratering events far from the central part of the nearside; i.e., they might have originated from the farside or rims. The lunar meteorite studies will provide us with useful information on the farside of the Moon, which has not been directly sampled.

SUMMARY

1. The pyroxene mineralogy and chemistry of granulitic clasts in feldspathic fragmental breccia 67016 and the Y86032-type meteorites are much alike.

2. Poikilitic clasts, which are dominant in regolith breccia 60019, are rare in feldspathic fragmental breccia 67016 and the Y86032-type meteorites.

3. The preservation of breccias such as 67016 on the nearside will yield useful information in interpreting the heterogeneity, origin, and history of the lunar meteorites.

4. The feldspathic fragmental breccia suite may have been part of a very old megaregolith that formed during or even before the basin-forming events.

5. Similarities in components, textures, and noble gas contents between the Y86032-group and 67016 suggest that formation processes of the feldspathic fragmental breccias are also important in the genesis of lunar meteorites.

6. Differences in matrix microtextures imply that 67016 and Y86032 breccia formation temperatures were different, however.

7. The feldspathic fragmental breccias, especially 67016, were probably never exposed to the solar wind because the high porosity and preservation of interstitial amorphous materials in the matrix indicate that shock melting could not have been intense enough to drive off solar wind gases.

8. Because the Y86032-type meteorites contain a small regolith component but are otherwise very much like feldspathic fragmental breccias, these lunar meteorites may represent samples intermediate between the feldspathic fragmental and the regolith breccia suites.

9. The preservation of Ca-zoning in olivines of a granulitic clast indicates that the original rocks of the granulite before metamorphic annealing may have had a relatively quickly cooled thermal history regardless of the subsequent thermal events.

Acknowledgments. We are indebted to National Institute of Polar Research for the samples, and Drs. M. M. Lindstrom, P. H. Warren, A. Treiman, J. Shervais, T. Ishii, and C. Koeberl for discussion and reviews. We thank Mr. E. Yoshida and Mr. O. Tachikawa for their help in microanalyses, and Mmes. H. Hatano and K. Hashimoto for their technical assistance. This work was supported in part by funds from Cooperative Program (No. 84134) provided by Ocean Research Institute, University of Tokyo, and by Grant-in-Aid for Scientific Research of the Japanese Ministry of Education, Science, and Culture, and by the Mitsubishi Foundation.

REFERENCES

Bischoff A., Palme H., Weber H. W., Stöffler D., Braun O., Spettel B., Begemann F., Wänke H., and Ostertag R. (1987) Petrography, shock history, chemical composition and noble gas content of the lunar meteorites Yamato-82192 and -82193. *Mem. Natl. Inst. Polar Res., Spec. Issue, 46,* 21-42.

Buening D. K. and Buseck P. R. (1973) Fe-Mg lattice diffusion in olivine. *J. Geophys. Res., 78,* 6852-6862.

Cremers C. J. and Hsia H. (1974) Thermal conductivity of Apollo 16 lunar fines. *Proc. Lunar Sci. Conf. 5th,* pp. 2703-2708.

Eugster O. and Niedermann S. (1988) Noble gases in lunar meteorites Yamato-82192 and -82193 and history of meteorites from the moon. *Earth Planet. Sci. Lett., 89,* 15-27.

Eugster O., Niedermann S., Burger M., Krähenbühl U., Weber H., Clayton R. N., and Mayeda T. K. (1988) Preliminary report on the Yamato-86032 lunar meteorite III: Ages, noble gas, isotopes, oxygen isotopes and chemical abundances. *Proc. NIPR Symp. Antarct. Meteorites, 2,* in press.

Horai K. and Winkler J. (1974) Thermal diffusivity of lunar rock sample 12002,85 (abstract). In *Lunar Science V,* pp. 354-356. The Lunar Science Institute, Houston.

Koeberl C., Warren P. H., Lindstrom M. M., Spettel B., and Fukuoka T. (1989) Preliminary examination on the Yamato-86032 lunar meteorite II: Major and trace element chemistry. *Proc. NIPR Symp. Antarct. Meteorites, 2,* in press.

Lindsley D. H. and Anderson D. J. (1983) A two-pyroxene thermometer. *Proc. Lunar Planet. Sci. Conf. 13th,* in *J. Geophys. Res., 88,* A887-A906.

Lindstrom M. M. and Lindstrom D. J. (1986) Lunar granulites and their precursor anorthositic norites of the early lunar crust. *Proc. Lunar Planet. Sci. Conf. 16th,* in *J. Geophys. Res., 91,* D263-D276.

Lindstrom M. M. and Salpas P. A. (1983) Geochemical studies of feldspathic fragmental breccias and the nature of North Ray Crater ejecta. *Proc. Lunar Planet. Sci. Conf. 13th,* in *J. Geophys. Res., 88,* A671-A683.

Lindstrom M. M., Lindstrom D. J., Korotev R. L., and Haskin L. A. (1986) Lunar meteorites Yamato 791197: A polymict anorthositic norite breccia. *Mem. Natl. Inst. Polar Res., Spec. Issue, 41,* 58-78.

Lucey P. G. and Hawke B. R. (1987) A remote mineralogic perspective on gabbroic units in the lunar highlands. *Proc. Lunar Planet. Sci. Conf. 18th,* pp. 355-363.

McKay D. S., Bogard D. D., Morris R. V., Korotev R. L., Johnson P., and Wentworth S. J. (1986) Apollo 16 regolith breccias: Characterization and evidence for early formation in the megaregolith. *Proc. Lunar Planet. Sci. Conf. 16th,* in *J. Geophys. Res., 91,* D277-D303.

Miyamoto M. (1987) Constraints on cooling histories of ordinary chondrites as inferred from chemical zoning of porphyritic olivine (abstract). In *Lunar and Planetary Science XVIII,* pp. 651-652. Lunar and Planetary Institute, Houston.

Miyamoto M., McKay D. S., McKay G. A., and Duke M. B. (1986) Chemical zoning and homogenization of olivines in ordinary chondrites and implications for thermal histories of chondrules. *J. Geophys. Res., 91,* 12804-12816.

Morioka M. (1981) Cation diffusion in olivine-II: Ni-Mg, Mn-Mg, Mg and Ca. *Geochim. Cosmochim. Acta, 45,* 1573-1580.

Nord G. L., Christie J. M., Heuer A. H., and Lally J. S. (1975) North Ray Crater breccias: An electron petrographic study. *Proc. Lunar Sci. Conf. 6th,* pp. 779-797.

Norman M. D. (1981) Petrology of suevitic lunar breccia 67016. *Proc. Lunar Planet. Sci. 12B,* pp. 235-252.

Ostertag R., Stöffler D., Bischoff A., Palme H., Schultz L., Spettel B., Weber H., Weckwerth G., and Wänke H. (1986) Lunar meteorite Yamato-791197: Petrography, shock history and chemical composition. *Mem. Natl. Inst. Polar Res., Spec. Issue, 41,* 17-44.

Reid A. M., Warner J., Ridley W. I., and Brown R. W. (1972) Major element composition of glasses in three Apollo 15 soils. *Meteoritics, 7,* 395-415.

Ryder G. and Norman M. (1978a) Catalog of pristine non-mare materials. Part 1. Non-anorthosites. *JSC Publ. No. 14565.* NASA Johnson Space Center, Houston. 146 pp.

Ryder G. and Norman M. (1978b) Catalog of pristine non-mare materials. Part 2. Anorthosites. *JSC Publ. No. 14603.* NASA Johnson Space Center, Houston. 86 pp.

Ryder G. and Norman M. (1980) Catalog of Apollo 16 Rocks. *Lunar Curatorial Facility Publ. No. 52.* NASA Johnson Space Center, Houston.

Ryder G. and Ostertag R. (1983) ALHA 81005: Moon, Mars, petrography, and Giordano Bruno. *Geophys. Res. Lett., 10,* 791-794.

Stöffler D., Knoll H.-D., and Maerz U. (1979) Terrestrial and lunar impact breccias and the classification of lunar highlands rocks. *Proc. Lunar Planet. Sci. Conf. 10th,* pp. 639-675.

Takeda H. and Miyamoto M. (1988) Mineralogical studies of lunar highland breccia 67016, an analog of the Yamato-82 lunar meteorites (abstract). In *Lunar and Planetary Science XIX,* pp. 1171-1172. Lunar and Planetary Institute, Houston.

Takeda H., Miyamoto M., and Ishii T. (1979) Pyroxenes in early crustal cumulate found in achondrites and lunar highland rocks. *Proc. Lunar Planet. Sci. Conf. 10th,* pp. 1095-1107.

Takeda H., Mori H., and Tagai T. (1986) Mineralogy of Antarctic lunar meteorites and differentiated products of the lunar crust. *Mem. Natl. Inst. Polar Res., Spec. Issue, 41,* 45-57.

Takeda H., Mori H., and Tagai T. (1987a) Mineralogy of lunar meteorites, Yamato-82192 and -82193 with reference to breccia in breccia. *Mem. Natl. Inst. Polar Res., Spec. Issue, 46,* 43-55.

Takeda H., Miyamoto M., Galindo C., and Ishii T. (1987b) Mineralogy of basaltic clast in lunar highland regolith breccia 60019. *Proc. Lunar Planet. Sci. Conf. 17th,* in *J. Geophys. Res., 92,* E462-E470.

Takeda H., Miyamoto M., Mori H., and Tagai T. (1988a) Mineralogical studies of clasts in lunar highland regolith breccia 60019 and in lunar meteorites Y82192. *Proc. Lunar Planet. Sci. Conf. 18th,* pp. 33-43.

Takeda H., Kojima H., Nishio F., Yanai K., and Lindstrom M. M. (1988b) Preliminary report on the Yamato-86032 lunar meteorite I: Recovery, sample descriptions, mineralogy and petrography. *Proc. NIPR Symp. Antarct. Meteorites, 2,* in press.

Treiman A. H. and Drake M. J. (1983) Origin of lunar meteorite ALHA 81005: Clues from the presence of terrae clasts and a very low-titanium basalt clast. *Geophys. Res. Lett., 10,* 783-786.

Warner J. L., Phinney W. C., Bickel C. E., and Simonds C. H. (1977) Feldspathic granulitic impactites and pre-final bombardment lunar evolution. *Proc. Lunar Sci. Conf. 8th,* pp. 2051-2066.

Warren P. H., Taylor G. J., and Keil K. (1983) Regolith breccia Allan Hills A81005: Evidence of lunar origin and petrography of pristine and nonpristine clasts. *Geophys. Res. Lett., 10,* 779-782.

Warren P. H., Jerde E. A., and Kallemeyn G. W. (1989) Lunar meteorites: Siderophile element contents and implications for the composition and origin of the Moon. *Earth Planet. Sci. Lett., 91,* 245-260.

Wentworth S. J. and McKay D. S. (1988b) Glasses in ancient and young Apollo 16 regolith breccias: Populations and Ultra Mg' glass. *Proc. Lunar Planet. Sci. Conf. 18th*, pp. 67-77.

Yanai K. and Kojima H. (1987) *Photographic Catalog of the Antarctic Meteorites*, pp. 197-199. Natl. Inst. of Polar Res., Tokyo.

Proceedings of the 20th Lunar and Planetary Science Conference, pp. 101-108
Lunar and Planetary Institute, Houston, 1990

Modeling of Lunar Basalt Petrogenesis: Sr Isotope Evidence from Apollo 14 High-Alumina Basalts

C. R. Neal and L. A. Taylor

Department of Geological Sciences, University of Tennessee, Knoxville, TN 37996

Trace-element modeling indicates that assimilation of KREEP by a fractionating parental magma [i.e., AFC (assimilation and fractional crystallization)] can account for the range in high-alumina basalt compositions at the Apollo 14 site. However, variability of magnesium number (MG#) with trace-element contents has been used to argue against such a hypothesis (Shih and Nyquist, 1989a,b), but these as well as Sr isotope data can be interpreted in terms of an AFC process at the Apollo 14 site. Positive linear correlations of $^{87}Sr/^{86}Sr$ with $^{87}Rb/^{86}Sr$ and Sr abundance (both leading to KREEP) are used as evidence of such an AFC process in the petrogenesis of Apollo 14 high-alumina basalts. Calculation of AFC paths on Sr isotope and MG# vs. La (ppm) diagrams has been undertaken. In the case of MG# vs. La (ppm), the effect of the KREEP assimilant upon the major element contents of the residual melt, the errors associated with the MG#, and the cyclicity of the AFC process were considered. Comparison of the amounts of AFC required to generate high-alumina basalts from both trace element and Sr isotope plots are similar. However, trace-element modeling suggests higher amounts of AFC are required to generate the "Group 1" basalt reported by Dasch et al. (1987) than in the Sr isotope modeling. This is probably due to the fact that the sample analyzed for Sr isotopes was a mixture of five small "Group 1" basalt clasts. The most trace-element enriched basalt should contain the most radiogenic initial $^{87}Sr/^{86}Sr$ ratio, if AFC was a single-stage process or the basalts were derived from a single source. This is not the case, as the "Group 3" or "intermediate" basalt contains the highest initial $^{87}Sr/^{86}Sr$ ratio. We conclude that the AFC process occurred more than once. This requires that each AFC cycle evolved essentially the same path, but as the Apollo 14 site has a "KREEP-rich" signature (e.g., Wood and Head, 1975; Etchegaray-Ramirez et al., 1983), this is a feasible supposition. Although subject to large errors, age determinations demonstrate that magmatism could have occurred on at least three occasions at the Apollo 14 site (4.3, 4.1, and 3.95 Ga). It is apparent that further precise Sr isotope analyses of Apollo 14 high-alumina basalts are required.

INTRODUCTION

Petrogenesis of Apollo 14 high-alumina basalts has been the subject of considerable study (e.g., *Hubbard et al.,* 1972; *Dickinson et al.,* 1985; *Shervais et al.,* 1985), particularly as the more evolved types appear to contain a KREEP signature. Breccia "pull-apart" efforts have dramatically increased the number of "new" Apollo 14 high-alumina basalt samples. Correspondingly, there has been an evolution in the models proposed for their petrogenesis as more data became available. *Dickinson et al.* (1985) and *Shervais et al.* (1985) concluded that there were five groups of Apollo 14 high-alumina basalts, some related by KREEP assimilation, but with at least three distinct source regions. *Dickinson et al.* (1985) suggested that these high-alumina basalts were formed by different degrees of partial melting of a common source or similar sources at different times. However, with the acquisition of more data, *Neal et al.* (1987, 1988, 1989) noted that rather than distinct groups, a continuum of high-alumina basalt trace-element compositions existed. This continuum was modeled by progressive assimilation of KREEP by a fractionating, Mg-rich parental magma [i.e., an "AFC" (assimilation and fractional crystallization) process] depleted in incompatible elements (*Neal et al.,* 1987, 1988, 1989).

In this paper we reexamine the limited Sr isotope data available for the Apollo 14 high-alumina basalts in light of the proposed AFC model. We demonstrate that these isotope data are entirely consistent with a petrogenesis for the Apollo 14 high-alumina basalts by AFC with KREEP. However, the small sample size of these clasts has precluded a comprehensive isotopic study of Apollo 14 high-alumina basalts. *Papanastassiou and Wasserburg* (1971) reported Sr isotopic data from basalt 14053 and a basalt clast from breccia 14321. Unfortunately, this latter sample was given no subnumber, so the whole-rock chemistry cannot be traced. *Taylor et al.* (1983) reported Sr isotope data from another Apollo 14 basalt clast (14305,122), noting that it yielded a pre-4.2 Ga crystallization age (4.23 ± 0.05 Ga), which made it the oldest dated basalt at the time. However, we will not include this sample in our study, because it is not a high-alumina basalt, containing only 4 wt.% Al_2O_3 (see Table 2 of *Taylor et al.,* 1983). *Dasch et al.* (1986, 1987) analyzed some of the largest basalt clasts from breccia 14321 for Sr isotopes and three of these for Nd isotopes. One of these ("14321") was a combination of five clasts defined by *Dickinson et al.* (1985) as being from "Group 1" of their classification. Also included in the study of *Dasch et al.* (1987) was the tridymite ferrobasalt described by *Shervais et al.* (1985). As demonstrated by *Neal et al.* (1988), this basalt is unrelated to the Apollo 14 high-alumina suite and, as such, also will not be discussed here.

GENERATION OF APOLLO 14 HIGH-ALUMINA BASALTS BY AFC

Inasmuch as the AFC model of *Neal et al.* (1987, 1988, 1989) will be extended to include Sr isotope data, it is deemed

TABLE 1. Selected trace-element abundances and Sr isotopic ratios from Apollo 14 high-alumina basalts.

Sample	"Group"	La	Hf	Sc/Sm	Sr	$^{87}Sr/^{86}Sr(P)$	$^{87}Sr/^{86}Sr(I)$	Age	Rb	$^{87}Rb/^{86}Sr$
"14321"	1	25.5	8.7	4.8	103.6	0.70398±5	0.69939±8	4.12±0.08	2.83	0.0791
9056	2	20.7	7.7	5.4	105.6	0.70414±4	0.69940±3	4.07±0.03	2.95	0.0809
1318/1394	2	20.8	8.2	5.7	102.3	0.70349±3	0.69934±14	4.12±0.15	2.54	0.0719
14053	3	13.0	4.7	8.5	97.3	0.70276±7	0.69948±6	3.96±0.04	2.39	0.0572
9059	5	3.9	2.3	25.4	77.7	0.70073±3	0.69913±5	4.33±0.13	0.685	0.0255
1161/1384	5	2.9	1.8	28.9	61.7	0.70117±5	0.69908±6	4.24±0.14	0.736	0.0345

Data from *Shervais et al.* (1985), *Dickinson et al.* (1985), *Neal et al.* (1988), *Papanastassiou and Wasserburg* (1971), and *Dasch et al.* (1987). Trace-element abundances are in ppm and 2 σ errors are given on isotope and age determinations.

necessary to give a brief description of parameters used in this petrogenetic model.

The AFC model was developed in order to generate the continuum of high-alumina basalt compositions and the similarity of REE patterns of the more evolved (i.e., incompatible trace-element rich) basalts to those of KREEP. The more evolved basalts, corresponding to "Groups 1 and 2" of *Dickinson et al.* (1985), require a maximum of 70% fractional crystallization and 15.4% KREEP assimilation, where r = 0.22 (r = mass assimilated/mass crystallized). The "IKFM" KREEP composition 15386 of *Vaniman and Papike* (1980) was taken as the assimilant. The most primitive high-alumina basalts (i.e., incompatible trace-element poor) correspond to "Group 5" of *Dickinson et al.* (1985) and were taken as parental. The basalts forming their "Groups 3 and 4" were described as "intermediate" by *Neal et al.* (1987, 1988, 1989). It should be emphasized that the categorizing of Apollo 14 high-alumina basalts into "groups" is not justified, as subsequent data have proven a continuum of basalt compositions at this site. However, the isotopic study of *Dasch et al.* (1986, 1987) adhered to this earlier classification.

Major-element modeling was undertaken on an Ol-Q-An pseudoternary, and this was used, as well as petrographic observations, to estimate the compositions and proportions of crystallizing phases. During the first 14% crystallization (i.e., F = 0.86), olivine (90%) and chromite (10%) were liquidus phases. Between 15% and 21% crystallization (i.e., F = 0.85-0.79), plagioclase (50%), olivine (40%), and chromite (10%) crystallized, followed by pyroxene (60%), plagioclase (30%), and ilmenite (10%) after 21% (i.e., F = 0.79) of the parental magma had crystallized. A parental magma composition was calculated on the basis of low incompatible trace-element abundances and position on the Ol-Q-An pseudoternary.

Sr ISOTOPES

Those Apollo 14 high-alumina basalts that have been analyzed for Sr isotopes range in crystallization ages from 3.96 to 4.33 Ga, and in initial $^{87}Sr/^{86}Sr$ from 0.69908 to 0.69948 (Table 1). Present-day $^{87}Sr/^{86}Sr$ ratios range from 0.70073 to 0.70414, and Sr and Rb abundances from 61.7 to 105.6 ppm and 0.685 to 2.95 ppm, respectively. As crystallization age becomes younger, initial $^{87}Sr/^{86}Sr$ ratios and abundances of Sr increase. Those basalts containing the most radiogenic present-day Sr isotope ratios also contain the highest abundances of incompatible trace elements (Fig. 1).

DISCUSSION

In their Sr isotope study, *Dasch et al.* (1987) analyzed Apollo 14 high-alumina basalts belonging to "Groups 1-5" of *Dickinson et al.* (1985). Whole-rock chemical data for the samples reported by *Dasch et al.* (1987) can be found in *Hubbard et al.* (1972) for 14053; *Shervais et al.* (1985) for 14321,1161/,1384; and *Dickinson et al.* (1985) for "14321", 14321,9056, and 14321,9059. However, the "Group 4" basalt (14321,1394) reported by *Dasch et al.* (1987) is more evolved than this classification suggests, as stated by these authors. The Sm-Nd data required this subsample to be from a basalt more enriched in the REE (*Dasch et al.,* 1987). Curatorial records describing the dissection of breccia 14321 revealed that the subsample from which the whole-rock composition of ,1394 was determined was not 14321,1149 as reported by *Dasch et al.* (1987). Rather, the whole-rock

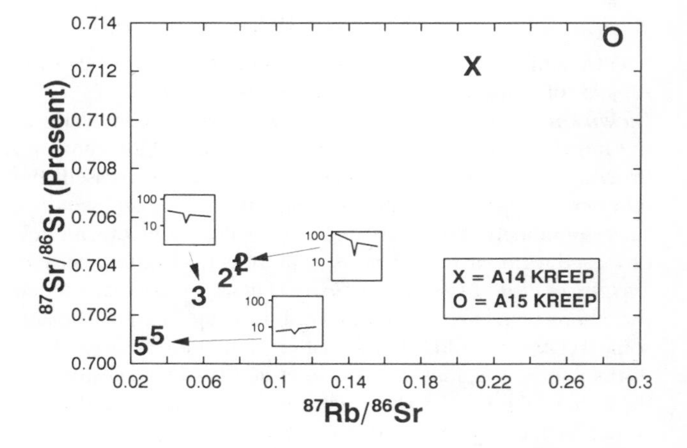

Fig. 1. Strontium isotope isochron plot for Apollo 14 high-alumina basalts and Apollo 14 and Apollo 15 KREEP. There is no age significance in this correlation due to different crystallization ages of the basalts plotted. Also plotted on this diagram are the relative REE profiles and abundances of the samples analyzed—those with a more radiogenic $^{87}Sr/^{86}Sr$ signature possess KREEP-like REE profiles. This diagram illustrates the possibility of a mixing relationship between a primitive, parental "Group 5" Apollo 14 high-alumina basalt and KREEP. Large, bold numbers refer to the basalt group. Data from *Papanastassiou and Wasserburg* (1971), *Dasch et al.* (1987), *McKay et al.* (1978, 1979), and *Nyquist et al.* (1973, 1975).

composition was reported as 14321,1318 by *Neal et al.* (1988) (JSC Curatorial Staff, personal communication, 1988). This is supported by the similarity of Sm and Nd abundances between ,1394 and ,1318 reported by *Dasch et al.* (1987) and *Neal et al.* (1988) (Sm = 11.2 and 10.4 ppm; Nd = 38.9 and 30 ± 4 ppm respectively). Therefore, this composition is *more* akin to a "Group 2" basalt (20.8 ppm La; 8.15 ppm Hf).

Dasch et al. (1987) suggested that the range in crystallization ages and initial $^{87}Sr/^{86}Sr$ ratios of the Apollo 14 high-alumina basalts negates a petrogenesis involving KREEP but could be resolved by assimilating urKREEP residuals (after *Binder,* 1982, 1985). These authors also supported the contention of *Dickinson et al.* (1985), that the high-alumina basalts evolved by different degrees of partial melting of a common source, or similar sources at different times early in the history of mare volcanism. Furthermore, the Sr isotope data indicated that most "groups" of Apollo 14 high-alumina basalts could have been erupted at different times from either a single source or from different sources containing the same Rb/Sr ratio, but with different absolute amounts of Rb and Sr (*Shih and Nyquist,* 1989a). Although on their own these Sr isotope data do not negate a partial melting model, such as that proposed by *Dickinson et al.* (1985), they cannot be used as absolute proof of such a petrogenesis for Apollo 14 high-alumina basalts.

Isotopic Evidence for AFC

Isotopic evidence for KREEP assimilation can be found in the positive correlation of the $^{87}Sr/^{86}Sr$ (present-day) ratios with $^{87}Rb/^{86}Sr$, and $^{87}Sr/^{86}Sr$ (initial) with Sr ppm (Figs. 1 and 2, Table 1). In Fig. 1, an isochronous relationship is displayed between the Apollo 14 high-alumina basalts. Note that Apollo 14 and 15 KREEPy basalts plot at the upper end of this correlation and that the REE profiles of the high-alumina basalts become more KREEP-like from "Group 5" to "Group 1."

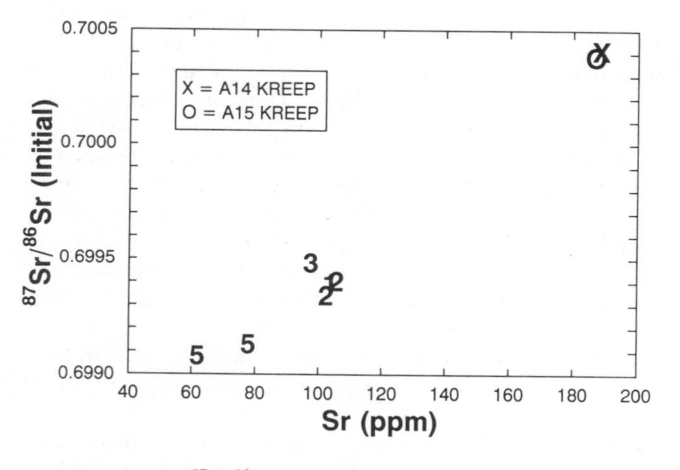

Fig. 2. Initial $^{87}Sr/^{86}Sr$ ratios plotted against Sr (ppm). Large, bold numbers refer to the basalt group. Data from *Papanastassiou and Wasserburg* (1971), *Dasch et al.* (1987), *McKay et al.* (1978, 1979), and *Nyquist et al.* (1973, 1975).

However, this type of plot can have no age significance because the Apollo 14 high-alumina basalts yield different crystallization ages (*Dasch et al.,* 1987). We conclude that this correlation represents a mixing line or AFC trend between "Group 5" basalts and KREEP.

This contention is borne out in Fig. 2 where a general positive correlation exists between initial $^{87}Sr/^{86}Sr$ ratios and Sr (ppm) for the Apollo 14 high-alumina basalts. The intermediate and evolved Apollo 14 high-alumina basalts contain progressively more radiogenic initial $^{87}Sr/^{86}Sr$ ratios and a greater abundance of Sr (Fig. 2). An exception to this is the intermediate "Group 3" basalt 14053, which contains the highest initial $^{87}Sr/^{86}Sr$ ratio (Fig. 2), inconsistent with the trace-element AFC modeling (*Neal et al.,* 1987, 1988, 1989). The significance of this observation will be discussed below. *Hawkesworth and Vollmer* (1979) demonstrated that positive correlations between $^{87}Sr/^{86}Sr$ and Sr abundance could be the result of two-component mixing. This can be extended to an AFC process, which in its simplest form is essentially the mixing of two components. Note that the initial Sr isotope ratio and Sr ppm increase from the most primitive "Group 5" basalts (i.e., LREE-depleted, low incompatible-/high compatible-element abundances, low SiO_2) to the more evolved basalts (Fig. 2), which exhibit KREEP-like REE profiles.

If the process was simply mixing (i.e., bulk assimilation with no fractional crystallization), a plot of initial $^{87}Sr/^{86}Sr$ vs. Sr (ppm) would yield a curved trend (*Langmuir et al.,* 1978; Fig. 5 of *Shih and Nyquist,* 1989b). However, a somewhat linear trend is portrayed (Fig. 2). This suggests something other than simple mixing or bulk assimilation is affecting the evolution of these basalts (i.e., fractional crystallization).

In order to interpret the Sr isotopic evolution of these samples, we have plotted crystallization age against initial $^{87}Sr/^{86}Sr$ for all Apollo 14 high-alumina basalt samples for which data are available (Fig. 3; Table 1). A negative correlation is apparent, which indicates a progressive increase in initial Sr isotopic ratio with younger basalts. *Dasch et al.* (1987) suggested that these data can be interpreted in terms of a single-stage evolution of several source regions from a whole Moon (estimated using the Sr isotope value of BABI), with a $^{87}Rb/^{86}Sr$ ratio of 0.06 (similar to the bulk Moon value of 0.05 proposed by *Nyquist,* 1977). Such an evolution line passes through the errors on age and initial $^{87}Sr/^{86}Sr$ determinations, except for one "Group 5" basalt (,1384). However, in light of trace-element modeling, the abundances of which can be adequately modeled by AFC (*Neal et al.,* 1987, 1988, 1989), we have placed a different interpretation upon these data.

Sr Isotope Modeling

The Sr isotope compositions of these basalts have been modeled by AFC (with KREEP) using equations (6a) and (15a) of *DePaolo* (1981). Unlike the trace elements (*Neal et al.,* 1987, 1988, 1989), modeling of the Sr isotopes requires that age relationships be taken into account. The Apollo 14 high-alumina basalts range in crystallization age from 3.96 to 4.33 Ga (*Papanastassiou and Wasserburg,* 1971; *Dasch et al.,* 1987; see Fig. 3, Table 1). Although errors on these ages are

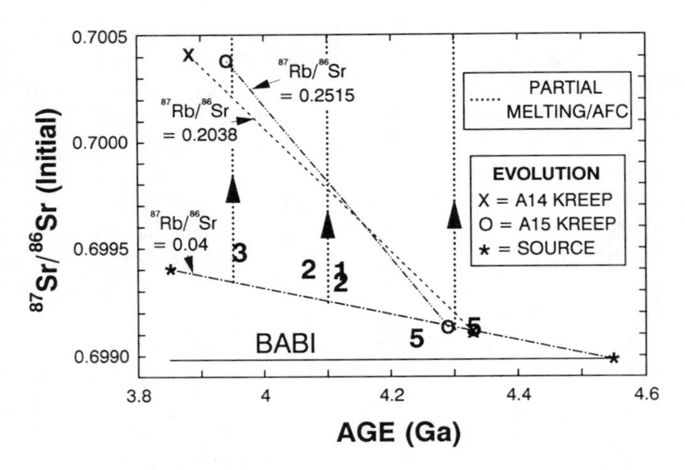

Fig. 3. Evolution diagram of the Apollo 14 high-alumina basalt source and resulting melts. The high-alumina basalt source is considered to have evolved along a single-stage evolution path similar to that of the bulk Moon. The evolution paths of the Apollo 14 and 15 KREEP components have been constructed by assuming crystallization of urKREEP at 4.29-4.33 Ga (i.e., the 'Sr model ages of Apollo 14 and 15 KREEP basalts) and subsequent evolution at a higher Rb/Sr ratio. The data points at the other end of the KREEP evolution paths correspond to crystallization ages and calculated initial ratios at the time of KREEP basalt extrusion. The different periods of magmatism and AFC are shown as vertical dotted lines. Large, bold numbers refer to the basalt group. Data from *Papanastassiou and Wasserburg* (1971) and *Dasch et al.* (1987), *McKay et al.* (1978, 1979), and *Nyquist et al.* (1973, 1975).

large (Table 1), it is evident that a single-stage AFC process cannot account for the distribution of the basalts in Fig. 3—such a process between two isotopically distinct end members would form a vertical trend on this plot. However, because the trace-element data define a single AFC trend (*Neal et al.,* 1988, 1989), each AFC cycle must have followed essentially the same path. This cyclical AFC process is not unreasonable, considering the Apollo 14 site has a "KREEP-rich" signature (e.g., *Wood and Head,* 1975; *Etchegaray-Ramirez et al.,* 1983).

As illustrated in Fig. 3, we consider that the source region for the high-alumina basalts evolved along an evolution line of ^{87}Rb/^{86}Sr 0.04 ($\cong$ whole Moon value proposed by *Nyquist,* 1977), similar to *Dasch et al.* (1987). However, at various points along the source evolution path, melting occurred, producing the primitive "Group 5" parental magma. From the clustering of the isotope data and taking into account the trace-element contents of these basalts, the first event occurred at $\cong$4.3 Ga, the second at 4.1 Ga, and the third at 3.95 Ga. It is the position of 14053, the "Group 3" basalt (classified as "intermediate" by *Neal et al.,* 1988), in Fig. 3 that indicates that single-stage evolution of a single source coupled with one AFC cycle are inadequate to generate all the observed Sr isotopes. If single source evolution or one AFC cycle was responsible for the range of high-alumina basalt compositions at the Apollo 14 site, this basalt should contain an intermediate initial ^{87}Sr/^{86}Sr ratio basalts of "Groups 1 and 2" and "Group 5." 14053 can be adequately generated by a later (i.e., $\cong$3.95 Ga) AFC cycle. However, more Sr isotope data from Apollo 14 high-alumina basalts are required to constrain more definitively the number and nature of the proposed AFC cycles.

The present-day isotopic signatures of the two "Group 5" basalts may be the result of a heterogeneous source with respect to absolute abundances of Rb and Sr (*Shih and Nyquist,* 1989a,b). The initial ratios of these basalts are similar (Fig. 3, Table 1), but as shown by *Dasch et al.* (1987), the error parallelograms for these two basalts are not overlapping. This heterogeneity, which may also have been induced by variable degrees of partial melting, is witnessed in a higher present-day ^{87}Sr/^{86}Sr ratio of ,1384, which has lower Sr abundances (Table 1).

In our previous trace-element studies (*Neal et al.,* 1987, 1988, 1989), 15386 "IKFM" KREEP basalt was taken as the assimilant. This representative KREEPy sample was also used by *Neal and Taylor* (1989) in a preliminary report on the Sr isotopic evidence for AFC in the petrogenesis of Apollo 14 high-alumina basalts. However, in the present study we have undertaken a more detailed approach to the determination of the Sr isotopic compositions of the assimilant. The initial ^{87}Sr/^{86}Sr ratio of 15386 (0.70038) and the average Sr abundance of all Apollo 15 KREEP basalts (186.7 ppm; *Nyquist et al.,* 1973, 1975) were utilized in this modeling. Furthermore, we have also used an average Apollo 14 KREEPy basalt composition (Table 2: data from *Papanastassiou and Wasserburg,* 1971; *McKay et al.,* 1978, 1979). The average Apollo 14 KREEP possesses an initial ^{87}Sr/^{86}Sr ratio of 0.70041 and 188.3 ppm Sr. Both Apollo 14 and Apollo 15 KREEPy compositions plot at the end of the positive correlation delineated by the high-alumina basalts in ^{87}Sr/^{86}Sr (present) vs. ^{87}Rb/^{86}Sr and ^{87}Sr/^{86}Sr (initial) vs. Sr (ppm) (Figs. 1 and 2).

The crystallization ages of KREEPy basalts are younger (Apollo 14 = 3.88 Ga; Apollo 15 = 3.94 Ga) than the Apollo 14 high-alumina basalts. However, model ages for Apollo 14 (*McKay et al.,* 1978, 1979; *Papanastassiou and Wasserburg,* 1971) and Apollo 15 KREEPy basalts (*Nyquist et al.,* 1973, 1975) range from 4.29-4.33 Ga. *Dowty et al.* (1976) suggested that the model ages represent the creation or crystallization of the "urKREEP" (pristine, unadulterated KREEP) reservoir from the "lunar magma ocean" (*Warren and Wasson,* 1979; *Warren,* 1985). This reservoir was subsequently remelted at $\cong$3.9 Ga and erupted as KREEPy basalts. These basalts are the best representatives of the proposed KREEP assimilant. In Fig. 3 we have plotted average Apollo 14 and Apollo 15 KREEPy basalts and drawn a tie line between their respective initial ratios at time of crystallization and the bulk Moon evolution path at 4.29 Ga (Apollo 15) and 4.33 Ga (Apollo 14). These tie lines trace the evolution of the urKREEP reservoir prior to KREEPy basalt magmatism. As such, the Sr isotopic composition of KREEP can be estimated at the inferred times of magmatism. We assume that Sr abundances do not change. Modeling parameters can be found in Table 2, and F values shown are those calculated from the major-element modeling (*Neal et al.,* 1988; see previous section).

The effect of KREEP assimilation by a parental "Group 5"

TABLE 2. Modeling parameters.

F Value	4.1 Ga $^{87}Sr/^{86}Sr$	Sr (ppm)	F	3.95 Ga $^{87}Sr/^{86}Sr$	Sr(ppm)
		Apollo 14 KREEP			
0.95	0.699260	75.4	0.95	0.699360	75.4
0.90	0.699278	82.6	0.90	0.699390	82.6
0.86	0.699292	88.9	0.86	0.699412	88.9
0.79	0.699312	89.9	0.79	0.699445	89.9
0.70	0.699336	95.7	0.70	0.699484	95.7
0.60	0.699363	103.0	0.60	0.699527	103.0
0.50	0.699389	111.4	0.50	0.699570	111.4
0.40	0.699415	121.1	0.40	0.699612	121.1
Parent	0.699240	69.0	Parent	0.699328	69.0
Apollo 14 KREEP	0.699780	188.3	Apollo 14 KREEP	0.700200	188.3
		Apollo 15 KREEP			
0.95	0.699260	75.4	0.95	0.699365	75.4
0.90	0.699279	82.5	0.90	0.699399	82.5
0.86	0.699294	88.8	0.86	0.699425	88.8
0.79	0.699315	89.8	0.79	0.699462	89.8
0.70	0.699340	95.6	0.70	0.699507	95.6
0.60	0.699367	102.8	0.60	0.699557	102.8
0.50	0.699395	111.1	0.50	0.699606	111.1
0.40	0.699422	120.8	0.40	0.699654	120.8
Parent	0.699240	69.0	Parent	0.699328	69.0
Apollo 15 KREEP	0.699800	186.7	Apollo 15 KREEP	0.700335	186.7

Mineral	Sr Kd	Reference
Olivine	0.02	(1)
Plagioclase	2.25	(2,3)
Pyroxene	0.02	(4)
Ilmenite	0.005	(5)
Chromite	0.005	Estimated

References: (1) *Hart and Brooks* (1974); (2) *Drake and Weill* (1975); (3) *Philpotts and Schnetzler* (1970); (4) *McKay and Weill* (1976); (5) *Binder* (1982).

magma has been evaluated at 4.3, 4.1, and 3.95 Ga. At 4.3 Ga the parental magma and KREEP assimilant will contain virtually identical $^{87}Sr/^{86}Sr$ ratios (Fig. 3). We would predict that all Apollo 14 high-alumina basalts erupted at 4.3 will contain initial $^{87}Sr/^{86}Sr$ ratios of between 0.6991 and 0.6992. Until more data is collected, it is uncertain whether basalts spanning the complete range of observed trace-element compositions were erupted at this time.

The one "Group 1" and two "Group 2" high-alumina basalts were erupted at ≅4.1 Ga. These basalts contain the most evolved trace-element compositions. According to the model of *Neal et al.* (1987, 1988, 1989), between 45-54% fractional crystallization of a parental "Group 5" magma (9.9-11.9% KREEP assimilation) is required to generate these compositions. We have calculated an AFC path between a parental "Group 5" basalt and Apollo 14 and 15 KREEP (Fig. 4). The $^{87}Sr/^{86}Sr$ ratios for both Apollo 14 and 15 KREEP and the parental magma were estimated from the evolution paths in Fig. 3. In our previous trace-element AFC modeling of Apollo 14 high-alumina basalts, a primitive "Group 5" basalt was taken as the parental magma (*Neal et al.,* 1988, 1989). Although Sr abundance and initial $^{87}Sr/^{86}Sr$ ratio of the "Group 5" basalt source region was heterogeneous (see previous section), we have for simplicity taken as our parental magma an average of the $^{87}Sr/^{86}Sr$ ratio and Sr (ppm) of the two "Group 5" basalts (Table 2). Modeling parameters are presented in Table 2. The initial Sr isotopic ratio of the "Group 3" basalt 14053 can only be generated by a separate AFC cycle (at 3.95 Ga), different from that which produced the "Group 1 and 2" basalts (Fig. 3). Therefore, an AFC path between a parental "Group 5" basalt and Apollo 14 and 15 KREEP at 3.95 Ga has also been calculated (the upper two paths in Fig. 4), in order to generate the "Group 3" basalt 14053.

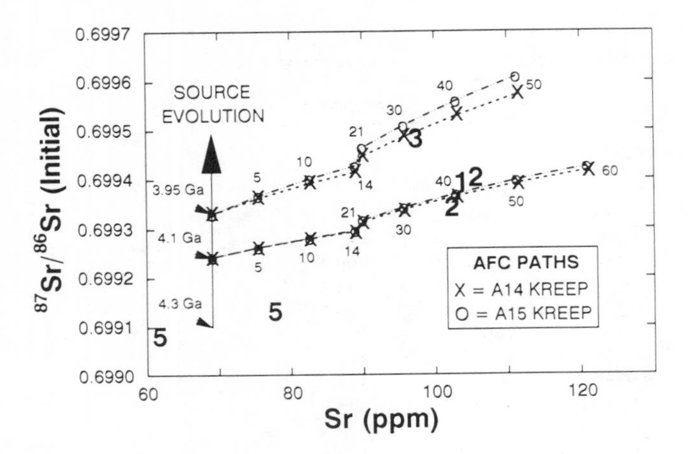

Fig. 4. AFC modeling of initial $^{87}Sr/^{86}Sr$ ratios and Sr (ppm) using equations (6a) and (15a) of *DePaolo* (1981) for Apollo 14 high-alumina basalts and Apollo 14 and 15 KREEP as assimilants. Large, bold numbers refer to the basalt group and small numbers refer to the percent crystallized. Data from *Papanastassiou and Wasserburg* (1971), *Dasch et al.* (1987), *McKay et al.* (1978, 1979), and *Nyquist et al.* (1973, 1975). The average parental basalt and KREEP assimilant compositions are estimated from Fig. 3 ($^{87}Sr/^{86}Sr$ ratio) and averaging the two analyzed "Group 5" basalts (Sr abundance).

The calculated AFC paths between a parental high-alumina basalt and KREEP indicate the applicability of this process in generating the Sr isotopic compositions of Apollo 14 high-alumina basalts (Fig. 4, Table 2). Note that the increase in Sr is slowed when plagioclase becomes a liquidus phase after 14% fractional crystallization (F = 0.86). Such results can be related to trace-element modeling by examining selected trace-element contents (Table 1). These elements have been plotted on definitive graphs (La vs. Hf and Sc/Sm vs. La) used by *Neal et al.* (1988, 1989) to illustrate the applicability of an AFC process to Apollo 14 high-alumina basalt petrogenesis (Figs. 5a,b). Also plotted are the calculated AFC paths (*Neal et al.*, 1988, 1989), from which can be estimated the amount of fractional crystallization and KREEP assimilation required to generate each sample that has been analyzed for Sr isotopes. The parental basalt used in these trace-element AFC calculations was the one defined by *Neal et al.* (1988) in order to be consistent with previously published models (*Neal et al.*, 1988, 1989). We consider the average Sr isotope value of the "Group 5" basalts reported by *Dasch et al.* (1987) to be representative of the parental magma.

The AFC calculations involving trace-elements are presented in Figs. 5a,b. For each basalt, the amount of required AFC is approximately the same, both for the trace elements and the Sr isotopes (Figs. 4 and 5, Table 3). However, the amount of AFC required for the "Group 1" ("14321") basalt is somewhat higher for trace-element modeling than that in the modeling of the Sr isotope data (Table 3). This may be due to the fact that this sample is a mixture of five different "Group 1" basalt clasts.

Shih and Nyquist (1989a,b) used a plot of MG# vs. La (ppm) (Fig. 6) to demonstrate the groupings of the Apollo 14 high-alumina basalts. These authors demonstrated that the AFC path of *Neal et al.* (1988) does not pass through the central portion of the data on such a plot. However, in the construction of this AFC path, the major elements were modeled by fractional crystallization alone (after the method of *DePaolo,* 1981), not allowing for the influence of the assimilant.

In Fig. 6, we have reevaluated the AFC process inasmuch as it affects the major elements. We have plotted the original AFC curve (dotted line) from *Neal et al.* (1988) along with two AFC curves that include the effect of the KREEP assimilant (IKFM 15386 KREEP from *Vaniman and Papike,* 1980) on the major element contents of the residual melt (solid lines). These curves are drawn from the parental "Group 5" basalt defined by *Neal et al.* (1988) and the "Group 5" basalt

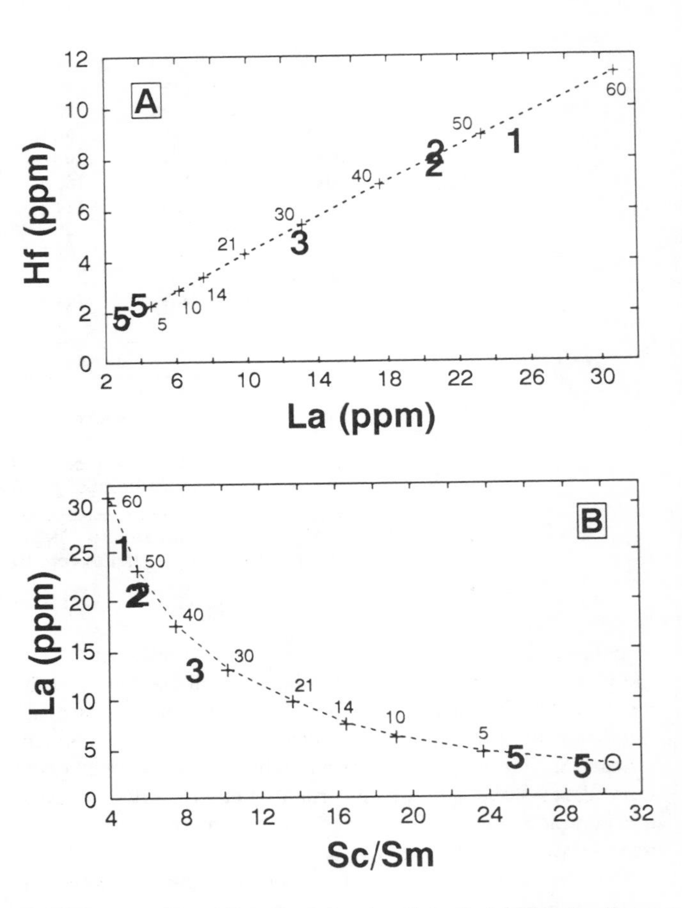

Fig. 5. Trace-element modeling for all Apollo 14 high-alumina basalts that have been analyzed for Sr isotopes. Large, bold numbers refer to the basalt group and small numbers refer to the percent crystallized. **(a)** La (ppm) vs. Hf (ppm); **(b)** Sc/Sm vs. La (ppm). Parental basalt composition from *Neal et al.* (1988). Apollo 14 high-alumina basalt trace-element data from *Shervais et al.* (1985), *Dickinson et al.* (1985), and *Neal et al.* (1988). AFC paths were calculated by *Neal et al.* (1988, 1989).

TABLE 3. Comparison of amounts of AFC required to generate the trace-element contents and isotopic signatures of Apollo 14 high-alumina basalts.

Sample	"Group"	La vs. Hf		La vs. Sc/Sm		Sr Isotopes	
		A	FC	A	FC	A	FC
"14321"	1	11.2-11.7	51-53	11.4-11.9	52-54	9.1-9.2	41-42
9056	2	9.9-10.3	45-47	10.3-10.8	47-49	9.7-9.9	44-45
1318/1394	2	10.1-10.6	46-48	10.3-10.8	47-49	8.5-8.6	38-39
14053	3	6.2-6.6	28-30	6.8-7.3	31-33	6.6-6.9	30-32

A = assimilation; FC = fractional crystallization.

containing the highest MG#. Drawing more than one AFC curve is feasible due to the cyclical nature of the AFC process at Apollo 14 (see above). Also shown on Fig. 6 are bulk mixing lines between the two parental "Group 5" basalts and our KREEP assimilant (dashed lines).

Like the original AFC curve (*Neal et al.,* 1988), the two mixing paths miss the bulk of the data, but these mixing lines do not rule out the possibility of post-KREEP-assimilation olivine fractionation as proposed by *Shih and Nyquist* (1989a,b). However, much of the horizontal scatter may be a function of the large errors associated with the MG# calculation. When the affect of the assimilant on the major elements, the magnitude of the errors associated with the MG# calculation and the cyclical nature of AFC at the Apollo 14 site are taken into consideration, this plot does not negate a petrogenesis for these high-alumina basalts by AFC with KREEP.

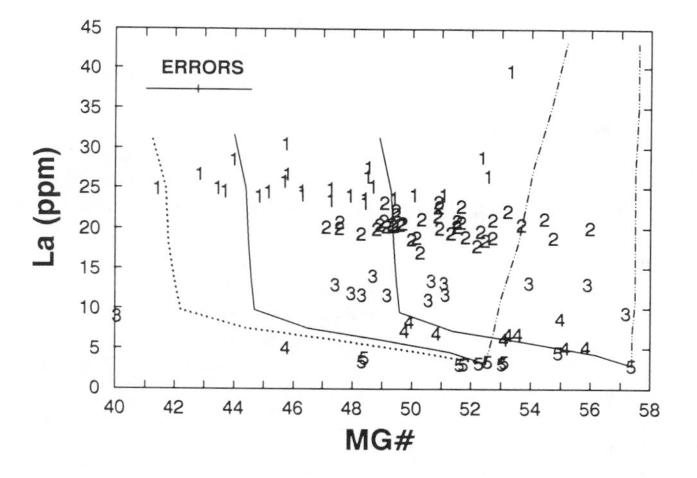

Fig. 6. A plot of MG# vs. La (ppm) for all Apollo 14 high-alumina basalts (data from *Hubbard et al.,* 1972; *Dickinson et al.,* 1985; *Shervais et al.,* 1985; *Neal et al.,* 1988, 1989). Numbers correspond to the different basalt "groups" defined by *Dickinson et al.* (1985). The dotted line represents the original AFC path reported by *Neal et al.* (1988). Solid lines represent AFC paths where the effect of the assimilant on the major elements has been taken into consideration. These curves have been derived from the parent defined by *Neal et al.* (1988) and the "Group 5" basalt with the highest MG#. Dashed lines represent bulk mixing lines between these parents and the KREEP assimilant (15386 "IKFM" of *Vaniman and Papike,* 1980). Maximum errors were calculated from INAA data.

GEOLOGICAL SETTING AND THE AFC PROCESS

When interpreted in terms of an AFC process, the existing Sr isotope data for Apollo 14 high-alumina basalts suggest that there were three AFC cycles. The primitive ("Group 5"-type) parental magma was generated from a source region that essentially underwent Sr isotope evolution along a bulk Moon trend. This source region was isotopically heterogeneous (*Dasch et al.,* 1987), and absolute abundances of Rb and Sr were variable (*Shih and Nyquist,* 1989a,b). The nature of the present-day Sr isotope ratios indicates this variability in Rb/Sr ratios (Figs. 1 and 2) either within the source or induced by slight variations in the degree of partial melting.

The holding chamber for the parental "Group 5" magma was situated in a KREEP-rich area, allowing the progressive incorporation of a KREEPy signature into the evolving melt. The initial AFC cycle at 4.3 Ga between the "Group 5" parent and KREEP would have produced basalts of limited range in initial $^{87}Sr/^{86}Sr$. This is because both KREEP and the "Group 5" parent source region contained similar Sr isotope ratios at this time (Fig. 3). It is unclear whether high-alumina basalts spanning the whole compositional range defined by *Neal et al.* (1988, 1989) were erupted at this time. Subsequent AFC cycles will produce a range in initial $^{87}Sr/^{86}Sr$ ratios because the evolution paths of the source and assimilant have diverged sufficiently (Fig. 3).

Although the range in ages and initial Sr isotopic composition of the five analyzed basalts requires three AFC cycles (at 4.3, 4.1, and 3.95 Ga), the trace-element AFC trends (Figs. 5a,b) are not sensitive to age relationships. These trends suggest that one AFC cycle generated all observed Apollo 14 basalt compositions. Only when the Sr isotopes are considered can the cyclicity of this process be demonstrated. ***Each cycle*** must follow essentially the same evolutionary path in order for the trace-element and isotopic data to form coherent trends. This study highlights the need for an increased number of more accurate Sr isotope data for Apollo 14 high-alumina basalts in order to thoroughly evaluate their petrogenesis.

Acknowledgments. This paper has greatly benefited from thoughtful reviews by J. Davidson, C.-Y. Shih, T. Dickinson, and D. Lindstrom. The research undertaken was supported by NASA grant NAG 9-62 (L.A.T.).

REFERENCES

Binder A. B. (1982) The mare basalt magma source region and mare basalt magma genesis. *Proc. Lunar Planet. Sci. Conf. 13th,* in *J. Geophys. Res., 87,* A37-A53.

Binder A. B. (1985) Mare basalt genesis: Modeling trace elements and isotopic ratios. *Proc. Lunar Planet. Sci. Conf. 16th*, in *J. Geophys. Res., 90,* D19-D30.

Dasch E. J., Shih C.-Y., Bansal B. M., Wiesmann H., and Nyquist L. E. (1986) Isotopic provenance of aluminous mare basalts from the Fra Mauro formation (abstract). In *Lunar and Planetary Science XVII,* pp. 150-151. Lunar and Planetary Institute, Houston.

Dasch E. J., Shih C.-Y., Bansal B. M., Wiesmann H., and Nyquist L. E. (1987) Isotopic analysis of basaltic fragments from lunar breccia 14321: Chronology and petrogenesis of pre-Imbrium mare volcanism. *Geochim. Cosmochim. Acta, 51,* 3241-3254.

DePaolo D. J. (1981) Trace element and isotopic effects of combined wallrock assimilation and fractional crystallization. *Earth Planet. Sci. Lett., 53,* 189-202.

Dickinson T., Taylor G. J., Keil K., Schmitt R. A., Hughes S. S., and Smith M. R. (1985) Apollo 14 aluminous mare basalts and their possible relationship to KREEP. *Proc. Lunar Planet. Sci. Conf. 15th,* in *J. Geophys. Res., 90,* C365-C374.

Dowty E., Keil K., Prinz M., Gros J., and Takahashi H. (1976) Meteorite-free Apollo 15 crystalline KREEP. *Proc. Lunar Sci. Conf. 7th,* pp. 1833-1844.

Drake M. J. and Weill D. F. (1975) Partitioning of Sr, Ba, Ca, Y, Eu^{2+}, and Eu^{3+}, and other REE between plagioclase and magmatic liquid: An experimental study. *Geochim. Cosmochim. Acta, 39,* 689-712.

Etchegaray-Ramirez M., Metzger A., Haines E. L., and Hawke B. R. (1983) Thorium concentrations in the lunar surface: IV. Deconvolution of the Mare Imbrium, Aristarchus, and adjacent regions. *Proc. Lunar Planet. Sci. Conf. 13th,* in *J. Geophys. Res., 88,* A529-A543.

Hart S. R. and Brooks C. (1974) Clinopyroxene-matrix partitioning of K, Rb, Cs, Sr, and Rb. *Geochim. Cosmochim. Acta, 38,* 1799-1806.

Hawkesworth C. J. and Vollmer R. (1979) Crustal contamination versus enriched mantle: $^{143}Nd/^{144}Nd$ and $^{87}Sr/^{86}Sr$ evidence from the Italian volcanics. *Contrib. Mineral. Petrol., 69,* 151-165.

Hubbard N. J., Gast P. W., Rhodes J. M., Bansal B. M., Wiesmann H., and Church S. E. (1972) Non-mare basalts: Part II. *Proc. Lunar Sci. Conf. 3rd,* pp. 1161-1179.

Langmuir C. H., Vocke R. D. Jr., Hanson G. N., and Hart S. R. (1978) A general mixing equation with applications to Icelandic basalts. *Earth Planet. Sci. Lett., 37,* 382-390.

McKay G. A. and Weill D. F. (1976) Petrogenesis of KREEP. *Proc. Lunar Sci. Conf. 7th,* pp. 2427-2447.

McKay G. A., Wiesmann H., Nyquist L. E., Wooden J. L., and Bansal B. M. (1978) Petrology, chemistry, and chronology of 14078: Chemical constraints on the origin of KREEP. *Proc. Lunar Planet. Sci. Conf. 9th,* pp. 661-687.

McKay G. A., Wiesmann H., Bansal B. M., and Shih C.-Y. (1979) Petrology, chemistry, and chronology of Apollo 14 KREEP basalts. *Proc. Lunar Planet. Sci. Conf. 10th,* pp. 181-205.

Neal C. R. and Taylor L. A. (1989) Apollo 14 high-alumina basalt petrogenesis: Isotopic evidence for assimilation and fractional crystallization (abstract). In *Lunar and Planetary Science XX,* pp. 768-769. Lunar and Planetary Institute, Houston.

Neal C. R., Taylor L. A., and Lindstrom M. M. (1987) Mare basalt evolution: The influence of KREEP-like components (abstract). In *Lunar and Planetary Science XVIII,* pp. 706-707. Lunar and Planetary Institute, Houston.

Neal C. R., Taylor L. A., and Lindstrom M. M. (1988) Apollo 14 mare basalt petrogenesis: Assimilation of KREEP-like components by a fractionating magma. *Proc. Lunar Planet. Sci. Conf. 18th,* pp. 139-153.

Neal C. R., Taylor L. A., Schmitt R. A., Hughes S. S., and Lindstrom M. M. (1989) High alumina (HA) and very high potassium (VHK) basalt clasts from Apollo 14 breccias, Part 2—Whole rock geochemistry: Further evidence for combined assimilation and fractional crystallization within the lunar crust. *Proc. Lunar Planet. Sci. Conf. 19th,* pp. 147-161.

Nyquist L. E. (1977) Lunar Rb-Sr chronology. *Phys. Chem. Earth, 10,* 103-142.

Nyquist L. E., Hubbard N. J., Gast P. W., Bansal B. M., and Wiesmann H. (1973) Rb-Sr systematics for chemically defined Apollo 15 and 16 materials. *Proc. Lunar Sci. Conf. 4th,* pp. 1823-1846.

Nyquist L. E., Bansal B. M., and Wiesmann H. (1975) Rb-Sr ages and initial $^{87}Sr/^{86}Sr$ for Apollo 17 basalts and KREEP basalt 15386. *Proc. Lunar Sci. Conf. 6th,* pp. 1445-1465.

Papanastassiou D. A. and Wasserburg G. J. (1971) Rb-Sr ages if igneous rocks from the Apollo 14 mission and the age of the Fra Mauro formation. *Earth Planet. Sci. Lett., 12,* 36-48.

Philpotts J. A. and Schnetzler C. C. (1970) Phenocryst-matrix partition coefficients for K, Rb, Sr, and Ba, with applications to anorthosite and basalt genesis. *Geochim. Cosmochim. Acta, 34,* 307-322.

Shervais J. W., Taylor L. A., and Lindstrom M. M. (1985) Apollo 14 mare basalts: Petrology and geochemistry of clasts from consortium breccia 14321. *Proc. Lunar Planet. Sci. Conf. 15th,* in *J. Geophys. Res., 90,* C375-C395.

Shih C.-Y. and Nyquist L. E. (1989a) Isotopic constraints on the petrogenesis of Apollo 14 igneous rocks (abstract). In *Workshop on Moon in Transition: Apollo 14, KREEP, and Evolved Lunar Rocks* (G. J. Taylor and P. H. Warren, eds.), pp. 128-136. LPI Tech. Rpt. 89-03, Lunar and Planetary Institute, Houston.

Shih C.-Y. and Nyquist L. E. (1989b) Isotopic and chemical constraints on models of aluminous mare basalt genesis (abstract). In *Lunar and Planetary Science XX,* pp. 1002-1003. Lunar and Planetary Institute, Houston.

Taylor L. A., Shervais J. W., Hunter R. H., Shih C.-Y., Bansal B. M., Wooden J., Nyquist L. E., and Laul J. C. (1983) Pre-4.2 AE mare-basalt volcanism in the lunar highlands. *Earth Planet. Sci. Lett., 66,* 33-47.

Vaniman D. T. and Papike J. J. (1980) Lunar highland melt rocks: Chemistry, petrology and silicate mineralogy. In *Proceedings of the Conference on the Lunar Highlands Crust* (J. J. Papike and R. B. Merrill, eds.), pp. 271-337. Pergamon, New York.

Warren P. H. (1985) The magma ocean concept and lunar evolution. *Annu. Rev. Earth Planet. Sci., 13,* 210-240.

Warren P. H. and Wasson J. T. (1979) The origin of KREEP. *Rev. Geophys. Space Phys., 17,* 73-88.

Wood C. A. and Head J. W. (1975) Geologic setting and provenance of spectrally distinct pre-mare material of possible volcanic origin (abstract). In *Papers Presented to the Conference on Origins of Mare Basalts and Their Implications for Lunar Evolution,* pp. 189-193. The Lunar Science Institute, Houston.

Proceedings of the 20th Lunar and Planetary Science Conference, pp. 109-126
Lunar and Planetary Institute, Houston, 1990

Chemical Differences Between Small Subsamples of Apollo 15 Olivine-Normative Basalts

J. W. Shervais

Department of Geological Sciences, University of South Carolina, Columbia, SC 29208

S. K. Vetter

Department of Geological Sciences, University of South Carolina, Columbia, SC 29208

M. M. Lindstrom

Planetary Science Branch, Mail Code SN2, NASA Johnson Space Center, Houston, TX 77058

Nine samples of Apollo 15 mare basalt have been analyzed to assess chemical and petrological variations within the mare basalt suite at this site. All nine (15536, 15537, 15538, 15546, 15547, 15548, 15598, 15605, and 15636) are low-silica olivine normative basalts (ONBs) that correlate with the ONB suite as defined in previous studies. Partial analyses have been published for two of these samples, but the other seven have not been analyzed previously. Five of these samples are part of a concurrent study by Schuraytz and Ryder (1988). The nine samples vary in texture and grain size from fine-grained, intergranular or subophitic basalts to coarse-grained, granular "microgabbros." Six of these samples are small (original sample weights 1.9 g to 27.8 g) but three are relatively large (original sample weights 135.7 g to 336.7 g). Two splits from each sample were analyzed separately to assess chemical differences between small subsamples of the same rock as a function of grain size and texture. Seven of the basalts have subsample pairs that are similar in composition and plot on olivine control lines with other Apollo 15 low-silica ONBs. Two of the basalts (15547 and 15636) have subsample pairs that differ significantly in composition from each other and from large subsamples of the same basalt. They also deviate from the overall trend of well-analyzed ONBs (Rhodes and Hubbard, 1973). These samples are coarse-grained microgabbros characterized by the heterogeneous distribution of late-forming mesostasis phases (glass, ilmenite, fayalite, whitlockite, troilite, Fe-metal) that are rich in FeO, TiO_2, and incompatible trace elements. Our data support the conclusions of Ryder and Steele (1988) that the ONB suite (referred to here as the low-silica ONB suite to distinguish it from high-silica ONBs parental to the quartz-normative basalt suite) represents a single chemical group related by fractional crystallization of olivine. Chemical variations observed within the low-silica ONB suite that did not form by olivine removal are probably the result of nonrepresentative sampling and analytical uncertainty. These problems are most acute for small subsamples of coarse-grained granular basalts with heterogeneously distributed mesostasis phases. These data can be applied to breccia clast studies to infer the extent to which small clasts can be considered representative of their parent rock. The high-silica ONBs of Vetter et al. (1988) fall on olivine control lines with primitive quartz-normative basalts and are probably related to the QNB suite by olivine fractionation. The high-silica ONBs do not exhibit variations of the same magnitude or type as the low-silica ONBs studied here, which are affected primarily by mesostasis enrichments or depletions. We conclude that the high-silica ONBs of Vetter et al. (1988) are indeed a distinct rock type unrelated to the more common low-silica ONBs at the Apollo 15 site.

INTRODUCTION

Early studies of the Apollo 15 mare basalt suite established the presence of two distinct groups: the olivine-normative basalts and quartz-normative basalts (e.g., *Rhodes and Hubbard,* 1973; *Chappell and Green,* 1973). Olivine normative basalts (ONBs) are the more common variety and are inferred to overlie quartz-normative basalts (QNBs) at the Apollo 15 site (*ALGIT,* 1972; *Ryder,* 1989). These groups cannot be related to one another by fractional crystallization, so at least two separate parental magmas are needed (*Rhodes and Hubbard,* 1973; *Chappell and Green,* 1973).

Recent studies of mare basalts at the Apollo 14 site have revealed a wide variety of previously unrecognized mare basalt types (e.g., *Shervais et al.,* 1985a,b; *Dickinson et al.,* 1985; *Neal et al.,* 1988a,b). The new basalts, which occur as clasts in breccias or as coarse-fine fragments, have considerably broadened our views on mare basalt petrogenesis. These studies show that the lunar crust is far more complex than suspected previously, and that processes such as magma-mixing and wall-rock assimilation were important in its petrogenesis (e.g., *Warren et al.,* 1983; *Shervais et al.,* 1985a,b; *Neal et al.,* 1988a,b; *Warren,* 1988). However, in many cases the samples may be only a few grain diameters across and weigh less than 50 mg. This can create problems in obtaining a representative sample. These problems are most acute for coarse-grained highland rocks, but can also cause considerable uncertainty in the analysis of mare basalt clasts.

Similar problems may arise when small subsamples of individual hand samples are allocated for analysis. *Mason et al.* (1972) discuss this problem in regard to the Apollo 15 mare basalt suite. They report that their analysis of 15085, which has an average grain size of about 3 mm, deviates significantly from the trend of other analyzed mare basalts, and that this deviation is not in the direction expected by normal igneous processes. *Rhodes and Hubbard* (1973) note differences in separate analyses of coarse-grained mare basalt 15555 by different laboratories that were too large to be caused by interlaboratory analytical error. They also find large differences in their own replicate analyses of 15076, a coarse-grained basalt or microgabbro. *Mason et al.,* (1972) and *Rhodes and Hubbard* (1973) both attribute these discrepancies to inadequate sampling of the coarse-grained rocks. *Helmke et al.* (1973) discuss sampling problems in their study of Apollo 15 basalts, which includes 24 "walnut" and "peanut" sized samples from the rake samples and coarse-fines. They note that the compatible elements display relatively little scatter, but that the incompatible elements are more sensitive to the distribution of late mesostasis phases.

Clanton and Fletcher (1976) developed a model for major element variations in small samples based on Monte Carlo simulations of mineral grain distribution, observed grain size, mode, and sample weight. The model is designed to convert uncertainties in the mineral grain distribution into uncertainties in the abundance of the major elements. It does not consider the distribution of mesostasis phases, however, and thus cannot explain variations in the incompatible trace elements.

Ryder and Steele (1988) and *Schuraytz and Ryder* (1988) have attempted to minimize this problem for the Apollo 15 ONB suite by analyzing splits taken from large, homogenized subsamples. *Ryder and Steele* (1988) homogenized subsamples weighing 200 to 500 mg for 12 of their analyses. *Schuraytz and Ryder* (1988) increased that by an order of magnitude, homogenizing 4 to 5 g per sample. This approach is impractical for breccia clast studies, where the weight of sample extracted from the breccia matrix may be 100 mg or less.

We report here new chemical and petrographic data on nine samples of Apollo 15 ONB. Seven of these samples have not been analyzed previously; one has been analyzed by INAA only (15605; *Ma et al.,* 1978), and another has been analyzed by XRF and INAA (15636; *Compston et al.,* 1972; *Fruchter et al.,* 1973). These basalts exhibit a range in average grain size from coarse to fine, and several display macroscopic heterogeneity. Our approach is to analyze subsamples similar in size to those commonly used in breccia clast studies (100 to 150 mg), and to analyze two separate rock fragments from each sample. In addition, we received subsamples from five of the large samples studied by *Schuraytz and Ryder* (1988). Our goal is to present new data on the Apollo 15 olivine basalt suite, and to assess the effects of small sample size on the apparent bulk chemistry of mare basalt samples. We extend our conclusions to analyses of small clasts in breccia 15498 reported by *Vetter et al.* (1988).

METHODS

Nine mare basalts were sampled in this study: 15536, 15537, 15538, 15546, 15547, 15548, 15598, 15605, and 15636. Existing thin sections and probe mounts of each sample were studied using standard petrographic techniques, and their constituent mineral phases analyzed using the SX-50 electron microprobe at the University of South Carolina. Minerals were analyzed at 15 kV with a 25 nA beam current and a 1-2-μ spot size; representative mineral analyses are reported in the Appendix, and complete analytical data are on file in the datapacks at NASA-JSC. A suite of natural and synthetic minerals provided by E. Jarosewich of the National Museum of Natural History at the Smithsonian Institution were used as standards. Modes were determined by point counts on one to three thin sections of each sample; these data are reported in Table 1. The number of sections counted for the modes was a function of the number of sections available for each sample, not their grain size. Grain size distributions were determined using color photomicrographs; for each sample, *all* of the mineral grains in one or more photos were measured.

Five of the nine samples studied here were received as two separate subsamples, each weighing approximately 150 mg, taken from different parts of the parent sample. The other four samples were received as single 200-mg samples (15536, 15598, 15605, 15636) and were split into two 100-mg subsamples prior to further processing. Each subsample was crushed to a fine powder in an agate mortar and further subdivided into two fractions: 35-50 mg for major element analysis using the fused bead-electron microprobe technique (*Brown,* 1977) and 60-100 mg for trace element analysis by instrumental neutron activation analysis (INAA). FeO, Na_2O, and Cr were also determined by INAA. Subsamples for fused

TABLE 1. Modal composition of Apollo 15 olivine normative basalts based on petrographic analysis.

Sample	15536 ,7	15536 ,8	15536 Avg.	15537 ,4	15538 ,4	15538 ,5	15538 Avg.	15546 ,6	15546 ,7	15546 ,8	15546 Avg.	15547 ,6	15547 ,7	15547 ,8	15547 Avg.
Points Counted	6750	6079	12829	831	532	563	1095	1211	1162	1398	3771	719	1218	1086	3023
Modal %															
Olivine	28.0	20.2	24.3	17.7	18.0	20.2	19.2	7.8	20.8	10.6	12.9	19.5	16.2	16.5	17.1
Pyroxene	35.9	40.2	37.9	41.2	41.0	35.7	38.3	51.6	39.1	53.6	48.5	47.6	45.4	49.0	47.2
Plagioclase	31.2	30.3	30.8	32.3	31.4	29.3	30.3	34.2	27.5	26.5	29.3	26.8	30.2	23.8	27.1
Opaques	3.1	5.4	4.2	6.5	0.8	2.3	1.6	3.7	5.8	5.1	4.9	1.0	5.1	2.2	3.1
Mesostasis	0.2	0.7	0.5	0.7	2.6	2.8	2.7	0.6	3.4	1.6	1.8	0.7	1.1	2.1	1.4
Fayalite	0.2	0.8	0.5	0.0	0.6	1.1	0.8	2.1	2.2	0.1	1.4	0.1	1.6	0.9	1.0
Cristobalite	1.4	2.4	1.9	1.7	5.6	8.5	7.1	0.0	1.2	2.5	1.3	4.3	0.3	5.4	3.1

TABLE 1. (continued).

Sample	15548 ,4	15598 ,10	15598 ,11	15598 ,12	15598 Avg.	15605 ,5	15605 ,6	15605 Avg.	15636 ,8	15636 ,9	15636 Avg.
Points Counted	532	1836	2300	2286	6422	1488	1480	2968	1058	858	1916
Modal %											
Olivine	22.4	15.3	14.7	12.7	14.2	16.3	15.3	15.8	16.2	20.3	18.0
Pyroxene	41.4	42.6	40.8	45.1	42.8	47.6	48.0	47.8	47.9	40.8	44.7
Plagioclase	28.2	32.2	35.0	32.7	33.4	27.0	23.9	25.4	24.0	29.1	26.3
Opaques	5.6	7.1	6.9	5.7	6.5	4.5	8.3	6.4	6.1	6.2	6.2
Mesostasis	0.9	2.1	0.8	2.4	1.7	1.2	1.7	1.4	1.6	1.4	1.5
Fayalite	0.2	0.6	1.6	1.3	1.2	2.8	2.2	2.5	2.6	1.0	1.9
Cristobalite	1.3	0.2	0.1	0.0	0.1	0.5	0.7	0.6	1.6	1.2	1.4

bead EMP analysis were fused in an electric strip furnace at the University of South Carolina, using Mo foil sample boats and a dry nitrogen gas atmosphere. Fusion time was generally 20 seconds or less. The fused beads were analyzed at 15 kV with a 15 nA beam current and a 15-20-μ-diameter spot size on a Cameca SX-50 electron microprobe at the University of South Carolina (Table 2). Natural and synthetic minerals and glasses provided by E. Jarosewich of the National Museum of Natural History at the Smithsonian Institution were used as standards.

In order to test this method, ten separate fractions of USGS basalt standard W-2 were fused into glass beads and analyzed using the same conditions and standards as the mare basalts. The results were averaged and a standard deviation for each element determined (Table 2). This estimate of analytical uncertainty includes counting error, sample preparation, and matrix correction effects, which are the most important sources of error inherent in the analytical technique.

The trace element subsamples were analyzed at NASA-JSC using the procedures outlined by *Lindstrom et al.* (1989); the results are reported in Table 2 along with the major element data. The analytical uncertainties listed in the rightmost column reflect both counting errors and 1% uncertainty for noncounting errors. The silica glass tube containing one sample, 15537B, broke during irradiation or transport, resulting in the loss of about one-third of the sample. The mass of sample remaining was obtained by comparing the apparent FeO in the INAA sample to the concentration of FeO determined by fused bead electron microprobe analysis of powder from the same split.

PETROGRAPHY AND MINERAL CHEMISTRY

All the samples studied here are low-SiO_2 ONBs typical of the Apollo 15 site. Six of these samples are medium- or coarse-grained granular basalts and probably best classified as olivine microgabbros; the other three (15536, 15548, and 15598) are olivine-phyric basalts with medium- or fine-grained groundmass. The grain size designations used here differ from those used by *Ryder* (1985): fine-grained <0.35 mm average, medium-grained = 0.35-0.5 mm average, and coarse-grained >0.5 mm average. The reason for these differences is that the average grain size determined by counting *all* the grains in a given area is smaller than that estimated from random measurements because very small grains tend to be ignored in random measurements. The average grain size here was determined by calculating the area of each measured grain, averaging the areas, and determining the equivalent length of the average grain. This results in a slightly larger average grain size than simply averaging the grain diameters. Photomicrographs of each sample, all made at the same magnification, are shown in Fig. 1.

Petrography

Basalt 15536. 15536 is a medium-grained olivine-phyric basalt, with olivine phenocrysts 0.5 to 1.5 mm in diameter (Fig. 1a). Olivine is less common than pyroxene in the groundmass, where both minerals range from 0.1 to 1.1 mm across (average about 0.4 mm). The mafic phases are enclosed by plagioclase laths up to 1.7 × 0.5 mm in size (0.5 × 0.2 mm average) that form a subophitic to poikilitic texture, with small pyroxene chadacrysts in plagioclase oikocrysts. Ilmenite, ulvospinel, fayalite, cristobalite, glass, and troilite all crystallized late and are interstitial to the mafic phases, but are not generally associated with plagioclase. *Ryder* (1985) noted that 15536 contains mafic and felsic bands in hand specimen, but these bands are not obvious in thin section.

Basalt 15537. 15537 is a medium-grained, vesicular olivine basalt with rounded olivine grains, 0.20 to 1.5 mm across, forming chadacrysts in large, subhedral pigeonite oikocrysts (Fig. 1b). Small chadacrysts of plagioclase in pyroxene are less common. Pigeonite forms blocky prisms up to 4 mm long, but most are 0.2 to 1 mm in length with subhedral to anhedral outlines (0.41 mm average). The pigeonite is surrounded by plagioclase laths 0.1 to 0.8 mm in length (about 1.5 mm maximum, 0.38 mm average) that enclose pyroxene subophitically, but some laths are partly enclosed by pyroxene. Late pyroxene-plagioclase intergrowths are finer-grained and have variolitic textures. Mesostasis phases, which include ilmenite, ulvospinel, chromite, glass, and fayalite, are relatively scarce and evenly distributed throughout the sections.

Basalt 15538. 15538 is medium-grained granular olivine basalt or microgabbro with a poikilitic, cumulus texture (Fig.

TABLE 2. Whole rock major and trace element analyses of Apollo 15 mare basalts.

	15536 A	15536 B	15536 A+B	15536 ,9003	15537 A	15537 B	15537 A+B	15538 A	15538 B	15538 A+B	15546 A	15546 B	15546 A+B	15546 ,9003	15547 A	15547 B	15547 A+B	15547 ,9003
Sample wt. (mg)	101	101	202		146	143	289	159	145	304	152	147	299		154	151	305	
SiO_2	44.6	44.7	44.6	44.4	44.6	44.5	44.5	44.8	45.5	45.2	45.3	45.1	45.2	45.2	42.5	45.2	43.8	44.3
TiO_2	2.05	2.23	2.14	2.30	2.18	2.31	2.25	1.82	2.00	1.91	1.90	2.07	1.98	2.41	3.21	1.73	2.47	2.22
Al_2O_3	7.46	7.57	7.52	7.66	7.28	7.17	7.22	8.90	8.78	8.84	9.54	8.03	8.78	8.15	5.48	9.38	7.43	8.03
FeO	23.56	23.02	23.29	23.24	22.31	23.07	22.69	21.67	21.44	21.56	20.72	21.44	21.08	22.54	28.21	20.68	24.44	22.65
MnO	0.29	0.29	0.29	0.31	0.310	0.320	0.32	0.30	0.300	0.30	0.29	0.27	0.28	0.30	0.34	0.31	0.32	0.28
MgO	11.89	11.36	11.63	11.11	13.00	12.96	12.98	11.50	11.01	11.26	10.96	11.85	11.41	10.75	10.68	11.78	11.23	11.46
CaO	9.23	9.42	9.32	9.53	9.09	8.77	8.93	9.65	10.16	9.90	10.05	9.78	9.91	9.83	8.52	9.79	9.15	9.40
Na_2O	0.22	0.21	0.21	0.25	0.200	0.210	0.21	0.26	0.250	0.26	0.29	0.25	0.27	0.23	0.20	0.28	0.24	0.21
K_2O	0.02	0.04	0.03	0.04	0.030	0.030	0.03	0.02	0.020	0.02	0.02	0.02	0.02	0.03	0.05	0.02	0.04	0.03
P_2O_5	0.04	0.04	0.04	0.04	0.050	0.060	0.06	0.06	0.030	0.04	0.05	0.03	0.04	0.04	0.11	0.03	0.07	0.06
Cr_2O_3	0.69	0.65	0.67	0.60	0.700	0.720	0.71	0.58	0.580	0.58	0.64	0.67	0.65	0.58	0.62	0.63	0.62	0.64
Total	100.02	99.48	99.75	99.47	99.78	100.07	99.93	99.57	100.12	99.85	99.76	99.52	99.64	100.04	99.88	99.87	99.87	99.33
Na_2O	0.234	0.255	0.244		0.225	0.224	0.225	0.274	0.246	0.260	0.282	0.28	0.281		0.183	0.274	0.229	
FeO	22.4	22	22.2		21.5	23.1	22.3	20.5	22.2	21.4	20.1	20.2	20.1		27.8	20.6	24.2	
Sc	40	41.1	40.5		41.3	39.7	40.5	39.8	40.9	40.3	40.9	42.3	41.6		45.1	38.7	41.9	
Cr	4700	4590	4645		5100	5414	5257	4300	3980	4140	4360	4180	4270		4170	4450	4310	
Co	55.9	55.4	55.6		56.7	61.9	59.3	53.3	55.8	54.5	50.4	51.6	51.0		58.7	56.5	57.6	
Ni	60	70	65		80	78	79	61	90	76	80	70	75		60	70	65	
Sr	90	120	105		90	93	92	124	130	127	120	120	120		130	80	105	
Ba	41	23	32		47	50	48	34	45	40	33	45	39		72	32	52	
La	3.82	4.27	4.04		5.26	5.15	5.20	3.25	5.38	4.31	2.64	2.64	2.64		7.89	2.79	5.34	
Ce	12.3	12.1	12.2		14.6	14.6	14.6	10.3	14	12.2	8.5	7.7	8.1		22.2	8.2	15.2	
Sm	2.87	3.07	2.97		3.64	3.59	3.62	2.37	3.74	3.06	2.02	1.99	2.00		5.54	2.11	3.83	
Eu	0.739	0.81	0.774		0.811	0.790	0.801	0.746	0.89	0.818	0.706	0.71	0.708		1.03	0.721	0.875	
Tb	0.64	0.74	0.69		0.78	0.78	0.78	0.52	0.84	0.68	0.46	0.46	0.46		1.2	0.51	0.85	
Yb	1.87	1.94	1.91		2.25	2.30	2.28	1.6	2.35	1.98	1.39	1.42	1.40		3.3	1.41	2.35	
Lu	0.255	0.273	0.264		0.308	0.314	0.311	0.212	0.311	0.262	0.199	0.187	0.193		0.468	0.196	0.332	
Zr	70	<190	70		80	<150	80	60	120	90	<150	<130	<150		110	110	110	
Hf	2.25	2.3	2.27		2.66	2.61	2.64	1.79	2.7	2.25	1.59	1.51	1.55		4.25	1.52	2.88	
Ta	0.31	0.32	0.32		0.352	0.376	0.36	0.237	0.36	0.30	0.218	0.19	0.20		0.62	0.2	0.41	
U	0.06	<0.2	0.06		0.24	0.12	0.18	<0.2	<0.3	<0.3	0.13	<0.3	0.13		0.15	<0.23	0.15	
Th	0.29	0.3	0.29		0.44	0.44	0.44	0.25	0.4	0.33	0.16	0.17	0.17		0.76	0.23	0.49	

TABLE 2. (continued).

	15548 A	15548 B	15548 A+B	15598 A	15598 B	15598 A+B	15598 ,9003	15605 A	15605 B	15605 A+B	15605 ,3	15636 A	15636 B	15636 A+B	15636 ,9003	15636 composite	W-2	One sigma
Sample wt. (mg)	158	145	303	100	101	201		93	91	184		101	100	201				
SiO_2	45.0	44.6	44.8	44.7	45.3	45.0	45.1	44.9	45.1	45.0		45.3	42.3	43.8	44.8	44.6	53.07	0.2
TiO_2	2.39	2.63	2.51	2.48	2.65	2.57	2.45	2.27	2.27	2.27	2.1	2.26	3.60	2.93	1.93	2.22	1.01	0.12
Al_2O_3	8.65	8.13	8.39	7.92	8.19	8.05	8.47	8.55	7.73	8.14	9.1	6.73	5.01	5.87	9.14	8.55	15.47	0.6
FeO	22.13	22.96	22.54	23.42	22.74	23.08	22.56	22.23	22.77	22.50	22.3	23.99	28.77	26.38	21.22	22.67	9.58	0.29
MnO	0.30	0.28	0.29	0.32	0.31	0.31	0.28	0.30	0.33	0.32		0.32	0.34	0.33	0.27		0.17	0.02
MgO	9.72	9.84	9.78	10.35	9.39	9.87	9.82	10.22	10.70	10.46	11	11.08	10.36	10.72	11.60	11.32	6.53	0.23
CaO	10.29	9.95	10.12	9.66	10.21	9.94	10.16	10.05	9.80	9.92	10.1	9.45	8.54	9.00	9.63	9.58	11.04	0.12
Na_2O	0.27	0.25	0.26	0.24	0.25	0.25	0.23	0.25	0.22	0.24	0.257	0.17	0.16	0.17	0.23	0.26	2.29	0.06
K_2O	0.03	0.04	0.04	0.03	0.03	0.03	0.03	0.04	0.03	0.03	0.051	0.03	0.05	0.04	0.03	0.04	0.59	0.04
P_2O_5	0.07	0.06	0.07	0.04	0.06	0.05	0.05	0.05	0.06	0.06		0.04	0.10	0.07	0.03	0.07	0.14	0.05
Cr_2O_3	0.60	0.55	0.57	0.66	0.49	0.58	0.54	0.64	0.60	0.62		0.59	0.59	0.59	0.61			
Total	99.43	99.32	99.37	99.79	99.65	99.72	99.72	99.49	99.62	99.55		99.95	99.82	99.89	99.55	99.29	99.89	
Na_2O	0.261	0.252	0.257	0.254	0.28	0.267		0.261	0.255	0.258		0.221	0.175	0.198				0.005
FeO	21.8	22.7	2.3	22.6	22	22.3		22.1	21.9	22.0		24.4	29.1	26.8				0.25
Sc	47.7	45.6	46.7	44.2	45.6	44.9		43.8	44	43.9	42	45.3	47	46.1				0.5
Cr	4170	4100	4135	4460	3710	4085		4530	4310	4420	4150	4290	4160	4225		3840		50
Co	48.6	50.5	49.5	53.6	47.2	50.4		49.9	51.3	50.6	51	54.1	57.1	55.6		52		0.6
Ni	<100	n.d.	<100	<90	<100	<100		50	60	55	40	<110	90	90				25
Sr	110	140	125	170	130	150		140	90	115		100	140	120				25
Ba	50	39	45	48	55	52		52	49	51	45	65	111	88				9
La	4.62	5.17	4.89	4.44	4.97	4.71		5.95	5.45	5.70	5.4	7.81	11.49	9.65		2.6		0.08
Ce	13.8	15	14.4	13.8	14.3	14.1		16.2	15.7	15.9		23.4	32.1	27.8				1
Sm	3.44	3.78	3.61	3.33	3.6	3.46		4.19	3.82	4.00	3.6	5.53	7.8	6.67		1.9		0.04
Eu	0.86	0.92	0.890	0.82	0.93	0.875		0.96	0.93	0.945	0.84	1.07	1.31	1.190		0.66		0.02
Tb	0.79	0.86	0.82	0.74	0.83	0.78		0.95	0.86	0.91		1.17	1.71	1.44				0.03
Yb	2.16	2.33	2.25	2.12	2.3	2.21		2.55	2.4	2.47	0.7	3.19	4.49	3.84		1.3		0.05
Lu	0.338	0.318	0.328	0.299	0.313	0.306		0.36	0.322	0.341	0.29	0.443	0.622	0.532		0.22		0.01
Zr	90	90	90	160	110	135		130	130	130		150	210	180		77		40
Hf	2.55	2.72	2.63	2.54	2.7	2.62		3.14	2.85	3.00	2.5	4.02	5.76	4.89		1.3		0.08
Ta	0.33	0.38	0.35	0.34	0.38	0.36		0.45	0.38	0.42	450	0.53	0.84	0.69				0.02
U	<0.27	<0.3	<0.3	<0.12	<0.3	<0.3		0.11	0.24	0.17		0.14	0.24	0.19				0.05
Th	0.37	0.39	0.38	0.414	0.41	0.41		0.48	0.42	0.45		0.71	0.99	0.85				0.04

Explanation: All ,9003 samples from large subsamples of *Schuraytz and Ryder* (1988). 15605,3 analysis by *Ma et al.,* 1978; 15636 composite from *Chappel and Green* (1973; majors) and *Fruchter et al.* (1973; most trace elements). Samples "A+B" are averages of the two small subsamples for each basalt analyzed here.

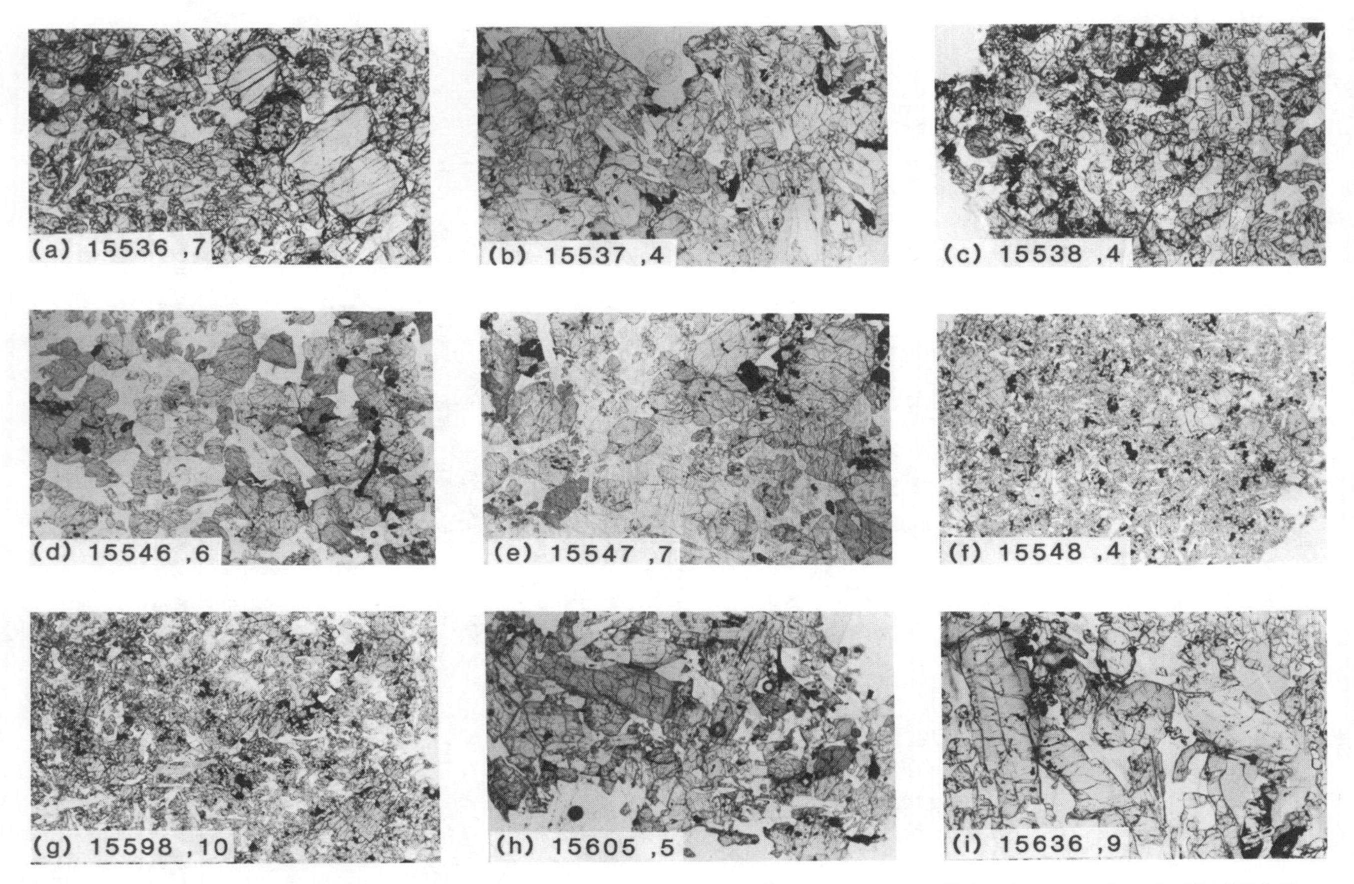

Fig. 1. Photomicrographs of the nine mare basalt samples studied here, all at the same scale (field of view = 5.6 mm). **(a)** 15536 ,7; **(b)** 15537 ,4; **(c)** 15538 ,4; **(d)** 15546 ,6; **(e)** 15547 ,7; **(f)** 15548 ,4; **(g)** 15598 ,10; **(h)** 15605 ,5; **(i)** 15636 ,9.

1c). Rounded olivine and pyroxene grains 0.1 to 0.8 mm across are poikilitically enclosed by blocky, lath-shaped plagioclase oikocrysts 1.0 to 2.5 mm long, and up to 1.0 mm wide. Smaller plagioclase laths are interstitial to both plagioclase and pyroxene. Magnesian olivine also occurs as small (0.3 mm average) rounded inclusions in pigeonite, which zones outward to augite. The pigeonites contain very thin exsolution lamellae of a pale brown phase, possibly spinel.

Half of thin section ,4 is mafic and the other half is plagioclase-rich; this banding was also noted in hand specimens by *Ryder* (1985). The mafic half contains abundant pigeonite, olivine, ulvospinel, and ilmenite surrounded by 25-30% interstitial plagioclase. The other half of the section contains larger lath-shaped plagioclase oikocrysts that comprise 50-60% of the basalt, with small, rounded chadacrysts of olivine and pigeonite. Mesostasis clots consisting of ilmenite, fayalite, cristobalite, and brown glass are common in the mafic half of the section, but are rare in the plagioclase-rich half.

Basalt 15546. 15546 is a medium- to coarse-grained, granular-textured olivine basalt (Fig. 1d). Olivine forms blocky or rounded crystals less than 1 mm across (0.4 mm average) that generally are either overgrown by pigeonite or form inclusions within single pigeonite grains. Pigeonite forms blocky, subhedral crystals, mostly 0.1 to 1.0 mm long, although some are as large as 2 mm (0.4 mm average). The mafic phases are surrounded by interstitial plagioclase laths approximately 0.8×0.2 mm in size that enclose pyroxene subophitically, but in some cases this relationship is reversed and pyroxene molds around plagioclase.

Except for a few chromite inclusions, opaque phases are generally interstitial to pyroxene and mold around pyroxene crystal faces. The opaques, ilmenite and ulvospinel, are usually small (0.1 to 0.2 mm) but a few ilmenite blades are as large as 1.2 x 0.4 mm. Late-forming mesostasis phases such as cristobalite, fayalite, and pale brown glass are commonly associated with the opaques, but these mesostasis-rich areas are small and evenly distributed throughout the section.

Basalt 15547. 15547 is a coarse-grained, granular-textured olivine basalt or microgabbro (Fig. 1e). Pyroxene and plagioclase form an interlocking network of subhedral to anhedral prisms and laths 0.5 to 2 mm long (0.58 mm average), with the plagioclase commonly enclosing pyroxene subophitically. The pyroxene (0.62 mm average) is strongly zoned from pigeonite cores to reddish-brown ferroaugite rims that are rich in opaque inclusions. The pigeonites commonly contain very thin exsolution lamellae of a pale brown phase that are too thin to analyze but which may be spinel; similar lamellae are found in 15605 and 15538. Olivine occurs as tiny

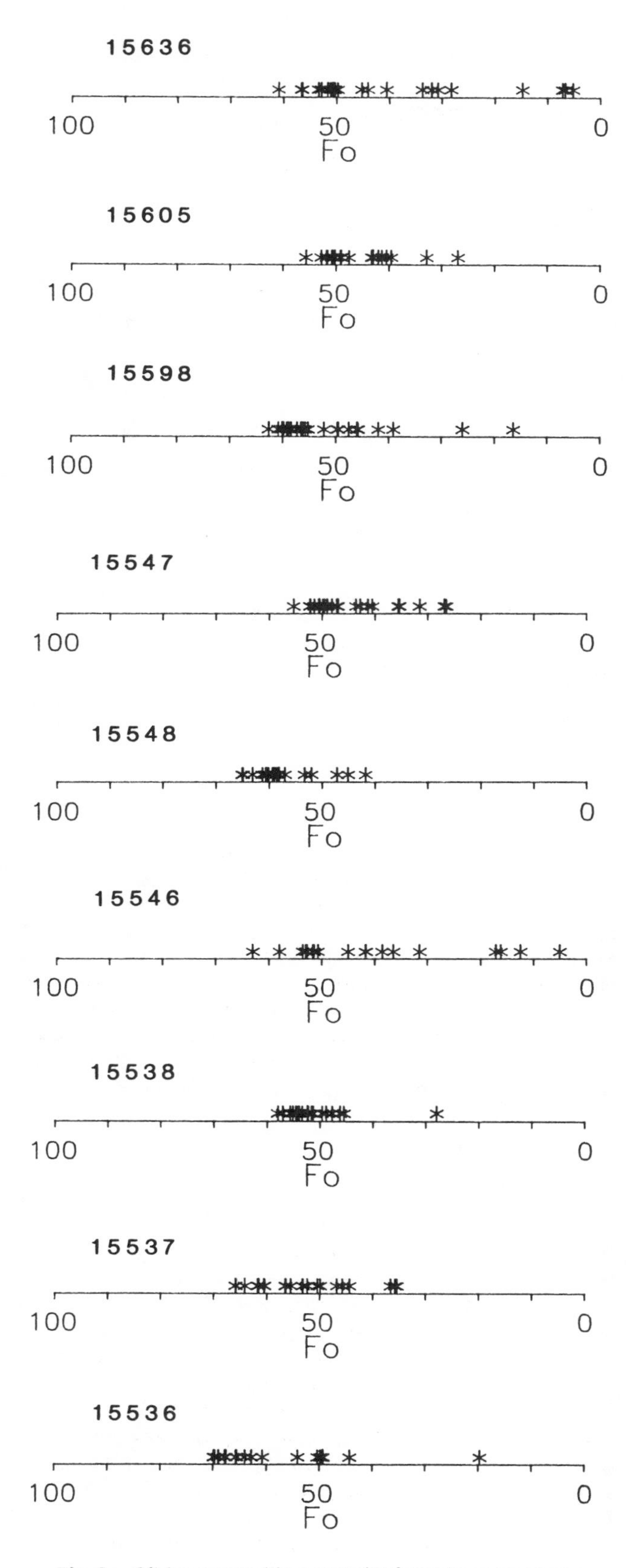

Fig. 2. Olivine compositions in mol.% forsterite component.

(0.1 to 0.5 mm) rounded inclusions in pigeonite and plagioclase. Ilmenite and ulvospinel occur interstitially either as individual grains or in clots with other mesostasis phases.

Mesostasis phases typically occur as clots up to 0.6 mm across that are irregularly distributed throughout the thin sections. The clots are scarce and small in sections ,6 and ,7, whereas in section ,8 they are abundant. The clots generally consist of fayalite, ilmenite, cristobalite, pale brown glass, ulvospinel, whitlockite, troilite, and Fe-metal; fayalite and glass are commonly associated in vermicular intergrowths.

Basalt 15548. 15548 is a very fine-grained, olivine-microphyric basalt with an intergranular to subophitic texture in which plagioclase laths partially enclose smaller pyroxene grains (Fig. 1f). The olivine microphenocrysts are 0.25 to 1.25 mm across (0.4 mm average) and have subhedral to nearly euhedral outlines with ragged grain boundaries. A few olivines are embayed and the embayment is filled with groundmass phases. The groundmass consists of a framework of plagioclase laths 0.05 to 0.55 mm long (0.23 mm average) with interstices that are filled with granular aggregates of pyroxene (0.17 mm average) and opaques (0.09 mm average). A few of the larger plagioclase grains are poikilitic and enclose small pyroxene grains, but partial molding of plagioclase around pyroxene is more common. The opaque phases, ilmenite and ulvospinel, are evenly distributed as small, anhedral grains that are interstitial to pyroxene; cristobalite and glass are scarce.

Basalt 15598. 15598 is a fine-grained, olivine-phyric basalt consisting of scattered olivine phenocrysts 0.6 to 1.3 mm across, set in an intergranular matrix of plagioclase, pyroxene, and opaques (Fig. 1g). The olivine phenocrysts are subhedral to anhedral in outline, with embayed rims and ragged or fritted grain boundaries (Fig. 1g). The groundmass consists of subhedral plagioclase laths 0.1 to 0.8 mm long (0.33 mm average) separated by a fine granular aggregate of pyroxene, ilmenite, and ulvospinel. Cristobalite, fayalite, and residual glass are distributed in residual patches, but these are generally small and evenly distributed throughout the sample. The groundmass pyroxene is generally only 0.1 to 0.4 mm in diameter (0.27 mm average), and forms a granular mosaic that fills in between plagioclase laths.

Basalt 15605. 15605 is a coarse-grained, vesicular olivine-bearing basalt with scarce, large, rounded olivines up to 1.25 mm across. Most olivines form small (<0.8 mm) rounded and embayed inclusions in pyroxene, but section ,5 contains a large skeletal olivine grain 3 mm long × 0.3 mm wide with a hollow core (*Ryder,* 1985). Pigeonite forms subhedral prisms up to 2.5 × 0.6 mm, but most are 0.2 to 2 mm in length (0.56 mm average). Chromite inclusions are common in both pigeonite and olivine. Many of the larger pigeonites have very thin exsolution lamellae of a pale brown phase—possibly spinel—that are too thin to analyze. Plagioclase forms laths up to 2 mm long that are interstitial to pyroxene and commonly mold around them to form a subophitic texture.

Mesostasis phases include fayalite, ilmenite, ulvospinel, cristobalite, brown glass, and a phosphate. These are generally concentrated in mesostasis-rich areas between the mafic silicate grains, but these areas are small and evenly distributed

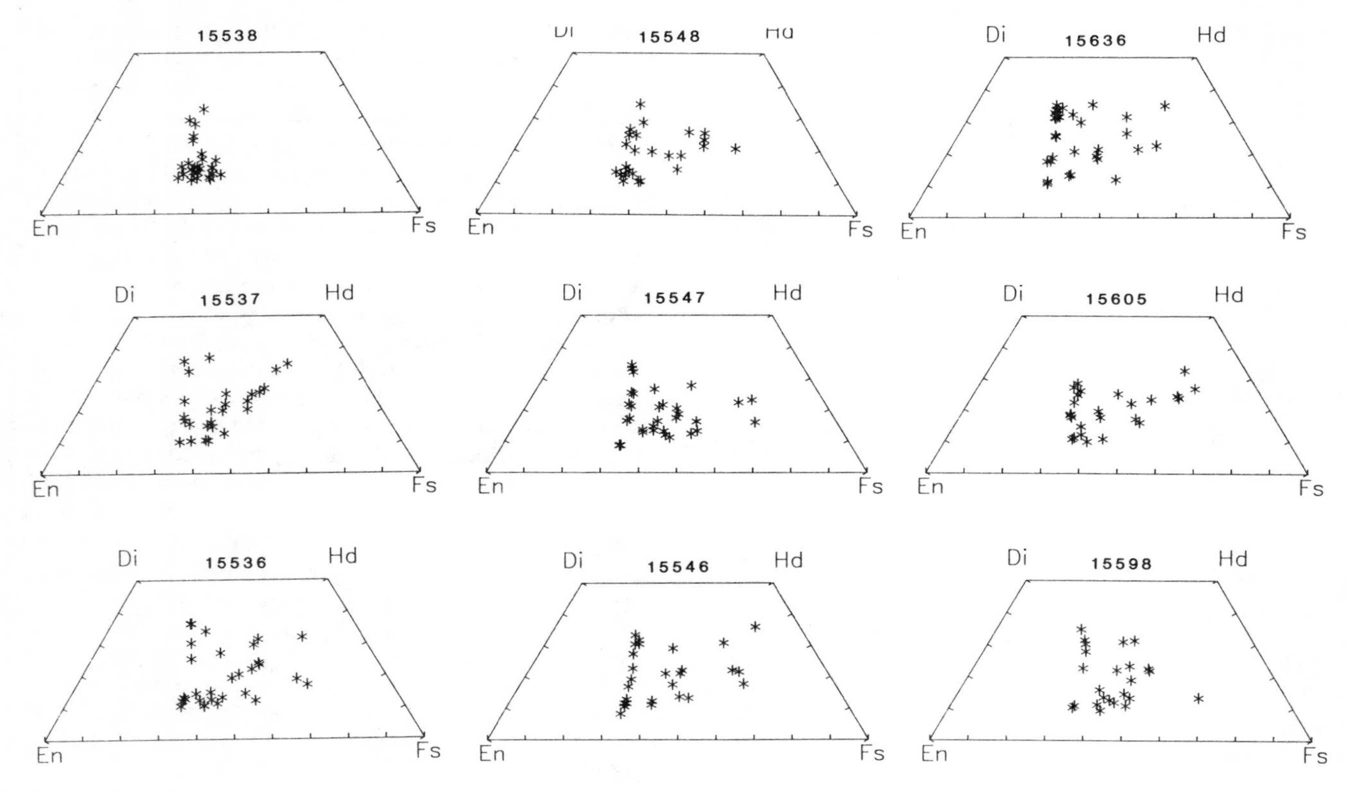

Fig. 3. Pyroxene compositions.

throughout the sample. One interesting textural feature found in section ,5 is a deformed pyroxene grain that has been curved around an adjacent olivine. The pyroxene has undulatory extinction where the cleavage is curved, and a sharp kink-band boundary with the undeformed termination of the crystal. Despite the strong curvature in the pyroxene, the adjacent plagioclase shows no indication of strain or deformation.

Basalt 15636. 15636 is a coarse-grained, granular-textured olivine basalt or microgabbro consisting of large subhedral to anhedral pyroxene and plagioclase laths up to 3.5 mm long (Fig. 1i). Pyroxene (0.74 mm average) is strongly zoned from pigeonite cores to reddish-brown, inclusion-rich ferroaugite rims. The pyroxenes are commonly twinned but no exsolution lamellae are evident. Blocky plagioclase laths (0.65 mm average) enclose pyroxenes subophitically, and also poikilitically enclose some of the smaller olivine and pyroxene grains. Olivine occurs as small (0.1 to 0.5 mm) subhedral to euhedral inclusions in plagioclase, or as rounded inclusions in pyroxene. Opaque phases include chromite, ilmenite, ulvospinel (which may contain ilmenite exsolution lamellae), troilite, and Fe-metal. Chromite generally occurs as inclusions in pigeonite or olivine; the other opaque phases are commonly associated with mesostasis clots that also include fayalite, cristobalite, brown glass, and whitlockite.

The mesostasis clots are up to 1.0 × 0.5 mm in size and are distributed irregularly throughout the sample. Their occurrence is generally restricted to plagioclase-free areas between pyroxene grains, and they are commonly intergrown with the ferroaugite rims. 15636 resembles 15538 in the irregular distribution of mafic phases and plagioclase: two-thirds of thin section ,8 consists of plagioclase-rich basalt (45-50% plagioclase) with rare opaques; the other one-third of ,8 contains only 25-30% plagioclase. The mafic portion of ,8 and most of thin section ,9 contain abundant mesostasis clots, but even in section ,9 (which is relatively mafic) the mesostasis-rich areas are clustered in three distinct regions. This highly irregular distribution of opaque and mesostasis phases is also evident in hand specimen.

Mineral Chemistry

Olivine. Olivine phenocrysts in the three olivine-phyric basalts (15536, 15548, 15598) range in composition from Fo50 to Fo71, with rims that are somewhat richer in iron—Fo43 to Fo63 (Fig. 2). The cores of the olivine phenocrysts are generally more magnesian than the resorbed olivine chadacrysts in the coarser-grained olivine basalts and olivine microgabbros, which generally range from Fo40 to Fo57, although chadacrysts in one sample (15537) are more magnesian (Fo60 to Fo66). Fayalitic olivines associated with other late mesostasis phases in the coarse-grained basalts are greenish in color, and range in composition from Fo37 to Fo15.

Pyroxene. Pyroxenes in all the samples have similar compositional ranges, regardless of texture or grain size (Fig.

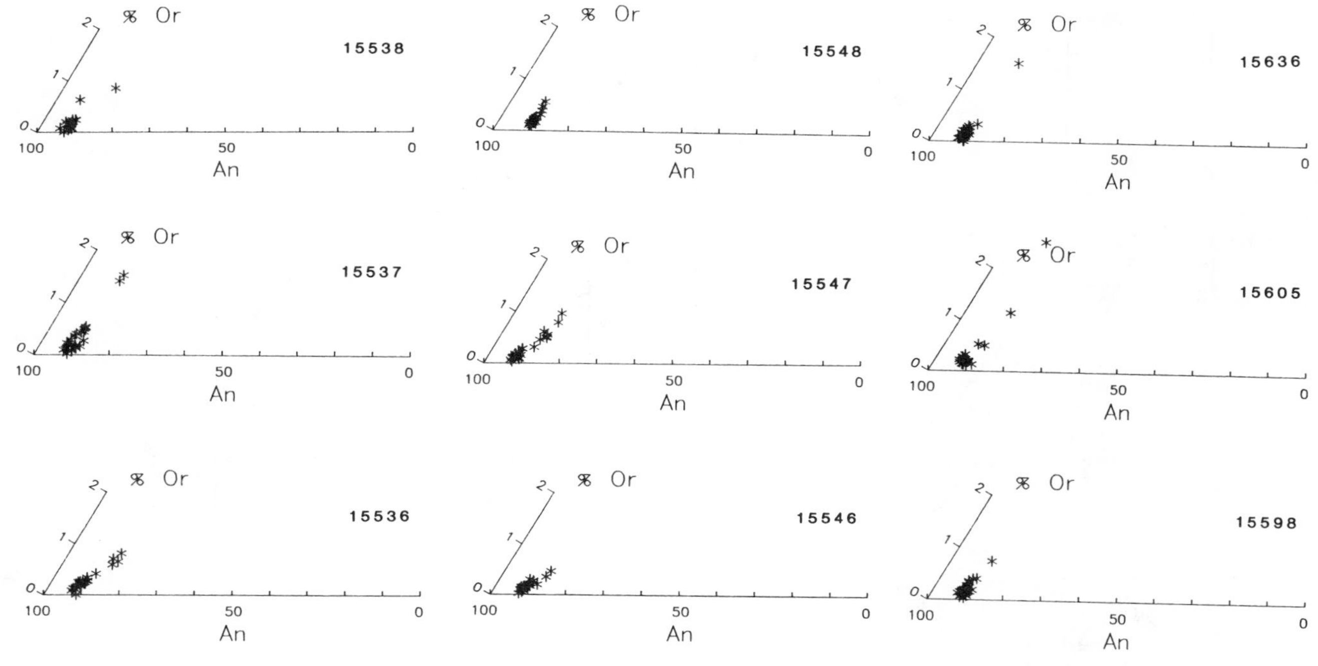

Fig. 4. Plagioclase compositions.

3). Groundmass pyroxenes in the olivine-phyric samples have 100*Mg/(Mg+Fe) ratios (MG#s) that range from 66 to 24. Large, subhedral pyroxenes in the coarse-grained olivine microgabbros have pigeonitic cores (MG#s = 67 to 46) that zone outwards toward augite or ferroaugite rims (MG#s = 63 to 24). Minor element concentrations are also similar, with Al_2O_3 = 0.8 to 3.1 and TiO_2 = 0.3 to 1.8 in all samples. In general, the highest minor element concentrations are associated with higher wollastonite components in the pyroxene.

Plagioclase. Plagioclase displays limited compositional variation, ranging from An93 to An85 and averaging about An91 in most samples (Fig. 4). One of the coarse-grained basalts (15547) seems to include a larger proportion of more sodic plagioclase, lowering the average composition to around An89 (Fig. 4). Potassium contents are uniformly low, with compositions ranging up to 2 mol.% Or maximum.

Opaques. Ilmenite and ulvospinel are the most common opaque phases in all samples; low-Ti chromite is much less abundant and generally occurs as small euhedral inclusions in olivine or pigeonite. Ilmenites display limited solid solution towards geikielite, with 0.1 to 1.5 wt.% MgO. Chromites include about 3 to 4 wt.% TiO_2, the ulvospinels 20 to 30 wt.% TiO_2.

Mesostasis glass. Mesostasis glass is scarce in the fine-grained samples, but is common in many of the coarser-grained rocks. This glass is rich in SiO_2 (74 to 79 wt.%), Al_2O_3 (10.7 to 12.9 wt.%), and K_2O (2.6 to 8.8 wt.%), with relatively minor TiO_2 (0.4 to 0.6 wt.%), FeO (0.9 to 2.15 wt.%), MgO (0 to 0.03 wt.%), CaO (0.6 to 4.0 wt.%), and Na_2O (0.12 to 0.3 wt.%).

WHOLE ROCK GEOCHEMISTRY

Whole rock geochemical data for the nine mare basalts studied here are presented in Table 2, along with additional data from the literature for samples 15605 and 15636 (*Ma et al.,* 1978; *Compston et al.,* 1972; *Fruchter et al.,* 1973). Table 2 also includes major element data that we determined for five of the basalts studied here, using subsamples from the 4 to 5 g homogenized powders prepared by *Schuraytz and Ryder* (1988). These data, along with the previously published results for samples 15605, provide our best estimate of "true" bulk rock compositions for comparison with our analyses of the small subsamples.

Major Elements

All the samples studied here have major element geochemical characteristics typical of the Apollo 15 olivine-normative basalt suite. Data for the 18 small subsamples (2 per basalt sample) and 6 large subsamples are plotted on MgO variation diagrams in Fig. 5. All of the large subsamples and 16 of the small subsamples plot within the same region as the other Apollo 15 ONBs. Two of our small aliqouts (15547A and 15636B) are enriched in FeO and TiO_2, and depleted in SiO_2 and CaO; these samples fall outside the area of typical Apollo 15 ONBs and far from the other samples studied here (Fig. 5).

Five of the nine samples studied here (15536, 15537, 15538, 15548, 15605) have subsample pairs that are essentially the same within analytical uncertainty (assumed to be about twice the standard deviation of replicate analyses, Table 2). Two other samples (15546, 15598) have subsample pairs that differ

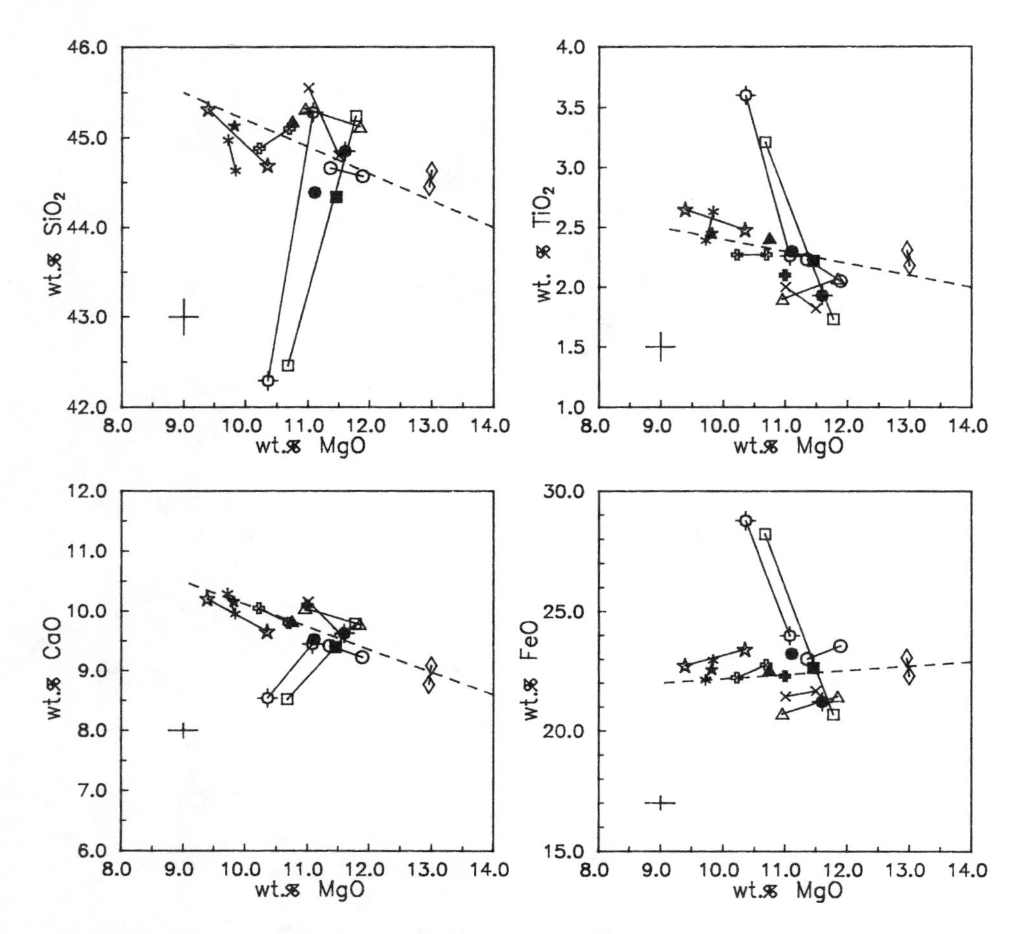

Fig. 5. MgO variation diagrams for SiO_2, TiO_2, FeO, and CaO. Open symbols connected by tie-lines are small subsamples analyzed here, closed symbols are large subsamples. Circles = 15536, diamonds = 15537, X = 15538, triangles = 15546, squares = 15547, asterisks = 15548, stars = 15598, cross = 15605, circle with cross = 15636. Dashed lines are olivine control lines projected from composition Fo70. Error bars in lower left corner of each diagram represent two-sigma analytical uncertainty (from Table 2).

by larger amounts, but still lie within the field of Apollo 15 ONBs. Of these seven samples, four have been analyzed as large samples (15536, 15546, 15598 analyzed here from the powders of *Schuraytz and Ryder,* 1988, and 15605 analyzed by *Ma et al.,* 1978). Three of these large samples are within analytical uncertainty of at least one corresponding subsample and within twice the analytical uncertainty of both subsamples (Fig. 5). The large sample of 15546 is much higher in FeO and TiO_2 than either corresponding subsample (Fig. 5).

Two of the nine samples studied here, 15547 and 15636, have subsamples that differ significantly from each other and from their corresponding large samples. Each of these has a subsample that deviates from the overall trend of Apollo 15 ONBs and does not lie along olivine control lines drawn through the other ONB data (Fig. 5). These large differences in composition affect all the important major elements—Fe, Ti, Si, Ca, Al, and Mg.

Trace Elements

Differences in trace element concentrations among the nine subsample pairs studied here are shown in Fig. 6 for six representative elements as a function of MgO content. MgO was chosen for the abscissa to facilitate comparison with the major element data in Fig. 5. Concentrations of compatible elements such as Cr and Co correlate positively with MgO overall, which probably reflects control by olivine (Co) and Cr-spinel (Cr). Several subsample pairs show distinct negative correlations within this overall trend (Fig. 6). Details of the Cr, Co, and MgO variations among the subsample pairs suggest some decoupling of olivine and chromite.

Incompatible trace elements such as La, Ta, Hf, and Sc correlate negatively with MgO overall (as expected) and most data scatter around simple olivine control lines (Fig. 6). Data for the small subsamples parallel that for the large samples but with somewhat greater scatter. However, subsamples that fall

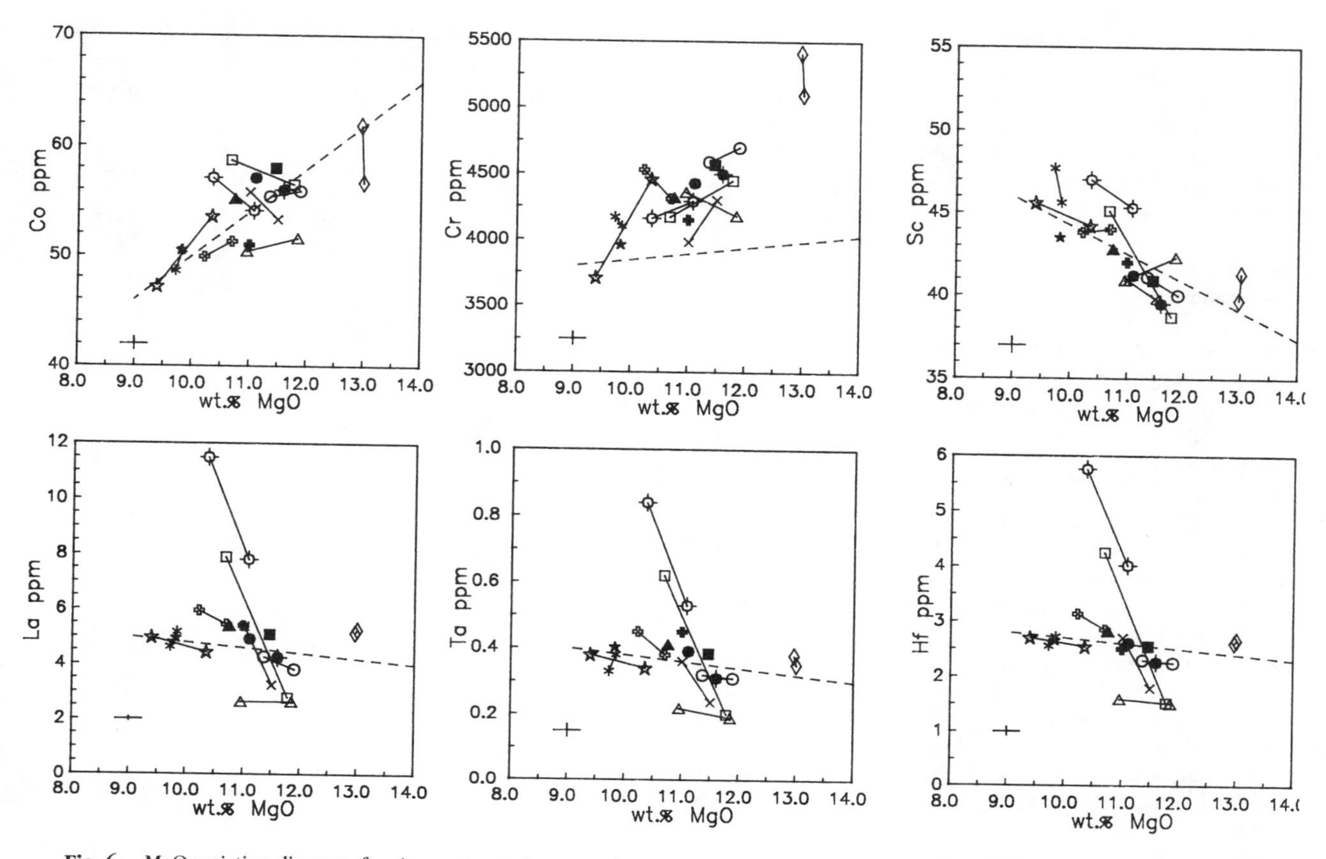

Fig. 6. MgO variation diagrams for six representative trace elements: Co, Cr, Sc, La, Ta, and Hf. Same symbols and conventions as Fig. 5.

outside the normal ONB fields on major element plots (Fig. 5) show strong enrichments in La, Ta, and Hf which are not consistent with subtraction of phenocrystic olivine or any other major silicate phase (Fig. 6).

DISCUSSION

Chemical Evolution of the Olivine Normative Basalt Suite

The chemical and petrologic data presented above show that all nine of the samples studied here are low-silica ONBs similar to those described earlier from the Apollo 15 site (e.g., *Rhodes and Hubbard,* 1973; *Chappell and Green,* 1973; *Helmke et al.,* 1973; *Dowty et al.,* 1973). Variation diagrams show that the overall trend of the data is consistent with the fractionation of olivine (plus minor Cr-spinel) from a high-MgO parent magma (Figs. 5 and 6). Olivine control lines (dashed) are generally parallel to the trend of the data; the exact position of the control line depends on the olivine composition used (Fo70 in this case) and the low MgO projection point.

One sample analyzed here, 15537, contains ~13.0 wt.% MgO and is the most primitive Apollo 15 low-silica ONB described yet. The two small subsamples analyzed here have compositions that are nearly identical, despite the fact that these subsamples were taken from different parts of the parent rock (see methods). As noted here and by *Ryder* (1985), olivine in 15537 occurs as small chadacrysts in pyroxene, not as phenocrysts (Fig. 1b). Thus, it seems unlikely that the high MgO in these analyses is due to excess (nonrepresentative) olivine in the subsamples. This is supported by the observation that 15537 is not deficient in incompatible elements.

Least square mixing calculations using 15537 as the parent magma indicate that about 12% olivine fractionation is needed to model the most evolved sample analyzed here (15598); about 17% olivine fractionation is needed to model the most evolved Apollo 15 ONB analyzed previously (15085). These results are consistent with previous suggestions of <15% olivine fractionation (*Rhodes and Hubbard,* 1973; *Chappell and Green,* 1973; *Helmke et al.,* 1973) when the more primitive nature of 15537 is considered.

Causes of Chemical Heterogeneity

The dispersion of chemical data in replicate analyses can be attributed to two basic causes: analytical error and nonrepresentative sampling. Deviations that exceed the expected analytical uncertainty may be due to sampling problems or to nonsystematic errors (e.g., human errors). Sampling problems can be reduced by analyzing samples that are large enough to contain representative proportions of the constituent mineral phases, but this may be impractical in many situations.

The mare basalts studied here vary from very fine to medium or coarse in grain size, and from intergranular or subophitic to poikilitic-granular in texture. One of the fine-grained, olivine-phyric samples, 15598, shows significant variations in the compatible elements Mg, Cr, and Co, suggesting that the olivine phenocrysts (and chromite inclusions within them) were not sampled representatively. Chemical variations created by the nonrepresentative sampling of olivine are common even in large samples of terrestrial rocks. However, in samples that are saturated with olivine-only (such as the Apollo 15 ONB suite), the addition or subtraction of olivine does not remove the rock from its normal liquid line of descent—it merely moves the rock toward or away from more primitive compositions.

One result that is somewhat surprising is that the small subsamples of the medium-grained basalts are consistently within the compositional range of Apollo 15 ONBs defined by larger samples (Figs. 5 and 6). Compositional differences between subsample pairs may exceed the analytical uncertainty of the major elements somewhat (generally by a factor of less than two), but credible fractionation trends can be drawn through the data, which tend to form linear arrays trending away from olivine. Differences between the subsample pairs and corresponding large samples show that caution must be exercised against trying to squeeze too much from the data (e.g., *Binder,* 1976).

Subsample pairs of the coarse-grained basalts 15547 and 15636 differ dramatically in both their major element and trace element concentrations. The differences exceed the expected analytical uncertainty significantly and result in subsamples that do not plot with other Apollo 15 ONBs. These samples are characterized by the heterogeneous distribution of late-forming mesostasis phases such as fayalite and ilmenite, and one sample (15636) is also heterogeneous with respect to plagioclase and pyroxene. The mesostasis phases (fayalite, ilmenite, glass, cristobalite, ulvospinel, and whitlockite) tend to occur together in the interstices between mafic phases (generally pigeonite) where these interstices are not already filled by plagioclase (which crystallizes before the mesostasis phases). This competition between the mesostasis phases and plagioclase for the interstitial areas between the early forming mafic silicates is graphically illustrated by the scarcity of mesostasis phases in the plagioclase-rich zones of samples like 15538.

The mesostasis phases contain much of the Fe (fayalite, ilmenite) and almost all of the TiO_2, Ta, and Hf (ilmenite, ulvospinel), P_2O_5 and REE (whitlockite), and U and Th (glass) found in these samples. This results in the positive correlation between FeO and TiO_2, P_2O_5, and La observed (Fig. 7). The trend of the data in Fig. 7 is at a high angle to possible olivine control lines and cannot be due to olivine subtraction or addition. Many of these phases are also low in SiO_2 (fayalite, ilmenite, spinel, whitlockite), resulting in a negative correlation between these elements and SiO_2. Apparently the silica-rich glass and cristobalite are not abundant enough in these clots to cause silica enrichment.

The importance of mesostasis phase distribution is illustrated by the medium-grained basalts that have subsample pairs that are essentially identical to one another. These samples are characterized by small mesostasis clots that are more or less

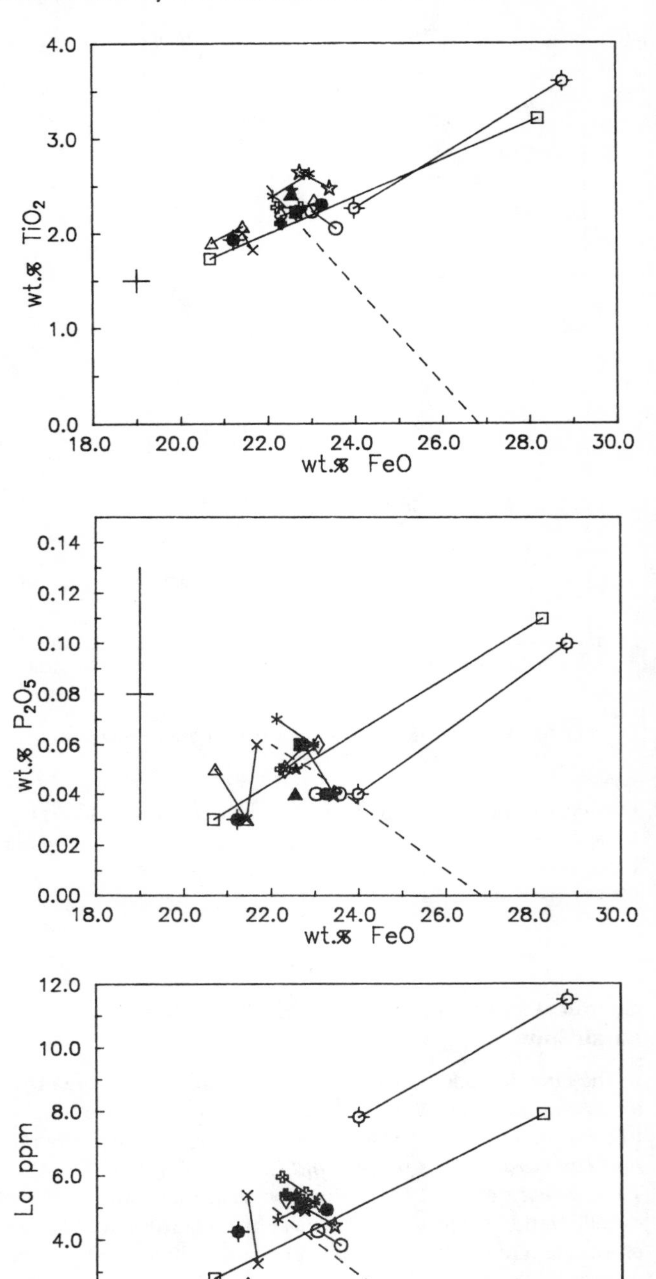

Fig. 7. FeO variation diagrams for TiO_2, P_2O_5, and La. Symbols and conventions same as Fig. 5. Note the strong positive correlation between these elements and FeO.

evenly distributed throughout the sample. The variation seen between some of these subsample pairs is not much larger than that observed by *Schuraytz and Ryder* (1988) between 4-or 5-g subsamples of 15555 and 15556. These samples can be used to model the geochemical evolution of the ONB suite despite the small size of the subsamples analyzed, as long as due consideration is given to the effects of analytical uncertainty on the results.

Haskin and Korotev (1977) were the first to document the effect of mesostasis distribution on bulk analyses of small mare basalt fragments. They found that small, 5- to 7-mg subsamples of basalt 70135 may differ by a factor of 20 or more in incompatible trace element concentrations, especially the REE, Ta, and Hf (*Haskin and Korotev,* 1977). They also found that separate subsamples of this basalt weighing 0.25 to 1.4 g may have trace element concentrations that vary by up to a factor of four—enough to significantly effect any trace element models based on these analyses (*Haskin and Korotev,* 1977). Haskin and Korotev attribute these chemical variations to differences in the proportions of mesostasis phases (primarily glass and ilmenite) present in each subsample.

These results are consistent with the so-called "short-range unmixing" model of *Lindstrom and Haskin* (1978, 1981). They proposed that chemical variations observed between hand samples from a single lava flow of Icelandic basalt, and within different suites of lunar mare basalt, could be explained by nonuniform distribution of the primary mineral phases and an incompatible-element-rich mesostasis. They show that nonrepresentative samples may be characterized by combinations of minerals and mesostasis that are not petrologically likely (e.g., olivine plus glass) and that the mesostasis could separate from the other phases at any stage during crystallization. *Lindstrom and Haskin* (1978, 1981) note that much of the incompatible element variation between samples from a single flow can be attributed to the nonuniform distribution of the mesostasis (*Lindstrom and Haskin,* 1981). Our results support this conclusion for mare basalt samples, where distribution of the mesostasis in small subsamples appears to be even more important than in terrestrial rocks.

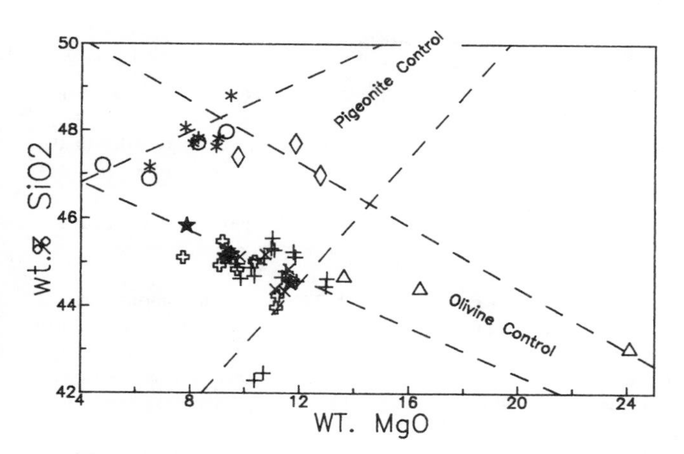

Fig. 8. MgO vs. SiO_2 plot for Apollo 15 mare basalts. Stars = low-silica ONB clasts in breccia 15498 (*Vetter et al.,* 1988); open crosses = low-silica ONBs (*Rhodes and Hubbard,* 1973); pluses = low-silica ONBs studied here (small subsamples); crosses = large subsamples of low-silica ONBs produced by *Schuraytz and Ryder* (1988), data from Table 2; open circles = QNB clasts in brecia 15498 (*Vetter et al.,* 1988); asterices = large QNB samples (*Rhodes and Hubbard,* 1973); diamonds = high-silica ONB clasts in breccia 15498 (*Vetter et al.,* 1988); triangles = olivine-pyroxene cumulate clasts in breccia 15498 (*Vetter et al.,* 1988). Dashed lines represent possible olivine control lines or pigeonite control lines based on high-Mg olivine and pigeonite phenocryst compositions.

Application to Breccia Clast Studies

One goal of this study is to evaluate the reliability of the small subsamples normally used in breccia clast studies. The primary limitation in these studies is imposed by the size of the clast, not the Lunar and Planetary Sample Team. Breccia clast samples allocated for chemical analysis are commonly in the 100-mg range, and in some cases even smaller. Despite the problems inherent in analyzing small samples, the wide array of potential samples not available in the large sample suite make these studies necessary and even desirable.

Vetter et al. (1988) investigated 25 mare basalt clasts extracted from lunar breccia 15498. Eleven of these clasts proved to be olivine normative, but three are chemically distinct from all previously studied ONB suite samples. These three clasts, which constitute the "high silica ONB" group of *Vetter et al.* (1988), are lower than normal ONBs in TiO_2 and FeO, and higher in SiO_2 (Fig. 8). *Vetter et al.* (1988) noted that these three clasts closely resemble basalts of the QNB suite, but are olivine normative because of their high MgO concentrations. They propose that the high-silica ONB clasts in 15498 represent primitive magmas that were parental to basalts of the QNB suite. This view is supported by a single large sample, 15065, which is grouped with the QNB suite even though it is slightly olivine-normative (*Rhodes and Hubbard,* 1973).

Another possible explanation for the high-silica ONBs is that they are nonrepresentative samples of normal ONB suite basalt. In this scenario, the high-silica ONBs are interpreted as spurious samples which result from the analysis of small subsamples of coarse-grained, heterogeneous, low-silica ONBs. The samples analyzed by *Vetter et al.* (1988) weighed only 100-150 mg, which is common for breccia clast studies but much smaller than the 4 to 5 g preferred for normal hand specimens (e.g., *Schuraytz and Ryder,* 1988).

Our results reported here can be used to test this hypothesis. The high-silica ONBs are medium-grained basalts similar to several of those studied here, and the subsamples analyzed by *Vetter et al.* (1988) were the same size as ours. Several lines of evidence support the conclusion of *Vetter et al.* that the high-silica ONBs do represent a distinct rock type that is *not* related to the low-silica ONB suite.

1. The most deviant samples of ONB studied here (15547, 15636) are characterized by the concentration of mesostasis-enriched areas in the subsample analyzed. This concentration of the mesostasis clots results in subsamples that are higher in FeO, TiO_2, and the incompatible trace elements such as La,

Yb, and Hf, and lower in CaO, Al_2O_3, MgO, and SiO_2. The high-silica ONBs are lower in FeO and TiO_2 than the low-silica ONBs, suggesting a depletion in mesostasis, but have incompatible element concentrations that are similar to both the low-silica ONB suite and the QNB suite. Unlike the most deviant samples studied here, the high-silica ONBs have major and trace element concentrations that fall on smooth trends with other Apollo 15 mare basalts; they are not depleted in incompatible elements as would be expected if they formed by subtraction of mesostasis clots from the bulk rock.

2. The deviant ONB samples studied here are characterized by heterogeneous distribution of mesostasis-rich clots containing fayalite, ilmenite, cristobalite, glass, and whitlockite. The high-silica ONBs do not contain any fayalitic olivine, glass, or whitlockite. Ilmenite and cristobalite do occur, but they are evenly distributed through the sections and do not form large mesostasis clots.

3. One olivine-phyric ONB studied here (15598) exhibits variations in Mg, Co, and Cr that suggest that olivine may be heterogeneously distributed. Although the high-silica ONB samples studied by *Vetter et al.* (1988) may be enriched or depleted in olivine, this will not move the samples off of their normal liquid line of descent because olivine is the primary liquidus phase in both the ONB suite and primitive members of the QNB suite.

4. A cursory examination of Fig. 8 suggests that enrichment of a low-silica ONB in pigeonite could create a high-silica ONB. This suggestion is wrong for several reasons. First, thin sections of high-silica ONB samples are not noticeably enriched in pigeonite. Second, sufficient enrichment of a low-silica ONB with pigeonite would lower the incompatible element concentration of the rock significantly; this is not observed. Third, pigeonite enrichment would also enrich Sc; this is not observed. Fourth, least squares mixing models are not consistent with this explanation for the other chemical elements not shown on Fig. 8.

5. Short-range unmixing calculations (*Vetter et al.,* 1988) show that compositional ranges within 15498 basalt suites cannot be due to short-range unmixing, but must represent fractional crystallization of the liquidus phases. Differences between suites require at least two separate parent magmas (*Vetter et al.,* 1988).

In summary, our data on small subsamples of mare basalt suggest that certain systematic variations can be expected if these subsamples deviate from the bulk composition of their parent rock because of nonrepresentative sampling. The high-silica ONBs of *Vetter et al.* (1988) exhibit variations from the normal, low-silica ONB group that are *not* consistent with those caused by nonrepresentative sampling. This supports the conclusion of *Vetter et al.* (1988) that the high-silica ONB group is a distinct rock type that is not related to the normal, low-silica Apollo 15 ONB suite.

CONCLUSIONS

The nine samples studied here are all normal low-silica olivine normative basalts similar to other Apollo 15 ONBs described previously (*Rhodes and Hubbard,* 1973; *Mason et al.,* 1972; *Chappell and Green,* 1973; *Helmke et al.,* 1973). One sample analyzed for this investigation (15537) is primitive in composition and may represent the ONB parent magma; other samples studied here can be related to it by up to 12% olivine fractionation. There is no evidence to suggest more than one low-silica ONB magma series.

Small subsamples of Apollo 15 ONBs may approximate the bulk rock if the late mesostasis phases are evenly distributed throughout the sample and do not form large clots that can be fractionated into or from the subsample during sampling. Where this does occur, systematic enrichment or depletion of the subsample in FeO, TiO_2, and the incompatible trace elements can result. Enrichment may also deplete the sample in SiO_2, Al_2O_3, and CaO; depletion will have the opposite effect. These trends are consistent with the short-range unmixing model of *Lindstrom and Haskin* (1978, 1981).

These results can be applied to breccia clast studies to show that a new group of high-silica ONBs described by *Vetter et al.* (1988) from breccia 15498 are a distinct rock type that is not related to the low-silica ONBs normally found at the Apollo 15 site. The high-silica ONBs have trace element concentrations that are in the range of Apollo 15 mare basalts, and so cannot represent low-silica ONBs that have been depleted in mesostasis-rich areas by the sampling process.

Acknowledgments. This work was supported by NASA's Planetary Materials and Geochemistry program through grant NAG9-169 to J. W. Shervais and RTOP 152-13-40-21 to M. M. Lindstrom. Curatorial work by C. Galindo of the JSC-LESC curatorial staff is greatly appreciated. This paper in its various incarnations was improved substantially by thorough reviews from R. Korotev, B. Schuraytz, D. Lindstrom, S. Hughes, and J. Jones.

REFERENCES

ALGIT (Apollo Lunar Geology Investigation Team) (1972) Geological setting of the Apollo 15 samples. *Science, 175,* 407-415.

Binder A. B. (1976) On the compositions and characteristics of the mare basalt magmas and their source regions. *The Moon, 16,* 115-150.

Brown R. W. (1977) A sample fusion technique for whole rock analysis with the electron microprobe. *Geochim. Cosmochim. Acta, 41,* 435-438.

Chappell B. W. and Green D. H. (1973) Chemical composition and petrogenetic relationships in the Apollo 15 mare basalts. *Earth Planet. Sci. Lett., 18,* 237-246.

Clanton U. S. and Fletcher C. R. (1976) Sample size and sampling errors as the source of dispersion in chemical analyses. *Proc. Lunar Sci. Conf. 7th,* pp. 1413-1428.

Compston W., de Laeter J. R., and Veron M. J. (1972) Strontium isotope geology of Apollo 15 basalts. In *The Apollo 15 Lunar Samples* (J. W. Chamberlain and C. Watkins, eds), pp. 347-351. The Lunar Science Institute, Houston.

Dickinson T., Taylor G. J., Keil K., Schmitt R. A., Hughes S. S., and Smith M. R. (1985) Apollo 14 aluminous mare basalts and their possible relationship to KREEP. *Proc. Lunar Planet. Sci. Conf. 15th,* in *J. Geophys. Res., 90,* C365-C374.

Dowty E., Prinz M., and Keil K. (1973) Composition, mineralogy, and petrology of 28 mare basalts from Apollo 15 rake samples. *Proc. Lunar Sci. Conf. 4th,* pp. 423-444.

Fruchter J. S., Stoeser J. W., Lindstrom M. M., and Goles G. G. (1973) Apollo 15 clastic materials and their relationship to geological features. *Proc. Lunar Sci. Conf. 4th,* pp. 1227-1237.

Haskin L. A. and Korotev R. L. (1977) Test of a model for trace element partition during closed system solidification of a silicate liquid. *Geochim. Cosmochim. Acta, 41,* 921-939.

Helmke P. A., Blanchard D. P., Haskin L. A., Telander K., Weiss C., and Jacobs J. W. (1973) Major and trace elements in igneous rocks from Apollo 15. *The Moon, 8,* 129-148.

Lindstrom M. M. and Haskin L. A. (1978) Causes of compositional variations within mare basalt suites. *Proc. Lunar Planet. Sci. Conf. 9th,* pp. 465-486.

Lindstrom M. M. and Haskin L. A. (1981) Compositional inhomogeneities in a single Icelandic tholeiite flow. *Geochim. Cosmochim. Acta, 45,* 15-31.

Lindstrom M. M., Marvin U. B., and Mittlefehldt, D. W. (1989) Apollo 15 Mg and Fe norites: a redefinition of the Mg-suite differentiation trend. *Proc. Lunar Planet. Sci. Conf. 19th,* pp. 245-254.

Ma M.-S., Schmitt R. A., Warner R. D., Taylor G. J., and Keil K. (1978) Genesis of Apollo 15 olivine normative basalts: Trace element correlations. *Proc. Lunar Sci. Conf. 7th,* pp. 523-533.

Mason B., Jaroesewich E., Melson W. G., and Thompson G. (1972) Mineralogy, petrology, and chemical composition of lunar samples 15085, 15256, 15271, 15471, 15475, 15476, 15535, and 15556. *Proc. Lunar Sci. Conf. 3rd,* pp. 785-796.

Neal C. R., Taylor, L. A., and Lindstrom, M. M. (1988a) Importance of lunar granite and KREEP in VHK basalt petrogenesis. *Proc. Lunar Planet. Sci. Conf. 18th,* pp. 121-137.

Neal C. R., Taylor L. A., and Lindstrom M. M. (1988b) Apollo 14 Mare Petrogenesis: Assimilation of KREEP-like components by a fractioning magma. *Proc. Lunar Planet. Sci. Conf. 18th,* pp. 139-153.

Rhodes J. M. and Hubbard N. J. (1973) Chemistry, classification, and petrogenesis of Apollo 15 mare basalts. *Proc. Lunar Sci. Conf. 4th,* pp. 1127-1148.

Ryder G. (1985) Catalog of Apollo 15 Rocks. *Curatorial Branch Publication 72, JSC 20787.* NASA-JSC, Houston. 1296 pp.

Ryder G. (1989) Mare basalts of the Apennine Front and the mare stratigraphy of the Apollo 15 site. *Proc. Lunar Planet. Sci. Conf. 19th,* pp. 43-50.

Ryder G. and Steele A. (1988) Chemical dispersion among Apollo 15 olivine- normative basalts. *Proc. Lunar Planet. Sci. Conf. 18th,* pp. 273-282.

Schuraytz B. and Ryder G. (1988) New petrochemical data base of Apollo 15 olivine normative mare basalts (abstract). In *Lunar and Planetary Science XIX,* pp. 1041-1042. Lunar and Planetary Institute, Houston.

Shervais J. W., Taylor L. A., and Lindstrom M. M. (1985a) Apollo 14 mare basalts: Petrology and geochemistry of clasts from consortium breccia 14321. *Proc. Lunar Planet. Sci. Conf. 15th,* in *J. Geophys. Res., 90,* C375-C395.

Shervais J. W., Taylor L. A., Laul J. C., Shih C.-Y., and Nyquist L. E. (1985b) Very high potassium (VHK) basalt: Complications in mare basalt petrogenesis. *Proc. Lunar Planet. Sci. Conf. 16th,* in *J. Geophys. Res., 90,* D3-D18.

Vetter S. K., Shervais J. W., and Lindstrom M. M. (1988) Petrolgoy and geochemistry of olivine-normative and quartz-normative basalts from regolith breccia 15498: new diversity in Apollo 15 mare basalts. *Proc. Lunar Planet. Sci. Conf. 18th,* pp. 255-271.

Warren P. H. (1988) The origin of pristine KREEP: effects of mixing between urKREEP and the magmas parental to the Mg-rich cumulates. *Proc. Lunar Planet. Sci. Conf. 18th,* pp. 233-241.

Warren P. H., Taylor G. J., Keil K., Kallemeyn G. W., Shirley D. S., and Wasson J. T. (1983) Seventh foray: whitlockite-rich lithologies, a diopside-bearing troctolitic anorthosite, ferroan anorthosites, and KREEP. *Proc. Lunar Planet. Sci. Conf. 14th,* in *J. Geophys. Res., 88,* B151-B164.

APPENDIX 1. Representative olivine compositions for Apollo 15 mare basalts.

	15536			15538			15537			15546			15547		
			Avg.			Avg.			Avg.						Avg.
SiO_2	37.7	31.7	38.2	35.8	34.3	36.3	37.3	33.3	35.7	37.2	29.9	34.0	35.2	32.8	34.3
FeO	26.80	59.06	30.76	35.66	43.60	38.03	29.24	49.90	36.63	31.39	66.47	46.62	36.82	49.37	42.56
MnO	0.25	0.58	0.32	0.39	0.47	0.41	0.30	0.37	0.36	0.32	0.74	0.45	0.35	0.50	0.42
MgO	35.31	8.65	29.20	27.95	21.35	24.58	33.38	16.90	27.32	31.44	2.35	17.96	26.87	15.91	22.11
CaO	0.28	0.52	1.58	0.32	0.27	0.74	0.21	0.39	0.31	0.21	0.50	0.67	0.35	0.38	0.33
Cr_2O_3	0.22	0.03	0.28	0.09	0.16	0.19	0.21	0.08	0.17	0.14	0.00	0.54	0.22	0.06	0.13
Total	100.59	100.65	100.75	100.29	100.22	100.40	100.68	101.03	100.56	100.75	100.27	100.43	99.83	99.12	99.94
Si	0.996	0.998	1.027	0.994	0.995	1.014	0.997	0.990	0.994	1.003	0.994	1.001	0.989	0.997	0.994
Fe	0.593	1.557	0.704	0.829	1.058	0.896	0.654	1.241	0.857	0.708	1.845	1.170	0.866	1.254	1.033
Mn	0.006	0.015	0.007	0.009	0.012	0.010	0.007	0.009	0.009	0.007	0.021	0.012	0.008	0.013	0.010
Mg	1.392	0.407	1.166	1.157	0.923	1.029	1.331	0.749	1.129	1.265	0.116	0.768	1.127	0.721	0.952
Ca	0.008	0.017	0.044	0.010	0.009	0.021	0.006	0.012	0.009	0.006	0.018	0.021	0.011	0.012	0.010
Cr	0.005	0.001	0.006	0.002	0.004	0.004	0.004	0.002	0.004	0.003	0.000	0.012	0.005	0.001	0.003
Sum	3.001	3.000	2.964	3.003	3.002	2.980	3.001	3.007	3.004	2.995	3.002	2.990	3.007	3.001	3.004
Fo	69.94	20.54	61.59	58.01	46.33	53.36	66.82	37.46	56.62	63.86	5.87	39.67	56.30	36.25	47.71
Fa	30.06	79.46	38.41	42.00	53.67	46.64	33.18	62.54	43.38	36.14	94.13	60.33	43.70	63.75	52.29

APPENDIX 1. (continued).

	15548			15598			15605			15636		
						Avg.						Avg.
SiO_2	37.2	33.9	35.8	36.6	34.6	37.0	35.1	31.7	34.3	35.8	30.7	34.5
FeO	29.89	46.13	35.36	33.64	43.62	37.31	38.68	55.54	44.17	35.91	61.77	42.73
MnO	0.25	0.35	0.34	0.33	0.47	0.37	0.37	0.54	0.43	0.32	0.60	0.44
MgO	32.84	19.37	28.14	29.67	21.60	24.84	25.49	12.09	21.29	27.39	6.48	22.42
CaO	0.27	0.29	0.28	0.26	0.34	0.84	0.29	0.32	0.34	0.28	0.46	0.37
Cr_2O_3	0.15	0.09	0.14	0.18	0.12	0.17	0.19	0.04	0.13	0.82	0.01	0.18
Total	100.67	100.30	100.10	100.52	100.82	100.79	100.22	100.38	100.71	100.64	100.17	100.76
Si	0.999	0.996	0.995	0.996	0.997	1.025	0.993	0.984	0.992	0.993	0.991	0.992
Fe	0.670	1.133	0.825	0.771	1.051	0.872	0.914	1.441	1.073	0.833	1.667	1.034
Mn	0.006	0.009	0.008	0.008	0.011	0.009	0.009	0.014	0.011	0.007	0.017	0.011
Mg	1.313	0.848	1.163	1.212	0.927	1.030	1.074	0.559	0.915	1.133	0.312	0.951
Ca	0.008	0.009	0.008	0.008	0.011	0.024	0.009	0.010	0.010	0.008	0.016	0.011
Cr	0.003	0.002	0.003	0.004	0.003	0.004	0.004	0.001	0.003	0.018	0.000	0.004
Sum	2.999	3.000	3.003	3.001	3.001	2.970	3.005	3.015	3.006	2.996	3.006	3.005
Fo	66.00	42.62	58.26	60.88	46.61	53.84	53.77	27.76	45.77	57.40	15.62	47.66
Fa	34.00	57.38	41.74	39.12	53.39	46.16	46.23	72.24	54.23	42.60	84.38	52.34

APPENDIX 2. Representative pyroxene compositions for Apollo 15 mare basalts.

	15536			15537			15538			15546			15547		
SiO_2	52.5	51.2	47.6	51.9	49.9	49.0	51.8	50.2	50.5	52.4	51.6	48.3	53.6	50.9	47.3
TiO_2	0.52	0.83	1.33	0.34	1.07	1.15	0.45	0.92	0.94	0.45	0.62	0.83	0.27	0.92	1.10
Al_2O_3	1.32	2.39	1.62	1.08	3.38	1.62	0.94	2.44	2.26	1.31	1.87	0.84	0.73	2.79	1.32
FeO	19.13	14.35	29.43	19.61	13.63	24.93	21.95	16.91	17.23	18.38	17.02	31.46	18.99	12.40	32.77
MnO	0.25	0.27	0.51	0.28	0.22	0.38	0.43	0.39	0.34	0.35	0.33	0.42	0.26	0.30	0.46
MgO	20.90	15.88	5.10	20.71	15.02	10.39	18.85	16.37	16.60	20.39	18.20	7.47	21.42	15.05	5.98
CaO	5.10	14.50	14.36	4.94	15.15	11.42	5.42	11.54	10.89	5.61	9.19	9.72	4.22	16.08	10.32
Na_2O	0.02	0.07	0.06	0.00	0.04	0.04	0.02	0.02	0.04	0.03	0.03	0.04	0.05	0.04	0.03
Cr_2O_3	0.78	0.86	0.17	0.51	0.88	0.27	0.38	0.82	0.74	0.76	0.85	0.11	0.43	1.05	0.14
Total	100.50	100.32	100.20	99.39	99.29	99.17	100.25	99.57	99.55	99.72	99.75	99.23	100.02	99.51	99.38
Si	1.945	1.914	1.926	1.951	1.886	1.936	1.954	1.902	1.913	1.956	1.936	1.962	1.985	1.912	1.936
Ti	0.015	0.023	0.040	0.010	0.031	0.034	0.013	0.026	0.027	0.013	0.017	0.025	0.008	0.026	0.034
Al	0.058	0.106	0.077	0.048	0.151	0.075	0.042	0.109	0.101	0.057	0.082	0.040	0.032	0.124	0.064
Fe	0.593	0.449	0.995	0.616	0.431	0.824	0.692	0.536	0.545	0.573	0.534	1.068	0.588	0.390	1.122
Mn	0.008	0.008	0.017	0.009	0.007	0.013	0.014	0.013	0.011	0.011	0.010	0.014	0.008	0.010	0.016
Mg	1.155	0.886	0.307	1.160	0.847	0.612	1.060	0.925	0.937	1.134	1.017	0.452	1.182	0.843	0.365
Ca	0.203	0.582	0.622	0.199	0.614	0.484	0.219	0.469	0.442	0.224	0.369	0.423	0.167	0.647	0.453
Na	0.002	0.005	0.004	0.000	0.003	0.003	0.001	0.001	0.003	0.002	0.002	0.003	0.004	0.003	0.002
Cr	0.023	0.025	0.006	0.015	0.026	0.008	0.011	0.024	0.022	0.022	0.025	0.004	0.013	0.031	0.005
Sum	4.000	3.999	3.995	4.008	3.996	3.989	4.007	4.005	4.000	3.992	3.993	3.992	3.987	3.986	3.997
Fe/Fe+Mg	34.2	34.0	76.7	35.0	34.1	57.8	40.0	37.2	37.3	34.0	34.8	70.5	33.5	32.1	75.7
Wo	10.3	30.2	32.0	10.0	32.3	25.0	11.0	24.1	22.8	11.5	19.1	21.6	8.6	34.3	23.2
En	59.0	46.0	15.8	58.5	44.6	31.7	53.4	47.6	48.4	58.4	52.7	23.1	60.7	44.6	18.7
Fs	30.7	23.8	52.1	31.5	23.1	43.3	35.6	28.3	28.8	30.1	28.2	55.3	30.6	21.1	58.2

APPENDIX 2. (continued).

	15548			15598			15605			15636		
SiO_2	52.6	50.4	48.2	52.4	51.2	47.7	52.8	50.4	47.9	52.9	51.6	48.6
TiO_2	0.43	0.88	0.79	0.41	0.78	0.94	0.41	0.87	0.85	0.40	0.80	1.12
Al_2O_3	1.29	2.50	0.87	1.19	2.33	1.20	1.08	2.64	0.81	1.04	2.20	1.38
FeO	20.33	16.81	33.23	19.96	17.18	36.29	20.36	16.41	32.07	19.09	15.33	31.03
MnO	0.39	0.30	0.40	0.39	0.33	0.43	0.34	0.28	0.41	0.31	0.29	0.40
MgO	19.69	17.00	7.12	20.51	16.71	7.44	19.91	15.91	5.19	20.70	16.94	7.89
CaO	5.09	10.31	9.37	5.07	10.81	5.99	5.18	12.33	11.88	5.22	12.33	10.29
Na_2O	0.00	0.07	0.02	0.00	0.03	0.02	0.04	0.05	0.04	0.03	0.02	0.02
Cr_2O_3	0.66	0.91	0.07	0.51	0.85	0.17	0.41	0.74	0.19	0.56	0.82	0.20
Total	100.48	99.20	100.04	100.44	100.25	100.16	100.55	99.64	99.36	100.23	100.35	100.93
Si	1.960	1.909	1.953	1.951	1.922	1.941	1.965	1.907	1.963	1.964	1.925	1.935
Ti	0.012	0.025	0.024	0.012	0.022	0.029	0.012	0.025	0.026	0.011	0.022	0.034
Al	0.056	0.111	0.041	0.052	0.103	0.058	0.047	0.118	0.039	0.045	0.097	0.065
Fe	0.633	0.533	1.127	0.622	0.539	1.236	0.634	0.519	1.099	0.593	0.478	1.033
Mn	0.012	0.010	0.014	0.012	30.011	0.015	0.011	0.009	0.014	0.010	0.009	0.014
Mg	1.093	0.960	0.430	1.138	0.934	0.452	1.104	0.897	0.317	1.146	0.942	0.468
Ca	0.203	0.419	0.407	0.202	0.434	0.261	0.206	0.500	0.521	0.208	0.493	0.439
Na	0.000	0.006	0.001	0.000	0.002	0.002	0.003	0.004	0.003	0.002	0.001	0.002
Cr	0.020	0.027	0.002	0.015	0.025	0.005	0.012	0.022	0.006	0.016	0.024	0.006
Sum	3.990	4.000	4.001	4.004	3.992	3.999	3.994	4.000	3.989	3.995	3.993	3.996
Fe/Fe+Mg	37.1	36.1	72.6	35.8	37.0	73.5	36.8	37.1	77.8	34.5	34.1	69.1
Wo	10.5	21.8	20.6	10.2	22.6	13.3	10.6	26.0	26.7	10.6	25.6	22.5
En	56.3	50.0	21.8	57.6	48.7	23.0	56.5	46.6	16.2	58.6	49.0	24.0
Fs	33.2	28.2	57.7	32.1	28.6	63.7	33.0	27.4	57.0	30.8	25.4	53.6

APPENDIX 3. Representative plagioclase compositions for Apollo 15 mare basalts.

	15536			15537			15538			15546			15547		
			Avg.			Avg.			Avg.			Avg.			Avg.
SiO_2	46.5	49.6	47.7	46.4	48.4	47.7	46.0	49.0	48.3	46.4	48.8	47.3	46.6	49.5	47.7
Al_2O_3	34.31	31.59	33.18	33.74	32.41	33.06	34.26	31.76	32.19	34.06	32.38	33.65	33.95	31.57	32.86
FeO	0.43	1.00	0.61	0.61	0.56	0.61	0.53	0.83	0.71	0.43	0.61	0.51	0.41	0.85	0.64
MgO	0.19	0.15	0.22	0.25	0.33	0.31	0.18	0.10	0.29	0.21	0.18	0.22	0.20	0.07	0.19
CaO	18.51	16.57	17.82	18.50	17.09	17.90	19.01	16.69	17.73	18.36	17.02	17.96	18.42	16.44	17.52
Na_2O	0.74	1.55	1.05	0.82	1.16	0.96	0.61	1.50	0.84	0.76	1.27	0.95	0.74	1.22	1.11
K_2O	0.02	0.11	0.05	0.02	0.04	0.03	0.01	0.14	0.03	0.03	0.06	0.03	0.03	0.50	0.10
Total	100.71	100.59	100.73	100.39	100.08	100.63	100.67	100.12	100.17	100.29	100.39	100.65	100.45	100.13	100.19
Si	2.125	2.261	2.180	2.134	2.217	2.179	2.112	2.247	2.212	2.132	2.228	2.161	2.138	2.266	2.190
Al	1.850	1.697	1.786	1.828	1.749	1.781	1.853	1.715	1.741	1.844	1.742	1.812	1.834	1.704	1.777
Fe	0.017	0.038	0.023	0.023	0.021	0.023	0.020	0.032	0.027	0.017	0.023	0.020	0.016	0.032	0.025
Mg	0.013	0.010	0.015	0.017	0.022	0.021	0.012	0.007	0.019	0.014	0.012	0.015	0.013	0.005	0.013
Ca	0.907	0.809	0.872	0.911	0.838	0.876	0.935	0.819	0.872	0.903	0.832	0.879	0.905	0.807	0.862
Na	0.066	0.137	0.093	0.073	0.103	0.085	0.054	0.133	0.075	0.067	0.112	0.084	0.066	0.108	0.099
K	0.001	0.006	0.003	0.001	0.003	0.002	0.001	0.008	0.002	0.001	0.003	0.002	0.002	0.029	0.006
Sum	4.981	4.960	4.973	4.989	4.958	4.971	4.988	4.964	4.952	4.980	4.957	4.975	4.976	4.951	4.972
Ab	6.8	14.4	9.6	7.4	10.9	8.9	5.5	13.9	7.8	6.9	11.9	8.7	6.8	11.5	10.2
Or	0.1	0.6	0.3	0.1	0.3	0.2	0.1	0.9	0.2	0.2	0.3	0.2	0.2	3.1	0.6
An	93.1	85.0	90.1	92.5	88.8	90.9	94.5	85.3	91.9	92.9	87.8	91.1	93.0	85.4	89.2

APPENDIX 3. (continued).

	15548			15598			15605			15636		
			Avg.			Avg.			Avg.			Avg.
SiO_2	47.4	47.6	47.6	46.7	48.1	47.2	46.6	48.8	47.2	46.5	47.6	47.1
Al_2O_3	33.16	32.12	32.81	34.17	32.70	33.65	34.02	32.14	33.38	33.76	33.51	33.49
FeO	0.69	0.87	0.78	0.56	0.60	0.61	0.51	1.11	0.67	0.48	0.64	0.53
MgO	0.30	0.28	0.26	0.17	0.36	0.23	0.20	0.07	0.20	0.25	0.20	0.25
CaO	18.13	17.50	17.85	18.27	17.61	18.14	18.41	17.08	18.06	18.33	17.72	18.13
Na_2O	0.95	1.04	1.01	0.83	1.06	0.94	0.70	1.07	0.91	0.78	1.11	0.94
K_2O	0.02	0.06	0.04	0.03	0.03	0.03	0.04	0.41	0.07	0.04	0.05	0.03
Total	100.72	99.58	100.43	100.77	100.51	100.87	100.47	100.79	100.54	100.16	100.88	100.55
Si	2.168	2.201	2.183	2.134	2.199	2.155	2.135	2.227	2.161	2.138	2.171	2.157
Al	1.787	1.749	1.772	1.841	1.761	1.811	1.838	1.730	1.803	1.831	1.801	1.807
Fe	0.027	0.034	0.030	0.021	0.023	0.023	0.019	0.043	0.026	0.019	0.024	0.020
Mg	0.020	0.019	0.018	0.012	0.025	0.015	0.013	0.005	0.014	0.017	0.013	0.017
Ca	0.888	0.866	0.877	0.895	0.862	0.887	0.905	0.836	0.887	0.904	0.866	0.890
Na	0.084	0.094	0.090	0.074	0.094	0.084	0.062	0.095	0.080	0.070	0.098	0.083
K	0.001	0.004	0.002	0.001	0.001	0.002	0.002	0.024	0.004	0.002	0.003	0.002
Sum	4.978	4.970	4.975	4.981	4.966	4.980	4.976	4.963	4.977	4.981	4.978	4.979
Ab	8.6	9.7	9.3	7.6	9.8	8.6	6.4	9.9	8.3	7.1	10.1	8.6
Or	0.1	0.4	0.2	0.2	0.2	0.2	0.2	2.5	0.4	0.2	0.3	0.2
An	91.3	89.9	90.5	92.3	90.0	91.2	93.4	87.6	91.3	92.7	89.5	91.3

Proceedings of the 20th Lunar and Planetary Science Conference, pp. 127-138
Lunar and Planetary Institute, Houston, 1990

Chemistries of Individual Mare Volcanic Glasses: Evidence for Distinct Regions of Hybridized Mantle and a KREEP Component in Apollo 14 Magmatic Sources

S. S. Hughes[1]
Departments of Chemistry and Geology and The Radiation Center, Oregon State University, Corvallis, OR 97331

J. W. Delano
Department of Geological Sciences, State University of New York, Albany, NY 12222

R. A. Schmitt[2]
Departments of Chemistry and Geology and The Radiation Center, Oregon State University, Corvallis, OR 97331

Major and trace element compositions have been determined by electron probe microanalysis (EPMA) and instrumental neutron activation analysis (INAA) on 21 individual mare volcanic glasses (MVGs) from Apollo 14, 15, 16, and 17, plus one basaltic impact glass from Apollo 17, in order to assess the impact of chemical hybridization in mantle sources. Chemical variations within each major group of picritic green, yellow/brown, and orange volcanic glasses are minor compared to distinct variations between each group. Each group of MVGs can be delineated with respect to the chemical signatures obtained in source regions and a comparison with mare basaltic compositions suggests that basaltic magma sources have experienced similar mechanisms of source evolution. Although mare basalts have experienced variable secondary processes of differentiation and assimilation, chemical distinctions among various MVG and mare basalt groups strongly support the hypothesis of hybridized lunar mantle regions from which all mare magmas have been generated. Apollo 14 VLT and green (type A) volcanic glass chemistries display a somewhat wider range in overall abundances, as well as a distinct segregation, relative to other MVG groups. Chemical signatures of Apollo 14 glasses depict a strong affinity to KREEP and indicate the presence of a KREEP-like component in their hybridized mare source regions. The KREEP signature in picritic glasses is consistent with a KREEP component observed in many Apollo 14 mare basalts, although the process of KREEP hybridization in mare sources is not evident in all Apollo 14 basalts. These relations suggest that processes of assimilation and/or hybridization were operative during source evolution, but similar processes were likely during the evolution of some primary basaltic magmas.

INTRODUCTION

Analyses of individual mare volcanic glasses (MVGs) indicate distinct segregations in major element chemistries (e.g., *Delano and Livi,* 1981; *Delano,* 1986) that are largely upheld in trace element abundances (*Ma et al.,* 1981; *Hughes et al.,* 1988, 1989). Chemical signatures in lunar mare basalts are equally divisible into separate groups that reflect a wide range of petrogenetic processes. Contrary to mare basalts, many MVGs have picritic compositions and therefore represent mantle-derived magmas that retain chemical signatures obtained in mare source regions. Generally, MVG compositions range, like mare basalts, from low-Ti (green) to intermediate-Ti (yellow/brown) to high-Ti (orange) to very high-Ti (red/black). At least six MVG varieties spanning this range are known to occur at the Apollo 14 landing site (*Delano,* 1986, 1988; *Delano et al.,* 1988), an observation that is commensurate with the large variety of mare basalt clasts in Apollo 14 breccias. A seventh group of Apollo 14 volcanic glass, a low alkali picrite (LAP), is proposed by *Shearer et al.* (1989); however, no glasses belonging to that LAP group were observed by *Delano* (1988) among the nearly 1200 glasses analyzed. Consequently, the LAP variety at Apollo 14 awaits verification due to its rarity.

Remelting of cumulate sources (e.g., *Philpotts and Schnetzler,* 1970; *Wood et al.,* 1970; *Taylor and Jakes,* 1974; *Shih and Schonfeld,* 1976) deposited during the fractionation of a primordial lunar magma ocean (LMO) is currently regarded as the fundamental process in mare petrogenesis. However, melting of simple cumulates is inadequate to produce many of the major and trace element signatures (e.g., *Ringwood and Kesson,* 1976) observed in mare basalts and glasses. Chemical variations in mare basalts, particularly the wide ranges in Ti contents and REE patterns, lead researchers to propose scenarios ranging from source hybridization to assimilation and fractionation of primary magmas (*Hubbard and Minear,* 1975, 1976; *Ringwood and Kesson,* 1976; *Binder,* 1980, 1982, 1985a,b; *Delano,* 1979, 1986). Regardless of these

[1]Now at Montana Bureau of Mines and Geology, Montana College of Mineral Science and Technology, Butte, MT 59701
[2]Also at College of Oceanography, Oregon State University, Corvallis, OR 97331

processes, which may have operated on a restricted number of sources or evolving magmas, it is likely that much of the chemical variety in mare basaltic magmas is source related.

Apollo 14 aluminous mare basalt clasts extracted from 14321 breccia (*Shervais et al.,* 1985; *Dickinson et al.,* 1985; *Neal et al.,* 1988) exhibit a substantial range in incompatible trace element abundances, although their bulk compositions are grossly similar. *Shervais et al.* (1985) suggested that the chemical variability requires at least three separate magmatic sources; however, they further argued that assimilation of a KREEP-like component is necessary to produce distinct types of 14321 mare basalts. Models calculated by *Dickinson et al.* (1985) for five separate aluminous mare basalt groups support the concept of KREEP assimilation to derive trace element variations in 14321 basalts, although this requirement is not apparent in major element systematics. Recently, *Neal et al.* (1988) presented additional chemical data that reveals a continuum of compositions among the 14321 high alumina (HA) basalts that is likely controlled by the combined effects of assimilation and fractional crystallization (AFC) at depth within the lunar crust. They propose a scenario in which a parental magma (14321, 1422; La = 3.2 ppm, MgO = 11%) evolved by AFC with KREEP such that crystallization of first olivine and chromite, which are subsequently joined by plagioclase, then co-precipitation of orthopyroxene, plagioclase, and ilmenite yielded the range in 14321 HA compositions. Whereas this scenario is suitable for crystalline mare basalts, which apparently do not represent samples of primary magma (e.g., *Longhi,* 1987), the six chemical varieties of pristine mare glass recognized to occur at the Apollo 14 site (MgO = 13.3-19.1%, *Delano,* 1988) are not controlled by assimilation and fractionation and thus reflect the chemical signatures of mantle source regions.

Variable trace element signatures in mare sources are exemplified by analyses of individual low-, intermediate-, and high-Ti MVG samples. Comparisons in trace element signatures of (1) Apollo 15 MVGs (15426 green and 15427 yellow/brown glasses) relative to average Apollo 12 and 15 olivine basalts (*Hughes et al.,* 1988) and (2) Apollo 17 MVGs (74220 orange glass) relative to Apollo 11 and 17 high-Ti basalts (*Hughes et al.,* 1989) indicate that distinct source regions are required to produce each MVG type. Similarities in trace element patterns between MVGs and their comparative basalts suggest that equivalent processes were operative on the sources of mare basalt primary magmas. These studies also show that trace element signatures are possibly controlled by minor amounts of evolved LMO components (late-stage cumulates and intercumulus liquid) having high, yet distinct, concentrations of trace elements that were involved in source hybridization with major proportions of early cumulate olivine ± orthopyroxene. Models presented by *Hughes et al.* (1989) for Apollo 11 and 17 magmas further demonstrate that source hybridization alone is inadequate to produce Apollo 11 intermediate- and high-K basalts, although appropriate models were obtained for Apollo 11 low- and low-intermediate-K and Apollo 17 basalts. Their models indicate that 10% KREEP assimilation into a magma initially having a low-K signature, combined with variable fractionation, is required to produce the intermediate-and high-K Apollo 11 basalts.

In light of the apparent requirement for source-related variations between the major categories of volcanic glasses and some mare basalts, the present study reports major and trace element abundances in a variety of MVG types. Major element systematics of at least 26 MVG types have been outlined in detail by extensive electron probe microanalysis (EPMA) research (e.g., *Delano,* 1979, 1981, 1986, 1988; *Delano and Livi,* 1981; *Delano and Lindsley,* 1983). Chemical differences among MVGs from various sites lead to the designation of six types of Apollo 14 volcanic glass (green A and B, green VLT, yellow, orange, and red/black glass; *Delano,* 1981, 1986, 1988) and five types of Apollo 15 green glass (green A-E; *Delano,* 1979, 1986). Arguments presented here will concentrate on trace element signatures necessary in delineating the effects of source hybridization and/or assimilative processes. These analyses are intended to evaluate further the amount of variation among the members of major MVG groups relative to potentially much larger variations between the major categories. Major and trace element chemistries are reported for three 14163 VLT (very low-Ti) green glasses (GGs), one 14163 low-Ti GG (type A), one 65501 GG, four 15301 GGs (types A and D), four 15301 yellow/brown glasses (YBGs), two 79221 orange glasses (OGs), six 74241 OGs, and a VHT (very high-Ti) basaltic glass from 74241 apparently derived by an impact into a mare lava flow. The Apollo 14 type A MVG has higher TiO_2 and FeO and lower Al_2O_3 relative to other green glasses (*Delano,* 1981). Both Apollo 15 GG types, A and D, which exhibit minor differences in ratios of CaO/MgO and CaO/FeO, represent the most primitive Mg/(Mg + Fe) ends of higher Mg/Si (types D and E) and lower Mg/Si (types A, B, and C) chemical trends (*Delano,* 1979). Samples were selected, according to availability as extractable beads, to determine (or update) the chemical signatures of Apollo 14 and 16 low-Ti glasses. In addition, Apollo 15 and 17 glasses were analyzed to discriminate further among the patterns previously reported and to obtain better analyses of individual Apollo 15 YBG glass using a technique updated since YBG analyses by *Hughes et al.* (1988).

VOLCANIC GLASS CHEMISTRIES

Individual glasses were handpicked from lunar soils and assessed chemically and petrographically using the criteria listed by *Delano* (1986). Major element abundances were determined by an EPMA procedure developed by J.W.D., the details of which are outlined in *Hughes et al.* (1988). Trace element abundances, as well as Cr, Na, and K contents, were obtained by instrumental neutron activation analysis (INAA) following the procedure described in *Hughes et al.* (1989). The INAA procedure involved activation for ~150 hr in the University of Missouri Research Reactor (neutron flux = 4.6×10^{14} n $cm^{-2}sec^{-1}$), an appropriate technique for samples weighing less than 100 μg. Sample masses ranged from ~1.3 μg to ~90 μg, except for one sample of 14163 VLT glass weighing 300 μg. Very low masses were estimated from ^{59}Fe activities, in which Fe is known from EPMA, using a reproducible gamma counting geometry and assuming negligible variations in neutron flux. Moreover, INAA data were normalized to FeO contents determined by EPMA in order to correct for variations in total neutron fluence, counting geometry, and errors in sample mass determinations.

Major Element Compositions

Bulk chemistries and CIPW normative mineralogies of low-Ti MVGs from Apollo 14, 15, and 16, listed in Table 1, are dominantly picritic with the exception of one 15301 YBG composition (CS-19) that is quartz normative. The low MgO content (7.5%) plus higher CaO and Al_2O_3 in this YBG relative to other YBG glasses is probably due to mafic mineral fractionation, known to occur in many MVG types (e.g., *Delano,* 1979; *Delano and Lindsley,* 1983), which places this sample outside the realm of primary MVG compositions (*Delano,* 1986). Olivine fractionation will have little effect on the incompatible trace element signature (other than overall increases in abundances) compared to major element variations, so the CS-19 trace element pattern might be considered to reflect the near-primary composition by accounting for this effect. However, additional complications are evident in significantly higher Cr_2O_3 contents (0.74%) in CS-19 relative to Cr_2O_3 = 0.43-0.65% in other low-Ti MVGs, whereas the reverse is true for TiO_2 contents. These relations, which are consistent with fractionation of (chromite-free) olivine + ilmenite or olivine + armalcolite from a picritic YBG parent to yield the CS-19 composition, argue against using CS-19 as a representative of primary magma. In the remaining low-Ti MVGs, normative olivine ranges from 14-23% in Apollo 14 VLT glasses, 24-27% in Apollo 15 YBGs, 30% in Apollo 14 GG, 33-35% in Apollo 15 GGs, and up to 41% in Apollo 16 GG. Normative augite (Di) contents are 16.7-19.5% in the picritic glasses, whereas normative Di = 24.1% in the CS-19 quartz-normative sample.

Compositions of eight high-Ti orange MVGs listed in Table 2 are also picritic with MgO = 12.2-15.1% and normative olivine = 15-27%. Relative to low-Ti MVGs, the overall lower values for SiO_2, Al_2O_3, and CaO coupled with higher Cr_2O_3, TiO_2, and Na_2O reflect a fundamental difference in source compositions between major MVG types. Much of this difference is due to the increased proportion of ilmenite required in source regions (e.g., *Green et al.,* 1975; *Hubbard and Minear,* 1976; *Binder,* 1982, 1985b), causing a mass-balance reduction in other components. It follows that ilmenite presence in MVG sources, as a late-stage LMO phase, could occur only after a significant amount of LMO crystallization. Experimental evidence from Ti-rich Apollo 17 mare compositions (*Green et al.,* 1975) shows that the parental magmas are not saturated in Fe-Ti oxide, such that ilmenite would have been depleted from source regions during magma genesis. The analysis of a 74241 IG (impact-generated glass), CS-30 (VHT-IG), is included with the high-Ti group. For reasons discussed below, this very high-Ti composition is believed to be the result of meteorite impact into a lava flow having characteristics similar to high-Ti Apollo 17 mare basalts.

Besides the anomalous CS-19 and CS-30 (VHT-IG) samples, all of the MVG compositions in Tables 1 and 2 fall within either array I (Apollo 14 VLT and Apollo 15 YBG) or II (Apollo 14, 15, and 16 green and Apollo 17 orange) of *Delano and Livi*

TABLE 1. Major element compositions of low- and intermediate-Ti mare volcanic glasses from Apollo 14, 15, and 16.

Apollo #	14163, 162			14163	65501	15301, 78				15301, 78			,30
	VLT	VLT	VLT	Green-A	Green	Green-A	Green-A	Green-D	Green-D	YBG	YBG	YBG	YBG
OSU #	CS-1	CS-4	CS-5	CS-2	CS-12	CS-14	CS-16	CS-15	CS-17	CS-18	CS-19	CS-20	CS-21
SiO_2	47.1	47.2	46.3	44.7	43.3	44.8	45.0	44.4	44.9	42.3	48.0	43.0	42.8
TiO_2	0.64	0.69	0.53	0.88	0.46	0.44	0.45	0.47	0.46	3.75	3.42	3.55	3.69
Al_2O_3	10.2	10.1	9.3	7.0	7.9	7.8	7.5	7.5	7.4	8.7	9.5	9.1	8.9
Cr_2O_3*	0.51	0.52	0.52	0.49	0.43	0.53	0.51	0.53	0.51	0.58	0.74	0.56	0.65
FeO	17.4	17.6	18.0	22.4	22.1	20.1	20.0	20.7	20.7	22.3	20.0	22.1	22.7
MnO	0.28	0.27	0.27	0.35	0.31	0.31	0.31	0.32	0.27	0.34	0.26	0.30	0.31
MgO	13.6	13.2	15.5	15.7	16.6	17.2	17.9	17.2	17.1	12.8	7.5	12.2	11.5
CaO	9.8	9.9	9.1	8.1	8.6	8.7	8.1	8.6	8.5	8.7	10.6	8.7	9.1
Na_2O*	0.37	0.34	0.30	0.19	0.25	0.15	0.17	0.17	0.19	0.48	0.44	0.57	0.54
K_2O*	0.11	0.06	0.04	0.05	0.03	0.02	0.02	0.02	0.02	0.07	0.06	0.08	0.08
CIPW normative compositions													
Q	—	—	—	—	—	—	—	—	—	—	3.48	—	—
Or	0.65	0.36	0.24	0.30	0.18	0.12	0.12	0.12	0.12	0.41	0.35	0.47	0.47
Ab	3.10	2.87	2.54	1.57	2.08	1.25	1.40	1.48	1.60	4.03	3.67	4.79	4.54
An	25.8	25.9	23.9	18.1	20.4	20.6	19.7	19.7	19.3	21.5	23.7	21.9	21.5
Di (cpx)	18.8	19.3	17.5	18.3	18.2	18.5	16.7	19.0	18.6	18.1	24.1	17.5	19.5
Hy (opx)	33.1	35.2	31.0	29.0	16.1	24.8	27.2	22.7	25.3	20.8	37.1	23.0	22.0
Ol	16.6	14.3	23.0	30.3	41.4	33.1	33.4	35.4	33.4	27.3	—	24.8	24.0
Il	1.22	1.31	1.01	1.67	0.87	0.84	0.86	0.89	0.87	7.13	6.47	6.74	7.00
Cm	0.74	0.76	0.77	0.72	0.63	0.78	0.75	0.79	0.76	0.85	1.09	0.83	0.96

Weight percent abundances by electron microprobe analysis, except (*)Cr, Na, and K analyses by neutron activation.

TABLE 2. Major element compositions of Apollo 17 high-Ti mare volcanic glasses and VHT basaltic impact glass.

Apollo #	79221, 86 Orange		74241, 143 Orange				74241, 171 Orange		74241 VHT-IG
OSU #	CS-6	CS-7	CS-22	CS-23	CS-24	CS-25	CS-27	CS-28	CS-30
SiO_2	39.2	39.1	38.8	39.0	38.7	38.3	40.4	39.1	33
TiO_2	9.2	9.3	8.7	8.7	9.8	8.8	8.6	9.6	15.0
Al_2O_3	5.9	5.8	5.8	5.8	6.5	5.9	7.8	6.3	9.4
Cr_2O_3*	0.63	0.64	0.66	0.65	0.67	0.66	0.63	0.66	0.20
FeO	22.3	22.5	22.9	22.8	23.1	22.9	21.3	22.6	24.2
MnO	0.29	0.26	0.26	0.27	0.28	0.28	0.3	0.31	0.44
MgO	14.5	14.7	14.9	15	12.2	15.1	12.2	12.8	6.9
CaO	7.4	7.2	7.4	7.3	8.3	7.6	8.4	8.1	10.2
Na_2O*	0.44	0.47	0.44	0.46	0.50	0.17	0.52	0.49	0.20
K_2O*	0.06	0.03	0.07	0.08	0.08	0.08	0.1	0.08	0.08
CIPW normative compositions									
Q	—	—	—	—	—	—	—	—	—
Or	0.36	0.18	0.41	0.47	0.47	0.47	0.59	0.47	0.48
Ab	3.73	4.01	3.74	3.85	4.23	1.43	4.35	4.11	0.61
An	13.93	13.59	13.67	13.51	15.25	14.99	18.67	14.82	25.20
Di (cpx)	18.66	18.09	18.91	18.68	21.28	18.63	18.64	20.92	21.41
Hy (opx)	23.36	23.32	19.14	19.71	20.38	20.96	25.35	21.96	12.97
Ol	21.55	22.19	26.55	26.32	18.87	25.79	15.18	18.53	10.39
Il	17.49	17.67	16.60	16.49	18.53	16.76	16.28	18.22	28.65
Cm	0.93	0.95	0.97	0.95	0.98	0.97	0.93	0.97	0.30

Weight percent abundances by electron microprobe, except (*)Cr, Na, and K analyses by neutron activation.

(1981). These compositions clearly place low-Ti picritic glasses within the experimentally determined olivine stability field at low pressures (e.g., *Kesson,* 1975). Figure 1, which shows the projection of MVG compositions on a pseudoternary olivine-plagioclase-silica diagram (*Walker et al.,* 1973), indicates that MVGs fall well within the olivine stability field at low pressures. The overall spread of MVG compositions in Fig. 1 also suggests variations controlled by residual olivine, although this effect would be minor within any of the MVG groups that tend to cluster, especially the GGs.

Binder (1982, 1985a) suggested, on the basis of olivine control, that primary mare magmas were derived from low pressure (5-10 kbars) regions in the lunar mantle where olivine alone remained in the residuum. However, multiple saturation would occur at greater depths such that the olivine-pyroxene cotectic in Fig. 1 would occur at a lower position on the pseudoternary plot. Therefore, a magma derived at pressures of ~20 kbars (~400 km depth) would require a residuum of both olivine and low-Ca orthopyroxene. Such relations lead experimental petrologists (e.g., *Green et al.,* 1975; *Chen et al.,* 1982; *Chen and Lindsley,* 1983; *Delano,* 1980) to argue for magma generation at, or around, 400 km. Models presented by *Hughes et al.* (1988, 1989) suggest that low-intermediate-Ti Apollo 12 and 15 mare magmas were ultimately derived from multiple saturation depths, whereas high-Ti Apollo 11 and 17 mare magmas were derived in shallower, olivine-dominated regions. Since eutectic or cotectic melting is thermodynamically predicted for magma genesis, then it is likely that high-Ti primary magmas segregated from their source regions at, or slightly above, the temperature of orthopyroxene consumption during the melting process. Compositions that require only olivine in the residuum were likely derived from a mantle region slightly above the multiple saturation point. Moreover, relatively minor differences in the overall incompatible element patterns will be obtained between resultant liquids derived from either olivine or olivine + orthopyroxene residuals.

Trace Element Covariations

Trace element abundances, compiled in Tables 3 and 4, cover a broad range of values within the entire suite that is contrasted by relative uniformity among MVGs of a given major type. The distinct signature of each type is useful in delineating a common mechanism of mare magma source evolution (e.g., *Hughes et al.,* 1988, 1989); however, some element covariations are scattered within a particular group. Such relations are evident in a Co-Sc plot (Fig. 2), which shows a general clustering among each group of GGs, YBGs, and OGs, whereas a slightly larger field exists for the Apollo 14 VLT glasses. The scattering is probably due, in part, to the small number of samples analyzed, but the range in values may also indicate additional complexities in source evolution among the Apollo 14 VLTs. Moreover, it is apparent from Fig. 2 that GGs obtained from three sites (Apollo 14, 15, and 16) exhibit nearly equal Co and Sc abundances, which attests to a regional simplicity in GG source petrology suggested by major elements (Fig. 1). Although the low-Ti characteristics of Apollo 14 VLT glasses imply a possible genetic relation to other GGs, an equivalent mechanism for trace element systematics is not maintained in the VLT glass sources.

The displaced position of the Apollo 15 quartz-normative YBG in Fig. 2 may not be source related, although a rough overall trend of decreasing Co with increasing Sc is apparent

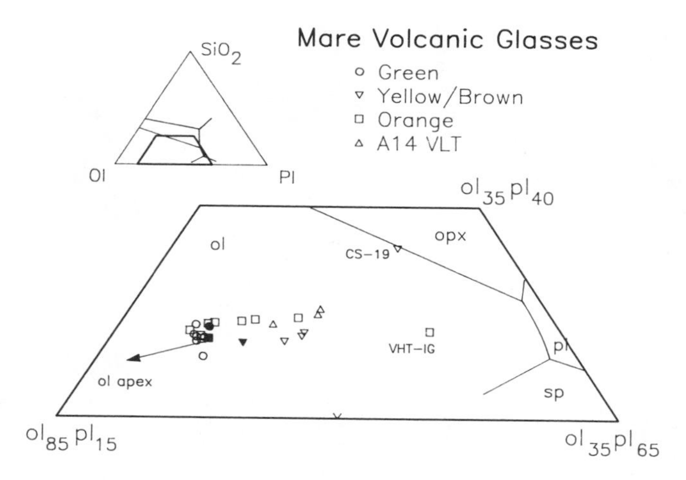

Fig. 1. Projection of mare volcanic glass major element compositions, recalculated on an ilmenite-free basis, on the Ol-Pl-SiO_2 pseudoternary diagram of *Walker et al.* (1973). Filled symbols represent average mare volcanic glass compositions obtained previously that include: Apollo 15 green (*Ma et al.,* 1981), Apollo 15 yellow/brown (*Hughes et al.,* 1988), and Apollo 17 orange (*Hughes et al.,* 1989) glasses. With the exception of the CS-19 and VHT-IG evolved compositions, the diagram illustrates a small range in volcanic glass compositions relative to a much larger field of mare basalt compositions (not shown) that ranges from the MVG field to the ol-opx and ol-pl cotectic boundaries.

in the pristine glass compositions. This negative trend is possibly due to variation in the amount of pyroxene (Sc compatible) relative to olivine (Co compatible) in source regions or, more appropriately, the proportion of pyroxene removed by melting. Also, the quartz-YBG plots within the region expected for many mare basalts, which have nonprimary compositions (e.g., *Longhi,* 1987). Therefore, the quartz-YBG position on Fig. 2 is likely due to differentiation that is unrelated to the source variations evident in the other MVGs.

Covariations in light rare earth elements (LREE) and the high-field strength element Hf are illustrated in La-Hf (Fig. 3) and La-La/Sm (Fig. 4) diagrams. The low abundances of all incompatible elements in Apollo 15 and 16 GGs produce a very tight cluster in Fig. 3, yet an elongate array is evident in Fig. 4 (Apollo 16 GG has highest La/Sm). By contrast, the Apollo 14 GG-A composition is set apart from other GGs and plots close to fields of YBGs and the Apollo 14 VLTs. In these two diagrams, as in Fig. 2, the limits of GG, OG, YBG, and VLT compositions are reasonably well defined, although some overlap in absolute Hf and La abundances exists between YBGs and VLTs. With reference to possible KREEP-like contamination, suggested for Apollo 14 mare basalts, the position of average KREEP (*Warren and Wasson,* 1979) is located substantially off each of these plots. Minor contamination by a KREEP-like composition would cause variations along the weakly converging lines shown. From these plots, Apollo 14 GG-A and VLTs might be interpreted as being related to either

TABLE 3. INAA trace element abundances (ppm) of individual low- and intermediate-Ti mare volcanic glasses from Apollo 14, 15, and 16.

Apollo #	14163, 162			14163	65501	15301, 78				15301, 78			,30
Type	VLT	VLT	VLT	Green-A	Green	Green-A	Green-A	Green-D	Green-D	YBG	YBG	YBG	YBG
OSU #	CS-1	CS-4	CS-5	CS-2	CS-12	CS-14	CS-16	CS-15	CS-17	CS-18	CS-19	CS-20	CS-21
Wgt (μg)	299.8	4.0*	89.3	22.5	1.4*	8.0*	10.5	30.5	37.5	6.3*	21.5	74.8	1.3*
Sc	39.6	35.2	38.4	34.2	32.6	37.3	33.6	37.3	35.4	43.2	56.6	47.8	43.4
Co	60	54	68	73	79	75	77	77	77	67	34	73	65
Ni	107	88	151	113	289	155	159	163	191	162	64	72	72
Sr	320	20	40	70	40	30	40	20	20	160	110	170	320
Cs	0.08	0.18	0.04	0.03	0.07	-nd-	0.05	0.04	0.03	0.24	0.09	0.15	0.08
Ba	177	119	153	95	51	38	26	8	39	142	116	173	129
La	13.0	13.2	9.5	8.6	2.3	1.3	1.7	1.5	1.3	8.7	11.7	8.9	10.0
Ce	30.8	31.6	24.0	21.5	5.8	4.9	3.6	3.4	3.4	25.6	36.2	25.5	25.6
Nd	19.8	21.5	16.2	15.2	3.2	3.3	2.9	2.4	3.5	18.6	23.3	18.0	16.0
Sm	6.1	6.0	5.0	4.5	0.91	0.77	0.79	0.81	0.84	6.4	9.0	7.1	6.7
Eu	0.74	0.57	0.56	0.49	0.31	0.26	0.26	0.26	0.29	1.61	1.68	1.89	1.60
Tb	1.54	1.19	1.11	0.78	0.22	0.16	0.20	0.21	0.22	1.58	2.02	1.64	1.58
Yb	4.2	4.2	3.3	2.8	0.86	0.88	0.91	0.85	1.01	4.2	7.3	4.0	4.6
Lu	0.74	0.47	0.42	0.35	0.10	0.14	0.13	0.16	0.17	0.49	0.93	0.59	0.55
Zr	180	150	210	120	-nd-	20	-nd-	15	-nd-	260	250	190	200
Hf	5.7	4.6	4.4	3.3	0.63	0.51	0.50	0.59	0.48	4.7	5.5	5.4	4.7
Ta	0.69	0.60	0.61	0.44	0.09	0.14	0.21	0.13	0.15	0.85	1.0	1.0	0.85
Th	2.4	2.5	1.7	1.3	0.29	0.10	0.13	0.20	0.16	1.11	0.86	1.11	0.95
U	0.42	0.37	0.36	0.35	0.08	-nd-	0.11	-nd-	-nd-	0.34	0.37	0.25	0.26

*Weights less than ~10 μg were calculated from ^{59}Fe activity level.
Statistical counting errors are: 1-2% Sc, Co, La, and Sm; 2-4% Eu, Yb, and Hf; 4-8% Tb, Lu, Ta, and Th; 10-20% Ni, Ba, Ce, Nd, and U; and 20-40% Sr, Cs, and Zr.
nd = not detected.

TABLE 4. INAA trace element abundances (ppm) of individual Apollo 17 high-Ti mare volcanic glasses.

Apollo #	79221, 86		74241, 143				74241, 171		VHT-IG
OSU #	CS-6	CS-7	CS-22	CS-23	CS-24	CS-25	CS-27	CS-28	CS-30
Wgt (μg)	15.5	12.6	2.8*	3.0*	3.7*	16	1.8*	2.3*	2.4*
Sc	45.4	46.3	47.6	46.7	49.8	48.2	47.8	49.2	89.4
Co	58	59	61	60	59	61	51	51	45
Ni	27	96	75	18	123	108	68	69	560
Sr	230	280	220	220	230	190	270	140	140
Cs	0.04	0.05	0.15	0.19	-nd-	0.06	0.05	0.14	0.07
Ba	120	59	161	31	80	129	79	124	83
La	5.3	5.6	5.7	6.2	5.2	6.6	7.9	6.3	5.1
Ce	16.7	15.2	16.2	18.4	18.6	18.7	21.2	18.4	17.1
Nd	14.9	16.8	17.5	15.4	21.7	19.5	17.2	19.5	23.1
Sm	6.9	6.7	6.9	6.8	7.4	6.8	7.5	7.8	8.4
Eu	1.72	1.74	1.83	2.04	2.14	1.91	1.81	2.18	1.54
Tb	1.42	1.49	1.42	1.65	2.02	1.77	1.50	1.89	2.15
Yb	3.5	3.4	3.6	3.6	4.1	3.7	4.4	4.4	8.1
Lu	0.39	0.48	0.50	0.47	0.49	0.54	0.46	0.60	1.05
Zr	160	170	150	190	280	150	220	210	330
Hf	5.5	5.9	5.8	5.5	6.5	5.8	6.3	6.5	9.4
Ta	0.96	1.1	1.2	-nd-	1.3	1.2	1.3	1.3	2.0
Th	0.30	0.30	0.35	0.04	0.48	0.22	0.65	0.77	0.98
U (ppm)	0.12	-nd-	0.22	0.22	0.34	0.28	0.12	-nd-	-nd-

Same parameters as Table 3.

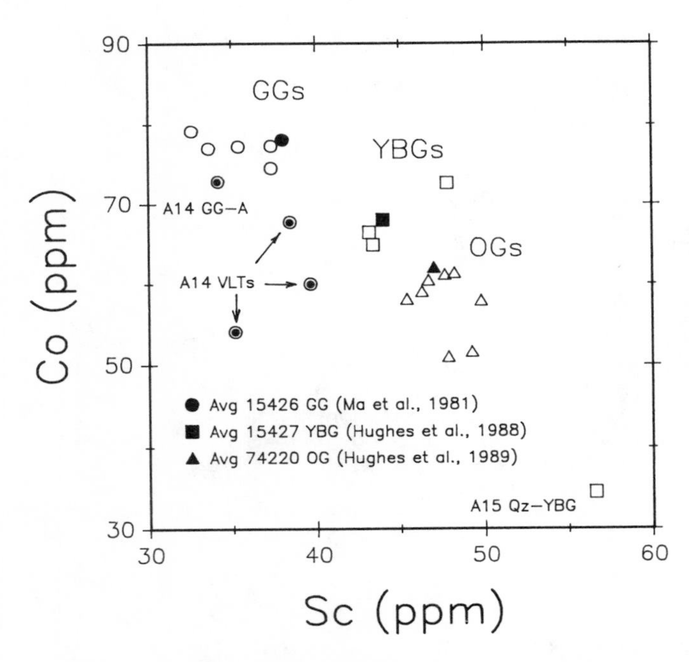

Fig. 2. Covariation diagram of Co vs. Sc abundances in pristine mare glasses. Open symbols correspond to analyses determined in the current study: circles = green glasses, dotted circles = Apollo 14 VLT and green-A glasses, squares = yellow/brown glasses, triangles = orange glasses. Closed symbols refer to the average compositions previously determined for similar types of glasses extracted from soil samples other than the ones used in this study.

assimilation of KREEP into a GG-like magma or to effects of a KREEP-like trapped liquid, obtained during late stages of LMO evolution, in hybridized source regions.

The Apollo 17 VHT-IG (impact melt of a lava flow) composition exhibits a striking similarity in LREE to the OGs (Fig. 4), but a definite segregation occurs in Fig. 3 due to a much higher Hf abundance. Moreover, the VHT-IG composition lies far outside the bounds in Fig. 2, and therefore the composition is likely related to an added component. Although the evidence is inconclusive in Figs. 2-4, the high Ni content in this glass compared to mare basalts and MVGs probably resulted from impactor contamination. It may be that CS-30 represents an impact-melted VHT mare basalt that became somewhat contaminated by an Fe-Ni bolide. Besides relatively high Ti and Fe, most other major and trace element abundances in the VHT-IG are not unlike Apollo 17 mare basalts. Ratios of Sr/Eu and Hf/La in the VHT-IG, which are nearly identical to those in Apollo 17 mare basalts, would not be significantly affected by contamination due to the low abundances of these elements in a primordial impactor. More important, the compositions of elements such as Co and Ni would be greatly altered while FeO contents would be increased an insignificant amount. Assuming that an Fe-Ni meteorite would have Fe, Ni, and Co abundances of ~90%, 8%, and 0.5%, respectively, the increase in Ni content from ~7 ppm in an average Apollo 17 mare basalt (*Hughes et al.,* 1989) to 560 ppm in VHT-IG would require ~0.7% addition of the meteoritic component to the basalt. The contents of Fe and Co would be increased by only ~0.8% FeO and by ~34 ppm Co, which is consistent with the higher Co (45 ppm) in VHT-

IG relative to Apollo 17 mare basalts (~20 ppm). The much higher FeO in VHT-IG (~24%) relative to mare basalts (~19%) is due to greater contribution of ilmenite in the VHT-IG source as evident in higher relative Ti and Ta abundances.

Multiple-element Patterns

Trace element abundances, shown normalized to the bulk Moon composition (*Taylor,* 1982) in Figs. 5-7, provide an effective means of defining the overall signatures of each MVG type. The relative differences in petrogeneses between major MVG types, as well as the evidence for uniformity among the members of each type, are substantiated in the use of five element groups that include alkali (K), divalent plagioclase compatible (Sr, Eu), large-ion-lithophile (Ba, Th, U, REE), pyroxene compatible (Sc), and high-field-strength quadrivalent (Hf, Ta, Ti). Plotted points represent analyses obtained with confidence, whereas a few erratic data points, having higher statistical errors due to nuances of INAA, are either left out or are labeled with error bars. Statistical errors are shown for the sample having either the highest abundance or most erratic value of the element.

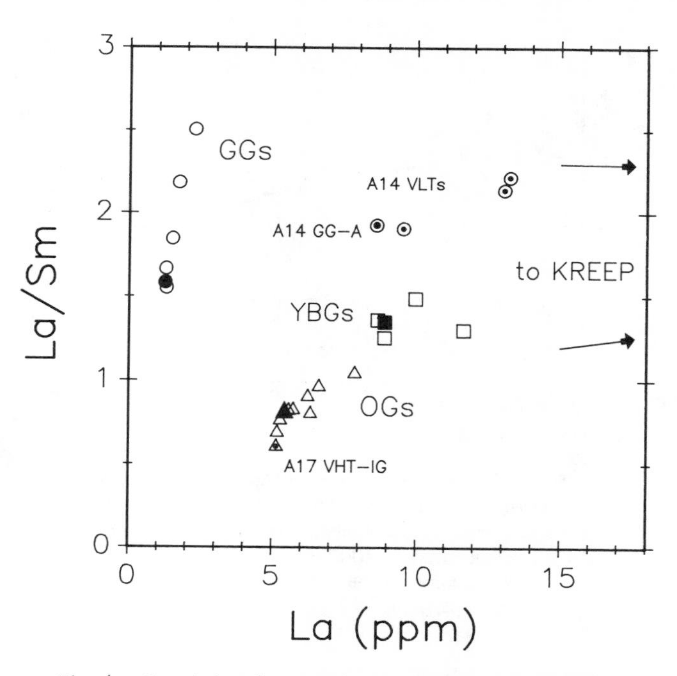

Fig. 4. Covariation diagram of La/Sm ratios vs. La abundances in mare volcanic glasses. Symbols are the same as in Fig. 2. Possible influence of KREEP in sources of Apollo 14 VLT and GG-A glasses is supported by the projection lines to the KREEP composition (*Warren and Wasson,* 1979).

Apollo 14 VLT and Green-A Glasses

Apollo 14 MVG compositions (Fig. 5) display a general pattern of REE, Sc, Hf, Ta, and Ti that is nearly equivalent, albeit lower, to that in the average group 2 aluminous mare basalt of *Dickinson et al.* (1985). The primary difference from the basalt pattern is a higher LILE (Ba, Th) signature relative to LREE. Average KREEP (*Warren and Wasson,* 1979), which has slightly higher bulk Moon-normalized abundances of Ba, Th, and U (as well as lower alkalis K and Rb) relative to LREE is also characterized by strongly negative anomalies in Sr, Eu, Sc, and Ti that were depleted during LMO evolution by the extraction of plagioclase, pyroxene, and ilmenite. [The Ba concentration in KREEP listed in Table 2 of *Warren and Wasson* (1979) should be 1200, not 1800, μg/g.] Similar depletions in these compatible elements are evident in the Apollo 14 VLT and GG-A patterns. Therefore, the overall pattern (i.e., relative elemental abundances) exhibited by the MVGs is quite similar to that of the average KREEP pattern. Average KREEP-normalized abundances (×100) of incompatible elements in Apollo 14 GG-A and VLTs are fairly uniform at Ba = 11.3, Th = 11.0, La = 10.2, Sm = 11.0, Yb = 10.1, and Hf = 12.2. Higher relative values are obtained for compatible elements Sr, Eu, Sc, and Ti (as well as slightly higher Hf and Ta) due to their low abundances in KREEP. Such higher values are consistent with mass-balance arguments for KREEP contamination in a cumulate or magma that would have initially contained higher relative abundances of these compatible elements. The KREEP signature in the Apollo 14 units, not observed in other MVG compositions, suggests a ubiquitous tendency of KREEP in the sources of Apollo 14 MVG liquids and the likelihood that hybridization with KREEP may also have affected the sources of primary basaltic magmas.

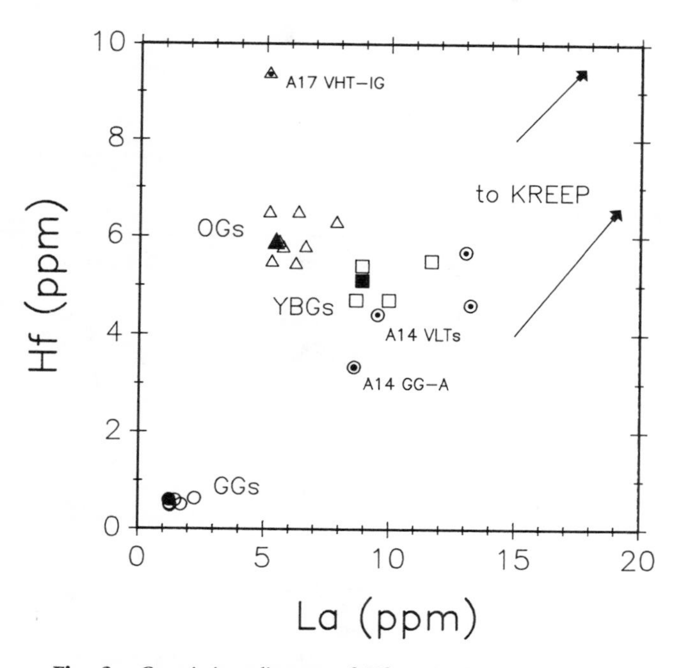

Fig. 3. Covariation diagram of Hf vs. La in mare volcanic glasses. Symbols are the same as in Fig. 2. Weakly convergent lines indicate the position of the KREEP component (*Warren and Wasson,* 1979; La = 110 ppm, Hf = 37 ppm) and a possible cause for KREEP signatures in Apollo 14 MVGs that is due to contamination of GG-like sources.

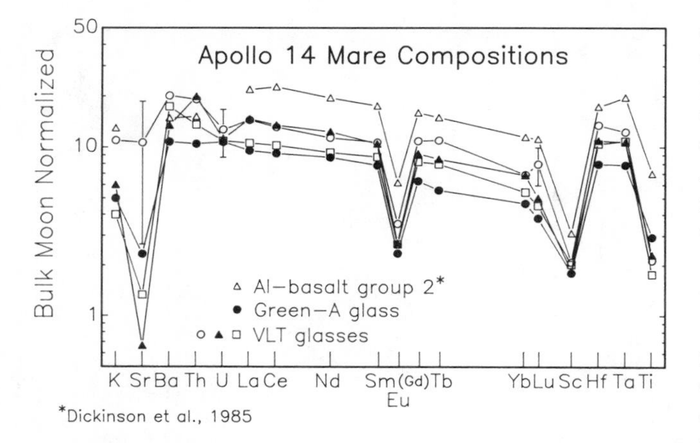

Fig. 5. Multiple-element plots of Apollo 14 compositions shown normalized to bulk Moon values (*Taylor,* 1982). The overall signatures of these units are equivalent to, albeit much lower than, the average KREEP pattern (not shown). The average group 2 aluminous mare basalt (*Dickinson et al.,* 1985) is shown for comparison.

Apollo 15 YBG and GG Glasses

Olivine-normative yellow/brown glasses from 15301 exhibit a narrow range of abundances in a well-defined pattern (Fig. 6) that is contrasted by the signature obtained for the quartz-normative MVG (CS-19), which has higher overall REE. Nearly identical, or somewhat lower, abundances of K, Sr, Ba, Th, and Ti in CS-19 relative to other YBGs do not support derivation of this composition by simple fractionation, or even AFC, of a parent magma having the picritic YBG signature. It is more likely that the quartz-normative YBG is a derivative of some as-yet-undefined parental mare basalt liquid that had gross similarities to the olivine-normative YBG magma, but was derived from a different hybridized source. Comparison of the 15301 picritic YBG pattern to the average 15427 YBG pattern (lower part of Fig. 6) illustrates a chemical uniformity among these types that argues for (1) a common source or even the same eruption, and (2) magma homogeneity at the scale of individual MVG volumes. The positive Ba signature affirms the inference made by earlier results (*Hughes et al.,* 1988; Fig. 1) that the overall pattern of YBGs is roughly equivalent to, although ~2× higher than, the average Apollo 15 olivine basalts, which have a correspondingly high Ba pattern. These comparisons, while not supportive of simple differentiation schemes, argue for analogous mechanisms of source evolution for both MVGs and mare basalts by variable influence of hybrid components.

Green glasses from 15301 also exhibit an overall uniformity in their trace element signatures (Fig. 6), although much lower abundances yield a dispersion of some elements (e.g., Sr, Th, Ce, and Nd) that have higher relative errors during INAA. Comparison with the average 15426 GG (*Ma et al.,* 1981) again supports a common magma or similarities in source evolution; however, the 65501 GG composition exhibits higher K, Sr, Ba, Th, U, and LREE. The relative enrichment of these incompatible elements in the Apollo 16 sources, while maintaining a similar pattern of more compatible elements, argues for differences in source regions that are not related to the cumulate mineralogy. Such relative differences are more likely due to the influence of late-stage trapped liquid, which would more strongly affect the signatures of incompatible elements. Moreover, the differences between Apollo 15 and Apollo 16 GG sources illustrates the distinction between mantle regions underlying each landing site and supports the concept of a heterogeneous lunar mantle.

Apollo 17 Orange and VHT Basaltic Glasses

Orange MVG from Shorty crater (74220) has proved from numerous analyses to be remarkably homogeneous in major element (*Delano and Livi,* 1981) and trace element (*Hughes et al.,* 1989) abundances. Except for sample CS-27, which has somewhat higher relative LREE abundances, the 74241 and 79221 orange MVGs (Tables 2 and 4) exhibit similar element ratios that yield comparable patterns. The analyses of these samples also agree with the analyses of orange/black soils (*Blanchard and Budahn,* 1978) obtained from various depths in cores 74001 and 74002. Figure 7 depicts the average trace element pattern obtained for the remaining seven Apollo 17 orange MVGs and the equivalent pattern of 74220 OG. The

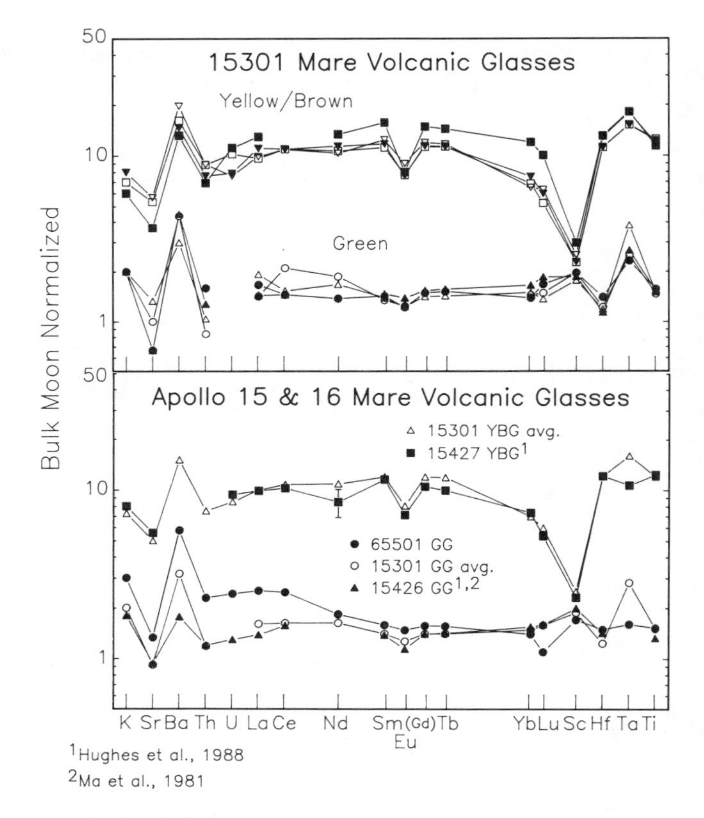

Fig. 6. Multiple-element patterns of 15301 MVGs and average Apollo 15 MVG compositions (same parameters as in Fig. 5). The general uniformity within each group of yellow/brown and green glasses is apparent, although the quartz-normative YBG (basaltic, not picritic, top half of figure in solid squares) and the Apollo 16 GG patterns are likely derived from source regions that have somewhat divergent petrogeneses.

similarity between the two patterns thus indicates a common eruption or magma genesis and strengthens the argument for an overall homogeneity of the fire-fountained lava in the vicinity of the Apollo 17 landing site.

The overall trace element signature in Apollo 17 OGs is unique and, compared to the Apollo 17 basaltic pattern shown in Fig. 7, has a steeper HREE (heavy REE) pattern and overall lower REE, Sc, Hf, Ta, and Ti abundances. Trace elements in the Apollo 17 VHT impact glass are largely indistinguishable from those in the average Apollo 17 mare basalt. Although the VHT-IG exhibits higher Ta and Ti, likely due to a higher relative proportion of ilmenite in the source region, the overall likeness in trace elements argues for parallel petrogeneses. Moreover, the distinction between MVG and basalt signatures in Apollo 17 compositions, as in the other Apollo compositions, supports petrogeneses from separate mantle source regions.

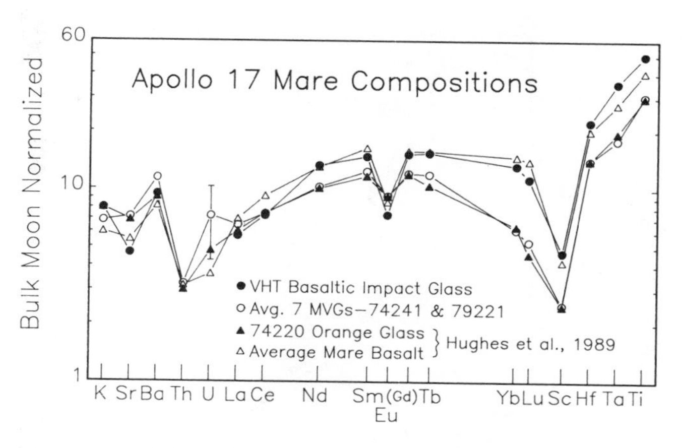

Fig. 7. Multiple element patterns of average Apollo 17 mare volcanic glasses and high-Ti basalts. A definite separation is apparent from these two distinct chemical signatures, which supports their derivation from magmatic sources that were unrelated, yet analogous in petrogenetic mechanism. The VHT basaltic impact glass has an equivalent pattern to the average Apollo 17 basalt composition, although relatively high Ni contents indicate contamination from an Fe-Ni meteorite.

DELINEATION OF MARE SOURCES

Covariations in trace elements shown in Figs. 2-4 and multiple-element diagrams depicted in Figs. 5-7 illustrate the effects of mantle evolution on the chemical signatures of primary mare magmas. The trace element variations among MVGs are too complex to be derived by simple fractionation processes or minor differences in dominant source mineralogies reflected in their major element compositions. Evolution of the lunar mantle likely involved numerous complexities during LMO crystallization (e.g., *Warren,* 1985), which produced mineralogically and chemically variable regions due to convective mass transfer (*Longhi,* 1978) and reassimilation of previously crystallized anorthositic crust (*Warren,* 1986). Four major groups of picritic MVGs (green, VLT, yellow/brown, and orange) display variations that are probably related to (1) the LMO stages of their respective source cumulate deposition and (2) the relative proportions of cumulate source minerals and trapped LMO liquid that have commingled into hybridized regions (e.g., *Ringwood and Kesson,* 1976) capable of yielding the observed signature upon remelting. Compatible and major element chemistries in MVGs are controlled by the dominant early cumulate minerals and their melting proportions; however, trace element signatures are controlled by late-stage cumulates and evolved (KREEP-like) trapped liquids (e.g., *Hughes et al.,* 1988, 1989). Such relations in picritic mare glasses are contrary to models proposed for basaltic magmas (e.g., *Hubbard and Minear,* 1976; *Ringwood and Kesson,* 1976; *Binder,* 1982, 1985b; *Shervais et al.,* 1985; *Neal et al.,* 1988) that argue for secondary effects of assimilation by a primary magma as well as variation in source regions. Perhaps more significant is the range of possibilities in primary mare sources, owing to the fact that a meager sampling during the Apollo and Luna programs has produced myriad complexities in major and trace element chemistries.

The delineation of presently known mare compositions that relate to their source regions is most evident in REE systematics that were controlled by late-stage components of LMO evolution. Assuming that sources were depleted of all but olivine and/or orthopyroxene, TiO_2 contents were controlled by the proportion of ilmenite in the source. Two covariant diagrams, shown in Figs. 8 and 9, indicate well-defined segregations between signatures obtained from various mare source regions. These plots were selected not only on the basis of distinct separation between signatures, but also because simple mafic mineral fractionation will yield negligible effects in compositions plotted in Fig. 8 while linear trends would be evident in Fig. 9.

Ratios of La/Sm relative to Sm/Yb (Fig. 8) delineate the four major MVG sources, as well as the likely sources of some mare basalts and the possible designation of an additional source for Apollo 16 GG. The position of Apollo 17 OGs reflects the strong "bow-shaped" appearance in the normalized diagrams (Fig. 7). The OG signature is possibly related to accessory apatite (or other phosphate) included with the late-stage cumulate mass (*Hughes et al.,* 1989), which because of high partitioning of middle REE (*Nagasawa,* 1970; *Watson and Green,* 1981; *Fujimaki,* 1986) would result in the observed pattern. The spread in Apollo 15 GG positions may be related to minor differences in the amounts of trapped LMO liquid in their source regions or to analytical uncertainty for low REE abundances in these samples.

Apollo 15 YBGs define a field in Fig. 8 that is close to the Apollo 15 (ol and qtz) basalts, and Apollo 15 quartz YBG falls in the field of average Apollo 12 (ol, pig, ilm) basalts (data compiled in *Hughes et al.,* 1988, Appendix). Average Apollo 11 intermediate- and high-K basalts (*Hughes et al.,* 1989) also occur within the YBG field, whereas the Apollo 17 basalts and Apollo 11 low- and low-intermediate-K basalts define a small cluster along with the VHT-IG composition. Hughes et al. argue that the two higher-K Apollo 11 basalts originally had a low-K basalt signature, but assimilated KREEP, which is consistent with their position in Fig. 8 between the Apollo 11-17 cluster and KREEP. In this regard, the position of the two higher-K Apollo 11 basalts in the field with Apollo 12 average compositions is probably fortuitous. The average Apollo 17 VLT basalt composition (*Hughes et al.,* 1988) has an early

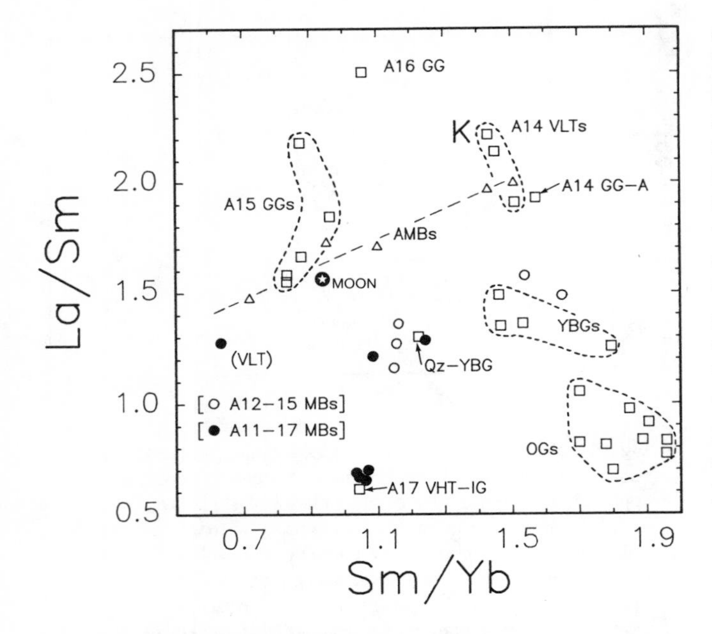

Fig. 8. Covariations in La/Sm ratios vs. Sm/Yb ratios provide a means of delineating variations in mare source chemistries that are due to the effects of late-stage LMO components rich in incompatible elements. Definite chemical segregations are shown to exist between mantle sources of major volcanic glass and mare basalt groups. The trend shown for Apollo 14 aluminous mare basalts, groups 1-5 (AMBs, *Dickinson et al.,* 1985), illustrates the variable involvement of a KREEP-like composition ("K," *Warren and Wasson,* 1979), either in source regions or as an assimilant in secondary processes of basalt petrogenesis. Apollo 14 VLT and green-A glasses exhibit a strong influence of KREEP that argues for KREEP contamination in source regions of some picritic magmas.

cumulate-dominated REE signature, placing it near the Apollo 15 GG field in a region that lacks significant contribution by late-stage LMO cumulates or trapped liquid.

Apollo 14 MVGs, which exhibit the expected close affinity to KREEP, define a field at one end of the aluminous mare basalt trend (averages for groups 1-5, *Dickinson et al.,* 1985) shown by the dashed line in Fig. 8. The line passes through the most primitive GG compositions, which supports the view that VLT Al-basalts are possibly related to GG compositions by variable KREEP involvement in otherwise similar source regions. This is evident especially for the Apollo 14 VLT and GG-A glass compositions in Figs. 3 and 4, which depict covariant trends toward KREEP from the GG field. Moreover, the compatible element Co-Ti variations in Fig. 9 show that Apollo 14 VLT and GG-A glasses have close affinity to GGs and trend toward a KREEP-like composition. However, the positions of Al-basalts in Fig. 9 do not support their derivation from sources that, without the KREEP involvement, would be similar to GG sources. The higher Co abundance in the A17 VHT-IG is very likely related to Co addition from an iron meteorite impact into an Apollo 17 VHT mare basalt flow.

Well-defined MVG regions in Fig. 9 strengthen the rationale for separate source regions that are defined by proportions of major cumulate minerals. Generally, an increase in TiO_2 abundance corresponds to source development with more late-stage LMO components, hence the increased likelihood of phosphate involvement. This is also consistent with Apollo 11 and 17 high-Ti sources being somewhat shallower (i.e., less than ~400 km) in the lunar mantle, although the difference in depth from low-Ti Apollo 12 and 15 sources may be only a few tens of kilometers. Figure 9 also indicates considerable variation for Apollo 11 and 17 basalts, whereas a much lower spread is evident for Apollo 12 and 15 compositions. A gross trend of decreasing TiO_2 with increasing Co is evident in all mare compositions; however, this trend is best defined by the GGs, YBGs, and OGs, which indicate the possibility of an "upper limit." Lunar mare compositions that lie above this line would likely be related to impact processes or some other mechanism of contamination. Normal igneous processes such as assimilation and fractionation would produce liquids that plot in the regions below the line. Unless other unique glass compositions are discovered to lie above the line in Fig. 9, the upper limit of Co-Ti variation in primary magmas is defined. Moreover, the fields portrayed in Figs. 8 and 9 support major element variations (*Longhi,* 1987) that argue for separate primary magmas for mare basalts and picritic glasses.

CONCLUSIONS

Major and trace element abundances in individual picritic mare volcanic glasses are controlled by the chemical signatures obtained in mantle source regions. Major groups of mare volcanic glasses can be chemically defined such that their designation reflects the variation in proportions of source components. Incompatible trace elements are largely con-

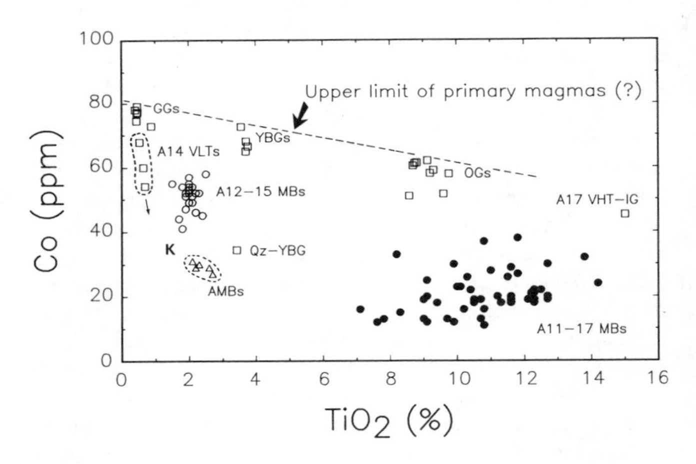

Fig. 9. Covariation of Co vs. TiO_2 abundances in mare compositions illustrate the effects of cumulate mineralogy. Tight clustering of volcanic glasses and relatively uniform Apollo 12 and 15 mare basalt compositions are contrasted by the wide range in Apollo 11 and 17 high-Ti mare basalts. The relation of Apollo 14 MVG and Al-basalts (AMBs, *Dickinson et al.,* 1985) to KREEP ("K," *Warren and Wasson,* 1979) is evident, but a potential relation between normal green glass sources and Al-basalt sources via KREEP contamination is not supported by their relative Co and TiO_2 abundances. Data source for mare basalts are *Ma et al.,* 1978, 1980; *Warner et al.,* 1979; and unpublished results from this laboratory.

trolled by minor amounts of late-stage LMO cumulates, accessory minerals, or trapped liquid that have relatively high trace element concentrations. Major and compatible trace elements are controlled by early cumulates that are proportionately dominant but relatively deficient in incompatible elements.

Picritic volcanic glasses are derived from sources that are separate from mare basalt primary magmatic sources. Current assessments allow the designation of at least four major volcanic glass sources that yielded distinct chemical signatures reflected in mare glass compositions: (1) Apollo 15 and 16 green glasses, (2) Apollo 15 yellow/brown glasses, (3) Apollo 17 orange glasses, and (4) Apollo 14 VLT and green glass. Distinct sources are also evident for low-Ti and high-Ti mare basalt types, although many basalts have experienced assimilation and fractionation that disqualify their use for strict interpretation of source regions. Apollo 14 volcanic glasses have some chemical affinity to green glasses, yet the trace element compositions are dominated by the influence of a KREEP-like component. KREEP involvement in Apollo 14 mare volcanic glass petrogenesis occurred in source regions and is unrelated to possible KREEP contamination as a secondary process noted for some mare basaltic compositions.

Acknowledgments. This research was conducted with valuable assistance by the crew of the University of Missouri Research Reactor, who assisted in neutron irradiations. We also thank D. Pratt and V. Golightly of the Oregon State University Radiation Center Health Physics Department for assistance in sample shipments. Reviews by C. R. Neal and an anonymous reviewer significantly improved the quality of arguments presented in this paper. Research was supported by NASA grants NAG 9-63 to R.A.S. and NAG 9-78 to J.W.D.

REFERENCES

Binder A. B. (1980) On the mare basalt source region. *Proc. Lunar Planet. Sci. Conf. 10th,* pp. 1-22.

Binder A. B. (1982) The mare basalt magma source region and mare basalt genesis. *Proc. Lunar Planet. Sci. Conf. 13th,* in *J. Geophys. Res., 87,* A37-A53.

Binder A. B. (1985a) The depths of the mare basalt source region. *Proc. Lunar Planet. Sci. Conf. 15th,* in *J. Geophys. Res., 89,* C396-C404.

Binder A. B. (1985b) Mare basalt genesis: Modeling trace elements and isotopic ratios. *Proc. Lunar Planet. Sci. Conf. 16th,* in *J. Geophys. Res., 90,* D19-D30.

Blanchard D. P. and Budahn J. R. (1978) Chemistry of orange/black soils from core 74001/2. *Proc. Lunar Planet. Sci. Conf. 9th,* pp. 1969-1980.

Chen H.-K. and Lindsley D. H. (1983) Apollo 14 very low titanium glasses: Melting experiments in iron-platinum alloy capsules. *Proc. Lunar Planet. Sci. Conf. 14th,* in *J. Geophys. Res., 88,* B335-B342.

Chen H.-K., Delano J. W., and Lindsley D. H. (1982) Chemistry and phase relations of VLT volcanic glasses from Apollo 14 and Apollo 17. *Proc. Lunar Planet. Sci. Conf. 13th,* in *J. Geophys. Res., 87,* A171-A181.

Delano J. W. (1979) Apollo 15 green glass: Chemistry and possible origin. *Proc. Lunar Planet. Sci. Conf. 10th,* pp. 275-300.

Delano J. W. (1980) Chemistry and liquidus phase relations of Apollo 15 red glass: Implications for the deep lunar interior. *Proc. Lunar Planet. Sci. Conf. 11th,* pp. 251-288.

Delano J. W. (1981) Major-element composition of volcanic green glasses from Apollo 14 (abstract). In *Lunar and Planetary Science XII,* pp. 217-219. Lunar and Planetary Institute, Houston.

Delano J. W. (1986) Pristine lunar glasses: Criteria, data, and implications. *Proc. Lunar Planet. Sci. Conf. 16th,* in *J. Geophys. Res., 91,* D201-D213.

Delano J. W. (1988) Apollo 14 regolith breccias: Different glass populations and their potential for charting space/time variations. *Proc. Lunar Planet. Sci. Conf. 18th,* pp. 59-65.

Delano J. W. and Livi K. (1981) Lunar volcanic glasses and their constraints on mare petrogenesis. *Geochim. Cosmochim. Acta, 45,* 2137-2149.

Delano J. W. and Lindsley D. H. (1983) Mare glasses from Apollo 17: Constraints on the moon's bulk composition. *Proc. Lunar Planet. Sci. Conf. 14th,* in *J. Geophys. Res., 88,* B3-B16.

Delano J. W., Hughes S. S., and Schmitt R. A. (1988) Apollo 14 pristine mare glasses. In *Workshop on Moon in Transition: Apollo 14, KREEP, and Evolved Lunar Rocks* (G. J. Taylor and P. H. Warren, eds.), pp. 34-37. Lunar and Planetary Institute, Houston.

Dickinson T., Taylor G. J., Keil K., Schmitt R. A., Hughes S. S., and Smith M. R. (1985) Apollo 14 aluminous mare basalts and their possible relationship to KREEP. *Proc. Lunar Planet. Sci. Conf. 15th,* in *J. Geophys. Res., 90,* C365-C374.

Fujimaki H. (1986) Partition coefficients of Hf, Zr, and REE between zircon, apatite, and liquid. *Contrib. Mineral. Petrol., 94,* 42-45.

Green D. H., Ringwood A. E., Hibberson W. O., and Ware N. G. (1975) Experimental petrology of Apollo 17 mare basalts. *Proc. Lunar Sci. Conf. 6th,* pp. 871-893.

Hubbard N. J. and Minear J. W. (1975) A physical and chemical model of early lunar history. *Proc. Lunar Sci. Conf. 6th,* pp. 1057-1085.

Hubbard N. J. and Minear J. W. (1976) Hybridization: An answer to the problem of heterogeneous source materials (abstract). In *Lunar Science VII,* pp. 393-395. The Lunar Science Institute, Houston.

Hughes S. S., Delano J. W., and Schmitt R. A. (1988) Apollo 15 yellow-brown volcanic glass: Chemistry and petrogenetic relations to green volcanic glass and olivine-normative mare basalts. *Geochim. Cosmochim. Acta, 52,* 2379-2391.

Hughes S. S., Delano J. W., and Schmitt R. A. (1989) Petrogenetic modeling of 74220 high-Ti orange volcanic glasses and the Apollo 11 and 17 high-Ti mare basalts. *Proc. Lunar Planet. Sci. Conf. 19th,* pp. 175-188.

Kesson S. E. (1975) Mare basalts: Melting experiments and petrogenetic interpretations. *Proc. Lunar Sci. Conf. 6th,* pp. 921-944.

Longhi J. (1978) Pyroxene stability and the composition of the lunar magma ocean. *Proc. Lunar Planet. Sci. Conf. 9th,* pp. 285-306.

Longhi J. (1987) On the connection between mare basalts and picritic volcanic glasses. *Proc. Lunar Planet. Sci. Conf. 17th,* in *J. Geophys. Res., 92,* E349-E360.

Ma M.-S., Schmitt R. A., Warner R. D., Taylor G. J., and Keil K. (1978) Genesis of Apollo 15 olivine normative mare basalts: Trace element correlations. *Proc. Lunar Planet. Sci. Conf. 9th,* pp. 523-533.

Ma M.-S., Schmitt R. A., Beaty D. W., and Albee A. L. (1980) The petrology and chemistry of basaltic fragments from the Apollo 11 soil: Drive tubes 10004 and 10005. *Proc. Lunar Planet. Sci. Conf. 11th,* pp. 37-47.

Ma M.-S., Liu Y.-G., and Schmitt R. A. (1981) A chemical study of individual green glasses and brown glasses from 15426: Implications for their petrogenesis. *Proc. Lunar Planet. Sci. 12B,* pp. 915-933.

Nagasawa H. (1970) Rare earth concentrations in zircons and apatites and their host dacites and granites. *Earth Planet. Sci. Lett., 9,* 359-364.

Neal C. R., Taylor L. A., and Lindstrom M. M. (1988) Apollo 14 mare basalt petrogenesis: Assimilation of KREEP-like components by a fractionating magma. *Proc. Lunar Planet. Sci. Conf. 18th,* pp. 139-153.

Philpotts J. A. and Schnetzler C. C. (1970) Apollo 11 lunar sample: K, Rb, Sr, Ba and rare-earth concentrations in some rocks and separated phases. *Proc. Apollo 11 Lunar Sci. Conf.*, pp. 1471-1486.

Ringwood A. E. and Kesson S. E. (1976) A dynamic model for mare basalt petrogenesis. *Proc. Lunar Sci. Conf. 7th*, pp. 1697-1722.

Shearer C. K., Papike J. J., Simon S. B., Shimizu N., Yurimoto H., and Sueno S. (1989) Ion microprobe studies of trace elements in Apollo 14 volcanic glass beads, comparisons to Apollo 14 mare basalts and petrogenesis of picritic magmas. *Geochim. Cosmochim. Acta,* in press.

Shervais J. W., Taylor L. A., and Lindstrom M. M. (1985) Apollo 14 mare basalts: Petrology and geochemistry of clasts from consortium breccia 14321. *Proc. Lunar Planet. Sci. Conf. 15th,* in *J. Geophys. Res., 90,* C375-C395.

Shih C.-Y. and Schonfeld E. (1976) Mare basalt genesis: A cumulate-remelting model. *Proc. Lunar Sci. Conf. 7th*, pp. 1757-1792.

Taylor S. R. (1982) *Planetary Science: A Lunar Perspective.* Lunar and Planetary Institute, Houston. 481 pp.

Taylor S. R. and Jakes P. (1974) The geochemical evolution of the moon. *Proc. Lunar Sci. Conf. 5th*, pp. 1287-1305.

Walker D. J., Grove T. L., Longhi J., Stolper E. M., and Hays J. F. (1973) Origin of lunar feldspathic rocks. *Earth Planet. Sci. Lett., 20,* 325-336.

Warner R. D., Taylor G. J., Conrad G. H., Northrop H. R., Barker S., Keil K., Ma M.-S., and Schmitt R. (1979) Apollo 17 high-Ti mare basalts: New bulk compositional data, magma types, and petrogenesis. *Proc. Lunar Planet. Sci. Conf. 10th*, pp. 225-247.

Warren P. H. (1985) The magma ocean concept and lunar evolution. *Annu. Rev. Earth Planet. Sci., 13,* 201-240.

Warren P. H. (1986) Anorthosite assimilation and the origin of the Mg/Fe-related bimodality of pristine Moon rocks: Support for the magmasphere hypothesis. *Proc. Lunar Planet. Sci. Conf. 16th,* in *J. Geophys. Res., 91,* D331-D343.

Warren P. H. and Wasson J. T. (1979) The origin of KREEP. *Rev. Geophys. Space Phys., 17,* 73-88.

Watson E. B. and Green T. H. (1981) Apatite/liquid partition coefficients for the rare earth elements and strontium. *Earth Planet. Sci. Lett., 56,* 405-421.

Wood J. A., Dickey J. S. Jr., Marvin U. B., and Powell B. N. (1970) Lunar anorthosites and a geophysical model of the Moon. *Proc. Apollo 11 Lunar Sci. Conf.*, pp. 965-988.

Proceedings of the 20th Lunar and Planetary Science Conference, pp. 139-143
Lunar and Planetary Institute, Houston, 1990

Parental Magmas of Mare Fecunditatis: Evidence from Pristine Glasses

Y. Jin and L. A. Taylor

Department of Geological Sciences, University of Tennessee, Knoxville, TN 37996

Glasses within the lunar soils can reveal significant aspects of both the impact process and the petrogeneses of mare basalts. In an attempt to understand the glass compositions and to identify parental magma types at Mare Fecunditatis, electron microprobe analyses have been performed on 116 glass particles from a new Luna 16 soil sample 21036,15 (200-500 micron size fraction). Four types of glass particles were identified: (1) basaltic glass, (2) anorthosite-norite-troctolite (ANT) suite glass, (3) maskelynite, and (4) high-alumina silica-poor (HASP) glass. The majority of the 116 glasses are of impact origin, with about 55% of all the glasses classed as ANT suite. Although it is to be expected that the compositions of the impact glasses will resemble the average soil, the compositions of these ANT glasses do not overlap that of the Luna 16 bulk soil. This would seem to indicate that these glasses are not indigenous and were transported to the Luna 16 site by impacts into the neighboring highlands. Alternatively, major impacting has melted and excavated highland materials from beneath the thin basalt flows. Fourteen grains with basaltic compositions are judged to be of volcanic origin. These pristine glasses are further divided into two groups (A and B) on the basis of MgO, FeO, Al_2O_3, and CaO contents. Compared to Apollo pristine glasses (e.g., *Delano,* 1986), they have higher CaO (>9% vs. 6-9.4%) and Al_2O_3 (>9% vs. 4-9.6%), along with lower MgO (<15% vs. 11.6-19.5%) and Ni (<10-97 vs. <22-188 ppm). At least two parental magmas are needed to explain the chemical variations among these glasses. The Group B glasses appear to represent primitive parental magma that evolved by olivine fractionation to the compositions of the Luna 16 aluminous mare basalts. The Group A volcanic glasses are high in CaO (11.6-15.1%) and Al_2O_3 (13.9-19.6%). This Group A may represent an unusual, new basalt magma type that contains a high plagioclase component. However, the possibility also exists that the glasses of this group were formed by meteorite impact.

INTRODUCTION

It has been well demonstrated by J. Delano and coworkers that volcanic glasses in lunar soil are a key to an understanding of magmatic activities, the petrogenesis of mare basalts, and the mantle geochemistry of the Moon. The nature and compositions of impact-formed glasses can reveal information concerning the various aspects of impacting processes and formation of the soils. Based upon studies of thousands of glasses from Apollo soils, Delano and his coworkers have identified 25 different magma types (e.g., *Delano and Livi,* 1981; *Delano,* 1986). However, their studies were restricted to sites visited by the Apollo missions. The samples reported on here are from Mare Fecunditatis, which was sampled only by the Soviet Luna 16 mission. It is reasonable to assume that this region of the Moon will possess evidence of additional magma types, perhaps quite different from those of other regions of the Moon.

Recently, we received a new Luna 16 core sample, from the 20-28 cm portion of horizon C (*Vinogradov,* 1971). A preliminary study of this sample was reported by *Jin and Taylor* (1988, 1989). A polished grain mount was prepared of the 200-500 micron size fraction (PM 21036,15) of this soil. An optical microscopic and electron microprobe survey has been performed on all the discrete glass particles within this grain mount. This was done in an attempt to understand the provenance and origin of glasses in Mare Fecunditatis regolith and to search for parental magmas of mare basalts at this locale. Detailed microprobe study, involving special attention to analysis for trace amounts of Ni, was performed on 30 potentially pristine glass beads and fragments, so classified based upon petrographic criteria. Applying other criteria of pristinity (e.g., *Delano,* 1986), the number of pristine glasses was narrowed down to 14. An emphasis of this paper is on the nature and characterization of these 14 pristine glasses from Luna 16. Previous data on lithic fragments and regoliths from the Luna 16 drill core (e.g., *Kurat et al.,* 1976) have been integrated to show the relationship between various components present in Mare Fecunditatis regolith. Comparisons are made between Luna 16 pristine glasses and the 25 types of volcanic glasses from Apollo samples (*Delano,* 1986), and the possibility is discussed of new types of volcanic glasses, representative of new basaltic parental magmas.

PETROGRAPHY

A total of 116 discrete glass beads and fragments ranging in size from 20-400 microns was microscopically identified in sample 21036,15. Most of the glasses are colorless (77 grains) or orange to greenish (38); one is opaque. Some (8) are partially devitrified, 11 have distinct schlieren, and many grains (>70%) contain minute opaque inclusions, especially at or near their outer margins.

ANALYTICAL PROCEDURES

Major- and minor-element compositions of the glasses were obtained with a fully automated Cameca SX-50 electron microprobe, utilizing ZAF correction procedures and natural and synthetic standards. The accelerating voltage was 15 KeV, beam current was 10-20 na, and beam diameter was 5 or 10 microns, depending upon grain size. Particular attention was given to ensure no detectable Na loss from the glass due to electron-beam heating. Up to 12 replicate analyses were performed on each grain. Those glasses judged to be "potentially pristine" were analyzed for Ni by counting 400 sec for each analysis (200 sec at peak and 100 sec each in either side of the peak for background counts). Up to 10 replicate analyses were performed on each particle utilizing 30 KeV accelerating potential, 50-na beam current, and a beam diameter of 5-20 microns, depending upon the grain size. Approximately 11 counts/ppm Ni were achieved. Based on replicate analyses on natural standards with trace amounts of Ni, the analytical precision is estimated to be 10 ppm, with a detection limit calculated to be 5-10 ppm.

RESULTS

Among the 116 glass particles analyzed, 104 grains are chemically homogeneous, with negligible intragranular variations of SiO_2, Al_2O_3, MgO, FeO, or CaO. However, as a group, the glasses have a wide range of compositions (e.g., SiO_2 30.7-53.7%, Al_2O_3 8.01-37.0%, CaO 9.36-21.9%, MgO 0.02-14.8%, FeO 0.2-25.9%, TiO_2 0.02-8.87%, and Ni $<$10-737 ppm). Based upon their MgO + FeO vs. CaO + Al_2O_3 contents (Fig. 1), they are divided into four types: (1) Basaltic glass, (2) anorthosite-norite-troctolite (ANT) suite glass, (3) maskelynite, and (4) high-alumina silica-poor (HASP; *Naney et al.,* 1976) glass.

Basaltic glasses are distinguished by relatively high FeO, MgO, and TiO_2 contents, and CaO/Al_2O_3 ratios (12.6-25.9%, 6.5-14.8%, 0.77-8.87%, and 0.66-1.18, respectively) and low Al_2O_3 and CaO (8.01-19.6% and 9.36-15.0%). Whole rock analyses of basaltic lithic fragments from Luna 16 (e.g., *Kurat et al.,* 1976) plot within this basaltic glass area; as expected for this mare region, average compositions of the Luna 16 regolith (soil) occur here as well (Fig. 1). *ANT suite glasses* are characterized by high Al_2O_3 and CaO (19.8-31.2% and 9.78-19.6%, respectively) and low FeO (1.11-8.78%). The MgO contents span a wide range from 0.07 to 15.8% but are mostly around 5-10%. CaO/Al_2O_3 ratios range from 0.56-0.78, mostly around 0.61. It is noteworthy that over half (55%) of the glasses examined are of this type, different from the bulk soil composition. These may represent glasses added to the soil from (1) distant highland impacts and/or (2) local deep impacts that melted and excavated highland materials from beneath the relatively thin basalt flows (cf. *Farrand,* 1987, 1988). This glass type is equivalent to the highland basaltic glass reported by *Jakes et al.* (1972). Five *maskelynite* grains were identified in this sample. Four of them retain plagioclase stoichiometry, with An values 89, 93, 97, and 98. One grain is particularly rich in Na_2O (3.31%; An68) and may represent a nonstoichiometric feldspar (cf. *Taylor et al.,* 1971). One of the glasses represents *HASP glass* (*Naney et al.,* 1976), whose unusual composition is reportedly the result of fractional vaporization due to impact melting.

Although the glass compositions do not strictly correlate with color, the nonbasaltic glasses are all colorless; most of the basaltic glasses, especially those with high TiO_2 contents are greenish to orange. The opaque glass bead is high in TiO_2 (6.64%) and FeO (20.3%) and contains minute ilmenite dendrites, much like the black glass equivalent of Apollo 17 orange glass.

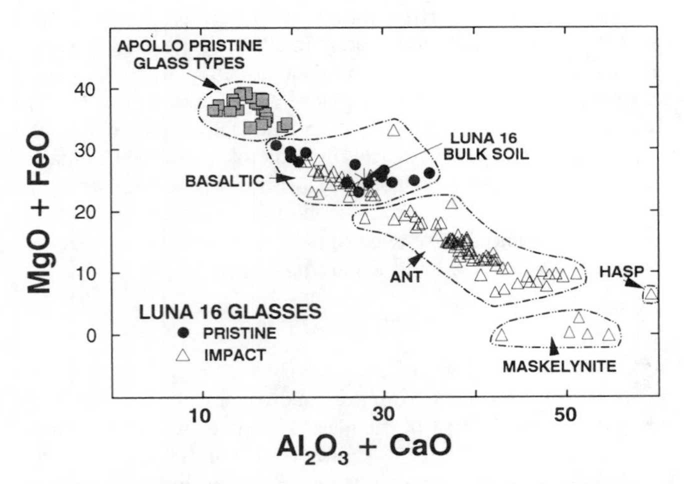

Fig. 1. MgO + FeO vs. Al_2O_3 + CaO plot showing the four major types of glass from horizon C of the Luna 16 core (PM 21036,15): basaltic glass, ANT suite (anorthosite-norite-troctolite) glass, maskelynite, and HASP. The average composition of Luna 16 regolith plot into the central area of basaltic type glass. For comparison, 25 Apollo pristine glass types (*Delano,* 1986) are also plotted, which are higher in MgO+FeO contents than all of the Luna 16 glasses.

PRISTINE MARE FECUNDITATIS GLASS

Criteria

Although the criteria for recognition of volcanic vs. impact glasses are not unambiguous (*Delano and Livi,* 1981; *Delano,* 1986), they do provide logical ways by which to determine their origins. Various criteria have been employed to identify the pristinity of the glasses among the Luna 16 glass particles. Pristine glasses must be chemically homogeneous without schlieren or exotic inclusions. However, specific chemical criteria also have been used to further evaluate possible volcanic melt products. Only those glasses with CaO/Al_2O_3 ratios greater than 0.75 are considered to have potential mare parentage (cf. *Naney et al.,* 1976). It is important that the Ni contents of the glass should correlate with MgO (*Delano,* 1986). As discussed below, the Luna 16 glasses do not strictly follow these criteria. But these were developed using only the Apollo samples as a database, and Mare Fecunditatis basalts are significantly different. Applying these modified criteria, 14 inclusion-free grains of the original 116 glass particles were considered to be of volcanic origin. All examples of the ANT suite glasses, maskelynite, HASP glass, and 32 of the 46 basaltic glasses are judged to be nonpristine.

The criteria of cluster analysis (*Delano,* 1986) are not entirely applicable to the present study, mainly because of limited number of glass particles in our populations. However,

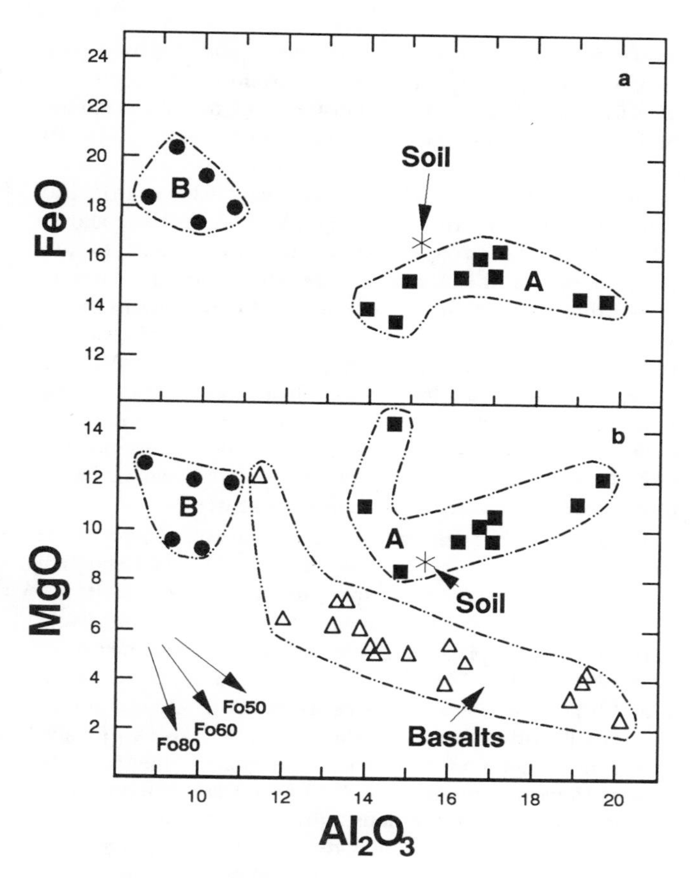

Fig. 2. FeO **(a)** and MgO **(b)** vs. Al_2O_3 showing subdivision of Luna 16 pristine glasses into two groups and their relationships with the basalts (from *Kurat et al.,* 1976) and the average composition of Luna 16 regolith (*Vinogradov,* 1971). Note that the Group B glasses can be related to Luna 16 basalts by fractional crystallization of olivine. Group A also seems to be related to the basalts by olivine fractionation; however,other element-element plots (e.g., CaO-Al_2O_3) negate this possibility. Group A glasses are unusual in their high plagioclase component, coupled with relatively low Mg#.

the five glasses in Group B form a restricted range of compositions (Fig. 2) similar to some of those of *Delano* (1986). Group A glasses, on the other hand, have considerable compositional variability and may represent two or more magma types. Considerably more glass particles will need to be examined in order to obtain sufficient statistics with which to further evaluate this statement.

Luna 16 Pristine Glass

As shown in Fig. 3, the Luna 16 pristine glasses are unusual in that they have much lower Ni contents (<10-97 ppm) than the majority of their Apollo glass counterparts (*Delano,* 1986). Rather, they have Ni abundances similar to those of Apollo mare basalts (*Delano,* 1986). The low MgO contents of these glasses (Table 1) also reflect their evolved nature. The Ni contents of the glass should correlate with MgO (*Delano,* 1986) demonstrating that Ni behaves as a lithophile element during magma evolution. Although the Luna 16 glasses do not entirely follow the Apollo glass trend (Fig. 3), all except one of the group A glasses do plot within an expansion of the volcanic glass Ni-MgO array of *Delano* (1986). Some of these glasses may represent magmas more evolved than those represented by the Apollo array of glasses. The Luna 16 basalts are more evolved than most Apollo basalts; therefore, it is to be expected that their parental magmas would be less primitive than at the Apollo locales.

The Luna 16 pristine glasses have compositional ranges of FeO, TiO_2, CaO, and Al_2O_3 similar to those of Luna 16 basalts (e.g., *Vinogradov,* 1971; *Albee et al.,* 1972; *Keil et al.,* 1972; *Kurat et al.,* 1976; *Ma et al.,* 1979). However, the majority of the pristine glasses have higher MgO contents (7-12%) and Mg/(Mg + Fe) ratios (0.45-0.65) than their lithic equivalents (2-7% and 0.28-0.37, respectively), except for one olivine basalt (12% and 0.49, respectively; Fig. 2).

Inspection of the pristine glass chemistry permits division into two groups (Fig. 2, Table 1). The Group A glass (nine particles) is characterized by the highest CaO (11.6-15.1%) and Al_2O_3 (13.9-19.6%), low FeO (13.5-16.4%), and generally

TABLE 1. Electron microprobe analyses of Luna 16 pristine glasses in sample 21036,15.

	Group A									Group B				
SiO_2	46.5	42.9	34.6	40.0	36.5	43.1	43.9	38.8	37.9	47.9	45.7	46.4	43.3	39.1
TiO_2	0.95	1.37	2.70	3.00	3.43	3.44	3.59	3.81	4.27	0.77	1.64	2.29	5.58	8.87
Al_2O_3	13.9	14.7	19.6	16.7	19.0	16.1	14.9	17.1	17.0	8.64	10.8	9.84	10.1	9.33
Cr_2O_3	0.39	0.29	0.19	0.25	0.23	0.26	0.31	0.20	0.22	0.77	0.45	0.37	0.22	0.41
MgO	11.0	14.5	12.2	10.2	11.1	9.65	8.51	9.56	10.5	12.6	11.9	12.0	9.17	9.61
CaO	11.6	11.8	15.1	12.7	13.9	11.8	11.9	13.6	12.7	9.27	10.4	9.70	10.3	9.96
MnO	0.3	0.22	0.18	0.24	0.22	0.21	0.23	0.24	0.22	0.26	0.25	0.30	0.30	0.24
FeO	14.0	13.5	14.4	16.0	14.5	15.2	15.1	15.5	16.4	18.3	17.9	17.3	19.3	20.4
Na_2O	0.03	<0.03	<0.03	0.04	<0.03	0.17	0.25	<0.03	<0.03	0.37	0.31	0.15	0.16	0.26
K_2O	<0.03	<0.03	<0.03	<0.03	<0.03	0.04	0.08	<0.03	<0.03	<0.03	0.03	0.03	0.04	0.12
Ni (ppm)	<10	35	15	17	30	97	<10	21	68	10	82	16	<10	53
Total	98.70	99.34	99.03	99.16	98.94	99.97	98.77	98.87	99.27	98.91	99.38	98.38	98.47	98.30
Mg#*	0.58	0.65	0.60	0.53	0.57	0.53	0.50	0.53	0.53	0.55	0.54	0.55	0.45	0.45

*Mg# = Mg/(Mg + Fe) atomic ratio.

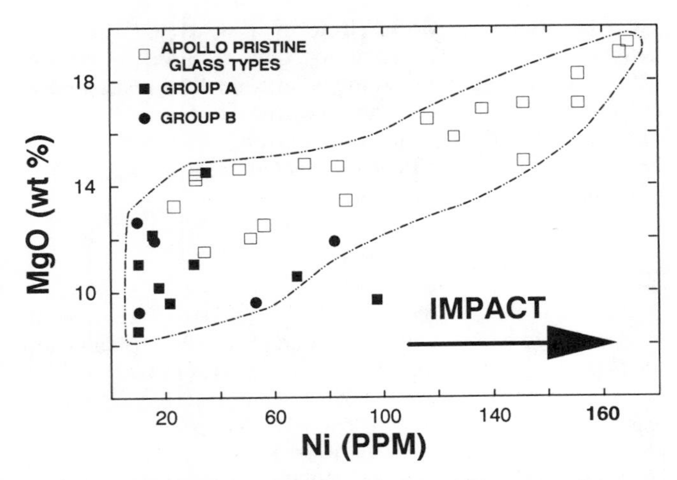

Fig. 3. MgO vs. Ni of the Luna 16 and Apollo pristine glasses (*Delano,* 1986). Although the Luna 16 glasses occupy a scattered area, they generally fall along the trend defined by the Apollo glasses. The impact glasses have Ni contents that are largely due to meteorite contamination. In such cases the Ni does not display the lithophile relationship with MgO as in the basalts and magmas.

high Mg# (0.53-0.65). The Na_2O contents of most of this glass group are below the detection limit (0.03%). Group B glasses (five particles) are recognized by low CaO (9.27-10.4%) and Al_2O_3 (8.64-10.8%), high FeO (17.3-20.4%), and relatively low Mg# (0.45-0.55). The Na_2O contents (0.15-0.37%) are generally higher than those of Group A glasses. Note that the average composition of the Luna 16 regolith (*Vinogradov,* 1971) plots within or near the compositional fields of the Group A glasses, as shown in Fig. 2.

New Pristine Glass Types

Delano (1986) has reported 25 different basaltic magma compositions from Apollo volcanic glasses. It would appear that Mare Fecunditatis glasses may represent additional magma types. Compared to the pristine glasses of *Delano* (1986), Luna 16 magmas have unusual compositions enriched in plagioclase components, being higher in Al_2O_3 (8.64-19.6% vs. 4.6-9.6%) and CaO (9.27-15.0% vs. 6.27-9.4%) and lower in MgO (8.51-14.4% vs. 12.1-19.5%). However, the Mg# (0.45-0.65) and Ca/(Ca + Na) (0.93-1.0) ratios of the Luna 16 pristine glasses overlap those of Apollo pristine glass types (0.47-0.66 and 0.93-1.0, respectively).

Figure 2 reveals that at least two magma types are responsible for the chemical variations observed among the Luna 16 pristine glasses. One is represented by Group A pristine glasses, while the Group B glasses comprise the other magma type. One Luna 16 olivine basalt (*Kurat et al.,* 1976) plots very close to our Group B glasses and may well be a crystalline equivalent. The Group B glasses can be related to the other Luna 16 basalts by fractional crystallization of 5-20% olivine (Fo_{80-50}) (Fig. 2b). It is probable that these glass compositions are representative of a magma that was parental to the Luna 16 basalts. This kind of fractionation trend is similar to that determined for Apollo magmatic glasses (see Fig. 5a, *Delano,* 1986). Although this group is represented by only five glass fragments, the clustering of the compositions is similar to many of the Apollo groups observed by *Delano* (1986).

The Group A glass population is distinctly different from the composition of Luna 16 basalts (Fig. 2). Although Fig. 2 allows for some of the Group A glasses to have evolved to the compositions of the Al-rich basalts, other elemental trends negate this. In addition, no continuum of compositions exists in the present data array between the Group A glasses and the Luna 16 basalts. Likewise, although it would appear that the Group A magma could have evolved by fractionation of olivine + pyroxene from the Array I or II magmas of *Delano and Livi* (1981), no intermediate compositions connect these populations (e.g., Fig. 1). Therefore, we suggest that the Group A magmas were generated from an unusual source.

Ma et al. (1979) demonstrated that, "the Luna 16 basalts . . . have high Al_2O_3 (>13 wt.%), low CaO/Al_2O_3 ($\lesssim$0.9), and high FeO/MgO (3-4) compared to most other mare basalts . . ." They proposed that Luna 16 aluminous mare basalts were generated from a plagioclase-rich source shallower than those of Apollo 11 and 17 basalts. As discussed above, both Groups A and B Luna 16 volcanic glasses are richer in Al_2O_3 and CaO than Apollo pristine glasses. These volcanic glasses probably represent samples of the parental magma of the Luna 16 basalts. However, the distinctly high plagioclase component of the Group A glass compositions remains to be explained.

An alternative origin for the Group A glasses is possible. The composition of the Luna 16 average soil plots within or near the Group A glass areas of Fig. 2, but toward the plagioclase (i.e., Al_2O_3) deficient region. The soil does possess evidence, in the presence of numerous ANT glasses (Fig. 1), for admixed highlands materials. Thus, it is possible that the Group A glasses are of impact origin and represent a mixture of a highlands plagioclase component and the indigenous regolith. In fact, the low SiO_2 contents (Table 1) of some of the Group A glasses may be the product of fractional vaporization of the impact melt. The criteria for pristinity are ambiguous. Although possessing the properties of pristine glasses of volcanic origin, we cannot rule out completely the distinct possibility of an impact genesis.

SUMMARY

Although the database of glass analyses from this study is relatively small, several significant observations can be made. Over half (55%) of the glass beads and fragments observed in the Luna 16 soil sample are "ANT suite glasses." These are high in CaO and Al_2O_3 and low in FeO and TiO_2 compared to the compositions of the basalts occurring at this locale. Indeed, the bulk soil composition does not plot in this array. It seems that these ANT impact glasses have been added to the soil by (1) transportation from adjacent highland areas by large impacts and/or (2) the melting and excavation of local, deep-penetrating impacts that sampled highland materials beneath the relatively thin basalt flows.

There appear to be at least two parental magma types represented by pristine glasses at Mare Fecunditatis. Both Groups A and B are more evolved than those of the Arrays I and II lunar pristine glasses (*Delano and Livi,* 1981). The Group B magma type seems to be parental to the Luna 16 aluminous basalts, which can be so derived by simple olivine fractionation. The Group A volcanic glasses, however, may represent an unusual magma type. Although some of these Group A glasses might be related to Arrays I and II of *Delano and Livi* (1981) by olivine and pyroxene fractionation, there is no continuum between them. The possibility exists that these Group A glasses represent impact products, mixtures of a highland component with the indigenous basalts. Further data are required to thoroughly evaluate these relationships.

Acknowledgments. The authors wish to thank J. Delano, C. Neal, and G. Wasserburg for invaluable suggestions. Likewise, the constructive comments and critical reviews of J. Delano, A. Basu, and D. J. Lindstrom are greatly appreciated. Assistance by A. Patchen with some of the electron microprobe analyses is acknowledged. The Cameca SX-50 electron microprobe was purchased with funds from NASA and NSF grants to L.A.T. In addition, the research presented in this paper was supported by NASA grant NAG 9-62.

REFERENCES

Albee A. L., Chodos A. A., Gancarz A. J., Haines E. L., Papanastassiou D. A., Ray L., Tera F., Wasserburg G. J., and Wen T. (1972) Mineralogy, petrology, and chemistry of a Luna 16 basaltic fragment, sample B-1. *Earth Planet Sci. Lett.,* 13, 353-367.

Delano J. W. (1986) Pristine lunar glasses: Criteria, data and implications. *Proc. Lunar Planet. Sci. Conf. 16th,* in *J. Geophys. Res., 91,* D201-213.

Delano J. W. and Livi K. (1981) Lunar volcanic glasses and their constraints on mare petrogenesis. *Geochim. Cosmochim. Acta, 45,* 2137-2149.

Farrand W. H. (1987) Vertical versus lateral mixing of highland materials and minimum basalt thickness in northern Mare Fecunditatis (abstract). In *Lunar and Planetary Science XVIII,* pp. 282-283. Lunar and Planetary Institute, Houston.

Farrand W. H. (1988) Highland contamination and minimum basalt thickness in northern Mare Fecunditatis. *Proc. Lunar Planet. Sci. Conf. 18th,* pp. 319-329.

Jakes P., Warner J., Ridley W. I., Reid A. M., Harmon R. S., Brett R., and Brown R. W. (1972) Petrology of a portion of the Mare Fecunditatis regolith. *Earth Planet. Sci. Lett., 13,* 257-271.

Jin Y.-Q. and Taylor L. A. (1988) Mineral chemistry and petrology of Luna 16 basalts: Sample 21036 (abstract). In *Lunar and Planetary Science XIX,* pp. 555-556. Lunar and Planetary Institute, Houston.

Jin Y.-Q. and Taylor L. A. (1989) Volcanic and impact glasses from Mare Fecunditatis (abstract). In *Lunar and Planetary Science XX,* pp. 466-467. Lunar and Planetary Institute, Houston.

Keil K., Kurat G., Prinz M., and Green J. (1972) Lithic fragments, glasses and chondrules from Luna 16 fines. *Earth Planet. Sci. Lett., 13,* 243-256.

Kurat G., Kracher A., Keil K., Warner R., and Prinz M. (1976) Composition and origin of Luna 16 aluminous mare basalts. *Proc. Lunar Sci. Conf. 7th,* pp. 1301-1321.

Ma M.-S., Schmitt R. A., Nielsen R. L., Warner R. D., Taylor G. J., and Keil K. (1979) Luna 16 basalts and breccias: New chemical and petrologic data (abstract). In *Lunar and Planetary Science X,* pp. 762-764. Lunar and Planetary Institute, Houston.

Naney M. T., Crowl D. M., and Papike J. J. (1976) The Apollo 16 drill core: Statistical analysis of glass chemistry and the characterization of a high alumina-silica poor (HASP) glass. *Proc. Lunar Sci. Conf. 7th,* pp. 155-184.

Taylor L. A., Kullerud G., and Bryan W. B. (1971) Opaque mineralogy and textural features of Apollo 12 samples and a comparison with Apollo 11 rocks. *Proc. Lunar Sci. Conf. 2nd,* pp. 855-871.

Vinogradov A. P. (1971) Preliminary data on lunar ground brought to Earth by the automatic probe "Luna-16." *Proc. Lunar Sci. Conf. 2nd,* pp. 1-16.

Geology of the Moon

Proceedings of the 20th Lunar and Planetary Science Conference, pp. 147-160
Lunar and Planetary Institute, Houston, 1990

The Relationship Between Orbital, Earth-based, and Sample Data for Lunar Landing Sites

P. E. Clark

Mail Stop 238/420, Jet Propulsion Laboratory, California Institute of Technology, Pasadena, CA 91109

B. R. Hawke

Planetary Geosciences Division, Hawaii Institute for Geophysics, University of Hawaii, Honolulu, HI 96822

A. Basu

Department of Geology, Indiana University, Bloomington, IN 47405

Reported in this paper are the results of a study involving a very detailed examination of remote-sensing data available for the Apollo lunar landing sites. Remote observations include Apollo orbital measurements of six major elements derived from X-ray fluorescence (XRF) and gamma-ray instruments and geochemical parameters derived from Earth-based spectral reflectivity data. Wherever orbital coverage exists for Apollo landing sites (for 12, 14, 15, 16, and 17), remote data are correlated with geochemical data derived from soil sample averages for major geological units and major rock components associated with these units. Typically, the samples were collected along meandering multikilometer traverses at the landing sites (except at the Apollo 12 site, where samples were obtained in the immediate vicinity of the lander). The effective fields of view for the orbital instruments, on the other hand, are 30 km on a side for the XRF instrument (Al and Mg data); 300 km on a side for gamma-ray-derived K, Fe, and Ti (and in some cases Mg and Al) data, and 60 km on a side for deconvoluted gamma-ray-derived Th data. At the Apollo 14, 15, and 17 sites, a greater KREEP component appears to be present in the larger area around the landing site than predicted on the basis of sample data. The area around the Apollo 12 site is more anorthositic than the site itself. The Apollo 16 site is less enriched in gabbroic anorthosite than the surrounding area. The area around Apollo 17 contains more material similar to the North Massif mixed with a ferroan anorthosite component than that represented at the site. The discrepancies between remote and soil-analysis elemental concentration data that we have observed are apparently due to the differences in the extent of exposure of geological units, and hence major rock components, in the area sampled. Typically, major geological units were represented and sampled at the landing sites but not necessarily in the same proportions in which these units exist in the areas around the sites. Differences in signal depths between various orbital experiments, which may provide a mechanism to explain differences between XRF and other landing site data, appear to have minimal influence on the observed discrepancies.

INTRODUCTION

Analyses of available lunar orbital geochemical data done during the last decade have led to a greater understanding of the distribution and relative abundances of major rock types on the lunar surface. The results of some of these ongoing studies, done at global, regional, and local scales, are summarized below. The most recent work, described in this paper, includes a detailed examination of remote-sensing data available for the Apollo landing sites. In the last decade, many new analyses of lunar soil samples have been made available and a comparison of remote-sensing and sample data is in order.

Adler and Trombka (1977) calculated the average concentrations of Fe, Mg, Th, and K, and the concentration ratios of Mg/Si and Al/Si for some regions of the Moon that included some landing sites; a very limited number of surface samples were used to provide "ground truth." A comparison between the average chemical composition of specific landing sites derived from lunar soil geochemistry and that derived from an integrated study of orbital X-ray fluorescence (XRF) and gamma-ray spectrometer (GRS) experiments for areas including the landing sites has not been done. Further, the nature of the relationship of these two independent datasets has not been previously discussed. It should be pointed out that the depths sampled (i.e., depths from which meaningful signals were received) by the XRF and GRS experiments are restricted to the upper 1 μm and about 10 cm, respectively, although signals from a depth of up to 1 m for naturally radioactive elements like Th may be measured. Therefore, only the uppermost regolith and virtually no bedrock was probably sampled (*Metzger et al.,* 1973; *Reedy et al.,* 1973; *Adler and Trombka,* 1977; *Reedy,* 1978; *Wilhelms,* 1987). Most of the soil samples also come from the uppermost regolith.

The purpose of this work is to compare and contrast orbital and soil-sample compositional data from the Apollo 12, Apollo 14, Apollo 15, Apollo 16, and Apollo 17 landing sites (Fig. 1) and to evaluate how well the orbital data may represent the "ground truth" of soil samples. In addition, a limited amount

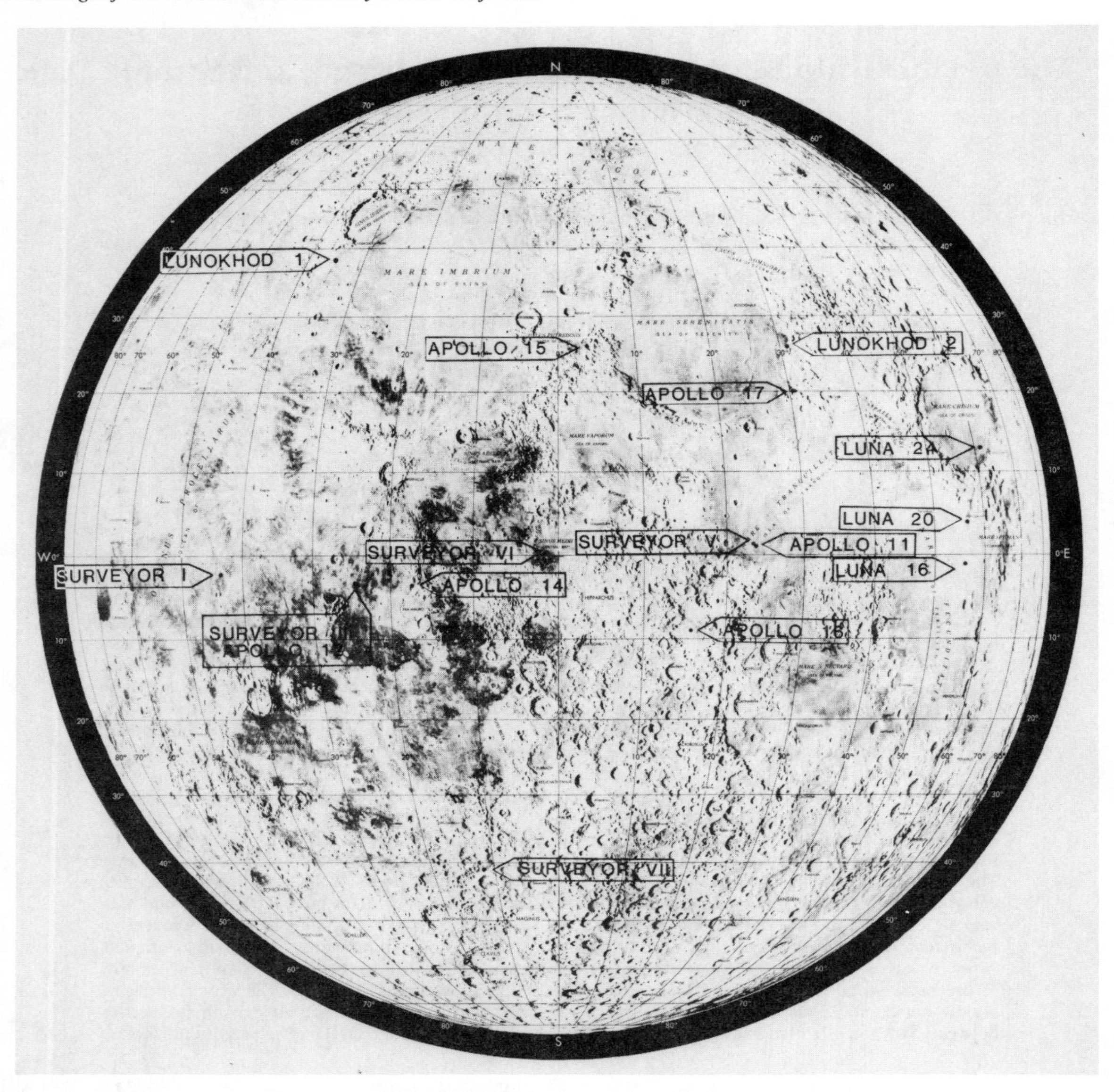

Fig. 1. Apollo landing sites plotted on a map of the lunar nearside.

of Earth-based spectral reflectance data is compared to laboratory measurements of lunar soils for selected landing sites. We restrict our work to direct measurements and deliberately stay away from comparing model-dependent soil compositions that may represent large areas of the Moon, or from finding a soil whose composition matches an orbital measurement exactly; these are important questions but are beyond the scope of this paper.

PREVIOUS WORK

Studies of the relationship among various chemical variables on a global scale have led to the characterization of the lunar surface in terms of geochemical terrains (*Clark et al*, 1978; *Clark and Hawke*, 1982; *Clark*, 1985). An early study (*Clark et al.*, 1978) demonstrated that although the highlands were uniformly high in aluminum, varying amounts of Fe, Ti, and different surface morphologies in different regions made the highlands quite heterogeneous. In a later study *Clark and Hawke* (1982) used multidimensional supervised and unsupervised classification techniques, and cluster analysis, to correlate orbital XRF Al/Si and Mg/Si intensity ratios (*Clark and Adler*, 1978; *Clark and Hawke*, 1981) and GRS Fe, Ti, and Th concentration data (*Davis*, 1980; *Metzger et al.*, 1977). Regional scale correlation resulted in the identification of major rock units, including four basalt types that consisted of a KREEP unit primarily associated with the Imbrium basin region and three basin-age correlated mare basalt types associated with nearside basins. Also identified were five highland units dominated by ANT suite material but obviously contaminated by varying amounts of KREEP-rich material and mare basalts.

TABLE 1. List of lunar soils used to calculate the average composition of surficial material at five Apollo landing sites.

Landing Site	Soil Samples	References
Apollo 12	12001; 12023; 12030; 12032; 12037; 12041; 12042; 12044; 12070	*Laul and Papike* (1980); *Warren and Wasson* (1978); *Woodcock and Pillinger* (1978); *Frondel et al.* (1971); *Schnetzler and Philpotts* (1971); *Cuttitta et al.* (1971); *Haskin et al.* (1971); *Wakita et al.* (1971); *Goles et al.* (1971)
Apollo 14	14003; 14148; 14149; 14156	*Rose et al.* (1972); *Lindstrom et al.* (1972); *Philpotts et al.* (1972)
Apollo 15	15012; 15013; 15020; 15030; 15040; 15070; 15080; 15470; 15500; 15530; 15600; 15090; 15100; 15210; 15221; 15230; 15250; 15270; 15290; 15300; 15400; 15410; 15430	*Apollo 15 PET* (1972); *Christian et al.* (1976); *Laul et al.* (1972); *Laul and Schmitt* (1973); *Morgan et al.* (1972); *Duncan et al.* (1975); *Wänke et al.* (1973); *Brunfelt et al.* (1972); *Carron et al.* (1972); *Korotev* (1987); *Laul and Papike* (1980); *Masuda et al.* (1972); *Rose et al.* (1975); *Taylor et al.* (1973); *Willis et al.* (1972)
Apollo 16	60050; 60500; 61140; 61161; 61180; 61220; 61240; 61500; 62240; 62280; 63320; 63340; 63500; 67460; 67480; 67600; 67700; 67710; 68500; 68820; 69920; 69940; 64420; 64500; 64800; 65500; 65700; 66040; 66080	*Apollo 16 PET* (1973); *Duncan et al.* (1973); *Korotev* (1981, 1982); *Rose et al.* (1973, 1975); *Simkin et al.* (1973); *Wänke et al.* (1973, 1975); *Taylor et al.* (1973); *Haskin et al.* (1973); *Boynton et al.* (1976); *Laul and Schmitt* (1973); *Brunfelt et al.* (1973); *Krähenbühl et al.* (1973); *Bansal et al.* (1972); *Compston et al.* (1973); *Mason et al.* (1973); *Finkelman et al.* (1975); *Fruchter et al.* (1974)
Apollo 17	70011; 70160; 70180; 71040; 71060; 71500; 72160; 75060; 75080; 79220; 79240; 72320; 72440; 72460; 72500; 73120; 73140; 73220; 73280; 74120; 76240; 76260; 76280; 76320; 76500; 77530; 78220; 78420; 78440; 78460; 78480	*Apollo 17 PET* (1973); *Rose et al.* (1974); *Wänke et al.* (1974); *Rhodes et al.* (1974); *Korotev* (1976); *Laul et al.* (1974); *Philpotts et al.* (1974); *Baedecker et al.* (1974); *Miller et al.* (1974)

Three lunar regions were studied in detail in order to investigate geochemical heterogeneity in the highlands (*Clark and Hawke,* 1981, 1987, 1989). Data from the Hadley-Apennine region are consistent with the presence of a mixture of ANT suite and Fra Mauro basalt components, dominated by KREEP basalt west of the Apennine Front (*Clark and Hawke,* 1981; *Hawke et al.,* 1985). The Apennine Bench is composed of KREEP basalt and is probably the source of KREEP in this region. A more noritic component, possibly low-K Fra Mauro basalt (*Clark and Hawke,* 1981), is concentrated along the northern Apennine crest and backslope. The region located between Crisium, Fecunditatis, and Smythii, and including the Balmer basin, has also been studied in detail (*Maxwell and Andre,* 1981; *Hawke et al.,* 1985). These studies indicated the presence of variable amounts of a basalt component associated with both mare and light plains deposits, in a region otherwise dominated by ANT suite material. A basaltic rock, intermediate in composition between mare basalt and KREEP basalt, has been proposed as a probable component of the plains associated with the Balmer basin. The highland region east of Smythii is dominated by very anorthositic material. However, a few geochemical anomalies indicating the presence of buried basalts have been found, particularly in the vicinity of Pasteur (*Clark and Hawke,* 1989; *Hawke et al.,* 1985). In a few places, particularly south and west of Pasteur and near Babcock crater, photogeological units mapped as plains deposits may have a KREEP-enriched basalt component. In two of the highland regions studied (*Clark and Hawke,* 1987, 1989), areas adjacent to Smythii were found to be enriched in a troctolitic component.

CURRENT WORK

In the study reported here, elemental concentrations derived from the analyses of landing site soil samples are compared with Apollo orbital geochemical measurements. The differences between the two datasets are likely to be the result of (1) discrepancies between local and regional distribution of major rock components (i.e., discrepancies between the abundances of rock types represented in the landing site soils and those present in the surface units that provide signals for orbital measurements); (2) differences in the quantitative detection of the composition of geological units due to different signal depths (1 μm to 1 m) associated with each experiment; this could also result in effective detection of only the finest soil fraction, including the micron to submicron dust coating larger fragments at the surface for very shallow XRF measurements; (3) nonrepresentative sampling of soils at landing sites; (4) an interpretive bias in the determination of average soil compositions of landing sites from a nonrepresentative geographic distribution of sample stations, or an interpretive bias in the mapping of morphological (lithological) features; or (5) systematic errors in detection efficiency by orbital instruments.

In this study each possibility is considered in the interpretation of the correlation between all measurement types. The last probability (systematic errors) may be ruled out in general, because the relationship between concentration of the same element in landing site soils and that measured by remote experiments is different for each landing site. We have weighted the averages of the Apollo soil compositions in approximate proportions of geologic units sampled at the landing sites, as described below, so that nonrepresentative sampling at landing sites should not be a problem. The distribution of major geological units as identified and mapped by photogeologic methods both at landing sites and in the surrounding areas seems adequate for explaining the discrepancies that are found. However, the second explanation (*Adler and Trombka,* 1977) could still be a contributing factor for the XRF data.

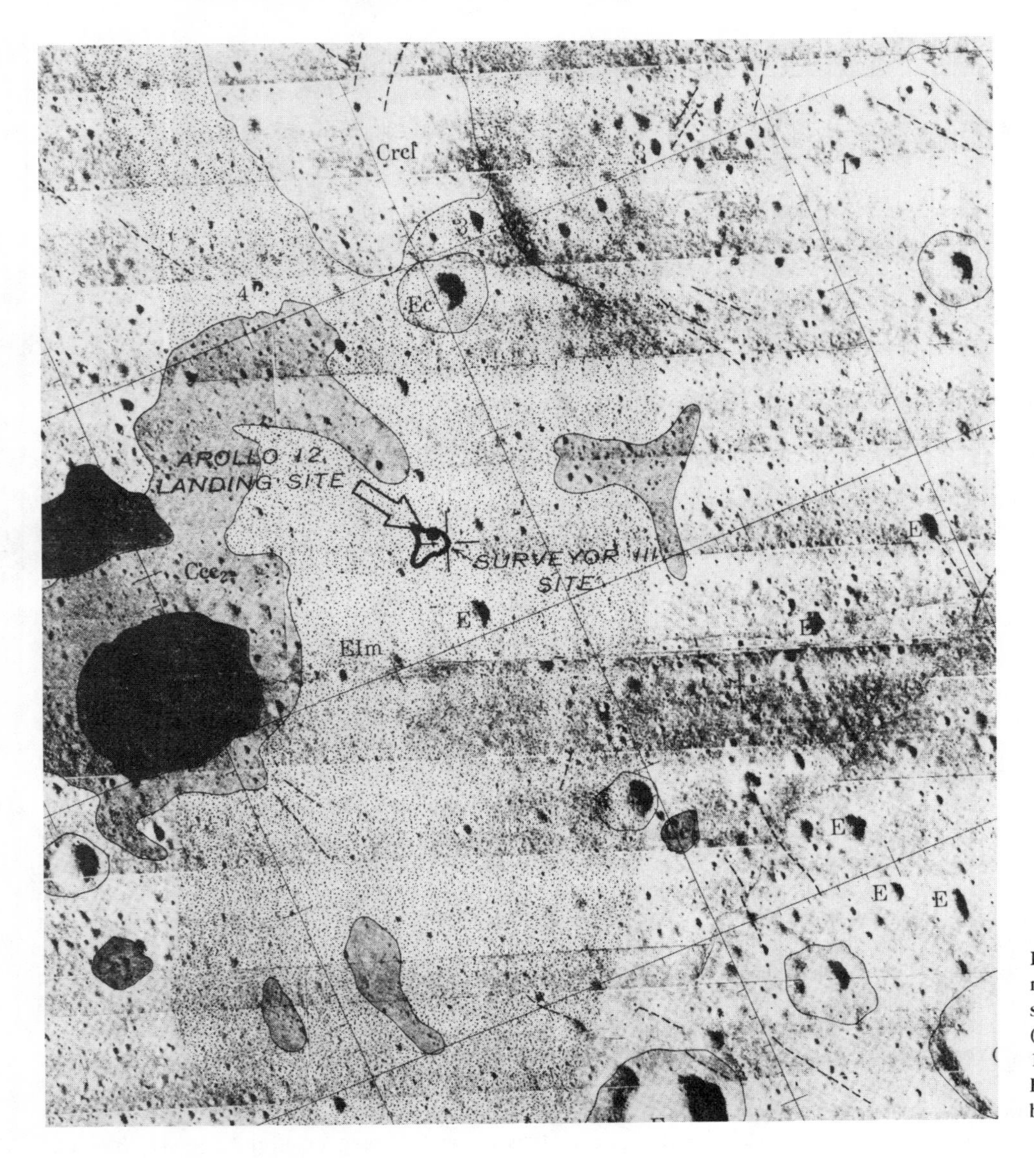

Fig. 2. Photograph of planning map for the Apollo 12 site with sample traverse superimposed (*Apollo 12 PET,* 1971; *Pohn,* 1971). Note major geological units Elm (mare basalt) and Cc (crater bottom material—anorthositic?).

PROCEDURE

We have compared elemental concentrations derived from different sources, including laboratory analysis of soil samples (XRF, INAA, and a few RNAA; *Morris et al.,* 1983), remote XRF and gamma-ray detector measurements in the case of orbital data (*Clark and Hawke,* 1981, 1989; *Davis,* 1980; *Metzger et al.,* 1977), and Earth-based telescopic measurements in the case of spectral reflectivity data (*McCord et al.,* 1972, 1981; *Adams et al.,* 1974). Analytical procedures and data reduction methods are well established and will not be discussed again here. Instead, we will focus on the procedure we used in the comparison of these datasets.

Average compositions of soil samples from different landing sites were computed from published analysis of individual soils analyzed by different groups (*Morris et al.,* 1983; *Korotev,* 1987, and personal communication, 1989). Soil samples used for averaging were selected to represent the largest area possible for the traverses made by the astronauts. This ensured that no single station would be overrepresented in the average. In addition, for the Apollo 15, 16, and 17 sites, the average composition of soils in different geological units were computed first, and then a weighted average of these mean values was calculated to be representative of the site. It is well

TABLE 2. Comparison of averaged geochemical data for the Apollo 12 site.

	Al_2O_3	MgO	FeO	TiO_2	K_2O	Th(ppm)
Compositions						
Orbital Data	13.0	8.3	15.6	2.7	0.33	6.0
Soil Samples	12.9	9.3	15.1	3.0	0.31	6.4
Ratios						
Orbital/Soil	1.01	0.89	1.03	0.90	1.06	0.94

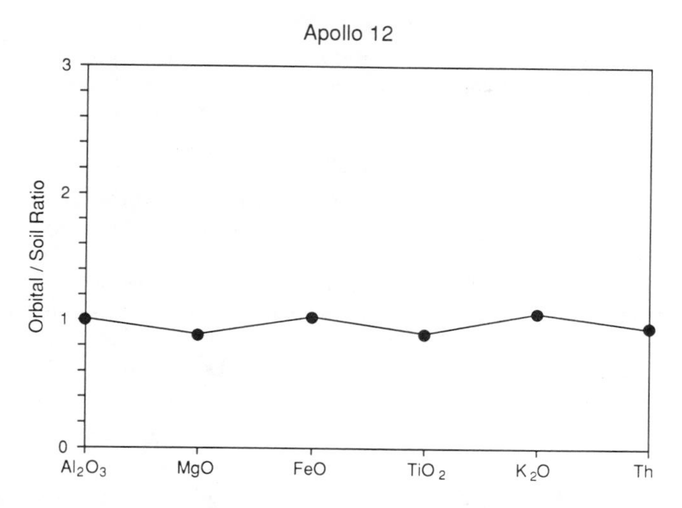

Fig. 3. Plot of orbital vs. sample concentration ratios at the Apollo 12 site for elements considered in this study.

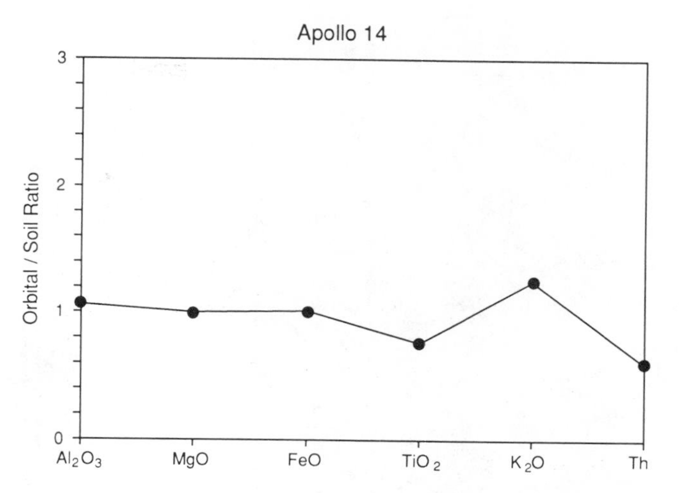

Fig. 4. Plot of orbital vs. sample concentration ratios at the Apollo 14 site for elements considered in this study.

established that different grain-size fractions of lunar soils have different chemical compositions (e.g., *Laul and Papike,* 1980). We used the chemical compositions of the <1 mm size fraction of the selected soils, which is the most abundant size fraction as well as the most commonly analyzed fraction. This ensured uniformity and internal consistency of sample data. The list of soils used for calculating averages is given in Table 1.

The methodology used to derive average elemental concentrations from orbital measurements varied with the source and resolution of the data. The Apollo 11 site had no direct coverage by any orbital experiment. For the Apollo 15, 16, and 17 sites, concentrations for Al and Mg were derived from XRF Al/Si and Mg/Si intensity ratios by averaging all available measurements within a $1° \times 1°$ area around the landing site corresponding to about $30\,km \times 30\,km$ spatial resolution of these data (*Clark and Hawke,* 1981, 1989). Gamma-ray spectrometer-derived Fe and Ti concentrations were taken from the published list of concentrations for these elements for the landings sites (*Davis,* 1980). In general, Al and Fe concentrations, as indicated by orbital measurements, are inversely proportional (*Miller,* 1974); Al concentrations for the Apollo 12 and 14 sites, where no coverage by the XRF experiment exists, were derived from Fe concentrations (*Davis,* 1980). Gamma-ray spectrometer-derived concentrations of Mg, K, and Th, as well as some calculated values for Fe and Mg, are all taken from published lists of areal averages

TABLE 3. Comparison of averaged geochemical data for the Apollo 14 site.

	Al_2O_3	MgO	FeO	TiO_2	K_2O	Th(ppm)
Compositions						
Orbital Data	18.4	9.4	10.5	1.3	0.69	8.2
Soils Samples	17.4	9.4	10.4	1.7	0.55	13.7
Ratios						
Orbital/Soil	1.06	1.00	1.01	0.76	1.25	0.6

that included landing sites (*Bielefeld et al.,* 1976; *Metzger et al.,* 1977). All of the gamma-ray-derived data have spatial resolution of 200 to 300 km except for Th concentrations, which have a maximum spatial resolution of about 60 km for deconvoluted data. A conservative estimate of errors on the averaged remote-sensing measurements would be usually in the range of 1-2% for Al, Mg, Fe, and Ti; about 0.07% for K; and about 0.3 ppm for Th. It should be noted, however, that counting statistics for raw data are not uniformly good for all measurements and detailed discussion of such uncertainties are to be found in *Metzger et al.* (1977), *Davis* (1980), and in *Clark and Hawke* (1981).

Another important consideration for orbital data is that although they all represent average concentrations of surficial material (presumably all soils) encountered within the field of view, effective depths of coverage vary according to the instrument flown. Errors in the instrument footprint position (FOV: field of view) resulting from uncertainties in spacecraft orientation would be small compared to the size of the footprint for all of the orbital geochemical experiments. The XRF-derived data represent concentration averages to an approximate depth of about 100 Å (*Adler et al.,* 1972), whereas GRS-derived naturally radioactive Th concentrations are for depths of up to 1 m, and the remaining gamma-ray-derived concentrations represent the surficial material to an approximate depth of 10 cm (*Reedy and Arnold,* 1972). Spectral reflectivity derived parameters represent soil properties to an average depth of 1 μm or less (*Morris,* 1985).

Whereas it is relatively easy to ratio the averages of elemental abundances derived from orbital measurements and sample analyses, and define any departure from unity as a discrepancy, explanations of discrepancies noted can be very subjective. We have therefore followed a systematic method to explain the data. We assumed that the photogeologic units identified in the planning maps represent suites or assemblages of rocks in the landing sites (*Carr et al.,* 1971; *Eggleton and Offield,* 1970;

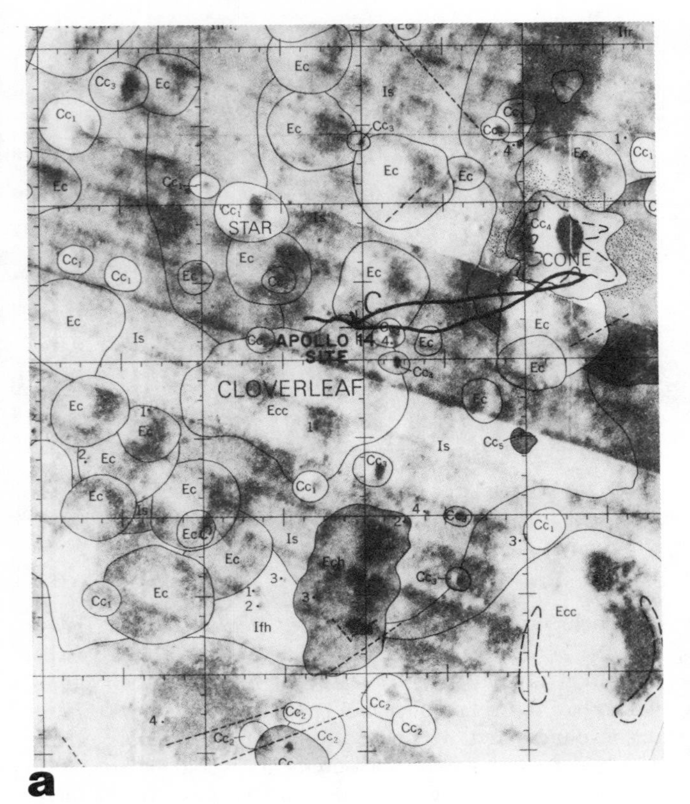

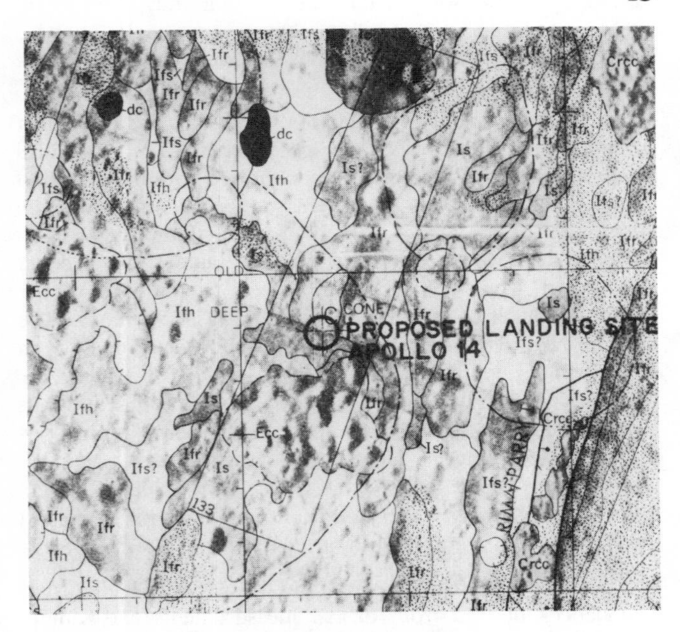

Fig. 5. Photographs of planning maps for the Apollo 14 site with **(a)** sample traverse superimposed to show the immediate vicinity of the landing site, and **(b)** an approximate area of about 30 km × 30 km around the landing site (*Apollo 14 PET,* 1971; *Eggleton and Offield,* 1970). Note major geological units Ifr (Fra Mauro formation) and Is (smooth-terrain material).

Milton and Hodges, 1972; *Pohn,* 1971; *Scott et al.,* 1972). Premission planning maps rather than postmission geologic maps were used because of higher resolution in the premission maps as well as for uniformity of mapping techniques between landing sites. We then qualitatively compared the relative abundances of these units to the abundances of the photogeologic units in the regions (approximately 30 km × 30 km corresponding to the field of view of orbital instruments) surrounding the landing sites. The differences in the abundances of map units, and therefore between rock assemblages present, should correspond to the discrepancies between sample and orbital data. Any lack of correspondence here could be due to differences in the relative signal depths and in the size fractions sampled by the different techniques.

DISCUSSION

We compare and contrast the geochemical data and the distribution of photogeologic map units at and around each of the five landing sites. In general, a discrepancy between orbital and sample data may mean that local samples are not representative of the general area in the field of view of remote detectors (e.g., *Metzger et al.,* 1977). Important for comparative purposes is the proportion of rock units in the field of view of the orbital instruments, as well as the presence of rock fragments of all of these units in the landing site. For example, the Apollo 12 landing and sampling site is contained within the photogeologic unit Elm (Fig. 2). But other units (e.g., Ec, Cc, Crcl, etc.) are also present in the region surrounding the Apollo 12 landing site, and these units may not be necessarily represented in the soil samples.

Analysis and Interpretation of Apollo 12 Landing Site

The Apollo 12 landing site in eastern Mare Procellarum (Fig. 1) consists primarily of mare basalt, as the chemical composition of soil samples suggest (Table 2). Samples were only collected in the immediate vicinity of the lander (Fig. 2; *Apollo 12 PET,* 1971; *Pohn,* 1971). Despite the small sampling

TABLE 4. Comparison of averaged geochemical data for the Apollo 15 site.

	Al_2O_3	MgO	FeO	TiO_2	K_2O	Th(ppm)
Compositions						
Orbital Data	14.0	10.9	12.7	1.7	0.42	4.2
Soils Overall	14.8	10.7	14.1	1.6	0.21	4.0
Soil Suites (areal averages)						
A: Mare	13.2	10.9	16.3	1.7	0.23	3.8
B: Apennine	17.1	10.5	11.7	1.4	0.20	4.4
C: Spur Crater	13.4	13.0	14.9	1.0	0.19	3.0
Ratios						
Orbital/All Soil	0.95	1.02	0.90	1.06	2.00	1.05
Orbital/Mare	1.06	1.00	0.78	1.00	1.83	1.10
Orbital/ Apennines	0.82	1.04	1.08	1.21	2.10	0.95
Orbital/Glass	1.37	0.84	0.85	1.70	2.21	1.40

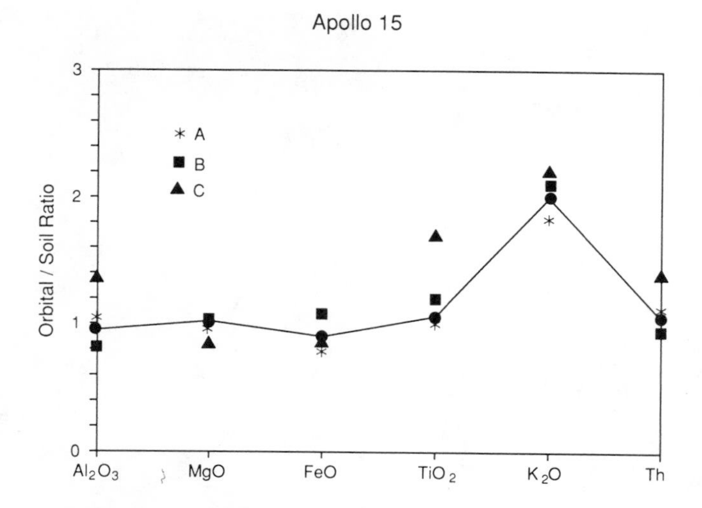

Fig. 6. Plot of orbital vs. sample concentration ratios at the Apollo 15 site for elements considered in this study. Such ratios are plotted not only for averages of all soil samples, but also for those from the mare region (A), the Apennine Front (B), and the Spur crater area (C) as well.

area at the site, orbital elemental concentrations closely correspond to the average concentration of soils at this site, as illustrated in Fig. 3. The small differences in elemental abundances can be accounted for on the basis of a somewhat larger highland component exposed in the area around the site (Fig. 2). Some of this material occurs at the site in the form of primary ejecta in a Copernicus ray. Significant amounts of terra material is also exposed within the 200-300 km field of view of the Apollo GRS experiment. Orbital K, Fe, and Ti concentrations may also be affected by compositional differences between the basalt flows to the west in Mare Procellarum and those present at the landing site (*Pieters et al.,* 1980).

Analysis and Interpretation of Apollo 14 Landing Site

The Apollo 14 landing site north of Fra Mauro crater (Fig. 1) consists primarily of breccias resembling a medium-K KREEP basalt in composition. Serious discrepancies exist between orbital and soil sample averages, as illustrated in Table 3 and Fig. 4. The area generally south of the landing site may have a larger KREEP component (*Apollo 14 PET,* 1971; *Eggleton and Offield,* 1970; Fig. 5). However, the soil samples are significantly enriched in Th relative to orbital data. Additionally, despite the presence of high-Al and high-K mare basalts in the sample suite (e.g., *Dasch et al.,* 1987; *Neal et al.,* 1989), the soils are depleted in Al and K relative to the orbital measurements. Higher orbital Al and K concentrations could indicate that a larger anorthositic component and perhaps some K-rich norite or some kind of "LKFM" melt rock have been sampled by orbital instruments in the area generally south of the site.

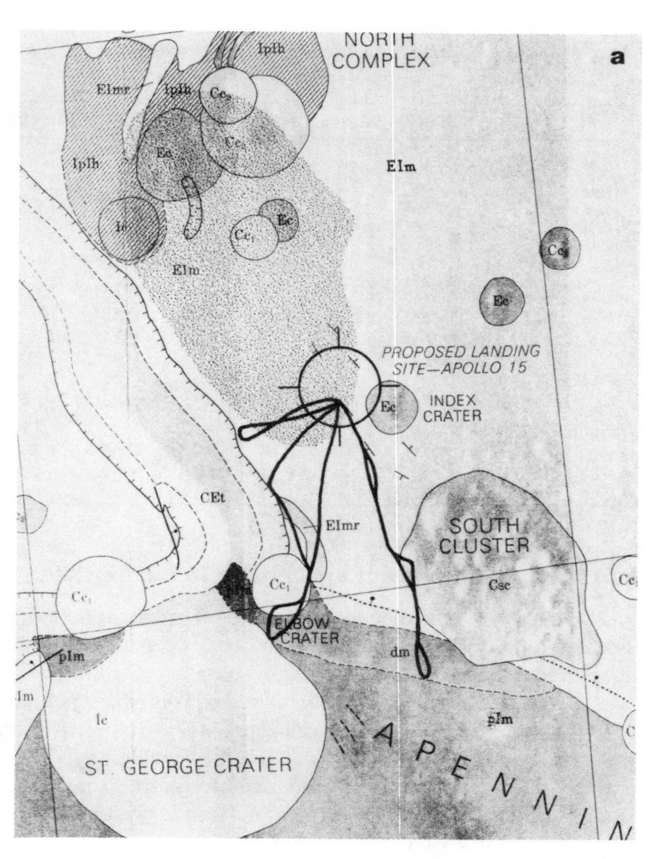

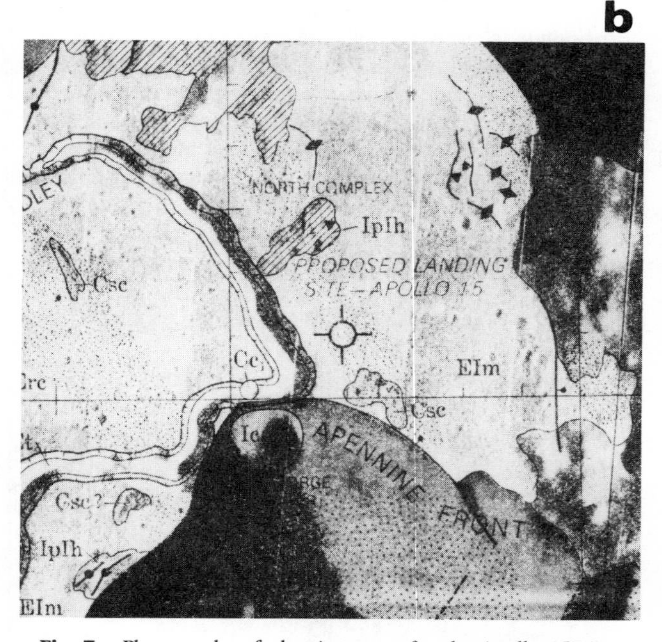

Fig. 7. Photographs of planning maps for the Apollo 15 site with **(a)** sample traverse superimposed to show the immediate vicinity of the landing site, and **(b)** an approximate area of about 30 km × 30 km around the landing site (*Apollo 15 PET,* 1972; *Carr et al.,* 1971). Note major geological units EIm (mare basalt) and pIm (massif material).

TABLE 5. Comparison of averaged geochemical data for the Apollo 16 site.

	Al_2O_3	MgO	FeO	TiO_2	K_2O	Th(ppm)
Compositions						
Orbital Data	26.7	6.8	5.6	0.6	0.15	2.0
Soils Overall	27.3	5.7	5.1	0.5	0.17	1.9
Soil Suites (areal averages)						
A: Cayley	27.1	5.8	5.2	0.56	0.13	1.8
B: North Ray	28.0	5.6	4.7	0.47	0.23	1.6
C: South Ray	26.8	5.7	5.4	0.60	0.14	2.2
Ratios						
Orbital/All Soil	0.98	1.19	1.10	1.20	0.88	1.05
Orbital/Cayley	0.99	1.17	1.08	1.07	1.15	1.11
Orbital/North Ray	0.95	1.21	1.19	1.28	0.65	1.25
Orbital/South Ray	1.00	1.19	1.04	1.00	1.07	0.91

Analysis and Interpretation of Apollo 15 Landing Site

The Apollo 15 landing site, located on a basalt unit near Hadley Rille (Fig. 1) sampled a variety of geologic units. Returned samples included mare basalts, KREEP basalt, volcanic glasses, and ANT suite material from the Apennine Front. Soil suite averages are calculated for three areas with different dominant components (Table 4). Significant differences exist between orbital and sample-analysis elemental concentrations (Table 4) as shown in Fig. 6 in which ratios of overall soil averages as well as soil suite averages are plotted. Orbitally derived elemental concentrations show the greatest correspondence to elemental concentrations for the mare suite soils mapped as EIm (Figs. 7a,b; *Apollo 15 PET,* 1972; *Carr et al.,* 1971). This suite shows the greatest KREEP enrichment of any suite because of the ubiquitous presence of KREEP basalt fragments. Thus, the geographic extent of the area mapped as a mare *basalt* unit and also dominated by fragments

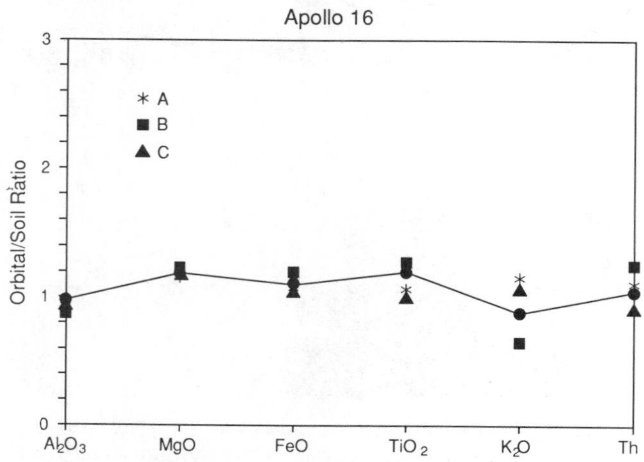

Fig. 8. Plot of orbital vs. sample concentration ratios at the Apollo 16 site for elements considered in this study. Such ratios are plotted not only for averages of all soil samples, but also for the Cayley (A), North Ray (B), and South Ray (C) soil suites.

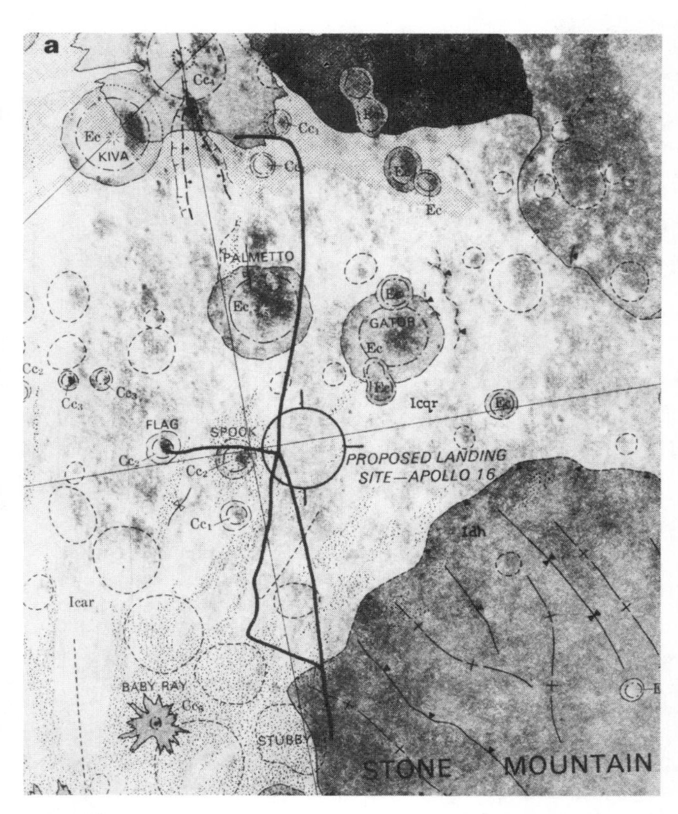

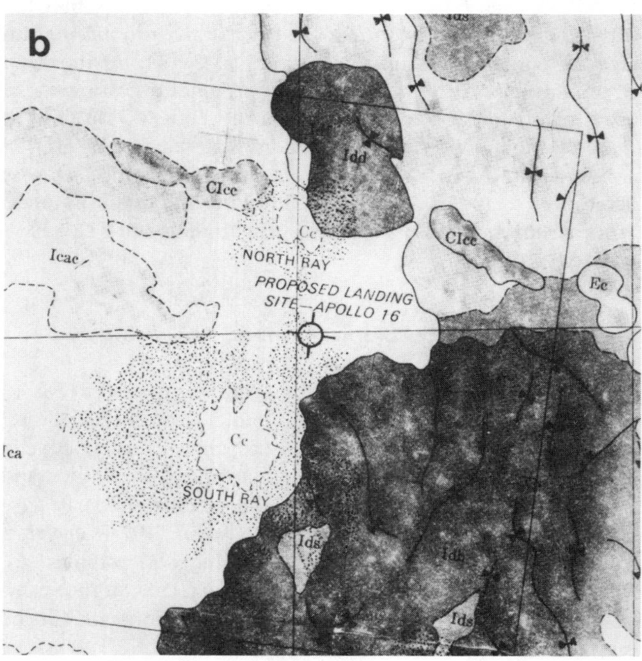

Fig. 9. Photographs of planning maps for the Apollo 16 site with **(a)** sample traverse superimposed to show the immediate vicinity of the landing site, and **(b)** an approximate area of about 30 km × 30 km around the landing site (*Apollo 16 PET,* 1972; *Milton and Hodges,* 1972). Note major geological units Icar (Cayley formation) and Cc (crater material), which were contributed to by North Ray and South Ray craters.

TABLE 6. Comparison of averaged geochemical data for the Apollo 17 site.

	Al_2O_3	MgO	FeO	TiO_2	K_2O	Th(ppm)
Compositions						
Orbital Data	16.9	9.1	13.8	6.0	0.17	1.9
Soils Overall	17.1	10.4	12.2	4.2	0.13	1.9
Soil Suites (areal averages)						
A: Mare	12.0	9.9	16.7	8.4	0.16	0.53
B: Light Mantle	20.7	9.8	8.8	1.7	0.16	2.7
C: North Massif	18.0	10.7	10.9	3.4	0.12	2.7
D: Sculptured Hills	17.4	11.1	12.2	3.5	0.09	1.5
Ratios						
Orbital/All Soil	0.99	0.88	1.13	1.43	1.31	1.00
Orbital/Mare	1.41	0.92	0.83	0.71	1.06	3.58
Orbital/Light Mantle	0.82	0.93	1.57	3.53	1.06	0.70
Orbital/North Massif	0.94	0.85	1.27	1.76	1.42	0.70
Orbital/Hills	0.97	0.82	1.13	1.71	1.89	1.27

of mare basalts still contain an orbital geochemical signature of the presence of KREEP basalt fragments. These KREEP basalt fragments were probably derived from the Apennine Bench formation below the mare basalts (*Spudis and Ryder,* 1986; *Basu and McKay,* 1979; *Spudis,* 1978; *Hawke and Head,* 1978). Orbital XRF-derived average Al and Mg concentrations are somewhat lower than the average concentrations of these elements in local soils, which is apparently due to the presence of a greater KREEP basalt component in the field of view of remote instruments (Fig. 6). Apennine Front material (mapped as pIm but probably dominated by Imbrium ejecta) apparently makes a larger contribution to soil composition at the site itself (Figs. 7a,b). The distribution of ANT suite soils in this region as indicated by the correlation of remote-sensing and major rock component data is discussed in detail by *Clark and Hawke* (1981). Relative to soils, regional KREEP enrichment is even more pronounced in the abundances of Fe, Ti, Th, and K as obtained from gamma-ray-derived elemental concentrations. Because the field of view of the GRS is much larger than that of XRF (see section entitled "Procedure"), KREEP-rich rocks must be inferred to be abundant in the region surrounding the landing site (i.e., the Apennine Bench formation).

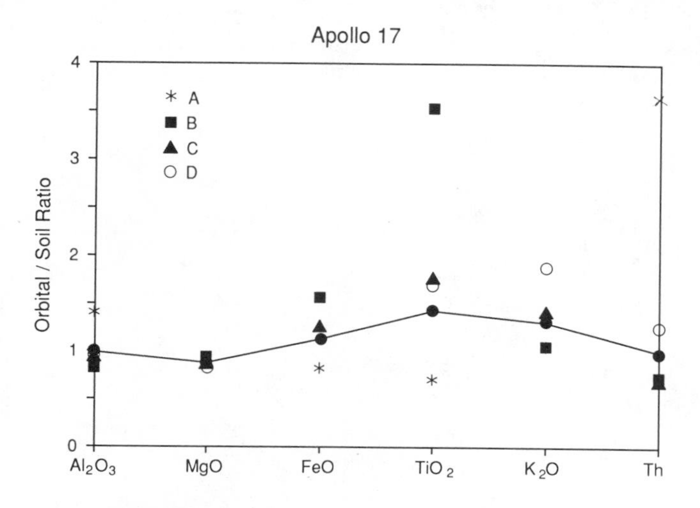

Fig. 10. Plot of orbital vs. sample concentration ratios at the Apollo 17 site for elements considered in this study. Such ratios are plotted not only for averages of all soil samples, but also for the mare (A), light mantle (B), North Massif (C), and Sculptured Hills (D) soil suites.

Analysis and Interpretation of Apollo 16 Landing Site

The Apollo 16 site in the Descartes region (Fig. 1) consists of a mixture of ANT suite components (*Wilhelms,* 1987). These are derived from the Cayley plains, North Ray and South Ray craters (soil suites A, B, and C in Table 5) and the Descartes mountains. Significant differences exist between orbital and sample-analysis elemental concentrations. Orbital chemistry is most closely approximated by the Cayley plain material except for K_2O (Fig. 8). Indications are that this material is made up of "mafic-rich crystalline melt breccias with KREEP- or very-high-alumina-basaltic affinities" (*Stöffler et al.,* 1981). In a sense this is more like anorthositic gabbro/norite in composition compared to other soil suites. The orbital elemental concentrations are least similar to those of North Ray crater. Apparently, North Ray samples may be overrepresented in the average composition of soils at this landing site than what is represented in the field of view of the remote instruments (Fig. 9a,b; *Apollo 16 PET,* 1972; *Milton and Hodges,* 1972). In fact, South Ray crater does have a much more extensive ejecta blanket in the area around the landing site and therefore has more influence on the orbital data.

Analysis and Interpretation of Apollo 17 Landing Site

The Apollo 17 site at Taurus Littrow (Fig. 1) consists of a number of geologic units. Some of these units have components with very different chemical compositions; thus, not surprisingly, the relationship between orbital measurements and sample analyses data is most difficult to interpret. Soil suites at this site include those from the mare plain, light mantle soils that are anorthositic in nature, along with a North Massif suite, and a Sculptured Hills suite (Soil suites A-D in Table 6). Elemental concentrations derived from orbital observations most closely approximate those of the North Massif soils (Fig. 10). More of the North Massif material (mapped as pltm) is seen in the field of view of the orbital instruments than what is represented at the landing site itself (Fig. 11a,b; *Apollo 17 PET,* 1973; *Scott et al.,* 1972). The addition of a small mare suite component to the North Massif suite comes close to approximating the chemistry detected by orbital measurements. One element, however, is the exception: Mg concentrations of all soil suite averages are greater than Mg values seen by the orbital instruments. Thus, an additional lower Mg component, i.e., some anorthositic material such as a ferroan anorthosite (e.g., *James et al.,* 1989), must be present in the larger field of view, possibly in the dominant Iplh unit as mapped (Fig. 11a).

Major geochemical units and a number of typical rocks and soils have been plotted in a MgO vs. Al_2O_3 diagram with each

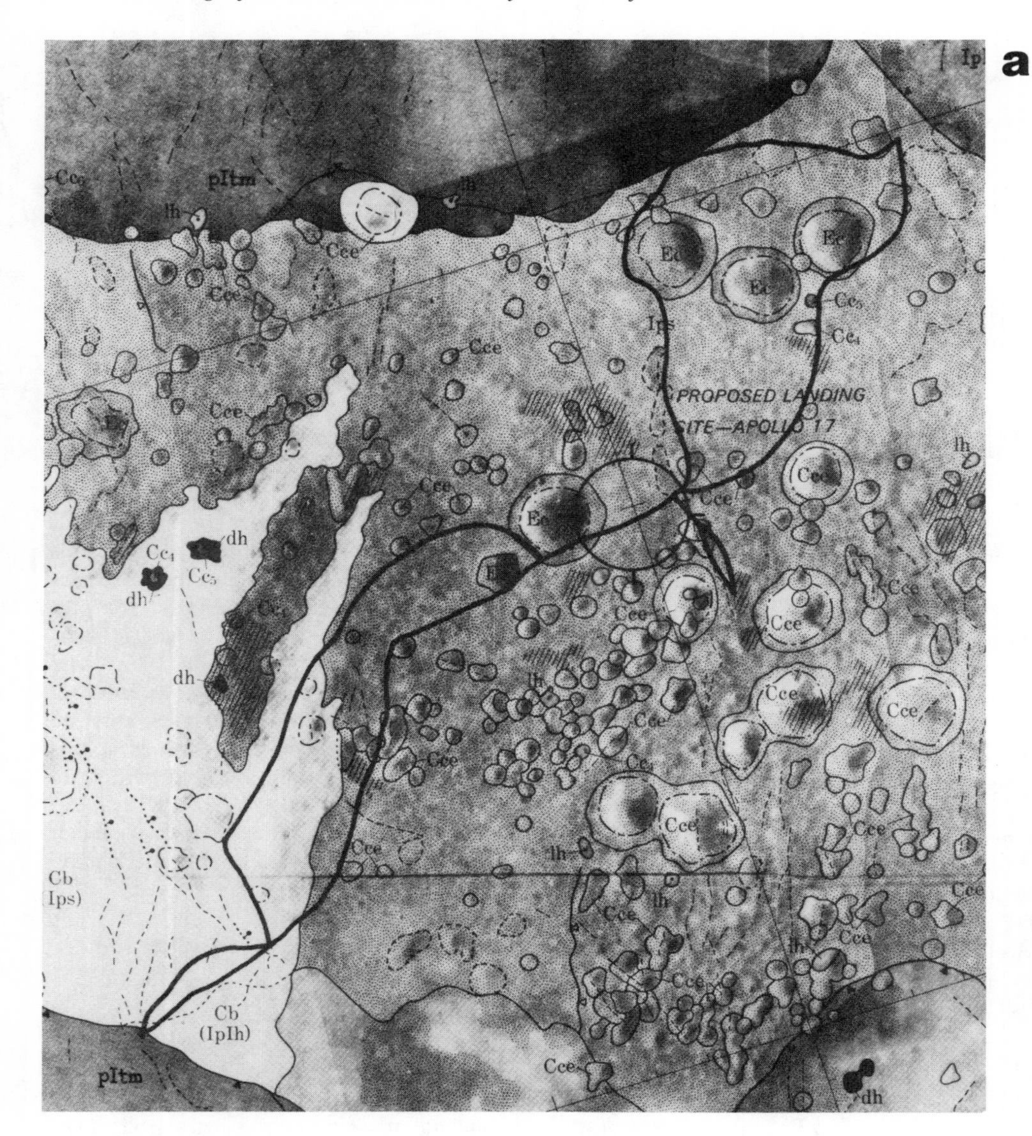

Fig. 11. Photographs of planning maps for the Apollo 17 site with **(a)** sample traverse superimposed to show the immediate vicinity of the landing site, and **(b)** an approximate area of about 30 km × 30 km around the landing site (*Apollo 17 PET,* 1973; *Scott et al.,* 1972). Note major geological units IpIh (unclassified highland material), Cd (dark mantle material), and pItm (massif material).

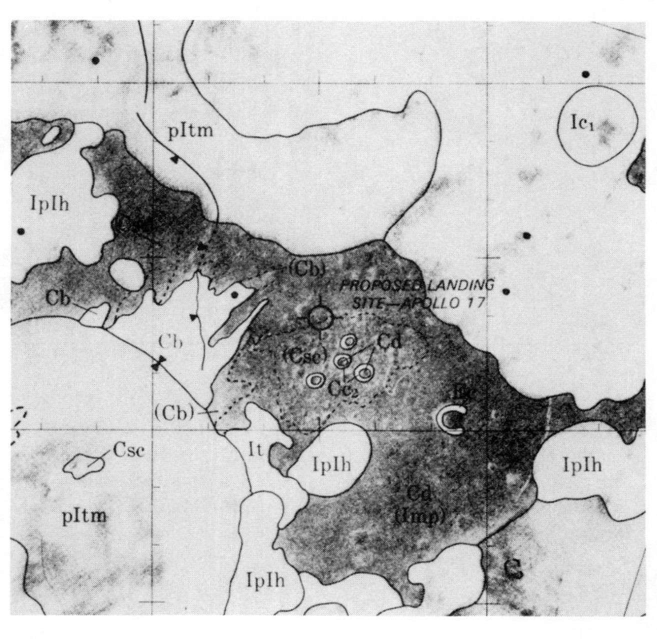

axis also keyed to XRF Mg/Si and XRF Al/Si ratios (*Clark and Hawke,* 1981, their Fig. 15). Averages of soil samples and orbital data are plotted on this diagram for a direct comparison (Fig. 12). We have used this figure to infer the distribution of major rock components in the field of view of orbital instruments, which are probably responsible for the discrepancies mentioned above. For example, the shift in the plot of Apollo 15 orbital composition away from that of Apollo 15 soil samples is directly toward the KREEP basaltic field in this diagram (Fig. 12) and supports the interpretation given above. Similarly, although both the soil sample average and orbital measurements of Apollo 16 site plot within the anorthositic field, the soils are closer to the gabbroic/noritic composition, whereas the orbital data are closer to anorthositic compositions (Fig. 12). This diagram, in which Apollo 17 sample and orbital compositions fall outside the marked fields, also illustrates that the Apollo 17 site is rather complicated and that the mixed material is not dominated by any single major geochemical unit.

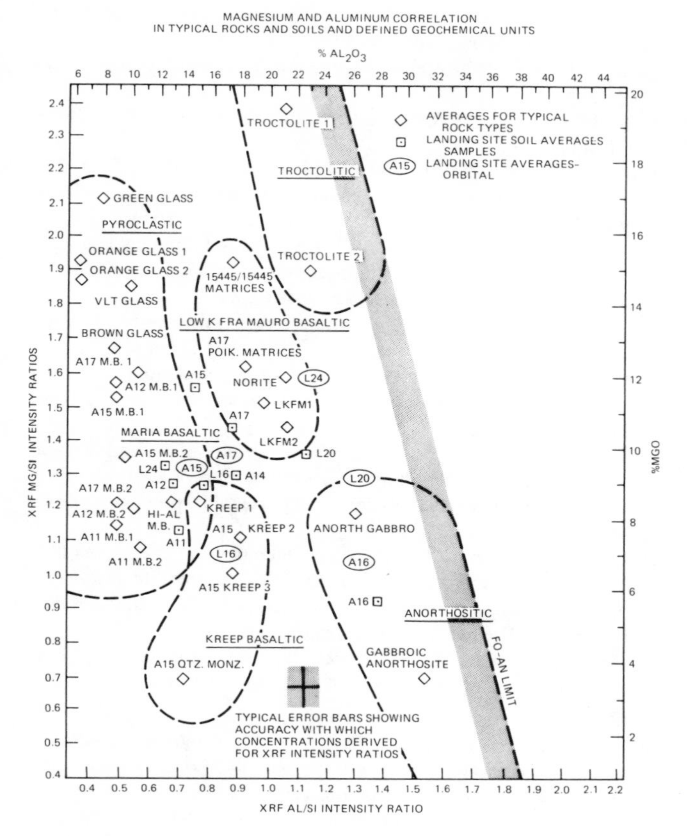

Fig. 12. Plot of Mg vs. Al concentrations showing typical rock compositions, average compositions of soil samples at each landing site, and orbital measurements of Mg/Si and Al/Si ratios in the landing site area (after *Clark and Hawke,* 1981). Note the similarity and dissimilarity between sample analyses and orbital observations as discussed in the text.

Spectral Contribution

McCord and coworkers established a program that related the optical properties of returned lunar samples measured in the laboratory with those of relatively small (2-20 km diameter) areas on the lunar surface obtained with Earth-based telescopes. Spectral reflectance measurements in the wavelength region of 0.3-2.5 μm have provided much compositional information concerning the lunar surface and the evolution of lunar soils (e.g., *McCord et al.,* 1972, 1981). The telescopic spectrum of a landing site and the spectrum of a representative mature soil from the same site agree within 1% or 2% as is seen Fig. 13. The spectra shown in Fig. 13 are scaled to unity at 0.56 μm to eliminate albedo effects and to emphasize only color characteristics. It is concluded that the telescope measurements refer mainly to the mature lunar soil rather than to rocks and breccias (*Adams et al.,* 1974).

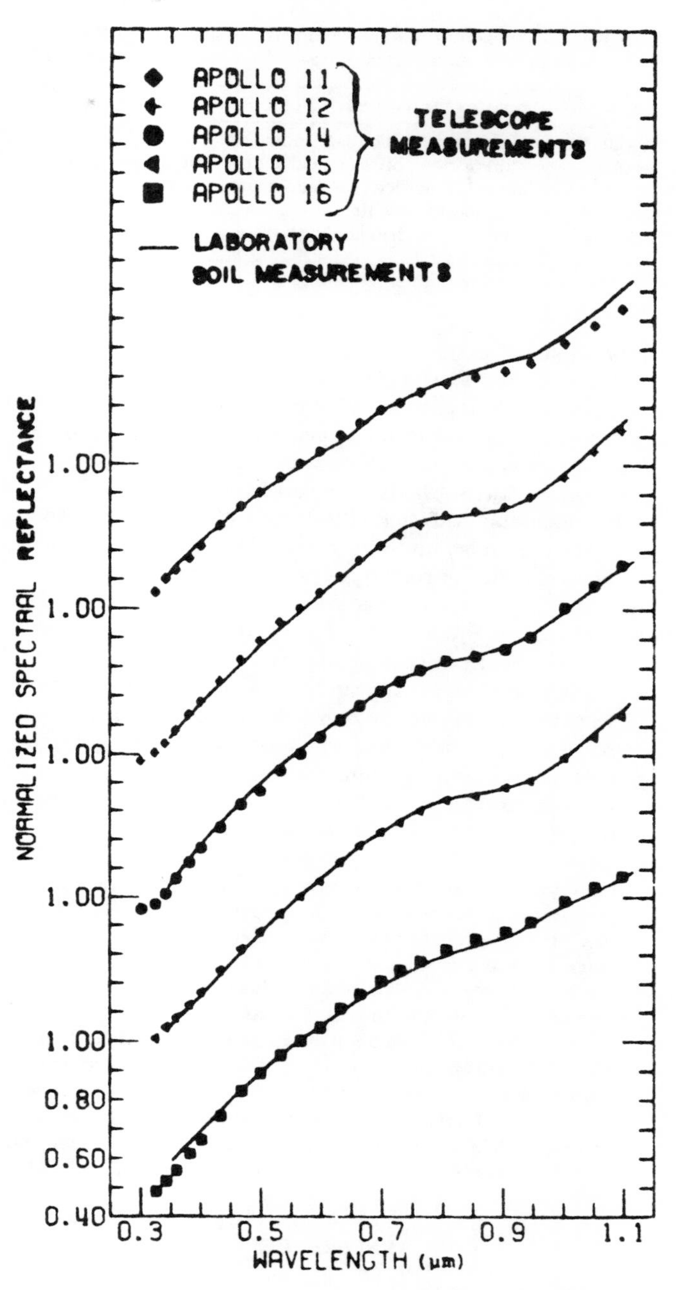

Fig. 13. Spectral reflectance measurements of areas centered on five Apollo landing sites compared to spectra obtained for representative mature soils returned from the respective sites. The curves are scaled to unity at 0.56 μm (after *Adams et al.,* 1974).

SUMMARY

The major findings from this study are summarized and listed by landing site in Table 7. At the Apollo 12 site, the greater abundance of highland material can be seen in the remote field of view than at the landing site. On the other hand, the greater contribution of KREEP in the larger area beyond the Apollo 15 site is certainly indicated. The area surrounding the Apollo 14 site is both more aluminous and enriched in Th, and probably more enriched in rare earth elements and in K (although this element still has relatively low concentrations here) as well, indicating a greater abundance of medium-K KREEP basalt or rocks of similar composition. The average soil composition of the Apollo 16 site does not reflect any significantly greater contribution of Cayley (or South Ray?) material than that seen

TABLE 7. A comparative summary of inferred compositions of remote fields of view and returned soil samples from landing sites.

Site	Nature of Lunar Surface in the Field of View*
Apollo 12	More anorthositic than landing site
Apollo 14	More K-rich noritic(?) and anorthositic than landing site
Apollo 15	More KREEP-rich than landing site
Apollo 16	Less anorthositic than landing site
Apollo 17	More noritic than landing site

*As represented by remote observations.

more extensively in the larger field of view of orbital instruments, the latter being more like anorthositic norite in composition. At the Apollo 17 site, the presence of a KREEP-enriched component in the larger field of view of the orbital instruments is indicated, along with a greater contribution of the more noritic North Massif material.

In conclusion, XRF- and GRS-derived elemental abundances (despite the errors involved in counting statistics) appear to be representative of rock types present on the surface (based on mapped geological units sampled at landing sites) within the respective fields of view of the two instruments. Differences in depth of sampling (i.e., effective signal depths) for either instrument appear to play minor roles; surficial observations are ostensibly representative of underlying units. If so, this is probably the result of an efficient gardening process of the lunar regolith. Telescopic reflectance spectral measurements refer primarily to the mature soils at the various landing sites and not to rocks and breccias.

Acknowledgments. We are most appreciative of the assistance of persons at the Regional Planetary Imaging Facilities at JPL, LPI, and Univ. of Hawaii, who helped us in compiling reference maps and drawings used in this study. We would like to thank M. Rudnyk, in particular, in this regard. One of the authors (P.E.C.) is also grateful for productive discussions with H. Schmidt, I. Ridley, J. Papike, and R. Reedy. M. Blom is thanked for technical assistance while J. Tolen, B. Hill, R. Hill, and M. LaRue assisted greatly in manuscript preparation. R. Korotev and J. Whitford-Stark helped with excellent reviews and C. Wood provided profound editorial comments. This work was partly carried out at JPL/Caltech under a NASA grant to T. Thompson, and other NASA grants NSG-7323 (B.R.H.), NAGW-237 (B.R.H.), NSG-9077 (A.B.).

REFERENCES

Adams J. B., Pieters C. M., and McCord T. B. (1974) Orange glass: Evidence for regional deposits of pyroclastic origin on the Moon. *Proc. Lunar Sci. Conf. 5th*, pp. 171-186.

Adler I. and Trombka J. I. (1977) Orbital chemistry—Lunar surface analysis from the X-ray and gamma ray remote sensing experiments. *Phys. Chem. Earth, 10*, 17-43.

Adler I., Gerard J., Trombka J., Lowman P., Blodgett H., Yin L., Eller E., and Lamothe R. (1972) The Apollo 15 X-ray fluorescence experiment. *Proc. Lunar Sci. Conf. 3rd*, pp. 2157-2178.

Apollo 12 PET (1971) Apollo 12 preliminary science report. *NASA SP-235*, 114 pp.

Apollo 14 PET (1971) Apollo 14 preliminary science report. *NASA SP-272*, 62 pp.

Apollo 15 PET (1972) Apollo 15 preliminary science report. *NASA SP-289*, pp. 5-45.

Apollo 16 PET (1972) The Apollo 16 lunar samples—Petrographic and chemical descriptions. *Science, 177*, 23-24.

Apollo 16 PET (1972) Apollo 16 preliminary science report. *NASA SP-315*, pp. 6-57.

Apollo 17 PET (1973) Apollo 17 preliminary science report. *NASA SP-330*, pp. 6-40.

Baedecker P. A., Chou C. L., Sundberg L. L., and Wasson J. T. (1974) Volatile and siderophilic trace elements in the soils and rocks of Taurus-Littrow. *Proc. Lunar Sci. Conf. 5th*, pp. 1625-1644.

Bansal B. M., Church S. E., Gast P. W., Hubbard N. J., Rhodes J. M., and Wiesmann H. (1972) Chemical composition of soil from Apollo-16 and Luna-20 sites. *Earth Planet. Sci. Lett., 17*, 29.

Basu A. and McKay D. S. (1979) Petrography and provenance of Apollo 15 soils. *Proc. Lunar Planet. Sci. Conf. 10th*, pp. 1413-1424.

Bielefeld M., Reedy R., Metzger A., Trombka J., and Arnold J. (1976) Surface chemistry of selected lunar regions. *Proc. Lunar Sci. Conf. 7th*, pp. 2661-2676.

Boynton W. V., Chou C. L., Bild R. W., Baedecker P. A., and Wasson J. T. (1976) Element distribution in size fractions of Apollo 16 soils: Evidence for element mobility during regolith processes. *Earth Planet. Sci. Lett., 29*, 21-33.

Brunfelt A. O., Heier K. S., Nilssen B., Steinnes E., and Sundvoll B. (1972) Elemental composition of Apollo 15 samples. In *The Apollo 15 Lunar Samples* (J. W. Chamberlain and C. Watkins, eds.), pp. 195-197. The Lunar Science Institute, Houston.

Brunfelt A. O., Heier K. S., Nilssen B., and Sundvoll B. (1973) Geochemistry of Apollo 15 and 16 materials. *Proc. Lunar Sci. Conf. 4th*, pp. 1209-1218.

Carr M. H., Howard K. A., and El-Baz F. (1971) Geologic map of part of the Apennine-Hadley region of the Moon—Apollo 15 premission map. *U.S. Geol. Surv. Misc. Geol. Inv. Map I-723*.

Carron M. K., Annell C. S., Christian R. P., Cuttitta F., Dwornik E. J., Ligon D. T., and Rose H. J. (1972) Elemental analyses of lunar soil samples from Apollo 15 mission. In *The Apollo 15 Lunar Samples* (J. W. Chamberlain and C. Watkins, eds.), pp. 198-201. The Lunar Science Institute, Houston.

Christian R. P., Berman S., Dwornik E. J., Rose H. J., and Schnepfe M. M. (1976) Composition of some Apollo 14, 15, and 16 lunar breccias and two Apollo 15 fines (abstract). In *Lunar Science VII*, pp. 138-140. The Lunar Science Institute, Houston.

Clark P. E. (1985) Geochemical differentiation of the maria on the early Moon (abstract). In *Lunar and Planetary Science XVI*, pp. 139-140. Lunar and Planetary Institute, Houston.

Clark P. E. and Adler I. (1978) Utilization of independent solar flux measurements to eliminate non-geochemical variation in X-ray fluorescence data. *Proc. Lunar Planet. Sci. Conf. 9th*, pp. 3029-3036.

Clark P. E. and Hawke B. R. (1981) Compositional variation in the Hadley Apennine region. *Proc. Lunar Planet. Sci. 12B*, pp. 727-749.

Clark P. E. and Hawke B. R. (1982) Geochemical classification of lunar terrain (abstract). *Eos Trans. AGU, 63*, 364.

Clark P. E. and Hawke B. R. (1987) The relationship between geology and geochemistry in the Undarum/Spumans/Balmer region. *Earth, Moon, and Planets, 38*, 97-112.

Clark P. E. and Hawke B. R. (1989) The lunar farside: The highlands east of Smythii. *Earth, Moon, and Planets*, in press.

Clark P. E., Eliason E., Andre C., and Adler I. (1978) A new color correlation method applied to XRF Al/Si ratios and other lunar remote sensing data. *Proc. Lunar Planet. Sci. Conf. 9th*, pp. 3015-3027.

Compston W., Vernon M. J., Chappell B. W., and Freeman R. (1973) Rb-Sr model ages and chemical composition of nine Apollo 16 soils (abstract). In *Lunar Science IV*, p. 158. The Lunar Science Institute, Houston.

Cuttitta F., Rose H. J., Annell C. S., Carron M. K., Christian R. P., Dwornik E. J., Greenland L. P., Helz A. W., and Ligon D. T. (1971) Elemental composition of some Apollo 12 lunar rocks and soils. *Proc. Lunar Sci. Conf. 2nd,* pp. 1217-1230.

Dasch E. J., Shih C.-Y., Bansal B. M., Wiesmann H., and Nyquist L. E. (1987) Isotopic analysis of basaltic fragments from lunar breccia 14321: Chronology and petrogenesis of pre-Imbrium mare volcanism. *Geochim. Cosmochim. Acta, 51,* 3241-3254.

Davis P. A. (1980) Iron and titanium distribution on the Moon from orbital gamma ray spectrometry with implications for crustal evolutionary models. *J. Geophys. Res., 85,* 3209-3224.

Duncan A. R., Erlank A. J., Willis J. P., and Ahrens L. H. (1973) Composition and inter-relationships of some Apollo 16 samples. *Proc. Lunar Sci. Conf. 4th,* pp. 1097-1114.

Duncan A. R., Sher M. K., Abraham Y. C., Erlank A. J., Willis J. P., and Ahrens L. H. (1975) Interpretation of the compositional variability of Apollo 15 soils. *Proc. Lunar Sci. Conf. 6th,* pp. 2309-2320.

Eggleton R. and Offield T. (1970) Geologic maps of the Fra Mauro region of the Moon (Apollo 14). *U.S. Geol. Surv. Map I-708.*

Finkelman R. B., Baedecker P. A., Christian R. P., Berman S., Schnepfe M. M., and Rose H. J. (1975) Trace-element chemistry and reducing capacity of size fractions from the Apollo 16 regolith. *Proc. Lunar Sci. Conf. 6th,* pp. 1385-1398.

Frondel C., Klein C. Jr., and Ito J. (1971) Mineralogical and chemical data on Apollo 12 lunar fines. *Proc. Lunar Sci. Conf. 2nd,* pp. 719-726.

Fruchter J. S., Kridelbaugh S. J., Robyn M. A., and Goles G. G. (1974) Breccia 66055 and related clastic materials from the Descartes region, Apollo 16. *Proc. Lunar Sci. Conf. 5th,* pp. 1035-1046.

Goles G. G., Duncan A. R., Lindstrom D. J., Martin M. R., Beyer R. L., Osawa M., Randle K., Meek L. T., Steinborn T. L., and McKay S. M. (1971) Analyses of Apollo 12 specimens—compositional variations, differentiation processes, and lunar soil mixing models. *Proc. Lunar Sci. Conf. 2nd,* pp. 1063-1082.

Haskin L. A., Helmke P. A., Allen R. O. Jr., Anderson M. R., Korotev R. L., and Zweifel K. A. (1971) Rare-earth elements in Apollo 12 lunar materials. *Proc. Lunar Sci. Conf. 2nd,* pp. 1307-1318.

Haskin L. A., Helmke P. A., Blanchard D. P., Jacobs J. W., and Telander K. (1973) Major and trace element abundances in samples from the lunar highlands. *Proc. Lunar Sci. Conf. 4th,* pp. 1275-1296.

Hawke B. R. and Head J. W. (1978) Lunar KREEP volcanism: Geologic evidence for history and mode of emplacement. *Proc. Lunar Planet. Sci. Conf. 9th,* pp. 3285-3309.

Hawke B. R., Spudis P., and Clark P. (1985) The origin of selected lunar geochemical anomalies: Implications for early volcanism and the formation of light plains. *Earth, Moon, and Planets, 32,* 257-273.

James O. B., Lindstrom M. M., and Flohr M. K. (1989) Ferroan anorthosite from lunar breccia 64435: Implications for the origin and history of lunar ferroan anorthosites. *Proc. Lunar Planet. Sci. Conf. 19th,* pp. 219-243.

Korotev R. L. (1976) Geochemistry of grain-size fractions of soils from the Taurus-Littrow valley floor. *Proc. Lunar Sci. Conf. 7th,* pp. 695-726.

Korotev R. L. (1981) Compositional trends in Apollo 16 soils. *Proc. Lunar Planet. Sci. 12B,* pp. 577-605.

Korotev R. L. (1982) Comparative geochemistry of Apollo 16 surface soils and samples from cores 64002 and 60002 through 60007. *Proc. Lunar Planet. Sci. Conf. 13th,* in *J. Geophys. Res., 87,* A269-A278.

Korotev R. L. (1987) Mixing levels, the Apennine Front soil component, and compositional trends in the Apollo 15 soils. *Proc. Lunar Planet. Sci. Conf. 17th,* in *J. Geophys Res., 92,* E411-E431.

Krähenbühl U., Ganapathy R., Morgan J. W., and Anders E. (1973) Volatile elements in Apollo 16 samples—Implications for highland volcanism and accretion history of the Moon. *Proc. Lunar Sci. Conf. 4th,* pp. 1325-1348.

Laul J. C. and Papike J. J. (1980) The lunar regolith: Comparative chemistry of the Apollo sites. *Proc. Lunar Planet. Sci. Conf. 11th,* pp. 1307-1340.

Laul J. C. and Schmitt R. A. (1973) Chemical composition of Apollo 15, 16, and 17 samples. *Proc. Lunar Sci. Conf. 4th,* pp. 1349-1368.

Laul J. C., Wakita H., Showalter D. L., Boynton W. V., and Schmitt R. A. (1972) Bulk, rare earth, and other trace elements in Apollo 14 and 15 and Luna 16 samples. *Proc. Lunar Sci. Conf. 3rd,* pp. 1181-1201.

Laul J. C., Hill D. W., and Schmitt R. A. (1974) Chemical studies of Apollo 16 and 17 samples. *Proc. Lunar Sci. Conf. 5th,* pp. 1047-1066.

Lindstrom M. M., Duncan A. R., Fruchter J. S., McKay S. M., Stoesser J. W., Goles G. G., and Lindstrom D. J. (1972) Compositional characteristics of some Apollo 14 clastic materials. *Proc. Lunar Sci. Conf. 3rd,* pp. 1201-1215.

Mason B., Simkin T., Norman A. F., Switzer G. S., Nelen J. A., Thompson G., and Melson W. G. (1973) Composition of Apollo 16 fines 60051, 60052, 64811, 67711, 67712, 68821 and 68822 (abstract). In *Lunar Science IV,* pp. 505-507. The Lunar Science Institute, Houston.

Masuda A., Nakamura N., Kurasawa H., and Tanaka T. (1972) Precise determination of rare-earth elements in the Apollo 14 and 15 samples. *Proc. Lunar Sci. Conf. 3rd,* pp. 1307-1315.

Maxwell T. and Andre C. (1981) The Balmer basin: Regional geology and geochemistry of an ancient lunar impact basin. *Proc. Lunar Planet. Sci. 12B,* pp. 715-725.

McCord T. B., Charette M. P., Johnson T. V., Lebofsky L. A., Pieters C. M., and Adams J. B. (1972) Lunar spectral types. *J. Geophys. Res., 77,* 1349-1359.

McCord T. B., Clark R. N., Hawke B. R., McFadden L. A., Owensby P. D., Pieters C. M., and Adams J. B. (1981) Moon: Near-infrared spectral reflectance, a good look. *J. Geophys. Res., 86,* 10,883-10,892.

Metzger A. E., Trombka J. I., Peterson L. E., Reedy R. C., and Arnold J. R. (1973) Lunar surface radioactivity: Preliminary results of the Apollo 15 and Apollo 16 gamma-ray spectrometer experiments. *Science, 179,* 800-803.

Metzger A. E., Haines E. L., Parker R. E., and Radocinski R. G. (1977) Thorium concentrations in the lunar surface I.: Regional values and crustal content. *Proc. Lunar Sci. Conf. 8th,* pp. 949-999.

Miller M. D., Pacer R. A., Ma M.-S., Hawke B. R., Lookhart G. L., and Ehmann W. D. (1974) Compositional studies of the lunar regolith at the Apollo 17 site. *Proc. Lunar Sci. Conf. 5th,* pp. 1079-1086.

Milton D. and Hodges C. (1972) Geologic maps of the Descartes region of the Moon (Apollo 16). *U.S. Geol. Surv. Map I-748.*

Morgan J. W., Krähenbühl U., Ganapathy R., and Anders E. (1972) Trace elements in Apollo 15 samples: Implications for meteorite influx and volatile depletion on the Moon. *Proc. Lunar Sci. Conf. 3rd,* pp. 1361-1377.

Morris R. V. (1985) Determination of optical penetration depths from reflectance and transmittance measurements on albite powders (abstract). In *Lunar and Planetary Science XVI,* pp. 581-582. Lunar and Planetary Institute, Houston.

Morris R. V., Score R., Dardano C., and Heiken G. (1983) Handbook of lunar soils. *NASA-JSC Planetary Materials Branch Publ. 67, JSC 19069.* 914 pp.

Neal C. R., Taylor L. A., and Patchen A. D. (1989) High alumina (HA) and very high potassium (VHK) basalt clasts from Apollo 14 breccias, part 1—Mineralogy and petrology: Evidence of crystallization from evolving magmas. *Proc. Lunar Planet. Sci. Conf. 19th,* pp. 137-145.

Philpotts J. A., Schnetzler C. C., Nava D. F., Bottino M. L., Fullagar P. D.,

Thomas H. H., Schumann S., and Kouns C. W. (1972) Apollo 14: Some geochemical aspects. *Proc. Lunar Sci. Conf. 3rd,* pp. 1293-1306.

Philpotts J. A., Schumann S., Kouns C. W., Lum R. K. L., and Winzer S. (1974) Origin of Apollo 17 rocks and soils. *Proc. Lunar Sci. Conf. 5th,* pp. 1255-1267.

Pieters C. M., Head J. W., Adams J. B., McCord T. B., Zisk S. H., and Whitford-Stark J. L. (1980) Late high-titanium basalts of the western maria: Geology of the Flamsteed region of Oceanus Procellarum. *J. Geophys. Res., 85,* 3913-3938.

Pohn H. (1971) Geologic map of the Lansberg P region of the Moon (Apollo 12). *U.S. Geol. Surv. Map I-627.*

Reedy R. C. (1978) Planetary gamma-ray spectroscopy. *Proc. Lunar Planet. Sci. Conf. 9th,* pp. 2961-2984.

Reedy R. and Arnold J. (1972) Interaction of solar and galactic cosmic-ray particles with the Moon. *J. Geophys. Res., 77,* 537-555.

Reedy R. C., Arnold J. R., and Trombka J. I. (1973) Expected γ ray emission spectra from the lunar surface as a function of chemical composition. *J. Geophys. Res., 78,* 5847-5866.

Rhodes J. M., Rodgers K. V., Shih C.-Y., Bansal B. M., Nyquist L. E., Wiesmann H., and Hubbard N. J. (1974) The relationships between geology and soil chemistry at Apollo 17 landing site. *Proc. Lunar Sci. Conf. 5th,* pp. 1097-1118.

Rose H. J. Jr., Cuttitta F., Annell C. S., Carron M. K., Christian R. P., Dwornik E. J., Greenland L. P., and Ligon D. T. Jr. (1972) Compositional data for twenty-one Fra Mauro lunar materials. *Proc. Lunar Sci. Conf. 3rd,* pp. 1215-1231.

Rose H. J. Jr., Cuttitta F., Berman S., Carron M. K., Christian R. P., Dwornik E. J., Greenland L. P., and Ligon D. T. Jr. (1973) Compositional data for twenty-two Apollo 16 samples. *Proc. Lunar Sci. Conf. 4th,* pp. 1149-1158.

Rose H. J. Jr., Cuttitta F., Berman S., Brown F. W., Carron M. K., Christian R. P., Dwornik E. J., and Greenland L. P. (1974) Chemical composition of rocks and soils at Taurus-Littrow. *Proc. Lunar Sci. Conf. 5th,* pp. 1119-1134.

Rose H. J. Jr., Baedecker P. A., Berman S., Christian R. P., Dwornik E. J., Finkelman R. B., and Schnepfe M. M. (1975) Chemical composition of rocks and soils returned by the Apollo 15, 16, and 17 missions. *Proc. Lunar Sci. Conf. 6th,* pp. 1363-1374.

Schnetzler C. C. and Philpotts J. A. (1971) Alkali, alkaline earth, and rare earth element concentrations in some Apollo 12 soils, rocks, and separated phases. *Proc. Lunar Sci. Conf. 2nd,* pp. 1101-1122.

Scott D., Lucchitta B., and Carr M. (1972) Geologic maps of the Taurus-Littrow region of the Moon (Apollo 17). *U.S. Geol. Surv. Map I-800.*

Simkin T., Noonan A., Switzer G. S., Mason B., Nelen J., Melson W. G., and Thompson G. (1973) Composition of Apollo 16 fines 60051, 60052, 64811, 64812, 68821, and 68822. *Proc. Lunar Sci. Conf. 4th,* pp. 279-290.

Spudis P. D. (1978) Composition and origin of the Apennine Bench formation. *Proc. Lunar Planet. Sci. Conf. 9th,* pp. 3379-3394.

Spudis P. D. and Ryder G. (1986) Geology and petrology of the Apollo 15 landing site: Past, present and future understanding. *Eos Trans. AGU, 66,* 721, 724-726.

Stöffler D., Ostertag R., Reimold W. U., Borchardt R., Malley J., and Rehfeldt A. (1981) Distribution and provenance of lunar highland rock types at North Ray crater, Apollo 16. *Proc. Lunar Planet. Sci. 12B,* pp. 185-207.

Taylor S. R., Gorton M. P., Muir P., Nance W., Rudowski W., and Ware N. G. (1973) Lunar highlands composition—Apennine front. *Proc. Lunar Sci. Conf. 4th,* pp. 1445-1460.

Wakita H., Rey P., and Schmitt R. A. (1971) Abundances of the 14 rare-earth elements and 12 other trace elements in Apollo 12 samples—five igneous and one breccia rocks and four soils. *Proc. Lunar Sci. Conf. 2nd,* pp. 1319-1330.

Wänke H., Baddenhausen H., Dreibus G., Jagoutz E., Kruse H., Palme H., Spettel B., and Teschke F. (1973) Multielement analyses of Apollo 15, 16, and 17 samples and the bulk composition of the Moon. *Proc. Lunar Sci. Conf. 4th,* pp. 1461-1481.

Wänke H., Palme H., Baddenhausen H., Dreibus G., Jagoutz E., Kruse H., Spettel B., Teschke F., and Thacker R. (1974) Chemistry of Apollo 16 and 17 samples: Bulk composition, late stage accumulation and early differentiation of the Moon. *Proc. Lunar Sci. Conf. 5th,* pp. 1307-1335.

Wänke H., Palme H., Baddenhausen H., Dreibus G., Teschke F., and Thacker R. (1975) New data on the chemistry of lunar samples: Primary matter in the lunar highlands and the bulk composition of the Moon. *Proc. Lunar Sci. Conf. 6th,* pp. 1313-1340.

Warren P. H. and Wasson J. T. (1978) Compositional-petrographic investigation of pristine nonmare rocks. *Proc. Lunar Planet. Sci. Conf. 9th,* pp. 185-217.

Wilhelms D. E. (1987) The geologic history of the Moon. *U.S. Geol. Surv. Prof. Pap. 1348.* 302 pp.

Willis J. P., Erlank A. J., Gurney J. J., and Ahrens L. H. (1972) Geochemical features of Apollo 15 materials. In *The Apollo 15 Lunar Samples* (J. W. Chamberlain and C. Watkins, eds.), pp. 268-271. The Lunar Science Institute, Houston.

Woodcock M. R. and Pillinger C. T. (1978) Major element chemistry of agglutinate size fractions. *Proc. Lunar Planet. Sci. Conf. 9th,* pp. 2195-2214.

Proceedings of the 20th Lunar and Planetary Science Conference, pp. 161-174
Lunar and Planetary Institute, Houston, 1990

The Alphonsus Region: A Geologic and Remote-Sensing Perspective

C. R. Coombs

Mail Code SN15, NASA Johnson Space Center, Houston, TX 77058

B. R. Hawke, P. G. Lucey, P. D. Owensby, and S. H. Zisk

Planetary Geosciences Division, Hawaii Institute of Geophysics, Honolulu, HI 96822

A number of interesting questions exist concerning the composition and origin of a number of geologic units in the west-central highlands crater Alphonsus. This paper utilizes a variety of newly obtained remote sensing data to address these questions and improve the current understanding of the geologic history of the Alphonsus region. Near-infrared and UV/VIS reflectance spectra were obtained and analyzed for many of the geologic features, the results of which are presented here. Spectra collected for pyroclastic debris associated with three endogenic dark halo crater complexes on the floor of Alphonsus indicate a basaltic assemblage rich in olivine. These pyroclastic deposits exhibit generally low returns on the depolarized 3.8 cm radar image. An exogenic dark halo crater was identified on the interior of Alphonsus and appears to be composed of a mixture of pyroclastic debris and highlands material. The light plains deposits and other highland units in the Alphonsus region generally exhibit a noritic composition. However, the Alphonsus central peak is composed of pure anorthosite. This composition is unique within the region.

INTRODUCTION

Alphonsus crater is located in the south-central lunar highlands (13°S, 4°W; Figs. 1 and 2), between the craters Ptolemaeus and Arzachel. The pre-Imbrian (>3.85 b.y.) aged Alphonsus is 118 km in diameter and is slightly elongate in shape. This crater is typical of many pre-Imbrium highland craters of this size, in that it has a relatively flat, cratered floor, a central peak, and a broad crater rim. The presence of numerous fractures, dark-halo craters, and a north-south trending central ridge on the floor, however (Figs. 3 and 4), distinguish Alphonsus from the other highland craters of this size.

To date, most of the work done on Alphonsus has centered around the nature and origin of the dark halo craters that occur around the perimeter of its floor (e.g., *Hartmann,* 1967; *Head,* 1974; *Peterfruend,* 1976; *Head and Wilson,* 1979; *Gaddis et al.,* 1985; *Coombs,* 1988; *Coombs and Hawke,* 1988a; *Hawke et al.,* 1989). Early mappers of Alphonsus suggested that these features may be either tephra deposits due to maars (e.g., *Howard and Masursky,* 1968), or pyroclastic deposits from a central vent (e.g., *Carr,* 1969). Vulcanian eruption models developed by *Head and Wilson* (1979) and *Wilson and Head* (1981) further support the pyroclastic origin of these dark halo craters. Later spectral studies of these deposits confirmed their basaltic composition (*Gaddis et al.,* 1985; *Coombs et al.,* 1987; *Hawke et al.,* 1989; *Coombs et al.,* 1989). This previous work has been instrumental in distinguishing lunar "endogenic" dark-halo craters from the "exogenic" ones, as well as defining their vulcanian eruption origin (e.g., *Head and Wilson,* 1979; *Bell and Hawke,* 1984).

Although the Alphonsus region has been the focus of several geologic and remote sensing studies (e.g., *Salisbury,* 1968; *Howard and Masursky,* 1968; *Carr,* 1969; *Zisk,* 1972; *Gaddis et al.,* 1985; *Hawke et al.,* 1989) problems such as (1) the determination of the composition of the dark mantle deposits associated with endogenic craters on the Alphonsus floor, (2) the nature and origin of the light plains on the crater floor, (3) the nature and origin of the central peak, (4) the origin of the north-south-trending ridge on the crater floor, (5) the processes responsible for the crater floor uplift, and (6) the effects of Imbrium basin impact/ejecta on Alphonsus, are not fully understood. This paper utilizes new remote-sensing data to address the above questions and present the current understanding of the geologic history of the Alphonsus region.

METHOD

A variety of available datasets were used during the course of this study. These include Apollo, Lunar Orbiter, and Ranger photopraphs, topographic and geologic maps, as well as Whitaker's color difference map of the region (in *Wilhelms,* 1987, p. 98). Also, several recently collected telescopic data sets were utilized. These include UV-VIS (0.3-1.1 μm) and near-infrared reflectance spectra (0.6-2.5 μm). In addition, 3.8-cm and new, very high resolution 3.0-cm radar images were utilized in this study.

Near-Infrared Observations

Near-infrared reflectance spectra were collected from a number of geologic units within the Alphonsus crater and vicinity including (from Alphonsus) (1) the Alphonsus central

Fig. 1. Full-Moon photograph showing location of study region. Box corresponds to map shown in Fig. 2.

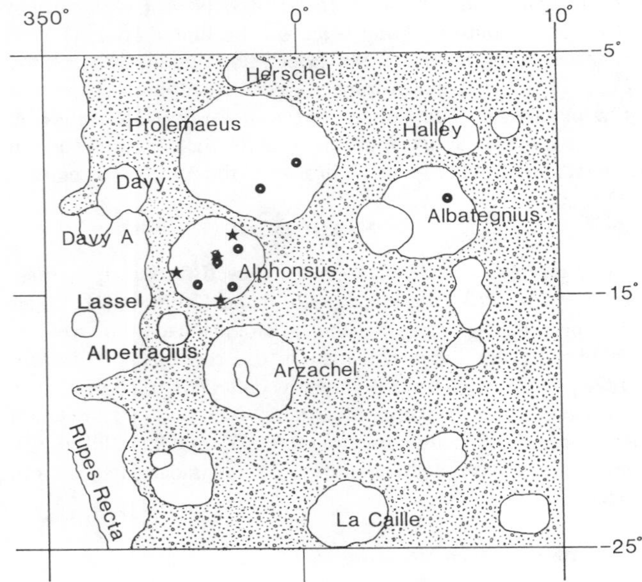

Fig. 2. Diagrammatic sketch map of the Alphonsus region. The black dots indicate the areas for which the near-infrared reflectance spectra were collected (see Fig. 5 for spectra). The stars show the locations of UV/VIS spectral collection sites (see Fig. 6 for spectra). North at top.

peak, (2) the light plains unit in the floor of Alphonsus, (3) the very fresh small impact crater on the floor of Alphonsus, and (4) three localized pyroclastic units surrounding the endogenic craters; (from surrounding region) (1) the Arzachel central peak, (2) the eastern floor of Ptolemaeus, (3) the southern floor of Ptolemaeus, and (4) the Albategnius east plains. Representative spectra are shown in Fig. 5a and b.

These near-IR reflectance spectra were collected at the UH Telescope facility on Mauna Kea, using the PGD indium antimonide circular-variable filter (CVF) spectrometer. The PGD spectrometer measures intensity in each of the 120 wavelength channels between 0.6-2.5 μm region through a rotating filter with a continuously variable band pass (*McCord et al.,* 1981; *Bell and Hawke,* 1984, 1987). To provide consistancy, each of the spots in the Alphonsus region was observed independently two or three times within an hour, with two successive scans being added together before the spectrum was written to tape. Each spectrum thus represents an average of four to six individual measurements.

During the observations, both the f/10 secondary and the higher magnification f/35 secondary mirrors were used. The f/10 secondary allows observation of an elliptical area 4.3 × 8.4 km in size with the 2.3 arcsec aperture in place. Smaller regions were also observed using the f/35 secondary mirror.

Fig. 3. Oblique photograph of Alphonsus showing the LDMD, central ridge, and floor fractures. Note position of dark mantle deposits peripheral to the roughly circular (endogenic) craters and floor fracures. View is looking southwest. Apollo 16 Metric Oblique 2477.

Generally, this oscillating mirror is used for thermal-infrared measurements; however, when used in its nonoscillating mode, and under optimum observing conditions, a 0.7 arcsec aperture can be used and a smaller elliptical region (1.6×3.1 km) may be observed on the lunar surface. Observing conditions varied depending upon the local weather and atmospheric conditions as well as lunar phase and position. Spectra were collected from various sites during a number of visits to Mauna Kea. All of these data were assembled and the best selected for analysis.

In order to monitor the atmospheric conditions throughout the observation period, frequent observations were also made of the Apollo 16 standard lunar reflectance area. Extinction corrections were made following the methods outlined by *McCord and Clark* (1979) and *Clark* (1980). This produces a reflectance ratio between the observed area and the standard area. Then, following the method of *Adams and McCord* (1971), the reflectance curve of the mature Apollo 16 soil sample 62231,1 was used to convert the relative reflectance spectra to spectral reflectance. For clarification and ease of comparison, the spectra have been scaled to unity at 1.02 μm and "stacked" relative to the 1.0 μm absorption band (Fig. 5a,b). In addition, to further investigate the individual nature of the 1.0 μm absorption band of each of the spots observed, a straight line continuum was removed from each spectrum (Fig. 5b). This enhances the absorption bands in the 1.0 μm region and facilitates the extraction of compositional information.

Further data reduction included a method developed by *Lucey et al.* (1986) that allows one to quantitatively classify the data. Following their method, four spectral parameters were derived for each of the observed locations: (1) the continuum slope, Δ/Δ_λ, defined as the slope of a straight line drawn through the peaks on either side of the 1 μm absorption band and measured as the change in the scaled reflectance over the change in wavelength; (2) the depth of the absorption, defined as 1 minus the reflectance at the absorption minimum relative to the continuum as defined above; (3) the wavelength of the band minimum, whereby parameters 2 and 3 were derived by fitting 15 channels on either side of the absorption minimum with a fourth order polynomial and using this equation to find the relative reflectance and wavelength of the relative minimum absorption, and (4) the width of the absorption band derived from the difference in wavelength between the intercepts of the spectrum under analysis and a line parallel to the continuum slope at a relative reflectance equal to half the absorption depth plus the reflectance at the absorption minimum. Selected spectra are shown in Fig. 5a and b. The spectral parameters for all of the near-IR reflectance spectra used in this study are presented in Table 1.

Visible Observations

The UV-VIS reflectance spectra were collected in November 1988 using the Planetary Geosciences Division (PGD) UV-VIS spectrometer at the University of Hawaii 2.2 m telescope on Mauna Kea. This instrument measures the spectra of the observed spots from 0.3-1.1 μm. Data collected with this instrument were reduced in a manner similar to that previously described for the near-infrared reflectance spectra. Four typical spectra are shown in Fig. 6. These spectra are displayed relative to the Apollo 16 reflectance standard. In addition, a small

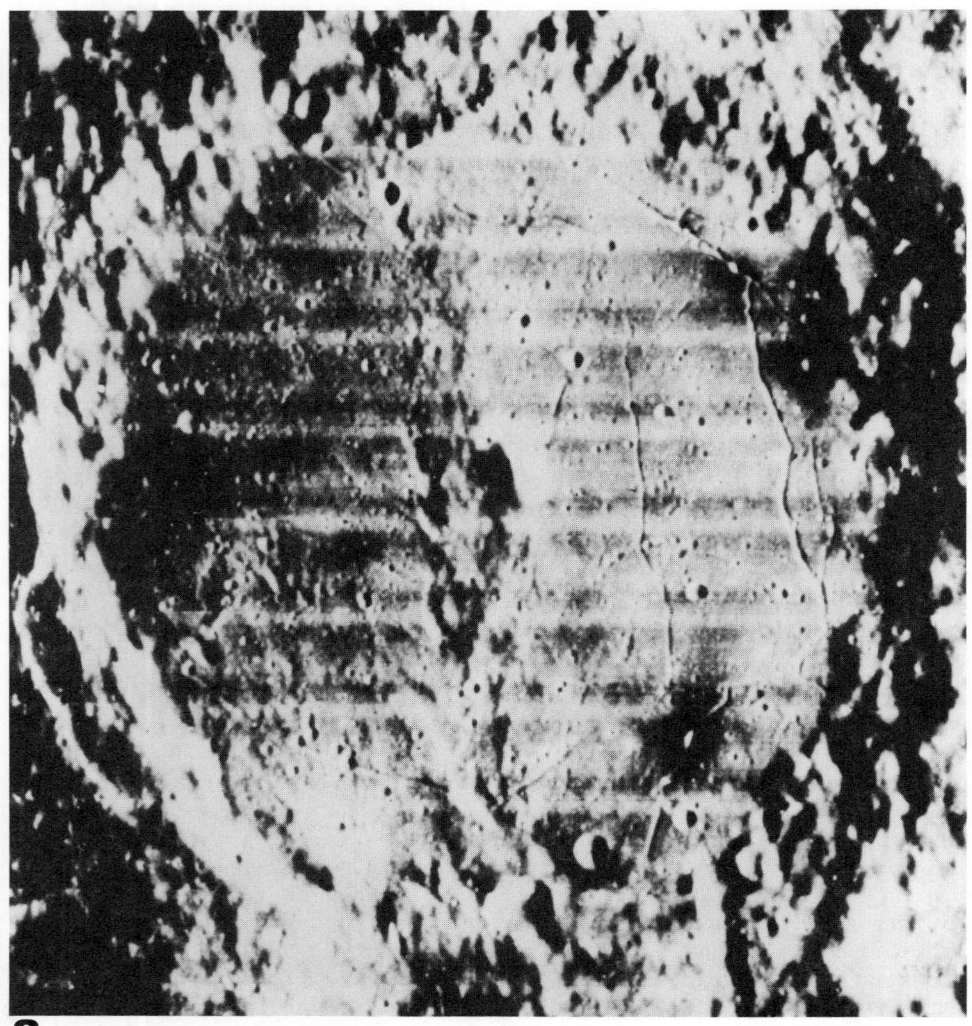

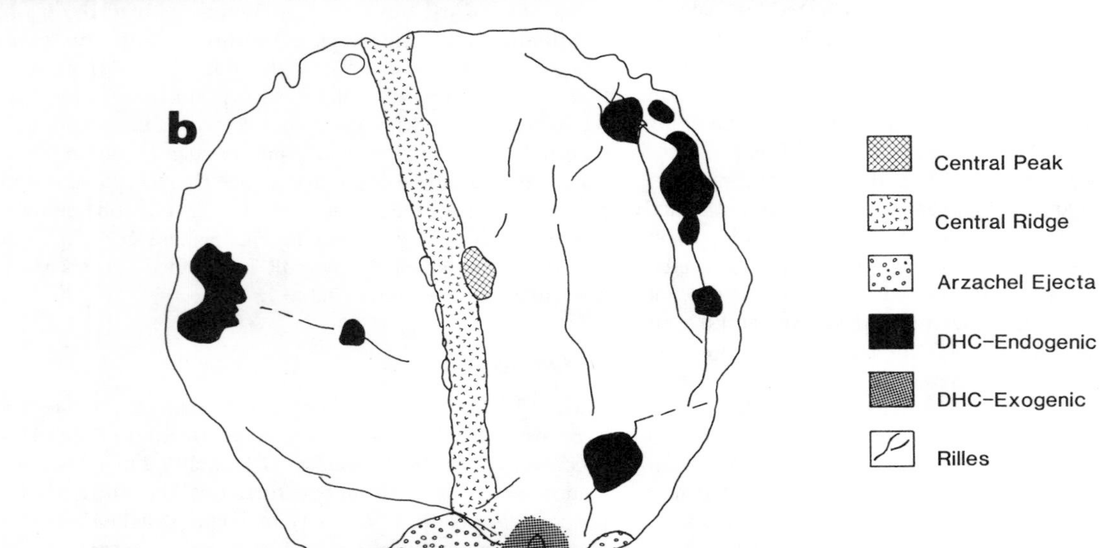

Fig. 4. (a) Near-vertical photograph of Alphonsus crater. Note structures along the central ridge and the exogenic dark-halo crater in southern portion of the crater, just east of the central ridge. Same view as map in **(b)**. North at top. (LO-108-H2.) **(b)** Geologic sketch map of Alphonsus crater (118 km). Refer to text for an in-depth discussion of the composition and nature of these features. North at top.

number of previously unpublished UV-VIS spectra obtained by T. B. McCord and coworkers during the early 1970s were examined for this study.

3.0 cm and 3.8 cm Radar

Previously obtained 3.8 cm radar images (*Zisk et al.,* 1974) were used in this study in addition to a pair of 3.0 cm radar maps covering two parts of the floor of Alphonsus crater (Figs. 7 and 8). These 3.0 cm data were collected very recently with MIT's Haystack Observatory radar system, located in Westford, Massachusetts. The radar maps are standard range-doppler resolved radar data, measured and analyzed using techniques that have been described by *Zisk et al.* (1974). These are maps of the "expected"-polarization echoes, or opposite-sense circular polarization. As presented here, each processed block comprises one map composed of 1024 pixels in the range direction and 8192 in the doppler, with a basic single-pixel resolution of about 15×75 m in the range and doppler directions respectively.

The images presented in Fig. 8, however, are preliminary and show only the central portion of the full data measurement. In addition, these images are single-look pictures and have been smoothed by a 2×2 spatial integration to reduce the speckle noise. Each image thus consists of a 512×1024 pixel block, with surface resolution (independent-pixel size) reduced to about 75×150 m by the integration and the effects of projection onto the lunar surface. The images have not been reprojected and thus are in the original range-doppler coordinate system. Because of the location of Alphonsus near the central lunar meridian, the vertical on the picture points only a few degrees off lunar North. Each picture spans a full 150 km diameter and extends roughly 38 km north to south. The darkening at the top of each picture is an aritifact caused by the anti-aliasing filter in the radar system.

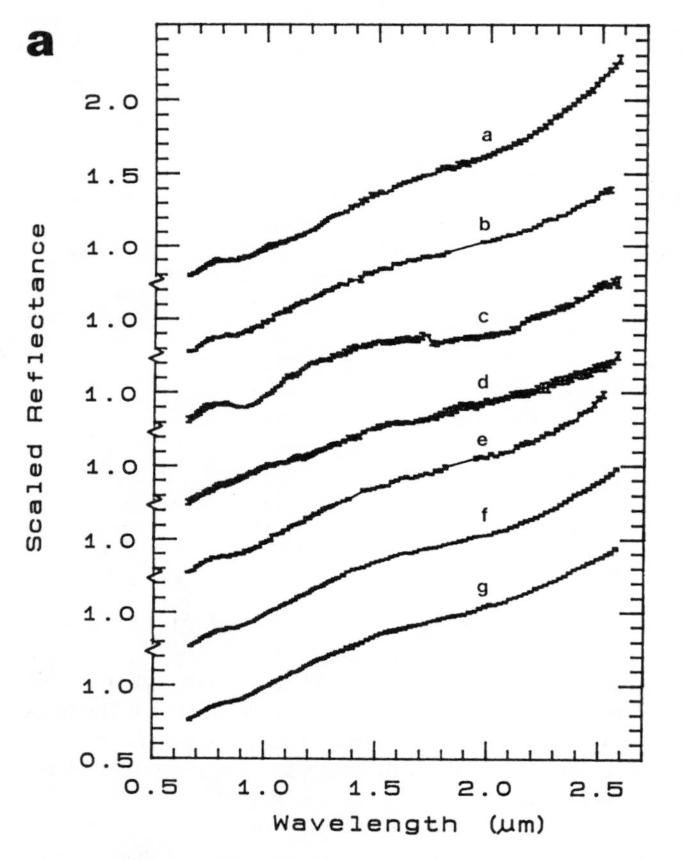

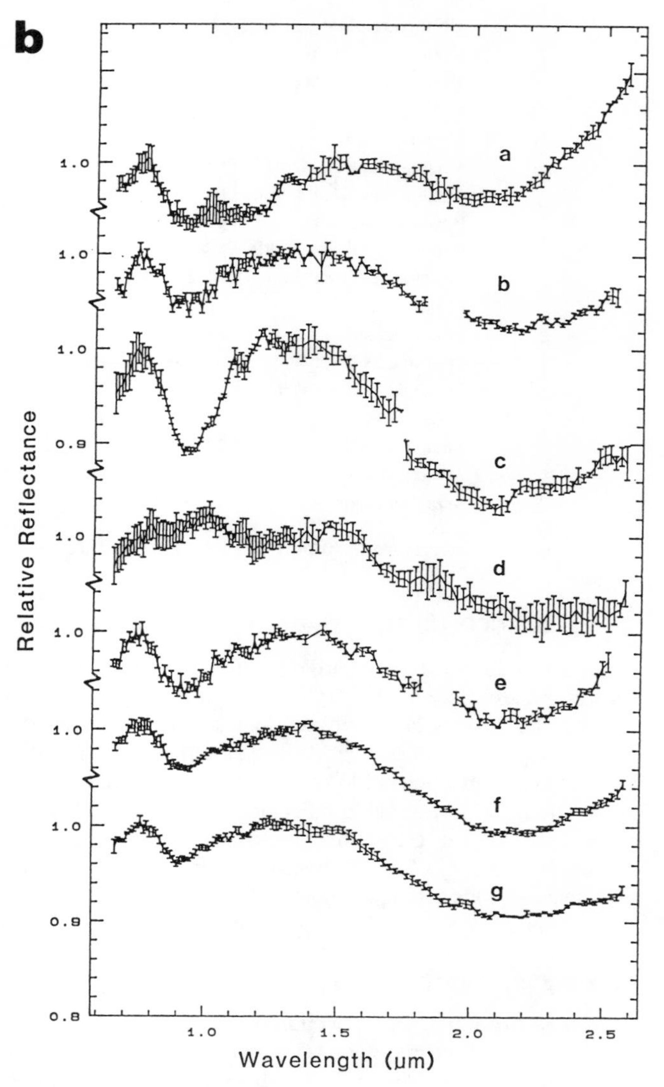

Fig. 5. (a) Near-infrared reflectance spectra of features in region shown as spectra/sun: a = Alphonsus DHC southeast, b = Alphonsus floor, c = Alphonsus crater south of central peak, d = Alphonsus central peak, e = Ptolemaeus floor, f = Ptolemaeus plains south, g = Albategnius east plains. **(b)** Continuum removed near-infrared reflectance spectra corresponding to those also shown in **(a)** as spectra/sun. a = Alphonsus DHC southeast, b = Alphonsus floor, c = Alphonsus crater south of central peak, d = Alphonsus central peak, e = Ptolemaeus floor, f = Ptolemaeus plains south, g = Albategnius east plains.

TABLE 1. Spectral characteristics determined for features in the Alphonsus region.

Location Name	Date	Telescope Aperture*	Relative Absorption Depth (%)	Relative Band Minimum (μm)	Continuum Slope (ΔR/λ)	Band Width at Half Height (μm)
Alphonsus Floor 1	6/12/87	2	5.3	0.915	0.626	0.231
Alphonsus Smooth Plains	9/7/82	1	5.0	0.914	0.674	0.222
Alphonsus: Small Crater South of Central Peak	9/13/84	1	10.8	0.925	0.588	0.207
Alphonsus: Light Plains on the NW floor	9/15/85	1	6.2	0.905	0.626	0.273
Alphonsus Peak	7/14/81	1	0.0	N.A.	0.596	N.A.
Alphonsus Peak	9/15/81	1	0.0	N.A.	0.530	N.A
Alphonsus Peak	9/7/82	1	0.0	N.A.	0.541	N.A.
Alphonsus Central Peak	9/13/84	1	0.0	N.A.	0.562	N.A.
Alphonsus G Crater	9/15/84	1	15.8	0.938	0.403	0.382
Alphonsus DHC East	10/25/83	1	4.1	1.080	0.662	0.434
Alphonsus DHC East	9/11/84	1	4.3	1.030	0.700	0.354
Alphonsus DHC East	9/13/84	1	6.0	0.953	0.652	0.432
Alphonsus DMM	9/7/82	1	8.6	0.963	0.676	0.303
Alphonsus DHC West	9/15/84	1	5.7	1.010	0.658	0.359
Alphonsus DHC West	9/11/84	1	6.0	0.979	0.681	0.401
Alphonsus R	8/16/81	1	6.0	0.951	0.661	0.410
Alphonsus CA	8/16/81	1	6.2	0.970	0.683	0.387
Albategnius East Plains	10/30/85	2	3.5	0.917	0.678	0.208
Albategnius Floor 1	10/25/83	2	3.9	0.914	0.699	0.248
Ptolemaeus Floor 1	6/12/87	2	5.9	0.918	0.668	0.241
Ptolemaeus Plains South	10/30/85	2	3.9	0.922	0.685	0.223
Ptolemaeus Floor	10/25/83	2	4.7	0.947	0.685	0.211
Ptolemaeus A	8/1/80	1	11.4	0.931	0.539	0.232
Ptolemaeus A	10/30/85	2	11.0	0.940	0.631	0.293
Ptolemaeus West Wall	8/1/80	1	5.8	0.935	0.549	0.247
Ptolemaeus East Wall	8/1/80	1	3.5	0.911	0.623	0.247
Arzachel Peak	7/14/81	1	6.6	0.924	0.634	0.261
ArzachelPeak	7/15/81	1	6.3	0.938	0.626	0.248
Arzachel Wall	7/12/81	1	5.0	0.916	0.648	0.304
Arzachel Floor	7/12/81	1	4.6	0.919	0.643	0.220

*Aperture 1 = 0.7 arcsec or 1.4 km; aperture 2 = 1.55 arcsec or 3.1 km.

RESULTS OF REMOTE-SENSING ANALYSES

Important information regarding the composition, physical nature, and areal extent of a particular deposit may be obtained through the analysis of remote-sensing data. In particular, the use of spectral reflectance data can aid in deciphering geochemical and compositional data from the lunar surface units. The near-IR and visible reflectance data collected in this study have helped greatly in unraveling the geologic and stratigraphic history of the Alphonsus region. Spectra collected from these various geologic units are discussed below.

Near-IR Reflectance Spectra

Dark-halo craters. Spectra collected for the LDMD, or dark-halo craters, on the floor of Alphonsus indicate a basaltic composition rich in olivine. These spectra fall into the Group 3 spectral class of lunar pyroclastic deposits described by *Hawke and Lucey* (1985), *Hawke et al.* (1989), *Coombs* (1988), and *Coombs and Hawke* (1988a). These spectra have a broad, moderately deep (~5-7%), asymmetrical absorption band in the 1.0 μm region (Table 2, a in Figs. 5a,b). The broad nature and asymmetrical shape of these bands are indicative of composite features with band centers at or beyond 1.0 μm. In lunar spectra, these band characteristics are most often produced by a mixture of olivine and pyroxene (*McCord et al.,* 1981; *Singer,* 1981). A detailed quantitative analysis of the "1.0 μm" band in the J. Herschel LDMD spectrum, another member of the Group 3 spectral class, was presented by *McCord et al.* (1981) and *Hawke et al.* (1989). It was determined that this feature could best be explained by a mixture of olivine and orthopyroxene. While in theory this band could be produced by the combination of clinopyroxene, orthopyroxene, and Fe^{2+}-bearing glass, the analyses presented by *McCord et al.* (1981) and *Hawke et al.* (1989) indicated that this is unlikely.

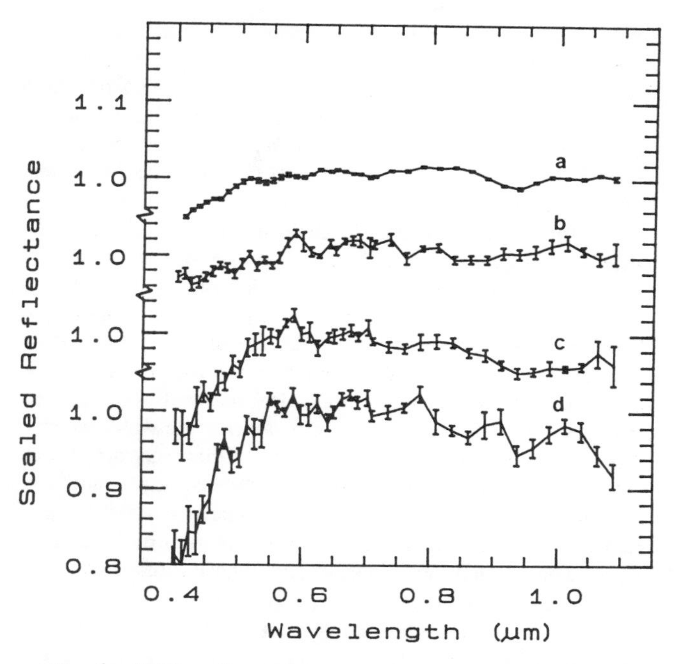

Fig. 6. Visible reflectance spectra of features in region, shown as spectra/sun: a = Alphonsus floor, b = exogenic DHC, c = DHC west, d = Alphonsus central peak.

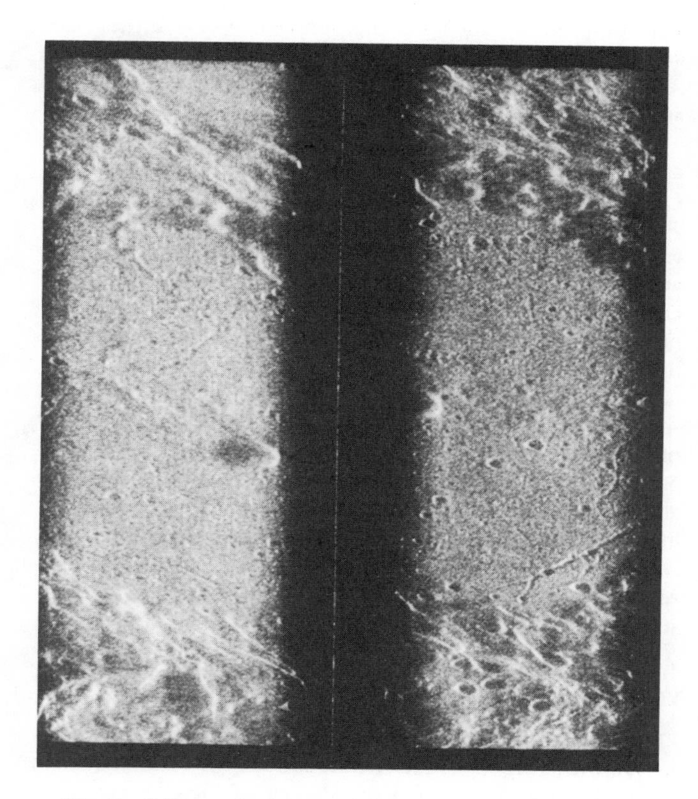

Fig. 7. 3.0 cm radar images of the Alphonsus crater: left half of the image is the western half of the crater floor; right half shows part of the eastern floor. Note prominent floor-fracture in lower right portion of image. North at top.

The Alphonsus endogenic DHC spectra appear to be dominated by a mixture of olivine and pyroxene. While the basaltic plug rock could have been the source of some of the olivine, the high olivine abundance indicated by the spectral data requires another source. No evidence for olivine was found in the spectra of highlands units in and around Alphonsus. Hence, the olivine was not emplaced with the highlands-rich wall rock. Rather, the bulk of the olivine was almost certainly emplaced with the juvenile material. The olivine in the juvenile material might occur in several forms: (1) the olivine may have existed in the form of phenocrysts in the magma and been emplaced with the juvenile material upon eruption, (2) the olivine now exists as devitrified glass in the localized pyroclastic deposits, and/or (3) olivine-rich mantle inclusions might have been present in the melt and now could be found as xenoliths in the pyroclastic deposits (*Coombs,* 1988; *Hawke et al.,* 1989).

The pyroxene in the LDMD on the floor of Alphonsus was probably derived from multiple sources. The bulk of the pyroxene was derived from the basaltic plug rock and the

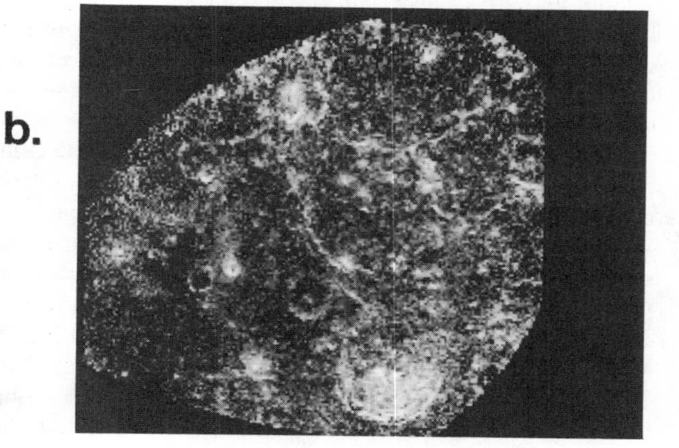

Fig. 8. 3.8 cm radar images of the Alphonsus crater region: **(a)** polarized (OC) signal **(b)** depolarized (SC) echo. Alphonsus crater is located in the middle of the images, between Ptolemaeus (top) and Arzachel (bottom) craters. See text for a discussion of these figures. North is at the top of the images.

TABLE 2. Characteristics of the "1.0 μm" bands of spectra obtained for localized pyroclastic deposits.

Group 1	center	0.93-0.95 μm
	depth	4-5%
	asymmetrical	checkmark-like shape
Group 2	center	0.96 μm
	depth	~7%
	more asymmetrical	
Group 3	center	1.0 μm region
	depth	5-7%
	asymmetrical	

From *Coombs* (1988).

juvenile material. However, *Head and Wilson* (1979) used morphometric data for the Alphonsus dark-halo craters to demonstrate that major amounts of nonjuvenile highlands material were incorporated into the pyroclastic deposits. Spectral data for the Alphonsus floor deposits indicate that the dominant mafic mineral in the highlands material is orthopyroxene. Therefore, some of the pyroxene in the LDMD is orthopyroxene derived from the highlands-rich wall rock that was eroded and emplaced during the explosive eruptions that produced the Alphonsus LDMD.

In summary, we interpret the Alphonsus LDMD, to be generally basaltic in composition and to be dominated by olivine and pyroxene. Almost all of the olivine and most of the pyroxene was emplaced with the juvenile material. However, some orthopyroxene was emplaced as a component in the highlands-rich wall rock. The basaltic plug rock could have contributed minor amounts of olivine and pyroxene. Some pyroclastic glass may also be present but no direct spectral evidence for the existence of a glass component was found.

Alphonsus light plains and other floor material. The floor of Alphonsus is largely mantled by a layer of relatively high albedo material. These plains deposits are relatively smooth and are of uncertain origin. Our spectra (Figs. 2 and b in Figs. 5a,b; Table 1) indicate that the Alphonsus light plains deposits are composed of noritic material. A small, very fresh crater on the southern portion of the Alphonsus crater floor also excavated noritic debris.

Alphonsus central peak. Spectra obtained for the central peak of Alphonsus indicate that it is composed of pure anorthosite. This important interpretation was first presented by *Pieters* (1986). We subsequently obtained a near-infrared spectrum for a slightly different portion of the central peak (d in Figs. 5a,b). Analysis of this spectrum supports the anorthosite interpretation. The spectra collected for the Alphonsus peak exhibit characteristics that are distinct (Table 1). No mafic absorption bands are detected and a distinct change of continuum slope is observed in these spectra near 1.6 μm. Similar spectra for the central peaks of Theophilus, Piccolomini, and Petavius craters (*Pieters,* 1986, 1989), the Inner Rook Mountains of the Orientale basin (*Hawke et al.,* 1984; *Spudis et al.,* 1984), and several craters associated with the various rings of Nectaris basin (*Hawke et al.,* 1985; *Spudis et al.,* 1989). Some of these spectra exhibit very shallow absorption features centered at about 1.25 μm that may be attributed to the presence of Fe^{2+}-bearing plagioclase feldspar. Other spectra for anorthosite deposits show no feldspar band at ~1.25 μm. The feldspar bands in these spectra have probably been destroyed by shock effects (*Pieters and Hörz,* 1985). Alternatively, the plagioclase in the Alphonsus peak has a very low abundance of Fe^{2+} iron.

No other major anorthosite deposits have been identified either in the Alphonsus region or the western central highlands. The nearest anorthosite deposit to the east is associated with Kant crater in the eastern central highlands. To the west, the nearest pure anorthosite units are located on the interior of the Orientale basin. The ultimate source of the Alphonsus anorthosite is uncertain. The material in the central peak was derived from a unit of pure anorthosite which existed ~10-12 km beneath the surface of the Alphonsus target site. The anorthosite exposed in the Alphonsus peak may have originated as a part of the primordial ferroan anorthosite upper crust that formed by plagioclase crystal flotation in the magma ocean. This ancient anorthosite could have been buried by impact ejecta, ancient volcanism, or some combination of these and other processes. Alternatively, the Alphonsus anorthosite may have been derived from the upper portion of an ancient pluton. It is interesting to note that anorthosite is not exposed in the central peak (or walls) of Arzachel, a 70-km-diameter impact crater immediately south of Alphonsus. If the anorthosite unit exists beneath the Arzachel target site, it was not sampled by the Arzachel impact event. Either the anorthosite layer does not extend beneath Arzachel or the Arzachel impact did not penetrate deep enough to expose the anorthosite.

Arzachel crater. Spectra for the Arzachel central peak and wall were presented by *Pieters* (1986). Both spectra (Table 1) exhibit moderately deep (5-6%) "1.0-μm" absorption bands centered shortward of 0.95 μm. The mafic assemblages in the areas for which these spectra were obtained are dominated by low-Ca orthopyroxene. The spectral characteristics indicate that the peak and wall are composed of anorthositic norite. A similar composition is indicated for the floor of Arzachel (Table 1).

Other highlands features in the region. Near-IR spectra were also obtained for light plains deposits on the floor of Ptolemaeus and Albategnius craters (Figs. 2 and e-g in Figs. 5a,b; Table 1). Spectral analysis indicated that these deposits are dominated by noritic material. A similar composition is indicated for the east and west walls of Ptolemaeus crater. In contrast, *Pieters* (1986) determined that the central peak of Alpetragius is dominated by high-Ca clinopyroxene and suggested a gabbroic composition.

Visible Reflectance Spectra

The UV-VIS spectra for features associated with Alphonsus are shown in Fig. 6. These spectra are shown relative to the Apollo 16 reflectance standard. All four of these spectra are "red" (low 0.40/0.56 μm values) relative to the Apollo 16 standard. The spectrum obtained for the northeastern portion of the Alphonsus floor is most similar to that of the Apollo

16 site. A similar highland composition is implied. The localized dark mantle deposit (LDMD) on the western portion of the floor is relatively "red." This is consistent with the results presented by *McCord* (1968) for the dark halo craters on the interior of Alphonsus. A basaltic composition is implied.

The UV-VIS spectrum obtained for the LDMD (Alphonsus R) on the southeastern portion of Alphonsus is different from the spectrum collected for the pyroclastic deposit on the western portion of the floor as well as the spectra presented by *McCord* (1968) for the western and eastern LDMD in that it is "blue" relative to the Apollo 16 standard. If this difference is real, it would imply that the composition of the pyroclastic deposit associated with Alphonsus R is distinct from that of the other pyroclastics on the floor of Alphonsus. Our confidence in the spectrum of the Alphonsus R pyroclastics is not high because of the large error bars associated with the spectrum and the fact that it was collected during twilight. However, an unpublished spectrum of this deposit obtained by McCord and co-workers also shows that the material is "blue" relative to the lunar reflectance standard MS-2. It will be necessary to repeat this observation in order to determine if the Alphonsus R pyroclastic deposit is indeed compositionally distinct.

The spectrum obtained for the central peak of Alphonsus shows that it is very "red." Most fresh lunar highland features are "blue" relative to the Apollo 16 standard in this wavelength range. Near-IR spectra show that the peak is composed of pure anorthosite (*Pieters,* 1986; *Coombs et al.,* 1989). Hence, it appears that the Alphonsus peak anorthosite deposit is "red" in the UV.

Finally, a spectrum was collected for the near-rim ejecta of Alphonsus A, a previously unidentified dark-haloed impact crater on the floor of Alphonsus (Figs. 3 and 4). This deposit may be a mixture of highland material and pyroclastic debris.

3.8 cm Radar

Figure 7 shows polarized and depolarized images of the 3.8 cm radar reflectivity of Alphonsus and its vicinity. At this low angle of incidence, or close proximity to the lunar sub-Earth point, the poloarized signal shows the usual strong quasirelief brightening. This strong signal is largely a function of local slopes. The depolarized signal, on the other hand, is dependent almost entirely on the inherent properties of the surface materials, with little or no slope effect. Thus, the depolarized map unambiguously depicts regions of differing surface roughness and dielectric constant (i.e., mean density).

The following discussion refers to just 3 of the 11 endogenic deposits on the floor of Alphonsus; the 2 coalesced deposits in the west and northeast, and the single deposit in the southeast. We omit the additional two smaller, less well-defined LDMDs on the floor and near the eastern wall (DHC numbers 6 and 8 from *Head and Wilson,* 1979) from this discussion.

In each of these three regions there appears to be strong radar-bright signatures near the center of the dark-mantled area. These bright features may indicate the presence of one or more rocky vents. The radar appearence of each of the mantling deposits in these areas is quite different, however. For the western LDMD there is a good correspondence of a very dark radar feature with the optically dark mantled area. This mantling deposit is clearly smooth enough that no surface scatter is observed as evidenced by the zero radar echo shown in several pixels. These deposits are thick enough and sufficiently rock-free that no subsurface echoes are produced. Assuming a somewhat higher absorption coefficient than the average lunar regolith material, we estimate that at least the upper 3 to 5 m of the deposit is essentially free of rocks larger than a few millimeters in size.

The southeast LDMD also shows a generally lower radar albedo than its surroundings. However, it is not as low as the western LDMD; that is, there are no zeros in the radar echo. The correspondence between radar and optically dark regions is not as good for this deposit. More specifically, the optical region is larger than the radar signature for this deposit.

Finally, for the northeast LDMD there is very little correlation between the dark mantle deposit itself and any radar feature. One radar-dark patch of material does coincide with about one third of the dark halo just south of the vent. However, the remainder of the dark mantle appears to have essentially the same radar signature as the rest of the crater floor. Our interpretation of this discrepancy is that the thickness of the mantling layer is not very great, especially north of the main source vent. It is also possible that the mantling unit contains enough small rocks to create a slightly brighter reflectance echo. However, it appears unlikely that the fraction of "reflector" rocks in this region is just enough to balance the darkness of the mantle.

The depolarized radar image appears to have sufficent sensitivity that any darkening clearly would be shown. For example, there are several concentric brightened rings on the southern part of the crater floor. These rings appear to be centered on the break in the crater wall through which the ejecta curtain from Arzachel crater came, and thus may be related to the Arzachel impact event. Alternatively, the rings may be caused by differences in surface roughness on the local slopes.

3.0 cm Radar Data

A comparison of the 3.0 cm radar images in Fig. 8 with the Lunar Orbiter picture in Fig. 3 immediately reveals one fundamental difference in the appearance of the two images. Several rilles appear on each data set that cannot be found (at least easily) on the other. This is because the radar illumination direction is roughly from the center of the Earth-facing hemisphere, or in this case, from the north. This northern *radar* illumination shows the east-west trending rilles and lineations very well, whereas the east-west *solar* illumination of the lunar surface preferentially highlights the north-south-trending rilles and lineaments.

Also, the well-known dark-halo craters are almost invisible in the 3.0 cm radar images. Evidently, the Fresnel reflectivity of the mantling material is not different enough from the surrounding crater floor that a contrast is evident in the "polarized" image. If, as expected, the mantled area shows up with a smoother texture than its surroundings at the 3.0 cm radar wavelength, as discussed above, then the cross-polarized image should contain the evidence for it. Additional analyses of these 3.0 cm images is underway.

GEOLOGY OF THE ALPHONSUS REGION

A number of geologic features are present within the Alphonsus crater region. Within the crater itself are the central peak, floor fractures, lineaments, a central ridge, light plains material, an uplifted floor, and numerous dark-halo craters.

Alphonsus Central Peak

The central peak of Alphonsus is located just right of center, midway along the north-south-trending central ridge (Figs. 3 and 4). The peak is approximately 9 km in diameter and is steeply sloped. The peak summit is approximately 1100 m above the crater floor. It may be considered a weak, single central peak after the classification of *Wood* (1968), based on its age and morphology. As discussed above, spectral data collected for this peak indicate that it is composed of pure anorthosite. This composition is unique within the Alphonsus region. Apparently, the impact that produced Alphonsus sampled an anorthosite unit at depth (Fig. 9).

Central Ridge

A prominent ridge runs north-south through the center of Alphonsus and butts up against the central peak. Faint lineaments crisscross the ridge, giving it a herringbone structure similar to that formed by secondary impact crater chains. A chain of craters marks most of the western boundary of the ridge with the crater floor. These appear to be secondary craters from the Arzachel impact event. Ideas as to the origin of the ridge vary. A possible volcanic origin was suggested by *Howard and Masursky* (1968). *Carr* (1969) suggested that it formed as a result of isostatic readjustment of the crater floor, and was further affected by the Imbrium impact.

A volcanic origin for this ridge seems unlikely because the albedo and multispectral imagery data indicate that the ridge is similar in composition to the wall and floor deposits in Alphonsus. While the ridge was uplifted along with the rest of the Alphonsus floor, uplift was not responsible for the origin of the ridge. Stratigraphic relations indicate that the ridge was formed prior to the emplacement of the light plains unit on the crater floor. Floor uplift occurred much later, after the deposition of Arzachel ejecta and during the flooding of the nearby Nubium basin. We suggest that the ridge was largely formed by the sculpting of Alphonsus floor features by Imbrium secondary projectiles. The resulting ridge-feature was immediately embayed by the Imbrium debris surge that formed the light plains unit (Fig. 9) on the floor of Alphonsus. Much later, the ridge was modified by the formation of Arzachel secondaries and uplifted along with the rest of the Alphonsus floor as a result of magma intrusion at depth.

Floor Fractures

Numerous floor fractures and lineaments are present on the floor of Alphonsus (*Howard and Masursky,* 1968; *Schultz,* 1976a,b; *Whitford-Stark,* 1982). The largest of these fractures are over 50 km long and more than a km wide and are aligned north-south, while the smaller, and more abundant fractures are predominately aligned in a west-northwest/east-southeast direction. The north-south-trending features parallel the central ridge of Alphonsus as well as the slight north-south elongation

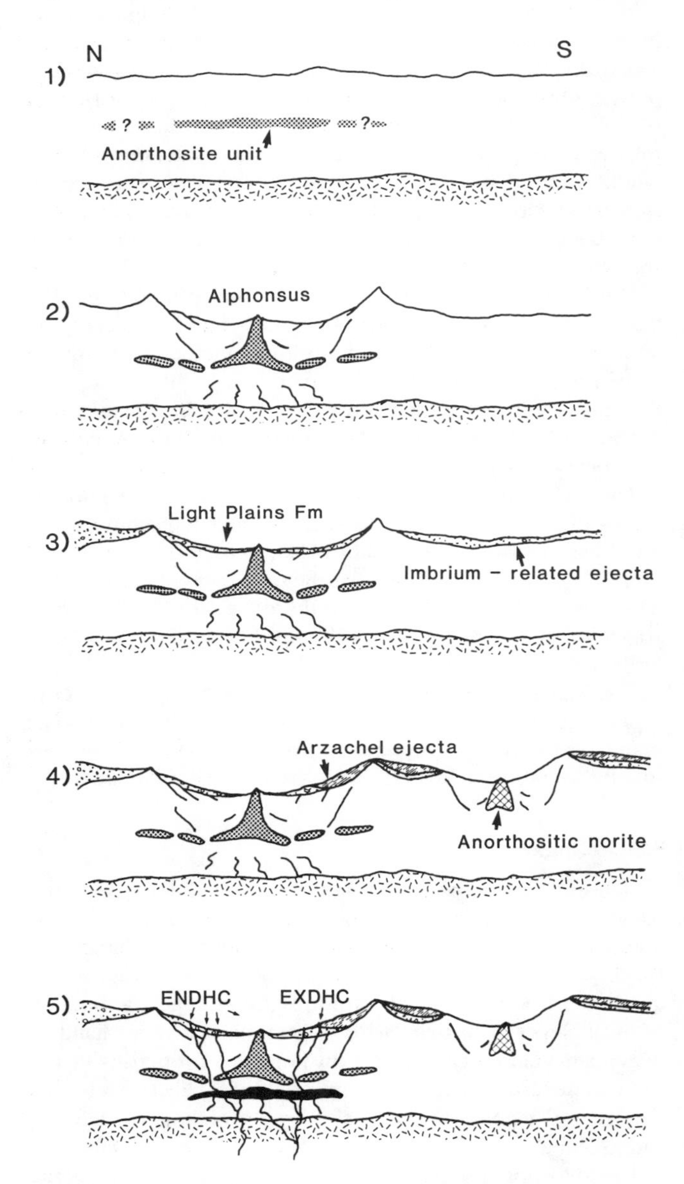

Fig. 9. Time-sequence cross-section of the Alphonsus region. Section is north to south. Stages: (1) pre-Alphonsus, discontinuous (?) anorthosite unit exists at depth, (2) Alphonsus impact, central peak formation, floor fracturing and brecciation of substrata, (3) Imbrium impact and deposition of Imbrium-related ejecta material, formation of light plains unit, (4) Arzachel impact, deposition of Arzachel ejecta on southern floor of Alphonsus, formation of anorthositic norite central peak, and (5) intrusion of magma at depth penecontemporaneous with nearby Mare Nubium infilling, vulcanian eruptions on floor of Alphonsus deposit pyroclastic material about *endogenic* craters (ENDHCs). Later, an impact directly over an ENDHC forms an *exogenic* dark-halo crater (EXDHC).

of the rim of Alphonsus crater. The floor fractures about the perimeter of the crater floor are associated with endongenic craters and pyroclastic deposits of various sizes. *Howard and Masursky* (1968), *Carr* (1969), and *McCauley* (1969) suggested that these floor fractures represent extension of the Alphonsus crater floor following the Imbrium and Arzachel impacts. Data currently available for this region indicate that the fractures on the floor of Alphonsus are generally graben bounded by normal faults. These graben formed during the uplift and resulting extension of the Alphonsus crater floor contemporaneous with magma intrusion in the local substratum.

Floor Uplift

Topographic data indicates that the floor of Alphonsus has been uplifted since its initial formation. A crater the diameter of Alphonsus (118 km) should be much deeper than the 1-3 km shown on the topographic maps (*Wu et al.,* 1972). This may be due to several factors: (1) infilling of the crater floor by material related to the Imbrium and other nearby impacts increased the floor elevation, or deceased the rim-to-floor depth of the crater; (2) erosion by Imbrium secondary-forming projectiles reduced the rim height of Alphonsus; and (3) floor uplift was caused by the differential intrusion of magma at depth beneath the crater floor at the same time as basalts were flooding the nearby Nubium basin. Floor fractures created in association with this floor uplift provided paths of preferred weakness for the magma to reach the surface in later localized vulcanian eruptions.

Light Plains

The floor of Alphonsus is mantled by a layer of moderately high albedo material. This unit was named the Cayley plains after similar deposits found around Julius Ceasar (*Morris and Wilhelms,* 1967). Light plains deposits are abundant in certain regions of the lunar highlands. They are distinguished by their moderately high albedoes and smooth surfaces. Among the origins that have been discussed for these light mantling deposits are (1) primary ejecta from large multiring basin and large crater formation (*Eggleton and Schaber,* 1972; *Hodges et al.,* 1973; *Chao et al.,* 1975), (2) highlands volcanism (*Hackman,* 1962), (3) debris surges resulting from local mixing by secondary craters (*Oberbeck,* 1974, 1975; *Oberbeck et al.,* 1975), (4) impact melt deposits (*Howard and Wilshire,* 1975; *Hawke and Head,* 1977), and (5) local crater and mass wasted material (*Head,* 1974). We suggest that the light plains deposits on the floor of Alphonsus were emplaced by the Imbrium debris surge that was produced by the impact of low-angle secondary-forming projectiles, and that as a result they contain a large amount of local material.

Adjacent Features

Prior to and following the Alphonsus impact event were several other sizeable impacts. Immediately to the north is the pre-Imbrian aged Ptolemaeus. This crater is 140 km in diameter, has a flat floor, and no central peak. Herschel crater debris overlie the northern rim and floor of Ptolemaeus. Immediately to the south of Alphonsus is the Imbrian-aged Arzachel crater. Arzachel is 70 km in diameter, has a hummocky floor, several floor fractures, and an elongate central peak. The majority of the rolling floor is covered with smooth impact melt material. Ninety kilometers to the northeast of Alphonsus are the Nectarian craters Albategnius (85-km-diameter) and Klein (40-km-diameter). Both craters have a small, irregular shaped central peak, and a relatively flat floor. Also, like Alphonsus, the floors of both Klein and Albategnius are mantled by light plains material, generally thought to have been emplaced as a result of the Imbrium impact event.

Dark-Halo Craters

Localized dark mantle deposits (LDMD) are small (typically $<$250-500 km^2), low albedo units of pyroclastic origin (e.g., *Coombs,* 1988; *Hawke et al.,* 1989) that are widely scattered across the lunar surface. The majority of these deposits are concentrated about the perimeters of the major lunar maria and in the floors of large Imbrian and pre-Imbrian impact structures. The smooth, block-free, low-albedo deposits, typified by those located in the floor of Alphonsus, are generally associated with small endogenic dark-halo craters that are commonly aligned along floor fractures. Endogenic dark-halo craters are characterized by their small size ($<$3 km), noncircular shapes, lack of rays, and alignment along crater floor fractures (*Head and Wilson,* 1979; *Coombs,* 1988; *Hawke et al.,* 1989). Other, less typical LDMD may be found as (1) isolated patches in highlands areas, (2) isolated patches on the maria, and (3) isolated patches in the highlands adjacent to mare deposits.

Sizes of the LDMD vary. Individual (from one source vent) LDMD are typically $<$100 km^2, while larger, coalesced, LDMD are in general $<$1000 km^2, with most falling between 250 and 500 km^2. Studies by *Head and Wilson* (1979) and *Wilson and Head* (1981) have suggested that these deposits are the result of vulcanian eruptions. In this style of eruption gas becomes trapped in a capped magma body, leading to explosive decompression and the subsequent emplacement of a pyroclastic deposit about the perimeter of an endogenic source crater.

Eleven endogenic dark-halo craters are present along the floor fractures within Alphonsus. These vents are all less than 3 km in diameter, and have noncircular shapes. Deposits peripheral to the vents are of different sizes; some are individual and some are coalesced. The largest coalesced deposit, along the western edge of the crater floor, covers an area of roughly 160 km^2, and the largest individual deposit in southeast Alphonsus covers roughly 45 km^2. These deposits are some of the youngest deposits on the floor of Alphonsus as they superpose other geologic units. They are composed of olivine-rich basaltic material derived in a vulcanian eruption (e.g., *Coombs,* 1988; *Coombs et al.,* 1988; *Hawke et al.,* 1989). These pyroclastic deposits may be of significance in the search for lunar resources such as Fe, Ti, He^3 and O_2. See *Coombs and Hawke* (1988a) and *Hawke et al.* (1990) for further reference.

Another crater, previously identified and mapped as an Eratosthenian-aged impact crater by *Howard and Masursky*

(1968), was recently identified as an exogenic (impact) dark-halo crater (*Coombs et al.,* 1989). This crater, Alphonsus A, is located on the Arzachel ejecta blanket in southern Alphonsus. It is 3.5 km in diameter and is surrounded by a 2-5-km-wide dark halo. The albedo of the dark halo is not as low as those of the LDMD elsewhere in the floor of Alphonsus. The southern portion of this dark halo butts up against the southern crater rim of Alphonsus. This exogenic dark-halo material may be composed of a mixture of pyroclastic material and highlands debris. The visible spectrum obtained for the dark halo (Fig. 6) is consistant with, but does not prove, this interpretation. An LDMD may have existed in the preimpact target site of Alphonsus A. Near-IR reflectance spectra will have to be obtained to resolve this problem.

GEOLOGIC HISTORY OF THE ALPHONSUS REGION

The Alphonsus region has undergone a complex history of formation. Beginning shortly after the lunar crust formed, it was bombarded by many impacting bolides. The first large basin to form within our study region was the Nubium basin. The main Nubium ring is located ~30 km southwest of Alphonsus. Following Nubium, other large impacts formed the basins known as South Procellarum, Cognitum, Crisium, Humorum, Nectaris, and Serentitatis. Each of these contributed more than one meter of ejecta to the Alphonsus region (*McGetchin et al.,* 1973; *Hawke and Head,* 1977).

Alphonsus formed in the early Nectarian Period, or more than 3.85 b.y. ago. Very little ejecta from Alphonsus is visible today due to the bombardment and mare infilling that took place within the region during the Imbrian, Eratosthenian, and Copernican periods. The formation of the Imbrium basin had a large effect on the Alphonsus region. Material emplaced as a result of the Imbrium impact was deposited throughout this area as evidenced by the smooth plains unit (Cayley Plains material) present on the crater floor and Imbrium sculpture about the region.

Arzachel crater, immediately to the south of Alphonsus, formed sometime in the Lower Imbrian (*Holt,* 1974) to Eratosthenian (*Howard and Masursky,* 1968). Its ejecta blanket is as wide as the crater cavity and blankets the southern rim and floor of Alphonsus. The crater floor is densely covered with small impact craters, and the presence of at least four prominent floor fractures suggest that it underwent a period of uplift following impact. The curved elongate central peak of Arzachel roughly parallels the curvature of the western crater rim.

Perhaps the most interesting feature in Alphonsus is the endogenic dark-halo craters scattered about the perimeter of the floor. The 11 endogenic dark-halo craters present along floor fractures and their associated dark mantling deposits formed late in the development of Alphonsus. These deposits mantle the light plains and cover the floor fractures as well as other floor lineaments. The nature and composition of the exogenic crater (Alphonsus A) on the southern rim of Alphonsus suggest that an endogenic DHC was located in the Alphonsus A target site, and that Alphonsus A excavated part of this material upon impact.

CONCLUSION

Alphonsus is host to a myriad of lunar geologic features. The 11 endogenic (volcanic) dark-halo craters associated with the floor fractures have become type examples for other localized pyroclastic deposits on the lunar nearside (e.g., *Coombs et al.,* 1987, 1988; *Coombs,* 1988; *Hawke et al.,* 1989). This study identified a previously unknown exogenic dark halo crater on the southern floor.

Visible and near-infrared reflectance spectra, coupled with the radar images of the region helped distinguish the composition of the various geologic units within the Alphonsus region. Overall, the Alphonsus area is noritic in composition. The majority of the sites analyzed are orthopyroxene-rich. The Alphonsus central peak, however, is composed solely of anorthosite, suggesting that a unit of the material must have been present at depth below the target site prior to impact. Similarly, the spectra for the Arzachel central peak suggest that anorthositic norite was present below that target site. These compositions, together with the noritic signatures from the crater rims and floors of Ptolemaeus and Albategnius, and the gabbroic composition of the Alpetragius plains suggest a heterogeneous composition of the crust in this area. The central ridge of Alphonsus is unusual in that no other crater of this size possesses such a feature. However, it is concluded in this study that the central ridge formed largely as a result of the Imbrium impact. The deposition of Imbrium primary and secondary ejecta aided in resurfacing the region, as evidenced by the Imbrium sculpture and light plains material present within Alphonsus. The Alphonsus impact occurred more than 3.85 b.y. ago. This event was followed by the Arzachel impact that deposited a blanket of ejecta over the southern rim of Alphonsus. Later localized vulcanian eruptions occurred contemporaneous with and/or just after the formation of the floor fractures that were produced as a result of floor uplift due to magma intrusion.

Acknowledgments. The authors would like to thank the telescope scheduling committees for allotting C.R.C. and B.R.H. time on the University of Hawaii 2.2-m Telescope on Mauna Kea, and S.H.Z. time at the MIT Haystack Observatory. Also, thanks are extended to L. Gaddis, J. Whitford-Stark, and R. Morris for their helpful reviews. This research was undertaken as part of a National Research Council Fellowship awarded to C.R.C. This work was supported by NASA grants NAGW-237 and NAGW-7323. This is PGD Publication No. 573, and Hawaii Institute of Geophysics Contribution No. 2194.

REFERENCES

Adams J. B. and McCord T. B. (1971) Optical properties of mineral separates, glass, and anorthositic fragments from Apollo mare samples. *Proc. Lunar Sci. Conf. 2nd,* pp. 2183-2195.

Bell J. F. and Hawke B. R.(1984) Lunar dark-haloed impact craters: Origin and implications for early mare volcanism. *J. Geophys. Res., 89,* 6899-6910.

Bell J. F. and Hawke B. R.(1987) Recent comet impacts on the Moon: The evidence form remote-sensing studies. *Publ. Astron. Soc. Pac., 99,* 862-867.

Carr M. H. (1969) Geologic map of the Alphonsus region of the Moon. *U.S. Geol. Survey Misc. Geol. Map I-599.*

Chao C., Boynton W. V., Sundberg L. L., and Wasson J. T. (1975) Volatiles on the surface of Apollo 15 green glass and trace element distributions amoung Apollo 15 soils. *Proc. Lunar Sci. Conf. 6th,* pp. 1701-1727.

Clark R. N. (1980) A large scale interactive one-dimensional array processing system. *Publ. Astron. Soc. Pac., 92,* 221-224.

Coombs C. R. (1988) Explosive volcanism on the Moon and the development of lunar sinuous rilles and their terrestrial analogs. Ph.D. dissertation, Univ. of Hawaii, Honolulu, 241 pp.

Coombs C. R. and Hawke B. R. (1988a) Explosive volcanism on the Moon: A review. *Proc. Kagoshima Intl. Conf. on Volcanoes,* Kagoshima, Japan, pp. 416-419.

Coombs C. R. and Hawke B. R. (1990) A Preliminary survey of lunar rilles: The search for intact lava tubes. In *Lunar Bases and Space Activities of the 21st Century.* Univelt, San Diego, in press.

Coombs C. R., Hawke B. R., and Gaddis L. R. (1987) Explosive volcanism on the Moon (abstract). *Lunar and Planetary Science XVIII,* pp. 197-198. Lunar and Planetary Institute, Houston.

Coombs C. R., Hawke B. R., and Owensby P. D. (1988) A recent survey of localized lunar dark mantle deposits (abstract). *Lunar and Planetary Science XIX,* pp. 209-210. Lunar and Planetary Institute, Houston.

Coombs C. R., Hawke B. R., Lucey P. G., and Head J. W. (1989) Geologic and remote sensing studies of the Alphonsus crater region (abstract). In *Lunar and Planetary Science XX,* pp. 185-186. Lunar and Planetary Institute, Houston.

Eggleton R. E. and Schaber G. G. (1972) Cayley Formation interpreted as basin ejecta. Pt. B of Photogeology, Chap. 29. In *Apollo 16 Preliminary Science Report,* pp. 29-7 to 29-16. NASA SP-315.

Gaddis L. R., Pieters C. M., Hawke B. R. (1985) Remote sensing of lunar pyroclastic mantling deposits. *Icarus, 61,* 461-489.

Hackman R. J. (1962) Geologic map and sections of the Kepler region of the Moon. *U.S.G.S. Map I-355,* (LAC-57).

Hartmann W. K. (1967) Lunar crater counts, I: Alphonsus. *Comm. Lunar Planet. Lab., Univ. Arizona, 6,* 31-38.

Hawke B. R. and Head J. W.(1977) Impact melt on lunar crater rims. In *Impact and Explosion Cratering* (D. J. Roddy, R. O. Pepin, and R. B. Merrill, eds.), pp. 815-841. Pergamon, New York.

Hawke B. R. and Lucey P. G. (1985) Spectral reflectance study of the Hadley-Apennine (Apollo 15) region. In *Reports of Planet. Geol. and Geophys. Prog.-1985,* pp. 523-525.

Hawke B. R., Cloutis E., Owensby P. D., Lucey P. G., Bell J. F., and Spudis P. D. (1984) Spectral reflectance studies of the Orientale region of the Moon: Preliminary results (abstract). In *Lunar and Planetary Science XV,* pp. 352-353. Lunar and Planetary Institute, Houston.

Hawke B. R., Lucey P. G., Bell J. F., and P. D. Owensby (1985) Spectral studies of the highlands around the Nectaris Basin: Preliminary results (abstract). In *Lunar and Planetary Science XVI,* pp. 329-330. Lunar and Planetary Institute, Houston.

Hawke B. R., Coombs C. R., Gaddis L. R., Lucey, P. G., Owensby P. D. (1989) Remote sensing and geologic studies of localized dark mantle deposits on the Moon. *Proc. Lunar Planet. Sci. Conf. 19th,* pp. 255-268.

Hawke B. R., Coombs C. R. and Clark B. (1990) Pyroclastic deposits: An ideal lunar resource. *Proc. Lunar Planet. Sci. Conf. 20th,* this volume.

Head J. W. (1974) Lunar dark-mantle deposits: Possible clues to the distribution of early mare deposits. *Proc. Lunar Sci. Conf. 5th,* pp. 207-222.

Head J. W. and Wilson L. (1979) Alphonsus-type dark halo craters: Morphology, morphometry and eruption conditions. *Proc. Lunar Planet. Sci. Conf. 10th,* pp. 2861-2897.

Hodges C. A., Muehlberger W. R., and Ulrich G. E. (1973) Geologic setting of Apollo 16. *Proc. Lunar Sci. Conf. 4th,* pp. 1-25.

Holt H. E. (1974) Geologic map of the Purbach quadrangle of the Moon. *U.S. Geol. Surv. Misc. Geol. Map I-822.*

Howard K. A. (1974) Fresh lunar impact craters: Review of variations with size. *Proc. Lunar Sci. Conf. 5th,* pp. 61-69.

Howard K. A. and Masursky H. (1968) Geologic map of the Ptolemaeus quadrangle of the Moon. *U.S. Geol. Surv. Misc. Geol. Inv. Map I-566.*

Howard K. A. and Wilshire H. G. (1975) Flows of impact melt at lunar craters. *U.S.G.S. J. Res., 3,* 237-257.

Lucey P. G., Hawke B. R., Pieters C. M., Head J. W., McCord T. B. (1986) A compositional study of the Aristarchus Region of the Moon using near-infrared reflectance spectroscopy. *Proc. Lunar Planet. Sci. Conf. 16th,* in *J. Geophys. Res., 91,* D344-D354.

McCauley J. E. (1969) Geologic map of the Alphonsus GA region of the Moon. *U.S.G.S. Surv. RLC Map 15, I-586.*

McCord T. B. (1968) Color differences on the lunar surface. Ph.D. dissertation, California Institute of Technology, Pasadena. 181 pp.

McCord T. B., Pieters C. M., and Feierberg M. A. (1976) Multispectral mapping of the lunar surface using ground-based telescopes. *Icarus, 29,* 1-34.

McCord T. B. and Clark R. N. (1979) Atmospheric extinction 0.65-2.50 μm above Mauna Kea. *Publ. Astron. Soc. Pac., 91,* 571-576.

McCord T. B., Clark R. N., Hawke B. R., McFadden L. A., and Owensby P. D. (1981) Moon: Near-infrared spectral reflectance, a first good look. *J. Geophys. Res., 86,* 10,883-10,892.

McGetchin T. R., Settle M., and Head J. W. (1973) Radial thickness variation in impact crater ejecta: Implications for lunar basin deposits. *Earth Planet. Sci. Lett., 20,* 226-236.

Morris E. C. and Wilhelms D. E. (1967) Geologic map of the Julius Caesar quadrangle of the Moon. *U.S. Geol. Surv. Misc. Geol. Inv. Map I-510.*

Oberbeck V. R. (1974) Smooth plains and continuous deposits of craters and basins. *Proc. Lunar Sci. Conf. 5th,* pp. 111-136.

Oberbeck V. R. (1975) The role of erosion and sedimentation in lunar stratigraphy. *Rev. Geophys. Space Phys., 13,* 337-362.

Oberbeck V. R., Morrison R. H., and Hörz F. (1975) Transport and emplacement of crater and basin deposits. *The Moon, 13,* 9-26.

Peterfreund A. R. (1976) Alphonsus dark-haloed craters: Examples of isolated dark mantle sources (abstract). *Eos Trans. AGU, 57,* 275.

Pieters C. M. (1986) Composition of the lunar highland crust from near-infrared spectroscopy. *Rev. Geophys., 24,* 557-578.

Pieters C. M. (1989) Compositional stratigraphy of the lunar highlands crust. (abstract). In *Lunar and Planetary Science XX,* pp. 848-849. Lunar and Planetary Institute, Houston.

Pieters C. M. and Hörz F. (1985) Spectral characteristics of experimental regoliths (abstract). *Lunar and Planetary Science XVI,* pp. 661-662. Lunar and Planetary Institute, Houston.

Salisbury J. W., Adler J. E. M., and Smalley V. G. (1968) Dark-haloed craters on the Moon. *Mon. Not. R. Astron. Soc., 138,* 245-249.

Schultz P. H. (1976a) *Moon Morphology.* Univ. of Texas, Austin. 626 pp.

Schultz P. H. (1976b) Floor-fractured lunar craters. *The Moon, 15,* 3-4, 241-273.

Singer R. B. (1981) Near-infrared spectral reflectance of mineral mixtures: Systematic combinations of pyroxenes, olivine, and iron oxides. *J. Geophys. Res., 86,* 7967-7982.

Spudis P. D., Hawke B. R., and Lucey P. G. (1984) Composition of Orientale basin deposits and implications for the lunar basin-forming process. *Proc. Lunar Planet. Sci. Conf. 15th,* in *J. Geophys. Res., 89,* C197-C210.

Spudis P. D., Hawke B. R., and Lucey P. G. (1989) The lunar Crisium Basin: Geology, rings and deposits (abstract). In *Lunar and Planetary Science XX,* pp. 1042-1043. Lunar and Planetary Institute, Houston.

Whitford-Stark J. W. (1982) A preliminary analysis of lunar extra-mare basalts: Distribution, compositions, ages, volumes, and eruption styles. *Moon and Planets, 26,* 323-338.

Wilhelms D. E. (1987) The geologic history of the Moon. *U.S.G.S. Prof. Paper 1348.*

Wilson L. and Head J. W. (1981) Ascent and eruption of basaltic magma on the Earth and Moon. *J. Geophys. Res., 78,* pp. 2971-3001.

Wood C. A. (1968) Statistics of central peaks in lunar craters. *Comm. Lunar Planet. Lab., Univ. of Arizona, 7,* pp. 157-160.

Wu S. S. C., Schafer F. J., Jordan R., and Nakata G. M. (1972) Photogrammetry using Apollo 16 orbital photography. In *Apollo 16 Preliminary Science Report,* pp. 30-5 to 30-10. NASA SP-315.

Zisk S. H. (1972) Lunar topography: First radar-interferometer measurements of the Alphonsus-Ptolemaeus-Arzachel region. *Science, 178,* 977-980.

Zisk S. H., Pettengill G. H., and Catuna G. W. (1974) High-resolution radar map of the lunar surface at 3.8-cm wavelength. *The Moon, 10,* 17-50.

Proceedings of the 20th Lunar and Planetary Science Conference, pp. 175-185
Lunar and Planetary Institute, Houston, 1990

The Volcanotectonic Evolution of Mare Frigoris

J. L. Whitford-Stark

Geology Department, Sul Ross State University, Alpine, TX 79832

The basalt fill within Mare Frigoris is thick (>400 m) and was emplaced in at least three major episodes within topographic lows overlying a thick lunar lithosphere. The oldest identifiable unit is dominated by titanium-poor lavas and was succeeded by basalts of intermediate and titanium-rich composition. The lavas appear to have been emplaced largely by flood-style eruptions. The topographic low of western Frigoris was probably created by the collapse of highland blocks into the Imbrium impact cavity. The line of separation may have been an Imbrium ring fracture or a ring fracture of the older Procellarum basin.

PREVIOUS WORK

Mare Frigoris is the fifth largest, but one of the least studied, of the lunar maria, covering approximately 320,000 km^2 of the nearside of the Moon between latitudes 50° and 60°N. Frigoris is elongate, having an average width of about 200 km and a length of about 1800 km (Fig. 1). Parts of Frigoris have been mapped by *Ulrich* (1969), *Lucchitta* (1972), and *M'Gonigle and Schleicher* (1972), and a small-scale geologic map of the entire mare has been made by *Lucchitta* (1978). A summary of the earlier spectral and age data for the Moon has been presented by *Wilhelms* (1987).

On the basis of shape, the lunar maria can be separated into regular and irregular types (*Head,* 1976a), the regular being circular in outline, representing infilled impact basins. Frigoris appears to be composite in being regular at its eastern end and irregular in the west. *Whitford-Stark and Fryer* (1975) hypothesized that the topographic low of Frigoris west of the crater Aristoteles originated through the horizontal separation of large crustal blocks. This was based on a computer reconstruction employing the match of surface geology on both sides of the mare. Subsequently, *Whitford-Stark* (1981a) suggested that the horizontal separation could have arisen through collapse of the south Frigoris highland block into the Imbrium impact cavity, perhaps along a ring fault concentric with the Imbrium impact cavity. The suggestion that a large, pre-Nectarian, Procellarum impact basin (*Whitaker,* 1981) exists offers the alternative possibility that the south Frigoris blocks separated along the line of a Procellarum ring fracture. Various authors (see summary in *Spudis et al.,* 1988) have attempted to relate the western Frigoris topographic low to the Imbrium impact structure based on the topographic relationships observed at the present day. It is the author's opinion that these models fail in that they employ improbable

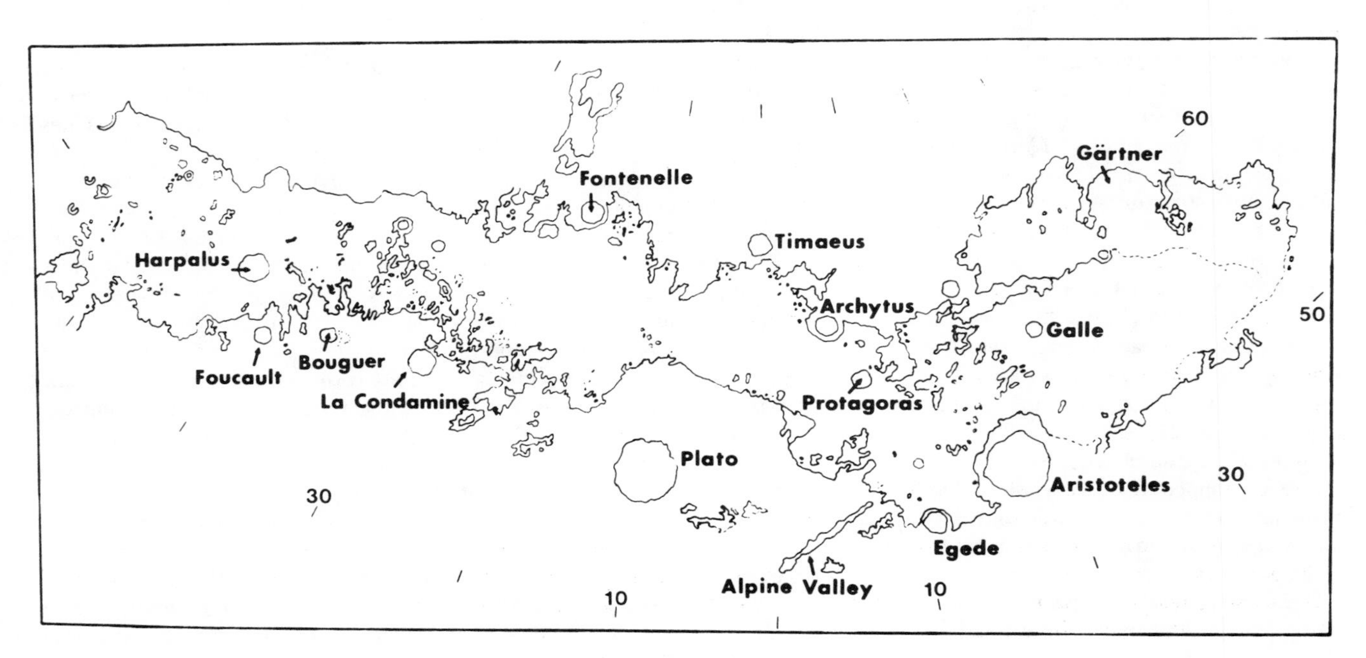

Fig. 1. Index map of Mare Frigoris on polar stereographic projection.

Fig. 2. Earth-based photograph of central Frigoris from the *Consolidated Lunar Atlas* (*Kuiper et al.,* 1967). The boundary between eastern and western Frigoris is obscured by the ejecta blanket of the crater Aristoteles (89-km diameter). Note also the flooded crater Egede to the left of Aristoteles and the northern end of the Alpine valley on the extreme left of the photo.

egg-shaped ring geometries, have "split" rings, or have rings that are in part topographic highs traced laterally into topographic lows.

A boundary between eastern and western Frigoris occurs at the location of the crater Aristoteles (Fig. 2) where there are changes in both surface compositions and tectonic features. The reason for the eastern Frigoris topographic low also is not obvious. *Spudis et al.* (1988) have illustrated two Imbrium-radial "megastructures" approximating the northern and southern boundaries of eastern Frigoris but do not discuss their significance. If these megastructures are faults, it is possible that eastern Frigoris is a large graben. Alternatively, it may represent an old, infilled impact basin heavily modified by subsequent impact events. There are no obvious rings marking the existence of such a basin, but there is an approximately circular area that encompasses eastern Frigoris and patches of the unit mapped as Ip2 by *Lucchitta* (1978) to the north of eastern Frigoris.

I have mapped the fill within Frigoris and the adjacent highlands using Lunar Orbiter and Earth-based photographs and multispectral images. There is no Apollo orbital coverage of Frigoris and the Lunar Orbiter photographs of eastern Frigoris are generally of poor quality. The high latitudes of Frigoris also allow only oblique Earth-based imagery of the mare.

THE FRIGORIS FILL

Age and Composition

The oldest recognizable unit within Frigoris is the light plains unit (Fig. 3) covering a large area of the northern portion of eastern Frigoris and occurring as isolated patches along the southern "shore." *Lucchitta* (1978) suggested that the plains might consist of a thin layer of ejecta from the Orientale impact event overlying mare basalts. The lack of dark-halo craters on the plains presents problems for such an interpretation since such craters might be expected if there were a subsurface composed of mare basalts (*Schultz and Spudis,* 1979; *Bell and Hawke,* 1984). However, the widespread distribution of ejecta from young, rayed craters in the highlands around eastern Frigoris may have obscured any dark halos (*Lucchitta,* 1978). At the wavelengths of the multispectral imagery, the surface of the plains unit is indistinguishable from that of the adjacent mare basalt, lending some support to this model. An alternative suggestion (*Lucchitta,* 1978) is that the thin Orientale ejecta covers earlier ejecta derived from the Imbrium and Humboltianum impact events. A geologically similar plains unit is located to the north of the crater Taruntius near Mare Fecunditatis. *Hawke and Spudis* (1980) were able to demonstrate that the Apollo orbital compositional data for the Taruntius region could be modeled by a mixture of 46% highlands material and 54% mare material. The nature of the plains unit in Frigoris is therefore enigmatic; all that can be added from the present study is that the surface of the plains unit is poor in titanium.

Spectra of the basalt units within Frigoris have been obtained by *Pieters and McCord* (1976). Additionally, a part of the Frigoris surface has been characterized on multispectral images by *Johnson et al.* (1977) and a simplified map of the major basalt types, based on a variety of spectral reflectance data, has been presented by *Pieters* (1978) and revised by *Wilhelms* (1987). A problem with spectra and multispectral imagery is that they only provide compositional data relating to the upper soil layer rather than the bedrock (see *Adams et al.,* 1981). This constraint is particularly important in Frigoris since the mare is extensively covered by rays from highland craters and from large craters within the mare (Harpalus and Aristoteles, Fig. 2 and 3). The extent to which these spectra are representative of the underlying basalt is undoubtedly dependent upon the percentage of highland material mixed with the locally derived soil and the extent to which impact gardening has taken place. Current estimates suggest that while vertical mixing is relatively efficient, lateral mixing is inefficient, with more than 50% of the regolith being derived from distances of less than 3 km (*Taylor,* 1982). Therefore, in those areas that have not suffered a recent influx of ejecta from a large impact event, the soil spectra is largely determined by the underlying bedrock (*Rhodes,* 1977). Another problem created by the large number of secondary craters within Frigoris is that the relative ages of the different units cannot be readily established by the crater-counting technique; a problem that is compounded by the poor images.

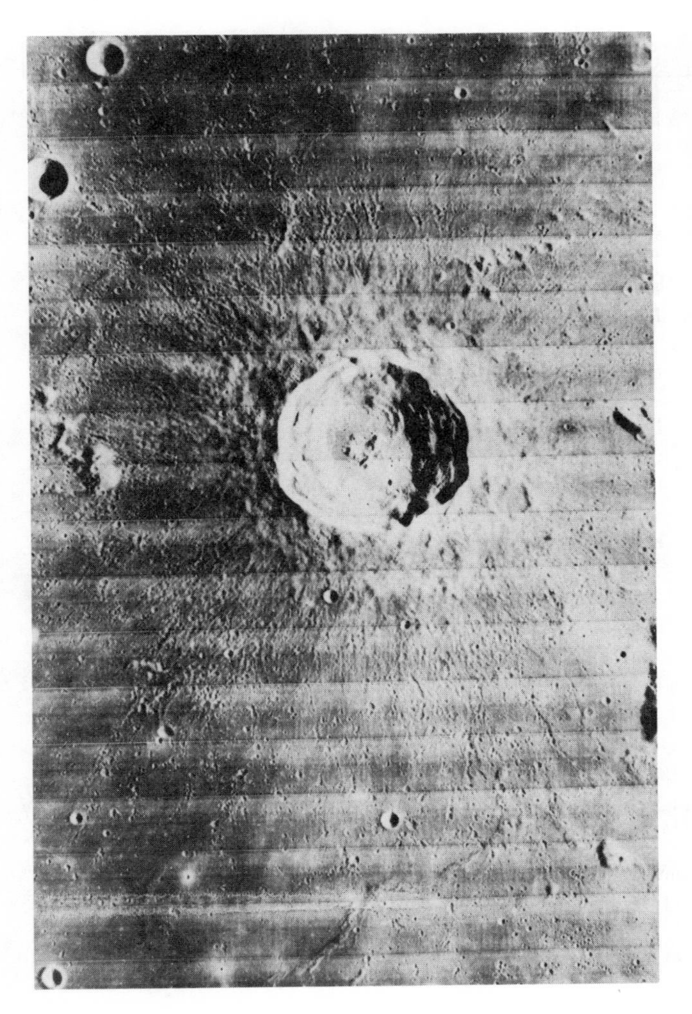

Fig. 3. Lunar Orbiter IV-163 H_3 photograph of the crater Harpalus (39 km diameter) with its extensive ejecta blanket covering much of the extreme western end of Frigoris.

The secondary craters have been employed, however, to determine the relative ages of the mare units on the basis of superposition relationships in a manner similar to that originally outlined by *Shoemaker and Hackman* (1962). While there are no lava flow fronts visible on available imagery, there are sharply defined boundaries on the multispectral images that are interpreted as approximating the boundaries of lava flows.

A geologic map of the mare basalt units is presented in Fig. 4. In this figure, red indicates titanium-poor basalts and blue, titanium-rich. Data that have been employed to characterize the unit are presented in Table 1. The units are described in what is interpreted as reverse stratigraphic order, since the younger units are the better exposed and least modified.

The youngest unit is titanium-rich, exposed in western Frigoris (Fig. 4). This unit is part of the Roris Basalt Member of the Sharp formation of Oceanus Procellarum (*Whitford-Stark and Head,* 1980). Part of the area demarcated as the young titanium-rich unit on Fig. 4 was designated as "LIS-" on the unit map presented by *Pieters* (1978). In this designation, L indicates a low ultraviolet/visible ratio, the I an intermediate normal albedo, the S a strong 1-μm band, and the - indicates an unknown 2-μm response. On both multispectral imagery and color difference photography (see *Adams et al.,* 1981, their Fig. 2.2.6), this unit appears blue; that is, it has a high rather than a low ultraviolet/visible ratio (Table 1). The reason for this discrepency is unclear. It is possible that the lower resolution of the telescopic spectra (4 to 20 km vs. 2 km for the multispectral imagery; *Head et al.,* 1981, their Table 5.4.1) resulted in the sampling of the extensive ejecta blanket of the large (39-km diameter), Copernican-aged crater, Harpalus (Fig. 3).

Another area of titanium-rich basalt is located at the eastern end of western Frigoris around 5°E, 53°N (Fig. 4). This unit is also considered to be young on the basis of its low crater density. It also has a low albedo.

The remaining intermediate- and low-titanium basalts are mapped as older than the titanium-rich basalts on the basis of their greater crater density and superposition relationships. The assignment of relative ages to the intermediate- and titanium-poor units proved to be a more difficult task than defining the relative age of the titanium-rich unit. The reasons for this are that greater modification has led to an equalization of their impact crater densities and degradation of any original topographic boundaries. In a detailed examination of the

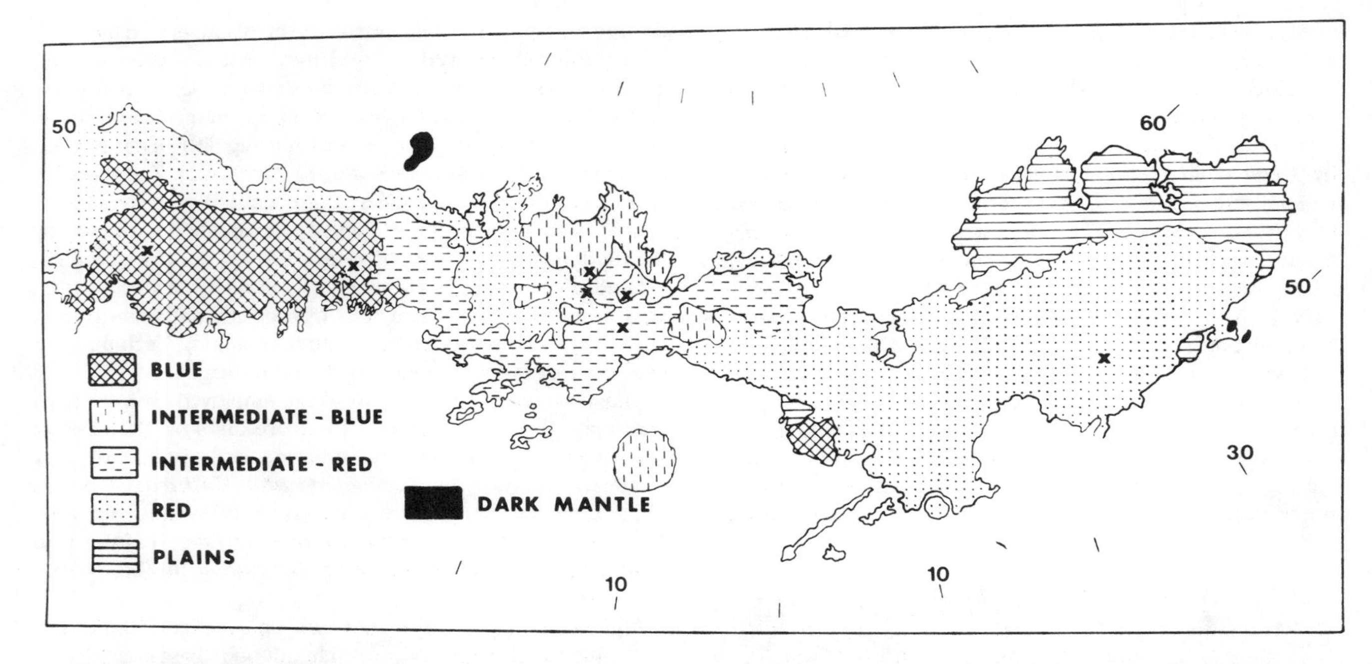

Fig. 4. Distribution of basalt units and plains unit within Mare Frigoris. The units are listed in approximate stratigraphic age (youngest at top). The red units are poor in titanium whereas the blue units are titanium-rich. The crosses indicated the approximate positions of telescopic spectra of *Pieters and McCord* (1976).

nearside of the Moon covered by high-resolution Apollo photography, *Wilhelms* (1980) demonstrated that a particular time span was dominated by, but not restricted to, basalts of a particular composition. Wilhelms' data show that basalts of intermediate composition were dominant in his Age 3, whereas red and very red basalts were predominant in the older Age 2. By analogy, therefore, it is inferred that the majority of the intermediate basalts within Frigoris are of younger age than the majority of titanium-poor basalts. Specific exceptions to this general pattern may include titanium-poor units (not illustrated on Fig. 4) to the east of the crater Timaeus and to the immediate north of the Alpine Valley. These areas appear to be less densely cratered relative to other areas of intermediate and titanium-poor composition. However, their small areal extents as well as the inadequate resolution of the available photography preclude a definitive determination.

By pursuing the analogy between the Frigoris sequence and the sequence within the adjacent Oceanus Procellarum (*Whitford-Stark and Head,* 1980) and the sequence outlined by *Wilhelms* (1980), the majority of the intermediate basalts are assigned an age of $3.3 \pm 0.3 \times 10^9$ years (Late Imbrian to Early Eratosthenian).

Parts of the areas herein mapped as intermediate in composition were described as mottled by *Pieters* (1978). The spectra of this area are complicated by the presence of partly flooded craters and island remnants of highland material. The difference between the two data sets is again probably a function of the lower resolution of the telescopic spectra.

The remaining titanium-poor unit covers the majority of eastern Frigoris and is preserved along the northern part of western Frigoris. The majority of the unit is considered to be of Age 2 of *Wilhelms'* (1980) scheme and is correlated with the Telemann Formation of Oceanus Procellarum (*Whitford-Stark and Head,* 1980) of Imbrian age.

Associated with, and probably the same age as, the titanium-poor basalts within the mare are volcanic dark-mantle deposits in the crater J. Herschel, located in the highlands to the north

TABLE 1. Characteristics of basalts within Mare Frigoris based on *Adams et al.* (1981), *Boyce* (1976), and *Pieters* (1978).

Units	UV/ Visible ratio	Color	Albedo	1 μm reflectance	TiO_2 content	Age	Other
Youngest	high	blue	low	strong	3- 11 wt.%	3.2 ± 0.2	Topographic boundaries
Medial	average	r/i/b	intermediate	average to strong	1-6 wt.%	3.3 ± 0.3	
Oldest	low	red	high	average to strong	<2 wt.%	3.6 ± 0.2	high crater density

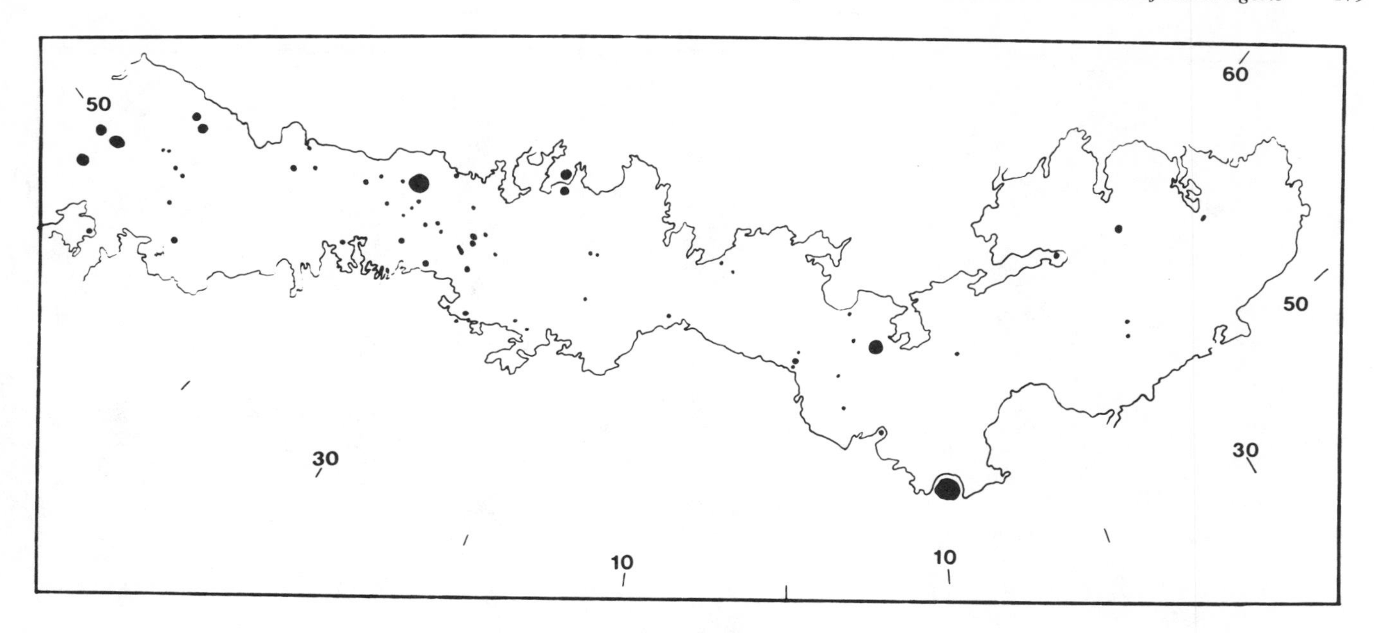

Fig. 5. Distribution of craters partially flooded by basalt within Mare Frigoris.

of western Frigoris [described in detail by *Hawke et al.* (1980)], and along the southern border of eastern Frigoris, east of Aristoteles A and B of *Hawke et al.* (1989). Both of these deposits are associated with fractures. *Gaddis et al.* (1985) illustrate possible mantling deposits surrounding the crater Plato.

There is no evidence at the surface of Frigoris for an early titanium-rich unit such as that found in Oceanus Procellarum, mapped by *Wilhelms* (1980) in equatorial regions, and returned as samples from the Apollo11 and 17 landing sites. If such a unit is present in Frigoris, it has been totally buried by younger deposits.

Fill Thickness

The amount of postimpact flooding has been estimated by the technique developed by *Whitford-Stark* (1980). Estimates by this method are believed to be accurate to within 30% of the derived value. Only three preflooding craters within

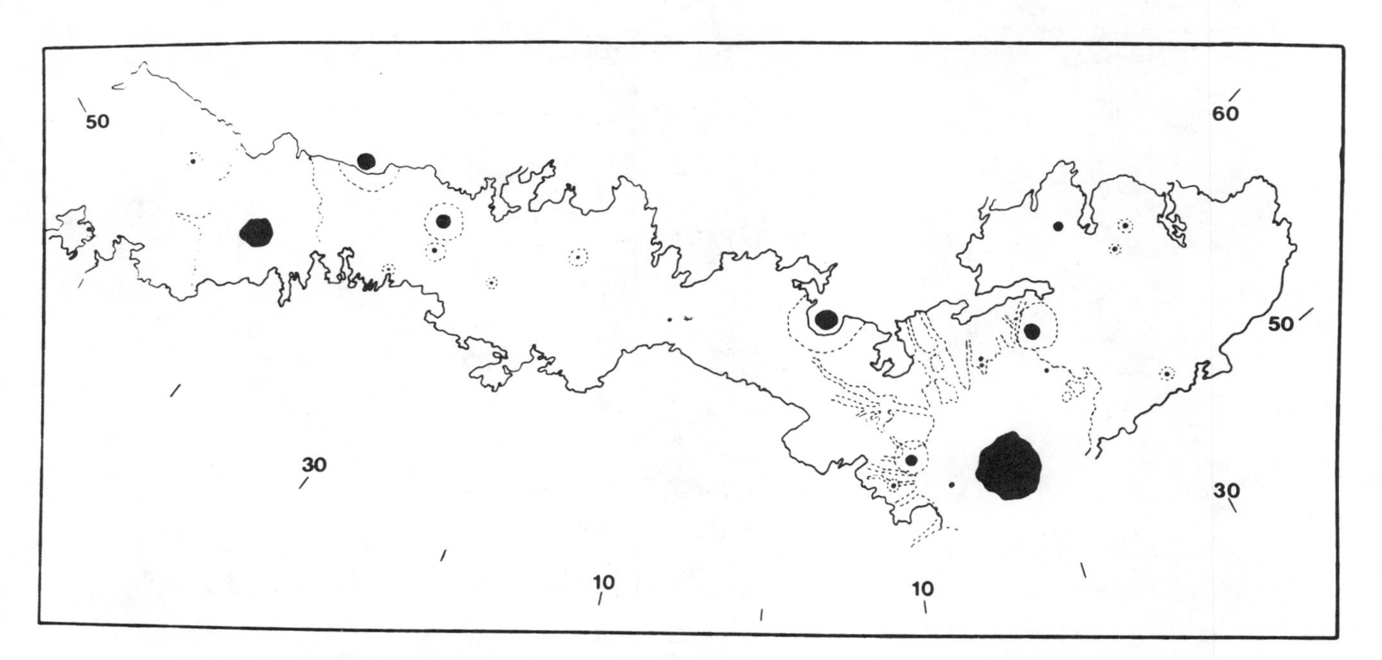

Fig. 6. Distribution of postmare craters within Mare Frigoris. Dashed lines indicate distribution of ejecta blankets.

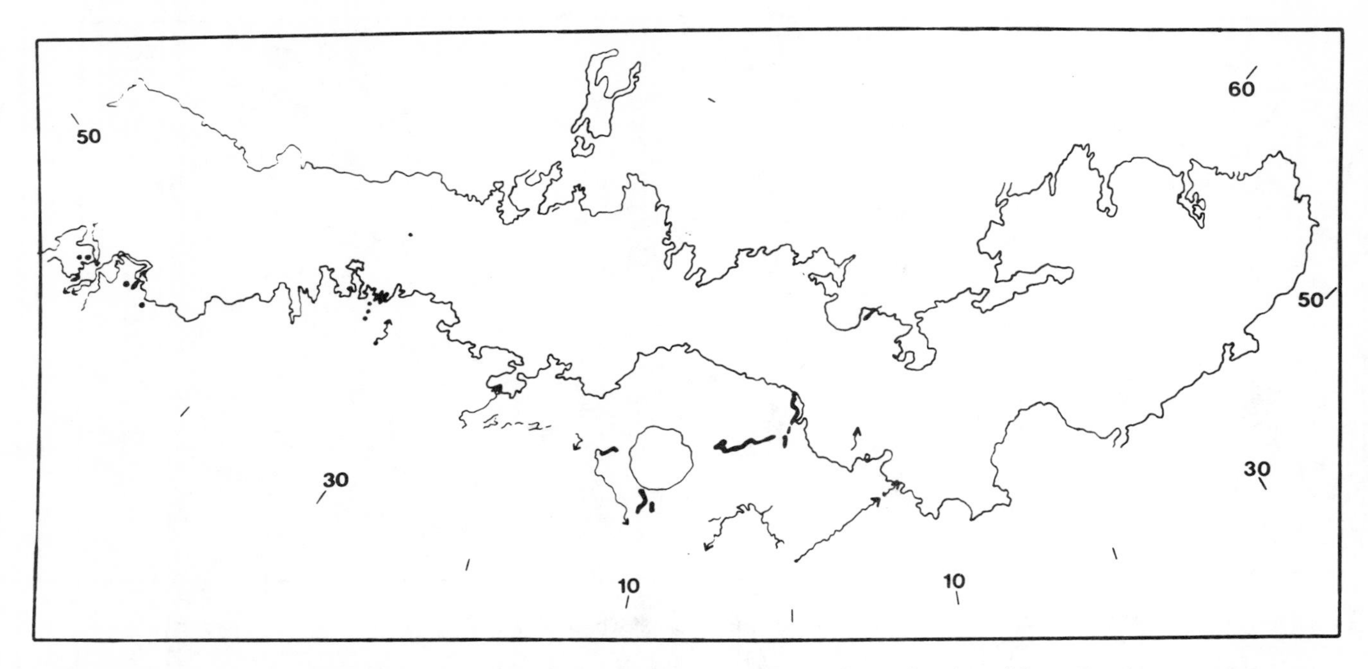

Fig. 7. Distribution of sinuous rilles within and around Mare Frigoris. Arrows indicate direction of flow. Also shown are locations of craters with peculiar floor morphologies (see text).

Frigoris are of sufficient size to enable reasonable estimates of fill thickness. Values of 430 m, 500 m, and 900 m were obtained, respectively, for the fill overlapping the craters J. Herschel F, Protagoras, and Egede. Since each of these craters is near the edge of the mare, they cannot be used to determine the thickness of the fill in the mare center.

Frigoris is somewhat peculiar in that there appears to be more, and larger, post-mare impact craters (eight greater than

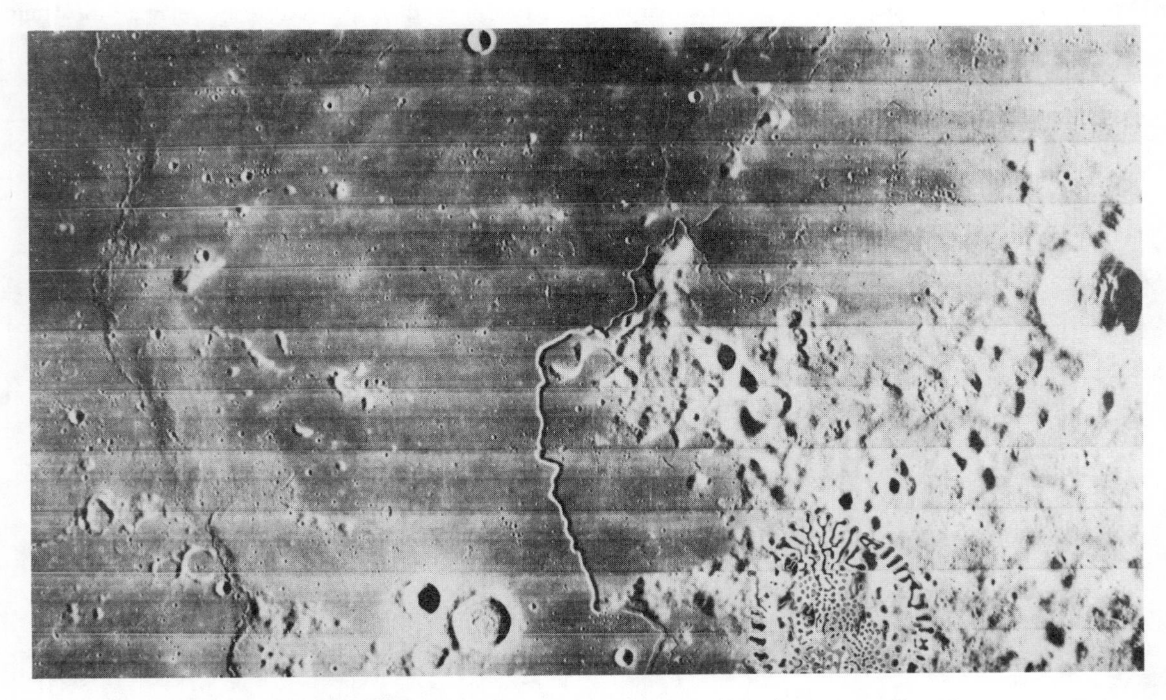

Fig. 8. Lunar Orbiter IV-163 H_3 photograph of western Mare Frigoris. Note the sinuous rille with a source crater close to the highland/mare contact. Note also toward the left side of the photo where the younger darker Frigoris flow has been restricted by an older, brighter mare ridge.

12.5-km diameter) than pre-mare craters (six greater than 12.5-km diameter; Figs. 5 and 6). There are several possible scenarios that could have led to this situation. First, an old surface with large pre-mare craters could be buried by a thick mare fill. Second, a young surface may not have been covered by large impact craters prior to lava flooding. This latter situation could have arisen if, as previously indicated (*Whitford-Stark,* 1981a), the Frigoris topographic low was produced by the collapse of the south Frigoris highlands into the Imbrium impact cavity. Third, premare craters may have been severely degraded by ejecta from the Imbrium impact event. In both the second and third scenarios, the premare floor of the topographic low would be only as old as the Imbrium impact event and would postdate the major phase of lunar meteorite bombardment. Until there is geophysical data for Frigoris, there is no way of substantiating any model. The tectonic features of Frigoris (see *Tectonics* section), however, would favor a thick basalt fill.

Eruption Style

In an analysis of the surface basalts within Mare Imbrium it was found (*Whitford-Stark,* 1980) that the lavas had three distinct morphologies: (1) those with flow lobes and channels, (2) those with sinuous rilles, and (3) those with neither of the previous characteristics. These three types were believed to represent three distinct eruption styles: (1) extremely rapid eruption rate, short duration; (2) lesser eruption rate, long duration flows on steep slopes; and (3) rapid eruption for long duration on shallow slopes. On the basis of available imagery, there are no flows of the first group within Frigoris. A sinuous rille flow is located at the extreme west of Frigoris (Figs. 7 and 8) within the young titanium-rich basalt. More typically, the sinuous rilles are located within the highlands and are preferentially developed on the southern side of Frigoris. These rilles terminate at, or before, the highland/mare boundary, probably as a result of a reduction in slope (*Whitford-Stark,* 1980). Four of the rilles were associated with the filling of the crater Plato. As Plato filled to near the rim with lava, it became easier for the magma to reach the lunar surface exterior to the crater rim. The sinuous rilles source craters are therefore concentrically arranged around the crater exterior. Two rilles are located within the Alpine Valley (Fig. 9). The southernmost rille has a source crater at the northern edge of the Montes Alpes and extends approximately 130 km to the north. This rille was partially infilled with lava from the younger sinuous rille flow located at the northern end of the valley. Other sinuous rilles are located in the topographically low highland separating Mare Frigoris from Mare Imbrium to the west of the crater Plato (Fig. 4).

The majority of the lavas within Frigoris do not have rilles, lobes or channels. Flows lacking such features may never have possessed them or those features may have been degraded or buried. Flows that originally lack sinuous rilles or channels are produced by flood-style eruptions, are thick, and tend to bury their own vents (*Greeley,* 1976). Alternatively, some flows, such as that at the northern end of the Alpine Valley, may have originated as sinuous rille-type flows and changed to featureless flows as the slope lessened.

Fig. 9. Lunar Orbiter IV-115 H_3 photograph of the Alpine Valley. Note sinuous rille source craters marked by solid arrows and area where rille has been infilled (hollow arrow). Valley approximately 200 km in length.

It is believed that the volcanic dark-halo craters, such as those within J. Herschel and in eastern Frigoris, were the product of a high gas-content, vulcanian eruption style (*Wilson and Head,* 1981; *Hawke et al.,* 1989). The dark-halo material within J. Herschel covers approximately 2500 km^2 and was probably erupted from four craters ranging in diameter from 4 to 8 km (*Hawke et al.,* 1980).

In the highlands near the western end of Frigoris, there are a number of small (<15-km diameter) craters with peculiar hummocky floor morphologies (Figure 10; see also *Schultz,* 1976a, Plate 23). It has been suggested (*Schultz and Mendenhall,* 1979) that this type of crater originated as secondaries produced by low-density, molten ejecta or by clustered impacts (*Schultz and Gault,* 1985). The close association of the craters with the mare, both in Frigoris and other maria (*Whitford-Stark,* 1981b) leads to the alternative suggestion that they may be the product of eruptions. The peculiar morphology could have arisen if the erupted material has a higher viscosity than typical mare basalts.

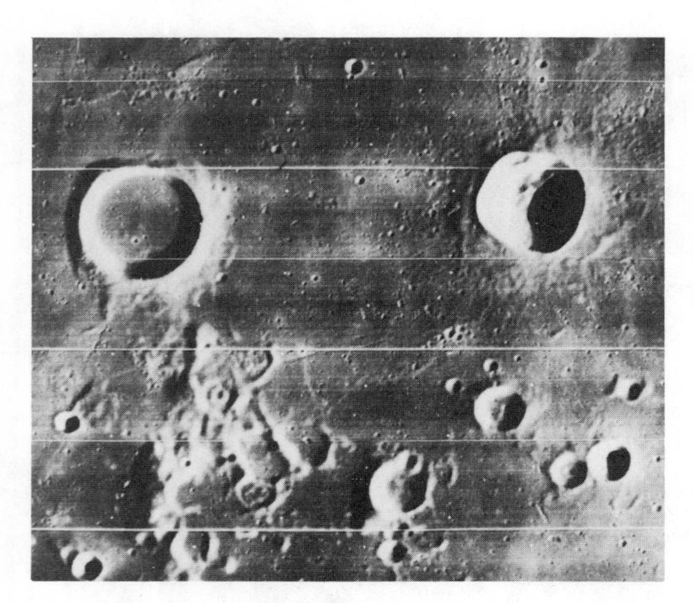

Fig. 10. Lunar Orbiter IV-152 H_1 of western Frigoris. The two larger craters are approximately 15 km in diameter. Note the hummocky topography within the smaller craters and the lava flooding of the left large crater.

TECTONICS

Rifts, Scarps, and Fractures

There are three rift-like structures located in the highlands adjacent to Frigoris (Fig. 11). These rifts have average widths of 10 km and trend at an angle (N55°E) to the shoreline of Frigoris. These features appear to have resulted from vertical crustal motion accompanying the formation of the Imbrium impact cavity since they are approximately radial to the center of Imbrium, postdate deposition of the Imbrium ejecta, and predate deposition of mare basalts.

Several fault scarps are present in the topographically low highlands to the east of the crater La Condamine (Fig. 11). These appear to postdate the basalt fill since they cut mare material and probably arose through subsidence induced by the weight of the mare fill (*Solomon and Head,* 1980).

There is a paucity of rifts within the highlands adjacent to western Frigoris. A small section of a partly buried rift does occur, however, on the southern side of the mare at approximately 1°E, 54.5°N. The formation of rifts follows the subsidence of mare basalts, which results in a tensional environment in the adjacent highlands (*Solomon and Head,* 1980). The lack of such features along western Frigoris could be ascribed to the mare basalts being of insufficient mass to have deformed the lithosphere and subside. This interpretation is, however, not supported by the presence of numerous, large mare ridges within western Frigoris (Fig. 12); features that also result from mare subsidence (*Solomon and Head,* 1980). *Lucchitta and Watkins* (1978) have shown that there was a global cessation of rift production at $3.6 \pm 0.2 \times 10^9$ years ago. A possible explanation for the lack of tectonic rilles peripheral to Frigoris is that a sufficient volume of fill to initiate subsidence was not introduced into the mare until after the global termination of rille production. Such an interpretation would be compatible with the age of $3.6 \pm 0.2 \times 10^9$ years assigned here to the apparently oldest basalts present at the surface of Frigoris.

Two sets of rifts are present within the light plains unit of northeastern Frigoris. The northernmost of these rilles is

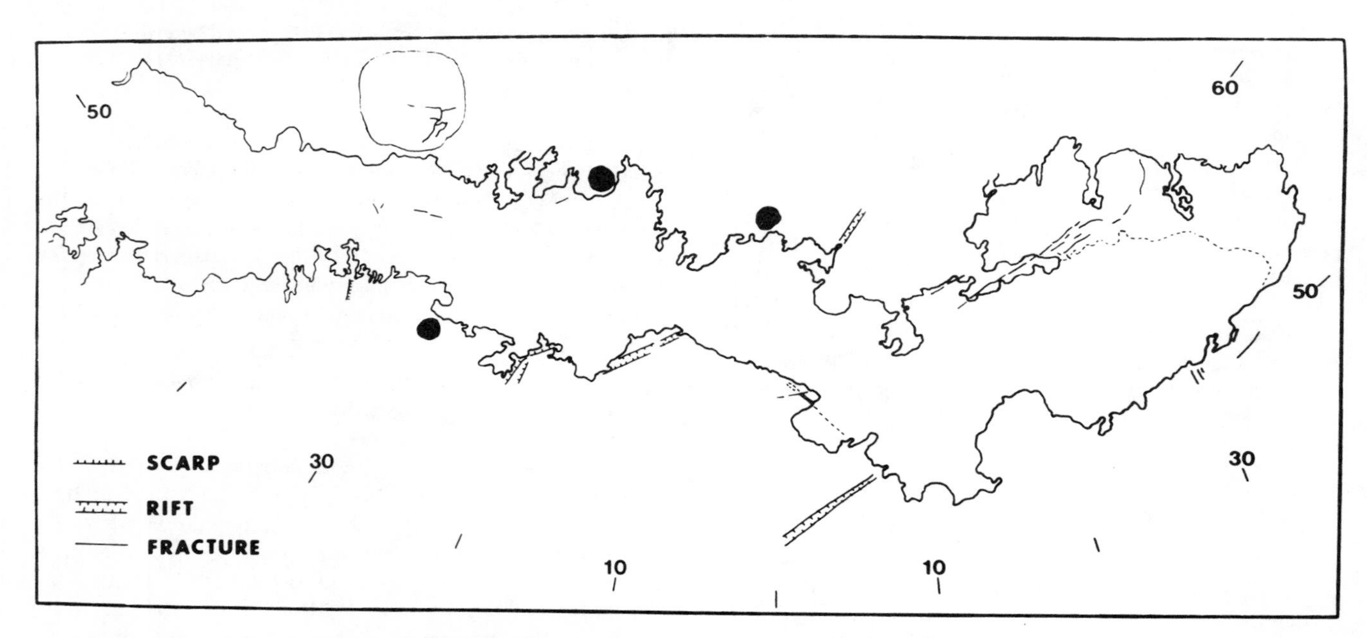

Fig. 11. Distribution of fractures within and adjacent to Mare Frigoris. Also in heavy shading are floor-fractured craters, excepting J. Herschel where the crater outline is indicated.

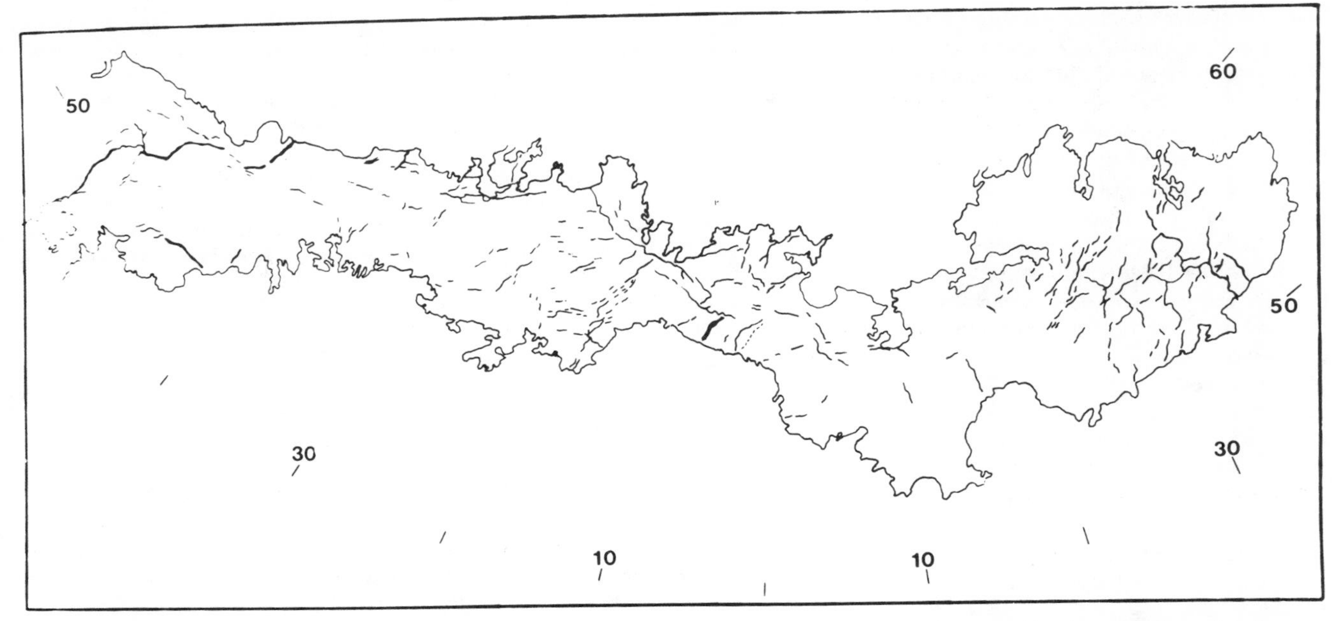

Fig. 12. Distribution of mare ridges in Mare Frigoris. Note preferential east-west alignment in west Frigoris and polygonal pattern in eastern Frigoris.

composed of three discrete features about 500 m in width and having a total length of about 200 km. These rilles are relatively small compared with similar features elsewhere on the Moon that typically have widths of the order of 2.5 km (e.g., *Golombek,* 1979). If, as suggested by *Golombek* (1979), the Frigoris fractures are conjugate fault planes with dip angles of approximately 60°, they would intersect at a depth of about 400 m. This is significantly less than the thickness of the lunar "megaregolith" (about 3 to 4 km; *Head,* 1976b) and could represent the thickness of the plains material. The presence of the features solely within the plains and not adjacent lavas suggests their formation at an early period of mare emplacement or that the plains have different mechanical properties than those of the mare basalts.

As is common around other lunar maria, floor-fracturing in craters adjacent to Mare Frigoris probably was caused by floor uplift subsequent to subfloor magmatic intrusion (e.g., *Schultz,* 1976b) that took place penecontemporaneously with basalt eruptions within the adjacent mare (*Whitford-Stark,* 1982).

Mare Ridges

The mare ridges within both western and eastern Frigoris are well developed (Fig. 12) but differ markedly in their orientations. In western Frigoris the major ridges trend approximately parallel with the shoreline and a subordinate set trends normal to the shoreline. In eastern Frigoris, however, the ridges form a polygonal pattern like that of northern Procellarum (*Whitford-Stark,* 1980). It is now widely believed (see *Solomon and Head,* 1980) that mare ridges are features of compressional tectonics resulting from the subsidence of the superisostatic load constituting the mare basalt fill. The trends of the ridges within other maria have been shown to be controlled by the shape of the basin and, in the case of young ridges, by a superposed global stress field that produced a preferential north-south alignment (*Fagin et al.,* 1978). These controls also appear to have been exerted in Frigoris; the strong east-west alignments in western Frigoris reflect the similarly aligned preflooding trough, whereas the polygonal pattern of eastern Frigoris reflect the amoeboid outline of the mare fill.

The major development of ridges appears to predate the emplacement of the youngest basalts within the mare. In western Frigoris a mare ridge acted as a topographic barrier, resulting in the diversion of flow of a young titanium-rich basalt about both sides of the ridge (Fig. 8). All of the basalt units at the surface of Frigoris contain mare ridges, and the ridges do not change orientation at unit boundaries nor do they crosscut one another. Both of these features indicate ridge formation within a uniform stress field over an unknown period of time. The preferential development of ridges toward the northern boundary of western Frigoris may reflect a thickness asymmetry; that is, the basalt fill thickens from the southern to the northern shore.

SUMMARY

The presence of well-developed mare ridges indicates a thick basalt fill within Mare Frigoris. The presence of craters that have been flooded by >400 m of postimpact materials and the absence of large premare flooded craters also support the existence of a thick mare fill. The absence of large flooded craters can also be explained by the western Frigoris topographic low being relatively uncratered at the time of basalt emplacement. That is, the time of formation of the trough postdated or was close to the younger limit for heavily cratered surfaces (about 3.8×10^9 years ago; *Taylor,* 1982). Such a situation could have arisen through the collapse of the south Frigoris highlands into the Imbrium impact cavity.

The mode of formation of the topographic low which is now eastern Frigoris is somewhat enigmatic. The paucity of rifts can be accounted for by the mare fill being of insufficient thickness to cause subsidence of the lithosphere at the time of cessation of global rift production. A large part of this fill is probably low density ejecta from the Imbrium impact event.

Three major basaltic units have been identified within Mare Frigoris. The youngest is a titanium-rich basalt with an age of $3.2 \pm 0.2 \times 10^9$ years. The other units are of intermediate and titanium-poor composition. By analogy with basaltic sequences developed elsewhere on the Moon, it is suggested that the majority of the intermediate composition basalts were erupted at $3.3 \pm 0.3 \times 10^9$ years ago and the majority of the titanium-poor basalts at $3.6 \pm 0.2 \times 10^9$ years ago. Most of the basalts appear to have been emplaced by flood-style eruptions.

Acknowledgments. The author would like to thank E. A. Whitaker for color composite and earth-based photography, B. R. Hawke (Hawaii Institute of Geophysics) for multispectral imagery and reviewing the manuscript, P. Spudis for an informed review, and S. Rudine for reproducing the figures. This work was made possible through the award of NASA Grant NAGW-394.

REFERENCES

Adams J. B., Pieters C. M., Metzger A. E., Adler I., McCord T. B., Chapman C. R., Johnson T. V., and Bielefeld J. (1981) Remote sensing of basalts in the solar system. In *Basaltic Volcanism on the Terrestrial Planets,* pp. 239-490. Pergamon, New York.

Bell J. F. and Hawke B. R. (1984) Lunar dark-haloed impact craters: Origins and implications for early mare volcanism. *J. Geophys. Res., 89,* 6899-6910.

Boyce J. M. (1976) Ages of flow units in lunar nearside maria based on Lunar Orbiter IV photographs. *Proc. Lunar Sci. Conf. 7th,* pp. 2717-2728.

Fagin S. W., Worrall D. M., and Muehlberger W. R. (1978) Lunar mare ridge orientations: Implications for lunar tectonic models. *Proc. Lunar Planet. Sci. Conf. 9th,* pp. 3743-3479.

Gaddis L. R., Pieters C. M., and Hawke B. R. (1985) Remote sensing of lunar pyroclastic mantling deposits. *Icarus, 61,* 461-489.

Golombek M. P. (1979) Structural analysis of lunar grabens and the shallow crustal structure of the Moon. *J. Geophys. Res., 84,* 4657-4666.

Greeley R. (1976) Modes of emplacement of basalt terrains and an analysis of mare volcanism in the Orientale basin. *Proc. Lunar Sci. Conf. 7th,* pp. 2747-2759.

Hawke B. R. and Spudis P. D. (1980) Geochemical anomalies on the eastern limb and farside of the Moon. In *Proceedings of the Conference on the Lunar Highlands Crust* (J. J. Papike and R. B. Merrill, eds.), pp. 467-481. Pergamon, New York.

Hawke B. R., McCord T. B., and Head J. W. (1980) Small lunar dark-mantle deposits of probable pyroclastic origin. *Reports of Planetary Geology Program—1980,* pp. 512-514. NASA TM-82385.

Hawke B. R., Coombs C. R., Gaddis L. R., Lucey P. G. and Owensby, P. D. (1989) Remote sensing and geologic studies of localised dark mantle deposits on the Moon. *Proc. Lunar Planet. Sci. Conf. 19th,* pp. 255-268.

Head J. W. (1976a) Lunar volcanism in space and time. *Rev. Geophys. Space Phys., 14,* pp. 265-300.

Head J. W. (1976b) The significance of substrate characteristics in determining morphology and morphometry of lunar craters. *Proc. Lunar Sci. Conf. 7th,* pp. 2913-2929.

Head J. W., Bryan W. B., Greeley R., Guest J. E., Schultz P. H., Sparks R. S. J., Walker G. P. L., Whitford-Stark J. L., Wood C. A., and Carr M. H. (1981) Distribution and morphology of basalt deposits on planets. In *Basaltic Volcanism on the Terrestrial Planets,* pp. 701-800. Pergamon, New York.

Johnson T. V., Mosher J. A., and Matson D. L. (1977) Lunar spectral units: A northern hemisphere mosaic. *Proc. Lunar Sci. Conf. 8th,* pp. 1013-1028.

Kuiper G. P., Strom R. G., Whitaker E. A., Fountain J. W., and Larson S. M. (1967) *Consolidated Lunar Atlas.* Contributions of the Lunar and Planetary Laboratory No. 4. Univ. of Arizona, Tucson.

Lucchitta B. K. (1972) Geologic Map of the Aristoteles Quadrangle of the Moon. *U.S. Geol. Surv. Misc. Invest. Ser. Map I-725.*

Lucchitta B. K. (1978) Geologic Map of the North Side of the Moon. *U.S. Geol. Surv. Misc. Invest. Ser. Map I-1062.*

Lucchitta B. K. and Watkins J. A. (1978) Age of graben systems on the Moon. *Proc. Lunar Planet. Sci. Conf. 9th,* pp. 3459-3472.

M'Gonigle J. W. and Schleicher D. (1972) Geologic Map of the Plato Quadrangle of the Moon. *U.S. Geol. Surv. Misc. Invest. Ser. Map I-701.*

Pieters C. M. (1978) Mare basalt types on the front side of the Moon: A summary of spectral reflectance data. *Proc. Lunar Planet. Sci. Conf. 9th,* pp. 2825-2849.

Pieters C. M. and McCord T. B. (1976) Characterization of lunar mare basalt types: 1. A remote sensing study using reflection spectroscopy of surface soils. *Proc. Lunar Sci. Conf. 7th*, pp. 2677-2690.

Rhodes J. M. (1977) Some compositional aspects of lunar regolith evolution. *Philos. Trans. R. Soc. London, Ser. A, 285,* 293-301.

Schultz P. H. (1976a) *Moon Morphology.* Univ. of Texas, Austin.

Schultz P. H. (1976b) Floor-fractured lunar craters. *The Moon, 15,* 241-273.

Schultz P. H. and Gault D. E. (1985) Clustered impacts: Experiments and implications. *J. Geophys. Res., 90,* 3701-3732.

Schultz P. H. and Mendenhall M. H. (1979) On the formation of basin secondary craters by ejecta complexes (abstract). In *Lunar and Planetary Science X,* pp. 1078-1080. The Lunar and Planetary Institute, Houston.

Schultz P. H. and Spudis P. D. (1979) Evidence for ancient mare volcanism. *Proc. Lunar Planet. Sci. Conf. 10th*, pp. 2899-2918.

Shoemaker E. H. and Hackman R. J. (1962) Stratigraphic basis for a lunar time scale. In *IAU Symposium Volume No. 14: The Moon* (Z. Kopal and Z. K. Mikhailov, eds.), pp. 289-300. Academic, New York.

Solomon S. C. and Head J. W. (1980) Lunar mascon basins: Lava filling, tectonics, and evolution of the lithosphere. *Rev. Geophys. Space Phys., 18,* 107-141.

Spudis P. D., Hawke B. R., and Lucey P. G. (1988) Materials and formation of the Imbrium basin. *Proc. Lunar Planet. Sci. Conf. 18th,* pp. 155-168.

Taylor S. R. (1982) *Planetary Science: A Lunar Perspective.* Lunar and Planetary Institute, Houston. 481 pp.

Ulrich G. E. (1969) Geologic map of the J. Herschel Quadrangle of the Moon. *U.S. Geol. Surv. Misc. Invest. Ser. Map I-604.*

Whitaker E. A. (1981) The lunar Procellarum basin. In *Multi-ring Basins, Proc. Lunar Planet Sci. 12A* (P. H. Schultz and R. B. Merrill, eds.), pp. 105-111. Pergamon, New York.

Whitford-Stark J. L. (1980) A Comparison of the Origin and Evolution of a Circular and an Irregular Lunar Mare. Ph.D. thesis, Brown Univ., Providence, Rhode Island, 360 pp. Reprinted in *Advances in Planetary Geology,* pp. 127-353, NASA TM-85630, 1983.

Whitford-Stark J. L. (1981a) Modification of multi-ring basins—the

Imbrium model. In *Multi-ring Basins, Proc. Lunar Planet. Sci. 12A,* (P. H. Schultz and R. B. Merrill, eds.), pp. 113-124. Pergamon, New York.

Whitford-Stark J. L. (1981b) The evolution of the lunar Nectaris multi-ring basin. *Icarus, 48,* 393-427.

Whitford-Stark J. L. (1982) A preliminary analysis of lunar extra-mare basalts: Distribution, compositions, ages, volumes, and eruption styles. *Moon and Planets, 26,* 323-338.

Whitford-Stark J. L. and Fryer R. J. (1975) Origin of Mare Frigoris. *Icarus, 26,* 231-242.

Whitford-Stark J. L. and Head J. W. (1980) Stratigraphy of Oceans Procellarum basalts: Sources and styles of emplacement. *J. Geophys. Res., 85,* 6579-6609.

Wilhelms D. E. (1980) Stratigraphy of part of the lunar near side. *U.S. Geol. Sur. Prof. Pap. 1046-A,* 71 pp.

Wilhelms D. E. (1987) The Geologic History of the Moon. *U.S. Geol. Surv. Prof. Pap. 1348.* 327 pp.

Wilson L. and Head J. W. (1981) Ascent and emplacement of basaltic magma on the Earth and Moon. *J. Geophys. Res., 86,* 2971-3001.

Proceedings of the 20th Lunar and Planetary Science Conference, pp. 187-194
Lunar and Planetary Institute, Houston, 1990

Near-Infrared Multispectral Images for Gassendi Crater: Mineralogical and Geological Inferences

S. Chevrel and P. Pinet

G.R.G.S. (Groupe de Recherches de Géodési Spatiale/O.M.P. (Observatoire Midi-Pyrénées)/C.N.E.S. (Centre National d'Etudes Spatiales), Toulouse, France

High-resolution near-infrared multispectral mapping of the interior of Gassendi crater has been performed by means of a charge-coupled device (CCD) camera placed at the focus of the 2-m telescope of Pic-du-Midi Observatory in France. The interpretation of these new data suggests that (1) the central part of the crater, including the peak complex, may have a more mafic composition with a higher pyroxene component than surrounding highlands; (2) extensive extrusive volcanism may have occurred within the eastern portion of the floor as indicated by the detection of a concentration of very low 0.97/0.56-μm-ratio spectral units, suggesting the presence of a significant pyroxene component that correlates with photogeologically identified volcanic features; and (3) the western part of the crater floor, away from the geometric continuation of the western edge of Mare Humorum, is composed of highlands-rich material. This apparent dichotomy between the western and eastern (adjacent to mare) side of the Gassendi floor-fractured crater may be strongly related to the early thermal history of Mare Humorum.

INTRODUCTION

Information on the mineralogy and chemical composition of the surfaces of planetary bodies is of importance in understanding planetary evolution. Earth-based spectral reflectance remote-sensing techniques have been developed to detect absorption features, characteristic of specific rock-forming minerals, occurring in the solar reflected light from planetary surfaces (*Adams,* 1974, 1975; *McCord,* 1988). Analysis of UV, visible, and near-infrared spectral reflectance data has proven to be the most powerful technique for determining surface composition and lithology.

Two significant accomplishments of solar system spectroscopy have been the studies of the lunar mare surfaces (e.g., *McCord et al.,* 1976; *Pieters,* 1978) and asteroids (e.g., *Gaffey et al.,* 1977). Compositional data of the lunar surface have been obtained from reflectance spectra in the 0.30-to 2.50-μm domain (e.g., *Pieters,* 1978, 1986; *Hawke et al.,* 1979; *McCord et al.,* 1981; *Gaddis et al.,* 1985; *Lucey et al.,* 1986; *Coombs et al.,* 1987) and from multispectral imaging using vidicon techniques (*McCord et al.,* 1976; *Johnson et al.,* 1977.) Individual reflectance spectra are helpful in characterizing selected spots, while the determination of the horizontal extent of identified spectral units is more easily done with the multispectral data. Regional geological investigations of the southern part of Oceanus Procellarum (*Pieters et al.,* 1980) and Mare Humorum (*Pieters et al.,* 1975) are examples of studies using both methods. Both of these studies concentrated on regional mapping and interpretation of the mare surfaces in relation to structural features and did not focus on the analysis of highland regions and complex morphological units such as impact craters. Such investigations were subsequently addressed by means of high-resolution (2-3-km-diameter areas) visible and near-infrared reflectance spectra (0.7-2.5 μm) within both mare and highland craters (*McCord et al.,* 1981; *Pieters,* 1982, 1986; *Pieters et al.,* 1985).

Rapid progress in the development of the charge-coupled device (CCD) imaging technique (*Vilas,* 1985; *Chevrel and Pinet,* 1988; *Jaumann and Neukum,* 1989; *Lucey and Hawke,* 1989) allows a complementary approach consisting of near-infrared multispectral mapping of complex morphological units. In this paper we present high spatial resolution (0.5 km/pixel) visible and near-infrared images, obtained with a CCD camera on the 2-m Pic-du-Midi telescope (*Chevrel and Pinet,* 1989), of Gassendi, a typical floor-fractured crater located at the edge of Mare Humorum. *Chevrel and Pinet* (1988) presented a set of lower resolution (3 km/pixel) images that had been taken during an earlier observational program.

DATA ACQUISITION AND PROCESSING

The multispectral imaging system comprises a CCD detector placed behind high-quality image interferential filters. This photometric detector allows one to obtain directly a numerical image of a particular target on the lunar surface. This system offers a combination of qualities: good quantum efficiency, excellent linearity over the whole spectral domain of sensitivity (from 0.40 to 1.10 μm), large dynamic range, low noise and dark current, and a good overall system stability. Images presented in this paper have been made using a CCD array with 576 × 384 pixels from Thomson-CSF (THX 31133). It was placed at the focus of the 2-m aperture telescope (F/D = 25) of the Pic-du-Midi Observatory in France (elevation: 2877 m). An evaluation of the performance of the CCD-Thomson detector (*Mellier et al.,* 1986) demonstrated a low read-out noise, a uniform response across the chip (leading to better photometric accuracy), and a high linearity (deviation <0.001) and storage capacity per pixel. With the whole chip illuminated when imaging the lunar surface, we did not exceed the maximum storage capacity limit for one pixel, an output signal of 60,000 electrons or 4500 ADU units (*Mellier et al.,* 1986). The optimal signal-to-noise ratio conditions are obtained by cooling the CCD to -80°C with liquid nitrogen. The high quality image interferential filters used are centered at 5585 Å and 9700 Å and have the following respective optical characteristics: bandwidth = 110 Å, 94 Å; transmission peak T (%) = 80%, 50%.

Given the optical system and the pixel dimension (a square 23 μm in size), the field of view is $54 \times 36''$ and corresponds to a target of 96×64 km on the lunar surface at the subterrestrial point and to a theoretical spatial resolution of 0.2 km/pixel.

The multispectral dataset for a given area on the Moon comprises a sequence of two sets of images, one in the near-infrared (0.97 μm) and the other in the visible (0.56 μm) domain. Within a set, images are taken at a rate of one per minute with a time interval of five minutes between two consecutive sets. Time integration, determined by an electronic shutter, is less than one second. Images are coded in 16-bit intensities and stored in FITS format on magnetic tape for subsequent processing.

The actual processing consists of (1) the selection of images and their calibration and normalization within a given set; and (2) the superposition and division of a near-infrared (IR) image by a visible (VIS) image, resulting in an image ratio IR/VIS. Selection of images in a set is made by dividing one image by each of the others in that set. The correlation coefficient and standard deviation (σ = r.m.s.) are calculated, and the pair of images that shows the best stability ($\sigma \sim$ 2-3%) is selected. Calibration of an image allows the removal of instrumental effects and a correction for the slight pixel-to-pixel nonuniformity of the chip for a given wavelength. For each raw image taken with a given filter, there is a corresponding flat field image, or one that exposes a uniformly lit field (twilight sky) through the telescope (with the same filter). A dark field image is also made, which is an exposure at the same integration time as the raw image and the flat field image, except the shutter is not opened. After the dark field image has been subtracted from both previous images, the raw image is divided by the flat field image resulting in a photometrically calibrated image.

To derive absolute reflectance values requires taking into account atmospheric corrections, instrumental sensitivity, flux calibrations, etc. However, accurate relative reflectance values can be obtained using relative reflectance produced during normalization. Consequently, calibrated images are normalized to a given area taken as a standard within the imaged region. The choice of the standard area is made on the basis of its spectral homogeneity at both wavelengths, evaluated by means of the standard deviations σ_{vis} and σ_{ir}. Normalization is accomplished by dividing the spectral images by the respective mean values of the standard area.

Because the repetition of the telescopic pointing on a given feature of the lunar surface is not better than a few pixels in accuracy, a pair of near-infrared and visible images have to be geometrically superposed before derivation of the spectral ratio image. One image is shifted onto the other and the best fit is reached by calculating the maximum correlation between the two images in a square 200 pixels in size, including "sharp" lunar features such as central peaks in a crater. Once the images are normalized and superposed, the IR/VIS image pixel-by-pixel ratio is calculated. The result is a spectral ratio image or multispectral image IR/VIS normalized to a standard area in the image. Independent ratio images are made using pairs of IR and VIS calibrated images having the best stability and the smallest time interval between them.

RESULTS

The selected zone of investigation is the 110 - km - wide Nectarian (*Carr,* 1984) impact crater Gassendi, situated along the northwestern border of Mare Humorum, a mare-filled multiringed basin located in the southwestern portion of the nearside of the Moon. As with many craters on the margins of large mare-filled basins, Gassendi displays a shallow, flat, fractured floor, with a central peak complex and a partial inundation by mare-like material (Fig. 1). Such craters have significance for studying the style of crater modification, the type of lunar volcanism, the sequence of inundation of the maria (possible interconnected sources responsible for partial inundation within the crater and of adjacent mare), and the thermal history of the Moon.

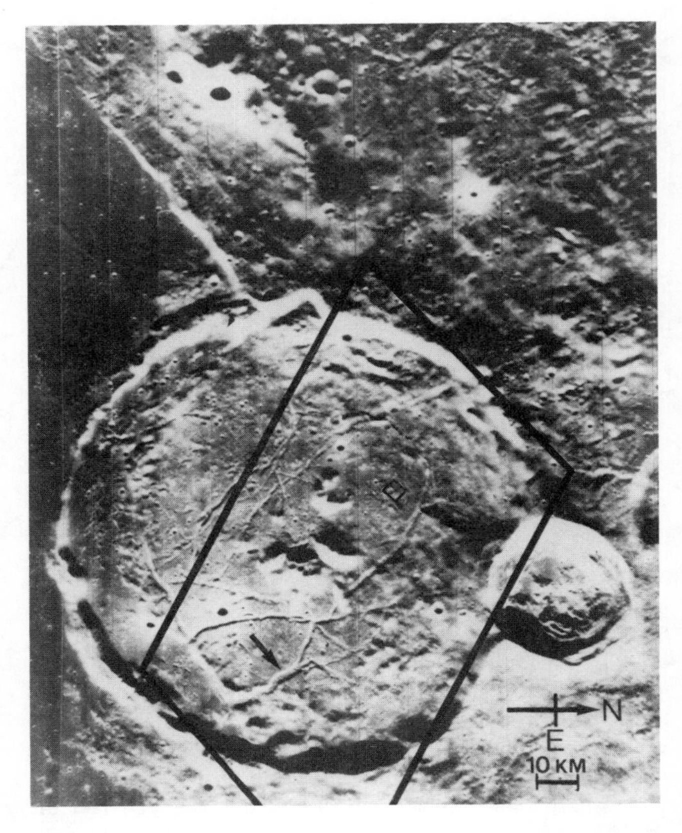

Fig. 1. Lunar Orbiter IV-143-H2 picture of the crater Gassendi, located on the northwestern edge of Mare Humorum (17°S; 40°W). Outlined area shows the imaged zone inside the crater (Figs. 2 and 3). Small square northwest of the central peak complex delimits the standard area NA (see text and Fig. 4). Arrow points to the ST unit (see text and Fig. 4) centered on the broad part of the fracture Rima Gassendi II.

Gassendi is classified as a class III crater (*Schultz,* 1976a). These craters are characterized by a wide annular depression, or "moat," between a wall scarp and a shallow plate-like floor that exhibits fractures and a central peak complex. Central peaks of craters presumably represent deep-seated material excavated during crater formation (*Grieve and Head,* 1983; *Grieve et al.,* 1981). The floor of Gassendi shows a subdued hummocky terrain and wide variations in local crater densities, reflecting the possible emplacement of extrusive units or, in the units where few craters exist, rapidly degraded surfaces.

Fractures on the floor of Gassendi are not well ordered (Fig. 1) and sometimes intersect previously existing features such as the central peaks. These fractures are probably due to the complex modification history of the crater. Major fractures form a northeast-trending scarp that cuts off the southeast portion of the crater floor (Rima Gassendi VII), a set of northwest-trending scarps located on both sides of central peaks (Rima Gassendi IV, V, VI), and a north-trending fracture in the western portion of the crater floor (see *NASA,* 1971). A partial inundation by a mare-like material occurred on the side adjacent to the mare, south of Rima Gassendi VII (Fig. 1), and several volcano-like structures have been photogeologically identified along the southern part of this fracture and the eastern portion of the floor (*Schultz,* 1976b).

The shallow, fractured floor and the presence of possible extrusive centers on the floor of Gassendi and similar craters along mare borders may be of relevance for modeling the crater evolution and its subsequent topographic modification (*Hall et al.,* 1981). Clearly, more information is required about the composition and distribution of the different types of materials present on the floor as well as their relevance to the tectonic features. At the present time no spectra are available from selected parts of Gassendi. Medium resolution vidicon images in three bands (0.40, 0.56, and 0.95 μm) of Mare Humorum (*Pieters et al.,* 1975) included crater Gassendi; however, no conclusive compositional interpretation has been carried out within the crater.

On September 29, 1988, an observational period on the 2-m telescope of the Pic-du-Midi Observatory was devoted to the multispectral mapping of Gassendi in the visible (0.56 μm) and the near-infrared (0.97 μm) wavelengths. As a result, we obtained 10 high-resolution images under stable weather and very good visibility conditions (0.3 to 0.5 arcsec). The internal consistency of the dataset, evaluated for each spectral range from the σ r.m.s. standard deviation, is better than 3.5% for the visible and 2.5% for the near-infrared set of calibrated images.

Figures 2 and 3, respectively, show a visible and a near-infrared high-resolution (better than 0.5 km/pixel) image of a

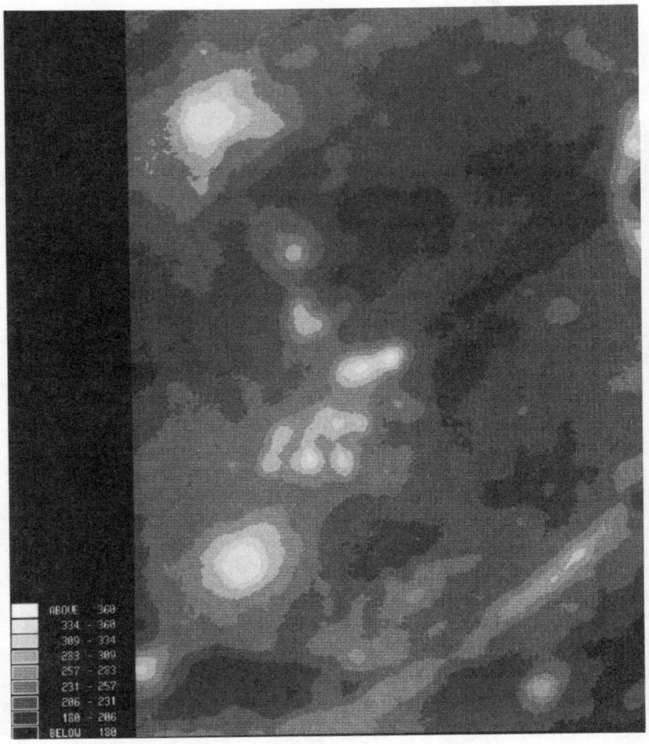

Fig. 2. Visible (0.56 μm) CCD image of the northern part of Gassendi. North is at right and east is at bottom. Scale: 9 mm = 10 km. White and dark, respectively, mean a bright and a low albedo with a gray tone interval class of 15% variation on the whole intensity range. Central peaks (bright) are close to the center of the image. Western and eastern parts of the rim are at the top and bottom of the image, respectively. Southern rim of the crater Gassendi A is at the upper right.

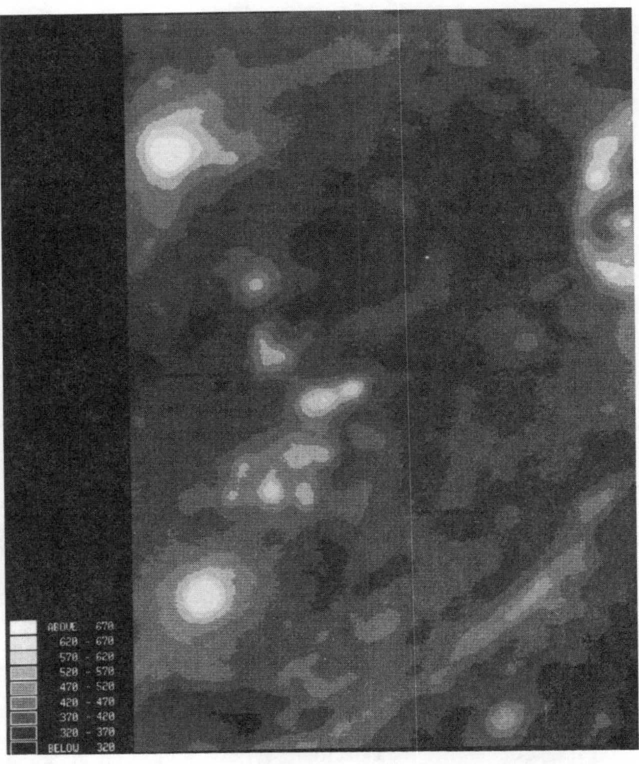

Fig. 3. Near-infrared (0.97 μm) CCD image of the northern part of Gassendi. For orientation, scale, and gray tone coding, see Fig. 2. Field of view is approximately the same as Fig. 2.

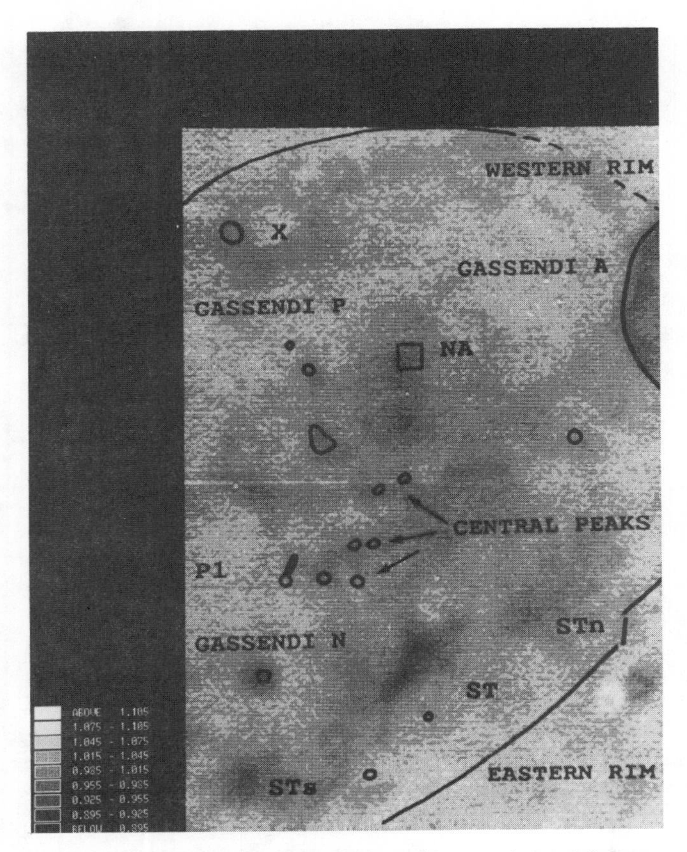

Fig. 4. A 0.97/0.56-μm spectral ratio image, obtained from images in Figs. 2 and 3 (same scale and orientation), normalized to the standard area NA (a square 20 × 20 pixels in size). Gray tone scale ranges from dark (10% below unity) to bright (10% above unity) with an interval class of 3% variation. P1 = peak 1; ST = spectral trough region; STn and STs = northern and southern extensions of the ST region; X = unnamed crater (see text for further description). Note that both white and dark patches near the eastern rim at bottom are artifacts in the image.

common region of the crater interior, corresponding to the outlined area in Fig. 1. Central peaks appear close to the center of the images. Northwestern and northeastern parts of the crater rim are at the top and the bottom, respectively. The southern part of the rim of Gassendi A, a crater located on the northern edge of Gassendi, appears at the upper right of the image. Figure 4 shows the spectral ratio image 0.97/0.56 μm of the two previous images displayed in Figs. 2 and 3. The spectral ratio is *relative* to the normalization area [NA; a small square, 20 pixels in size (Figs. 1 and 4)] in which σ_{vis} = 2.9%, σ_{ir} = 2.1%, and the standard deviation σ_r of the spectral ratio is 3.3%.

The 0.97/0.56-μm multispectral measurement accuracy is estimated to be 3% on the basis of the evaluation of the mean internal consistency of the dataset. The σ_r value, within NA, is on the order of the measurement accuracy and indicates that NA is spectrally very homogeneous. However, a 3% accuracy in the multispectral observations may correspond to important mineralogical composition differences (e.g., *Pieters,* 1978; *Hawke et al.,* 1979). Consequently, a 3% interval class is used in Fig. 4, and only significant compositional variations across Gassendi crater are detected.

Despite the high degree of correlation between the visible and the near-infrared images (r = 0.967), extreme variations in the range of ±10% relative to the standard zone NA are seen in the ratio image.

A critical assessment of the reproducibility of the multispectral information is performed by comparing two ratio images derived from two independent couples of visible/near-infrared images. This test demonstrates the excellent similarity of the two ratio images.

Inside Gassendi the floor globally displays a high 0.97/0.56-μm ratio (4% to 6% above NA), except for the following specific features: (1) regions related to fractures, (2) fresh craters, and (3) the central peaks. These exceptions exhibit a spectral ratio less than or equal to unity.

The most significant multispectral feature identified in the image (Fig. 4) is a triangle-shaped region, marked ST ("spectral trough"). Located in the northeastern portion of the crater floor, this feature exhibits a mean spectral ratio depleted by 8%, lower than that exhibited by NA, ranging down to 10-11% in its central part. The ST region is 10 km long and lies along Rima Gassendi II.

This Rima is a 2-km-wide fracture that bifurcates into two branches in this area (cf. Fig. 1). The ST region extends laterally over 5 km and thus overlaps and spreads out of the fracture zone. On the eastern side of the fracture, there is a plain-forming unit with very few craters on its relatively flat surface. Scarps to the north of this unit may indicate ancient levels of lava terraces, and a probable volcanic cone lies very close to the fracture on its western side (cf. plate 37, p. 85, in *Schultz,* 1976b). All these volcanic features are included in the very low 0.97/0.56-μm-ratio ST patch. North of the ST zone, in its straight continuation, there is another area spectrally depleted by 5-6% (STn in Fig. 4). Although it is not directly related to a fracture, it also corresponds to a portion of the floor where volcanic processes seem to have occurred, as evidenced by the presence of lava terraces from photogeological analysis (*Schultz,* 1976b). South of the ST zone, also almost in its continuation, is a low spectral ratio zone, depleted by 7% to 8% and noted STs in Fig. 4, which corresponds to the northern part of the mare-like material lying in the southeastern portion of the floor of the crater (Fig. 1).

Finally, an areally restricted zone, located in the eastern vicinity of the standard zone NA, has a 3% to 4% depleted ratio relative to NA, but there is no clear floor feature associated with it. One must notice, however, that a 3% depletion is at the limit of the spectral accuracy estimate and must be carefully considered.

The craters Gassendi A [Cc_1 material (*Wilhelms and McCauley,* 1971)] and Gassendi N also show a low multispectral ratio value, about 5-6% less than NA. Their spectral properties, considered as characteristic of fresh craters, may represent excavated immature material, relatively unmasked by agglutinates resulting from subsequent soil maturation processes (e.g., *Pieters,* 1977). Central peaks have the same ratio as the normalization area except for one peak, P1 in Fig. 4, which has the same ratio as the surrounding crater-floor material.

A last striking observation is that outside Gassendi, in the external vicinity of the northeastern rim, there exists a broad zone (bottom right part of Fig. 4) with an average spectral ratio of around 8% above NA, locally exceeding 10%.

INTERPRETATION AND CONCLUSIONS

The high spatial resolution 0.97/0.56-μm multispectral image of the northern portion of Gassendi presented in Fig. 4 shows horizontal spectral variations on the floor of the crater. The sketch map (Fig. 5) shows these variations in relation to topographic, morphological, and geological features, after the ratio image (Fig. 4) has been rectified to match the high-resolution orbital picture given in Fig. 1.

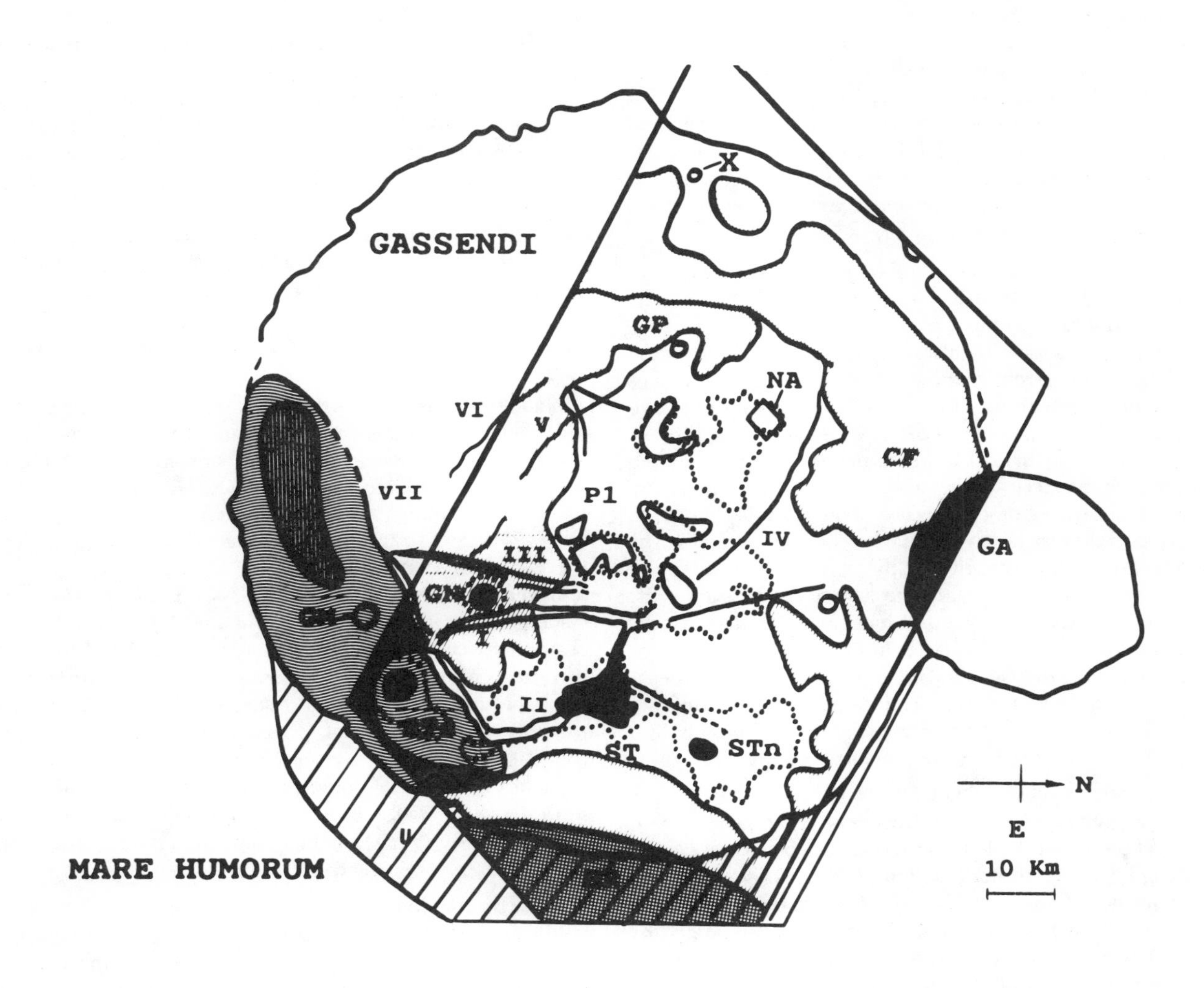

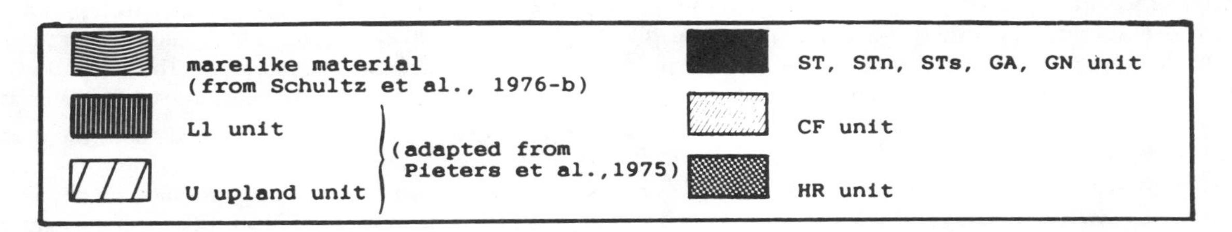

Fig. 5. Sketch map showing the spectral units identified from the 0.97/0.56-μm-ratio values of Fig. 4. Topographic features are adapted from the IV-143-H2 Lunar Orbiter Picture (Fig. 1). I to VII = Rimae Gassendi; NA = normalization area; P1 = Peak 1; GA = Gassendi A; GM =Gassendi M; GN =Gassendi N; GP = Gassendi P; X =unnamed crater.

Principal observations are (Figs. 4 and 5) (1) a high spectral ratio (5-6%) relative to NA for the major part of the crater floor (CF unit), (2) a low mean spectral ratio (4% to 8% less than NA) for relatively dark albedo local regions in the southeastern portion of the floor adjacent to Mare Humorum (ST zone with possible north and south extensions STn, STs) and for the fresh crater units GA and GN, (3) a uniform response of the central peaks identical to the standard area NA, and (4) a high spectral ratio zone (HR unit) located outside the northeastern rim of Gassendi.

Because of the lack of spectra in Gassendi, there is no way to refer to a standard area spectrally classified in terms of composition. As mentioned in the previous section, NA has been chosen on the basis of its high spectral homogeneity ($\sigma_r = 3.3\%$), but its mineralogical composition is unknown. Consequently, the interpretation of observed spectral variations on the floor of Gassendi in terms of spectral absorption related to composition is not straightforward.

However, an indirect piece of information relative to NA can be derived from the fact that the HR unit, which exhibits the highest spectral ratio within the image (8% to 10% above NA), lies within an extensive highland region identified by *Pieters et al.* (1975). The spectrophotometry of such highland regions shows them in general to have a weak 1-μm absorption band (*McCord et al.,* 1972; *Pieters et al.,* 1980; *Hawke et al.,* 1989). On this basis, high 0.97/0.56-μm ratio areas correspond to highlands-rich material units, whereas low ratio areas are generally indicative of more mare-like material units. However, this simple trend may be obscured by soil maturity effects. The 0.97/0.56-μm ratio is sensitive to the abundance and type of mafic mineral present as well as the degree of maturity of the local soil; relatively low ratio values may be due to the freshness of the surface material, as in the case of fresh craters. As a consequence, the NA standard zone and the central peaks (except for P1), which spectrally respond the same and are ratio-depleted by 8% to 10% in respect to HR, must have a somewhat 0.97-μm-absorbing component.

The NA area is about 7 km long $\times$ 7 km wide and has a relatively low albedo in both the visible and the near-infrared domains (Figs. 2 and 3). It is located northwest of the central peak complex, close to the junction of Rima Gassendi IV and a major north-trending fracture in the western part of the floor (Figs. 4 and 5). It lies within a gentle depression bounded by a local 5000-m elevation contour (see *NASA,* 1971), tentatively interpreted as being due to endogenetic modification of the crater floor (*Schultz,* 1976b). Alternatively, as a result of important degradations undergone by old craters (*Head,* 1975), this area might consist of ejecta material from nearby craters such as Gassendi A or from a more distant event. Nevertheless, the fact that it has the same spectral ratio as the central peaks argues against this possibility.

In strong contrast with the HR unit, the identification of areally restricted low albedo units (ST, STn, STs) displaying the lowest 0.97/0.56-μm ratios (Fig. 4) and their association with morphological volcanic and tectonic features (Figs. 1 and 5) on the basis of photogeological evidences (*Schultz,* 1976a,b) call for the presence of a mafic component within these units. Although more information concerning the reflectance properties of these materials in the near-infrared spectral domain is needed for mineral identification, this low ratio value may be indicative of a higher abundance of pyroxene such as is found for the mafic mare basalts (*Adams,* 1974; *Pieters,* 1986).

In the study of Mare Humorum by *Pieters et al.* (1975), a part of the mare-like material, located on the floor of Gassendi and adjacent to the mare (see L1 unit, Fig. 5), has been identified as being similar to the major basalt unit located in Mare Humorum. *Pieters et al.* (1975) determined that this Mare Humorum unit is composed of a low titanium basalt (2% TiO_2) with a major clinopyroxene mafic component.

An important point is that the spectral units, L1, ST, STn, and STs, are located directly or in the vicinity of volcanic features such as domes, cones, and lava terraces. These photogeologically identified (*Schultz,* 1976a,b) features suggest that extrusive activity occurred within the crater. Although the extensive mare-like material unit (Fig. 5) may be the result of partial lava flooding of Gassendi from Mare Humorum, it may also be a direct consequence of the extrusive volcanic activity, depending on its importance and its connection to the mare-related magma chamber through deep-seated fractures.

In addition, it is noteworthy that the crater Gassendi N, 4 km in diameter, is spectrally different from other equal-sized small craters such as Gassendi P and an unnamed crater located along the western rim of Gassendi (X; Figs. 4 and 5). These small fresh craters may be considered as near-subsurface probes (*Pieters,* 1977) that excavated material from a depth of a few hundred meters (*Gault,* 1973). These differences may be linked to their location within Gassendi. Whereas Gassendi P and X lie in the western part of the floor, Gassendi N is located between two fractures, Rima Gassendi I and III, on the mare-adjacent eastern side of Gassendi. The rather low spectral ratio GN unit (Fig. 5) may be due to the presence of mafic material with a significant pyroxene component, excavated from a shallow depth. Alternatively, the relatively low 0.97/0.56-μm value for GN could be due to the relative freshness of this crater. Radar backscatter measurements at 3.8 cm (*Pieters et al.,* 1975) are high for this crater, indicating a blocky or immature surface.

Similarly, the low-ratio GA unit (Figs. 4 and 5) located within the 33-km-diameter Gassendi A crater may be mafic in composition. If so, as a direct consequence of the diameter difference (*Settle and Head,* 1979; *Pike,* 1980) between Gassendi A and Gassendi N, the GA unit material may have originated from a depth inferred to be an order of magnitude greater than in the case of Gassendi N. This low-ratio GA unit may also be due to the relative immaturity of the crater walls.

The CF unit appears to be distributed about the perimeter of the crater interior (Fig. 5). This unit is spectrally intermediate between the NA and HR units and has a spectral ratio only 2% to 4% lower than HR. It is thought to be composed of highlands-rich material.

The interpretation of the spectral data, presented in this paper, suggests that the central part of Gassendi, inclusive of the central peak complex and those areas within the dotted lines of Fig. 5, may have a mafic pyroxene composition similar to the subsurface information derived from the Gassendi A spectral response. Furthermore, a concentration of very low 0.97/0.56-μm ratio units (ST, STn, STs, GN) located in the eastern side of the crater floor occurs in relation to several volcanic features and a dense interconnected tectonic pattern. While the western side of the crater is away from the edge of Mare Humorum, the eastern side may be directly above the pattern of fractures and thus may have experienced extrusive volcanic activity through scattered separated conduits, pipes, and fractures that lead to a partial inundation of the crater floor. This dichotomy, revealed by both photogeological (Fig. 1) and spectral (Figs. 4 and 5) observations, between the western and eastern (adjacent to mare) sides of the Gassendi floor-fractured crater may be strongly related to the thermal history of Mare Humorum.

Gassendi is classified as a class III crater (*Schultz,* 1976a) and presents a substantial floor uplift. Its stratigraphic age is PIc_3 (*Wilhelms and McCauley,* 1971) or Nc (*Carr,* 1984), corresponding to the terminal phase of Period I (*Head,* 1975) (i.e., 3.95-3.85 b.y. old). The high rate of modification of craters in Period I (*Pike,* 1968) suggests that the residence time of a crater on the lunar surface during this period may be relatively short and that the vast majority of crater degradation takes place nearly coincident with its time of formation (*Head,* 1975). Besides, the formation of Gassendi is contemporaneous with the formation and/or the early partial infilling of Mare Humorum (Nb age; *Carr,* 1984) within a time interval on the order of 300 m.y. or less. Locally elevated temperatures may have occurred about the periphery of Humorum basin, as a result of heat remaining from the basin formation and/or the subsequent mare volcanism, and these higher temperatures may have lowered the effective ancient viscosity of the lunar interior. In this case, both the rapid temporal evolution of the crater topography and its preserved perennial shape may be explained simply by a viscous relaxation model using an elastic-viscous, two-layered substrate (*Pinet,* 1985; *Pinet and Souriau,* 1989). With such a model the crater topography may have been modified within a relatively short time interval, on the order of 30-300 m.y., as a result of the substantial viscosity lowering of the lunar interior [to 10^{22}-10^{23} P (*Pinet and Souriau,* 1989)]. Fractures of the crater floor may have been caused by stresses in excess of the elastic strength of the thin, brittle lithosphere. In this context the spectral clues of volcanic activity within Gassendi (very low 0.97/0.56-μm-ratio ST, STn, and STs units), which reinforce *Schultz's* (1976b) previous photogeological analysis, and possible the L1 unit (Fig. 5), might well be indicative of the earliest infilling of Mare Humorum.

A future valuable task should be to complete the near-infrared spectral mapping of Gassendi, and, secondly, to map other craters around the edge of Mare Humorum in the same fashion. These craters include Doppelmayer, Vitello, Mersenius D, Mersenius G, and Liebig. These new maps will allow us to look at the spatial distribution of such low 0.97/0.56-μm-spectral ratio patches in relation with the main concentric scarps and faults of Mare Humorum. The acquisition of additional spectral bands within the 0.9-1.05-μm domain should help distinguish a larger variety of surface units in this region of the Moon.

Acknowledgments. This research was supported by the French A.T.P. Planetology Program (contracts Nos. 57-62, 67-15, and 37-11) of the Institut National des Sciences de l'Univers (I.N.S.U.) and the French Centre National d'Etudes Spatiales (C.N.E.S.). The authors acknowledge the support of the Pic-du-Midi Observatory staff during observations with the 2-m telescope and thank J. W. Head, in charge of the Planetary Geological Sciences Department of Brown University, for his very helpful support while the authors were guests in March 1989. We are grateful to P. H. Schultz and C. M. Pieters for the fruitful discussions from which a part of the proposed interpretations arose. We gratefully acknowledge the constructive reviews by C. R. Coombs, C. M. Pieters, and C. A. Wood.

REFERENCES

Adams J. B. (1974) Visible and near-infrared diffuse reflectance spectra of pyroxenes as applied to remote sensing of solid objects in the solar system. *J. Geophys. Res., 79,* 4829-4836.

Adams J. B. (1975) Interpretation of visible and near-infrared diffuse reflectance spectra of pyroxenes and other rock-forming minerals. In *Infrared and Raman Spectroscopy of Lunar and Terrestrial Materials* (C. Karr, ed.), pp. 91-116. Academic, New York.

Carr M. H., ed. (1984) The geology of the terrestrial planets. *NASA SP-269.* 317 pp.

Chevrel S. and Pinet P. C. (1988) Remote sensing on the Moon by CCD imaging (abstract). *Eos Trans AGU, 69,* 393.

Chevrel S. and Pinet P. C. (1989) Lunar mare-highland horizontal spectral variations from telescopic CCD-imaging (abstract). In *Lunar and Planetary Science XX,* pp. 153-154. Lunar and Planetary Institute, Houston.

Coombs C. R., Hawke B. R., and Wilson L. (1987) Geologic and remote sensing studies of Rima Mozart. *Proc. Lunar Planet. Sci. Conf. 18th,* pp. 339-353.

Gaddis L. C., Pieters C. M., and Hawke B. R. (1985) Remote sensing of lunar pyroclastic mantling deposits. *Icarus, 61,* 461-489.

Gaffey M. J. and McCord T. B. (1977) Asteroid surface materials: Mineralogical characterizations and cosmological implications. *Proc. Lunar Sci. Conf. 8th,* pp. 113-143.

Gault D. (1973) Displaced mass, depth, diameter, and effects of oblique trajectories for impact craters formed in dense crystalline rock. *The Moon, 6,* 32-44.

Grieve R. A. F. and Head J. W. (1983) The Manicougan impact structure: An analysis of its original dimensions and form. *Proc. Lunar Planet. Sci. Conf. 14th,* in *J. Geophys. Res., 88,* A867-A818.

Grieve R. A. F., Robertson P. B., and Dence M. R. (1981) Constraints on the formation of ring impact structures based on terrestrial data. *Proc. Lunar Planet. Sci. 12B,* pp. 665-678.

Hall J. L., Solomon S. C., and Head J. W. (1981) Lunar floor-fractured craters: Evidence for viscous relaxation of crater topography. *J. Geophys. Res., 86,* 9537-9552.

Hawke B. R., MacLaskey D., McCord T. B., Head J. W., Pieters C. M., and Zisk S. H. (1979) Multispectral mapping of the Apollo 15—Apennine region: The identification and distribution of regional pyroclastics. *Proc. Lunar Planet. Sci. Conf. 10th,* pp. 2995-3015.

Hawke B. R., Lucey P. G., Spudis P. D., and Owensby P. D. (1989) Impact structures as crustal probes: A summary of recent progress (abstract). In *Lunar and Planetary Science XX,* pp. 391. Lunar and Planetary Institute, Houston.

Head J. W. (1975) Processes of lunar crater degradation: Changes in style with geologic time. *The Moon, 12,* 299-329.

Jaumann R. and Neukum G. (1989) Spectrophotometric analysis of the lunar Plinius/Apollo 17 region (abstract). In *Lunar and Planetary Science XX,* pp. 456-457. Lunar and Planetary Institute, Houston.

Johnson T. V., Mosher J. A., and Matson D. L. (1977) Lunar spectral units: A northern hemisphere mosaic. *Proc. Lunar Sci. Conf. 8th,* pp. 1013-1028.

Lucey P. G. and Hawke B. R. (1989) Imaging spectroscopy of the central highland from .7 to 1.00 μm (abstract). In *Lunar and Planetary Science XX,* pp. 594-595. Lunar and Planetary Institute, Houston.

Lucey P. G., Hawke B. R., Pieters C. M., Head J. W., and McCord T. B. (1986) A compositional study of the Aristarchus region of the Moon using near-infrared reflectance spectroscopy. *Proc. Lunar Planet. Sci. Conf. 16th,* in *J. Geophys. Res., 91,* D344-D354.

McCord T. B., ed. (1988) Reflectance spectroscopy in planetary science: Review and strategy for the future. *NASA SP-493.* 45 pp.

McCord T. B., Charette M. P., Johnson T. V., Lebofsky L. A., Pieters C. M., and Adams J. B. (1972) Lunar spectral types. *J. Geophys. Res., 77,* 1349-1359.

McCord T. B., Pieters C. M., and Feierberg M. A. (1976) Multispectral mapping of the lunar surface using ground-based telescopes. *Icarus, 29,* 1-34.

McCord T. B., Clark R. N., Hawke B. R., McFadden L. A., Owensby P. D., Pieters C. M., and Adams J. B. (1981) Moon: Near-infrared spectral reflectance, a first good look. *J. Geophys. Res., 86,* 10,883-10,892.

Mellier Y., Cailloux M., Dupin J. P., Fort B., Lours C., Picat J. P., and Tilloles P. (1986) Evaluation of the performance of the 576384 Thomson CCD for astronomical use. *Astron. Astrophys., 157,* 96-100.

NASA (1971) Lunar topographic photomap, Gassendi, sheets A and B, Lunar Orbiter-V-Site 43.2, 1/250 000.

Pieters C. M. (1977) Characterization of lunar mare basalt types—II: Spectral classification of fresh mare craters. *Proc. Lunar Sci. Conf. 8th,* pp. 1037-1048.

Pieters C. M. (1978) Mare basalt types on the front side of the Moon: A summary of spectral reflectance data. *Proc. Lunar Planet. Sci. Conf. 9th,* pp. 2845-2849.

Pieters C. M. (1982) Copernicus crater central peak: Lunar mountain of unique composition. *Science, 215,* 59-61.

Pieters C. M. (1986) Composition of the lunar highland crust from near-infrared spectroscopy. *Rev. Geophys., 24,* 557-578.

Pieters C. M., Head J. W., McCord T. B., Adams J. B., and Zisk S. (1975) Geochemical and geological units of Mare Humorum: Definition using remote sensing and lunar sample information. *Proc. Lunar Sci. Conf. 6th,* pp. 2689-2710.

Pieters C. M., Head J. W., Adams J. B., McCord T. B., Zisk S. H., and Whitford-Stark J. L. (1980) Late high-titanium basalts of the western maria: Geology of the Flamsteed region of Oceanus Procellarum. *J. Geophys. Res., 85,* 3913-3938.

Pieters C. M., Adams J. B., Mouginis-Mark P. J., Zisk S H., Smith M. O., Head J. W., and McCord T. B. (1985) The nature of crater rays: The Copernicus example. *J. Geophys. Res., 90,* 12,393-12,413.

Pike R. J. (1968) Meteoritic and Consequent Endogenic Modification of Large Lunar Craters—A Study in Analytical Geomorphology. Ph.D. dissertation, Univ. of Michigan, Ann Arbor. 404 pp.

Pike R. J. (1980) Geometric interpretation of lunar craters. *U.S. Geol. Surv. Prof. Pap. 1046-C,* C1-C77.

Pinet P. C. (1985) Cratérisation lunaire: distribution et moyen d'étude rhéologique. Ph.D. dissertation, Toulouse Univ., France. 158 pp.

Pinet P. C. and Souriau M. (1989) Structural evolution of large lunar craters (abstract). In *28th International Geological Congress,* in press.

Schultz P. H. (1976a) Floor-fractured lunar craters. *The Moon, 15,* 241-273.

Schultz P. H., ed. (1976b) *Moon Morphology.* Univ. of Texas, Austin. 826 pp.

Settle M. and Head J. W. (1979) The role of rim slumping in the modification of lunar impact craters. *J. Geophys. Res., 84,* 3081-3096.

Vilas F. (1985) Mercury: Absence of crystalline Fe^{2+} in the regolith. *Icarus, 64,* 133-138.

Wilhelms D. E. and McCauley J. F. (1971) Geologic map of the nearside of the Moon. *U.S. Geol. Surv. Map I-703.* Washington, DC.

Proceedings of the 20th Lunar and Planetary Science Conference, pp. 195-206
Lunar and Planetary Institute, Houston, 1990

Terrestrial Analogs to Lunar Sinuous Rilles: Kauhako Crater and Channel, Kalaupapa, Molokai, and Other Hawaiian Lava Conduit Systems

C. R. Coombs[1], B. R. Hawke, and L. Wilson[2]

Planetary Geosciences Divison, Hawaii Institute of Geophysics, Honolulu, HI 96822

Two source vents, one explosive (Pu'u 'Ula) and one effusive (Kauhako Crater), erupted to form a cinder cone and low lava shield (Pu'u 'Uao) that together compose the Kalaupapa peninsula of Molokai, Hawaii. A 50-100-m-wide channel/tube system extends 2.3 km northward from Kauhako crater in the center of the shield. Based on modeling, a volume of up to ~0.2 km^3 of lava erupted at a rate of 260 m^{3}sec^{-1} to flow through the Kauhako conduit system in one of the last eruptive episodes on the peninsula. Channel downcutting by thermal erosion occurred at a rate of ~10 μmsec^{-1} to help form the 30-m-deep conduit. Two smaller, secondary tube systems formed east of the main lava channel/tube. Several other lava conduit systems on the islands of Oahu and Hawaii were also compared to the Kauhako and lunar sinuous rille systems. These other lava conduits include Whittington, Kupaianaha, and Mauna Ulu lava tubes. Morphologically, the Hawaiian tube systems studied are very similar to lunar sinuous rilles in that they have deep head craters, sinuous channels, and gentle slopes. Thermal erosion is postulated to be an important factor in the formation of these terrestrial channel systems and by analogy is inferred to be an important process involved in the formation of lunar sinuous rilles.

INTRODUCTION

Despite recent advances, the origin of lunar sinuous rilles remains controversial (*Coombs et al.,* 1987; *Spudis et al.,* 1987). Mechanisms believed feasible for their origin include: (1) thermal erosion by lava (*Hulme,* 1973; *Head and Wilson,* 1980, 1981; *Coombs et al.,* 1987), (2) structural control of lava flows (*Spudis et al.,* 1987), (3) formation as constructional features (*Kuiper et al.,* 1966; *Greeley,* 1971), (4) the result of lava lake drainage (*Howard et al.,* 1972), or (5) some combination of the above. Due to the current inaccessibility of the Moon for fieldwork, we have studied possible terrestrial analogs as part of a larger study of the nature and origin of lunar sinuous rilles. The lava tubes and channels associated with both recent and historic Hawaiian basalt flows have some broad morphological similarities to the rilles found on the lunar surface (e.g., *Greeley,* 1971; *Cruikshank and Wood,* 1972). These previous studies outlined the basic similarities between the lunar rilles and terrestrial lava channels and their associated tubes (i.e., basaltic, sinuous, tube-forming). In this current study we investigate the basic processes responsible for the formation of lunar rilles as well as the factors that control their development. During this investigation we identified a volcanic complex on the peninsula of Kalaupapa, Molokai, that has many of the characteristics of lunar sinuous rilles and appears to be a viable analog.

Recent photographic and map reconnaissances have shown that, morphologically, Kauhako crater/channel is very similar to many lunar sinuous rilles (*Coombs and Hawke,* 1988). Both the Kauhako and lunar channels formed as a result of basaltic volcanism, have deep source craters, exhibit some degree of tube formation, and follow sinuous paths. Also, according to the models presented by *Hulme* (1973) and *Wilson and Head* (1980, 1981), thermal erosion may have played a role in the formation of many lunar sinuous rilles and some terrestrial channels. This paper describes the geology, morphology, and volcanic history of Kauhako crater and channel and offers a thermal erosion model for its formation. Data used in this study include topographic and geologic maps, aerial photographs, and measurements taken in the field. In addition, preliminary studies of the geology and origin of several other lava conduit systems on the islands of Hawaii and Oahu were conducted.

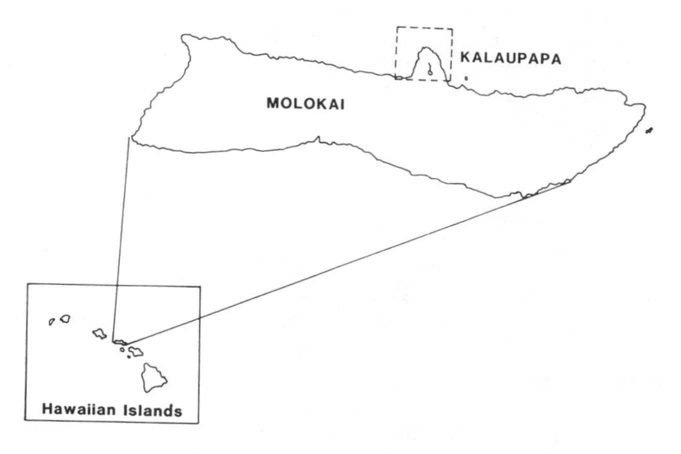

Fig. 1. Location map of Kalaupapa peninsula, a 100-km^2 shield volcano on the north shore of Molokai.

[1]Now at Mail Code SN15, NASA Johnson Space Center, Houston, TX 77058

[2]Also at Environmental Science Division, Institute of Environmental and Biological Sciences, University of Lancaster, Lancaster, LA1 4YQ, England

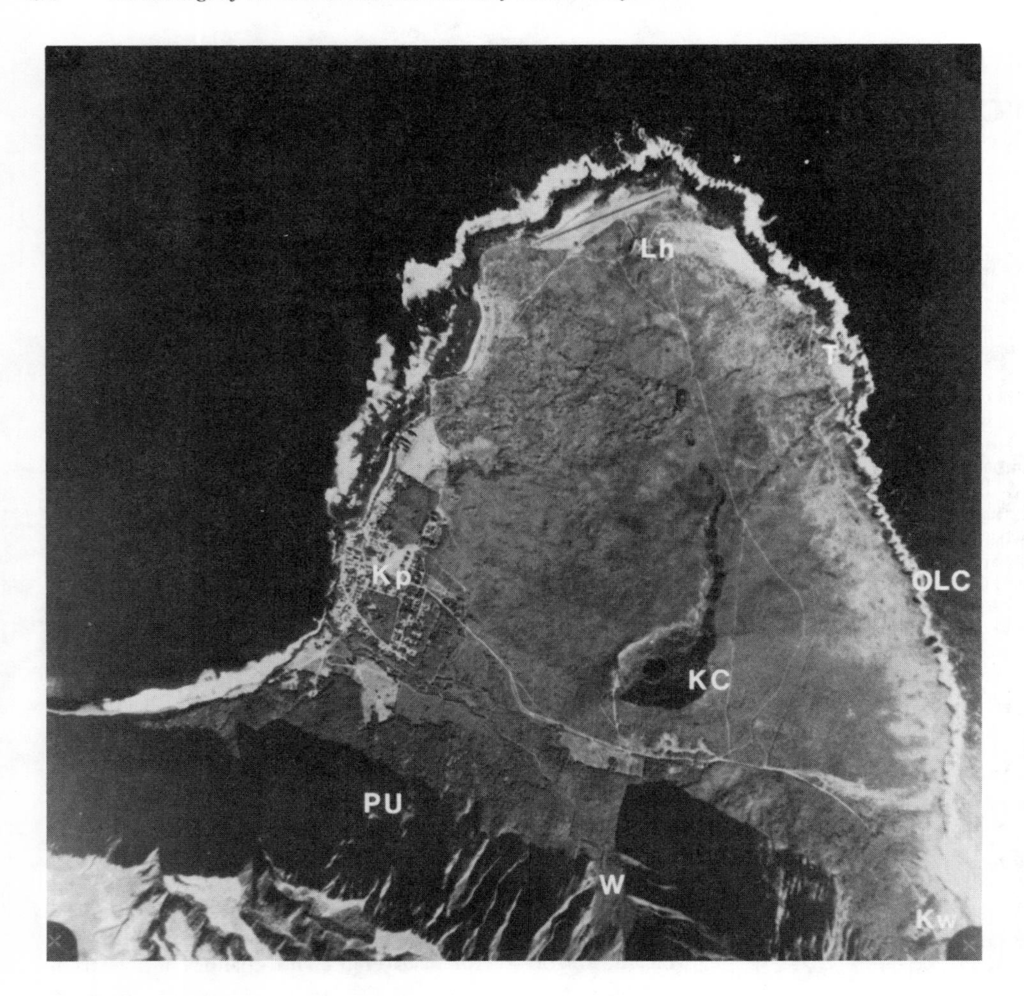

Fig. 2. Air photograph of the Kalaupapa peninsula. Kp - Kalaupapa, Kw - Kalawao, Lh - Lighthouse, KC - Kauhako crater/channel, T - Tumuli, PU - Pu'u 'Ula, W - Waihanau Stream, OLC-Old Lady Cave. North is at the top. (Photograph courtesy of the R. M. Towill Corporation, Honolulu, HI.)

GEOLOGIC SETTING

The Kalaupapa (*the flat plain*) peninsula is located along the northern cliffline of Molokai, the fifth largest island in the Hawaiian Archipelago (Fig. 1). The pali (*cliffs*) bounding the southern end of Kalaupapa locally reach 1100 m and are among the tallest sea cliffs in the world. Molokai is composed of two very large shield volcanoes and two posterosional stage volcanoes. West Molokai volcano, the oldest, rises 421 m above sea level, whereas the younger Wailau (East Molokai) volcano stands 1515 m above sea level. Both large shield volcanoes are capped by alkalic rocks from their post caldera eruptive phase and have been highly eroded (*Clague et al.,* 1982; *Compton,* 1985).

Pu'u 'Uao, or the Kalaupapa shield (Figs. 2 and 3), formed during the posterosional, or rejuvenation, stage of the Hawaiian islands. Access to the lava channel in the center of the Kalaupapa peninsula is severely restricted due to the presence of the Kalaupapa Settlement for those with Hansen's Disease. Consequently, very little is known about the geology and eruptive history of the Kalaupapa shield. Petrologic work done at the turn of the century by *Möhle* (1902) showed the Molokai lavas to be typical basalts. A much more comprehensive petrologic analysis of the Kalaupapa basalts was made by *Stearns and Macdonald* (1947), who also constructed the only geologic map of Molokai to date. More recently, a detailed investigation of the petrology, mineralogy, and geochemistry of the East Molokai volcanic series was completed by *Beeson* (1976).

Naughton et al. (1980) published a K-Ar age of 1.24 ± 0.57 m.y. for a single sample of basalt collected from Kauhako crater in the center of the peninsula. *Clague et al.* (1982) determined K-Ar ages of 0.34 to 0.57 m.y. for the Kalaupapa basalts. Compared with the 1.5-1.53-m.y. age determined for the Wailau volcano (*Naughton et al.,* 1980), the ages determined by *Clague et al.* (1982) indicate an erosional period of nearly 1 Ma years for the pali to form.

A volcanological reconnaissance of the Kalaupapa peninsula indicates that Kalaupapa may be a monogenetic volcano (G.P.L. Walker and C. R. Coombs, unpublished data, 1988). As such, it may be the largest documented single eruptive unit in the Hawaiian chain. A total erupted volume of 3 km^3 for the entire peninsula, only 22% of which is visible above sea level, has been calculated. A second, explosive, source vent 1.4 km southwest of the main crater, has also been identified.

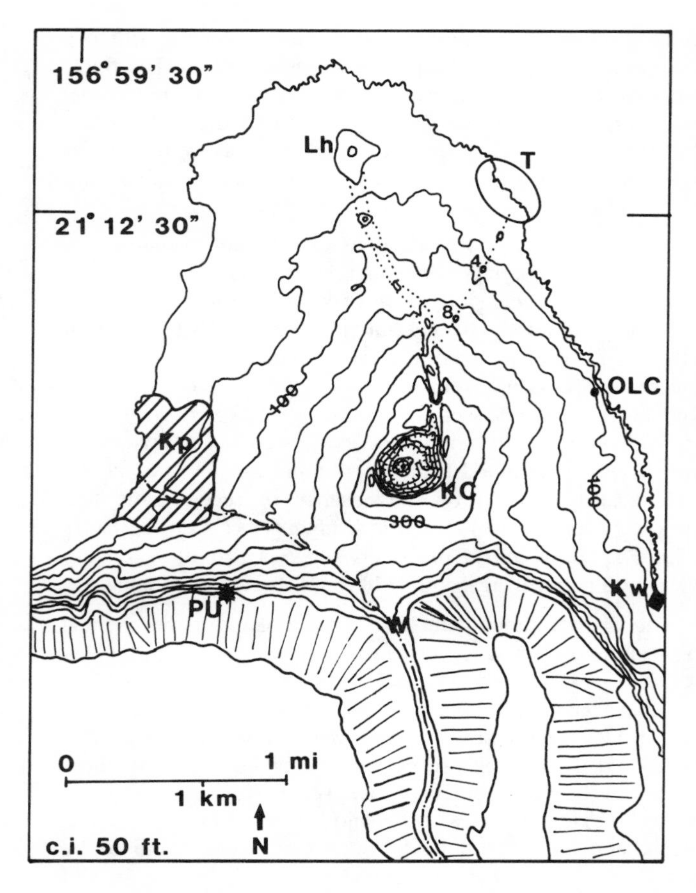

Fig. 3. Sketch map of the Kalaupapa peninsula. Topographic contours from Kaunakakai Quadrangle. Letters refer to locations described in Fig. 2 text. Double dotted line indicates the trend of the Kauhako channel/tube; single dotted line indicates the trend of the secondary tube system and the location of skylights #4 and #8 described in text.

FIELD STUDIES OF KAUHAKO CRATER AND CHANNEL

Mapping, measurements of the channel stratigraphy and morphology, and sample collecting were completed during eight days in February and August 1988. Progress in the field was often hampered by (1) the extremely dense foliage, (2) restricted access due to the presence of archeological/historical sites, (3) abundant wild game, and (4) extremely adverse weather.

Kalaupapa is a 10 km^2 lava shield volcano that rises to 135 m above sea level at Pu'u 'Uao (Peacemaker Hill, Figs. 2-5). It is relatively untouched by erosion, and the lava surfaces appear quite fresh (*Macdonald et al.,* 1986). A terrace composed of stream-laid conglomerate located 30 m above sea level on the western edge of the peninsula, near the site of the old Kalawao Leprosarium, was interpreted by *Macdonald et al.* (1986) to indicate a higher stand of the sea at one time. The inferred age of this graded stream terrace is in agreement with the K-Ar ages presented by *Clague et al.* (1982) and indicate a late Pleistocene formation of the peninsula.

Kauhako Crater

The Kalaupapa shield is capped by Kauhako crater (Figs. 2 and 3), measuring 500 m by 650 m in diameter, with a rim elevation of 135 m above sea level (a.s.l.). The crater forms a funnel-like pit with one circum-crater terrace and a lake in the bottom (Fig. 4). The terrace is at 40 m a.s.l. and is ~150 m wide. Kauhako Lake, at the bottom of the crater, is 50 m wide and 248 m deep (*Maciolek,* 1975). This depth has subsequently been confirmed by the U.S. Navy (Petty Officer Tano, personal communication, 1988). Also, surveys made in April 1988 by the U.S. Navy showed that no terraces are located under the water, but rather the conduit constricts to 20 m in diameter at a depth of 100 m on the southern side of the lake. Studies by *Maciolek* (1975, 1982) revealed some tidal fluctuations within the lake in 1973; none has been recorded since, however.

A second, smaller pit about 22 m wide by 10 m deep is located on the northeastern part of the crater terrace, near the mouth of the lava channel. This pit intersects a north-south trending lava tube at its base that measures 7 m wide by 5 m high and continues for an unknown distance northward. An avalanche has covered the northern entrance to the tube and sufficiently blocked the southern exposure to safe entry; thus, the tube was not examined.

Kauhako Channel System

A sinuous lava channel/tube extends northward from the northeast side of Kauhako crater (Figs. 2-4). The discontinuous channel is 1.0 km long, up to 30 m deep, and varies in width from 100 m to 150 m. Tumuli, formed by the squeeze-up of magma from an underlying tube system, extend the line of the channel another 1.3 km to the north-northwest. These tumuli presumably mark the course of a lava tube that once fed them. This tube appears to be an extension of the channel. The largest of these tumuli is 25 m high and is now the location of the Kalaupapa lighthouse (Figs. 2-4).

The walls of the crater and main channel/tube system are composed of two stratigraphic sequences: one a composite of

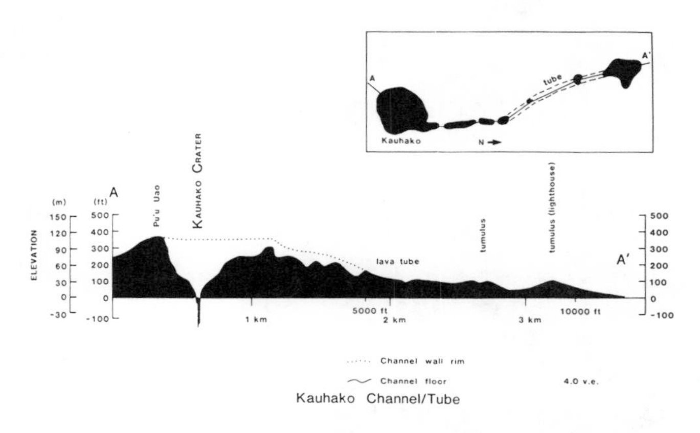

Fig. 4. Two-dimensional longitudinal section of the Kauhako channel-tube system based on the topographic map of the Kaunakakai Quadrangle.

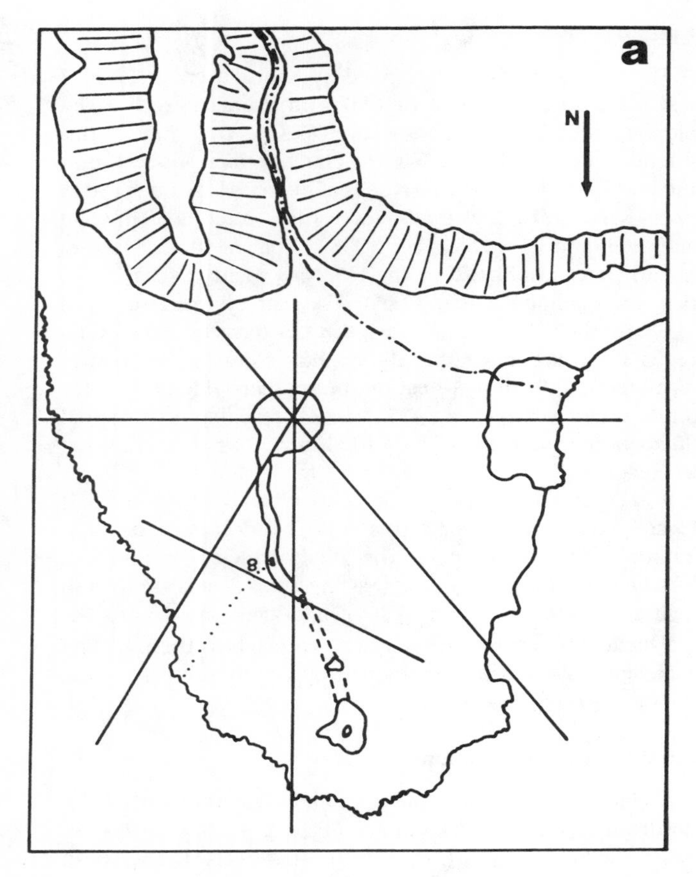

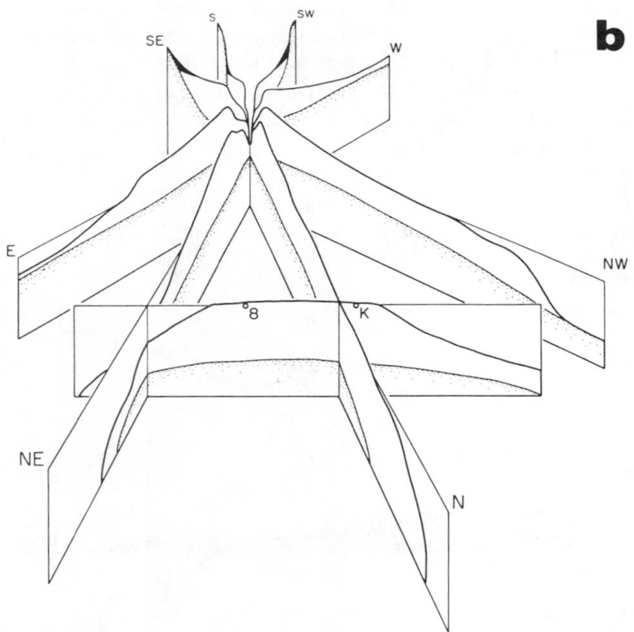

Fig. 5. **(a)** Reference sketch map for **(b)**. **(b)** Three-dimensional cross-section of the Kalaupapa peninsula. Heavy line indicates the ground surface. Stippled line indicates approximate boundary for the underlying stratigraphic unit. K indicates the location of Kauhako channel and 8 indicates the location of skylight #8 in the secondary tube system.

relatively thick and massive flow units, and the other a composite of very thin flows units (Figs. 6 and 7). The thick, massive sequence is exposed at the very base of the channel wall near the crater and around the base of the crater walls. Individual layers in it are 1-3 m thick. This massive unit meets the water surface of Kauhako Lake and continues down into the depths.

The crater rim and channel/tube walls are composed of multiple thin (average thickness is 10 cm) pahoehoe lava layers. A recent field study of an exposure of the west channel wall near the crater-channel junction revealed that these thin layers are not of pyroclastic origin, but rather that each thin layer is an individual lava flow (Figs. 6 and 7). More than 200 lava flows, or layers, were estimated to be present in the 25-m-thick section. The fact that these layers are not of pyroclastic origin was determined from the 5-6 m lateral extent of the individual layers and by the absence of recognizable lava droplets (Pele's tears) or scoria fragments between the layers (Fig. 7). The width of the layers suggests that they are thin lava flows rather than spatter balls that flattened upon impact.

The discontinuous nature and morphology of the channel as well as the presence of three land bridges along it (Figs. 2-5) suggest that the channel is almost certainly a partially collapsed lava tube. Efforts to inspect the two largest of the land bridges failed, however, to reveal a tube. This failure was attributed to the collapse of the original tube roof in these areas and the extremely dense vegetation and extensive reworking on both sides of the land bridges. Where accessible, the channel floor and land bridge wall were found to be extensively terraced and reworked, presumably by the ancient Polynesians, into housing platforms, walls, and burial sites.

Many blocks of the thin layered "rim" material were found on the channel floor, however, suggesting that the lava once formed a roof or crust that extended across the channel. This material may have subsided or collapsed when the magma supply diminished and the lava drained out. Much of the fallen material may have been incorporated in and/or carried away by the active flow. Several large blocks of material found on the crater floor show evidence for some thermal erosion. Their otherwise uniformly layered surface is contorted into abstract forms and several small "tubes" extend inward from the surface. The sheer size and morphology of this channel/tube suggest that it may have been thermally eroded, as is discussed and modeled below.

A second prominent tube system extends north-northwest from an elbow in Kauhako channel to the eastern coast (Figs. 2 and 3). A series of tumuli 5-10 m high occurs along the coastline and may have been associated with this tube. The tube itself is discontinuous along its trend, but eight skylights were found in the field, six of which allowed entrance into the lava tube. The original glassy surface and stalactites that developed during its formation are still intact along most of the tube. Several smaller tubes radiate outward from the main tube forming a small lava distributary system. Seven stairs leading down into the tube and a 1.5-m-high stone wall at the entrance indicate that this portion of the tube was once used for living and/or storage quarters. The flat floor is covered with

Fig. 6. Photograph of Kauhako channel wall. Photo shows the upper part of the measured section described in the text. Photo was taken from the landbridge closest to channel.

a layer of dirt. This dirt was most likely laid down by the early settlers and/or by runoff and drainage into the tube. It is unlikely that the original floor was observed.

A third tube, Old Lady cave (*Kaupikiawa*), is located on the eastern shore of the peninsula (Figs. 2 and 3). This tube can be entered from a skylight located 100 m inland from the edge of the sea cliff. The tube is 570 cm wide by 280 cm high at the edge of the cliff and stands 15 m a.s.l. (Fig. 8). This tube exhibits some of its original glassy formation surface and stalactites, but a great deal of the roof has given way as a result of small collapses. No other skylights were found on this tube.

EXPLOSIVE ACTIVITY AT PU'U 'ULA

A second, previously unknown, source vent was identified while in the field. This second vent is located near the base of the pali bounding the southern side of the peninsula, 1.4 km southwest of Kauhako crater (Fig. 3). The pyroclastic cone, for which the name Pu'u 'Ula (*Red Hill*) is proposed, is well bedded and dips steeply toward the sea. Lithics, poorly consolidated scoria, bombs, and Pele's tears are abundant. The upper layers contain bombs and have weathered to a noticeable bright red. The cone is extremely eroded due to its friable composition, the steep slope, and its location at the base of a water-eroded gully.

That this cone is associated with the Kauhako eruption has not been firmly established. However, the fact that these two vents formed during the posterosional stage of volcanism and that they are aligned along the same trend as the Koko fissure and other alignments associated with the rejuventation-stage volcanism on Oahu suggest that they may be related. Also, the presence of this pyroclastic cone explains the absence of any pyroclastic material around the main crater rim and lava channel. We suggest that degassing occurred at this vent. An analogy may be drawn between this Pu'u 'Uao-Kauhako system and the currently active Pu'u 'O'o-Kupaianaha system on Hawaii where, since July 1986, lavas erupted on Kilauea have issued quietly and almost continuously from the Kupaianaha lava pond, while degassing occurred at the Pu'u 'O'o cone 1.7 km away. Similar explosive activity is thought to have occurred with the formation of some lunar sinuous rilles (e.g., Rima Mozart; *Coombs et al.,* 1987).

Fig. 7. Photograph of a single flow unit (vesiculation unit) composing the channel wall. Vesicle size increases toward the top, and layers are separated by a gas blister in the center. This individual unit is 14 cm thick.

Fig. 8. Photograph of Old Lady cave on the eastern coast of Kalaupapa. View is looking toward the west. Figure in photo is 1.8 m tall.

THERMAL EROSION OF THE KAUHAKO CONDUIT

Several models have been developed for the formation of lunar sinuous rilles in conjuction with thermal erosion (*Hulme,* 1973; *Carr,* 1974; *Wilson and Head,* 1980, 1981; *Coombs et al.,* 1987). *Carr* (1974) argued that thermal erosion could occur if the eruption lasted long enough and if the lava moved in a laminar manner. *Hulme* (1973), on the other hand, showed that turbulent flow would lead to faster and more efficient thermal erosion. These models may also be utilized in modeling terrestrial eruptions despite the differences in the dimensions of the flows. This variation in dimension is due to the physical differences between the Earth and Moon: (1) lunar eruptions occur in a vacuum making the lava outgas much more rapidly and completely than on Earth (*Hulme,* 1973); (2) lunar gravity field is one-sixth that of the Earth, allowing the lava to travel more slowly on similar slopes and to form basaltic flows almost twice as deep from similar eruption rates on Earth (*Hulme,* 1973; *Hulme and Fielder,* 1977; *Wilson and Head,* 1983); and (3) lunar basaltic lavas are less viscous than terrestrial basaltic lavas (*Hulme,* 1973). These physical differences are accounted for in the calculations of thermal erosion, allowing for comparison between the two bodies.

To summarize briefly the differences between laminar and turbulent flow in a channel: in a laminar flow each portion of lava flows along a path roughly parallel to the ground, while in a turbulent flow, material from different depths is constantly being mixed. Heat is transferred by conduction perpendicular to the flow direction in a laminar flow and by convection in a turbulent flow. Heat loss is much more rapid in the turbulent flow since hot fresh lava is constantly brought into contact with the flow boundaries and the surface temperature is somewhat higher than in a laminar flow, where heat is lost by radiation from the surface and conduction into the underlying rock. The lava at the base of the flow is likely to be above its solidus temperature and may also be above the solidus temperature of the underlying rock (*Hulme,* 1973). These elevated temperatures along with the turbulent flow from above allow the flowing lava to melt and pluck away the underlying substrate. Although there is a mechanical aspect to the process (ie., plucking), it is minor, and the entire process is generally referred to as thermal erosion.

Thus, in order to form a rille, or lava channel, where thermal erosion plays a significant role, an initial flow is required where a high flow velocity (favored by a high effusion rate) creates turbulent flow. This maximizes the efficiency of heat transfer to the basal layer and leads to the heating of the substrate rocks above their solidus temperature (*Head and Wilson,* 1981). Then, as partial melting and mechanical erosion excavate the underlying terrain, the flow is progressively lowered. The vertical erosion rate is always greatest nearest the vent where the lava is hottest and is always zero at some finite distance downstream (*Hulme,* 1973). Beyond this zero point, the rille ceases to excavate by thermal erosion and a leveed lava flow continues across the surface.

In the following analysis of Kauhako crater and conduit, we have used the available topographic data in conjunction with

TABLE 1. Definition of variables.

Symbol	Quantity	Value
k	Thermal conductivity of lava	$1\ Wm^{-1}K^{-1}$
c	Specific heat of lava and of ground	$730\ Jkg^{-1}K^{-1}$
ρ_l	Density of lava and of ground	$1 \times 10^3\ kgm^{-3}$
T_{mg}	Melting temperature of ground	1360 K
L	Latent heat of fusion of ground	$5 \times 10^5\ Jkg^{-1}$
σ	Stefan-Boltzmann constant	$5.6703 \times 10^{-8}\ Wm^{-2}K^{-4}$
ϵ	Emissivity of the lava surface	1
Q	Volume rate of flow per unit width of channel	$5.2\ m^2sec^{-1}$
η	Viscosity of lava	$30\ kgm^{-3}sec^{-1}$
ρ	Density(ρ_l - of lava, ρ_g - of ground)	$1000\ kgm^{-1}$
u	Mean flow velocity	$6.8\ msec^{-1}$
h	Heat transfer coefficient at interface	$74.6\ Wm^{-2}k^{-1}$
E	Erosion rate	$10.5\ \mu msec^{-1}$

Hulme's (1973) model for thermal erosion of a lunar sinuous rille. By changing the variables to terrestrial values, this model should provide an adequate picture of the thermal erosive history of Kauhako crater/conduit. Here, we present a synopsis of Hulme's model rather than the entire method. For a complete presentation of this model we refer the reader to *Hulme* (1973). Table 1 lists the values and variables used in these calculations.

The volume rate of flow per unit width of channel, Q, was determined by

$$Q = \frac{3\,\sigma\epsilon\,X_m}{\rho_1 c\left(\dfrac{1}{T_{mg}^3} - \dfrac{1}{T_{1_0}^3}\right)} \qquad (1)$$

where σ is the Stefan-Boltzman constant, ρ_l is the density of the lava, c the specific heat, ϵ the emissivity (for the lava surface), X_m the length of the rille, T_{1_0} the temperature of the lava at the source, and T_{mg} the solidus temperature of the ground.

The depth, d, of the flowing lava, which in general does not fill the rille once sufficient erosion has taken place, is given by

$$d = Q^{2/3}\left(\frac{f}{2\,g\,\alpha}\right)^{1/3} \qquad (2)$$

where f is the friction factor, α is the slope of the ground, and g the acceleration due to gravity. Depth (d) then can be used to calculate the mean flow velocity in the channel, u.

$$u = \left(\frac{2\,g\,\alpha\,d}{f}\right)^{1/2} \qquad (3a)$$

Or, more simply

$$u = \frac{Q}{d} \qquad (3b)$$

If the lava viscosity is η, the Reynolds number for the flow will be

$$Re = \rho_1 \frac{Q}{\eta} \qquad (4)$$

If Re is of the order of 10^3 or greater, the flow will be turbulent, the optimum condition for thermal erosion to occur.

Using the standard engineering treatment of heat transfer from a fluid in a pipe and converting to an open channel geometry (after *Hulme,* 1973)

$$h = 0.017k^{0.6}\left(\frac{c}{\eta}\right)^{0.4}\rho_1^{0.8}Q^{0.13}\left(\frac{2\,g\,\alpha}{f}\right)^{0.33} \qquad (5)$$

where h is the heat transfer coefficient at the head of the stream, k the thermal conductivity.

The actual thermal erosion rate, E, is given by

$$E = h\,\frac{(T_{l_0} - T_{mg})}{\rho_g L}\cdot\frac{N}{(1+N)} \qquad (6)$$

where ρ_g and L are the density and latent heat of fusion of the ground, respectively, and N is defined by

$$N = \frac{L}{c(T_{mg} - T_0)} \qquad (7)$$

and T_0 is the initial temperature of the ground.

The flow parameters for Kauhako channel/tube were calculated as follows: Assuming values of T_{mg} and T_{1_0} of 1343 K and 1478 K, respectively; Q was calculated to be 5.2 m^2sec^{-1} for the visible channel/tube length (X_m) of 2.3 km.

The viscosity of the erupted lava, calculated and derived from the field samples, was close to 30 $kgm^{-1}sec^{-1}$. Using a lava density of 1000 kgm^{-3} (estimated as typical of the vesicular field samples) and the above value of Q, equation (4) then implies that the value of Re for this event was 173. This value implies that the flow was not in fact fully turbulent (Re $\gtrsim$2000); the motion was, however, well into the transition region between laminar and turbulent flow. If a more typical density value for terrestrial basalts is used, such as 2500 kgm^{-3}, the resulting Reynold's number is 433 and the motion of the flow would be considered more turbulent. Using data from *McAdams* (1954), a value of Re = 173 for a Newtonian fluid moving in what we assume to have been a relatively smooth channel implies a friction factor, f, of 0.017.

The overall gradient of the channel floor, measured from the U.S. Geological Survey Kaunakakai topographic quadrangle, is near 3° or 0.052 radians. Using this value, together with f and Q in equation (2) leads to a flowing lava depth of 0.77 m; equation (3b) then gives the lava velocity as 6.8 $msec^{-1}$.

The value of h in equation (5) can now be calculated as 74.6 $Wm^{-2}K^{-1}$, and insertion of this into equation (6) leads to a value for the thermal erosion rate. It should be noted that L, the latent heat of fusion of the ground required in equation (6), is close to 5×10^5 Jkg^{-1}; however, following *Hulme* (1973), we assume that only 40% of the ground needs to be melted before mechanical forces sweep away the remainder, and so modify L to 2×10^5 Jkg^{-1}. Using this value, the thermal erosion rate E is found to be 10.5 $\mu msec^{-1}$.

Taking into consideration the amount of time since the Kalaupapa peninsula formed, natural erosion, and the reworking of the channel walls and floor, it is difficult to determine the original channel/tube dimensions. Thus, based on field measurements, average values were used. For these calculations, it was estimated that 5 to 10 m of the total 30-m-deep channel was downcut by thermal erosion. Also, a mean channel-bottom floor width of 50 m was used, rather than the 100-150-m widths measured along the top of the channel walls.

The period of eruption is then taken to be the time required to erode 5 m to 10 m at a rate of 10.5 $\mu msec^{-1}$, or 0.5-0.9 $\times$ 10^6 sec (5 to 10 days). The total volume flux through the channel, or eruption rate, is 260 m^3sec^{-1}. The amount of material that flowed through the channel during the last eruption is then estimated to be 0.1 to 0.2 km^3.

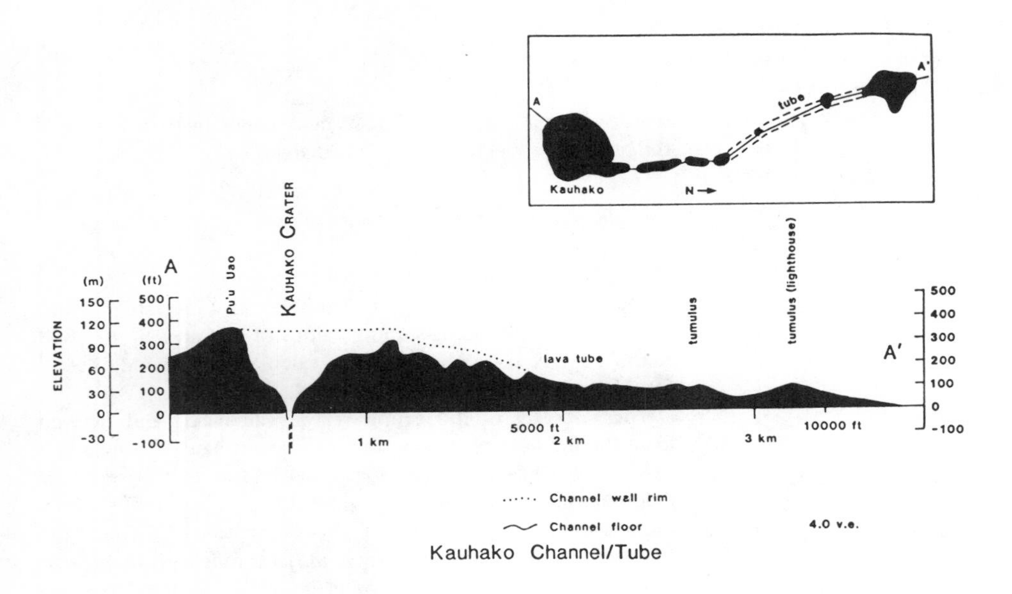

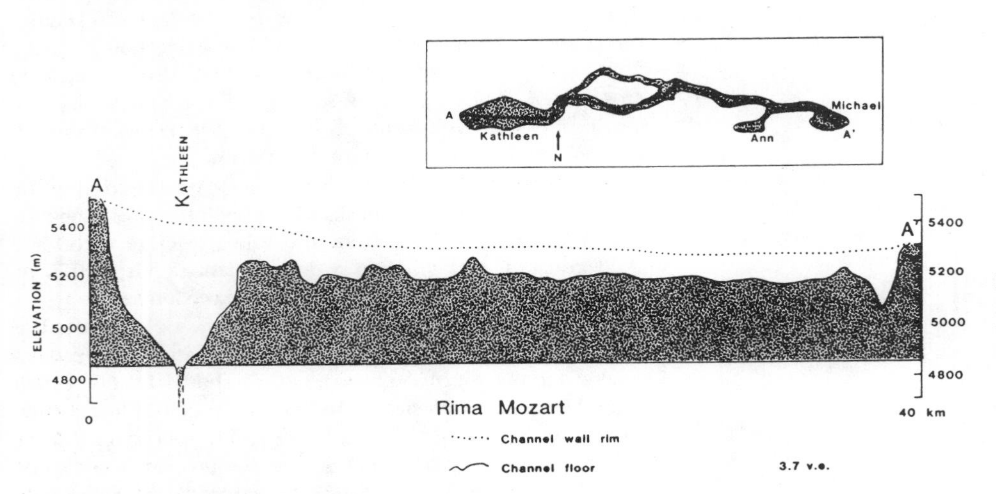

Fig. 9. Comparative figure showing the morphologic similarity between Kauhako channel and Rima Mozart. Note that the vertical exaggeration is about the same, but the Rima Mozart section is 10 times longer than Kauhako channel.

The calculated volume of 0.2 km^3 that flowed through the channel/tube is only a small fraction of the total volume (3.0 km^3) of material that comprises the Kalaupapa peninsula, estimated by Walker and Coombs (G.P.L. Walker and C. R. Coombs, unpublished data, 1988), and represents the last eruption sequence to have flowed through the channel. The remainder of the peninsula was formed by tube-fed and surface flows radiating outward from the main vent at Kauhako, as evidenced by the presence of lava tubes along the eastern coastline and in the floor of the crater.

HAWAIIAN ANALOGS TO LUNAR SINUOUS RILLES

Kauhako Conduit

Studies by *Greeley* (1971) and *Cruikshank and Wood* (1972) showed that terrestrial basaltic lava tubes and their associated features can be considered as valid analogs to lunar sinuous rilles. *Greeley* (1971) made an in-depth comparison of the rilles in the Marius Hills region in the northwestern part of the lunar nearside to terrestrial basaltic lava tubes in Oregon, California, and Washington. *Cruikshank and Wood* (1972), watched the eruptive activity of Mauna Ulu in Hawaii, and were able to observe several lava tubes in formation. A similar comparison is made here between the structure of the Kauhako conduit system on Kalaupapa, Molokai, to that of Rima Mozart, a lunar sinuous rille.

Both Kauhako channel and Rima Mozart originated at very deep source vents and evolved into large sinuous channel/tube systems (Fig. 9), were built up by overspilling of hot, low-viscosity, basaltic magma, and had a significant amount of thermal erosion involved in their downcutting. Kauhako channel/tube system originates in a very deep, noncircular crater (Figs. 4, 5 and 9) located atop a gently sloping (1.5-6°) shield. The channel walls are composed of many thin lava flows that formed a tube roof along several segments, three of which are still intact. The distal portion of the tube ends in a long lava tube segment, over which lie several large tumuli. Similarly, Rima Mozart originates at a very deep, elongate source crater on very gently sloping (0.1°) terrain (*Coombs et al.,* 1987). A sinuous channel formed from the erupted basalt (Fig. 9). The distal end of Rima Mozart terminates at a sink crater, presumably draining into an underground fissure system. The distal end of Kauhako channel is marked by a lava tube whose trend is marked by several tumuli.

Another feature common to both is the presence of a pyroclastic vent and the deposit of pyroclastic material associated with the channel/tube system. Pyroclastic activity at Pu'u 'Ula near Kalaupapa formed a cinder cone. Pyroclastic activity at Rima Mozart did not form a cinder cone, but rather formed noncircular deposits of what is interpreted to consist of spatter about the secondary source vents at Ann, as well as flat, asymmetrical deposits of pyroclastic debris about Kathleen (*Coombs et al.,* 1987).

Other Hawaiian lava conduit systems may also be considered analogs as they followed sinuous paths, have deep source craters, and show evidence of thermal erosion. These analog tube systems include the currently active Kupaianaha system, the recently active Mauna Ulu system, the older Whittington and Makapu'u lava tubes, and the prehistoric Thurston lava tube. Each of these conduits is briefly described below.

Mauna Ulu/Kupaianaha

During the course of the 1969-1974 eruption of Mauna Ulu, *Cruikshank and Wood* (1972) and *Peterson and Swanson* (1974) observed several channel/tube systems in formation. *Cruikshank and Wood* (1972) concluded that fluvial-like processes of meandering, bank cutting, channel capture, and channel deepening occur in active lava tubes and that degradation takes place by melting and plucking of country rock (thermal erosion) instead of by abrasion or dissolution (mechanical erosion). These same processes have been observed in the formation of the Kupaianaha lava flows on Kilauea's southeast rift zone by the present authors. Here, a 10-30-m wide by >100 m long extension has channeled lava since July 1986 from the southeast side of Kupaianaha lava pond into the master tube system that supplies lava to the Kalapana coast 3 km away.

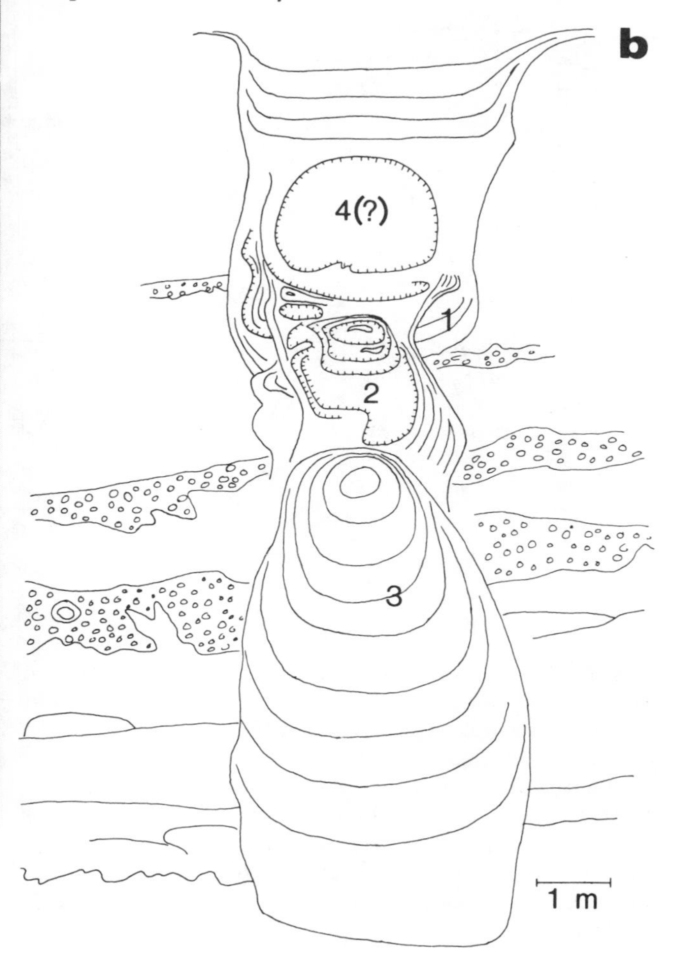

Fig. 10. (a) Photograph of Makapu'u lava tube. **(b)** Diagrammatic sketch map of Makapu'u lava tube illustrating the four (?) episodes of eruption and tube-forming phases. (Sketch by G. P. L. Walker)

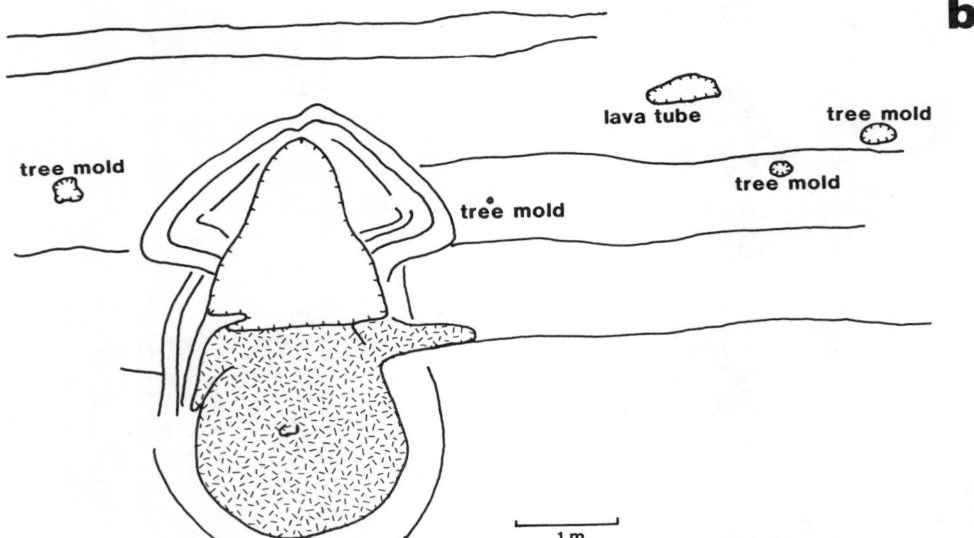

Fig. 11. **(a)** Photograph of Whittington lava tube, Whittington State Park, Hawaii. The tube is partially infilled: round section in bottom of tube. **(b)** Diagrammatic sketch of Whittington lava tube.

During its active periods, this tube has varied from an open channel to a crusted tube to a half channel-half tube feature. Collapse pits and skylights have also developed along its length allowing one to look in and observe the flow. Terraces and small ledges above the flow within the tube suggest that some thermal erosion may be occurring. That the active flow is below the ledges suggests that thermal erosion may be downcutting the base of the tube, and that the low level one observes is not necessarily due to a decrease in the volumetric lava flux.

Thurston Tube

Other Hawaiian channel/tube systems that may also be analogs to lunar sinuous rilles but were not observed during their formation include Thurston lava tube on Kilauea, Makapu'u lava tube on Oahu's eastern shore, and a lava tube along the southern coastline of the Big Island in Whittington State Park. Each of these lava tubes cuts across accreted layers of relatively thin lava flows and exhibits evidence for varying degrees of downcutting by thermal erosion.

Thurston lava tube is located along the Crater Rim Road in Hawaii Volcanoes National Park. It is prehistoric in age (~500 yr B.P.). The tube averages 5 m wide by 3 m high and still retains some of its original stalactites and glassy surface. Terraces or benches along the tube walls indicate several lava levels that may reflect the passage of several pulses of lava or pauses in the drawdown of lava through the tube as well as periods of thermal erosion activity. An examination of the tube entrance revealed that it cuts across multiple thin flows that make up the adjacent walls, much like the Kalaupapa lava tubes. Also, this lava tube is associated with a deep source crater, as are Kalaupapa, Mauna Ulu, Kupaianaha, and such lunar sinuous rille systems as Rima Mozart and Rima Hadley.

Makapu'u Tube

Both the Whittington and Makapu'u lava tubes (Figs. 10a,b and 11a,b) show evidence of thermal erosive activity in their formational histories. The Makapu'u tube, now completely filled in (Figs. 10a,b), was formed in several stages as illustrated in Fig. 10b. The first stage cut the main channel and is the conduit down which all the others flowed. A short period of thermal downcutting, or ablation, is thought to have been involved in this first stage. The erosion was then followed by a period of deposition when the lava flowing through the tube cooled and/or the supply rate decreased. Phase number two formed the jumbled section of the tube. Lavas flowing through this section of the tube cut through the preexisting layers formed in the first phase. Later this lava cooled and deposited several thin layers of its own. A third tube later formed at the base of this second tube. The third tube-forming phase eroded down through approximately 6.5-7 m of host material to form a long cylindrical-shaped tube system. This phase too ended in deposition of lava to form semiconcentric rings about the active flow. The last section to cool was the smallest circle seen at the top of this tube segment. It is unclear whether or not there was a fourth tube-forming phase. The uppermost section of the outcrop may have been either the fourth tube-building phase or is an upper extension of the second phase. In order for thermal erosion to have occurred, as is indicated by the evident cross-cutting relationships, the lava flowing through the tube must have been sufficiently hot for a relatively long period of time and flowed rapidly. A baked zone, or metamorphic aureole, around the margin of the tube also indicates that the material flowing through the tube was quite hot. Viscosity estimates are difficult to make at this outcrop since the lava lacks large mafic crystals.

Whittington Tube

Located on the southern shore of the Big Island, in Whittington St. Park, the Whittington tube (Figs. 11a,b) measures 1.9 m wide by 1.9 m high at its seaward exposure, and expands to 3.4 m wide by 4.6 m high just inside. Several phases of flow are evident within this tube, the last of which left the round exposure in the lower-central portion of the tube (Figs. 11a,b). This tube cut through several layers (0.9-1.2 m) of lava flows. Contorted layers adjacent to the open tube (Figs. 11a,b) suggest that the tube had a more complex history of formation than several of the smaller lava tubes (<1-m diameter) located nearby.

Evidence for thermal erosion in terrestrial lava tubes is difficult to find and to prove. It can, however, be demonstrated that thermal erosion played an important role in the formation of this tube. Several tree molds were identified at the base of two older, succesive lava flows adjacent to the tube, which are separated by a thin weathered paleosol. The presence of these features indicates that some time interval occurred between the eruption and deposition of the two flows, as the trees needed time to grow and the paleosol to form. Estimates for the difference in ages of these two flows is several hundred years, based on the size of the tree molds and the thickness of the paleosol. As illustrated in Fig. 11a and b, the tube cuts through these two layers, indicating that thermal erosion was involved in the formation of this tube. While it may be argued that the lava followed a preexisting fault plane or stream channel, no evidence was found in the field for the preexistence of either feature. The field evidence, rather, indicates that the tube did not form contemporaneously with the lava flows adjacent to it and supports a thermal erosion model for the downcutting of the tube into the previously formed stratigraphic layers. This lava tube is an unequivocal example of the involvement of thermal erosion in the formation of terrestrial lava tubes.

The actual whereabouts of the source vent for this tube is unknown, so proper calculations of the volume flux and erosion rate have not been made. Since thermal erosion is inferred to have occurred, however, it may be assumed that the lava flowing through the tube was very hot, had a low viscosity, a high density, and was moving quite rapidly.

CONCLUSION

Kalaupapa formed as a result of two types of volcanic activity, one pyroclastic at Pu'u 'Ula, and one effusive at Kauhako. The last eruption within the channel lasted for a period of 5 to 10 days and erupted a maximum volume of 0.2 km^3 at a rate of 260 m^3sec^{-1} that flowed through the channel/tube system. Also, based on the comparative analysis between the Kauhako system, other Hawaiian lava conduit systems and lunar sinuous rilles the following conclusions may be made:

1. The morphology and eruptive history of this volcano make it a good analog to lunar sinuous rilles (i.e., Rima Mozart). Morphologically, the Rima Mozart and Kauhako lava channels are very similar. Their channel floors are hummocky, their paths are sinuous and they begin at deep source craters. Volcanologically, the two are also very similar: both channel systems were carved out of local substrate by basaltic magma and were associated with a pyroclastic eruption phase (*Coombs et al.,* 1987).

2. The morphology and apparent eruptive history of Kauhako channel/tube make it a good analog to the Mauna Ulu and Kupaianaha eruptions on Kilauea, as well as several prehistoric eruptions on the Big Island of Hawaii and Oahu (i.e., the Thurston and Whittington and Makapu'u lava tubes).

3. Thermal erosion was an important process in the formation of the Kauhako channel/tube system. The hot, turbulent magma erupted from Kauhako crater helped to carve 5 m to 10 m of the sinuous, 30-m-deep channel/tube system at a rate of 10.5 μm sec^{-1} across the center of the Kalaupapa peninsula.

4. Thermal erosion also played an important role in the formation of several other terrestrial basaltic lava channels elsewhere in Hawaii: Whittington, Makapu'u, Mauna Ulu, and Kupaianaha.

5. Thermal erosion is an important process involved in the formation of lunar sinuous rilles.

Acknowledgments. The authors wish to express their sincere thanks to the residents of the Kalaupapa Settlement for those with Hansen's Disease, the Hawaii State Department of Health, and to N. Borgemeyer and H. Law of the Kalaupapa National Historical Park for sponsoring our visits and enabling us to conduct this research in a restricted community. Thanks are also gratefully extended to G. P. L. Walker for his field expertise and insightful discussions throughout this project, to G. Coombs and L. Parfitt for their help in the field, and to the R. M. Towill Corporation of Honolulu for providing air photographs of the peninsula. For their helpful reviews of the manuscript we would like to thank J. Whitford-Stark and P. Francis. This work was supported by a Harold T. Stearn's Fellowship awarded to C.R.C. and NASA grant numbers NAGW-237 and NAGW-437. This is PGD Publication No. 572 and HIG Contribution No.2193.

REFERENCES

Beeson M. H. (1976) Petrology, mineralogy, and geochemistry of the East Molokai Volcanic Series. *U.S. Geol. Surv. Prof. Pap. 961,* 53 pp.

Carr M. H. (1974) The role of lava erosion in the formation of lunar rilles and martian channels. *Icarus, 22,* 1-23.

Clague D. A., Dao-gong C., Murnane R., Beeson M. H., Lanphere M. A., Dalrymple G. B., Friesen W., and Holcomb R. (1982) Age and petrology of the Kalaupapa basalt, Molokai, Hawaii. *Pacific Sci., 36,* 411-420.

Compton R. R. (1985) The caldera of East Molokai volcano, Hawaiian Islands. *Res. Rept. Natl. Geogr. Soc., 21,* 81-87.

Coombs C.R. and Hawke B.R. (1988) Kauhako crater and channel, Kalaupapa, Molokai: A preliminary look at a possible analog to lunar sinuous rilles (abstract). In *Lunar and Planetary Science XIX,* pp. 207-208. Lunar and Planetary Institute, Houston.

Coombs C. R., Hawke B. R., and Wilson L. (1987) Geologic and remote sensing studies of Rima Mozart. *Proc. Lunar Planet. Sci. Conf. 18th,* pp. 339-353.

Cruikshank D. P. and Wood C. A. (1972) Lunar rilles and Hawaiian volcanic features: Possible analogs. *The Moon, 3,* 412-447.

Greeley R. (1971) Lunar Hadley Rille: Considerations of its origin. *Science, 172,* 722-725.

Head J. W. and Wilson L. (1980) The formation of eroded depressions around the sources of lunar sinuous rilles: Observations (abstract). In *Lunar and Planetary Science XI,* pp. 426-428. Lunar and Planetary Institute, Houston.

Head J. W. and Wilson L. (1981) Lunar sinuous rille formation by thermal erosion: Eruption conditions, rates and durations (abstract). In *Lunar and Planetary Science XII,* pp. 427-429. Lunar and Planetary Institute, Houston.

Howard K. A., Head J. W., and Swann G. A. (1972) Geology of Hadley Rille. *Proc. Lunar Sci. Conf. 3rd,* pp. 1-14.

Hulme G. (1973) Turbulent lava flow and the formation of lunar sinuous rilles. *Mod. Geol., 4,* 107-117.

Hulme G. (1982) A review of lava flow processes related to the formation of lunar sinuous rilles. *Geophys. Surv., 5,* 245-279.

Hulme G. and Fielder G. (1977) Effusion rates and rheology of lunar lavas. *Phil. Trans. R. Soc. London, A, 85,* 227-234.

Kaunakakai Quadrangle (1967) *U.S.G.S. Topographic Map of the Kaunakakai Quadrangle, Molokai Island, Hawaii - Maui Co.*

Kuiper G. P., Strom R. G., and LePoole R. S. (1966) Interpretation of the Ranger Records. *JPL Tech. Rept. 32-800,* 35-248.

Macdonald G. A., Abbott A. T., and Peterson F. L. (1986) *Volcanoes in the Sea: The Geology of Hawaii,* 2nd ed. Univ. of Hawaii Press, Honolulu. 517 pp.

Maciolek J. A. (1975) Limnological ecosystems and Hawaii's preservational planning. *Intl. Ver. Theor. Angew. Limnol. Verh., 19,* 1461-1467.

Maciolek J. A. (1982) Lakes and lake-like waters of the Hawaiian Archipelago. *Occasional Papers of Bernice P. Bishop Museum, XXV,* 1-14.

McAdams W. H. (1954) *Heat Transmission.* McGraw Hill, New York. 54 pp.

Möhle F. (1902) Beitrag zur Petrographie der Sandwich und Samoz Inseln. *Neues Jahrb. Mineral. Abh., 15,* 65-104.

Naughton J. J., Macdonald G. A., and Greenberg V. A. (1980) Some additional potassium-argon ages of Hawaiian rocks: the Maui volcanic complex of Molokai, Maui, Lanai, and Kahoolawe. *J. Volcanol. Geotherm. Res., 7,* 339-355.

Peterson D. W. and Swanson D. A. (1974) Observed formation of lava tubes. *Stud. Speleol., 2,* 209-222.

Spudis P. D., Swann G. A., and Greeley R. (1987) The formation of Hadley Rille and implications for the Geology of the Apollo 15 region. *Proc. Lunar Planet. Sci. Conf. 18th,* pp. 243-254.

Stearns H. T. and Macdonald G. A. (1947) Geology and groundwater resources of the island of Molokai. *Hawaii Hydrogr. Div. Bull., 11,* 113 pp.

Wilson L. and Head J. W. (1980) The formation of eroded depressions around the sources of lunar sinuous rilles: Theory (abstract). In *Lunar and Planetary Science XI,* pp. 1260-1262. Lunar and Planetary Institute, Houston.

Wilson L. and Head J. W. (1981) Ascent and eruption of basaltic magma on the Earth and Moon. *J. Geophys. Res. 86,* 2971-3001.

Wilson L. and Head J. W. (1983) A comparison of volcanic eruption process on Earth, Moon, Mars, Io and Venus. *Nature, 302,* 663-669.

Lunar Regolith Processes and Resources

Proceedings of the 20th Lunar and Planetary Science Conference, pp. 209-217
Lunar and Planetary Institute, Houston, 1990

Glass Variants and Multiple HASP Trends in Apollo 14 Regolith Breccias

D. T. Vaniman

Geology and Geochemistry, Mail Stop D462, Los Alamos National Laboratory, Los Alamos, NM 87545

Three distinctive samples can be recognized by analysis of all glass types, including devitrified glasses, in a suite of 27 thin sections from 26 Apollo 14 regolith breccias. These distinctive samples include (1) the well-studied sample 14315, which has an abundance of anorthositic gabbro glasses and devitrified glasses; (2) 14004,77, which contains only glasses of medium-K Fra Mauro (MKFM) composition similar to the local soil; and (3) 14076,5, which contains no glasses similar to the local soil or to low-K Fra Mauro (LKFM) compositions. Sample 14076,5 is clearly exotic, for it contains devitrified glasses of anorthositic composition and of a silica-volatilized (HASP) trend that stems from anorthosite; these silica-volatilized glasses have devitrified to form a new Al-rich, Si-poor mineral with the formula $Ca_{8-(x/2)}\square_{(x/2)}Al_{16-x}Si_xO_{32}$. The HASP glasses in this exotic sample and HASP glass spheres that stem from the Apollo 14 soil composition differ greatly from the HASP glasses at the Apollo 16 site. The various HASP glasses can be just as useful as nonvolatilized glasses in searching for major crustal or regolith compositions.

INTRODUCTION

Regolith breccias provide individual coherent soil samples that may or may not have formed from the surrounding regolith. Regolith breccias may be samples of regoliths that are much older than the unconsolidated, surrounding soil (*McKay et al.,* 1986) or of regoliths that are exotic to their collection locality (E. A. Jerde et al., in preparation, 1989). If old or exotic, the fragments within regolith breccias can provide clues to the compositions of regoliths that are not otherwise represented among known lunar samples, or of "fossil" regoliths that no longer exist on the Moon.

Glasses in regoliths and in regolith breccias can provide abundant information about the impact melt and volcanic lithologies that contributed to these samples. *Delano* (1988) studied glasses from five large (140 g to 1360 g) regolith breccias from the Apollo 14 site. His study showed that at least six types of pristine mare glasses occur in these samples, and he suggested that relative ages for the regolith breccias may be inferred from the different pristine glass populations of the samples (younger regolith breccias should have more types of pristine mare glasses, since they sample a longer history of mare volcanism). The present study summarizes data for a larger number of smaller (mostly <40 g) regolith breccia samples from Apollo 14. Because of the small sizes of the subsamples analyzed (most are <0.5 cm in diameter), it was generally rare to find more than 10 to 20 glasses in a single thin section.

One goal of this effort was to provide a comparison of the glass populations, in 27 thin sections, with bulk chemistry and I_s/FeO data from comparable splits analyzed at UCLA (P. H. Warren and E. A. Jerde) and at NASA/JSC (R. V. Morris). The purpose of the chemical study was to search for exotic regolith breccias that may provide new information not previously extracted from the Fra Mauro regolith. It was hoped that the glass fragments, which represent a variety of rock and soil types, might include correspondingly exotic compositions in those regolith breccias that had distinctive bulk chemical compositions. Indeed, one very unusual sample (representing half of the 2-g regolith breccia 14076) stands out not only for its very anorthositic bulk composition (E. A. Jerde et al., in preparation, 1989) but also for the presence of an exotic set of devitrified glass fragments. These fragments were formed by impact in an almost "pure" anorthosite terrain with extreme SiO_2 volatilization from the impact melt, and they contain a new Al-rich, Si-poor mineral with the formula $Ca_{8-(x/2)}\square_{(x/2)}Al_{16-x}Si_xO_{32}$, the first new mineral unique to the lunar highlands (*Vaniman et al.,* 1989; D. T. Vaniman and D. L. Bish, in preparation, 1989).

Because this study of glass fragments was intended to be a comparison with bulk chemistry, it was decided to analyze not only the small glass spheres but also to analyze glasses that had undergone obvious devitrification, glasses with vesicles and small inclusions, agglutinates, matrix glasses, and ropy glasses. Although these analyses might be argued to include features that are too varied for uniform interpretation in one study, it would be unacceptable to ignore large but "messy" glass types in a search for glass-based reflections of chemical variation between bulk samples. Thus, a wide variety of glass types is represented here, and the results include a consideration of the relations between glass type, size, and chemical composition.

Impact glasses are believed to (1) originate from impact melting events that sample regolith rather than rocks (*Meyer,* 1978), and (2) often have compositions that are skewed to lower K, Na, and Si content because of fractional vaporization from impact melts. *Naney et al.* (1976) documented the effects of SiO_2 vaporization from impact melt glasses, and *Delano et al.* (1981) have used refractory-element ratios to "see through" this effect and recover part of the original soil chemistry. In this report as much emphasis will be given to the effects of fractional vaporization as to the decipherment of original constituents.

METHODS

Electron microprobe analyses were obtained on a Cameca Camebax instrument with wavelength-dispersive detectors. Microprobe automation is described in *Chambers* (1985); data reduction using modified Bence-Albee empirical correction factors is described elsewhere (W. F. Chambers, in preparation, 1989). A beam current of 15 nA was used. Analyses were obtained either in rastered areas that covered large portions of homogeneous glasses and completely devitrified glasses, or with focused beams to avoid vesicles, inclusions, and local devitrification. Glass forms were classified and measured by petrographic microscopy.

Of the 27 thin sections studied, 16 represent individual regolith breccia "rocklets" (4-10-mm fragments) that were

TABLE 1. Glass compositions in Apollo 14 regolith breccias, compared with soil 14259.

Thin Section	NAA/EP split	"HASP"-like	Plag. Glass	Gabb. Anorth.	Anorth. Gabb.	LKFM	MKFM	Rhyol.	Mixed Mare	Pristine Mare
14004,77	(,55)*						8			
14004,78	(,56)*				1	1	3			2†
14004,79	(,57)*						4		1	
14004,80	(,58)*				1		5			
14004,81	(,59)*			2	1	3	8			1
14004,82	(,60)*			1	1	3	10			1†
14076,5‡	(,1)§	5¶	6	5	1			3		
14076,11	(,6)§		1	2	1	2	4			
14160,144	(,113)*	1	2	2	1	3	4	1	2	3
14160,145	(,114)*		1		3	8			1	
14160,147	(,116)*					1				
14160,148	(,117)*	2			1	1			2	
14160,149	(,118)*	1**	1		4	1	4			
14160,150	(,119)*			4	1	7			1	
14194,5	(,1)§		1	2	2	2	10			
14250,8	(,5)§	1			1	1	12		5	2
14251,5	(,2)§			4	9	3	3		3	1
14252,5	(,2)§	1			7	3	18		1	1
14263,23	(,8)*	1		1	1		4			1
14263,24	(,9)*			1	1		2			1
14263,25	(,10)*	1			5		4			1
14263,26	(,11)*					1	1	1		
14281,20	(,17)§				1	5	1	1		
14282,5	(,2)§				5	1	11		2	2
14309,10	(,8)§		1	1			18			2
14315,26	(,24)*			3	7		1			
14316,16	(,13)§	1		2	4	5		4		
Σ(regolith breccias)	n=	14	13	30	59	51	135	10	18	18
	%=	(4.0)	(3.7)	(8.6)	(17.0)	(14.7)	(38.8)	(2.9)	(5.2)	(5.2)
Soil 14259††	n=	1	1	8	100	48	155	6	23	15
	%=	(0.3)	(0.3)	(2.2)	(28.0)	(13.4)	(43.4)	(1.7)	(6.4)	(4.2)

*NAA/EP (neutron activation/electron microprobe) analysis reported in *Jerde et al.* (1987).
†The pristine mare glasses in 14004,78 and ,82 have exceptionally high TiO_2 contents (16-17% TiO_2); they are similar to the Ti-rich mare glasses from Apollo 14 described by *Delano* (1988).
‡All of the glasses in 14076,5 are devitrified.
§NAA/EP analysis reported by E. A. Jerde et al. (in preparation, 1989).
¶HASP glasses in 14076,5 are Al-rich and have devitrified to form a new Al-rich, Si-poor mineral ($Ca_{8-(x/2)}\square_{(x/2)}Al_{16-x}Si_xO_{32}$).
**The HASP glass in 14160,149 is the only glass found in this study that is similar to Apollo 16 HASP (see Fig. 2).
††Reclassification of glass analyses listed in *Brown et al.* (1971) to fit the classification used in this paper.

analyzed by *Jerde et al.* (1987); these small regolith breccias are from soil samples 14004, 14160, and 14263 (see Table 1 for individual split identifications.) The other 11 thin sections are from larger (>1 cm) parent samples: 14076 (2.0 g), 14194 (4.3 g), 14250 (4.1 g), 14251 (1.5 g), 14252 (0.9 g), 14281 (12.0 g), 14282 (1.9 g), 14309 (42.4 g), 14315 (115.0 g), and 14316 (38.2 g). If only one thin section were studied from each sample, the total would be 26. An additional thin section was made from sample 14076 because of this sample's notable heterogeneity in color and texture (E. A. Jerde et al., in preparation, 1989); data for the two very different parts of this sample are listed in Table 1. Only 1 of the 26 regolith breccias analyzed (14315) has been studied in detail before; the goal of our informal consortium was to look at the unstudied small regolith breccias for new information. An excellent summary of glass data from the larger Apollo 14 regolith breccias can be found in *Simon et al.* (1989).

CLASSIFICATION

Glasses were classified by color, form, special features, and size prior to analysis. Colorless, yellow, green, orange, black, and brown color classifications were used. Glass forms were classified as spheres (or sphere fragments), angular, splash coats, agglutinates, ropy glasses, and matrix glasses (those glass bodies with diffuse edges that penetrate into a porous matrix). Special features were defined for those fragments that were not clear glass; these features were (1) devitrification avoided in microprobe analysis, (2) pervasive devitrification that was analyzed by broad-beam methods, (3) edge or rim crystallization avoided in microprobe analysis, (4) inclusions or quench crystals avoided in analysis, and (5) vesicles.

For systematic chemical classification, highland glass types were categorized according to a modified scheme based on major-element composition (*Naney et al.*, 1976, 1977). The Naney scheme was used as is to classify *plagioclase* (or

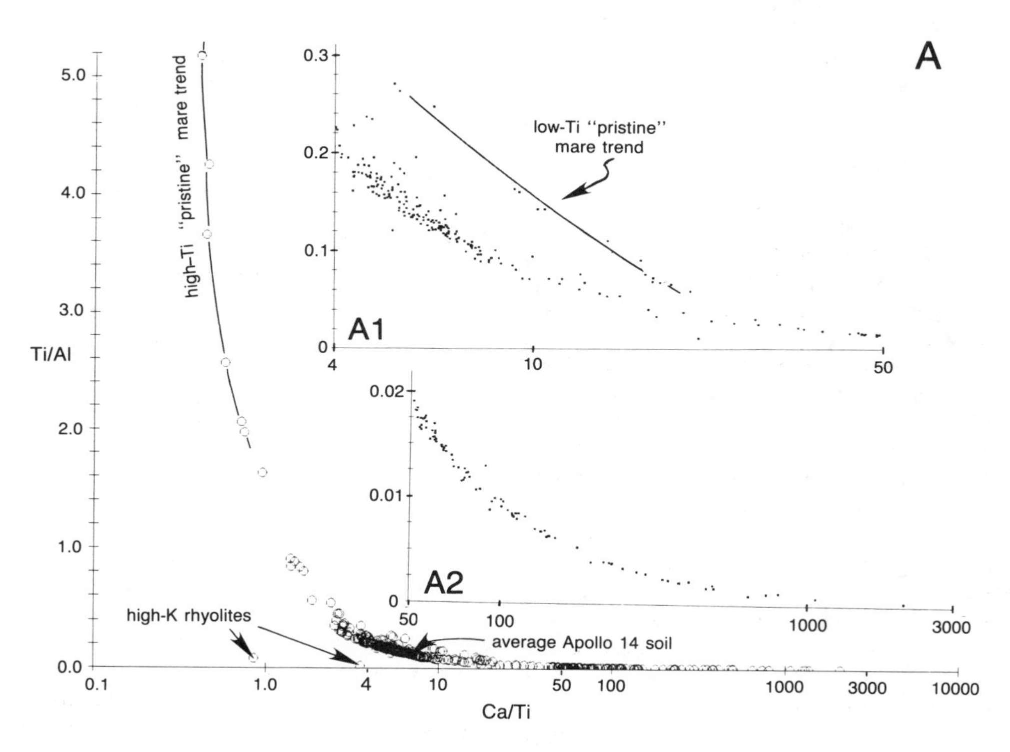

Fig. 1. (a) Plot of Ti/Al vs. Ca/Ti (weight ratios) for all glasses and devitrified glasses analyzed in this study. A semilog plot is used to accommodate the high Ca/Ti ratios (>100) found in plagioclase-like (anorthositic) glasses as well as in some gabbroic anorthosite glasses, in a few anorthositic gabbro glasses, and in three high-Ca rhyolite glasses from 14076,5. Two high-K rhyolite glasses with >7% K_2O (columns 1 and 2 in Table 2) plot below the general mixing hyperbola. Expanded inset plot (A1) shows that the portion of the curve from 4<(Ca/Ti)<20 contains a higher Ti/Al trend defined by low-Ti "pristine" mare glasses that are not part of the general mixing hyperbola, and inset plot (A2) shows that the portion of the entire curve ranging from 50<(Ca/Ti)<3000 is not compressed into the x-axis. The location of the average Apollo 14 soil composition is based on data in *Rose et al.* (1972). **(b)** Plot of Mg/Al vs. Ca/Ti for all glasses and devitrified glasses. Absence of a common element between plotted ratios generates greater scatter than in **(a)**, but similar limits on mixing are observed. The high-K rhyolites and low-Ti "pristine" mare glasses still plot off of the range of main mixing hyperbolas. Two possible mixing hyperbolas are shown, ranging from high-Ti "pristine" mare to either the main anorthositic gabbro glass cluster or to the high Ca/Ti glasses.

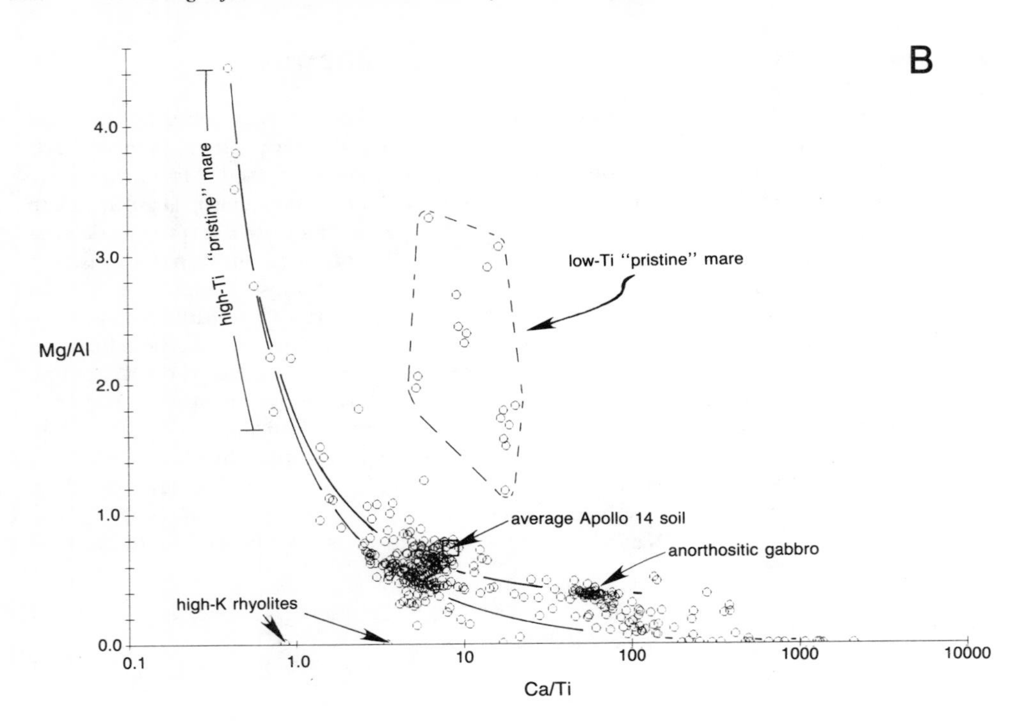

Fig. 1. (continued).

anorthositic) *glass* (~34% Al_2O_3), *gabbroic anorthosite* (~31% Al_2O_3), *anorthositic gabbro* (~26% Al_2O_3), *LKFM* (low-K Fra Mauro, ~22% Al_2O_3), and *MKFM* (medium-K Fra Mauro, ~16% Al_2O_3, the glass composition most similar to that of the Apollo 14 regolith). In addition, *mixed mare* glasses were defined as those that have CaO/Al_2O_3 weight ratios above 0.75 but do not fit the "pristine" Apollo 14 glass classifications of *Delano* (1988); *pristine mare* glasses are those that do fit his criteria. *Rhyolitic* glasses were defined as those with SiO_2 weight contents greater than 60%. Feldspathic highland glasses with >16% Al_2O_3 and <40% SiO_2 are classified as *"HASP"* (high-alumina, silica-poor), in a modification of the use of this acronym by *Naney et al.* (1976). Although Naney et al. used this term for glasses derived by SiO_2 volatilization from anorthositic gabbro (~26% Al_2O_3), the genetic implications of the term (high-Al_2O_3 residue due to SiO_2 volatilization) make it applicable to a broad range of impact-volatilized glasses. The HASP compositions in Apollo 14 regolith breccias are especially complex, since volatilization from both Fra Mauro and anorthositic sources is observed. This complexity, however, allows a general comparison of the fractional vaporization process.

The results of chemical classification, applied to all of the glass and devitrified glass constituents analyzed, are listed in Table 1. Table 1 also shows the application of this classification scheme to the 357 glass analyses from soil sample 14259 listed in *Brown et al.* (1971). Table 1 summarizes only the chemical data obtained in this study; a complete listing of color, form, special features, size, and chemical composition for each of the 348 regolith breccia glasses or devitrified glasses analyzed can be obtained by writing to the author.

DISCUSSION

Systematic Chemical Variation

Refractory-element ratios. All of the glasses and devitrified glasses analyzed can be considered as a group or as separate glass types. As a group, the effects of impact volatilization can be suppressed by examining the variation among refractory elements. In this manner, *Delano et al.* (1981) examined glasses from all of the Apollo sites except Apollo 14 and 17. They did not rely on the volatile constituents (K_2O, Na_2O, SiO_2, Fe) but instead developed hyperbolic mixing curves based on refractory-element ratios (Ti/Al vs. Ca/Ti, and Mg/Al vs. Ca/Ti). These curves were shown to match the larger-scale variation in soils due to mixing at each site, even though many of the glasses analyzed had suffered volatile-element losses. Similar curves are shown in Fig. 1 for all of the glasses analyzed in this study.

The plot of Ti/Al vs. Ca/Ti in Fig. 1a indicates a general hyperbolic mixing curve for all of the glasses and devitrified glasses analyzed. The alignment of compositions along this curve shows a mixing relationship (or several families of mixing relationships) that extends from high-Ti basalts (Ti/Al > 1.8) to a variety of glasses (anorthositic, gabbroic anorthosite, and anorthositic gabbro, plus three high-Ca rhyolites in 14076,5) with high Ca and low Ti contents (Ca/Ti > 100). This alignment passes through the average Apollo 14 (Fra Mauro) soil composition. Those glass compositions that deviate greatly cannot be major contributors to the Fra Mauro soils and regolith breccias. Figure 1a, and the upper inset plot (A1), indicate two exotic glass types: high-K (>7% K_2O) rhyolites that plot below the mixing curve, and low-Ti

TABLE 2. Distinctive glass sphere, ropy glass, and devitrified glass compositions in Apollo 14 regolith breccias.

	High-K Rhyolites		Pristine Low-Ti Mare		Apollo 14 HASP		High-Al HASP*	
	14160,144 (Sphere)	14281,20 (Ropy Glass)	14160,144 (Sphere)	14160,145 (Sphere)	14263,23 (Sphere)	14160,148 (Sphere)	14076,5 (Devitrified)	
	1	2	3	4	5	6	7	8
SiO_2	75.7	72.6	46.8	46.3	38.9	35.6	27.3	19.3
TiO_2	0.86	0.27	1.92	0.52	2.51	2.61	0.04	0.11
Al_2O_3	12.6	14.5	8.01	9.75	21.5	23.7	45.8	52.4
FeO	0.41	0.30	20.3	17.9	11.0	10.4	0.02	0.02
MgO	0.07	0.00	13.8	15.6	12.5	12.3	0.20	0.21
CaO	0.61	0.79	8.50	8.99	13.5	14.5	27.2	27.6
Na_2O	1.00	1.62	0.10	0.24	0.00	0.00	0.18	0.00
K_2O	7.77	7.20	0.04	0.07	0.01	0.00	0.00	0.00
BaO	0.00	2.86	n.a.†	n.a.	n.a.	n.a.	n.a.	n.a.
Σ	99.1	100.1	99.5	99.4	99.9	99.1	100.7	99.6
Ca/Ti	5.41	3.49	5.28	20.6	6.42	6.65	810	299
Ti/Al	0.077	0.021	0.271	0.060	0.132	0.125	0.001	0.002
Mg/Al	0.006	0.000	1.97	1.82	0.659	1.076	0.005	0.005

*The devitrified HASP in 14076,5 consists entirely of a new mineral with formula $Ca_{8-(x/2)}\square_{(x/2)}Al_{16-x}Si_xO_{32}$ (D. T. Vaniman and D. L. Bish, in preparation, 1989).
†n.a. = not analyzed.

(<2% TiO_2) pristine mare glasses that follow a separate trend above the mixing curve. The high-K rhyolite compositions, and end members for the apparent low-Ti pristine mare trend, are listed in Table 2. High-K rhyolite occurs both as a single clear glass sphere and as a ropy glass.

Refractory-element mixing variations form a more diffuse hyperbola where an element is not held in common between the ratios plotted. Nevertheless, the plot of Mg/Al vs. Ca/Ti in Fig. 1b shows the same distinctive separation of high-K rhyolites below the main trend, and low-Ti "pristine" mare glasses above the trend. The other glass/devitrified glass types scatter along the trend but indicate complex groupings and multiple mixing relationships.

In both Figs. 1a and 1b, the best-fit hyperbolic curves through the major glass clusters are those that tie high-Ti mare compositions to a variety of feldspathic highland compositions. The importance of high-Ti basalts in mixing may well be due to their greater age: *Delano* (1988) proposed that the high-Ti "pristine" glasses in Apollo 14 regolith breccias are older than the other pristine mare glasses. This inference was based on the occurrence of high-Ti compositions as the only mare glass in some regolith breccias, whereas the low-Ti compositions are often mixed with the high-Ti types (thus suggesting that there was an early episode of regolith breccia formation when only the high-Ti types were present). A greater age for the high-Ti mare composition would allow an earlier (perhaps more intense) and longer period of impact mixing to incorporate this component into glasses, regolith, and regolith breccias among the Apollo 14 samples.

In Fig. 1b, the scatter of glass compositions other than high-K rhyolite or low-Ti pristine mare allows a wide range of hyperbolic mixing curves between the pristine high-Ti mare compositions and a range of highland compositions. Two such curves are shown. One passes through the high Mg/Al part of the major glass cluster, just below the average Apollo 14 soil composition, and into the main anorthositic gabbro cluster. The other curve is of lower Mg/Al and passes closer to the x-axis toward the high Ca/Ti highland glasses. Clearly, most of the glasses with Ca/Ti >100 do not fall along the mixing trend that includes most anorthositic gabbros. *Delano et al.* (1981) proposed that the refractory-element ratio curves may be used to constrain the original lithologic components in a regolith sample; for the glasses shown in Fig. 1b, separate mixing trends for the typical and relatively mafic anorthositic gabbro (high Mg/Al) and for the very high Ca/Ti highland glasses (with low Mg/Al) can be perceived, although the highland end members themselves are impact mixtures rather than pristine igneous rocks. Notable in this figure is the displacement of the average Apollo 14 soil composition from the glass mixing curves; this displacement is apparently due to the addition of low-Ti "pristine" mare compositions to the regolith and is further evidence of the relative lack of this particular mare component in the glass mixing trends established prior to regolith breccia formation.

Volatile-element variation. Volatilization of constituents such as K_2O, Na_2O, SiO_2, and Fe modifies the chemical trends caused by impact mixing (*Delano et al.*, 1981). However, the volatilization process itself may be a measure of how severe the impact history has been for specific glass groups. *Naney et al.* (1976) related HASP glasses in Apollo 16 regolith to an anorthositic gabbro (~26% Al_2O_3) parent composition. Figure 2 plots a major refractory oxide (Al_2O_3) against relatively volatile SiO_2. Samples with more than 40% SiO_2 may have suffered some SiO_2 volatile loss, but silica volatilization from

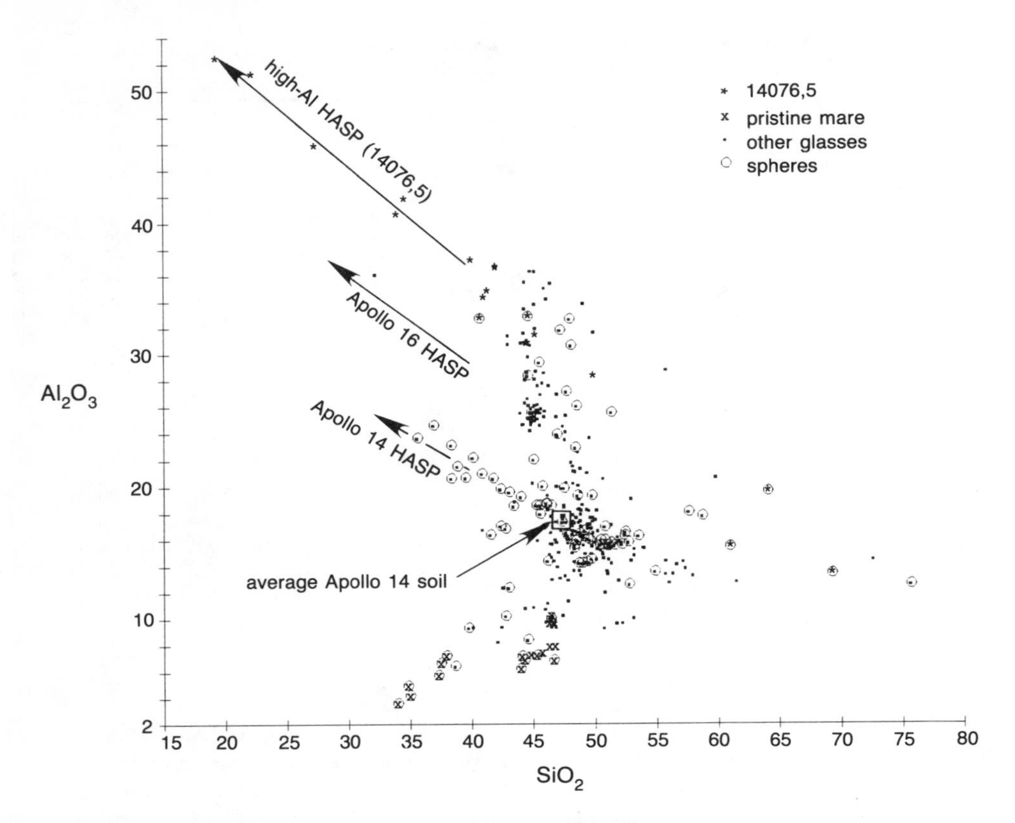

Fig. 2. Plot of Al_2O_3 vs. SiO_2; "x" symbols are used to distinguish the "pristine" mare glasses from all other glasses and devitrified glasses, and stars mark the odd range from high-Al HASP glasses to Ca-rich rhyolitic glasses (all of which are devitrified) in regolith breccia 14076,5. Circles are drawn around those glasses that occur as spheres or as fragments of spheres. All compositions with more than 16% Al_2O_3 and less than 40% SiO_2 are classified as "HASP-like" in this paper. The Apollo 14 HASP trend and the 14076,5 high-Al HASP trend are based on this study; the Apollo 16 HASP trend is based on information in *Naney et al.* (1976). Only one small (<0.001 mm^2) angular fragment of glass, in 14160,149, falls close to the Apollo 16 HASP trend. The location of the average Apollo 14 soil composition is based on data in *Rose et al.* (1972).

these samples is generally limited. Feldspathic (>16% Al_2O_3) highland glasses with low silica contents (<40% SiO_2) are attributable to some amount of SiO_2 volatilization (note that mafic highland glasses from dunitic sources and Ti-rich mare glasses, both of which have <16% Al_2O_3, are compositions that can have SiO_2 contents <40% without SiO_2 volatilization).

In Fig. 2 there are three HASP trends. One is common to a broad range of regolith breccias (columns 5 and 6 in Table 2) and can be tied to a Fra Mauro-type (16-18% Al_2O_3) source composition; this trend is considered to be the "Apollo 14 HASP" trend. Another trend occurs only in regolith breccia 14076,5 and can be tied to an anorthositic (~34% Al_2O_3) source. These two trends are far different from the third, the Apollo 16 HASP described by *Naney et al.* (1976). The recognition of at least three separate highland HASP trends indicates that there are no compositional constraints on the SiO_2-volatilization process.

When the term HASP was first employed (*Naney et al.*, 1976), it was used to describe a range of Si-volatilized glasses at Apollo 16 related to an anorthositic gabbro parent composition. The local soils at Apollo 16 have compositions between those of anorthositic gabbro and gabbroic anorthosite (about 25% to 30% Al_2O_3); most have Al_2O_3 contents >26%, and thus the soils themselves are not especially appropriate parent material for the Apollo 16 HASP. Materials of appropriate composition (~26% Al_2O_3), however, are represented by the Apollo 16 dimict breccias and by one group of impact-melt splashes (*See et al.*, 1986). The Apollo 16 HASP may have origins related to these two lithologic types, perhaps originating as silica-volatilized droplets derived from impact melts equivalent to the veined breccias formed beneath large impact craters.

Figure 2 shows that only one of the HASP-like glasses or devitrified glasses in Apollo 14 regolith breccias falls near the Apollo 16 HASP trend; this sample is a small, angular glass fragment in sample 14160,149. In contrast, the Apollo 14 HASP trend is common to many Apollo 14 regolith breccias. There is no significant difference between the composition of Apollo 14 regolith and the projected parent composition for the Apollo 14 HASP trend. If the Apollo 14 HASP formed in a manner similar to the Apollo 16 HASP, it is possible that at the Apollo 14 HASP source the impact regime equivalent to that of Apollo 16 dimict breccia formation occurred entirely within a thick regolith sequence of uniform Fra Mauro regolith composition.

The high-Al HASP trend occurs only in the sample 14076,5. The projected parent composition for this HASP trend is anorthositic, with ~34% Al_2O_3. The implications of this trend are far-reaching, because the regoliths collected from Apollo and Luna sampling sites do not consist of such pure anorthositic material.

Maps generated from Apollo orbital geochemical data reveal extensive terrains of anorthositic regolith on the lunar limbs and farside (*Davis and Spudis,* 1987). Although the closest of these terrains to Apollo 14 is near the crater Grimaldi, similar ancient anorthositic terrains may be buried closer to the site. The thin section 14076,5, and the related bulk sample 14076,1 (E. A. Jerde et al., in preparation, 1989), may be a product of impact into such terrains.

In addition to its distinctive composition, the high-Al HASP trend in 14076,5 is distinct from other HASP trends because (1) it occurs in devitrified rather than glassy fragments, (2) all of the fragments are angular rather than spheres, and (3) the devitrification is complete, and the sole devitrification product takes on the original glass composition to make a metastable mineral with composition $Ca_{8-(x/2)}\square_{(x/2)}Al_{16-x}Si_xO_{32}$ (D. T. Vaniman and D. L. Bish, in preparation, 1989). This new mineral is discussed below, in the more detailed description of 14076,5. Other HASP compositions are glassy spheres or fragments of spheres; this is so for the Apollo 16 HASP (*Naney et al.,* 1976) and for the Apollo 14 HASP (note the circles around all Apollo 14 HASP analyses in Fig. 2, indicating that the samples are either spheres or sphere fragments). The angular shapes and the devitrification of high-Al HASP samples may reflect their very Ca,Al-rich composition (Table 2, columns 7 and 8) or a different method of formation. Many of the answers to questions about these puzzling high-Al HASP fragments probably will not be answered until the anorthositic lunar highland terrains are explored.

Effects of size. The largest glass constituents in the regolith breccias are of MKFM (similar to the local Apollo 14 regolith) or LKFM composition. Four glasses were found with areas greater than 0.9 mm^2; the largest of these was an 8-mm^2 splash coat draping sample 14160,147 (this is the only glass analyzed in this sample and is thus the only data point shown for 14160,147 in Table 1). The other three are splash coats (in 14160,149 and 14282,5) and an agglutinate (in 14160,145).

In the size range of 0.2 to 0.4 mm^2, a greater variety of compositions and glass types is observed. Splash coats, ropy glasses, matrix-soaking glasses, and angular fragments of MKFM or LKFM composition are common. In addition, angular devitrified fragments of anorthositic gabbro glass in this size range occur in 14250,8, 14251,5, and 14315,26. The exotic regolith breccia 14076,5, which includes the high-Al HASP fragments, also has a large (0.32 mm^2) fragment of devitrified plagioclase-like (anorthositic) glass.

The largest glassy sphere found (in 14004,78) has an area of 0.2 mm^2 and is a vesicular, inclusion-rich brown glass of MKFM composition. The largest inclusion-free and nonvesicular glass sphere found is also of MKFM composition (0.06-mm^2 yellow glass in 14004,77). Over 90% of the glass spheres found and analyzed were in the size range 0.001-0.01 mm^2.

Ropy glasses. Ropy glasses in the regolith breccia samples have a wide range of compositions, from an anorthositic (33% Al_2O_3) ropy glass in 14076,11 to the high-K rhyolitic ropy glass in 14281,20 (column 2 of Table 2). Regolith breccias 14160,150 and 14282,5 also contain ropy glasses with anorthositic gabbro composition (~26% Al_2O_3). However, most of the ropy glasses analyzed have MKFM compositions (similar to the Apollo 14 soil).

The MKFM composition is also found in fragments of angular glass, glass beads, agglutinates, splash coats, and matrix glasses. Ropy glasses of the MKFM type differ slightly from the other MKFM glasses by having a higher average SiO_2 content ($51 \pm 3\%$ SiO_2 vs. $48 \pm 2\%$ SiO_2). Although the 1-σ ranges in these silica contents overlap, the largest MKFM ropy glass (0.35 mm^2; 14281,20) has the high-silica average composition (50.8% SiO_2). The MKFM ropy glass compositional range is also stretched toward compositions of lower alumina than the other MKFM glasses (13% vs. 16% Al_2O_3). In general, a higher proportion of rhyolitic components is present in the ropy MKFM glasses than in other MKFM glasses.

Glasses in Distinctive Regolith Breccias

The glass constituents in soil 14259 (Table 1) reflect the predominance of anorthositic gabbro, LKFM, and MKFM (recycled soil?) compositions in Apollo 14 soils. The MKFM composition is well represented, but the glass type next in abundance is not LKFM but anorthositic gabbro. Similar distributions of glass types are seen in the regolith breccias, but there are notably more HASP, plagioclase-like (anorthositic), and gabbroic anorthosite glasses in the regolith breccias than in soil 14259—even when the data from the exotic Al-rich regolith breccia 14076,5 are disregarded. The greater abundance of these three aluminous glass types in the regolith breccias (~16%, with or without 14076,5) than in the soil (~3%) suggests a greater sampling of aluminous terrains prior to the time of regolith breccia formation (a similar conclusion was reached by *Simon et al.,* 1989). It is also significant that the sole HASP fragment found by *Brown et al.* (1971) in soil 14259 is similar to Apollo 16 HASP; they found no Apollo 14 HASP, which accounts for all but one of nine HASP spheres (exclusive of the Al-rich HASP in 14076,5) listed in Table 1. It is probable that the impacts that formed HASP glasses from the Apollo 14 soil occurred before or during the episode of regolith breccia formation and were not an important part of bulk regolith maturation.

Several regolith breccias stand out because of particular glass types (e.g., the devitrified high-Al HASP in 14076,5). Glasses often have compositions that reflect the compositions of chemically deviate regolith breccias in which they occur. These glasses help to clarify the origins of regolith breccias with distinctive compositions.

14004,77. This breccia, rather than being exotic, has only MKFM glasses that mimic the local soil. Although its glass abundance is small, the absence of glasses other than those formed in the immediate vicinity may indicate derivation from a regolith of almost pure Fra Mauro deposits containing very little exotic material.

14076,5. This is by far the most exotic of the Apollo 14 regolith breccias studied. The regolith breccia from which this sample comes consists of two very different portions; E. A. Jerde et al. (in preparation, 1989) describe the portion represented by thin section 14076,5 as anomalously aluminous (30% Al_2O_3) and clearly exotic to the Apollo 14 sampling area. The other part (represented by thin section 14076,11) has the typical Apollo 14 regolith breccia composition.

The exotic portion (14076,5) contains a minor high-Ca rhyolite constituent as devitrified glass spheres; a significant fraction of its devitrified glasses consists of the high-Al HASP glass described above. This type of HASP has been found only in this sample. This HASP composition occurs not only as devitrified glasses, but also as single crystals up to 250 μm long in 14076,5. X-ray diffraction analysis of mineral separates from the equivalent sample powder (14076,1) has confirmed that these devitrified glasses and the single crystals form a new mineral (*Vaniman et al.,* 1989; D. T. Vaniman and D. L. Bish, in preparation, 1989). The new mineral is hexagonal, with a nepheline-like structure, and formula $Ca_{8-(x/2)}\square_{(x/2)}Al_{16-x}Si_xO_{32}$ where $\square$ represents a vacant lattice site. This mineral forms metastably in glasses equivalent to Si-depleted anorthite that have devitrified at about 950°-1200°C. Vaniman and Bish suggest that the new mineral forms in impact glasses from relatively "pure" anorthositic terrains. The occurrence of this aluminous regolith breccia with the new Al-rich mineral hints at large impacts into anorthositic terrains and indicates that anorthositic terrains are more important in the lunar highlands than the "typical" Apollo and Luna regolith samples suggest.

The extremely low I_s/FeO of 14076,5 (0.03; E. A. Jerde et al., in preparation, 1989) and the absence of agglutinates in this sample might be construed to cast some doubt on whether it is indeed a regolith breccia. However, 5 of the 20 devitrified glass fragments larger than 0.001 mm^2 in this sample are spheres or sphere fragments (*Vaniman et al.,* 1988). Smaller spheres can be found throughout the matrix. This sample's nonfriable nature and relatively low porosity (~15%) suggest that it is an exceptionally well-sintered regolith breccia; the lack of agglutinates also suggests that the parent regolith was immature. The very low I_s/FeO may be caused by both extreme sintering and immaturity.

14315,26. This sample is the "chondrule"-bearing regolith breccia studied recently by *Simon et al.* (1989) and *Jerde et al.* (1987), and earlier by other researchers (e.g., *Chao et al.,* 1972; *Simonds et al.,* 1977). The "chondrules" in 14315,26 have an anorthositic gabbro composition that accounts for its anomalously Al-rich composition (22% Al_2O_3) contrasted with the average Apollo 14 regolith breccia (17.6% Al_2O_3; *Fruland,* 1983). Although less abundant than the "chondrules," the glasses in this sample also have a predominantly anorthositic gabbro composition. It is noteworthy that next to the Fra Mauro glasses, the most abundant glass constituent in Apollo 14 regolith is anorthositic gabbro (soil 14259, Table 1). The regolith breccia 14004,77 and the chondrules in 14315,26 may provide examples of relatively "unmixed" MKFM and anorthositic gabbro soil types, respectively, preserved in regolith breccias at the Apollo 14 site.

CONCLUSIONS

Among Apollo 14 regolith breccias, compositions that deviate from that of the Apollo 14 soil are rare (*Fruland,* 1983). *Simonds et al.* (1977) analyzed 14 regolith breccias and, with the exception of 14315, found that all clustered around the composition of Apollo 14 soil. *Jerde et al.* (1987) reanalyzed 14315 and studied 18 additional small (4-10 mm) Apollo 14 regolith breccias; they found only one other anomalous sample, 14004,55 (our thin section 14004,77), which stands apart from the Apollo 14 regolith in terms of its higher (~1.4×) incompatible element contents. In the completion of their study, E. A. Jerde et al. (in preparation, 1989) analyzed another 9 small Apollo 14 regolith breccias and added 1 portion of 14076 (14076,1; our thin section 14076,5) to the final list of 3 deviate regolith breccias out of 41.

Glass compositions, including devitrified glass compositions, match the bulk chemical anomalies in these three odd regolith breccias. Sample 14315 stands apart because the glasses and "chondrules" it contains are largely of anorthositic gabbro composition; it has a notable paucity of glasses that match the local soil (Table 1). Sample 14004,77, on the other hand, is distinctive in that it contains only glasses (ropy glasses, spheres, angular glass fragments, and matrix glasses) that match the local soil; its lack of other more aluminous glasses is reflected in the higher incompatible element composition found by *Jerde et al.* (1987). Sample 14076,5 is by far the most exotic. This sample is not only very aluminous in bulk composition (30% Al_2O_3; E. A. Jerde et al., in preparation, 1989), it is the only sample with no glasses that match either the MKFM (similar to local soil) or LKFM compositions (Table 1). Moreover, all of the glasses in this subsample of 14076 are devitrified and are dominated by anorthositic gabbro, anorthosite, and high-Al HASP compositions—the latter devitrified to form a new Al-rich, Si-poor mineral ($Ca_{8-(x/2)}\square_{(x/2)}Al_{16-x}Si_xO_{32}$). This part of 14076 is clearly exotic. If it formed from impacts near the Apollo 14 site, then the target area is presently hidden. Our only sighting of appropriate highland terrains for this sample is in the far distant anorthositic surfaces mapped by *Davis and Spudis* (1987).

The high-Al HASP in 14076,5 and the Apollo 14 HASP found in many of the other regolith breccias broaden our understanding of silica volatilization processes. At both Apollo 14 and Apollo 16, the silica volatilization trends stem from compositions that match either the local soil or the dimict breccias that may arise from impacts through the local regolith. In the exotic regolith breccia 14076,5, the HASP trend implies a comparable origin but in a regolith of essentially pure anorthosite. Regolith of this type was not sampled by the Apollo or Luna explorations; the devitrified glasses in regolith breccia 14076,5 are our only samples. The suggestion of *Delano et al.* (1981), that surveys of glass compositions in grab-samples are a powerful tool to explore the chemistry of crustal lithologies, is amply borne out by the data from 14076,5.

Acknowledgments. This study was supported in part by the Institute of Geophysics and Planetary Physics of the University of California, grant #86-129. The work was done under the auspices of the U.S. Department of Energy. I owe much to other members of an informal consortium (G. Heiken, P. Warren, E. Jerde, and R. Morris) for the study of Apollo 14 regolith breccias. Reviews by G. Heiken, A. W. Laughlin, J. Delano, S. Wentworth, S. Simon, and D. Lindstrom helped to improve and correct the original manuscript.

REFERENCES

Brown R. W., Reid A. M., Ridley W. I., Warner J. L., Jakeš P., Butler P., Williams R. J., and Anderson D. H. (1971) Microprobe analyses of glasses and minerals from Apollo 14 sample 14259. *NASA TM X-58080.* 89 pp.

Chambers W. F. (1985) SANDIA TASK8: A subroutined electron microprobe automation system. *Sandia National Laboratory Research Report SAND85-2037.* 115 pp.

Chao E. C. T., Minkin J. A., and Best J. B. (1972) Apollo 14 breccias: General characteristics and classification. *Proc. Lunar Sci. Conf. 3rd,* pp. 645-660.

Davis P. A. and Spudis P. D. (1987) Global petrologic variations on the Moon: A ternary-diagram approach. *Proc. Lunar Planet. Sci. Conf. 17th,* in *J. Geophys. Res., 92,* E387-E395.

Delano J. W. (1988) Apollo 14 regolith breccias: Different glass populations and their potential for charting space/time variations. *Proc. Lunar Planet. Sci. Conf. 18th,* pp. 59-65.

Delano J. W., Lindsley D. H., and Rudowski R. (1981) Glasses of impact origin from Apollo 11, 12, 15, and 16: Evidence for fractional vaporization and mare/highland mixing. *Proc. Lunar Planet. Sci. 12B,* pp. 339-370.

Fruland R. M. (1983) Regolith breccia workbook. *NASA Johnson Space Center Planetary Materials Branch Publ. No. 66, JSC 19045.* 269 pp.

Jerde E. A., Warren P. H., Morris R. V., Heiken G. H., and Vaniman D. T. (1987) A potpourri of regolith breccias: "New" samples from the Apollo 14, 16, and 17 landing sites. *Proc. Lunar Planet. Sci. Conf. 17th,* in *J. Geophys. Res., 92,* E526-E536.

McKay D. S., Bogard D. D., Morris R. V., Korotev R. L., Johnson P., and Wentworth S. J. (1986) Apollo 16 regolith breccias: Characterization and evidence for early formation in the mega-regolith. *Proc. Lunar Planet. Sci. Conf. 16th,* in *J. Geophys. Res., 91,* D277-D303.

Meyer C. Jr. (1978) Ion microprobe analyses of aluminous lunar glasses: A test of the "rock type" hypothesis. *Proc. Lunar Planet. Sci. Conf. 9th,* pp. 1551-1570.

Naney M. T., Crowl D. M., and Papike J. J. (1976) The Apollo 16 drill core: Statistical analysis of glass chemistry and the characterization of a high alumina-silica poor (HASP) glass. *Proc. Lunar Sci. Conf. 7th,* pp. 155-184.

Naney M. T., Papike J. J., and Vaniman D. T. (1977) Lunar highland melt rocks: Major element chemistry and comparisons with lunar highland glass groups (abstract). In *Lunar Science VIII,* pp. 717-719. The Lunar Science Institute, Houston.

Rose H. J. Jr., Cuttitta F., Annell C. S., Carron M. K., Christian R. P., Dwornik E. J., Greenland L. P., and Ligon D. T. Jr. (1972) Compositional data for twenty-one Fra Mauro lunar materials. *Proc. Lunar Sci. Conf. 3rd,* pp. 1215-1229.

See T. H., Hörz F., and Morris R. V. (1986) Apollo 16 impact-melt splashes: Petrography and major-element composition. *Proc. Lunar Planet. Sci. Conf. 17th,* in *J. Geophys. Res., 91,* E3-E20.

Simon S. B., Papike J. J., Shearer C. K., Hughes S. S., and Schmitt R. A. (1989) Petrology of Apollo 14 regolith breccias and ion microprobe studies of glass beads. *Proc. Lunar Planet. Sci. Conf. 19th,* pp. 1-17.

Simonds C. H., Phinney W. C., Warner J. L., McGee P. E., Geeslin J., Brown R. W., and Rhodes J. M. (1977) Apollo 14 revisited, or breccias aren't so bad after all. *Proc. Lunar Sci. Conf. 8th,* pp. 1869-1893.

Vaniman D., Heiken G., Warren P., and Jerde E. (1988) Glasses and a "HASP"-mimicking mineral or mineral intergrowth in Apollo 14 regolith breccias (abstract). In *Lunar and Planetary Science XIX,* pp. 1215-1216. Lunar and Planetary Institute, Houston.

Vaniman D. T., Bish D. L., and Chipera S. J. (1989) A new Ca,Al-silicate mineral from the Moon (abstract). In *Lunar and Planetary Science XX,* pp. 1150-1151. Lunar and Planetary Institute, Houston.

Proceedings of the 20th Lunar and Planetary Science Conference, pp. 219-230
Lunar and Planetary Institute, Houston, 1990

Petrology and Chemistry of Apollo 17 Regolith Breccias: A History of Mixing of Highland and Mare Regolith

S. B. Simon[1,2], J. J. Papike[1], D. C. Gosselin[3,4], J. C. Laul[3], S. S. Hughes[5,6], and R. A. Schmitt[5]

Detailed petrologic and chemical study of regolith breccias and comparison with soils provides a way to gain insight into the history of the regolith of the area. Results for a suite of ten Apollo 17 breccias shows that two (74115 and 76565) consist predominantly of highland material, seven (70019, 70295, 74246, 78546, 79035, 79135, and 79175) are mare-dominated, and one (70175) is a welded volcanic glass deposit. All were formed at or near the Apollo 17 site and all contain both mare and highland components. Comparison with the soils shows that the regolith has become more mature since the formation of the breccias. However, the highland component in the basaltic valley floor soils has not increased significantly since that time. This observation, combined with the thorough distribution of orange glass throughout the site, suggests that vertical mixing has been dominant over lateral transport, especially if the orange glass originated in several extensive layers interbedded with basalt flows. Entry of the orange glass into the regolith predates the formation of the breccias, but with continued mixing and dilution the glass is now slightly less abundant than in the regolith breccia source regoliths. A previously unreported volcanic glass type may be present in breccia 74246. If so, this glass would provide important information regarding the igneous petrology and regolith evolution at the Apollo 17 site.

INTRODUCTION

Although the Apollo 17 soils have been thoroughly studied, the Apollo 17 regolith breccias have not (*Fruland,* 1983). They were overlooked at the time of sample return because of the greater interest in the Apollo 17 orange glass, high-Ti basalts, highland impact melt rocks, and the extensive soil suite, as well as the anorthositic rocks from the Apollo 16 mission. The only recent work on Apollo 17 regolith breccias is that of *Jerde et al.* (1987), who analyzed eight small (10-4 mm) regolith breccias via INAA (instrumental neutron activation analysis).

We have conducted the first complete chemical and petrographic study of the ten Apollo 17 members of the regolith breccia reference suite (*Fruland,* 1983): 70019, 70175, 70295, 74115, 74246, 76565, 78546, 79035, 79135, and 79175. We compare the new data for these breccias with our Apollo 17 soil data (*Vaniman et al.,* 1979; *Simon et al.,* 1981; *Laul et al.,* 1981) to improve our understanding of regolith evolution at the Apollo 17 site. As is the case for Apollo 15 samples, Apollo 17 regolith materials provide an opportunity to study regolith mixing in the immediate vicinity of a highland/mare lithologic contact. Comparison of regolith breccias, which have been "closed" to mixing for some time, with the present-day soils, which continue to be "open" to gain and loss of lithologic components, allows us to evaluate the extent of mixing and soil maturation that has occurred since the formation of the breccias. This insight can help us interpret the regolith evolution of the site, as we showed in our Apollo 15 study (*Simon et al.,* 1986).

This paper is the last in a series of studies of regolith breccia-soil suites from each of the Apollo sites (*Simon et al.,* 1984, 1985, 1986, 1988, 1989). Here we report the results of our Apollo 17 study. The results of our entire investigation of lunar regolith breccias are summarized in a review paper presented elsewhere (Simon and Papike, in preparation, 1990).

ANALYTICAL METHODS

Modal data were optically collected with a Zeiss photomicroscope, using reflected and transmitted light and an automated point-counter. For each breccia, 2000 points were counted, 1000 on each of two thin sections. Pyroxene/olivine ratios were determined by electron microprobe (EDS). Glass and mineral compositions were determined with a MAC-5 automated electron probe operated at 15 keV. Data were reduced according to the method of *Bence and Albee* (1968). For bulk chemistry samples were irradiated in the Oregon State University TRIGA reactor. Counting was done at Oregon State and at Battelle using procedures described by *Laul* (1979) and *Simon et al.* (1989).

MODAL PETROLOGY OF THE REGOLITH BRECCIAS

Clast-size distributions in the 10 breccias we studied are summarized in Fig. 1. All the breccias have fairly high matrix (<20 μm fraction and pore space) contents, giving this suite the highest average matrix content (59 vol.%) of all the breccia

[1]Institute for the Study of Mineral Deposits, South Dakota School of Mines and Technology, Rapid City, SD 57701
[2]Now at Department of the Geophysical Sciences, 5734 South Ellis Avenue, University of Chicago, Chicago, IL 60637
[3]Chemical Sciences Department, Battelle Pacific Northwest Laboratories, Richland, WA 99352
[4]Now at University of Nebraska—Lincoln, 700 W. Benjamin Avenue, Norfolk, NE 68702
[5]Departments of Chemistry and Geology and the Radiation Center, Oregon State University, Corvallis, OR 97331
[6]Now at Montana Bureau of Mines and Geology, Montana College of Mineral Science and Technology, Butte, MT 59701

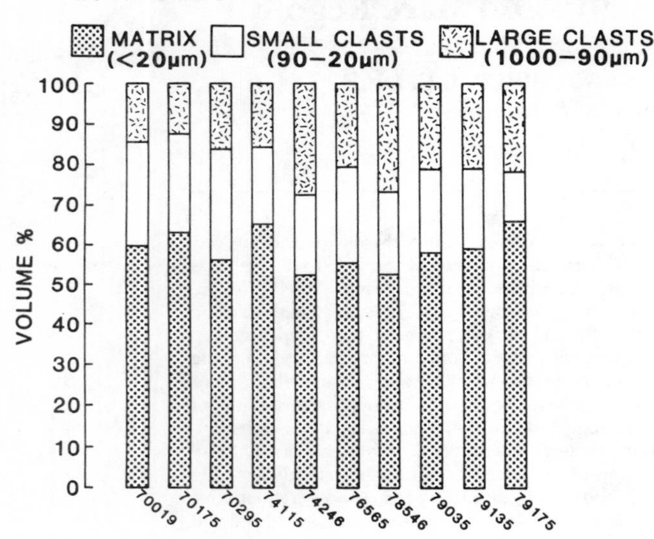

Fig. 1. Abundances of the three clast sizes used in this paper. Samples are arranged by sample number.

suites we have studied (*Simon et al.,* 1989). The mature soils of the Apollo 17 drill core also average ~60 vol.% matrix (*Vaniman et al.,* 1979), but those samples have higher porosities and agglutinate contents than the regolith breccias.

The complete modal petrologies of the breccias are given in Table 1, and matrix-free modes are illustrated in Fig. 2. Most of the breccias have very low agglutinate contents, and even the one with highest content, 79035, would qualify as an immature soil. It has been known for some time (e.g., *McKay et al.,* 1984) that regolith breccias have low fused-soil abundances. With this study now completed, we can state that this is a feature we have observed in the breccia suites from each Apollo site, and we conclude that for most breccias the low fused-soil abundances reflect formation of the breccias from immature regolith (*Simon et al.,* 1984; Simon and Papike, in preparation, 1990). This statement is based on observations of matrix porosity and the presence of agglutinates with circular, undeformed vesicles in thin sections of breccias, which demonstrates that agglutinates can survive breccia formation.

Partially due to the relative immaturity of their sources (lesser amount of recycling into agglutinates), the breccias have higher lithic and monomineralic components than soils (Simon and Papike, in preparation, 1990), and the Apollo 17

TABLE 1. Modal petrology of Apollo 17 regolith breccias.

	70019		70175		70295		74115		74246	
	S	L	S	L	S	L	S	L	S	L
Lithic Fragments										
Mare Component										
Mare Basalt	1.0	6.0	—	0.6	1.0	6.5	0.5	1.0	1.7	15.7
Highland Component										
Plutonic	0.2	0.2	—	0.1	0.4	0.6	0.4	1.5	0.2	1.7
Feld. Frag. Breccia	0.1	0.2	—	—	—	—	—	0.2	—	—
Feld. Basalt	—	—	—	—	—	—	—	—	—	—
Granulite/Poik.	0.2	0.3	—	—	—	0.6	0.2	1.0	0.1	0.3
Impact Melt	0.2	0.5	—	—	0.3	0.8	0.5	1.5	0.1	0.2
Fused Soil Component										
Regolith Breccia	0.2	0.4	—	—	0.3	0.6	0.3	2.5	—	0.2
Agglutinate	3.8	1.4	0.2	0.3	2.6	1.5	3.4	2.8	0.2	0.1
Mineral Fragments										
Pyroxene	10.3	2.5	—	—	7.9	0.9	3.7	1.1	6.5	3.3
Olivine	0.5	0.1	0.7	0.2	1.3	0.2	1.5	0.4	1.9	0.9
Plagioclase	3.1	1.0	—	—	3.3	1.1	3.8	1.1	2.6	1.5
Opaque	2.1	0.1	—	—	3.1	0.5	0.4	—	5.0	0.6
Glass Fragments										
Orange/Black	1.9	0.6	15.0	2.8	3.5	0.6	1.2	0.6	0.2	0.2
Yellow/Green	0.4	0.2	0.1	—	0.2	0.4	0.6	0.2	0.2	—
Colorless	0.3	0.1	tr	—	0.5	—	0.3	0.2	1.4	0.4
Brown	—	—	—	—	0.1	—	0.3	—	—	—
Miscellaneous										
Devit. Glass	1.5	0.7	8.4	8.5	3.1	1.7	1.9	1.4	0.6	1.1
Other	—	0.2	—	—	0.1	0.1	—	0.2	0.2	0.2
Total	25.8	14.5	24.4	12.5	27.7	16.1	19.0	15.7	20.9	26.4
Matrix	59.7		63.1		56.2		65.3		52.7	

TABLE 1. (continued).

	76565		78546		79035		79135		79175	
	S	L	S	L	S	L	S	L	S	L
Lithic Fragments										
Mare Component										
Mare Basalt	0.4	2.2	0.1	3.7	1.1	5.0	0.4	1.9	0.5	3.3
Highland Component										
Plutonic	0.5	2.5	0.1	1.8	0.1	0.1	0.1	0.6	0.3	3.0
Feld. Frag. Breccia	0.1	—	—	0.1	—	0.1	—	0.5	—	0.2
Feld. Basalt	—	—	—	0.1	—	—	0.1	—	—	—
Granulite/Poik.	0.5	1.9	0.2	0.5	—	—	0.2	0.6	0.1	0.2
Impact Melt	0.5	0.4	0.5	1.5	0.7	1.2	0.5	1.5	0.4	1.2
Fused Soil Component										
Regolith Brecc.	0.1	1.5	0.1	2.7	0.3	0.3	0.2	0.8	0.1	2.0
Agglutinate	1.6	1.5	0.7	3.5	5.0	8.6	0.9	5.0	0.6	4.0
Mineral Fragments										
Pyroxene	3.8	1.6	3.8	1.2	3.8	1.7	3.2	1.7	2.9	1.7
Olivine	2.4	1.0	1.4	0.4	0.8	0.4	0.9	0.5	0.6	0.3
Plagioclase	7.8	5.6	4.2	3.4	2.2	0.8	3.4	1.6	1.7	1.4
Opaque	1.5	0.3	0.8	—	1.7	0.2	1.9	0.3	1.5	0.6
Glass Fragments										
Orange/Black	1.1	—	2.5	0.5	1.1	—	2.7	0.7	0.7	0.3
Yellow/Green	0.6	0.6	1.5	0.4	0.8	0.4	1.2	1.1	0.2	0.1
Colorless	1.0	0.1	0.8	—	0.6	1.0	1.3	1.1	0.2	0.3
Brown	0.2	—	0.2	—	0.2	—	0.1	—	0.2	0.1
Miscellaneous										
Devit. Glass	1.7	1.2	3.7	6.8	2.2	1.4	3.0	2.7	2.0	3.1
Other	0.2	0.2	—	0.2	0.1	—	—	0.4	0.1	0.2
Total	24.0	20.6	20.6	26.8	20.7	21.2	20.1	21.0	12.1	22.0
Matrix	55.4		52.6		58.1		58.9		65.9	

Matrix = <20μm; S = small clasts (90 - 20μm); L = large clasts (1000 - 90μm); tr = trace.

breccias are no exception (Fig. 2). With the exception of 70175, the breccia modes show the effect of moderate highland/mare mixing. All the mare breccias contain at least 5 vol.% highland lithic clasts, and the highland breccias (74115 and 76565) contain ~5 vol.% mare lithic clasts. Sample 70175 is essentially a compacted orange/black volcanic glass deposit. It consists of orange glass beads and rare lithic and olivine fragments in a "matrix" of black devitrified glass. *Jerde et al.* (1987) apparently analyzed a similar sample.

The anomalous petrology of 70175 is clearly illustrated in Fig. 3, a summary of the monomineralic components in the breccias. Its small mineral fragment population consists of only olivine, whereas all the other breccias have either pyroxene or plagioclase as the dominant monomineralic clast type. Several of the basalt-dominated breccias have fairly high opaque (mostly ilmenite) contents, and the two highland breccias are relatively rich in plagioclase and olivine. Note that several of the mare regolith breccias have roughly equal proportions of feldspar and pyroxene. This would not be expected in a "pure" mare breccia, and it shows the effect of mixing of highland and mare rocks and minerals in the source regoliths.

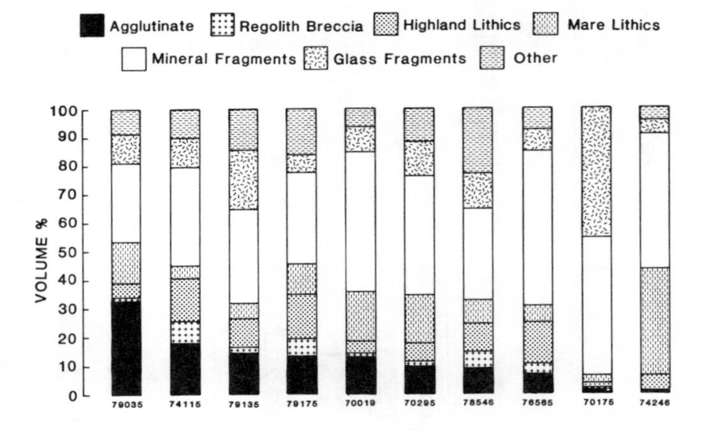

Fig. 2. Modal petrologies of the 1000-20 μm (i.e., matrix-free) fraction of the regolith breccias, in order of decreasing agglutinate content. "Other" includes devitrified glass and unidentifiable clasts.

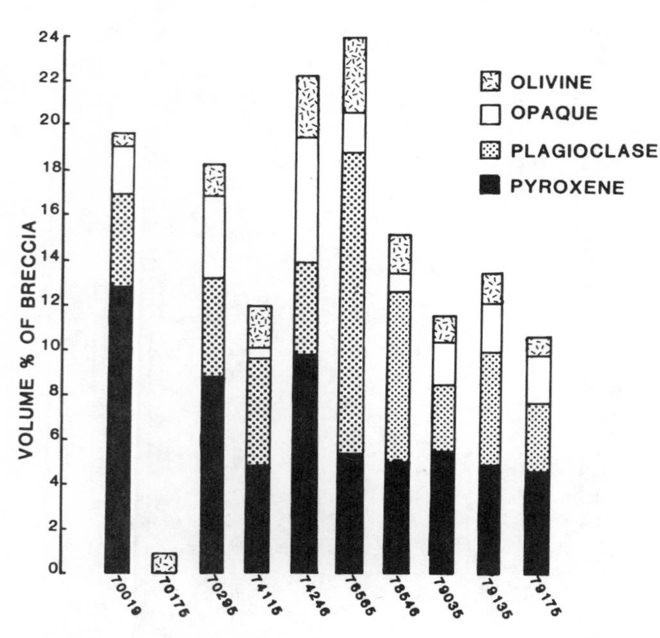

Fig. 3. Mineral fragment abundances in the breccias. Pyroxene/olivine ratios were determined for each sample by electron microprobe.

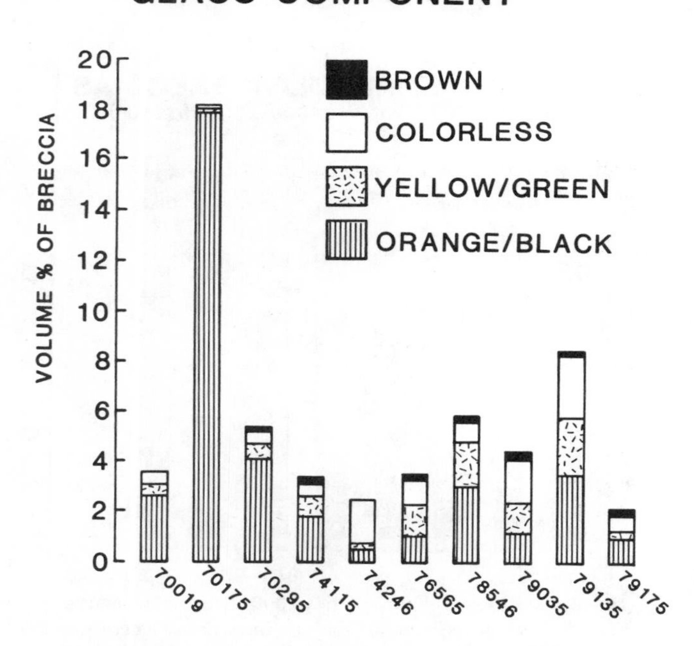

Fig. 4. Abundances of glass color groups in the regolith breccias.

Abundances of the different glass types, optically identified by color, are illustrated in Fig. 4. Except for 70175, the breccias have 3-8 vol.% glass. All the breccias contain Apollo 17 orange/black volcanic glass, evidence that the breccias are local to the site and that breccia formation postdated glass eruption and distribution around the site. This suggests that some time elapsed between the fire-fountaining event that formed the glasses and the formation of the breccias, because time is required for distribution of the glasses and mixing into the regolith. Sample 70175 formed from a highly concentrated glass deposit, but because rather pure deposits are still present (e.g., 74220) we cannot conclude that this breccia formed soon after glass formation and deposition.

MODAL COMPARISON OF APOLLO 17 REGOLITH BRECCIAS AND SOILS

Breccia modes are compared with data for Apollo 17 surface soils in Fig. 5. The breccias have lower fused-soil contents than the soils, a feature of all breccia-soil suites (*McKay et al.,* 1984; Simon and Papike, in preparation, 1990). In the right half of Fig. 5, the data have been renormalized to a fused soil-free base, to compare the mineral, lithic and glass components. The highland soils have slightly more highland lithic clasts than the highland breccias. This may be because the soils come from the bases of massifs, areas that are probably still accumulating material from higher up on the hillsides.

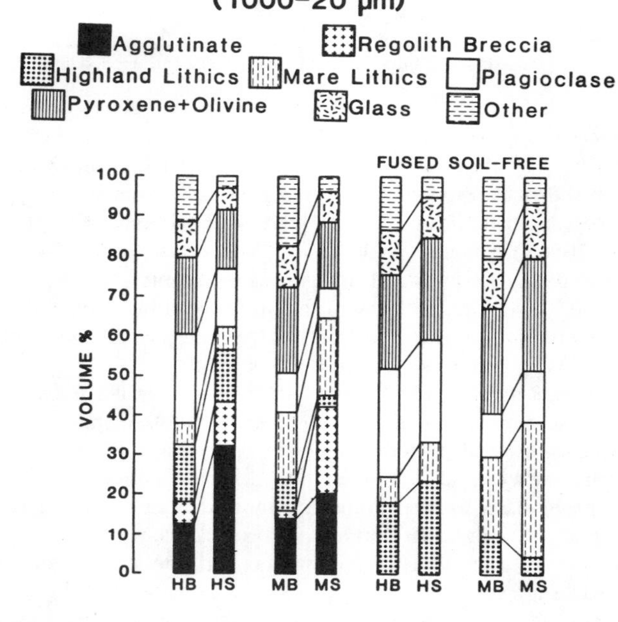

Fig. 5. Comparison of the 1000-20 μm fractions of Apollo 17 regolith breccias and soils. Highland breccias (HB) 74115 and 76565 are compared with highland soils (HS) 72501, 76501, and 78221. Mare breccias (MB) are compared with mare soils (MS).

Figure 5 also indicates that the mare soils have a higher proportion of basalt fragments than the mare breccias. This suggests that over the time since breccia formation derivation of basalt fragments from the bedrock has been a more efficient process than the breakdown of the fragments into their constituent minerals.

Note also the higher "other" content of the breccias compared to the soils. This relationship is also observed in the Apollo 11, Apollo 14, and Apollo 16 suites (Simon and Papike, in preparation, 1990). This is mostly a reflection of higher devitrified glass contents in the breccias, probably caused by heating of the glasses during breccia formation.

MINERAL AND GLASS CHEMISTRY

Compositions of individual pyroxene grains in the breccias are plotted in Fig. 6. In this and the following mineral and glass plots, the highland breccia data are plotted separately from the mare breccia data. Both plots have Mg-rich orthopyroxenes that plot within the highland pyroxene field of *Vaniman et al.* (1979). Such pyroxenes are relatively rare in the Apollo 17 soils having stronger mare affinities, such as the upper part of the deep drill core (*Vaniman et al.,* 1979). Pyroxene populations in the highland breccias are similar to those of the highland soils 72501, 76501, and 78221 (*Simon et al.,* 1981). Magnesium-rich orthopyroxene is most abundant, with lesser amounts of magnesian and more ferroan clinopyroxene. The mare breccias and soils differ from the highland regolith in having higher proportions of pigeonite and augite, which are typical compositions in Apollo 17 mare basalts (*Longhi et al.,* 1974; *Papike et al.,* 1974).

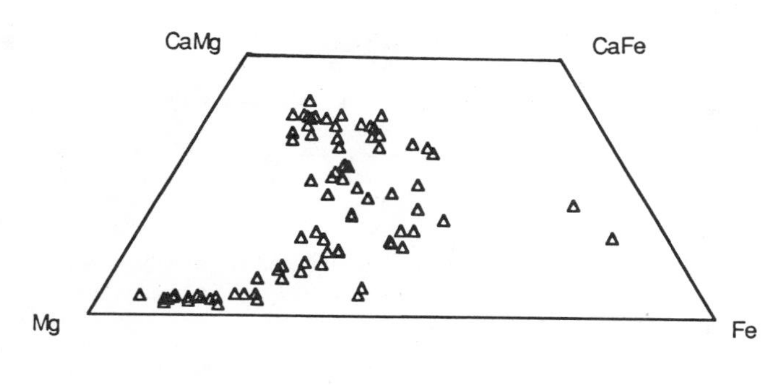

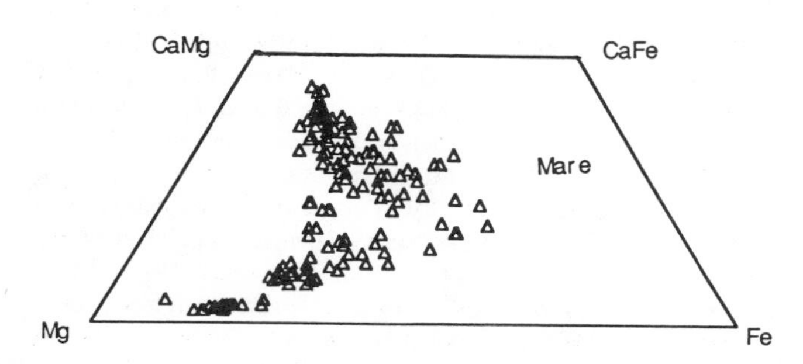

Fig. 6. Compositions of pyroxene grains in Apollo 17 highland regolith breccias (top) and mare regolith breccias.

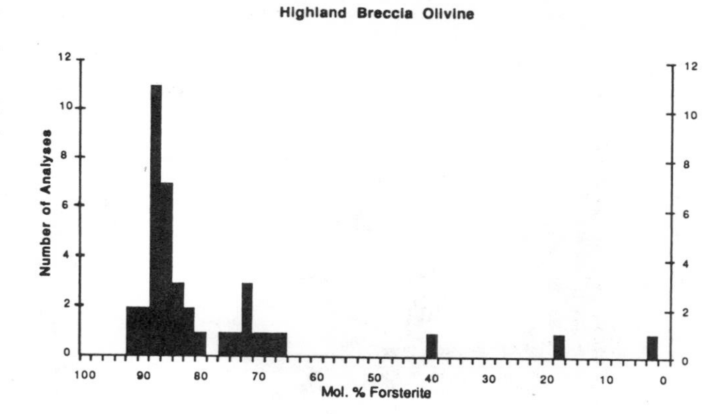

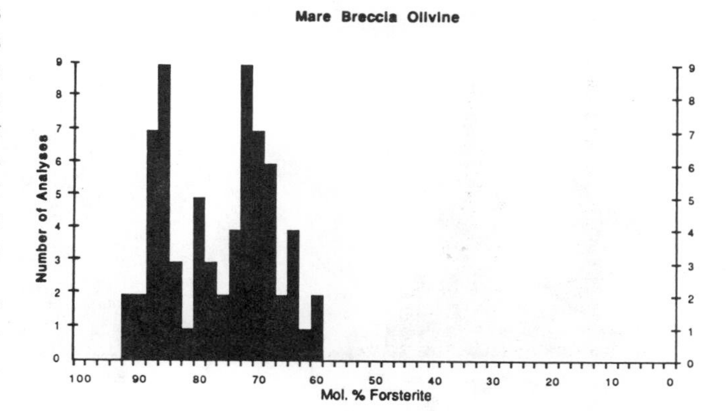

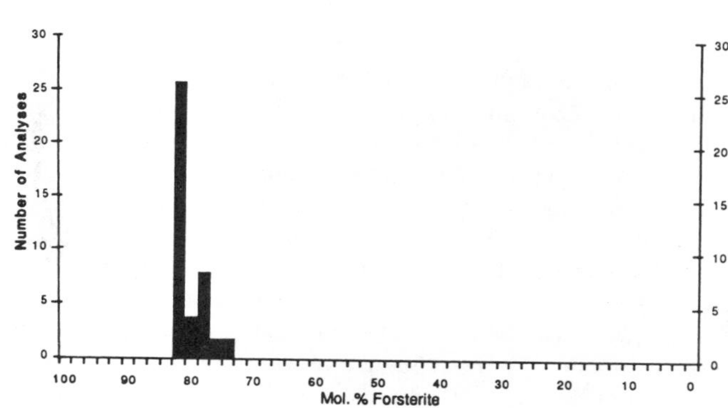

Fig. 7. Compositions of olivine grains in Apollo 17 regolith breccias. Note the bimodal distributions.

The highland and the mare breccias clearly have bimodal distributions of olivine with gaps at about Fo_{80} (Fig. 7). The relatively forsteritic olivines ($>Fo_{80}$) are probably of highland origin, and those less forsteritic than Fo_{80} are mostly from basalts. The olivines in 70175 have a narrow range, Fo_{75-82}, and their compositions are consistent with derivation from Apollo 17 orange volcanic glass.

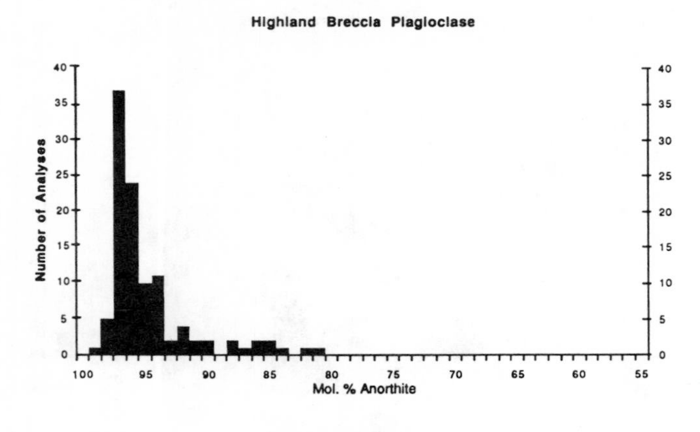

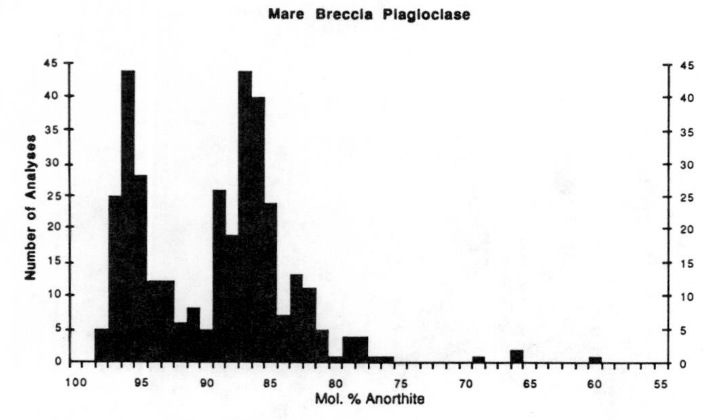

Fig. 8. Plagioclase compositions in the regolith breccias. Note the bimodal distribution in the mare breccias.

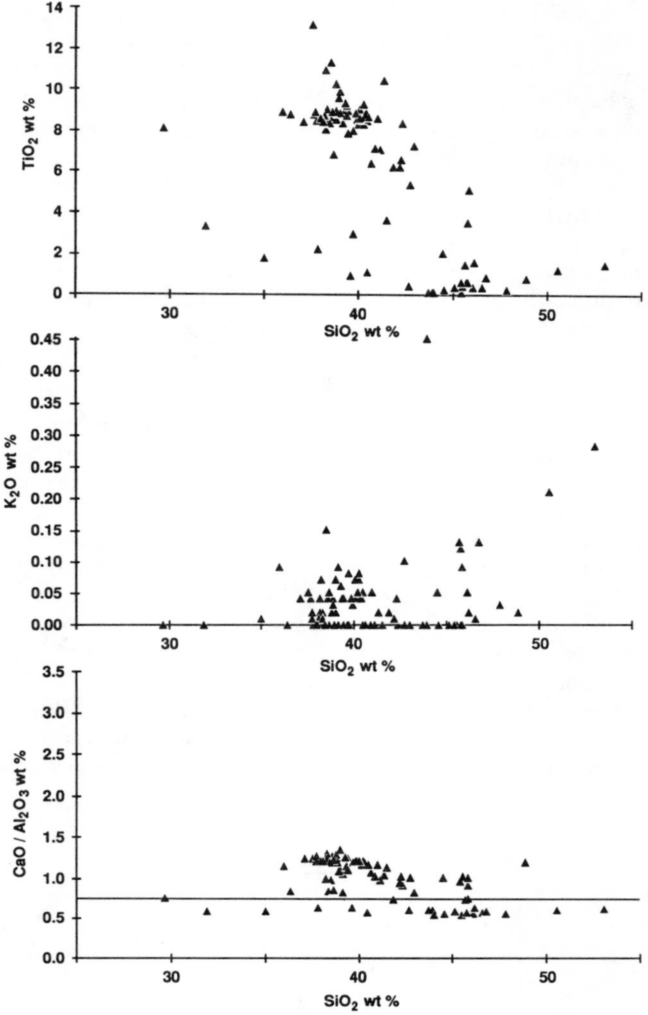

Fig. 9. Glass compositions in the highland regolith breccias. Most mare glasses plot above the $CaO/Al_2O_3 = 0.75$ line, and most highland glasses plot below the line.

The olivine populations in the breccias are remarkably similar to those in comparable soils. As in the highland breccias, most of the olivine grains in Apollo 17 highland soils have compositions of Fo_{80} to Fo_{90} (*Simon et al.,* 1981). The olivine population in the Apollo 17 deep drill core has a large peak at $\sim Fo_{70}$, and smaller ones at $\sim Fo_{84-90}$ and $\sim Fo_{78-80}$ (*Vaniman et al.,* 1979), much like the mare breccias. The latter peak may be due to olivine derived from devitrified orange glass. The breccia olivine populations strongly support a local (Apollo 17) origin for all of the breccias.

The breccia plagioclase populations are also similar to those of the corresponding soils. Figure 8 shows that almost all of the plagioclase in the highland breccias is more calcic than An_{90}, like the soils analyzed by *Simon et al.* (1981). This indicates that almost all of the feldspar in these samples is of highland origin. In contrast, the mare breccia plagioclase exhibits a bimodal distribution, and, as is the case for olivine, the gap falls at the approximate division between highland and mare mineral compositions.

GLASS COMPOSITIONS IN APOLLO 17 REGOLITH BRECCIAS

Approximately 50 glass beads and fragments per breccia were analyzed. The glass compositions are summarized in terms of their TiO_2, K_2O, and SiO_2 contents and their CaO/Al_2O_3 ratios in Figs. 9 through 11.

Glass compositions in the highland regolith breccias are shown in Fig. 9. The plot of TiO_2 vs. SiO_2 shows two clusters, one at ~9 wt.% TiO_2 (orange glass), and one at very low TiO_2 (highland glass). In between are glasses that represent a wide range of TiO_2 contents. These are mostly impact glasses of mare origin. The plot of K_2O vs. SiO_2 shows that only a few of these glasses are rich enough in Si and K to be considered KREEP. The CaO/Al_2O_3 ratio is useful as an indicator of mare vs. highland affinity. Highland glasses tend to have CaO/Al_2O_3 ratios ≤ 0.65 (*Naney et al.,* 1976), whereas mare basalts and glasses have ratios >0.75. Therefore, a high proportion of the glasses in the highland breccias are of mare origin (Fig. 9). This is consistent with their TiO_2 contents and with the reported orange and black glass abundances in South Massif and North Massif soils (*Heiken and McKay,* 1974; *Simon et al.,* 1981).

The mare breccias (Fig. 10) appear to contain essentially the same glasses as the highland breccias. Note the small cluster at ~14 wt.% TiO_2, which may be represented by one glass in the highland breccias. Although these occur in more than one breccia, sample 74246 contains an unusually high proportion of them. This may be a new orange volcanic glass group; representative analyses are given in Table 2. Further study, including a search for more beads, is required.

The K_2O plot shows that K_2O contents in glasses in the mare breccias extend to higher values than in the highland breccia glasses, and that these K-rich glasses have fairly high SiO_2 contents. Like the other CaO/Al_2O_3 plots, the one in Fig. 10 shows the tendency of highland glasses to approach a lower limit of 0.5, approximately the ratio in pure anorthite.

Figure 11 shows that although 70175 is essentially a welded orange/black glass deposit, the source material remained unconsolidated long enough to collect some highland glasses. They are colorless and not KREEPy.

BULK COMPOSITIONS OF APOLLO 17 REGOLITH BRECCIAS AND SOILS

Compositions of the Apollo 17 breccias analyzed via INAA for this study are given in Table 3. Variations in several of the major elements in Apollo 17 breccias and soils are illustrated in Fig. 12, in which TiO_2, MgO, and FeO are plotted against Al_2O_3. Stations 2, 3, 4, 6, and 8 are highland stations, and samples from these localities have higher Al_2O_3 contents than samples from mare stations. An exception is breccia 74246. Although it is from station 4, which is within the "white mantle" avalanche deposit of highland material onto the valley floor, it consists of mare regolith from the rim of Shorty crater (*Fruland,* 1983). This crater penetrated the "white mantle," so it is not surprising to have a mare breccia from this highland station.

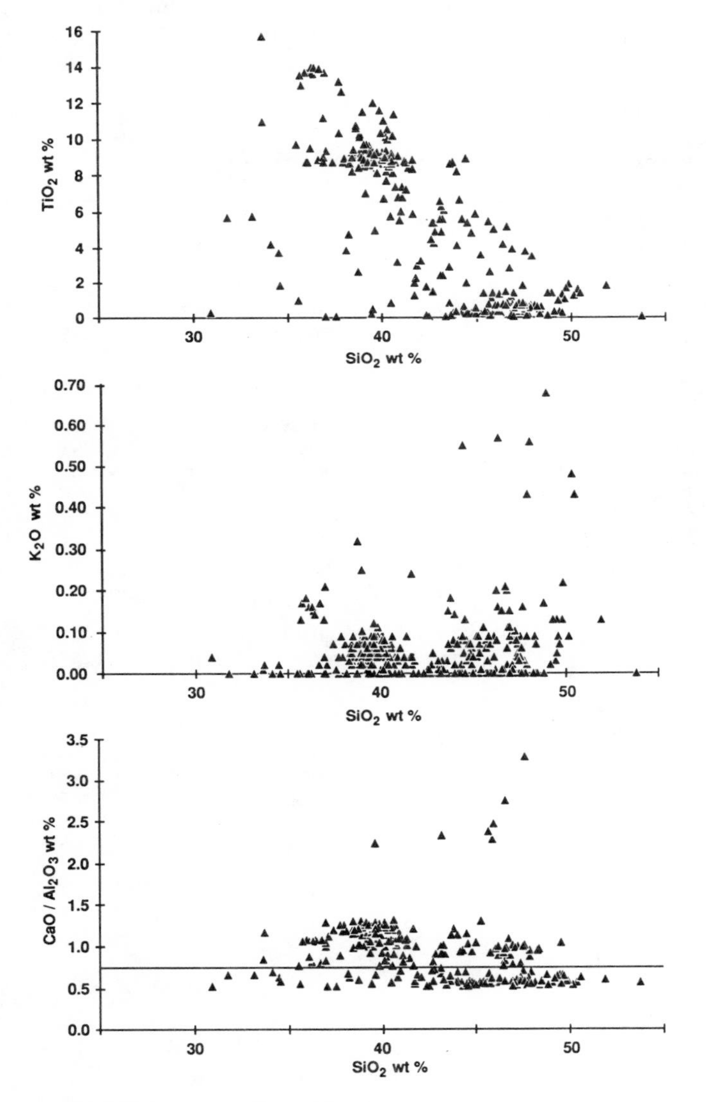

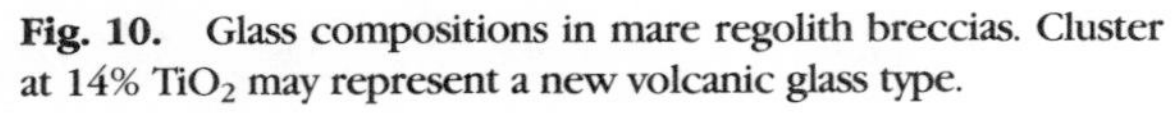

Fig. 10. Glass compositions in mare regolith breccias. Cluster at 14% TiO_2 may represent a new volcanic glass type.

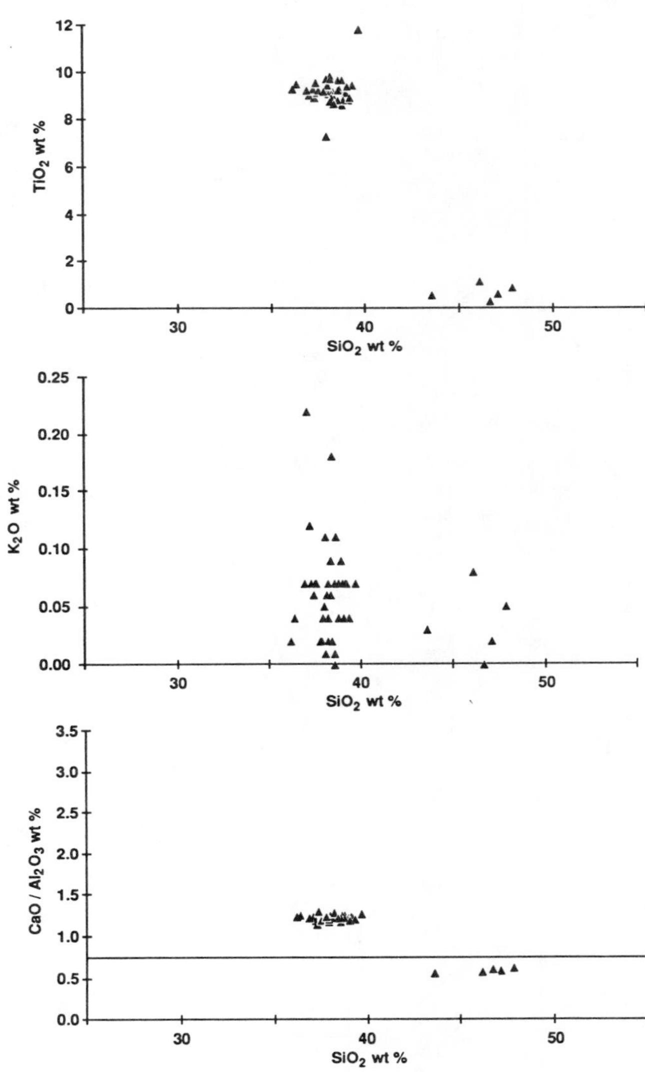

Fig. 11. Glass compositions in 70175. A few highland glasses are present.

TABLE 2. Representative analyses of very high-Ti glass in breccia 74246.

	1	2	3
SiO_2	36.43	36.70	36.95
TiO_2	14.01	13.95	13.74
Al_2O_3	7.77	8.15	8.19
Cr_2O_3	0.57	0.47	0.51
FeO	21.03	20.39	20.45
MnO	0.36	0.27	0.36
MgO	10.63	9.23	9.46
CaO	8.30	8.89	8.59
Na_2O	0.68	0.70	0.64
K_2O	0.14	0.17	0.21
Total	99.92	98.92	99.10

TiO_2 and FeO contents of Apollo 17 regolith materials vary inversely with Al_2O_3 contents, as shown in Fig. 12 and in the early studies of Apollo 17 samples (e.g., *LSPET,* 1973; *Rhodes,* 1973; *Rhodes et al.,* 1974). This relationship shows the effect of mixing between Ti-, Fe-rich components (basalts, volcanic glasses) and Al-rich highland lithologies. Note, however, that such a relationship does not exist for MgO vs. Al_2O_3 (Fig. 12). This is because the highland components are relatively magnesian noritic, gabbroic, and troctolitic rocks. The trends shown in Fig. 12 therefore also show that ferroan anorthosite (Al-rich, Mg-, Fe-poor) is not an important factor in these samples. If ferroan anorthosite was a significant component, the endmember Apollo 17 highland soils would have higher Al_2O_3 and lower MgO contents than are observed.

Sample 70175, collected near the LM, does not plot on the compositional trends defined by the other samples because its composition is essentially that of orange glass, as shown by its low Al_2O_3 content.

Chondrite-normalized REE abundances in the Apollo 17 breccias and soils are illustrated in Fig. 13. Except for breccia 74115, all the breccias plot within the range of analyzed Apollo 17 soils (*Morris et al.,* 1983). The dominant features of the patterns are: abundances at 20×-40× chondrite, positive LREE slope, negative Eu anomaly, and a negative HREE slope that is flatter than the LREE slope. As shown by *Rhodes et al.*

TABLE 3. Bulk compositions of Apollo 17 regolith breccias.

Sample No.	70019,104	70175,10	70295,8	74115,9	74246,6	76565,13	78546,10	79035,135	79135,153	79175,32
wt (mg)	157.6	176.1	146.4	164.8	166.4	140.3	141.5	163.4	133.2	162.3
TiO_2* (%)	7.91	8.65	7.83	2.65	8.16	4.57	4.33	6.58	5.75	5.64
Al_2O_3*	12.7	5.83	12.0	18.4	10.5	16.1	13.9	13.0	13.6	13.7
FeO*	15.9	22.9	17.3	10.5	17.1	12.4	13.6	15.2	14.8	14.5
MnO*	0.219	0.275	0.218	0.133	0.250	0.160	0.184	0.202	0.202	0.197
MgO*	9.0	13.9	9.6	9.4	7.3	10.3	10.6	9.8	10.3	10.0
CaO*	11.8	7.2	10.9	13.3	11.2	12.0	11.5	11.4	10.7	10.5
Na_2O*	0.41	0.39	0.42	0.46	0.38	0.41	0.47	0.41	0.5	0.49
K_2O*	0.083	0.071	0.079	0.15	0.067	0.089	0.11	0.072	0.116	0.107
Sc* (ppm)	59.6	47.0	58.0	27.2	74.3	38.7	40.0	49.2	41.1	42.0
V	87	105	92	55	105	66	78	87	85	79
Cr*	2870	4850	3050	1840	3000	2220	2840	2550	2930	2930
Co*	33.0	63.5	36.0	31.5	20.3	30.0	37.1	31.0	36.7	35.4
Ni	140	70	150	220	20	130	100	110	140	170
Zn	15	190	40	30	—	35	60	25	80	53
Rb	2.9	3.3	6.4	10.6	9.6	4.7	10.8	2.1	7.6	9.5
Sr*	170	200	170	160	120	160	150	150	185	130
Cs	0.42	0.22	0.15	0.15	0.40	0.14	0.13	0.86	0.12	0.16
Ba*	97	80	90	165	80	105	110	105	120	120
La*	7.90	5.90	8.07	14.9	7.14	8.55	8.62	8.20	10.1	9.16
Ce*	23.0	17.5	23.3	37.0	19.5	22.5	22.7	22.1	27.0	24.0
Nd*	18.2	15.3	20.4	26.0	18.0	18.7	17.8	18.2	22.3	20.0
Sm*	7.61	6.52	7.43	7.53	7.57	6.04	5.80	6.84	7.25	6.84
Eu*	1.60	1.85	1.80	1.35	1.40	1.38	1.40	1.40	1.55	1.58
Gd	12.0	7.7	11.3	8.8	10.4	7.4	7.3	10.0	10.1	8.8
Tb*	2.00	1.53	1.90	1.66	1.95	1.40	1.30	1.62	1.72	1.57
Dy*	12.7	8.9	11.3	10.1	13.1	8.7	8.6	11.1	10.9	9.4
Tm	1.0	0.61	0.98	0.86	1.0	0.81	0.71	0.87	0.82	0.77
Yb*	6.12	3.77	6.30	5.88	6.74	4.78	4.42	5.45	5.07	5.10
Lu*	0.89	0.50	0.97	0.81	1.02	0.72	0.66	0.79	0.70	0.75
Zr	135	120	130	140	120	120	110	120	130	130
Hf*	6.40	5.40	6.30	6.10	5.70	5.00	4.70	5.40	5.70	5.35
Ta	1.20	1.00	1.20	0.93	1.00	0.86	0.76	1.05	1.00	0.95
Th*	0.80	0.33	0.81	2.10	0.51	1.12	1.15	0.90	1.00	1.02
U (ppm)	0.20	0.08	0.18	0.62	0.06	0.32	0.33	0.18	0.38	0.14
Ir (ppb)	4.7	<1	5.0	5.7	<1	4.5	4.5	5.5	6.4	4.7
Au (ppb)	3.3	2.8	6.3	7.2	5.1	2.0	6.0	6.7	3.6	7.3

*Used in mixing model calculations.

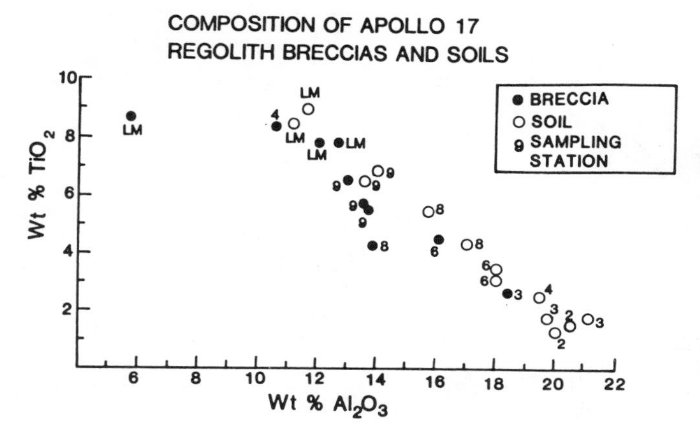

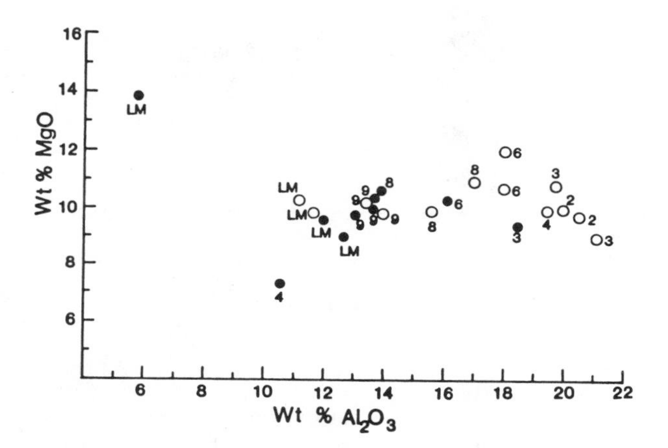

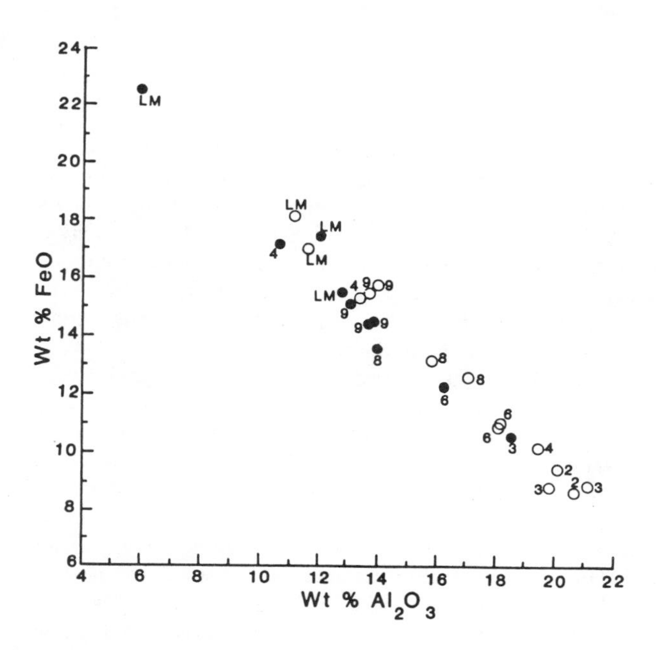

Fig. 12. Summary of TiO_2, MgO, FeO, and Al_2O_3 variations in Apollo 17 regolith breccias and soils.

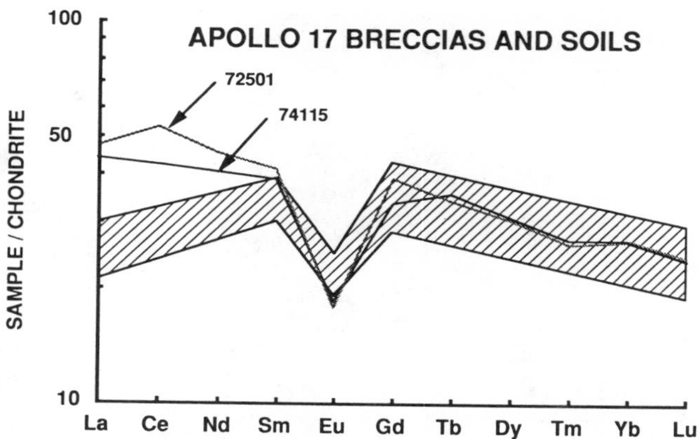

Fig. 13. Chondrite-normalized REE abundances in Apollo 17 regolith breccias and soils. All plot within the indicated range except for some South Massif samples, which are enriched in the light REE, as shown by 72501 and 74115.

(1974), these features are typical of Apollo 17 basalts, valley floor soils and North Massif soils. However, South Massif soils tend to have LREE at ~50× chondrite, with a slight negative slope (*Rhodes et al.,* 1974). The patterns for 72501 (*Laul et al.,* 1981) and 74115 show this characteristic, probably reflecting a higher noritic breccia component (*Phinney et al.,* 1974; *Rhodes et al.,* 1974; *Laul et al.,* 1979) relative to the other samples.

CHEMICAL MIXING MODELS

In addition to the data and observations discussed above, chemical mixing calculations provide a straightforward way to compare the breccias and soils. If the same lithologic components that work for the soils can be used to model the breccia compositions as well, we then have an objective way to compare the breccias and soils in terms of their possible source lithologies. Furthermore, this is strong evidence for local origin of the breccias. This is the case for Apollo 11 (*Simon et al.,* 1984), Apollo 15 (*Simon et al.,* 1986), and Apollo 17 regolith breccias and soils.

We modeled the Apollo 17 breccias with the same components used by *Laul et al.* (1979, 1981) to model Apollo 17 soil compositions: low-K KREEP (noritic breccia), anorthositic gabbro, high-Ti basalt, and orange glass. The least-squares method of *Boynton et al.* (1975, 1976) was used. The elements used in the modeling are indicated in Table 3. A weighting factor of 0.1 was used for the major elements and 0.2 was used for the trace elements. The goodness of fit of the model is indicated by the value of the reduced chi-square (χ^2), which is defined as the sum of the squares of the residuals for each element divided by the number of degrees of freedom, in this case the number of elements used minus the number of components. For a good fit, a reduced χ^2 should be $\leq$1. In the present study, for each breccia the reduced χ^2 was between 0.23 and 0.77.

The results of the mixing calculations are illustrated in Fig. 14. For comparison models for soils 70009 (*Laul et al.,* 1979), 72501, 76501, and 78221 (*Laul et al.,* 1981) are also shown. The result for 70175 is not shown because its composition is that of orange glass.

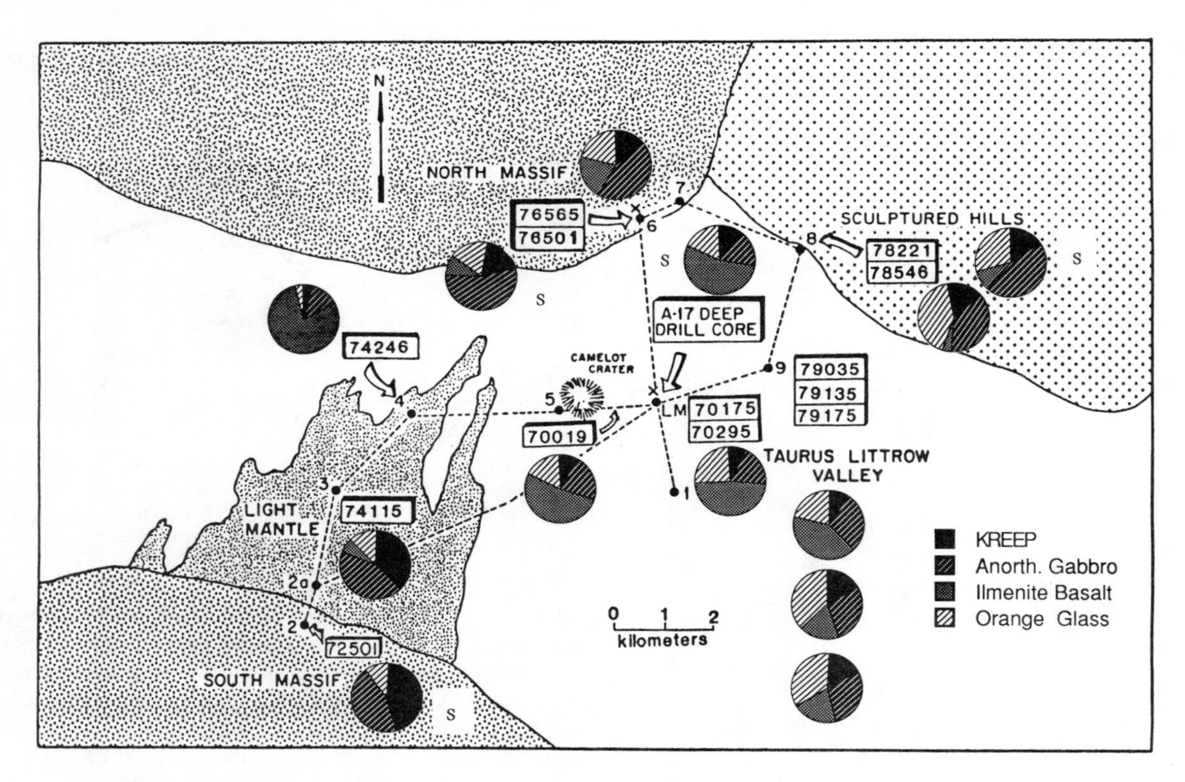

Fig. 14. Mixing model results superimposed on a map of the Apollo 17 site, showing the compositional variation with sampling locality. Results for 70175 are not shown because it has orange glass composition. The Apollo 17 drill core locality is represented by the top of 70009 (*Laul et al.,* 1979). Other soil data (72501, 76501, 78221) are from *Laul et al.* (1981) and soil diagrams are indicated by "S."

As might be expected from their REE patterns, 72501 and 74115 are enriched in noritic breccia (low-K KREEP) relative to the other samples. This is characteristic of South Massif samples (*Rhodes et al.,* 1974). Soil 72501 has no calculated basalt component, whereas station 4 breccia 74246 has ~75%. This reflects its high TiO_2 content, at least part of which is actually present in glass. Because our basalt end member has a higher TiO_2 content than the glass component (12.4% vs. 9.4%) the model probably overestimated the basalt component and underestimated the orange glass component. Its basaltic affinity is understandable because it was collected from the rim of Shorty crater. It therefore likely formed from the mare regolith beneath the light mantle.

The other valley floor samples also have high mare components (~60-75%), although the glass:basalt proportions vary. Breccias 79135 and 79175, from station 9, have very similar compositions and high glass:basalt ratios (~2:1) compared to 79035 and the samples from near the LM, which have ratios of about 1:2. The highland component in the valley floors has not increased significantly since the breccias formed.

The mixing models also indicate that the samples from the bases of North Massif and Sculptured Hills have anorthositic gabbro as their dominant lithology and more basalt and less KREEP than the South Massif samples. Although the breccias and soils "agree" in these respects, there are some differences as well. At both station 6 and station 8 the soils are enriched in anorthositic gabbro relative to the breccias. Both breccias are enriched in orange glass relative to the soils, possibly because of lithification prior to the eventual gardening and dilution of the glasses.

DISCUSSION

When the fractionated compositions of the finest soil fractions relative to bulk soil were first recognized (*Murthy et al.,* 1972; *Evensen et al.,* 1973) they were originally attributed to a widely distributed "exotic" component (*Murthy et al.,* 1973; *Evensen et al.,* 1974). However, the observation of sharp geologic contacts and the detection of numerous basalt units by remote-sensing techniques (e.g., *Head et al.,* 1978; *Pieters et al.,* 1980) argue against efficient lateral transport. Impact comminution of rocks, which has been experimentally shown to concentrate feldspar and mesostasis into the finest size fractions of the debris (*Haskin and Korotev,* 1977; *See et al.,* 1984; *Hörz et al.,* 1984) is now favored over lateral transport of a uniform ultrafine fraction.

Of course, some lateral transport does occur, and the samples considered in this study provide some insight into that process and into regolith mixing in general. Although mixing has occurred to the extent that none of the samples represents pure highland or mare regolith, we observe differences between the North Massif and South Massif highland soils. Also, the valley floor soils retain their strong mare affinity despite the adjacent highland massifs. *Rhodes et al.* (1974) noted that the Apollo 17 valley soils have highland components comparable to that of the Apollo 11 soil, which is ~50 km from the nearest highland outcrop. From this, *Rhodes et al.* (1974) concluded that lateral mixing is inefficient and proposed that the highland materials at Apollo 11 entered the regolith via vertical mixing. We find slightly higher highland components in the Apollo 17 valley regolith than in the Apollo 11 regolith (35-40 vs. ~25%). If the analogy used by *Rhodes et al.* (1974) is valid, then the difference in highland components can be attributed to local lateral mixing at the Apollo 17 site. The assumption is that for a highland component of 25% or less, transport from local highlands is not required, as illustrated by the Apollo 11 regolith (*Wood et al.,* 1970; *Simon et al.,* 1984). Lack of efficient lateral transport is also demonstrated by the similar magnitude of the highland components in the valley floor breccias and soils. Efficient transport from the massifs should have increased the highland component in the soils in the time since the formation of the breccias.

The distribution of orange glass around the site, even into the highland samples, is an additional indicator of material transport. The glass may have always been well distributed around the site, if it was deposited in extensive beds interlayered with basalt (*Heiken et al.,* 1974). If this is true, its presence in all the Apollo 17 soils is evidence for vertical mixing in the mare soils and lateral mixing with the highland soils (especially North Massif and Sculptured Hills). Generally, the breccias contain more orange glass than the soils (*Rhodes et al.,* 1974; *Laul et al.,* 1979) due to continued dilution of the glasses in the regolith after the formation of the breccias.

The orange glass (74220) has been dated at 3.71 ± 0.06 b.y. (*Schaeffer and Husain,* 1973), and it would be tempting to assign this age to 70175 and 72504,10 (*Jerde et al.,* 1987). However, because the concentrated glass deposits are still present at the Apollo 17 site, we can only say that their source glass deposits were exposed at the surface only briefly, then buried while still rather pure. The source for 70175 was unconsolidated and exposed long enough to acquire a small component of lithic fragments and highland glasses.

On the other hand, the very high-Ti glass in 74246 represents a type that had a very limited distribution and/or became thoroughly mixed into the regolith before the other breccias formed. The latter case would imply that 74246 is older than the other breccias. However, a complete absence of this glass from some of the breccias would suggest that these breccias predate the glass and are older than 74246 (*Delano,* 1988). Further study of this glass is needed to determine whether it is volcanic, whether it can be related to 74220, and whether it is present in all or some of the breccias.

CONCLUSIONS

1. The Apollo 17 regolith breccias have low agglutinate abundances, typical of lunar regolith breccias and indicative of formation from immature source regolith.

2. The breccias formed locally after the eruption of basalt and orange glass at the site. Sample 70175 is similar to 72504,10 and is a fragment of a welded orange/black glass deposit.

3. Since the formation of the breccias, the regolith at the Apollo 17 site has become more mature, and the orange glass abundance has been somewhat decreased by mixing. However, lateral transport has not been efficient enough to increase the highland component in the valley floor soils, much of which could be accounted for by vertical mixing from beneath the basalt.

4. Breccia 74246 may contain a previously unreported volcanic glass type. Further study of this glass is needed to determine if it is volcanic and to determine its abundance in other breccias.

Acknowledgments. We thank the OSU TRIGA reactor crew for technical assistance. M. Lum typed the manuscipt with L. Grossman's kind permission. Reviews by M. Lindstrom, D. Lindstrom, and G. Heiken led to improvements in the text. M. Spilde, I. Palaz, and P. Papike assisted in the figure preparation, and their efforts are appreciated. This work was funded by NASA grants NAG 9-22 (JJP), NAG 9-63 (RAS), and NAS 9-15357 (JCL).

REFERENCES

Bence A. E. and Albee A. L. (1968) Empirical correction factors for the electron microanalysis of silicates and oxides. *J. Geol., 76,* 382-403.

Boynton W. V., Baedecker P. A., Chou C.-L., Robinson K.L., and Wasson J.T. (1975) Mixing and transport of lunar surface material: Evidence obtained by the determination of lithophile, siderophile, and volatile elements. *Proc. Lunar Sci. Conf. 6th,* pp. 2241-2259.

Boynton W. V., Chou C.-L., Robinson K. L., Warren P. H., and Wasson J. T. (1976) Lithophile, siderophiles and volatiles in Apollo 16 soils and rocks. *Proc. Lunar Sci. Conf. 7th,* pp. 727-742.

Delano J. W. (1988) Apollo 14 regolith breccias: Different glass populations and their potential for charting space/time variations. *Proc. Lunar Planet. Sci. Conf. 18th,* pp. 59-65.

Evensen N. M., Murthy V. R., and Coscio M. R. Jr. (1973) Rb-Sr ages of some mare basalts and the isotopic and trace element systematics in lunar fines. *Proc. Lunar Sci. Conf. 4th,* pp. 1707-1724.

Evensen N. M., Murthy V. R., and Coscio M. R. Jr. (1974) Provenance of KREEP and the exotic component: Elemental and isotopic studies of grain size fractions in lunar soils. *Proc. Lunar. Sci. Conf. 5th,* pp. 1401-1417.

Fruland R. M. (1983) *Regolith Breccia Workbook.* Planetary Materials Branch Publ. No. 66, JSC 19045, NASA, Houston. 269 pp.

Haskin L. A. and Korotev R. L. (1977) Test of a model for trace element partition during closed-system solidification of a silicate liquid. *Geochim. Cosmochim. Acta, 41,* 921-939.

Head J. W., Pieters C., McCord T., Adams, J., and Zisk S. (1978) Definition and detailed characterization of lunar surface units using remote observations. *Icarus, 33,* 145-172.

Heiken G. H. and McKay D. S. (1974) Petrography of Apollo 17 soils. *Proc. Lunar Sci. Conf. 5th,* pp. 843-860.

Heiken G. H., McKay D. S., and Brown R. W. (1974) Lunar deposits of possible pyroclastic origin. *Geochim. Cosmochim. Acta, 38,* 1703-1718.

Hörz F., Cintala M. J., See T. H., Cardenas F., and Thompson T. D. (1984) Grain size evolution and fractionation trends in an experimental regolith. *Proc. Lunar Planet. Sci. Conf. 15th*, in *J. Geophys. Res., 89,* C183-C196.

Jerde E. A., Warren P. H., Morris R. V., Heiken G. H., and Vaniman D. T. (1987) A potpourri of regolith breccias: "New" samples from the Apollo 14, 16 and 17 landing sites. *Proc. Lunar Planet. Sci. Conf. 17th*, in *J. Geophys. Res., 92,* E526-E536.

Laul J. C. (1979) Neutron activation analysis of geologic materials. *Atom. Energy Rev., 17,* 603-695.

Laul J. C., Lepel E. A., Vaniman D. T., and Papike J. J. (1979) The Apollo 17 drill core: Chemical systematics of grain size fractions. *Proc. Lunar Planet. Sci. Conf. 10th*, pp. 1269-1298.

Laul J. C., Papike J. J., and Simon S. B. (1981) The lunar regolith: Comparative studies of the Apollo and Luna sites. Chemistry of soils from Apollo 17, Luna 16, 20 and 24. *Proc. Lunar Planet. Sci. 12B*, pp. 389-407.

Longhi J., Walker D., Grove T. L., Stolper E. M., and Hays J. F. (1974) The petrology of the Apollo 17 mare basalts. *Proc. Lunar Sci. Conf. 5th*, pp. 447-469.

LSPET (Lunar Sample Preliminary Examination Team) (1973) Apollo 17 lunar samples: Chemical and petrographic description. *Science, 182,* 659-672.

McKay D. S., Morris R. V., and Wentworth S. J. (1984) Maturity of regolith breccias as revealed by ferromagnetic and petrographic indices (abstract). In *Lunar and Planetary Science XV*, pp. 530-531. Lunar and Planetary Institute, Houston.

Morris R. V., Score R., Dardano C., and Heiken G. (1983) *Handbook on Lunar Soils.* Planetary Materials Branch Publ. No. 67, JSC 19069, NASA, Houston. 914 pp.

Murthy V. R., Evensen N. M., Jahn B., and Coscio M. R. Jr. (1972) Apollo 14 and 15 samples: Rb-Sr ages, trace elements, and lunar evolution. *Proc. Lunar Sci. Conf. 3rd*, pp. 1503-1514.

Murthy V. R., Evensen N. M., and Coscio M. R. Jr. (1973) Episodic lunacy—IV: Ages, trace elements and delphic speculations (abstract). In *Lunar Science IV*, pp. 549-551. The Lunar Science Institute, Houston.

Naney M. T., Crowl D. M., and Papike J. J. (1976) The Apollo 16 drill core: Statistical analysis of glass chemistry and the characterization of a high alumina-silica poor (HASP) glass. *Proc. Lunar Sci. Conf. 7th*, pp. 155-184.

Papike J. J., Bence A. E., and Lindsley D. H. (1974) Mare basalts from the Taurus-Littrow region of the Moon. *Proc. Lunar Sci. Conf. 5th*, pp. 471-504.

Phinney W.C., Anders E., Bogard D., Butler P., Gibson E., Gose W., Heiken G., Hohenberg C., Nyquist L., Pearce W., Rhodes J. M., Silver L., Simonds C., Strangeway D., Turner G., Walker R., Warner J., and Yuhas D. (1974) Progress report: Apollo 17 Station 6 Boulder Consortium (abstract). In *Lunar Science V, Suppl. A*, pp. 7-13. The Lunar Science Institute, Houston.

Pieters C. M., Head J. W., Adams, J. B., McCord T. B., Zisk S. H., and Whitford-Stark J. L. (1980) Late high-titanium basalts of the Western maria: Geology of the Flamsteed region of Oceanus Procellarum. *J. Geophys. Res., 85,* 3913-3938.

Rhodes J. M. (1973) Major and trace element chemistry of Apollo 17 samples (abstract). *Eos Trans. AGU, 54,* 609-610.

Rhodes J. M., Rodgers K. V., Shih C., Bansal B. M., Nyquist L. E., Wiesmann H., and Hubbard N. J. (1974) The relationship between geology and soil chemistry at the Apollo 17 landing site. *Proc. Lunar Sci. Conf. 5th*, pp. 1097-1117.

Schaeffer O. A. and Husain L. (1973) Isotopic ages of Apollo 17 lunar material (abstract). *Eos Trans. AGU, 54,* 614.

See T. H., Hörz F., and Cintala M. J. (1984) Regolith evolution experiments II: Modal and chemical analyses (abstract). In *Lunar and Planetary Science XV*, pp. 742-743. Lunar and Planetary Institute, Houston.

Simon S. B., Papike J. J., and Laul J. C. (1981) The lunar regolith: Comparative studies of the Apollo and Luna sites. Petrology of soils from Apollo 17, Luna 16, 20 and 24. *Proc. Lunar Planet. Sci. 12B*, pp. 371-388.

Simon S. B., Papike J. J., and Shearer C. K. (1984) Petrology of Apollo 11 regolith breccias. *Proc. Lunar Planet. Sci. Conf. 15th*, in *J. Geophys. Res., 89,* C109-C132.

Simon S. B., Papike J. J., Gosselin D. C., and Laul J. C. (1985) Petrology and chemistry of Apollo 12 regolith breccias. *Proc. Lunar Planet. Sci. Conf. 16th*, in *J. Geophys. Res., 90,* D75-D86.

Simon S. B., Papike J. J., Gosselin D. C., and Laul J. C. (1986) Petrology, chemistry, and origin of Apollo 15 regolith breccias. *Geochim. Cosmochim. Acta, 50,* 2675-2691.

Simon S. B., Papike J. J., Laul J. C., Hughes S. S., and Schmitt R. A. (1988) Apollo 16 regolith breccias and soils: Recorders of exotic component addition to the Descartes region of the Moon. *Earth Planet. Sci. Lett., 89,* 147-162.

Simon S. B., Papike J. J., Shearer C. K., Hughes S. S., and Schmitt R. A. (1989) Petrology of Apollo 14 regolith breccias and ion microprobe studies of glass beads. *Proc. Lunar Planet. Sci. Conf. 19th*, pp. 1-17.

Vaniman D. T., Labotka T. C., Papike J. J., Simon S. B., and Laul J. C. (1979) The Apollo 17 drill core: Petrologic systematics and the identification of a possible Tycho component. *Proc. Lunar Planet. Sci. Conf. 10th*, pp. 1185-1227.

Wood J. A., Dickey J. S. Jr., Marvin U. B., and Powell B. N. (1970) Lunar anorthosites and a geophysical model of the moon. *Proc. Apollo 11 Lunar Sci. Conf.*, pp. 965-988.

Proceedings of the 20th Lunar and Planetary Science Conference, pp. 231-238
Lunar and Planetary Institute, Houston, 1990

Recycled Grains in Lunar Soils as an Additional, Necessary, Regolith Evolution Parameter

A. Basu

Department of Geology, Indiana University, Bloomington, IN 47405

Agglutinate abundances in lunar soils, or parameters that increase in value with increasing maturity, are not sufficient to solve petrologic models of regolith evolution. Other empirical data, e.g., amounts of freshly added crystalline material in a soil, are necessary. It is proposed that measuring the amounts of recycled grains in lunar soils will provide an additional, necessary parameter. A statistical method, used to estimate the amount of recycled detrital grains in terrestrial sandstones, has been adapted to estimate the amounts of recycled crystalline grains in lunar soils. Optical data from 12 soils in the Apollo 16 core 64001/2 have been collected to estimate the proportion (W) of recycled crystalline grains in each of these soils. The values (W) show a correspondence with other independently derived parameters and the history of the core soils. Hence, W appears to be a valid soil evolution parameter. It is suggested that W varies directly with the relative exposure ages of soils, and that uniform values of W in a soil column probably imply uniform conditions during its evolution. A knowledge of W allows calculating estimated proportions of fresh and recycled fractions of soil components, which provides an independent means of testing if a soil is in a steady state, or if it is evolving along Path I (*in situ* reworking dominant) or Path II (mixing dominant). Finally, it appears that given (1) the estimated amounts of recycled grains in several lunar soils, (2) the measured amounts of agglutinates in these soils, and (3) the exposure ages of the same soils if determined from ferromagnetic resonance (FMR), rare gas, or neutron fluence studies, it may be possible to solve equations describing lunar regolith evolution models. Because W is a newly determined property of lunar soils and because the empirical data have been gathered only from 12 samples of one regolith core, any promise of its applicability in investigations of lunar soil processes i. preliminary. Modifications and refinements of interpretations will be necessary after testing the validity of some predictions made here.

INTRODUCTION

Lunar regolith dynamics are driven by several geologic agents. On a large and an overall long timescale, larger impacts excavate and move considerable amounts of regolith; gravitational creep down a slope such as a crater wall or a rille margin is another example. On a much smaller scale and presumably at a much shorter timescale, the regolith is gardened by frequent micrometeoritic impacts (e.g., *McKay et al.,* 1972; *Arnold,* 1975; *Langevin,* 1982; etc.). The process is perhaps also aided by electrostatic levitation of very fine dust at terminator regions (*Criswell,* 1972; *Rennilson and Criswell,* 1974; *McCoy,* 1976). Solar wind elements are implanted in lunar soil grains that are exposed at the surface; lunar soils thus retain a record of ancient solar activity (*Kerridge,* 1975; *Walker,* 1980; *Pepin et al.,* 1980). Small-scale gardening significantly affects this record as do other processes that mix unevenly matured soils. Unless gardening and other processes are well understood, interpretations of lunar soil data for deciphering solar history, for example, remain difficult. It is therefore important to continue to investigate lunar surface processes in order to better understand lunar soil evolution, including rates of lunar soil turnover.

The purpose of this paper is to show that it is possible to estimate the proportion of recycled material in lunar soils, which is an important but unknown parameter in lunar regolith evolution models. Recycled grains are defined as those soil grains that have been a part of either regolith breccias or agglutinates. Thus, mineral grains, rock fragments, older agglutinates, volcanic glass spherules, etc., if dislodged from an agglutinate or a regolith breccia, would all qualify as recycled grains. These grains can come from local agglutinates or regolith breccias or may be transported from elsewhere.

REGOLITH GARDENING (EVOLUTION) MODELS

Basic Scenario

All regolith evolution models build on the well-known scenario originally deduced by *McKay et al.* (1972, 1974). The basic idea is that as micrometeorites bombard the lunar surface, soil grains are comminuted to smaller sizes, reducing the mean grain size of the soils. Agglutinates are formed of the very fine dust by melting and agglomeration, providing a mechanism that tends to increase the grain size (or slow down its rate of decrease) as well as increase the agglutinate content of the soils. Comminuted agglutinates become finer as they are broken up but do not change their petrologic identity. Occasional larger impacts excavate fresh material from below the surface layer and either add fresh material to the soil at the surface or bury the surface layer. With continued exposure and increasing maturity, a steady state is reached in which the mean grain size and the agglutinate content of a soil do not change any more until interrupted by an unusual event that may bury the soil. Part or all of a buried soil may be reexcavated by a subsequent event and recycled back in bulk to the surface to undergo further maturation. Recycling of individual soil grains is not considered in this basic scenario.

Solar Wind Elements and FMR

Eberhardt et al. (1965) showed that solar wind implants ions in grains exposed to the sun on meteorite parent bodies and that the abundances of solar wind elements normalized to specific surface areas of soil grains in meteoritic regolith is proportional to their exposure ages. The same is true for the lunar regolith. Several models of regolith evolution using solar wind elements as the reference parameter have been proposed. *DesMarais et al.* (1973) show how recycling of fine-grained particles into coarser sizes via agglutination would affect the distribution of solar-wind-implanted elements in lunar soils. *Bogard* (1977) showed how the fraction of recycled solar wind elements in a lunar soil may be isolated and taken into account for modeling purposes. However, an estimate of the actual amount of soil grains that may have been recycled into a soil could not be obtained.

The process of agglutination involves melting of the target soil, which contains solar-wind-implanted hydrogen in proportion to the exposure age of the soil. Iron-bearing minerals and glass, upon melting in the presence of hydrogen, produce single domain, fine-grained Fe° (<300 Å), the amount of which may be measured by the intensity of ferromagnetic resonance (FMR) (*Housley et al.,* 1972, 1973, 1975). *Morris* (1978) has shown that FMR of a soil normalized to its FeO content (I_s/FeO) is an excellent measure of the maturity, i.e., relative exposure age of the soil. All other maturity parameters of lunar soils, if plotted against I_s/FeO, show an asymptotic distribution at higher values. This observation suggests that I_s/FeO does not seem to a reach a steady state with maturity. According to *Morris* (1977, 1978), I_s/FeO may be used in lieu of duration of exposure, i.e., in lieu of time. Note, however, that I_s/FeO only measures cumulative exposure age and does not isolate the preirradiated or recycled population of soil grains.

Mendell and McKay

The regolith evolution model of *Mendell and McKay* (1975) elegantly combines grain size and grain type, agglutinate in this case, and shows how maturities of different lunar soils with known exposure ages could be used to infer some rate constants of surface processes on the Moon. The model allows very fine-grained comminuted agglutinates to be reclassified as "fine" soil and therefore implicitly allows for recycling of grains. However, no reason is enunciated to separate fine-grained agglutinate-hash (not considered as agglutinates any more) and other, or "normal," agglutinates. In this model, the agglutinate population is taken to represent *in situ* growth and no allowance is made for a population of possible reagglutinated agglutinate-hash. It is now known that even large agglutinates could be made up of recycled grains, including fragments of older agglutinates (*Basu and Meinschein,* 1976, their Fig. 1). Some regolith breccias, e.g., 10064 (*Fruland,* 1983) are extremely rich in agglutinates; comminution products of these breccias will add to the population of agglutinates in a soil and not to the population of fresh, coarse-grained particles. Therefore, in practice it is not yet easy to collect the kind of data necessary to make this model very useful.

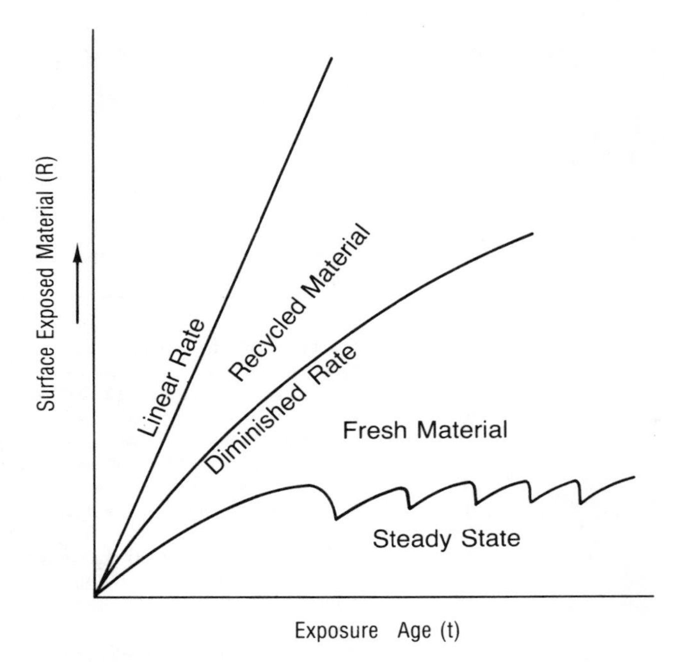

Fig. 1. Sketch to show the possible modes of growth of surface exposed material (R), e.g., agglutinates, in the lunar regolith. A linear rate of growth is obtained if a fixed amount of R is added per unit time. This rate would be diminished if part of R recycles back into the regolith. The area between the linear rate and the diminished rate curves is the amount of recycled material. This diminished rate may be balanced on a long timescale as fresh unexposed material (e.g. crystalline particles) is added to the regolith. The area between the diminished rate curve and the steady state curve is the amount of fresh material (after *McKay and Basu,* 1983 and *Basu,* 1977).

McKay and Basu

McKay and Basu (1983) and *Basu* (1977) explicitly consider populations of recycled grains ("recycled agglutinates" and "secondary nonagglutinates" in their terminology) in their models of agglutinate production in lunar soils. Some important features of these models are shown in Fig. 1. Both of these models invoke import of fresh particles into a maturing soil at the surface of the Moon and recycling of old particles within the same soil to collectively balance agglutinate production to obtain a steady state (cf. Path I of *McKay et al.,* 1974). Their equations cannot be solved with available data on lunar soils, which consist primarily of (1) total agglutinate content and (2) exposure ages determined independently from FMR, rare gas, or neutron fluence studies (e.g., decay of ^{53}Mn; *Nishiizumi and Imamura,* 1979). Further, in *Basu's* (1977) database the population of all "fresh-looking" grains is considered not to contain any recycled grain, which is an oversimplification. Yet these are the only two models that explicitly trace agglutinate production rates, i.e., petrologic evolution of the lunar regolith in terms of actual physical entities in lunar soils. It is also clear from Fig. 1 that data on amounts of surface exposed material in lunar soils of different maturities are necessary to model regolith evolution.

The object of this study is to establish a method for

estimating the population of recycled grains and the amount of fresh material in lunar soils. These parameters will provide the additional necessary data for use in these regolith evolution models and should lead to a better understanding of regolith evolution processes.

RECYCLED GRAINS

It is well established that lunar soils have recycled over time. They may have been buried and reexcavated; some soil grains may have become incorporated into agglutinates and/or regolith breccias, to be dislodged later by impacts and cycled into soils at the surface. Recycled soil grains may also retain a memory of their preirradiation at the surface (e.g., *McKay et al.,* 1974; *Basu and Meinschein,* 1976; *Becker,* 1980; *Bogard,* 1977; *Curtis and Wasserburg,* 1977; *Signer et al.,* 1977; *Morris,* 1978). A recycled grain could be part of an excavated soil that is mixed into another soil at the surface *a la* Path II of the *McKay et al.* (1974) model. A recycled grain may also be the product of *in situ* comminution of grains residing on the lunar surface. Thus, recycled grains are likely to be present in both Path I and Path II soils. Although this was not perhaps realized at the time regolith evolution models were proposed by *Basu* (1977) and *McKay and Basu* (1983), their models are equipped to handle this concept but require a knowledge of the population of recycled grains. Although it may be relatively simple to realize the importance of recycled grains, in practice it is quite difficult, if not impossible, to identify each and every recycled grain under a microscope. In fact, there are no data on the modal amounts of recycled grains in any lunar soil except for one preliminary report (*Basu and Bangs,* 1988).

In general, three broad classes of grains have been recognized in all modal analyses of lunar soils. They are (1) fused soil components (e.g., agglutinates, regolith breccias), (2) glass (e.g., crystal and clast-free, devitrified, etc.), and (3) crystalline grains (e.g., monomineralic, lithic, recrystallized breccias) as elaborated by *Papike et al.* (1982). All of these broad grain classes could and most likely do contain some recycled grains. For example, a soil grain consisting of a single crystal of feldspar (97%) and a little agglutinitic glass (3%) adhering to its margin would be classified and counted as a monomineralic feldspar grain (Fig. 2a). The margin of tolerance of exotic coating on grains varies between laboratories but is usually <10% of the total area observed under a microscope. As a matter of fact, many grains counted as monomineralic or crystalline lithic particles in lunar soils show minor patchy coatings of brown glassy material similar to the matrix of regolith breccias or the cement in agglutinates, which is best studied under a scanning electron microscope (SEM), although reflected light optical microscopy is also very revealing. This patchy glassy coating may be the residue of a regolith breccia matrix or a glassy agglutinate matrix. These are then recycled grains. However, a simple count of partially coated grains will not provide an estimate of recycled grains. The reason is that a random section of grain mounts will not always intersect a thin sliver of grain coating. The probability of intersection will depend on the relative size of the grain-cap and on the *completeness* of the inherited coating that a grain may preserve. Because comminution of a regolith breccia or an agglutinate may also break a clastic grain apart, and because a clastic grain may be attached to the rest of an agglutinate or a breccia only at certain points, there is no reason to expect the presence of complete coating in a recycled grain. Further, a grain dislodged from a regolith breccia or an agglutinate may undergo additional fracturing and lose much, if not all, of its inherited coating. Thus, a grain that "looks" absolutely fresh without any visible coating (Fig. 2b) may still be a recycled grain. It is therefore necessary to devise a method to obtain a statistical estimate of the total amount of recycled grains from what is observable under a microscope. Note that brown glass coating is easy to recognize on a crystalline grain, but it may be nearly impossible to recognize a brown glass coating on a brown glassy agglutinate. Therefore, the amount of recycled grains in a lunar soil has to be derived from an extrapolation of the estimated amount of recycled crystalline grains in the same soil.

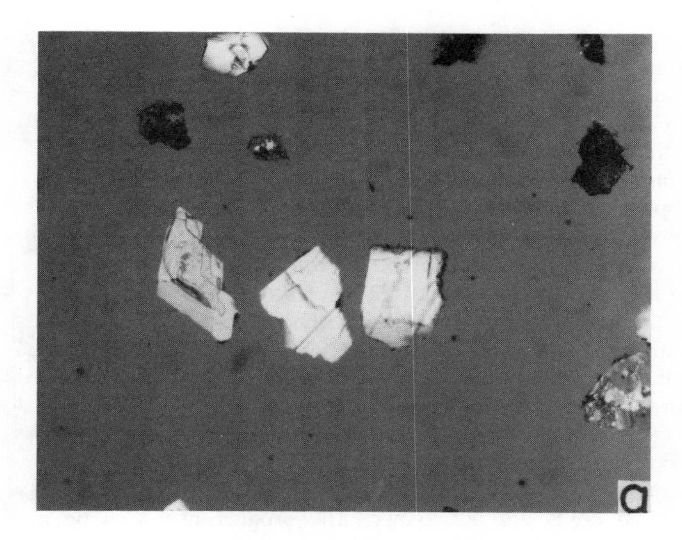

Fig. 2. Photomicrographs of monomineralic crystalline grains **(a)** with minor coating, and **(b)** without any coating.

We assume that coatings of broken, raggedy, brown glass and dust on a crystalline grain indicate that the grain is recycled subject to the provision below. Any brown glass coating on a crystalline grain probably, but not necessarily, indicates recycling. An impact excavating fresh crystalline material may produce a jet of melt that may splash on a fresh grain. Such splash glass is likely to have smooth, surface tension-dominated exterior morphology; such grains cannot be considered to be recycled grains. However, if these grains are later recycled, their splash-glass coatings will likely be broken. Therefore, a criterion to distinguish fresh crystalline grains from splash-glass and recycled grains is to check if either the glass coating has an entire or nearly entire margin with a smooth exterior or if the coating is shattered. A survey of crystalline grains under a SEM may determine the proportion of grains with splash-glass coating. The processes of interaction of freshly excavated 100-μm-size crystalline grains with even finer dust and melt cloud in an ejecta are not known. The product may well be a population of coated grains. We have no way of quantifying this fraction of fresh crystalline soil grains.

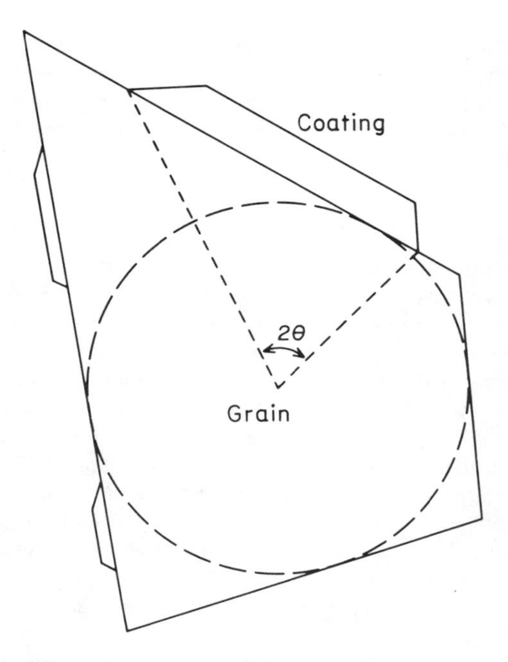

Fig. 3. Schematic diagram to show the details of measuring the angular distance of coating on a clastic grain. The center of the largest inscribed circle within a grain is taken to be the center of the grain. The angle subtended at this center by the length of the longest coating is the angular distance 2θ measured for the grain. Other shorter coatings, if present (like those on the left edge of the schematic grain), are ignored.

METHODOLOGY AND CALCULATIONS

Sanderson (1984) has devised a method to solve the problem of estimating the proportion of recycled quartz grains (i.e., those with coatings of overgrowths that are not necessarily seen in a thin section) on the basis of analytical geometry, experimental work, and empirical data on terrestrial sandstones. Briefly, the procedure consists of: (1) counting the number of grains with some coating (c) in a population of "n" grains to obtain the observed proportion (c/n) of grains with coatings in a sample; (2) measuring the length of coating on *individual grains* (in terms of angular distance), i.e., the angle (2θ by definition) that the coating on each grain subtends at the center of the grain (Fig. 3); and (3) averaging all the 2θ values (one value of 2θ per grain) to obtain a single 2θ for the sample. The statistically predicted proportion of coated grains calculated entirely from 2θ is given by

$$c/n_{(predicted)} = (\pi/4)(1 - \theta/180)\sin\theta + (1 - \cos\theta)/2 \quad (1)$$

[See *Sanderson* (1984) for the long mathematical derivation of this equation.] The estimated whole rock proportion (W) of grains with partial coating, i.e., *the estimated fraction of recycled grains in the population of seemingly fresh grains* is given by

$$W = (c/n_{observed})/(c/n_{predicted}) \quad (2)$$

Thus, with these two albeit painstaking but simple measurements (equations (1) and (2) above), it is possible to estimate the *"real"* proportion of coated, i.e., recycled grains.

Although the theoretical derivation was tested with empirical data on terrestrial sandstones (*Sanderson,* 1984), the method is applicable for estimating the real population of partially coated grains in any sample. Just as the detrital grains in sandstones are hardly ever fully coated by an overgrowth or cement (there are many pores and many grain-to-grain contacts), clastic grains in agglutinates and regolith breccias are also not fully coated by glass. Detrital grains dislodged from a terrestrial sandstone or clastic grains dislodged from a lunar agglutinate or a lunar regolith breccia will retain only a part of their cement; some may lose all. A thin section of such grains may not always show the coat of cement. Note that the method only estimates the proportion of grains with partial coatings (quartz overgrowth, glass splash, etc.). Whether such coated grains are recycled from a previous deposit or not must be determined from a geological evaluation of the properties of material concerned. If so, Sanderson's method estimates the total proportion of such grains reworked in a younger population such as a younger sandstone or, similarly, a reworked lunar soil.

Twelve soils from the Apollo 16 core 64001/2 were selected for the study. This core has been extensively studied for soil maturation, mixing, and provenance characteristics (*Korotev et al.,* 1984; *Basu and McKay,* 1984; *Houck,* 1982) and is suitable for this study. Following the standard procedure in lunar soil petrography (e.g., *Heiken,* 1975; *Heiken and McKay,* 1974; *Heiken et al.,* 1973), polished grain mounts of the 90-150-μm fraction of the soils were selected. A sandstone petrologist, at that time totally unfamiliar with the lunar literature and lunar science, was requested to perform the measurements. This assured unbiased data gathering. The operator (CLB) counted 250 random monomineralic, crystalline lithic, and recrystallized breccia grains to obtain $c/n_{observed}$. In the standard practice of modal analysis none of these grains would have been counted as either an agglutinate or a regolith breccia (Fig. 2). The operator then measured 2θ on 100 of these grains to calculate W (i.e., the recycled proportion of crystalline fresh-looking grains) in each of the soils (Table 1).

As mentioned above, it is possible to recognize recycled crystalline grains from the brown coating on them; this criterion fails for agglutinates and regolith breccias. Therefore, the value of W for crystalline grains in a soil has to be assumed for all grains in the soil. For example, the value of W is 0.49 for the crystalline grains in soil 64002,261 (Table 1), which forces the assumption that 49% of all the agglutinates and regolith breccias observed in this soil are also recycled. This is an unsubstantiated but not necessarily weak assumption, although different proportions of crystalline grains relative to agglutinates or regolith breccias *may* be excavated from the substrate. Unfortunately, there is no other means to estimate the recycled proportion of agglutinates or regolith breccias in a lunar soil. However, given this assumption, it is possible to recalculate the "real" proportion of agglutinates and regolith breccias in each soil by redistributing (i.e., adding) the recycled amount of crystalline grains to the observed modal proportions of agglutinates and regolith breccias. The formulae used are

$$A_n = C_r[A_o/(A_o+R_o)] + A_o \quad (3)$$

$$R_n = C_r[R_o/(A_o+R_o)] + R_o \quad (4)$$

where A_n = recalculated abundance of agglutinates, R_n = recalculated abundance of regolith breccias, A_o = observed abundance of agglutinates, R_o = observed abundance of regolith breccias, and C_r = estimated abundance of recycled crystalline grains. Mass balance is maintained because $C_r = (A_n - A_o) + (R_n - R_o)$.

Fifty random crystalline grains with some minor coatings in each of the six thin sections from the lower half of the core were observed under an SEM at a magnification of about ×10,000 to identify splash glass by an independent operator (SJW). Coated crystalline grains that could be identified as having a splash-glass coating (smooth outer coating, etc., as described above) and not the brown glass coating of an agglutinate or a regolith breccia numbered from zero to only two. This indicates that no more than 5% of the grains identified as recycled could be fresh but adorned with a jet of impact melt.

RESULTS AND DISCUSSION

Modal abundances of crystalline grains, agglutinates, and regolith breccias make up about 90% of the 64001/2 core soils (*Houck,* 1982). These abundances were multiplied by recycling factors (W) determined by optical microscopic observations to obtain the amounts of recycled material present in each of these classes of grains (C_r, A_r, R_r; Table 1). Because the proportion of splash-glass-coated grains is found to be <5%, the recycled populations were not corrected further. As described above, the recycled fraction of crystalline grains were redistributed into both agglutinates and regolith breccias to obtain more realistic abundances of these fused soil components. Note that we have used the 90-150-μm size fraction for our measurements, which has been the standard grain size range for nearly all comparative petrologic studies of lunar soils. The degree to which our results actually apply to lunar soil science depends on the degree to which the 90-150-μm size fraction represents lunar soils.

Steady State Considerations

It is seen that even though more than 50% (±2%) of the core soils consist of agglutinates, only about 10% (±1.5%) are fresh (A_f), i.e., newly produced (Table 1). If the core soils as

TABLE 1. Estimates of recycled proportion (W) of crystalline particles in the 90-150-μm fraction of the double drive core 64001/2 soils.

Sample No.	Depth	W	Crystalline Particles			Agglutinates				Regolith Breccias			
			Observed	Calculated		Observed	Calculated			Observed	Calculated		
64002/1	(cm)			Fresh	Recycled		Fresh	Recycled	Estimated		Fresh	Recycled	Estimated
			C_o	C_f	C_r	A_o	A_f	A_r	A_n	R_o	R_f	R_r	R_n
261	0.5	0.49	34.6	17.6	17.0	41.1	21.0	20.1	53.8	13.9	7.1	6.8	18.2
262	5.5	0.68	36.1	11.6	24.5	36.2	11.6	24.6	52.2	19.2	6.1	13.1	27.7
263	10.5	0.61	39.0	15.2	23.8	39.7	15.5	24.2	57.0	14.9	5.8	9.1	21.4
264	15.0	0.44	53.8	30.1	23.7	23.6	13.2	10.4	38.6	13.7	7.7	6.0	22.4
265	18.5	0.37	52.7	33.2	19.5	25.9	16.3	9.6	39.6	11.2	7.1	4.1	17.0
266	24.5	0.48	39.8	20.7	19.1	35.0	18.2	16.8	48.7	13.9	7.2	6.7	19.3
370	28.0	0.80	45.0	9.0	36.0	29.0	5.8	23.2	54.0	12.8	2.6	10.2	23.8
371	35.5	0.75	43.5	10.9	32.6	26.8	6.7	20.1	47.1	16.2	4.0	12.2	28.5
372	42.5	0.80	46.2	9.2	37.0	25.6	5.1	20.5	47.0	18.6	3.7	14.9	34.2
373	47.0	0.88	43.4	5.2	38.2	30.2	3.6	26.6	55.8	14.9	1.8	13.1	27.5
374	52.5	0.68	39.3	12.6	26.7	38.0	12.2	25.8	59.6	9.0	2.9	6.1	14.1
375	59.5	0.79	40.0	8.4	31.6	38.6	8.1	30.5	63.6	10.3	2.2	8.1	16.9
Average		0.65	42.8	15.3	27.5	32.5	11.4	21.1	51.7	14.1	4.9	9.2	22.4

W has been applied to obtain the amounts of fresh and recycled crystalline particles in these soils, and extrapolated to agglutinates and regolith breccias as well. Estimated amounts of agglutinates and regolith breccias include the redistributed fraction of recycled crystalline particles. Original modal data of *Houck* (1982) are used for observed values.

a whole were to be in a steady state, there would be about 10% fresh crystalline material (C_f) to balance the production of new fresh agglutinates in the soil. However, the amount of fresh crystalline material is estimated to be 15%, i.e., about 50% more than that required to maintain a steady state. Therefore, it is likely that this column of regolith has had a flux of fresh material mixed into it and has not been able to reach equilibrium before being buried. The topmost soil (64002,261; depth of 0.5 cm), however, contains 21% fresh agglutinates and about 17% fresh crystalline material. Thus, the soil 64002,261 is also not in a steady state. Rather, agglutinate production is more rapid than the influx of fresh crystalline material into this soil. This situation, opposite of that of the bulk core described above, is expected because agglutinate production occurs at the very surface of the regolith (*McKay et al.,* 1974). Soils 64002,262 and 64002,263, at depths of 5.5 cm and 10.5 cm, respectively, contain nearly equal amounts of fresh agglutinates and fresh crystalline material suggesting that these two soils should be in a steady state ($A_f = C_f$). Thus, W appears to provide an additional test as to whether a soil is in a steady state as defined by *McKay et al.* (1974).

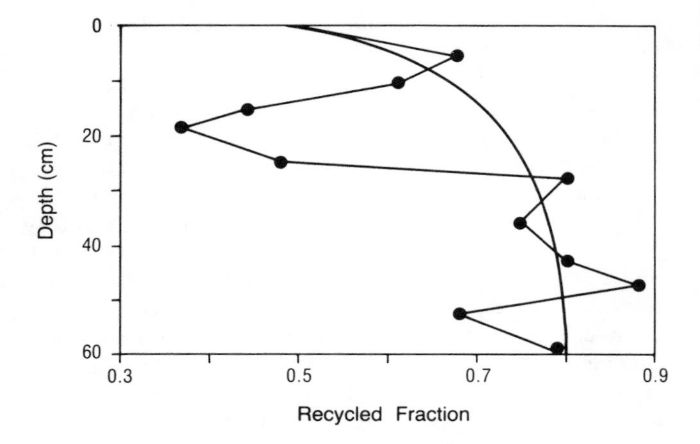

Fig. 4. Plot of the fraction of recycled crystalline particles in 12 soils of the core 64001/2 against depth. Note that a schematic idealized curve reaches a uniform value of about 0.8 in the lower part of the core. The deviation from the ideal around a depth of 18 cm is consistent with independent inferences that this represents an influx of fresh material presumably excavated from the South Ray crater.

Evolution of Soil Profile

Other data, independent of W, have shown that an influx of material, presumably from the South Ray crater, has produced a disturbed layer enriched in immature crystalline components approximately at an 18-cm depth of this core (e.g., *Nishiizumi et al.,* 1983; *Houck,* 1982; *Korotev et al.,* 1984). A plot of the recycled fraction (W) against depth for each of the 12 soils in this core shows a remarkable deviation around 18 cm (Fig. 4). Soils at 15- and 18-cm depths also have fresh crystalline material and are more than 100% enriched over fresh agglutinates (Table 1). This correspondence between W and other independent parameters confirms that W is a valid soil evolution parameter. Indeed, crystalline and lithic grains from more mature soils tend to have more glass and "crud" on their surfaces relative to those from less mature soils. Although dislodgement from an agglutinate or a regolith breccia will leave a coating on spalled grains, the source of the glass and the "crud" is not really known. Presence of glass and "crud" coatings does not guarantee that a grain is recycled because the mechanism of attaching "crud" on grains is not well understood. How useful W might turn out to be will remain unknown until many more measurements are made on different soils and cores.

The value of W increases rapidly with depth at Station 4 (Apollo 16), except for the excursion noted above, before reaching an approximately constant value of 0.8 toward the bottom of the core (Fig. 4). In the absence of measurements from other cores, this value cannot be compared to anything else. However, it is likely that the apparently constant value of W in the bottom half of the core may reflect the relative maturity of the soil profile. The more mature a soil is, the higher the amount of recycled material in that soil should be. This interpretation is completely compatible with that of *McKay et al.* (1974, 1980) in which the relative maturity of a soil column was predicted to correlate with its thickness. Their argument was that the thicker the soil column, the less likely would be the influx of fresh crystalline substance from the bedrock below. This soil column at Station 4 of Apollo 16 (site of core 64001/2) has a high value (0.8) of W and is considered to be fairly mature. *McKay et al.* (1980) showed that the soil column at Station 9 of Apollo 15 (core 15010/11), being adjacent to Hadley Rille where the soil column is thin, was immature and undergoing active replenishment. If W is an indicator of relative age of soil columns (i.e., the residence time of a column as a whole), then the core 15010/11 should have a low value of W. In addition, the apparently constant value of W in the lower part of 64001/2 suggests that this particular soil column below gardening and local mixing depths (i.e., depths up to which the soil column is affected by the South Ray crater event) may reflect soil evolution under uniform conditions. If uniform conditions of regolith evolution are related to relatively flat subregolith surfaces, then conditions might be nonuniform on slopes. If so, nonuniform values of W are expected along the profile of core 15007/8 taken on the slope of Apennine Front. These two predictions about the expected values of W should be good tests of its usefulness.

Agglutinate Production

Current petrologic models allow calculating the production rate of agglutinates in the lunar regolith if the amounts of three petrologic types of grains and the exposure ages of a few soils are known. The discrete function model of *Basu* (1977) uses an equation of the form

$$R = 1 - (1-k)^t - rt \qquad (5)$$

and the continuous function model of *McKay and Basu* (1983) uses an equation of the form

$$R = 1 - (1/k)[(k-r)e^{-kt} + r] \qquad (6)$$

where R = abundance of surface exposed material in a soil (e.g., agglutinates), k = rate constant for agglutinate production, r = rate constant for replenishment of fresh crystalline material, and t = exposure age (time).

Values of three variables are necessary to solve either of the above equations. Whereas the rate constants cannot be measured, one could determine the amounts of agglutinates in several soils, the exposure ages of these soils (presumably from FMR estimates or rare gas or neutron fluence models), and the amounts of replenished material in these soils using the methodology described in this paper. Such data would allow the equations of soil evolution models to be solved numerically, provided that many soils have been studied.

CONCLUSIONS

The purpose of this paper is to present a rationale for using the abundances of recycled grains as an additional parameter for tracing lunar regolith evolution. Therefore, data have been collected from only one soil profile, and the conclusions given below are preliminary in nature.

1. The methodology of *Sanderson* (1984), devised for estimating the population of recycled detrital grains (W) in terrestrial sediments, may be used to estimate the proportion of recycled grains in lunar soils if it is assumed that the glass coatings on grains reflect recycling of grains resident in agglutinates or regolith breccias. At the present time measurements may be made only on crystalline grains.

2. W in 12 soils from lunar core 64001/2 shows a depth profile similar to those obtained from other soil evolution parameters. Hence, W appears to be another valid soil evolution parameter.

3. As a soil column becomes more mature, more and more grains are expected to recycle; therefore, W in more mature soil columns should increase. If so, W may be an indicator of the relative age of a soil column as a whole. A test of this hypothesis would be to find a low value of W for the core soils in 15010/11, which are creeping down the edge of Hadley Rille and are thus kept young and immature.

4. Determination of the amounts of fresh crystalline material (C_f) and fresh agglutinates (A_f) in a soil provides a test to find out whether a soil is in a steady state ($A_f = C_f$), or it is evolving primarily along Path I ($A_f > C_f$), or it has evolved primarily along Path II ($C_f > A_f$) of *McKay et al.* (1974).

5. The value of W in the lower part of core 64001/2 appears to become uniform. This may indicate that this part of the soil column evolved under uniform conditions, albeit not in a steady state. If subregolith slope determines the uniformity of such (unknown) conditions, then the value of W should not be uniform in the coresoils of 15007/8, which was taken from the slopes of the Apennine Front.

6. Determination of W provides a way to estimate the amounts of freshly added crystalline material, freshly produced agglutinates,and the recycled amounts of other components in a soil. Together with the measured amounts of agglutinates in several soils, the exposure ages of which may be determined from rare gas, neutron fluence, or FMR studies, it may be possible to solve equations describing lunar regolith evolution models.

7. This study has shown W to be a viable new parameter to trace regolith evolution on the Moon. However, additional data are necessary not only to test the several predictions made, but also to establish a wider database to better understand lunar soil evolution in general.

Acknowledgments. This research was supported in part by NASA grant NSG-9077, which is gratefully acknowledged. R. Hill, J. Tolen, K. Sowder, and L. Zinn are thanked for their assistance in manuscript preparation. The author is extremely grateful to C. L. Bangs and S. J. Wentworth for collecting the optical and scanning electron microscopic data without which this research could not have been done. D. S. McKay very graciously allowed the use of his laboratory, loaned the polished grain mounts of sieved soils, and along with S. J. Wentworth and D. Vaniman kindly reviewed the manuscript, improved the English, critiqued the rationale, and raised many issues to be cautious about. Expert editorial handling by J. Shervais is greatly appreciated.

REFERENCES

Arnold J. R. (1975) A Monte Carlo model for the gardening of the lunar regolith. *The Moon, 13,* 159-172.

Basu A. (1977) Steady state, exposure age and growth of agglutinates in lunar soils. *Proc. Lunar Sci. Conf. 8th,* pp. 3617-3632.

Basu A. and Bangs C. (1988) Estimation of recycled proportion of monomineralic and crystalline lithic particles in lunar soils (abstract). In *Lunar and Planetary Science XIX,* pp. 47-48. Lunar and Planetary Institute, Houston.

Basu A. and McKay D. S. (1984) Petrologic profile of Apollo 16 regolith at Station 4. *Proc. Lunar Planet. Sci. Conf. 15th,* in *J. Geophys. Res., 89,* C133-C143.

Basu A. and Meinschein W. G. (1976) Agglutinates and carbon accumulation in Apollo 17 lunar soils. *Proc. Lunar Sci. Conf. 7th,* pp. 337-349.

Becker R. H. (1980) Evidence for a secular variation in the $^{13}C/^{12}C$ ratio of carbon implanted in lunar soils. *Earth Planet. Sci. Lett., 50,* 189-196.

Bogard D. D. (1977) Effects of soil maturation on grain size-dependence of trapped solar gases. *Proc. Lunar Sci. Conf. 8th,* pp. 3705-3718.

Criswell D. (1972) Lunar dust motion. *Proc. Lunar Sci. Conf. 3rd,* pp. 2671-2680.

Curtis D. B. and Wasserburg G. J. (1977) Stratigraphic processes in the lunar regolith—Additional insight from neutron fluence measurements on bulk soils and lithic fragments from the deep drill cores. *Proc. Lunar Sci. Conf. 8th,* pp. 3575-3593.

DesMarais D. J., Hayes J. M., and Meinschein W. G. (1973) Accumulation of carbon in lunar soils. *Nature, 246,* 65-68.

Eberhardt P., Geiss J., and Grögler N. (1965) Further evidence on the origin of trapped gases in the meteorite Khor Temiki. *J. Geophys. Res., 70,* 4375-4378.

Fruland R. M. (1983) Regolith breccia workbook. *NASA-JSC Planetary Materials Branch Publ. 66.* 269 pp.

Heiken G. (1975) Petrology of lunar soils. *Rev. Geophys. Space Phys., 13,* 567-587.

Heiken G. H. and McKay D. S. (1974) Petrography of Apollo 17 soils. *Proc. Lunar Sci. Conf. 5th,* pp. 843-860.

Heiken G. H., McKay D. S., and Fruland R. M. (1973) Apollo 16 soils: Grain size analyses and petrography. *Proc. Lunar Sci. Conf. 4th,* pp. 251-265.

Houck K. J. (1982) Modal petrology of six soils from Apollo 16 double drive tube core 64002. *Proc. Lunar Planet. Sci. Conf. 13th,* in *J. Geophys. Res., 87,* A210-A220.

Housley R. M., Grant R. W., and Abdel-Gawad M. (1972) Study of excess Fe metal in the lunar fines by magnetic separation, Mossbauer spectroscopy, and microscopic examination. *Proc. Lunar Sci. Conf. 3rd,* pp. 1065-1076.

Housley R. M., Grant R. W., and Paton N. E. (1973) Origin and characteristics of excess Fe metal in lunar glass welded aggregates. *Proc. Lunar Sci. Conf. 4th,* pp. 2737-2749.

Housley R. M., Cirlin E. H., Goldberg I. B., Crowe H., Weeks R. A., and Perhac R. (1975) Ferromagnetic resonance as a method of studying the micrometeorite bombardment history of the lunar surface. *Proc. Lunar Sci. Conf. 6th,* pp. 3173-3186.

Kerridge J. F. (1975) Solar nitrogen: Evidence for secular change in the ratio of nitrogen-15 to nitrogen-14. *Science, 188,* 162-164.

Korotev R., Morris R. V., and Lauer H. V. Jr. (1984) Stratigraphy and geochemistry of the Stone Mountain core (64001/2). *Proc. Lunar Planet. Sci. Conf. 15th,* in *J. Geophys. Res., 89,* C143-C160.

Langevin Y. (1982) Evolution of an asteroidal regolith: Granulometry, mixing and maturity (abstract). In *Workshop on Lunar Breccias and Soils and Their Meteoritic Analogs* (G. J. Taylor and L. L. Wilkening, eds.), pp. 87-93. LPI Tech. Rpt. 82-02, Lunar and Planetary Institute, Houston.

McCoy J. E. (1976) Photometric studies of light scattering above the lunar terminator from Apollo solar corona photography. *Proc. Lunar Sci. Conf. 7th,* pp. 1082-1112.

McKay D. S. and Basu A. (1983) The production curve for agglutinates in planetary regoliths. *Proc. Lunar Planet. Sci. Conf. 14th,* in *J. Geophys. Res., 88,* B193-B199.

McKay D. S., Heiken G. H., Taylor R. M., Clanton U. S., Morrison D. A., and Ladle G. H. (1972) Apollo 14 soils: Size distribution and particle types. *Proc. Lunar Sci. Conf. 3rd,* pp. 983-995.

McKay D. S., Fruland R. M., and Heiken G. H. (1974) Grain size and evolution of lunar soils. *Proc. Lunar Sci. Conf. 5th,* pp. 887-906.

McKay D. S., Basu A., and Nace G. (1980) Lunar core 15010/11: Grain size, petrography and implications for regolith dynamics. *Proc. Lunar Planet. Sci. Conf. 11th,* pp. 1531-1550.

Mendell W. W. and McKay D. S. (1975) A lunar soil evolution model. *The Moon, 13,* 285-292.

Morris R. V. (1977) Origin and evolution of the grain-size dependence of the concentration of fine-grained metal in lunar soils: The maturation of lunar soils to a steady state stage. *Proc. Lunar Sci. Conf. 8th,* pp. 3719-3748.

Morris R. V. (1978) The surface exposure (maturity) of lunar soils: Some concepts and I_s/FeO compilation. *Proc. Lunar Planet. Sci. Conf. 9th,* pp. 2287-2297.

Nishiizumi K. and Imamura M. (1979) The extent of lunar regolith mixing. *Earth Planet. Sci. Lett., 44,* 409-419.

Nishiizumi K., Murrell M. T., and Arnold J. R. (1983) ^{53}Mn profiles in four Apollo surface cores. *Proc. Lunar Planet. Sci. Conf. 14th,* in *J. Geophys. Res., 88,* B211-B219.

Papike J. J., Simon S. B., and Laul J. C. (1982) The lunar regolith: Chemistry, mineralogy, and petrology. *Rev. Geophys. Space Phys., 20,* 761-826.

Pepin R. O., Eddy J. A., and Merrill R. B., eds. (1980) *The Ancient Sun:*
Fossil Record in the Earth, Moon and Meteorites. Pergamon, New York. 580 pp.

Rennilson J. J. and Criswell C. R. (1974) Surveyor observations of lunar horizon-glow. *The Moon, 10,* 121-142.

Sanderson I. D. (1984) Recognition and significance of inherited quartz overgrowths in quartz arenites. *J. Sediment. Petrol., 54,* 473-486.

Signer P., Baur H., Derksen U., Etique Ph., Funk H., Horn P., and Wieler R. (1977) Helium, neon, and argon records of lunar soil evolution. *Proc. Lunar Sci. Conf. 8th,* pp. 3657-3684.

Walker R. M. (1980) Nature of the fossil evidence: Moon and meteorites. In *The Ancient Sun: Fossil Record in the Earth, Moon and Meteorites* (R. O. Pepin, J. A. Eddy, and R. B. Merrill, eds.), pp. 11-28. Pergamon, New York.

Proceedings of the 20th Lunar and Planetary Science Conference, pp. 239-247
Lunar and Planetary Institute, Houston, 1990

Characterization of Lunar Ilmenite Resources

G. H. Heiken and D. T. Vaniman

Los Alamos National Laboratory, Earth and Environmental Sciences Division, Los Alamos, NM 87544

Ilmenite will be an important lunar resource, to be used mainly for oxygen production but also as a source of iron. Ilmenite abundances in high-Ti basaltic lavas are higher (9-19 vol.%) than in high-Ti mare soils (mostly <10 vol.%). This factor alone may make crushed high-Ti basaltic lavas most attractive as a target for ilmenite extraction. Concentration of ilmenite from either a crushed basalt or regolith requires size sorting to avoid polycrystalline fragments. In coarse-grained high-Ti basaltic lavas, about 60-80% of the ilmenite will consist of relatively "clean" single crystals if the rocks are crushed to a size of 0.2 mm. Fine-grained high-Ti basalts, with thin skeletal or hopper-shaped ilmenites, would produce essentially no free or "clean" ilmenite grains even if crushed to 0.15 mm and only ~7% free ilmenite if crushed to 0.05 mm. Data from the 2.8-m-thick regolith sampled by coring at the Apollo 17 site show that in even the most basalt-clast-rich and least mature stratigraphic intervals, free ilmenite grains make up less than 2% of the 0.02- to 0.2-mm size fraction and a mere 0.3% of the 0.2- to 2-mm size fraction.

INTRODUCTION

The high-Ti mare basaltic lavas and pyroclastic deposits, which are rich in ilmenite ($FeTiO_3$), chromite ($FeCr_2O_4$), armalcolite [(Fe, Mg) Ti_2O_5) and troilite (FeS), are important to the development of self-sufficient lunar colonies. Ilmenite is the major resource for oxygen; other products that may be obtained from these minerals include sulfur, iron, and chromium. *Vaniman et al.* (1990) outline potential uses for sulfur, relatively abundant in high-Ti lunar basalts (0.16-0.27 wt.%), which will be an important byproduct of oxygen production.

On the lunar nearside, the maria cover an area of 6.4×10^6 km^2, which is about 17% of the Moon's surface (*Head,* 1976). The total volume of lunar basaltic lavas and associated pyroclastic rocks is estimated at between 10^6 and 10^7 km^3 (dependent on models used for thicknesses of basalt in the ring-basins). High-Ti lavas and pyroclastic materials make up approximately 20% of the visible mare units (*BVSP,* 1981). Spectral data and samples collected during the Apollo 11 and 17 missions indicate that the Tranquillitatis and Procellarum basins in particular contain abundant near-surface high-Ti basalts (*Pieters,* 1978). High-Ti basalts are thus a vast resource, with a likely minimum volume of 2×10^5 km^3.

McKay and Williams (1979) determined that regolith developed on high-Ti basalts may have as much as 5% ilmenite that is readily available as a resource. This is, however, for an immature soil with brief exposure at the lunar surface. By comparison, much higher ilmenite contents (9-19%) are locked in the high-Ti basaltic rocks.

Beneficiation of ilmenites will be based on the volume of single ilmenite crystals that can be extracted with the least amount of energy needed to crush the source material. The efficiency of this operation will depend upon (1) the grain size of ilmenite, (2) the grain shape, and (3) the degree to which the rock serving as feedstock has already been comminuted by brittle fracture of cooling and solidifying flow surfaces and by the breakup of lava flows by ongoing impact processes at the lunar surface.

The purpose of this study is to quantitatively characterize the sizes and shapes of ilmenite crystals within a variety of lunar samples collected at the Apollo 17 and Apollo 11 sites (located on high-Ti maria) and to evaluate their utility as the raw material for oxygen production. Basic descriptions of ilmenite shapes, volumes, and chemical compositions appeared in the earlier Lunar Science Conference Proceedings volumes, when investigators were getting their first looks at the lunar rocks (e.g., *Brown et al.,* 1970; *Cameron,* 1970; *El Goresy et al.,* 1974). During later missions and after the end of the Apollo program, very few basic, systematic rock descriptions were published. Exceptions to this lack of descriptive data include the reports by *Warner et al.* (1976, 1978), who catalogued the Apollo 17 basalt samples that were collected by raking lithic clasts from the regolith.

For this study, we have reviewed the data on ilmenites for 109 samples from the Apollo 11 and Apollo 17 landing sites, including lavas, regolith breccias, and pyroclastic deposits. Using polished thin sections, we collected size, shape, and chemical data for ilmenites from 30 of these samples. The chemical data were collected using an electron microprobe and an energy-dispersive X-ray detector on the scanning electron microscope (SEM), but are not presented here.

ANALYTICAL PROCEDURE

For this work, we have examined polished thin sections in a SEM. After acquiring a high-contrast backscattered electron image, the image is processed using a Tracor Northern Vista image processing system.

To process the image gray levels are segmented to separate the oxide minerals from iron and troilite, with high average atomic numbers per unit volume, and from the common silicate minerals, with low average atomic numbers per unit volume. In developing our identification technique, identifications were verified with X-ray spectra and electron microprobe analyses. A binary image was constructed from the segmented gray level spectra and used for size and shape analysis. The area filled by the binary image provided an accurate measurement of percent oxide minerals in the sample. Although armalcolite and chromite are included with ilmenite in this process, all three minerals are potential oxygen sources. Ilmenite predominates among these oxide minerals, and we use *ilmenite* throughout for convenience in describing the oxide mineral group.

Between 250 and 800 ilmenite grains were analyzed per sample with the sizing program. This program provided measurements for each ilmenite of area, perimeter, average diameter, length, width, shape factor [perimeter^2/(4π area)], aspect ratio (length/width), and orientation. These data were transferred to a computer file from which the graphs and table in this paper were prepared.

ILMENITE IN HIGH-TI LAVAS

Width

The best basalt textures to seek for ilmenite beneficiation are coarse grained, with ilmenite grains having low aspect ratios and low shape factors. Width is the most significant of the measurements, for it controls the size to which the rock must be crushed to extract clean ilmenite grains. Widths of ilmenite grains ranged from <10-850 μm (Fig. 1). Mean widths in the samples studied ranged from 17-131 μm (Fig. 2) and correlate reasonably well with the ilmenite abundance (9.0-19.4%) in the samples.

Based upon width information alone, ilmenites in high-Ti lavas fall into three general categories (Fig. 1, Table 1): (1) fine grained, with mean widths of 17-35 μm, which plot as steep curves having standard deviations of 19-68 μm; (2) medium grained, with mean widths of 40-56 μm and standard deviations of 31-108 μm; and (3) coarse grained, with mean widths ranging from 105-131 μm and standard deviations of 108-142 μm. There are limitations when working with very coarse-grained basalts on the scale of petrographic thin sections. For example, the cumulative curves for total ilmenite vs. grain width (Fig. 1) are nearly identical for basaltic lavas 78505 and 78506; however, the mean grain widths are very different (Table 1). The differences in Table 1 between these two samples are an artifact of small sample size.

Shape Factor and Aspect Ratio

In published descriptions of lunar ilmenites, shape descriptions are subjective and include the terms "blocky," "skeletal," "acicular," "laths," "tabular," and "feathery," but it is difficult to quantitatively compare grains or crystals using these terms. The sizing program used here determines a shape factor, which is [perimeter^2/(4π area)], and an aspect ratio (length/width) for each grain. The shape factor is 1.00 for any section through a sphere and 1.27 for a centered section through a cube. Thus, the shape factor correlates with an increase in surface area over volume.

Ilmenite shape factors for simple, blocky ilmenites show a positive correlation with increasing grain size for individual samples (Figs. 3a, 4a, and 4c). Ilmenites with quench crystal

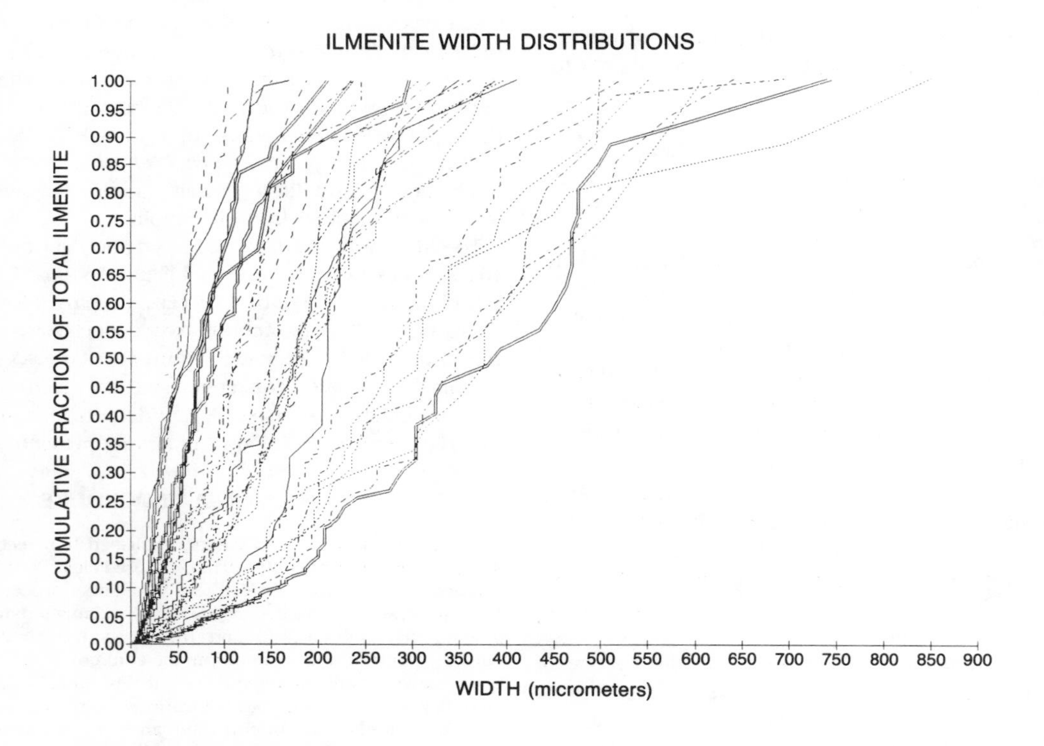

Fig. 1. Cumulative curves of fraction of total ilmenite vs. width of ilmenite grains in high-Ti basalts from the Apollo 11 and 17 landing sites. See Table 1 for sample listing and mean widths for each sample. The curves shown are from fine-grained rocks with quench-texture ilmenites on the left (steeper curves) to coarse-grained rocks with equant ilmenites on the right. Lines are too close for labeling, but are presented here on one graph to show the clustering of finer- and coarser-grained lavas. A series of labeled figures, one for each sample, can be obtained by writing to the authors.

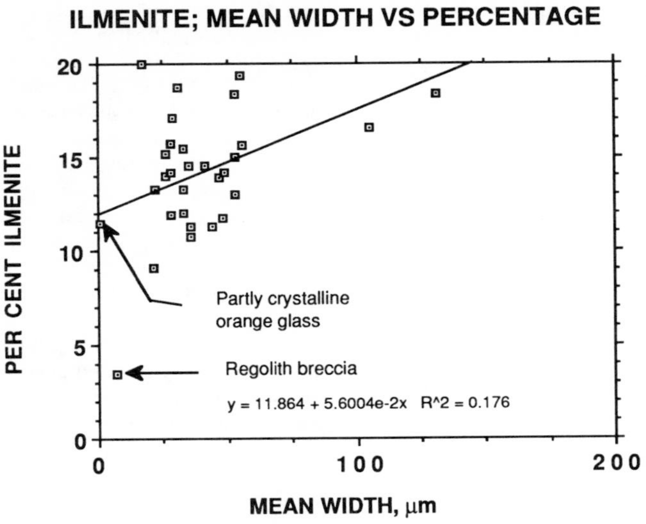

ILMENITE; MEAN WIDTH VS PERCENTAGE

PER CENT ILMENITE

MEAN WIDTH, μm

Partly crystalline orange glass

Regolith breccia

y = 11.864 + 5.6004e-2x R^2 = 0.176

Fig. 2. Percent ilmenite within each sample of high-Ti basalt vs. median width of ilmenite.

textures (dendritic, skeletal; Figs. 3a, 4b, and 4d), with grain size of less than 200 μm, show a similar but less pronounced correlation. However, when comparing mean shape factors between samples, there is no correlation between mean width or area with shape factors (Fig. 3b).

Aspect ratios measured within individual lava samples are mostly less than 3.0 for blocky, coarse-grained ilmenites but show a broad variation (1.0 to 9.5) for dendritic or skeletal ilmenites. Between samples there is a correlation of increasing mean aspect ratio with decreasing mean width (Table 1); the crystals change from blocky to lathlike and acicular with decreasing width.

ILMENITE IN PYROCLASTS AND REGOLITH BRECCIAS

Nearly all ilmenites in the partly crystalline orange glass droplets from the Apollo 17 landing site are fine grained and have dendritic shapes (*Heiken and McKay,* 1977). In the sample examined here (72504), which contains 11.5 vol.% ilmenite, the mean area of ilmenite grains is 1.5 μm^2, the mean width is 0.9 μm^2, and the mean aspect ratio is 2.47. Grain widths range from 0.2 to 7.5 μm (Fig. 5).

TABLE 1. Samples used in this study, showing means, skewness, and standard deviation for ilmenite widths, and means and standard deviation for ilmenite shape factors and aspect ratios.

Sample No.	Mean Width, μm	Std. Dev., Width	Skewness, Width	Mean Shape Factor	Std Dev. Shape Factor	Mean Aspect Ratio	Std. Dev. Aspect Ratio	Ilmenite %
				Lavas				
10017	41	31	1.15	3.09	2.44	1.98	0.79	14.5
10020	29	33	3.69	4.23	5.48	2.71	1.32	17.1
10050	33	68	4.20	1.98	4.73	2.02	0.60	13.3
10057	28	21	1.90	4.39	3.09	2.37	1.07	15.7
70017	105	108	1.64	3.26	2.92	1.86	0.69	16.5
70035	53	101	3.01	2.43	5.00	2.06	0.77	15.0
70135	55	98	2.57	6.13	17.04	2.09	0.96	19.4
70185	53	68	2.16	3.75	5.32	1.96	0.86	13.0
70215	17	19	3.66	4.10	3.93	2.57	1.41	20.0
70255	22	29	2.75	2.94	4.13	2.33	1.55	13.3
70275	21	24	4.81	3.67	3.65	2.65	1.32	9.1
70315	33	56	3.01	3.04	5.50	2.25	0.93	15.5
71035	49	59	2.17	2.98	3.88	2.11	1.01	14.2
71055	47	64	2.55	3.39	5.50	2.26	1.51	13.9
71135	48	59	2.43	3.43	4.78	2.31	0.99	11.7
71136	28	36	3.72	2.37	3.37	2.17	1.05	11.9
71155	36	45	3.08	4.42	6.94	2.35	1.26	11.3
71175	56	92	3.92	3.12	5.54	2.08	0.83	15.6
72155	26	34	2.92	2.42	3.67	2.28	0.97	15.2
74255	44	44	1.88	3.12	3.09	2.09	3.09	11.3
74275	28	18	1.61	4.32	3.97	2.22	1.28	14.2
75035	33	60	4.06	3.06	1.08	2.37	1.08	12.0
75055	36	57	2.84	2.95	4.36	2.06	0.83	10.7
75075	35	65	2.75	2.86	5.40	2.14	0.94	14.5
76136	26	30	2.66	4.85	6.97	2.37	1.02	14.0
78135	53	63	1.94	7.54	10.56	2.14	0.96	18.4
78505	131	142	1.73	4.56	5.17	2.04	0.77	18.7
78506	31	78	4.54	3.43	12.10	2.17	0.89	18.5
				Pyroclastic Deposit				
72504	0.94	0.72	4.06	2.69	2.36	2.47	1.08	11.5
				Regolith Breccia				
10046	6.81	5.16	2.47	2.23	1.72	1.87	0.81	3.5

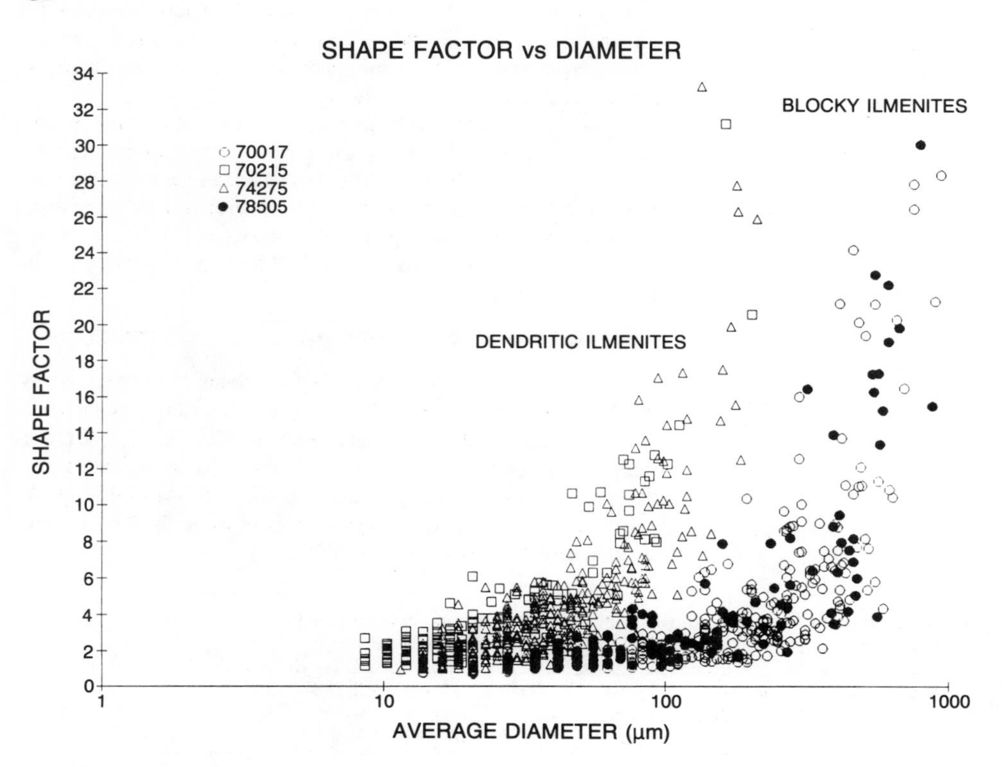

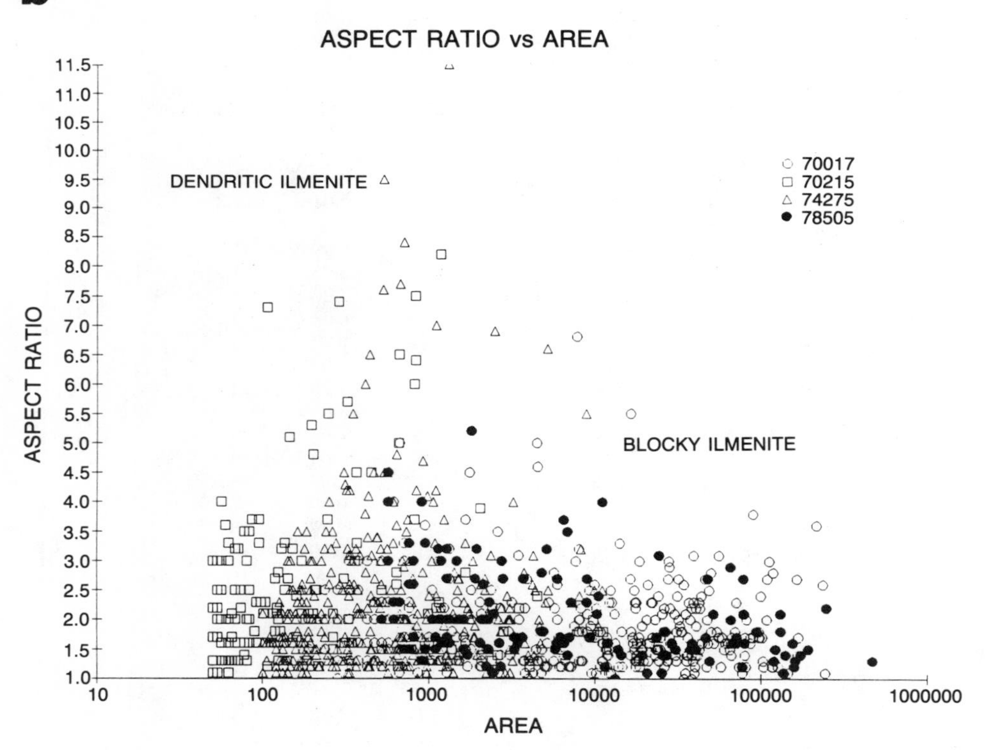

Fig. 3. **(a)** Ilmenite shape factors (1 = circular) plotted against average ilmenite grain diameter. Quickly cooled rocks with small, dendritic ilmenite grains show a large variation in shape factor within a small size range, whereas coarser-grained rocks show a rough correlation between size and shape factor. **(b)** Ilmenite aspect ratio vs. ilmenite grain size (grain area in μm^2). As with shape factor, there is a broad range of aspect ratios (from 1 to 11.5) for rocks with a dendritic texture; coarser-grained rocks contain ilmenites with aspect ratios of only 1 to 3.5.

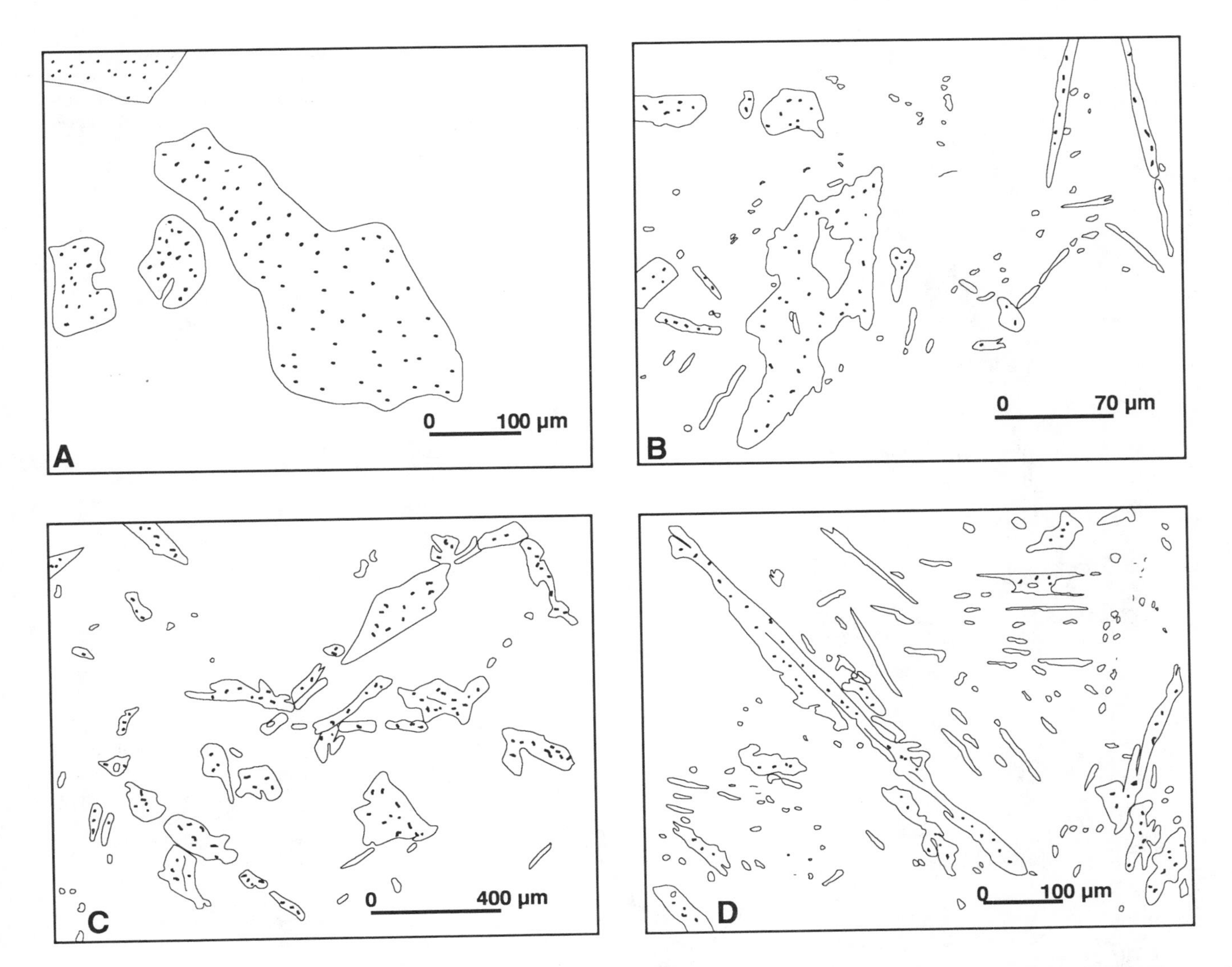

Fig. 4. Drawings of ilmenites in samples of high-Ti basaltic lavas. **(a)** 70017,117; mean width (mw) = 105 μm, mean shape factor (sf) = 3.26, and mean aspect ratio (ar) = 1.86. **(b)** 74275,84; mw = 28 μm, sf = 4.32, and ar = 2.22. **(c)** 75035,78; mw = 60 μm, sf = 3.06, and ar = 2.37. **(d)** 70215,89; mw = 19 μm, sf = 4.1, and ar = 2.57.

Ilmenite grains in regolith breccias are coarser grained than those in orange glass droplets but are still not an important component. A typical sample is 10046, in which the range of ilmenite widths is 11 to 105 μm, but the volume of ilmenite in the sample is only 3.5% (Fig. 6).

ILMENITE IN THE LUNAR REGOLITH

The first evaluations of ilmenite as a resource were made for ilmenite from the lunar regolith (*McKay and Williams,* 1979). These authors concluded that regoliths developed on high-Ti mare basalts may have as much as 5% ilmenite. A soil collected from near Station 1 at the Apollo 17 site (71060) is one of the most ilmenite-rich soils in the sample collection. It is an immature soil (I_s/FeO = 14; mean grain size = 114 μm; 27% agglutinates) that could be processed for free ilmenite grains (*McKay and Williams,* 1979; for a discussion of lunar soil maturity index, see *McKay et al.,* 1974). Even so, at grain sizes of <250 μm, free ilmenite makes up <2% of the sample. In the >250-μm size fraction, all ilmenite is bound up in basalt clasts, which must be crushed for processing.

The deepest sampled section of lunar regolith was collected at the Apollo 17 site (samples 70001 to 70009). Variation of opaque minerals (mostly ilmenite) with depth in this regolith corresponds with variation in maturity; immature soils contain the most ilmenite, whereas mature soils do not—the ilmenites are mostly bound up in agglutinates. Abundances of free opaque oxides in the core range from 0% to 4.8%. The highest abundances are in the less-than-0.2-μm fraction of immature soils (Fig. 7; data from *Vaniman and Papike,* 1977; *Vaniman et al.,* 1979; *Taylor et al.,* 1977; *Taylor et al.,* 1979).

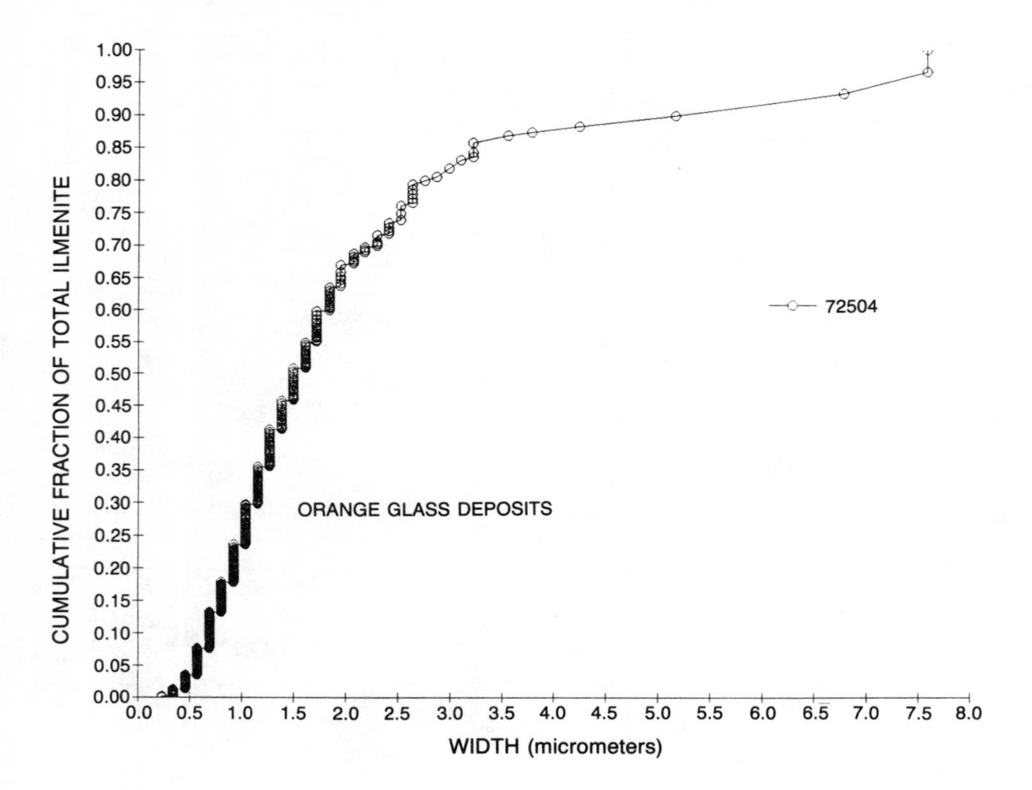

Fig. 5. Cumulative distribution of ilmenite widths, for a sample of Apollo 17 orange glass sample 72504. Note the x-axis; the ilmenites are very fine grained.

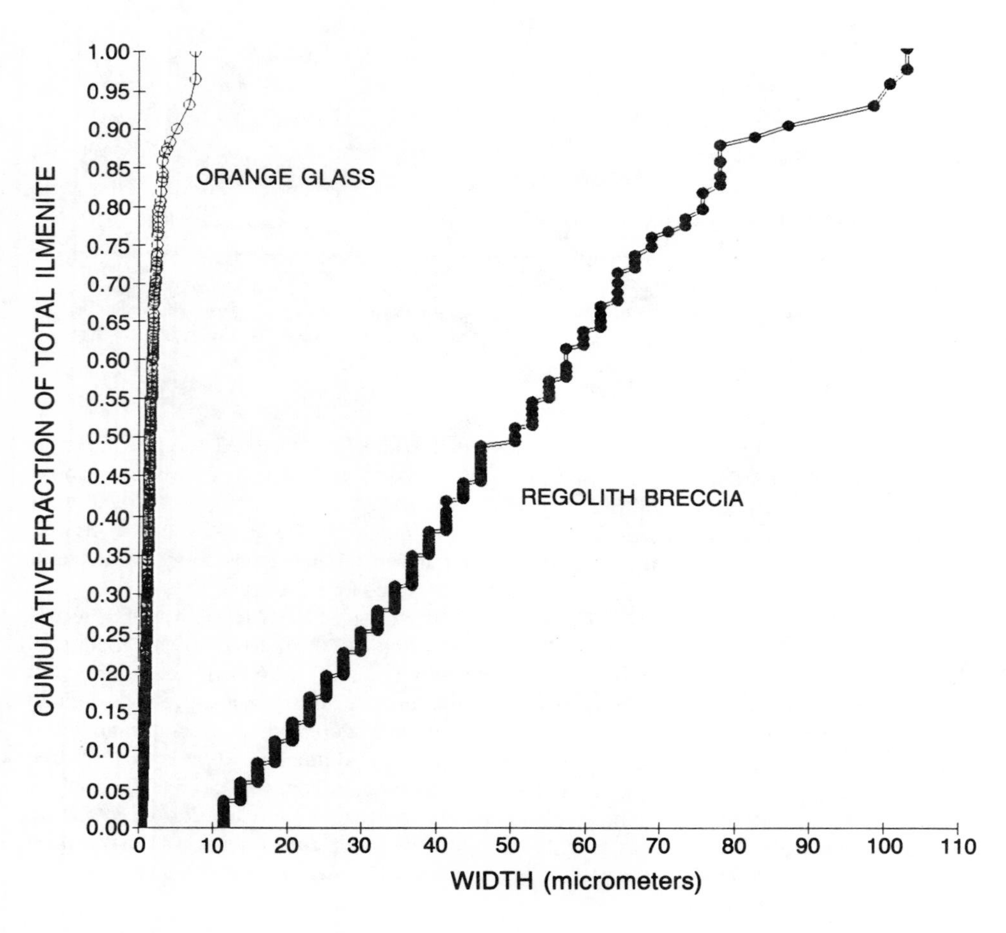

Fig. 6. Cumulative distribution of ilmenite widths for samples of Apollo 17 orange glass sample 72504 (11.5% ilmenite) and Apollo 11 regolith breccia 10046 (3.5% ilmenite).

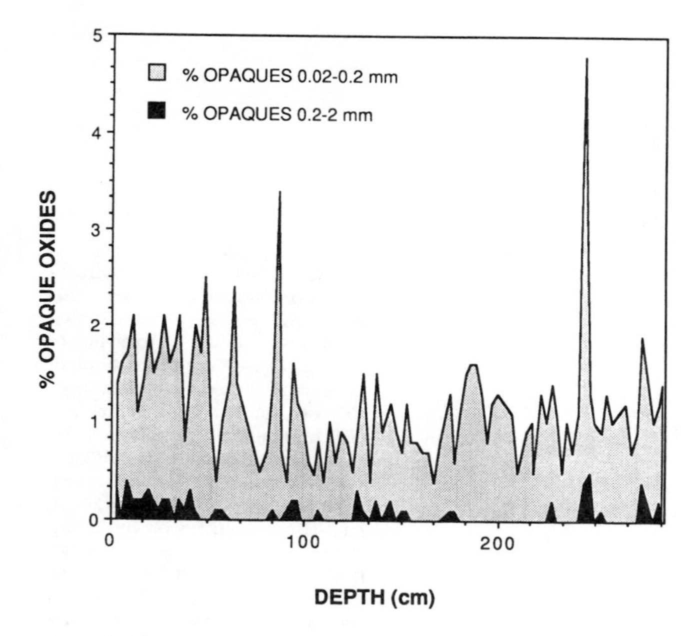

Fig. 7. Variation, with depth, of opaque oxides within the Apollo 17 regolith drill core (70001-70009). Data from *Vaniman and Papike* (1977), *Vaniman et al.* (1979), and *Taylor et al.* (1977, 1979).

BENEFICIATION OF ILMENITE FROM LUNAR ROCKS AND REGOLITH

The concentration of ilmenite from regolith, regolith breccia, or a high-Ti basaltic lava requires sizing to avoid polycrystalline fragments. Key factors in the most efficient beneficiation of ilmenite are grain size, grain width, and shape factors. Figure 8 combines the cumulative curves of Figs. 1 and 6 (fractions of total ilmenite vs. width) with percent ilmenite for each sample. The resulting curves show the rising percentage of "clean" ilmenite obtained with decreasing minimum crush size (fragment width). For example, if a coarse-grained high-Ti basalt containing 19.5% blocky ilmenites is crushed to 200 μm, then 14-16% relatively clean single ilmenite crystals can be freed from the lavas.

If the purpose of the beneficiation is only to extract ilmenite, then the ideal rocks to mine are coarse-grained high-Ti basalts. As can be seen from Fig. 8, little or no free ilmenite can be extracted from lavas with finer grained, quench-texture ilmenites (such as the one in Fig. 4d), from regolith breccias, or from Apollo 17 orange glasses. In fact, microprocessing would be required to extract even 1% of clean ilmenite from these deposits, because they would need to be crushed to a grain size of $<20\ \mu$m. In regolith breccias similar to the high-Ti mare soils, very little or no clean ilmenite will be obtained (Fig. 8).

Regoliths as sources of ilmenite have both advantages and disadvantages. If the regolith is to serve multiple uses, including a source of solar-wind-implanted hydrogen, free iron, and feedstock for glass-making, then the 1-2% ilmenite that might be obtained by quarrying and beneficiation is a bonus. If the goal is to obtain the maximum amount of clean ilmenite with the minimum mass movement and handling, then high-Ti mare lavas are the obvious choice for quarrying.

WHERE TO LOOK FOR ILMENITE RESOURCES

Textures of lunar high-Ti basalts have been used to interpret the spectrum of cooling rates for these lavas, to evaluate the role of crystal settling within the lava flows, and to estimate the flow thicknesses (e.g., *Usselman et al,* 1975; *Brett,* 1975). From these studies and from photogeologic interpretations of the lunar surface, we have constructed a hypothetical lunar high-Ti lava flow (Fig. 9).

Usselman et al. (1975) determined that the ilmenite shapes represented in the Apollo 17 lava samples indicate cooling rates ranging from 210°C/hr (curved dendrites) to 0.1°C/hr (subhedral, equant, and tabular). This range in cooling rates can be explained only if flows are a few meters to a few tens of meters thick. The lavas sampled at the Apollo 17 landing site are not like those of pooled lavas or very thick flows.

Although not high-Ti basalts, the extensive lava flows of Mare Imbrium have lobate flow scarps with heights ranging from 10 m to 63 m (average = 35 m; *Schaber,* 1973; *Gifford and El-Baz,* 1978). These thicknesses seem to be fairly constant over large distances (as much as 1200 km). *Brett* (1975) estimated, on the basis of petrologic models, that basalt flows at the Apollo 11, 12, and 15 sites were at least 10 m thick.

In the walls of Hadley Rille at the Apollo 15 site, at least three basalt flow units were observed in the upper 60 m. Supporting evidence from photogeologic studies of this site demonstrate that most individual flows in this area are about 20 m thick and that their cumulative thickness in Palus Putredinus is about 50 m (*Howard et al.,* 1972; *Gifford and El-Baz,* 1978; *Spudis and Ryder,* 1985; *Spudis et al.,* 1987).

Physical variations within a terrestrial plateau basalt flow, which may be similar in many ways to the mare flows of the Moon, show that the lower half of these flows usually consists of massive, dense holocrystalline basalt, whereas the upper half consists of massive basalt with quench textures, grading up into a vesicular top broken by cooling joints (*Lutton,* 1969; *Swanson and Wright,* 1978; *Arndt et al.,* 1977).

Our model Apollo 17 high-Ti lava flow (Fig. 9) consists of a dense, holocrystalline lower half (~12 m thick); this part of the flow contains coarse-grained ilmenites, which may have also been concentrated somewhat by crystal settling. The upper half of the flow contains mostly quench-textured ilmenites, with complex fine-grained phases that would be difficult to extract by crushing.

The thickness of holocrystalline, coarse-grained basalt in our hypothetical lava flow perhaps should be even greater. The immature ejecta in the upper part of the Apollo 17 drill core, which is inferred to be from Camelot crater (a 650-m-diameter crater on the Taurus-Littrow valley floor), consists of 80% equigranular basalt clasts (with coarse, tabular ilmenite) and 20% finer grained basalt clasts with dendritic or skeletal

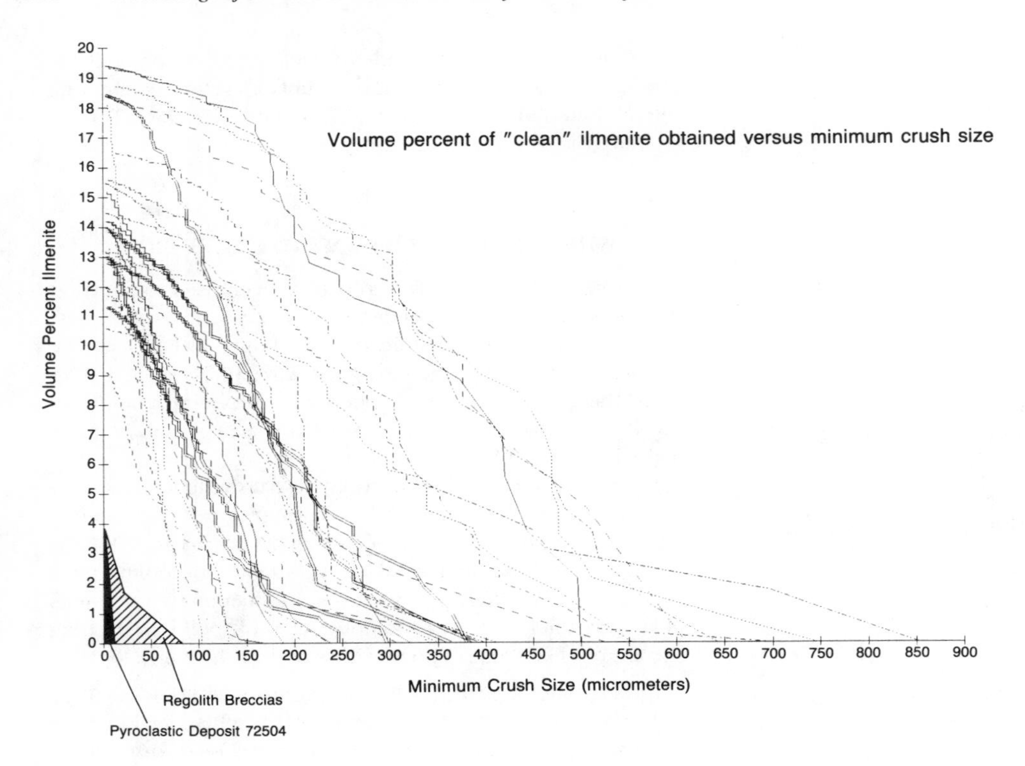

Fig. 8. The volume percent of "clean" ilmenite obtained by crushing vs. minimum crush size. This graph was created by combining cumulative fraction of total ilmenite (Fig. 1) with the percent ilmenite in each sample. Each curve rises to show the increasing percentage of "clean" ilmenite obtained if the size of fragmentation is reduced. The obvious choices as an ilmenite resource are the coarse-grained, high-Ti lavas (curves on the upper right). To obtain "clean" ilmenite from either regolith breccias or orange glass deposits would require crushing to less than 10 μm to obtain less than 2%. A series of labeled figures for each sample can be obtained by writing to the authors.

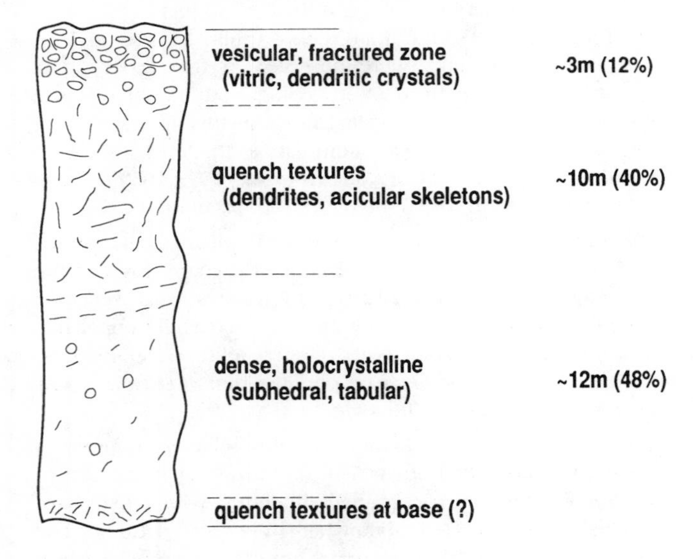

Fig. 9. Model high-Ti basaltic lava flow. The source for coarse-grained, blocky ilmenite crystals is within the lower half of this flow. Quarrying is possible, but if flows such as these are penetrated by impact craters, then the coarse-grained ejecta around crater rims would be the ideal source for ilmenite-rich lavas that have already been partly comminuted.

ilmenite clasts (*Vaniman and Papike,* 1977). If these clasts represent the ratio of basalt textures represented in flows penetrated by Camelot crater, then the coarse-grained ilmenite basalt resource at the Apollo 17 landing site is larger than predicted in our model.

Siting a mine for ilmenite would require information on regolith and lava flow thicknesses and coring of the flow(s) to verify the textural models. If these flows have been penetrated by impacts, then the quarry should be sited on the crater rims, where the basalt has already been partly excavated and may be collected as blocky material around the crater rim.

In early studies it appeared that the lunar regolith that developed on high-Ti basaltic lava flows and lunar pyroclastic deposits are good sources for ilmenite. This is not the case; regolith and pyroclasts may be excellent resources for other purposes, but they are not good sources of free or "clean" ilmenite grains.

Acknowledgments. This work is funded by the National Aeronautics and Space Administration under the auspices of the U.S. Department of Energy. The study was possible through support and encouragement by D. S. McKay (NASA Johnson Space Center). We appreciate the time and thought given to making this a better paper by the reviewers: B. Schuraytz, S. K. Kawatra, S. B. Simon, W. S. Baldridge, and D. J. Lindstrom.

REFERENCES

Arndt N. T., Naldrett A. J., and Pyke D. R. (1977) Komatiitic and iron-rich tholeiitic lavas of Munro Township, northeast Ontario. *J. Petrol., 18,* 319-369.

Brett R. (1975) Thickness of some lunar mare basalt flows and ejecta blankets based on chemical kinetic data. *Geochim. Cosmochim. Acta, 39,* 1125-1141.

Brown G. M., Emeleus C. H., Holland J. G., and Phillips R. (1970) Mineralogical, chemical, and petrological features of Apollo 11 rocks and their relationship to igneous processes. *Proc. Apollo 11 Lunar Sci. Conf.,* pp. 195-219.

BVSP (Basaltic Volcanism Study Project) (1981) *Basaltic Volcanism on the Terrestrial Planets.* Pergamon, New York. 1286 pp.

Cameron E. N. (1970) Opaque minerals in certain lunar rocks from Apollo 12. *Proc. Lunar Sci. Conf. 2nd,* pp. 193-206.

El Goresy A., Ramdohr P., Medenbach O., and Bernhardt H.-J. (1974) Taurus-Littrow TiO_2-rich basalts: Opaque mineralogy and geochemistry. *Proc. Lunar Sci. Conf. 5th,* pp. 627-652.

Gifford A. W. and El-Baz F. (1981) Thicknesses of lunar mare flow fronts. *Moon and Planets, 24,* 391-398.

Head J. W. (1976) Lunar volcanism in space and time. *Rev. Geophys. Space Phys., 14,* 265-300.

Heiken G. and McKay D. S. (1977) A model for eruption behavior of a volcanic vent in eastern Mare Serenitatis. *Proc. Lunar Sci. Conf. 8th,* pp. 3243-3255.

Howard K. A., Head J. W., and Swann G. A. (1972) Geology of Hadley Rille. *Proc. Lunar Sci. Conf. 3rd,* pp. 1-14.

Lutton R. J. (1969) Internal structure of the Buckboard Mesa Basalt. *Bull. Volcanol., 33,* 579-593.

McKay D. S. and Williams R. J. (1979) A geologic assessment of potential lunar ores. In *Space Resources and Space Settlements* (J. Billingham and W. Gilbreath, eds.), pp. 243-255. NASA SP-428.

McKay D. S., Fruland R. M., and Heiken G. H. (1974) Grain size and the evolution of lunar soils. *Proc. Lunar Sci. Conf. 5th,* pp. 887-906.

Pieters C. M. (1978) Mare basalt types on the front side of the Moon: A summary of spectral reflectance data. *Proc. Lunar Planet. Sci. Conf. 9th,* pp. 2825-2849.

Schaber G. G. (1973) Lava flows in Mare Imbrium: Geologic evidence from Apollo orbital photography. *Proc. Lunar Sci. Conf. 4th,* pp. 73-92.

Spudis P. D. and Ryder G. (1985) Geology and petrology of the Apollo 15 landing site: Past, present, and future understanding. *Eos Trans. AGU, 66,* 721-726.

Spudis P. D., Swann G. A., and Greeley R. (1987) The formation of Hadley Rille and implications for the geology of the Apollo 15 region. *Proc. Lunar Planet. Sci. Conf. 18th,* pp. 243-254.

Swanson D. A. and Wright T. L. (1978) Bedrock geology of the northern Columbia Plateau and adjacent areas. In *The Channeled Scabland* (V. R. Baker and D. Nummedal, eds.), pp. 35-57. National Aeronautics and Space Administration, Washington, DC.

Taylor G. J., Keil K., and Warner R. D. (1977) Petrology of Apollo 17 deep drill core—I: Depositional history based on modal analyses of 70009, 70008, and 70007. *Proc. Lunar Sci. Conf. 8th,* pp. 3195-3222.

Taylor G. J., Warner R. D., and Keil K. (1979) Stratigraphy and depositional history of the Apollo 17 drill core. *Proc. Lunar Planet. Sci. Conf. 10th,* pp. 1159-1184.

Usselman T. M., Lofgren G. E., Donaldson C. H., and Williams R. J. (1975) Experimentally reproduced textures and mineral chemistries of high-titanium mare basalts. *Proc. Lunar Sci. Conf. 6th,* pp. 997-1020.

Vaniman D. and Papike J. (1977) The Apollo 17 drill core: Characterization of the mineral and lithic component (sections 70007, 70008, 70009). *Proc. Lunar Sci. Conf. 8th,* pp. 3123-3159.

Vaniman D. T., Labotka T. C., Papike J. J., Simon S. B., and Laul J. C. (1979) The Apollo 17 drill core: Petrologic systematics and the identification of a possible Tycho component. *Proc. Lunar Planet. Sci. Conf. 10th,* pp. 1185-1227.

Vaniman D., Pettit D., and Heiken G. (1990) Uses of lunar sulfur. In *Lunar Bases and Space Activities of the 21st Century* (W. W. Mendell, ed.). Univelt, San Diego, in press.

Warner R. D., Berkley J. L., Mansker W. L., Warren R. G., and Keil K. (1976) Electron microprobe analyses of spinel, Fe-Ti oxides, and metal from Apollo 17 rake sample mare basalts. *Univ. of New Mexico Inst. of Meteoritics Spec. Publ. 16,* Albuquerque. 114 pp.

Warner R. D., Keil K., Nehru C. E., and Taylor G. J. (1978) Catalogue of Apollo 17 rake samples from stations 1A, 2, 7, and 8. *Univ. of New Mexico Inst. of Meteoritics Spec. Publ. 18,* Albuquerque. 87 pp.

Proceedings of the 20th Lunar and Planetary Science Conference, pp. 249-258
Lunar and Planetary Institute, Houston, 1990

Ilmenite-rich Pyroclastic Deposits: An Ideal Lunar Resource

B. R. Hawke, C. R. Coombs[1], and B. Clark

Planetary Geosciences Division, Hawaii Institute of Geophysics, University of Hawaii, Honolulu, HI 96822

It seems likely that a permanent lunar base will be established early in the next century. While the initial return to the Moon may be for the purpose of constructing a science outpost, a viable permanent settlement must return major economic benefits to the near-Earth space infrastructure. Attention has recently been focused on the production of oxygen for use as propellant as well as helium-3 for nuclear fusion fuel. Ilmenite-rich lunar material is generally preferred for the production of these substances. We have concluded that ilmenite-rich pyroclastic deposits would be excellent sites for the establishment of a permanent lunar base for the production of oxygen and He-3. A wide variety of potentially useful by-products could be produced. These include Fe, Ti, H_2, N, C, S, Cu, Zn, Cd, Bi, and Pb. Several ilmenite-rich pyroclastic deposits of regional extent have been identified on the lunar surface. These extensive, deep deposits of ilmenite-rich pyroclastic material are block-free and uncontaminated by vertical mixing or lateral transport. They could be easily excavated and would be ideal for lunar mining operations. These deep, loose pyroclastic deposits would be ideal for rapidly covering base modules with an adequate thickness of shielding.

INTRODUCTION

In recent years, interest in establishing a permanent settlement on the surface of the Moon has greatly increased. It now seems likely that a Permanent Lunar Base (PLB) will be established by some nation or group of nations early in the 21st century. While the initial return to the Moon may be for the purpose of constructing a science outpost for astronomical, geological, or other basic research purposes, a viable permanent settlement must return major benefits to the near-Earth space infrastructure. Because of the great expense, it is clear that a permanent lunar base will be established for economic reasons.

Early in the post-Apollo era, several workers suggested that titanium production might be a profitable economic activity for a PLB. Later, it was suggested that lunar material would be useful for shielding space habitats or military facilities in orbit. In recent years, attention has been focused on the production of oxygen propellant and helium-3 as nuclear fusion fuel (e.g., *Kulcinski et al.,* 1986; *Kulcinski,* 1988; *Gibson and Knudsen,* 1985; *Simon,* 1985). Ilmenite-rich material is generally preferred for the production of these substances (*Mendell,* 1985; *Hawke et al.,* 1989a). To date, efforts to locate ilmenite-rich deposits have been concentrated on the high-titanium mare basalts present in various lunar maria (e.g., *Cameron,* 1988; *Busarev and Shevchenko,* 1988; *Shevchenko and Busarev,* 1988). However, in this paper, we propose that ilmenite-rich pyroclastic deposits would be excellent sites for mining operations for the production of O_2 and He-3 as well as the establishment of permanent lunar bases. A wide variety of very useful by-products could be produced, and the extensive, deep unconsolidated deposits of pyroclastic debris that are relatively block-free and uncontaminated would be ideal for lunar mining operations and covering base modules with an adequate thickness of shielding.

The purposes of this paper include the following: (1) to demonstrate that ilmenite-rich pyroclastic deposits would be excellent sources of a wide variety of valuable elements; (2) to show that several ilmenite-rich pyroclastic deposits of regional extent exist on the lunar surface; and (3) to investigate the suitability of regional pyroclastic deposits for lunar mining operations, construction activities, and the establishment of permanent lunar settlements.

ECONOMIC RESOURCES DERIVED FROM ILMENITE-RICH DEPOSITS

One lunar product with a potentially large market yet requiring minimal processing is liquid oxygen for spacecraft propellant (*Mendell,* 1985). In any future settlement of the Moon, oxygen is clearly one of the most important materials to be supplied (*Mendell,* 1985; *Gibson and Knudsen,* 1985). It is required both for life support and propulsion. The production of oxygen propellant is particularly attractive because it immediately relieves some of the burden on the transportation system for lunar operations (*Mendell,* 1985).

Lunar oxygen may prove to be a viable export to sustain operations in low-Earth orbit (LEO) and elsewhere in near-Earth space. It has been predicted (e.g., *Mendell,* 1985) that several national, international, and private space stations will be in operation early in the 21st century and that a variety of commercial activities will be underway. These commercial activities may include large-scale space manufacturing and microgravity materials processing. Oxygen and other lunar resources will be needed to establish and sustain this near-Earth space infrastructure. Oxygen will be needed for use as a propellant to transport material from LEO to GEO and elsewhere in near-Earth space. Fuel will also be needed for proposed planetary missions.

A variety of methods for producing oxygen from lunar rocks and soils has been investigated (e.g., *Mendell,* 1985; *Briggs,* 1988). However, the reduction of ilmenite has been extensively studied and seems to be preferred by most workers (*Mendell,* 1985; *Gibson and Knudsen,* 1985; *Briggs,* 1988;

---

[1]Now at Mail Code SN15, NASA Johnson Space Center, Houston, TX 77058.

Cutler and Krag, 1985). Ilmenite is relatively abundant in many lunar soils, has been shown to be extractable from the soil, and can be stripped of its oxygen by relatively simple means (e.g., *Briggs,* 1988; *Agosto,* 1985). In addition, potentially useful by-products are produced by ilmenite reduction. Iron, almost 40% of the original mass of the ilmenite, remains behind in a TiO_2 matrix (*Briggs,* 1988). Ultimately, iron may be as useful as oxygen since it can be utilized for the construction of space structures. Titanium can also be extracted from ilmenite if enough energy is applied to the system. Titanium would also be a very useful material for space construction. Eventually, Ti might be exported to Earth if an inexpensive means of transportation (e.g., mass drivers) was available.

In recent years, another possible lunar resource has received considerable attention. The exploitation of lunar helium-3 as nuclear fusion fuel could dramatically improve our energy future (*Wittenberg et al.,* 1986; *Li and Wittenberg,* 1988; *Kulcinski,* 1988). He-3 could be used in D-He-3 fusion reactions and is not found in significant quantities on Earth. The energy is released in a form that can be converted directly to electricity with efficiencies of at least 70%. In comparison, current electric power plants fueled by coal, oil, nuclear fission, or other energy sources have efficiencies of 30% to 40% (*Kulcinski,* 1988). Preliminary estimates show that the available quantities of He-3 on the lunar surface could provide the equivalent of 40,000 years of the U.S. electrical power generation demand recorded in 1985 (*Kulcinski et al.,* 1986). In addition, it has been pointed out by *Kulcinski* (1988) and others that D-He-3 reactor operation is inherently very safe; the process itself simply cannot lead to a devastating accident. The radioactivity levels of the reactors are so low that the facilities can be dismantled at the end of their useful lifetimes and disposed of as low-level waste. Many other advantages of the utilization of He-3 as fusion fuel have been described by Kulcinski and coworkers in a series of publications (*Kulcinski et al.,* 1986, 1988; *Kulcinski,* 1988; *Wittenberg et al.,* 1986; *Kulcinski and Schmitt,* 1987; *Santarius,* 1988). Included among these benefits is the use of He-3 for fusion-powered spacecraft that could shorten trip times and increase payload capacities to other solar system objects (*Kulcinski,* 1988).

Lunar He-3 supplies originate from solar wind materials that are embedded in the near-surface regions of fine-grained regolith particles. Helium and other solar wind gases can be easily extracted from the lunar regolith by the relatively simple procedure of heating the soil to temperatures of 700°C or higher (*Haskin,* 1989; *Des Marais et al.,* 1974; *Bustin et al.,* 1984). Several factors were cited by *Cameron* (1987, 1988) as being important in controlling the helium abundance of the regolith. He noted that the helium content of a mature regolith is a function of its composition. In particular, mare regoliths rich in titanium (and hence ilmenite) exhibit high helium contents. Helium is concentrated in the finer regolith fractions (-50-μm size fraction, *Cameron,* 1988). Hence, an ilmenite-rich, mature regolith would be the best source of lunar helium (*Cameron,* 1988; *Hawke et al.,* 1989a).

During the course of operations to extract He-3 from the regolith, other valuable volatiles implanted by the solar wind could be collected with relatively small mass and power penalties (*Crabb and Jacobs,* 1988, 1989; *Haskin,* 1989). These other constituents of the solar wind (H, C, N, noble gases, etc.) can be used to meet the needs of such lunar base subsystems as transportion, life support, and agriculture (*Crabb and Jacobs,* 1988). *Kulcinski* (1988) has noted that these solar wind-derived elements and compounds are vital to human existence in space and will permit the growth of food, the establishment and maintenance of contained atmospheres, and provide a water supply. These by-products will have crucial impacts on the size of space settlements that can be supported.

In a recent study, *Haskin* (1989) noted that a typical cubic meter of lunar soil has a mass of 1.6-2 tonnes and contains ~100 g of hydrogen (*Des Marais et al.,* 1974; *Bustin et al.,* 1984), 200 g of carbon (*Chang et al.,* 1974), and 100 g of nitrogen (*Chang et al.,* 1974; *Muller,* 1973). *Haskin* (1989) determined that the upper 2 m of the lunar regolith contain approximately 8×10^9 tonnes of hydrogen, 1.5×10^{10} tonnes of carbon, and 8×10^9 tonnes of nitrogen. Adequate quantities of these materials are present in the lunar regolith, alleviating the necessity to transport these materials to the surface of the Moon from Earth or elsewhere in space. In fact, hydrogen and other materials might be exported to other space facilities for use in propulsion, life support, or other systems. In this regard, it should be noted that the hydrogen abundance in the regolith also increases with ilmenite content and maturity.

ILMENITE-RICH PYROCLASTIC DEPOSITS

To date, efforts at selection and evaluation of sites for the production of oxygen and He-3 have focused on high-titanium maria (e.g., *Cameron,* 1988; *Shevchenko and Busarev,* 1988; *Busarev and Shevchenko,* 1988). Both direct sampling by the Apollo and Lunar missions and remote-sensing studies have indicated that major deposits of high-titanium mare basalt exist on the surface of the Moon (e.g., *Cameron,* 1988; *Pieters,* 1978; *Johnson et al.,* 1977). Spectral reflectance and orbital geochemical data indicate that Mare Tranquillitatis contains the largest expanse of high-titanium mare basalt on the eastern nearside. In addition, large areas of relatively titanium-rich mare basalt have also been identified in Oceanus Procellarum and elsewhere on western nearside (e.g., *Pieters,* 1978; *Pieters et al.,* 1980; *Shevchenko and Busarev,* 1988). However, major problems may exist in mining and processing high-titanium mare basalts and regoliths for certain purposes. For example, *Cameron* (1988) noted that while the high-Ti regolith sampled by the Apollo 17 mission in the Taurus-Littrow region is probably extensive, the area may be too heterogeneous for large-scale helium mining. *Cameron* (1988) also pointed out that because of the concentration of helium in the finer soil fractions and considerations of ease of mining, the mining areas must be as block-free as possible. Unfortunately, the mare regolith is not block-free; impact cratering has excavated and emplaced fragments of basalt in the upper 2 m of the regolith. Other possible difficulties are mentioned by *Cameron* (1988). As an alternative to high-Ti mare basalt flows, we propose that ilmenite-rich pyroclastic mantling deposits of regional extent (Fig. 1) would be superior sites for mining operations and the establishment of a permanent lunar base (*Hawke et al.,* 1989a). These regional "dark mantle" deposits are extensive,

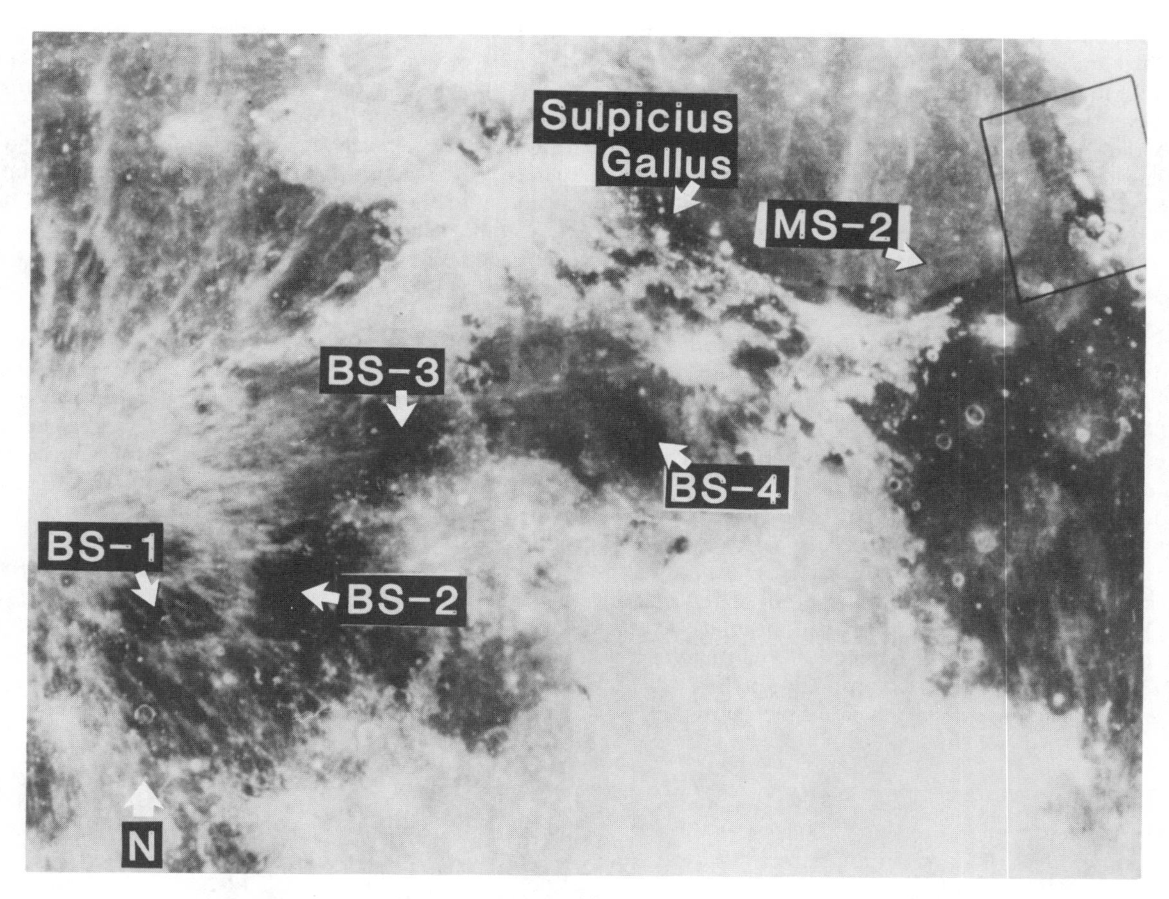

Fig. 1. Full Moon photograph showing the major pyroclastic mantling deposits in the north central portion of the lunar nearside. The lunar standard area MS-2 is shown. The Taurus-Littrow region is enclosed in solid lines. BS-1 and BS-2 comprise the Southern Sinus Aestuum deposit. BS-3 is near Rima Bode and BS-4 is the Southern Mare Vaporum deposit. The Sulpicius Gallus pyroclastic deposit appears not be dominated by ilmenite-rich debris. Reflectance spectra for these deposits are shown in Figs. 5 and 6. After *Adams et al.* (1974).

thick, and relatively numerous on the lunar nearside (*Pieters et al.,* 1973; *Gaddis et al.,* 1985; *Hawke et al.,* 1989a).

The material that comprises a major regional pyroclastic deposit was sampled at the Apollo 17 landing site (Fig. 2). An important objective of the Apollo 17 mission was to characterize the "dark mantle" unit at the site; premission analysis had indicated that this was a relatively young deposit of pyroclastic origin (e.g., *Lucchitta,* 1973; *Pieters et al.,* 1974; *Adams et al.,* 1974; *Gaddis et al.,* 1985). Immediately after the mission, it was considered surprising that no obvious evidence was found of a young dark mantling material component in the samples (*Pieters et al.,* 1974). The soils collected from the low albedo unit on the valley floor are composed largely of ancient (~3.76 b.y.) high-titanium mare basalt (*Taylor,* 1982; *Pieters et al.,* 1974; *Adams et al.,* 1974). However, subsequent work identified the orange glass droplets and the chemically equivalent partially crystallized black spheres as pyroclastic components in the Apollo 17 regolith (e.g., *Heiken et al.,* 1974; *Pieters et al.,* 1974; *Adams et al.,* 1974). The chemical compositions of the orange and black spheres are indistinguishable, the only difference being that the black spheres are largely crystallized. The quench-crystallized black spheres are rich in TiO_2 (9-10%), and ilmenite and they are similar, though not identical, in composition to the Apollo 17 mare basalts. The pyroclastic debris is slightly younger (~3.6 b.y.) than the mare basalts at the site (~3.7 b.y., *Taylor,* 1982).

The glasses and black spheres sampled at Shorty crater and elsewhere at the Apollo 17 site are fine-grained (median grain size = 40 μm) droplets. Particular shapes vary from simple glass spheres to compound droplets with quench textures (very fine-grained dendrites to coarser grained subequant skeletons). Quench crystals include mostly olivine (Fo_{61-80}) and ilmenite with minor amounts of chrome spinel (*Heiken and McKay,* 1977, 1978). Although opaque in optical microscopes, compound droplets with quench crystals all have glassy groundmasses. Rare subequant olivine phenocrysts have compositions of Fo_{79-82}. Within the cored deposit at Shorty crater samples contain between 0.2% and 2.9% olivine phenocrysts. Vesicles are uncommon, occurring in 0.2% to 6% of orange glass samples.

Nearly all ilmenites in the partly crystalline orange glass droplets from the Apollo 17 landing site are fine grained and have dendritic shapes (*Heiken and McKay,* 1977). In the sample examined in detail for this study (72504; 11.5%

ilmenite) the mean area of ilmenite grains is 1.5 μm^2, the mean width is 0.9 μm^2, and the mean aspect ratio is 2.47. Grain widths range from 0.2 to 7.5 μm.

Lunar pyroclastic materials that have been found at all of the Apollo landing sites, range from low-TiO_2 (0.4 wt.%) to high-TiO_2 ultramafic compositions (e.g., *Heiken et al.,* 1974; *Delano and Livi,* 1981; *Delano and Lindsley,* 1983; *Delano,* 1986).

One of the most unique features of the Apollo 17 and other lunar pyroclasts are the sublimates that coat grain surfaces. Micromounds, usually less than 1 μm thick, are composed of globular and crystalline clusters, visible only with a scanning electron microscope. *Meyer et al.* (1975) and *Butler and Meyer* (1976) determined that these clusters consist of sulfur compounds and include elements such as Zn, K, Cl, Na, Ga, Cu, and Pb. It has been inferred that the sublimates were deposited on pyroclast surfaces during lava fountaining associated with the eruption of mare lavas.

Although the orange and black pyroclastic spheres are not abundant at the Apollo 17 site, there is a major regional pyroclastic deposit (Taurus-Littrow dark mantle deposit) just over 50 km west of the site. A comparison of reflectance spectra obtained for the Taurus-Littrow dark mantling deposit (DMD) with laboratory and telescopic reflectance measurements (Fig. 3) have demonstrated that the Apollo 17 black spheres are the characteristic ingredient of the Taurus-Littrow pyroclastic mantling deposit (*Pieters et al.,* 1973, 1974; *Adams et al.,* 1974; *Gaddis et al.,* 1985; *Hawke et al.,* 1989a). The thickest part of the Taurus-Littrow DMD has an areal extent of just over 4000 km^2. However, the thinner portions of the deposit blanket a much larger area. Both geological and radar studies of the Taurus-Littrow DMD indicate that the core portion of the deposit has a thickness of many tens of meters. The deposit exhibits very weak to nonexistent echoes on the depolarized 3.8-cm radar maps produced by *Zisk et al.* (1974). These very low depolarized radar returns are generally attributed to the lack of scatterers (1-50 cm) on the smooth surface of the pyroclastic mantling deposits (*Pieters et al.,* 1973; *Zisk et al.,* 1974; *Gaddis et al.,* 1985). A very low degree of small-scale surface roughness and a relatively block-free surface are indicated. Areas of enhanced-return, higher-albedo material exposed by small impact craters that have penetrated this pyroclastic mantling unit are very rare. The 70-cm radar data obtained by *Thompson* (1979) shows that the Taurus-Littrow DMD is deficient in large (~1-2 m) blocks.

Several other major occurrences of regional pyroclastic mantling deposits have been documented (e.g., *Wilhelms and McCauley,* 1971; *Gaddis et al.,* 1985) including those at the following locations: Rima Bode, Aristarchus Plateau, Sulpicius Gallus, Montes Harbinger, southern Sinus Aestuum, southwestern Mare Humorum, and southern Mare Vaporum (see Fig. 4

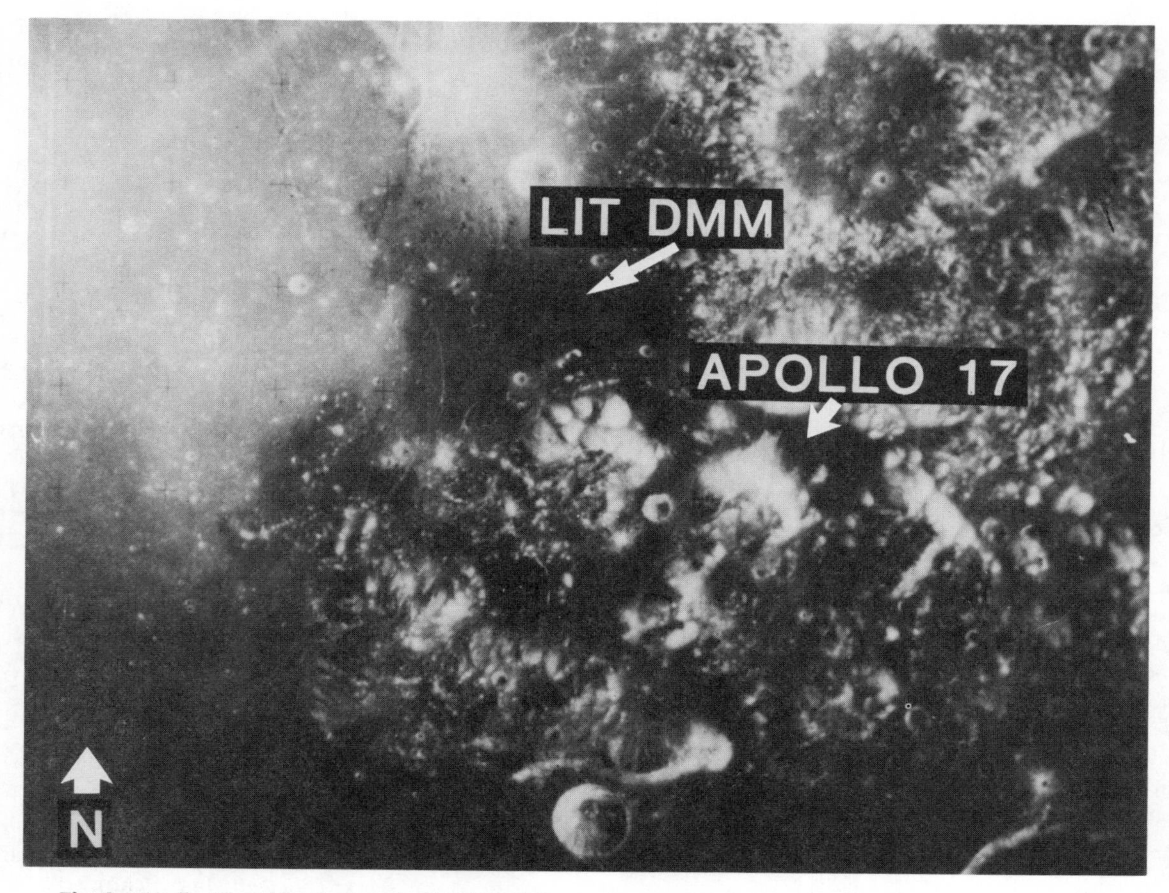

Fig. 2. Apollo 15 metric photograph (AS15 - 0972) of the Taurus-Littrow region. LIT DMM indicates the major regional pyroclastic deposit NW of the Apollo 17 landing site.

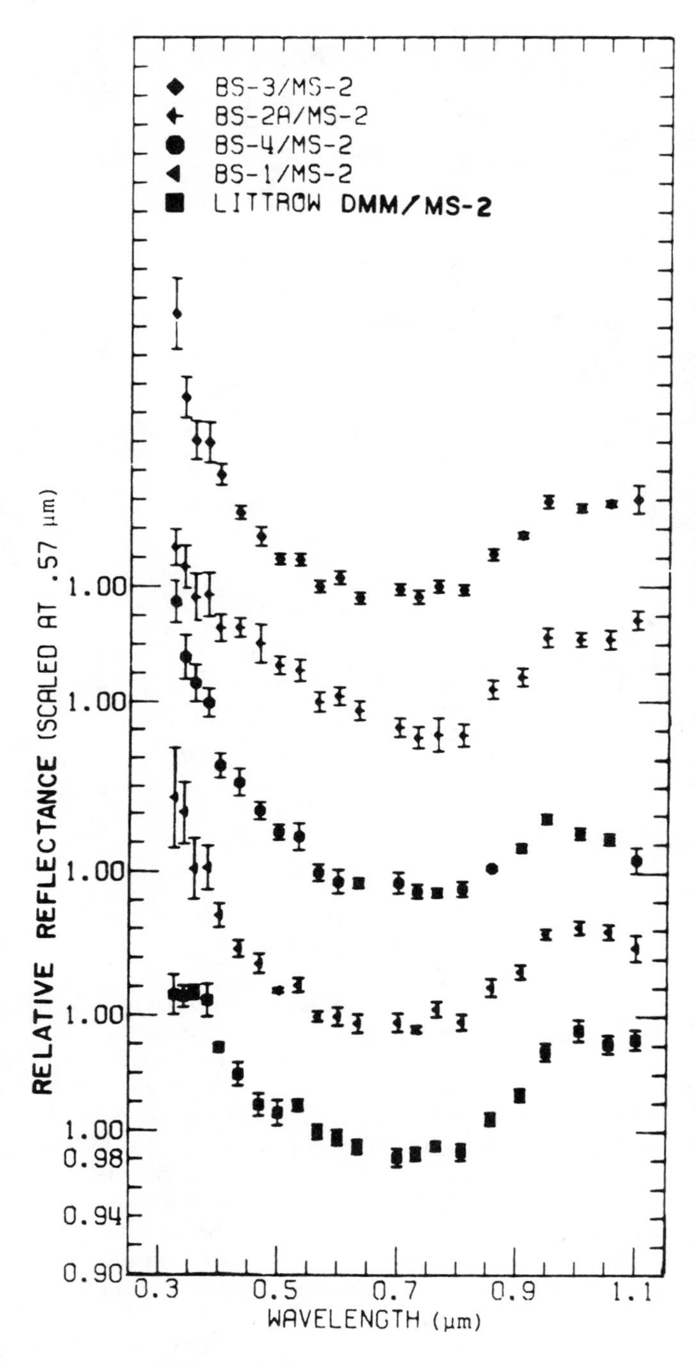

Fig. 3. Relative reflectance spectra for three lunar samples and four lunar areas. All spectra have been scaled to unity at 0.57 μm and divided by the spectrum for the standard lunar area, MS-2. The scale for sample 74001 has been reduced by a factor of ten to allow comparison with other relative spectra. After *Adams et al.* (1974).

and Table 1). These units have been characterized as extensive deposits of low albedo (0.079-0.096) material that appear to subdue and mantle underlying terrain. Photographs and visual observations made during the Apollo missions indicated that the surfaces of pyroclastic mantling deposits are relatively fine textured and exhibit a smooth, velvety appearance (see *Gaddis et al.,* 1985). The low returns on Earth-based 3.8-cm depolarized radar backscatter maps confirm these observations of mantled areas and indicate an absence of surface scatters in the 1 to 50-cm size range (*Pieters et al.,* 1973; *Zisk et al.,* 1974, 1977; *Gaddis et al.,* 1985). The surfaces of the regional pyroclastic deposits are rock-free.

Spectral reflectance studies have provided important information concerning the composition of regional pyroclastic (DMD) deposits (e.g., *Lucey et al.,* 1986; *Gaddis et al.,* 1985). Near-infrared spectra for a number of regional pyroclastic deposits are shown in Figs. 5a,b. These spectra were obtained at the 2.2-m University of Hawaii telescope facility on Mauna Kea, Hawaii and processed according to the methods outlined by *McCord et al.* (1981). As discussed above, the Taurus-Littrow DMD (Littrow 1 in Fig. 5) is dominated by ilmenite-rich black spheres. The spectra obtained for the Rima Bode pyroclastic deposits (Sinus Aestuum 2 in Fig. 5) are almost identical to those collected for various positions of the Taurus-Littrow deposit. A similar composition is indicated. The reflectance and continuum-removed spectra obtained for the DMD on the Aristarchus Plateau and southwest of Mare Humorum (Aristarchus Plateau 1 and Mare Humorum MM in Fig. 5) exhibit broader, longer-wavelength absorption bands than those that can be attributed to pyroxenes alone in highland and mare soils. As noted by *Gaddis et al.* (1985) and *Lucey et al.* (1986), there must be additional Fe-bearing soil components which have both modified the "1 μm band" and maintained the low albedo of the materials observed. The presence of Fe^{2+}-bearing volcanic glasses in the mantling deposits from which the spectra were obtained is most consistent with the available spectral evidence (*Gaddis et al.,* 1985; *Lucey et al.,* 1986; *Hawke et al.,* 1989a).

The spectral data presented by *Pieters et al.* (1973, 1974), *Adams et al.* (1974), and *Gaddis et al.* (1985) have demonstrated that several DMD on the lunar nearside are composed of ilmenite-rich spheres of pyroclastic origin. These include the following deposits: (1) Taurus-Littrow, (2) Rima Bode, (3) Southern Mare Vaporum, and (4) Southern Sinus Aestuum. As discussed above, the near-infrared reflectance spectra for the Taurus-Littrow and Rima Bode pyroclastic deposits are nearly identical and indicate the presence of large amounts of ilmenite-rich pyroclastic debris. Figure 6 presents relative UV-VIS spectra (0.3-1.1 μm) for several major regional pyroclastic deposits. These spectra have been divided by the spectrum for the lunar standard area, MS-2 and scaled to unity

TABLE 1. Major ilmenite-rich pyroclastic mantling deposits of regional extent.

Name	Location	Areal Extent (km^2)
Taurus-Littrow	30°E, 20°N	4000
Rima Bode	30°W, 13°N	10,000
Southern Sinus Aestuum	7°W, 5°N	30,000
Southern Mare Vaporum	7°E, 10°N	10,000

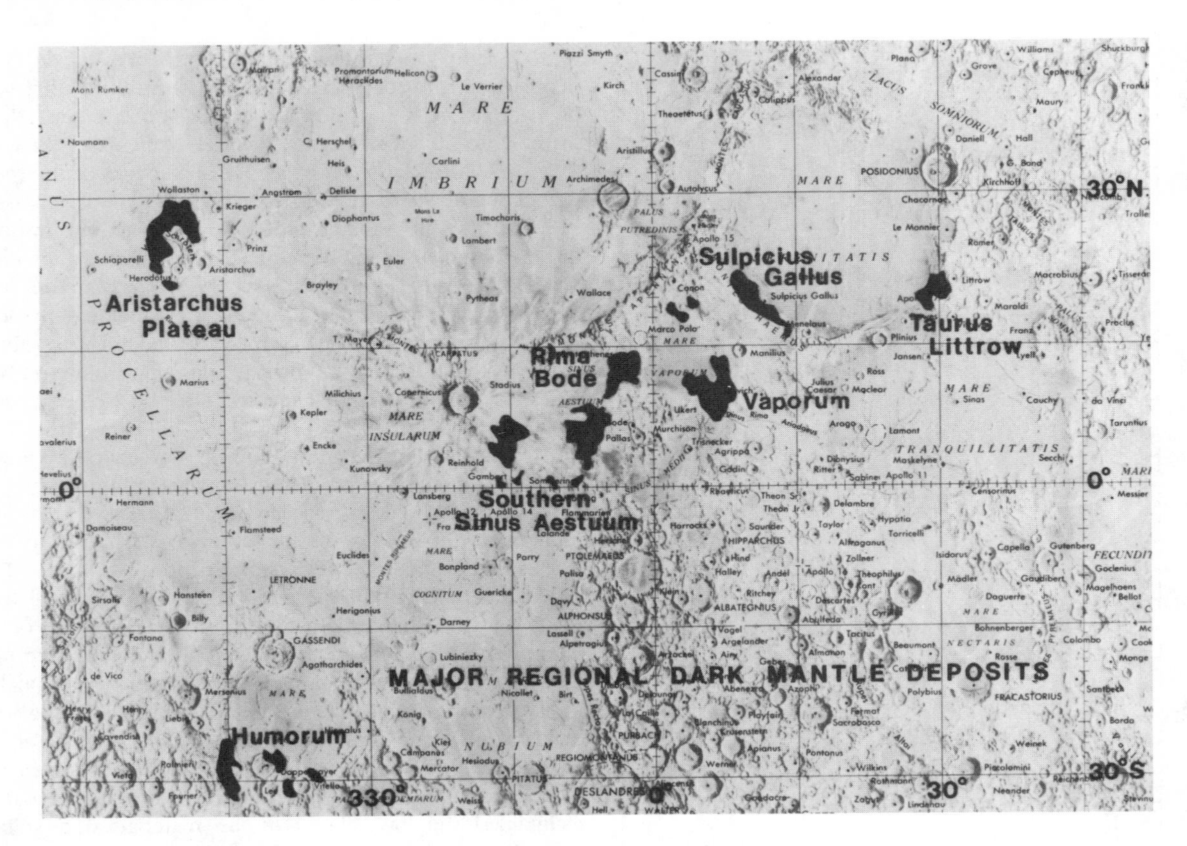

Fig. 4. The distribution of major lunar dark mantle deposits (after *Head,* 1974 and *Gaddis et al.,* 1985).

at 0.56 μm. All of these spectra exhibit similar characteristics. For example, they exhibit very high 0.40/0.56 μm values and the deposits appear "blue" in 0.40/0.56 μm multispectral ratio images presented by *Pieters et al.* (1974) and *McCord et al.* (1976). *Adams* (1974) and other workers have interpreted the spectra shown in Fig. 6 as indicating the presence of ilmenite-rich pyroclastic debris similar to the material returned from the Apollo 17 landing site.

As mentioned previously, all of the regional pyroclastic deposits exhibit very low values on depolarized 3.8-cm radar images (*Pieters et al.,* 1973; *Zisk et al.,* 1974; *Gaddis et al.,* 1985). Low values are also seen on 70-cm radar images (*Thompson,* 1979). 3.8-cm and 70-cm radar demonstrate that the regional pyroclastic deposits surfaces are generally free of rock fragments and blocks. The pyroclastic debris is apparently unwelded or annealed. Otherwise, small craters in these dark mantle deposits would have excavated welded blocks.

The locations and areal extents of the ilmenite-rich pyroclastic deposits are given in Table 1. The deposits cover very large areas (up to 30,000 km^2) and both geologic and radar data indicate that they have thicknesses of tens of meters. Most of the regional pyroclastic deposits overlie highlands terrain but a few (e.g., Taurus-Littrow) were emplaced on a mare substrate.

ADDITIONAL FACTORS FAVORING PYROCLASTIC DEPOSITS AS PLB SITES

The large areal extents of the lunar ilmenite-rich pyroclastic deposits, as well as their relatively great thicknesses (tens of meters), are major factors in favor of their selection as sites for permanent lunar bases. The recovery of significant amounts of He-3 will require the processing of thousands and eventually tens of thousands of square kilometers of regolith. The regional deposits are this large (see Table 1). For example, the areal extent of the Rima Bode DMD is 10,000 km^2 and the pyroclastics in southern Aestuum cover 30,000 km^2. The thickness of the ilmenite-rich pyroclastics becomes an important factor if the deposits are mined for ilmenite to be used in oxygen production.

It is clear that site selection must be directed toward the identification of large individual areas suitable for mining. *Cameron* (1988) has pointed out a variety of factors that must be considered in site selection and evaluation. These include regolith uniformity and ease of mining. For excavation and processing purposes, it is important that the deposit be as block-free as possible, that the regolith exhibit a high degree of uniformity, and that the deposit not be contaminated with low-Ti material derived by vertical mixing or lateral transport.

Both near-infrared spectra and multispectral images indicate that the surfaces of the ilmenite-rich regional pyroclastic deposits are relatively uniform and exhibit a low amount of contamination (*Pieters et al.,* 1974; *McCord et al.,* 1976, 1979; *Gaddis et al.,* 1981, 1985; *Hawke et al.,* 1989a). Work done by *Head and Wilson* (1979) and *Wilson and Head* (1981) suggests that the regional pyroclastic deposits were emplaced by strombolian or continuous eruption mechanisms and that it is unlikely that these extensive dark mantle deposits were produced by coalescing deposits of localized pyroclastics (i.e., Alphonsus-type dark-halo craters; *Gaddis et al.,* 1985; *Coombs and Hawke,* 1989; *Coombs et al.,* 1990; Hawke et al., 1989b). The results of calculations presented by *Head and Wilson* (1979) and *Wilson and Head* (1981) indicate that the lunar equivalent of strombolian activity is likely to lead to dispersal of pyroclasts over a wide area, with extreme sorting of particles. While some vent erosion undoubtedly occurred, the eruption mechanisms proposed for regional pyroclastics are those that would not lead to the emplacement of a high percentage of wall rock. Hence, little initial contamination by low-Ti wallrock would be expected.

Other factors have been important in keeping contamination to a minimum. The large areal extents of the various regional pyroclastic deposits tend to limit the effectiveness of lateral transport by impact. The relatively large thickness (at least tens of meters) of the regional pyroclastics is also a positive factor. The regional deposits are deep enough that small impact craters tens to hundreds of meters in diameter have generally not penetrated the ilmenite-rich pyroclastic debris and ejected subjacent blocky low-Ti material. This has helped keep surface contamination by low-Ti debris to a minimum.

As discussed in detail above, 3.8-cm radar backscatter images show that the regional pyroclastic deposits are deficient in surface scatterers in the 1-50-cm size range. The surfaces of the regional deposits are smooth and block-free. The near-surface areas of the pyroclastics are also deficient in fragments and blocks. The regional pyroclastic deposits are apparently composed of loose, unwelded particles. Most small craters excavated only incoherent debris. These extensive, deep deposits of loose ilmenite-rich pyroclastic material that are relatively rock-free and uncontaminated by vertical mixing or lateral transport would be ideal for lunar mining.

As discussed above, lunar pyroclastic material is enriched in certain surface-correlated volatile elements (*Butler and Myer,* 1976; *Butler,* 1978). These include S, Zn, Cu, Pb, Cl, Cd, and Bi. These elements are very rare on the Moon and are generally

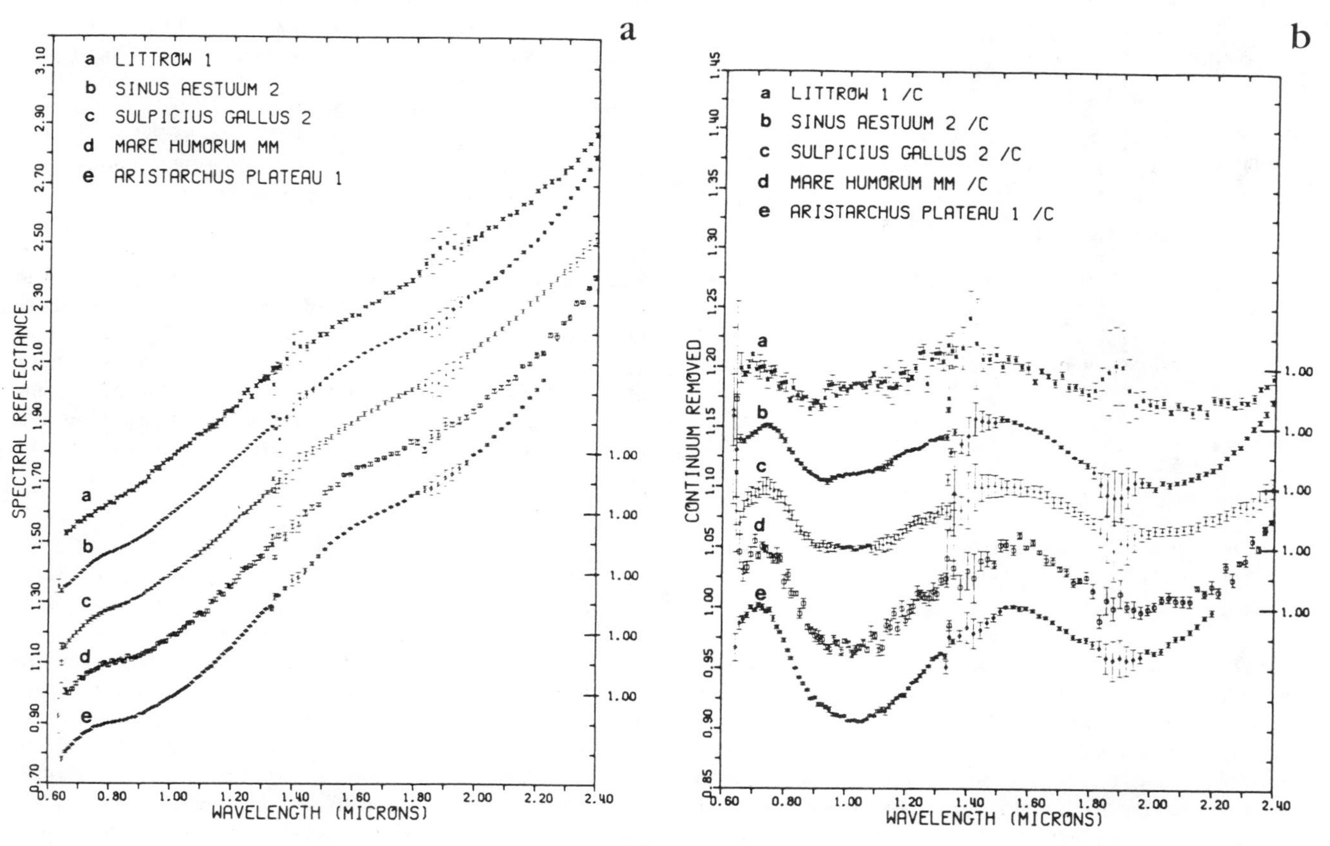

Fig. 5. (a) Scaled spectral reflectance measurements of several lunar regional pyroclastic deposits. The spectra are scaled to unity at 1.01 μm. **(b)** Residual absorption features for the same spectral measurements after continuum removal (after *Gaddis et al.,* 1985).

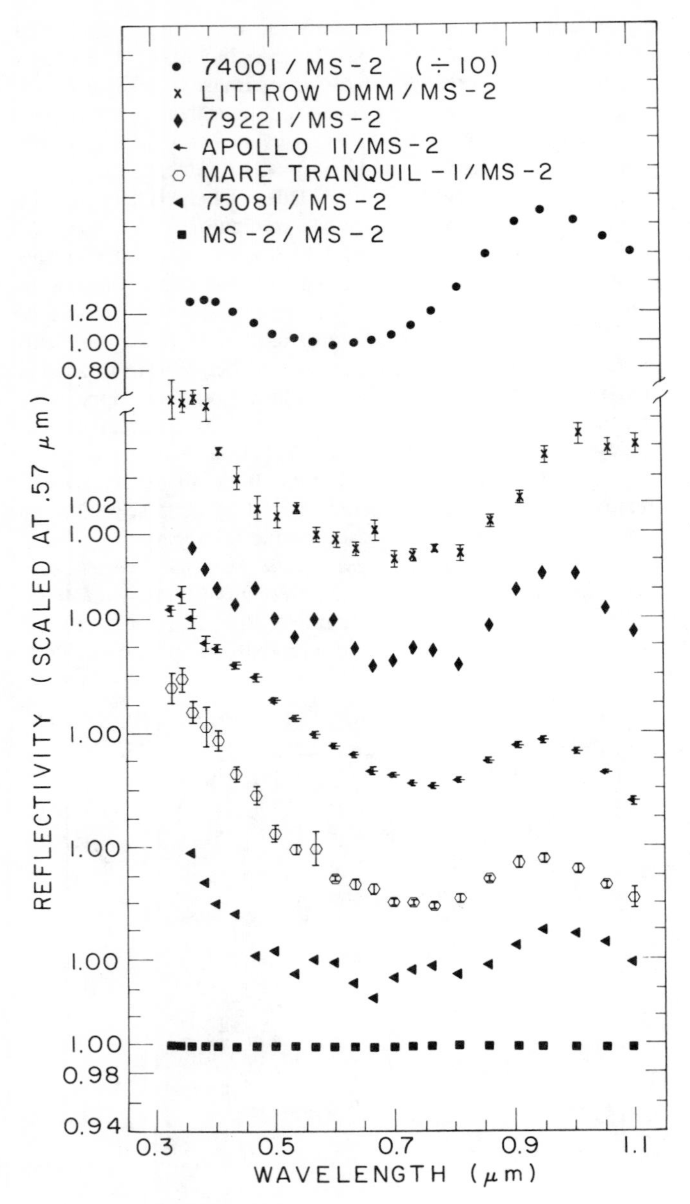

Fig. 6. Relative reflectance spectra for the regional pyroclastic deposits shown in Fig. 1 (see *Adams et al.,* 1974).

depleted in most common lunar rocks and soils. Since these volatile elements are concentrated on the surfaces of the pyroclastic spheres, they could be separated fairly easily by heating. The collection of these elements as a by-product of other processing techniques would provide an important source of material for use in lunar base developments and possible export to other structures in space. In particular, sulfur is a very important resource. It has multiple uses, especially in the waterless lunar environment (*Vaniman et al.,* 1989).

Glass may also prove to be an important lunar resource. Pure glass spheres, easily separated from partly crystalline droplets, may provide feedstock to be used in forming construction materials such as fiberglass and structural glass. In high vacuum, glasses may have tensile and compressive strengths greater than steel (*Blacic,* 1985).

As noted earlier, it has been suggested that lunar material would be useful as shielding material for space structures. Several factors favor the use of black pyroclastic spheres as shielding material. These include the following: (1) ease of mining, (2) handling characteristics, and (3) relatively high density.

The unconsolidated nature of the regional pyroclastic deposits is also important from the standpoint of lunar base construction. Protection from cosmic rays and meteorite impacts will be needed from the earliest stages of lunar base development. Recent studies have shown that greater thicknesses (4-5 m) of shielding material will be required than had previously been thought. Hence, base planners are very concerned about the ability to cover modules with adequate thicknesses of shielding. The deep, loose pyroclastic deposits would be ideal for quickly covering base modules with shielding material.

Binder (1988) has detailed the advantages of establishing a lunar base on or near KREEP-rich terrain. Several elements contained in KREEP would be very useful in lunar base development and expansion. A regional pyroclastic deposit composed of ilmenite-rich spheres on a KREEP-rich substrate would be a particularly good site for lunar base development.

Although the emphasis in this paper has been on economic resources for lunar base development, it is clear that a wide variety of scientific studies will also be conducted at the PLB. Lunar geological investigation will be important among these studies (e.g., *Cintala et al.,* 1985). In this connection, it should be noted that the ilminite-rich regional pyroclastic deposits (Fig. 1) are located in geologically interesting areas of the lunar surface. The location of a lunar base at one of these regional deposits would allow numerous lunar science questions to be addressed. These include the following: (1) the composition and mode of emplacement of Imbrium basin ejecta, (2) the impact cratering process, (3) the nature and origin of crater rays, (4) the formation of sinuous rilles, and (5) the compositions and eruption styles of a variety of mare basalt units.

CONCLUSIONS

1. Ilmenite-rich pyroclastic deposits would be excellent sites for the establishment of one or more permanent lunar bases for the production of oxygen and helium-3.

2. A wide variety of potentially useful by-products could be produced during the course of operations to extract oxygen and/or helium-3. Iron and titanium could be obtained as by-products of oxygen production from ilmenite. During the course of operations to extract He-3 from the regolith, other valuable volatiles implanted by the solar wind (H_2, N_2, CO_2, etc.) could be collected with relatively small mass and power penalties. In addition, lunar pyroclastic deposits are enriched in certain surface-correlated volatile elements (S, Cu, Pb, Zn, Cd, etc.). The collection of these elements as by-products of other processing techniques would provide an important source of material for use in lunar base development and expansion.

3. Numerous relatively thick pyroclastic deposits of regional extent have been identified on the lunar surface. Several of these regional deposits are rich in ilmenite and are located in geologically interesting areas.

4. These extensive, deep deposits of ilmenite-rich pyroclastics are unconsolidated and block-free. Their surfaces are relatively uncontaminated with low-Ti debris introduced by vertical mixing or lateral transport. These fine-grained, block-free, uncontaminated deposits would be ideal for lunar mining operations.

5. The deep, loose pyroclastic debris would be ideal for quickly covering lunar base modules with an adequate thickness of shielding material for protection from meteorite impact and space radiation. Eventually, this may prove to be the strongest argument for siting a lunar base at a regional pyroclastic deposit.

6. Finally, we conclude that if a lunar base is established for any of the economic reasons that have been seriously suggested to date, it will be located at a regional pyroclastic deposit.

Acknowledgments. This research was supported by NASA grant NAGW-237. The authors would like to thank J. Delano, R. Colson, and L. Gaddis for their very helpful reviews of the manuscript. We would also like to thank A. Binder, A. Basu, D. McKay, A. Cutler, and W. Agosto for useful discussions, and special thanks are due to G. Heiken and D. Vaniman for providing previously unpublished data and very helpful discussions. This is PGD No. 576 and HIG Contribution No. 2195.

REFERENCES

Adams J. B., Pieters C. M., and McCord T. B. (1974) Orange glass: Evidence for regional deposits of pyroclastic origin on the Moon. *Proc. Lunar Sci. Conf. 5th,* pp. 171-186.

Agosto W. (1985) Electrostatic concentration of lunar soil minerals. In *Lunar Bases and Space Activities of the 21st Century* (W. W. Mendell, ed.), pp. 453-464. Lunar and Planetary Institute, Houston.

Binder A. B. (1988) Lunar base site selection: Lunar resource criteria (abstract). *Papers Presented to the 1988 Symposium on Lunar Bases and Space Activities of the 21st Century,* p. 25. Lunar and Planetary Institute, Houston.

Blacic J. D. (1985) Mechanical properties of lunar materials under anhydrous, hard vacuum conditions: Applications of lunar glass structure components. In *Lunar Bases and Space Activities of the 21st Century* (W. W. Mendell, ed.), pp. 487-495. Lunar and Planetary Institute, Houston.

Briggs R. A. (1988) Oxidation/reduction of ilmenite and the design of an oxygen production facility for use on the Moon. *The High Frontier Newsletter, Vol. XIV, Issue 6,* Space Studies Institute, Princeton.

Busarev V. V. and Shevchenko V. V. (1988) TiO_2 in crater Le Monnier (abstract). In *Papers Presented to the 1988 Symposium on Lunar Bases and Space Activities of the 21st Century,* pp. 44-45. Lunar and Planetary Institute, Houston.

Bustin R., Kotra R. K., Gibson E. K., Nace G. A., and McKay D. S. (1984) Hydrogen abundances in lunar soils (abstract). In *Lunar and Planetary Science XV,* pp. 112-113. Lunar and Planetary Institute, Houston.

Butler P. Jr. (1978) Recognition of lunar glass droplets produced directly from endogenous liquids: The evidence from S-Zn coating. *Proc. Lunar Planet. Sci. Conf. 9th,* pp. 1459-1471.

Butler P. Jr. and Meyer C. Jr. (1976) Sulfur prevails in coatings on glass droplets: Apollo 15 green and brown glasses and Apollo 17 orange and black (devitalized) glasses. *Proc. Lunar Sci. Conf. 7th,* pp. 1561-1581.

Cameron E. N. (1987) *WCSAR-TR-AR3-8708.* Wisconsin Center for Space Automation and Robotics (WCSAR), Madison.

Cameron E. N. (1988) Mining for helium-site selection and evaluation (abstract). In *Papers Presented to the 1988 Symposium on Lunar Bases and Space Activities of the 21st Century,* p. 47. Lunar and Planetary Institute, Houston.

Chang S., Lennon K., and Gibson E. K. (1974) Abundances of C, N, H, He, and S in Apollo 17 soils from stations 3 and 4: Implications for solar wind exposure ages and regolith evolution (abstract). In *Lunar Science V,* pp. 106-108. The Lunar Science Institute, Houston.

Cintala M. J., Spudis P. D., and Hawke B. R. (1985) Advanced geologic exploration supported by a lunar base: A traverse across the Imbrium-Procellarum region of the Moon. In *Lunar Bases and Space Activities of the 21st Century* (W. W. Mendell, ed.), pp. 223-237. Lunar and Planetary Institute, Houston.

Coombs C. R. and Hawke B. R. (1989) Explosive volcanism on the Moon: A review. *Proc. Kagoshima Intl. Conf. on Volcanoes,* Kagoshima, Japan, pp. 416-419.

Coombs C. R. and Hawke B. R., Lucey P. G., Owensby P. D., and Zisk S. H. (1990) The Alphonsus region: A geologic and remote sensing perspective. *Proc. Lunar Planet. Sci. Conf. 20th,* this volume.

Crabb T. M. and Jacobs M. K. (1988) Synergism of He-3 acquisition with lunar base evolution (abstract). In *Papers Presented to the 1988 Symposium on Lunar Bases and Space Activities of the 21st Century,* p. 62. Lunar and Planetary Institute, Houston.

Crabb T. M. and Jacobs M. K. (1990) Synergism of He-3 acquisition with lunar base evolution. In *Lunar Bases and Space Activities of the 21st Century* (W. W. Mendell, ed.). Univelt, San Diego, in press.

Cutler A. H. and Krag P. (1985) A carbothermal scheme for lunar oxygen production. In *Lunar Bases and Space Activities of the 21st Century* (W. W. Mendell, ed.), pp. 559-569. Lunar and Planetary Institute, Houston.

Delano J. W. (1986) Pristine lunar glasses: Criteria, data, and implications. *Proc. Lunar Planet. Sci. Conf. 16th,* in *J. Geophys. Res., 91,* D201-D213.

Delano J. W. and Lindsley D. H. (1983) Mare volcanic glasses from Apollo 17 (abstract). In *Lunar and Planetary Science XIV,* pp. 156-157. Lunar and Planetary Institute, Houston.

Delano J. W. and Livi K. (1981) Lunar volcanic glasses and their constraints on mare petrogenesis. *Geochim. Cosmochim. Acta, 45,* 2137-2149.

DesMarais D. J., Hayes J. M., and Meinschein W. G. (1974) The distribution in lunar soil of hydrogen released by pyrolysis. *Proc. Lunar Sci. Conf. 8th,* pp. 1811-1822.

Gaddis L. R., Adams J. B., Hawke B. R., Head J. W., McCord T. B., Pieters C. M., and Zisk S. H. (1981) Characterization and distribution of pyroclastic units in the Rima Bode region of the Moon (abstract). In *Lunar and Planetary Science XII,* pp. 318-320. Lunar and Planetary Institute, Houston.

Gaddis L. R., Pieters C. M., and Hawke B. R. (1985) Remote sensing of lunar pyroclastic mantling deposits. *Icarus, 61,* 461-489.

Gibson M. A. and Knudsen C. W. (1985) Lunar oxygen production from ilmenite. In *Lunar Bases and Space Activities of the 21st Century* (W. W. Mendell, ed.), pp. 543-550. Lunar and Planetary Institute, Houston.

Haskin L. A. (1989) The Moon as a practical source of hydrogen and other volatile elements (abstract). In *Lunar and Planetary Science XX,* pp. 387-388. Lunar and Planetary Institute, Houston.

Hawke B. R., Coombs C. R., Clark B. (1989a) Pyroclastic deposits: An ideal lunar resource (abstract). In *Lunar and Planetary Science XX,* pp. 389-390. Lunar and Planetary Institute, Houston.

Hawke B. R., Coombs C. R., Gaddis L. R., Lucey P. G., and Owensby P. D. (1989b) Remote sensing and geologic studies of localized dark mantle deposits on the Moon. *Proc. Lunar Planet. Sci. Conf. 19th,* pp. 255-268.

Head J. W. (1974) Lunar dark-mantle deposits: Possible clues to the distribution of early mare deposits. *Proc. Lunar Sci. Conf. 5th,* pp. 207-222.

Head J. W. and Wilson L. (1979) Alphonsus-type dark-halo craters: Morphology, morphometry and eruption conditions. *Proc. Lunar Planet. Sci. Conf. 10th,* pp. 2861-2897.

Heiken G. H. and McKay D. S. (1977) A model for eruption behavior of a volcanic vent in eastern Mare Serenitatis. *Proc. Lunar Sci. Conf. 8th,* pp. 3243-3255.

Heiken G. H. and McKay D. S. (1978) Petrology of a sequence of pyroclastic rocks from the valley of Taurus-Littrow (Apollo 17 landing site). *Proc. Lunar Planet. Sci. Conf. 9th,* pp. 1933-1943.

Heiken G. H. and McKay D. S., and Brown R. W. (1974) Lunar deposits of possible pyroclastic origin. *Geochim. Cosmochim, Acta, 38,* 1703-1718.

Johnson T. V., Mosher J. A., and Matson D. L. (1977) Lunar spectral units: A northern hemispheric mosaic. *Proc. Lunar Sci. Conf. 8th,* pp. 1013-1028.

Kulcinski G. L., ed. (1988) *Astrofuel for the 21st Century.* College or Engineering, Univ. of Wisconsin, Madison. 20 pp.

Kulcinski G. L. and Schmitt H. H. (1987) The Moon: An abundant source of safe fusion fuel for the 21st century. Presented at the 11th International Scientific Forum on Fueling the 21st Century, 2; 9 September-6 October 1987, Moscow, USSR.

Kulcinski G. L., Santarius J. F., and Wittenberg L. J. (1986) Presentation at the 1st Lunar Development Symposium, September, 1986, Atlantic City, New Jersey.

Kulcinski G. L., Sviatolslavsky I. N., Santarius J. F., and Wittenberg L. J. (1988) Fusion energy from the Moon for the 21st century (abstract). In *Papers Presented to the 1988 Symposium on Lunar Bases and Space Activities of the 21st Century,* p. 147. Lunar and Planetary Institute, Houston.

Li Y. T. and Wittenberg L. J. (1988) Lunar surface mining for automated acquisition of helium-3: Methods, processes, and equipment (abstract). In *Papers Presented to the 1988 Symposium on Lunar Bases and Space Activities of the 21st Century,* p. 158. Lunar and Planetary Institute, Houston.

Lucchitta B. K. (1973) Photogeology of the dark material in the Taurus-Littrow region of the Moon. *Proc. Lunar Sci. Conf. 4th,* pp. 149-162.

Lucey P. G., Hawke B. R., Pieters C. M., Head J. W., and McCord T. B. (1986) A compositional study of the Aristarchus region of the Moon using near-infrared reflectance spectroscopy. *Proc. Lunar Planet. Sci. Conf. 16th,* in *J. Geophys Res., 91,* D344-D354.

McCord T. B., Pieters C. M. and Feierberg M. A. (1976) Multispectral mapping of the lunar surface using ground-based telescopes. *Icarus, 29,* 1-34.

McCord T. B., Grabow M. A., Feierberg M. A., MacLaskey D., and Pieters C. M. (1979) Lunar multispectral maps: Part II of the lunar nearside. *Icarus, 37,* 1-28.

McCord T. B., Clark R. N., Hawke B. R., McFadden L. A. Owensby P. D., Pieters C. M., and Adams J. B. (1981) Moon: Near-infrared spectral reflectance, a first good look. *J. Geophys. Res., 86,* 10,883-10,892.

Mendell W. W., ed. (1985) *Lunar Bases and Space Activities of the 21st Century.* Lunar and Planetary Institute, Houston. 866 pp.

Meyers C., Jr., McKay D. S., Anderson D. H., and Butler P. (1975) The source of sublimates on the Apollo 17 orange glass samples. *Proc. Lunar Sci. Conf. 6th,* pp. 1673-1699.

Müller O. (1973) Chemically bound nitrogen contents of Apollo 16 and Apollo 15 lunar fines. *Proc. Lunar Sci. Conf. 4th,* pp. 1625-1634.

Pieters C. M. (1978) Mare basalt types on the front side of the moon: A summary of spectral reflectance data. *Proc. Lunar Planet. Sci. Conf. 9th,* pp. 2825-2849.

Pieters C. M., McCord T. B., Zisk S. H., and Adams J. B. (1973) Lunar black spots and the nature of the Apollo 17 landing area. *J. Geophys. Res, 78,* 5867-5875.

Pieters C. M., McCord T. B., Charette M. P., and Adams J. B. (1974) Lunar surface: Identification of the dark mantling material in the Apollo 17 soil samples. *Science, 183,* 1191-1194.

Pieters C. M., Head J. W., Adams J. B., McCord T. B., Zisk S. H., and Whitford-Stark J. (1980) Late high titanium basalts of the western maria: Geology of the Flamsteed region of Oceanus Procellarum. *J. Geophys. Res., 85,* 3913-3938.

Santarius J. F. (1988) Lunar 3He, fusion propulsion, and space development (abstract). In *Papers Presented to the 1988 Symposium on Lunar Bases and Space Activities of the 21st Century,* p. 212. Lunar and Planetary Institute, Houston.

Shevchenko V. V. and Busarev V. V. (1988) Ilmenites in Oceanus Procellarum (abstract). In *Papers Presented to the 1988 Symposium on Lunar Bases and Space Activities of the 21st Century,* p. 219. Lunar and Planetary Institute, Houston.

Simon M. C. (1985) A parametric analysis of lunar oxygen production. In *Lunar Bases and Space Activities of the 21st Century* (W. W. Mendell, ed.), pp. 531-541. Lunar and Planetary Institute, Houston.

Taylor S. R. (1982) *Planetary Science: A Lunar Perspective.* Lunar and Planetary Institute, Houston. 481 pp.

Thompson T. W. (1979) A review of Earth-based radar mapping of the Moon. *Moon and Planets,* 20, 179-198.

Vaniman D., Pettit D., and Heiken G. (1990) Uses of lunar sulfur. In *Lunar Bases and Space Activities of the 21st Century* (W. W. Mendell, ed.). Univelt, San Diego, in press.

Wilhelms D. E. and McCauley J. F. (1971) Geologic map of the nearside of the Moon. *U.S. Geol. Surv. Misc. Map I-703.*

Wilson L. and Head J. W. (1981) Ascent and eruption of basaltic magma on the Earth and Moon. *J. Geophys. Res, 78,* 2971-3001.

Wittenberg L. J., Santarius J. F., and Kulcinski G.L. (1986) Lunar source of He-3 for commercial fusion power. *Fusion Technology, 10,* 167-178.

Zisk S. H., Pettengill G. H., and Catuna G. W. (1974) High-resolution radar map of the lunar surface at 3.8-cm wavelength. *The Moon, 10,* 17-50.

Zisk S. H., Hodges C. A., Moore H. J., Shorthill R. W., Thompson T. W., Whitaker E. A., and Wilhelms D. E. (1977) Th Aristarchus-Harbinger region of the Moon: Surface geology and history from recent remote-sensing observations. *The Moon, 17,* 59-99.

Proceedings of the 20th Lunar and Planetary Science Conference, pp. 259-269
Lunar and Planetary Institute, Houston, 1990

Sources and Subsurface Reservoirs of Lunar Volatiles

B. L. Cooper

The University of Texas at Dallas, Department of Geosciences FO 2.1, P.O. Box 830688, Richardson, TX 75083-0688

Estimates are presented for volumes and compositions of lunar volatiles in proposed subsurface traps, and possibilities are discussed for such shallow gas reservoirs. Crater fillings could serve as porous and permeable gas reservoirs, as could magma chambers and conduits. Terrestrial-analog structural domes covered by a less porous and less permeable cap might exist, as well as stratigraphic "pinch-out" traps. Likewise, fault-related traps might also exist. There are three sources for the gas that might be found in such reservoirs. Firstly, solar wind implanted H, He and other noble gases, as well as N, C, and O, are known to exist in significant quantities in the lunar regolith. Secondly, volcanic outgassing could produce CO, CO_2, CH_4, He, N_2, NH_3, H_2O, and H_2S. Finally, radioactive decay in the lunar interior is thought to produce He, Rn, and Ar. The geologic systems affecting regolith outgassing and trapping may be of practical importance for in situ lunar resource utilization.

INTRODUCTION

Although the Moon is often considered "dry" and "airless," it would be more correct to say that the Moon is dry and airless *in comparison to the Earth*. In fact, there are enough volatiles in a cubic meter of lunar soil to fabricate a lunch for two of cheese sandwiches, fruit, and soda pop (*Haskin*, 1989).

Volatiles are understood in this context to be atomic or molecular species lightly bound (through ion implantation or surface adsorption) to the surfaces of grains (J. Jordan, personal communication, 1989). These include H, He, C, N, O, Ne, Na, P, S, Cl, Ar, Kr, Xe, Rn, and could also include many other elements depending upon the kinetic energy of the particle (atom, molecule, or ion) in which the atomic nucleus in question resides.

The problem of the "dry" and "airless" Moon is not so much one of short supply of volatiles, but rather where these elements reside. They are bound to the surfaces of regolith particles or inside the crystal lattices of minerals, and therefore they must be mined. In previous studies, mining the Moon for volatiles involves either surface strip mining or tunnelling to extract ore that is then hauled to a processing facility for treatment or refining (*Christiansen*, 1988). This is a labor-intensive process involving not only EVA time but also IVA time. Furthermore, mass and power requirements to transport mining equipment and operate it on the Moon make such an operation expensive. Finally, the problem of maintaining a pressure seal in a dusty environment presents engineering obstacles. Extracting volatiles such as hydrogen and oxygen from lunar regolith or lunar basalt requires tremendous heat energy or large amounts of dangerous reagents, and sometimes both. Also, the most promising methods of accomplishing this are largely untested. Of the 13 proposed processes studied by *Christiansen* (1988) for extracting oxygen from lunar rock or soil, only a few have experimental support for the techniques discussed.

WHY LOOK FOR SUBSURFACE VOLATILE RESOURCES?

If gas reservoirs exist below the lunar surface, they could provide much reduction in the complexity of obtaining O_2, H_2, He and other useful volatiles. As mentioned above, the pressure vessel manipulation problem is a fundamental obstacle to processing lunar regolith for volatiles. It is necessary to place the regolith material inside a pressure vessel where it can be heated and the volatiles immediately collected. Volatiles trapped below the surface may not need to undergo such a process, as they are trapped naturally. In order for 1 tonne of hydrogen to be extracted from a lunar mine, an average of 16,600 tonnes of hydrogen-bearing lunar regolith must be beneficiated, and 3800 tonnes of this concentrate must be moved inside the pressure vessel (*Carter,* 1985). After processing, almost the entire mass (minus the extracted volatiles) must then be removed from the pressure vessel. This technology would be the equivalent of maintaining a vacuum chamber on Earth at 10^{-8} torr, while intermittently opening and closing the high-vacuum seals to transport dust through the chamber.

While the technology for performing the above operation has not yet been fully developed (*Howe et al.,* 1989), some preliminary steps have been taken with the dual feed-hopper design that allows semicontinuous feedstock conveyance with minimal losses (e.g., *Christiansen,* 1988). Furthermore, samples have been degassed and collected in light- and rare-gas labs for some time. Yet these laboratory-scale methods are not cost-effective for commercial-scale operations, nor do they deal with the problems of flowing dust.

The entire purpose for moving the regolith and sealing it against the lunar vacuum is to contain the emitted volatiles. Thus, the possibility of utilizing a natural system that duplicates this function is of interest. Several such gas-trapping systems may exist on the Moon in the form of structural straps and natural cavities, and also as stratigraphic traps such as porosity pinch-outs.

EVIDENCE FOR LUNAR SUBSURFACE GASES

Several Apollo experiments detected gases near the Moon's surface (e.g., *Hodges et al.,* 1972; *Gorenstein et al.,* 1973; *Hoffman and Hodges,* 1975). A surface mass spectrometer placed on the Moon during the Apollo 17 mission recorded a daily variation in the amounts and kinds of gases present. It was learned that daytime gas concentrations were much higher than expected and that the instrument saturated out so that it had to be turned off during the day to prevent damage (J. H. Hoffman, personal communication, 1987). Thirty-seven different species with mass numbers from 1 to 94 were detected; however, only a few, He, Ne, Ar, CO_2, and O_2, could be verified by other Apollo experiments (*Hoffman and Hodges,* 1975). The Apollo orbital mass spectrometer detected the same mass peaks, also with a diurnal variation in concentration (*Hodges et al.,* 1972). Furthermore, variations in the orbital measurements of Rn and Ar could be correlated with certain features on the lunar surface (*Gorenstien et al.,* 1973).

The various experiments conducted during the Apollo missions gave evidence not only for a rarefied atmosphere of H_2, He, CH_4, NH_3, H_2O, CO, O_2, and CO_2 (*Hoffman et al.,* 1975), but also evidence that some gas concentrations were localized near the reported sites of lunar transient phenomena (*Gorenstein et al.,* 1973). The Aristarchus region is one of those shown to have unusually high Rn readings, and it is also a frequently reported site of lunar transient phenomena (LTP). These LTP are temporary brightenings and darkenings that are occasionally observed on the Moon's surface, and have been reported since the 1600s (*Kozyrev,* 1959; *Geake and Mills,* 1977; *Middlehurst,* 1977). They are believed to be evidence of periodic lunar outgassing events (*Cameron,* 1977). If these events are indeed outgassing phenomena, then we may infer that (1) volcanic gases exist in the lunar interior and (2) periodic overpressuring, perhaps triggered by tidal effects, causes these gases as well as radiogenic gases to be released from wherever they have been collecting. Such phenomena lend credence to the idea that gas reservoirs may exist in the lunar subsurface, occasionally becoming overpressured and "blowing out." This suggests a level of geologic activity within the Moon that, while minor compared to the Earth's, could nevertheless be significant in its effects.

By virtue of the fact that the Earth is about 81 times as massive as the Moon, the Earth's gravitational effect on the Moon is 81 times stronger than that of the Moon upon the Earth (*Weast,* 1971). Tidal effects alone are thus a major source of geologic stress and strain, and may be the cause of the high-frequency teleseismic (HFT) events reported by *Meissner et al.* (1970). These events are thought to be associated with thermal and tidal stresses acting on the deep fractures surrounding the multiring basins. These HFT are the most energetic of moonquakes, and they suggest the occurrence of gas volcanism (*Meissner et al.,* 1970).

Volatiles emitted at mare-filled basin edges may originate from deep within the lunar lithosphere, reaching the surface through fault systems that ring the mare borders (*Friesen,* 1975). Alternatively, I suggest that these volatiles may be the result of lateral gas migration from regions in the interior of the basin.

TERRESTRIAL ANALOGS FOR GAS TRAPPING

Natural occurrences of He, N_2, and CO_2 in the subsurface are known on Earth (*Keesom,* 1942). Hydrocarbon gas wells on Earth often contain He (*Elworthy,* 1926; *Cook,* 1961; *Mattill,* 1977), whereas CO_2 and N_2 often occur together in wells that do not contain much hydrocarbon (*Keesom,* 1942). The He is thought to occur because of radioactive decay of U or Ra (*Cook,* 1961).

The relatively high concentrations of He around the Amarillo, Texas, area have never been satisfactorily explained. If there are no local sources of He, then the gas must have migrated from some distance below the surface. They are probably radiogenic in nature, and while sources of ocean-ridge He are conceptually understood, it is not known why over half of the world's known He reserves accumulated in just a few places in the Earth's continental crust, especially when He is so diffusive owing to its small size. This fact suggests that some similar type of selective concentration might occur on the Moon.

Although the exact origin of these terrestrial nonhydrocarbon gases remains uncertain, the nature of the structural and stratigraphic traps is fairly well understood. A permeability barrier is required, and many geometries of reservoir rock can exist as long as gas escape is inhibited to some degree by a cap rock or other permeability barrier. Because He atoms are smaller than hydrocarbon molecules, some rocks might be able to trap hydrocarbons while allowing He to pass through. Furthermore, on Earth, and probably on the Moon as well, there is no such thing as a "perfect" trap. In other words, the flow of gas or liquid can only be slowed down; it cannot be entirely stopped. Some remote-sensing hydrocarbon exploration techniques are based upon this principle. Terrestrial gas "leaks" can be detected from orbit, and it should also be possible to detect lunar gas "leaks" from lunar orbit. Furthermore, because of the greatly reduced "background" noise values on the Moon, in comparison to Earth, these leaks—if they exist—should be much easier to detect by remote sensing.

Various types of terrestrial gas-traps exist. In one type, for example, a dome allows immiscible gas or oil to remain relatively immobile in its top while subsurface waters flow freely below the hydrocarbon-rich zone. Another kind of structural trap may occur when a porous and permeable bed is truncated against a less porous and less permeable bed on its up-dip end. Stratigraphic traps occur when a change in porosity or permeability within the formation itself prevents further upward migration of fluids. Pinch-outs are an example of this kind of trap, in which a porous and permeable bed gradually thins and disappears, thereby trapping the gas at the juncture.

MECHANISMS FOR TRAPPING GASES IN THE LUNAR SUBSURFACE

The pressure differential that might exist below the lunar surface remains uncertain. Because there is no denser medium such as water to hold gases in place in the domes of structures on the Moon, it might be argued that no gas-trap on the Moon could possibly be effective. However, a positive temperature gradient with depth might act as a vertical barrier causing gases to preferentially collect in a cooler zone (J. L. Carter, personal communication, 1988). A fairly linear temperature gradient exists from the surface to a depth of 200 km (*Taylor,* 1982), which indicates that at a depth of from 40 to 80 km,

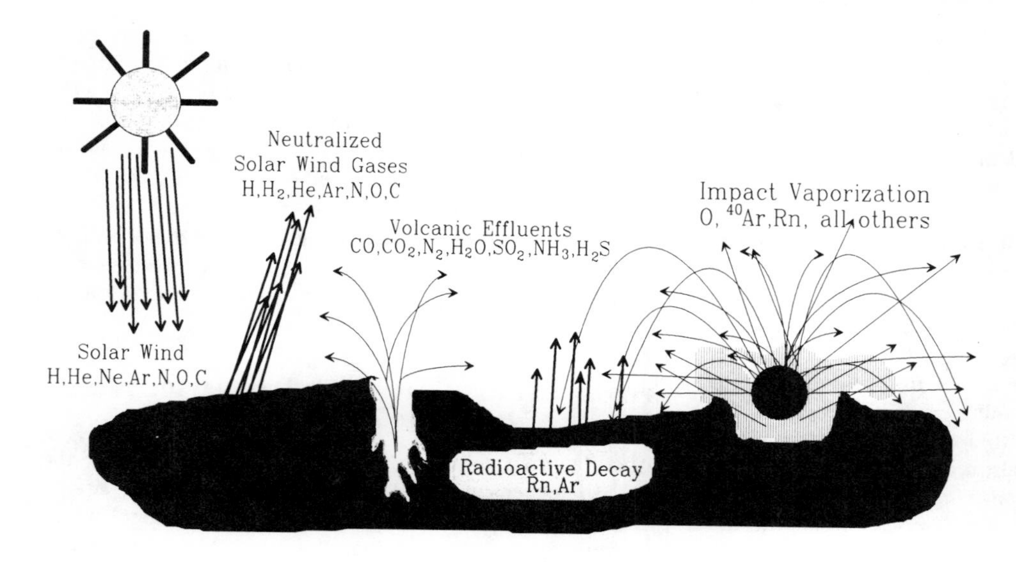

Fig. 1. Sources of lunar atmospheric gases. Although solar-wind implantation is believed to be the major source of volatiles in the lunar regolith, there are also several other sources. The lunar atmosphere may be likewise affected by volcanic effluents and radioactive decay products. Impact vaporization may volatilize any of the elements that exist in the lunar soil, thus redistributing them, and the impactor itself may also contribute volatiles to the system (after J. H. Hoffman, personal communication, 1988).

temperature would rise by 100°C; a 1° gradient over 400 to 800 m. In-situ measurements will be necessary to learn whether this temperature gradient produces a significant concentration of gas in subsurface traps. Direct measurements are also needed to quantify the effectiveness of any potential lunar gas-trap, and to refine volume and pressure calculations presented here.

Apollo orbital data showed elevated Rn readings over the Aristarchus plateau, which, again, is a frequent location of reported lunar transient phenomena. If indeed these LTPs are caused by escaping gases, then this lends some support to the suggestion that gas reservoirs exist in the lunar subsurface; however, the possibility also exists that this may be purely an effect of Ra decay of KREEP basalt. Gas trapping could occur in a reservoir that has permeability barriers on all sides, such as a porous lens encapsulated between layers of impermeable basalt. For example, a crater ray might be emplaced on a fresh basalt, and subsequently buried by another unit of basalt in a later eruption. The breccias formed by the impact event would be greatly enriched in gases (*Heymann and Yaniv,* 1971). If the ray material remained exposed to the solar wind for 20 m.y. or so it would be a "mature" soil (e.g., *Blanford,* 1986) and as such would be considered an ore-grade regolith from the viewpoint of extracting solar-wind volatiles. Thus, the ray would be enriched in gases from both the impact event and from solar-wind implantation. The heat of the overlying basalt might liberate many of the surface-correlated volatiles, and burial at depth could cause extraction of the breccia pore gases by crushing in a fashion similar to the extraction technique used by *Heymann and Yaniv* (1971).

Because the solar wind implants ions of H, He, Ne, Ar, N, O, and C into the outermost part of regolith grains (e.g., *Zeller and Dreschoff,* 1970; *Hoffman and Hodges,* 1975), a mature lunar soil may have economically significant quantities of these elements (*Carter,* 1985). Volcanic effluents such as CO, CO_2, N_2, H_2O, SO_2, NH_3, and H_2S have been released sporadically into the lunar atmosphere (*Hoffman and Hodges,* 1975; *Hoffman et al.,* 1975). Once in the atmosphere these volatiles may enter cold traps and recondense into the regolith. Furthermore, if volcanic outgassing occurs in the subsurface, these volatiles may be retrapped very close to where they were released in a subsurface reservoir. Finally, radioactive decay of Ra produces Rn and He, and decay of K produces Ar. These gases may remain in subsurface traps or make their way to the surface to become part of the atmosphere. Figure 1 shows the various sources of lunar atmosphere and thus of lunar volatiles that could become trapped in the subsurface.

Several sources have been suggested for gas venting from depth to the lunar surface. Deep fractures may have resulted from basin-forming impacts, or from the formation of smaller craters above a previously overpressured zone. Although these fractures are normally "closed," gas pressures could temporarily force them open during deep moonquakes and HFTs, allowing the gases to make their way to the surface (*Middlehurst,* 1977). The question remains as to whether elements can exist in the gas phase at the temperatures and pressures in the lunar lithosphere where the HFTs originate. Because gases such as H and He are small compared to other elements, even "solid" rock is permeable to them to some extent. It is also unknown how much gas would be liberated by a moonquake or an impact event.

Volatiles may be produced by radioactive decay within the interior and/or from magma crystallization. If the escaping gases arrive on the lunar night surface, the condensible gases (CH_4, NH_3, CO_2, and Ar, for example) will adsorb to the surface/subsurface grains. If they arrive upon the daylight side, they may be released immediately into space (*Friesen,* 1975). Gases that are adsorbed to the regolith during the lunar night will be heated and mobilized upon the crossing of the dawn terminator. At daybreak, the adsorbed gases will either escape into space, or assume ballistic trajectories which will carry them to a new cold trap, either on the night side or at the poles (see Fig. 1) (*Hoffman et al.,* 1975).

The lunar regolith has also acquired volatiles as a result of solar-wind bombardment (e.g., *Zeller and Dreschoff,* 1970). These volatiles include H, He, C, N, O, Ne, Ar, Kr, and many others in trace amounts. Calculations done by *Carter* (1985) further showed that solar wind hydrogen may have significant resource potential. The "gardening" effect that occurs during meteorite bombardment reworks the regolith layers, thereby burying older soils under fresh debris (e.g., *Taylor,* 1982). As a result, gas-rich particles may exist to an unknown depth in the lunar subsurface.

The Apollo drill core shows an irregularly varying amount of solar-wind-implanted volatiles with increasing depth rather than monotonic change with depth (*Heymann et al.,* 1978; J. Jordan, personal communication, 1989). We do not yet know the vertical extent of this gardening effect, which mixes immature, gas-poor soils with more mature, gas-rich soils to create the uneven pattern. It could extend throughout the entire thickness of the regolith, to a depth of 20 m or more, and it might also be found in paleoregoliths buried to depths of hundreds of meters (e.g., *Sharpton and Head,* 1982).

Volcanic outpourings of lava were very common in the Moon's early history, with vast lava plains now covering roughly 17% of its surface (*Taylor,* 1982). While most of these lavas are believed to range in age from 3.85 to 2.5 b.y., it has been shown that some flows may be as young as 1.0 b.y. and others older than 4.0 b.y. (*Schultz and Spudis,* 1983). It is possible that mature regoliths were covered by these basalts. If so, this suggests the possibility of source and reservoir beds in the lunar subsurface that have been capped by less porous and less permeable mare basalts. The thickness of these mare basalts is estimated to range from tens of meters to 2.5 km (*Taylor,* 1982).

It has been established that temperatures of 600°C or higher can cause outgassing of a substantial amount of the H and He from lunar regoliths (*Pepin et al.,* 1970; *Gibson and Johnson,* 1971; *Gibson and Hubbard,* 1972; *Gibson et al.,* 1974; *Carter,* 1985). Thus a lunar lava flowing across the surface could easily cause significant outgassing of the regolith below it, while at the same time preventing the gases from escaping into space by acting as a crust (vertical barrier) between the regolith and the lunar atmosphere. Eruption rates of 8×10^4 m^3/sec were estimated for the Rima Mozart flows, with a duration of about 900 days (*Coombs et al.,* 1988). While the gas might initially bubble through a thin lava layer, the formation of a crust will retard heat loss and will also impede further loss of volatiles. Thus the effect of ancient lunar lava flows and hot impact ejecta blankets on regolith outgassing may be of interest in resource exploration (J. L. Carter, personal communication, 1989).

POSSIBLE STRUCTURAL AND STRATIGRAPHIC TRAPS IN THE LUNAR SUBSURFACE

Coombs et al. (1988) produced a fence diagram for the Rima Mozart region of the Moon, which is near the Apollo 15 landing site (Fig. 2). Rima Mozart is a lava channel which cuts through a diverse geological setting. Mare basalts and pyroclastics are superposed on units of the Apennine Bench formation, Imbrium ejecta, and Serenitatis ejecta. Construction of Rima Mozart began with an explosive eruption at Kathleen and turbulent lava began carving the rille to the southeast. Explosive eruptions at Ann also deposited pyroclastic material, and resulted in the formation of a shallow, secondary rille that is joined to the main channel near Michael, a sink crater. The distal end of Rima Mozart is embayed by "Lacus Mozart," a mare pond (*Coombs et al.,* 1988). The pyroclastic activity associated with the formation of Rima Mozart supports the idea that volatiles might be present in the region.

An undulating topography is shown in cross-section in Fig. 2. The region consists of the Apennine Bench formation overlain by a less porous and less permeable mare basalt, and underlain by Imbrium ejecta material that may be more porous (e.g., *Coombs et al.,* 1988). If variations in porosity or permeability

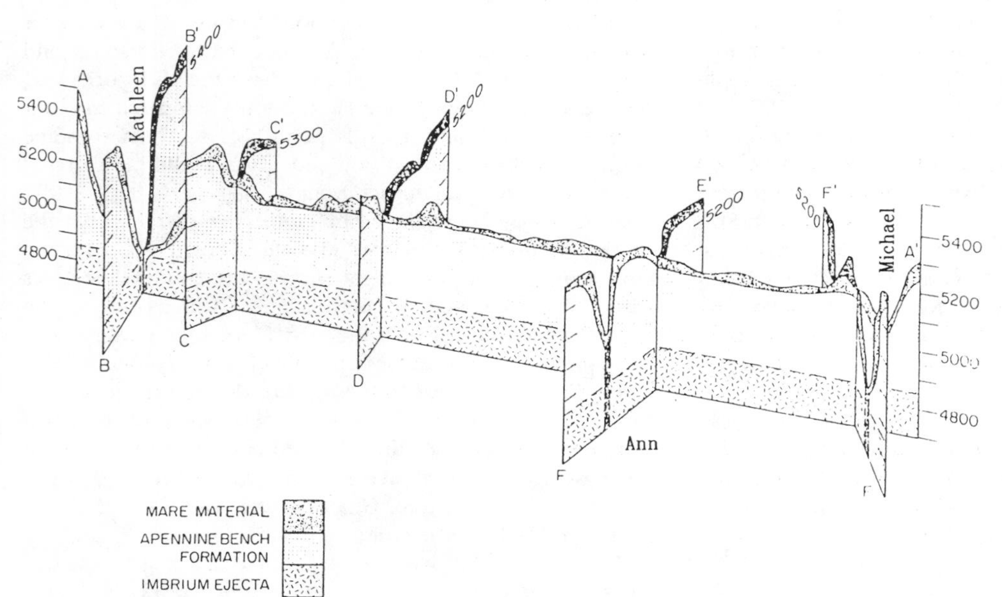

Fig. 2. Fence diagram of Rima Mozart structure, suggesting significant geologic/topographic relief in this area of the Moon near the Apollo 15 landing site. The entire region is covered by a layer of mare basalt that varies in thickness from 0 to 50 m. Vertical exaggeration is roughly 10× (from *Coombs et al.,* 1988).

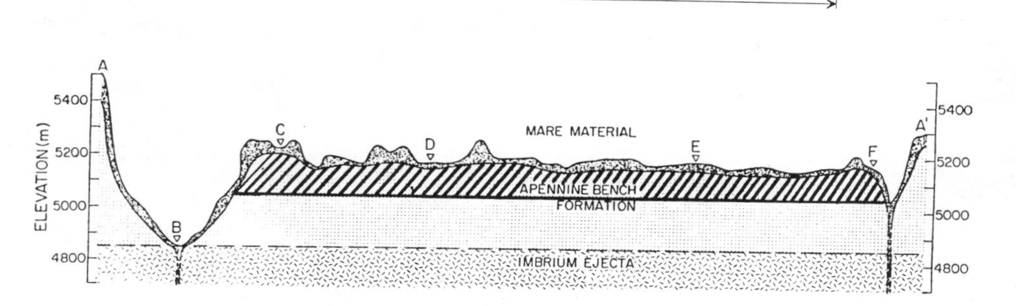

Fig. 3. Cross-section of Rima Mozart structure (modified from *Coombs et al.,* 1988). Cross-hatched area represents a possible gas-trap configuration resulting from structural closure by relatively impermeable mare basalt superposed on the feature. This section was chosen because it gives an idea of the scale of structures that might be expected on the Moon.

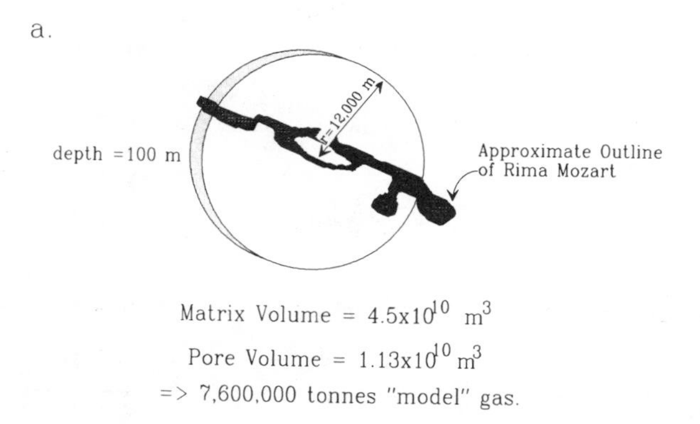

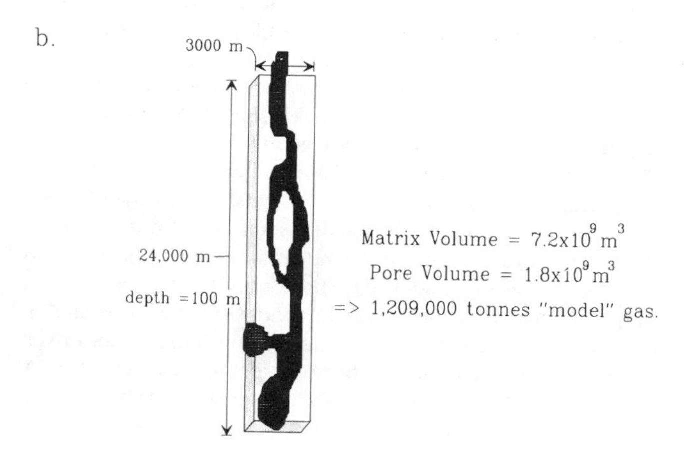

Fig. 4. Volumes of disk- and tablet-shaped reservoirs on the Rima Mozart structure shown in Fig. 3. **(a)** A best-case example for estimating reservoir volume; **(b)** a worst case example for estimating reservoir volume in a gas-trap based on this scenario. In **(a)**, a disk is shown with diameter equal to roughly half the diameter of Rima Mozart, from Kathleen to Michael (see Fig. 2). The irregular gray shape indicates the general outline and size of the part of Rima Mozart shown in the previous cross-section (Fig. 2). The thickness of the disk, 100 m, is based on the thickness of the cross-hatched zone from Fig. 2, the conjectured gas reservoir. In **(b)**, the volume calculation for an area with the same length, but narrowly restricted in width, is shown. The thickness, 100 m, is held constant.

exist within the Apennine Bench formation or the Imbrium ejecta, this would seem to be an ideal area for exploration for volatile accumulations. A sizable gas reservoir could exist, for example, in a portion of the Apennine Bench formation as shown by the cross-hatched area in Fig. 3. This 24-km section represents a topographic low along Rima Mozart. Nevertheless, its geometry is used for the reservoir volume calculations that follow because it expresses the scale of possible gas-bearing structures that might be found in similar areas on the Moon.

Figure 4 shows the result of a simplified volume calculation for two possible reservoir geometries of the Rima Mozart structure. Table 1 expresses the result of these and other volume calculations done in this paper. A porosity value of 25% (averaged from *McKay et al.,* 1986) was used to calculate the available pore volume for gases in a buried soil zone or in buried rubble blankets. Partial pressures for He in pore spaces of a lunar breccia were estimated at 3 atmospheres (*Heymann and Yaniv,* 1971) and could be much higher (J. Jordan, personal communication, 1989); however, other lunar-soils (*Heymann et al.,* 1978) yielded an average He partial pressure of 1/13 atm (58.5 torr) in calculations based upon 25% pore volume in the rock. Using these and other low-end values for solar wind gases such as the smaller of many measurements reported by *Epstein and Taylor* (1970, 1973), *Kothari and Goel* (1973), and *Heymann et al.* (1978), a "model" subsurface gas composition was calculated (Table 2). The model assumes that 6.72×10^{-4} tonne/m^3 of solar-wind volatiles are contained within one cubic meter of regolith. The simplifying assumption was then made that if some subsurface volume were devoid of regolith (i.e., in a cavern), then the associated volatile mass would neither increase nor decrease. Table 3 shows the results of tonnage calculations of individual gas species of this "model" gas for the various structures discussed in this paper.

Treating the Rima Mozart reservoir within the Apennine Bench Formation as a disk shape of radius 12,000 m and thickness 100 m results in a reservoir size of 4.52×10^{10} m^3 (Fig. 4a). Treating the reservoir as a rectangular volume 24,000 m $\times$ 3000 m $\times$ 100 m (Fig. 4b) yields a smaller volume of 7.20×10^9 m^3. If one quarter of this smaller estimated volume were pore space, this would represent a gas volume on the order of 1.80×10^9 m^3, or, 1.21×10^6 tonnes "model" gas. This amounts to roughly 3.42×10^5 tonnes O_2, 3.96×10^5

TABLE 1. Reservoir volumes and "model" gas tonnages for various structures.

Structure Discussed	Shape	Size	Formula Used	Volume of Reservoir	Estimated Porosity	Total Volume of Pore Space	Tonnes* "Model" Gas	Drilling Depth
Rima Mozart Magma Chamber		r = 2500 m	$(4/3)\pi r^3$	6.54×10^{10} m³	100%	6.54×10^{10} m³	4.37×10^7	600-800 m
Rima Mozart Magma Conduit		r = 250 m l = 1000 m	$\pi r^2 l$	2.16×10^9 m³	100%	2.16×10^9 m³	1.48×10^6	600-800 m
Lava Tube		r = 340 m l = 750 m	$\pi r^2 l$	2.72×10^8 m³	100%	2.72×10^8 m³	1.83×10^5	65-200 m
Rima Mozart Disk-Shaped Closure		r = 12000 m d = 100 m	$\pi r^2 d$	4.52×10^{10} m³	25%	1.13×10^{10} m³	7.59×10^6	50-100 m
Rima Mozart Tabular Closure		l = 24000 m w = 3000 m d = 100 m	lwd	7.20×10^9 m³	25%	1.80×10^9 m³	1.21×10^6	50-100 m
Flynn Creek Fold Pocket		r = 5 m l = 350 m	$\frac{\pi r^2 l}{2}$	1.37×10^4 m³	100%	1.37×10^4 m³	9.24	100-200 m
Buried Regolith Reservoir		l = 50000 m w = 50000 m d = 2 m	lwd	5.0×10^9 m³	25%	1.25×10^9 m³	8.40×10^5	200-2000 m
Postulated Crater Fill Reservoir		r = 5000 m d_i = 1000 m d_o = 2000 m	$\frac{(4/3)\pi r^2[d_o - d_i]}{2}$	5.23×10^{10} m³	25%	1.31×10^{10} m³	3.51×10^7	50-500 m

*Tonnage calculation, "model" gas: pore volume $\times 6.72 \times 10^{-4}$ mt/m³ = tonnage.

tonnes H_2, 2.52×10^4 tonnes He, 1.58×10^5 tonnes N_2, and 2.88×10^5 tonnes C. The Imbrium ejecta, below the Apennine Bench formation, might provide a more porous and permeable reservoir rock, as shown in Fig. 5, but closure is more difficult to imagine for that formation. Nevertheless, very little is known about the porosity/permeability variations of either formation, thus no immediate conclusions can be drawn.

Figure 5 also shows another possibility for a gas reservoir in the Rima Mozart area. Magma chambers and interconnected conduits are believed to exist below the volcanic vents Kathleen and Ann (*Coombs et al.,* 1988). Assuming a radius of 2500 m, and assuming that the chambers have not been refilled with magma or rock, these postulated chambers would have a total volume of 6.5×10^{10} m³. Results of simple tonnage calculations made for the chamber and conduit are shown in Table 1. Such gas reservoirs were postulated by *Friesen* (1975, 1985).

TABLE 2. "Model" gas calculations for worst-case reservoir volumes.

Element	Source	Values*
He	*Heymann et al.* (1978)	1.4×10^{-5} mt/m³
H_2	*Epstein and Taylor* (1970)	2.2×10^{-4} mt/m³
C	*Epstein and Taylor* (1973)	8.8×10^{-5} mt/m³
N_2	*Kothari and Goel* (1973)	1.6×10^{-4} mt/m³
O_2	Estimated to correspond to total C content based on data given by *Gibson et al.* (1974)	1.9×10^{-4} mt/m³
Total		6.72×10^{-4} mt/m³

*Values used in this calculation represent the lowest values reported by the authors listed.

Lava tubes may exist in the lunar subsurface and may also be reservoir structures. Intact lava tubes may be buried at depths of 65 m or greater, with widths varying from 450 to 970 m (*Coombs and Hawke,* 1988). Intact surface segments have lengths of 500 to 1000 m (*Coombs and Hawke,* 1988). Table 1 shows a calculation for one such postulated enclosure of average size. A tube this size could contain as much as 183,000 tonnes of "model" gas, including 51,700 tonnes of O_2. This amounts to 34,367% of the yearly baseline production goal of 150 tonnes of O_2 per year for the initial years of the lunar outpost (D. S. McKay, personal communication, 1989). It would be necessary to find and successfully drill only one such structure to meet the total O_2 goal, if indeed such a sealed underground chamber exists, and if it also contains gases in the pressures supposed here.

It has been shown that 5 MHz radar data combined with geological data can be used to model the relatively shallow (<1000 m) lunar stratigraphy, and this technique was used in the Mare Serenitatis area to suggest the existence of two buried regolith layers, each at least 2 m thick (*Sharpton and Head,* 1982). A cross-section of the region was created that showed the possible existence of porosity/permeability "pinch-outs" that might act as gas-traps (Fig. 6). A closure of areal extent 50,000 m × 50,000 m would represent a pore volume of 1.25×10^9 m³, containing a total of 2.38×10^5 tonnes O_2, 2.75×10^5 tonnes H_2, 1.75×10^4 tonnes He, 1.1×10^5 tonnes

TABLE 3. "Model" gas species tonnages for various structures.

Structure Discussed	Shape	Total Volume of Pore Space	Tonnes* Oxygen	Tonnes* Hydrogen	Tonnes* Helium	Tonnes* Nitrogen	Tonnes* Carbon	Total Tonnes "Model"	Drilling Depth
Rima Mozart Magma Chamber	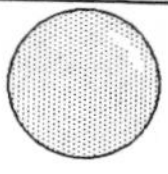	6.54×10^{10} m^3	1.24×10^7	1.43×10^7	9.10×10^5	5.72×10^6	1.04×10^7	4.37×10^7	600-800 m
Rima Mozart Magma Conduit	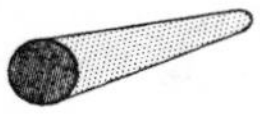	2.16×10^9 m^3	4.18×10^5	4.84×10^5	3.08×10^4	1.94×10^5	3.52×10^5	1.48×10^6	600-800 m
Lava Tube	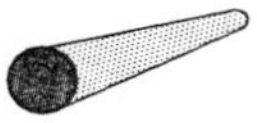	2.72×10^8 m^3	5.17×10^4	5.98×10^4	3.81×10^3	2.39×10^4	4.35×10^4	1.83×10^5	65-200 m
Rima Mozart Disk-Shaped Closure		1.13×10^{10} m^3	2.15×10^6	2.49×10^6	1.58×10^5	9.94×10^5	1.81×10^6	7.59×10^6	50-100 m
Rima Mozart Tabular Closure		1.80×10^9 m^3	3.42×10^5	3.96×10^5	2.52×10^4	1.58×10^5	2.88×10^5	1.21×10^6	50-100 m
Flynn Creek Fold Pocket	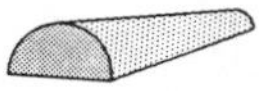	1.37×10^4 m^3	2.61	3.02	1.92×10^{-1}	1.21	2.20	9.24	100-200 m
Buried Regolith Reservoir		1.25×10^9 m^3	2.38×10^5	2.75×10^5	1.75×10^4	1.10×10^5	2.00×10^5	8.40×10^5	200-2000 m
Postulated Crater Fill Reservoir		1.31×10^{10} m^3	9.94×10^6	1.15×10^7	7.32×10^5	4.60×10^6	8.37×10^6	3.51×10^7	50-500 m

*Tonnage calculation based on data from Table 2.

N_2, and 2.00×10^5 tonnes C. Furthermore, these tonnage values could be lowered by three orders of magnitude and still represent a significant and possibly economic quantity of subsurface gas.

We must have a reservoir of sufficient size before it is of interest. Drilling in terrestrial impact craters has revealed much about the structures associated with these features, including the existence of gas cavities (*Hörz et al.,* 1977; *Roddy,* 1979). Figure 7 shows a view of the eastern crater wall of the Flynn Creek impact crater in Tennessee, and the location of a cave that was formed by impact deformation. If the volume of this cave is modeled as a half-cylinder of radius 5 m and length 0.1 radian (Table 1), it would yield a volume of 13,700 m^3. While this is quite small by terrestrial standards, it nevertheless amounts to 1.7% of the annual O_2 baseline production goal for the lunar outpost (D. S. McKay, personal communication, 1989). However, considering that it would be necessary to find and successfully drill about 58 such structures per year to meet the total O_2 goal (*McKay,* 1989), this does not seem to be a likely candidate for further study. The existence of a cavity does not necessarily imply that significant quantities of volatiles are contained within it.

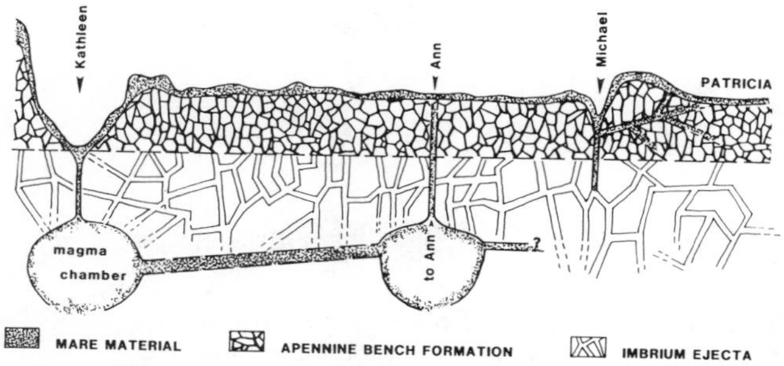

Fig. 5. Magma chambers, magma conduits, and lava tubes may also serve as reservoirs. If a large-scale closure existed, then any of these structures might serve as a large-volume gas-trap. See Table 1 for calculations. (From *Coombs et al.,* 1988.)

Other kinds of reservoirs on the Moon can be imagined. Figure 8 shows a sequence of events that might lead to a gas reservoir within a crater. After a crater is formed, normal crater degradational processes will form a regolith in the crater basin (*Head,* 1975). This regolith would also experience implantation of volatiles due to the solar wind. At some later time, a volcanic eruption could bury the regolith under a layer of nonporous and nonpermeable basalt. At that point, the regolith, acting as a source and reservoir bed, would be virtually enclosed on all sides. Gases that had been liberated by the volcanic heating would have no easy escape, and gases generated at depth by the crystallization of magma might also be emplaced in the reservoir. For reservoirs less than 20 km deep, these conduits might remain open on a continuous or semicontinuous bases.

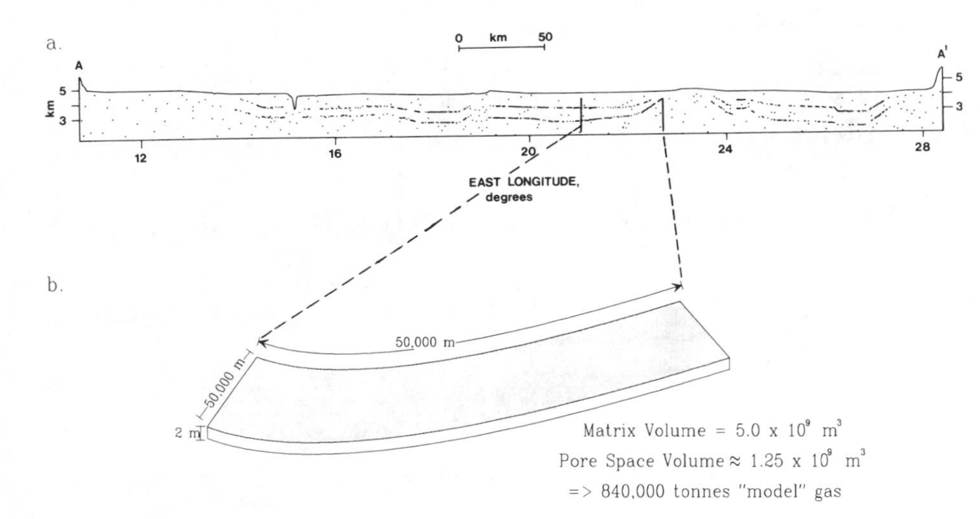

Fig. 6. A possible stratigraphic "pinch-out" gas-trap in the southern Mare Serenitatis region. The subsurface structure from 400 m to 1000 m has been reconstructed based on 5 MHz radar data (*Sharpton and Head,* 1982). The enhanced radar-reflective layers (darker colored zones in this sketch) are probably paleoregoliths of at least 2 m in thickness (*Peeples et al.,* 1978).

TECHNOLOGY READINESS/FURTHER STUDY REQUIRED

Major technology requirements for use of these postulated lunar subsurface volatiles are (1) a better understanding of the geology of the lunar subsurface, (2) an appropriate lunar prospecting and exploration method, and (3) production technology appropriate for the depth, pressure, and other unusual drilling conditions that may be encountered both on and below the lunar surface.

A positive temperature gradient with depth might act as a vertical barrier causing gases to preferentially collect in a cooler zone (J. L. Carter, personal communication, 1989). Calculations need to be done to determine if the effect of a vertical temperature gradient could cause gases to accumulate in significant quantities to be of practical interest.

It has been shown that cavities in the Earth's subsurface may be detected by seismic resonance phenomena (*Watkins et al.,*

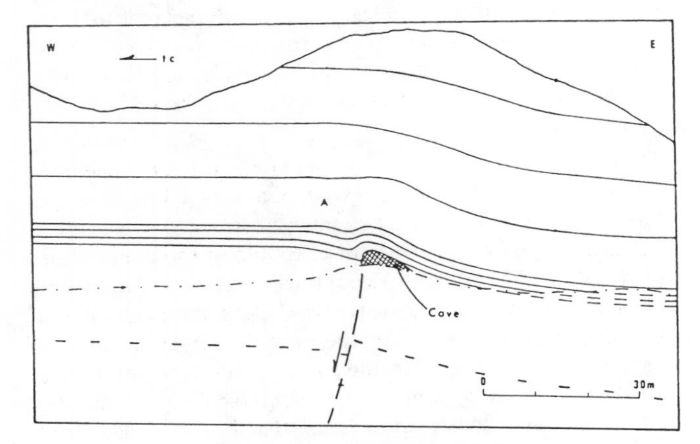

Fig. 7. Cave in wall of Flynn Creek impact crater, Tennessee. In this cross-section of a terrestrial impact crater, a cavity has been formed due to overfolding of adjacent rock layers. Many possibilities exist on Earth for cavern formation due to thrust faulting and to shock deformation that may also exist on the Moon. See text of *Roddy* (1979) for further discussion.

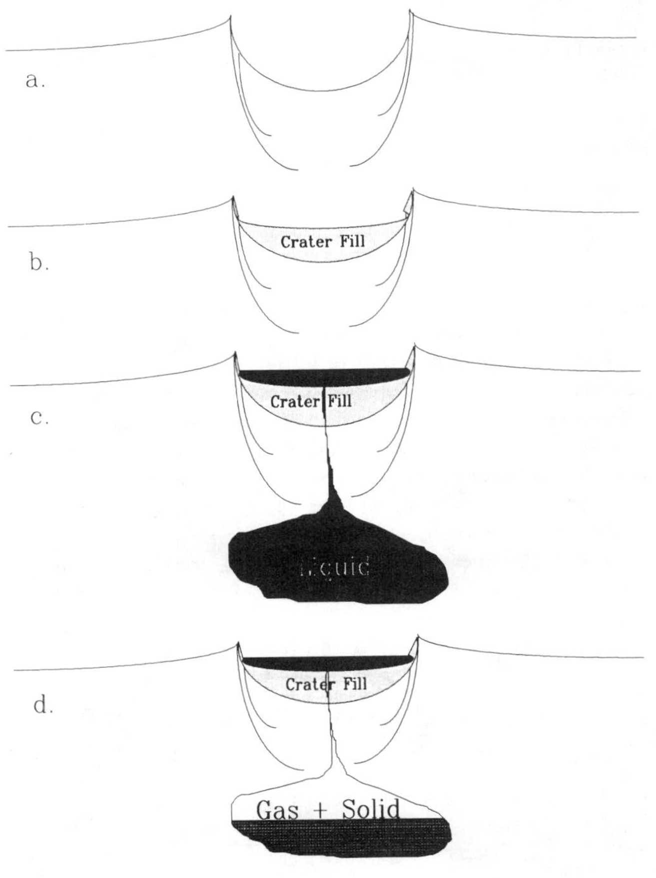

Fig. 8. A possible scenario for the formation of large basin reservoirs. **(a)** A bowl-shaped crater less than 15 km in diameter is formed. **(b)** The crater is partially filled with regolith. **(c)** Volcanic eruption covers the area with flood basalts and heats regolith below. **(d)** Gases from buried regolith and from magma crystallization may be trapped within crater fill material or within the magma chamber. Faults generated by basin formation serve as natural gas conduits.

1967). This suggests the same calculations could be applied to search for lunar cavities. However, seismic reflections on the Moon are significantly different in form and appearance from those on Earth (*Meissner et al.,* 1970). To my knowledge, appropriate lunar seismic models have not been determined that would allow the detection of relatively shallow (<1000 m) and relatively small (<10-km-diameter) subsurface structures. This step is critical for exploration planning, and addresses points 1 and 2 above.

It has been noted that the Moon exhibits a "ringing" phenomenon. Lunar seismic signals reverberate for a much longer duration than events of similar magnitude on Earth (*Meissner et al.,* 1970). The explanation for this has been that the highly fractured nature of the regolith and upper crust causes a great deal of seismic scattering (*Kovach and Watkins,* 1973). It is also thought that the absence of water and volatile constituents allows the seismic signals to propagate with little or no attenuation (*Kovach and Watkins,* 1973). Thus, gas reservoirs should be easily detectable by seismic exploration, as they would show different attenuation properties. The variation in measured Q (internal friction quality factor) should be an almost direct reading of the amount of entrapped water or hydrocarbon gas in a rock, whereas Q is unaffected by the presence of H_2, He, N_2, O_2, CO, or CO_2 (*Tittman et al.,* 1973). Velocity variations indicate the presence or lack of substantial interstitial pressure between rock fragments (*Taylor,* 1982). High-velocity materials have been observed on the Moon below basalt layers (*Kovach and Watkins,* 1973) that might be an indication of trapped volatiles. Further understanding should be gained about the possible variation in lunar seismic signals of gas-pressured reservoirs vs. dense, nonporous rock.

The planned Lunar Orbiter is expected to carry a radar altimeter and possibly other radio science instruments (*Solar System Exploration Committee,* 1986). Considering the usefulness of the 5 MHz radar data in subsurface investigations (e.g., *Sharpton and Head,* 1982), it would be very useful if radar coverage near the 5 MHz band were extended across the lunar surface.

The Apollo drill cores show an irregular amount of volatiles at various depths as opposed to a monotonic change with depth (*Heymann et al.,* 1978; J. Jordan, personal communication, 1989). This is probably due to gardening and different surface residence times for different soils (*Heymann et al.,* 1978). A drilling exploration program needs to be developed for a lunar-outpost precursor mission to establish the true depth to which this effect exists. Such drilling should be integrated with a plan to learn more about the variation in subsurface structure across the nearside. Coring would thus be preferred over drilling in the early research phases.

Rowley and Neudecker (1985) have developed an excavation technology based upon melting rock and soil. Prototype hardware has been fabricated, and laboratory and field tests have been performed. The excavation method is relatively insensitive to soil or rock types and can form in situ glass linings or casings in the walls of boreholes. If coring were required, then a different type of bit would be used. It is likely that penetration speed varies with bit design in this system because bit design has a large impact on penetration rate with conventional rotary drilling systems (*Roepke,* 1975). Electrical energy is used for resistive heaters for the melting penetrators, although direct heating by nuclear power is possible for larger equipment. This exploration and production technology should be carefully compared and contrasted to conventional drilling techniques. Such a calculation could rate one technique against the other for technology readiness, risk assessment, cost, energy, and astronaut-tending requirements as well as survivability in the dusty lunar vacuum.

A multidisciplinary exploration approach is envisioned, in which the first step would be orbital surveying (e.g., *Friesen,* 1985) and combined interpretation of all available data. Later this would give way to surface exploration, such as a traverse over an area of interest (e.g., *Cintala et al.,* 1985). Eventually, surface exploration would give way to subsurface exploration, i.e., drilling.

Before exploration can begin, calculations need to be made regarding the logistics and operation of a drilling program on the Moon. For example, what depth, on average, would need to be attained before we could feel reasonably confident that no realistic opportunity had been overlooked? Secondly, how much time and energy would it take to achieve that depth in an average case? Also, how could interstitial gas be detected from the surface? And, what pressure-sealing technology could be used to block the hole and prevent blowouts?

Finally, a number of geological questions will have to be answered before a thorough exploration program can be planned. It will be important to know what regional structural features should be considered as areas of interest for more detailed exploration. The size and type of structure sought should be more clearly understood. An estimate could then be made of how many features of that size potentially exist. Criteria could also be developed for narrowing the search to the most promising areas based on knowledge gained thus far. These and other specific calculationas and measurements would aid in the planning of an exploration program.

CONCLUSIONS

Although the direct evidence is not overwhelming, the presence of structural and stratigraphic gas-traps on the Moon is a logical inference from the facts at hand. Our knowledge of terrestrial and structural gas-traps on Earth suggests that the basic occurrence of a reservoir rock overlain by a nonporous and nonpermeable cap-rock with a source for those gases, should result in gas-traps. Reservoir rocks, porosity/permeability barriers, and sources for volatiles all exist on the Moon, and further exploration should be done to characterize these possibilities.

Acknowledgments. This work has benefited greatly by numerous discussions with J. Carter, M. Cintala, J. Dailey, and J. Jordan. E. Bayley also provided valuable insights. This work was funded in part by Brown and Root, U.S.A. Thanks also to C. Coombs and J. Jordan for their reviews of this paper. This paper is Contribution No. 639 of the Geosciences Program at the University of Texas at Dallas.

REFERENCES

Blanford G. E. (1986) Particle track measurements in lunar regolith breccias (abstract). In *Workshop on Past and Present Solar Radiation: The Record in Meteoritic and Lunar Regolith Material* (R. O. Pepin and D. S. McKay, eds.), pp. 14-17. LPI Tech. Rpt. 86-02, Lunar and Planetary Institute, Houston.

Cameron W. S. (1977) Lunar transient phenomena (LTP): Manifestations, site distribution, correlations and possible causes. *Phys. Earth Planet. Inter., 14,* 194-216.

Carter J. L. (1985) Lunar regolith fines: A source of hydrogen. In *Lunar Bases and Space Activities of the 21st Century* (W. W. Mendell, ed.), pp. 571-581. Lunar and Planetary Institute, Houston.

Christiansen E. L. (1988) Conceptual design of a lunar oxygen pilot plant. *NASA NAS9-17878,* EEI Report 88-182, Eagle Engineering, Houston. 260 pp.

Cintala M. J., Spudis P. D., and Hawke B. R. (1985) Advanced geologic exploration supported by a lunar base: A traverse across the Imbrium-Procellarum region of the Moon. In *Lunar Bases and Space Activities of the 21st Century* (W. W. Mendell, ed.), pp. 223-237. Lunar and Planetary Institute, Houston.

Cook G. A. (1961) *Argon, Helium and the Rare Gases Volume 1: The Elements of the Helium Group.* Interscience, New York. 393 pp.

Coombs C. R. and Hawke B. R. (1990) A survey of lunar rilles: The search for intact lava tubes (abstract). In *Papers Presented to the 1988 Symposium on Lunar Bases and Space Activities of the 21st Century,* p. 58. Lunar and Planetary Institute, Houston.

Coombs C. R., Hawke B. R., and Wilson L. (1988) Geologic and remote sensing studies of Rima Mozart. *Proc. Lunar Planet. Sci. Conf. 18th,* pp. 339-353.

Elworthy R. T. (1926) *Helium in Canada.* Canada Department of Mines, Ottawa. 64 pp.

Epstein S. and Taylor H. P. Jr. (1970) The concentration and isotopic composition of hydrogen, carbon and silicon in Apollo 11 lunar rocks and minerals. *Proc. Apollo 11 Lunar Sci. Conf.,* pp. 1085-1096.

Epstein S. and Taylor H. P. Jr. (1973) The isotopic composition of water, hydrogen, and carbon in some Apollo 15 and 16 soils and in the Apollo 17 orange soil. *Proc. Lunar Sci. Conf. 4th,* pp. 1559-1575.

Friesen L. J. (1975) Volatile emission on the Moon: Possible sources and release mechanisms. *The Moon, 13,* 425-430.

Friesen L. J. (1985) Search for volatiles and geologic activity from a lunar base. In *Lunar Bases and Space Activities of the 21st Century* (W. W. Mendell, ed.), pp. 239-244. Lunar and Planetary Institute, Houston.

Geake J. E. and Mills A. A. (1977) Possible physical processes causing transient lunar events. *Phys. Earth Planet. Inter., 14,* 299-320.

Gibson E. K. Jr. and Hubbard N. J. (1972) Volatile element depletion investigations on Apollo 11 and 12 lunar basalts via thermal volatilization (abstract). In *Lunar Science III,* pp. 303-305. The Lunar Science Institute, Houston.

Gibson E. K. Jr. and Johnson S. M. (1971) Thermal analysis-inorganic gas release studies of lunar samples. *Proc. Lunar Sci. Conf. 2nd,* pp. 1351-1366.

Gibson E. K. Jr., Moore G. W., and Johnson S. M. (1974) *Summary of analytical data from gas release investigations, volatilization experiments, elemental abundance measurements on lunar samples, meteorites, minerals, volcanic ashes and basalts.* NASA Johnson Space Center, Houston.

Gorenstein P., Golub L., and Bjorkholm P. (1973) Radon emanation from the Moon, spatial and temporal variability. *The Moon, 9,* 129-140.

Haskin L. A. (1989) The Moon as a practical source of hydrogen and other volatile elements (abstract). In *Lunar and Planetary Science XX,* pp. 387-388. Lunar and Planetary Institute, Houston.

Head J. W. (1975) Processes of lunar crater degradation: changes in style with geologic time. *The Moon, 12,* 299-329.

Heymann D. and Yaniv A. (1971) Breccia 10065: Release of inert gases by vacuum crushing at room temperature. *Proc. Lunar Sci. Conf. 2nd,* pp. 1681-1692.

Heymann D., Jordan J. L., Walker A., Dziczkaniec M., Ray J., and Palma R. (1978) Inert gas measurements in the Apollo-16 drill core and an evaluation of the stratigraphy and depositional history of this core. *Proc. Lunar Planet Sci. Conf. 9th,* pp. 1885-1912.

Hodges R. R. Jr., Hoffman J. H., Yeh T. T. J., and Chang G. K. (1972) Orbital search for lunar volcanism. *J. Geophys. Res., 77,* 4079-4085.

Hoffman J. H. and Hodges R. R. Jr. (1975) Molecular gas species in the lunar atmosphere. *The Moon, 14,* 159-167.

Hoffman J. H., Hodges R. R. Jr., and Johnson F. S. (1975) Lunar atmospheric composition results from Apollo 17. *Proc. Lunar Sci. Conf. 4th,* pp. 2865-2875.

Hörz, F., Gall H., Huttner R., and Oberbeck V. R. (1977) Shallow drilling in the 'Bunte Breccia' impact deposits, Ries Crater, Germany. In *Impact and Explosion Cratering* (D. J. Roddy, R. O. Pepin, and R. B. Merrill, eds.), pp. 425-448. Pergamon, New York.

Howe S. D., Hughes K. J., Johnson S. W., Leigh G. G., and Leonard R. S., eds. (1989) *A Center for Extraterrestrial Engineering and Construction (CETEC).* BDM Corporation, Albuquerque. 126 pp.

Keesom W. H. (1942) *Helium.* Elsevier, New York. 494 pp.

Kothari B. K. and Goel P. S. (1973) Nitrogen in lunar samples. *Proc. Lunar Sci. Conf. 4th,* pp. 1587-1596.

Kovach R. L. and Watkins J. S. (1973) Apollo 17 seismic profiling: Probing the lunar crust. *Science, 180,* 1063-1064.

Kozyrev M. A. (1959) Observation of a volcanic process on the Moon. *Sky and Telescope, 18,* 184-186.

Mattill J. (1977) The short and the long of helium. *Technol. Rev., 89,* 77.

McKay D. S., Bogard D. D., Morris R. V., Korotev R. L., Johnson P., and Wentworth S. J. (1986) Apollo 16 regolith breccias: Characterization and evidence for early formation in the mega-regolith. *Proc. Lunar Planet Sci. Conf. 16th,* in *J. Geophys. Res., 91,* D277-D303.

Meissner R., Sutton G. H., and Duennebier F. K. (1970) Moonquakes (abstract). *Eos Trans. AGU, 51,* 776.

Middlehurst B. M. (1977) A survey of lunar transient phenomena. *Phys. Earth Planet. Inter., 14,* 185-193.

Peeples W. J., Sill W. R., May T. W., Ward S. H., Phillips R. J., Jordan R. L., Abbott E. A., and Kilpatrick T. J. (1978) Orbital radar evidence for lunar subsurface layering in Maria Serenitatis and Crisium. *J. Geophys. Res., 83,* 3459-3470.

Pepin R. O., Nyquist L. E., Phinney D., and Black D. C. (1970) Rare gases in Apollo 11 lunar material. *Proc. Apollo 11 Lunar Sci. Conf.,* pp. 1435-1454.

Roddy D. J. (1979) Structural deformation at the Flynn Creek impact crater, Tennessee: A preliminary report on deep drilling. *Proc. Lunar Planet Sci. Conf. 10th,* pp. 2519-2534.

Roepke W. W. (1975) Experimental system for drilling simulated lunar rock in ultrahigh vacuum. NASA Contract #R-09-040-001, U.S. Department of the Interior, Bureau of Mines, Minneapolis. 12 pp.

Rowley J. C. and Neudecker J. W. (1985) In-situ rock melting applied to lunar base construction and for exploration drilling and coring on the Moon. In *Lunar Bases and Space Activities of the 21st Century* (W. W. Mendell, ed.), pp. 465-477. Lunar and Planetary Institute, Houston.

Schultz P. H. and Spudis P. D. (1983) Beginning and end of lunar mare volcanism. *Nature, 302,* 233-236.

Sharpton V. L. and Head J. W. III (1982) Stratigraphy and structural evolution of southern Mare Serenitatis: A reinterpretation based on Apollo lunar sounder experiment data. *J. Geophys. Res., 87,* 10,983-10,998.

Solar System Exploration Committee of the NASA Advisory Council (1986) *Planetary Exploration through the Year 2000: An Augmented Program.* U.S. Govt. Printing Office, Washington, DC. 239 pp.

Taylor S. R. (1982) *Planetary Science: A Lunar Perspective.* Lunar and Planetary Institute, Houston. 481 pp.

Tittman B. R., Housley R. M., and Cirlin E. H. (1973) Internal friction of rocks and volatiles on the moon. *Proc. Lunar Sci. Conf. 4th,* pp. 2631-2637.

Watkins J. S., Godson R. H., and Watson K. (1967) Seismic detection of near-surface cavities. In *Contributions to Astrogeology,* U.S. Geol. Surv. Prof. Pap. #599-A. U.S. Govt. Printing Office, Washington, DC. 16 pp.

Weast R. D., ed. (1971) *Handbook of Chemistry and Physics.* CRC, Cleveland. 2042 pp.

Zeller E. J. and Dreschoff G. (1970) Chemical alterations resulting from proton irradiation of the lunar surface. *Mod. Geol., 1,* 141-148.

Petrology and Geochemistry of Achondrites

Proceedings of the 20th Lunar and Planetary Science Conference, pp. 273-280
Lunar and Planetary Institute, Houston, 1990

Complex Petrogenesis of the Nakhla (SNC) Meteorite: Evidence from Petrography and Mineral Chemistry

A. H. Treiman

Geology Department, Boston University, Boston, MA 02215

The Nakhla (SNC) meteorite, of putative martian origin, consists of euhedra and subhedra of augite and olivine (0.5-4 mm dia.) in a fine-grained basaltic mesostasis. The augites have been interpreted as phenocrysts; the olivines have been interpreted variously. Both mineral species have strong and similar compositional zonings, concentric from centers to edges. Euhedral to anhedral cores of both minerals are of relatively constant Mg/Fe, with irregularly variable minor element contents. Olivine core compositions vary, but average Mg/Fe = 0.65; augite cores average Mg/Fe = 1.7. Rim zones are of smoothly varying compositions, becoming enriched outward in more incompatible elements (like Fe). Most of the rim zoning is consistent with growth during Rayleigh fractionation. Overgrowths beyond the rims are of constant, evolved composition. Some augites contain ferroan replacement zones that transsect the other zones. Their similar zoning patterns suggest that the olivine and augite grains had similar chemical and thermal histories, and so preclude hypotheses that require different origins for the two minerals. However, the compositions of the augites' and olivines' cores are not near chemical equilibrium under magmatic conditions; only if the cores grew at gross disequilibrium is it possible that they were comagmatic. The Mg/Fe distribution between the cores is consistent with subsolidus equilibration, so the cores may represent xenocrysts from a partially equilibrated source rock. It is also possible that the cores represent phenocrysts from separate magmas, which mixed before the formation of Nakhla. All of the permitted hypotheses for the formation of Nakhla involve complex processes, so Nakhla can not be treated as a simple, closed-system igneous rock.

INTRODUCTION

The Nakhla meteorite, an igneous rock rich in augite and olivine, is one of the "SNC" meteorites of probable martian origin (*Wood and Ashwal*, 1981; *Bogard et al.*, 1984). Nakhla is generally considered to be a quickly cooled cumulate rock (*Bunch and Reid*, 1975; *Reid and Bunch*, 1975; *Berkley et al.*, 1980; *Nakamura et al.*, 1982; *Treiman*, 1986, 1989a; *Longhi and Pan*, 1989), possibly emplaced in a thick flow or shallow sill (*Treiman*, 1987). But the formation of olivine in Nakhla has been unclear; it has been argued that the olivines are all postcumulus (*Reid and Bunch*, 1975; *Berkley et al.*, 1980), xenocrysts with postcumulus overgrowths (*Treiman*, 1986), phenocrysts with postcumulus overgrowths (*Treiman*, 1989a), and partially re-equilibrated phenocrysts with overgrowths (*Nakamura et al.*, 1982; *Longhi and Pan*, 1989).

Understanding the origin of the olivine is important for understanding Nakhla itself, and I have collected petrographic and mineral chemical data to help distinguish among these hypotheses. To anticipate the conclusions, the pyroxene "phenocrysts" and olivine "megacrysts" in Nakhla experienced very similar chemical and magmatic events, but their cores could not have approached equilibrium under magmatic conditions. The best available hypothesis is that the cores of both the olivines and augites are xenocrysts although it is possible that both cores are phenocrysts, grown either at gross disequilibrium or in different magmas.

METHODS

Chemical analyses were obtained by electron microprobe using wavelength dispersive X-ray analysis. Most of the analyses were obtained with the JEOL 733 "Superprobe" in the laboratory of Dr. J. Wood, Harvard-Smithsonian Astrophysical Observatory, Cambridge, MA. Some olivine analyses were obtained with the ARL SEMQ in the laboratory of Dr. M. Drake, Lunar and Planetary Laboratory, University of Arizona, Tucson, AZ. All analyses were with a focused electron beam, current of 20 namp in a Faraday cup, at a potential of 15 kV. For each element, X-rays were collected for 35 to 60 sec, and K-ratios reduced using "ZAF" correction procedures. Standards were well-documented minerals and synthetic glasses.

PETROGRAPHY OF NAKHLA

Only a synopsis of Nakhla's petrography is needed; more complete data are in *Bunch and Reid* (1975), *Reid and Bunch* (1975), *Berkley et al.* (1980), and *Treiman* (1986). Nakhla is a clino-pyroxenite, containing ~80% subcalcic augite, ~5% olivine, and ~15% fine-grained "mesostasis" material of thin radiating laths of plagioclase, pyroxenes, olivine, magnetite, and other species (Figs. 1 and 2; *Bunch and Reid*, 1975; *Berkley et al.*, 1980). Shock effects are minor, but see *Lambert* (1987) for an alternate interpretation.

Clinopyroxene, here subcalcic augite, is present as euhedra and subhedra set in mesostasis material and poikilitic olivine grains (Fig. 1a). Augite grains average 0.5 mm across (up to 1 mm), and have a weak preferred orientation (*Berkley et al.*, 1980). Cores zones, euhedral to subhedral in outline, are relatively homogeneous (see below and *Berkley et al.*, 1980) and contain scattered inclusions of glass. Rim zones, present on most augites, are enriched in iron.

Olivine is present in three textures: subhedral grains larger than 0.5 mm across, poikilitic grains enclosing cumulus augites, and minute (<0.05 mm) equant grains in the mesostasis (*Bunch and Reid*, 1975; *Reid and Bunch*, 1975; *Berkley et al.*,

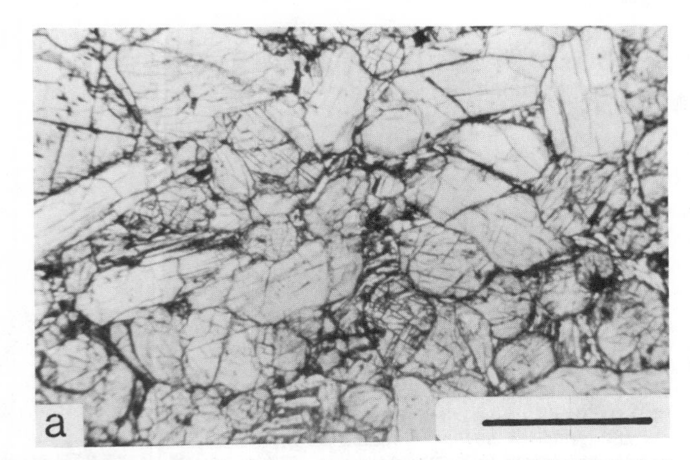

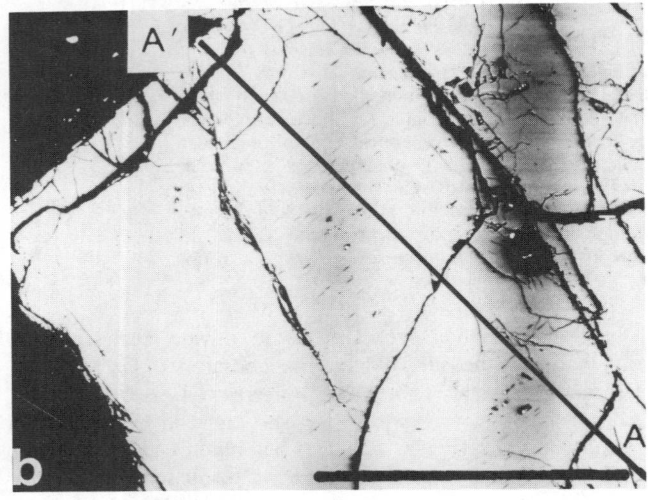

Fig. 1. Nakhla in thin section. **(a)** Typical textures: euhedral augites with overgrowths among dark mesostasis and poikilitic olivine. Plane light, scale bar 1 mm. **(b)** Euhedral olivine; #2 of Harvard Univ. thin section #1. Backscattered electron image, vertical bands are artifacts; scale bar 0.5 mm. A-A′ is location of analytical traverse, Fig. 3.

1980; *Treiman,* 1986). Only the larger subhedral grains are controversial; the poikilitic and mesostasis grains grew after accumulation of the augite crystals.

The larger subhedral olivines are commonly 2-3 times as long as the augites (Fig. 1b; ***Bunch and Reid,*** 1975; ***Nakamura et al.,*** 1982; *Treiman,* 1986). Rare grains are larger still; up to 4 mm by 3 mm (Fig. 2). In thin section, these subhedral olivines form rude squares, rectangles, or rhombs from crystal faces in the [001] zone and the {021} or {101} terminations. Where the olivines abut mesostasis, the crystal faces are sharp (Fig. 1b); olivines become subhedral where they abut augite, and may enclose the augite (Fig. 2). The poikilitic olivines may be the edges of such grains.

Scattered within the olivines are dark inclusions of euhedral augite, Fe $\pm$ Ti oxides and olivine in silica-rich glass (Fig. 2), similar to those in the Chassigny meteorite (*Floran et al.,* 1978). The inclusions are round or elliptical and up to 150 μm across. Within each large olivine crystal, inclusions smaller than 10 μm across are concentrated in a thin zone, commonly delimiting a euhedral to subhedral "ghost crystal" with faces parallel to those of the full crystal (Fig. 1b of *Treiman,* 1986) or in a rounded anhedral zone.

COMPOSITIONAL ZONING: DESCRIPTION

All of the hypotheses about Nakhla's olivine rely critically on interpretation of its chemical zoning, first noted by B. Noyes (*Reid and Bunch,* 1975) and briefly described in *Treiman* (1986). Controversy has centered on whether the zoning was real, was of primary magmatic origin, or reflected late- and postmagmatic diffusional reequilibration.

Choice of properly oriented grains is critical in analysis of chemical zoning, because the apparent zoning pattern depends on the grain's orientation and position relative to the thin section plane (*Pearce,* 1984). Ideally, the grain's center should be in the plane of the section, and the grain's crystal faces should be perpendicular to the thin section. Only for these grains will a measured zoning profile be an undistorted representation of the true zoning pattern.

For olivine, properly oriented grains were identified by their rectangular outlines and large sizes; two such olivines have been analyzed: HU #1.2 (Fig. 1), and USNM426-2.1 (Fig. 2 of *Treiman,* 1986). For augite, properly oriented grains were recognized as having cleavage traces at right angles (thin section perpendicular to crystallographic "c") and regular octagonal cross-sections (not near a crystal termination).

Chemical zoning in mineral grains is usually presented as graphs of a compositional parameter vs. distance, but other spatial parameters may be more useful. In igneous systems, crystal volume may be better than distance because chemical effects of many igneous processes depend on the volumes of phases produced or removed, e.g., crystal growth during Rayleigh fractionation (*Pearce,* 1987). In this case the compositon of crystallizing material (the edge of the growing crystal) is a function of the volume of the system already crystallized. In Rayleigh fractionation growth, abundance ratios

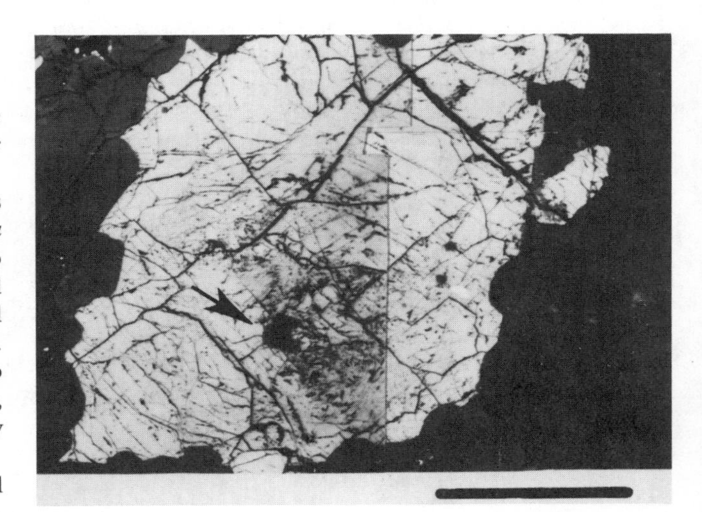

Fig. 2. Largest olivine; Harvard Univ. thin section #2, olivine 2. Composite of backscattered electron images, scale bar 1 mm. Arrow points to magmatic inclusion.

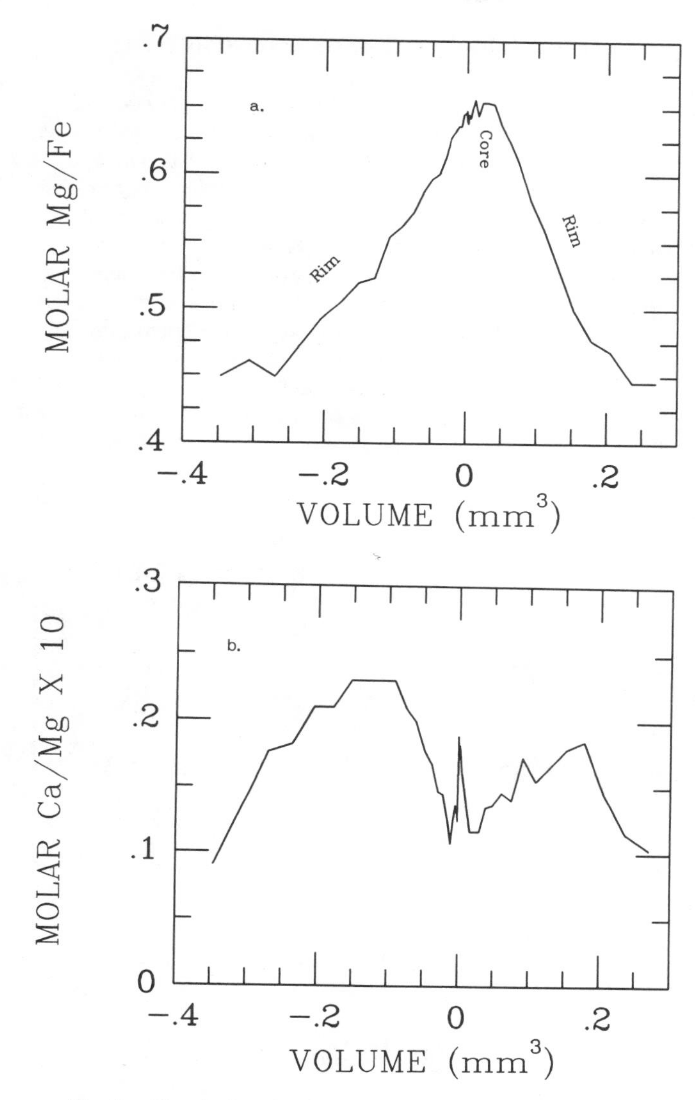

Fig. 3. Chemical zoning in olivine; from electron microprobe traverse shown in Fig. 1b. Molar ratios vs. volume from grain's core (2·distance^3). **(a)** Mg/Fe; **(b)** Ca/Mg.

of relatively compatible elements divided by relatively incompatible elements will be nearly linear functions of crystal volume (*Pearce,* 1987). If chemical zoning results from solid-state diffusion, the zoning will not be linear in crystal volume (*Crank,* 1975).

Olivine

Figure 3 shows compositional zoning of olivine #1.2 (Fig. 1), which is representative of those in all analyzed olivines (Table 1; unpublished data; *Treiman,* 1986). The zoning is mostly normal, from cores richer in Mg to edges richer in Fe. The variation is not smooth, and can be divided into zones concentric around the grain centers.

Core zones are relatively homogeneous with moderate Mg/Fe ratio (Fig. 3a). Mg/Fe in the cores is irregularly variable, more so than can be ascribed to analytical error. Similar though greater variability is also seen in Ca/Mg and Mg/Al ratios. Fe/Mn is essentially constant.

The core zones of the larger subhedral olivines are not all of the same composition. Analyses in *Treiman* (1986) show two grains with different core compositions, and this is confirmed by further analyses. The variations are probably not all from effects of crystal orientation; the two best-oriented olivines have different core compositions [HU #1.2 has Mg/Fe = 0.64; USNM-436-2.1 has Mg/Fe = 0.82 (*Treiman,* 1986)].

Next are *rim* zones, of continuously decreasing Mg/Fe ratio. The Mg/Fe vs. volume slopes for rim zones are different for different crystals, and even for different sides of a single crystal. Mg/Al generally follows Mg/Fe, but Ca/Mg shows first an increase and then a decrease in the rim zone.

Beyond the rim zones of some crystals are strongly ferroan *overgrowth* zones, of constant low Mg/Fe ratios, but not necessarily constant Ca/Mg or Mg/Al. Not all crystals have overgrowth zones, nor are they necessarily of identical composition on all sides of a single crystal (Fig. 3a).

Augite

Figure 4 shows the compositional zoning of a typical augite crystal in Nakhla; its pattern is similar to that in the olivines (Fig. 3, Table 2; *Treiman,* 1989a): magnesian core zones, rim zones of decreasing Mg/Fe, and variably developed overgrowth zones (pigeonite) of constant low Mg/Fe.

As in the olivine, the core zones of the augites have fairly constant abundances of Fe and Mg. However, abundances of minor and trace elements vary widely (Fig. 4); Ti content varies by a factor of four, Al content by a factor of six, and Cr content by a factor of three. This variation is apparently not symmetric around the crystal core, but randomly disposed. The cores may appear to be reversely zoned, where both sides

TABLE 1. Representative olivines from Nakhla.

Chemical Analyses							
Anal.#	58	62	68	72	75	79	81
Zone*	c	c	c	r	r	r	o
SiO_2	33.51	33.21	33.55	33.20	33.09	32.48	32.55
TiO_2	0.03	0.04	0.05	0.05	0.03	0.03	0.07
Al_2O_3	0.07	0.05	0.08	0.05	0.02	0.12	0.04
Cr_2O_3	0.02	0.07	0.03	0.00	0.00	0.01	0.02
FeO	48.42	48.33	48.19	48.90	50.66	52.15	53.39
MnO	0.93	0.88	0.96	0.97	0.98	1.14	1.08
MgO	17.44	17.13	17.50	16.98	15.77	13.58	13.21
CaO	0.38	0.44	0.29	0.34	0.34	0.28	0.19
Na_2O	0.05	0.00	0.05	0.06	0.01	0.01	0.01
Total	100.85	100.15	100.70	100.55	100.90	99.70	100.56
Normalizations to 3 cations							
Mg	0.773	0.771	0.778	0.764	0.709	0.625	0.607
Fe	1.196	1.196	1.191	1.222	1.264	1.335	1.362
Mn	0.023	0.023	0.024	0.024	0.025	0.030	0.028
Ca	0.012	0.014	0.009	0.011	0.011	0.009	0.006
Al	0.002	0.002	0.002	0.002	0.001	0.004	0.001
Si	0.993	0.992	0.994	0.996	0.992	0.997	0.996

*c = core, r = rim, o = overgrowth.
Electron microprobe analyses as described in text. All analyses from traverse of Fig. 1b.

of the core are (by chance) relatively magnesian. Abundances of Ti and Al (i.e., Mg/Ti and Mg/Al) are roughly correlated, as expected if they are in coupled substitution. However, their abundances are essentially independent of Cr or Na abundances.

Also, as in the olivine, the augites' rims include regions where compositional ratios are nearly linear in crystal volume (Fig. 4). Also present are nonlinear regions, such as in the left side of Fig. 4. Near the edges of some augites (not shown in Fig. 4) are *replacement zones* of constant moderate Fe/Mg that abruptly transsect and erase the other zones (*Treiman*, 1989b). The augites are also zoned with respect to Ca content (Fig. 5), and overall follow a normal trend ending at ferroan pigeonite, similar to the minute grains in the mesostasis (*Berkley et al.*, 1980).

COMPOSITIONAL ZONING: INTERPRETATION

The chemical zoning patterns in Nakhla's olivines and augites (Figs. 3, 4, and 5) are crucial in understanding the meteorite. The interpretation here follows *Pearce's* (1984, 1987) quantitative analysis of mineral zoning in igneous systems. In an equilibrium fractionating magma, the composition of the melt is a function of the masses or volumes of crystal species removed, and the mineral/melt distribution coefficients for elements in the system. Important also is the volume of melt that an individual crystal can equilibrate with chemically, i.e., its reservoir. For elements with constant mineral/melt distribution coefficients, *Pearce* (1987) showed that crystal growth during Rayleigh fractionation yields nearly linear relationships between crystal volume and abundance ratios of

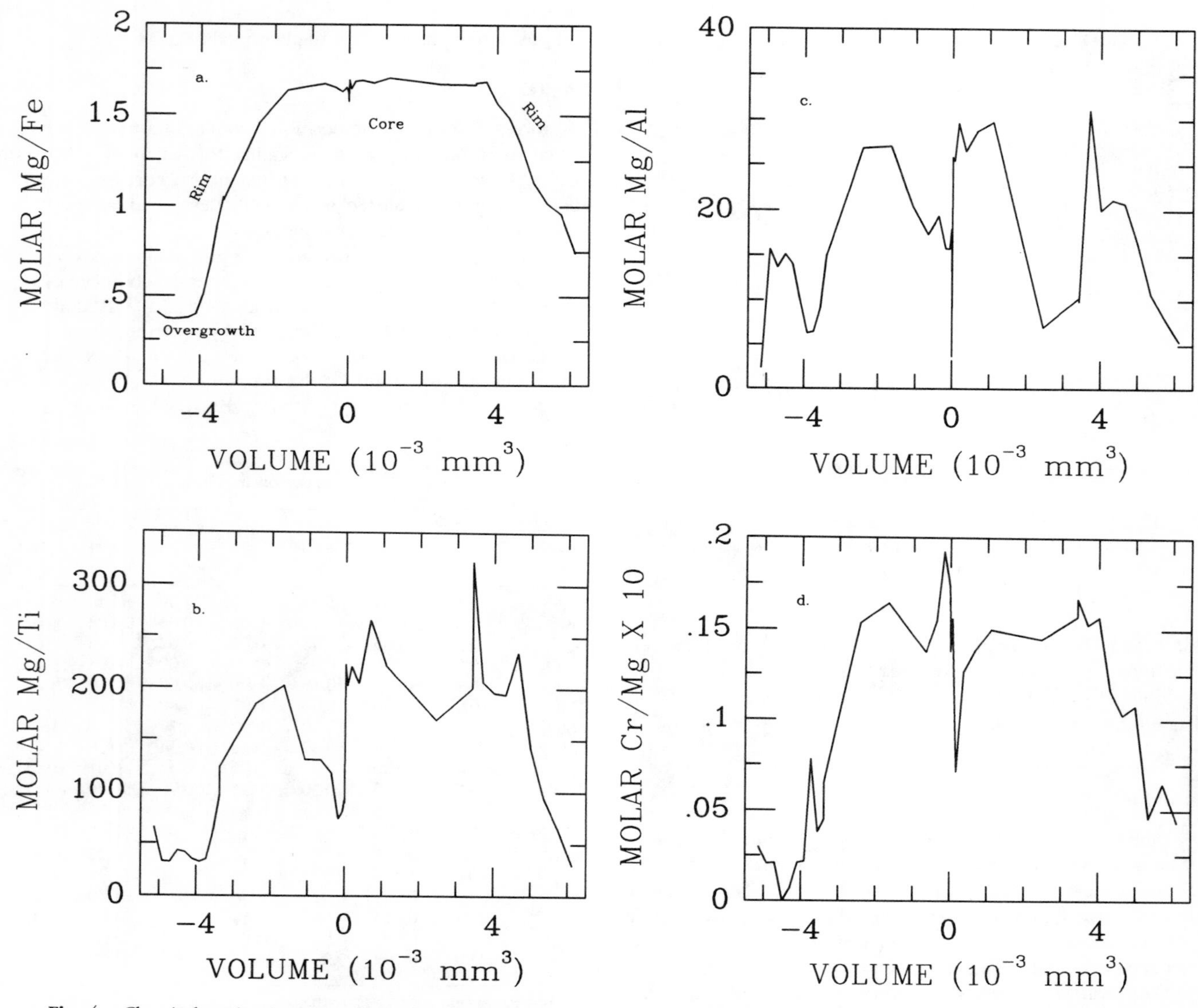

Fig. 4. Chemical zoning in augite; from electron microprobe traverse across grain AU6, Harvard Univ. thin section #2. Molar ratios vs. volume from grain's core (2·distance^3). **(a)** Mg/Fe; **(b)** Mg/Al; **(c)** Mg/Ti; **(d)** Cr/Mg.

TABLE 2. Representative pyroxenes from Nakhla.

Anal.#	153	146	141	137	133	132	131	128
Zone*	c	c	c	c	r	r	r	o
SiO_2	50.59	53.26	52.68	53.12	51.14	50.25	49.82	49.64
TiO_2	0.16	0.12	0.34	0.13	0.32	0.46	0.44	0.37
Al_2O_3	2.31	0.67	1.03	0.63	1.35	1.59	1.45	0.69
Cr_2O_3	0.36	0.35	0.47	0.41	0.08	0.11	0.03	0.01
FeO	13.68	14.18	14.18	14.43	19.37	21.80	25.57	36.17
MnO	0.45	0.45	0.44	0.48	0.61	0.64	0.71	1.03
MgO	12.84	13.48	12.97	13.26	9.83	8.19	7.27	7.73
CaO	17.43	17.72	18.35	18.10	17.18	16.70	14.86	5.15
Na_2O	0.19	0.22	0.22	0.18	0.25	0.23	0.23	0.12
Total	97.99	100.45	100.67	100.73	100.49	100.14	100.38	100.91
Normalizations to 4 cations								
Na	0.013	0.016	0.016	0.013	0.019	0.018	0.017	0.002
Ca	0.720	0.715	0.740	0.730	0.710	0.709	0.630	0.228
Mg	0.742	0.762	0.732	0.748	0.569	0.482	0.432	0.467
Fe	0.440	0.445	0.445	0.453	0.635	0.712	0.843	1.225
Mn	0.014	0.014	0.014	0.015	0.020	0.021	0.024	0.035
Ti	0.004	0.003	0.010	0.004	0.009	0.014	0.013	0.021
Cr	0.013	0.011	0.013	0.012	0.002	0.003	0.001	0.000
Al	0.105	0.020	0.046	0.028	0.062	0.073	0.068	0.031
Si	1.950	2.005	1.983	1.999	1.974	1.970	1.972	2.005

*c = core, r = rim, o = overgrowth.
Electron microprobe analyses as described in text. All analyses of augite HU2AU6.

more compatible to less compatible elements. Crystal growth from a large melt reservoir yields element ratio vs. volume graphs of shallow slope; growth from a small reservoir yields steep slopes as the compatible elements in the melt reservoir are rapidly depleted.

Olivine

The most easily interpretable facet of the olivines' zonings is the rim, where Mg/Fe is a decreasing linear function of crystal volume. This zoning pattern is consistent with growth during Rayleigh fractionation from a small melt reservoir (*Pearce,* 1987). The rims could not have been caused by simple solid-state diffusion, because their compositions are not linear in the co-error function of distance from the crystal wall (*Crank,* 1975). As noted, the linear core regions may not have the same slopes. Differing slopes could suggest different distribution coefficients or different reservoir volumes on opposite sides of a crystal, or adjacent to different crystals. Different distribution coefficients might suggest different phase assemblages in the magma pockets at different sides of the crystal(s). Different reservoir volumes might suggest different sized pores on different sides of the crystal(s). But both explanations imply that there was little chemical communication between different crystals or different sides of a single crystal, a state easily realized during crystal growth in a tight cumulate pile.

If the linear zones represent igneous crystal growth within the cumulate, chemical zoning in the olivine might reflect the cocrystallization of other minerals. This effect may be seen in the Ca/Mg ratio, which increases and then decreases in the rim zone (Fig. 3). If olivine alone were growing, Ca/Mg ought to increase monotonically with crystallization because Ca is relatively incompatible in olivine. The succeeding outward decrease in Ca/Mg may be ascribed to crystallization of a Ca-rich phase, probably augite. That Ca/Mg doesn't decrease through the whole rim zone could mean that part of the rim formed before crystal accumulation, far away from augite crystals. Or, the whole rim could have formed in the cumulate pile, but the melt reservoir for the inner rims were quite small and the growing olivines were not in chemical contact with nearby augite crystals.

Toward the olivines' edges Mg/Fe slopes in the rim zones may become shallower, and finally flat in the overgrowth zone. Following *Pearce* (1987), the flattening might represent increasing reservoir volume or changing partition coefficients. One could justify increasing reservoir volume by invoking circulation of intercumulus magma (e.g., *Berkeley et al.,* 1980; *Irvine,* 1980). However, magma circulation is seemingly inconsistent with the continuing changes of Mg/Ca and Mg/Al in the olivines (Fig. 5). Changes in bulk partition coefficients are more easily explained by the nucleation of new phases, like Fe ± Ti oxides.

Augite

Grossly, the compositional zonations in Nakhla's augites are similar to those in olivines (Fig. 4). Both species have core zones with irregularly variable, but nearly constant Mg/Fe, rims with rapidly varying compositions, and variably developed overgrowth zones of homogeneous evolved composition. On some augites, replacement zones of constant compositions transsect and erase the other zones. These zones are apparent in all of the elements analyzed.

As in the olivines, the rim zones contain segments that have linear relationships between crystal volume and compatible/incompatible element abundance ratio. This linearity may be consistent with growth during Rayleigh fractionation, if there was little change in temperature during formation of the rim zone (i.e., *Nielsen and Drake,* 1979). Unlike the zonings observed so far in olivine, the zoning in augite rarely inflects through a curve from flat core to linear rim (left side, Fig. 4). Within the model of *Pearce* (1987), this could represent growth during which the volume of the magma reservoir was

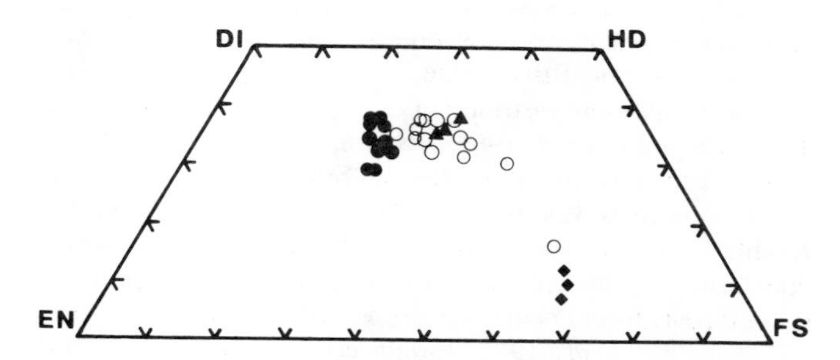

Fig. 5. Pyroxene quadrilateral projection of selected augite analyses, projected by the scheme of *Lindsley and Andersen* (1983). Filled circles are from cores, open circles from rims, diamonds from overgrowths, and filled triangles from replacement zones.

rapidly dropping, or could represent growth while temperature and partition coefficients were changing rapidly. Or it could merely represent a compositional irregularity at the edge of the core.

Similar Histories

That Nakhla's olivines and augites have similar chemical zoning patterns (core, mantle, rim) suggests that grains of both minerals experienced similar chemical and thermal histories. The core zones, relatively homogeneous in Mg/Fe (especially in olivine), but variable in other elements, imply considerable chemical equilibration of Mg and Fe, but not of Ca and monovalent, trivalent or tetravalent elements. Incomplete chemical equilibration is also consistent with the range of core zone compositions in different grains. There are several possible explanations for the core zones. They could be igneous grains grown at disequilibrium, perhaps as filled dendritic crystals. They could be irregularly zoned phenocrysts that have incompletely equilibrated with their host magma. Or they could be xenocrysts from a partially equilibrated rock, possibly an augite-olivine cumulate.

Rim zones with compositions linearly dependent on volume suggest crystal growth from a restricted reservoir, as in Rayleigh fractionation growth from a restricted volume of magma. During at least part of this Rayleigh fractionation time, the olivine and augite were in chemical communication, and so were probably growing together in the cumulate. Next outward, overgrowth zones of constant Mg/Fe and Mg/Ti ratios suggest growth in buffered reservoirs, perhaps influenced by the presence of Fe and Fe-Ti oxides.

CORE COMPOSITIONS: INTERPRETATION

Since Nakhla's olivines and augites share similar thermal histories, they probably formed in similar environments. It is reasonable that the rims and overgrowths reflect crystal growth within the cumulate. But petrography and mineral zoning do not explain the olivines' and augites' cores: what processes could have produced grains relatively homogeneous in Fe and Mg, and irregularly zoned in other elements? Could the cores have approached chemical equilibrium with each other, as the similarity of history may suggest? If the cores approached chemical equilibrium, could they also have equilibrated with a magma? An approach to these questions is through the distribution of elements between the augite and olivine cores.

It has been noted that Nakhla's augites and olivines do not have an equilibrium distribution of Mg and Fe (*Berkley et al.,* 1980; *Nakamura et al.,* 1982; *Treiman,* 1986; *Longhi and Pan,* 1989), and this apparent disequilibrium has driven most theories of Nakhla's origins. The Fe/Mg distribution for Nakhla's olivine and augite, $D^{Fe/Mg}_{Aug/Ol}$ is near 0.4, while equilibrium coefficient for phenocrysts in basaltic systems is 0.7-1.0 (data from *Roeder and Emslie,* 1970; *Grove and Bence,* 1977; *Walker et al.,* 1977; *Longhi et al.,* 1978; *Nielsen and Drake,* 1979; *Walker et al.,* 1979). In addition, the work of *Longhi and Pan* (1989), for which $D^{Fe/Mg}_{Aug/Ol}$ = 0.68-0.92, is most significant because the augites they grew are similar to Nakhla's, and the bulk compositions are like those inferred as parental to Nakhla (*Treiman,* 1986). It seems reasonable that the augite and olivine cores did not approach magmatic equilibrium.

It remains barely possible that the augite and olivine cores represent comagmatic phenocrysts grown at gross disequilibrium. Such growth would be consistent with the many magmatic inclusions in the olivines, but still presents problems for the distribution of Fe and Mg. The problem is that the Fe/Mg distributions between augite, olivine, and melt do not vary greatly with cooling rate (*Grove and Bence,* 1979; *Gamble and Taylor,* 1980), certainly not enough to yield the observed Fe/Mg distribution between the cores of Nakhla's olivines and augites. So there is little evidence in favor of disequilibrium growth, but I hesitate to exclude it entirely.

However, the core compositions are close to subsolidus equilibrium. The apparent distribution coefficient $D^{Fe/Mg}_{Aug/Ol}$ (~0.4) for the cores is in the range for olivine and augite equilibrated below the solidus, either in metamorphosed rocks, or in rocks with evidence of extensive subsolidus chemical equilibration (e.g., *Boctor et al.,* 1976; *Ashwal,* 1982; *Naslund,* 1984; *Treiman and Drake,* 1985; *Kozul and Hewins,* 1988; *Kolker and Lindsley,* 1989).

In summary, it seems that the cores of Nakhla's augites and olivines are within the equilibrium distribution of Fe and Mg for subsolidus equilibration. It is clear that the augite and olivine cores did not approach chemical equilibrium for magmatic conditions, but it is barely possible that they represent disequilibrium comagmatic phenocrysts.

CONCLUSIONS

Portions of Nakhla's petrogenesis are not in dispute. It is generally agreed that Nakhla is a cumulate igneous rock, consisting of accumulated grains of augite ± olivine, that these grains were coated with postcumulus overgrowths, and that the remaining basaltic magma crystallized to a fine-grained, radiate-textured mesostasis among the cumulus grains and postcumulus overgrowths. What has not been clear is the origin of the olivine and augite cores.

Although many hypotheses for Nakhla's olivines and augite have been published, all are inconsistent with the meteorite's petrography and mineral zonings. Two other, less intuitive, hypotheses remain: that the cores of the olivines and augites are both xenocrystic; and that the cores are both phenocrystic, but from different magmas. Both hypotheses imply that Nakhla did not form as a closed magmatic system.

The first published hypothesis, that Nakhla's olivine is all postcumulus (*Reid and Bunch,* 1975; *Berkley et al.,* 1980), is inconsistent with petrography and mineral zoning. It cannot explain the sizes of the large subhedral olivines (Fig. 2; *Treiman,* 1986), nor their euhedral, constant composition cores and inclusion-rich zones.

The second hypothesis, that the olivines' cores are xenocrysts while the augites' cores are phenocrysts (*Treiman,* 1986), is also difficult to reconcile with the data here. The olivine xenocrysts must be of relatively constant composition (chemically equilibrated); this presents no problem. But the augites' cores must also be of relatively constant composition, have irregular zoning like those in the olivine, and have grown from a magma. Thus, the hypothesis seems to require two separate processes to produce the very similar core zones, and *prima facie* fails the test of Occam's razor.

The third hypothesis, that both the olivines and augites are phenocrysts (*Treiman,* 1989a), is consistent with petrographic and mineral-chemical data, but is quite inconsistent with equilibrium between the cores at magmatic temperatures.

The fourth hypothesis, that olivine in Nakhla is partially re-equilibrated with the parental magma (*Nakamura et al.,* 1982; *Longhi and Pan,* 1989), is inconsistent with the similarities of, and slopes of, the olivines' and augites' zoning patterns. First, solid-state chemical diffusion cannot produce both the olivines' and augites' zonings, because diffusion in pyroxene is orders of magnitude slower than in olivine (e.g., *Buening and Buseck,* 1973; *Brady and McCallister,* 1983). The time-at-temperature needed to produce zoning like that of Nakhla's olivines would have had essentially no effect on the augite. Second, solid-state diffusion should yield compositional profiles that are linear in the co-error function of the distance from the crystal wall (*Crank,* 1975), not the observed profiles linear in distance from the crystal center cubed (Figs. 3 and 4).

New Hypotheses

A new hypothesis suggested by the data here is that the cores of augites and olivines are both xenocrysts from a partially equilibrated source rock. This idea is apparently consistent with all available petrographic and mineral chemical data, though possible sources for the xenocrysts are severely limited. The strongest evidence in favor of this hypothesis is the the distribution of Fe and Mg between the species' cores; that distribution is consistent with subsolidus equilibration. Notably, the equilibrated nakhlite Lafayette has augite and olivine of compositions very similar to Nakhhla's (*Boctor et al.,* 1976). Compositional variability within and among cores are reasonable if they are xenocrysts from a partially equilibrated rock. The euhedral to subhedral shapes of the cores are also consistent, as some chemically equilibrated rocks retain magmatic crystal forms; a notable example is the nakhlite meteorite Lafayette (*Boctor et al.,* 1976; *Berkley et al.,* 1980). The magmatic inclusions in the olivines and augites also present little problem; relict inclusions may be present in chemically equilibrated rocks (*Floran et al.,* 1978), and the inclusions may remelt when their host crystal is heated by new magma.

If the cores of augites and olivines are xenocrysts, the character of their source rock is strongly constrained. The source rock must have consisted primarily of olivine and augite, like (but less equilibrated than) those in the Lafayette meteorite (*Boctor et al.,* 1976). The source rock must have had euhedral to subhedral grains, as does Lafayette, and must have been essentially contemporaneous with Nakhla's magma and of essentially the same initial isotope ratios, or Nakhla's isotope systematics would be disturbed (i.e., *Gale et al.,* 1975, *Nakamura et al.,* 1982). In sum, the source of xenocrysts must have been much like the Lafayette meteorite, or Nakhla itself.

Two other hypotheses are also possible, but apparently less likely. First is that the cores of the augites and olivines are comagmatic phenocrysts grown under gross disequilibrium. This hypothesis has little support at present, but may be possible. Second is that the cores of the olivines and augites represent phenocrysts, but from different magmas. The distribution of Fe and Mg show that the cores could not have been comagmatic, and noncomagmatic phenocrysts could result if Nakhla formed from a mixture of pyroxene-phyric and olivine-phyric basalts. There is little evidence for or against magma mixing, but the inclusion-rich zones in Nakhla's olivines are like those formed during magma mixing (*Hibbard,* 1981).

Complex Petrogenesis

The petrography and mineral chemistry of the Nakhla meteorite seem to preclude all simple petrogenetic hypotheses, such as those discussed above. The remaining possibilities all require some sort of petrogenetic complexity: gross disequilibrium, abundant xenocrysts, or magma mixing. Under these constraints, attempts to retrieve a magma composition from core augite compositions (*Longhi and Pan,* 1989) seem problematic. Attempts to retrieve a magma composition by mass balance (*Treiman,* 1986) may still be fruitful, but it is not clear whether the composition would be that of a primary magma, a magma contaminated by assimilation of xenocrysts, or a mixture of two other magmas.

Much work will be required to unravel Nakhla's petrogenesis, if indeed it can be done. Detailed mapping of the chemical zoning in the augites' and olivines' cores may clarify whether they are xenocrysts or phenocrysts grown under unusual circumstances. The magmatic inclusions in the olivines and augites hold great promise, for they are probably the best available relicts of the magmas from which the grains grew. Finally, it will be important to compare Nakhla with the other nakhlites, in an extension of the seminal work of *Berkley et al.* (1980); single samples are often of limited utility in deciphering complex magmatic processes.

Acknowledgments. The reviews of H. Naslund, J. Delano, and J. Shervais are greatly appreciated. Discussions with J. Berkley, J. Jones, and J. Longhi have been helpful. I am grateful to J. Wood and M. Drake for use of microprobes; technical support from J. Paque, A. McGuire, and B. Holmberg is gratefully acknowledged. This work was supported by NASA grant NAG 9-168.

REFERENCES

Ashwal L. D. (1982) Mineralogy of the mafic and Fe-Ti oxide-rich differentiates of the Marcy anorthosite massif, Adirondaks, New York. *Am. Mineral., 67,* 14-27.

Berkley J. L., Keil K., and Prinz M. (1980) Comparitive petrology and origin of Governador Valadares and other nakhlites. *Proc. Lunar Planet. Sci. Conf. 11th,* pp. 1089-1102.

Bogard D. D., Nyquist L. E., and Johnson, P. (1984) Noble gas contents of shergottites and implications for the martian origin of SNC meteorites. *Geochim. Cosmochim. Acta, 48,* 1723-1739.

Boctor N. Z., Meyer H. O. A., and Kullerud G. (1976) Lafayette meteorite: Petrology and opaque mineralogy. *Earth Planet. Sci. Lett., 32,* 69-76.

Brady J. B. and McCallister R. H. (1983) Diffusion data for clinopyroxenes from homogenization and self-diffusion experiments. *Am. Mineral., 68,* 95-105.

Buening D. K. and Buseck P. R. (1973) Fe-Mg lattice diffusion in olivine. *J. Geophys. Res., 78,* 6852-6862.

Bunch T. E. and Reid A. M. (1975) The nakhlites, Part I: Petrography and mineral chemistry. *Meteoritics, 10,* 303-315.

Crank J. (1975) *The Mathematics of Diffusion.* Oxford Univ., Oxford. 414 pp.

Floran R. J., Prinz M., Hlava P. F., Keil K., Nehru C. E., and Hinthorne J. R. (1978) The Chassigny meteorite: A cumulate dunite with hydrous amphibole-bearing melt inclusions. *Geochim. Cosmochim. Acta, 42,* 1213-1229.

Gale N. H., Arden J. W., and Hutchison R. (1975) The chronology of the Nakhla achondritic meteorite. *Earth Planet. Sci. Lett., 26,* 195-206.

Gamble R. P. and Taylor L. A. (1980) Crystal/liquid partitioning in augite: Effects of cooling rate. *Earth Planet. Sci. Lett., 47,* 21-33.

Grove T. L. and Bence A. E. (1977) Experimental study of pyroxene-liquid interaction in quartz-normative basalt 15597. *Proc. Lunar Sci. Conf. 8th,* pp. 1549-1579.

Grove T. L. and Bence A. E. (1979) Crystallization kinetics in a multiply saturated basalt magma: An experimental study of Luna 24 ferrobasalt. *Proc. Lunar Planet. Sci. Conf. 10th,* pp. 439-478.

Hibbard M. J. (1981) The magma mixing origin of mantled feldspars. *Contrib. Mineral. Petrol., 76,* 158-170.

Irvine T. N. (1980) Magmatic infiltration metasomatism, double-diffusive fractional crystallization, and adcumulus growth in the Muskox and other layered intrusions. In *Physics of Magmatic Processes* (R. B. Hargraves, ed.), pp. 325-383. Princeton Univ., Princeton, New Jersey.

Kolker A. and Lindsley D. H. (1989) Geochemical evolution of the Maloin Ranch pluton, Laramie Anorthosite Complex, Wyoming: Petrology and mixing relations. *Am. Mineral., 74,* 307-324.

Kozul J. and Hewins R. H. (1988) Mafic mineral compositions of igneous clasts in the Lewis Cliffs 85300, 85302, 85303 polymict eucrites (abstract). *Meteoritics, 23,* 281-282.

Lambert P. (1987) SNC meteorites: The metamorphic record (abstract). In *Lunar and Planetary Science XVIII,* pp. 529-530. Lunar and Planetary Institute, Houston.

Lindsley D. H. and Andersen D. J. (1983) A two-pyroxene thermometer. *Proc. Lunar Planet. Sci. Conf. 13th,* in *J. Geophys. Res., 88,* A887-A906.

Longhi J. and Pan V. (1989) The parental magmas of the SNC meteorites. *Proc. Lunar Planet. Sci. Conf. 19th,* pp. 451-464.

Longhi J., Walker D., and Hayes J. F. (1978) The distribution of Fe and Mg between olivine and lunar basaltic liquids. *Geochim. Cosmochim. Acta, 42,* 1545-1558.

Nakamura N., Unruh D. M., Tatsumoto M., and Hutchison R. (1982) Origin and evolution of the Nakhla meteorite inferred from the Sm-Nd and U-Pb systematics and REE, Ba, Sr, Rb, and K abundances. *Geochim. Cosmochim. Acta, 46,* 1555-1573.

Naslund H. R. (1984) Petrology of the upper border series of the Skaergaard intrusion. *J. Petrol., 25,* 185-214.

Nielsen R. L. and Drake M. J. (1979) Pyroxene-melt equilibria. *Geochim. Cosmochim. Acta, 43,* 1259-1273.

Pearce T. H. (1984) The analysis of zoning in magmatic crystals with emphasis on olivine. *Contrib. Mineral. Petrol., 86,* 149-154.

Pearce T. H. (1987) The theory of zoning patterns in magmatic minerals using olivine as an example. *Contrib. Mineral. Petrol., 97,* 451-459.

Reid A. M. and Bunch T. E. (1975) The nakhlites, Part II: Where, when and how. *Meteoritics, 10,* 317-324.

Roeder P. L. and Emslie R. F. (1970) Olivine-liquid equilibrium. *Contrib. Mineral. Petrol., 19,* 275-289.

Treiman A. H. (1986) The parental magma of the Nakhla achondrite: Ultrabasic volcanism on the shergottite parent body. *Geochim. Cosmochim. Acta, 50,* 1061-1070.

Treiman A. H. (1987) Geology of the nakhlite meteorites: Cumulate rocks from flows and shallow intrusions (abstract). In *Lunar and Planetary Science XVIII,* pp. 1022-1023. Lunar and Planetary Institute, Houston.

Treiman A. H. (1989a) Origin of olivine in the Nakhla achondrite, with implications for distribution of Mg/Fe between olivine and augite (abstract). In *Lunar and Planetary Science XX,* pp. 1130-1131. Lunar and Planetary Institute, Houston.

Treiman A. H. (1989b) Post-cumulus processes in a volcanic cumulate rock (abstract). *Geol. Soc. Am. Abstr. with Progr., 21,* 71.

Treiman A. H. and Drake M. J. (1985) Basaltic volcanism on the eucrite parent body: Petrology and chemistry of the polymict eucrite ALHA80102. *Proc. Lunar Planet. Sci. Conf. 15th,* in *J. Geophys. Res., 90,* C619-C628.

Walker D., Longhi J., Lasaga A. C., Stolper E. M., Grove T. L., and Hayes J. F. (1977) Slowly cooled microgabbros 15555 and 15065. *Proc. Lunar Sci. Conf. 8th,* pp. 1521-1547.

Walker D., Shibata T., and DeLong S. E. (1979) Abyssal tholeiites from the Oceanographer fracture zone II. Phase equilibria and mixing. *Contrib. Mineral. Petrol., 70,* 111-125.

Wood C. A. and Ashwal L. D. (1981) SNC meteorites: igneous rocks from Mars? *Proc. Lunar Planet. Sci. 12B,* pp. 1359-1375.

Proceedings of the 20th Lunar and Planetary Science Conference, pp. 281-297
Lunar and Planetary Institute, Houston, 1990

Pomozdino: An Anomalous, High-MgO/FeO, yet REE-rich Eucrite

P. H. Warren and E. A. Jerde

Institute of Geophysics and Planetary Physics, University of California, Los Angeles, CA 90024

L. F. Migdisova

V. I. Vernadsky Institute of Geochemistry and Analytical Chemistry, USSR Academy of Sciences, Moscow, USSR

A. A. Yaroshevsky

V. I. Vernadsky Institute of Geochemistry and Analytical Chemistry, and Geological Department, Moscow State University, Moscow, USSR

We report new chemical analyses and petrographic data for the Pomozdino basaltic achondrite. Our data confirm earlier indications that Pomozdino is a eucrite (based on ratios such as Ga/Al, P/Sm, and Mn/Fe), and that it is a monomict breccia with an anomalous, REE-rich yet high-mg′ [=MgO/(MgO+FeO)] bulk composition. Pomozdino is texturally heterogeneous, with coarse-grained clasts set amidst a fine-grained matrix; some of this textural heterogeneity probably predates the brecciation. However, for the bulk-rock composition we find an mg′ ratio only slightly lower than that reported by Kvasha and Dyakonova (1972) and REE concentrations that are even higher (typically by a factor of ~1.8) than reported by Kolesov (1976). The pyroxenes are also relatively magnesian, and only slightly more Fs-rich than the pyroxenes of the lowest-mg′ recognized cumulate eucrite (Moore County). The pyroxenes have exsolved into compositionally uniform orthopyroxene and augite, but the original assemblage comprised two distinct varieties of primary pyroxene. The most abundant variety had a primary Wo_9 composition, similar to the primary pyroxene of Moore County. However, some of the primary pyroxenes had much higher Wo contents (as high as Wo_{39}), with pronounced Ca-zonation, resembling pyroxenes from the eucrites Stannern and Bouvante. Chromite and ilmenite also have mg′ ratios intermediate between the ranges generally found for cumulate and noncumulate eucrites. Our preferred model for the origin of this unique meteorite is as a partial cumulate, with an uncommonly high content of trapped liquid (roughly 59-80 wt.%). If so, the parent melt is inferred to have been roughly similar in composition to Stannern. Alternatively, Pomozdino might be a primary partial melt, derived from a source region far more magnesian than generally envisaged for the sources of primary eucritic partial melts. Barring great heterogeneity in the mantle of the eucrite asteroid, the primary partial melt model would imply a higher mg′ for the bulk asteroid than has generally been assumed for the eucrite parent body. In either case, we can infer that at least some Stannern-like (REE-rich and moderate-mg′) eucrites were involved in fractional crystallization and thus do not represent primary partial melts.

INTRODUCTION

The Pomozdino meteorite is an essentially unweathered find, originally 327 g. Previous studies of this meteorite comprised a wet-chemical bulk analysis and basic petrography (*Kvasha and Dyakonova,* 1972); an INAA study that determined bulk-rock concentrations of REE, Sc, Cr, Co, and Ta (*Kolesov,* 1976); and a study of ^{10}Be and ^{26}Al contents (*Aylmer et al.,* 1988). *Mason et al.* (1979) also gave a brief (one sentence) petrographic description. The studies of *Kvasha and Dyakonova* (1972) and *Kolesov* (1976) indicated that Pomozdino is a brecciated but monomict eucrite (pyroxene-plagioclase achondrite) with an extraordinary, high-MgO/FeO yet REE-rich, bulk composition. In an effort to clarify the nature and origin of this unique meteorite, we have undertaken to study its geochemistry and petrology as part of a consortium that also includes oxygen isotopic studies (*Petaev et al.,* 1988).

ANALYTICAL PROCEDURES

In Los Angeles, a single 470-mg chip (derived from specimen N2530; sawn surfaces were evident on two sides of the chip) was studied, but it was initially split into two subequal pieces, thereafter treated as separate samples. These samples were analyzed by the techniques of *Warren and Jerde* (1987). Thus, Mg, Al, Si, Ca, Ti, and Fe were determined by electron probe analysis of fused beads (EPFBA); instrumental neutron activation analysis (INAA) was used to determine 32 elements (including backup determinations for all of the fused bead elements except Si); and the trace siderophile elements Ni, Ge, Re, Os, Ir, and Au were determined by radiochemical NAA (RNAA).

In Moscow, a 48 mm^2 thin section was used for detailed analysis, along with selected regions of a second thin section. On the Camebax-Microbeam electron microprobe of the

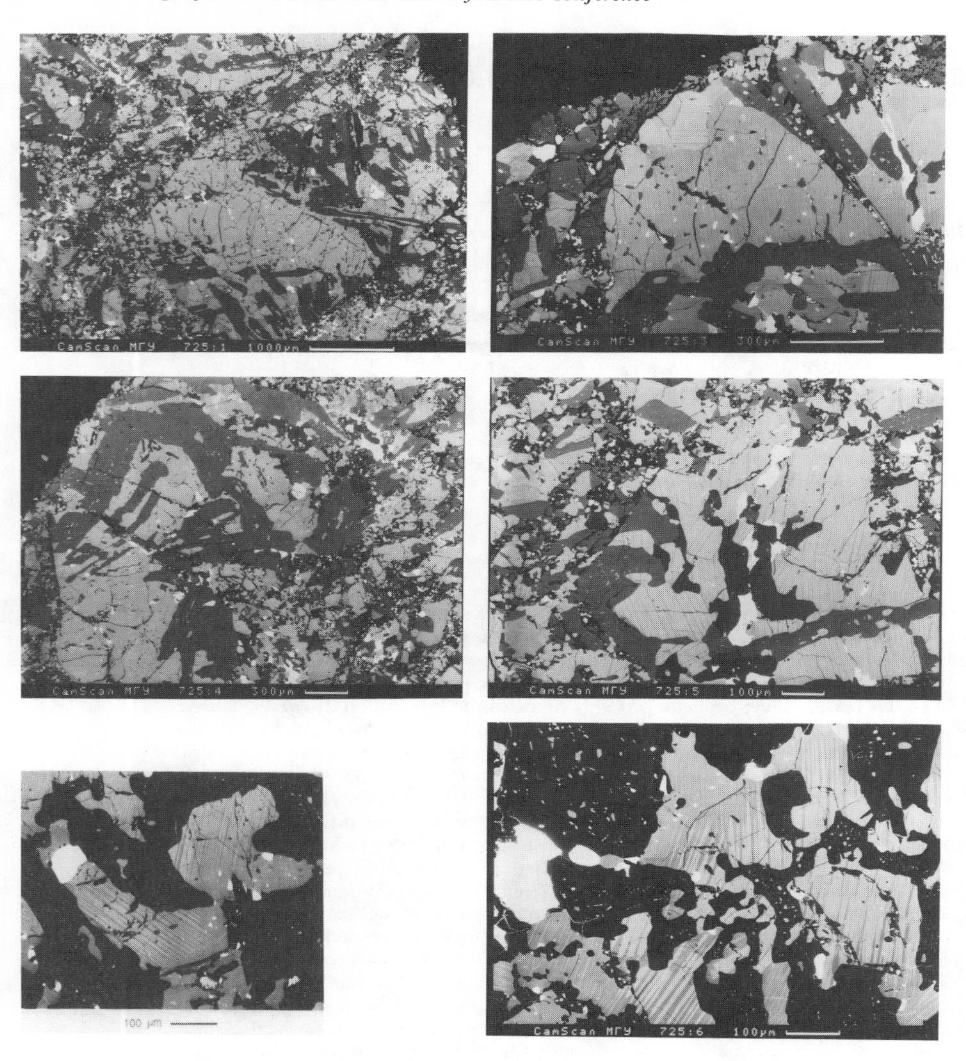

Fig. 1. Backscattered-electron views of the Pomozdino thin section that was used for detailed study. Light greys are orthopyroxene with slightly darker lamellae of augite, medium-dark greys are plagioclase, darkest greys (blacks) are tridymite, whites are opaque minerals. In sequence clockwise from upper left, the views are centered around clasts: 1, 6, 11, 10, 31, and 8 (clast 31 is in a separate thin section from the others). Note abundance and complexity of exsolution lamellae in the Ca-zoned pyroxenes of clasts 10 and 31.

Fig. 2. View of the ophitic texture preserved in an essentially unbrecciated portion of clast 1 (cf. right-central area of the clast 1 view in Fig. 1), in reflected light (left), and in transmitted light (right). The field of view is 780×1000 μm. The large opaque grain in the lower right is an ilmenite.

Vernadsky Institute, individual phases were analyzed using a beam diameter of 2 μm; and bulk compositions of clasts and different matrix regions were determined by defocused beam analysis (DBA), using a 100-μm-diameter beam. All electron probe analyses were run at 15 keV, and were corrected with standard ZAF procedures. The DBA analyses were further corrected using the *Nazarov et al.* (1982) refinement of the *Albee et al.* (1977) DBA correction procedure.

RESULTS

Petrography and Mineral Chemistry

As noted by *Kvasha and Dyakonova* (1972), Pomozdino is a monomict breccia, composed dominantly of pyroxene and plagioclase. Pyroxenes range up to 3 mm across, commonly in spindle-shaped forms (Figs. 1 and 2). In the 48 mm^2 thin section used for our detailed study, 25 intact polymineralic clasts, ranging in size from 0.8 × 0.8 mm to 4.8 × 2.4 mm, together compose 52% of the area. The remaining 48% is a thoroughly brecciated matrix of fine (<0.3 mm), mostly monomineralic, fragments. Seven of the largest clasts were chosen for particularly detailed study. Their modal compositions cover wide ranges of pyroxene/plagioclase ratio and SiO_2 content (Table 1).

The diversity among the modes of the various clasts is not surprising, considering the coarse granularity of much of the prebrecciation lithology. Note that the modal extremes tend to occur among the smallest clasts. The opaque minerals are clearly more abundant in the fine-grained matrix portion of the meteorite (Table 1, cf. *Kvasha and Dyakonova,* 1972). However, opaque minerals may have been concentrated in the finer-grained portions of the meteorite even before brecciation (see following paragraph). Only mild clast-to-clast mineral-composition differences were found, and no significant mineral-compositional heterogeneity was found that appeared related to differences between the aggregate clasts and the matrix. Thus, we concur with *Kvasha and Dyakonova* (1972) that the meteorite is probably monomict.

The clasts are usually irregular in shape, but some are angular or rounded, and a few are almost circular. Grain sizes, even excluding comminuted (matrix) regions, are diverse. On the basis of grain size and overall texture, the clasts studied may be divided into two groups. The first group (clasts 1, 5, 6, 8, and 9) includes irregular coarse-grained lithologies with ophitic, poikilophitic, and in some cases doleritic textures. The largest clast (No. 1) contains glomeroporphyritic aggregates of prismatic pyroxene crystals, from 1.6 × 0.8 to 1.1 × 0.7 mm in size. Plagioclase occurs as laths, typically 1-2 mm long, or as euhedral and subhedral platelets, intergrown and poikilitically enclosed by pyroxene. The second group (clasts 10 and 11) includes fine-grained lithologies with anhedral-granular textures. Grain sizes in these clasts range from 0.02 to 0.2 mm. The plagioclase occurs as subhedral laths. It appears that the opaque minerals were already associated mainly with fine-grained areas, before the brecciation. Note in Table 1 that, on average, clasts 10 and 11 have far higher opaque mineral contents than the coarse-grained (group one) clasts.

Pomozdino's pyroxene is uncommonly magnesian, for a eucrite, and exhibits an extraordinary range in primary Wo (Ca/[Ca+Mg+Fe]) ratio. The current (postexsolution) pyroxene assemblage comprises orthopyroxene, $Wo_{1.8-3.4}En_{42-47}Fs_{55-50}$, and augite, $Wo_{40-44}En_{33-36}Fs_{26-22}$. Average compositions

TABLE 1. Modes of clasts, matrix regions, and overall meteorite.

	Clasts								Matrix				Meteorite
	1	5	6	8	9	10	11	Avg	1	2	3	Avg	Avg
Mode, vol%													
Pyroxene	61.5	39.3	70.6	47.4	69.0	57.7	65.7	59.8	51.7	52.3	35.1	46.4	53.3
Plagioclase	32.4	58.4	27.6	48.2	28.9	37.9	16.0	35.2	42.5	39.0	42.9	41.4	38.2
SiO_2	5.2	0.5	0.2	2.6	1.1	3.4	15.0	3.8	0.4	5.0	11.2	5.5	4.6
Ilmenite	0.62	1.77	0.45	1.13	0.27	0.28	0.16	0.66	0.94	1.09	1.63	1.22	0.93
Chromite	0.16	—	—	—	0.13	0.38	0.16	0.12	0.47	2.63	—	1.03	0.56
Metal+ troilite	0.2	—	1.0	0.65	0.5	0.4	3.0	0.46	4.0	—	8.2	4.4	2.4
Area, mm²	8.2	0.5	1.4	3.4	2.4	0.9	0.5	17.3	0.7	0.4	0.4	1.5	18.8
*Composition, wt%**													
SiO_2	51.0	47.5	48.5	48.2	49.3	49.7	51.9	49.9	43.8	48.5	41.9	44.6	47.3
TiO_2	0.72	1.70	0.64	1.16	0.50	0.61	0.42	0.77	0.94	1.29	1.32	1.18	0.97
Al_2O_3	9.38	17.1	7.86	13.9	8.17	11.0	4.60	10.0	11.7	11.5	11.3	11.5	10.7
Cr_2O_3	0.29	0.14	0.24	0.18	0.35	0.47	0.33	0.28	0.44	2.12	0.07	0.85	0.56
V_2O_3	0.004	0.007	0.002	0.007	0.016	0.004	0.002	0.006	0.008	0.032	—	0.014	0.010
FeO	19.80	12.4	20.5	15.8	21.3	14.3	19.4	18.9	15.2	17.3	10.5	14.3	16.6
MnO	0.64	0.38	0.64	0.47	0.67	0.45	0.64	0.60	0.50	0.55	0.34	0.46	0.53
MgO	9.38	6.5	10.9	7.9	10.6	8.4	9.8	9.3	8.0	8.5	5.3	7.2	8.3
CaO	7.84	13.3	8.2	10.1	7.3	13.1	5.8	8.6	9.6	9.6	8.2	9.1	8.8
Na_2O	0.50	0.91	0.44	0.72	0.49	0.48	0.23	0.53	0.61	0.59	0.59	0.60	0.56
K_2O	0.055	0.094	0.047	0.076	0.051	0.043	0.062	0.058	0.068	0.065	0.086	0.076	0.067
Fe-metal + FeS	0.38	—	2.25	1.60	1.26	0.89	6.8	1.13	9.3	—	20.5	10.2	5.6

*Bulk compositions implied by combinations of modes with electron microprobe analyses of individual phases.

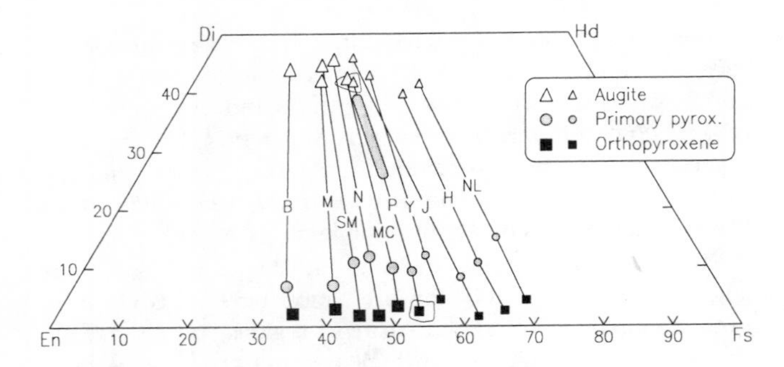

Fig. 3. Compositions of pyroxenes in Pomozdino, with literature results for other eucrites included for comparison: B = Binda, M = Moama, SM = Serra de Magé, N = Nagaria, MC = Moore County, P = Pomozdino, Y = Y-791195, J = Juvinas, H = Haraiya, and NL = Nuevo Laredo (*Lovering,* 1975; *Takeda et al.,* 1976, 1988; *Takeda and Mori,* 1985; *Harlow et al.,* 1979; *Mason et al.,* 1979). Larger symbols denote cumulate eucrites.

are shown in Fig. 3 and Tables 2 and 3. The compositions of all analyzed pyroxenes plot close to a single orthopyroxene-augite tie line, similar to that for pyroxenes of the Moore County cumulate eucrite, except with a few mol.% higher Fs (Fig. 3). However, optical microscopy and calcium K_α X-ray images indicate that two or three distinct types of primary pyroxene originally existed.

Most pyroxene grains consist of ~85 vol.% orthopyroxene and 15 vol.% exsolution lamellae of augite, implying a primary composition of $Wo_9En_{43}Fs_{48}$ (Table 3); cf. the primary pigeonite composition of the Moore County cumulate eucrite (*Harlow et al.,* 1979; *Takeda et al.,* 1988). Within these pyroxenes, optical-microscopic study reveals two generations of augite exsolution: first with lamellae parallel to (001), and later parallel to (100). The widths of the (001) and the (100) lamellae are 0.4-3.0 μm and <1 μm, respectively.

In clasts 5 and 10, the dominant primary pyroxene appears to have been strongly zoned with respect to Ca; in general, it corresponds in composition to subcalcic augite. Clasts 1, 6, and 8 also contain minor proportions of this type of pyroxene. These pyroxene grains consist of 45-65 vol.% augite lamellae and only 55-35 vol.% opx, implying that the primary compositions ranged from $Wo_{19}En_{40}Fs_{41}$ to $Wo_{39}En_{34}Fs_{27}$ (Table 3). At their rims these zoned grains are typically almost pure augite with only thin lamellae (too thin to permit electron microprobe analysis) of orthopyroxene. Typically associated with these grains are small, anhedral individual grains, similar in composition to the rims of the large zoned grains, along with relatively high abundances of tridymite and opaque minerals. In the large zoned grains, the width of the augite lamellae, which are parallel to (001), ranges from 0.5 to 14 μm, and the distance between them ranges from 0.9 to 9 μm. However, the predominant widths and abundances of the augite lamellae are heterogeneous not only from grain to grain, but also within individual grains, due to Ca-zoning in the primary crystals. For example, the augite content in a grain of clast 10 varies from 38 to 70 vol.%. This zoning apparently engendered diverse inclinations for the augite lamellae even within grains, albeit in a symmetrical pattern if traversed from one edge to the opposite one. Similar Ca-zonation of pyroxenes was noted by *Takeda et al.* (1983) in Stannern, along with roughly comparable zonation in Nuevo Laredo and in basaltic clasts from polymict eucrite Yamato-75011. In one pyroxene grain from Stannern, the average bulk composition in the outermost 150 μm is $Wo_{26}En_{31}Fs_{43}$, yet the core of the grain is $Wo_{1.7}En_{36.5}Fs_{61.8}$. *Christophe Michel-Lévy et al.* (1987) found similar Ca-zonation in the pyroxenes of Bouvante. In all of these eucrites except for the Y-75011 clasts, despite extensive Ca zonation, there is little or no zonation in Mg/Fe ratio other than slight zonation along the appropriate augite-opx tie lines.

The average composition of the plagioclase is An_{85} (Table 4). Mostly it is An_{85-87}, but the largest grains tend to

TABLE 2. Average chemical composition (wt.%) and molecular norms (mol.%) of the pyroxene grains in Pomozdino.

	Matrix		Clast 1					Clast 5			Clast 6		
				Bulk					Bulk			Bulk	
	opx	Augite Lamellae	opx	pig-1*	pig-2†	Low-Ca Augite†	Augite Lamellae	opx	pig†	Augite Lamellae	opx	pig*	Augite Lamellae
SiO_2	50.6	50.9	51.2	51.1	51.2	51.4	51.6	50.9	50.9	50.4	52.3	50.5	52.6
TiO_2	0.25	0.59	0.28	0.32	0.40	0.45	0.57	0.34	0.46	0.65	0.26	0.38	0.60
Al_2O_3	0.29	0.93	0.27	0.43	0.51	0.61	0.83	0.26	0.51	0.92	0.26	0.61	0.94
Cr_2O_3	0.16	0.16	0.15	0.24	0.23	0.26	0.34	0.26	0.28	0.31	0.19	0.32	0.18
FeO	31.2	14.0	31.9	28.9	24.4	21.5	14.7	30.9	24.3	13.2	32.0	28.1	14.9
MnO	1.02	0.52	0.94	0.95	0.74	0.67	0.49	0.98	0.78	0.43	1.03	0.88	0.47
MgO	15.7	12.1	14.7	14.0	13.3	12.7	11.5	15.4	14.2	12.1	15.5	14.5	12.4
CaO	1.15	20.4	1.18	4.42	9.15	12.3	19.7	1.79	8.51	19.5	0.9	4.12	19.7
Na_2O	0.03	0.14	0.02	0.04	0.05	0.06	0.09	0.00	0.03	0.07	0.00	0.07	0.10
K_2O	0.02	0.00	0.01	0.01	0.01	0.01	0.01	0.02	0.02	0.02	0.01	0.01	0.01
Total	100.45	99.75	100.74	100.47	99.99	100.00	99.81	100.89	100.01	97.63	102.40	99.55	101.82
Wo	2.4	42.3	2.5	9.5	19.6	26.3	41.8	3.8	18.0	41.8	1.80	8.9	40.6
En	46.1	34.9	44.0	42.0	39.6	37.9	33.9	45.2	41.8	36.1	45.5	43.6	35.3
Fs	51.5	22.7	53.5	48.5	40.8	35.8	24.3	51.0	40.2	22.1	52.7	47.5	24.1
n	10	2	6	20	1	1	2	1	1	2	1	6	1

TABLE 2. (continued).

	Clast 8			Clast 9		Clast 10			Clast 11		Clast 31		
		Bulk			Bulk		Bulk			Bulk		Bulk	
	opx	pig*	Augite Lamella	opx	pig*	opx	Low-Ca Augite†	Augite Lamellae	opx	pig*	Host Augite	Low-Ca Augite*	opx Lamella
SiO_2	51.2	49.8	53.5	50.9	51.0	50.1	50.6	50.8	51.5	50.8	53.4	51.3	53.9
TiO_2	0.32	0.44	0.67	0.28	0.37	0.32	0.55	0.72	0.34	0.42	0.36	0.48	0.26
Al_2O_3	0.30	0.50	0.32	0.27	0.45	0.29	0.68	0.96	0.31	0.54	1.06	0.76	0.33
Cr_2O_3	0.20	0.33	0.23	0.13	0.33	0.12	0.27	0.37	0.20	0.31	0.38	0.26	0.13
FeO	32.6	28.6	14.7	31.7	28.6	32.1	21.6	14.0	31.5	28.1	14.1	22.2	31.5
MnO	1.02	0.85	0.43	1.06	0.90	0.95	0.68	0.49	1.03	0.92	0.46	0.69	1.02
MgO	14.9	14.8	11.3	15.1	14.4	14.9	13.2	12.0	15.3	14.4	11.6	12.6	14.7
CaO	0.91	3.96	20.0	1.07	4.29	1.31	12.3	20.2	1.30	5.02	20.2	12.5	1.00
Na_2O	0.10	0.02	0.21	0.03	0.10	0.02	0.06	0.09	0.02	0.03	0.13	0.10	0.04
K_2O	0.01	0.01	0.01	0.01	0.01	0.01	0.00	0.00	0.01	0.03	0.01	0.01	0.00
Total	101.60	99.35	101.37	100.56	100.51	100.16	99.99	99.69	101.59	100.57	101.69	100.77	102.84
Wo	1.93	8.4	42.4	2.2	9.2	2.8	25.9	42.3	2.8	10.7	42.7	26.2	2.2
En	43.9	44.0	33.4	44.9	43.0	44.1	38.7	34.9	45.1	42.6	34.1	36.6	44.5
Fs	54.1	47.6	24.2	52.8	47.8	53.1	35.4	22.8	52.2	46.7	23.3	36.2	53.3
n	3	4	1	3	11	4	1	5	7	7	3	7	1

Abbreviations: opx = orthopyroxene, pig = pigeonite, n = number of analyses, Wo = Ca/(Ca+Mg+Fe), En = Mg/(Ca+Mg+Fe), Fs = Fe/(Ca+Mg+Fe).
* Estimated bulk-grain composition, derived by averaging defocused beam analyses.
† Estimated bulk-grain composition, based on modal proportions of orthopyroxene and augite lamellae.

show mild zoning, with rim compositions of An_{80-82} (Fig. 4). The largest grains also tend to contain abundant microinclusions, which tend to be systematically oriented, and apparently formed by exsolution of silica and pyroxene out of the plagioclase. The average composition is close to the calcic end of the range for noncumulate eucrites, but distinctly sodic compared to recognized cumulate eucrites (Fig. 4, *Dodd,* 1981).

TABLE 3. Average chemical composition (wt.%) and molecular norms (mol.%) of the pyroxene phases in Pomozdino.

	Host Hypersthene		Augite Lamellae		Primary Pigeonite*		Primary Subcalcic Augite*		Primary Augite*	
	$\bar{x}$	s	$\bar{x}$	s	$\bar{x}$	s	$\bar{x}$	s	$\bar{x}$	s
SiO_2	50.6	1.0	51.4	1.1	50.7	1.3	51.0	0.5	51.2	0.5
TiO_2	0.30	0.05	0.65	0.08	0.36	0.08	0.50	0.07	0.60	0.06
Al_2O_3	0.29	0.04	0.87	0.17	0.47	0.11	0.64	0.05	1.07	0.15
Cr_2O_3	0.16	0.05	0.30	0.09	0.29	0.12	0.26	0.01	0.38	0.05
FeO	31.3	0.9	14.2	0.5	28.5	1.1	21.6	0.1	16.5	0.1
MnO	0.98	0.07	0.48	0.05	0.92	0.08	0.68	0.01	0.52	0.02
MgO	15.2	0.5	12.0	0.3	14.3	0.6	13.0	0.4	11.4	0.1
CaO	1.19	0.04	20.0	0.5	4.39	0.81	12.3	0.0	18.3	0.5
Na_2O	0.04	0.04	0.10	0.08	0.05	0.05	0.06	0.01	0.12	0.01
K_2O	0.01	0.01	0.01	0.01	0.01	0.01	0.01	0.01	0.01	0.01
Total	100.0		100.0		100.0		100.0		100.0	
Wo	2.6		41.9		9.4		26.1		38.9	
En	45.1		34.9		42.7		38.3		33.7	
Fs	52.3		23.2		47.9		35.6		27.4	
n	43		15		48		2		13	

Abbreviations: $\bar{x}$ = mean, s = standard deviation, n = number of analyses, Wo = Ca/(Ca+Mg+Fe), En = Mg/(Ca+Mg+Fe), Fs = Fe/(Ca+Mg+Fe).
* Averages of defocused beam analyses.

Opaque minerals tend to be clustered, mainly in fine-grained areas of the meteorite. *Kvasha and Dyakonova* (1972) reported unusually high modal Fe-metal (0.4 vol.%) and troilite (0.8 vol.%); and even higher values are implied by their statement (p. 114) that the overall mode for the meteorite includes 5 vol.% opaque minerals. Their bulk-rock analysis indicated extraordinarily high reduced Fe and S (Fig. 5), implying ~3.0 vol.% for combined Fe-metal + troilite. We have not measured modal troilite and Fe-metal separately, but their combined proportion is found to be 2.4 vol.% in the meteorite as a whole, and 4.4 vol.% in the matrix. The Fe-metal is extremely Ni-poor, averaging 99.76 wt.% Fe, 0.12 wt.% Co, and 0.025 wt.% Ni (based on eight analyses of matrix metals plus two of metal from clast 1).

Chromite and ilmenite are found as intergrowths and as individual grains up to 0.5 mm across, situated mainly near, and commonly within, pyroxenes. Based on wide variations in TiO_2 content, three compositional types of chromite were recognized (Table 5, Fig. 6). The chromite with the highest TiO_2 content (15.4 wt.%) is a grain enclosed by a subcalcic augite variety of primary pyroxene. The most abundant type of chromite, which is relatively TiO_2-poor, has an average MgO content of 1.15 wt.%. For comparison, among noncumulate eucrites chromite MgO contents are consistently <0.74 wt.%, and generally <0.3 wt.%; for three cumulate eucrites means of 1.30, 1.31, and 2.18 wt.% have been reported (*Bunch and Keil,* 1971; *Lovering,* 1975). The MgO contents of the ilmenites (Table 6) are likewise high compared to those of noncumulate eucrites, and near the lower end of the range for cumulate eucrites (*Bunch and Keil,* 1971).

The silica phase appears to be tridymite. It tends to occur as relatively fine (7-30 μm), roughly lath-shaped grains, clustered inhomogeneously throughout the meteorite, and in

TABLE 4. Average chemical composition (wt.%) and molecular norms (mol.%) of plagioclase in Pomozdino.

	Matrix		Clast 1		Clast 5			Clast 8		Clast 9				Meteorite	
	plag-1	plag-2	plag-1	plag-2	plag-1	plag-2	Clast 6	plag-1	plag-2	plag-1	plag-2	Clast 10	Clast 11	plag -1	plag-2
SiO_2	46.6	47.7	47.1	47.9	46.6	48.2	46.4	46.8	47.0	47.1	48.2	47.0	45.1	46.5	47.7
TiO_2	0.01	0.02	0.01	0.02	0.01	0.02	0.02	0.04	0.07	0.01	0.02	0.03	0.02	0.02	0.02
Al_2O_3	33.7	33.4	33.8	34.0	33.1	33.1	33.4	33.0	33.3	34.0	33.9	33.3	32.2	33.2	33.2
Cr_2O_3	0.02	0.03	0.03	0.01	0.03	0.00	0.01	0.01	0.00	0.01	0.00	0.01	0.02	0.02	0.01
FeO	0.43	0.37	0.39	0.33	0.27	0.31	0.40	0.40	0.20	0.33	0.50	0.44	0.42	0.32	0.30
MnO	0.02	0.02	0.00	0.02	0.00	0.00	0.03	0.03	0.00	0.02	0.02	0.03	0.03	0.02	0.02
MgO	0.10	0.02	0.03	0.02	0.01	0.01	0.05	0.01	0.00	0.02	0.00	0.02	0.01	0.03	0.01
CaO	17.8	17.1	18.1	17.5	18.7	17.5	18.5	18.4	18.1	18.3	17.7	18.6	18.5	18.3	17.6
Na_2O	1.54	1.92	1.62	1.89	1.61	2.17	1.61	1.62	2.02	1.68	1.92	1.41	1.62	1.57	1.96
K_2O	0.15	0.21	0.14	0.17	0.15	0.21	0.15	0.15	0.21	0.15	0.21	0.12	0.15	0.14	0.21
Total	100.32	100.78	101.19	101.81	100.47	101.56	100.58	100.51	100.86	101.68	102.47	100.97	98.10	100.15	101.00
An	85.69	82.07	85.34	82.87	85.81	80.76	85.69	85.54	82.24	85.05	82.64	87.39	85.60	85.87	82.29
Ab	13.44	16.72	13.86	16.18	13.37	18.09	13.48	13.63	16.64	14.12	16.22	11.95	13.57	13.34	16.53
Or	0.87	1.21	0.80	0.95	0.82	1.16	0.83	0.83	1.12	0.83	1.14	0.66	0.83	0.79	1.18
n	5	4	9	5	3	1	3	4	1	5	1	3	5	50	16

Abbreviations: plag = plagioclase, n = number of analyses, An = Ca/(Ca+Na+K), Ab = Na/(Ca+Na+K), Or = K/(Ca+Na+K).

some cases associated with clusters of opaques. It contains impurities of FeO (0.4-1.1 wt.%), Al_2O_3 (0.4 wt.%), K_2O (0.25 wt.%), CaO (0.1 wt.%), Na_2O (0.03 wt.%), and MgO (0.03 wt.%). Rare phosphates are found enclosed within plagioclase laths. One such phosphate was determined to have a low-chlorine apatite composition. Phosphates have previously been found in the eucrites Ibitira (*BVSP,* 1981) and Bouvante (*Christophe Michel-Lévy et al.,* 1987).

The original igneous texture of Pomozdino has been partly obscured by brecciation. Direct textural evidence of cumulate origin is not strong, being limited to the coarse, blocky nature of some of the pyroxenes. The moderately extensive exsolution within the pyroxene, into high- and low-Ca extremes, attests to relatively slow cooling, and slow cooling suggests an origin in a setting favorable for crystal accumulation. Previously described eucrites with exsolution lamellae as coarse as those of Pomozdino are exclusively cumulates, except for Y-791195, which shows an ambiguous, granular to microgabbroic texture (*Takeda et al.,* 1988). However, the extensive primary Ca-zonation of some of the pyroxenes and the elongated-lath shapes of most of the plagioclase grains are more suggestive of a normal, extrusive (or shallow hypabyssal) eucrite. The tendency for the opaque minerals to be concentrated into fine-grained areas suggests that they crystallized after most of the pyroxene. However, both the pyroxene and the opaque oxides have high, cumulate-like MgO contents. It appears that

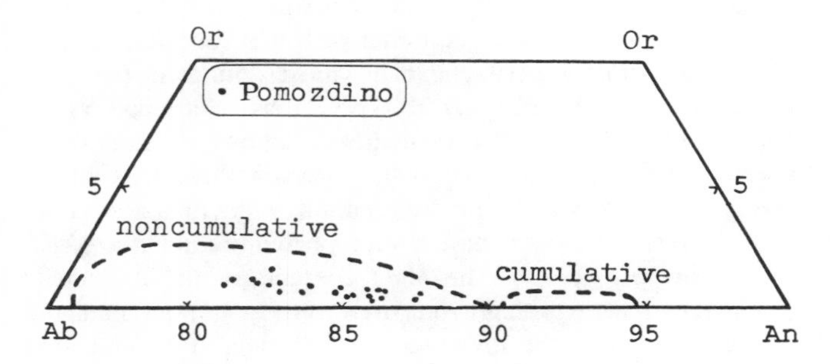

Fig. 4. Plagioclase compositions in Pomozdino are within the range observed for noncumulate eucrites and more sodic than the plagioclase of recognized cumulate eucrites.

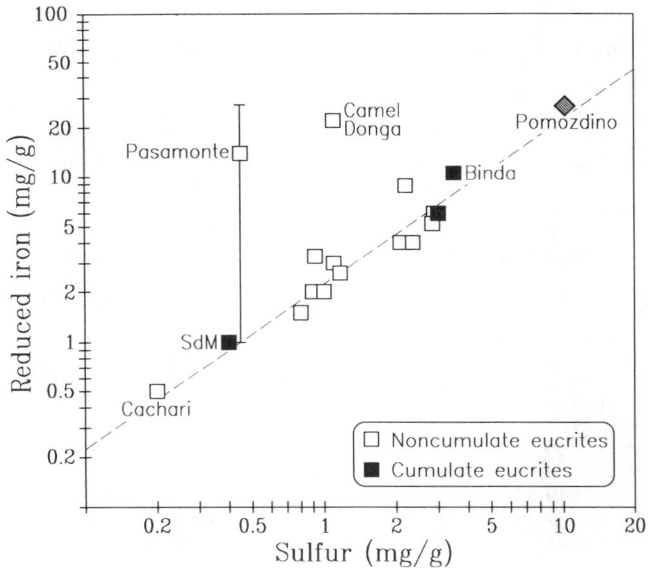

Fig. 5. Bulk-rock data for S vs. reduced iron (i.e., the sum of Fe present as Fe^0 and Fe present as FeS) in eucrites. Data are from *Duke and Silver* (1967), *Kvasha and Dyakonova* (1976), *Kharitonova and Barsukova* (1982), *Cleverly et al.* (1987), *Palme et al.* (1988), and a variety of older sources cited by *Urey and Craig* (1953). Dashed line shows Fe/S ratio of the eutectic in the system Fe-S (*Hansen and Anderko,* 1958).

TABLE 5. Chemical composition (wt%) of chromite in Pomozdino.

	Matrix	Clast 1		Clast 9		Clast 10	Clast 11	Meteorite		
Chromite Type*	1	1	3	1	2	1	1	1	2	3
TiO_2	5.49	4.20	14.8	3.32	10.3	4.78	3.57	4.52	9.9	14.8
Al_2O_3	7.0	7.5	4.50	7.7	6.30	7.1	7.5	7.3	5.89	4.50
Cr_2O_3	49.8	49.4	33.5	51.4	45.1	49.6	50.0	49.8	42.9	33.5
V_2O_3	0.70	0.72	0.58	0.54	0.58	0.51	0.67	0.68	0.55	0.58
FeO	34.3	35.2	44.8	31.8	34.2	33.2	34.1	34.2	37.7	44.8
MnO	0.66	0.63	0.67	0.61	0.62	0.68	0.80	0.67	0.64	0.67
MgO	1.30	0.98	0.88	1.55	1.59	0.84	1.06	1.15	1.23	0.88
ZnO	0.03	0.00	0.02	0.07	0.00	0.00	0.03	0.02	0.01	0.02
Total	99.27	98.66	99.67	97.02	98.73	96.88	97.71	98.33	98.85	99.67
n	5	3	3	1	1	1	2	16	2	3

Abbreviation: n = number of analyses averaged.
* Compositional classification of the Pomozdino chromites based on their TiO_2 contents (see text).

Pomozdino has some cumulate affinities, but judging from the textural and mineralogical heterogeneity it seems that only a slight degree of purification of crystals apart from the parent melt occurred before the crystal-melt mush became a closed system, isolated from the main mass of melt.

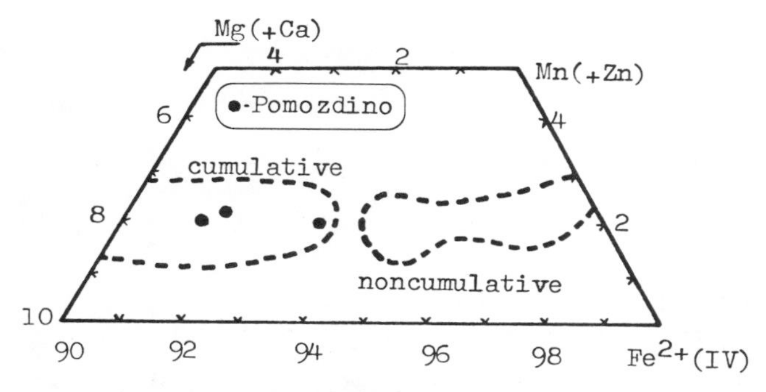

Fig. 6. Three compositional varieties of chromite are found in Pomozdino, but all have MgO/(MgO+FeO) ratios comparable to the chromites of cumulate eucrites and more magnesian than virtually any chromite from noncumulate eucrites.

Bulk-Rock Chemistry

Results for bulk-rock chemistry are shown in Table 7. The modal recombination analysis (No. 5) is taken from the Table 1, which shows the bulk compositions of various clasts and matrix regions, as well as the calculated bulk composition of the meteorite. The compositions in Table 1 were calculated based on the modes for the regions considered, the corresponding mineral compositions shown in Tables 2-6, and the average tridymite composition cited above. The modes were converted to weight proportions based on the densities of the various phases (e.g., *Deer et al.,* 1965). The DBA composition in Table 7 is based on a mean for 345 defocused beam analyses, including clast and matrix lithologies, with the mean "weighted" so as to preserve the modal clast/matrix ratio. *Kvasha and Dyakonova* (1972) reported an uncommonly high pyroxene/plagioclase ratio (63.5 vol.% to ~28.5 vol.%). However, our data for the mode (Table 1), and all of the bulk-rock Al results in Table 7, indicate that the pyroxene/plagioclase ratio is scarcely different from the average for noncumulate eucrites (e.g., *Dodd,* 1981).

The analyses in Table 7 are generally in fair-to-good agreement, but Pomozdino is coarse-grained and heterogene-

TABLE 6. Chemical composition (wt.%) of ilmenite in Pomozdino.

	Matrix	Clasts							Meteorite
		1	5	6	8	9	10	11	
TiO_2	54.2	54.6	54.6	54.9	54.4	54.9	53.8	53.7	54.3
Al_2O_3	0.08	0.07	0.11	0.08	0.08	0.04	0.08	0.07	0.07
Cr_2O_3	0.22	0.13	0.13	0.07	0.24	0.24	0.44	0.18	0.22
V_2O_3	0.21	0.26	0.24	0.25	0.27	0.25	0.23	0.16	0.23
FeO	43.4	43.2	42.3	44.3	44.5	42.4	43.2	43.9	43.5
MnO	1.05	0.93	1.06	1.03	0.97	1.06	1.23	1.47	1.10
MgO	1.73	1.50	1.53	1.63	1.43	1.94	1.57	1.35	1.56
ZnO	0.02	0.02	0.02	0.00	0.05	0.00	0.03	0.00	0.02
Total	100.98	100.61	99.93	102.21	101.93	100.75	100.53	100.87	101.01
n	9	12	1	2	2	1	2	1	33

Abbreviation: n = number of analyses averaged.

TABLE 7. Bulk-rock concentrations of 39 elements in Pomozdino, and literature results for it and other eucrites.

	Na mg/g	Mg mg/g	Al* mg/g	Si mg/g	P mg/g	K μg/g	Ca mg/g	Ti mg/g	Cr mg/g	Mn mg/g	Fe mg/g	Co μg/g	mg′ × 100	mg* × 100
Pomozdino														
(1) Literatue 1	3.0	60.1	59.9	224.2	1.13	0.2	76.9	4.4	0.7	4.1	140.3	10	54.9†	49.6
(2) Literatue 2	—	—	—	—	—	—	—	—	3.9	—	—	9.0	—	—
(3) INAA/FBA-1	3.67	52.8	60.7	233.2	—	0.40	69.3	6.3	2.96	3.98	135.3	5.0	—	47.3
(4) INAA/FBA-2	3.23	53.0	55.3	232.3	—	0.33	65.9	5.5	3.66	4.27	162.8	13.6	—	42.8
(5) Modal recom.	4.2	50.0	56.8	221.1	—	0.6	63.0	5.8	3.8	4.1	169.5	—	47.0†	40.4
(6) DBA	4.0	52.0	59.1	237.9	0.65	0.5	73.5	3.1	2.5	4.0	131.3‡	—	—	47.6‡
Bouvante lit. avg.	3.80	39.0	55.6	234.8	—	0.52	74.5	6.0	2.1	4.10	151.0	6.0	38	37.2
Stannern lit. avg.	4.43	41.1	65.0	228.2	0.54	0.67	75.0	5.7	2.22	4.03	147.5	5.1	39.9	39.0
Avg. 25 Mnc eucrites	3.27	39.9	66.3	230.9	0.36	0.36	73.9	4.4	2.33	4.14	147.6	5.6	39.2	38.3
Avg. 6 Mc eucrites	2.22	68.6	75.5	224.8	0.17	0.23	69.6	1.33	3.73	3.58	114.5	11.3	59	57.9
Avg. 31 M eucrites	3.07	45.6	68.2	229.5	0.32	0.33	73.0	3.7	2.60	4.05	141.2	6.9	43.6	42.6

	Sc μg/g	V μg/g	Ni μg/g	Zn μg/g	Ga μg/g	Ge ng/g	Sr μg/g	Zr μg/g	Ba μg/g	La μg/g	Ce μg/g	Nd μg/g	Sm μg/g	Eu μg/g
Pomozdino (2)	23	—	—	—	—	—	—	32	—	2.6	9	7	2.3	0.66
Pomozdino (3)	28.8	72	1.12	13	2.1	134	117	136	57	6.0	14.6	10.8	3.3	0.91
Pomozdino (4)	30.0	72	3.1	28	1.13	370	108	<172	62	5.2	13.2	9.1	2.8	0.78
Bouvante lit. avg.	30.5	52	<15	<50	2.1	—	94	86	57	5.8	15.7	11.3	3.6	0.82
Stannern lit. avg.	31.1	65	3	1.7	1.65	106	91	84	54	5.2	13.36	10.4	3.2	0.80
Avg. 25 Mnc eucrites	30.8	69	10.5	4.7	1.63	55	78	56	40	3.4	9.295	7.1	2.06	0.65
Avg. 6 Mc eucrites	20.6	96	11.0	4.4	1.40	13.5	51	16.3	12.8	0.53	1.38	1.19	0.37	0.34
Avg. 31 M eucrites	28.3	74	10.6	4.6	1.57	41	74	49	35	2.7	7.4	5.4	1.66	0.59

TABLE 7. (continued).

	Tb μg/g	Dy μg/g	Ho μg/g	Tm μg/g	Yb μg/g	Lu μg/g	Hf μg/g	Ta μg/g	Re pg/g	Os pg/g	Ir pg/g	Au ng/g	Th μg/g	U μg/g
Pomozdino (2)	0.4	—	0.6	—	1.3	0.20	0.20	0.20	—	—	—	—	—	—
Pomozdino (3)	0.78	5.0	—	0.45	2.84	0.42	2.46	0.3	<3	<13	8.7	0.37	0.71	0.20
Pomozdino (4)	0.78	<10	1.03	0.45	2.60	0.39	1.96	0.27	3.2	9	62	0.88	0.68	0.18
Bouvante lit. avg.	0.85	5.7	1.25	0.50	3.14	0.46	2.70	0.41	—	—	—	0.4	0.67	0.19
Stannern lit. avg.	0.75	4.9	1.09	0.47	2.71	0.41	2.54	0.41	—	—	130	1.0	0.60	0.17
Avg. 25 Mnc eucrites	0.51	3.42	0.82	0.38	2.02	0.32	1.55	0.22	6.2	49	185.8	2.57	0.43	0.12
Avg. 6 Mc eucrites	0.07	0.71	0.10	0.05	0.48	0.09	0.42	0.07	1.5	—	201.7	2.11	0.04	0.02
Avg. 31 M eucrites	0.45	2.75	0.69	0.30	1.68	0.27	1.36	0.20	4.3	49	190.1	2.44	0.35	0.09

Abbreviations: M = monomict, Mnc = monomict noncumulate, Mc = monomict cumulate, avg. = average, mg′ = molar MgO/(MgO+FeO), mg* = molar Mg/(Mg+Fe).

Pomozdino data are from: (1) *Kvasha and Dyakonova* (1972), wet-chemical analysis; (2) *Kolesov* (1976), INAA; (3) and (4) INAA /EPFBA at Los Angeles; (5) modal recombination at Moscow (Table 1); and (6) DBA at Moscow (see text).

Bouvante average is based on *Palme and Ramensee* (1981) and *Christophe Michel-Lévy et al.* (1988). Stannern and eucrite-group averages (which exclude Pomozdino) are based on many sources, mostly cited by *Warren and Jerde* (1987).

*In addition, *Aylmer et al.* (1988) report an Al concentration for Pomozdino of 57.8 mg/g.

†Analyses (1) and (5) determined FeO separately from total Fe. For these analyses, the molar MgO/(MgO+FeO) ratio is shown as mg′. The mg′ ratio is estimated for the other eucrites by assuming that virtually all of the Fe occurs as FeO. Most of the Pomozdino analyses determined only total Fe. The molar Mg/(Mg+Fe) or mg* ratio is considerably lower than the molar MgO/(MgO+FeO) or mg′ ratio.

‡The DBA result for Fe is actually a lower limit, as it is based on a correction procedure assuming all Fe is in silicates or oxides and none in Fe-metals or sulfides (and consequently the DBA result for mg* is actually an upper limit).

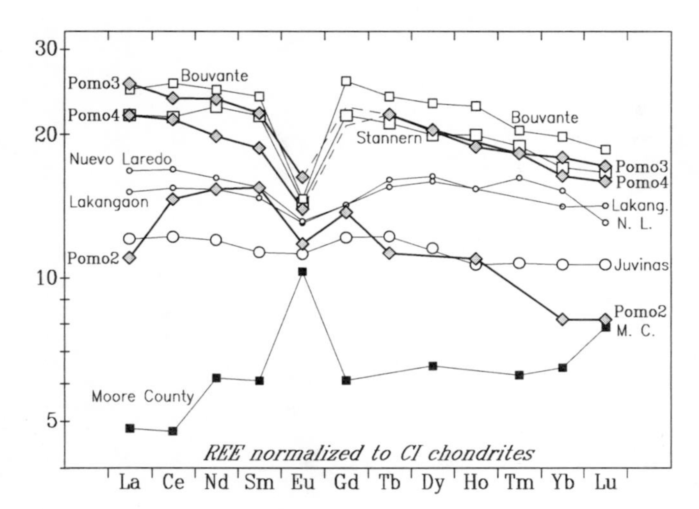

Fig. 7. Concentrations of rare earth elements (REE) normalized to CI chondrites. Shown for comparison are various relatively REE-rich eucrites, as well as a more typical noncumulate eucrite (Juvinas) and the most REE-rich of the recognized cumulate eucrites (Moore County). Pomozdino data are from Table 7; other sources include *Jerome* (1970), *Philpotts and Schnetzler* (1970), *Palme and Ramensee* (1981), *Christophe Michel-Lévy et al.* (1987), *Warren and Jerde* (1987), and *Palme et al.* (1988). See also *Warren and Jerde's* (1987) compilation of a large number of references for compositional data on eucrites.

ous (especially for Fe-metal and troilite) at the scale of our samples, and sample heterogeneities are no doubt responsible for much of the disparity among the various analyses. The fine-grained matrix appears far more homogeneous (at the roughly 1-mm scale involved in distinguishing the 10 regions featured in Table 1) than the coarse-grained, uncomminuted clast portion of the meteorite. As discussed above, the matrix appears to have significantly higher contents of opaque minerals (especially Fe-metal and FeS) than the average clast, and this disparity most likely preserves an anticorrelation between opaque abundance and grain size from the prebrecciation texture of the meteorite. The Cr result of *Kvasha and Dyakonova* (1972) and the Hf result of *Kolesov* (1976) appear spuriously low. The general tendency for incompatible elements (e.g., REE, Fig. 7) to be lower in Kolesov's analysis than in our two analyses presumably reflects real heterogeneity among the samples analyzed. *Kolesov's* (1976) anomalously low La result (see the resultant La/Ce, La/Nd, and La/Sm ratios; Fig. 7) might be an analytical error. However, it might simply be another manifestation of Pomozdino's heterogeneity. The Moore County cumulate eucrite shows a comparably low La/Sm ratio (Fig. 7). Without further constraints, it is difficult to judge between these alternatives.

Results for siderophile trace elements (Co, Ni, Ge, Re, Os, Ir, and Au) are consistently, with the possible exception of Os, higher for INAA/FBA-2 than for INAA/FBA-1. Osmium is reported as an upper limit for the INAA/FB-1 sample because during Os radiochemistry a centrifuge tube broke, and the sample had to be salvaged from the bottom of the centrifuge, which probably resulted in at least minor "cross-sample" contamination. Presumably, the concentration of Fe-metal was higher in the INAA/FBA-2 sample than in the INAA/FBA-1 sample. Note that the concentration of total Fe is also relatively high for INAA/FBA-2—even though its Mg content is essentially the same as that of INAA/FBA-1. Subsequent discussion and figures will be based on an average of all the bulk-rock data for Pomozdino, including (except as noted) the earlier analyses compiled in Table 7.

DISCUSSION

Confirmation that Pomozdino Is a Eucrite

Given the unusual composition and petrology of this meteorite, an obvious question arises as to whether it is truly derived from the eucrite parent body. The most powerful tool for resolving such a question is probably the three-isotope plot for oxygen isotopes (*Clayton et al.,* 1976), and especially the ratio $\Delta^{17}O$ ($= \delta^{17}O - 0.52\ \delta^{18}O$). Preliminary results from *Petaev et al.* (1988) are shown in Fig. 8. One of four whole-rock samples plots at surprisingly low $\delta^{18}O$, and two other samples (a whole-rock and a plagioclase separate) plot slightly above the terrestrial fractionation line. The reason that one of the whole-rock samples yielded such an unusually low $\delta^{18}O$ remains obscure; and the uncertainty in the Petaev et al. data (quoted as 0.1‰) is somewhat greater than that achieved among recent analyses from the long-developed University of Chicago lab (e.g., *Mayeda and Clayton,* 1989). However, the mean $\Delta^{17}O$ of the seven Pomozdino samples is -0.15 (±0.16), which is in good agreement with the mean of 46 analyses reported from the Chicago lab for meteorites of the eucrite-howardite-diogenite association: -0.232 ± 0.119 (Fig. 8). In contrast, except for a few early analyses of Shergotty (*Clayton et al.,* 1976), all SNC meteorite analyses plot well above the terrestrial line. The mean $\Delta^{17}O$ of 13 SNC analyses reported by *Clayton and Mayeda* (1983) is +0.249 ± 0.091. Thus, Fig. 8 strongly suggests that Pomozdino is not of SNC affinity.

Further evidence of the distinction between Pomozdino and SNC meteorites comes from comparison of ratios involving element pairs that are geochemically similar, except for their volatilities. For example, Fig. 9 shows the plagiophile/refractory element Al plotted vs. the plagiophile/volatile element Ga. The SNC meteorites consistently have Ga/Al in the range $2.1\text{-}4.5 \times 10^{-4}$, whereas eucrites have Ga/Al in the range $1.5\text{-}4.0 \times 10^{-5}$, and Pomozdino has Ga/Al $\sim 2.7 \times 10^{-5}$. This parameter also effectively distinguishes eucrites from two other varieties of achondrite that might conceivably be related to Pomozdino: the brachinites (Ga/Al range $0.7\text{-}2.3 \times 10^{-3}$) and the angrites (Ga/Al $\sim 5 \times 10^{-6}$). A plot of incompatible/refractory Sm vs. incompatible/volatile P (Fig. 10) is roughly analogous, except angrites are not so well resolved from eucrites. For further confirmation, Fig. 11 shows Fe plotted vs. the volatile, FeO-affiliated element Mn. Although eucrites and SNC meteorites show considerable overlap on this plot, brachinites and angrites clearly have Fe/Mn far too high to be plausibly linked with eucrites, including Pomozdino. The Fe/Mn ratio also clearly distinguishes Pomozdino from lunar materials, which have Fe/Mn ratios of about 75±5 (*Warren,* 1990).

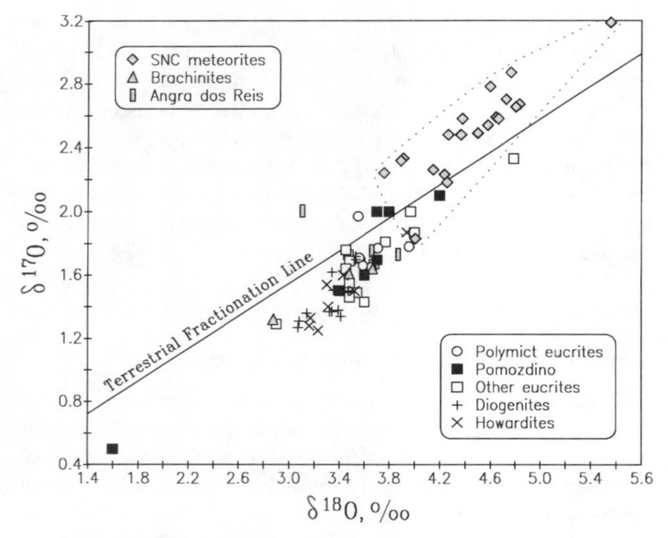

Fig. 8. Oxygen isotopic data for various samples of Pomozdino, from *Petaev et al.* (1988), except for two additional unpublished analyses (M. I. Petaev, personal communication, 1989). Shown for comparison are results for a variety of other achondrites (mineral separates as well as bulk-rock samples), from the Chicago laboratory (*Clayton et al.,* 1976, 1979; *Clayton and Mayeda,* 1983; *Mayeda and Clayton,* 1989; *Mayeda et al.,* 1987; and R. N. Clayton, personal communication, 1988, for the ALH84025 brachinite).

Conceivably, Pomozdino might be a portion of a former silicate enclave in a mesosiderite. However, data from ***Rubin and Jerde*** (1988) show that even samples from deep within 5-cm silicate "pebbles" from the Vaca Muerta mesosiderite have consistently high siderophile element concentrations. For example, among the 10 Vaca Muerta pebbles studied by ***Rubin and Jerde*** (1988), Au ranges from 4.5 to 19 ng/g, and Ni ranges from 290 to 1690 μg/g. ***Mittlefehldt*** (1979) obtained similar results for silicate enclaves from seven other mesosiderites. The contrast with Au and Ni in Pomozdino (Table 7) is stark. In summary, all available evidence indicates that Pomozdino is a eucrite.

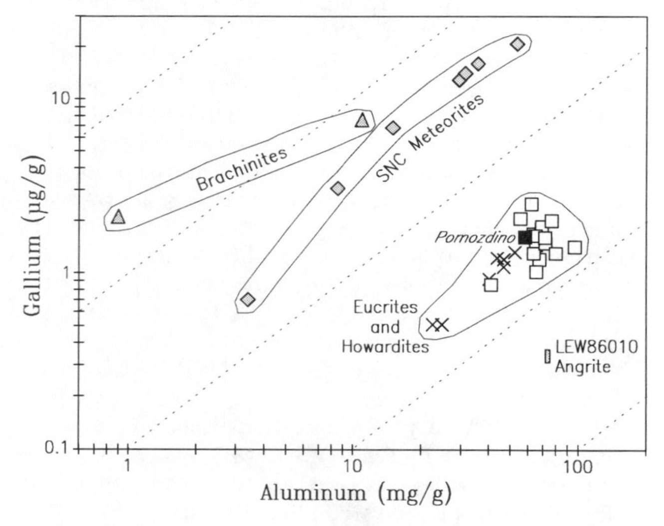

Fig. 9. Bulk-rock Al vs. Ga for monomict eucrites and a variety of other achondrites, showing that Pomozdino has a characteristically eucritic Ga/Al ratio (diagonal lines represent various fixed Ga/Al ratios). Data are from Table 7 and a variety of literature sources, nearly all cited by *Warren and Jerde* (1987); see also *Kallemeyn and Warren* (1989) and *Warren and Kallemeyn* (1989).

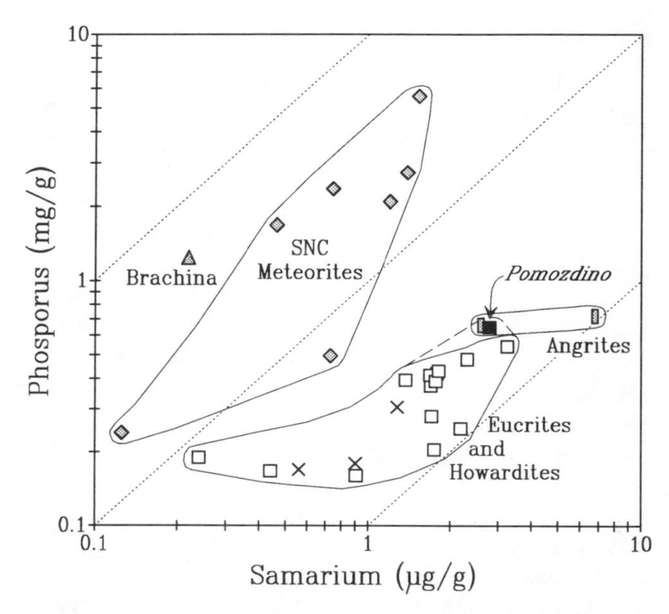

Fig. 10. Bulk-rock Sm vs. P for monomict eucrites, analogous to Fig. 9.

Relationship to Other Types of Eucrite

The geochemical uniqueness of Pomozdino is exemplified by a plot of the MgO/(MgO+FeO) ratio vs. a representative incompatible element, such as Sm (Fig. 12). Note that the

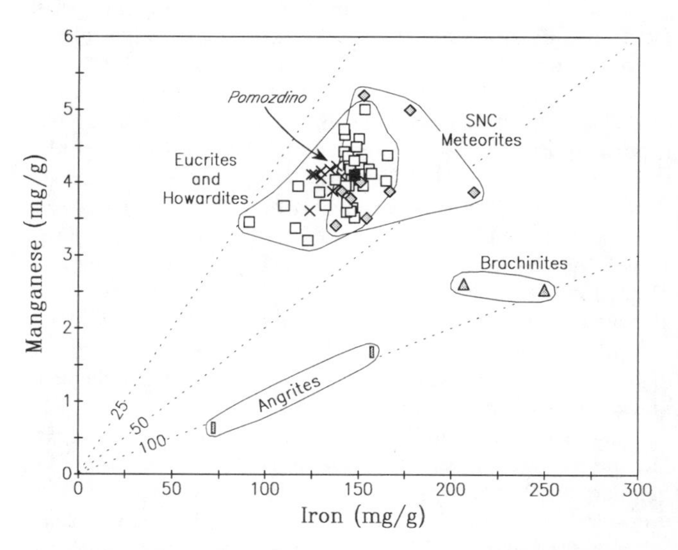

Fig. 11. Bulk-rock Fe vs. Mn for monomict eucrites, analogous to Fig. 9.

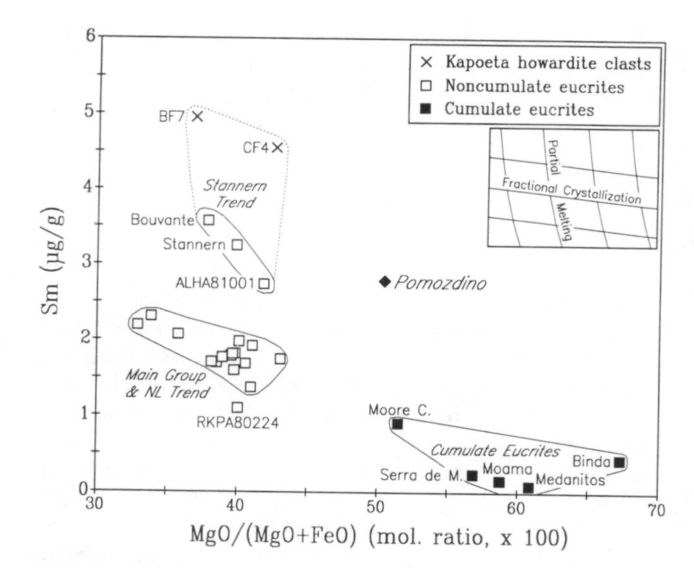

Fig. 12. Bulk-rock MgO/(MgO+FeO) vs. Sm, showing anomalous composition of Pomozdino in relation to other monomict eucrites. Data are from Table 7 and a variety of literature sources, nearly all cited by *Warren and Jerde* (1987). Also shown are two uncommonly REE-rich clasts from the Kapoeta howardite (*Smith,* 1982).

plotted ratio is not simply Mg/(Mg+Fe). For most eucrites, the difference between MgO/(MgO+FeO) and Mg/(Mg+Fe) is negligible. In the cases of eucrites for which only total-Fe data are available, the Fe^0/FeO ratio is assumed, for plotting purposes, to be the same as the average for all eucrites.

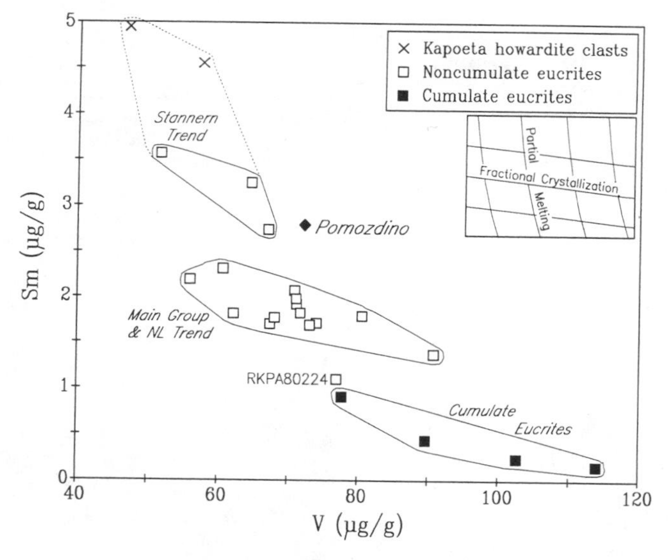

Fig. 13. Bulk-rock V vs. Sm for monomict eucrites, analogous to Fig. 12—including data sources; except note that some of the oldest V data are problematical. Data from the CalTech lab (e.g., *Duke,* 1963) show unaccepatble scatter (mostly toward high values) and hence have been ignored. Data from *Jerome* (1970) appear to be precise, but systematically high, and have been recalibrated (reduced) by a factor of 0.75.

However, for Pomozdino, which has unusually high contents of Fe-metal and FeS (Fig. 5), the difference is quite significant. Hereafter, these ratios will be abbreviated as mg′ and mg*, for MgO/(MgO+FeO) and Mg/(Mg+Fe), respectively.

Previous work (e.g., *Stolper,* 1977; *Reid et al.,* 1979; *BVSP,* 1981; *Warren and Jerde,* 1987) has suggested that the noncumulate eucrites can be classified into two or three types, on the basis of diagrams such as Fig. 12: (1) a "main Group" of eucrites with mg′ = roughly 40 and Sm ~1.7 µg/g; (2) a "Stannern trend" with similar mg′, but Sm extending upward, as if reflecting lower and lower degrees of partial melting; and (3) a "Nuevo Laredo Trend" extending toward higher Sm but also lower mg′ from the main group, as if reflecting a fractional crystallization residue sequence. Also shown in Fig. 12 are two unusual clast lithologies from the Kapoeta howardite (*Smith,* 1982) that appear to represent extensions of the Stannern trend. The main group eucrites may actually be related to one (or both) of the two trends, but available constraints do not resolve a clear hiatus between either of the trends and the main group. Apart from these two to three types of noncumulate eucrites, the cumulate eucrites are distinguished by far higher mg′ ratios and far lower Sm contents. Pomozdino's composition does not fit into any of these classes. Its mg′ ratio is too high for any of the three types of noncumulate eucrites. Its Sm content is far higher than expected for a cumulate eucrite and would only be expected for an extremely low-mg′ extension of the Nuevo Laredo trend.

Pomozdino also falls in an analogous position on a plot of V vs. Sm (Fig. 13); which is not surprising because, like the mg′ ratio, V tends to be higher in crystals than in their equilibrium melt counterparts (*Warren and Jerde,* 1987). Even more closely analogous would be plots of mg′ vs. any of a number of other incompatible elements (e.g., Ba, Hf, and Ti).

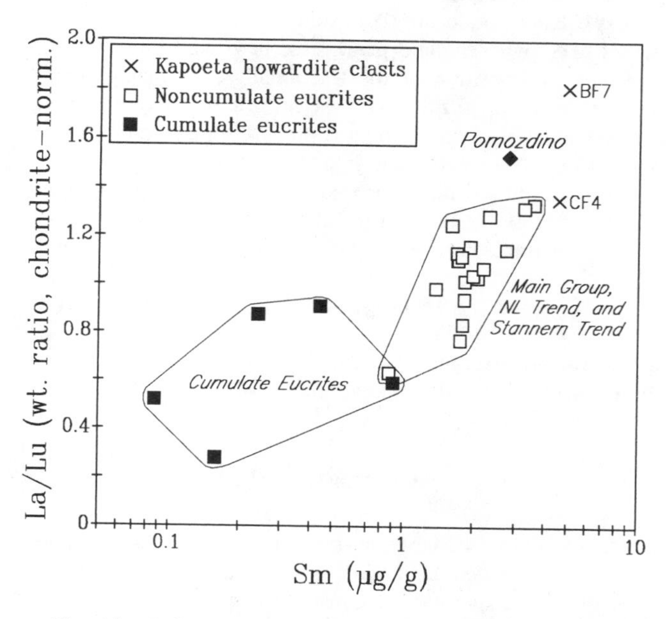

Fig. 14. Bulk-rock Sm vs. La/Lu for monomict eucrites and REE-rich Kapoeta clasts, showing that Pomozdino has an uncommonly high La/Lu ratio.

However, except for Ti, fewer eucrites have been analyzed for these elements. Perhaps because Ti is partially compatible, it does not separate the Stannern and Nuevo Laredo trends as clearly as Sm.

Sampling "errors" linked to the heterogeneity of Pomozdino do *not* plausibly account for these anomalous aspects of its composition. Table 7 shows five separate measurements for the mg^* ratio, and two separate measurements for the mg' ratio. The two mg' ratios are obviously high compared to all except cumulate eucrites (see also Fig. 12). The mg^* ratios are superficially more normal, but considering the abundant evidence for an uncommonly high Fe^0+FeS content (e.g., Fig. 5), even the lowest among the mg^* ratios implies an anomalously high mg' ratio. Of course, the pyroxene and opaque oxide compositions (Figs. 3 and 6) also imply a bulk-rock mg' virtually as high as that of Moore County. However, the modal recombination analysis in Table 7 is based on these same mineral-composition data, and note that the modal recombination analysis has the *lowest* mg^* of all the analyses in Table 7. The high Sm content is not the only "evolved" aspect of the REE composition. Among eucrites in general, the ratio La/Lu tends to correlate with Sm (Fig. 14). Pomozdino's La/Lu is the highest of any known eucrite (notwithstanding the strangely low La/Ce ratio of *Kolesov's* (1976) analysis).

Origin of Pomozdino

Pomozdino clearly does not fit into any of the previously recognized subgroupings of eucrites, but it appears to have closer affinities to the Stannern trend than to either of the other two types of noncumulate eucrites. The relatively high mg' ratio, along with the textural features discussed above, suggest that Pomozdino has some cumulate affinities, albeit its origin seems to have involved only a slight degree of purification of crystals apart from the parent melt. Consider a mass-balance model for Pomozdino as mixture of cumulus matter (i.e., pyroxene and plagioclase in equilibrium with the transitory composition of the fractionally crystallizing parent melt at the point in time where Pomozdino's matter became isolated from the parent melt), plus "trapped" melt. If the proportion of trapped melt is abbreviated as TM, then the fraction of cumulus matter is equal to 1-TM. For any assumed parent melt MgO/FeO ratio, the MgO/FeO ratio of the cumulus matter is implied by the relationship

$$K_D = ([FeO]_{xtl}/[FeO]_{liq})/([MgO]_{xtl}/[MgO]_{liq}) \qquad (1)$$

where the square brackets denote molar concentrations, and the subscripts are xtl = crystal and liq = liquid. Another way of writing equation (1) is

$$mg'_{xtl} = 1/(1+K_D/mg'_{liq}-K_D) \qquad (2)$$

Essentially all of the Mg+Fe in the cumulus matter is contained in pyroxene (pigeonite and/or low-Ca augite), for which a large body of experimental constraints implies that K_D ~0.30 (*Warren,* 1986). Given this precise relationship between assumed parent melt MgO/FeO and implied cumulus matter MgO/FeO, and knowing the MgO/FeO ratio of the mixture (Pomozdino), a trivial mass-balance calculation can be used to infer the TM implied by any given parent melt MgO/FeO. The Sm concentration of the parent melt, S_{pa} can then be calculated, because mass balance implies that

$$S_{pa} = S_{mix}/(D+TM(1-D)) \qquad (3)$$

where S_{mix} is the Sm concentration of the mixture (Pomozdino), and D is the bulk crystal/liquid distribution coefficient for Sm. Based on experimental results of *McKay* (1982a,b), the value of D is almost negligibly small, ~0.025 for a mixture of low-Ca pyroxene and plagioclase in 6:4 wt. proportions. The results from these mass balance calculations are curves for the potential compositions of a Pomozdino parent melt, and for the cumulus component within Pomozdino (Fig. 15). Despite the great uncertainty associated with TM (textural evidence only implies that TM is definitely greater than 0, and probably also significantly less than 1), Fig. 15 indicates that the range of potential parent melts obliquely crosses the general region of the Stannern trend, and remains well to the high-Sm side of any plausible extension of the Nuevo Laredo trend.

Analogous calculations for the recognized cumulate eucrites lead to curves for their potential parent melts (not shown in the figure) that pass slightly to the low-MgO/FeO side of the most Sm-rich members of the Nuevo Laredo trend and far to the low-MgO-FeO side of the Stannern trend, except for Binda. The curve for potential parent melts of Binda crosses through the Stannern trend, but it also crosses the high-MgO/FeO end of the main group - Nuevo Laredo trend region.

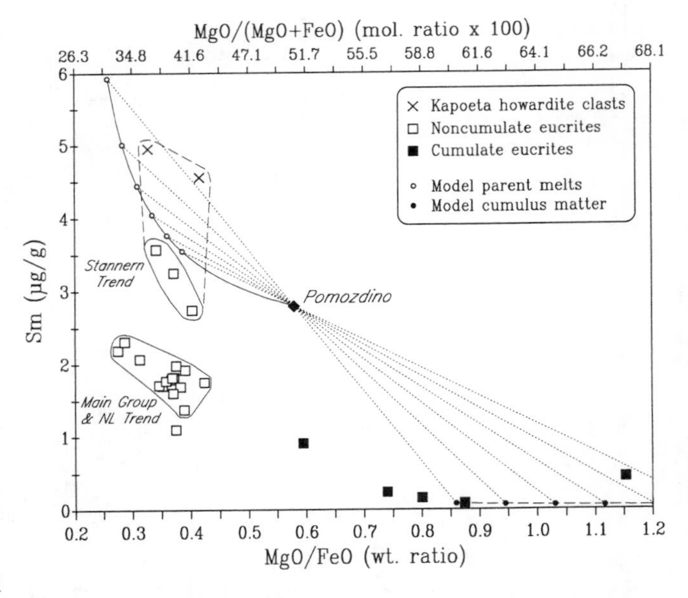

Fig. 15. Bulk-rock MgO/FeO vs. Sm for monomict eucrites and REE-rich Kapoeta clasts, with results from models for Pomozdino as a mixture of high-MgO/FeO, REE-poor cumulus matter, and low-MgO/FeO, REE-rich trapped liquid.

Conversely, assuming that Pomozdino is a product of a melt roughly along the Stannern trend, its TM can be estimated by inversion of the same technique. For example, if the range of MgO/FeO wt. ratios for the intercept between the potential parent melt curve (Fig. 15) with the Stannern trend is assumed to be between 0.3 and 0.4, then Pomozdino's TM is implied to be between 59 and 80 wt.%. A problem with this model is that there is no obvious mechanism for production of "cumulates" with TM higher than ~45 vol.%, the approximate TM that results from simple settling, without any compaction, of crystals (*Wager et al.,* 1960). However, mechanical settling and compaction are not necessarily the only processes for replacement of "trapped" liquid with compositionally "cumulus" matter (*Campbell,* 1978; *Irvine,* 1982); e.g., simple *in situ* growth of crystals, if accompanied by continued compositional diffusion between the interstitial melt and the main mass of melt, may lead to compositionally similar results. As discussed by *Warren and Jerde* (1987, p. 720), data for the Palisades Sill (*Shirley,* 1986) indicate that diabases with roughly Pomozdino-like textures [i.e., elongate grains, especially plagioclases; little indications of any "cumulus framework" (*Irvine,* 1982)] may contain ~50 wt.% TM. Conceivably, in the far lower gravity field of the eucrite parent asteroid, an otherwise similar environment engendered slightly higher TM as Pomozdino formed.

Of course, the mass balance calculations used to construct Fig. 15 involve considerable uncertainties, mainly associated with the composition of Pomozdino. The possibility cannot be rigidly excluded that Pomozdino's anomalous composition is simply that of a quenched melt (TM = 1). If Pomozdino does represent a melt, equation (2) implies (assuming K_D ~0.30) that the equilibrium mafic silicates would have mg′ ~0.78. Pyroxenes this magnesian are common among the diogenites, which probably come from the same parent body as the eucrites. Although diogenites are invariably plagioclase- and REE-poor, the nature of their parent melts is still poorly constrained (e.g., *Mittlefehldt and Lindstrom,* 1988b). In any case, the conclusion that the parent melt of Pomozdino was more Stannern-like than Nuevo Laredo-like appears at least as firm as the hypothesis (e.g., *Stolper,* 1977; *Reid et al.,* 1979; *BVSP,* 1981; *Warren and Jerde,* 1987) that the Stannern and Nuevo Laredo "trends" formed by fundamentally different processes.

The unusual combination of Ca-zonation with minimal mg′ zonation within the primary pyroxenes also hints at kinship with Stannern and Bouvante, which represent the best examples of similar zonation among other eucrites (*Takeda et al.,* 1983). Analogous zonation in the case of Nuevo Laredo does not appear to be nearly so well defined (*Takeda et al.,* 1983). Bouvante also resembles Pomozdino in texture, being an apparently monomict breccia (or possibly a genomict breccia), with coarse, subophitic enclaves set amidst a much finer grained, mostly brecciated matrix (*Delaney et al.,* 1984; *Christophe Michel-Lévy et al.,* 1987; *Boctor et al.,* 1989).

A geochemical distinction among howardite clasts has been postulated on the basis of plots of the En (or mg′) ratio in pyroxene vs. the An or Ca/(Ca+Na+K) ratio in plagioclase (*Delaney et al.,* 1981; *Ikeda and Takeda,* 1985). A wide range of An is observed for any given En, and *Delaney et al.* (1981) postulated that two subparallel trends, a high-An "peritectic" trend and a low-An "evolved" trend, are resolved. The average mineral compositions of Pomozdino (Tables 3 and 4) place it in the low-An or "evolved" field on this diagram; i.e., the same field as Stannern. *Hewins and Newsom* (1988) regard the two postulated trends on the En vs. An diagram as a more significant manifestation of geochemical and petrogenetic diversity than the distinction between the Stannern trend and the main group + Nuevo Laredo trend eucrites (Fig. 12). However, the bulk-rock Ca/(Ca+Na+K) ratio of Pomozdino is less "evolved" than those of many typical, moderate-Sm, and presumably "peritectic," eucrites.

Hewins and Newsom (1988) suggest that the most primitive member of the "evolved" trend may be the clast "ρ" from Kapoeta (*Dymek et al.,* 1976). This clast has sodic plagioclase (average An_{79}), a moderately high bulk Ti content (4.7 mg/g), and yet a bulk mg* ratio of 0.490. Thus, superficially at least, clast ρ resembles Pomozdino. However, clast ρ has a well-preserved porphyritic texture with ~40% pyroxene phenocrysts. *Dymek et al.* (1976) discussed the possibility that ρ nonetheless closely resembles the composition of its parent melt [some porphyritic lunar basalts are thought to be compositionally almost identical to their parent melts (*Grove and Walker,* 1977)]. However, we find this interpretation strongly at odds with the clast's noncotectic pyroxene/plagioclase ratio (2.05 by volume), especially considering the supposedly evolved nature of the lithology. Barring remelting of magma ocean cumulates (as may have occurred in the case of the lunar mare basalts), it is difficult to conceive of ways to generate noncotectic melts on a small body such as an asteroid. *Dymek et al.* (1976) also cited mineralogic evidence suggesting that "a portion of the pyroxene phenocrysts are cumulate in origin." Thus, we infer that the parent melt of clast ρ probably had a far lower mg* ratio than ρ itself, possibly even as low as the mg* of Stannern.

The texture of Pomozdino does not rule out an origin as a porphyry. However, its normal-eucritic pyroxene/plagioclase ratio is difficult to reconcile with entrainment of a large component of phenocrysts prior to the main stage of cooling and crystallization, unless both pyroxene and plagioclase phenocrysts were present, and in roughly equal proportions. In any case, the distinction between a slowly cooled porphyry and a high-TM cumulate is a subtle one, especially for products of a small parent body where the source region and the emplacement region of a partial melt cannot be more than a few kilometers apart.

Origin of the High Fe-metal and S Contents

The high Fe^0+FeS content of Pomozdino (Fig. 5, Table 1) is not a component of foreign matter introduced by impact-mixing. We find extremely low contents of all highly siderophile elements (Ni, Ge, Re, Os, Ir, and Au) in both of the splits analyzed (Table 7). Derivation of the several weight percent of Fe^0+FeS in Pomozdino from a chondrite or iron meteorite would yield gross enrichments of the siderophile elements relative to uncontaminated eucrites. In fact, for most of these elements, with the exception of Ge and possibly Os, the average of our two Pomozdino bulk-rock analyses is lower than the literature average for eucrites. This conclusion holds, even if the literature database is restricted to recent, more reliable analyses, as recommended by *Warren and Jerde* (1987). Based on comparison between our microprobe result for average Ni content of the Fe-metal (250 $\mu g/g$) and our bulk-rock results (Table 7), it appears that the bulk-rock samples contained an average of 0.8 wt.% Fe-metal (assuming negligible Ni in phases other than Fe-metal). Thus, the explanation for the low siderophile results is probably not that our samples were grossly unrepresentative.

The eucrite Camel Donga (*Palme et al.,* 1988) contains about as much Fe^0 as does Pomozdino (Fig. 5). *Palme et al.* (1988) infer that Camel Donga originally crystallized with a fairly normal Fe^0 content and later became enriched in Fe-metal at the expense of FeO and FeS, during heating, with loss of S_2 and SO_2, caused by intense impact shock. The main source of the FeO was probably pyroxene. Thus, on Fig. 5, brecciation moved Camel Donga diagonally upward and to the left. In the case of Pomozdino, the high Fe^0 content apparently dates from the initial igneous crystallization, because the Fe^0/FeS ratio is quite normal for a eucrite (Fig. 5). The significance of these high Fe-metal and FeS contents is unclear. They may merely reflect vagaries of eucrite genesis that have little to do with other distinctions among eucrites. However, the two REE-rich Kapoeta clasts, BF7 and CF4 (*Smith,* 1982), also have relatively high contents of Fe-metal and FeS (*Warren and Taylor,* 1982). If these three lithologies all were derived (more or less) directly from primary partial melts of the eucrite parent body (the conventional interpretation of the Stannern trend), it does not seem surprising that they have higher Fe-metal and FeS contents than eucrites derived by extensive fractional crystallization (i.e., the Nuevo Laredo trend, and possibly most of the main group), given the likelihood that fractional crystallization was a protracted process accompanied by extensive volatile loss (*Ikeda and Takeda,* 1985; *Mittlefehldt,* 1987) and settling of Fe-metal. However, neither Stannern nor Bouvante (*Christophe Michel-Lévy et al.,* 1987) appears to be unusually rich in Fe-metal and FeS.

In geochemical processing of planetary matter, siderophile elements tend to follow Fe-metal. Pomozdino is uncommonly rich in Fe-metal, and yet it has even lower concentrations of most siderophile elements than an average eucrite. Apparently, the Fe-metal enrichment in Pomozdino formed by complicated, multistage processing of the ultimate parental (presumably roughly chondritic) material. It is tempting to speculate that the high Fe-metal content reflects reduction of FeO within a parent magma initially along the Stannern trend (i.e., with melt mg′ ~0.40). This model would simultaneously help account for the anomalously high mg′. However, the mg* ratio is also anomalously high, albeit less dramatically so (Table 7). Moreover, Pomozdino is also enriched in FeS, and has a normal eucritic Fe^0/FeS ratio (Fig. 5), both of which would be unlikely if the Fe-metal were derived mainly by reduction of FeO from a normal parent magma.

On the other hand, the extreme depletions of the siderophile elements in Pomozdino, as well as in the other REE-rich eucrites, including Stannern and Bouvante (e.g., *Chou et al.,* 1976; *Palme et al.,* 1988; *Christophe Michel-Lévy et al.,* 1987), are difficult to reconcile with models for production of any of these eucrites as primary partial melts of primitive, little-differentiated sources. Genesis of the high REE contents (etc.) would require an exceptionally low degree of partial melting. However, as pointed out by *Hewins and Newsom* (1988), low siderophile element levels imply that virtually complete segregation of metal apart from silicate occurred before the fractionation events that produced the REE-rich magmas; yet relatively high degrees of partial melting probably have to be sustained before a silicate-metal mush becomes sufficiently lubricated to render complete silicate/metal segregation (cf. *Taylor,* 1989).

Other Implications Regarding the Eucrite Parent Body

Interpretation of Pomozdino in terms of the composition and evolution of the eucrite parent body hinges on whether it is interpreted as a partial cumulate (Fig. 16a) or as a quenched melt (Fig. 16b). If Pomozdino is essentially a quenched melt, then the disparity in mg′ between it and the Stannern trend calls into question the hypothesis (*Stolper,* 1977) that the Stannern trend formed as a series of low-degree primary partial melts of primitive, little-differentiated source material. Except at high degrees of melting, the melt mg′ should remain relatively uniform as equilibrium partial melting proceeds (*Warren and Jerde,* 1987). The partial melting hypothesis could be salvaged by invoking a surprising degree of heterogeneity for the mantle of the eucrite asteroid (à la *Smith,* 1982; see also *Hewins and Newsom,* 1988) and/or by invoking a complex, multistage genesis for the source material of Pomozdino. However, invoking such a complex history for Pomozdino would tend to undermine the presumption that the source regions for the roughly similar Stannern trend eucrites were little differentiated. Conceivably, Pomozdino is the only case among known eucrites of a low-degree primary (or near-primary) partial melt of primitive, little-differentiated source material and the Stannern trend actually comprises residual products of fractional crystallization of REE-rich, yet magnesian melts similar to Pomozdino (Fig. 16b). If so, and if the primitive mantle of the parent asteroid was roughly uniform in mg′, then the main group cluster of eucrites would also have to be fractional crystallization products, except derived from far higher-degree primary melts than Pomozdino. The early stages of crystallization of these melts might have engendered near-monomineralic pyroxene cumulates—the diogenites. The narrow range of mg′ and REE contents among the main group eucrites might seem difficult to engender by such a model. However, the diversity of both the degree of partial melting (which would the main factor governing the REE) and the degree of subsequent fractional crystallization (which would be the main factor governing the mg′ ratio) would tend to be dampened by rheological effects. The degree of melting prior to melt/solid separation would tend to cluster near a "critical melt fraction" beyond which the solid matrix of the rock breaks down (*van der Molen and Paterson,* 1979). During fractional crystallization, the onset of crystallization of

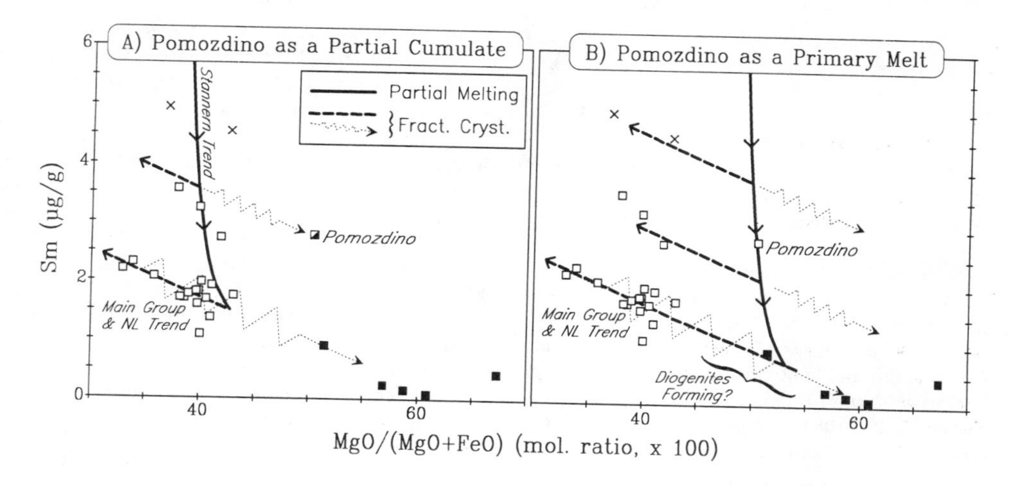

Fig. 16. Two models to account for the anomalous position of Pomozdino on the mg′ vs. Sm diagram. **(a)** According to our preferred endmember model, Pomozdino is a partial cumulate from a melt along the Stannern trend. **(b)** In an alternative endmember model that cannot be rigorously excluded, Pomozdino is a primary partial melt, in which case the Stannern trend eucrites might be residues from fractional crystallization of Pomozdino-like melts. A scenario effectively intermediate between **(a)** and **(b)**, i.e., with Pomozdino being "not quite" primary, may be most realistic.

plagioclase, which leads to an aggregate density for the crystallizing phases scarcely different from the density of the melt, would tend to leave the magma choked with crystals and thus would tend to dampen all subsequent fractional crystallization (*Warren,* 1985).

If, as seems more likely, Pomozdino is a partial cumulate (Fig. 16a), then it still constitutes circumstantial evidence that the Stannern trend does not comprise strictly primary partial melts. The only plausible parent melt for Pomozdino, viewed as a partial cumulate, is approximately Stannern-like (Fig. 14). Thus, this model would imply that evolved, REE-rich eucritic melts, and not just REE-poor, main group and Nuevo Laredo-like eucritic melts, were differentiated by fractional crystallization. The textures of Stannern (*Duke and Silver,* 1967), Kapoeta-CF4 (*Warren and Taylor,* 1982), and especially Bouvante (*Christophe Michel-Lévy et al.,* 1987) are not sufficiently fine grained to prove that they formed as quenched melts; moreover, even a quenched melt might have had a prior history of fractional crystallization. A recently discovered clast from the LEW85003 polymict eucrite that contains fayalitic olivine (*Kozul and Hewins,* 1988) and has a bulk Sm concentration of ~5.6 µg/g (*Mittlefehldt and Lindstrom,* 1988a) appears to have formed by the fractional crystallization of a Stannern-like melt (*Kozul and Hewins,* 1989).

The primary melt model (Fig. 16b) implies a relatively high mg′ ratio for the eucrite parent body, ~0.76, based on the model of *Warren and Jerde* (1987). The bulk mg′ ratio of the eucrite parent body has generally been estimated at ~0.67 (e.g., *Consolmagno and Drake,* 1977; *Stolper,* 1977; *Morgan et al.,* 1978), which is appropriate to account for the Stannern trend eucrites as primary, moderate-degree partial melts and would also be consistent with derivation of Pomozdino as a partial cumulate from a lower-mg′ parent melt (Fig. 16a). However, *Dreibus and Wänke* (1980) advocate a bulk-asteroid composition with mg′ = 0.79. A similarly high mg′ ratio would probably be necessary to account for the diogenites and the Nuevo Laredo trend eucrites as a single fractional crystallization series (*Warren and Jerde,* 1987). Of course, the two models shown in Fig. 16 are essentially endmember scenarios, and the most accurate model is likely to be a hybrid of these two; i.e., it may be that *none* of the known eucrites is an absolutely unfractionated primary basalt. However, if forced to choose between the two endmember models, we would have to favor the partial cumulate (Fig. 16a) model. We are confident that the present uncertainties regarding the composition and evolution of the eucrite parent asteroid can be greatly reduced through further investigations of eucrites, howardites, and diogenites, particularly samples that are compositionally evolved (e.g., low-mg′ or REE-rich) and not severely brecciated.

Acknowledgments. The UCLA portion of this research was supported by NASA grant NAG 9-87. We thank R. N. Clayton and M. I. Petaev for permission to cite unpublished oxygen-isotopic data, M. A. Nazarov and A. E. Rubin for extremely valuable discussions, G. W. Kallemeyn, F. T. Kyte, and L. Zhou for technical assistance with INAA and RNAA, and the Committee on Meteorites of the USSR Academy of Sciences for the samples of Pomozdino.

REFERENCES

Albee A. L., Quick J. E., and Chodos A. A. (1977) Source and magnitude of errors in "broad beam analysis" (DBA) with the electron probe (abstract). In *Lunar Science VIII,* pp. 7-9. The Lunar Science Institute, Houston.

Aylmer D., Herzog G. F., Lein J., and Mittleton R. (1988) ^{10}Be and ^{26}Al contents of eucrites: Implications for production rates and exposure ages. *Geochim. Cosmochim. Acta, 52,* 1691-1698.

Boctor N. Z., Carlson R. W., and Tera F. (1989) Petrology of vitric and basaltic clasts from the Bouvante eucrite (abstract). *Meteoritics, 24,* in press.

Bunch T. T. and Keil K. (1971) Chromite and ilmenite in nonchondritic meteorites. *Am. Mineral., 56,* 146-157.

BVSP (Basaltic Volcanism Study Project) (1981) *Basaltic Volcanism on the Terrestrial Planets.* Pergamon, New York. 1286 pp.

Campbell I. H. (1978) Some problems with the cumulus theory. *Lithos, 11,* 311-323.

Chou C.-L., Boynton W. V., Bild R. W., Kimberlin J., and Wasson J. T. (1976) Trace element evidence regarding a chondritic component in howardite meteorites. *Proc. Lunar Sci. Conf. 7th,* pp. 3501-3518.

Christophe Michel-Lévy M., Bourot-Denise M., Palme H., Spettel B., and Wänke H. (1987) L'eucrite de Bouvante: chimie, pètrologie et mineralogie. *Bull. Minéral., 110,* 449-458.

Clayton R. N. and Mayeda T. K. (1983) Oxygen isotopes in eucrites, nakhlites, and chassignites. *Earth Planet. Sci. Lett., 62,* 1-6.

Clayton R. N., Onuma N., and Mayeda T. K. (1976) A classification of meteorites based on oxygen isotopes. *Earth Planet. Sci. Lett., 30,* 10-18.

Clayton R. N., Mayeda T. K., and Onuma N. (1979) Oxygen isotopic compositions of some Antarctic meteorites (abstract). In *Lunar and Planetary Science X,* pp. 221-223. Lunar and Planetary Institute, Houston.

Cleverly W. H., Jarosewich E., and Mason B. (1987) Camel Donga meteorite, a new eucrite from the Nullarbor Plain, West Australia. *Meteoritics, 21,* 263-269.

Consolmagno G. J. and Drake M. J. (1977) Composition and evolution of the eucrite parent body: Evidence from rare earth elements. *Geochim. Cosmochim. Acta, 41,* 1271-1282.

Deer W. A., Howie R. A., and Zussman J. (1965) *An Introduction to the Rock-forming Minerals.* Longman, London. 528 pp.

Delaney J. S., Prinz M., Nehru C. E., and Harlow G. E. (1981) A new basalt group from howardites: Mineral chemistry and relationships with basaltic achondrites (abstract). In *Lunar and Planetary Science XII,* pp. 211-213. Lunar and Planetary Institute, Houston.

Delaney J. S., O'Neill C., and Prinz M. (1984) Two magma types in the eucrite Bouvante (abstract). In *Lunar and Planetary Science XV,* pp. 210-211. Lunar and Planetary Institute, Houston.

Dodd R. T. (1981) *Meteorites: A Petrologic-Chemical Synthesis.* Cambridge Univ., New York. 368 pp.

Dreibus G. and Wänke H. (1980) The bulk composition of the eucrite parent asteroid and its bearing on planetary evolution. *Z. Naturforsch., 35a,* 204-216.

Duke M. B. (1963) Petrology of basaltic achondrites. Ph.D. dissertation, California Institute of Technology, Pasadena. 362 pp.

Duke M. B. and Silver L. T. (1967) Petrology of eucrites, howardites and mesosiderites. *Geochim. Cosmochim. Acta, 31,* 1637-1665.

Dymek R. F., Albee A. L., Chodos A. A., and Wasserburg G. J. (1976) Petrography of isotopically-dated clasts in the Kapoeta howardite and petrologic constraints on the evolution of the parent body. *Geochim. Cosmochim. Acta, 40,* 1115-1130.

Grove T. L. and Walker D. (1977) Cooling histories of Apollo 15 quartz-normative basalts. *Proc. Lunar Sci. Conf. 8th,* pp. 1501-1520.

Hansen M. and Anderko K. (1958) *Constitution of the Binary Alloys.* McGraw-Hill, New York. 1305 pp.

Harlow G. E., Nehru C. E., Prinz M., Taylor G. J., and Keil K. (1979) Pyroxenes in Serra de Magé: Cooling history in comparison with Moama and Moore County. *Earth Planet. Sci. Lett., 43,* 173-181.

Hewins R. H. and Newsom H. E. (1988) Igneous activity in the early solar system. In *Meteorites and the Early Solar System* (J. F. Kerridge and M. S. Matthews, eds.), pp. 73-101. Univ. of Arizona, Tucson.

Ikeda Y. and Takeda H. (1985) A model for the origin of basaltic achondrites based on the Yamato 7308 howardite. *Proc. Lunar Planet. Sci. Conf. 15th,* in *J. Geophys. Res., 90,* C649-C663.

Irvine T. N. (1982) Terminology for layered intrusions. *J. Petrol., 23,* 127-162.

Jerome D. Y. (1970) Composition and origin of some achondritic meteorites. Ph.D. dissertation, Univ. of Oregon, Eugene. 166 pp.

Kallemeyn G. W. and Warren P. H. (1989) Geochemistry of the LEW86010 angrite (abstract). In *Lunar and Planetary Science XX,* pp. 496-497. Lunar and Planetary Institute, Houston.

Kharitonova V. Ya. and Barsukova L. D. (1982) The chemical composition of meteorites Aliskerovo, Aprel'sky, Vetluga and Kuleshovka. *Meteoritika, 40,* 41-44.

Kolesov G. M. (1976) Determination of some trace and rare-earth elements in achondrites and tektites by instrumental neutron activation analysis. *Meteoritika, 35,* 59-66.

Kozul J. and Hewins R. H. (1988) Mafic mineral compositions of igneous clasts in the Lewis Cliffs 85000, 85002, 85003 polymict eucrites (abstract). *Meteoritics, 23,* 281-282.

Kozul J. M. and Hewins R. H. (1989) Fayalite-bearing eucrites and the origins of HED magmas (abstract). *Meteoritics, 24,* in press.

Kvasha L. G. and Dyakonova M. I. (1972) The Pomozdino eucrite. *Meteoritika, 31,* 109-115.

Lovering J. F. (1975) The Moama eucrite—a pyroxene-plagioclase adcumulate. *Meteoritics, 10,* 101-114.

Mason B., Jarosewich E., and Nelen J. A. (1979) The pyroxene-plagioclase achondrites. *Smithsonian Contrib. Earth Sci., 22,* 27-45.

Mayeda T. K. and Clayton R. N. (1989) Oxygen isotopes in the Bholghati howardite (abstract). In *Lunar and Planetary Science XX,* p. 648. Lunar and Planetary Institute, Houston.

Mayeda T. K., Clayton R. N., and Yanai K. (1987) Oxygen isotopic compositions of several Antarctic meteorites. *Mem. Natl. Inst. Polar Res., Spec. Issue 46,* pp. 144-150. National Institute of Polar Research, Tokyo.

McKay G. A. (1982a) Partitioning of REE between olivine, plagioclase, and synthetic basaltic melts: implications for the origin of lunar anorthosites (abstract). In *Lunar and Planetary Science XIII,* pp. 493-494. Lunar and Planetary Institute, Houston.

McKay G. A. (1982b) Experimental REE partitioning between pigeonite and Apollo 12 olivine basaltic liquids (abstract). *Eos Trans. AGU, 63,* 1142.

Mittlefehldt D. W. (1979) Petrographic and chemical characterization of igneous lithic clasts from mesosiderites and howardites and comparison with eucrites and diogenites. *Geochim. Cosmochim. Acta, 43,* 1917-1935.

Mittlefehldt D. W. (1987) Volatile degassing of basaltic achondrite parent bodies: Evidence from alkali elements and phosphorus. *Geochim. Cosmochim. Acta, 51,* 267-278.

Mittlefehldt D. W. and Lindstrom M. M. (1988a) Geochemistry of diverse lithologies in Antarctic eucrites (abstract). In *Lunar and Planetary Science XIX,* pp. 790-791. Lunar and Planetary Institute, Houston.

Mittlefehldt D. W. and Lindstrom M. M. (1988b) HED petrogenesis: View from the diogenite end (abstract). *Meteoritics, 23,* 290.

Morgan J. W., Higuchi H., Takahashi H., and Hertogen J. (1978) A "chondritic" eucrite parent body: inference from trace elements. *Geochim. Cosmochim. Acta, 42,* 27-38.

Nazarov M. A., Ignatenko K. I., and Shevaleevsky I. D. (1982) Source of errors in defocussed beam analysis with the electron probe, revisited (abstract). In *Lunar and Planetary Science XIII,* pp. 582-583. Lunar and Planetary Institute, Houston.

Palme H. and Rammensee W. (1981) The significance of W in planetary differentiation processes: Evidence from new data on eucrites. *Proc. Lunar Planet. Sci. 12B*, pp. 949-964.

Palme H., Wlotzka F., Spettal B., Dreibus G., and Weber H. (1988) Camel Donga: A eucrite with high metal content. *Meteoritics, 23*, 49-57.

Petaev M. I., Ustinov V. I., Zaslavskaya N. I., Gavrilov E. Ya., and Shukolyukov Yu. A. (1988) Oxygen isotopes in Pomozdino meteorite (abstract). In *Lunar and Planetary Science XIX*, pp. 917-918. Lunar and Planetary Institute, Houston.

Philpotts J. A. and Schnetzler C. C. (1970) Apollo 11 lunar samples: K, Rb, Sr, Ba and rare-earth concentrations in some rocks and separated phases. *Proc. Apollo 11 Lunar Sci. Conf.*, pp. 1471-1486.

Reid A. M., Duncan A. R., and Le Roex A. (1979) Petrogenetic models for eucrite genesis (abstract). In *Lunar and Planetary Science X*, pp. 1022-1024. Lunar and Planetary Institute, Houston.

Rubin A. E. and Jerde E. A. (1988) Compositional differences between basaltic and gabbroic clasts in mesosiderites. *Earth Planet. Sci. Lett., 87*, 485-490.

Shirley D. N. (1986) Differentiation and compaction in the Palisades Sill, New Jersey. *J. Petrol., 28*, 835-865.

Smith M. R. (1982) A chemical and petrologic study of igneous lithic clasts from the Kapoeta howardite. Ph.D. dissertation, Oregon State University, Corvallis. 194 pp.

Stolper E. (1977) Experimental petrology of eucritic meteorites. *Geochim. Cosmochim. Acta, 41*, 587-611.

Takeda H. and Mori H. (1985) The eucrite-diogenite links and the crystallization history of a crust of their parent body. *Proc. Lunar Planet. Sci. Conf. 15th*, in *J. Geophys. Res., 90*, C636-C648.

Takeda H., Miyamoto M., Iishi T., and Reid A. M. (1976) Characterization of crust formation on a parent body of achondrites and the Moon by pyroxene crystallography and chemistry. *Proc. Lunar Sci. Conf. 7th*, pp. 3535-3548.

Takeda H., Mori H., Delaney J. S., Prinz M., Harlow G. E., and Ishii T. (1983) Mineralogical comparison of Antarctic and non-Antacrtic HED (howardites-eucrites-diogenites) achondrites. *Mem. Natl. Inst. Polar Res., Spec. Issue 30*, pp. 181-205. National Institute of Polar Research, Tokyo.

Takeda H., Tagai T., and Graham A. (1988) Mineralogy of slowly cooled eucrites and thermal histories of the HED parent body (abstract). In *Thirteenth Symposium on Antarctic Meteorites*, pp. 142-144. National Institute of Polar Research, Tokyo.

Taylor G. J. (1989) Metal segregation in asteroids (abstract). In *Lunar and Planetary Science XX*, pp. 1109-1110. Lunar and Planetary Institute, Houston.

Urey H. C. and Craig H. (1953) The composition of the stone meteorites and the origin of the meteorites. *Geochim. Cosmochim. Acta, 4*, 36-82.

van der Molen I. and Paterson M. S. (1979) Experimental deformation of partially-melted granite. *Contrib. Mineral. Petrol., 70*, 299-318.

Wager L. R., Brown G. M., and Wadsworth W. J. (1960) Types of igneous cumulates. *J. Petrol., 1*, 73-85.

Warren P. H. (1985) Origin of howardites, diogenites and eucrites: A mass balance constraint. *Geochim. Cosmochim. Acta, 49*, 577-586.

Warren P. H. (1986) The bulk-Moon MgO/FeO ratio: A highlands perspective. In *Origin of the Moon* (W. K. Hartmann, R. J. Phillips, and G. J. Taylor, eds.), pp. 279-310. Lunar and Planetary Institute, Houston.

Warren P. H. (1990) Miscellaneous minor elements. In *Lunar Sourcebook* (G. H. Heiken, D. T. Vaniman, and B. M. French, eds.), Chapter 8.6. Cambridge Univ., New York.

Warren P. H. and Jerde E. (1987) Composition and origin of Nuevo Laredo trend eucrites. *Geochim. Cosmochim. Acta, 51*, 713-725.

Warren P. H. and Kallemeyn G. W. (1989) Allan Hills 84025: The second brachinite, far more differentiated than Brachina, and an ultramafic achondritic clast from L chondrite Yamato 75097. *Proc. Lunar Planet. Sci. Conf. 19th*, pp. 475-486.

Warren P. H. and Taylor G. J. (1982) The diversity of igneous rocks of the eucrite parent body, as explored via clasts from Kapoeta (abstract). *Meteoritics, 17*, 293-294.

Proceedings of the 20th Lunar and Planetary Science Conference, pp. 299-308
Lunar and Planetary Institute, Houston, 1990

New High P-T Experimental Results on Orthopyroxene-Chrome Spinel Equilibrium and a Revised Orthopyroxene-Spinel Cosmothermometer

A. B. Mukherjee

Department of Geology and Geophysics, Indian Institute of Technology, Kharagpur — 721 302, West Bengal, India

V. Bulatov

Institute of Lithosphere, 22 Staromonetny Per., 109180 Moscow, USSR

A. Kotelnikov

Institute of Experimental Mineralogy, Chernogolovka 142 432, Moscow District, USSR

Equilibrium experiments between 800° and 1200°C and 2.5 and 22 kbar were carried out with a set of natural high-Mg orthopyroxene/high-Cr spinel assemblages employing both cold seal hydrothermal and solid media piston-cylinder techniques. The run products were analyzed by electron microprobe and the orthopyroxene and spinel compositional data were used to supplement Mukherjee and Viswanath's (1987) analysis of the compositional relationships of equilibrated orthopyroxene and spinel from wide-ranging P-T regimes. The new data fill the gap in the 1200°C regression of Mg-Fe^{2+} orthopyroxene-spinel distribution coefficients against the Cr mole fraction of the Cr-rich spinel, and yield an improved formulation of the orthopyroxene-spinel thermometer. The revised thermometric expression is as follows

$$T(K) = \frac{1175.549 + 2580.473\ Y_{Cr}^{Sp}}{\ln K_D^{\circ} + 0.282\ Y_{Cr}^{Sp} + 0.328}$$

where

$$K_D^{\circ} = \frac{X_{Mg}^{Opx} \cdot X_{Fe}^{Sp}}{X_{Fe}^{Opx} \cdot X_{Mg}^{Sp}},$$ the ferric iron-free distribution coefficient,

$$Y_{Cr}^{Sp} = \frac{Cr}{Cr + Al + Fe^{3+}}$$

and Opx and Sp represent orthopyroxene and spinel, respectively. The new calibration gives reasonable temperatures for terrestrial ultramafic rocks that agree, on the whole, with olivine-spinel temperatures (Fabriès, 1979) and are well within 0.7% to 6% of the estimates obtained from the earlier calibration of Mukherjee and Viswanath (1987). The revised estimates for the high temperature diogenites, i.e., Yamato-74013, Yamato-74136, and Yamato-6902, are in the range 1000°-1100°C, which still sets them apart from the slowly cooled, deep-seated diogenites, e.g., Johnstown and Yamato-75032, with equilibration temperatures in the 700°-800°C range. The new calibration is consistent with a range of acceptable values for symmetric or nearly symmetric Cr-Al mixing, i.e., with $W_{Al\text{-}Cr}^{Sp}$ = 2.3-3.4 kcal/gm atom, in the spinel phase. The revised cosmothermometer should be useful in delineating the thermal history of a wide range of orthopyroxene-spinel parageneses, including achondrites with histories of shock heating, remelting, and recrystallization.

INTRODUCTION

The need for an orthopyroxene-spinel thermometer arises in attempting to determine thermal histories of the widely occurring orthopyroxenite units of ultramafic complexes, e.g., the orthopyroxenite dikes of the Red Mt. peridotites of New Zealand (*Sinton,* 1977), the orthopyroxenite members of the eastern part of the Bushveld complex (*Cameron and Emerson,* 1959), or the orthopyroxenite intrusive of the Sukinda ultramafic complex of eastern India (*Viswanath,* 1983). The need exists also for the orthopyroxene-spinel achondrites (the diogenites), which have an important role in the modeling of the layered HED (howardite-eucrite-diogenite) parent bodies (*Takeda et al.,* 1976, 1979, 1983). *Mukherjee and Viswanath* (1987) proposed an empirical orthopyroxene-spinel thermometer essentially based on Mg-Fe^{2+} exchange equilibrium. However, this formulation suffers from a lack of suitable reference equilibrium data at 1200°C for assemblages with relatively low-Mg orthopyroxene and high-Cr spinel. Since the Mg-Fe^{2+} distribution coefficient between orthopyroxene and spinel at a particular temperature, similar to that between olivine and spinel (*Fabriès,* 1979), is strongly dependent on the Cr content of the spinel, it is important that the reference isotherms be based on data points spread out over a wide range of orthopyroxene-spinel Mg-Fe^{2+} distribution coefficient ($K_{D, Mg\text{-}Fe^{2+}}^{Sp\text{-}Opx}$) and Y_{Cr}^{Sp}.

In this paper we present the result of high P-T equilibrium experiments on some natural orthopyroxene-spinel assemblages in which ($K_{D, Mg\text{-}Fe^{2+}}^{Sp\text{-}Opx}$) and Y_{Cr}^{Sp} are approximately 1.6 and 0.8, respectively. From these experiments, we obtain a new 1200°C isotherm for orthopyroxene-spinel equilibrium and formulate a revised orthopyroxene-spinel thermometer.

For the equilibrium experiments, we have used samples from the orthopyroxenites of the Sukinda ultramafic complex, Orisa, India (*Banerjee,* 1972; *Viswanath,* 1983). The Sukinda complex is a late Precambrian, roughly 40 km^2 alpine-type body occupying the core of a regional, plunging syncline of the iron-ore group quartzites within the Bihar-Orissa craton of eastern India. Serpentinites, talc-tremolite-chlorite schists, chromitites, and orthopyroxenites are the major rock types of the complex. The orthopyroxenites occur as a 6-km-long, 300-m-wide tabular body of nearly uniform petrography and composition and are believed to represent the last intrusive phase of the Sukinda complex (*Banerjee,* 1972; *Viswanath,* 1983).

EXPERIMENTAL METHODS

Starting Materials

Four samples from the orthopyroxenites of Sukinda, numbered S-24, S-26, S-30, and S-32, were used as the starting materials of the high P-T experiments. All these samples are coarse-grained, pale green to greyish-green rocks with nearly 98% modal orthopyroxene, the rest being disseminated chrome spinel and traces of serpentine and/or talc. In thin section, chrome spinel euhedra (0.08-0.6 mm, most often square shaped, and deep wine red in color in transmitted plane polarized light) occur as discrete, disseminated grains at the boundaries of coarsely interlocked, xenomorphic granular orthopyroxene (0.6-10.0 mm). The orthopyroxenes are colorless in thin section, do not show exsolution lamellae, and are commonly fractured. A small amount of felted talc-serpentine intergrowth can be seen, mostly as secondary fracture-filling.

Small fragments of the samples were crushed and ground in an agate mortar under acetone to ~5-μm grain size for the experimental runs. No attempt was made to increase artificially the proportion of spinel in the powdered samples. Compositions of orthopyroxene and spinel in each sample are given in Table 1. The orthopyroxenes are essentially homogeneous. The spinels show compositional gradients in Mg, Fe, Al, and Cr across the original grains in the thin sections. In spinel, Fe and Mn increase from core to rim, while Mg and Al decrease. Chromium remains nearly the same or decreases very slightly toward the core. The rim and core compositions of the spinels in Table 1 are from the original thin sections. In the starting material mixes, the original zoning of the spinel can be detected in the form of grain-to-grain compositional variation.

High P-T Experiments

Runs were carried out using both piston-cylinder and cold seal hydrothermal equipment. CaF_2 cells with graphite heaters

TABLE 1. Microprobe analyses of minerals of the rock samples used in the experiments.

	S-24			S-26			S-30			S-32		
	Orthopyroxene	Spinel Core	Spinel Rim	Orthopyroxene	Spinel Core	Spinel Rim	Orthopyroxene	Spinel Core	Spinel Rim	Orthopyroxene	Spinel Core	Spinel Rim
SiO_2	57.26	—	—	57.50	—	—	57.48	—	—	56.99	—	—
Al_2O_3	0.85	8.52	7.72	0.79	8.26	7.69	0.82	8.01	7.58	0.93	8.38	8.01
Cr_2O_3	0.66	60.17	61.90	0.49	60.95	61.52	0.52	61.32	60.62	0.40	61.25	60.75
FeO*	6.39	21.52	26.10	7.01	22.78	25.58	6.12	22.32	25.92	6.50	22.62	25.03
MnO	0.41	0.27	0.35	0.29	0.25	0.34	0.32	0.28	0.34	0.38	0.26	0.38
MgO	33.06	7.86	3.23	33.27	7.85	3.92	33.85	7.84	3.82	33.91	6.32	4.25
CaO	0.65	—	—	0.50	—	—	0.58	—	—	0.64	—	—
Total	99.28	98.34	99.30	99.85	100.09	99.05	99.68	99.77	98.28	99.75	98.83	98.42
Oxygen	6	4	4	6	4	4	6	4	4	6	4	4
Si	1.99	—	—	1.99	—	—	1.99	—	—	1.98	—	—
Al	0.04	0.35	0.32	0.03	0.33	0.32	0.03	0.32	0.32	0.03	0.34	0.33
Cr	0.02	1.64	1.70	0.01	1.64	1.71	0.01	1.65	1.70	0.01	1.67	1.69
Fe^{3+}	—	0.01	—	—	0.03	—	—	0.03	—	—	—	—
Fe^{2+}	0.19	0.59	0.76	0.20	0.58	0.75	0.18	0.58	0.77	0.19	0.65	0.73
Mn	0.01	0.01	0.01	0.01	0.01	0.01	0.01	0.01	0.01	0.01	0.01	0.01
Mg	1.71	0.40	0.17	1.72	0.40	0.21	1.75	0.40	0.20	1.75	0.33	0.22
Ca	0.02	—	—	0.02	—	—	0.02	—	—	0.02	—	—
Total	3.98	3.0	2.96	3.98	2.99	3.0	3.99	2.99	3.0	3.99	3.0	2.98
X_{Mg}	0.86	0.40	0.17	0.86	0.40	0.21	0.88	0.40	0.20	0.88	0.33	0.22
Y_{Cr}^{Sp}	—	0.82	0.85	—	0.82	0.86	—	0.83	0.85	—	0.84	0.85

*Total Fe.
Spinel Fe^{3+}/Fe^{2+} was calculated assuming ideal stoichiometry.

TABLE 2. Experimental results.

Run No.	Run Type*	T (°C)	P (kb)	Duration (hr)	Starting composition	Product Composition: Spinel X_{Mg}	Spinel X_{Al}	Spinel X_{Cr}	Orthopyroxene X_{Mg}
1	P	1200	22	24	Product of run no. 6	0.72	0.19	0.78	0.83
2	P	1200	20	24	S-24	0.64	0.18	0.80	0.88
3	P	1200	20	24	Product of run no. 7	0.68	0.19	0.79	0.88
4	P	1200	15	24	S-30	0.70	0.20	0.77	0.86
5	H	800	2.5	4	S-32	0.21	0.16	0.84	0.88
6	H	800	3	22	S-32	0.23	0.17	0.85	0.88
7	H	800	3	96	S-32	0.30	0.17	0.84	0.89
8	H	800	3	20	S-26	0.22	0.16	0.85	0.86

*P = piston cylinder.
H = cold seal hydrothermal. For all H-type runs, the spinel compositions correspond to grains with the lowest X_{Mg} value.

were used in a ½-inch piston-cylinder apparatus (*Mirwald et al.,* 1975, p. 1520). Temperature was measured using W3Re-W25Re thermocouples and the recorded temperatures were controlled to within ±5°C of the set point. A correction of -10% of the nominal load pressure was employed in the piston-in position. The charge was put in an inner capsule of W with flanged lids on both ends and this was enclosed within a sealed Pt outer capsule. This proved effective in preventing Fe loss from the sample. The f_{O_2} of the piston-cylinder experiments was indeterminate, the inherent f_{O_2} (bar) of the apparatus material at 1000°C lying possibly between ~10^{-8}[Ni-NiO buffered f_{O_2} of Renè-41 (*Popp et al.,* 1984)] and ~10^{-12} [wüstite-iron; *Myers and Eugster* (1983)]. The hydrothermal runs were carried out in conventional cold seal equipment, using chromel alumel thermocouples, gold capsules (sealed during the experiments), a small amount of water along with the charge, and hematite-magnetite buffer put in a separate, smaller diameter unsealed gold capsule within the main outer capsule. Run conditions are given in Table 2. Run products were examined optically and by the JEOL JXA-35 electron-probe microanalyzer at IIT, Kharagpur. The analyses were carried out employing wavelength dispersive spectrometry, an accelerating voltage of 15 kV, specimen current between 20-50 nA, and a beam diameter not larger than 2 μm. Reduction of the microprobe data was carried out using the Bence-Albee method (*Bence and Albee,* 1968).

TABLE 3. Microprobe analyses of minerals of the run products.

Run No.	1		2		3		4		5		6		7		8	
	Opx	Sp	Opx	Sp	Opx	Sp	Opx	Sp	Opx	Sp	Opx	Sp	Opx	Sp	Opx	Sp
SiO_2	58.25	—	57.56	—	57.15	—	58.01	—	57.12	—	57.02	—	57.03	—	57.47	—
Al_2O_3	0.98	9.55	0.83	9.19	0.90	9.56	0.90	10.01	0.93	7.80	0.92	8.04	0.87	8.20	0.74	7.70
Cr_2O_3	0.66	60.37	0.60	61.23	0.38	60.82	0.66	59.56	0.40	60.35	0.42	60.82	0.36	61.01	0.50	61.42
FeO*	5.28	14.14	5.26	15.40	5.54	14.28	5.12	14.86	6.56	25.72	6.48	24.75	5.86	22.98	6.92	25.46
MnO	0.39	0.12	0.32	0.21	0.35	0.17	0.30	0.14	0.38	0.42	0.35	0.36	0.28	0.29	0.29	0.34
MgO	32.00	14.82	34.12	12.96	34.09	13.90	33.45	14.32	33.90	4.08	33.89	4.48	34.40	5.80	33.31	4.20
CaO	0.83	—	0.70	—	0.65	—	0.60	—	0.66	—	0.65	—	0.60	—	0.49	—
Total	98.39	99.00	99.39	98.99	99.06	98.73	99.04	98.89	99.95	98.37	99.73	98.45	99.40	98.28	99.72	99.12
Oxygen	6	4	6	4	6	4	6	4	6	4	6	4	6	4	6	4
Si	2.03	—	1.99	—	1.98	—	2.01	—	1.98	—	1.98	—	1.98	—	1.99	—
Al	0.04	0.37	0.03	0.36	0.04	0.37	0.04	0.39	0.04	0.32	0.04	0.33	0.04	0.34	0.05	0.32
Cr	0.02	1.56	0.02	1.60	0.01	1.58	0.02	1.54	0.01	1.68	0.01	1.69	0.01	1.68	0.02	1.70
Fe^{3+}	—	0.07	—	0.04	—	0.05	—	0.07	—	—	—	—	—	—	—	—
Fe^{2+}	0.15	0.32	0.15	0.38	0.16	0.34	0.15	0.34	0.19	0.76	0.19	0.73	0.17	0.67	0.20	0.75
Mn	0.01	0.003	0.01	0.006	0.01	0.005	0.01	0.004	0.01	0.01	0.01	0.01	0.01	0.01	0.01	0.01
Mg	1.66	0.72	1.76	0.64	1.76	0.68	1.72	0.70	1.76	0.21	1.75	0.23	1.78	0.30	1.72	0.22
Ca	0.03	—	0.03	—	0.02	—	0.02	—	0.02	—	0.02	—	0.02	—	0.02	—
Total	3.94	3.04	3.99	3.03	3.98	3.03	3.97	3.04	4.01	2.98	4.00	2.99	4.01	3.00	4.01	3.00
X_{Mg}	0.83	0.72	0.88	0.64	0.88	0.68	0.86	0.70	0.88	0.21	0.88	0.23	0.89	0.30	0.86	0.22
X_{Cr}^{Sp}		0.78		0.80		0.79		0.77		0.84		0.85		0.84		0.85

*Total Fe.
Opx = orthopyroxene; Sp = spinel. Spinel Fe^{3+}/Fe^{2+} was calculated assuming ideal stoichiometry.

EXPERIMENTAL RESULTS

Mineral compositions of the run products are given in Table 3. No other minerals were present or could be detected in the run products except for orthopyroxene and spinel. In all hydrothermal runs at 800°C, the compositional spread of the spinels inherited from the original zoning was preserved, while the orthopyroxenes were homogeneous. The products of the piston-cylinder runs at 1200°C were all homogeneous within the limits of precision of the microprobe. The experiments, however, caused a marked resetting of the Mg-Fe^{2+} distribution between the spinel and the orthopyroxene. This is shown in Fig. 1, through the X_{Mg} tie line dispositions between spinel and orthopyroxene before and after the experiments. The original spinel core and rim compositions are indicated by the largest and the smallest X_{Mg} values for the spinels of the four samples. These values were obtained from a microprobe examination of as many spinel grains as possible from each sample. The range of this variation decreased with increasing run duration of the hydrothermal experiments. For clarity, the results for sample S-32 only after a 4, 22, and 96 hr run at 800°C are shown in Fig. 1. The smallest X_{Mg} value for this sample changed from 0.21 to 0.30 with the increase of the hydrothermal run duration from 4 to 96 hr.

The products of the piston cylinder runs at 1200°C show a rather tight clustering of the spinel-orthopyroxene X_{Mg} tie lines within a narrow range of 0.64-0.72 for spinel and 0.83-0.88 for orthopyroxene. Two piston cylinder runs (nos. 1 and 3) used the products of the hydrothermal run nos. 6 and 7, which had sample no. S-32, as the starting material. While the X_{Mg} and X_{Al} of the spinels increased with run temperature and run duration, the orthopyroxene compositions changed only slightly within narrow limits.

The hydrothermal experiments were clearly not equilibrated. However, the purpose of carrying out and reporting these experiments along with the piston-cylinder experiments, using some common samples, was to show that with increasing temperature and run duration the spinel compositions changed in the *right* direction and possibly to the *right* extent relative to the orthopyroxenes. In ultramafic igneous rocks, reequilibrated to lower temperatures, Mg/(Mg+Fe^{2+}) of the spinel (modal volume 2-3%) decreases appreciably, while the change in olivine composition is often nearly insignificant (*Henry and Medaris,* 1980, pp. 227-228). *Arai* (1980, Table 5, Fig. 13) has estimated that in the dunites and harzburgites of Sangun-Yamaguchi zone, in western Japan cooling from about 1200°C to 700°C reset the spinel (modal volume 1-2%) Mg/(Mg+Fe^{2+}) from a narrow 0.78-0.80 range to a wide scatter from 0.492 to 0.619, while the olivine Mg/(Mg+Fe^{2+}) remained nearly constant within 0.91-0.92. Arai also believed that the Mg-Fe^{2+} distribution relation of the orthopyroxenes with spinels in these rocks was similar to that of the olivines. This rotation, rather than displacement, of the Mg-Fe silicate-spinel X_{Mg} tie lines during reequilibration is a consequence of the "lever rule" (*Irvine,* 1965; *Dick and Bullen,* 1984, p. 65). We believe that the overall disposition of the X_{Mg} tie lines between spinel and orthopyroxenes in Fig. 1 is consistent with an approach to equilibrium from low to high temperature.

Our piston-cylinder run duration of 24 hr at 20 kbar and 1200°C is believed to be adequate for achieving a fast enough

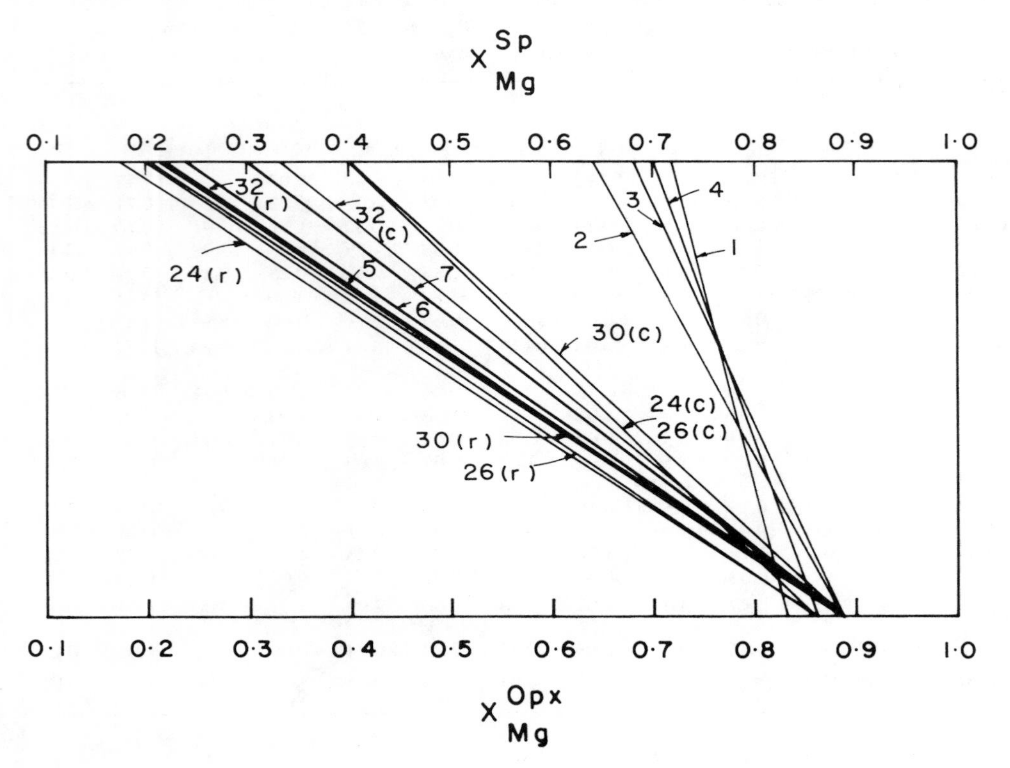

Fig. 1. X_{Mg} tie lines between spinel and orthopyroxene. The starting compositions are indicated as 24 (r) for S-24 (spinel rim), 26 (c) for S-26 (spinel core), etc. The hydrothermal experiments with S-32 for 4, 22, and 96 hr are labeled 5, 6, and 7 respectively. The piston-cylinder experiments are labeled 1, 2, 3, and 4.

reaction rate to approach equilibrium in ultramafic mineral mixtures (see *Danckwerth and Newton,* 1978; *Gasparik,* 1984). The piston-cylinder run products, orthopyroxene and spinel are essentially homogenous and the tight clustering of their tie lines is consistent with attainment of equilibrium of the piston-cylinder experiments. We have no good explanation for the crossing of some of the orthopyroxene-spinel tie lines, albeit within a narrow range. For the piston-cylinder runs, the tie-line crossing may be due to a small amount of nonuniform, disequilibrium partial melting that went undetected in the recovered run products.

REVISED CALIBRATION OF THE ORTHOPYROXENE-SPINEL THERMOMETER

$Mg\text{-}Fe^{2+}$ exchange equilibrium between chrome spinel and orthopyroxene can be written as

$$FeSiO_3 + Y^{Sp}_{Cr} \cdot MgCr_2O_4 + Y^{Sp}_{Al} \cdot MgAl_2O_4 + Y^{Sp}_{Fe^{3+}} \cdot MgFe_2O_4 = MgSiO_3 + Y^{Sp}_{Cr} \cdot FeCr_2O_4 + Y^{Sp}_{Al} \cdot FeAl_2O_4 + Y^{Sp}_{Fe^{3+}} \cdot Fe_3O_4 \quad (1)$$

For this reaction we may write (see *Irvine,* 1965; *Fabriès,* 1979; *Mukherjee and Viswanath,* 1987)

$$\ln K_D = \ln \frac{X^{Opx}_{Mg} \cdot X^{Sp}_{Fe}}{X^{Opx}_{Fe} \cdot X^{Sp}_{Mg}} = \ln K_3 + Y^{Sp}_{Cr} \cdot \ln K_4 + Y^{Sp}_{Fe^{3+}} \cdot \ln K_5 \quad (2)$$

where K_3, K_4, and K_5 are, respectively, the equilibrium constants of the reactions

$$FeSiO_3 + MgAl_2O_4 = MgSiO_3 + FeAl_2O_4 \quad (3)$$

$$MgCr_2O_4 + FeAl_2O_4 = MgAl_2O_4 + FeCr_2O_4 \quad (4)$$

$$MgFe_2O_4 + FeAl_2O_4 = MgAl_2O_4 + Fe_3O_4 \quad (5)$$

For Fe^{3+}-poor terrestrial ultramafic rocks ($Y^{Sp}_{Fe^{3+}}$ being generally less than 0.05 (*Fabriès,* 1979) and achondrites [($Y^{Sp}_{Fe^{3+}}$ being less than 0.04 (*Bunch and Keil,* 1971)], the ferric iron-free distribution coefficient can be expressed as

$$\ln K^\circ_D = \ln K_3 + Y^{Sp}_{Cr} \cdot \ln K_4 \quad (6)$$

where $\ln K^\circ_D$ can be calculated as follows, assuming the value of −4 for $\ln K_5$ after *Irvine* (1965), *Fabriès* (1970) and *Fujii* (1978)

$$\ln K^\circ_D = \ln K_D - 4\, Y^{Sp}_{Fe^{3+}} \quad (7)$$

Mukherjee and Viswanath (1987) solved equation (6) for $\ln K_3$ and $\ln K_4$ from linear regression of two datasets of orthopyroxene-spinel $\ln K^\circ_D$ - Y^{Sp}_{Cr} with independent temperature estimates. Combining these solutions they obtained the thermometric expression

$$T\,(K) = \frac{1662.97}{\ln K^\circ_D - 2.37\, Y^{Sp}_{Cr} + 0.829} \quad (8)$$

The 1200°C isotherm used in this formulation was based on three experimental data points (G, M, and F in Fig. 2) covering a Y^{Sp}_{Cr} range of only 0.10-0.27. The calibration, therefore, suffered from relatively high uncertainty for assemblages having spinel with Y^{Sp}_{Cr} in the region of 0.7 or above and orthopyroxene with X_{Mg} in the range 0.6-0.7. We have now redetermined the 1200°C isotherm from linear regression of the original three points (G, M, and F in Fig. 2) plus the four new points from our piston-cylinder experiments. The estimated error brackets for $\ln K^\circ_D$ and Y^{Sp}_{Cr} determined from the piston-cylinder run products are indicated in Fig. 2. The error brackets span 2σ and were calculated by propagating the compositional uncertainties of orthopyroxene and spinel, following standard methods (*Le Maitre,* 1982, pp. 178-199). The revised regression equation for the 1200°C isotherm is

$$\ln K^\circ_D = 0.47 + 1.47\, Y^{Sp}_{Cr} \quad (9)$$

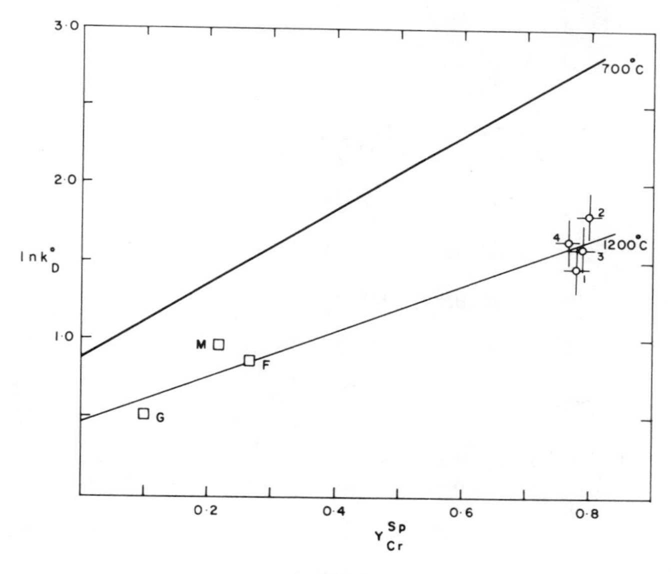

Fig. 2. Magnesium-iron distribuition coefficient between orthopyroxene and spinel as a function of Y^{Sp}_{Cr} and isotherms drawn by regression. Symbols: open squares = G [lunar basalt 14310 (*Green et al.,* 1972)], M [spinel ilherzolite 2L equilibrated at 1200°C and 16 kbar (*Mori,* 1977)], and F [olivine-orthopyroxene-spinel mixture equilibrated at 1200°C and 15 kbar (*Fujii,* 1978)]; 1, 2, 3, and 4 for piston-cylinder experiments of this work. The 700°C isotherm is reproduced from *Mukherjee and Viswanath* (1987).

Combining this with the previously determined 700°C isotherm (Fig. 2) for the metaperiodotites of the Seiad and Finero complexes (*Medaris,* 1975; *Mukherjee and Viswanath,* 1987) we obtain the revised thermometric expression

$$T\,(K) = \frac{1175.549 + 2580.473\ Y_{Cr}^{Sp}}{\ln K_D^{\circ} + 0.282\ Y_{Cr}^{Sp} + 0.328} \qquad (10)$$

TABLE 4. Mineral compositions and calculated temperatures of some orthopyroxene + spinel ± olivine ± clinopyroxene parageneses.

Sample	Olivine	Orthopyroxene			Clinopyroxene		Spinel			T°C	T°C
	X_{Mg}	X_{Mg}^{M1}	X_{Mg}^{M2}	X_{Al}^{M1}	X_{Al}^{M1}	X_{Mg}^{M2}	X_{Mg}	X_{Al}	X_{Cr}	(this work)	(other calibrations)*
Alpine peridotites											
Lizard (Green, 1964)											
No. 90683	0.895	0.776	0.834	0.107	0.767	0.152	0.771	0.757	0.192	1042	1040(1), 1115(2), 1393(3), 1299(4)
No. 90681	0.895	0.777	0.835	0.100	0.773	0.190	0.771	0.751	0.194	1119	1059(1), 1175(2), 1335(3), 1242(4)
Ronda (Obata, 1980)											
R 123 (primary)	0.878	0.785	0.856	0.099	0.687	0.079	0.754	0.890	0.100	868	863(1), 923(2), 1110(3), 1039(4)
R 123 (neoblast)	0.878	0.825	0.880	0.059	0.687	0.079	0.754	0.890	0.100	872	863(1), 908(2), 842(3), 775(4)
R 196 (neoblast)	0.096	0.832	0.892	0.066	0.742	0.076	0.806	0.853	0.122	1117	960(1), 919(2), 943(3), 838(4)
R 243 (primary)	0.898	0.786	0.865	0.107	0.725	0.096	0.779	0.892	0.098	897	814(1), 975(2), 1153(3), 1099(4)
R 243 (matrix)	0.898	0.852	0.909	0.042	0.725	0.096	0.779	0.892	0.098	880	814(1), 948(2), 830(3), 676(4)
Red Mt.(Sinton, 1977)											
No. 37010 (rim)	0.915	0.905	0.905	0.005	0.900	0.055	0.461	0.338	0.610	702	734(1), 881(2), 520(3), 481(4)
No. 37014 (rim)	0.907	0.857	0.894	0.046	0.849	0.052	0.696	0.695	0.289	775	772(1), 869(2), 833(3), 765(4)
No.37140 (rim)	0.913	0.883	0.905	0.024	0.918	0.025	0.674	0.671	0.315	735	737(1), 757(2), 715(3), 596(4)
SW Oregon (Medaris, 1972)											
Vondergreen (augen)	0.895	0.802	0.878	0.088	0.740	0.066	0.794	0.837	0.133	1099	954(1), 899(2), 1047(3), 1015(4)
Vondergreen (matrix)	0.895	0.866	0.894	0.034	0.789	0.065	0.794	0.837	0.133	1069	954(1), 892(2), 742(3), 640(4)
Carpenterville (augen)	0.897	0.808	0.894	0.082	0.817	0.077	0.788	0.843	0.136	1020	903(1), 944(2), 1151(3), 965(4)
Carpenterville (matrix)	0.897	0.852	0.894	0.044	0.839	0.077	0.788	0.843	0.136	1006	903(1), 939(2), 931(3), 701(4)
Signal Butte (augen)	0.904	0.857	0.885	0.036	0.801	0.073	0.757	0.795	0.186	908	840(1), 919(2), 913(3), 664(4)
Signal Butte (matrix)	0.904	0.881	0.887	0.014	0.826	0.076	0.757	0.795	0.186	878	840(1), 929(2), 801(3), 528(4)
Balmuccia (Shervais, 1979)											
No. 76B - 388	0.913	0.809	0.876	0.092	0.742	0.049	0.764	0.887	0.084	854	801(1), 844(2), 950(3), 1009(4)
No. 75B - 197	0.917	0.833	0.881	0.068	0.763	0.064	0.701	0.788	0.178	743	721(1), 891(2), 900(3), 927(4)
No. 76B - 517	0.903	0.808	0.881	0.083	0.714	0.051	0.736	0.880	0.097	800	746(1), 839(2), 973(3), 964(4)
No. 76B - 524	0.902	0.812	0.878	0.082	0.743	0.062	0.725	0.887	0.096	739	709(1), 882(2), 975(3), 965(4)
Chugoku dt. (Arai, 1975)											
No. AS11	0.917	0.899	0.912	0.000	0.889	0.090	0.605	0.514	0.465	744	724(1), 974(2), 696(3), 453(4)
No. AS5 (average cpx)	0.913	0.837	0.927	0.004	0.917	0.063	0.546	0.451	0.533	707	723(1), 912(2), 274(3), 477(4)
No. 2503 (average cpx and sp)	0.914	0.894	0.875	0.000	0.885	0.083	0.582	0.510	0.478	703	717(1), 967(2), 716(3), 453(4)

TABLE 4. (continued).

Sample	Olivine	Orthopyroxene			Clinopyroxene		Spinel			T°C	T°C
	X_{Mg}	X^{M1}_{Mg}	X^{M2}_{Mg}	X^{M1}_{Al}	X^{M1}_{Mg}	X^{M2}_{Mg}	X_{Mg}	X_{Al}	X_{Cr}	(this work)	(other calibrations)*
Ultramafic nodules											
Tasmania (Mori, 1977)											
No. LJ7	0.893	0.821	0.882	0.081	0.753	0.098	0.798	0.859	0.119	1004	932(1), 975(2), 937(3), 932(4)
No. SC4	0.887	0.830	0.885	0.071	0.767	0.072	0.784	0.885	0.103	891	876(1), 912(2), 841(3), 848(4)
Hoggar (Girod et al., 1981)											
No. ATK2	0.901	0.840	0.889	0.065	0.804	0.083	0.794	0.845	0.126	1018	903(1), 950(2), 866(3), 838(4)
No. TAH 48 (augen)	0.901	0.851	0.906	0.051	0.747	0.103	0.780	0.854	0.112	956	908(1), 973(2), 849(3), 743(4)
No. TAH 48 (neoblast)	0.900	0.834	0.900	0.066	0.774	0.088	0.780	0.854	0.112	987	898(1), 951(2), 923(3), 845(4)
No. ATK 88 (augen)	0.887	0.826	0.862	0.074	0.733	0.088	0.764	0.807	0.144	983	902(1), 951(2), 1096(3),933(4)
No. ATK 88 (neoblast)	0.886	0.816	0.864	0.074	0.733	0.088	0.764	0.807	0.144	1044	910(1), 950(2), 923(3), 938(4)
No. ADN 91	0.922	0.861	0.903	0.043	0.810	0.095	0.786	0.849	0.122	983	862(1), 972(2), 910(3), 696(4)
No. ADN 3	0.902	0.810	0.894	0.093	0.733	0.071	0.801	0.890	0.096	953	870(1), 908(2), 992(3), 991(4)
Diogenites											
Johnstown diogenite (Floran et al., 1981)	—	0.728	0.712	0.005	—	—	0.154	0.221	0.750	693	
Y-74013 diogenite (Mukherjee and Viswanath, 1986)											
Core compositions	—	0.693	0.684	0.000	—	—	0.259	0.158	0.822	1096	
Rim compositions	—	0.715	0.685	0.000	—	—	0.261	0.245	0.731	1020	
Y-6902 diogenite (Takeda et al., 1975)	—	0.739	0.713	0.003	—	—	0.297	0.161	0.812	1086	
Y-74136 (M) diogenite (Takeda et al., 1978)	—	0.723	0.718	0.000	—	—	0.314	0.270	0.696	1058	
Y-75032 diogenite (Takeda et al., 1978)	—	0.646	0.647	0.000	—	—	0.097	0.152	0.771	789	

*Numbers in parentheses refer to the following sources:(1) *Fabries* (1979): Olivine-spinel; (2) *Wells* (1977): Orthopyroxene-clinopyroxene; (3) *Gasparik and Newton* (1984): Olivine-orthopyroxene-spinel [using Gasparik and Newton's formulations for $(X^{M1}_{Mg})_{Opx}$ and $(X^{M1}_{Al})_{Opx}$]; (4) *Gasparik and Newton* (1984): Olivine-orthopyroxene-spinel [using $(X^{M1}_{Mg})_{Opx}$ and $(X^{M1}_{Al})_{Opx}$ formulations of *Wood and Banno* (1973) and *Powell* (1978)].

Mg and Al mole fractions in ortho- and clinopyroxenes were calculated following *Powell* (1978). All Gasparik-Newton calculations were done for 10 kbar pressure. The Gasparik-Newton (3) temperatures for the Lizard, Oregon, Ronda, and Hoggar assemblages are reproduced from *Gasparik and Newton* (1984, Table 5). Opx = orthopyroxene, cpx = clinopyroxene, sp = spinel.

The 700°C isotherm is in good agreement with the empirical 700°C isotherm of *Evans and Frost* (1975) and Engi's experimental reversals of olivine-spinel equilibria at 700°C (*Engi,* 1978; *Henry and Medaris,* 1980, pp. 222-223). The location of the isotherm has been generally accepted (see *Arai,* 1980, p. 156; *Obata,* 1980, p. 561).

APPLICATION AND EVALUATION

Revised Temperature Estimates

Calculated orthopyroxene-spinel equilibrium temperatures of 38 terrestrial and meteoritic parageneses (23 alpine-type periodotites, 9 ultramafic nodule suites, 6 diogenites) using the revised calibration are given in Table 4. Excepting the diogenites Yamato-74013, Yamato-74136, and Yamato-6902, the calculated temperatures for all other assemblages differ only slightly from the previous estimates of *Mukherjee and Viswanath* (1987). The revised thermometer is able to distinguish between the high temperature peridotites, e.g., of Lizard (1081°C, average of two determinations; *Green*, 1964), the moderately high temperature ultramafic nodules, e.g., of Hoggar (993°C, average of seven determinations; *Girod et al.*, 1981), and the lower temperatures of some alpine peridotites, e.g., Red Mt. (737°C, average of three determinations; *Sinton*, 1977), or Balmuccia (784°C, average of four determinations; *Shervais*, 1979).

Among the meteorites, the Johnstown and Yamato-75032 diogenites give 693°C and 789°C, respectively, reaffirming their similarity with slowly cooled, deep-seated, terrestrial plutonic bodies. The Yamato-74013, Yamato-74136, and Yamato-6902 diogenites, all having relatively low $\ln K_D^o$ and high Y_{Cr}^{Sp}, show the largest differences from the previous temperature estimates of *Mukherjee and Viswanath* (1987). The revised estimates are between 1020°C and 1096°C, while the previous estimates were all above 1500°C. The high-temperature history of this diogenite group, including shock-heating, remelting, and recrystallization (*Takeda et al.*, 1981), is possibly reflected in the calculated temperatures, although these temperatures need no longer be considered as "anomalously high" (*Mukherjee and Viswanath*, 1987, p. 213).

For comparison, temperatures were calculated using three other calibrations, olivine-spinel thermometer of *Fabriès* (1979), orthopyroxene-clinopyroxene thermometer of *Wells* (1977), and olivine-aluminous orthopyroxene-spinel thermometer of *Gasparik and Newton* (1984). The Gasparik-Newton computations were done using $(X_{Mg}^{M1})_{orthopyroxene}$ and $(X_{Al}^{M1})_{orthopyroxene}$ formulations as suggested by *Gasparik and Newton* (1984, p. 193) and also by *Wood and Banno* (1973) and *Powell* (1978). These two separate results have been indicated by (3) and (4) in Table 4. The entire set of calculations (Table 4) was done using the THERM1.BAS computer program.

The differences in the calculated temperatures for each assemblage are generally smaller in the nodules than in the Alpine peridotites. The average temperature differences for the 9 nodule suites are 84°C (Fabriès vs. this work), 39°C (Wells vs. this work), 87°C (Gasparik-Newton (3) vs. this work, 74°C (Gasparik-Newton (3) vs. Wells), 60°C (Gasparik-Newton (3) vs. Fabriès), 53°C (Fabriès vs. Wells). For the 22 alpine peridotites (omitting Ronda R 196 (neoblast), which has much higher Fe^{3+} in the spinel than other samples of this group), the average temperature differences are 48°C (Fabriès vs. this work), 104°C (Wells vs. this work), 147°C (Gasparik-Newton (3) vs. this work), 163°C (Gasparik-Newton (3) vs. Wells), 160°C (Gasparik-Newton (3) vs. Fabriès), 108°C (Fabriès vs. Wells). Taking the entire data set (alpine peridotites and nodules), the average temperature differences are 59°C (Fabriès vs. this work), 85°C (Wells vs. this work), 129°C (Gasparik vs. Newton (3) vs. this work), 137°C (Gasparik-Newton (3) vs. Wells), 131°C (Gasparik-Newton (3) vs. Fabriès), 92°C (Fabriès vs. Wells).

Differences in the calculated temperatures, using different geothermometers, are due to the inherent uncertainties and limitations of each calibration as well as the differences in the kinetics and the blocking temperatures of the reactions on which these calibrations are based. In general, methods based on olivine-pyroxene-spinel equilibria give the highest temperatures and the olivine-spinel equilibria the lowest (*Henry and Medaris*, 1980, p. 223). Temperatures calculated from the two pyroxene equilibria commonly occupy the intermediate postiion. This is presumably due to the progressively lower blocking temperatures of the olivine-pyroxene-spinel, orthopyroxene-clinopyroxene, and olivine (or orthopyroxene)-spinel equilibria (*Henry and Medaris*, 1980, p. 226; *Shervais*, 1979, p. 808). Moderate temperature differences between our calibration and that of *Fabries* (1979) and progressively larger differences with respect to that of *Wells* (1977) and of *Gasparik and Newton* (1984) are, on the whole, in conformity with the general experience in the thermometry of ultramafic rocks.

Al-Cr Mixing in Spinel and the Revised Calibration

Although the revised calibration has not explicitly taken note of the effect of possible nonidealities of the solid solutions, we may test it against current information in this regard. Figure 3 shows the RT $\ln K_D^o$ vs. Y_{Cr}^{Sp} plots of all the assemblages in Table 4 for which equilibrium temperatures have been calculated using the revised thermometer. The plots give a nearly perfectly linear correlation with a positive slope, with only the diogenites Yamato-74013, Yamato-74136, and Yamato-6902 somewhat off the straight line.

Magnesium-iron mixing in most major rock-forming oxides and silicates is generally treated as ideal for practical thermometry. For instance, *Harley* (1984) has shown that Mg-Fe interaction in orthopyroxene on the M2 site, Mg-Fe-Al interaction of the M1 site, and the reciprocal interactions across the sites give rise to only a very small compositional

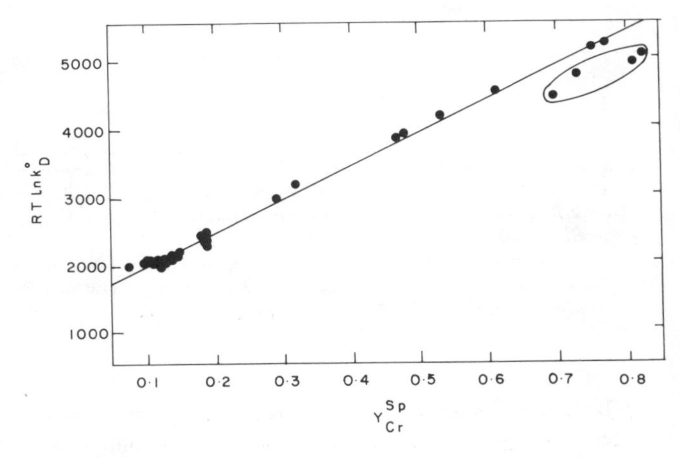

Fig. 3. Solid circles are plots of RT $\ln K_D^o$ as a function of Y_{Cr}^{Sp}. The temperatures are calculated for the natural paragenesis given in Table 4 using the revised calibration. The enclosed points are of Yamato-74013 (core and rim), Yamato-74136, and Yamato-6902.

dependency of the orthopyroxene-garnet Mg-Fe distribution coefficient, which may be neglected. In spinel solid solutions Mg-Fe mixing is treated as ideal (*Wood and Nicholls,* 1978, p. 395; *Webb and Wood,* 1986, p. 476). The slope of the RT ln K_D^o - Y_{Cr}^{Sp} line (Fig. 3) in principle depends on the nonideality of both spinel and orthopyroxene solid solutions. Assuming Mg-Fe mixing in both orthopyroxene and spinel to be ideal, we may obtain from equation (6) (see *Webb and Wood,* 1986, p. 474)

$$RT\left(\frac{\partial \ln K_D^o}{\partial Y_{Cr}^{Sp}}\right)_{P,T} \cong 2\,W_{Al-Cr}^{Sp} \qquad (11)$$

where $W_{Al\text{-}Cr}^{Sp}$ is the symmetric Al-Cr interaction parameter in a regular solution model for spinel. Equation (11) implies that we may roughly estimate $W_{Al\text{-}Cr}^{Sp}$ from the slope of the RT ln K_D^o - Y_{Cr}^{Sp} line in Fig. 3. The slope, after excluding the three Yamato samples (enclosed in a loop in Fig. 3), yields a value of 2.45 kcal for $W_{Al\text{-}Cr}^{Sp}$. This is in good agreement with several recent estimates, e.g., the symmetric value of 2.7±0.5 kcal (*Webb and Wood,* 1986), the symmetric value of 3.35 kcal/gm atom (*O'Neill and Navrotsky,* 1984), the asymmetric values of $W_{Al\text{-}Cr}^{Sp}$ = 3.4 kcal/gm atom and $W_{Al\text{-}Cr}^{Sp}$ = 2.3 kcal/gm atom (*Sack,* 1982), and asymmetric values of $W_{Al\text{-}Cr}^{Sp}$ = 3.33 kcal/gm atom and $W_{Al\text{-}Cr}^{Sp}$ = 2.48 kcal/gm atom (*Oka et al.,* 1984). The nearly perfectly collinear RT ln K_D^o - Y_{Cr}^{Sp} relationship of Fig. 3 is also a strong supportive evidence for the essential soundness of the revised calibration.

CONCLUSIONS

New high P-T equilibrium experiments have been used to formulate an impoved orthopyroxene-spinel cosmothermometer suitable for terrestrial and meteoritic orthopyroxene-spinel parageneses of a wide range of Mg-Fe distribution coefficients and Cr-Al ratios of the spinel. The calculated temperatures are reasonable and realistic and are broadly in accord with the inferred thermal histories of the specific assemblages. Although empirically calibrated, the thermometer is consistent with a range of acceptable values for symmetric or nearly symmetric Al-Cr mixing in the spinel phase.

Acknowledgments. One of us (A.B.M.) is deeply indebted to V. A. Zharikov and the USSR Academy of Sciences for much help and assistance for the experimental part of this work. The hydrothermal equipment and the microprobe at IIT, Kharagpur, were made available through a generous grant of the Department of Science and Technology, Government of India. The first author also wishes to thank R. A. Ishbulatov and A. V. Kosyakov for helpful advice on the piston-cylinder work.

REFERENCES

Arai S. (1980) Dunite-harzburgite-chromitite complexes as refactory residue in the Sangun-Yamaguchi zone, Western Japan. *J. Petrol., 21,* 141-165.

Banerjee P. K. (1972) Geology and Geochemistry of the Sukinda ultramafic field, Cuttack district, Orissa. *Mem. Geol. Surv. India, 103.* 171 pp.

Bence A. E. and Albee A. L. (1968) Empirical correction factors for the electron microanalysis of silicates and oxides. *J. Geol., 76,* 382-403.

Bunch T. E. and Keil K. (1971) Chromite and ilmenite in nonchondritic meteorites. *Am. Mineral., 56,* 146-157.

Cameron E. N. and Emerson M. E. (1959) The origin of certain chromite deposits in the eastern part of Bushveld complex. *Econ. Geol., 54,* 1151-1213.

Danckwerth P. A. and Newton R. C. (1978) Experimental determination of the spinel peridotite to garnet peridotite reaction in the system $MgO\text{-}Al_2O_3\text{-}SiO_2$ in the range 900°-1100°C and Al_2O_3 isopleths of enstatite in the spinel field. *Contrib. Mineral. Petrol., 66,* 189-201.

Dick H. J. B. and Bullen T. (1984) Chromian spinel as a petrogenetic indicator in abyssal and alpine-type peridotites. *Contrib. Mineral. Petrol., 86,* 54-76.

Engi M. (1978) Mg-Fe exchange equilibria among Al-Cr spinel, olivine, orthopyroxene and cordierite. Ph.D. thesis, E.T.H., Zürich.

Evans B. W. and Frost B. R. (1975) Chrome-spinel in progressive metamorphism—a preliminary analysis. *Geochim. Cosmochim. Acta, 39,* 959-972.

Fabriès J. (1979) Spinel-olivine geothermometry in peridotites from ultramafic complexes. *Contrib. Mineral. Petrol., 69,* 329-336.

Floran R. J., Prinz M. Hlava P. F., Keil K. Spettel B., and Wänke H. (1981) Mineralogy, petrology and trace element geochemsitry of the Johnstown meteorite: a brecciated orthopyroxenite with sidferophile and REE-rich components. *Geochim. Cosmochim. Acta, 45,* 2385-2391.

Fujii T. (1978) Fe-Mg partitioning between olivine and spinel. *Carnegie Inst. Wash. Yearb., 76,* 563-569.

Gasparik T. (1984) Two-pyroxene thermobarometry with new experimental data in the system $CaO\text{-}MgO\text{-}Al_2O_3\text{-}SiO_2$. *Contrib. Mineral. Petrol., 87,* 87-97.

Gasparik T. and Newton R. C. (1984) The reversed alumina contents of orthopyroxene in equilibrium with spinel and forsterite in the system $MgO\text{-}Al_2O_3\text{-}SiO_2$. *Contrib. Mineral. Petrol., 85,* 186-196.

Girod M., Dautria J. M., and de Giovanni R. (1981) A first insight into the constitution of the upper mantle undere the Hoggar area (Southern Algeria): the lherzolite xenoliths in the alkali basalts. *Contrib. Mineral. Petrol., 77,* 66-73.

Green D. H. (1964) The petrogenesis of the high-temperature peridotite intrusion in the Lizard area. *Cornwall. J. Petrol. 5,* 134-188.

Green D. H., Ringwood A. E., Ware N. G., and Hibberson W. O. (1972) Experimental petrology and petrogenesis of Apollo 14 basalt. *Proc. Lunar Sci. Conf. 3rd,* pp. 197-206.

Harley S. L. (1984) An experimental study of the partitioning of Fe and Mg between garnet and orthopyroxene. *Contrib. Mineral. Petrol., 86,* 359-373.

Henry D. J. and Medaris L. G. (1980) Application of pyroxene an olivine-spinel geothermometers to spinel peridotites in Southwestern Oregon. *Am J. Sci., 280-A,* 211-231.

Irvine T. N. (1965) Chromian spinel as petrogenetic indicator. Part I. Theory. *Canad. J. Earth Sci., 2,* 648-672.

Le Maitre R. W. (1982) *Numerical Petrology.* Elsevier, New York. 281 pp.

Medaris L. G. (1972) High pressure peridotite in Southwestern Oregon. *Bull. Geol. Soc. Am., 83,* 41-58.

Medaris L. G. (1975) Coexisting spinel and silicates in alpine peridotites of the granulite facies. *Geochim. Cosmochim. Acta, 39,* 947-958.

Mori T. (1977) Geothermometry of spinel lherzolites. *Contrib. Mineral. Petrol., 59,* 261-279.

Mirwald P. W., Gatting I. C., and Kennedy G. C. (1975) Low friction cell for piston cylinder high pressure apparatus. *J. Geophys. Res., 80,* 1519-1525.

Mukherjee A. and Viswanath T. A. (1986) On the thermal and redox history of the Yamato diogenite Y-74013. In *Proceedings of the Indian Academy of Sciences: Earth and Planetary Sciences, 95,* 381-395.

Mukherjee A. and Viswanath T. A. (1987) Thermometry of diogenites. *Mem. Natl. Inst. Polar Res., Spec. Issue 46,* pp. 205-215.

Myers J. and Eugster H. P. (1983) The system Fe-Si-O: Oxygen buffer calibrations to 1500 K. *Contrib. Mineral. Petrol., 82,* 75-90.

Obata M. (1980) The Ronda peridotite: garnet-, spinel-, and plagioclase-ilherzolite facies and the P-T trajectories of a high temperature mantle instrusion. *J. Petrol., 21,* 533-572.

Oka Y., Steinke P., and Chatterjee N. D. (1984) Thermodynamic mixing properties of $Mg(Al,Cr)_2O_4$ spinel crystalline solution at high temperatures and pressures. *Contrib. Mineral. Petrol., 87,* 196-204.

O'Neill H. St. C. and Navrotsky A. (1984) Cation distributions and thermodynamic properties of binary spinel solid solutions. *Am. Mineral., 69,* 733-753.

Popp R. K., Nagy K. L., and Hajash A. (1984) Semiquantitative control of hydrogen fugacity in rapid-quench hydrothermal vessels. *Am. Mineral., 69,* 557-562.

Powell R. (1978) *Equilibrium Thermodynamics in Petrology.* Harper and Row, New York. 284 pp.

Sack R. O. (1982) Spinels as petrogenetic indicators: activity-composition relations at low pressure. *Contrib. Mineral. Petrol., 71,* 169-186.

Shervais J. W. (1979) Thermal emplacement model for the alpine lherzolite massif at Balmucci, Italy. *J. Petrol., 20,* 795-820.

Sinton J. M. (1977) Equilibrium history of the basal alpine type peridotite, Red Mountain, New Zealand. *J. Petrol., 18,* 216-246.

Takeda H., Reid A. M., and Yamanaka T. (1975) Crystallographic and chemical studies of a bronzite and chromite in the Yamato (b) achondrite. *Mem. Natl. Inst. Polar Res., Spec. Issue 5,* pp. 83-90.

Takeda H., Miyamoto H., Ishii T., and Reid A. M. (1976) Characterization of crust formation on a parent body of achondrites and the moon by pyroxene crystallography and chemistry. *Proc. Lunar Sci. Conf. 7th,* pp. 3535-3548.

Takeda H., Miyamoto M., Yanai K., and Haramura H. (1978) A preliminary mineralogical examination of the Yamato-74 achondrites. *Mem. Natl. Inst. Polar Res., Spec. Issue 8,* pp. 170-184.

Takeda H., Miyamoto M., Ishii T., Yanai K., and Matsumoto Y. (1979) Mineralogical examination of the Yamato-75 achondrites and their layered crust model. *Mem. Natl. Inst. Polar Res., Spec. Issue 12,* pp. 82-108.

Takeda H., Mori H., and Yanai K. (1981) Mineralogy of the Yamato diogenites as possible pieces of a single fall. *Mem. Natl. Inst. Polar Res., Spec. Issue 20,* pp. 81-99.

Takeda H., Mori H., Delaney J. S., Prinz M., Harlow G. E., and Ishii T. (1983) Mineralogical comparison of Aantarctic and Non-Antarctic HED (howardites-eucrites-diogenites) achondrites. *Mem. Natl. Inst. Polar Res., Spec. Issue 30,* pp. 181-205.

Viswanath T. A. (1983) A study of orthopyroxene-chrome spinel equilibrium with application to the orthopyroxenite band of the Sukinda ultramafic complex of Orissa, India. Ph. D. thesis. Indian Institute of Technology, Kharagpur. 139 pp.

Webb S. A. C. and Wood B. J. (1986) Spinel pyroxene garnet relationships and their dependence on Cr/Al ratio. *Contrib. Mineral. Petrol., 92,* 471-480.

Wells P. R. A. (1977) Pyroxene thermometry in simple and complex systems. *Contrib. Mineral. Petrol., 62,* 129-139.

Wood B. J. and Banno S. (1973) Garnet-orthopyroxene and orthopyroxene-clinopyroxene relationships in simple and complex systems. *Contrib. Mineral. Petrol., 42,* 109-124.

Wood B. J. and Nicholls J. (1978) The thermodynamic properties of reciprocal solid solutions. *Contrib. Mineral. Petrol., 66,* 389-400.

Proceedings of the 20th Lunar and Planetary Science Conference, pp. 309-319
Lunar and Planetary Institute, Houston, 1990

U-Th-Pb Systematics of the Estherville Mesosiderite

M. Brouxel and M. Tatsumoto

U.S. Geological Survey, Branch of Isotope Geology, MS 963, P.O. Box 25046, Denver, CO 80225

A detailed U-Pb systematic study that includes stepwise leaching experiments has revealed that the 4288 ± 85 Ma $^{206}Pb/^{204}Pb$ vs. $^{207}Pb/^{204}Pb$ "isochron" obtained for the Estherville mesosiderite is a mixing line. Lead-lead internal isochrons for the Estherville mesosiderite yield ages of 4556 ± 35 Ma, 4506 ± 75 Ma, and 4422 ± 50 Ma. These ages show that the silicate fraction of the Estherville mesosiderite is very heterogeneous and was formed early in the solar system history. The 4556 ± 35 Ma age (the oldest reported yet for a mesosiderite) is similar to eucrite ages whereas the youngest Pb-Pb age obtained in this study (4422 ± 50 Ma) compares with cumulate eucrite ages. Eucrite and cumulate eucrite clasts are commonly described in mesosiderites, confirming the very heterogeneous nature of this meteorite group. On a modified concordia diagram, nearly all the points plot to the left of concordia, defining chords with upper-intercept ages either in agreement with the Pb-Pb ages (4571 ± 18 Ma and 4437 ± 11 Ma) or completely discordant (4082 ± 210 Ma) due to differential leaching of U and Pb during our experimental procedure. Lower-intercept ages are poorly defined, and are either close to 0 Ma (-40 ± 56 Ma and 144 ± 110 Ma), or close to the age of a heating event recorded by Ar-Ar data of ≈ 3.6-3.8 Ga for most of the mesosiderites (Bogard et al., 1988).

INTRODUCTION

Mesosiderites form a relatively small group of meteorites (≈ 20 samples recovered). They contain approximately equal parts of iron-nickel metal and silicates by weight (*Prior,* 1920), and show complex petrologic features that record both magmatic and metamorphic events. These events include magmatic differentiation, brecciation, and thermal metamorphism due to meteoritic impacts, mixing with metal, and removal from the mesosiderite parent body by a catastrophic collision (*Powell,* 1969, 1971; *Floran,* 1978; *Wasson and Rubin,* 1985; *Bogard et al.,* 1988). Most of the geochronological data published on mesosiderites give rather young ages with significant variation. Potassium-argon and Ar-Ar ages range between 3.23 and 4.25 Ga (*Megrue,* 1966; *Begemann et al.,* 1976; *Bogard et al.,* 1988) and fission tracks ages between 2.6 and 4.2 Ga (*Crozaz and Tasker,* 1981). Some mesosiderites show young Rb-Sr model ages, likely indicating disturbances (3.6 Ga, *Mittelfehldt et al.,* 1986), but older internal isochron ages have been obtained for Estherville (Rb-Sr: 4.24 ± 0.03; *Murthy et al.,* 1978) and Morristown (Sm-Nd: 4.51 ± 0.12 Ga; *Prinzhofer et al.,* 1989).

Our study has been conducted on Estherville, which is one of the best-studied mesosiderites. Reported ages on Estherville include 2.6 Ga (Fission track; *Crozaz and Tasker,* 1981), 3.6 Ga (Ar-Ar with no well defined plateau; *Murthy et al.,* 1978), 4.0 Ga (K-Ar; *Megrue,* 1966), and 4.24 ± 0.03 Ga (Rb-Sr; *Murthy et al.,* 1978).

Uranium-thorium-lead data have not been published yet for any mesosiderite. Because two radioactive-decay schemes are involved in the U-Pb system, the system has an advantage that internal isochrons may be obtained by measuring only the Pb isotopic composition. Measuring the U concentration also, however, provides not only U-Pb ages, but often provides information regarding later disturbances. Uranium-Pb ages may be more easily disturbed than Rb-Sr or Sm-Nd ages (*Unruh et al.,* 1977), but the $^{207}Pb/^{206}Pb$ ages are not sensitive to recent metamorphic events (*Tatsumoto et al.,* 1976; *Manhes and Allègre,* 1978). We present in this detailed U-Pb systematic study an example that clearly identifies a Pb-Pb isochron obtained for leaches of Estherville as a mixing line, although differential extraction of U and Pb during the leaching may have destroyed some U-Pb ages.

SAMPLE DESCRIPTION AND ANALYTICAL TECHNIQUES

Estherville (Emmet County, Iowa, U.S.A.), which fell on the 10th of May, 1879, is a heterogeneous polymict breccia. It has been classified as subgroup 1 [mildly recrystallized with brecciated angular silicate lithic fragments (*Powell,* 1971)], as subgroup 4 [igneous, impact melts (*Floran,* 1978)], or as subgroup 3 [highly recrystallized with many subrounded clasts (*Floran,* 1978)]. However, these authors studied different fragments of the Estherville mesosiderite, one from the U.S. National Museum (*Powell,* 1971) and the other from the American Museum of Natural History (*Floran,* 1978), illustrating that a unique classification is not always possible for some brecciated and heterogeneous mesosiderites when only a small fraction of the meteorite is examined (*Hewins,* 1984).

Two clasts of the Estherville mesosiderite, totaling ≈ 29 g, were kindly provided to us by Dr. V. R. Murthy from a slice kept at the University of Minnesota. The other portion is located now at the U.S. National Museum. Even though Estherville is a fall, the silicate fraction, and especially the fine-grained silicate clasts, are highly oxidized. This oxidation and the well-mixed silicate and metallic phases were a major problem in the mineral separation procedures. One of the clasts containing mostly iron was not used in this study. The second clast (≈ 13 g) was first washed for one minute using 1N HCl in an ultrasonic bath and then rinsed repeatedly in distilled water. Crushing, mineral separation techniques, and leaching procedures are summarized in the Appendix.

The study was conducted on nine fractions (fine-grained, orthopyroxene, brecciated fragments, $\rho > 3.3$, $\rho > 3.3$—magnetic, $2.95 < \rho < 3.3$, $2.95 < \rho < 3.3$—magnetic, $2.58 < \rho < 2.95$, and $\rho < 2.58$). Our density separates were clearly not pure monomineralic fractions and even the Opx fraction separated by hand-picking from the largest grains was probably not pure. Some duplicate analyses have been made (fine-grained fraction and $2.58 < \rho < 2.95$) for which the leaching technique has been slightly modified (see Appendix). The chemical procedures used for elemental separations were similar to those described by *Premo et al.* (1989).

All isotopic measurements were performed on an NBS-type two-stage mass spectrometer using an ion pulse counter at the end of the second stage. The analyses were corrected for laboratory U, Th, and Pb contamination (blank Pb had an isotopic composition of $^{206}Pb/^{204}Pb = 18.80 \pm 0.6$, $^{207}Pb/^{204}Pb = 15.65 \pm 0.2$, and $^{208}Pb/^{204}Pb = 38.65 \pm 0.75$; *Premo et al.,* 1989) and for the initial Pb using the Cañon Diablo Troilite primordial Pb isotopic composition of *Tatsumoto et al.* (1973). A blank correction of 50 to 100 pg for Pb and 2 pg for U and Th were applied. Thorium and U decay constants used in this paper are respectively from *Leroux and Glendenin* (1963) and *Jaffey et al.* (1971).

RESULTS

The U, Th, and Pb analytical data are given in Tables 1a, b, and c. For the residues, the Pb concentrations range from 25 to 310 ppb, with the highest concentrations in the whole-rock (310 ppb) and in the lowest density fraction (250 ppb). The lowest Pb concentration (25 ppb) was measured in the high density fraction (mainly Opx). Significant variations in the Pb concentration exist in the two heaviest density fractions; the

TABLE 1a. U-Th-Pb analytical data for the Estherville mesosiderite (residues, and acetic acid and dilute HBr leaches).

Sample	Weight (mg)	Pb* (ppb)	U* (ppb)	Th* (ppb)	$^{206}Pb/^{204}Pb^*$	$^{207}Pb/^{204}Pb^*$	$^{208}Pb/^{204}Pb^*$	$^{238}U/^{204}Pb^*$	$^{232}Th/^{238}U^*$	$^{206}Pb/^{238}U^{*§}$	$^{207}Pb/^{235}U^{*§}$	$^{207}Pb/^{206}Pb^{*§}$	$^{208}Pb/^{232}Th^{*§}$
Residues													
R-WR	78.1	307	2.90	4.95	19.13	16.70	38.61	0.616	1.765	15.93	1434	0.653	8.401
					(0.10)†	(0.15)	(0.20)	(0.45)	(0.25)	(1.2)	(1.7)	(0.80)	(1.2)
R-Opx	65.4	71.5	1.59	4.92	18.48	15.50	37.91	1.402	3.196	6.537	512.0	0.568	1.881
					(0.11)	(0.17)	(0.22)	(0.54)	(0.24)	(1.4)	(2.1)	(1.0)	(1.4)
R-Breccia	28.7	105	4.93	15.8	19.03	15.94	37.88	2.979	3.321	3.263	261.3	0.581	0.850
					(0.12)	(0.17)	(0.21)	(0.44)	(0.16)	(1.3)	(1.92)	(0.92)	(1.4)
R-ρ>3.3	70.4	25.2	1.74	4.64	17.54	14.93	36.98	3.796	2.745	2.169	168.5	0.563	0.738
					(0.14)	(0.17)	(0.25)	(0.77)	(0.60)	(1.6)	(2.4)	(1.2)	(1.4)
R-ρ>3.3-mag.	88.1	109	3.10	9.42	21.01	17.85	40.98	1.985	3.136	5.897	525.0	0.646	1.849
					(0.35)	(0.32)	(0.38)	(2.2)	(1.6)	(1.0)	(1.4)	(0.67)	(0.88)
R-2.95<ρ<3.3	89.2	46.6	2.99	6.17	18.78	17.01	38.24	4.095	2.134	2.314	226.1	0.709	1.003
					(0.11)	(0.16)	(0.21)	(0.50)	(0.34)	(1.3)	(1.6)	(0.69)	(1.5)
R-2.95<ρ<3.3-mag.	66.0	59.4	2.82	7.29	18.80	16.52	38.76	3.084	2.670	3.078	278.4	0.656	1.127
					(0.30)	(0.35)	(0.38)	(2.8)	(3.7)	(3.3)	(3.8)	(1.1)	(1.2)
R-2.58<ρ<2.95	75.5	132	1.32	4.29	12.53	13.40	32.65	1.322	3.245	2.437	324.2	0.965	0.739
					(0.11)	(0.17)	(0.22)	(1.1)	(1.6)	(3.6)	(3.3)	(1.6)	(3.6)
R-ρ<2.58	25.2	248	3.62	7.76	12.93	12.70	32.50	0.717	2.211	5.047	463.3	0.666	0.472
					(0.10)	(0.16)	(0.22)	(0.73)	(0.50)	(2.9)	(4.3)	(2.1)	(3.5)
Acetic acid leach (duplicate)													
A-2.58<ρ<2.95 (d)	137.6	20.0‡	0.94	6.10	22.47	17.80	42.01	3.383	6.713	3.890	306.1	0.571	0.552
					(0.14)	(0.18)	(0.22)	(1.7)	(1.4)	(4.1)	(4.6)	(0.99)	(1.3)
Very dilute HBr (0.005N) leaches (duplicate)													
B-WR (d)	101.3	3.11	0.31	1.78	20.14	17.21	40.34	6.757	5.958	1.602	141.1	0.639	0.270
					(0.58)	(0.37)	(0.42)	(4.1)	(3.8)	(4.3)	(4.3)	(1.0)	(3.7)
B-2.58<ρ<2.95 (d)	137.6	0.63	0.04	13.5	22.77	18.31	42.31	4.052	397.8	3.321	272.9	0.596	0.008
					(0.50)	(0.41)	(0.45)	(2.9)	(2.7)	(5.5)	(5.6)	(0.70)	(1.0)
Very dilute HBr (0.05N) leaches (duplicate)													
C-WR (d)	101.3	88.1	11.6	37.7	25.31	19.47	45.56	10.36	3.371	1.546	122.2	0.573	0.461
					(0.12)	(0.16)	(0.22)	(1.4)	(2.3)	(5.8)	(5.2)	(1.2)	(1.8)
C-2.58<ρ<2.95 (d)	137.6	29.1	2.08	19.1	29.04	22.71	48.42	6.270	9.475	3.147	272.9	0.629	0.319
					(0.15)	(0.18)	(0.21)	(0.44)	(0.34)	(0.77)	(0.98)	(0.41)	(0.65)

*Data corrected for mass fractionation and laboratory blank (*Ludwig,* 1985a; see text for details).
†2 sigma errors (in percent).
‡Pb, U, and Th concentration of leaches are shown against the starting weight.
§Data corrected for initial Pb (see text for details).

magnetic split is much more Pb-rich than the nonmagnetic split, leading us to speculate that troilite was enriched in the magnetic split.

Uranium (1.3 to 4.9 ppb) and Th (4.3 to 15.8 ppb) concentrations of the residues show smaller variations than Pb. The highest U and Th concentrations were measured in the brecciated fragments.

In most cases, U and Pb show the same type of variations during the leaching procedure: rather high concentrations in the first HBr leaching (0.1 N) and decreasing during the HBr (0.5 N), HCl, and HF+HNO_3 leaching (Fig. 1a). Except for the brecciated fragments, concentrations in the residues are always higher than in the leaches (Table 1).

In some leaches, large amounts of Pb relative to U were observed (Fig. 1b). In another fraction, a large Pb peak could also be related to the abnormally high U peak (Fig. 1c). A large amount of U relative to Pb may also be observed in some fractions (Fig. 1d).

Approximately 50% of the U and Pb was leached during the entire procedure, whereas a surprisingly high percentage of the Th (65% and sometimes up to 90%) was measured in the leaches.

The Pb concentration of the acetic acid leach of the fine-grained fraction (duplicate analysis) is rather low (20 ppb) compared to the concentration of the residue (310 ppb), suggesting that only a small percentage of sample Pb was removed during our "cleaning" process. More Pb was removed from this fraction by the dilute HBr (0.05 N) leach (90 ppb). On the other hand, an insignificant amount of Pb was removed by the very dilute HBr (0.005 N) leach (0.6 and 3 ppb, Table

TABLE 1b. U-Th-Pb analytical data for the Estherville mesosiderite (HBr leaches).

Sample	Weight (mg)	Pb* (ppb)	U* (ppb)	Th* (ppb)	$^{206}Pb/^{204}Pb^*$	$^{207}Pb/^{204}Pb^*$	$^{208}Pb/^{204}Pb^*$	$^{238}U/^{204}Pb^*$	$^{232}Th/^{238}U^*$	$^{206}Pb/^{238}U^{*§}$	$^{207}Pb/^{235}U^{*§}$	$^{207}Pb/^{206}Pb^{*§}$	$^{208}Pb/^{232}Th^{*§}$
Dilute HBr (0.1N) leaches													
D-WR	100.0	53.9‡	0.61	8.61	29.77	21.84	49.35	0.997	14.59	20.52	1597	0.565	1.366
					(0.13)†	(0.17)	(0.22)	(0.78)	(0.28)	(1.4)	(1.5)	(0.49)	(0.63)
D-Opx	68.7	48.2	0.55	6.45	17.25	15.56	36.78	0.699	12.05	11.35	1038	0.663	0.867
					(0.11)	(0.18)	(0.21)	(1.2)	(0.38)	(2.3)	(2.7)	(1.0)	(1.6)
D-Brec.	34.1	141	8.42	49.1	19.21	15.92	38.70	3.874	6.030	2.555	200.3	0.568	0.395
					(0.11)	(0.15)	(0.21)	(0.44)	(0.26)	(1.2)	(1.9)	(0.99)	(1.2)
D-$\rho>3.3$	96.1	15.7	0.40	1.88	19.97	16.41	39.50	1.698	4.857	6.281	496.6	0.573	1.216
					(0.16)	(0.17)	(0.23)	(1.5)	(0.86)	(2.3)	(2.6)	(0.89)	(1.4)
D-$\rho>3.3$-mag.	94.5	19.1	0.31	1.61	19.63	16.66	39.56	1.068	5.422	9.670	821.5	0.616	1.743
					(0.14)	(0.17)	(0.22)	(1.7)	(0.78)	(2.7)	(3.0)	(0.85)	(1.4)
D-$2.95<\rho<3.3$	96.5	16.4	4.77	14.4	20.38	17.22	40.13	19.79	3.110	0.560	48.29	0.626	0.173
					(0.16)	(0.18)	(0.22)	(0.80)	(0.72)	(1.2)	(1.7)	(0.78)	(1.3)
D-$2.95<\rho<3.3$-mag.	95.3	10.2	0.22	1.73	19.98	17.05	40.27	1.457	8.107	7.325	639.0	0.633	0.914
					(0.21)	(0.20)	(0.24)	(2.5)	(1.3)	(3.7)	(3.9)	(0.83)	(1.6)
D-$2.58<\rho<2.95$	95.4	44.9	0.23	1.00	27.66	22.03	47.73	0.443	4.399	41.38	3648	0.639	9.356
					(0.14)	(0.17)	(0.21)	(4.0)	(0.81)	(3.2)	(3.3)	(0.44)	(1.0)
D-$\rho<2.58$	34.8	41.2	3.00	1.00	20.01	16.42	39.68	4.850	0.343	2.207	174.2	0.572	6.138
					(0.16)	(0.19)	(0.23)	(4.1)	(4.6)	(3.7)	(3.9)	(0.91)	(4.8)
HBr (0.5) leaches													
E-WR	100.0	18.3	4.47	23.2	26.00	20.53	45.26	19.53	5.371	0.854	72.24	0.613	0.150
					(0.21)	(0.21)	(0.23)	(0.76)	(0.67)	(0.81)	(1.2)	(0.53)	(0.89)
E-Opx	68.7	3.28	0.77	0.34	17.67	15.62	37.30	14.44	0.457	0.579	50.85	0.637	1.186
					(0.73)	(0.35)	(0.5)	(5.9)	(6.1)	(6.1)	(6.0)	(1.5)	(6.8)
E-Brec.	34.1	24.7	0.70	2.09	19.80	16.20	39.10	1.861	3.091	5.637	437.3	0.563	1.674
					(0.24)	(0.18)	(0.25)	(2.0)	(1.0)	(3.5)	(3.8)	(1.0)	(1.5)
E-$\rho>3.3$	96.1	2.65	0.13	1.24	20.11	17.19	40.01	3.260	10.04	3.315	291.7	0.638	0.322
					(0.71)	(0.46)	(0.49)	(6.1)	(4.8)	(7.3)	(7.3)	(1.2)	(4.7)
E-$\rho>3.3$-mag.	94.5	4.09	0.14	3.23	19.83	17.12	39.82	2.213	24.75	4.753	425.3	0.649	0.189
					(0.46)	(0.31)	(0.36)	(4.7)	(3.0)	(6.3)	(6.3)	(0.95)	(3.1)
E-$2.95<\rho<3.3$	96.5	8.04	0.10	4.32	19.28	16.68	39.11	0.828	44.12	12.05	1063	0.640	0.264
					(0.24)	(0.20)	(0.26)	(4.4)	(1.5)	(7.4)	(7.5)	(0.86)	(1.9)
E-$2.95<\rho<3.3$-mag.	95.3	4.30	0.26	3.38	18.87	17.09	39.38	4.020	13.30	2.379	233.0	0.710	0.185
					(0.47)	(0.34)	(0.37)	(4.2)	(3.3)	(5.2)	(5.1)	(1.0)	(3.50)
E-$2.58<\rho<2.95$	95.4	18.2	0.39	9.50	19.80	17.06	39.34	1.424	25.21	7.367	655.2	0.645	0.275
					(0.15)	(0.17)	(0.22)	(1.4)	(0.66)	(2.3)	(2.5)	(0.77)	(1.3)
E-$\rho<2.58$	34.8	10.9	0.95	6.18	19.31	16.73	39.05	5.733	6.740	1.744	154.9	0.644	0.248
					(0.45)	(0.30)	(0.66)	(3.6)	(3.2)	(4.0)	(4.1)	(1.0)	(3.4)

*Data corrected for mass fractionation and laboratory blank (*Ludwig*, 1985a; see text for details).
†2 sigma errors (in percent).
‡Pb, U, and Th concentration of leaches are shown against the starting weight.
§Data corrected for initial Pb (see text for details).

1a). Uranium was not preferentially leached during this procedure; concentrations never exceed 2.1 ppb in the acetic acid and HBr (0.005 N and 0.05 N) leaches.

All lead isotopic data obtained in this study (50 analyses) are depicted in a $^{206}Pb/^{204}Pb$ vs. $^{207}Pb/^{204}Pb$ diagram (Fig. 2). The HBr leaches contained Pb with the most-radiogenic isotopic compositions ($^{206}Pb/^{204}Pb$ ratios = 17.3-29.8). The highest $^{206}Pb/^{204}Pb$ values were from the fine-grained fraction and the 2.58-2.95 density fraction, suggesting some dissolution of a very radiogenic phase (such as a phosphate) during the HBr leaching. Because we smelled H_2S during HBr leaching of virtually all fractions, we infer the presence of troilite. Therefore, we speculate that radiogenic Pb was masked by nonradiogenic Pb derived from the troilite. For the second set of leaching experiments, higher isotopic compositions were not measured in the very dilute HBr leaches, contrary to our expectation that only a very radiogenic phase would have been dissolved ($^{206}Pb/^{204}Pb$ ratios range between 20.1 and 29.0). Likewise, radiogenic Pb was not measured in the acetic acid leach of the fine-grained groundmass ($^{206}Pb/^{204}Pb$ ratio of 20.0). The HCl leaches exhibit small variations of their $^{206}Pb/^{204}Pb$ ratios (17.5 to 20.8), very close to the blank isotopic composition ($^{206}Pb/^{204}Pb$ = 18.80). The lowest $^{206}Pb/^{204}Pb$ ratios have been measured in the HF+HNO_3 leaches ($^{206}Pb/^{204}Pb$ range between 11.9 and 14.1). The residues show the largest variation of Pb isotopic composition with $^{206}Pb/^{204}Pb$ ratios ranging between 12.5 and 21.0.

These Pb data, which show significant scatter, define a linear array whose slope corresponds to an age of 4288 ± 85 Ma. The relatively nonradiogenic nature of the Pb explains in part why the data plot only on a poorly defined "isochron."

The U-Pb data are plotted in Fig. 3 on the modified

TABLE 1c. U-Th-Pb analytical data for the Estherville mesosiderite (HCl and HF+HNO_3 leaches).

Sample	Weight (mg)	Pb* (ppb)	U* (ppb)	Th* (ppb)	$^{206}Pb/^{204}Pb^*$	$^{207}Pb/^{204}Pb^*$	$^{208}Pb/^{204}Pb^*$	$^{238}U/^{204}Pb^*$	$^{232}Th/^{238}U^*$	$^{206}Pb/^{238}U^{*§}$	$^{207}Pb/^{235}U^{*§}$	$^{207}Pb/^{206}Pb^{*§}$	$^{208}Pb/^{232}Th^{*§}$
HCl (2N) leaches													
F-WR	100.0	16.1‡	1.34	6.43	20.76	18.05	41.55	5.837	4.966	1.961	183.3	0.678	0.416
					(0.11)†	(0.15)	(0.20)	(0.54)	(0.27)	(1.2)	(1.6)	(0.65)	(1.1)
F-Opx	68.7	30.4	1.48	0.67	17.74	15.54	37.54	3.006	0.470	2.806	240.5	0.622	5.671
					(0.10)	(0.15)	(0.20)	(2.3)	(1.6)	(3.2)	(3.5)	(1.0)	(2.1)
F-Brec.	34.1	24.8	10.8	1.71	19.12	15.95	38.93	28.26	0.163	0.347	27.60	0.576	2.050
					(0.16)	(0.17)	(0.21)	(0.89)	(1.2)	(1.2)	(1.8)	(0.93)	(1.7)
F-ρ>3.3	96.1	79.8	0.20	0.79	18.25	15.43	38.01	0.157	4.083	57.10	4519	0.574	13.35
					(0.11)	(0.16)	(0.22)	(2.4)	(0.94)	(4.0)	(4.3)	(1.1)	(1.6)
F-ρ>3.3-mag.	94.5	5.76	0.30	2.29	19.12	17.39	38.69	3.463	7.810	2.833	282.5	0.723	0.341
					(0.35)	(0.34)	(0.30)	(3.1)	(2.4)	(3.8)	(3.7)	(0.86)	(2.6)
F-2.95<ρ<3.3	96.5	8.03	0.39	1.30	17.50	16.21	37.00	2.996	3.454	2.734	272.0	0.722	0.727
					(0.31)	(0.36)	(0.40)	(3.7)	(0.60)	(7.0)	(7.0)	(0.56)	(1.3)
F-2.95<ρ<3.3-mag.	95.3	26.8	0.17	1.38	18.53	16.20	38.13	0.393	8.630	23.45	2071	0.640	2.551
					(0.12)	(0.17)	(0.22)	(4.1)	(0.55)	(4.6)	(4.8)	(0.89)	(1.4)
F-2.58<ρ<2.95	95.4	8.10	0.41	1.42	17.70	16.36	38.00	3.175	3.607	2.644	263.6	0.723	0.745
					(0.20)	(0.25)	(0.31)	(3.5)	(0.57)	(6.5)	(6.6)	(0.62)	(1.4)
F-ρ<2.58	34.8	17.0	1.84	23.5	18.21	16.75	38.19	6.791	13.16	1.311	131.0	0.725	0.097
					(0.47)	(0.49)	(0.50)	(1.3)	(1.2)	(1.8)	(2.2)	(0.80)	(2.7)
HF(1N) + HNO_3(2N) leaches													
G-WR	100.0	31.9	0.36	0.73	13.60	13.74	33.53	0.602	2.103	7.125	788.0	0.802	3.199
					(0.24)	(0.26)	(0.31)	(1.6)	(1.0)	(3.2)	(3.6)	(1.3)	(2.9)
G-Opx	68.7	6.22	0.08	0.34	13.18	12.81	32.95	0.688	4.258	5.630	503.3	0.648	1.187
					(0.30)	(0.35)	(0.38)	(7.6)	(3.1)	(12.6)	(13.0)	(2.0)	(4.4)
G-Brec.	34.1	31.7	0.25	3.93	12.96	13.33	32.95	0.408	9.020	8.951	1026	0.832	0.520
					(0.32)	(0.20)	(0.25)	(2.7)	(1.2)	(5.3)	(5.6)	(1.7)	(4.0)
G-ρ>3.3	96.1	7.31	0.36	0.36	14.09	14.31	34.73	2.717	1.030	1.761	203.8	0.840	1.740
					(0.31)	(0.35)	(0.42)	(3.7)	(2.1)	(5.6)	(5.3)	(1.1)	(3.0)
G-ρ>3.3-mag.	94.5	10.2	0.38	0.68	13.10	13.73	32.84	1.976	5.604	1.920	239.8	0.906	0.926
					(0.33)	(0.37)	(0.42)	(2.7)	(1.1)	(5.1)	(5.3)	(1.3)	(3.6)
G-2.95<ρ<3.3	96.5	7.15	0.18	0.33	12.93	13.75	32.88	1.322	1.886	2.742	360.9	0.954	1.364
					(0.32)	(0.36)	(0.41)	(5.3)	(2.2)	(8.8)	(8.7)	(1.2)	(3.9)
G-2.95<ρ<3.3-mag.	95.3	15.7	0.19	0.80	11.98	13.05	32.06	0.606	4.360	4.409	626.4	1.031	0.975
					(0.21)	(0.26)	(0.31)	(2.4)	(0.93)	(5.4)	(5.4)	(1.7)	(4.4)
G-2.58<ρ<2.95	95.4	23.6	0.38	0.94	12.24	13.23	32.25	0.815	2.557	3.590	496.3	1.003	1.329
					(0.19)	(0.17)	(0.22)	(1.7)	(0.94)	(4.4)	(4.3)	(1.6)	(4.2)
G-ρ<2.58	34.8	39.2	0.60	2.01	11.94	12.98	31.79	0.768	1.701	3.431	482.3	1.019	0.876
					(0.19)	(0.20)	(0.25)	(2.2)	(1.3)	(5.2)	(5.0)	(1.9)	(5.3)

*Data corrected for mass fractionation and laboratory blank (*Ludwig*, 1985a; see text for details).
†2 sigma errors (in percent).
‡Pb, U, and Th concentration of leaches are shown against the starting weight.
§Data corrected for initial Pb (see text for details).

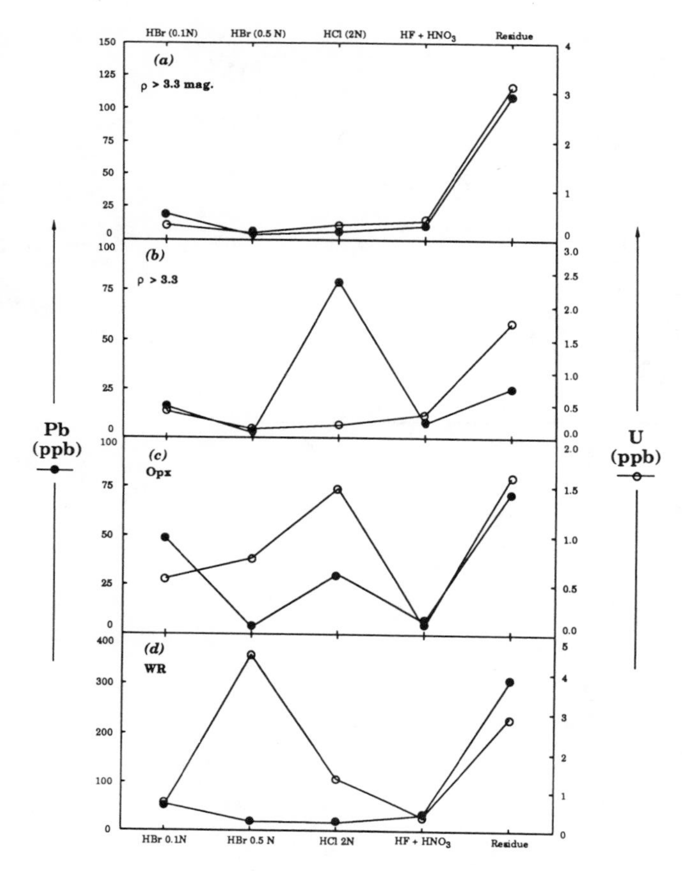

Fig. 1. Variation of U and Pb concentrations during our leaching procedure.

concordia diagram of *Tera and Wasserburg* (1972). The data are corrected using Cañon Diablo Troilite Pb (*Tatsumoto et al.,* 1973) as the assumed initial Pb isotopic composition. All but four (one HCl and three HBr leaches) data points plot to the left of the concordia curve, indicating an excess of radiogenic Pb.

It can be seen from Fig. 3 (insert) that most HBr leaches have $^{207}Pb/^{206}Pb$ ratios ranging between 0.56 and 0.66 whereas HCl leaches (0.57-0.73) and HF+HNO_3 leaches (0.64-1.02) exhibit higher $^{207}Pb/^{206}Pb$ isotopic ratios. Residues show variable $^{207}Pb/^{206}Pb$ isotopic ratios (0.56-0.97).

Despite the rather wide scatter, our data can be divided into four groups that define chords intersecting the concordia curve at different ages.

Group 1. Twenty-one analyses (mainly HBr leaches and residues) define a chord intersecting concordia at 4571 ± 18 Ma and 144 ± 110 Ma (Fig. 3). On a $^{206}Pb/^{204}Pb$ vs. $^{207}Pb/^{204}Pb$ diagram, the same data define a linear array with a slope that corresponds to an age of 4556 ± 35 Ma, in agreement with the U-Pb age (Fig. 4).

Group 2. Thirteen analyses (also mainly HBr leaches and residues) define an almost parallel trend intersecting concordia at 4437 ± 11 Ma and -40 ± 56 Ma. Once again, this U-Pb age is in agreement with the Pb-Pb age obtained from the same data (4422 ± 50 Ma, Fig. 5).

In contrast with the first two groups, which yield chords with lower-intercept ages close to the origin, all other data define chords with much older lower-intercept ages.

Group 3. Eight analyses (mainly HCl and HF+HNO_3 leaches) define a chord intersecting concordia at 4082 ± 210 Ma and 2842 ± 380 Ma. However, on the $^{206}Pb/^{204}Pb$ vs. $^{207}Pb/^{204}Pb$ diagram, these data plot on a 4557 ± 45 Ma isochron (Fig. 6a).

Group 4. Eight other analyses (mainly HF+HNO_3 leaches) define a chord that intersects concordia at 4480 ± 160 Ma and 3210 ± 340 Ma. On the $^{206}Pb/^{204}Pb$ vs. $^{207}Pb/^{204}Pb$ diagram, they define a linear array corresponding to an age of 4496 ± 80 Ma (Fig. 6b).

DISCUSSION

Uranium-Lead Systematics

As one of us has argued previously regarding lunar samples, when three or more stages of Pb evolution are involved, the age for the beginning of the second stage is not definable in the U-Pb system (*Tatsumoto and Unruh,* 1976). Moreover, differential leaching of U and Pb in successive leaches (Fig. 1) may have altered the $^{238}U/^{206}Pb$ ratios. So it is likely that some U-Pb ages do not correspond to true events. Those data that exhibit high and variable $^{207}Pb/^{206}Pb$ values are likely affected. As can be seen in Fig. 3, they are mostly HCl and HF+HNO_3 leaches (groups 3 and 4). This may explain the scatter of those data on the $^{238}U/^{206}Pb$ vs. $^{207}Pb/^{206}Pb$ diagram, the resulting large errors on the ages (between 160 and

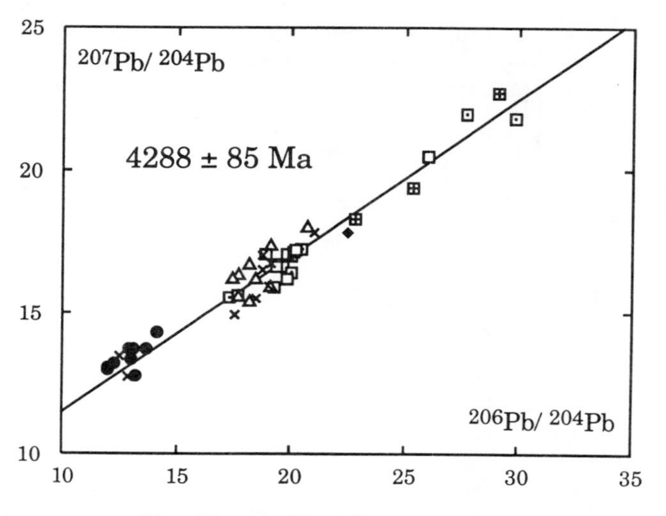

Fig. 2. $^{206}Pb/^{204}Pb$ vs. $^{207}Pb/^{204}Pb$ diagram of all the data obtained on the Estherville mesosiderite leaches and residues. Diamond: acetic acid leach; squares: HBr leaches (with pluses: dilute HBr; with points: 0.1 N HBr; open: 0.5 N HBr); triangles: HCl leaches; circles: HF+HNO_3 leaches; crosses: residues. All ages were determined using the algorithms of *Ludwig* (1980, 1985b), which use the regression approach of *York* (1969). All U-Pb and Pb-Pb regressions shown in this study present significant excess scatter with MSWD >> 1. Uncertainties are reported at the 95% confidence level.

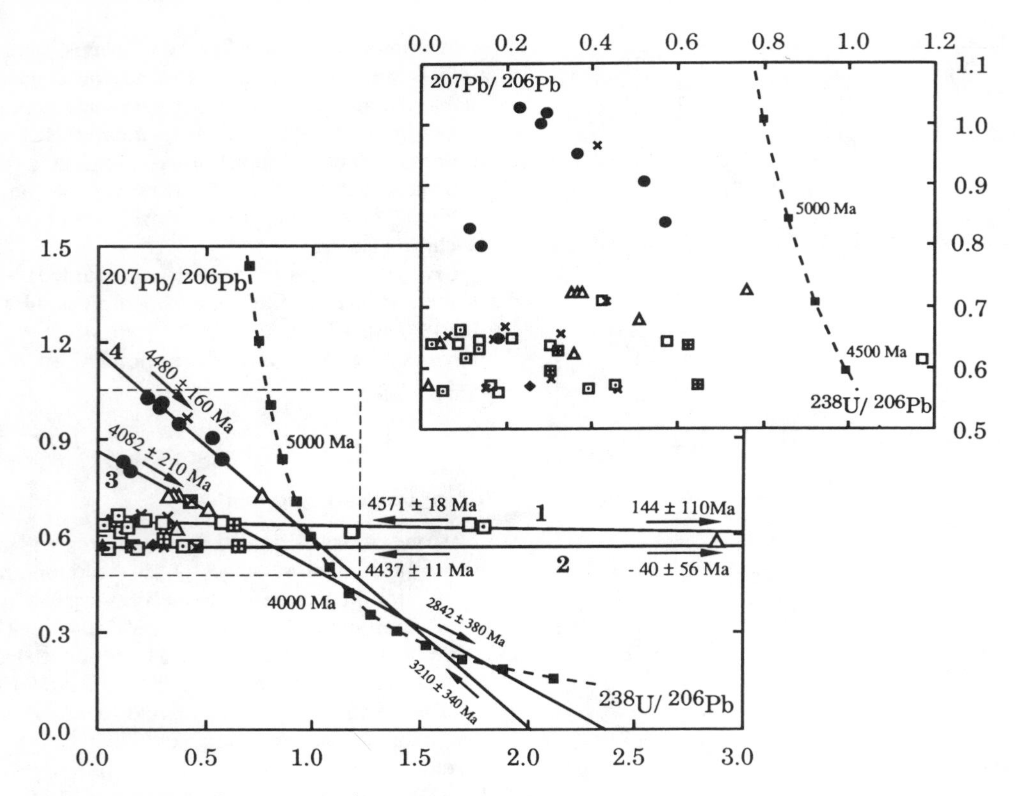

Fig. 3. $^{238}U/^{206}Pb$ vs. $^{207}Pb/^{206}Pb$ modified concordia diagram after *Tera and Wasserburg* (1972). Same symbols as in Fig. 2. Concordia is represented by a dotted line with a filled square each 250 Ma. The numbers on the chords refer to the four groups defined in the text. Arrows indicate the location of intercepts (upper and lower). The location of the insert diagram is indicated by the dotted rectangle.

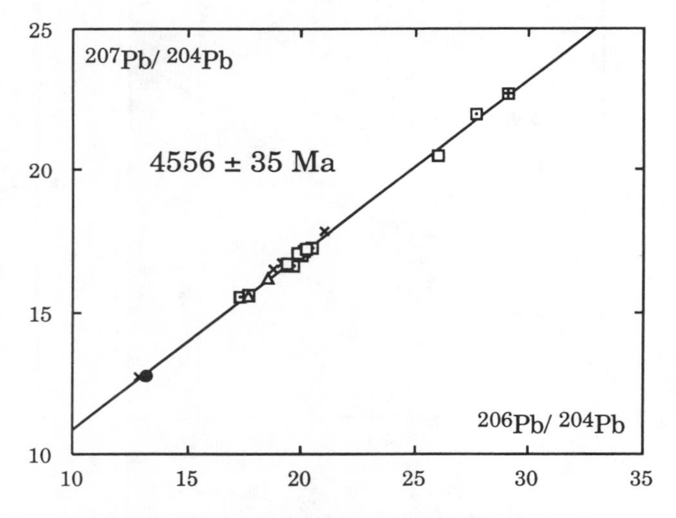

Fig. 4. $^{206}Pb/^{204}Pb$ vs. $^{207}Pb/^{204}Pb$ isochron diagram for 21 data of the Estherville mesosiderite (group 1: mainly HBr leaches and residues). Data involved in this diagram are E-ρ>3.3; D-ρ>3.3mag.; E-ρ>3.3mag.; R-ρ>3.3.mag.; D-2.95<ρ<3.3; E-2.95<ρ<3.3; D-2.95<ρ<3.3mag.;F-2.95<ρ<3.3mag.; R-2.95<ρ<3.3mag.; D-2.58<ρ<2.95; E-2.58<ρ<2.95; E-ρ<2.58; R-ρ<2.58; D-OPX; E-OPX; F-OPX; G-OPX; E-WR; R-WR; C-2.58<ρ<2.95 (d); B-WR (d).

380 Ma), and the differences observed between the Pb-Pb and U-Pb ages (Table 2). However, we do not suspect that our leaching procedure has affected those data that exhibit rather similar $^{207}Pb/^{206}Pb$ ratios, and that plot on nearly horizontal chords with lower-intercept ages close to the origin (groups 1 and 2; Fig. 3). Those chords show upper-intercept ages with small errors in agreement with the Pb-Pb ages (Table 2).

Data of group 1 and 2 that plot close to the Y-axis (Pb-rich components) in Fig. 3 may indicate "real" ages or, at least, the existence of different age components. The data that plot far away from the Y-axis may not have any real meaning, and grouping the data is actually guesswork. It must be mentioned, nonetheless, that the U-Pb ages of these groups are in good agreement with the Pb-Pb ages. Because our technique leached U-Pb from specific mineral phases, but without differentially extracting U and Pb from every mineral phase in the density fractions (for example, the acetic acid leached the iron coating and fine inclusions as well as terrestrial contamination, the dilute HBr dissolved troilite and phosphate minerals, and the stronger HBr and HCl dissolved the phosphate minerals), the U-Pb and Pb-Pb ages of the leaches may match (e.g., group 4), although we do not deny that some of the agreement may be fortuitous. One may consider the $^{207}Pb/^{206}Pb$ ages of the leaches as having a similar meaning to the ages produced by stepwise heating during $^{40}Ar/^{39}Ar$ dating, because chemical

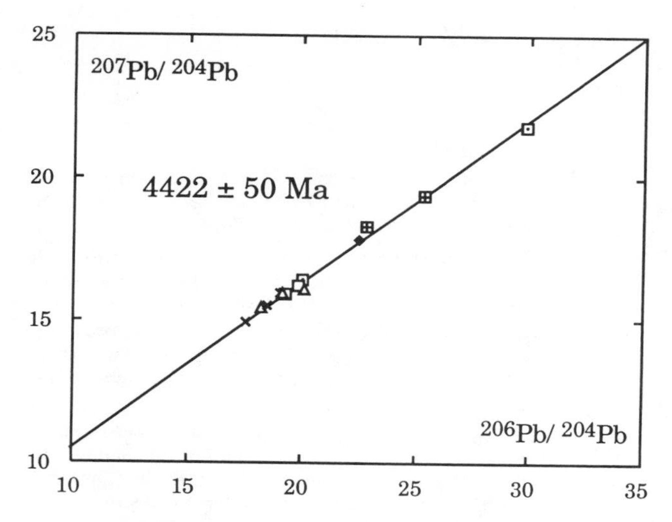

Fig. 5. $^{206}Pb/^{204}Pb$. vs. $^{207}Pb/^{204}Pb$ isochron diagram for 13 data of the Estherville mesosiderite (group 2: mainly HBr leaches and residues). Data involved in this diagram are D-ρ>3.3; F-ρ>3.3; R-ρ>3.3; D-ρ<2.58; R-OPX; D-Brec.; E-Brec.; F-Brec.; R-Breccia; D-WR; A-2.58<ρ<2.95 (d); B-2.58<ρ<2.95 (d); C-WR (d).

leaching cannot fractionate isotopes. Although the $^{207}Pb/^{206}Pb$ values measured on polymict breccias do not correspond to single phases, the corresponding Pb-Pb ages may be very close to the true ages of components.

We emphasize that a U-Pb systematics study involving a multileaching technique can distinguish ages of several components in polymict breccias for which the Pb isotope measurements alone could not distinguish between an isochron and a mixing line.

Ages of the Estherville Mesosiderite

Mesosiderites, which are considered to be aggregated fragments of differentiated meteorite parent bodies, have been affected by a series of magmatic and metamorphic events: brecciation by meteoritic impacts (sometimes several generations of brecciation; *Wasson and Rubin,* 1985), mixing with metal (either liquid or under extreme plastic deformation in the solid state; *Powell,* 1971), and a heating event (catastrophic collision of two asteroids; *Hewins,* 1983; *Bogard et al.,* 1988). All these events may have affected the different isotopic systems usually used to determine ages of meteorites.

A heating event probably occurred 3.6 to 3.8 Ga ago (Ar-Ar ages on seven mesosiderites; *Bogard et al.,* 1988). For Estherville, *Murthy et al.* (1978) reported an Ar-Ar age of 3.6 Ga (poorly defined plateau age) in agreement with ages from *Bogard et al.* (1988). This event, which reset the Ar-Ar isotopic system apparently partially disturbed the Rb-Sr isotopic system as well as the U-Pb system. Estherville, like many achondrites, contains high Rb/Sr mesostasis phases (*Duke and Silver,* 1967; *Powell,* 1971; *Allègre et al.,* 1975; *Murthy et al.,* 1977) that are very sensitive to perturbations. The 4.24 ± 0.03 Ga Rb-Sr age obtained by *Murthy et al.* (1978) on Estherville is, moreover, controlled only by the mesostatis phase (*Murthy et al.,* 1977). Without this phase, an age of 4.53 ± 0.26 Ga was obtained by these authors. So it is possibly a coincidence that the Rb-Sr age of 4.24 ± 0.03 Ga is, within

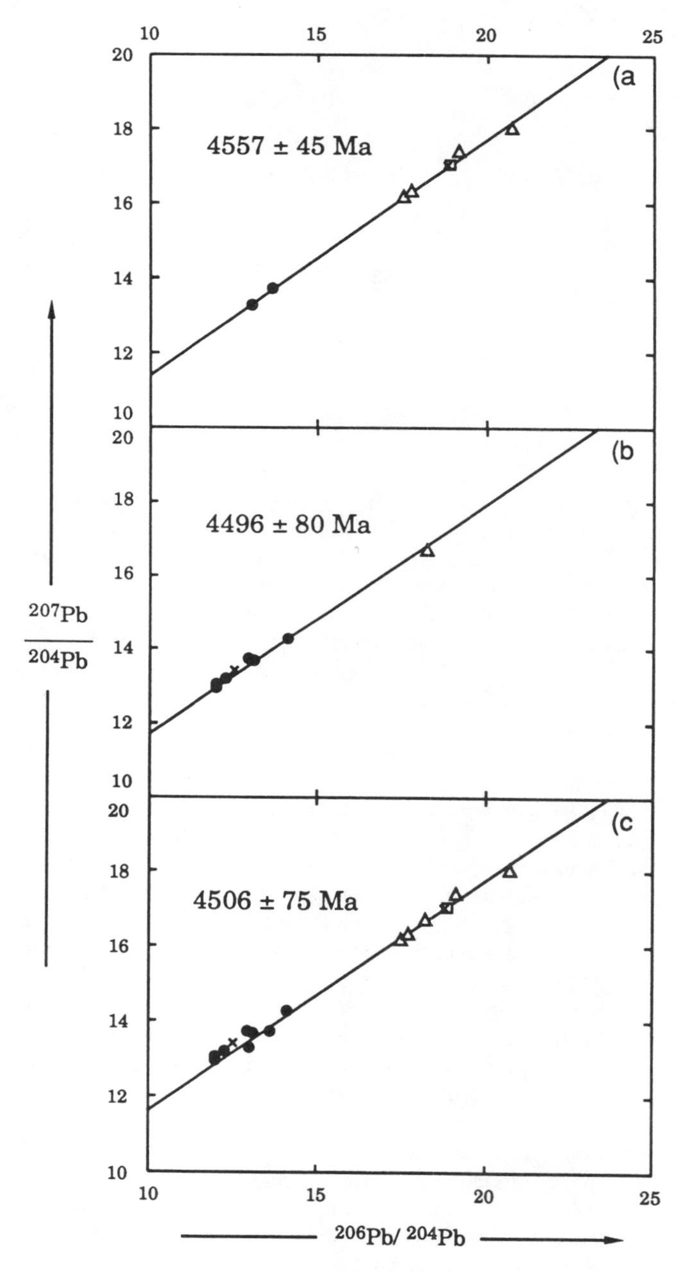

Fig. 6. $^{206}Pb/^{204}Pb$ vs. $^{207}Pb/^{204}Pb$ isochron diagram for the 16 other data of the Estherville mesosiderite. **(a)** Group 3: mainly HCl and $HF+HNO_3$ leaches. Data involved in this diagram are F-ρ>3.3mag.; F-2.95<ρ<3.3; R-2.95<ρ<3.3; E-2.95<ρ<3.3mag.; F-2.58<ρ<2.95; G-Brec.; F-WR; G-WR. **(b)** Group 4: mainly $HF+HNO_3$ leaches. Data involved in this diagram are G-ρ>3.3; G-ρ>3.3mag.; G-2.95<ρ<3.3; G-2.95<ρ<3.3mag.; G-2.58<ρ<2.95; R-2.58<ρ<2.95; F-ρ<2.58; G-ρ<2.58. **(c)** All the 16 data noted above.

TABLE 2. Summary of the Pb-Pb and U-Pb ages obtained on the Estherville mesosiderite.

Pb-Pb age	U-Pb upper-intercept age	U-Pb lower-intercept age	Data	Remarks
4288 ± 85 Ma	—	—	All	Mixing isochron.
4556 ± 35 Ma	4571 ± 18 Ma	144 ± 110 Ma	Group 1	U-Pb lower-intercept age is close to the origin. U-Pb upper-intercept age and Pb-Pb age are similar. *This age likely has geochronological significance.*
4422 ± 50 Ma	4437 ± 11 Ma	-40 ± 56 Ma	Group 2	U-Pb lower intercept age is close to the origin. U-Pb upper-intercept age and Pb-Pb are similar. *This age likely has geochronological significance.*
4557 ± 45 Ma	4082 ± 210 Ma	2842 ± 380 Ma	Group 3	Lower-intercept age is very different from the origin. U-Pb upper-intercept age and Pb-Pb age are different. U and Pb have probably been fractionated during the leaching procedure.
4496 ± 80 Ma	4480 ± 160 Ma	3210 ± 340 Ma	Group 4	Lower-intercept age is very different from the origin. U-Pb upper-intercept age and Pb-Pb age should be different (likely a coincidence). U and Pb have probably been fractionated during the leaching procedure.
4506 ± 75 Ma	—	—	Group 3 & 4	Group 3 and 4 cannot be differentiated on the basis of their U-Pb data.

error, similar to the Pb-Pb age of 4.28 ± 0.085 Ga obtained by us from all data (Fig. 2). Indeed, this Pb-Pb isochron appears to be a mixing line and not a "true" isochron. We have already shown that 4 Pb-Pb isochrons with ages ranging between 4.42 and 4.56 Ga can be drawn with this set of data (Figs. 4, 5, and 6a,b), and four chords with ages ranging between 4.08 and 4.57 Ga on a concordia plot (Fig. 3). We suspect that the Pb-Pb and U-Pb systems are probably more easily reset than the Rb-Sr system because Pb appears to be more mobile than Sr (for example, the closing temperatures of Ar, Pb, and Sr in apatite are, respectively, ≈350°C, ≈550°C, and 420°-700°C; *Bogard et al.,* 1988; *Watson et al.,* 1985; *Farver,* 1988). Such a disturbance in the U-Pb system can be identified by the lower-intercept of a chord on a U-Pb diagram (Fig. 3). However, some lower-intercept ages obtained in this study may not always correspond to disturbance events because of the possible differential extraction of U and Pb during our leaching experiments.

It is likely that at least part of the silicate fraction of the Estherville mesosiderite was formed at ≈4.56 Ga. This statement appears to be supported by the Pb-Pb age of 4.56 Ga obtained on 21 analyses (group 1, Fig. 4). On a concordia diagram, this set of data plot on an almost horizontal chord that has a lower intercept near the origin (144 ± 110 Ma). Therefore, differential leaching of U and Pb will have very little effect on the U-Pb upper-intercept age. Moreover, because the Pb-Pb and U-Pb upper-intercept ages are similar (Table 2), we believe that this 4.56 Ga age has geochronological significance. This age, which is the oldest reported yet for a mesosiderite, may be compared to some eucrite ages (Juvinas: 4.56 by Rb-Sr and Pb-Pb, *Allègre et al.,* 1975; *Tatsumoto et al.,* 1973; Ibitira: 4.55 by Pb-Pb, *Chen and Wasserburg,* 1985). Eucrites and mesosiderites do exhibit some petrological similarities (*Duke and Silver,* 1967; *Powell,* 1971). However, eucrite clasts, described in nine other mesosiderites, have not been recognized in Estherville, although diogenite clasts have been described (*Floran,* 1978). Very little geochronologic work has been published on diogenites, but *Birck and Allègre* (1978; 1981) have shown that it was not possible to distinguish them from eucrites on the basis of their Rb-Sr isotopic compositions.

Estherville also contains cumulate eucrite clasts (*Floran,* 1978) that have been shown to be younger than noncumulate eucrites (*Tera et al.,* 1989). The Pb-Pb age of 4.42 ± 0.05 Ga obtained on some data from Estherville (group 2) compare very well with cumulate eucrite ages (4.41 to 4.46 Ga: Moama, Serra de Mage, Moore County; *Lugmair et al.,* 1977; *Jacobsen and Wasserburg,* 1984; *Tera et al.,* 1987, 1989). As for group 1, this set of data plots on an almost horizontal chord that has a lower intercept close to the origin (-44 ± 56 Ma; Fig. 3). Because once again the U-Pb upper-intercept age and the Pb-Pb age match (Table 2), we believe that this age has a geochronological significance.

The other Pb-Pb ages obtained from HCl and HF+HNO_3 leaches also range between 4.4 and 4.5 Ga (4557 ± 45 Ma for group 3 and 4496 ± 80 Ma for group 4, Figs. 6a,b). However, it is unclear, as discussed earlier, if these data can be separated on the basis of their U-Pb isotopic characteristics. If not, they can be combined to form a 4506 ± 75 Ma in a Pb-Pb diagram (Fig. 6c).

The different Pb-Pb ages obtained on Estherville lead us to suggest that this mesosiderite contains a heterogeneous silicate fraction that was formed early in the solar system history (between 4.42 and 4.56 Ga) and mixed together later. This mixing is likely related to the brecciation of the silicates by meteoritic impacts. Several generations of breccia formation recorded on some meteorites by *Wasson and Rubin* (1985) are in agreement with the heterogeneity of clasts and ages from Estherville. The technique used in this study (density separates and leaching with progressively stronger acid) allows us to describe this heterogeneity. Measurements of the Pb isotopic composition of only a few residues would have given a mixing age between 4.55 and 4.42 Ga, or even a 4.28 Ga age, as obtained from all the data. This may explain why only young ages have been obtained thus far on mesosiderites.

The U-Pb lower-intercept ages obtained on Estherville are, unfortunately, poorly constrained, mainly because most of the data plot to the left of concordia and also because of the differential leaching of U and Pb (for groups 3 and 4).

It is, however, useful to point out that some of the lower-intercept ages coincide with young ages measured with other methods. For group 1, the 144 ± 110 Ma lower-intercept age, which is not distinguishable from 0 because of its large error, may be compared to the 62 Ma cosmic ray exposure age measured by *Begeman et al.* (1976), recording the final breakdown of the mesosiderite before reaching Earth. For group 2, the slightly negative lower-intercept age (possibly owing to overcorrection for initial Pb?) is, within error, close to 0. Lower-intercept ages for groups 3 and 4 are clearly speculative. It may be observed that the lower-intercept age of 3.21 ± 0.34 Ga for group 4 is, within error, similar to the Ar-Ar age of 3.6 Ga (*Murthy et al.,* 1978), interpreted as resulting from a major heating event (collision of two asteroids?).

History of the Estherville Mesosiderite

The following scenario is proposed as a possible explanation of the petrological and geochronological data obtained on the Estherville mesosiderite.

1. Magmatic differentiation of the silicates could have occurred between 4.56 and 4.42 Ga.

2. The brecciation and mixing of the different silicate clasts, likely by meteoritic impacts, occurred early in solar system history, and was followed by the mixing with metal (*Hewins,* 1983; *Floran,* 1978; *Wasson and Rubin,* 1985). This mixing with metal is one major problem concerning the origin of mesosiderites and, as noted by *Hewins* (1983), no single model appears to be satisfactory (mainly impact vs. internal processes; *Powell,* 1971; *Delaney et al.,* 1981; *Chapman and Greenberg,* 1981; *Hewins,* 1983; *Wasson and Rubin,* 1985). Our results indicate that the mixing with metal, which supposedly happened after brecciation of the silicates (*Powell,* 1971), occurred after 4.42 Ga.

3. The Ar release at relatively high extraction temperatures around 3.6-3.8 Ga (*Bogard et al.,* 1988) can be explained either by very slow cooling from 4.55 to 3.6 Ga or by a later thermal event. Since cooling rates as low as 0.1°C/Ma (*Powell,* 1971) have been ruled out and corrected by at least a factor of 5 (*Crozaz and Tasker,* 1981), the heating event appears to be the likely explanation and is interpreted as a catastrophic collision of two asteroids (*Hewins,* 1983; *Bogard et al.,* 1988). This collision reset the Ar-Ar isotopic system and partially disturbed the Rb-Sr clock (mesostasis phase) as well as the Pb-Pb and U-Pb systematics. For Estherville, closure of the Ar-Ar isotopic system occurred around 3.6 Ga (*Murthy et al.,* 1978), suggesting a temperature as low as 370°C (*Turner et al.,* 1978). It is puzzling to us that a young Pb-Pb isochron age was not obtained in our study. The only explanation that can be given at present is that the silicate fraction is dominated by excess Pb that masked a young component, and/or, if we could separate a mesostatis phase that was the key component for the young Rb-Sr age, we might be able to find a younger Pb-Pb age.

4. The mesosiderite cooled down slowly (0.3°C/Ma) until ≈ 2.6 Ga (fission track age recording temperatures as low as 100°C; *Crozaz and Tasker,* 1981).

5. The mesosiderite was exposed to cosmic rays since 62 Ma (*Begemann et al.,* 1976).

CONCLUSIONS

1. Estherville contains a heterogeneous silicate fraction probably formed early in solar system history. The Pb-Pb ages range between 4556 ± 35 and 4422 ± 50 Ma, and may be compared with eucrite, diogenite, and cumulate eucrite ages.

2. Some U-Pb upper-intercept ages are in agreement with the Pb-Pb ages suggesting that these ages are true. Some discordance between U-Pb and Pb-Pb ages are probably due to differential leaching of U and Pb during our experimental procedure.

3. All younger ages measured on Estherville (Rb-Sr and Ar-Ar) are likely due to a disturbance (major heating event such as a collision between two asteroids?) around 3.6-3.8 Ga.

4. The U-Pb lower-intercept ages can be divided into two groups: (a) around 3 Ga (3.2 and 2.8), and likely related to the 3.6 Ga heating event, but with important errors and probably disturbed by the differential leaching of U and Pb; (b) close to 0 Ma (-44 ± 68 Ma) and to 62 Ma (cosmic rays exposure age, 144 ± 110 Ma).

Acknowledgments. We thank Dr. V. R. Murthy for providing the sample. K. R. Ludwig and D. M. Unruh gave helpful reviews of the first draft. Critical comments by J. Chen and D. Mittlefehldt are appreciated. We would like to thank E. E. Foord who kindly performed mineral identification by X-ray diffraction and SEM techniques, X. Li for his assistance during chemistry, and W. R. Premo, who had a hard time improving our "franco-japonais" English. This study was supported by NASA Interagency Transfer order T-783-H.

REFERENCES

Allègre C. J., Birck J. L., Fourcade S., and Semet M. (1975) ^{87}Rb-^{87}Sr age of the Juvinas basaltic achondrite and early igneous activity of the solar system. *Science, 197,* 436-438.

Begemann F., Weber H. W., Vilcsek E., and Hintenberger H. (1976) Rare gas and ^{36}Cl in stony-iron meteorites: Cosmogenic elemental production rates, exposures ages, diffusion losses, and thermal histories. *Geochim. Cosmochim. Acta, 40,* 353-368.

Birck J. L. and Allègre C. J. (1978) Chronology and chemical history of the parent body of basaltic achondrites studied by the ^{87}Rb-^{87}Sr method. *Earth Planet. Sci. Lett., 39,* 37-51.

Birck J. L. and Allègre C. J. (1981) ^{87}Rb-^{87}Sr study of diogenites. *Earth Planet. Sci. Lett., 55,* 116-122.

Bogard D., Mittlefehldt D., and Jordan J. (1988) ^{39}Ar-^{40}Ar dating of mesosiderites: A case for major parent body disruption less than 4.0 Gy ago? (abstract). In *Lunar and Planetary Science XIX,* pp. 112-113. Lunar and Planetary Institute, Houston.

Chapman C. R. and Greenberg R. (1981) Meteorites from the asteroid belt: The stony-iron connection (abstract). In *Lunar and Planetary Science XII,* pp. 129-131. Lunar and Planetary Institute, Houston.

Chen J. H. and Wasserburg G. J. (1985) U-Th-Pb isotopic studies on meteorite ALHA 81005 and Ibitira (abstract). In *Lunar and Planetary Science XVI,* pp. 119-120. Lunar and Planetary Institute, Houston.

Crozaz G. and Tasker D. R. (1981) Thermal history of mesosiderites revisited. *Geochim. Cosmochim. Acta., 45,* 2037-2046.

Delaney J. S., Nehru C. E., Prinz M., and Harlow G. E. (1981) Metamorphism in mesosiderites. *Proc. Lunar Planet. Sci. 12B,* pp. 1315-1342.

Duke M. B. and Silver L. T. (1967) Petrology of eucrites, howardites and mesosiderites. *Geochim. Cosmochim. Acta, 31,* 1637-1665.

Farver J. R. (1988) The diffusive kinetics of oxygen in diopside, oxygen and strontium in apatite, and applications to thermal histories and igneous and metamorphic rocks. Ph.D. thesis, Brown University, Providence, Rhode Island. 159 pp.

Floran R. J. (1978) Silicate petrography, classification and origin of the mesosiderites: Review and new observations. *Proc. Lunar Planet. Sci. Conf. 9th,* pp. 1053-1081.

Hewins R. H. (1983) Impact versus internal origins for mesosiderites. *Proc. Lunar Planet. Sci. Conf. 14th,* in *J. Geophys. Res., 88,* B257-B266.

Hewins R. H. (1984) The case for a melt matrix in plagioclase-POIK mesosiderites. *Proc. Lunar Planet. Sci. Conf. 15th,* in *J. Geophys. Res., 89,* C289-C297.

Jacobsen S. B. and Wasserburg G. J. (1984) Sm-Nd isotopic evolution of chondrites and achondrites, II. *Earth Planet. Sci. Lett., 67,* 137-150.

Jaffey A. H. Flynn K. F., Glendenin L. E., Bentley W. C., and Essling A. M. (1971) Precision measurements of half-lives and specific activities of ^{234}U, and ^{238}U. *Phys. Rev., C4,* 1889-1906.

LeRoux L. J. and Glendenin L. E. (1963) Half-life of thorium 232. *Proc. Natl. Mtg. on Nucl. Energy,* pp. 83-94. Pretoria, S. Africa.

Ludwig K. R. (1980) Calculation of uncertainties of U-Pb isotope data. *Earth Planet. Sci. Lett., 46,* 212-220.

Ludwig K. R. (1985a) PBDAT200: A computer program for processing raw Pb-U-Th isotope data. *U.S. Geol. Surv. Open-File Rept., 85-547.* 54 pp.

Ludwig K. R. (1985b) ISOPLOT200: A plotting and regression program for isotope geochemists, for use with HP series 200 computers. *U.S. Geol. Surv. Open-File Rept., 85-513.* 102 pp.

Lugmair G. W., Scheinin N. B., and Carlson R. W. (1977) Sm-Nd systematics of the Serra de Mage eucrite (abstract). *Meteoritics, 12,* 300-301.

Manhes G. and Allègre C. J. (1978) Time differences as determined from the ratio of lead 207 to lead 206 in concordant meteorites. *Meteoritics, 13,* 543-548.

Megrue G. H. (1966) Rare gas chronology of calcium-rich achondrites. *J. Geophys. Res., 71,* 4021-4027.

Mittlefehldt D. W., Bansal B. M., Shih C. Y., Wiesmann H., and Nyquist L. E. (1986) Petrology, chronology and chemistry of basaltic clasts from mesosiderites (abstract). In *Lunar and Planetary Science XVII,* pp. 553-554. Lunar and Planetary Institute, Houston.

Murthy V. R., Coscio M. R., and Sabelin T. (1977) Rb-Sr internal insochron and the initial ^{87}Sr/^{86}Sr for the Estherville mesosiderite. *Proc. Lunar Sci. Conf. 8th,* pp. 117-186.

Murthy V. R., Alexander E. C. Jr., and Saito K. (1978) Rb-Sr and ^{40}Ar-^{39}Ar systematics of the Estherville mesosiderite (abstract). In *Lunar and Planetary Science XIV,* pp. 781-810. Lunar and Planetary Institute, Houston.

Powell B. N. (1969) Petrology and chemistry of mesosiderites—I. Textures and composition of nickel-iron. *Geochim. Cosmochim. Acta, 33,* 789-810.

Powell B. N. (1971) Petrology and chemistry of mesosiderites—II. Silicate textures and composition of metal-silicate relationships. *Geochim. Cosmochim. Acta, 35,* 5-34.

Premo W. R., Tatsumoto M., and Wang J. W. (1989) Pb isotopes in anorthositic breccias 67075 and 62237: A search for primitive lunar lead. *Proc. Lunar Planet. Sci. Conf. 19th,* pp. 61-71.

Prinzhofer A., Papanastassiou D. A., and Wasserburg G. J. (1989) Sm-Nd chronology of differentiation of small planets (abstract). In *Lunar and Planetary Science XX,* pp. 872-873. Lunar and Planetary Institute, Houston.

Prior G. T. (1920) The classification of meteorites. *Mineral. Mag., 19,* 51-63.

Tatsumoto M. and Unruh D. M. (1976) KREEP basalt age: Grain by grain U-Th-Pb systematics study of the quartz monzodiorite clast 15405,88. *Proc. Lunar Sci. Conf. 7th,* pp. 2107-2129.

Tatsumoto M., Knight R. J., and Allègre C. J. (1973) Time differences in the formation of meteorites as determined by the ^{207}Pb/^{206}Pb. *Science, 180,* 1279-1283.

Tatsumoto M., Unruh D. M., and Desborough G. A. (1976) U-Th-Pb and Rb-Sr systematics of Allende and U-Th-Pb systematics of Orgueil. *Geochim. Cosmochim. Acta, 40,* 617-634.

Tera F. and Wasserburg G. J. (1972) U-Th-Pb systematics in three Apollo 14 basalts and the problem of initial Pb in lunar rocks. *Earth Planet. Sci. Lett., 14,* 281-304.

Tera F., Carlson R. W., and Boctor N. Z. (1987) Isotopic and petrologic investigation of the eucrites Cachari, Moore County, and Stannern (abstract). In *Lunar and Planetary Science XVIII,* pp. 1004-1005. Lunar and Planetary Institute, Houston.

Tera F., Carlson R. W., and Boctor N. Z. (1989) Contrasting Pb-Pb ages of the cumulate and non-cumulate eucrites (abstract). In *Lunar and Planetary Science XX,* pp. 1111-1112. Lunar and Planetary Institute, Houston.

Turner G., Enright M. C., and Cadogan P. H. (1978) The early history of choncrite parent bodies inferred from ^{40}Ar-^{39}Ar ages. *Proc. Lunar Planet. Sci. Conf. 9th,* pp. 989-1025.

Unruh D. M., Nakamura N., and Tatsumoto M. (1977) History of the Pasamonte achondrite: relative susceptibility of the Sm-Nd, Rb-Sr, and U-Pb systems to metamorphic events. *Earth Planet. Sci. Lett., 37,* 1-12.

Wasson J. T. and Rubin A. E. (1985) Formation of mesosiderites by low velocity impacts as a natural consequence of planet formation. *Nature, 318,* 168-170.

Watson E. B., Harrison T. M., and Ryerson F. J. (1985) Diffusion of Sm, Sr, and Pb in fluorapatite. *Geochim. Cosmochim. Acta, 49,* 1813-1323.

York D. K. (1969) Least squares fitting of a straight line with correlated errors. *Earth Planet. Sci. Lett., 5,* 320-324.

APPENDIX

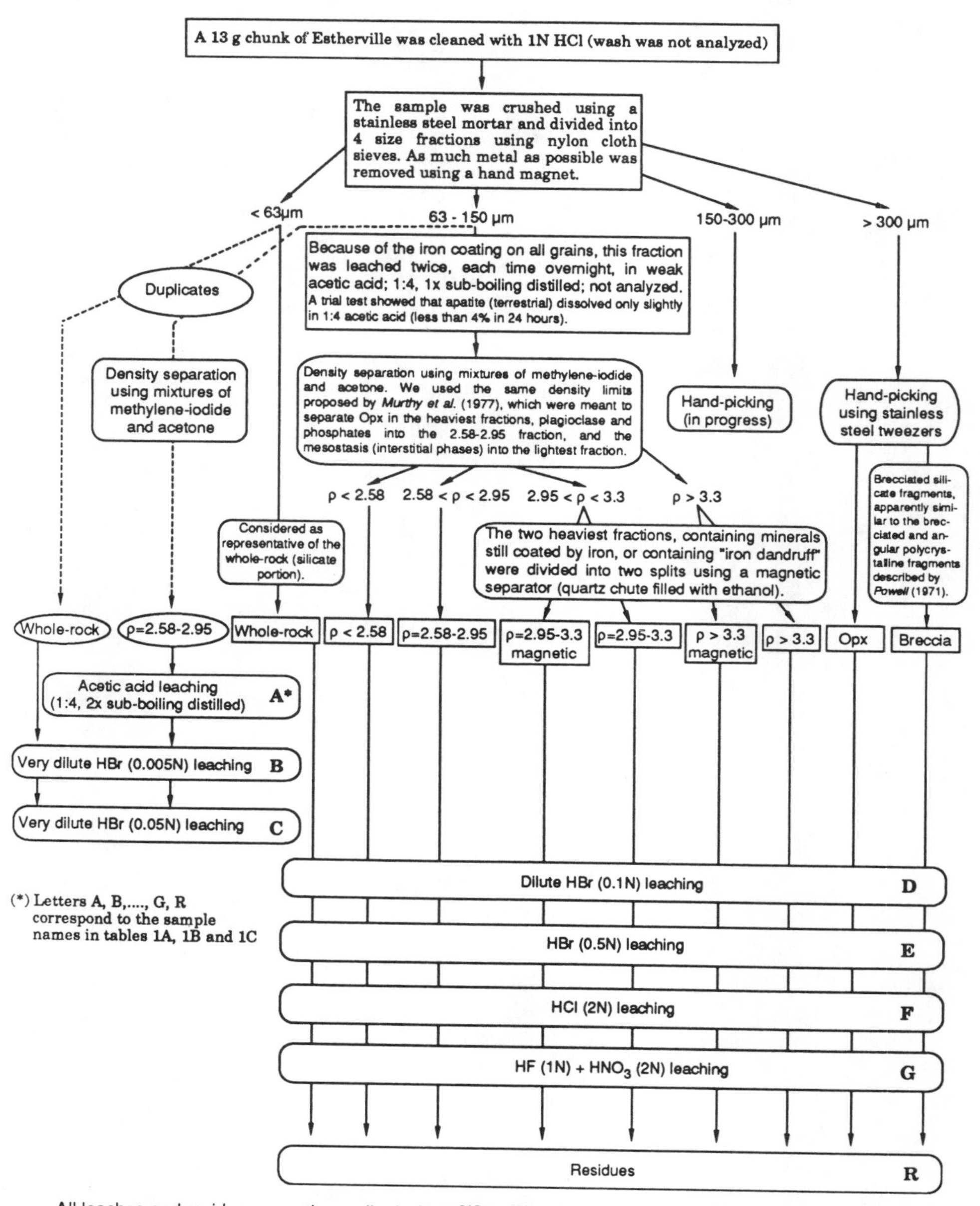

All leaches and residues were then spiked with a ^{205}Pb-^{233}U-^{236}U-^{230}Th mixed tracer solution and analyzed.

Comets and Interplanetary Dust

Proceedings of the 20th Lunar and Planetary Science Conference, pp. 323-333
Lunar and Planetary Institute, Houston, 1990

Titanium Oxide Magnéli Phases in Four Chondritic Porous Interplanetary Dust Particles

F. J. M. Rietmeijer

Department of Geology, University of New Mexico, Albuquerque, NM 87131

I. D. R. Mackinnon

Electron Microscope Centre, University of Queensland, St. Lucia, Brisbane, Queensland 4067, Australia

Detailed analytical electron microscope analyses of four fine-grained chondritic porous interplanetary dust particles (IDPs) reveal the presence of titanium oxide Magnéli phases, Ti_nO_{2n-1} (n = 4, 5, 6), and rare Ti-metal. The titanium minerals are indigenous to these chondritic IDPs. The association of Magnéli phases, Ti-metal, and carbonaceous material in chondritic IDPs, along with the grain size distributions of Ti-metal and Magnéli phases and equilibrium dissociation pressures for these oxygen-deficient Ti-oxides, support *in situ* solid carbon gasification in these extraterrestrial particles. The active catalyst in this process is titanium metal that we infer may be of interstellar origin. This favorable catalysis uniquely leads to the formation of Magnéli phases. As chondritic IDPs may be solid debris of short-period comets, our data indicate that nuclei of active short-period comets may show distinctive chemical reactions that lead to Ti-mineral assemblages that typically include Magnéli phases. The proposed model provides a plausible mechanism to explain the higher solid carbon content of chondritic IDPs relative to bulk carbon abundances typical for carbonaceous chondrite matrices that represent another type of more evolved, that is, metamorphosed, undifferentiated solar system bodies.

INTRODUCTION

Equilibrium vapor phase condensation models predict that various titanium oxides may condense in a cooling solar nebula gas and in supernova ejecta (*Lattimer and Grossman,* 1978). Observed depletions in the interstellar gas phase suggest that some fraction of titanium is indeed present in interstellar dust grains (*Morton,* 1978; *Morton and Hu,* 1975; *Snow,* 1976). Titanium isotopic anomalies of nucleosynthetic origin were carried into the early solar system in refractory interstellar dust, some of which is presently incorporated in carbonaceous chondrites (*Fahey et al.,* 1987). The presence of Ti-bearing phases at the locations of comet accretion in the early solar system is also inferred from spectral analyses of sun-grazing comet Ikeya Seki 1965f (*Delsemme,* 1982) and the P/comet Halley dust experiments (*Jessberger et al.,* 1988; *Hsiung and Kissel,* 1987). While anatase (TiO_2) has been observed in a carbonaceous chondrite acid residue (*Wopenka and Swan,* 1985), Ti-oxides are seemingly rare in primitive, fine-grained solar system materials such as carbonaceous chondrites and unequilibrated ordinary chondrites, while Ti-metal has not yet been identified in these meteorites (*Mason,* 1979).

Chondritic interplanetary dust particles (IDPs) are fine-grained extraterrestrial materials that are collected in the Earth's stratosphere and that sample small solar system bodies, such as comets and asteroids (*Brownlee,* 1985a; *Mackinnon and Rietmeijer,* 1987). It is probable that some fraction of chondritic IDPs are solid debris from short-period comets (*Bradley and Brownlee,* 1986; *Rietmeijer,* 1989a). Titanium oxides have been observed in four different chondritic IDPs out of a total of ~35 individual IDPs studied in detail using analytical electron microscopy (AEM) in addition to a chondritic IDP reported by *Flynn et al.* (1978). The presence of Ti-oxides in chondritic IDPs has been attributed to *in situ* low-temperature diagenetic processes, but this interpretation required the *ad hoc* postulate of (unobserved) $BaTiO_3$ (*Rietmeijer and Mackinnon,* 1984).

We report here new observations of Ti-minerals in additional chondritic porous IDPs. Our current model for the presence of Ti-minerals in these IDPs combines aspects of the previous interpretation and incorporates all observations of Ti-rich grains in these extraterrestrial materials. We propose that Ti-metal reacted to various Ti-oxide Magnéli phases during *in situ* catalytic gasification of solid carbon in these chondritic porous IDPs.

EXPERIMENTAL

Small particles ($\lesssim 50\,\mu m$ in diameter) are routinely collected in the Earth's stratosphere between 18 km and 20 km altitude (*Mackinnon and Rietmeijer,* 1987 and references therein for details on collection and curation procedures). A subset of these particles has approximately chondritic elemental abundance patterns in conjunction with a fluffy morphology. To a first approximation, the low bulk density and elemental abundance pattern are reliable determinants for an extraterrestrial origin, while noble gas contents, large D/H fractionation ratios, and solar flare tracks in constituent minerals uniquely point to an extraterestrial origin for porous, chondritic IDPs. The articles by *Brownlee* (1985a), *Mackinnon and Rietmeijer* (1987), and *Bradley et al.* (1988) provide current reviews of these observations and interpretations on the extraterrestrial nature of certain stratospheric particles.

Sample allocations of chondritic porous (CP) IDPs W7029C1 (*Rietmeijer and Mackinnon,* 1985a) and

W7010*A2 (*Rietmeijer and McKay,* 1986) were transferred to holey carbon films supported by a Be mesh grid. During transfer, the samples broke apart onto the holey carbon films. Thus, individual particles from each allocation were suitably dispersed on a sample substrate and accessible to study using a JEOL 100CX analytical electron microscope (AEM) equipped with a Princeton Gamma Tech (PGT) System IV energy dispersive spectrometer (EDS) and a Philips 420T equipped with an EDAX windowless EDS (WEDS) (see *Rietmeijer and Mackinnon,* 1985a,b, for details on analytical procedures). Selected Ti-grains were also analyzed using a JSM-35CF scanning electron microscope operating at an accelerating voltage of 10 keV and fitted with PGT WEDS for analysis of elements with atomic number $Z < 11$. Allocations of samples from CP IDPs U2011C2, U2034C12, and W7029C1 were embedded in epoxy and prepared for ultramicrotome thin sectioning. Thin sections were analyzed using a JEOL 2000FX AEM operating at an accelerating voltage of 200 keV that was equipped with a Tracor-Northern TN 5500 EDS. The SAED data obtained with the JEOL 100CX AEM have a relative error <2%; for those obtained with the JEOL 2000FX AEM the relative error is <1.5% (*Rietmeijer and Mackinnon,* 1985a, 1987a). In most cases, diffraction data were obtained from fine-grained aggregates or clusters that gave rise to polycrystalline (or "ring") diffraction patterns. For Ti-grains in ultramicrotomed thin sections, identifications are generally supported by single crystal electron diffraction data. Mineral identification is possible by a combination of electron microscope imaging, EDS and WEDS analysis, and electron diffraction. However, the crystallographic similarity of Ti-oxide Magnéli phases may interfere with unambiguous identification of a particular Magnéli phase, and we cannot unequivocally exclude the possibility of other Magnéli phases in these CP IDPs in addition to those identified in this paper.

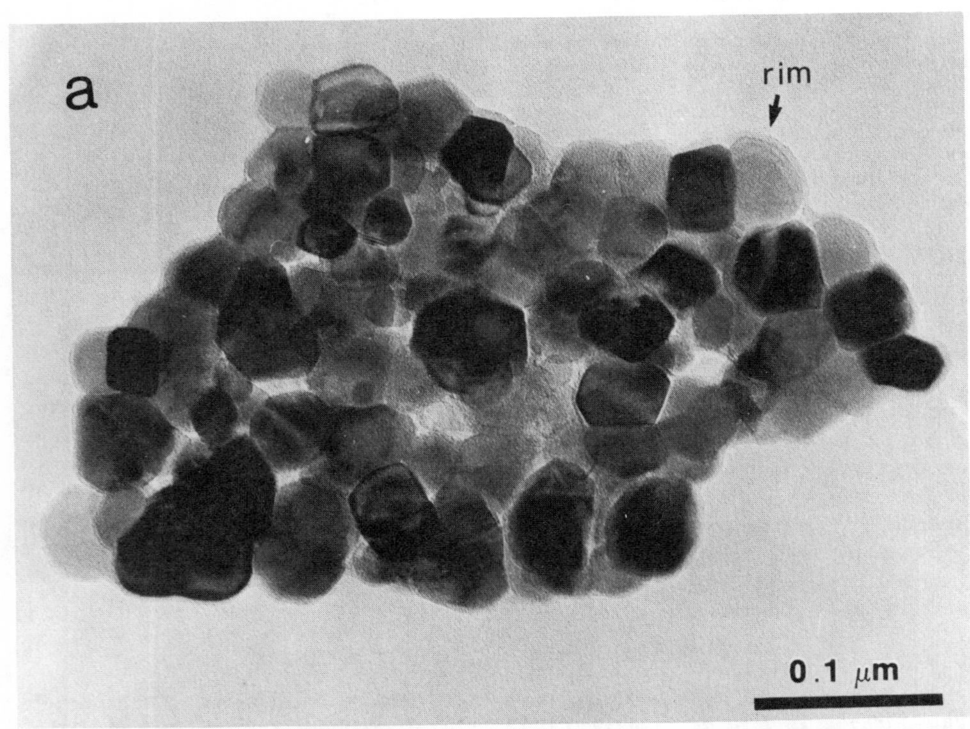

Fig. 1. Clusters of Ti-oxide grains in chondritic porous IDP U2011C2. **(a)** Large, open cluster. Carbonaceous rims visible on some individual grains (arrows). Insets show polycrystalline SAED patterns obtained with the cluster in two different orientations relative to the incident electron beam. **(b)** Cluster of Ti-oxide grains attached to carbonaceous material. Narrow rims visible around individual grains. **(c)** Cluster of subhedral and rounded Ti-oxide grains with carbonaceous rims clearly visible on each grain. **(d)** Subhedral, single crystal, platey Ti_6O_{11} grain in chondritic porous IDP U2011C2 with dislocation structures. Inset: single crystal electron diffraction pattern; zone axis $\bar{2}3\bar{1}$.

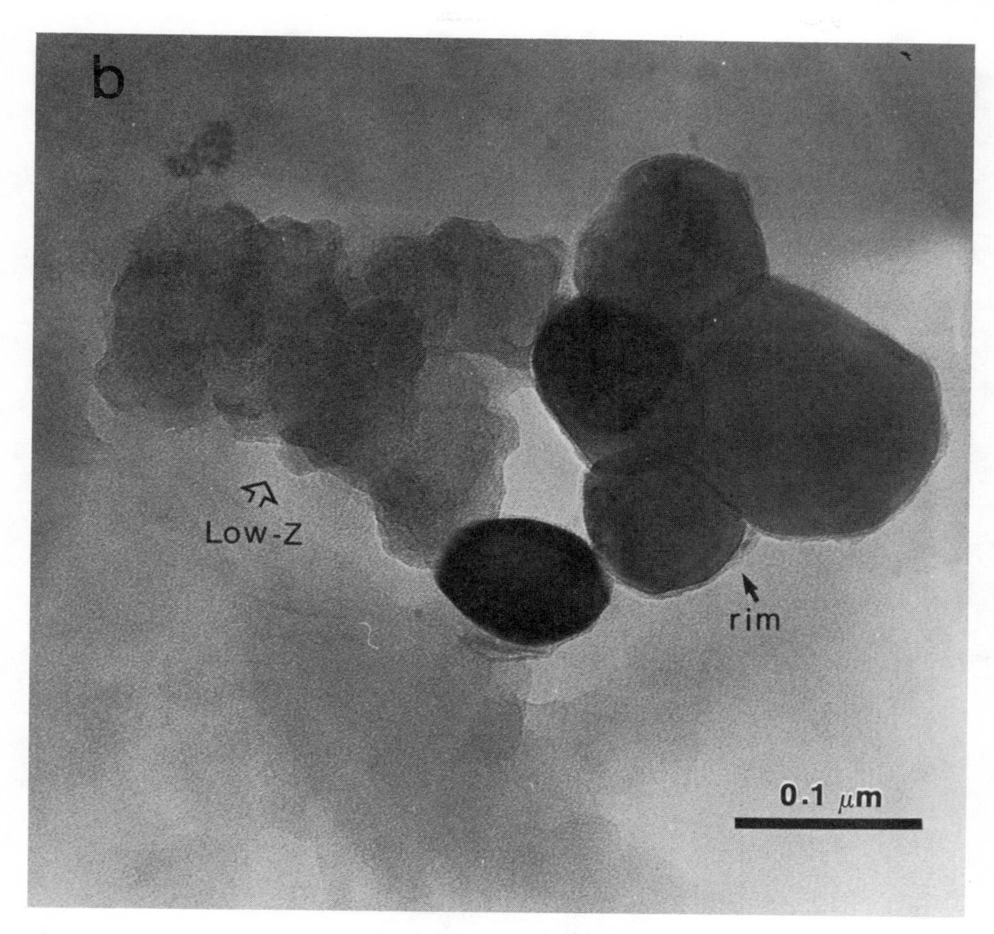

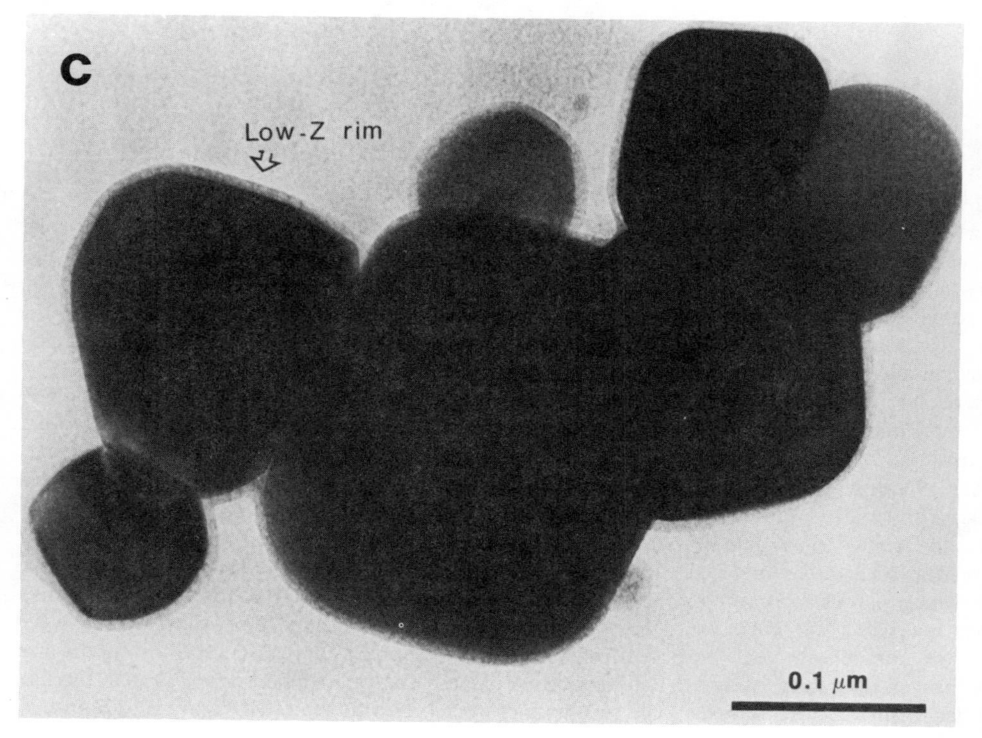

Fig. 1. (continued).

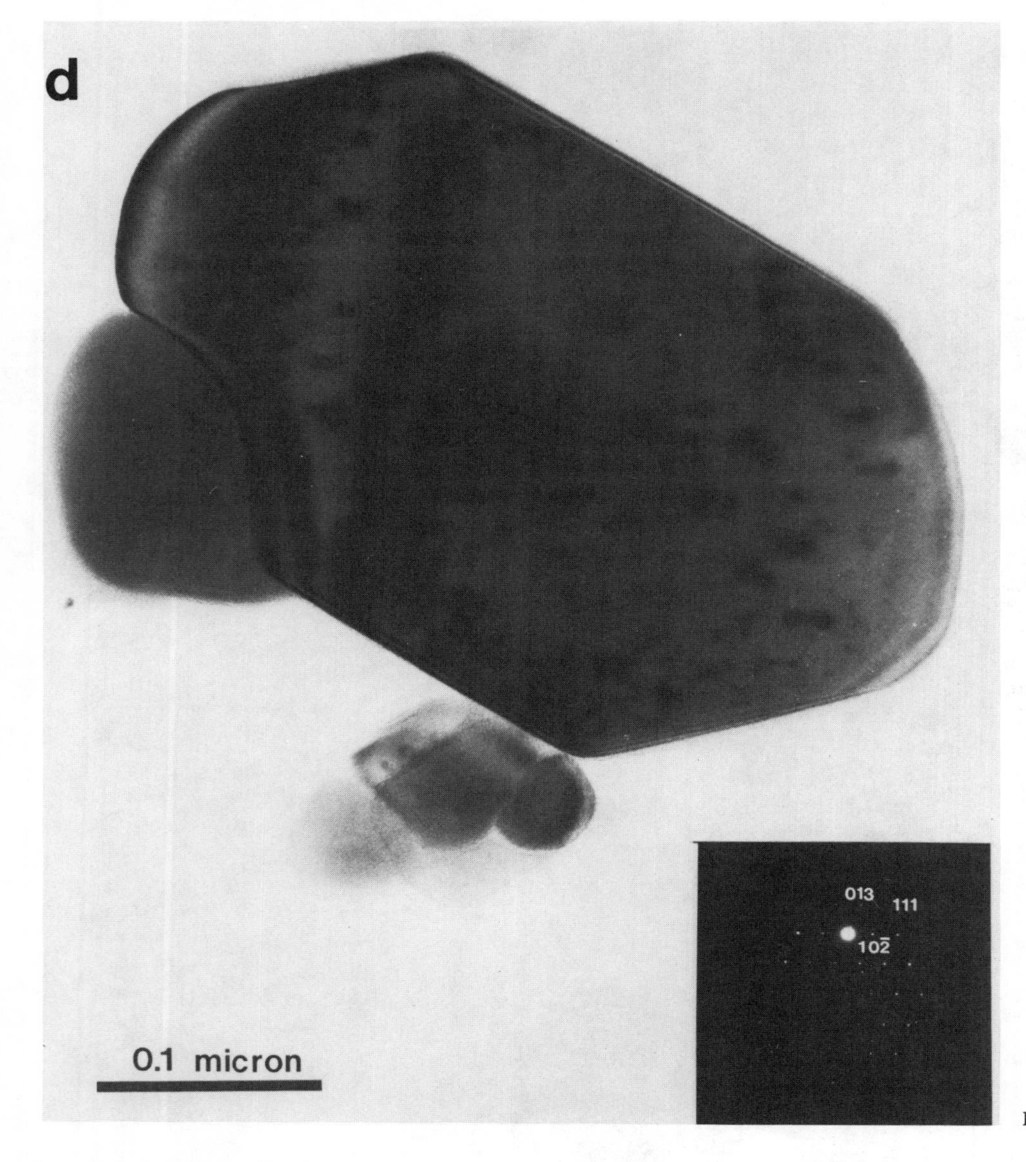

Fig. 1. (continued).

Assessment of Contamination

As noted in previous publications, all microanalytical studies of fine-grained extraterrestrial materials may be subject to minor levels of contamination. However, collection and curation procedures, as well as electron microscope laboratory procedures for stratospheric particles, are continuously monitored for contamination problems. Thus, contamination of our IDP samples by terrestrial materials is negligible and recognizable in the rare instances when present and are reviewed by *Mackinnon and Rietmeijer* (1987). Particularly relevant to the present study are the results of a detailed inventory of small-sized particles in the NASA/JSC Curatorial Facility that did not reveal Ti-containing particles present in this facility (*Cosmic Dust Courier,* Issue 8, 1989, NASA-JSC Curatorial Branch). Also, the use of ultramicrotomed thin sections considerably reduces the risk of laboratory contamination. In addition, we report on data from allocated samples that were handled in two different laboratories: at the Solar System Exploration Division of the NASA Johnson Space Center (W7029C1 and W7010*A2) and at the Electron Microbeam Analysis Facility within the Department of Geology at the University of New Mexico (W7029C1, U2011C2, and U2034C12).

Previous discussions on contamination of chondritic IDPs have not considered the possibility that IDP contamination with anthropogenic materials may occur in near-Earth space. The IDPs used in the present study were collected from the stratosphere between September 1981 and August 1985. During this period the Solar Maximum satellite was in orbit at about 570 km altitude. Parts of the satellite's surface have been examined for impacted orbital debris and micrometeoroids. The results from these examinations show that between February 1980 and April 1984 the submicrometer-size anthropogenic debris flux was significantly lower than the flux of natural micrometeoroids in the same size range (*Barrett et al.,* 1988). The former includes a significant number of secondary impacts caused by spacecraft paint particles (*Bernhard and McKay,* 1988). Spacecraft paint particles on surfaces from the Solar Max Satellite have been analyzed using scanning electron microscopy and WEDS analysis. A prime property of these paint particles is the presence of Ti, Si, Zn, C, and O in major amounts with minor amounts of K, S, Cl, Fe, and Al (*Barrett et al.,* 1988) and, typically, individual particles contain *at least* two of these major elements (*Rietmeijer et al.,* 1986). Individual submicrometer-size Ti-oxide grains are liberated due to atomic oxygen erosion of spacecraft paints in low-Earth orbit (*Bernhard and McKay,* 1988) but an X-ray diffraction analysis of these spacecraft paints shows that the Ti-oxides are rutile (*Zolensky et al.,* 1988). While we recognize the possibility of contamination, our observations on Magnéli phases and Ti-metal in four chondritic IDPs are inconsistent with an origin of these phases by contamination in near-Earth space, in the Earth's atmosphere, or in the laboratory.

OBSERVATIONS

In samples of CP IDPs W7029C1 and W7010*A2 several platey, angular Ti-grains display a typical, often noncoherent, polygonal texture (*Mackinnon and Rietmeijer,* 1987, Fig. 6). Direct lattice imaging using tilted beam illumination and SAED data are consistent with Ti-metal. The polygonal texture is interpreted as the result of nucleation and growth during the Tiβ-Tiα phase transformation that occurs between ~1155 K and ~1075 K as a function of cooling rate (*Cormier and Claisse,* 1974). The angular Ti-mental grains are up to 0.6×0.4 μm in size and have a narrow (~15 nm) rim of carbonaceous material similar to rims that commonly surround Fe,Ni grains (*Bradley et al.,* 1984) and Bi_2O_3 grains (*Mackinnon and Rietmeijer,* 1984) in chondritic IDPs.

In all CP IDP allocations reported in this study, the Ti-minerals typically occur as oxides in which the presence of oxygen, at least in randomly selected grains, was confirmed by WEDS analysis. The Ti-oxide grains form thin (~1.5 nm) platey (1) (sub-) rounded grains, (2) slightly elongated sub- to euhedral grains, and (3) rare, angular grains. These grains generally form clusters ($\gtrsim$0.4 μm in diameter) (Figs. 1a, b) but isolated grains also occur. Individual Ti-oxide grains within the clusters are single crystal grains. Isolated Ti-oxide grains have a 2-230 nm wide rim of carbonaceous material (Fig. 1c) and occasionally display dislocation substructures (Fig. 1d). Adjoining grains have straight intergranular boundaries and equi-angular triple junctions typical of solid-state recrystallization.

In CP IDPs W7010*A2 and U2011C2, Ti-oxide grains are embedded in amorphous carbon that also contains platey grains of Al_2O_3, SiO_2, a Si,Al layer silicate, and unidentified Ca- and Fe-minerals. These platey grains are typically ~0.1 μm in diameter but may be up to 0.2×0.4 μm in size (*Rietmeijer and Mackinnon,* 1987a). In CP IDPs W7029C1 and U2011C2, rare Ti-grains are intimately associated with angular Bi_2O_3 grains and a layer silicate (Fig. 2).

Electron diffraction analysis of individual Ti-oxide crystals show that these grains are not stoichiometric TiO_2 polymorphs, that is, rutile, anatase, or brookite. The Ti-oxides in these four CP IDPs are oxygen-deficient Ti-oxides of the homologous Magnéli series, Ti_nO_{2n-1} (n = 4-10) (*Andersson et al.,* 1957). Selected area electron diffraction data for

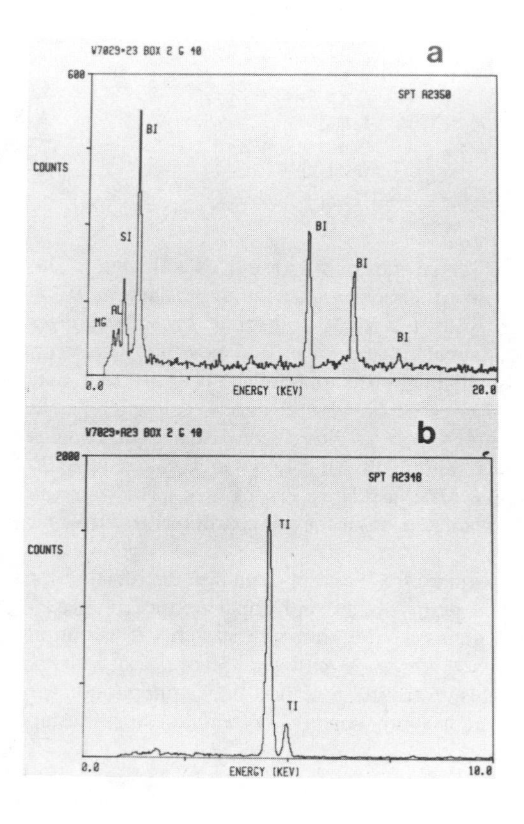

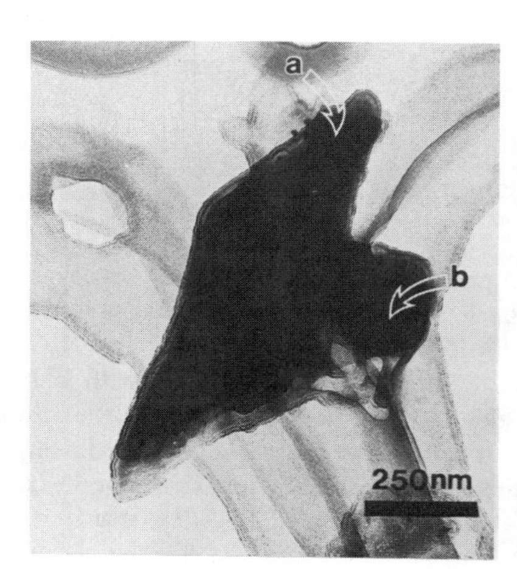

Fig. 2. Intimately associated grains of **(a)** bismuth oxide, **(b)** titanium oxide, and a Mg-Al-(Fe) layer silicate (dark gray grain) in chondritic porous IDP W7029Cl (allocation W7029*A23). The EDS spectra for the Ti and Bi grains are shown on the lefthand side.

TABLE 1. Observed selected area electron diffraction data for Ti grains from chondritic porous IDPs W7029C1, W7010*A2, and U2011C2 compared with selected interplanar spacings for Magnéli phases, Ti_nO_{2n-1} (n = 4, 5, 6), calculated using unit cell dimensions for Magnéli phases from *Andersson et al.* (1957).

W7029C1		W7010*A2				U2011C2					Ti_4O_7	Ti_5O_9	Ti_6O_{11}
6.04											6.16		
						5.31	5.31				5.26		5.37
5.07											5.07		
								4.98					4.96
		4.75											4.77
			4.71								4.70	4.72	
			4.67									4.67	4.66
	4.65										4.63		4.64
								4.29			4.27		
									4.10		4.10		
				4.03								4.04	4.00
						3.92	3.92		3.92	3.92	3.94	3.93	
								3.88			3.85	3.89	3.89
										3.67			3.67
3.60											3.60		3.59
					3.55								3.55
				3.50								3.50	
		3.45									3.45		3.43
3.40												3.42	3.40
		3.34									3.34	3.35	3.36
									3.20		3.22	3.23	3.20
										3.03	3.02		
										2.96	2.94	2.95	2.97
	2.88										2.91	2.91	2.86
								2.73	2.73			2.73	2.73
	2.71											2.71	
						2.67					2.66	2.67	2.67
						2.65	2.65	2.65				2.65	2.64
				2.63							2.62	2.63	2.63
									2.60		2.60	2.60	2.60
			2.58									2.58	2.58
								2.57				2.57	2.57
				2.55							2.55		
2.53											2.53	2.53	2.53
							2.43				2.43		2.43
					2.42						2.42	2.42	2.42
2.40			2.40								2.41	2.40	2.41

polycrystalline Ti-oxide clusters are presented in Table 1 and compared with calculated interplanar spacings of Magnéli phases (n = 4, 5, 6) using unit cell data reported by *Andersson et al.* (1957). Compared with X-ray diffraction, the sensitivity of electron diffraction will allow detection of low-intensity reflections that will not show in X-ray diffraction listings such as the ASTM Powder diffraction files. The SAED data for Ti-oxides in the IDPs are consistent with Ti_4O_7, Ti_5O_9, and Ti_6O_{11}. The identification of Ti_5O_9 in the polycrystalline clusters is based on the presence of interplanar spacing, d = 3.50 Å, which uniquely fits this Magnéli phase (Table 1). In addition, identifications of Ti_4O_7 and Ti_6O_{11} are supported by single crystal electron diffraction for grains in ultramicrotomed thin sections of allocation U2011C2 (Fig. 1d). Typically, single crystal diffraction data have been obtained at two different orientations of the grain relative to the incident electron beam. Single crystal diffraction pattern identification relies on a combination of observed interplanar distances and angles. The single crystal electron diffraction data for three random Ti_4O_7 and Ti_6O_{11} grains shown in Table 2 illustrate that a unique fit with individual Magnéli phases is indeed possible. However, these results also highlight the potential ambiguity that may be involved in Magnéli phase identifications. For example, in rare instances, combinations of interplanar distances and angles that resulted in a unique fit for Ti_4O_7 or Ti_6O_{11} (Table 2) involve assuming, *ad hoc*, a relative error of ~4% between the measured and calculated values for a particular interplanar distance.

The root-mean-square $[(x^2+y^2)^{1/2}]$ grain size distribution for individual Ti-metal grains in the polygonal texture of platey, angular Ti-metal grains is <150 nm whereas this distribution for isolated Ti-oxide grains is between 50 nm and 500 nm (Fig. 3). While this distribution is not fully understood, the sharp boundary at 100 nm suggests a reaction relationship

TABLE 2. Single crystal data for three randomly selected Magnéli phases in chondritic porous IDP U2011C2, and for two grains, in two different orientations of the grain relative to the incident electron beam.

Ti_4O_7				Ti_6O_{11}	
1		2		3	
d (Å)	hkl	d (Å)	hkl	d (Å)	hkl
5.31	$10\bar{1}$	3.30	$1\bar{2}1$	5.31	$10\bar{2}$
3.95	$00\bar{3}$	2.66	$1\bar{1}\bar{4}$	3.92	013
2.65	$20\bar{2}$	2.41	$01\bar{5}$	2.71	111
1.90	$20\bar{5}$			2.65	$20\bar{4}$
zone axis					
$0\bar{3}0$		951		$\bar{2}3\bar{1}$	
3.95	$1\bar{1}\bar{2}$			4.24	$1\bar{1}1$
1.90	$\bar{2}2\bar{4}$			3.95	$10\bar{6}$
1.65	$3\bar{3}\bar{2}$			2.71	$2\bar{1}\bar{5}$
zone axis					
$\bar{4}\bar{4}0$				$\bar{6}\bar{7}\bar{1}$	

hkl refers to the Miller indices.

between Ti-metal and Ti-oxides in which the surface energy of participating solids significantly contributes to the stability of submicron-sized Magnéli phases (*Rietmeijer and Mackinnon,* 1989).

DISCUSSION

The mineralogy (*Mackinnon and Rietmeijer,* 1987), infrared (IR) signatures (*Sandford and Walker,* 1985), and Fe-Mg-Si distribution patterns (*Bradley,* 1988) are used to classify chondritic IDPs as (1) nominally anhydrous IDPs, that is, IDPs characterized by olivines, or pyroxenes plus carbonaceous material and (2) hydrated IDPs dominated by layer silicates plus carbonaceous material. In this broadly-based classification scheme, the CP IDPs in this study are nominally anhydrous particles, but CP IDP W7029C1 has properties intermediate between these two major IDP classes (*Rietmeijer and Mackinnon,* 1985a,b; *Carey et al.,* 1987). Anhydrous CP IDPs typically consist of an ultrafine-grained matrix and their unique Fe/(Mg+Fe) distributions provide a distinct link with cometary dust (*Bradley,* 1988; *Rietmeijer,* 1989a). We note that currently it is difficult to unambiguously establish a cometary origin for chondritic IDPs, although P/comet Halley dust shows a unique spectral feature indicative for small crystalline olivines (*Campins and Ryan,* 1989). If CP IDPs are solid debris from short-period comets, an important condition to be considered is that accretion of these chondritic IDPs probably occurred in a part of the solar nebula that largely escaped the effects of evaporation typical of the inner solar nebula.

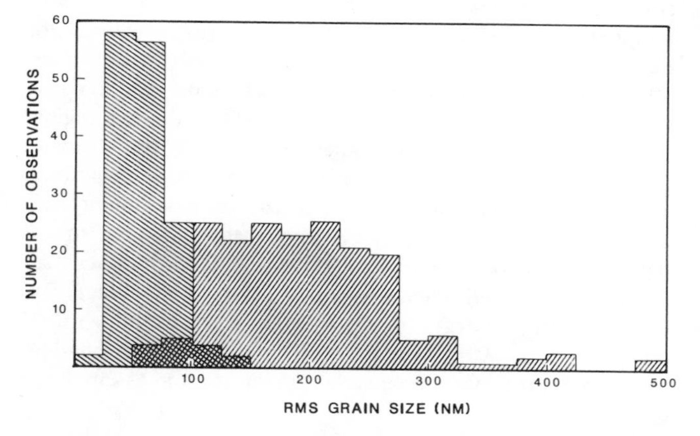

Fig. 3. Root-mean-square grain size distribution of individual Ti-metal (northwest-southeast hatching) and Ti-oxide (northeast-southwest hatching) grains in chondritic porous IDPs W7029C1, W7010*A2, U2011C2, and U2034C12.

Recently, *Blake et al.* (1989) have reported Ti-bearing grains in a nonchondritic, i.e., "low-atomic number," IDP that displays a D/H isotope enrichment consistent with an extraterrestrial origin. Indirect support for an extraterrestrial origin of Magnéli phases may be gleaned from a study by *Zolensky et al.* (1989) that reported on titania phases from >100,000-year-old Antarctic ice. Citing these observations and considering the unlikelihood of IDP contamination with Magnéli phases, which are well-characterized from industrially and laboratory prepared sources, we conclude that Magnéli phases described in this paper are indigenous to chondritic IDPs. Although *Flynn et al.* (1978) observed considerable submicron heterogeneity of Ti-bearing phases in a chondritic IDP, normalized Ti abundances in these IDPs are similar to abundances observed in the fine-grained matrix of carbonaceous chondrites (*Rietmeijer,* 1987). It is possible to broadly constrain silicon normalized elemental abundances in chondritic IDPs using AEM observations, although errors in these calculations are uncertain and may be considerable (*Rietmeijer,* 1989b). With these caveats, *Rietmeijer and Mackinnon* (1989) calculated Ti/Si abundances in CP IDP U2034C12 well within the range for bulk chondritic IDPs (*Fraundorf et al.,* 1982a).

Model for Ti-phase Formation

A model for the origin of Ti-minerals in chondritic IDPs must explain (1) the intimate association of Ti-grains with carbonaceous material, (2) the presence of isolated Ti-oxides as well as clusters of Ti-oxide grains, (3) the association of Ti-phases with other fine-grained minerals, and (4) the spatially close association of Ti-metal and oxygen deficient Ti-oxides. All of these observations on Ti-rich phases in IDPs have been reported either in this study or in previous publications (*Flynn et al.,* 1978; *Mackinnon and Rietmeijer,* 1983; *Rietmeijer and Mackinnon,* 1984, 1987b).

Clues from previous research. A useful clue to an explanation for Ti-mineral formation can be derived from equilibrium condensation models. An assumption of these models is that gas phase molecules preferentially condense as simple solids rather than as chemically complex solids. Thus, Ti-metal, TiO, TiO_2, and Ti_nO_{2n-1} phases are the most likely Ti-minerals to form in a cooling solar nebular gas or in supernova shells as a function of C/O ratio or location (*Lattimer and Grossman,* 1978). In these models, Ti-metal only condenses in supernova (Si) shells between 1620 K and 1075 K as a function of gas pressure. These temperatures exceed the Tiβ-Tiα transformation temperature.

A second clue is provided by the mode of Ti-oxide occurrence, i.e., isolated grains and clusters of grains in IDP W7029C1 and isolated grains embedded in carbonaceous material in IDPs W7010*A2, U2011C2, and U2034C12, which suggest that parent CP IDP history may affect the distribution of Ti-phases. Thus, we consider that layer silicates, carbon-2H and poorly graphitized carbons (PGCs) in W7029C1 support a thermal regime up to ~500 K while the pauacity of layer silicates and PGCs in the other IDPs suggest lower thermal regimes in these IDPs (*Rietmeijer and Mackinnon,* 1985a,b, 1987a,c).

Catalysis reactions/low-temperature catalysis. Catalytic reactions have previously been proposed for the formation of carbon compounds (*Bradley et al.,* 1984) and epsilon carbide (*Christoffersen and Buseck,* 1983) in chondritic IDPs. In the low (<100 K) thermal regime of the solar system at the loci of IDP accretion, catalytically activated disproportionation of CO may be an important reaction that forms solid carbons such as graphite (*Hayatsu et al.,* 1980).

Another type of reaction in which Ti-metal can be an active catalyst is gasification of solid elemental carbon in the binary systems C-CO_2, C-O_2, or C-H_2O (*Thomas,* 1965; *Walker et al.,* 1968). Drawing upon a review paper by *Walker et al.* (1968) we summarize the following properties for a catalytically active metal in solid carbon gasification.

1. The transition metal can form intermediate oxides before reaching its highest oxidized state when the catalyst becomes deactivated, i.e., its catalytic properties have ceased. Preferably, but not necessarily, the intermediate oxides are stoichiometrically deficient oxides.

2. The intermediate metal and its oxides are able to enter a reduction-oxidation loop that allows reactivation of the oxidized catalyst either by changing the composition of the ambient gas phase or the temperature. However, the extent of catalyst reactivation or deactivation depends on the metal/oxide thermodynamic equilibrium stability and is a function of the ambient gas phase composition and temperature.

3. Catalysis is a surface reaction; thus, the catalyst particle size will have a strong effect on its catalytic efficiency. For the industrial purposes reviewed by *Walker et al.* (1968) catalyst grain sizes are between ~3 and 20 μm at temperatures typically >1050 K. In the case of the CP IDPs, the root-mean-square size of Ti-metal grains is <100 nm.

4. The catalytic activity involves the weakening of neighboring C-bonds. As a result, catalytically activated gasification of crystalline carbons (i.e., graphite) is strongly crystallographically controlled.

Ti-oxide stability. Oxides in the Ti-TiO_2 binary system (*DeVries and Roy,* 1954; *Wahlbeck and Gilles,* 1966) have properties consistent with points 1 and 2 above. Thermodynamic data of *Robie et al.* (1978) can be used to calculate the equilibrium dissociation pressures (RTlnfO_2 in kJ) of titanium oxides as a function of temperature (K). The results of these calculations are shown in Fig. 4 and indicate that (1) TiO, Ti_2O_3, and Ti_4O_7 are stable phases intermediate between Ti-metal and Ti_5O_9, Ti_6O_{11}, and TiO_2 (anatase) and (2) the Magnéli phases (n = 4 – 6) and anatase have similar equilibrium dissociation pressures. However, low-temperature formation of specific submicron-sized Magnéli phases (Ti_nO_{2n-1}; n = 4, 5, or 6) is kinetically controlled and conceivably relates to subtle differences in surface energy of these oxides relative to Ti-metal as indicated by the rms-grain size distributions (Fig. 3).

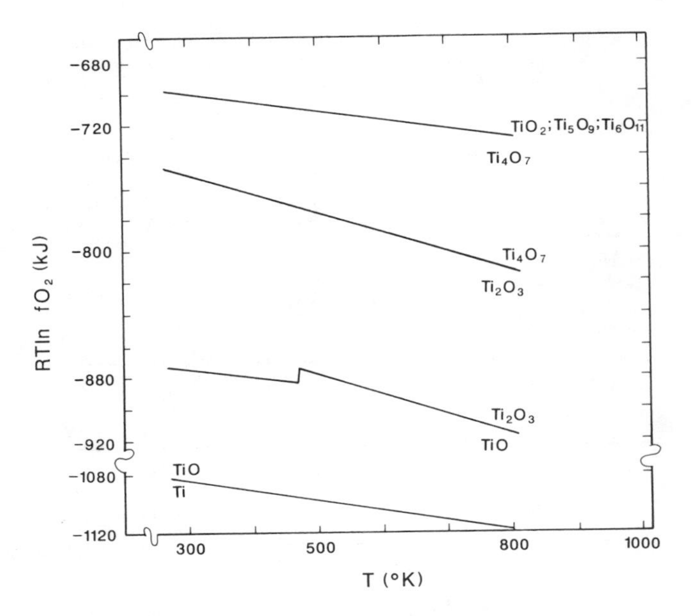

Fig. 4. Equilibrium dissociation pressures (RTlnfO_2 (kJ)) for various titanium oxides as a function of temperature (K).

We propose that a plausible mechanism to explain the presence of Magnéli phases in chondritic IDPs is low-temperature catalysis of elemental carbons using a Ti-metal support. The proposed catalytically activated carbon gasification probably takes place below ~500K. While these temperatures are well below typical thermal regimes for experimental or industrial graphite gasification (*Walker et al.,* 1968), we note that the proposed gasification reaction may still be an effective process in the assumed low thermal regime of chondritic IDP parent bodies because (1) the Ti-metal particle size in CP IDPs is at least two orders of magnitude smaller than typical catalyst particle sizes in experimental and industrial gasification, and (2) carbonaceous material in CP IDPs predominantly occurs in a noncrystalline form rather than in the form of well-ordered graphite (*Rietmeijer,* 1988). The C-C bonding energy in amorphous carbon is ~3.5 eV compared to ~6.9 eV in graphite. This difference in bonding energy will probably enable carbon gasification in assemblages typical of chondritic IDPs.

Location for catalysis reaction. Various investigators have commented upon a continuum of chemical and physical properties between carbonaceous chondrites and chondritic IDPs (*Brownlee,* 1985b; *Brownlee et al.,* 1977; *Rietmeijer,*

1985a) although important mineralogical differences between these chondritic materials have been emphasized by *Mackinnon and Rietmeijer* (1987). For example, a higher bulk carbon content for chondritic IDPs (*Schramm et al.,* 1989) compared to carbonaceous chondrites is consistent with a continuum concept in which carbonaceous chondrites experiences a higher degree of metamorphism. In this context, catalyzed carbon gasification is an attractive process because it relates the solid carbon content of primitive fine-grained chondritic solar system materials to their thermal history. Citing theoretical and observational constraints *Rietmeijer* (1985b,c), *Rietmeijer and Mackinnon* (1985a, 1987b,c), and *Rietmeijer et al.* (1989) have suggested that ambient conditions in a protoplanetary body, or comet nucleus, are conducive to chemical and mineralogical evolution. While our current models for these ambient conditions may not readily support the existence of a gas phase equilibrium $(pCO/pCO_2) \cdot (pH_2O/pH_2) = \text{Constant}_{(T)}$, the proposed catalytically supported solid carbon gasification is not inconsistent with proposed models for (proto-)planetary body evolution (*DuFresne and Anders,* 1962). For example, catalytically activated carbon gasification ceases with increased oxygen fugacity that deactivates the catalytic metals in this process. This condition may not be pervasive in extraterrestrial environments. Our observations, as well as previously reported anatase in an acid residue of the Leoville carbonaceous chondrite (*Wopenka and Swan,* 1985), suggest the Magnéli phases, and other Ti-oxides, will occur in all primitive, fine-grained extraterrestrial materials.

CONCLUSIONS

Nanometer-size Magnéli phases, Ti_nO_{2n-1} (n = 4, 5, 6), and Ti-metal associated with carbonaceous material in some chondritic IDPs support a model for formation of oxygen deficient Ti-oxides during *in situ* catalytically activated gasification of carbonaceous material in which Ti-metal was the active catalyst, some of which still occurs in these IDPs. Consistent with their relative equilibrium dissociation pressures, the oxides formed in this manner typically include Magnéli phases. So far, we have looked in vain for solar flare tracks in Magnéli phases. Solar flare tracks occur in olivine and pyroxene grains in chondritic IDPs (*Bradley et al.,* 1988). However, pulse-heating experiments have shown that tracks in these silicates rapidly anneal between 773 K and 873 K (*Fraundorf et al.,* 1982b). Given the fact that activation energies for diffusion in Ti-oxides are considerably lower compared to these energies in silicates, it is conceivable that tracks in Ti-oxides may be erased even during mild atmospheric entry heating of chondritic IDPs.

While it is difficult to unambiguously establish an origin for chondritic IDPs, some similarities between IDPs and cometary dust suggest that a fraction of all chondritic IDPs studied in the laboratory represents cometary dust. Using equilibrium vapor phase condensation models for stellar environments and probable loci of comet nucleus accretion, we imply that Ti-metal in CP IDPs may have an interstellar origin. The presence of Ti-oxide Magnéli phases in chondritic IDPs supports the notion that protoplanetary bodies and comet nuclei show distinct chemical and mineralogical activity. Catalytically activated gasification provides a plausible mechanism with which to explain the higher solid carbon content of chondritic IDPs compared to carbonaceous chondrites.

Acknowledgments. This study was initiated while both authors were affiliated with the Solar System Exploration Division at NASA Johnson Space Center. This paper benefited from constructive criticism by Drs. D. E. Blake and R. M. Walker. I.D.R.M. is grateful to R. McConnville and J. Fahey of the Applications Laboratory, Philips Electronic Instruments Inc. We acknowledge D. McKay for use of the electron microscope facilities at NASA/JSC and we appreciate assistance from L. Schramm and K. Thomas (Lockheed EMSCO at NASA/JSC) with the WEDS analyses. This study was continued at the Electron Microbeam Analysis Facility at the Department of Geology (University of New Mexico) where we are grateful for technical support by S. Kaser and F. Rietmeijer-Engelsman. This work is supported by NASA grant NAG9-160.

REFERENCES

Andersson S., Collén B., Kruuse G., Kuylenstierna U., Magnéli A., Pestmalis H., and Åsbrink S. (1957) Identification of titanium oxides by X ray powder patterns. *Acta Chim. Scand., 11,* 1653-1657.

Barrett R. A., Bernhard R. P., and McKay D. S. (1988) Impact holes and impact flux on returned Solar Max louver material (abstract). In *Lunar and Planetary Science XIX,* pp. 39-40. Lunar and Planetary Institute, Houston.

Bernhard R. P. and McKay D. S. (1988) Micrometer-sized impact craters on the Solar Maximum Satellite: The hazards of secondary ejecta (abstract). In *Lunar and Planetary Science XIX,* pp. 65-66. Lunar and Planetary Institute, Houston.

Blake D., Fleming R. H., and Bunch T. E. (1989) Identification and characterization of a carbonaceous, titanium containing interplanetary dust particle (abstract). In *Lunar and Planetary Science XX,* pp. 84-85. Lunar and Planetary Institute, Houston.

Bradley J. P. (1988) Analysis of chondritic interplanetary dust thin-sections. *Geochim. Cosmochim. Acta, 52,* 889-990.

Bradley J. P. and Brownlee D. E. (1986) Cometary particles: Thin sectioning and electron beam analysis. *Science, 231,* 1542-1544.

Bradley J. P., Brownlee D. E., and Fraundorf P. (1984) Carbon compounds in interplanetary dust: Evidence for formation by heterogeneous catalysis. *Science, 223,* 56-58.

Bradley J. P., Sandford S. A., and Walker R. M. (1988) Interplanetary dust particles. In *Meteorites and the Early Solar System* (J. F. Kerridge and M. S. Matthews, eds.), pp. 861-898. Univ. of Arizona, Tucson.

Brownlee D. E. (1985a) Cosmic dust: Collection and research. *Annu. Rev. Earth Planet. Sci., 13,* 147-173.

Brownlee D. E. (1985b) Electron microbeam analysis of cosmic dust. In *Microbeam Analysis—1985* (J. T. Armstrong, ed.), pp. 269-272. San Francisco Press, San Francisco.

Brownlee D. E., Rajan R. S., and Tomandl D. A. (1977) A chemical and textural comparison between carbonaceous chondrites and interplanetary dust. In *Comets, Asteroids, Meteorites—Interrelations, Evolution, Origins* (A. H. Delsemme, ed.), pp. 137-141. Univ. of Toledo, Toledo.

Campins H. and Ryan E. V. (1989) The identification of crystalline olivine in cometary silicates. *Astrophys. J., 341,* 1059-1066.

Carey W. C., Walker R. M., and Bradley J. P. (1987) Interplanetary dust particles and comet Halley: A comparative study. *Meteoritics, 22,* 348-349.

Christoffersen R. and Buseck P. R. (1983) Epsilon carbide: A low-temperature component of interplanetary dust particles. *Science, 222,* 1327-1329.

Cormier M. and Claisse F. (1974) Beta-alpha phase transformation in Ti and Ti-O alloys. *J. Less-common Metals, 34,* 181-189.

Delsemme A. H. (1982) Chemical composition of comet nuclei. In *Comets* (L. L. Wilkening, ed.), pp. 85-130. Univ. of Arizona, Tucson.

DeVries R. C. and Roy R. (1954) A phase diagram for the system Ti-TiO_2 constructed from data in the literature. *Ceramic Bull., 12,* 370-372.

DuFresne E. R. and Anders E. (1962) On the chemical evolution of the carbonaceous chondrites. *Geochim. Cosmochim. Acta, 26,* 1085-1114.

Fahey A. J., Goswami J. N., McKeegan K. D., and Zinner E. K. (1987) ^{16}O excess in Murchison and Murray hibonites: A case against a late supernova injection origin of isotopic anomalies in O, Mg, Ca, and Ti. *Astrophys. J., 323,* L91-L95.

Flynn G., Fraundorf P., Shirck J., and Walker R. M. (1978) Chemical and structural studies of "Brownlee" particles (abstract). In *Lunar and Planetary Science IX,* pp. 338-340. Lunar and Planetary Institute, Houston.

Fraundorf P., Brownlee D. E., and Walker R. M. (1982a) Laboratory studies of interplanetary dust. In *Comets* (L. L. Wilkening, ed.), pp. 383-412. Univ. of Arizona, Tucson.

Fraundorf P., Lyons T., and Schubert P. (1982b) The survival of solar flare tracks in interplanetary dust silicates on deceleration in the Earth's atmosphere. *Proc. Lunar Planet. Sci. Conf. 13th,* in *J. Geophys. Res., 87,* A409-A412.

Hayatsu R., Scott R. G., Studier M. H., Lewis R. S., and Anders E. (1980) Carbynes in meteorites: Detection, low-temperature origin, and implications for interstellar molecules. *Science, 209,* 1515-1518.

Hsiung P. and Kissel J. (1987) Composition of cometary dust particles. *ESA SP-278,* pp. 355-358.

Jessberger E. K., Christoforidis A., and Kissel J. (1988) Aspects of the major element composition of Halley's comet. *Nature, 332,* 691-695.

Lattimer J. M. and Grossman L. (1978) Chemical condensation sequences in supernova ejecta. *Moon and Planets, 19,* 169-184.

Mackinnon I. D. R. and Rietmeijer F. J. M. (1983) Layer silicates and a bismuth phase in chondritic aggregate W7029*A. *Meteoritics, 18,* 343-344.

Mackinnon I. D. R. and Rietmeijer F. J. M. (1984) Bismuth in interplanetary dust. *Nature, 311,* 135-138.

Mackinnon I. D. R. and Rietmeijer F. J. M. (1987) Mineralogy of chondritic interplanetary dust particles. *Rev. Geophys., 25,* 1527-1553.

Mason B. (1979) *Data of Geochemistry* (M. Fleischer, ed.,), 6th ed. Geol. Survey Prof. Paper 440-B-1. U.S. Govt. Printing Office, Washington, DC. 132 pp.

Morton D. C. (1978) Interstellar absorption lines in Zeta Puppis. *Astrophys. J., 222,* 863-880.

Morton D. C. and Hu E. M. (1975) Interstellar absorption lines toward Gamma Area. *Astrophys. J., 202,* 838-849.

Rietmeijer F. J. M. (1985a) On the continuum between chondritic interplanetary dust and CI and CM carbonaceous chondrites: A petrological approach (abstract). In *Lunar and Planetary Science XVI,* pp. 698-699. Lunar and Planetary Institute, Houston.

Rietmeijer F. J. M. (1985b) A model for diagenesis in proto-planetary bodies. *Nature, 313,* 293-294.

Rietmeijer F. J. M. (1985c) Low-temperature aqueous and hydrothermal activity in a proto-planetary body: Goethite, opal-CT, gibbsite, and anatase in chondritic porous aggregate W7029*A (abstract). In *Lunar and Planetary Science XVI,* pp. 696-697. Lunar and Planetary Institute, Houston.

Rietmeijer F. J. M. (1987) Chondritic interplanetary dust and primitive chondrite matrices: The search for chemically pristine solids in the Solar System (abstract). In *Lunar and Planetary Science XVIII,* pp. 832-833. Lunar and Planetary Institute, Houston.

Rietmeijer F. J. M. (1988) On graphite in primitive meteorites, chondritic interplanetary dust, and interstellar dust. *Icarus, 74,* 446-453.

Rietmeijer F. J. M. (1989a) Ultrafine-grained mineralogy and matrix chemistry of olivine-rich chondritic interplanetary dust particles. *Proc. Lunar Planet. Sci. Conf. 19th,* pp. 513-521.

Rietmeijer F. J. M. (1989b) Tin in a chondritic interplanetary dust particle. *Meteoritics, 24,* 43-48.

Rietmeijer F. J. M. and Mackinnon I. D. R. (1984) Diagenesis in interplanetary dust: Chondritic porous aggregate W7029*A. *Meteoritics, 19,* 301.

Rietmeijer F. J. M. and Mackinnon I. D. R. (1985a) Layer silicates in a chondritic porous interplanetary dust particle. *Proc. Lunar Planet. Sci. Conf. 16th,* in *J. Geophys. Res., 90,* D149-D155.

Rietmeijer F. J. M. and Mackinnon I. D. R. (1985b) Poorly graphitized carbon as a new cosmothermometer for primitive extraterrestrial materials. *Nature, 316,* 733-736.

Rietmeijer F. J. M. and Mackinnon I. D. R. (1987a) Interstellar titanium-oxides in interplanetary dust. *Meteoritics, 22,* 490-491.

Rietmeijer F. J. M. and Mackinnon I. D. R. (1987b) Cometary evolution: Clues from chondritic interplanetary dust particles. *ESA SP-278,* pp. 363-367.

Rietmeijer F. J. M. and Mackinnon I. D. R. (1987c) Metastable carbon in two chondritic porous interplanetary dust particles. *Nature, 326,* 162-165.

Rietmeijer F. J. M. and Mackinnon I. D. R. (1989) Grain size distributions of Magnéli phases and metallic titanium in chondritic porous interplanetary dust particles (abstract). In *Lunar and Planetary Science XX,* pp. 902-903. Lunar and Planetary Institute, Houston.

Rietmeijer F. J. M. and McKay D. S. (1986) Fine-grained silicates in a chondritic interplanetary dust particle are evidence for annealing in the early history of the Solar System (abstract). In *Lunar and Planetary Science XVII,* pp. 710-711. Lunar and Planetary Institute, Houston.

Rietmeijer F. J. M., Schramm L. S., Barrett R. A., McKay D. S., and Zook H. A. (1986) An inadvertent capture cell for orbital debris and micrometeorites; The main electronics box thermal blanket of the Solar Maximum satellite. *Adv. Space Sci., 6,* 145-149.

Rietmeijer F. J. M., Mukhin L. M., Fomenkova M. N., and Evlanov E. N. (1989) Layer silicate chemistry in P/comet Halley from PUMA-2 data (abstract). In *Lunar and Planetary Science XX,* pp. 904-905. Lunar and Planetary Institute, Houston.

Robie R. A., Hemingway B. S., and Fisher J. R. (1978) Thermodynamic properties of minerals and related substances at 298.15 K and 1 bar (10^5 Pascals) pressure and at higher temperature. *U.S. Geol. Survey Bull., 1452.* U.S. Govt. Printing Office, Washington, DC. 456 pp.

Sandford S. A. and Walker R. M. (1985) Laboratory infrared transmission spectra of individual interplanetary dust particles from 2.5 to 25 microns. *Astrophys. J., 291,* 838-851.

Schramm L. S., Brownlee D. E., and Wheelock M. M. (1989) The elemental composition of interplanetary dust. *Meteoritics, 24,* 99-112.

Snow T. P. (1976) Interstellar material toward Omicron Persei. *Astrophys. J., 204,* 759-774.

Thomas J. M. (1965) Microscopic studies of graphite oxidation. In *Chemistry and Physics of Carbon, 1* (P. L. Walker Jr., ed.), pp. 122-203. Dekker, New York.

Wahlbeck P. G. and Gilles P. W. (1966) Reinvestigation of the phase diagram for the system titanium-oxygen. *J. Amer. Ceramic Soc., 49,* 180-183.

Walker P. L., Shelef M., and Anderson R. A. (1968) Catalysis of carbon gasification. In *Chemistry and Physics of Carbon, 4* (P. L. Walker Jr., ed.), pp. 287-384. Dekker, New York.

Wopenka B. and Swan P. D. (1985) Identification of micron sized phases in meteorites by laser Raman microprobe spectroscopy. *Meteoritics, 20,* 788-789.

Zolensky M. E., McKay D. S., and Kaczor L. A. (1988) A ten-fold increase in the abundance of large solid particles in the stratosphere, as measured over the period 1976-1984. *J. Geophys. Res., 94,* 1047-1056.

Zolensky M. E., Pun A., and Thomas K. L. (1989) Titanium carbide and titania phases in Antarctic ice particles of probable extraterrestrial origin. *Proc. Lunar Planet. Sci. Conf. 19th,* pp. 505-511.

Proceedings of the 20th Lunar and Planetary Science Conference, pp. 335-342
Lunar and Planetary Institute, Houston, 1990

Synchrotron X-ray Fluorescence Analyses of Stratospheric Cosmic Dust: New Results for Chondritic and Low-Nickel Particles

G. J. Flynn

Department of Physics, SUNY-Plattsburgh, Plattsburgh, NY 12901

S. R. Sutton

Department of the Geophysical Sciences and Center for Advanced Radiation Sources, The University of Chicago, Chicago, IL 60637 and Department of Applied Science, Brookhaven National Laboratory, Upton NY 11973

Trace element abundance determinations were performed using synchrotron X-ray fluorescence on nine particles collected from the stratosphere and classified as "cosmic." Improvements to the synchrotron X-ray fluorescence microprobe at the National Synchrotron Light Source allowed the detection of all elements between Cr and Mo, with the exceptions of Co and As, in our largest particle (W7027C5; 30 μm). Minor and trace element concentrations in one particle, W7013A11, were within a factor of 2 of CI concentrations for all elements detected except Br. Five other particles exhibited chondritic abundance patterns, though each deviated by more than a factor of 2 from CI for one or more elements other than Br. The remaining three particles exhibited extremely low Ni concentrations, with Fe/Ni ratios similar to Earth or lunar crustal materials. Abundance patterns of two of the particles with low nickel were remarkably similar to that of basalt. Bromine enrichments in the nine particles ranged from 1.3 to 40× the CI concentration. Bromine concentrations were uncorrelated with particle size. An inverse correlation would be expected for a surface contamination acquired from the stratosphere.

INTRODUCTION

Micrometeorites that are small enough (<100 μm) to survive atmospheric entry without melting are of great importance in studying the nature of the interplanetary dust cloud and primitive bodies, such as comets and some asteroids. After deceleration in the upper atmosphere, these particles descend into the stratosphere where they are concentrated because of their low settling rates. NASA aircraft collect samples of the stratospheric dust, which are mixtures of cosmic dust, natural terrestrial dust, and man-made debris. The Cosmic Dust Curatorial Facility, NASA Johnson Space Center (JSC), curates the collection (*Zolensky et al.,* 1984).

One subset of the stratospheric dust exhibits major element abundances similar to those of the chondritic meteorites (*Brownlee et al.,* 1977). Extensive studies on particles selected from this chondritic subset have shown the presence of solar flare tracks (*Bradley et al.,* 1984), solar-wind-implanted noble gases (*Rajan et al.,* 1977; *Hudson et al.,* 1981), and non-terrestrial isotopic ratios (*Esat et al.,* 1979; *Zinner et al.,* 1983), all of which are indicative of an extraterrestrial origin. Prior analyses have demonstrated that the cosmic particle type is subdivided into three groups—so-called layer-lattice silicates, olivines, and pyroxenes—based on the dominant 10-μm silicate feature in the infrared absorption spectrum of each particle (*Sandford and Walker,* 1985). Solar flare tracks demonstrating space exposure have been reported in particles of all three groups, with *Bradley* (1988) indicating that such tracks are seen in more than 90% of the anhydrous cosmic-dust particles and in all smectite particles that also contain crystals suitable for track imaging in the transmission electron microscope. These observations have led to the working assumption that all particles with a "chondritic" major element chemistry are extraterrestrial, and most research has concentrated on stratospheric dust of the chondritic type.

The determination of chemical compositions is of fundamental interest in understanding the histories of the particles and their relationships with conventional meteorites. Major elements are identified qualitatively using electron beam energy dispersive X-ray (EDX) analysis and these results together with morphological characteristics are used by JSC to classify individual particles. We have focused our attention on the particles classified as cosmic, or "C-type," in the JSC catalogs, which are so classified based on the presence of Mg, Si, and Fe in major proportions and minor contents of Al, Ca, S, and/or Ni. The bremsstrahlung background and short analysis times limit the JSC analyses to the identification of elements present at the percent level or higher.

Several techniques have been employed to measure minor and trace element contents. Proton-induced X-ray emission (PIXE) reduces bremsstrahlung interference, permitting analyses of some elements down to the 10-ppm level (*van der Stap et al.,* 1986). However, severe heating can occur during analysis causing volatilization of elements such as Cl, Se, and Br (*Wallenwein et al.,* 1989). Instrumental neutron activation analysis (INAA) allows analyses for elements uniquely suited for this technique (e.g., Ir, Au, Co, Sc), but many elements are detectable in only large particles >100 ng (*Zolensky et al.,* 1989a).

We have been developing synchrotron X-ray fluorescence techniques for this purpose. One of the main advantages is the negligible degree of heating by the synchrotron radiation, several orders of magnitude less power density than for charged particle techniques. Because of the nondestructive nature of the analytical technique, particles are preserved for subsequent investigation by TEM, ion microprobe, infrared transmission spectroscopy, and Raman spectroscopy. Element sensitivities below 10 ppm have been achieved for 20-μm particles.

In this paper, we report quantitative minor and trace element abundances for nine new C-type particles using the synchrotron X-ray microprobe (XRM) at the National Synchrotron Light Source, Brookhaven National Laboratory. Results for three additional C-type particles have been previously reported (*Sutton and Flynn,* 1988). Improvements to the XRM have increased sensitivity and decreased the required analysis time since our first measurements.

EXPERIMENTAL METHOD

Samples

Nine C-type stratospheric particles were analyzed in the present experiment. The particle characteristics as determined at the Cosmic Dust Curatorial Facility at JSC are provided in Table 1. Five of these particles were selected because their JSC EDX spectra were typical of chondritic particles but had variable Fe/Si ratios: U2022B2, U2022G2, W7013A11, W7027C5, and W7013H17. Four other particles were selected because of one or more elemental enrichments apparent in the JSC EDX spectra: U2022C18 (high Ti), U2022G17 (high Ca), W7066*A4 (high Ca and Ti), and U2001B6 (high K).

In addition, aluminum-oxide spheres (AOS) and natural terrestrial particles (TCN) were included in our analyses in an attempt to better understand contamination acquired during stratospheric residence, collection, curation, and analysis.

TABLE 1. Particle description.

Particle	Size (μm)	Class*	EDX Peaks†
W7027C5	30	C	Mg,Al,**Si**,S,Ca,Cr,**Fe**,Ni
U2022G17	12	C	**Mg**,Al,**Si**,**S**,**Ca**,**Fe**,Ni
W7013H17	12	C	Mg,Al,**Si**,**S**,Ca,Cr,**Fe**,Ni,Cu§
W7013A11	20	C	Mg,Al,**Si**,S,Cr,**Fe**,Ni,Cu§
U2022G2	15	C	Mg,Al,**Si**,**S**,Ca,Fe,Ni
U2022B2	15	C	Mg,Al,**Si**,S,Ca,Cr,**Fe**
U2022C18	12	C	Mg,Al,**Si**,S,Ti,**Fe**
W7066*A4	7	C	Mg,Al,**Si**,S,**Ca**,Ti,Fe
U2001B6	25	C?	Mg,Al,**Si**,S,K,Ca,**Fe**
W7013A1	20	TCN	**Al**,**Si**,K,Ca,Fe,Cu§
W7013B1	25	TCN	Al,**Si**,S,K,Ca,Ti,Fe,Cu§
W7013B11	20	TCN	**Na**,Mg,Al,**Si**,**S**,**K**,Ca,Cu§
AOS 87-B1	10	AOS	‡
AOS 87-B2	10	AOS	‡

*JSC classes: C = cosmic dust, AOS = aluminum oxide sphere, TCN = natural terrestrial, ? indicates uncertainty in the JSC classification.
†X-ray peaks in JSC EDX spectrum; most intense peak(s) in boldface.
‡Supplied by D. Brownlee, no EDX spectrum available.
§Cu in some JSC spectra is an artifact of the analysis technique.

To minimize contamination from handling, each particle provided by JSC was transferred directly from its JSC glass-slide shipping container to an individual 7.5-μm-thick Kapton foil upon which it was analyzed. No attempt was made to remove the silicone oil in which the particles were shipped since previous control experiments demonstrated that interferences from elements in the oil and Kapton were negligible for all elements analyzed with the possible exception of Zn (*Sutton and Flynn,* 1988).

The two AOS particles provided by D. Brownlee had been mounted on a carbon film suspended on a Be TEM grid and were analyzed in this configuration.

Apparatus

The XRM at the National Synchrotron Light Source, Brookhaven National Laboratory, uses synchrotron radiation as the excitation source for X-ray fluorescence analyses of trace elements with high spatial resolution. A tungsten, double-pinhole collimator (courtesy of A. Krieger) was used to produce an X-ray beam 40×18 μm in size. The beam could be reduced in size down to about 15×15 μm by rotating the collimator in the horizontal and/or vertical planes, thereby adjusting its apparent aperture. A Klinger translation and rotation stage allowed 1-μm specimen movements. Samples were analyzed in air with a Si(Li) energy dispersive detector at 90° to the incident beam. A 170 μm Al filter was used on the detector for Fe-rich particles to suppress the intense Fe-K fluorescence. A major improvement was the incorporation of a Nikon Optiphot petrographic microscope with transmitted and reflected light capabilities that allowed easy positioning of the 10-μm particles in the synchrotron beam. In addition, the XRM has now been installed on beamline X26A, a recently completed beamline whose experimental area is twice as close to the source as that of the previous beamline used (X26C). The resulting increase in flux improved elemental sensitivities by about a factor of 2. Acquisition times were typically 20-30 minutes.

Analytical procedures were similar to those used previously for stratospheric particles (*Sutton and Flynn,* 1988). X-ray spectra were reduced using STRIP and SPCALC. Elemental concentrations for "cosmic" particles were calculated using the NRLXRF routine modified to perform standardless analyses (*Lu et al.,* 1989) based on the known concentration of the major element Fe determined from JSC EDX spectra. Concentrations for the terrestrial particles used the chondritic particle spectra as standards with corrections for mass and major element composition obtained from the prediction mode of NRLXRF.

RESULTS

The elemental contents for the nine C-type particles are reported in Table 2. The enhanced sensitivity has allowed us for the first time to determine the concentrations of Rb, Sr, Y, Zr, Nb, and Mo in a cosmic dust particle. All elements between Cr and Mo (Z = 24 to 42), except Co and As were detected in the largest particle, W7027C5 (Fig. 1).

TABLE 2. Elemental contents and CI-normalized abundances.

Particle	Cr	Mn	Fe*	Ni	Cu	Zn	Ga	Ge	Se	Br	Rb	Sr
W7027C5												
(pg)	15.4	11.6	1200	117	0.81	2.3	0.05	0.16	0.14	0.11	0.03	0.11
	±3.2	±2.3	±240	±23	±0.17	±0.47	±0.01	±0.03	±0.03	±0.02	±0.01	±0.02
(ppm)	0.24%	0.18%	18.5%	1.8%	125	360	8.0	25	22	17	4	17
(CI norm)	0.89	0.91	1.0	1.6	1.1	1.2	0.80	0.78	1.2	4.2	2	2.2
U2022G17												
(pg)	0.17	0.98	23	1.1	0.08	0.11	<0.005	<0.005	<0.005	0.02	<0.005	<0.018
	±0.03	±0.20	±5	±0.2	±0.02	±0.02	-	-	-	±0.004	-	
(ppm)	0.16%	0.89%	21.0%	1.0%	686	1010	<50	<50	<50	150	<50	<104
(CI norm)	0.59	4.6	1.1	0.92	6.1	3.3	<5	<2	<3	40	<20	<20
W7013H17												
(pg)	3.6	3.8	190	9.2	0.22	1.1	0.03	0.06	0.05	0.01	0.017	0.033
	±0.7	±0.8	±40	±1.8	±0.04	±0.22	±0.01	±0.01	±0.01	±0.003	±0.007	±0.009
(ppm)	0.30%	0.31%	16%	0.75%	180	880	28	46	40	5	14	27
(CI norm)	1.1	1.6	0.84	0.68	1.6	2.8	2.8	1.4	2.2	1.3	6.0	3.4
W7013A11												
(pg)	7.3	6.5	350	25	0.56	1.2	0.06	0.19	0.16	0.21	<0.018	0.039
	±1.5	±1.3	±70	±5.0	±0.11	±0.24	±0.01	±0.04	±0.03	±0.04	-	±0.017
(ppm)	0.21%	0.19%	10%	0.71%	160	340	18	53	45	60	<5	11
(CI norm)	0.79	0.95	0.54	0.64	1.4	1.1	1.8	1.7	2	15	<2	1.4
U2022G2												
(pg)	<4.0	3.8	260	20	0.10	0.33	<0.016	0.04	0.02	0.07	<0.014	0.02
	-	±0.8	±50	±4.0	±0.02	±0.07	-	±0.008	±0.005	±0.014	-	±0.006
(ppm)	<0.20%	0.19%	13%	1.0%	50	170	<8	18	10	34	<7	9
(CI norm)	<0.75	0.95	0.70	0.92	0.44	0.54	<0.8	0.6	0.6	9	<3	1
U2022C18												
(pg)	<5.0	0.88	330	14	0.30	0.06	<0.010	0.02	0.004	0.03	<0.017	0.01
	-	±.22	±70	±3	±0.06	±0.01	-	±0.005	±0.003	±0.007	-	±0.004
(ppm)	<0.15%	0.10%	37%	1.6%	330	64	<11	24	5	36	<19	7
(CI norm)	<0.55	0.50	2.0	1.5	3.0	0.21	<1	0.75	0.3	9.0	<8	.9
U2022B2												
(pg)	6.8	6.1	260	1.1	0.002	0.04	0.003	0.004	<0.009	0.01	0.004	0.01
	±1.4	±1.2	±52	±0.22	±0.002	±0.01	±0.002	±0.002	-	±0.003	±0.003	±0.004
(ppm)	0.51%	0.46%	19.5%	0.081%	2	29	2.	3.	<7	8	3	7
(CI norm)	1.9	2.3	1.1	.079	0.02	0.09	0.20	0.09	<0.4	2	1	0.9
U2001B6												
(pg)	0.96	4.8	320	0.33	0.46	0.48	0.07	<0.029	0.01	0.43	0.07	0.13
	±0.40	±0.97	±64	±0.068	±0.09	±0.096	±0.015	-	±0.004	±0.086	±0.014	±0.027
(ppm)	0.02%	0.10%	6.7%	0.007%	96	100	15	<6	3	90	14	28
(CI norm)	0.11	0.58	0.36	0.006	0.93	0.36	0.17	<0.20	0.22	23	6.5	3.6
*W7066*A4*												
(pg)	0.25	0.38	36	0.10	0.10	0.21	<0.012	<0.003	<0.0017	0.02	0.01	0.25
	±0.051	±0.076	±7.2	±0.02	±0.02	±0.042	-	-	-	±0.004	±0.002	±0.051
(ppm)	0.06%	0.09%	8.5%	0.024%	240	500	<29	<6	<4	43	16	600
(CI norm)	0.22	0.46	0.46	0.02	2.1	1.7	<2.9	<0.20	<0.20	11	7.4	77

†Fe abundance is determined from JSC EDX spectrum as described in text.
Additional concentrations (CI-normalized abundance in parentheses): W7027C5: Y = 4 ppm (3), Zr = 4 ppm (1), Nb = 8 ppm (30), Mo = 6 ppm (7); U2001B6: Zr = 30 ppm (8); W7066*A4: Zr = 45 ppm (12), Pb = 900 ppm.

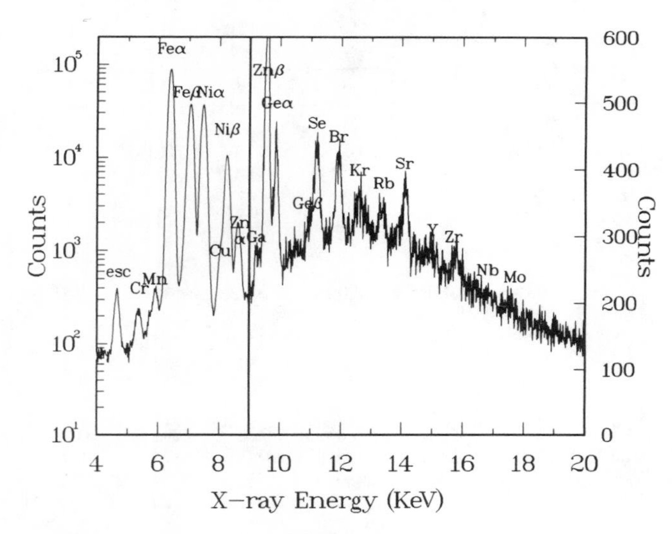

Fig. 1. Synchrotron X-ray fluorescence spectrum of W7027C5, a 30 μm chondritic particle, with a 30-minute acquisition time. All elements from Cr to Mo, except Co and As, were detected in this particle. Intensities for energies less than 9 keV are plotted logarithmic, while those above 9 keV are linear. Krypton fluorescence derives from air along the beam path. The Fe K_α escape peak is labelled "esc."

Chondritic Particles

Six of the particles (W7013A11, W7013H17, U2022C18, U2022G2, U2022G17, and W7027C5) exhibited minor and trace element abundance patterns generally similar to those of the CI meteorites. The CI-normalized abundances for these particles are given in Table 2 and plotted in Figs. 2a,b.

The four cosmic particles selected because their major element abundance patterns were similar to CI, exhibited minor and trace element abundance patterns generally similar to CI. All detected elements were within a factor of 2 of CI concentrations in W7027C5, with the exceptions of Br (4.2× CI), Sr (2.2× CI), Y (3× CI), Nb (30× CI) and Mo (7× CI). Only Br (15× CI) deviated by more than a factor of 2 from CI in W7013A11. In U2022G2, all detected elements except Br (9× CI) and Cu (0.44× CI) were within a factor of 2 of CI. W7013H17, which is high in S, showed enrichments in Zn, Ga, Se, Rb and Sr.

The cosmic particles that were selected because one or more major elements were substantially different from CI showed more unusual minor and trace element patterns (Figs. 2a,b). Particle U2022C18, which is high in Ti, also exhibited Cu and Br enrichments, 3 and 9× CI, respectively, and Zn and Se depletions, of 0.2 and 0.3× CI. The high-Ca particle, U2022G17, showed major enrichments of Mn, Cu, Zn, and Br, by factors of 4.6, 6.1, 3.3, and 40.

Thus, the general conclusion is that particles that are "chondritic" in their major elements exhibit minor and trace element patterns similar to CI. However, even in this group, substantial deviations from CI are usually seen for Br and one or more other elements. Those particles that show deviations from CI in one or more major elements have minor and trace element patterns that deviate more substantially from CI. This variation may be due to heterogeneity of a single parent and/or sampling of multiple parents.

Low-nickel Particles

Three particles, U2022B2, U2001B6, and W7066*A4, exhibited low-Ni contents with CI-normalized abundances of 0.08, 0.02, and 0.006, respectively. *Brownlee et al.* (1980) previously reported that 1 of 57 chondritic cosmic dust particles analyzed by EDX had a substantial Ni depletion (0.1× CI). In contrast, 3 of the 12 cosmic particles we have measured to date have low Ni. The apparent discrepancy in the abundance of low-Ni particles between these two studies may at least in part be due to the fact that we purposely selected samples with unusual major-element enrichments.

The Fe/Ni ratios of the three low Ni particles, U2022B2, U2001B6, and W7066*A4, were 240, 960, and 350, respectively, quite distinct from the CI value of 17 (Fig. 3). Particle U2022B2 appeared to be a "typical" chondritic particle based on the JSC spectrum, while W7066*A4 exhibited high Ca and Ti in its JSC spectrum. Particle U2001B6 was categorized as "C?" because of its high K.

One interpretation of the low-nickel particles is that they are terrestrial despite their major element chondritic compositions. Based on the low densities of these three particles (0.6, 0.8, and 1.7 g/cm^3; *Sutton and Flynn,* 1989) the possibility that these are simply single mineral grains can be ruled out. Alternatively, each particle may be extraterrestrial, in which case there are two obvious explanations for its low nickel: (1) the particle is a chondritic micrometeorite and the

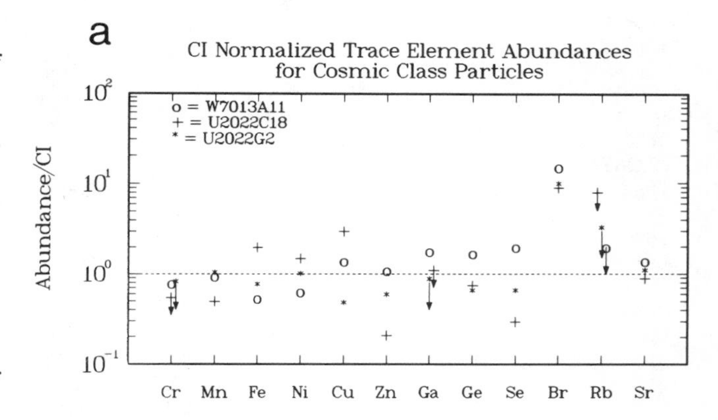

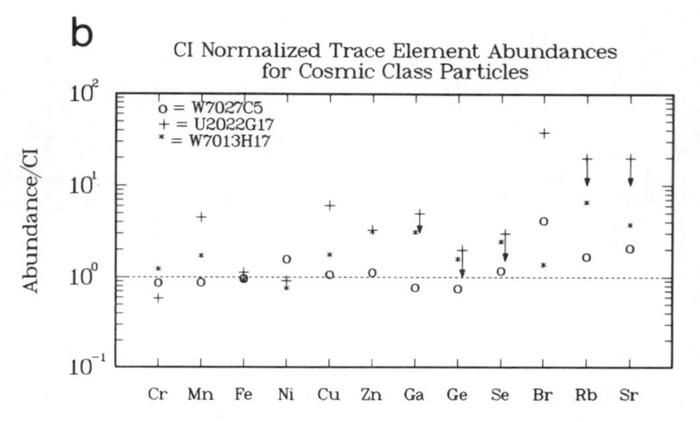

Fig. 2. (a and b) CI-normalized abundances for six chondritic particles. Most elements are within a factor of 2 of CI abundances. Enriched elements include Mn, Cu, Zn, Br, and Nb.

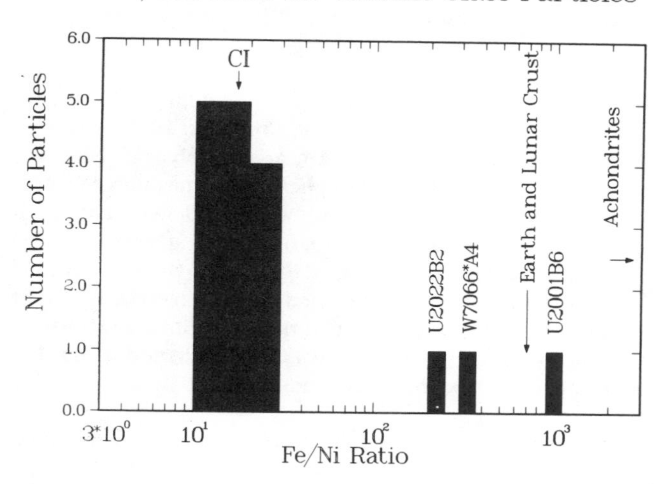

Fig. 3. Fe/Ni ratio histogram for all 12 cosmic class particles analyzed to date using the synchrotron X-ray fluorescence method. The three low-nickel particles are clearly distinct from the chondritic cluster between 10 and 30.

observed discrepant composition represents a statistical artifact due to the small particle size, or (2) the particle derives from a chemically evolved parent body.

Expected statistical fluctuations in Fe/Ni due to heterogeneities depend on the specific mineral phases in which these elements are concentrated. However, an analysis of twelve 30-μm fragments of Orgueil (CI) matrix revealed only one fragment (S-rich) that deviated by more than a factor of 3 from the CI Fe/Ni ratio (*Flynn et al.,* 1978). The relatively large grain size of the Orgueil matrix suggests that large statistical fluctuations would occur less frequently in cosmic dust.

Clearly, extraterrestrial material can have nonchondritic Fe/Ni ratios, e.g., achondrites and lunar samples. *Golenetskii et al.* (1978) have inferred the composition of the object responsible for the Tunguska event of 1908 by subtraction of bulk peat composition from that of the peat layer enriched in Tunguska material. They obtain an Fe/Ni = 163 for the Tunguska object. *Ganapathy* (1981) has analyzed submillimeter-sized metallic spheres associated with the Tunguska event and found that five of eight exhibited Fe/Ni well above CI, the highest with Fe/Ni = 2110. The Fe and Ni compositions of meteors derived by analysis of the spectral emission lines (*Harvey,* 1973) from a sporadic, a Taurid, and a Geminid were consistent with CI. However, Ni was not detected in the spectrum of a Perseid meteor (associated with Comet Swift-Tuttle), suggesting a higher Fe/Ni ratio than in the other three cases.

Most of the chondritic stony spheres (300 μm or larger) recovered from deep-sea sediments, presumed to be melt products produced on atmospheric entry of extraterrestrial material, are depleted in Ni and Cr (*Brownlee,* 1985). The Ni depletion was explained by inertial separation of a Ni-depleted stony phase from a Ni-rich metallic phase during atmospheric entry heating (*Brownlee,* 1985). *Sutton et al.* (1988) found coupled loss of Ni, Cu, and Zn in deep-sea spheres. However, this mechanism seems unlikely for the low-nickel stratospheric particles. One of our Ni-depleted particles, U2022B2, exhibits a small Cr enrichment (1.9× CI) but depletions in Cu and Zn. The second, U2001B6, shows normal CI Cu but substantial depletions in Cr and Zn. The third, W7066*A4, shows depleted Cr but enriched Cu and Zn.

The more likely possibility is that these low-Ni particles are samples of a differentiated parent [i.e., the Earth, Moon, eucrite parent body (EPB), etc.] since igneous differentiation leads to Fe-Ni fractionation. Thus, the Fe/Ni ratio of an unknown sample has frequently been used to identify primitive, unequilibrated extraterrestrial material. Six of the C-type particles reported here as well as the three particles we previously analyzed have Fe/Ni ratios ranging from 10.3 to 27.7, consistent with the CI value of 17. The high Fe to Ni abundance ratios (Fe/Ni = 240, 350, and 960 for U2022B2, W7066*A4, and U2001B6, respectively) in the three anomalous particles indicate that these samples may have experienced igneous fractionation.

A comparison of the abundance patterns of two of the particles (U2001B6 and W7066*A4) with that of terrestrial basalt reveals remarkable similarities (Fig. 4a). These abundance patterns are characterized by lithophile enrichments and siderophile depletions relative to CI. Volatile elements are notably different, however, generally enriched in the particles relative to mafic igneous materials. The pattern for U2022B2, a nickel-depleted particle with a typical chondritic major element EDX spectrum, is more nearly chondritic in trace elements than the other two Ni-depleted particles showing mainly anomalies in minor and trace siderophiles (Fig. 4b). The interpretation of these patterns is complicated by the fact that volatile elements, including S observed in the JSC spectrum of each of these three particles, are enriched relative to terrestrial and lunar basalts. Particle W7066*A4 contains 900 ppm of Pb. Some form of stratospheric or collection contamination seems required to explain the high contents of these volatile elements. The TEM analyses for solar flare tracks and mineralogy will be required to better understand the source of these interesting particles.

Bromine

Bromine enrichments of a factor of 30× the CI meteorite concentration were first reported in three chondritic cosmic dust particles, all from the same collector, by *van der Stap et al.* (1986). Subsequently, we have analyzed 12 cosmic-class particles from three different collectors and found Br enrichments by factors of 1.3 to 40× CI. *Fegley and Lewis* (1979) have performed extensive thermodynamic calculations on the formation of compounds of volatile elements, including Br, in a condensing solar nebula. In their homogeneous accretion model, Br was the last element of those studied (Na, K, F, Cl, Br, and P) to begin condensation, forming bromapatite, $Ca_5(PO_4)_3Br$, at about 350 K. This low-Br condensation temperature indicates it should be a very sensitive probe of nebula conditions in the dust formation region, since incomplete Br condensation in certain regions of the nebula would give rise to a Br-rich gas phase. Thus, widely varying Br abundances might be expected for materials condensing at different times or locations in the nebula.

In *van der Stap et al.* (1986) the authors suggested the Br enrichment was part of a general trend of volatile enrichment

a

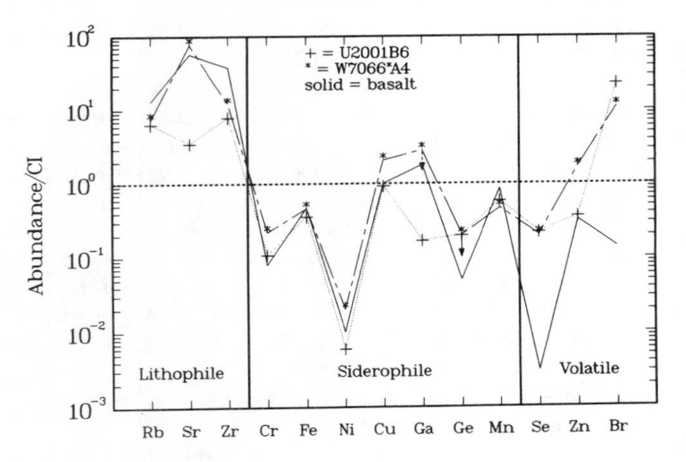

b

CI Normalized Trace Element Abundances
for a Low–Nickel Cosmic Class Particle

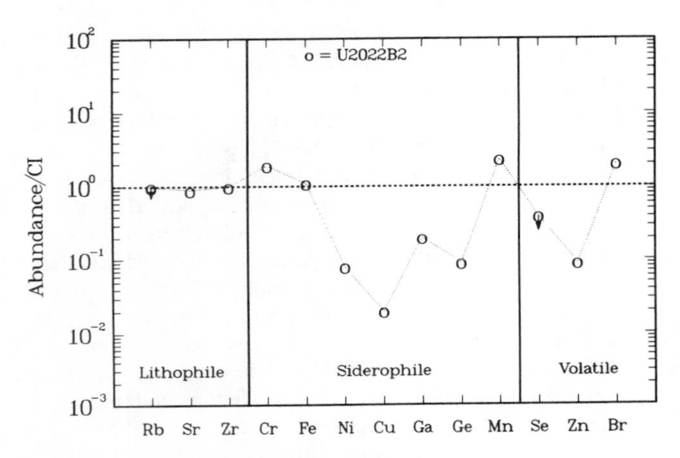

Fig. 4. (a and b) CI-normalized abundance patterns for the three Ni-depleted particles and terrestrial basalt (*Krauskopf,* 1979). The horizontal line represents the composition of CI carbonaceous chondrites.

in chondritic particles, consistent with the particles being late-stage nebula condensates. In this study we find that U2022G17, the particle most enriched in Br, also exhibits the largest enrichments in the volatile elements Cu and Zn, by factors of 6.1 and 3.3 over CI, respectively. Particle W7013H17, with Br only 1.3× CI, also exhibits lesser enrichments in Cu and Zn, by factors of 1.6 and 2.8× CI, respectively. When all the particles are considered, however, no clear correlation between Cu and Zn enrichments and Br enrichment is observed.

If this Br enrichment is preterrestrial, the ubiquity of the effect suggests a large fraction of the chondritic cosmic dust is derived from a parent body (bodies) that formed under different nebula conditions (location or time) than the meteorite parent bodies. However, the possibility that the Br could be a contaminant must be considered since the light halogens are present in trace quantities in the Earth's stratosphere: Cl = 2×10^{-9} g/g and Br = 5×10^{-11} g/g (*Cicerone,* 1981). In addition, *Mackinnon and Mogk* (1985) have provided direct evidence that aluminum oxide spheres (AOS) (particles of rocket exhaust; *Brownlee,* 1976) collected from the stratosphere contain a near-surface enrichment of S and Cl, presumably from stratospheric contamination. We have previously discussed the circumstantial evidence bearing on the question of Br contamination during stratospheric residence (*Sutton and Flynn,* 1988). Here we will examine the direct evidence, the expected inverse correlation of Br enrichment with particle size and the degree of contamination present on terrestrial particles also collected from the stratosphere and curated in the same manner as the chondritic particles.

Two mechanisms have been proposed for the incorporation of stratospheric Br into cosmic-dust particles. First, stratospheric aerosols, potentially rich in Br, could impact the particles during stratospheric residence. The incorporation of a significant amount of Br by this mechanism is inconsistent with the collision rate calculated by *Zolensky et al.* (1986,1989b), but *Rietmeijer* (1986) has invoked this mechanism to explain the presence of silica-rich glass fragments in cosmic dust. The number of collisions, and thus the mass of accreted Br, would depend on the cross-sectional area of the particle. A second mechanism is the incorporation of Br-rich stratospheric aerosols on the collector and subsequent particle impact into this contamination layer (D. E. Brownlee, personal communication, 1987). In this case the mass of contaminant Br would also vary with the area of the particle. For both mechanisms the Br concentration would be expected to correlate inversely with particle size. However, no obvious inverse correlation of Br concentration with particle dimension exists (Fig. 5). To the contrary, the second largest particle thus far examined, U2022G1, has one of the largest Br concentrations (37× CI) reported, and one of the smallest particles, W7013H17, shows the smallest Br concentration. The lack of a size dependence of observed Br concentrations argues against a stratospheric origin for the Br.

We have also attempted to assess the magnitude of stratospheric Br contamination in the chondritic particles by measuring the Br concentrations in terrestrial particles also collected from the stratosphere. According to C. C. A. H. van der Stap (personal communication, 1986) the lack of detectable Br in an iron-sulfur-nickel (FSN) sphere collected from the stratosphere limits the degree of stratospheric contaminant. However, since FSN particles have a much higher density and thus shorter stratospheric residence time, they might accumulate less stratospheric contaminant than chondritic particles.

We analyzed two AOS particles, AOS 87-B1 and AOS 87-B2, supplied to us by D. Brownlee. No Br was detectable in either sphere with an upper limit on their Br contents of 1 ppm. These spheres are about 10 μm in diameter, comparable in size to the smaller chondritic particles. However, the upper limits on their Br contents are two orders of magnitude below the Br abundances in the most Br-enriched particles: 150 ppm in U2022G17 and 90 ppm in U2001B6. All of the chondritic

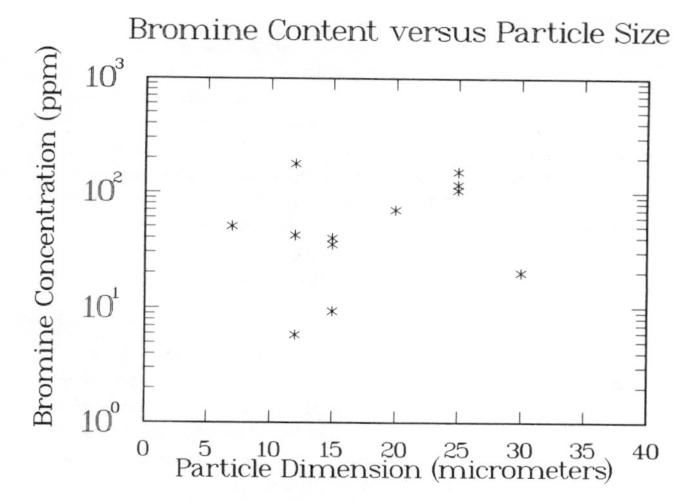

Fig. 5. Bromine enrichment vs. the particle dimension for the chondritic particles reported in this paper and the three previously reported in *Sutton and Flynn* (1988). An inverse correlation would be expected if the excess Br were a result of stratospheric contamination.

particles measured thus far have Br abundances of 5 ppm or higher. Thus, if the Br in the chondritic particles is a stratospheric contaminant, either it must be added more efficiently to chondritic particles than to AOS particles, or it must be more efficiently removed from the AOS particles during curation. The difference in surface texture of the AOS and chondritic particles might favor this effect.

We have also analyzed three particles classified as natural terrestrial (TCN): W7013A1, W7013B2, and W7013B11. These particles have surface textures more similar to some of the chondritic particles. The first two are alumino-silicates while the third gave comparable levels of Na, Si, S, and K in the JSC spectrum. A low Br abundance in these would be a clear indicator of the lack of significant stratospheric Br contamination. The Br contents of W7013A1 and W7013B2 were 10 ppm and 50 ppm, respectively, comparable to levels seen in all but the most Br rich of the chondritic particles. No Br was detected in the third particle, with a sensitivity of 10 ppm. Since the natural abundance of Br in the TCN particles prior to stratospheric injection is unknown, the high observed abundances do not necessarily signal stratospheric contamination as the source. Volcanic eruptions are known to be a significant source for terrestrial material in the Earth's stratosphere and *Kotra et al.* (1983) report 100 ppm Br concentrations in ash from the El Chichon volcano. The TCN particles of volcanic origin might exhibit Br concentrations of the magnitude we have measured.

Thus, Br measurements on terrestrial particles from the stratospheric collectors seem unable to resolve the question of the extent of contamination. Direct measurement of the Br concentrations on particles collected in space by Solar Max, the Long Duration Exposure Facility, or the proposed Space Station Cosmic Dust Collection Facility should resolve the question, since these particles receive no stratospheric exposure. Alternatively, identification of the Br carrier phase(s) in the cosmic dust collected from the stratosphere might provide a useful indicator of source, particularly if the Br were concentrated in phases of likely terrestrial origin or in phases suggested by *Fegley and Lewis* (1979) to be likely Br-rich solar nebula condensates.

CONCLUSIONS

Nine cosmic-class stratospheric particles were analyzed by synchrotron X-ray fluorescence. Five of those selected had typical "chondritic" major element EDX spectra. Four of these (W7027C5, W7013H17, W7013A11, and U2022G2) exhibited minor and trace element abundances generally within a factor of 2 of the CI values, although each particle had apparently unique enrichments and/or depletions in some elements. TEM observations of mineralogy are required to determine if there are correlations of these enrichments or depletions with particle type. The fifth particle (U2022B2) showed marked depletions in Ni, Cu, Zn, Ga, Ge, and Se. The remaining four particles were purposely selected on the basis of clearly nonchondritic peak heights in their EDX spectra (Ti, Ca, or K visible in their JSC spectra). Two of these (U2022G17 and U2022C18) showed generally CI trace element abundances. The remaining two unusual particles (U2001B6 and W7066*A4) exhibited minor and trace element abundance patterns grossly different from CI abundances and remarkably similar to those of mafic igneous rocks.

Bromine, enriched in all the cosmic-dust particles by factors ranging from 1.3 to 40, does not exhibit an inverse correlation with particle size that would be expected for a surface-correlated component acquired from stratosphere aerosols. No Br was detected in two aluminum-oxide spheres, also collected from the stratosphere, but was found in natural terrestrial particles from the JSC collectors in concentrations similar to those in the cosmic particles. Bromine analyses on terrestrial particles collected from the stratosphere seem unable to resolve the question of stratospheric contamination.

These results emphasize the value of the synchrotron X-ray fluorescence measurements in identifying and characterizing nonchondritic, extraterrestrial particles. Notably, Ni was not observed in the JSC EDX spectra of U2022C18 and W7066*A4, yet the Fe/Ni ratios and minor and trace element abundance patterns measured here clearly show that the former is chondritic and the latter is nonchondritic.

Acknowledgments. We are grateful to K. W. Jones, head of the Atomic and Applied Physics Division, Department of Applied Science, Brookhaven National Laboratory, who is responsible for development and operation of the SXRF facility and to the staff of the National Synchrotron Light Source for providing beam time for this research. Invaluable technical assistance was provided by M. Rivers (University of Chicago/BNL). We thank the Cosmic Dust Curatorial Facility (JSC) and especially M. Zolensky (JSC) for supplying samples for these measurements. D. Brownlee kindly provided the aluminum-oxide spheres. This paper benefited from the thorough reviews by F. Rietmeijer, D. Lindstrom, and M. Zolensky. This work was supported by the National Science Foundation (grant No. EAR-8313683) and the National Aeronautics and Space Administration (grant Nos. NAG 9-106 and NAG 9-257). The Brookhaven facilities are also supported by the U.S. Department of Energy, Division of Materials Sciences and Division of Chemical Sciences (contract No. DE-ACO2-76CH00016).

REFERENCES

Bradley J. (1988) Analysis of chondritic interplanetary dust thin-sections. *Geochim. Cosmochim. Acta, 52,* 889-900.

Bradley J. P., Brownlee D. E., and Fraundorf P. (1984) Discovery of nuclear tracks in interplanetary dust. *Science, 226,* 1432.

Brownlee D. E. (1985) Cosmic dust. *Annu. Rev. Earth Planet. Sci., 13,* 147-173.

Brownlee D. E., Ferry G. V., and Tomandl E. (1976) Stratospheric aluminum oxide. *Science, 191,* 1270-1721.

Brownlee D. E., Tomandl D. A., and Olsewski E. (1977) Interplanetary dust; a new source of extraterrestrial material for laboratory studies. *Proc. Lunar Sci. Conf. 8th,* pp. 149-160.

Brownlee D. E., Pilachowski L., Olszewski E., and Hodge P. W. (1980) Analysis of interplanetary dust collections. In *Solid Particles in the Solar System* (I. Halliday and B. A. McIntosh, eds.), pp. 333-342. IAU Symposium 90, Reidel.

Cicerone R. J. (1981) Halogens in the atmosphere. *Rev. Geophys. Space Phys., 19,* 123-139.

Esat T. M., Brownlee D. E., Papanastassiou D. A., and Wasserburg G. J. (1979) The Mg composition of interplanetary dust particles. *Science, 206,* 190-192.

Fegley B. Jr. and Lewis J. S. (1979) Volatile element chemistry in the solar nebula: Na, K, F, Cl, Br and P. *Icarus, 41,* 439-455.

Flynn G. J., Fraundorf P., Shirck J., and Walker R. M. (1978) The chemical and structural study of "Brownlee" particles. *Proc. Lunar Planet. Sci. Conf. 9th,* pp. 1187-1208.

Ganapathy R. (1981) The Tunguska explosion of 1908: Discovery of meteoritic debris near the explosion site and at the South Pole. *Science, 220,* 1158-1161.

Golenetskii S. P., Stepanok V. V., Kolesnikov E. M., and Murashov D. A. (1978) The question about the chemical composition and nature of the Tunguska cosmic body. *Solar Sys. Res., 11,* 103-113.

Harvey G. A. (1973) Spectral analysis of four meteors in evolutionary and physical properties of meteoroids. *NASA SP-319,* pp. 103-129.

Hudson B., Flynn G. J., Hohenberg C. M., and Shirck J. (1981) Noble gases in stratospheric dust particles: Confirmation of extraterrestrial origin. *Science, 211,* 383-386.

Kotra J. P., Finnegan D. L., Zoller W. H., Hart M. A., and Moyers J. L. (1983) El Chichon: Composition of plume gases and particles. *Science, 222,* 1018-1021.

Krauskopf K. B. (1979) *Introduction to Geochemistry.* McGraw-Hill, New York. 617 pp.

Lu F.-Q., Smith J. V., Sutton S. R., Rivers M. L., and Davis A. M. (1989) Synchrotron X-ray fluorescence analysis of rock-forming minerals. I. Comparison with other techniques. II. White-beam energy dispersive procedure for feldspars. *Chem. Geol., 75,* 123-143.

Mackinnon I. D. R. and Mogk D. W. (1985) Surface sulphur measurements on stratospheric particles. *Geophys. Res. Lett., 12,* 93-96.

Rajan R. S., Brownlee D. E., Tomandl D., Hodge P. W., Farrar H. N., and Bitten R. A. (1977) Detection of ^{4}He in stratospheric particles gives evidence of extraterrestrial origin. *Nature, 267,* 133-134.

Rietmeijer F. J. M. (1986) Silica-rich glass and tridymite in a chondritic porous aggregate: Evidence for stratospheric contamination of interplanetary dust particles (abstract). In *Lunar and Planetary Science XVII,* pp. 708-709. Lunar and Planetary Institute, Houston.

Sandford S. A. and Walker R. M. (1985) Laboratory infrared transmission of individual interplanetary dust particles from 2.5 to 25 microns. *Astrophys. J., 291,* 838-851.

Sutton S. R. and Flynn G. J. (1988) Stratospheric particles: Synchrotron X-ray fluorescence determination of trace element contents. *Proc. Lunar Planet. Sci. Conf. 18th,* pp. 607-614.

Sutton S. R. and Flynn G. J. (1989) Density estimates for eleven cosmic dust particles based on synchrotron X-ray fluorescence analyses (abstract). In *Abstracts for the 52nd Annual Meeting of the Meteoritical Society,* p. 235. Lunar and Planetary Institute, Houston.

Sutton S. R., Herzog G., and Hewins R. (1988) Chemical fractionation trends in deep sea spheres. *Meteoritics, 23,* 304.

van der Stap C. C. A. H., Vis R. D., and Verheul H. (1986) Interplanetary dust: Arguments in favour of a late stage nebular origin (abstract). In *Lunar and Planetary Science XVII,* pp. 1013-1014. Lunar and Planetary Institute, Houston.

Wallenwein R., Antz Ch., Jessberger E. K., Buttewitz A., Knochel A., Traxel K., and Bavdaz M. (1989) Multielement analyses of interplanetary dust with PIXES and SYXFA (abstract). In *Lunar and Planetary Science XX,* pp. 1172-1173. Lunar and Planetary Institute, Houston.

Zinner E. K., McKeegan K. D., and Walker R. M. (1983) Laboratory measurements of D/H ratios in interplanetary dust. *Nature, 305,* 119-121.

Zolensky, M. E. and Kaczor L. (1986) Stratospheric particle abundance and variations over the last decade (abstract). In *Lunar and Planetary Science XVII,* pp. 969-970. Lunar and Planetary Institute, Houston.

Zolensky M. E., Lindstrom D. J., Thomas K. L., Lindstrom R. M., and Lindstrom M. M. (1989a) Trace element compositions of six "chondritic" stratospheric dust particles. In *Lunar and Planetary Science XX,* pp. 1255-1256. Lunar and Planetary Institute, Houston.

Zolensky M. E., McKay D. S., and Kaczor L. A. (1989b) A tenfold increase in the abundance of large solid particles in the stratosphere as measured over the period 1976-1984. *J. Geophys. Res., 94,* 1047-1056.

Zolensky M. E., Mackinnon I. D. R., and McKay D. A. (1984) Towards a complete inventory of stratospheric dust with implications for their classification (abstract). In *Lunar and Planetary Science XV,* pp. 963-964. Lunar and Planetary Institute, Houston.

Proceedings of the 20th Lunar and Planetary Science Conference, pp. 343-355
Lunar and Planetary Institute, Houston, 1990

In Situ Extraction and Analysis of Volatiles and Simple Molecules in Interplanetary Dust Particles, Contaminants, and Silica Aerogel

C. P. Hartmetz and E. K. Gibson Jr.[1]

Mail Code SN2, Planetary Sciences Branch, NASA Johnson Space Center, Houston, TX 77058

G. E. Blanford

University of Houston—Clear Lake, 2700 Bay Area Boulevard, Houston, TX 77058

Eight "chondritic" interplanetary dust particles (IDPs) have been analyzed using a laser microprobe/mass spectrometer. This technique is capable of extracting volatile elements and molecules from particles that are $\geq 10\ \mu m$. Three compounds used during collection and curation [silicone oil ($(C_2H_5)_3SiOH$), freon ($C_2Cl_3F_3$), and hexane (C_6H_{14})] were analyzed by the same methods as the IDPs. The volatile species from contaminants are always present in the spectra of IDPs resulting in difficulties with the interpretation of the data. Analysis of U2022F5, U2034D1, U2015B20, and U2015F20 revealed mostly species that can be attributed to contaminants. Some of these species may be indigenous to the IDP; however, separation of the contamination signal and IDP signal is problematic. U2022F5 and U2015F20 released OH that may be associated with hydroxide-bearing minerals or hydrates. U2015B20 released C and CS_2 while U2034D1 released CO_2 and sulfur-related components (SO and COS, respectively). Analysis of U2022G13, U2034D7, U2017A4, and U2017A5 revealed sulfur-related components (typically S, H_2S, SO, COS, SO_2, S_2, and CS_2) that are either related to sulfur aerosols or sulfides (possibly sulfate related when SO and SO_2 are present), which is consistent with energy dispersive X-ray (EDX) data that indicate these IDPs contain significant amounts of sulfur. U2017A4 contained a carbon-bearing phase, and U2034D7 may contain carbonate components (major amounts of CO_2 and its fragment CO) and hydroxide (from hydrates or hydroxide-bearing minerals). Silica aerogel (0.06 and 0.12 g/cm^3), which is a proposed IDP collection medium, was also examined for volatile contaminants. Our studies show that aerogel contains large amounts of volatile contaminants that are from tetramethoxysilane components and solvents used in its preparation. The 0.06 g/cm^3 aerogel contains 1.74 ± 0.1 times more volatiles than the 0.12 g/cm^3 aerogel. The volatile content of aerogel can be reduced 50% by heating to 250°C under vacuum for ~72 hr. However, the remaining volatile abundances were still greater (i.e., more organic species at higher abundances) than what is introduced to IDPs by methods currently used in our stratospheric collection program, indicating that silica aerogel is not an ideal IDP capture medium.

INTRODUCTION

Interplanetary dust particles (IDPs) are samples of primitive materials probably derived from small solar system bodies, such as comets and asteroids, although an undetermined fraction may be from interstellar media. Because comets and asteroids escaped capture by planets, it is believed that they have suffered minimal heating and alteration during the accretional and postaccretional phases of planetesimal formation (*Brownlee,* 1978). Therefore, information obtained from these materials enhances our understanding of primordial solar system conditions. Our interest in IDPs stems in part from their contribution to the elucidation of the cosmic history of the elements H, C, N, O, S, and P (often referred to as the organogenic or biogenic elements *without* implying that these elements, and molecules made up of these elements, were derived from lifeforms). Three important issues involved in the study of these particles are (1) the chemical, physical, and geological phenomena involved in the path taken by these elements from their incorporation as compounds and minerals in primitive solar system bodies (i.e., what are they made of, where do they come from, and how did they form?); (2) the contribution made by these particles to our carbon-rich biosphere (*Thomas et al.,* 1989); and (3) the properties of these materials that influenced processes in the origin and early evolution of the solar system (*Wood and Chang,* 1985).

Recent encounters with Comet Halley showed that high abundances of carbon-bearing compounds exist in the dust released from Halley (*Kissel and Krueger,* 1987). Studies using the PUMA time-of-flight mass spectrometer data have suggested that the organic component in dust from Halley consists predominantly of chondritic cores (layer silicates are a possible candidate) with an organic mantle that is composed mainly of highly unsaturated hydrocarbons. Carbonaceous chondrites also contain highly unsaturated hydrocarbons, and it is suspected that IDPs do as well. Contamination makes determinations of the hydrocarbon content of IDPs difficult. However, evidence from the infrared spectra of IDPs indicates that a portion of the spectral features may be due to hydrocarbons in the particles (*Bradley et al.,* 1988). Raman spectra also indicate that carbonaceous material exists in IDPs (*Wopenka,* 1987).

Laboratory studies show that typical IDPs are fine-grained, black aggregates of approximately chondritic elemental abundance [for 10 to 12 of the most abundant elements in chondrites (*Brownlee,* 1978)]. *Blanford et al.* (1988) compared the composition of chondritic IDPs (based upon

---

[1]Author to whom correspondence should be addressed

their microbeam analysis of major elements, including carbon and oxygen) with that of Comet Halley dust, CI meteorites, and the solar photosphere. They found good agreement between the composition of these different objects. *Hartmetz et al.* (1988) also found similarities in the volatile data from IDPs, Comet Halley, and carbonaceous chondrites. *Blanford et al.* (1988) reported carbon contents for chondritic IDPs that generally range from 8-16% by weight, with one exceptional particle having 49% carbon. From these results as well as various analytical studies (i.e., Raman and infrared spectrometry), it is apparent that the carbon abundances for IDPs are quite variable and reflect their sources and past histories including collection and curation procedures that contaminate the particles (*Rietmeijer,* 1987).

The study of IDPs thus far indicates that they are complex heterogeneous assemblages of high-temperature and low-temperature crystalline phases and amorphous components (*Mackinnon and Rietmeijer,* 1987). These components may have originated as stellar or nebular condensates, interstellar dust, or products of parent body processes (aqueous alteration, and shock or thermal metamorphism). The degree of success in distinguishing between these sources rests largely on our ability to carry out compositional analysis for elemental, isotopic, molecular, and mineralogical characterization; physical determination of grain size, density, and morphology; and petrographic studies. One of the major obstacles in obtaining these results is the persistent contamination problem associated with the collection and curation of the IDPs (*Rietmeijer,* 1987). This contamination problem has put into question most published volatile element analyses of IDPs. To fully understand IDPs we must use analytical techniques that

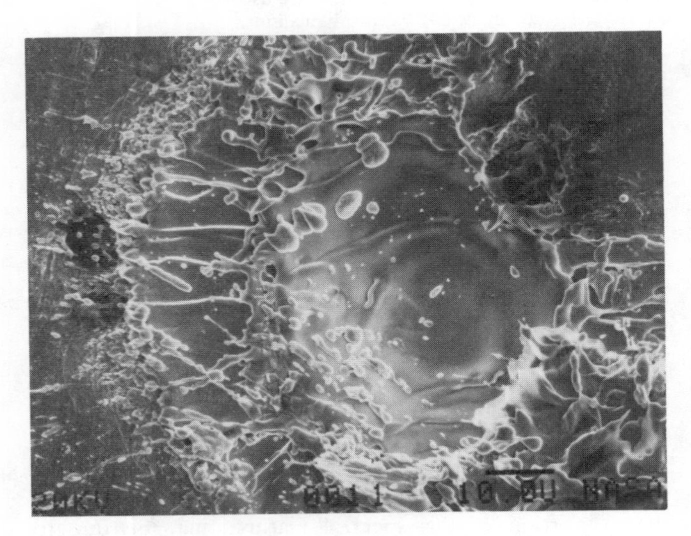

Fig. 1. SEM photomicrograph of U2015B20 that was missed by the laser pulse (IDP approximately at one o'clock in relation to pit). Clearly this IDP has been melted by the nearby laser pulse. The dark spray that emanates from the side of the particle away from the laser zap appears red under transmitted light and has only been seen around the laser zap pits that were performed in silicone oil (i.e., silicone oil was still present in this particle). The scale bar is 10 μm in length.

can accurately determine the elemental and molecular compositions of these particles free of contamination. Hopefully, adequate collection facilities will be devised for the capture of uncontaminated IDPs aboard space station Freedom.

Several questions concerning the volatile content of IDPs remain unanswered:

1. How complex were the volatile phases and the organic chemistry of the gas phases associated with the formation of IDPs?
2. What are the abundances, distributions, and carbon isotopic compositions of the volatiles and associated organic components in IDPs?
3. What relationships exist between the organic compounds, carbonaceous grains, and hydrous phases of IDPs, comets, and the parent bodies of carbonaceous or unequilibrated ordinary chondrites?
4. How do the nature, abundance, and distribution of the organogenic elements in IDPs, comets, and asteroids impose bounds on solar system formation?

To answer some of these questions we have undertaken a study of the volatile elements (H, C, N, O, and S) and their molecular species present in IDPs and the volatiles associated with contaminants. This investigation utilizes a laser microprobe interfaced with a quadrupole mass spectrometer and represents the first study of its kind.

EXPERIMENTAL METHODS

Interplanetary Dust Particles

One of the most critical steps in the analysis of IDPs is sample collection and preparation. Collection of the IDPs usually occurs in the stratosphere (~20 km) on collector plates coated with a 20:1 mixture of silicone oil and freon. Particles are removed from the collectors and placed on scanning electron microscope (SEM) grids for SEM/EDX characterization. After characterization, selected samples were rinsed with hexane and embedded in specially prepared softened gold. In the past we analyzed particles on beryllium; however, the particles fragmented upon laser interaction, causing incomplete vaporization of the particle. Gold, on the other hand, is soft enough to allow a particle to be embedded with a minimum of alteration. We have also analyzed the volatiles released from many substrates: aluminum, alumina, beryllium, tungsten, tantalum, fused silica, silica aerogel, borosilicate glass, metallized kapton, indium, gold, and Torr seal (*Blanford and Gibson,* 1988). The most volatiles were released from silica aerogel, the least volatiles were released from aluminum, beryllium, and gold. Gold was chosen because of its low contaminant content and its soft nature that prevents the IDP fragmentation problems noted above.

The gold was prepared by annealing from ~900°C. The softened gold was then cleaned with a commercial detergent and rinsed with distilled water, methanol, and freon. Then the gold was placed in a drying oven at ~130°C before it was oxygen-plasma-etched in two 15-minute cycles. The EDX spectra indicate that the 30 minutes of plasma etching considerably reduced the carbon (low-Z) contamination on the surface of the softened gold. The SEM photomicrographs of the embedded particles (Fig. 1) were taken using a JEOL-35CF SEM both before and after the particles were analyzed with the laser microprobe system. After the particles were photographed for the first time, they were placed in the sample chamber and "baked out" (~130°C) under vacuum (~1×10^{-7} torr). Several practice laser (pulsed neodymium glass) shots into the outer edges of the 0.5-cm^2 piece of gold were performed to aid in the aiming of the laser. The sample was vaporized by the laser at its lowest power setting, and then several laser shots were performed into the gold near the sample location to determine the average background contribution to the signal from contaminants adsorbed on the gold. The average amount of contaminants, inherent in all vacuum lines, was also determined before each laser shot into substrate or particle. This vacuum line background spectra was also subtracted from the particle's spectra. A Hewlett Packard quadrupole mass spectrometer was utilized to analyze the gases released by the laser. The analytical system is described in detail in *Gibson and Carr* (1989).

For each particle the ion current from each individual peak was multiplied by 1000 and divided by all eight particles' average total ion current. This normalization allows the comparison of spectra from the individual IDPs. For freon, hexane, silicone oil, and the treated Allende matrix sample the denominator in the normalization was the total ion current for freon on gold minus background; this value was very close to the average total ion current for all four contaminants. The only peaks that we attempted to interpret were peaks that were >1 σ above the gold background. This ensures a 67% probability that the peak is actually from the particle; unfortunately, this also includes contaminants that were adsorbed on the surface of the IDP. Silica aerogel was not analyzed with the laser while embedded in gold as the IDPs were. Therefore, its normalization was handled as we handle meteorite data (*Hartmetz et al.,* 1989). The ion current for a peak is divided by the total ion current of that sample and multiplied by 10,000.

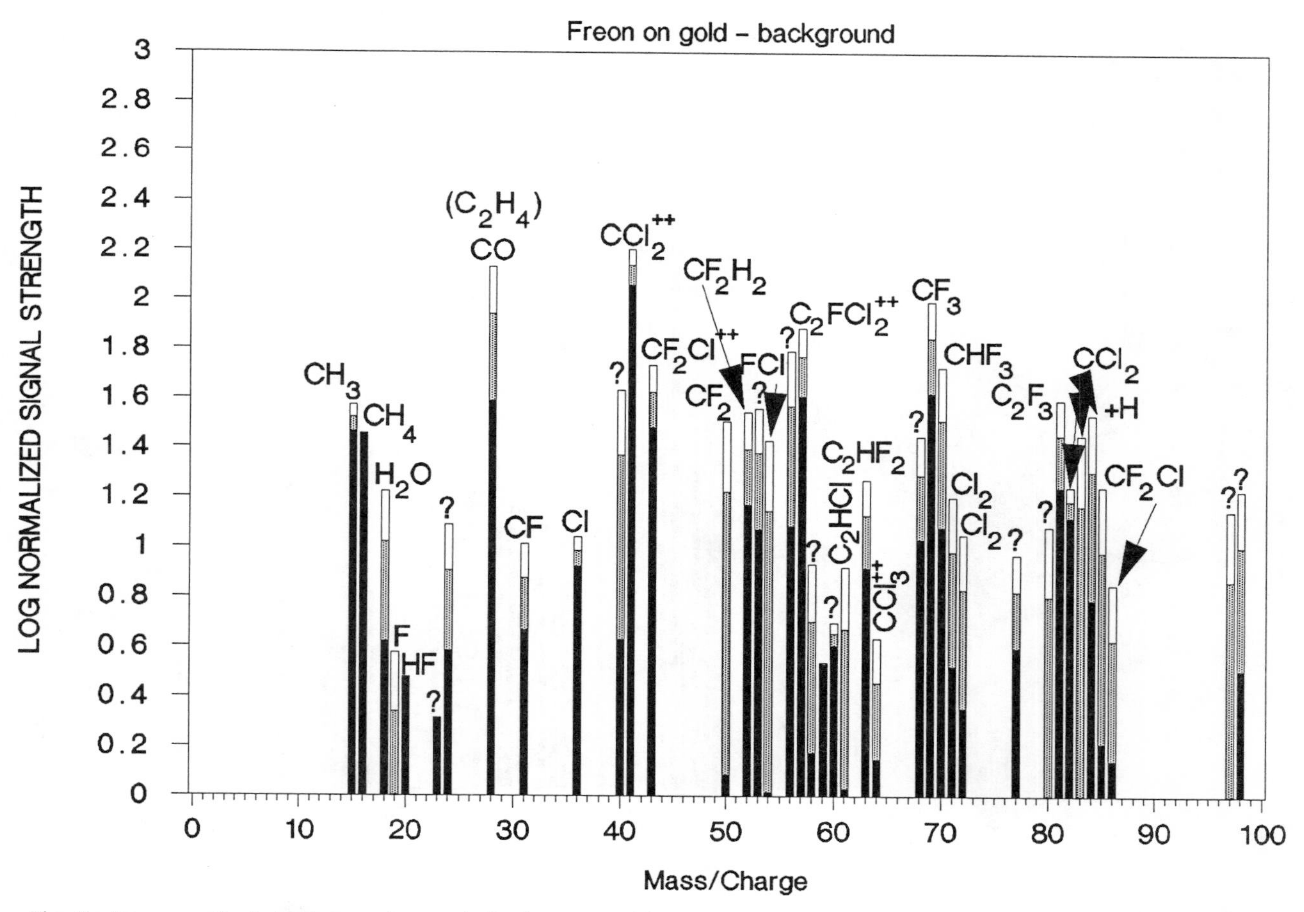

Fig. 2. Log normalized signal strength vs. m/z for freon on gold. Average background present in the vacuum line before the laser zap and the average signal from four zaps into the same piece of gold were subtracted from the IDP signal. Peaks shown are all 1 σ above the average gold background. The top of the shaded bar is the mean while the top of the filled and unfilled bars represent $\pm 1\ \sigma$. This spectra is extremely complex and at m/z = 82, 83, and 84 there was not room for each peak's formula; m/z = 82 (CCl_2) is given, but m/z = 83 and 84 are denoted by arrows and a +H symbol because their formulas are CCl_2H and CCl_2H_2, respectively.

Contaminant Analyses

Collection of IDPs in the stratosphere occurs on quartz plates coated with a 20:1 silicone oil/freon mixture, and the subsequent rinsing of the IDPs with hexane and handling prior to analysis introduce volatiles that are not indigenous to the IDPs. These contaminants were examined by coating the gold with the contaminant and analyzing them in the same manner as the IDPs.

Silica aerogel is a candidate for collection media on the Cosmic Dust Collection Facility (CDCF), which is under consideration as an experiment aboard the space station Freedom. Therefore, the volatiles associated with 0.06 and 0.12 g/cm^3 silica aerogel were also examined. Analyses were performed after the aerogel was heated overnight at 70°C, after two days at 150°C, and after three days at 250°C. All heating was performed under vacuum ($\sim 1 \times 10^{-7}$ torr).

We also analyzed small (<100 μm) pieces of Allende matrix that had been treated with 2M H_2SO_4 to determine the effect of sulfuric acid aerosols (*Mackinnon and Mogk,* 1985), which are present in the stratosphere, on an IDP. The same analysis procedures employed in the IDP studies were used starting after hexane rinsing (i.e., the Allende particle was mounted without coming in contact with any of the contaminants).

RESULTS

Contaminant Analyses

The relative abundances and error limits of volatile species released from silicone oil, hexane, and freon placed on gold are available on request (from correspondence author). The analysis of freon revealed a complex array of volatiles. Major cations detected include m/z = 15 (CH_3), 16 (CH_4 and/or O), 28 (CO and C_2H_4), 40 (?), 41 (CCl_2^{++}), 43 ($CH_2Cl_2^{++}$), 50 (CF_2), 52 (CH_2F_2), 53 (?), 54 (FCl), 56 (?), 57 ($C_2FCl_2^{++}$), 63 (C_2HF_2), 68 ($C_2F_2Cl_2^{++}$), 69 (CF_3), 70 (CHF_3), 81 ($C_2H_3F_3$), and 84 (CCl_2). The less abundant species include m/z = 19 (F), 20 (HF), 31 (CF), 36 (Cl), 85 ($CHCl_2$), and 98 ($C_2H_2Cl_2$) (Fig. 2). Analysis of silicone oil showed that certain species are released in high abundances and that silicone oil released the largest amount of total volatiles of any

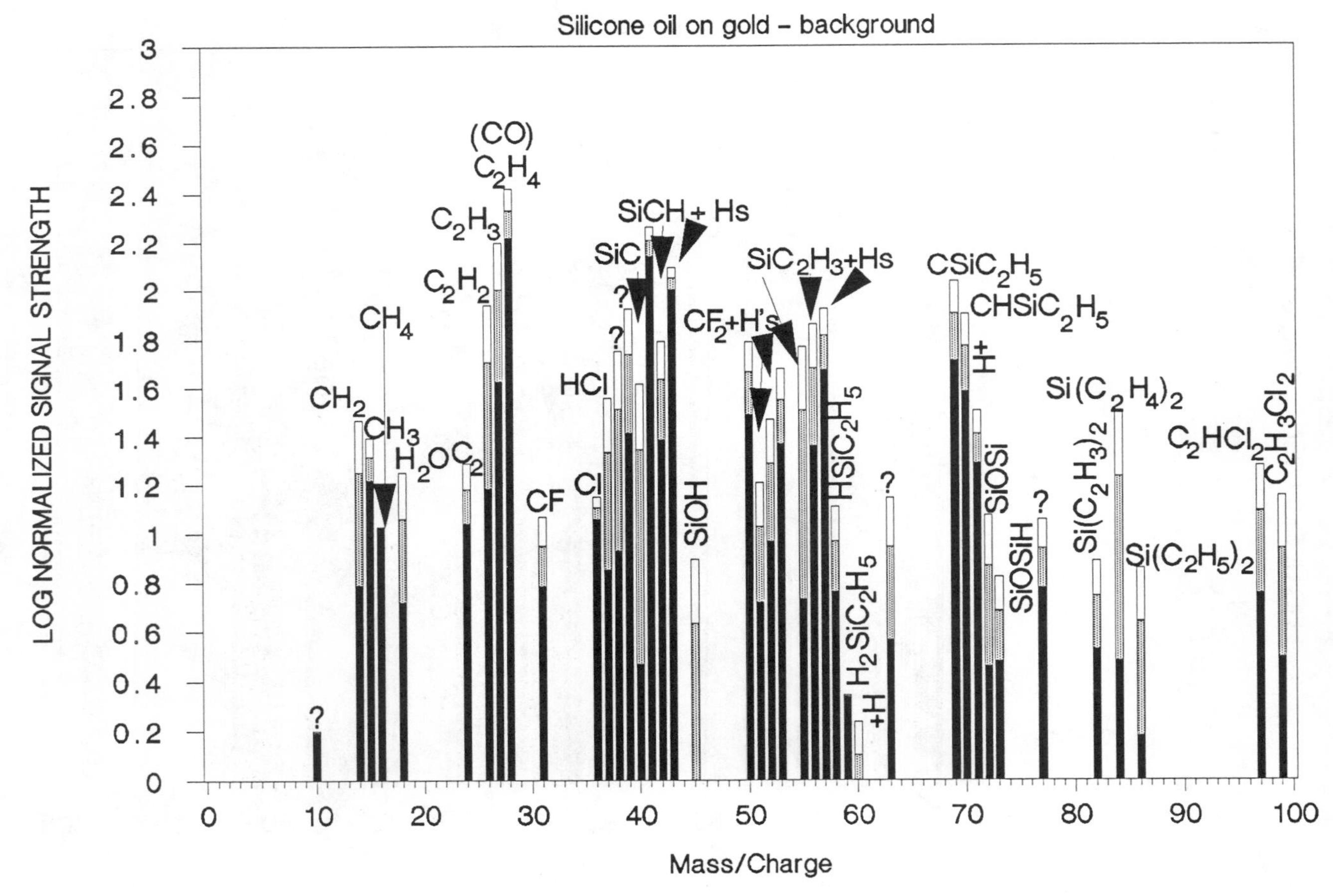

Fig. 3. Log normalized signal strength vs. m/z for silicone oil on gold. Background subtractions and bars are the same as in Fig. 2. Peaks shown are all 1 σ above the average gold background. The complexity of the spectra has caused us to identify groups of peaks that are different only in the number of hydrogen atoms by a +Hs and usually an arrow (e.g., SiCH); m/z = 41 is shown but m/z = 42 and 43 are indicated by arrows as their formulas are $SiCH_2$ and $SiCH_3$, respectively. M/z = 60 and 71 are simply the formula of m/z = 59 and 70, respectively, plus one hydrogen as indicated by the +H symbol (no arrow).

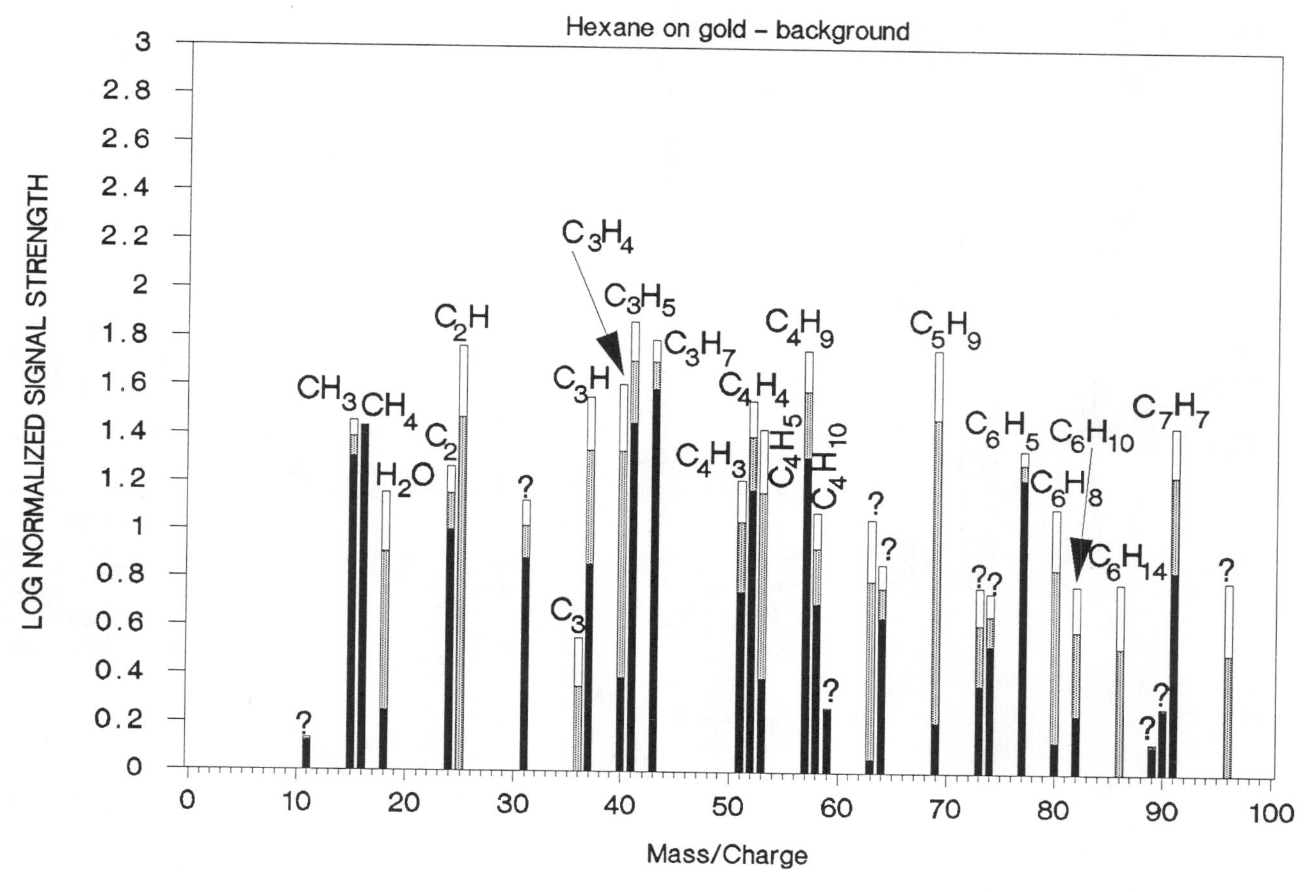

Fig. 4. Log normalized signal strength vs. m/z for hexane on gold. Background subtractions and bars are the same as in Fig. 3. Peaks shown are all 1 σ above the average gold background. All peaks in this spectra are from the fragmentation of hexane.

of the contaminants. Major volatiles released include m/z = 26 (C_2H_2), 27 (C_2H_3), 28 (C_2H_4 and CO), 41 (SiCH), 42 ($SiCH_2$), 43 ($SiCH_3$), 57 (SiC_2H_5), 69 ($CSiC_2H_5$ or CF3), and 70 ($CHSiC_2H_5$ or CHF_3). Less abundant species include m/z = 14 (CH_2), 15 (CH_3), 18 (H_2O), 24 (C_2), 31 (CF), 36 (Cl), 37 (HCl), 40 (SiC), 52 (CH_2F_2), 53 (SiC_2H), 55 (SiC_2H_3), 71 ($CH_3SiC_2H_4$), and 84 (CCl_2) (Fig. 3). The analysis of hexane released far fewer volatiles than freon or silicone oil, and the individual cations are generally found at lower abundances. Volatile species that were released include m/z = 15 (CH_3), 16 (CH_4), 24 (C_2), 25 (C_2H), 37 (C_3H), 40 (C_3H_4), 41 (C_3H_5), 43 (C_3H_7), 52 (C_4H_4), 53 (C_4H_5), 57 (C_4H_9), 69 (C_5H_9), 77 (C_6H_5), and 91 (C_7H_7) (Fig. 4). The Allende matrix particle that was treated with 2M H_2O_4 released sulfur-related components at m/z = 32 (S), 48 (SO), and 64 (SO_2). The latter two volatile species were released in greater abundance (Fig. 5).

Interplanetary Dust Particles

Traces of the molecules and ions associated with freon, hexane, and silicone oil have been observed in the spectra for all eight IDPs. Clearly some of these peaks are superimposed on the signal from the particle. Four of the analyzed IDPs (U2015B20, U2015F20, U2022F5, and U2034D1) released mostly species that may be interpreted as contaminants (silicone oil, freon, and hexane). Some of these peaks may include signal from the IDP, but this signal cannot easily be separated from the contamination signal [relative abundance data and error limits are available upon request (from correspondence author)]. Some species released from these IDPs were determined to be indigenous to the IDPs. U2022F5 and U2015F20 released m/z = 17 (hydroxide) that may be associated with hydroxide-bearing minerals or hydrates (Figs. 6 and 7, respectively). U2015B20 (Fig. 8) released carbon while U2034D1 (Fig. 9) released carbon dioxide and some sulfur-related components (m/z = 48 and 60, SO and COS, respectively). Four IDPs (U2022G13, U2017A4, U2017A5, and U2034D7) released a wider variety of indigenous volatile species at greater abundances than the other four IDPs. U2022G13 released sulfur-related species (H_2S, SO, COS, and SO_2) (Fig. 10). U2017A4 released large abundances of many sulfur-related components (H_2S, SO, and COS) (Fig. 11). This particle's spectra also contains a large release of CO that is interpreted as evidence for a carbonaceous phase. U2017A5 released large abundances of sulfur-related components (SO

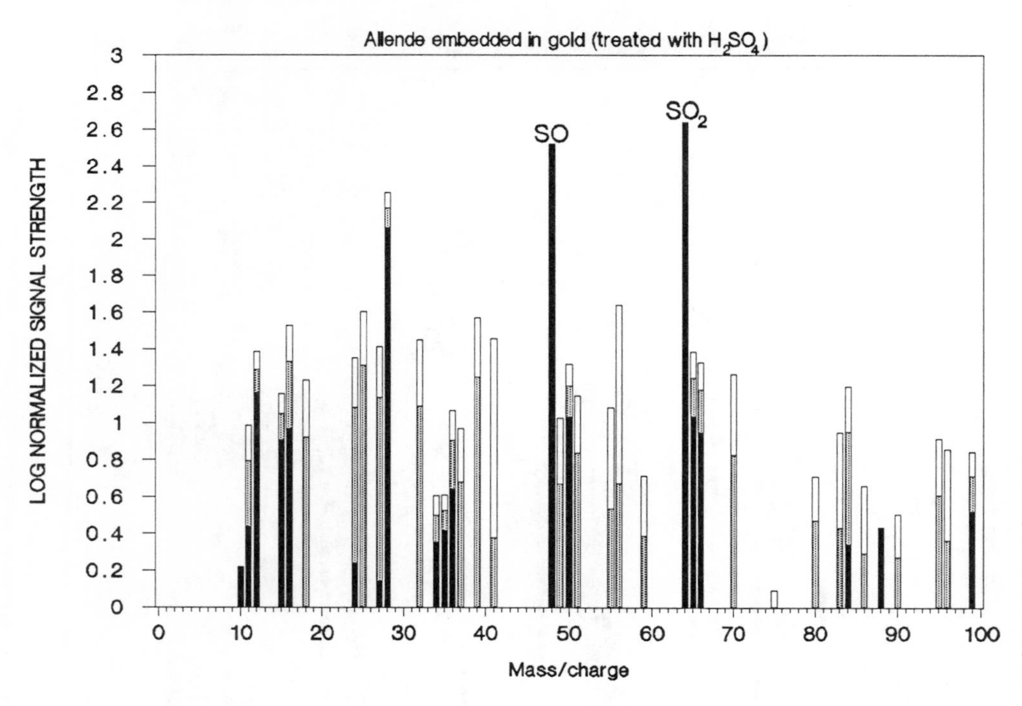

Fig. 5. Plot of log normalized signal strength vs. m/z for an Allende matrix particle treated with 2M H_2SO_4. Background subtractions and bars are the same as in Fig. 3. Peaks shown are all 1 σ above the average gold background. Note the large release of m/z = 48 (SO) and 64 (SO_2).

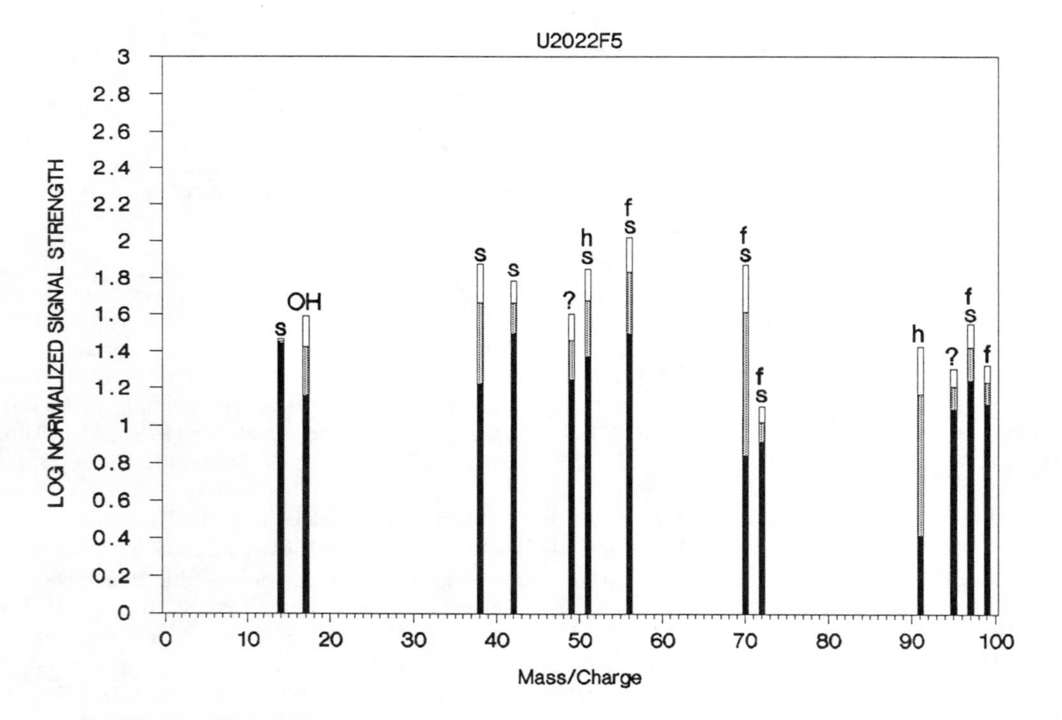

Fig. 6. Log normalized signal strength vs. m/z for U2022F5, a porous chondritic IDP (EDX indicates the presence of little or no volatiles). Average background present in the vacuum line before the laser zap and the average signal from four zaps into the same piece of gold were subtracted from the IDP signal. Peaks shown in Figs. 7-13 are all 1 σ above the average gold background. In Figs. 7-13 the top of the shaded bar is the mean while the tops of the filled and unfilled bars represent $\pm 1\ \sigma$. In Figs. 7-13 the symbols f, h, and s indicate contaminants from freon, hexane, and silicone oil, respectively. Peaks labeled with a chemical formula are from the IDP. Note the lack of overall volatiles, with m/z = 17 (OH) as the only peak that can be interpreted as coming from the IDP.

and CS_2) (Fig. 12). The most interesting particle studied (the particle with the largest release of indigenous species) was U2034D7, which was a porous IDP (Fig. 13). This particle also released large abundances of CO and CO_2 that could be from a carbonate phase within the particle. U2034D7 also released sulfur-related components (CS_2) and hydroxide (OH) from a hydroxide-bearing mineral or hydrate. Our data from these four IDPs (U2022G13, U2017A4, U2017A5, and U2034D7) concur with the results from EDX data that indicated these IDPs contained significant amounts of sulfur. It is possible that a portion of these sulfur species are from aerosol contamination. Large abundances of m/z = 48 and 64 (SO and SO_2, respectively) were released from the Allende particle after treatment with H_2SO_4. *MacKinnon and Mogk* (1985) indicate that sulfur aerosol-related species other than H_2SO_4 are possible (CS_2, COS, and H_2S).

Silica Aerogel

The 0.06 g/cm^3 silica aerogel reportedly contains 3 wt.% volatiles, while the 0.12 g/cm^3 aerogel contains 1% added during manufacturing (J. Williams, Los Alamos National Lab, personal communication, 1989). We found that the 0.06 g/cm^3 silica aerogel contains 1.74 ± 0.1 times more total volatiles than the 0.12 g/cm^3 silica aerogel. The volatiles released include every m/z value between 10 and 180; many of these species were released in high abundances. As the aerogel was heated to higher temperatures, the total volatile content was lowered. After being heated under vacuum at 250°C for three days, the volatile content of both samples was 50% less than the starting materials. The spectra of the 0.12 g/cm^3 aerogel that had experienced 250°C for three days released the fewest total volatiles and is shown in Fig. 14. The compound that is used in the manufacturing of silica aerogel (tetramethoxysilane, m/z = 152) (J. Williams, Los Alamos National Lab, personal communication, 1989) and its fragments appear in high abundances in the spectra of silica aerogel that has not experienced heating >150°C.

DISCUSSION AND CONCLUSIONS

Contaminant Analyses

The laser interaction with freon residue on gold released essentially every component possible from the breakup of freon. Recombination reactions (addition of hydrogen atoms to many of the fragments) apparently also occur prior to detection by the mass spectrometer. This particular molecule (CF_3CCl_3) does not completely evaporate from the gold surface or particles it comes in contact with. This occurs despite heating to ~130°C under vacuum (~1 × 10^{-7} torr) for

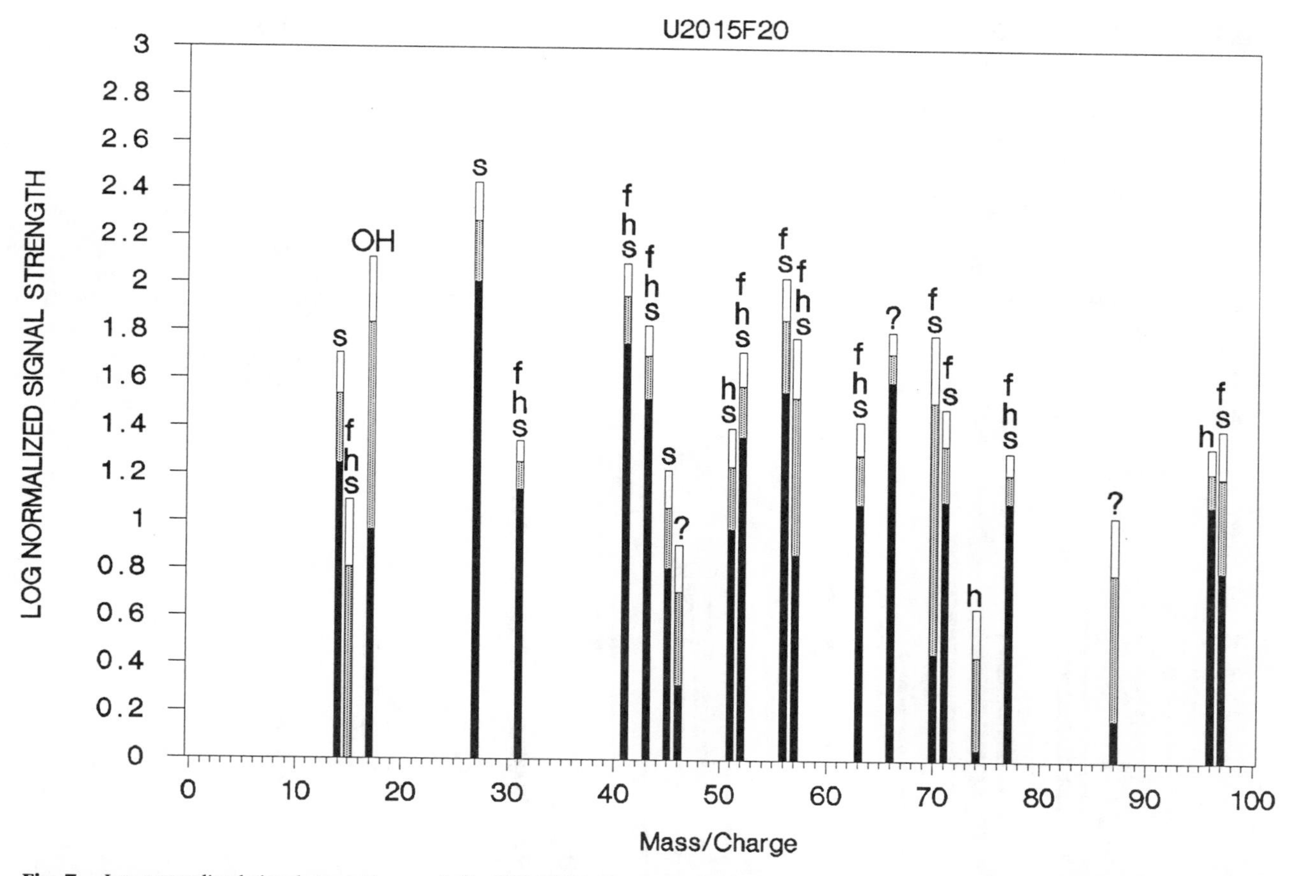

Fig. 7. Log normalized signal strength vs. m/z for U2015F20, a porous chondritic IDP (EDX indicates small quantities of sulfur). Note the absence of sulfur-related components; the only peak that can be attributed to the IDP is the hydroxyl (m/z = 17) peak.

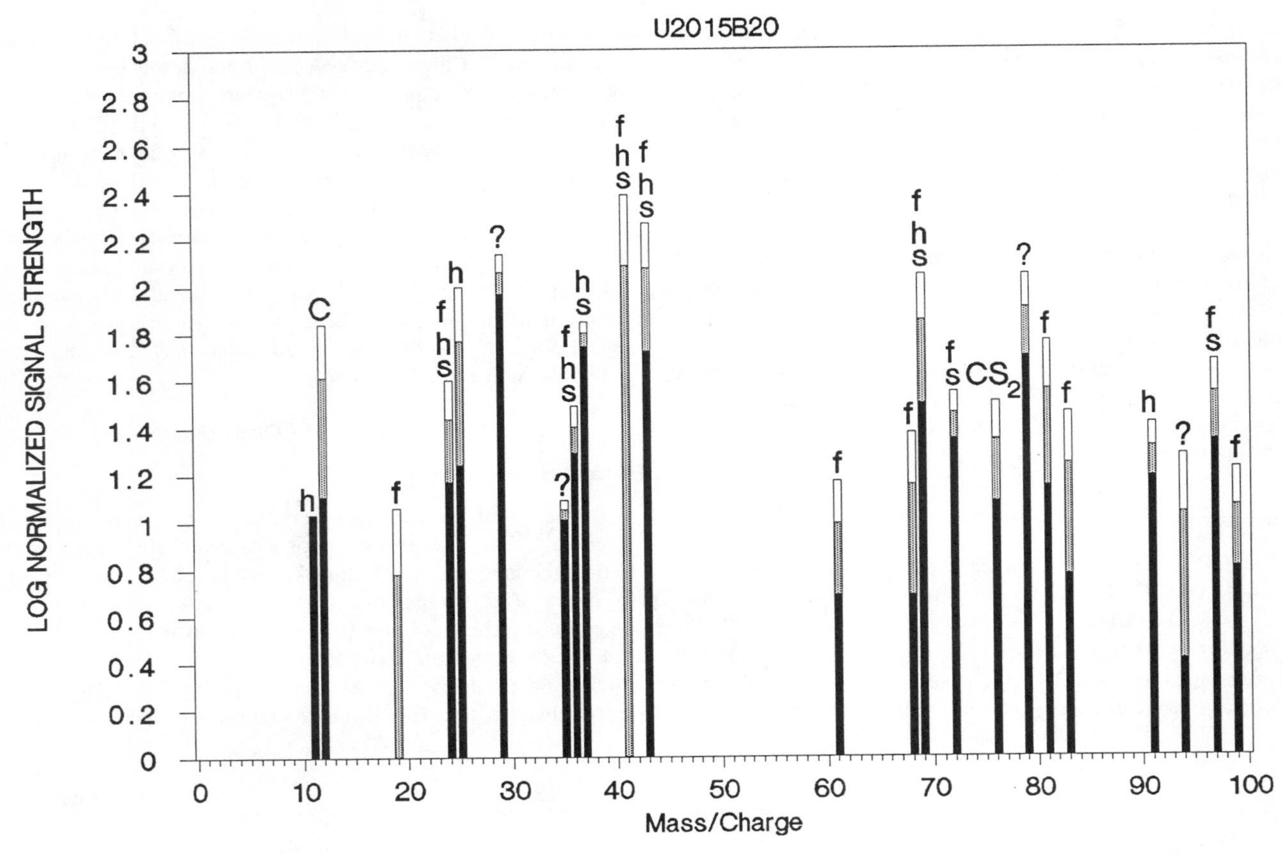

Fig. 8. Log normalized signal strength vs. m/z for U2015B20, a porous chondritic IDP (EDX indicates small quantities of sulfur). Note the presence of sulfide-related component (CS_2) and carbon.

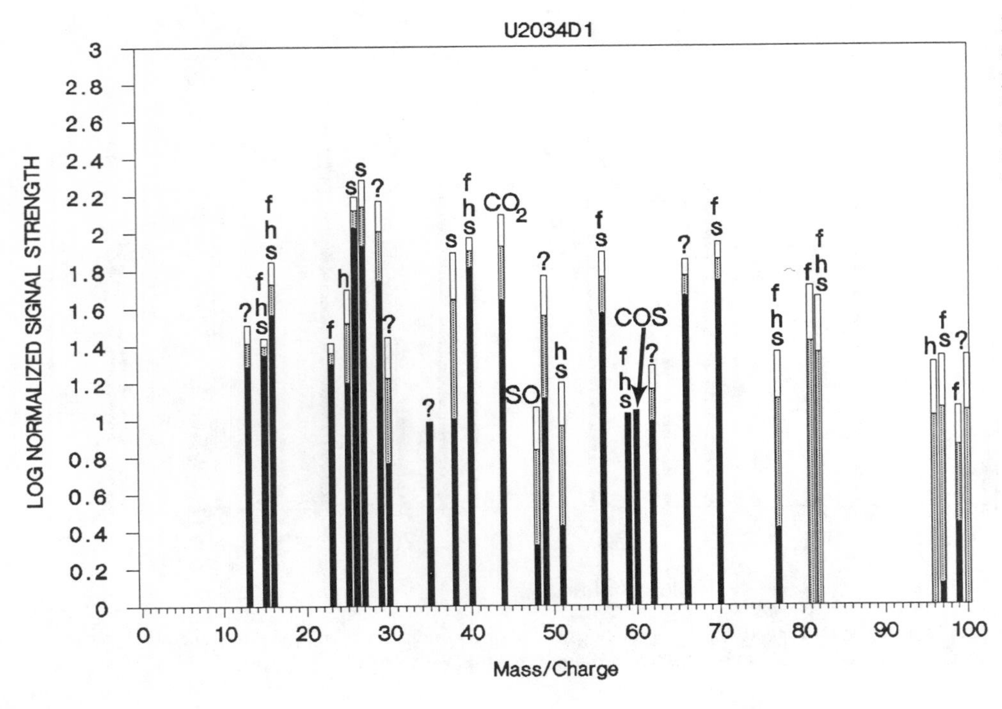

Fig. 9. Log normalized signal strength vs. m/z for U2034D1, a porous chondritic IDP (EDX indicates very small quantities of sulfur). Note the presence of the sulfur-related components (SO and COS) and a carbon dioxide peak.

several days. The high electronegativity of F and Cl atoms results in favorable conditions for the adsorption of freon on many surfaces including gold and IDPs.

Hexane evaporates more readily from the gold than does freon (laser interaction of hexane residue on gold released far fewer fragments despite an identical "bake-out" procedure used on freon) and presumably from the IDPs as well. Many unsaturated fragments are observed in the spectra, because carbon-hydrogen bonds are easily broken. The released hydrogen is free to recombine with other molecular fragments present prior to the encounter with the mass spectrometer's detector. Evidence for recombination reactions is suggested by the presence of deprotonated toluene (C_7H_7), which can only form from the recombination of hexane fragments. Because the hexane used by the curatorial facility is ultrahigh purity grade, no toluene contamination is expected. The released hydrogen can also flow through the mass spectrometer undetected because m/z = 1-3 are not recorded with this quadrupole.

The 20:1 mixture of silicone oil and freon that is used on stratospheric collection plates released many fragments at high abundances upon laser interaction. Again many of the species are unsaturated because of the easily broken carbon-hydrogen bonds. The presence of H_2O (m/z = 18) and SiOSi (m/z = 72) species indicate that a hydrogenation reaction may be occurring [i.e., $R_3Si\text{-}OH + H\text{-}O\text{-}SiR_3 \rightarrow H_2O + R_3Si\text{-}O\text{-}SiR_3$ (where R = methyl group, ethyl group, or a Si atom)]. This is confirmed by the almost total absence of RSiOH species; the only species released of this type was a low abundance of SiOH.

The Allende matrix treated with 2M H_2SO_4 released large abundances of m/z = 48 and 64 (SO and SO_2 or S_2, respectively) and lesser amounts of m/z = 32 (S). While these three species may be observed in particles that have been affected by aerosols, they are not the only species to be expected from stratospheric aerosols. Several other stratospheric sulfur-rich aerosols or gases include COS, SO_2, and CS_2 (*Turco et al.,* 1982). Other possible aerosols are H_2O_2 and HCl (*Baldwin and Golden,* 1979; *Stolarski and Cicerone,* 1975) and NO_2 (*Farlow et al.,* 1978; *Molina et al.,* 1987). The amounts of aerosol that can interact with the particle during a single encounter are small, as the aerosols are ≤ 0.1 μm (*Mackinnon and Mogk,* 1985). Therefore, the abundance of aerosol species will be considerably lower than that seen in this test case, which was designed solely to see what molecular species might be adsorbed on an IDP, not how much.

Interplanetary Dust Particles

As this study illustrates, the interpretation of the data from the analysis for indigenous volatile species in IDPs is very

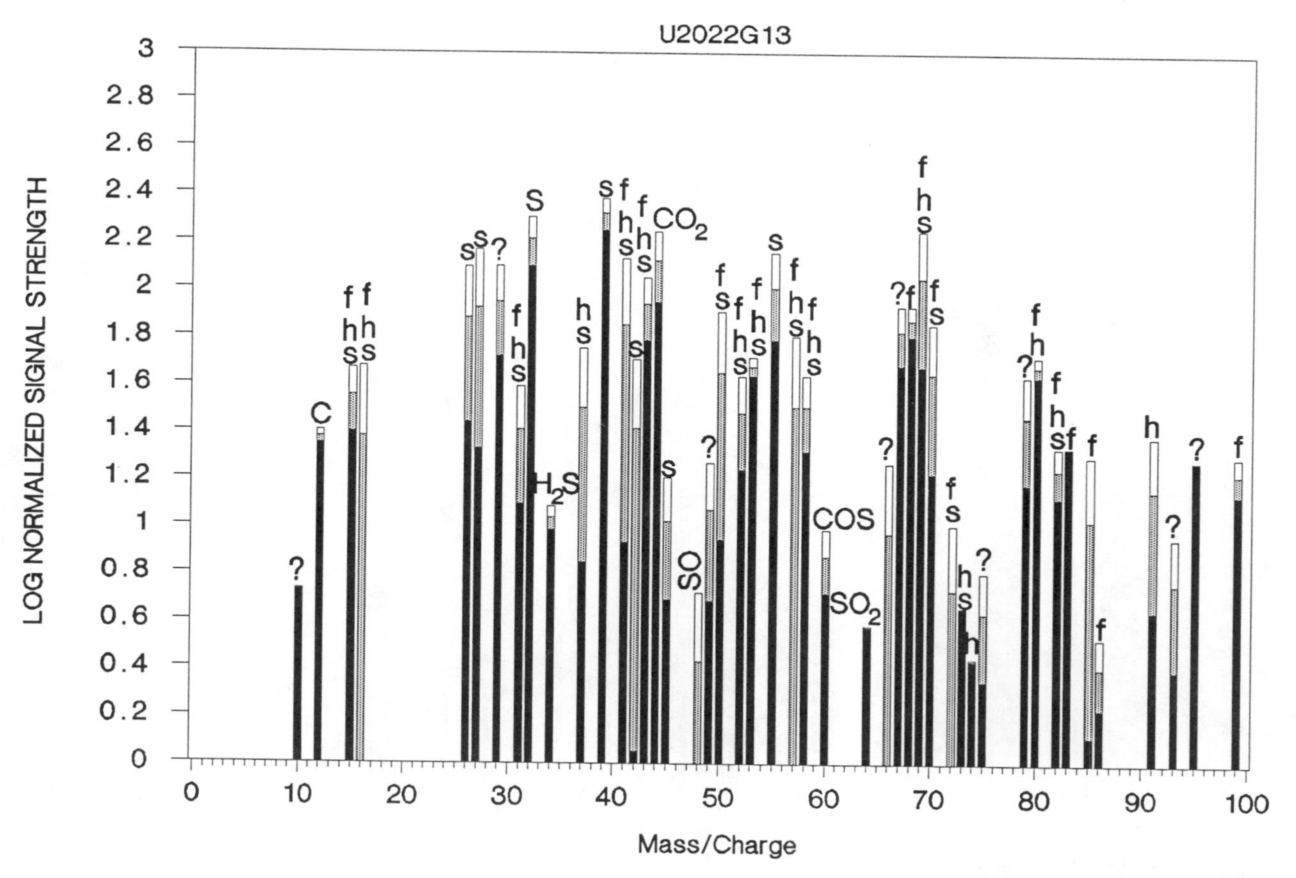

Fig. 10. Log normalized signal strength vs. m/z for U2022G13, a porous chondritic IDP (EDX indicates presence of sulfur). Note the presence of sulfur-related components.

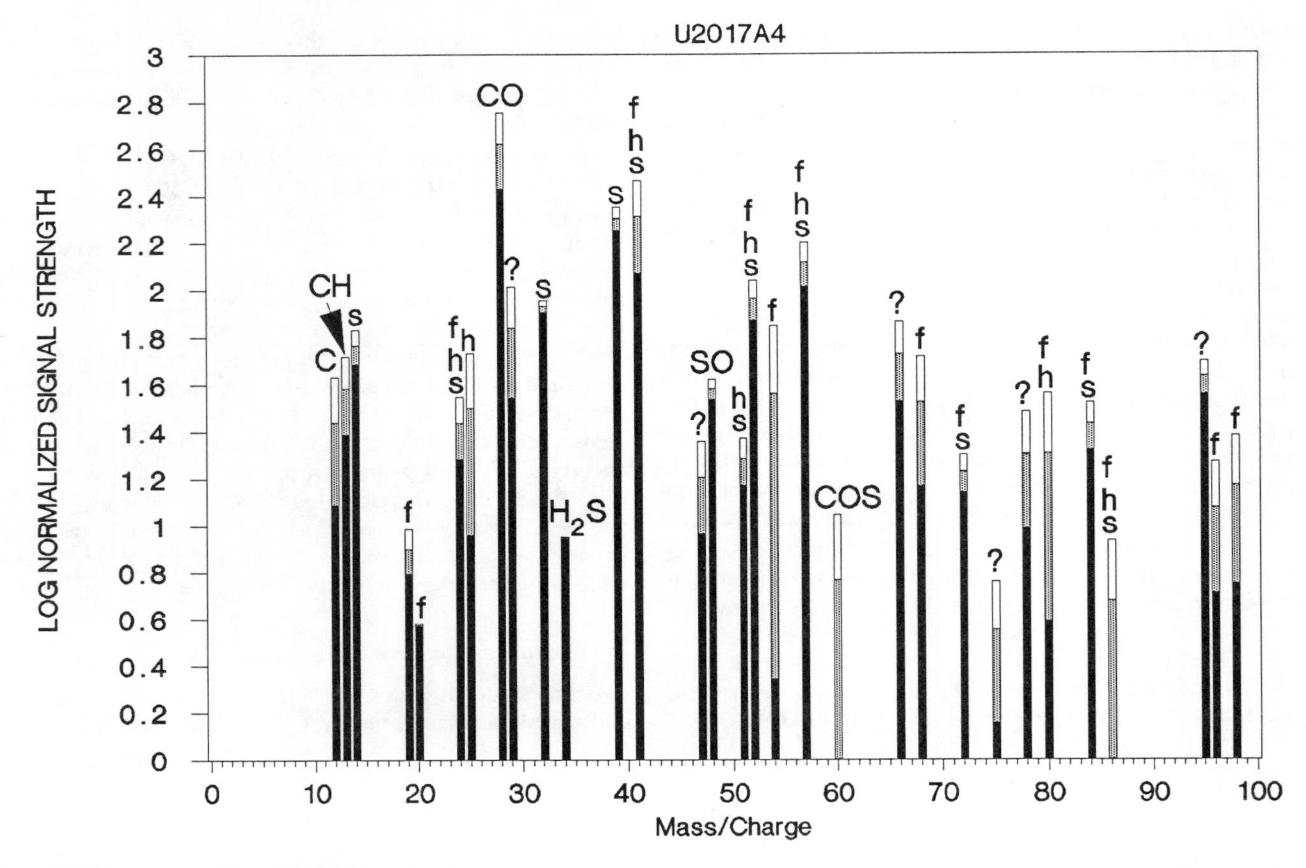

Fig. 11. Log normalized signal strength vs. m/z for U2017A4, a porous chondritic IDP (EDX indicates large quantities of sulfur). Note the presence of sulfur-related components and a carbonyl peak.

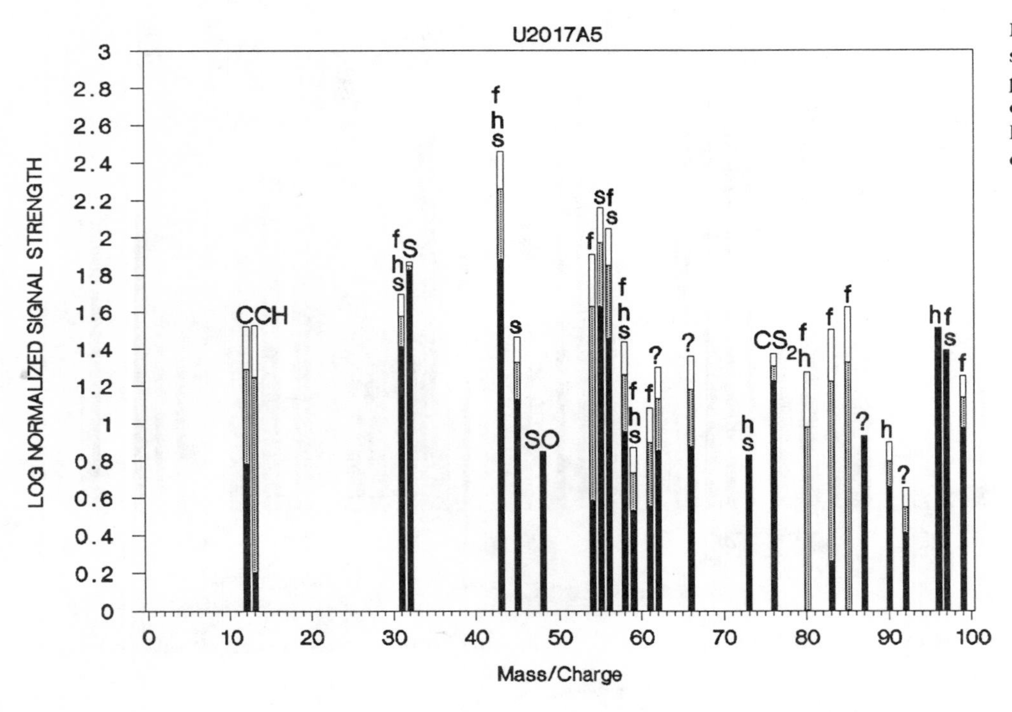

Fig. 12. Log normalized signal strength vs. m/z for U2017A5, a porous chondritic IDP (EDX indicates large quantities of sulfur). Note the presence of sulfur-related components.

difficult. Most of the signal from the IDPs can be interpreted as contamination, and as yet no satisfactory means of subtracting out this contamination signal has been devised. In fact, this is the first study that has attempted to characterize and account for all three possible contaminants associated with IDP collection and curation. Some workers in the field claim the ability to determine the amount of silicone oil (contamination) that is adsorbed and/or coats an IDP without considering hexane and freon. Presumably the wide variety of sample densities, porosities, and chemical affinities of individual IDPs can also be accounted for. We would like any attempt to subtract the effects of contaminants to take into account both the hexane used to rinse the 20:1 mixture of silicone oil and freon and any residual silicone oil/freon left after rinsing. Even if all the contaminants can be taken into account, it is problematic to translate the amount of contaminant adsorbed on a particle into quantitative amounts of individual fragments produced by laser interaction with the contaminants.

Despite these contamination problems several indigenous molecular species have been identified. These include m/z = 17 (OH) from U2034D7, U2015F20, and U2022F5; m/z = 28 (CO_2 or C_2H_4) (found in much greater relative abundance than can be attributed solely to contamination) in U2017A4 and U2034D7; m/z = 12 (C) and 76 (CS_2) (carbonaceous phases and sulfide phases are indicated, respectively) in U2015B20; m/z = 44 (CO_2) along with 28 (CO) possibly indicates the presence of carbonate in U2034D7, while just CO_2 in U2022G13 is not fully understood; m/z = 32 (S), 34 (H_2S), 48 (SO), 60 (COS or contamination), 64 (SO_2 or contamination), and 76 (CS_2), all sulfur-related components are observed in many of the particles. The sulfur components can be attributed to aerosols in some cases. However, in U2022G13 the presence of H_2S, SO, COS, and SO_2 indicates the possible presence of elemental sulfur. The most unique particle studied was U2034D7, which was a porous IDP. As mentioned earlier, the particle may have contained carbonates (m/z = 28 and 44 released in large quantities is consistent with terrestrial carbonate release patterns). Sulfide-related species (CS_2) were also released (CS_2 is the primary sulfur-bearing fragment released from both terrestrial sulfides and elemental sulfur). The presence of OH indicates that there are hydroxide-bearing minerals and possibly hydrates. All of these species are released from the groundmass of CI and CM chondrites, with higher quantities of CS_2 being released from CI chondrites (*Hartmetz et al.,* 1988). Many have made comparisons between the compositions of CI chondrites, IDPs, and dust from Comet Halley (*Blanford and Gibson,* 1988; *Hartmetz et al.,* 1988; *Rietmeijer et al.,* 1989). However, both volatiles (H, C, N, O, and S) and nonvolatiles were analyzed by the PUMA

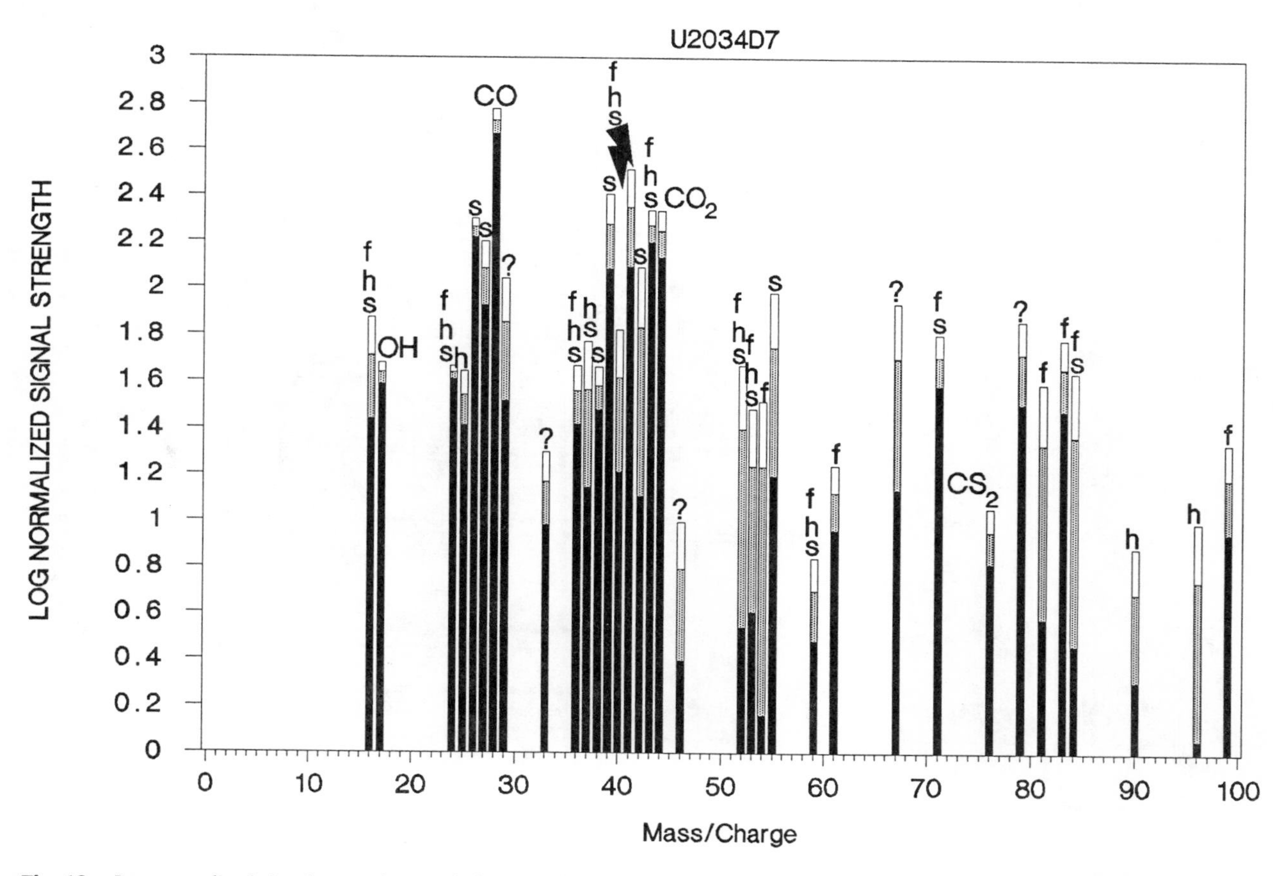

Fig. 13. Log normalized signal strength vs. m/z for U2034D7, a porous chondritic IDP. Carbonate-related components may be present, along with an hydroxide-bearing mineral, as well as sulfide and/or elemental sulfur-related components.

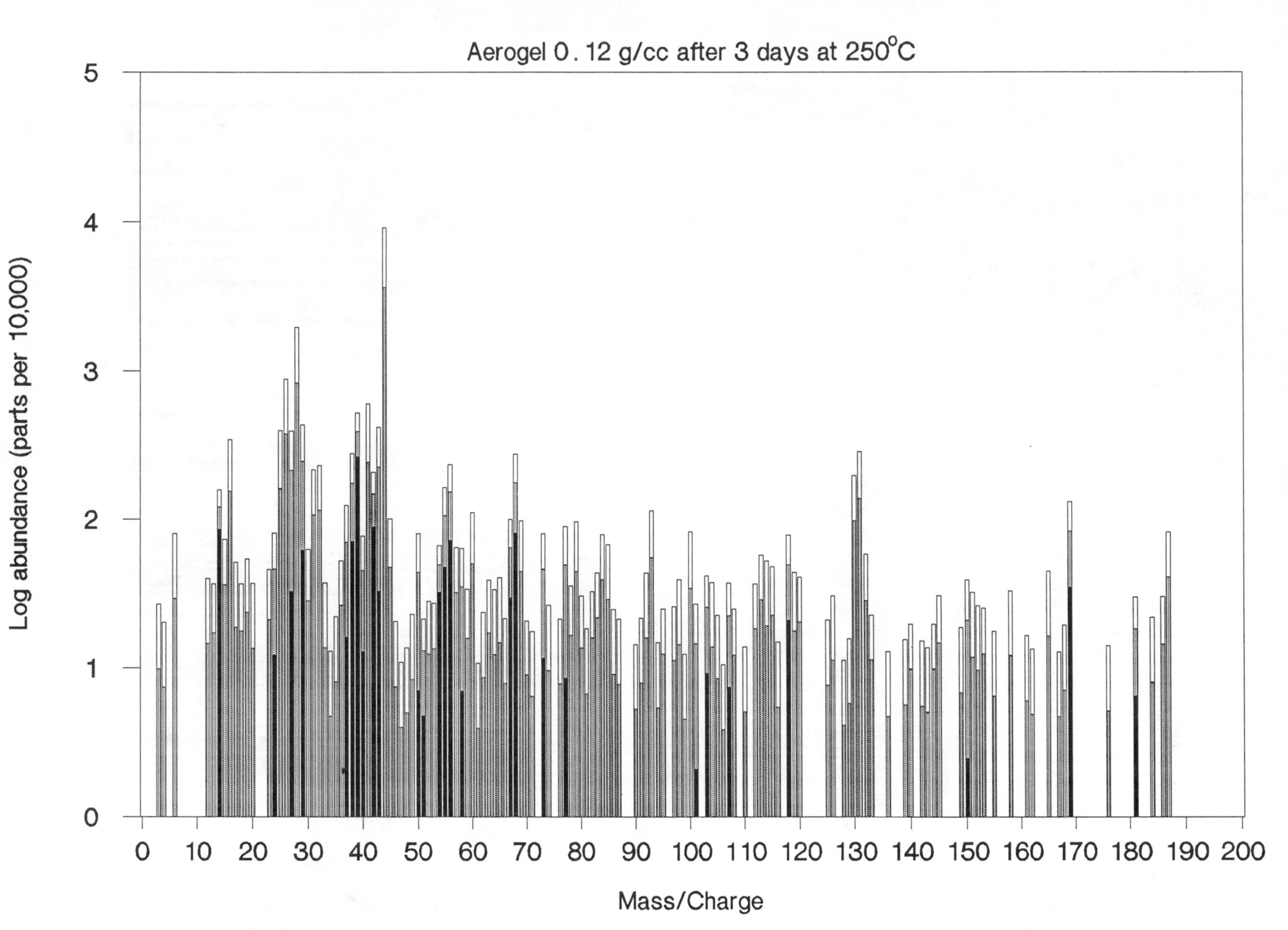

Fig. 14. Log normalized signal strength vs. m/z for 0.12 g/cm^3 silica aerogel that had been heated to 250°C for several days. The bars are the same as in Fig. 3. Only the vacuum line contaminants (background) are subtracted from the aerogel spectra. This is the aerogel spectra that contained the least amount of total volatiles of any of the six average spectra taken from aerogel. It is clear that the volatile contamination that will be introduced into the IDPs collected in this media will create impossible conditions for volatile element analyses of these particles.

instrument, making comparisons difficult. Further complications in this comparison include the very small size and amounts of dust analyzed by PUMA and the manner in which IDPs are collected (contaminated). Until particles are collected in an uncontaminated manner, these types of comparisons and questions will remain a mystery.

We intend to carefully explore the possibilities of using different cleaning solvents (other than freon, e.g,. methyl ethyl ketone) and longer rinsing times in hexane. Another possibility is using an oxygen-plasma etch technique on the IDPs. These clean-up procedures have to be done as long as our present and future collection methods contaminate the IDPs. We have an opportunity with the Cosmic Dust Collection Facility to employ methods that do not contaminate or lose the volatile contents of these important particles. Our studies indicate that using silica aerogel as the sole capture media would *not* be advisable.

Acknowledgments. We would like to thank K. Thomas and G. Flynn for their helpful thoughts and ideas, and J. Warren (Curatorial Staff) for his expertise in handling, rinsing, and mounting of the IDPs. We are also grateful for reviews from F. J .M. Rietmeijer, B. Clark, H. Karlsson, M. Zolensky, and R. Socki. One of us (C.P.H.) would like to thank the National Academy of Sciences/National Research Council's Postdoctoral Program for support during this work.

REFERENCES

Baldwin A. C. and Golden D. M. (1979) Heterogeneous atmospheric reactions: Sulfuric acid aerosols as tropospheric sinks. *Science, 206,* 562-563.

Blanford G. E. and Gibson E. K. Jr. (1988) Laser extracted volatiles from chondritic interplanetary dust particles (abstract). In *Lunar and Planetary Science XIX,* pp. 100-101. Lunar and Planetary Institute, Houston.

Blanford G. E., Thomas K. L., and McKay D. S. (1988) Microbeam analysis of chondritic interplanetary dust particles for carbon, oxygen, and major elements (abstract). In *Lunar and Planetary Science XIX,* pp. 102-103. Lunar and Planetary Institute, Houston.

Bradley J. P., Sandford S. A., and Walker R. M. (1988) Interplanetary dust particles. In *Meteorites and the Early Solar System* (J. F. Kerridge and M. S. Matthews, eds.), pp. 861-895. Univ. of Arizona, Tucson.

Brownlee D. E. (1978) Interplanetary dust: Possible implications for comets and presolar interstellar grains. In *Protostars and Planets* (T. Gehrels, ed.), pp. 134-150. Univ. of Arizona, Tucson.

Farlow N. H., Snetsinger K. G., Hayes D. M., Lem H. Y., and Tooper B. M. (1978) Nitrogen-sulfur compounds in stratospheric aerosols. *J. Geophys. Res., 83,* 6207-6211.

Gibson E. K. Jr. and Carr R. H. (1989) Laser microprobe-quadrupole mass spectrometer system for the analysis of gases and volatiles from geologic materials. In *New Frontiers in Stable Isotope Research: Laser Probes, Ion Probes, and Small Sample Analysis,* pp. 35-49. U.S. Geological Survey Bulletin No. 1890, Reston, Virginia.

Gibson E. K. Jr., Hartmetz C. P., and Blanford G. E. (1989) Analysis of interplanetary dust particles for volatiles and simple molecules (abstract). In *Lunar and Planetary Science XX,* pp. 382-383. Lunar and Planetary Institute, Houston.

Hartmetz C. P., Gibson E. K. Jr., and Blanford G. E. (1988) Comparison of volatiles released from carbonaceous chondrites, and IDPs with Halley cometary volatiles (abstract). *Meteoritics, 23,* 273.

Hartmetz C. P., Gibson E. K. Jr., and Blanford G. E. (1989) In situ analysis of volatile elements and molecules in carbonaceous chondrites (abstract). In *Lunar and Planetary Science XX,* pp. 381-382. Lunar and Planetary Institute, Houston.

Kissel J. and Krueger F. R. (1987) The organic component in dust from Comet Halley as measured by the PUMA mass spectrometer on board Vega 1. *Nature, 326,* 755-761.

Mackinnon I. D. R. and Mogk D. W. (1985) Surface sulfur measurements on stratospheric particles. *Geophys. Res. Lett., 12,* 93-96.

Mackinnon I. D. R. and Rietmeijer F. J. M. (1987) Mineralogy of chondritic interplanetary dust particles. *Rev. Geophys., 25,* 1527-1553.

Molina M. J., Tso T.-L., Molina L. T., and Wang F. C.-Y. (1987) Antarctic stratospheric chemistry of chlorine nitrate, hydrogen chloride, and ice: Release of active chlorine. *Science, 238,* 1253-1257.

Rietmeijer F. J. M. (1987) Silicone oil: A persistent contaminant in chemical and spectral micro-analyses of interplanetary dust particles (abstract). In *Lunar and Planetary Science XVIII,* pp. 836-837. Lunar and Planetary Institute, Houston.

Rietmeijer F. J. M., Mukhin L. M., Fomenkova M. N., and Evlanov E. N. (1989) Layer silicate chemistry in P/comet Halley from PUMA-2 data (abstract). In *Lunar and Planetary Science XX,* pp. 904-905. Lunar and Planetary Institute, Houston.

Stolarski R. S. and Cicerone R. J. (1975) Chlorine in the stratosphere. In *Proc. Conf. Clim. Impact Assess. Program, 4th* (T. M. Hard and A. J. Broderick, eds.), pp. 280-285. National Technical Information, Springfield, Virginia.

Thomas P. J., Chyba C. F., Brookshaw L., and Sagan C. (1989) Impact delivery of organic molecules to the early Earth and implications for the terrestrial origins of life (abstract). In *Lunar and Planetary Science XX,* pp. 1117-1118. Lunar and Planetary Institute, Houston.

Turco R. P., Whitten R. C., and Toon O. B. (1982) Stratospheric aerosols: Observation and theory. *Rev. Geophys. Space Phys., 20,* 233-279.

Wopenka B. (1987) Raman observations of individual interplanetary dust particles (abstract). In *Lunar and Planetary Science XVIII,* pp. 1102-1103. Lunar and Planetary Institute, Houston.

Wood J. A. and Chang S. (1985) The cosmic history of the biogenic elements and compounds. *NASA SP-476.* 80 pp.

Proceedings of the 20th Lunar and Planetary Science Conference, pp. 357-361
Lunar and Planetary Institute, Houston, 1990

Extraterrestrial Halogen and Sulfur Contents of the Stratosphere

S. R. Sutton

Department of the Geophysical Sciences and Center for Advanced Radiation Sources, The University of Chicago, Chicago, IL 60637 and Department of Applied Science, Brookhaven National Laboratory, Upton, NY 11973

G. J. Flynn

Department of Physics, SUNY-Plattsburgh, Plattsburgh, NY 12901

Interplanetary dust represents a potential source of environmentally important chemical species in the Earth's atmosphere. Previous studies have used computational models of atmospheric evolution of meteor debris to conclude that the steady-state stratospheric component of extraterrestrial matter is a small fraction of the total aerosol load. Observational data suggest such calculations may underestimate stratospheric residence times and, thus, concentrations. Two computational methods were employed here to obtain reasonable limits for the stratospheric contents of halogens and sulfur from extraterrestrial sources. The lower limit was based on the total stratospheric aerosol load and the relative influxes from interplanetary dust and tropospheric sources. The upper limit was obtained using a viscous settling method. These results suggest that the steady-state extraterrestrial influxes of halogens are minor compared to tropospheric sources but the sulfur input may be comparable to the present observed stratospheric content. Temporal enhancements in the meteoroid flux, such as passage through comet debris lanes or impact by large bodies, may produce significant chemical perturbations in the atmosphere.

INTRODUCTION

Aerosols (*Pollack et al.,* 1976; *Turco et al.,* 1980) and trace gases (*Ramanathan et al.,* 1985) play important roles in modulating the delicate atmospheric energy balance. Sulfuric acid aerosols of volcanic and anthropogenic origin modify the atmosphere's transmissivity to incoming solar radiation and outgoing radiated energy (*Toon and Farlow,* 1981). Coupled photochemical reactions between chlorinated and brominated molecules have been implicated in the scavenging of ozone (*Prather et al.,* 1984; *Yung et al.,* 1980). Repeated heterogeneous reactions on ice particle surfaces appear to be important in the creation of the Antarctic ozone hole (*Krüger et al.,* 1987; *McElroy et al.,* 1986). Some climatic changes depend on the altitude, e.g., troposphere vs. stratosphere, at which compositional modifications take place (*Wang et al.,* 1980).

Trace gases and aerosols deposited in the atmosphere by vaporization of extraterrestrial materials have been largely dismissed as sources of climatically significant species, a conclusion that is essentially based on the theoretical results of ablation evolution models. There are two reasons that the stratospheric content of these species may be greater than previously inferred. First, direct observations of particulate stratospheric residence times tend to be substantially longer than would be predicted by the models. Second, recent experimental evidence suggests micrometeorites are enriched in volatile elements relative to the compositions of hand specimen-sized meteorites. Consequently, in this paper, we discuss the atmospheric influx of extraterrestrial volatiles, specifically, halogens and sulfur, in the light of these new data.

PREVIOUS STUDIES

Attempts to measure directly the amount of extraterrestrial matter present in the atmosphere are rare. *Shedlovsky and Paisley* (1966) used neutron activation analysis to measure the composition of aerosol-bearing filters collected at an altitude of 20 km. The Fe/Na ratio, 1.2 ± 1.2, was interpreted as being consistent with that of average crustal material implying <10% meteoritic component but the large blank corrections leave this conclusion in some doubt. However, they measured a Co/Fe ratio of 0.006 that is actually closer to the carbonaceous chondrite CI value of 0.003 than to the terrestrial crustal value of 0.0004. In addition, optical scattering data (*Newkirk and Eddy,* 1964) and direct collection of stratospheric particles (*Zolensky et al.,* 1989b) suggest a major meteoric component for particles greater than 1 μm above 20 km. Thus, details of the fate of extraterrestrial material in the atmosphere are poorly understood.

Mathematical models have been developed to predict various properties of meteor debris. *Hunten et al.* (1980) and *Hunten* (1981) examined theoretically the evolution of ablation products including recondensation and vertical transport and concluded that the equilibrium vertical mass distribution implies a dominant meteoritic component in the upper stratosphere and a maximum in particle density near 80 km. These results are supported by condensation nuclei counts, which tend to show particle densities increasing with height above 20 km, consistent with a significant source from the top of the stratosphere (*Käselau et al.,* 1974). The atmospheric chemistry effects of metal vapors (Fe, Ni, Na, K,

Mg, Ca) and residual meteoric particles from large meteorites were examined by *Turco et al.* (1981) who concluded that (1) meteoric debris may dominate the <0.01 μm aerosols above 20 km, (2) chemical reactions are likely to occur on the surfaces of stratospheric particles, and (3) metallic elements derived from ablating meteorites may significantly neutralize sulfuric acid in the stratosphere.

A direct test of the *Hunten et al.* (1980) model of meteor ablation was provided by the 1964 reentry of a SNAP-9A nuclear power generator containing ^{238}Pu. Although the ablation model would suggest a relatively short stratospheric residence time (approximately one month) for this quasimeteoric debris, concentrations of ^{238}Pu in the atmosphere above 12 km were measurable (at levels of the order of 5% of the initial activity) as late as the end of 1970 (six years later) (*Hardy et al.*, 1973). Further evidence that model calculations may underestimate the stratospheric residence of particulate material is the collection of a large, 34 μm radioactive particle five days after the fourteenth Chinese nuclear test explosion in 1972 (*Moore et al.*, 1973). Furthermore, *Clemesha et al.* (1981) have pointed out that experimentally determined sodium ion densities are in poor agreement with those calculated by *Hunten* (1981).

COMPUTATIONAL APPROACH

Based on the apparent disagreement between debris evolutionary models and direct observations, it is clear that the greatest uncertainty in calculations of the extraterrestrial halogen and sulfur contents of the stratosphere concerns the stratospheric residence time of meteor debris. We therefore employed two semi-independent computational approaches to determine reasonable limits for the steady-state stratospheric content of extraterrestrial matter. The lower limit was derived from the equilibrium total aerosol mass predicted by the model of *Hunten et al.* (1980) after folding in the ratio of extraterrestrial to terrestrial influxes to determine a model cosmic stratospheric component. The upper limit was obtained assuming atmospheric settling of smoke-sized particles by a viscous, "Stokes' type," process only, and used the settling time from the vaporization zone to the aerosol layer to estimate the total inventory. These results were then multiplied by the apparent concentrations of halogens and sulfur in interplanetary dust particles (IDPs) and compared with the total observed stratospheric mass fractions of each element.

EXTRATERRESTRIAL MATERIAL INFLUX

The mass distribution of extraterrestrial objects in the vicinity of the Earth-Moon system has been determined from satellite observations for 10^{-12}grams < particle mass < 10^{-6} grams, radio data for 10^{-6}grams < particle mass < 10^{-2} grams and visual observations for m > 10^{-2} grams (*Hughes,* 1978). A total integrated mass influx of 1.6×10^{10} grams/year was used for the present purpose. An important observation is that the particle mass distribution exhibits a sharply peaked profile with the maximum occurring at about 10^{-5} grams, i.e., sand-sized particles. Seventy-five percent of the extraterrestrial mass is contributed by particles between about 10^{-6} and 10^{-2} grams (*Hughes,* 1978).

The fraction of this mass that is sufficiently heated to drive off volatile elements must be considered. The recovery of apparently unmelted micrometeorites from the stratosphere (*Brownlee,* 1985) and Greenland (*Maurette et al.,* 1986) demonstrates that most particles <10 μm survive atmospheric entry as well as some larger particles, presumably those with unusually oblique trajectories and low velocities. The entry heating and survival problem has been studied from a theoretical basis (*Fraundorf,* 1980). Recently, Flynn (G. J. Flynn, unpublished data, 1989) included a more accurate physical model of the atmosphere and considered the effect of particle density. These results indicate that micrometeoroids in the size range represented by the maximum in the mass distribution will be very strongly heated. For example, assuming that these small particles have the same velocity distribution as that measured for radar meteors by *Southworth and Sekanina* (1973), 90% of the 20-μm particles of density 1.0 g/cm^3 will be heated in excess of 800°C. Laboratory heating experiments on meteorites and stratospheric particles indicate that such thermal spikes are sufficiently intense to mobilize the volatile elements considered here (*Fraundorf et al.,* 1982a). This conclusion is supported by chemical analyses on melted micrometeorites collected from the sea sediments (*Bates,* 1986; *Sutton et al.,* 1988). These analyses show that the deep-sea spheres have very low S contents but suggest that they were originally chondritic particles, i.e., S-rich.

The precise magnitude of the heating pulse is sensitive to particle density. Unfortunately, the density of large IDPs is presently the subject of considerable debate. Synchrotron XRF measurements (*Sutton and Flynn,* 1988) and direct weighings (*Fraundorf et al.,* 1982b) suggest densities between 0.7 and 2.2 g/cm^{-3}. Lunar crater morphologies are consistent with densities between 1 and 7 g/cm^3 and reveal no low density events. However, estimates as low as 0.1 g/cm^3 are suggested by radar meteor data (*Verniani,* 1969) and porosity estimates of microtomed sections (*Mackinnon et al.,* 1987). The important point in the present context is that the small surviving particles <10 μm can certainly be no more massive than 10^{-8} g and therefore represent a negligible fraction of the total mass influx. The mass influx of unmelted larger particles with unusual orbital parameters is difficult to estimate. *Maurette et al.* (1986) estimate that about 25-30% of the Greenland chondritic particles up to 300 μm are "unmelted." Ablation of ice coatings is one possible mechanism for reducing atmospheric heating. However, it is likely that even these have been sufficiently heated to drive off a significant fraction of the highly volatile elements considered here. Based on these considerations, we estimate that volatile elements from about 10^{10} grams of extraterrestrial matter are currently deposited into the atmosphere each year.

HALOGEN AND SULFUR CONTENTS OF INTERPLANETARY DUST

Recent advances in microanalytical techniques, laboratory, and spacecraft-supported instruments have resulted in direct measurements of the composition of interplanetary dust and cometary debris. Stratospheric micrometeorites (≈10 μm in size) have been collected by high-flying aircraft (*Brownlee,* 1985) and analyzed individually by electron microprobe

(*Schramm et al.,* 1989), neuron activation (INAA; *Ganapathy and Brownlee,* 1979; *Zolensky et al.,* 1989a), particle-induced X-ray emission (PIXE; *van der Strap et al.,* 1986) and synchrotron X-ray fluorescence (SXRF; *Sutton and Flynn,* 1988; *Flynn and Sutton,* 1990). The first-order conclusion is that interplanetary dust particles have compositions similar to those of CI or CM carbonaceous meteorites (*Schramm et al.,* 1989) as given by *Anders and Ebihara* (1982). This conclusion is supported by the inferred, CI-chondrite-like meteoritic component of lunar soil (*Anders et al.,* 1973). Sulfur contents of IDPs are about 30% lower than the CI value (i.e., 4.0%; *Schramm et al.,* 1989). However, other volatile elements are the notably enriched. *Van der Stap et al.* (1986) observed about twofold enrichments relative to CI in Ge, Zn, and Se, while *Sutton and Flynn* (1988) found comparable enrichments in these elements plus Ga. Zinc was also enriched in two particles studied by INAA. Most significantly, all 14 particles analyzed for Br were enriched in this element by factors of 1.3 to 40. Although the possibility remains that the high volatile contents of stratospheric cosmic dust have a terrestrial origin (stratospheric or collection/handling contamination), it is more likely that these are inherent cosmochemical signatures. Indirect evidence supporting a preterrestrial source includes the high Br content observed in peat layers associated with the Tunguska event (*Golenetskii et al.,* 1977) and the large volatile component (so-called CHON particles) detected in the Comet Halley debris by the Giotto and Vega mass spectrometers (*Brownlee et al.,* 1987). Chlorine detected by one Vega instrument could indicate enriched halogens in cometary material but instrumental contamination has also been suggested.

PIXE and SXRF (*Wallenwein et al.,* 1989) analyses on four chondritic IDPs, Bounce, Zodiac, SP-85, and SP-87, argue against a general halogen enrichment indicating CI-equivalent Cl contents but large enrichments in Br.

No data exist for fluorine and iodine in individual micrometeorites from the stratosphere. The best analog materials are carbonaceous chondrites. Two sets of fluorine measurements, one using the reaction $^{19}F(p,\alpha\gamma)^{16}O$ (*Goldberg et al.,* 1974) and the other using RNAA (*Dreibus et al.,* 1979), suggest a concentration of about 60 ppm. Iodine contents in carbonaceous meteorites vary from 0.58 ppm (Orgueil) to 1.73 ppm (Ivuna), presumably reflecting alteration (*Anders and Ebihara,* 1982). We adopt a mean I value of 1 ppm for the present purpose.

STRATOSPHERIC CONTENT OF EXTRATERRESTRIAL MATTER

Vaporization products probably recondense into "smoke" particles with a mean size of about 1 nm at an altitude of about 80 km (*Hunten et al.,* 1980). Thus, the deposition of meteoric debris occurs well above the main concentration layer of stratospheric aerosols, 20 km. The effective descent rate of these products depends on the importance of physical processes such as coagulation and eddy diffusion that are difficult to characterize quantitatively. Because of the large uncertainties in the magnitudes of these effects, we employed two methods, one based on model aerosol load and relative tropospheric and extraterrestrial influxes and the second assuming viscous, "Stokes' type," settling only.

Relative Aerosol Influx Method

Total aerosol load in the stratosphere reaches a maximum at about 20 km altitude where the mass fraction is about 1.2 ppb (*Hunten et al.,* 1980). The extraterrestrial fraction of this material is approximately 0.05 based on the relative mass influxes from extraterrestrial and tropospheric sources [1×10^{-16} g/cm^2 sec^{-1}, see above; and 2×10^{-15} g/cm^2 sec^{-1} (*Hunten et al.,* 1980), respectively]. The present total meteoric mass fraction is therefore about 60 ppt.

Viscous Settling Method

The largest estimated inventory results when coagulation and atmospheric stirring by turbulence are negligible since, in this case, settling obeys Stokes' Law and the descent times are the longest. Descent velocities for spherical aerosols have been tabulated by *Kasten* (1968) as functions of altitude and particle radius. Plots of the logarithm of falling speed vs. altitude are very nearly linear. Velocity w for 1 nm particles of density 1.0 g/cm^3 are given approximately by

$$w\ (km/sec) = 10^{(z-176)/16.2}$$

for altitude z in kilometers. Total descent times to reach 40, 30, and 20 km, found by integrating the inverse of velocity down to each altitude, are 50, 250, and 1000 years, respectively. This approach, which assumes negligible coagulation for 1-nm spheres, predicts that the lower stratosphere contains about 10^3 years accumulated inventory of extraterrestrial material. The corresponding mass fraction is 30 ppb, 500 times greater than the estimate above. Viscous settling times for real smoke particles may be much longer (i.e., larger stratospheric contribution) than those for the spheres considered here since such particles produced in the laboratory are porous, low density, and exhibit fractal structures (*Donn,* 1986).

STRATOSPHERIC CONTENTS OF EXTRATERRESTRIAL HALOGENS AND SULFUR

The estimated steady-state halogen and sulfur contents of the lower stratosphere from meteors are given in Table 1 for these two methods assuming the compositions above. Comparison with observed mass fractions for these elements shows that the extraterrestrial halogen components are at the few percent level by the viscous settling method and in the range 10 to 100 ppm by the relative influx method. The large differences between the results of the two approaches emphasizes the sensitivity of these estimations to the magnitudes of evolutionary processes such as coagulation and turbulence.

The results suggest that extraterrestrial sources of stratospheric halogens are presently minor compared to the tropospheric influx. The sulfur result is perhaps the most significant since the estimated content from extraterrestrial sources is roughly equivalent to the observed stratospheric content. The atmospheric injection of sulfur in this way may account for the unexpectedly high atmospheric content of

TABLE 1. Extraterrestrial halogen and sulfur contents of the stratosphere.

Element	Method	Steady State Mass (grams)	Extraterrestrial Fractional Mass*	Observed Fractional Mass†	Extraterrestrial/ Observed
F	Viscous‡	6×10^{8}	2×10^{-12}	6×10^{-11}	3×10^{-2}
	Rel. Influx§	1×10^{6}	4×10^{-15}		7×10^{-5}
Cl	Viscous‡	7×10^{9}	2×10^{-11}	2×10^{-9}	1×10^{-2}
	Rel. Influx§	1×10^{7}	4×10^{-14}		2×10^{-5}
Br	Viscous‡	1×10^{9}	3×10^{-12}	5×10^{-11}	6×10^{-2}
	Rel. Influx§	2×10^{6}	6×10^{-15}		1×10^{-4}
I	Viscous‡	4×10^{6}	1×10^{-14}	1×10^{-11}	1×10^{-3}
	Rel. Influx§	1×10^{4}	3×10^{-17}		3×10^{-6}
S	Viscous‡	4×10^{11}	2×10^{-9}	7×10^{-10}	2
	Rel. Influx§	1×10^{9}	3×10^{-12}		4×10^{-3}

*Mass of stratosphere = 3.3×10^{20} grams.
†*Cicerone* (1981); *McElroy et al.* (1986).
‡Assumes Stokes' type settling only.
§Based on model aerosol load (*Hunten et al.,* 1980) and relative extraterrestrial and tropospheric influxes.

sulfur dioxide, a very short residence time species (*Ramanathan et al.,* 1985).

Temporal enhancements in meteoroid influx will increase the stratospheric content of these elements in the same manner as volcanic eruptions. Short-term excursions, such as passage of the Earth through cometary debris lanes, are likely to temporarily increase the halogen and sulfur content of the stratosphere but long-lived perturbations are also possible. We have assumed the present day extraterrestrial mass influx of 10^{10} g/yr. However, the long term average flux inferred from the Ir abundance of sea sediments is about fivefold greater. The present total halogen content of the stratosphere is equivalent to that contained in a single carbonaceous chondritic body about 100 m in size, comparable to the estimated size of the Tunguska meteoroid (*Ganapathy,* 1981). Terrestrial impacts of moderately-sized bodies may, therefore, have profound effects on the chemistry of the Earth's atmosphere from the standpoints of acid rain production and increased contents of greenhouse gases and ozone scavengers. Direct measurements of halogens and sulfur in the upper stratosphere are needed emphasizing the <1-μm size fraction.

Acknowledgments. This work was supported by National Aeronautics and Space Administration grants NAG 9-106 and NAG 9-257.

REFERENCES

Anders E. and Ebihara M. (1982) Solar system abundances of the elements. *Geochim. Cosmochim. Acta, 46,* 2363-2380.

Anders E., Ganapathy R., Krähenbühl U., and Morgan J. (1973) Meteoritic material on the moon. *Moon, 8,* 3-24.

Bates B. A. (1986) The elemental composition of stony extraterrestrial particles from the ocean floor. Ph.D. thesis, Univ. Washington, Seattle. 199 pp.

Brownlee D. (1985) Cosmic dust: collection and research. In *Annu. Rev. Earth Planet. Sci., 13,* 147-173.

Brownlee D. E., Wheelock M. M., Temple S., Bradley J. P., and Kissel J. (1987) A quantitative comparison of Comet Halley and carbonaceous chondrites at the submicron level (abstract). In *Lunar and Planetary Science XVIII,* pp. 133-134. Lunar and Planetary Institute, Houston.

Cicerone R. J. (1981) Halogens in the atmosphere. *Rev. Geophys. Space Phys., 19,* 123-139.

Clemesha B. R., Kirchhoff V. W. J. H., and Simonich D. M. (1981) Comments on "A meteor-ablation model of the sodium and potassium layers" by D. M. Hunten. *Geophys. Res. Lett., 8,* 1023-1025.

Donn B. (1986) Grain formation and accretion: Initial stages (abstract). In *Lunar and Planetary Science XVIII,* pp. 243-244. Lunar and Planetary Institute, Houston.

Dreibus G., Spettel B., and Wänke H. (1979) Halogens in meteorites and their primordial abundances. In *Origin and Distribution of the Elements* (L. H. Ahrens, ed.), pp. 33-38. Pergamon, New York.

Flynn G. J. and Sutton S. R. (1990) Synchrotron x-ray fluorescence analyses of stratospheric cosmic dust: New results for chondritic and low nickel particles. *Proc. Lunar Planet. Sci. Conf. 20th,* this volume.

Fraundorf P. (1980) The distribution of temperature maxima for micrometeorites in the Earth's atmosphere without melting. *Geophys. Res. Lett., 10,* 765-768.

Fraundorf P., Brownlee D. E., and Walker R. M. (1982a) Laboratory studies of interplanetary dust. In *Comets* (L. L. Wilkening, ed.), pp. 393-409. Univ. of Arizona, Tucson.

Fraundorf P., Hintz C. Lowry O., McKeegan K. D., and Sandford S. A. (1982b) Determinations of the mass, surface density and volume of interplanetary dust particles (abstract). In *Lunar and Planetary Science XIII,* pp. 225-226. Lunar and Planetary Institute, Houston.

Ganapathy R. (1981) The Tunguska explosion of 1908: Discovery of meteoritic debris near the explosion site and at the South Pole. *Science, 220,* 1158.

Ganapathy R. and Brownlee D. E. (1979) Interplanetary dust: trace element analyses of individual particles by neutron activation. *Science, 206,* 1075-1077.

Goldberg R. H., Burnett D. S., Furst M. J., and Tombrello T. A. (1974) Fluorine concentrations in carbonaceous chondrites. *Meteoritics, 9,* 347-348.

Golenetskii S. P., Stepanok V. V., Kolesnikov E. M., and Murashov D. A. (1977) The question about the chemical composition and nature of the Tunguska cosmic body. *Sol. Syst. Res., 11,* 103-113.

Hardy E. P., Krey P. W., and Volchok H. L. (1973) Global inventory and distribution of fallout plutonium. *Nature, 241,* 444-445.

Hughes D. (1978) Meteors. In *Cosmic Dust* (J. A. M. McDonnell, ed.), pp. 123-185. Wiley, New York.

Hunten D. M. (1981) A meteor-ablation model of the sodium and potassium layers. *Geophys. Res. Lett., 8,* 369-372.

Hunten D. M., Turco R. P., and Toon O. B. (1980) Smoke and dust particles of meteoric origin in the mesosphere and stratosphere. *J. Atmos. Sci., 37,* 1342-1357.

Käselau K. H., Fabian P., and Röhrs H. (1974) Measurements of aerosol concentration up to a height of 27 km. *Pure Appl. Geophys., 112,* 877-885.

Kasten F. (1968) Falling speed of aerosol particles. *J. Appl. Meteor., 7,* 944-947.

Krüger B. C., Wang G., and Fabian P. (1987) The Antarctic ozone depletion caused by heterogeneous photolysis of halogenated hydrocarbons. *Geophys. Res. Lett., 14,* 523-526.

Mackinnon I. D. R., Lindsay C., Bradley J. P., and Yatchmenoff B. (1987) Porosity of serially sectioned interplanetary dust particles. *Meteoritics, 22,* 450-451.

Maurette M., Hammer C., Brownlee D. E., Reeh N., and Thomsen H. H. (1986) Placers of cosmic dust in the blue ice lakes of Greenland. *Science, 233,* 869-872.

McElroy M. B., Salawitch R. J., Wofsy S. C. and Logan J. A. (1986) Reductions of Antarctic ozone due to synergistic interactions of chlorine and bromine. *Nature, 321,* 759-762.

Moore D. T., Beck J. N., Miller D. K., and Kuroda P. K. (1973) Radioactive hot particles from the recent Chinese nuclear weapons tests. *J. Geophys. Res., 78,* 7039-7050.

Newkirk G. Jr. and Eddy J. A. (1964) Light scattering by particles in the upper atmosphere. *J. Atmos. Sci., 21,* 35-60.

Pollack J. B., Toon O. B., Sagan C., Summers A., Baldwin B., Van Camp W. (1976) Volcanic explosions and climatic change: A theoretical assessment. *J. Geophys. Res., 81,* 1071-1083.

Prather M. J., McElroy M. B., and Wofsy S. C. (1984) Reductions in ozone at high concentrations of stratospheric halogens. *Nature, 312,* 227-231.

Ramanathan V., Cicerone R. J., Singh H. B., and Kiehl J. T. (1985) Trace gas trends and their potential role in climate change. *J. Geophys. Res., 90,* 5547-5566.

Schramm L. S., Brownlee D. E., and Wheelock M. M. (1989) Major element composition of stratospheric micrometeorites. *Meteoritics, 24,* 99-112.

Shedlovsky J. P. and Paisley S. (1966) On the meteoritic component of stratospheric aerosols. *Tellus, XVIII,* 501-503.

Southworth R. B. and Sekanina Z. (1973) Physical and dynamical studies of meteors. *NASA CR-2316.*

Sutton S. R. and Flynn G. J. (1988) Stratospheric particles: Synchrotron X-ray fluorescence analysis of trace element contents. *Proc. Lunar Planet. Sci. Conf. 18th,* pp. 607-614.

Sutton S.R. and Flynn G.J. (1989) Density estimates for eleven cosmic dust particles based on synchrotron x-ray fluorescence analyses. In *Abstracts for the 52nd Annual Meeting of the Meteoritical Society,* p. 235. Lunar and Planetary Institute, Houston.

Sutton S. R., Herzog G., and Hewins R. (1988). Chemical fractionation trends in deep sea spheres. *Meteoritics, 23,* 304.

Toon O. B. and Farlow N. H. (1981) Particles above the tropopause. In *Annu. Rev. Earth Planet. Sci., 9,* 19-58.

Turco R. P., Whitten R. C., Toon O. B., Pollack J. B., and Hamill P. (1980) OCS, stratospheric aerosols and climate. *Nature, 283,* 283-286.

Turco R. P., Toon O. B., Hamill P., and Whitten R. C. (1981) Effects of meteoric debris on stratospheric aerosols and gases. *J. Geophys. Res., 86,* 1113-1128.

van der Stap C. C. A. H., Vis R. D., and Verheul H. (1986) Interplanetary dust: arguments in favour of a late stage nebular origin (abstract). In *Lunar and Planetary Science XVII,* pp. 1013-1014. Lunar and Planetary Institute, Houston.

Verniani F. (1969) Structure and fragmentation of meteoroids. *Space Sci. Rev., 10,* 230-261.

Wallenwein R., Antz Ch., Jessberger E. K., Buttewitz A., Knöchel A., Traxel K., and Bavdaz M. (1989). Multielement analyses of interplanetary dust with PIXES and SYXFA (abstract). In *Lunar and Planetary Science XX,* pp. 1171-1172. Lunar and Planetary Institute, Houston.

Wang W. C., Pinto J. P., and Yung Y. L. (1980) Climatic effects due to halogenated compounds in the Earth's atmosphere. *J. Atmos. Sci., 37,* 333-338.

Yung Y. L., Pinto J. P., Watson R. T., and Sander S. P. (1980) Atmospheric bromine and ozone perturbations in the lower stratosphere. *J. Atmos. Sci., 37,* 339-353.

Zolensky M. E., Lindstrom D. J., Thomas K. L., Lindstrom R. M., and Lindstrom M. M. (1989a) Trace element compositions of six "chondritic" stratospheric dust particles (abstract). In *Lunar and Planetary Science XX,* pp. 1255-1256. Lunar and Planetary Institute, Houston.

Zolensky M. E., McKay D. S., and Kaczor L. A. (1989b) A tenfold increase in the abundance of large solid particles in the stratosphere, as measured over the period 1976-1984. *J. Geophys. Res., 94,* 1047-1056.

Proceedings of the 20th Lunar and Planetary Science Conference, pp. 363-371
Lunar and Planetary Institute, Houston, 1990

The Near-Earth Enhancement of Asteroidal over Cometary Dust

G. J. Flynn

Department of Physics, State University of New York-Plattsburgh, Plattsburgh, NY 12901

The cosmic dust particles collected from the stratosphere in the NASA Cosmic Dust Collection Program are samples of the zodiacal cloud. Whipple's (1967) suggestion that interplanetary dust is derived mainly from comets, and particularly from an earlier active phase of Comet Encke, coupled with the assumption that stratospheric cosmic dust is an unbiased sample of the interplanetary dust at 1 A.U., has been used to suggest that the cosmic dust particles collected from the stratosphere are samples of comets. Gravitational focusing, which is dependent on the velocity of the incoming particles, is shown to bias all near-Earth collections in favor of those components of the interplanetary dust having low geocentric velocity. Atmospheric entry heating further biases the stratospheric dust sample, since dust having a high atmospheric entry velocity is more likely to be melted or volatilized. These two effects bias the stratospheric collections in favor of dust collected from nearly circular heliocentric orbits of low inclination. These are the orbits expected for dust derived from mainbelt asteroidal parents and evolving to Earth under Poynting-Robertson drag. The ratio of the penetration rates of micrometeoroid detectors near the Earth and near the Moon provides a measure of the gravitational focusing and a constraint on the velocity distribution of the cosmic dust particles. The measured ratio is shown to be inconsistent with the majority of the stratospheric cosmic dust particles being samples of Comet Encke, or of other comets with perihelia <1.2 A.U. The observed near-Earth enhancement of micrometeoroids indicates that a low-velocity component of the interplanetary dust exists, and that this component is substantially over-represented in near-Earth collections, both from the stratosphere and on impact collectors in Earth orbit.

INTRODUCTION

The zodiacal light cloud is composed of small particles orbiting about the sun. Because of their small size, Poynting-Robertson drag (PR drag) causes the orbits of these particles to decay into the sun on time scales short compared to the age of the solar system (*Dohnanyi,* 1978). Thus, sources active in the recent past are required to resupply the zodiacal cloud. In recent years, the NASA Cosmic Dust Collection Program has provided samples of the zodiacal cloud. These cosmic dust particles are micrometeorites small enough (generally 5 to 100 μm in diameter) to decelerate in the upper atmosphere without being heated to their melting temperature (see *Whipple,* 1950). The particles are then collected from the stratosphere, where their low settling rate causes them to be concentrated. The atmospheric entry process, collection procedure, and particle properties are reviewed by *Fraundorf et al.* (1982a), *Brownlee* (1985), *Sandford* (1987), and *Bradley et al.* (1988).

Those cosmic dust particles that exhibit major element abundances similar to the primitive CI meteorites have been termed "chondritic." Although similar in major element composition to the CI meteorites, many chondritic cosmic dust particles exhibit one or more of the following features: (1) enrichments in the volatile elements above CI (*van der Stap et al.,* 1986), (2) unmetamorphosed mineral assemblages (*Fraundorf,* 1981), (3) large deuterium to hydrogen isotopic fractionations (*Zinner et al.,* 1983), and (4) bulk M_g^{26} isotopic excesses (*Esat et al.,* 1979).

All these observations suggest the cosmic dust particles are more primitive samples of the early solar system than are the CI meteorites. Thus, some of the parent bodies of the cosmic dust particles must themselves be composed of very primitive materials. The parent bodies of these particles are unknown.

In principle, almost every solar system object is able to contribute to the interplanetary dust environment by outgassing, ejection due to cratering events, collisional fragmentation, tidal disruption, or volcanic activity. It has been proposed that comets are the major contributor of dust to the inner solar system (*Whipple,* 1967; *Millman,* 1972; *Delsemme,* 1976; *Dohnanyi,* 1976). However, *Kresak* (1980) has shown that less than 2% of the dust required to maintain the observed zodiacal cloud can be supplied by presently active comets. *Whipple* (1967) suggests that Comet Encke may have been more active in the past, supplying the required dust during that extremely active phase. Thus, the extent of the cometary contribution to the zodiacal cloud is, at present, uncertain.

The Infrared Astronomical Satellite (IRAS) has identified a second major source of dust in the solar system: the main asteroid belt (*Low et al.,* 1984). The dust bands in the mainbelt are suggested to result from collisional fragmentation among the asteroids (*Sykes and Greenberg,* 1986). These collisions are likely to produce particles in the cosmic dust size range that would contribute to the zodiacal cloud. *Zook and McKay* (1986) suggest that main belt asteroids may contribute a major fraction of the interplanetary dust smaller than 50 μm in size.

Whipple's (1967) suggestion that the interplanetary dust cloud is mainly cometary has been used to suggest a cometary origin for most of the stratospheric cosmic dust (e.g., *Brownlee et al.,* 1976; *Fraundorf et al.,* 1982a). This follows from the assumption that the stratospheric cosmic dust is an unbiased sample of the interplanetary dust cloud at 1 A.U. If true, then the analysis of the cosmic dust particles collected from the stratosphere would provide information on the composition and properties of the comets themselves.

However, the stratospheric cosmic dust is a biased sample of the zodiacal cloud. Factors contributing to this bias include

(1) gravitational focusing, which significantly increases the near-Earth collection probability for dust of low geocentric velocity over that for dust of higher geocentric velocity (*Opik,* 1951), (2) atmospheric entry heating, which removes some of the dust with high entry velocity by melting or volatilization (*Whipple,* 1950), (3) atmospheric fragmentation, which may selectively remove the most fragile dust at any given entry velocity, and (4) stratospheric settling, which will preferentially concentrate dust of lower density and/or higher drag at the collection altitude.

The first two factors each bias the stratospheric collections in favor of the low velocity component of the interplanetary dust. Other near-Earth collections, such as micrometeorite impact collectors on the Long Duration Exposure Facility and the proposed Space Station Cosmic Dust Collection Facility, are also biased by the gravitational focusing effect. Thus, the stratospheric cosmic dust particles, as well as the particles collected by impact on Earth-orbiting satellites, preferentially sample particles collected from nearly circular heliocentric orbits of low inclination.

Dust derived from mainbelt asteroidal parent bodies will evolve, under PR drag, into low eccentricity orbits of low inclination at the Earth collection opportunity (*Flynn,* 1989). *Zook* (1980) has calculated that a pseudo-PR drag resulting from interactions between the dust particles and solar wind protons reduces the time scales by 30% from those calculated with only PR drag, but the orbital parameters evolve in the same manner. As a result, the geocentric velocity of the particles derived from main belt asteroids will be low, and proportion of this material will be enhanced on near-Earth collections over its in-space proportion.

Particles derived from comets also evolve, under PR drag, from their initial orbit to an orbit from which Earth collection is possible. However, the initial orbit of most cometary dust is more eccentric than the main belt asteroidal particles. Though the effect of PR drag is to reduce the eccentricity as the orbit of the particle decays toward the sun, most cometary particles will still retain considerable eccentricity at their Earth collection opportunity. This gives rise to a higher geocentric velocity and a lower near-Earth gravitational enhancement for the cometary particles than for the mainbelt asteroidal particles (*Flynn,* 1989).

The extent of the near-Earth enhancement, as determined by satellite dust detectors orbiting the Earth and the Moon, will be used to constrain the velocity distribution of the interplanetary dust at 1 A.U. This observed enhancement will also provide indications of the proportions of asteroidal and cometary material in the stratospheric and Earth-orbiting collections, and in the zodiacal cloud.

GRAVITATIONAL FOCUSING

A model to determine the collection probability for a small particle on an intersecting orbit with a planet was developed by *Opik* (1951). The dominant factor in determining the collection probability is gravitational focusing, which increases the effective Earth collection cross-section for particles arriving with lower geocentric velocities. This cross-section, σ, is given by *Opik* (1951) as

$$\sigma = \pi R_e^2 \, [1 + (V_e/V)^2] \qquad (1)$$

where R_e is the radius of the Earth, V_e is the escape velocity of the Earth, and V is the geocentric velocity of the dust particle at the collection opportunity but prior to gravitational infall acceleration (i.e., at "infinity" relative to the Earth's gravitational field).

Since πR_e^2 is the physical cross-section of the Earth, we can define the gravitational enhancement factor as the ratio $\sigma/\pi R_e^2$. Values of the gravitational enhancement factor for Earth are given in Table 1 for various geocentric velocities. The gravitational enhancement factor for particles of extremely low velocity can be more than a factor of 100 greater than for high-velocity particles.

TABLE 1. Near-Earth gravitational enhancements.

Geocentric Velocity	Enhancement Factor
1 km/sec	126
2 km/sec	32
3 km/sec	15
4 km/sec	9
5 km/sec	6
8 km/sec	3
10 km/sec	2
15 km/sec	1.5
20 km/sec	1.3
30 km/sec	1.1
50 km/sec	1.0

Since some meteor streams are identified with presently active comets, they are generally accepted to be debris given off by comets, and they provide an indication of the gravitational enhancement expected for material from cometary sources. The particles in these streams have evolved from the orbit of the parent comet to an Earth-intersecting orbit from which Earth collection is possible. Under the influence of PR drag, the smaller cosmic dust particles will arrive at Earth with encounter geometries that are often not greatly different from larger meteors from the same initial orbit. The smaller particles will, however, evolve to that collection orbit more rapidly than the larger particles. Thus, the measured atmospheric entry velocities of the stream meteors should be generally representative of the smaller cosmic dust particles from those same sources.

Radiation pressure is, however, increasingly effective in altering the initial orbit of a particle ejected from its parent comet as the particle size decreases (*Burns et al.,* 1979; *Kresak,* 1980). Thus, inclusion of the radiation pressure effect results in the initial orbits of the smaller cosmic dust particles being somewhat more eccentric than those of the larger stream meteors. Orbital evolution, under PR drag, then results

TABLE 2. Near-Earth enhancement for particles from meteor streams.

Meteor Shower/Comet	Geocentric Velocity*	Near-Earth Enhancement
Orionids/Halley	66.4 km/sec	1.03
April Lyrids/Thatcher	47.6 km/sec	1.05
τ Herculids/Schwassmann-Wachmann 3	15 km/sec	1.55
Northern Taurids/Encke	29.2 km/sec	1.15
Leonids/Temple-Tuttle	70.7 km/sec	1.03
October Draconids/Giacobini-Zinner	20.4 km/sec	1.30
Andromedids/Biela	16.5 km/sec	1.46
o Draconids/Metcalf	23.6 km/sec	1.22
June Bootids/Pons-Winnecke	13.9 km/sec	1.65
Pegasids/Blanplain	11.2 km/sec	2.00

*Data from *Cook* (1973).

in the cosmic dust particles from the meteor stream sources being collected from orbits of similar, but slightly larger, geocentric velocity than the larger stream meteors. The effects of ejection velocity, which may be size dependent, have not been considered; however, the existence of meteor streams indicates that many particles are ejected from comets with low relative velocities.

To illustrate the magnitude of the expected near-Earth gravitational enhancement for cosmic dust particles from the meteor stream parent bodies, the gravitational enhancement factors for particles collected from the orbits of a variety of meteor streams are given in Table 2. The parent comet associated with each stream is also indicated. If the comets responsible for the meteor streams were the major suppliers of dust to the interplanetary medium, the near-Earth gravitational enhancement of the dust collected at Earth would be less than 2.0, the maximum enhancement seen for a meteor stream. If as proposed by *Whipple* (1967), the major source of interplanetary dust was Comet Encke, then the near Earth enhancement would be only 1.15, the value for the Taurid stream associated with Comet Encke.

NEAR-EARTH ENHANCEMENT

No direct measurement of the near-Earth enhancement of the micrometeorite flux over its in-space at 1 A.U. value is available. However, the near-Earth and near-Moon fluxes can be compared in the cosmic dust size range.

Micrometeorite detectors of similar design were flown on two Earth-orbiting satellites, Explorer XVI and Explorer XXIII, as well as on the five Lunar Orbiters. Each satellite carried pressure-cell penetration detectors, having 25-μm-thick walls. These detectors were sensitive to particles from 2×10^{-9} g upward. Because of the steeply decreasing number of micrometeorites with increasing mass (*Hughes,* 1978), most of the penetration events were probably produced by particles within an order of magnitude of the minimum sensitive mass. Thus, the satellite detectors provided flux measurements for particles in the cosmic dust mass range.

While gravitational focusing significantly enhances the near-Earth flux of particles with low geocentric velocities, the near-Moon flux is almost unaltered, except for particles with velocities below about 3 km/sec, because of the lower lunar escape velocity. Thus, a comparison of the near-Earth and near-Moon fluxes provides an indication of the near-Earth enhancement over the interplanetary flux at 1 A.U.

The penetration rates for the near-Earth satellites, corrected for Earth and satellite shielding, are 0.445 events/m^2·day for Explorer XVI and 0.526 events/m^2·day for Explorer XXIII (*Naumann,* 1966). The average near-Moon flux from the five Lunar Orbiters, also corrected for shielding, is 0.19 events/m^2·day (*Grew and Gurtler,* 1971). A comparison of the flux measurements gives a near-Earth enhancement factor of 2.3 times the lunar flux for the ratio of the Explorer XVI to Lunar Orbiter penetration rates or 2.8 for Explorer XXIII ratioed to Lunar Orbiter rates. Averaging the two Explorer Satellites gives a mean near-Earth enhancement over the lunar flux of 2.6.

The actual enhancement could be larger since the Lunar Orbiter data may include events from lunar impact ejecta, which can briefly enhance the particle flux hundreds of kilometers above an impact event by factors of 10^2 to 10^3 over the interplanetary flux (*McDonnell,* 1978). Significant dust flux enhancements correlated with known meteor showers were detected near the Moon by a sensor with a 5-picogram minimum detection limit on Lunar Explorer 35 (*Alexander et al.,* 1970; *Alexander and Hyde,* 1989). Estimates of the flux of secondary meteoroids indicate that it should drop off sharply with altitude (*Gault et al.,* 1963). *Grew and Gurtler* (1971) examined the Lunar Orbiter penetration rate as a function of altitude above the lunar surface and found no detectable variation, suggesting the effect of secondary meteoroids on the total flux in the size range measured by Lunar Orbiter is small.

The statistical uncertainty in the near-Earth enhancement is significant. Only 22 penetration events were detected by the Lunar Orbiters, giving rise to a 95% confidence limit range on the shielding corrected penetration rate of from 0.12 to 0.32 events/m^2·sec (*Grew and Gurtler,* 1971). The Earth orbiting satellites recorded a total of 94 penetration events, 44 by Explorer XVI and 50 by Explorer XXIII (*Naumann,* 1966), giving a statistical uncertainty of approximately 30% in the near-Earth flux. Because of the low counting statistics, the near-Earth enhancement could vary from 1.2 to 5.0.

Grew and Gurtler (1971) have considered a variety of factors including counting statistics, errors in calculation of the shielding factors, and possible temporal variations in the micrometeoroid flux. They concluded that the most plausible explanation for the difference between the near-Earth and near-Moon flux measurements was the gravitational enhancement effect. *Grew and Gurtler* (1971) did not consider the increase in the near-Earth penetration rate that results from the higher mean velocity, due to gravitational infall acceleration, which allows somewhat smaller particles to penetrate the near-Earth puncture detectors than similar detectors in lunar orbit. However, it seems unlikely that this effect could be responsible for the large near-Earth enhancement they report.

To produce a near-Earth gravitational enhancement over the

lunar flux of 2.6, a substantial low-velocity component of the interplanetary dust is required. If the velocity distribution of the interplanetary dust is such that the near-Earth enhancement over the lunar flux is of 2.6, there will also be small (probably of order 5% to 10%) near-Moon enhancement over the interplanetary flux. The actual lunar enhancement can only be determined if the micrometeorite velocity distribution is known, but it will be much smaller than the Earth enhancement because of the much lower lunar escape velocity. Estimating the near-Moon enhancement at 8% raises the near-Earth enhancement over the interplanetary space flux to 2.5 for Explorer XVI and 3.0 for Explorer XXIII, giving a mean enhancement near Earth over the interplanetary flux of 2.8.

If, as suggested by *Whipple* (1967), most of the interplanetary dust came from the single source, Comet Encke, then the geocentric velocity at Earth collection would be similar to that of the Encke-associated meteor stream, the Taurids. In this case, the near-Earth gravitational enhancement factor would be 1.15. The large difference between the inferred enhancement factor of approximately 2.8 and the enhancement expected for material from Comet Encke suggests that Encke is not the sole contributor to the stratospheric or the zodiacal cosmic dust. Even when the large uncertainty in the near-Earth enhancement due to counting statistics is considered, the satellite penetration results are inconsistent with the low enhancement factor expected for material from Comet Encke.

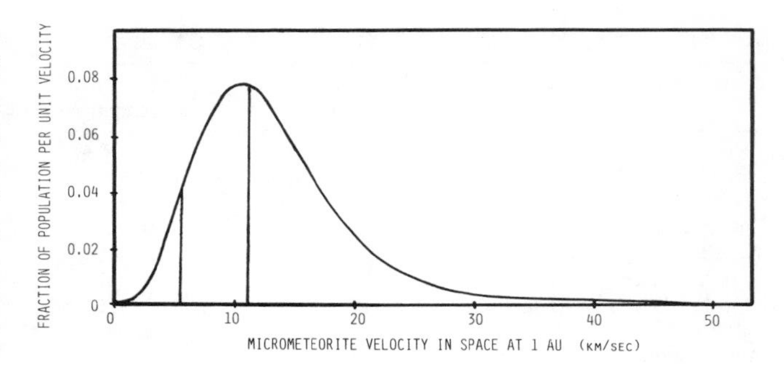

Fig. 2. The micrometeorite velocity distribution in interplanetary space was derived from Fig. 1 by reducing each velocity to account for gravitational infall and by removing the near-Earth gravitational enhancement. The vertical lines at 5.5 km/sec and 11 km/sec divide the particles into the same source types indicated in Fig. 1 after removal of the Earth infall acceleration effect. (Data from H. A. Zook, personal communication, 1988.)

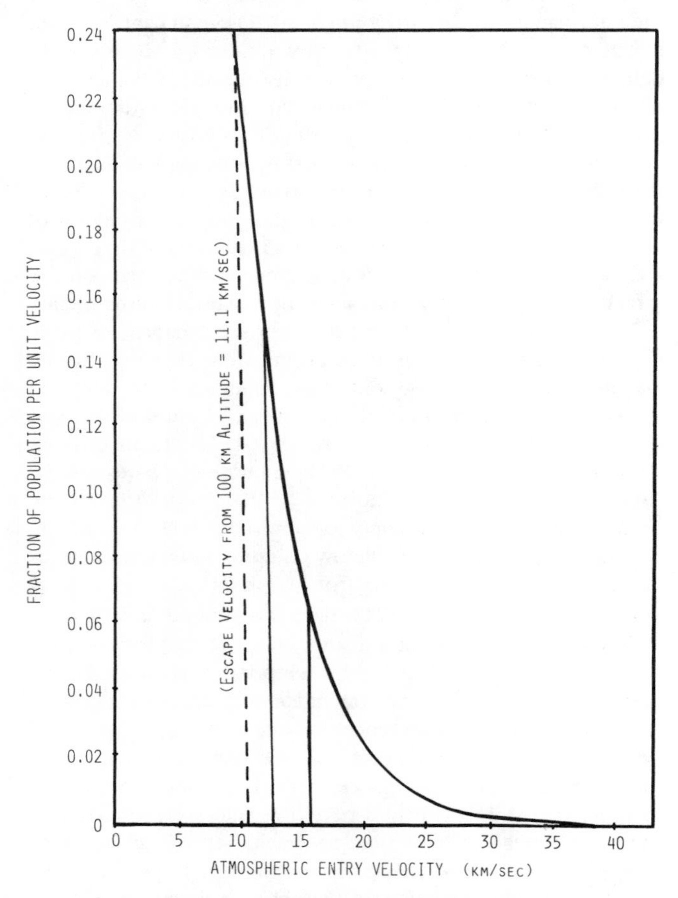

Fig. 1. Atmospheric entry velocity distribution for radar meteors at the top of the atmosphere as determined by *Southworth and Sekanina* (1973). The vertical lines at 11.1 km/sec, 12.5 km/sec, and 15.6 km/sec divide the particles into source types as described in the text.

If the velocity distribution of the cosmic dust collected at Earth were known, then the hypothesis that Encke was the major contributor to the interplanetary dust cloud could be tested in a second way. For the radar meteors, which are only slightly larger in size than the cosmic dust, the arrival velocity distribution has been established for 14,000 meteors by *Southworth and Sekanina* (1973). Their atmospheric entry velocity distribution, given in Fig. 1, shows no noticeable spike near the 29 km/sec velocity of the Taurid meteors. Even when the near-Earth gravitational enhancement effect, which biases the Southworth and Sekanina data toward the low-velocity component of the interplanetary dust, is removed by correcting the near-Earth abundances back to the in space at 1 A.U. distribution and removing the Earth infall contribution to the velocity (as described by *Zook,* 1975), no peak near the Taurid meteor stream velocity is observed. This velocity distribution is shown in Fig. 2. This also suggests that Comet Encke is not the dominant contributor to dust in the radar meteor size range.

Measurements on cosmic dust particles recovered from the stratosphere provide independent evidence of the velocity distribution of the dust particles. *Sandford* (1986) showed that a layer-lattice silicate cosmic dust particle, pulse heated to simulate atmospheric entry heating, showed pronounced changes in the depth of the 3.0-μm water band in its infrared absorption spectrum after the 560°C temperature step. Only 2 of the 12 layer-lattice silicate particles he examined appeared to have been heated above 700°C on entry. Using the atmospheric entry heating model of *Fraundorf* (1980), *Sandford* (1986) concluded the mean atmospheric entry velocity of the layer-lattice silicate particles was around 11.5 km/sec, implying a geocentric velocity before Earth infall acceleration of less than 3 km/sec. *Flynn* (1989) and *Sandford and Bradley* (1989) have used additional internal thermometers in the cosmic dust particles, including the presence of volatile elements, minerals of low stability temperature, and solar flare ion "tracks" that are annealed on heating (*Fraundorf et al.,* 1982b), to constrain the atmospheric entry velocities for chondritic cosmic dust particles. *Flynn* (1989) suggests the degree of heating experienced on atmospheric entry by chondritic cosmic dust collected from the stratosphere is inconsistent with a geocentric velocity at the collection opportunity characteristic of particles from Comet Encke.

The observed enhancement of about 2.8 clearly excludes a micrometeorite population composed solely of particles from the cometary sources represented by the stream meteors. Particles from these sources would arrive at Earth with geocentric velocities generally greater than 15 km/sec. Enhancement factors of 2.0 or less would be expected for particles from these sources (see Table 2). Thus, the near-Earth enhancement demonstrates that a low-velocity component is present in the interplanetary dust.

Realistically the interplanetary dust must consist of particles from a variety of different sources. The simplest mixture to model is a two-component mixture containing dust from the parent bodies of the meteor stream particles, which have low near-Earth enhancements, and a second dust component of lower geocentric velocity. The proportions of the two components can be selected to match the observed near-Earth enhancement. To illustrate the magnitude of the collection bias between the in-space and near-Earth dust samples, I will consider a mixture of high-velocity dust from an Encke-like orbit having an enhancement of 1.2 and a second monovelocity component having a higher enhancement. If the high-velocity component is assumed to contribute 99% of the dust in interplanetary space at 1 A.U., then low-velocity dust, representing only 1% of the in-space dust, must be enhanced by a factor of 160 near-Earth in order to provide an overall enhancement of 2.8. In this case the proportions of low- and high-velocity material near Earth would be 57% and 43%, respectively—a dramatic reversal of 1% and 99% interplanetary proportions. To produce this enhancement, the geocentric velocity of the second component would have to be 0.9 km/sec. These calculations are repeated in Table 3 for a variety

TABLE 3. Near-Earth enhancement.

High-Velocity Fraction*	Low-Velocity Fraction	Low-Velocity Enhancement	Characteristic Geocentric Velocity (km/sec)	Near-Earth Low-/High-Velocity Fraction
0.99	0.01	160	0.9	57/43
0.90	0.10	17	2.8	61/39
0.75	0.25	7.6	4.4	68/32
0.50	0.50	4.4	6.1	79/21
0.25	0.75	3.3	7.3	90/10
0.10	0.90	3.0	8.0	96/4
0.01	0.99	2.8	8.3	99/1

*The high-velocity fraction has an assumed enhancement factor of 1.2, corresponding to Encke dust.

of different starting fractions of the high-velocity dust. As seen in Table 3, no matter what proportion of this high-velocity dust component is assumed in interplanetary space, the low-velocity dust component dominates in the near-Earth collections if a near-Earth enhancement factor of 2.8 is required.

The analysis is more complex if a large number of sources, all giving different geocentric velocities at the collection opportunity, are assumed, but the same principle applies. To match the observed near-Earth enhancement, the fraction of low-velocity dust in the near-Earth collections will be significantly increased over the fraction in interplanetary space.

ASSESSMENT OF THE SOURCES

The Infrared Astronomy Satellite has identified two types of solar system dust sources: mainbelt asteroids (*Low et al.,* 1984) and comets (*Sykes et al.,* 1986). Since the orbits of these parent bodies and the dust particles they generate are generally not Earth-intersecting, the dust particles cannot be collected until their orbits have evolved to be Earth-intersecting. Poynting-Robertson drag is the most effective mechanism for orbital perturbation on particles in the cosmic dust size range (*Dohnanyi,* 1978). The effect of PR drag is to rapidly decrease the aphelion of the orbit of the particle with a much slower decrease in the perihelion. This causes the ellipticity of the orbit to decrease as the orbit decays into the sun.

Examination of the orbital evolution, under the influence of only PR drag, suggests three distinct ranges of geocentric velocity at the Earth collection opportunity (as described in *Flynn,* 1989) for dust derived from mainbelt asteroidal and cometary sources. These source types are (1) mainbelt asteroids, which generate particles collected from near-circular orbits of low inclination, having geocentric velocities $\leq$5.5 km/sec, (2) comets with perihelia >1.2 A.U., which generate particles collected from orbits of small ellipticity having velocities in the range from 5.5 to 11 km/sec, and (3) comets with perihelia <1.2 A.U., which generate particles collected from orbits of larger ellipticity having geocentric velocities >11 km/sec.

Although these groups are clearly separated if PR drag is the only perturbing force, planetary gravitational perturbations, as described by *Gustafson and Misconi* (1986), may cause some mixing among the groups. However, *Gustafson et al.* (1987) indicate that gravitational perturbations generally cause the eccentricity to decrease less rapidly than calculated from PR drag alone. Thus, some particles from mainbelt asteroids may arrive at Earth with velocities characteristic of cometary particles. However, gravitational perturbation is unlikely to transfer a large fraction of the cometary particles into the velocity range characteristic of mainbelt asteroidal particles.

Particle collisions can also cause some mixing between the groups. However, for particles in the 10- to 20-μm-diameter size range being considered, the time required to fall into the sun is short compared to the catastrophic collision lifetime (*Dohnanyi,* 1978). Thus, these small particles are unlikely to experience a significant orbital alteration due to collision during their lifetimes. Larger particles from either cometary or asteroidal sources may, however, be fragmented in collisions to produce 10- to 20-μm particles. The extent of this contribution has not been assessed.

Comets with perihelia greater than 1.2 A.U. could also contribute to the very-low-velocity group if the sources had low ellipticity and low orbital inclination. But among the presently active comets tabulated by *Porter* (1963), Comet Kopff has the smallest inclination (4.8°) of any short-period comet with perihelion greater than 1.5 A.U. Even with this favorable combination of low inclination and large perihelion, the geocentric velocity at Earth collection under PR drag orbital evolution was calculated to be 6 km/sec (*Flynn,* 1989). This velocity is mainly due to the residual eccentricity (0.2 for Comet Kopff) at the collection opportunity. Despite the eccentricity reduction during Poynting-Robertson orbital evolution, particles emitted in the highly elliptical orbits that characterize most presently active comets still have a moderate eccentricity at the Earth collection opportunity. Only if comets in near-circular orbits, such as Schwassman-Wachmann I, were a dominant source of the interplanetary dust would the cometary contribution to the low-velocity group be significant. Thus, if the velocity distribution of the cosmic dust particles arriving at Earth were known, the fraction of particles from each of these three types of sources could be estimated.

For larger meteors, the entry velocity distributions have been measured directly (e.g., *Southworth and Sekanina,* 1973; *Erickson,* 1968). However, the cosmic dust particles, which do not melt on atmospheric entry, leave no ionized or luminous trails that can be measured. Determination of the velocities of these particles is proposed for the Space Station Cosmic Dust Collection Facility (CDCF) (*Carey and Walker,* 1986; *Zook,* 1986) that may be launched in the 1990s. Until then, the observed near-Earth gravitational enhancement can be used to constrain the velocity distribution.

The atmospheric entry velocity distribution for the radar meteors, only slightly larger in size than the cosmic dust, was derived by *Southworth and Sekanina* (1973). The near-Earth gravitational enhancement over the lunar flux for the Southworth and Sekanina velocity distribution is 2.4 (*Zook,* 1975). This enhancement is consistent with the near-Earth over lunar enhancement of about 2.6 observed by the micrometeorite puncture detectors sensitive in the cosmic dust size range. There is independent evidence, based on the degree of heating experienced by the cosmic dust particles on atmospheric entry (*Sandford,* 1986; *Flynn,* 1989; *Sandford and Bradley,* 1989), to suggest that the actual cosmic dust velocity distribution has a slightly lower mean than the Southworth and Sekanina distribution. However, the actual velocity distribution cannot be derived from the atmospheric entry heating observations. If we take the Southworth and Sekanina velocity distribution as representative of the cosmic dust velocity distribution, then we can estimate the fraction of the cosmic dust in each of the three velocity ranges.

TABLE 4. Near-Earth enhancement from the *Southworth and Sekanina* (1973) distribution.

Source	Fraction in Space	Fraction Near-Earth	Surviving Atmospheric Entry
Asteroids (V_{space} <5.5 km/sec)	7%	30%	32%
Comets (perihelion >1.2 A.U.) (5.5 km/sec <V_{space} <11 km/sec)	33%	37%	40%
Comets (perihelion <1.2 A.U.) (V_{space} >11 km/sec)	60%	34%	28%

The geocentric velocity ranges for the three source types must be adjusted for gravitational infall acceleration in order to give the atmospheric entry velocities. By energy conservation, the atmospheric entry velocity, v_a, is related to the geocentric velocity, v_g, by

$$v_a = (v_g^2 + v_e^2)^{1/2} \qquad (2)$$

where v_e is the Earth escape velocity.

When adjusted for gravitational infall, the respective velocity ranges are 11.1 km/sec to 12.5 km/sec for mainbelt asteroids, 12.5 km/sec to 15.6 km/sec for large perihelion (>1.2 A.U.) comets, and over 15.6 km/sec for small perihelion (<1.2 A.U.) comets. The percentage of particles in each velocity range in the Southworth and Sekanina atmospheric entry velocity distribution (shown in Fig. 1) are 30% below 12.5 km/sec, 37% in the 12.5 km/sec to 15.6 km/sec velocity range, and 33% above 15.6 km/sec on atmospheric entry (Table 4). This would suggest that the proportions of asteroidal, large perihelion cometary, and small perihelion cometary material entering the Earth's atmosphere or on impact collection surfaces on orbiting spacecraft are about 30:37:33.

The stratospheric collections are further altered since a significant fraction of the small perihelion cometary material melts on atmospheric entry and thus is lost from the set of unaltered particles that are most frequently studied.

The Southworth and Sekanina velocity distribution at the top of the atmosphere can be transformed to the distribution in interplanetary space. The procedure, described by *Zook* (1975), is to use equation (2) to calculate the geocentric velocity corresponding to each atmospheric entry velocity. The fraction of material in each velocity increment is then corrected by removing the near-Earth gravitational enhancement. A detailed mathematical description of the procedure is given by *Morgan et al.* (1988). This distribution is given in Fig. 2 (H. A. Zook, personal communication, 1988). Using his curve, the proportions of particles in each category in interplanetary space can be established using the geocentric velocity ranges appropriate for each type of source. The in-space distribution gives relative abundances of 7% asteroidal, 33% large perihelion cometary, and 60% small perihelion cometary particles. Thus, using the Southworth and Sekinina velocity distribution, low-velocity (most likely asteroidal) material goes from 7% of the in space particles to 30% of the material entering the atmosphere.

As discussed earlier, planetary gravitational scattering can perturb the orbital evolution from that calculated for purely PR drag. Generally, the effect of planetary gravitational perturbations is to hinder the orbital circularization caused by

PR drag (*Gustafson et al.,* 1987). This would increase the geocentric velocity of the particles at the collection opportunity, causing an underestimation of the asteroidal fraction of the stratospheric and zodiacal dust in this model. However, Zook (personal communication, 1988) indicates that for particles in the 10- to 20-μm diameter range being considered, the PR optical evolution is so rapid that planetary gravitational scattering is a small effect.

It is possible that some comets, active in an earlier era and having very large perihelia or very low eccentricities as well as low orbital inclinations, could have contributed cometary dust to the low-velocity group. However, the absence of streams of larger particles with such orbits in the IRAS data suggests the contribution from these potential sources is small, since larger particles would persist in the streams for times much longer than the solar infall time for cosmic dust size particles.

ENTRY HEATING LOSS

Atmospheric entry heating will introduce a further bias for particles collected after atmospheric deceleration. Particles with high entry velocities are preferentially removed by melting or volatilization. Using the micrometeoritic entry heating model developed by *Whipple* (1950) and extended by *Fraundorf* (1980), the extent of the entry heating loss can be estimated.

The peak temperature experienced on atmospheric entry depends on the entry velocity, the angle between the particle trajectory and the normal to the Earth's surface, and physical properties of the particle. *Fraundorf* (1980) extended the model of micrometeorite entry heating developed by *Whipple* (1950) to predict the fraction of an isotropically incident particle flux of any given velocity that would be heated above any specific temperature. The Fraundorf model was applied to the Southworth and Sekanina velocity distribution, and those particles heated above 1500K (the melting point of fayalitic olivine) were removed. For 20-μm diameter particles, about 20% of the particles in the high-velocity range are removed, while none of the particles in the two lower velocity ranges are heated above this temperature. The last column of Table 4 shows the fraction of particles in each velocity range that are likely to survive atmospheric entry.

The combined effects of gravitational focusing and melting or volatilization on atmospheric entry serve to increase the low-velocity component in the stratospheric collections to 32%, an increase by a factor of 4.6 over the in-space fraction in this velocity range. The proportions of high-velocity material are reduced from 60% in-space to only 28% on the stratospheric collectors. If, as suggested in *Flynn* (1989), the low-velocity dust is mainly asteroidal while the high-velocity dust is from comets in orbits like Encke, the relative proportions of material from these two source types are altered by an order of magnitude from the interplanetary value to that on the stratospheric collectors. Thus, the proportions of the different types stratospheric cosmic dust are not necessarily representative of the zodiacal cloud.

This assessment has taken the *Southworth and Sekanina* (1973) radar meteor velocity distribution as representative of the smaller cosmic dust particles. This was done because of the consistency of the near-Earth enhancement determined by the satellite puncture detectors with that calculated for *Southworth and Sekanina* (1973) distribution. However, the low counting statistics on the Lunar Orbiter puncture detectors would permit some deviation from this velocity distribution, and thus from the inferred source proportions. *Delcourt* (1980) has analyzed the progressive changes in the orbital characteristics from photographic to radar meteors. He concluded that as mass decreases, the fraction of particles with small values of eccentricity increases. If this trend continues as the mass decreases to the cosmic dust range, then these particles would be collected from more nearly circular orbits than the radar meteors and would have a lower mean velocity than the Southworth and Sekanina distribution. That would increase the inferred asteroidal component in both the zodiacal and near-Earth categories. Determination of the velocity distribution of the micrometeorites, as suggested for the CDCF, will allow the extent of the correction to be assessed.

CONCLUSIONS

The stratospheric cosmic dust, as well as dust sampled by Earth-orbiting impact collectors, is not representative of the true composition of the zodiacal cloud. A substantial near-Earth collection bias enhances the low-velocity (most likely asteroidal) component in stratospheric and Earth-orbiting collectors. This arises from (1) gravitational focusing, which substantially biases all near-Earth micrometeorite collections in favor of the low-velocity component of the interplanetary dust, and (2) atmospheric entry heating, which further biases the stratospheric cosmic dust in favor of the low-velocity component of the interplanetary dust.

In addition to these two effects, the collision probability between a dust particle and a nongravitating planet increases as the orbits of the two objects become similar in inclination and aphelion or perihelion distance. This effect has not been considered because the detailed orbital properties of the cosmic dust particles are unknown; however, it will bias the near-Earth collections farther in favor of dust with lower geocentric velocity at the collection opportunity. Direct assessment of the magnitude of this effect should be possible if dust orbits are determined.

These selection effects are consistent with the observation by *Flynn* (1989) and *Sandford and Bradley* (1989) that most of the stratospheric cosmic dust exhibits evidence of minimal heating ($<800°C$) on atmospheric entry, suggesting it was collected from orbits of low geocentric velocity at collection.

If, as proposed, the CDCF has the capability of determining the velocity of incoming dust particles, corrections can be made to infer the proportions of various particle types in the zodiacal cloud.

The near-Earth enhancement observed by micrometeoroid detectors in Earth and lunar orbit are inconsistent with the hypothesis that a great majority of the stratospheric cosmic dust particles are from Comet Encke, or from other comets in orbits with perihelia <1.2 A.U.

Acknowledgments. I would like to thank H. Zook for providing unpublished calculations of the Southworth and Sekanina velocity distribution corrected to interplanetary space (Fig. 2), as well as for many valuable discussions on this topic, and S. Sandford and M. Zolensky for very constructive reviews of this paper. This work was supported by NASA Grant NAG 9-257.

REFERENCES

Alexander W. M. and Hyde T. W. (1989) Micron and submicron particle flux enhancement within the Earth's magnetosphere I: In-situ and laboratory source data (abstract). *Lunar and Planetary Science XX,* pp. 9-10. Lunar and Planetary Institute, Houston.

Alexander W. M., Arthur C. W., Corbin J. D., and Bohn J. L. (1970) Picogram dust particle flux: 1967-1968 measurements in selenocentric, cislunar, and interplanetary space. *Space Research, X,* North Holland, 252-253.

Bradley J. P., Sandford S. A., and Walker R. M. (1988) Interplanetary dust particles. In *Meteorites in the Early Solar System* (J. F. Kerridge and M. S. Matthews, eds.), pp. 861-898. Univ. of Arizona, Tucson, in press.

Brownlee D. E. (1985) Cosmic dust: collection and research. *Annu. Rev. Earth Planet. Sci., 13,* 147-173.

Brownlee D. E., Hörz F., Tomandl D. A., and Hodge P. W. (1976) Physical properties of interplanetary grains. In *The Study of Comets* (B. Donn et al., eds.), pp. 962-982. NASA SP-393.

Burns J. A., Lamy P. L., and Soter S. (1979) Radiation forces on small particles in the solar system. *Icarus, 40,* 1-48.

Carey W. C. and Walker R. M. (1986) Prospects for an orbital determination and capture cell experiment. In *Trajectory Determinations and Collection of Micrometeorites on the Space Station* (F. Hörz, ed.), pp. 49-51. LPI Tech. Rpt. 86-05. Lunar and Planetary Institute, Houston.

Cook A. F. (1973) A working list of meteor streams. In *Evolutionary and Physical Properties of Meteoroids* (C. Hemmenway, P. Millman, and A. F. Cook, eds.), pp. 183-191. NASA SP-319.

Delcourt J. (1980) Experimental and theoretical study of radiometeors. In *Solid Particles in the Solar System, IAU Symposium No. 90* (I. Halliday and B. A. McIntosh, eds.), pp. 133-136. D. Reidel, Dordrecht, Holland.

Delsemme A. H. (1976) The production rate of dust by comets. In *Interplanetary Dust and the Zodiacal Light* (H. Elsässer and H. Fechtig, eds.), p. 314. Springer-Verlag, Heidelberg.

Dohnanyi J. S. (1976) Sources of interplanetary dust: Asteroids. *Lect. Notes Phys., 48,* 187-205.

Dohnanyi J. S. (1978) Particle dynamics. In *Cosmic Dust* (J. A .M. McDonnell, ed.), pp. 527-605. Wiley, New York.

Erickson J. E. (1968) Velocity distribution of sporadic photographic meteors. *J. Geophys. Res., 73,* 3721-3726.

Esat T. M., Brownlee D. E., Papanastassiou D. A., and Wasserburg G. J. (1979) The Mg isotopic composition of interplanetary dust particles. *Science, 206,* 190-192.

Flynn G. J. (1989) Atmospheric entry heating: A criterion to distinguish between asteroidal and cometary sources of interplanetary dust. *Icarus, 77,* 287-310.

Fraundorf P. (1980) The distribution of temperature maxima for micrometeorites in the Earth's atmosphere without melting. *Geophys. Res. Lett., 10,* 765-768.

Fraundorf P. (1981) Interplanetary dust in the transmission electron microscope: Diverse materials from early solar system. *Geochim. Cosmochim. Acta, 45,* 915-943.

Fraundorf P., Brownlee D. E., and Walker R. M. (1982a) Laboratory studies of interplanetary dust. In *Comets* (L. Wilkening, ed.), pp. 383-408. Univ. of Arizona, Tucson.

Fraundorf P., Lyons T., Sandford S. A., and Schubert P. (1982b). The survival of solar flare tracks in interplanetary dust silicates on deceleration in the Earth's atmosphere. *Proc. Lunar Planet. Sci. Conf. 13th* in *J. Geophys. Res., 87,* A409-A412.

Gault D. E., Shoemaker E. M., and Moore H. J. (1963) Spray ejected from the lunar surface by meteoroid impact. *NASA TND-1767,* pp. 1-22.

Grew G. W. and Gurtler C. A. (1971) The lunar orbiter meteoroid experiments. *NASA TND-6266,* pp. 1-44.

Gustafson B. A. S. and Misconi N. Y. (1986) Interplanetary dust dynamics: I. Long term gravitational effects of the inner planets on zodiacal dust. *Icarus, 66,* 280-287.

Gustafson B. A. S., Misconi N. Y., and Rusk E. T. (1987) Interplanetary dust dynamics: II. Poynting-Robertson drag and planetary perturbations on cometary dust. *Icarus, 72,* 568-581.

Hughes D. W. (1978) Meteors. In *Cosmic Dust* (J. A. M. McDonnell, ed.), pp. 123-185. Wiley, New York.

Kresak L. (1980) Sources of interplanetary dust. In *Solid Particles in the Solar System* (I. Halliday and B. A. McIntosh, eds.), pp. 211-222. IAU Symposium No. 90, Reidel, Dordrecht.

Low F. J., Beinema D. A., Gautier T. N., Gillette F. C., Beichman C. A., Neugebauer G., Young E., Aumann H. H., Boggess N., Emerson J. P., Habing H. J., Heuser M. G., Houck J. R., Rowan-Robinson M., Soifer B. T., Walker R. G., and Wesselius P. R. (1984) Infrared cirrus: New components of the extended infrared emission. *Astrophys. J., 278,* L19.

McDonnell J. A. M. (1978) Microparticle studies by space instrumentation. In *Cosmic Dust* (J. A. M. McDonnell, ed.), pp. 337-426. Wiley, New York.

Morgan T. H., Zook H. A., and Potter A. E. (1988) Impact-driven supply of sodium and potassium to the atmosphere of Mercury. *Icarus, 75,* 156-170.

Millman P. M. (1972) Cometary meteoroids. In *Nobel Symposium No. 21: From Plasma to Planet* (A. Elvinus, ed.), pp. 157-168. Wiley, New York.

Naumann R. J. (1966) The near-Earth meteoroid environment. *NASA TND-3717,* pp. 1-38.

Opik E. J. (1951) Collision probabilities with the planets and the distribution of interplanetary matter. *Proc. R. Irish Acad., 54,* 165-199.

Porter J. G. (1963) The statistics of comet orbits. In *Moon, Meteorites, and Comets* (B. M. Middelhurst and G. P. Kuiper, eds.), pp. 550-572. Univ. of Chicago, Chicago.

Sandford S. A. (1986) Laboratory heating of an interplanetary dust particle: Comparisons with an atmospheric entry model (abstract). In *Lunar and Planetary Science XVII,* pp. 754-755. Lunar and Planetary Institute, Houston.

Sandford S. A. (1987) The collection and analysis of extraterrestrial dust particles. *Fund. Cosmic Phys., 12,* 1.

Sandford S. A. and Bradley J. P. (1989) Interplanetary dust particles collected in the stratosphere: Observations of atmospheric heating and constraints on their interrelationships and sources. *Icarus,* in press.

Southworth R. B. and Sekanina Z. (1973) Physical and dynamical studies of meteors. *NASA CR-2316,* pp. 1-108.

Sykes M. V. and Greenberg R. (1986) The formation and origin of the IRAS zodiacal dust bands as a consequence of single collisions between asteroids. *Icarus, 65,* 51-69.

Sykes M. V., Lebofsky L. A., Hunten D. M., and Low F. (1986) The discovery of dust trails in the orbits of periodic comets. *Science, 232,* 1115-1117.

van der Stap C. C. A. H., Vis R. D., and Verheul H. (1986) Interplanetary dust, arguments in favor of a late stage nebular origin (abstract). In *Lunar and Planetary Science XVII,* pp 1013-1014. Lunar and Planetary Institute, Houston.

Whipple F. L. (1950) The theory of micro-meteorites. Part I: In an isothermal atmosphere. *Proceedings of the National Academy of Sciences, 36,* 687-695.

Whipple F. L. (1967) On maintaining the meteoritic complex. In *The Zodiacal Light and the Interplanetary Medium,* p. 409. NASA SP-150.

Zinner E., McKeegan K. D., and Walker R. M. (1983) Laboratory measurements of D/H ratios in interplanetary dust. *Nature, 305,* 119-121.

Zook H. A. (1975) The state of meteoritic material on the Moon. *Proc. Lunar Sci. Conf. 6th,* pp. 1653-1672.

Zook H. A. (1986) Precision requirements on cosmic dust trajectory measurements. In *Trajectory Determinations and Collection of Micrometeorites on the Space Station* (F. Hörz, ed.), pp. 97-99. LPI Tech. Rpt. 86-05. Lunar and Planetary Institute, Houston.

Zook H. A. and McKay D. S. (1986) On the asteroidal component of cosmic dust (abstract). In *Lunar and Planetary Science XVII,* pp. 977-978. Lunar and Planetary Institute, Houston.

Proceedings of the 20th Lunar and Planetary Science Conference, pp. 373-378
Lunar and Planetary Institute, Houston, 1990

The Comet Nucleus: Ice and Dust Morphological Balances in a Production Surface of Comet P/Halley

J. A. M. McDonnell, G. S. Pankiewicz, P. N. W. Birchley, S. F. Green, and C. H. Perry

Unit for Space Sciences, University of Kent, Canterbury, Kent CT2 7NR, England

Dust-to-gas mass ratios are presented based on in-situ measurements of the coma dust and gas distributions of comet P/Halley by the DIDSY, PIA, and NMS experiments aboard the Giotto spacecraft. A relative excess, compared with preflight expectations, of particulates with masses $>10^{-9}$ kg is observed; if this is representative of the coma as a whole, the dust-to-gas ratio, μ, within the nucleus must be about 2. This compares with a value of μ ~0.2 for a model distribution without a large particle excess. The implied mass and area distributions of particulates in the parent nucleus material is presented for observed and modeled coma distributions. Some basic properties expected of the surface material are identified that are relevant to nucleus modeling and to the remote sensing of nucleus material during cometary rendezvous.

INTRODUCTION

The close approach of Giotto to comet P/Halley during its encounter on the 13th and 14th of March 1986 offered a unique opportunity to study the distribution of particulates in the inner coma. The large effective area (~2 m^2 for masses $>10^{-9}$ kg) of the dust impact detection system DIDSY (*McDonnell,* 1987) combined with the highly sensitive front end channels of the mass spectrometer PIA (*Kissel,* 1986) provided a coverage of the mass distribution from 10^{-19} kg (~0.02 μm radius) to some 30 mg (~2 mm radius) and can be inferred for masses up to ~1 g (the largest particle to hit the spacecraft) by using the total spacecraft deceleration causing a velocity change of 23.05 cm sec^{-1} (*Edenhofer et al.,* 1987). No particles of masses larger than ~30 mg were measured by DIDSY because the discrete data for which individual masses were determined were only a subset of the impacts detected by the threshold counters, the shield did not cover the whole cross-sectional area of the spacecraft, and the period of highest flux occurred during loss of telemetry at encounter.

This paper presents the results of transforming the measured coma distribution back to the nucleus via well-established modeling to give the flux and size distribution of particles leaving the surface. These are then used to calculate the dust-to-gas mass ratio and grain mass and area distributions in the nucleus matrix. The results may be applied to the task of remote sensing of a cometary nucleus to locate active source emission areas in cometary rendezvous. In a comet nucleus sample return mission such as CNSR-ROSETTA, one of ESA's four cornerstone missions and studied jointly with NASA (*ESA,* 1986), the identification of a fresh production surface by reflectance or emission properties at wavelengths from optical to radar depends upon the nature of the absorbing material and the detectability of ice in the matrix. The dust-to-ice ratio is critical to the prediction of these properties.

DUST MASS DISTRIBUTIONS IN THE COMA AND NUCLEUS

The calculation of dust flux rates based on the Giotto DIDSY and PIA data is described by *McDonnell et al.* (1987). Figure 1 shows the measured fluence F(m) (representing the time-integrated number of particles of mass >m impacting the spacecraft per m^2). This is shown in the coma for the pre-encounter period -120 to -43 sec (Fig. 1a) and for the post-encounter +40 to +120 sec (Fig. 1b). These results are based on a detailed analysis of all available data for impacts in these periods when there is unbiased sampling of all large impacts. An "excess" of particles with mass $>10^{-9}$ kg is apparent in both data sets relative to a uniform power law distribution of constant mass index. The asymmetry between the pre- and post-encounter data is consistent with the dynamical modeling of large grains (*Fertig and Schwehm,* 1984) for a rotating nucleus. The dotted line (*Divine et al.,* 1986, MODEL) indicates the constant mass index distribution. This would be contrary to the observed excess at large masses and would be sustained only if it were shown that Giotto data were not representative of the average properties of the coma. Distributions of this model form (*Divine et al.,* 1986) have generally been assumed from interpretation of remote observations and were used for preflight modeling with ***differential size*** distribution index at large masses of U = 3.7 (where $N(a)da = C a^{-U} da$) in the coma (corresponding to the ***cumulative mass*** distribution index α = 0.9 where $N(>m) = C m^{-\alpha}$).

Figure 2 shows the derived total fluence distribution over the whole of encounter. The solid line (IN-SITU) represents the distribution as measured after correction for telementry loss and constrained by the total spacecraft deceleration (*Edenhofer et al.,* 1987) and indicates the average coma distribution along the Giotto trajectory. The distribution for masses greater than 30 mg was determined by extrapolating

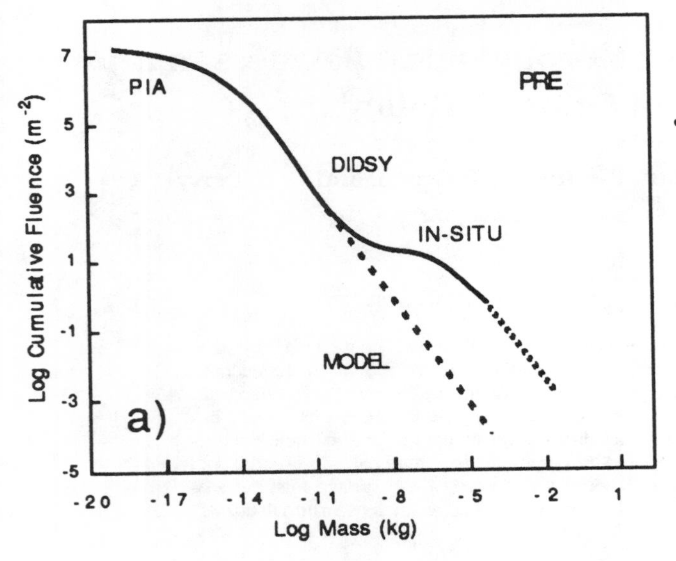

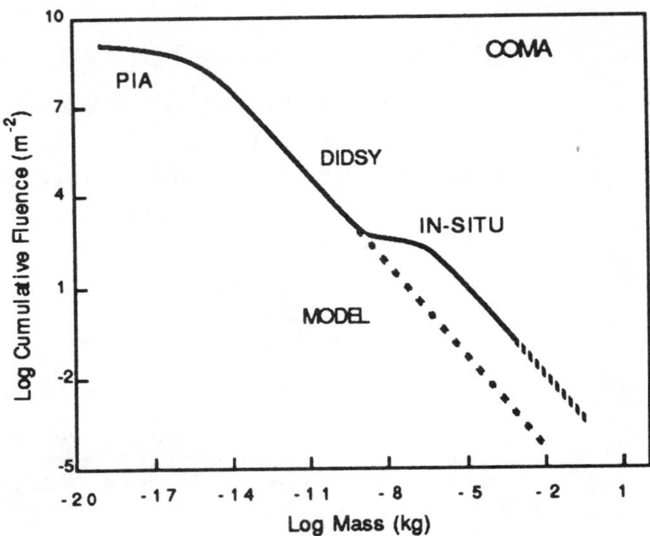

Fig. 2. Cumulative fluence distribution measured over the whole of encounter. For IN-SITU data (solid line) and MODEL distribution (dotted line).

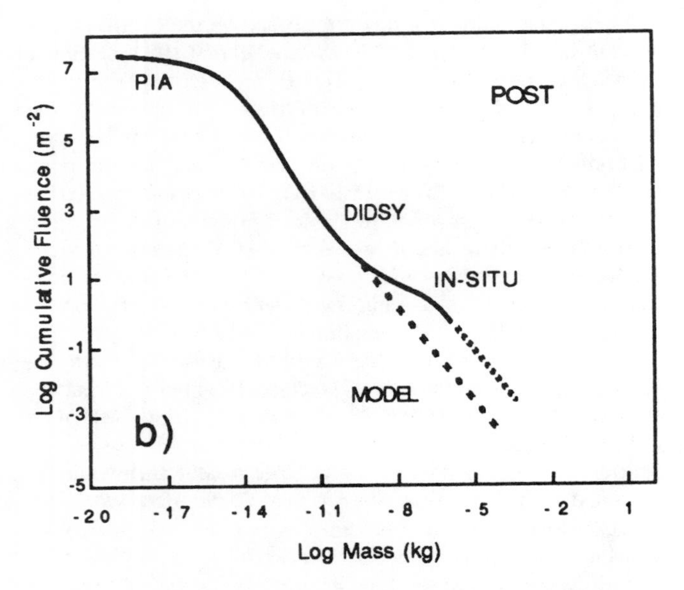

Fig. 1. Cumulative fluence distributions in the coma: **(a)** fluence -120 to -43s, (PRE); **(b)** fluence +40 to +120s, (POST). The solid curve is for Giotto data (this paper) and the dotted line for a model distribution (*Divine et al.,* 1986) with uniform mass index at large masses.

the observed distribution to the mass at which the fluence and spacecraft area product equals one (corresponds to the single largest impact ~1 g). The largest particles that penetrated the front shield produced less momentum enhancement (ϵ) from secondary ejecta than nonpenetrating particles ($\epsilon = 11$). A consistent solution for the total deceleration and DIDSY data was obtained with a "momentum enhancement derating exponent, γ" (*McDonnell et al.,* 1987) of 0.4.

The cumulative dust flux $\Phi_n(m)$ (the number of particles with mass >m leaving 1 m² of the nucleus per second) is given by

$$\Phi_n(m) = \int_m^{m_{max}} n_n(m')\, d\log m' = \int_m^{m_{max}} \frac{F'(m')}{\pi R_{miss}}\, v(a) \left(\frac{R_{miss}}{R_n}\right)^2 d\log m' \qquad (1)$$

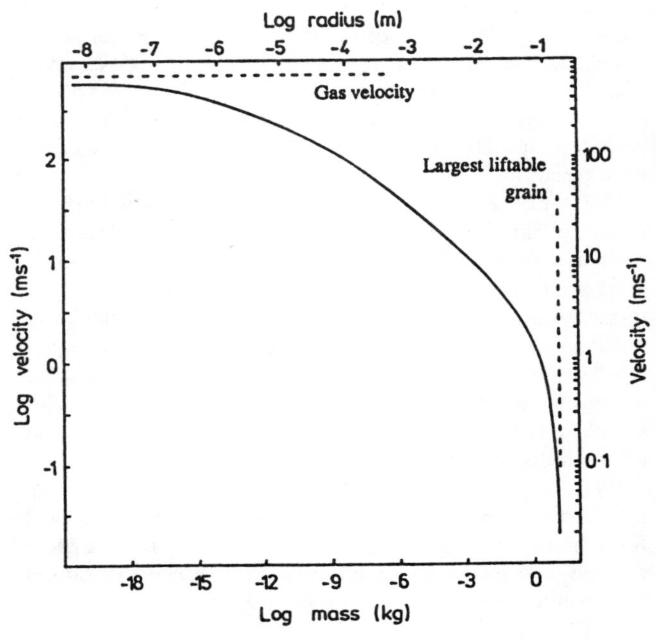

Fig. 3. Grain velocity distribution from *Divine* (1981). More recent data by *Gombosi* (1986) yields a similar result.

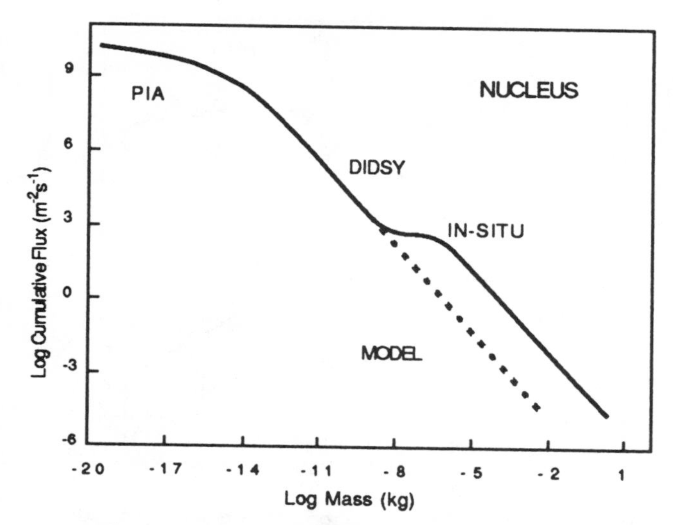

Fig. 4. Cumulative flux distribution at the nucleus derived from the total fluence. Solid line: IN-SITU distribution; dotted line: MODEL distribution.

assuming radial trajectories and where the grain velocity v(a) is taken from Fig. 3, R_n = 5.2 km is the effective nucleus radius (derived from the model of *Sagdeev et al.,* 1986), R_{miss} is the Giotto miss distance of 600 km, $n_n(m)$ is the differential flux at the nucleus, and F′(m) is the differential fluence measured over the whole encounter period. Using this relation we derive the cumulative flux distribution (flux of all particles with mass >m) at the nucleus, shown in Fig. 4. Dust velocities are calculated using the approximation of *Divine* (1981) with a maximum liftable mass of radius 15 cm (assuming a nucleus density of 800 kg m^{-3} and an active fraction of the nucleus of 10% of the total surface; *Keller et al.,* 1987). The grain density is assumed to take the form

$$\rho(a) = 3000 - 2200 \left(\frac{a}{a + a_0}\right) \text{ kg m}^{-3} \qquad (2)$$

(*Divine et al.,* 1986) where $a_o = 2 \times 10^{-6}$m.

DUST-TO-GAS MASS RATIO

The resultant fluxes are then integrated over all masses and over the active cometary surface and used to obtain a dust production rate that may be directly compared to the measured gas production rate of 2.55×10^4 kg sec^{-1} (*Krankowsky et al.,* 1986) to yield a dust-to-gas mass ratio as a function of the maximum mass of included particles (Fig. 5) where

$$\mu(m) = \frac{4\pi R_n^2}{Q_g M} \int_0^m n_n(m') \, m' \, d\log m' \qquad (3)$$

This gives a likely value of μ(m) shown by the solid line from the IN-SITU results whereas the dotted line (MODEL distribution) may be compared with previously derived results from remote sensing. Q_g is the gas production rate = 6.9×10^{29} mol sec^{-1} (*Krankowsky et al.,* 1986) and M is the mean molecular mass = 3.7×10^{-26} kg (*Divine et al.,* 1986). Although the nucleus is known to comprise localized active areas on a predominantly inactive surface, the gas production rate is an average value for the data measured during the whole encounter, i.e., assuming the whole surface has uniform activity. Likewise, the DIDSY and PIA fluence represents an average along the entire Giotto trajectory and hence is directly comparable to the gas measurements. Since the value of μ is a ratio of the production rates the actual active area is not required.

The MODEL distribution produces a bulk dust-to-gas ratio of ~0.2, namely classifying comet P/Halley as a gas rich comet, but Fig. 5 clearly shows how the measured large mass distribution enhances the total mass of dust to produce a bulk ratio of ~2; in both the IN-SITU and the MODEL cases the resultant dust-to-gas ratio is largely independent of the maximum liftable grain size although very different values of μ(m) result from the two cases. The tighter constraint on the dust-to-gas ratio now provided over that reported earlier by *McDonnell et al.* (1987, 1989) is due to the full analysis of large particle impacts that indicate an increase in the mass distribution index for masses above $\sim 5 \times 10^{-7}$ kg.

There is considerable evidence for an abundance of millimeter- to centimeter-sized grains in cometary comae. A number of comets observed by IRAS exhibited trails of material close to their orbital paths that are composed of grains with a lower size limit in the submillimeter range (*Eaton et al.,* 1984; *Sykes et al.,* 1986). Radar measurements of comet IRAS-Araki-Alcock by *Harmon et al.* (1989) showed a radar cross-section that was larger than expected from the nucleus alone and that was attributed to grains in excess of 1 cm. This interpretation was also applied by *Campbell et al.* (1989) to measurements of comet P/Halley that similarly exhibited a larger than expected radar cross-section. However, ground-based infrared observations of silicate emission (e.g., *Hanner et al.,* 1987) indicate that for much of the time grains <20 μm in size dominate the coma.

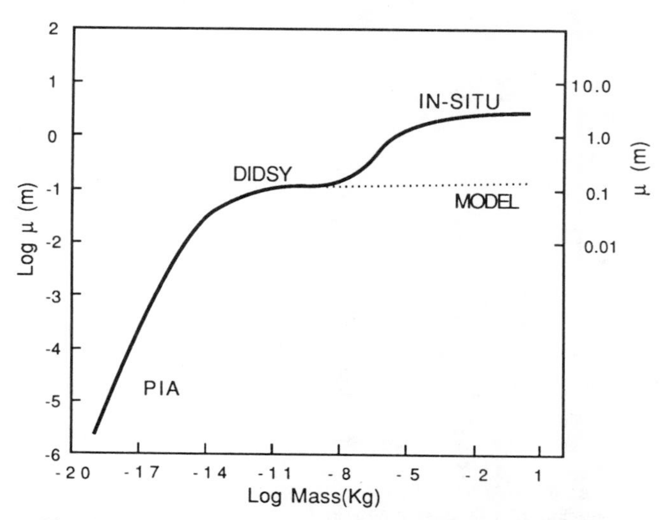

Fig. 5. Dust-to-gas mass ratios μ as a function of the largest grains included (m).

The analysis above is based on the assumption of a dust ejection velocity distribution of the form described by *Divine* (1981) and *Gombosi* (1986). In particular, the mass of the largest ejectable grain is very uncertain. In extreme cases, up to splitting of cometary nuclei, there is sufficient energy to lift extremely large masses from the surface. However, the analysis above indicates that the very largest "grains" do not dominate the total mass emitted by comet Halley.

AREA AND MASS DISTRIBUTIONS

Using the above distributions, the numbers of dust particles embedded in any volume of interest may be calculated. Figures 6 and 7 display the number, area, and mass distributions in 1 m^3 in the coma (600 km from the nucleus) and in the nucleus matrix itself. The differential distribuions $\mathfrak{n}_c(m)$, $A_c(m)$, and $\mathfrak{m}_c(m)$ are shown for the IN-SITU distribution (solid line) and the MODEL distribution (dotted line) in Fig. 6. The number of particles per m^3 per log mass interval

$$\mathfrak{n}_c(m)\ \mathrm{dlogm} = (n_c(m)/v_s)\ \mathrm{dlogm} \tag{4}$$

the area of particles per m^3 per log mass interval

$$A_c(m)\ \mathrm{dlogm} = (n_c(m)\pi a^2/v_s)\ \mathrm{dlogm} \tag{5}$$

and the mass of particles per m^3 per log mass interval

$$\mathfrak{m}_c(m)\ \mathrm{dlogm} = (n_c(m)m/v_s)\ \mathrm{dlogm} \tag{6}$$

where v_s = 68.4 km sec^{-1} and is the velocity of the spacecraft relative to the comet and n_c is the differential flux (number of particles impacting the spacecraft per m^2 per second) measured in the coma. Although the majority of grains are of mass $<10^{-9}$ kg, the total grain *mass* is dominated by the largest grains. Remote-sensing observations, which are dependent on the cross-sectional *area,* indicate dominant grain masses $\sim10^{-14}$ kg. The in situ DIDSY data indicate, however, that a significant contribution to the cross-sectional area could come from large grains. Figure 7 shows similar differential distributions for the nucleus material where the number of particles per m^3 per log mass interval is

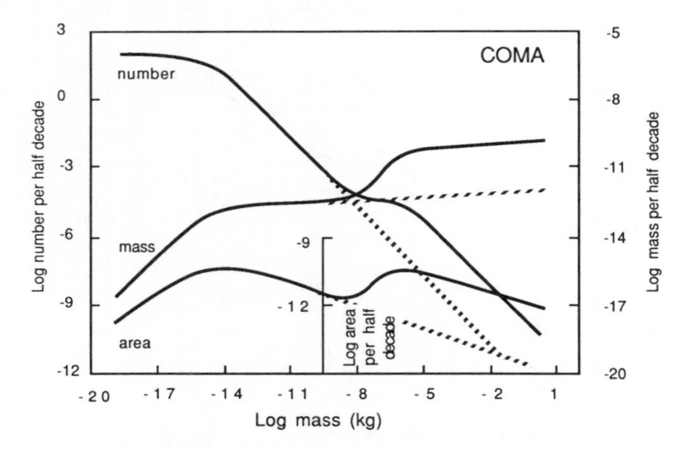

Fig. 6. Differential number, cross-sectional area, and mass distributions of dust grains in the coma. Solid line: IN-SITU distribution; dotted line: MODEL distribution.

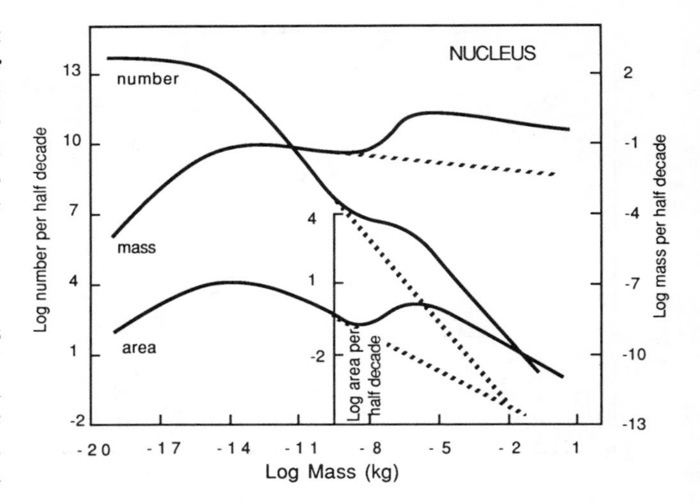

Fig. 7. Differential number, cross-sectional area, and mass distributions of dust grains on the nucleus. Solid line: IN-SITU distribution; dotted line: MODEL distribution.

$$\mathfrak{n}_n(m)\ \mathrm{dlogm} = (n_n(m)/V)\ \mathrm{dlogm} \tag{7}$$

The area of particles per m^3 per log mass interval is

$$A_n(m)\ \mathrm{dlogm} = (n_n(m)\pi a^2/V)\ \mathrm{dlogm} \tag{8}$$

and the mass of particles per m^3 per log mass interval

$$\mathfrak{m}_n(m)\ \mathrm{dlogm} = (n_n(m)m/V)\ \mathrm{dlogm} \tag{9}$$

where V is the total volume of material ejected per m^2 per second from the nucleus. $\mathfrak{n}_n(m)$ is calculated from the dust flux at the nucleus surface $n_n(m)$

$$\mathfrak{n}_n(m) = \frac{n_n(m)}{\displaystyle\int_0^{m_{max}} n_n(m)\frac{m}{\rho(m)_{dust}}\left(1+\frac{\rho(m)_{dust}}{\mu(m_{max})\rho_{ice}}\right)\mathrm{dlogm}} \tag{10}$$

where the denominator is the volume occupied by gas and dust. The dust-to-gas mass ratio $\mu(m_{max})$ is taken from Fig. 5 at m = 10 kg and ρ_{ice} assumed to be 200 kg m^{-3} for this work. The true value of ρ_{ice} may be as high as 800 kg m^{-3} resulting in an increase in $\mathfrak{n}_n(m)$ by factors of 2 and 3.5 for the IN-SITU and MODEL distributions respectively.

The nucleus mass distribution is dominated by grains in the milligram size range. The maximum mass plotted is the calculated largest liftable mass $\sim$10 kg. The total cross-sectional area of grains at the nucleus for the MODEL distribution is dominated by grains $\sim10^{-14}$ kg.

NATURE OF THE NUCLEUS SURFACE

The large masses measured by DIDSY (IN-SITU) are found to significantly enhance the nucleus area and mass distributions above $\sim 10^{-9}$ kg. The nucleus surface, however, appears more "transparent" than for the MODEL distribution if particles are absorbing. The MODEL distribution is optically thick (optical depth $\tau = 1$; 63% attenuation) at a depth of 0.1 mm with grains of size <50 μm dominating. The presence of large mass grains in the IN-SITU distribution (Fig. 5) reduces the optical depth ($\tau = 1$ at a depth of 3.3 mm), allowing incoming radiation to penetrate further into the nucleus. This can be understood by the fact that the distributions $n_n(m)$ apply to the flux leaving the surface *per second* whereas the nature of the surface depends on the cross-sectional area of material *per cubic meter*, $A_n(m)$. The IN-SITU flux distribution has more large grains and more mass leaving the surface per second than for the MODEL case. However, the bulk material of the nucleus [as defined by $n_n(m)$, $A_n(m)$ and $m_n(m)$ for the IN-SITU case] has the same volume, more large grains, but less small ones per m^3 than the MODEL distribution. Since the large grains have a smaller cross-sectional area-to-volume ratio than the smaller ones, the material will be less opaque. The grain extinction efficiencies will depend on the composition and structure of grains but would be expected to be greatest for the micron-sized grains, thus enhancing the effect. Figure 8 shows a schematic representation of the nucleus material for the IN-SITU (Fig. 8a) and MODEL (Fig. 8b) distributions; each box has sides of 1 cm and a depth corresponding to $\tau = 1$. Only particles of radius >10 μm are shown but the contribution of smaller particulates can be inferred from the total attenuation due to all sizes of 63%.

CONCLUSIONS

The dust-to-ice ratio in the nucleus matrix, and hence the probable nature of the nuclear surface, is seen to depend critically on the true distribution of large mass grains at the nucleus. Giotto data indicate an excess of particles up to m ~1 g along its trajectory but gives no information for larger grains or for the coma as a whole. We have shown, however, by other data and by the form of the result that many significant conclusions can be drawn without recourse to exact knowledge of larger particles. Indeed, any particles measured above those encountered by the spacecraft would increase the dust-to-gas mass ratio above a value of 2. The observed excess of large grains over preflight models implies a less opaque surface with greater penetration of radiation to lower levels. In reality, the distribution of dust at the surface (crust of large grains or homogeneous?) and the structure and density of the volatile component will determine the appearance of the surface.

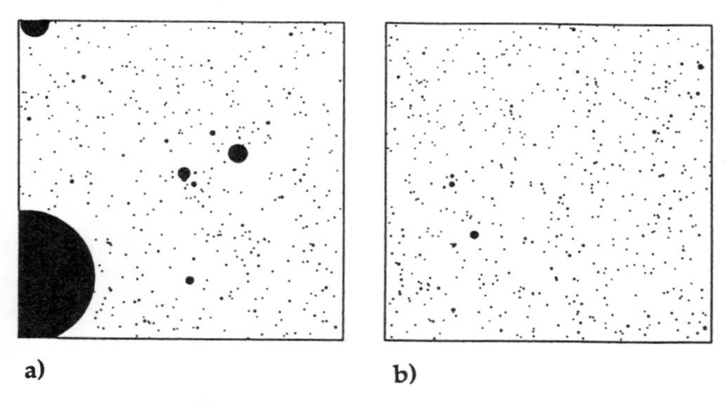

Fig. 8. Schematic representation of nucleus material for IN-SITU and MODEL distributions. In both cases the depth of the box is chosen such that the attenuation due to all particles is 63% ($\tau = 1$); however, only particles with radius >10 μm are shown. **(a)** IN-SITU: 1 cm × 1 cm × 0.33 cm; **(b)** 1 cm × 1 cm × 0.01 cm.

Acknowledgments. We acknowledge the U.K. Science and Engineering Research Council for financial support and provision of research studentships, and thank H. Zook for his constructive comments in reviewing the manuscript.

REFERENCES

Campbell D. B., Harmon J. K., and Shapiro I. I. (1989) Radar observations of comet Halley. *Astrophys. J., 338,* 1094-1105.

Divine N. (1981) Numerical models for Halley dust environments. *ESA SP-174,* pp. 25-30.

Divine N., Fechtig H., Gombosi T. I. Hanner M. S., Keller H. U., Larson S. M., Mendis D. A., Newburn R. L., Rheinhard R., Sekanina Z., and Yeomans D. A. (1986) The Comet Halley dust and gas environment. *Space Sci. Rev., 43,* 1-104.

Eaton N., Davies J. K., and Green S. F. (1984) The anomalous dust tail of comet P/Tempel 2. *Mon. Not. R. Astr. Soc., 211,* 15-19.

Edenhofer P., Bird M. K., Brenkle J. P., Buschert H., Kursinski E. R., Mottinger N. A., Porsche H., and Stelzried C. T. (1987) Dust distribution of comet P/Halley's inner coma determined from the Giotto Radio-Science experiment. *Astron. Astrophys., 187,* 712-718.

ESA (1986) *Comet Nucleus Sample Return.* ESA SP-249.

Fertig J. and Schwehm G. H. (1984) Dust environment models for comet P/Halley: Support for targeting of the GIOTTO S/C. *Adv. Space Res., 4,* 213-216.

Gombosi T. I. (1986) A heuristic model of the comet Halley dust size distribution. *ESA SP-250, Vol. II,* pp. 167-172.

Hanner M. S., Tokunaga A. T., Golisch W. F., Griep D. M., and Kaminski C. D. (1987) Infrared emission from P/Halley's dust coma during March 1986. *Astron. Astrophys., 187,* 653-660.

Harmon J. K., Campbell D. B., Hine A. A., Shapiro I. I., and Marsden B. G. (1989) Radar observations of comet IRAS-Araki-Alcock (1983d). *Astrophys. J., 338,* 1071-1093.

Keller H. U., Delamere W. A., Huebner W. F., Reitsema H. J., Kramm R., Thomas N., Arpigny C., Barbieri C., Bonnet R. M., Cazes S., Coradini M., Cosmovici C. B., Hughes D. W., Jamar C., Malaise D., Schmidt K., Schmidt W. K. H., and Siege P. (1987) Comet P/Halley's nucleus and its activity. *Astron. Astrophys., 187,* 807-823.

Kissel J. (1986) The Giotto particulate impact analyser. *ESA SP-1077,* pp. 67-83.

Krankowsky D., Lammerzahl P., Herrwerth I., Woweries J., Eberhardt P., Dolder U., Hermann U., Schulte W., Berthellier J. J., Iliano J. M., Hodges R. R., and Hoffman J. H. (1986) In situ gas and ion measurements of comet Halley. *Nature, 321,* 326-330.

McDonnell J. A. M. (1987) The Giotto dust impact detection system. *J. Phys. E.: Sci. Instrum., 20,* 741-758.

McDonnell J. A. M., Alexander W. M., Burton W. M., Bussoletti E., Evans G. C., Evans S. T., Firth J. G., Grard R. J. L., Green S. F., Grün E., Hanner

M. S., Hughes D. W., Igenburgs E., Kissel J., Kuczera H., Lindblad B. A., Langevin Y., Mandeville J. C., Nappo S., Pankiewicz G. S. A., Perry C. H., Schwehm G. H. Sekanina Z., Stevenson T. J., Turner R. F., Weishaupt U., Wallis M. K., and Zarnecki J. C. (1987) The dust distribution within the inner coma of comet P/Halley 1982i: encounter by Giotto's impact detectors. *Astrophys J., 187,* 719-741.

McDonnell J. A. M., Pankiewicz G. S., Birchley P. N., Green S. F., and Perry C. .H. (1989) The in-situ cometary particulate size distribution measured for one comet: P/Halley. In *Proceedings of Workshop on Analysis of Returned Comet Nucleus Samples,* NASA Ames Research Center, in press.

Sagdeev R. .Z., Krasikov V. A., Shamis V. A., Tarnopolsi V. I., Szego K., Toth I., Smith B., Larson S., and Merenyi E. (1986) Rotation period and spin axis of comet Halley. *ESA SP-250, Vol. II,* pp. 335-338.

Sykes M. V., Lebofsky L. A., Hunten D. M., and Low F. (1986) The discovery of dust trails in the orbits of periodic comets. *Science, 232,* 115-1117.

Proceedings of the 20th Lunar and Planetary Science Conference, pp. 379-388
Lunar and Planetary Institute, Houston, 1990

A Model Comet Made from Mineral Dust and H_2O-CO_2 Ice: Sample Preparation Development

K. Roessler[1], P. Hsiung[1], H. Kochan[2], H. Hellmann[2], H. Düren[3], K. Thiel[4], and G. Kölzer[4]

Sample analogs for comet nucleus simulation experiments KOSI are prepared by spraying aqueous suspensions of mineral dust into liquid nitrogen. In KOSI 3 (November 1988), CO_2 was introduced for the first time in that series as a representative for volatile components. It is incorporated into the sample by using CO_2 as a propellant gas. Several spraying parameters, such as the flow rate of the mineral suspension through the nozzle and the propellant gas pressure, are varied and the properties of the different materials obtained are studied. The sample material is characterized in particular by visual inspection of structure and texture ("mud"- or "snow"-like material), determination of CO_2 and mineral content, density, porosity, material strength, etc. Especially interesting are studies on the effect of insolation with an artificial light source on dust emission activity in the small simulation chamber. The material finally used for KOSI 3 in the space simulator of DLR Köln-Wahn contained 77.9 wt.% H_2O-ice, 13.8 wt.% CO_2-ice, 8.3 wt.% mineral dust (olivine/montmorillonite 9:1), and 0.08 wt.% carbon (for albedo adjustment) and consisted of fine grains with an appearance in between "mud"- and "snow"-like material. The changes of the sample after 41 hr insolation with 1.3-1.4 SC intensity include material loss by ejection and sublimation, formation of a dry dust mantle at the surface, build-up of a porous ice crust, and condensation of CO_2 in the interior of the sample near or on the cold backplate. The results are discussed in view of improvements for KOSI 4 and following experiments and in view of cometary relevance.

INTRODUCTION

The comet simulation experiments KOSI performed in the space simulator of DLR Köln-Wahn in cooperation with 11 scientific institutions are aimed at yielding information on the properties and dynamics of cometary nuclei, to better interpret the data obtained by the Halley swarm and to plan future sample return missions. Three KOSI experiments have been performed in the course of the last two years; a new one, KOSI 4, was scheduled for April 1989 (*Grün et al.,* 1987, 1988; *Bischoff and Stöffler,* 1988; *Klinger et al.,* 1989; *Kochan et al.,* 1989; *Roessler et al.,* 1988, 1989; *Spohn et al.,* 1989; *Thiel et al.,* 1989). KOSI 1 and 2 experiments used icy sample material prepared from aqueous mineral dust suspensions by spraying it into liquid nitrogen following a method described by *Saunders et al.* (1986) and *Storrs et al.* (1988). In KOSI 3 (29 November - 2 December 1988) CO_2-ice was introduced for the first time as a representative of volatile compounds. According to a proposal of J. Klinger (Grenoble), CO_2 was applied as propellant gas in the spraying procedure instead of N_2 gas used in KOSI 1 and 2.

In a different approach, the group at Dushanbe, USSR, reported the preparation of material containing dust and H_2O ice by simple freezing of suspensions. Samples containing H_2O and CO_2 ice were formed by co-condensation in cold cylindrical cells (*Ibadinov et al.,* 1987; *Ibadinov and Aliev,* 1987). These methods, however, do not allow the incorporation of both CO_2 and dust into the H_2O ice.

The task of the KOSI team was to produce material with the typical composition and properties listed in Table 1. The underlying idea is to build up the analog of a simplified and rather young comet at the beginning of the KOSI studies and to increase the complexity in future experiments. This is one of the reasons to use a relatively low mineral content as compared to the 20 to 70 wt.% discussed for comet P/Halley. Another more technical reason was that suspensions with

TABLE 1. Parameters aimed for KOSI 3 sample.

Composition	
H_2O-ice	75-78 wt.%
CO_2-ice	12-15 wt.%
Mineral dust	10 wt.% (olivine-montmorillonite 9:1)
Carbon	0.08 wt.%
Density	0.3-0.5 g/cm^3 ("fluffy")
Albedo	0.03-0.15
Grain size	100 μm to mm
Mixture	H_2O- and CO_2-ice adherent to dust (layers)
Size	30 cm ϕ, 15 cm height (~11 liters)
Temperature	77 to 150 K

more than 10% of minerals are difficult to control and to spray. Also, the problem of aggregation of mineral grains increases with the concentration. In later experiments (KOSI 6 and following) the mineral content will be changed in quantity and quality. From KOSI 5 on, other components will be added such as methanol, ammonia, formaldehyde, POM and eventually tars. In KOSI 3 the addition of carbon for adjusting the albedo was restricted in order to avoid the simulation of material that was too "old." It has to be mentioned here that the energy input to the sample was not only regulated by the reflectance, but also by the geometry of the sample and the strength of the artificial sun (1.3 to 1.4 SC).

[1]Institut für Chemie 1 der Kernforschungsanlage Jülich, Postfach 1913, D-5170 Jülich, Federal Republic of Germany
[2]Institut für Raumsimulation der DLR Köln-Wahn, Postfach 906058, D-5000 Köln, Federal Republic of Germany
[3]Institut für Planetologie der Universität Münster, Wilhelm-Klemm-Strasse 10, D-4400 Münster, Federal Republic of Germany
[4]Abteilung Nuklearchemie der Universität zu Köln, Zülpicher Strasse 47, D-5000 Köln, Federal Republic of Germany

Fig. 1. Equipment of the small simulation chamber of DLR Köln-Wahn with liquid-nitrogen-cooled sample container ($10 \times 7 \times 6$ cm), drilling device for material strength determination, and active (time dependent) dust collector.

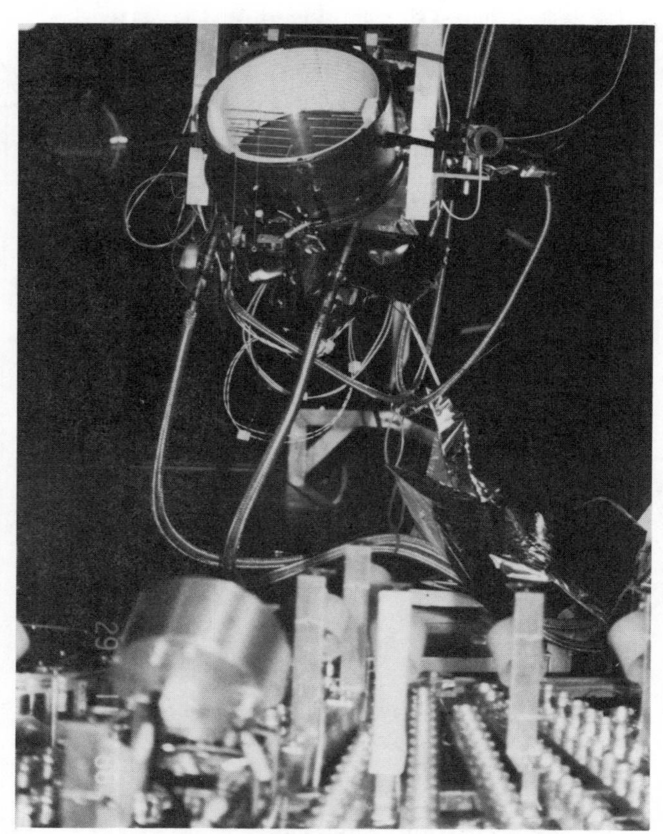

Fig. 2. Sample container of KOSI 3 experiment (30 cm ϕ, 15 cm height) with thermocouples and liquid-nitrogen cooling devices inside the space simulator of DLR Köln-Wahn. In the foreground are passive dust collectors and piezo-detectors.

An important problem that had to be solved for preparative experiments in the small simulation chamber and also for KOSI 3 in the space simulator was to prevent sublimation of CO_2 that in the vacuum becomes vigorous at $T > 150$ K. It was thus necessary to construct new sample containers with effective liquid nitrogen circulation for both chambers (cf. Figs. 1 and 2).

The analytical methods applied to characterize the sample material are listed in Table 2: visual inspection of structure and texture, determination of CO_2 and mineral dust content, and measurement of density and porosity. These analyses were performed (1) on the starting material, (2) after filling the sample containers of the small simulation chamber ($10 \times 7 \times 6$ cm, cf. *Grün et al.,* 1987; *Kochan et al.,* 1989) and space simulator (cylinder of 30 cm diameter and 15 cm height), and (3) after insolation experiments in both chambers. Other analytical methods such as determination of dust emission activity by video cameras, piezo impact detectors, and passive or active (time dependent) dust collection, measurement of material strength by drilling boreholes, albedo determination via optical reflexion spectroscopy, and mineralogical studies on bore kernels were applied to small chamber experiments and/or runs in the space simulator. For reviews of analytical methods applied, including glove box techniques, see *Kochan et al.* (1989), *Roessler et al.* (1989), and *Thiel et al.* (1989). This paper describes sample preparation development by evaluating the effect of various spraying parameters on composition and properties of the material, the selection of conditions for preparation of KOSI 3 material, and finally first results obtained by that experiment. Special emphasis is given to general structure and texture and CO_2 content. Recommendations for improvements are discussed for KOSI 4 and following experiments.

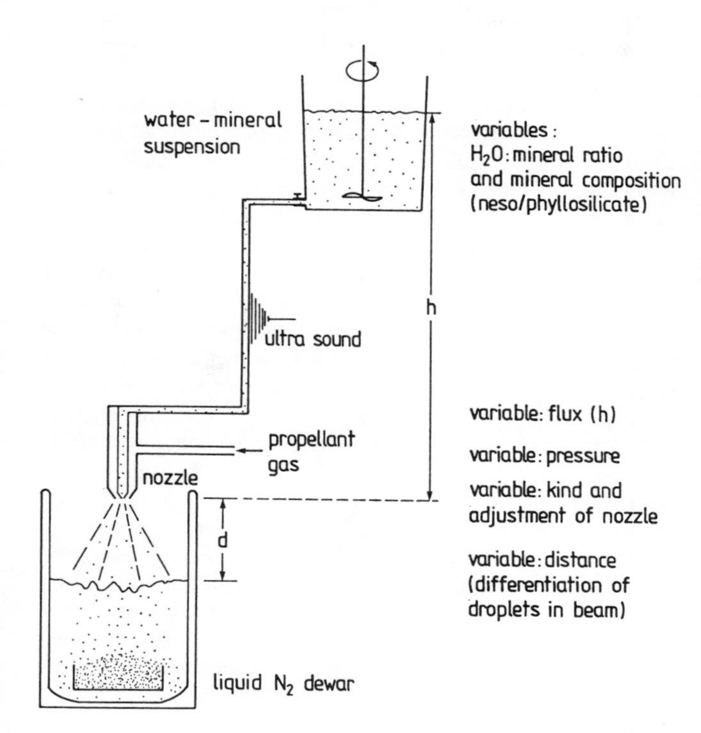

Fig. 3. Schematic drawing of the spraying device for sample preparation. Parameters influencing the composition and quality of the material are indicated at the right side.

SAMPLE PREPARATION

Mixtures of olivine and montmorillonite with diameters ranging from 0.02 to 50 μm and a median distribution of 7.3 to 8.0 μm in the weight ratios 7:3 and 9:1 were suspended in water in a 1:10 weight ratio at the Institut für Planetologie der Universität Münster. All samples contained 0.08 wt.% carbon in the form of an emulsion of finely divided soot (Derussol, Fa. Degussa) in order to adjust low albedos. After transport to DLR Köln-Wahn the suspensions were agitated up to 12 hr by rotating the containers before usage. The device for spraying into liquid nitrogen is shown in Fig. 3. The suspension is stored in a tank and agitated by a rotor in order to prevent precipitation of solid components. On the way to the nozzle, the suspension passes an ultrasound device that is aimed to destroy aggregates of mineral grains. The height of the level of the suspension over the nozzle determines the suspension flow rate. Other parameters that influence the flow rate are suspension composition and grain size distribution. The CO_2 propellant gas pressure varied from 1.8 to 2.5 bar, and the flow rate from 1 to 4 ml/sec^{-1}. Important parameters were the type and the adjustment of the nozzle, which were kept constant through all experiments (Walther Pilot IV from Fa. Rothert, Osnabrück).

During the experiment ice condensation was induced in the nozzle by the cold N_2 gas evaporating from the liquid nitrogen dewar. In order to keep conditions constant, the spraying was interrupted when the flow rate decreased and the nozzles were heated up and cleaned. The position of the nozzle with respect to the liquid nitrogen surface was kept constant by refilling the dewar, since droplets and aerosols coming out of the nozzle can vary their size with increasing distance to the cooling liquid. Another parameter, the radial distribution of material inside the dewar, which was first thought to play a role, proved to be of no significance. By the method described above, it was possible to prepare kilogram amounts of icy material for small chamber experiments within ½ to 1 hr and 10-15 kg for KOSI 3 in approximately 5 hr. The ice-mineral grains showed diameters ranging from 100 μm to some millimeters. The material was put into the sample container and a "dummy." The latter consisted of a pot of the same dimensions as the sample but which was not subjected to vacuum and insolation. The cooling geometry of the dummy inside the glove box was the same as that of the sample inside the space simulator. The dummy was prepared, stored, and analyzed in a dry, cold nitrogen atmosphere (<200 K). After removal from the space simulator the sample was analyzed under the same conditions as the dummy.

DIAGNOSIS OF THE MATERIAL AND INFLUENCE OF SPRAYING PARAMETERS

Table 2 reports schematically the diagnostic methods applied. Two of the methods were of primary importance for a quick check of sample quality:

1. *The visual inspection of structure and texture.* This method indicated two basic kinds of material to be formed. At one extreme were small crystalline samples of some 100-μm diameter that did not show any aggregation. Together with the residual liquid nitrogen they formed somewhat like a "mud" in which the individual grains were barely visible (cf. Fig. 4a). At the other extreme was coarser crystalline material with grain sizes up to 2-mm diameter with a more "snow"-like character. The individual crystals can easily be seen in Fig. 4b. The petrographic analysis of the samples of KOSI 2 (water ice-minerals) by the Institut für Planetologie of the Universität

TABLE 2. Methods of material analysis used in the preparation phase and the KOSI 3 experiment itself.

Methods	Starting Material	Small Chamber Experiments	KOSI 3 in Space Simulator
Visual inspection of structure and texture	+	+	+
CO_2 content	+	+	+
Density	+	+	+
Porosity	+	+	+
Dust content	+	+	+
"Dust activity"		+	+
Material strength by drilling		+	+
Albedo determination via reflexion spectroscopy	+		+
Mineralogy, texture, etc.			+

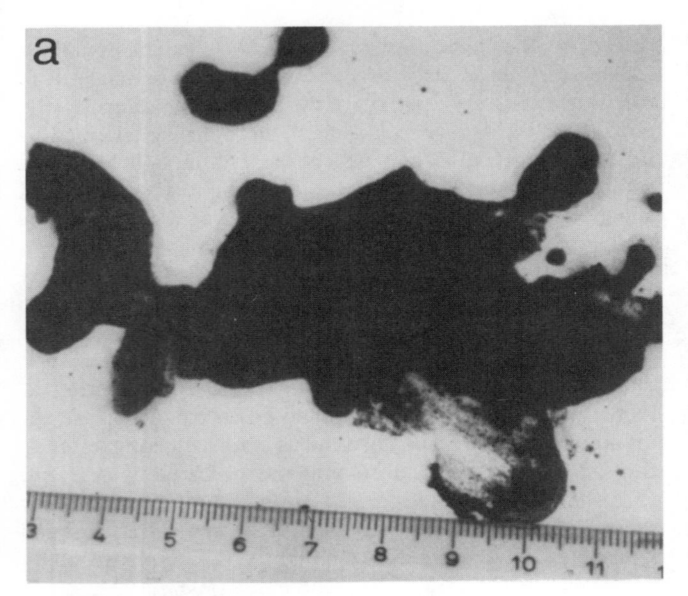

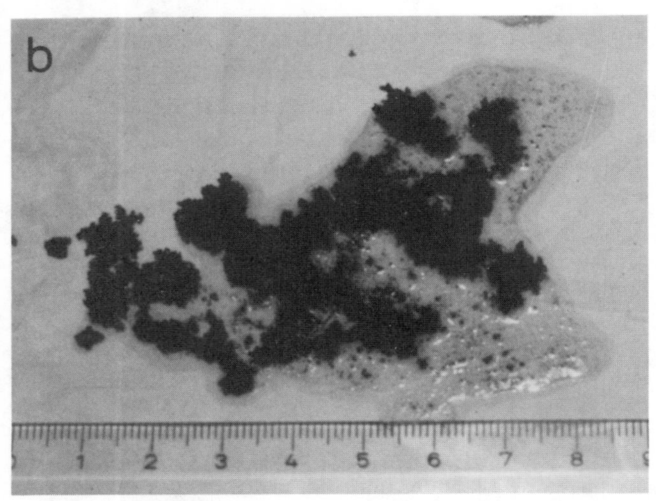

Fig. 4. Typical examples of **(a)** "mud"-like and **(b)** "snow"-like material. Spraying conditions: suspension flow 2.5 ml/sec; propellant gas pressure 2.5 bar (mud), 1.5 bar (snow). The scale is in centimeters. The fine mud crystals in **(a)** are barely visible in the liquid nitrogen blob.

Münster (Prof. D. Stöffler, Frau H. Düren) showed big ice crystals and the dust grains at their surface. It has to be mentioned that the mineral grains form aggregates when in solution. They were, however, segregated to a large extent by the ultrasound treatment before spraying. The petrographic analysis did not show large clusters of minerals in the ice. Larger blobs of H_2O ice that fell from the nozzle or the walls were relatively rare and were manually removed from the sample to a large extent. The samples studied in this work were quite uniform. For the consistency of the material after insolation, see *Thiel et al.* (1989).

2. *The analysis of the CO_2 content.* This was determined by taking samples from the starting material, the small chamber experiments, and KOSI 3 material with precooled tools inside the glove box (*Roessler et al.,* 1989) and putting them into 77 K cold teflon or stainless steel containers of 16 and 13 ml content, which were closed by means of Leybold flanges with a cover part provided with a valve. These samples were stored under liquid nitrogen until analysis. Then the valve was opened, and the sample was heated slowly to room temperature. The gases evolved (N_2 and CO_2) were collected in water-filled cylindrical glass recipients with three septum ports to allow gas samples to be drawn by syringes. Gas chromatography was carried out in a Siemens L 350 chromatograph on 3-m Porapak Q columns with a heat conductivity detector (Fig. 5). The peaks of N_2, CO_2, and H_2O were reasonably separated. From the total amount of gas and its composition on the one hand, and weighing of the aqueous mineral dust remnants in the containers on the other, it was possible to determine the weight percent of CO_2 in the sample. The technique was recently improved by using an aqueous solution containing 5 wt.% H_2SO_4 + 20 wt.% Na_2SO_4 as the liquid in the recipients, which led to a much lower degree of CO_2 absorption (Fig. 6). This avoids corrections for the absorption in H_2O that otherwise would be necessary. Furthermore, the gases were not bubbled through the water but introduced from the top of the recipient. The contact with the water surface was furthermore minimized by using cork swimmers (Fig. 7). Corrections had to be made for small CO_2 contents (small gas amounts) and the variation of the volume with the remaining water column in the recipient.

Figure 8 shows the effect of the suspension flow rate on the CO_2 content for a constant propellant gas pressure of 1.8 bar. It can be seen that the CO_2 content increases up to 15 wt.% with the flow rate decreasing to 1 ml/sec. For higher propellant gas pressures of 2.2 and 2.5 bar the 15 wt.% margin is reached also at higher flow rates (2-3 ml/sec). The scattering of the points is characteristic in particular for the variation of the product composition with geometry of the nozzle (see above) and a still inhomogeneous suspension, at least at the moment of reaching the nozzle.

Some general tendencies were observed. The higher the suspension flow and the lower the CO_2 propellant gas pressure, the lower the CO_2 content, and the more "snow"-like the material. In general this material shows a somewhat higher activity in dust emission in small chamber experiments. This, however, is also a function of CO_2 content, thus rendering unequivocal statements difficult. A higher content of CO_2 (lower suspension flow, higher CO_2 pressure) yields more "mud"-like samples with separate, fine CO_2 crystals. Furthermore, a higher content of nesosilicates (olivine) seems in general to favor "mud"-like material.

Some properties of materials prepared for small chamber experiments, determined in addition to general structure and texture and CO_2 content by the methods mentioned in Table 2, are compiled in Table 3 and include the results obtained for KOSI 3.

The small chamber experiments were carried out at an average pressure of $\approx 10^{-5}$ mbar, the KOSI 3 run at approximately 2×10^{-6} mbar. Temperatures of the samples ranged from 80 K at the backplate to 200 to 210 K at the surface for both small and big chamber experiments. The temperatures were measured with thermocouples. PT-100 resistors showed the same results. The measurement is relatively reliable since the temperature changes proceed slowly. The insolation of KOSI 3 was performed in four steps: a first period of 10 hr at

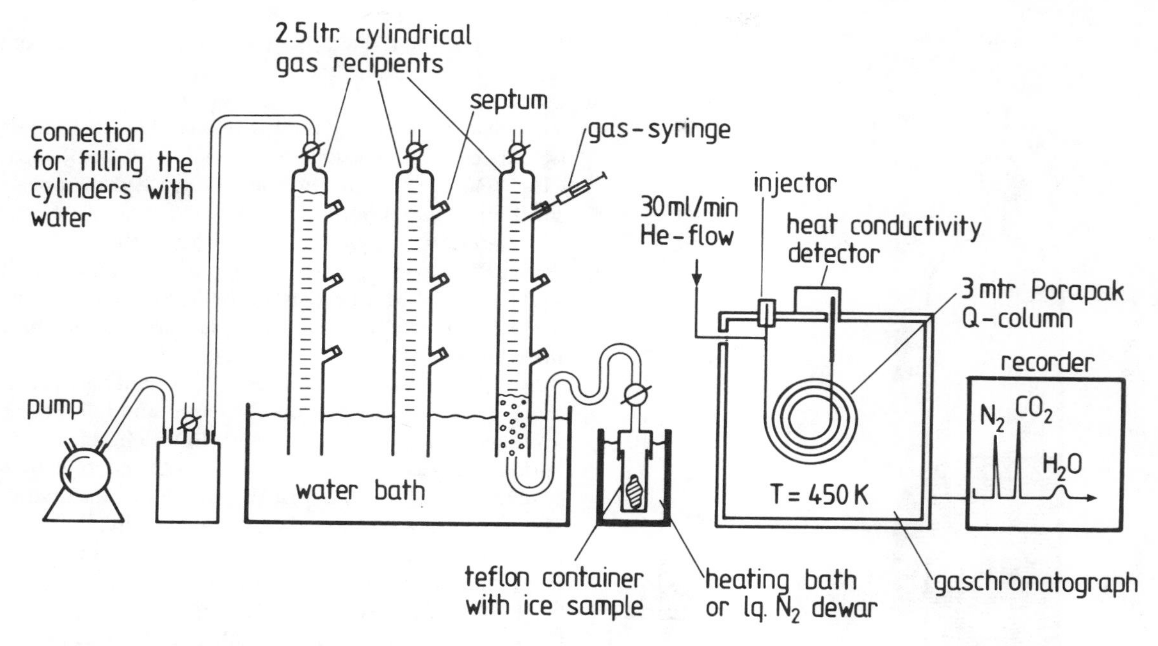

Fig. 5. Array for determination of CO_2 content including sample holder with valve, glass recipients for N_2 and CO_2 gas inside a bath, and gas chromatography of the samples on 3-m Poropak Q columns.

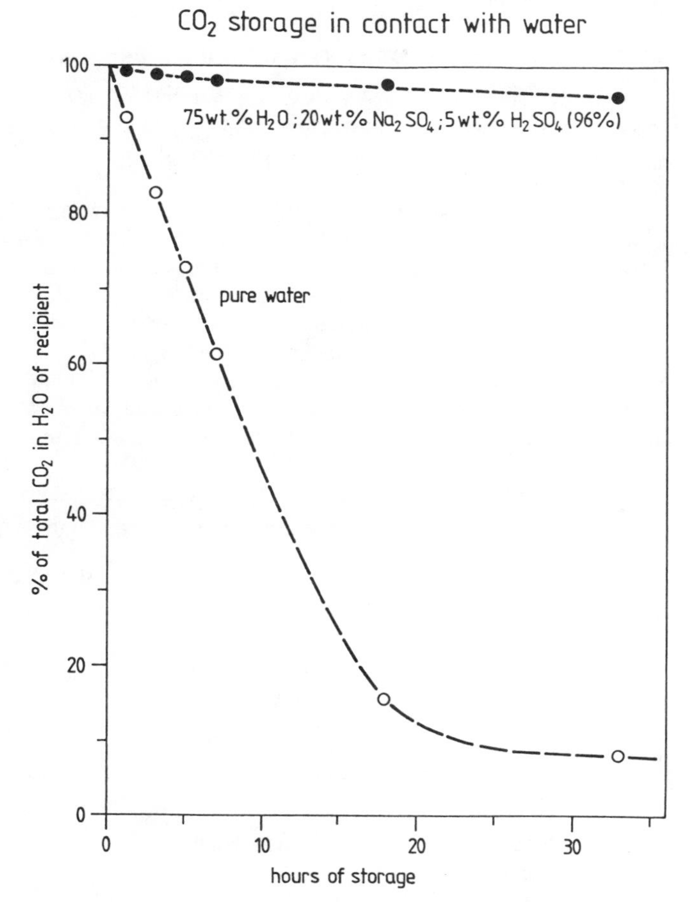

Fig. 6. The effect of addition of H_2SO_4 and sulfate salts on the dissolution of CO_2 in water.

1.3 SC, a dark phase of 6 hr, a second insolation period of 20 hr at 1.3 SC, and a final phase of 2 hr at 1.4 SC. The insolation of the small chamber samples was performed without interruption. For further details on small and big chamber experimental setup, see *Grün et al.* (1987, 1989) and *Kochan et al.* (1989).

From Table 3 it can be seen that the interference of some of the spraying parameters, in particular the geometry of the nozzle, can lead to unexpected results. However, the demand of densities between 0.3 and 0.5 g/cm^{-3} and CO_2 content of about 14% could easily be fulfilled. The albedos were in general sufficiently low (<0.2). During solar irradiation, crusts were formed whose thicknesses and strengths were measured via the borehole drilling facility (ESA-ESTEC, G. Schwehm, Nordwijkerhout; cf. also *Thiel et al.,* 1989). From the above results the preparation conditions for KOSI 3 material were derived: suspension flow rate, 2.0 ml/sec; CO_2 propellant gas pressure, 1.6 to 1.7 bar; distance from nozzle to liquid nitrogen surface, 10-12 cm; and olivine/montmorillonite ratio 9:1. During the actual preparation it became necessary to increase the CO_2 pressure up to 1.9 bar, since the CO_2 content remained relatively low. Hence, the material obtained showed an appearance in between "mud" and "snow" (cf. Table 3).

THE KOSI 3 RESULTS IN VIEW OF MATERIAL CHANGES

The starting material contained 77.9 wt.% H_2O ice, 13.8 wt.% CO_2, 8.3 wt.% mineral dust, and 0.08 wt.% carbon. The rather low mineral content was due primarily to the dilution with CO_2 ice, but also to co-condensation of atmospheric moisture and CO_2 that were dragged by the nozzle beam into the liquid nitrogen (J. Knölker, Münster). The density (of the dummy) was 0.48 g/cm^3. Figure 9 shows the KOSI 3 sample inside the space simulator of DLR Köln-Wahn before, and Fig. 10 shows

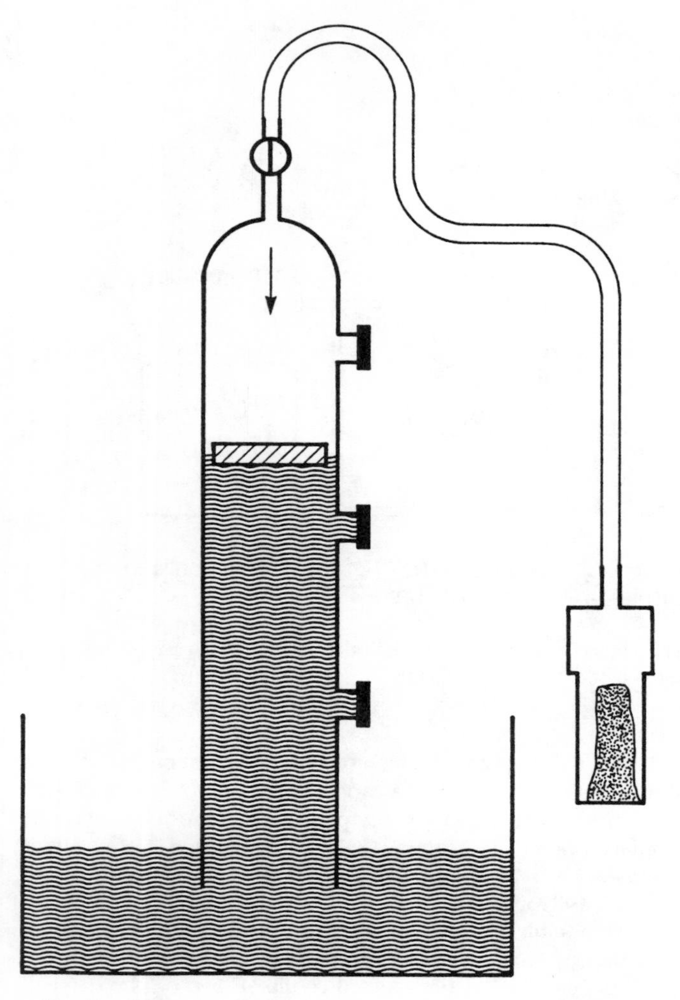

Fig. 7. New array for avoiding contact of CO_2 to the water of the recipient.

the same sample after, 41 hr insolation with 1.3 to 1.4 SC intensity. The albedo (500-2000 nm) changed from 0.192 before to 0.148 after irradiation (A. Oehler, DLR Oberpfaffenhoren; cf. also *Roessler et al.,* 1989). A layer of 2 to 5 cm of the material had been removed by evaporation of the volatiles. The surface had become rougher than before. A 1-3-cm-thick mantle of dry dust had formed asymmetrically because of mechanical motion along the tilted sample surface (inclination 45° toward the sun normal and 30° toward the vertical plane on the sun normal). Below the dust mantle was a 3- to 5-cm-thick porous ice crust. Figure 11 summarizes the visual inspections. The mineral content of the crust proved to be somewhat higher than for the underlying material, which yielded almost no visual changes. In the region near the 77 K cold backplate and on the plate itself a white layer was observed.

The CO_2 analysis of the dummy material in Fig. 12 (left) showed that the filling operation in the nitrogen-flushed glove box with the dummy container inside a liquid nitrogen bath (*Roessler et al.,* 1989) did not give rise to detrimental loss of CO_2. Only a few weight percent are lacking at the surface. It should be mentioned that the actual filling conditions of the proper sample inside the space simulator and the cooling of the surface were less optimal than in the glove box. Thus, a somewhat higher loss than that observed for the dummy could be expected. After insolation (41 hr, 1.3-1.4 SC) most of the CO_2 (630 to 650 g of the total 770 g) had disappeared. The small amounts found near or at the surface are certainly due to secondary condensation of atmospheric CO_2 during demounting of the sample in the space simulator and transport to the glove box for analysis. In the deeper layers of the sample some 50-70 g of CO_2 remained, in particular in the form of the white crust near or on the backplate.

The CO_2 contents given in Fig. 12 are not very precise since the material was not removed by layers, and CO_2-poor material from above may have mixed with CO_2-rich material from the bottom. At least one value shows a CO_2 content of 20 wt.%, significantly higher than that of the original material (13.8%). The observations prove a significant inward diffusion of the CO_2. The loss of CO_2 goes well with the finding of a smooth evaporation of H_2O and CO_2 in molar ratios between 4:1 and 6:1 by mass spectrometry of the gas phase over the sample (D. Krankowsky, P. Lämmerzahl, and D. Hesselbarth, MPI-Kernphysik, Heidelberg). Last but not least, the temperature profiles within the sample show a steep rise at a depth where the CO_2 depleted (upper) zone begins (T. Spohn and J. Benkhoff, Münster).

RECOMMENDATIONS FOR FURTHER KOSI EXPERIMENTS

From the successful KOSI 3, with its findings of a dust mantle, crust formation, inward diffusion of CO_2, dust and gas emission, albedo changes, thermal properties, etc., some improvements for further KOSI experiments can be derived. First, it seems necessary to modify the spraying apparatus in a way that avoids demixing of the suspension in the pipes leading to the nozzle, e.g., by loops circulating the suspension between reservoir and outlet to the nozzle. Since the addition of CO_2 in weight percentages of 13-15 created some drawbacks as to the sample homogeneity, and since the CO_2 was present in the sample mostly as isolated small ($\approx$100-μm diameter) crystals anyhow, returning to the spraying of pure aqueous suspensions with N_2-propellant gas and adding the

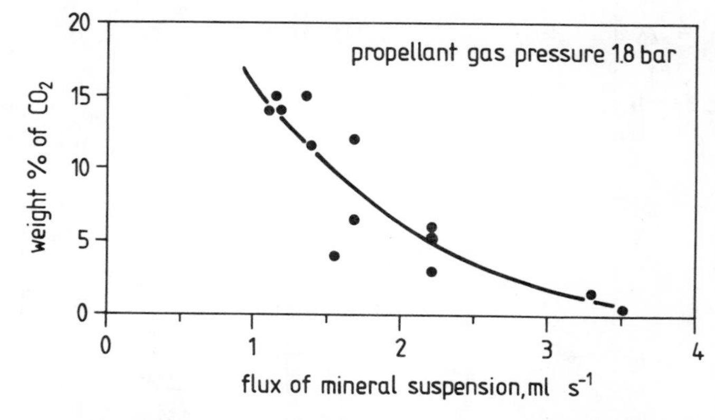

Fig. 8. CO_2 content in weight percent of mineral dust H_2O-CO_2-ice mixtures as a function of the suspension flow rate in the spraying device (10 wt.%; olivine/montmorillonite 9:1).

TABLE 3. Typical parameters of small chamber experiments 8-12 (October 18, 19, 20, 25, and 28, 1988) and the big chamber experiment KOSI 3 (November 29-December 2, 1988).

Experiment	8	9	10	11	12	KOSI 3
Material Before Insolation						
Mineral composition (olivine:montmorillonite)	7:3	9:1	7:3	7:3	9:1	9:1
Spraying pressure (atm.)	1.8	1.8	1.8	1.8	2.5	1.6-1.9
Spraying flow (ml/sec)	1.7-2.0	~2.2	~1.6	1.3-1.4	2.9-3.3	~2.0
Content of CO_2-ice (wt.%)	6.5	4.0	5.0	15.0	13.0	13.8
Density (g/cm^3)	0.41	0.43	0.38	0.36	0.41	0.48
Porosity (%)	—	—	59	63	60	57
Texture	mud	snow	snow	mud	snow	mud/snow
Intensity of irradiation	2.0-2.4	2.0	2.0-2.4	2.0-2.4	2.0-2.4	1.3-1.4
Period of irradiation (hr:min)	3:38	2:12	2:15	2:45	3:38	41:10
T_i, T_f (K) 2 cm below surface*	149-196	143-209	151-218	163-211	164-222	100-~150
Max. dust activity (μg/cm^2 min)	—	132	19	607	173	—
Material After Insolation						
Thickness of crust (mm)	20-25	16-18	20-24	10-26	20-42	28-70
Strength of crust (MPa)	0.30-1.3	0.43-0.55	0.15-0.19	0.75-1.10	0.35-0.88	1.3-5.1
Strength below crust (MPa)	0.1-0.3	0.04-0.06	0.03-0.05	0.08-0.24	0.03-0.08	0.2-0.5
No. of strength measurements	4	7	9	5	6	7

*T_i, T_f = initial and final temperature of the sample during the experiment.

CO_2-ice separately by mixing perhaps should be considered. In order to avoid the dragging of atmospheric H_2O, CO_2, Ar, etc. into the liquid nitrogen, the path from nozzle to nitrogen surface should be better isolated from the surrounding air by shieldings. The mineral-to-volatile ratio obtained seems to be too low with respect to the approximately 1:1 ratios discussed at present for comet P/Halley. Suspensions containing a 15 to 20 wt.% mineral dust component could still be processed in the spraying apparatus. Keeping the mineral content as high as possible and avoiding its dilution by the dragging of atmospheric gases should be attempted in future experiments. The determination of CO_2 content could be greatly improved by removing the sample in centimeter layers and analyzing each of them. Thus, a proper CO_2 profile could be obtained, maybe also an indication of an enrichment zone in deeper layers, such as already observed qualitatively in KOSI 3. The inward diffusion of volatiles may depend on their partial pressure in a way that the lighter ones are condensed in deeper (colder) regions of the sample. This may lead to separated enrichment zones in a kind of "inward thermochromatography." In order to observe this effect, it would be necessary to reduce the total thermal fluence to preserve interesting profiles in the middle of the sample.

The preparation of KOSI materials by spraying suspensions into liquid nitrogen has a major drawback: It includes liquid water in its reaction with the grains, which is rather untypical

Fig. 9. KOSI 3 sample in space simulator of DLR Köln-Wahn before the experiment.

Fig. 10. KOSI 3 sample in space simulator of DLR Köln-Wahn after 41 hr insolation with 1.3 to 1.4 SC intensity.

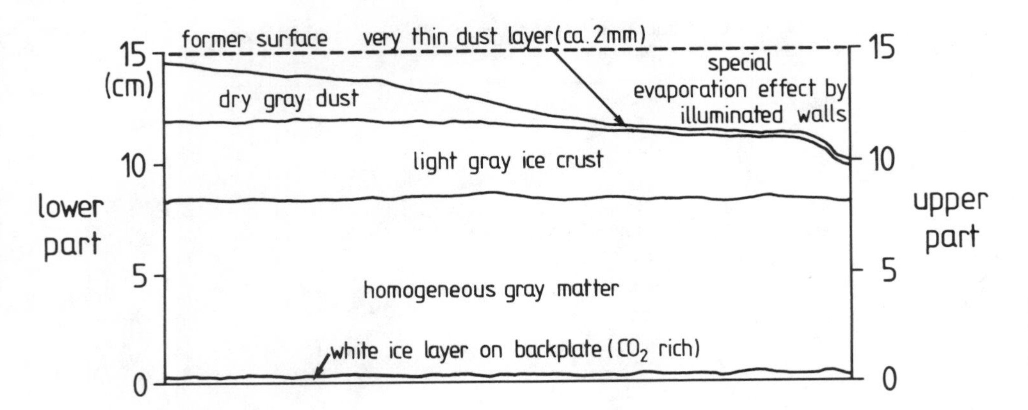

Fig. 11. Schematic view of KOSI 3 sample after the experiment (material loss, dust mantle, crust, CO_2 layers near backplate).

for the space environment. Furthermore, the samples always contain a certain amount of liquid or solid nitrogen (when bringing them into vacuum), the sudden evaporation of which can be responsible for special dust ejection events at the beginning of the insolation period. Thus, the material for KOSI 6 and following experiments will be prepared from the gas phase. Figure 13 shows the apparatus for preliminary test runs that was developed in Jülich. Precooled dust is transported by a cold-gas jet from a liquid-He dewar into a He-purged glove box. The dust beam is crossed by water gas beams. The majority of condensation events take place on the surface of the grains; some H_2O condenses together with the dust on the liquid-N_2-cooled catcher plate. It is possible at present to prepare some 100 g of very fluffy ($g \leq 0.1 g/cm^3$) icy material within an hour. The dust-to-water ratio can be easily modified in the range of 2:1 to 1:2. This procedure will probably *not* yield amorphous ices, since the heat coupling via the He-gas (1 bar) and the warm H_2O vapor causes the condensation to take place at temperatures >130 K (cooling rates <10^5 K/sec). Amorphous ice would be an excellent material to study primary effects of young comets. It can be prepared by condensation in the vacuum, a method that is at present developed by the Jülich and Tel-Aviv (Prof. Bar-Nun) groups. However, the production rates are rather small and the competent incorporation of dust is not yet achieved. Thus, the method of condensation in the He-gas phase, presented here, seems to be at present a reasonable compromise.

RELATION TO REAL COMETS

The residue after insolation showed some of the typical features discussed for comets: (1) a dry dust mantle at the surface; (2) a relatively thick and strong crust of crystalline ice and minerals that, however, is porous and allows gases to pass through; and (3) an almost unchanged region underneath, even if the CO_2 is lacking. The fact that the CO_2 leaves the sample without destroying the overall morphology is one of the interesting results of KOSI 3. CO_2 (and other volatiles likewise) seems to behave independently from the bulk H_2O ice. Even more important is the finding of inward diffusion of CO_2 and its condensation at the bottom plate. For a real comet, this may have the consequence that the concentration of frozen volatiles becomes higher in greater depths from revolution to revolution, until a point is reached where an individual heat transfer induced sudden evaporation, a catastrophic event needed for the emission of gas and dust jets. The coma composition thus cannot be a good clue for that

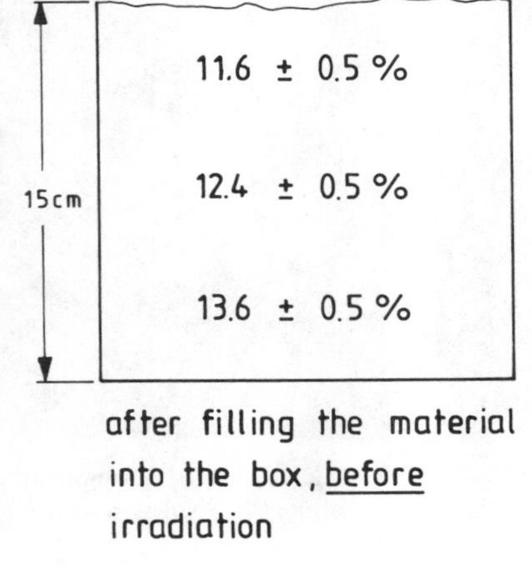

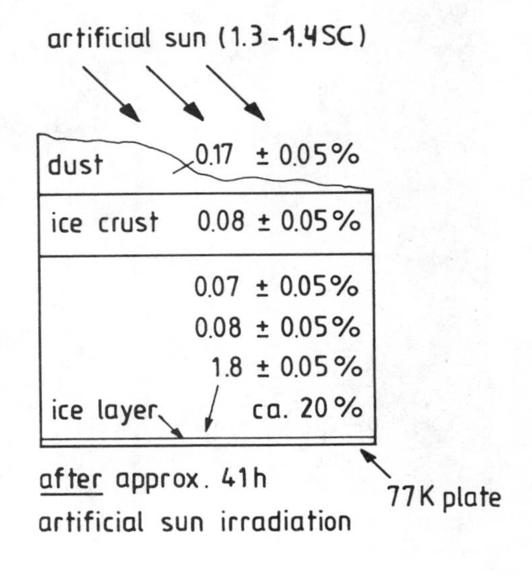

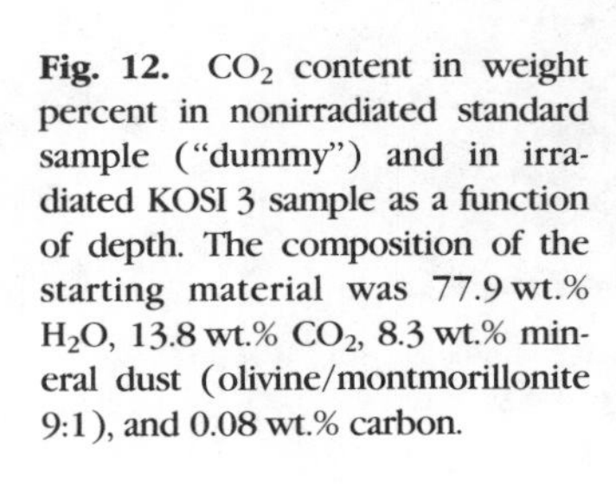

Fig. 12. CO_2 content in weight percent in nonirradiated standard sample ("dummy") and in irradiated KOSI 3 sample as a function of depth. The composition of the starting material was 77.9 wt.% H_2O, 13.8 wt.% CO_2, 8.3 wt.% mineral dust (olivine/montmorillonite 9:1), and 0.08 wt.% carbon.

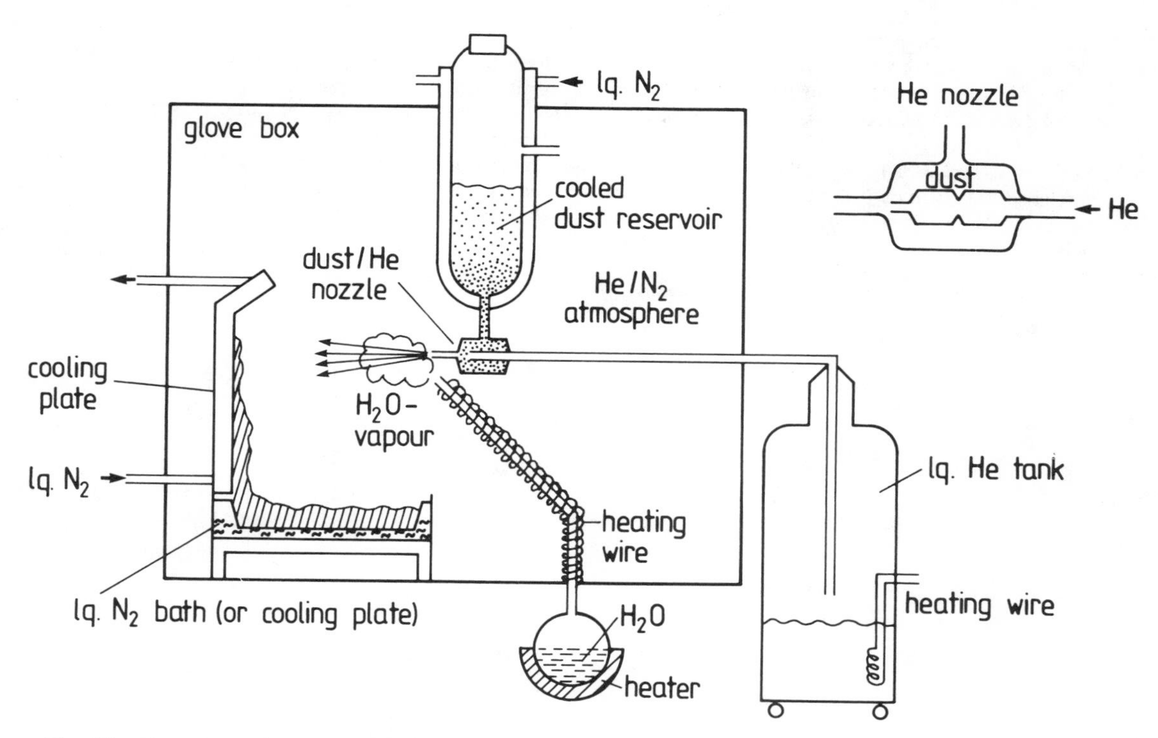

Fig. 13. New approach for condensation of water vapor on cold mineral dust grains carried by a cold He jet in a He-purged glove box. This method, developed in Jülich, avoids the contact of mineral grains with the liquid phase and allows for a wide variation of the dust-to-volatile ratio.

of the nucleus. These results are of a general kind and do not depend too much on the individual material studied.

One of the primary requests for analog materials was that the mineral grains were not too connected to each other in the beginning. The method of preparation presented here was able to fulfill this demand. As to the size of the individual particles, nothing is known for the cometary nucleus. The size distribution measured in the coma at $\geq$ 900 km distance has no bearing on the size in the real comet. The 7-μm mineral grains of the suspensions used in this experiment and the 100-μm to 2-mm ice-mineral grains formed by spraying may be somewhat large in the general cometary context. The thermal dynamics studied in these grains can nevertheless give interesting hints on the behavior of a real comet.

Acknowledgments. This study is financially supported by the Deutsche Forschungsgemeinschaft, Bonn, within its Schwerpunktprogramm "Kleine Körper im Sonnensystem." The authors thank the members of the Institut für Raumsimulation of DLR Köln-Wahn for running the small simulation chamber and the space simulator. Particular gratitude is expressed to G. Eich and M. Heyl (Jülich), W. Feibig (Köln), Prof. D. Stöffler and J. Knölker (Münster), and A. Oehler (Oberpfaffenhofen).

REFERENCES

Bischoff A. and Stöffler D. (1988) Comet nucleus simulation experiments: Mineralogical aspects of sample preparation and analysis (abstract). In *Lunar and Planetary Science XIX*, pp. 90-91. Lunar and Planetary Institute, Houston.

Grün E., Kochan H., Roessler K., and Stöffler D. (1987) Simulation of cometary nuclei. *ESA SP-278*, pp. 501-508.

Grün E., Kochan H., Roessler K., and Stöffler D. (1988) Initial comet simulation experiments at DFVLR. In *Experiments on Cosmic Dust Analogues* (E. Bussoletti et al., eds.), pp. 17-23. Kluwer Academic Publ., Dordrecht.

Grün E., Bar-Nun A., Benkhoff J., Bischoff A., Düren H., Hellmann H., Hesselbarth P., Hsiung P., Keller H. U., Klinger J., Knölker J., Kochan H., Kohl H., Kölzer G., Krankowsky D., Lämmerzahl P., Mauersberger K., Neukum G., Oehler A., Ratke L., Roessler K., Spohn T., Stöffler D., and Thiel K. (1989) Laboratory simulation of cometary processes. *Proc. Comets in the Post-Halley Era*, Bamberg, 24-28 April 1989, in press.

Ibadinov Kh. I. and Aliev S. A. (1987) Sublimation characteristics of H_2O nucleus with CO_2 impurities. *ESA SP-278*, pp. 717-719.

Ibadinov Kh. I., Aliev S. A., and Rahmonov A. A. (1987) Physical-mechanical properties of matrixes on the comet nuclei surface models. *ESA SP-278*, pp. 713-716.

Klinger J., Benkhoff J., Espinasse S., Grün E., Ip W., Joó F., Keller H. U., Kochan H., Kohl H., Roessler K., Seboldt W., Spohn T., and Thiel K. (1989) How far do results of recent simulation experiments fit current models of cometary nuclei? *Proc. Lunar Planet. Sci. Conf. 19th*, pp. 493-497.

Kochan H., Benkhoff J., Bischoff A., Fechtig H., Feuerbacher B., Grün E., Joó F., Klinger J., Kohl H., Krankowsky D., Roessler K., Seboldt W., Thiel K., Schwehm G., and Weishaupt U. (1989) Laboratory simulation of a cometary nucleus: Experimental setup and first results. *Proc. Lunar Planet. Sci. Conf. 19th*, pp. 487-492.

Roessler K., Bischoff A., Eich G., Grün E., Fechtig H., Joó F., Klinger J., Kochan H., Stöffler D., and Thiel K. (1988) Cometary matter in observation and simulation experiments (abstract). In *Lunar and Planetary Science XIX*, pp. 996-997. Lunar and Planetary Institute, Houston.

Roessler K., Hsiung P., Heyl M., Neukum G., Oehler A., and Kochan H. (1989) Handling and analysis of ices in cryostats and glove boxes in view of cometary samples . In *Proceedings of the Workshop on Analysis of Returned Comet Nucleus Samples.* NASA Ames Research Center, Moffett Field, in press.

Saunders R. S., Fanale F. P., Parker T. J., Stephens J. B., and Sutton S. (1986) Properties of filamentary sublimation residues from dispersion of clay in ice. *Icarus, 47,* 94-104.

Spohn T., Benkhoff J., Klinger J., Grün E., and Kochan H. (1989) Thermal histories of the samples of two KOSI comet nucleus simulation experiments. In *Proceedings of the Workshop on Analysis of Returned Comet Nucleus Samples.* NASA Ames Research Center, Moffett Field, in press.

Storrs A. D., Fanale F. P., Saunders, R. S., and Stephens J. B. (1988) The formation of filamentary sublimate residues (FSR) from mineral grains. *Icarus, 76,* 493-512.

Thiel K., Kochan H., Roessler K., Grün E., Schwehm G., Hellmann H., Hsiung P., and Kölzer G. (1989) Mechanical and SEM analysis of artificial comet nucleus samples. In *Proceedings of the Workshop on Analysis of Returned Comet Nucleus Samples.* NASA Ames Research Center, Moffett Field, in press.

Proceedings of the 20th Lunar and Planetary Science Conference, pp. 389-399
Lunar and Planetary Institute, Houston, 1990

Crustal Evolution and Dust Emission of Artificial Cometary Nuclei

K. Thiel[1], G. Kölzer[1], H. Kochan[2], E. Grün[3], H. Kohl[3], and H. Hellmann[2]

A series of experiments of the Kometensimulation Project (KOSI) yielded an improved understanding of near-surface processes of comet nucleus analogues under artificial insolation. Irradiation-induced sublimation of superficial ice leads to the development of a loose, fluffy dust mantle and a solidified but porous dust/ice crust within the upper region of the sample material. The thickening of the dust mantle during irradiation increasingly quenches the dust emission activity and reduces the emission of larger particles. A variety of dust-detecting and collecting devices allows the study of angle, velocity, and size-frequency distributions of emitted dust particles as a function of the main experimental parameters (sample material, irradiation intensity, surface temperature, chamber pressure, exposure time). The angular distribution of emitted particles turned out to be velocity and size dependent: large, low-speed particles are ejected in a steep forward orientation along the surface normal of the model "comet"; small high-speed grains show a broader emission maximum. The greatest observed velocity of a particle >100 μm was 5.2 m/sec. The relevance of the experiments to current comet observations is discussed.

1. INTRODUCTION

For more than two decades, the experimental simulation of cometary processes has been carried out in small (ce centimeter-sized) laboratory devices mainly by groups in the Soviet Union and the United States (*Kajmakov and Sharkov,* 1967a,b; *Donn and Urey,* 1956; *Saunders et al.,* 1986). A comprehensive list of publications on comet simulation is given in *Grün et al.* (1987).

With the beginning of the Kometensimulation (KOSI) Project in 1987 (*Grün et al.,* 1987), comet simulation was extended to larger facility dimensions (target diameter, 30 cm; target depth, 13-14 cm; chamber dimensions, 4.5 × 5.2 m), opening a wealth of new possibilities (*Grün et al.,* 1988; *Kochan et al.,* 1989a,b; *Roessler et al.,* 1989; *Klinger et al.,* 1989) mainly based on the utilization of larger scale lengths inside and outside the sample and newly developed diagnostics to study the relevant parameters. Our knowledge to ask more precise questions about comets after the comet P/Halley missions and the demand for more reliable data on soil mechanical parameters of cometary nuclei in view of the planned CNSR/Rosetta-missions stimulated new efforts in this project.

The present paper summarizes recent results on crust formation and dust emission of comet nucleus analogues obtained in the simulation chambers (cf. *Kochan et al.,* 1989a) at DLR/Köln.

[1]Abteilung Nuklearchemie der Universität zu Köln, Zülpicher Strasse 47, D-5000 Köln 1, F. R. Germany
[2]Institut für Raumsimulation, DLR, WB-RS, Postfach 906058, D-5000 Köln 90, F. R. Germany
[3]Max-Planck-Institut für Kernphysik, Postfach 103980, D-6900 Heidelberg 1, F. R. Germany

2. DIAGNOSTIC DEVICES

2.1. Video Observation

High-speed-shutter video cameras are used for optical observations either within the Space Simulator, in special containments, or outside the small simulation chamber monitoring the experiments via an optical window flange. The cameras are equipped with telemacro lenses to record the sample surface at high (8×-25×) magnifications or with wide-angle lenses to survey the trajectories of flying dust particles in the near-surface space.

This video equipment allows on-line recording of dust emission activity, angle, velocity, and size distributions of particles with diameters >100 μm as well as the observation of dynamic processes of surface features on a submillimeter scale.

2.2. Recording of Microseismic Events

By means of sensitive accelerometers mounted on the sample container, microseismic events occurring in the target material can be detected in well-defined frequency windows between 10 kHz and 2 MHz. The system (Brüel & Kjaer, SE-accelerometer type 8314, resonance frequency ≈800 kHz) is capable of recording faint high-frequency cracks that indicate mechanical stress relaxation during "active comet" phases of the experiments. These diagnostics are employed to reveal a possible time correlation between microcracks and particle emission.

2.3. Piezo Dust Detectors

By means of eight Piezo dust detectors in front of the model comet, ice-containing particles that are large enough (>250 μm) to induce reliable signals can be analysed with respect to size, mass, and velocity (*Kohl,* 1989). The sensitive part of the detectors consists of a 50-mm-diameter ceramic plate of 5-mm thickness. An impacting particle produces a voltage pulse the height U and the rise time t_r of which is related to its diameter $d = t_r \cdot c_p/(2n)$ and its momentum $p = const \cdot U \cdot t_r$. The symbols c_p and n stand for the sound velocity in the projectile and the number of complete passages of the impact pulse through the projectile (back *and* forth).

The calibration of the detectors was carried out by means of a low-velocity particle accelerator under vacuum conditions. Well-defined spherical and irregularly shaped particles of ices, minerals, ice-mineral mixtures, and porous dust residues similar to those produced in KOSI experiments were used in a velocity range between 0.8 and 8.9 m/sec to produce the calibration signals (*Kohl et al.,* 1989). Since the Piezo detectors only register particles that are hard enough (ice containing) and large enough (>250 μm in diameter) to produce data that can be evaluated, more sensitive particle diagnostics is presently being developed.

TABLE 1. Basic parameters for the experiments KOSI 1 to KOSI 4.

Experiment Big Space Simulator	KOSI 1 May 1987	KOSI 2 Apr. 1988	KOSI 3 Nov. 1988	KOSI 4 May 1989
Composition (wt.%)				
H_2O ice	90	89.9	77.9	77.6
CO_2 ice	—	—	13.8	13.8
Mineral dust	10	10.1	8.3	8.6
Dust composition (Total dust = 100 wt.%)				
Olivine	—	90	90	90
Montmorillonite	—	10	10	10
Kaolinite	100	—	—	—
Carbon	1	0.083	0.083	0.083
Density (g/cm³)	0.39-0.46	0.48-0.59	0.48	0.51
Porosity (%)	~60	40-60	50-60	50-60
Texture	"snow"	"snow"	"mud/snow"	"snow"
Penetration strength (MPa)	n.d.	0.2	0.2	0.1
Irradiation sequence time (h)/Intensity (SC)	2.2/0.7 12.1/1-1.6	17.4/1.6 4/0 4/1.6 4/0 4/1.6 4/0 0.25h each: 0.4; 0.8; 1.2; 1.6SC 1/2.9-3.0	10.3/1.3 6/0 28.8/1.3 2/1.6	0.5h each: 0.1; 0.2; 0.5 SC 20.0/0.6 6.3/0 12.0/0.6 4.4/0.9 1.5/0.2-0.5
Thickness of crust (mm)	~20-50	50-70	28-70	24-40
Strength of crust (MPa)	n.d.	n.d.	1.3-5.1	0.5-1.0
Strength below crust (MPa)	n.d.	n.d.	0.2-0.5	0.03-0.1

The texture designates the appearance of the sample material when stored in liquid nitrogen: "Snow" material looks fluffy and highly porous and forms coarse-grained agglomerates due to irregularly shaped ice/dust particles hooked together in a snowy sponge. "Mud" material shows a muddy appearance (cf. Table 2). The irradiation intensity is given in units of 1.4 kW/m²(SC).

TABLE 2. Basic parameters for selected small chamber experiments.

Experiment Small Chamber	KOSI SC 16-08-88	KOSI SC 17-08-88	KOSI SC 20-10-88	KOSI SC 25-10-88	KOSI SC 28-10-88
Composition (wt.%)					
H_2O ice	90.0	90.0	85.5	77.4	78.3
CO_2 ice	—	—	5.0	14.0	13.0
Mineral dust	10.0	10.0	9.5	8.6	8.7
Dust composition (Total dust = 100 wt.%)					
Olivine	70	90	70	70	90
Montmorillonite	30	10	30	30	10
Carbon	0.083	0.083	0.083	0.083	0.083
Density (g/cm³	0.40	0.40	0.38	0.36	0.41
Porosity (%)	55	55	59	63	60
Texture	"snow"	"snow"	"snow"	"mud"	"snow"
Period of insolation (h:min)	2:40	2:40	2:15	2:45	3:38
Intensity of insolation (SC)	2.0-2.4	2.0-2.4	2.0-2.4	2.0-2.4	2.0-2.4
Thickness of crust (mm)	n.d.	11-20	20-24	10-26	20-42
Strength of crust (MPa)	n.d.	0.10-0.21	0.15-0.19	0.75-1.10	0.35-0.88
Strength below crust (MPa)	n.d.	0.04-0.09	0.03-0.05	0.08-0.24	0.03-0.08

The texture designates the appearance of the sample material when stored in liquid nitrogen: "Snow" material looks snowy (cf. Table 1), "mud" material looks muddy and fine grained, the particles lying loosely together forming only small pores. The irradiation intensity is given in units of 1.4 kW/m²(SC).

2.4. Dust Collectors

For off-line dust studies an array of 264 passive collectors (Al-cans) on 13 radially arranged Al-rails (1-m length; 5°-10° spacing) are placed in front of the model comet to gather the dust residuals of the emitted particles. The collector cans are 30 mm in diameter and 30 mm in height to obtain material at sufficient local resolution.

Time-dependent information on the dust-particle flux is obtained with a step-motor driven and computer-controlled rotating sampler. This sampler consists of a turntable 30 cm in diameter, carrying 20 collector cans that are shielded by a fixed cover with a single hole, allowing the exposure of only 1 can at a time to the particle flux. For mass budget considerations, particles that miss the collectors are gathered on the cover of the rotating sampler and on a plastic foil below the collector and sampler array.

2.5. Optical and SEM Microscopy

For size measurements and characterization of texture and structure, the dust residuals are studied by means of conventional optical and scanning electron microscopy (SEM) (80×-50,000×). The emitted grains as well as the material of the target itself are compared with stratospheric Brownlee particles of the "fluffy" (presumably cometary) type with respect to their outer appearance (texture, porosity, shape), and this information in turn is used to improve the sample preparation procedure.

2.6. Sample Drilling

To obtain the basic soil mechanical parameters of the irradiated target material after the experiments, a special drilling device allows to record the mechanical stress vs. depth during the penetration of a test rod of 5-mm thickness into the model comet. The stress-depth profile reveals details of internal layering and gives the strength data of the different layers as a funciton of sample composition and duration and intenstiy of insolation.

3. RESULTS

To simplify comparison and discussion, the results will be presented following the physics of phenomena rather than the sequence of the diagnostics described.

3.1. Soil Mechanical Parameters

Insolation of the sample produces a distinct layering of the material. All experiments performed so far yielded a thin (<1 mm to ~5 mm), fluffy, and porous dust mantle that is almost free of volatiles. Below this dust mantle a solidified dust-ice crust of several centimeters thickness and considerable hardness is formed (for data, cf. Tables 1 and 2). The crust in turn is overlaying the original, essentially unaltered material. Depending on the experimental parameters, further alteration of the sample may occur near the liquid-nitrogen-cooled backplate (*Roessler et al.,* 1989, 1990).

Mechanical drilling (cf. section 2.6) before and after insolation is mainly aimed to characterize crust formation. Figure 1 summarizes the results of a series of stress-depth profiles recorded along the sample radius in the KOSI 4 experiment. Despite the lateral inhomogeneity with respect to the sample hardness, four well-defined layers can be distinguished (dashed lines in Fig. 1): (1) an upper crust of 12-35-mm thickness and a mechanical strength ranging from 0.2 to 0.95 MPa; (2) a lower crust of 5-20-mm thickness and decreasing strength, forming a transition region between crust and original material; (3) a mechanically weak part with a consistency similar to the original material, with a depth extension of 25-40 mm and a strength of only 0.03-0.09 MPa; and finally (4) solidified CO_2-rich material of 0.45-1.0 MPa strength covering a depth interval of 50-52 mm at the bottom of the target body. The increase of strength near the cooled backplate is very pronounced in KOSI 4 (30 hours insolation with 0.6 solar constants intensity, 2 hours with 0.8 solar constants) and was not observed in KOSI 3 (42 hours insolation with 1.3-1.6 solar constants intensity). It is ascribed to the inward migration and recondensation of CO_2-gas or to a compression of material caused by the test rod of the force meter (cf. *Roessler et al.,* 1989, 1990). The data of the crust in the KOSI 3 experiment (strength 1.3-5.1 MPa, thickness 28-70 mm) strikingly differs from KOSI 4, obviously due to the much higher total energy input of KOSI 3.

3.2. Mantle Formation and Dust Emission

When the virgin material is exposed to the light source, ice sublimation immediately starts and the emanating water vapor (and CO_2-gas in case of admixed CO_2-ice) tears off loosely bound particles from the uppermost surface layer. Gases incorporated during sample preparation (air, nitrogen) may also contribute to the first outburst of dust emission activity. Figure 2 shows the particle flux during the KOSI 3 experiment

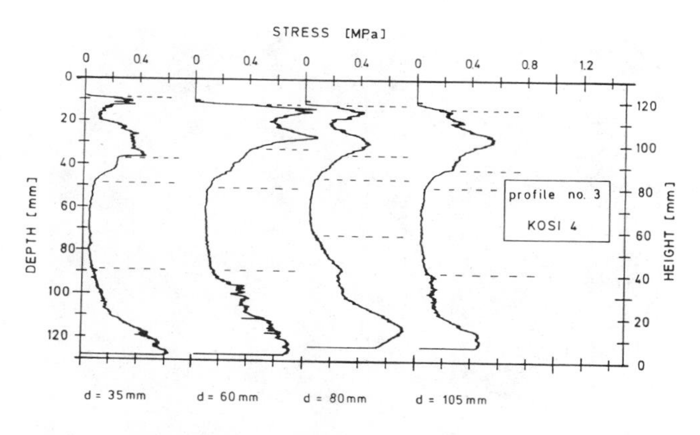

Fig. 1. Stress-depth profiles of the sample KOSI 4 measured with the ESA/ESTEC-drilling device at different radial distances d (35 mm, 60 mm, 80 mm, 105 mm) from the center of the target. Zero depth refers to the original surface, zero height refers to the bottom plate. Dashed lines mark layer interfaces of upper and lower crust, central and bottom material.

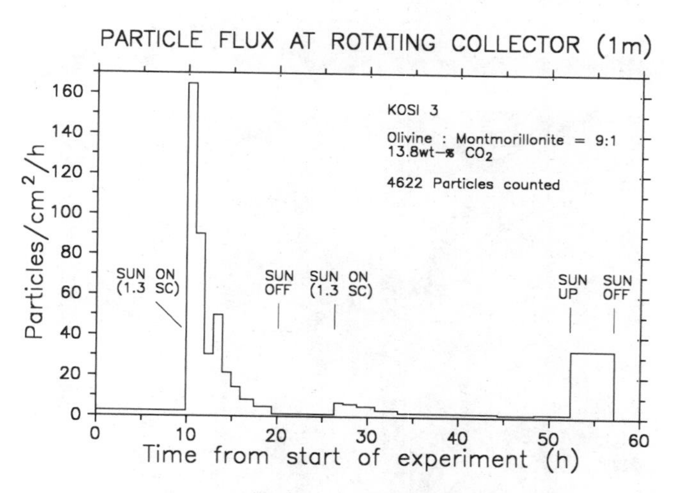

Fig. 2. Dust particle flux vs. time during the KOSI 3 experiment as recorded by the rotating sampler. The measurement does not include pure ice particles that are destroyed by sublimation in the (heated) collectors during the experiment. In the final insolation phase, the irradiation intensity was increased to 1.6 solar constants.

as measured by means of the rotating sampler ~1 m in front of the model comet. During the first 10 hours of insolation, a steep decrease of the dust-emission activity was observed. At the beginning of the second irradiation phase, the particle flux starts approximately at the activity level of the end of the first insolation period, decreasing very slowly with time. After 52 hours the irradiation intensity was increased from 1.3 to 1.6 solar constants leading again to a considerable enhancement of the dust emission. The dust emission activity shows a prositive correlation with the total gas emission rate (D. Krankowsky, P. Lämmerzahl, and P. Hesselbarth, personal communication, 1989), indicating that the overall gas emanation of the sample material is the main propellant for the particle ejection.

Video scanning of the sample surface during the experiments using magnifications of 8×-25× shows that particle emission obviously occurs in two different regimes: (1) predominantly gravity controlled (a particle lying loosely on top of the surface may be easily blown away by the emanating gas stream if the size and mass of the particle is so small that the particle gravity is overcome by the gas drag) and (2) predominantly cohesive force controlled (particles that are sticking to the surface via ice bonds or because of their hooked shape can only be lifted after the cohesion has been removed). The most frequent way to weaken cohesion of surface grains is sublimation of the ice bonds. A particle that has only a minor surface of contact may start moving in the emanating gas stream, thus gradually getting rid of the bonding, e.g., by the formation of fatigue fractures in contact "bridges." Movement of a loosely fixed particle may also influence the adhesion of adjacent particles causing their ejection. Material science aspects of these microerosion processes are more comprehensively considered in *Kochan et al.* (1990).

Sublimation of the superficial ice leads to the formation of a dry dust mantle made up of particles that are either too heavy or too tightly adhering to other large dust residuals to be carried away by the gas stream. This dust layer gradually quenches further sublimation of ice and reduces the gas

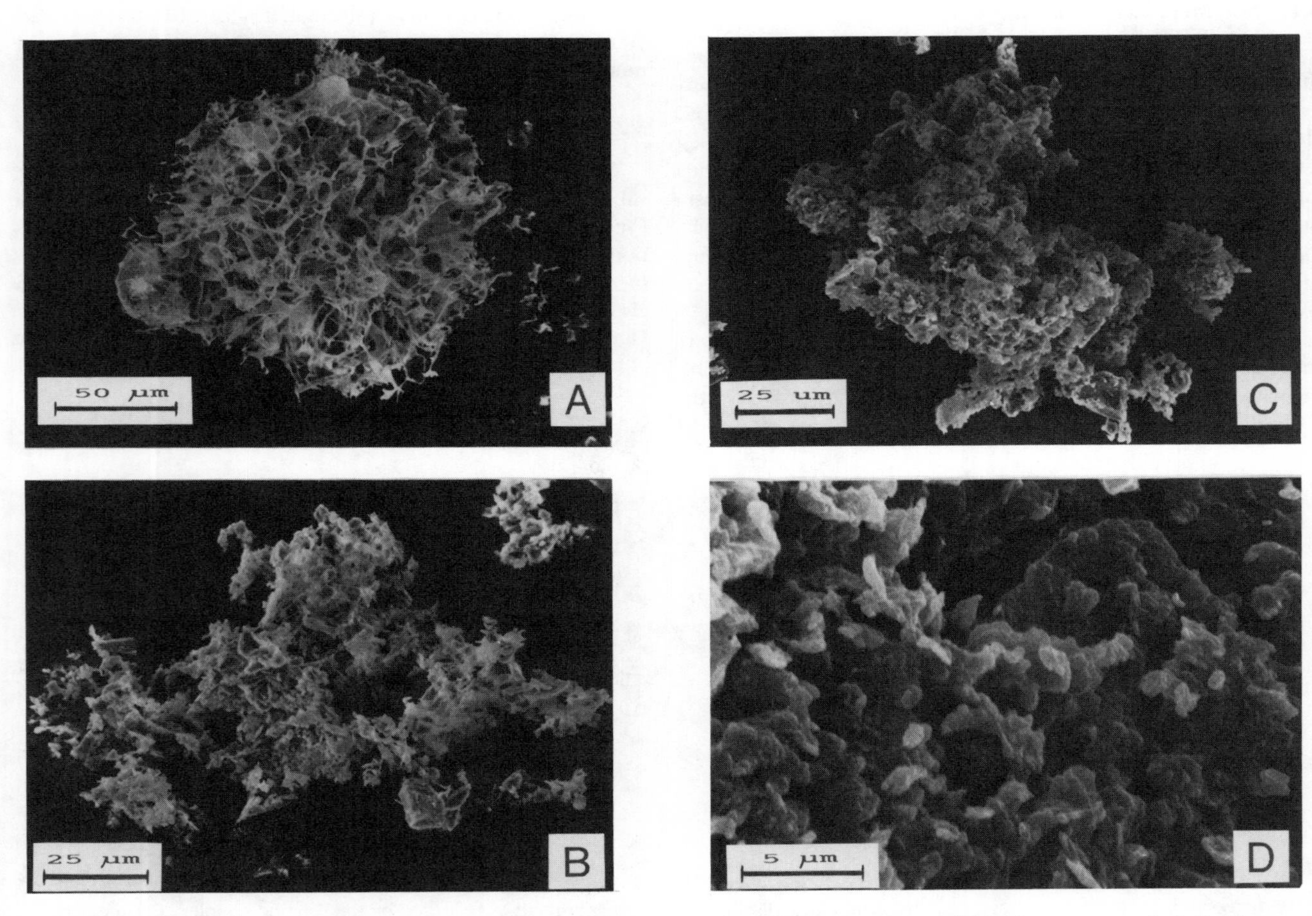

Fig. 3. Scanning-electron micrographs of typical dust residuals obtained in KOSI experiments. **(a)** Higher content of phyllosilicates (montmorillonite) yields fluffy dust particles of high, sometimes regular porosity, containing roundish vesicular pores with thin walls formed of phyllosilicate platelets. **(b-d)** Higher content of nesosilicates (olivine) leads to less fluffy aggregates of hooked to roundish mineral grains. **(a)** Residual containing olivine and montmorillonite at a ratio of 7:3. The particle has been collected in experiment KOSI SC 16-08-88 (cf. Table 2). **(b-c)** Residuals obtained in the KOSI 2 experiment (olivine:montmorillonite ratio 9:1; cf. Table 1). **(d)** Close-up view of a KOSI 2 particle.

emission and thus also the dust emission activity (cf. Fig. 2). Recent temperature measurements during the KOSI 4 experiment using an IR-camera revealed that the outer skin of the dust mantle is heated to temperatures exceeding 350 K. At the same time temperatures around 140-150 K were recorded by a thermocouple approxiamtely 15 mm below the surface, indicating extremely low heat conductivity of the dust mantle. The dry dust therefore is an effective heat shield preventing deeper ice layers from rapid sublimation. Unless considerable pressure buildup below the dust cover leads to a removal of part of the mantle and to the exposure of fresh icy material to irradiation, the sample shows no noticeable activity (Fig. 2). Pressure-induced detaching of larger cohering parts of the dust mantle has been observed only with low porosity samples of "mud"-like appearance and with massive ice-dust blocks (*Thiel et al.,* 1989a,b; *Grün et al.,* 1989).

3.3. Dust Residuals

The investigation of particles after complete ice sublimation (dust residuals) by means of SEM techniques is aiming at a better simulation of the cometary dust component (cf. section 2.5). Taking the structure and texture of "fluffy"-type stratospheric particles of presumably cometary origin as a reference, the parameters of the sample preparation procedure are varied in order to obtain dust residuals of roughly similar appearance. This is especially important, since sample preparation in KOSI experiments is presently still based on the freezing of aqueous suspensions of mineral dust and not on the solidification of the ices from the vapor phase (which is under development).

Typical examples of residual material are shown in Fig. 3. There are mainly two parameters ruling the phenomenology of the residuals: (1) the content of phyllosilicates in the mineral dust component, and (2) the gas pressure of the spray injection into liquid nitrogen (cf. *Roessler et al.,* 1990). Phyllosilicate concentrations in the order of 30% of the dust content favor the formation of aggregates containing pores with thin smooth separating walls. This is obviously due to surface forces causing the agglomeration of the phyllosilicate platelets predominantly parallel to their cleavage planes. High nesosilicate (olivine) contents (70-100%) favor irregular agglomeration of the dust grains, yielding particles that resemble in structure and texture certain stratospheric particles except for the *scale* of the microscopic structure (Fig. 3). Sublimation experiments with cometary analogues of the KOSI type indicate that a given dust agglomerate embedded in the ice/dust mixture does not necessarily keep its original shape after ice removal. Shrinkage of the sample volume of several percent during ice sublimation causes separate mineral grains that are embedded in ice to contact each other to form a new aggregate of dry dust.

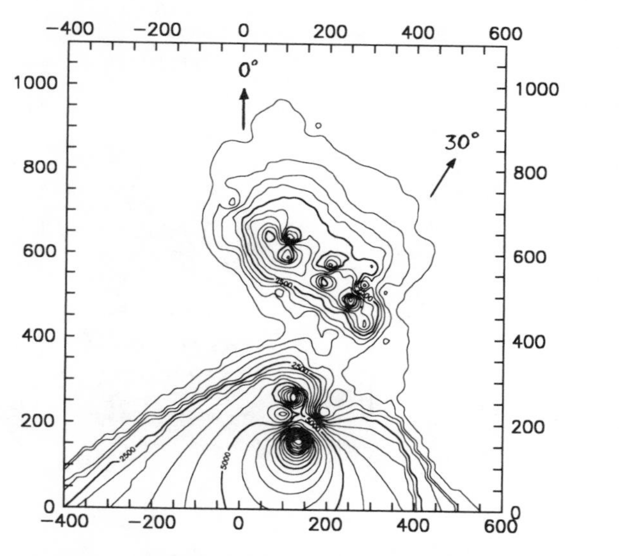

Fig. 4. Topography of dust deposition in front of the KOSI 2 target as measured by the passive collector array. The axes correspond to the distance from the chamber axis (x) and the distance from the target center (y), both in mm. The position of the sample (30 cm in diameter, 15 cm high) is located at (0,0), approximately 70 cm above the plane of the horizontal collector array. The sample is tilted by 45° toward the light source, which is situated at (0,1500), and the (tilted) sample is turned clockwise by an azimuth angle of 30° off the direction of the incident light. The dust deposition level interval is 500 μg. There is a broad concentration in the dust deposition near (200,600). The "mountain" at (130,150) near the sample (0,0) reflects material that has fallen down due to the 45°-tilt of the sample.

3.4. Dynamical Parameters of Particle Emission

3.4.1. Spatial distribution of emitted dust. The amount of residuals deposited in the 264 passive dust collectors in front of the target (cf. section 2.4) allows us to establish the spatial mass distribution of emitted dust with a fair resolution. Figure 4 is a plot of the lines of constant dust deposition (by mass) obtained from the KOSI 2 experiment. The tilted and turned orientation of the sample relative to the light source leads to an asymmetric dust emission (pattern

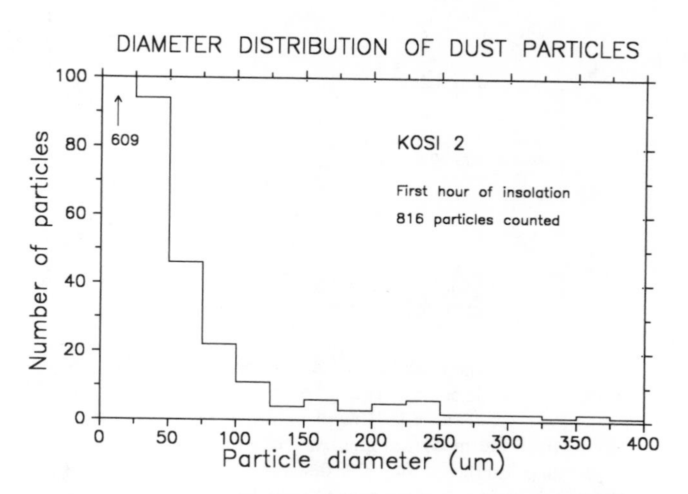

Fig. 5. Size-frequency distribution of emitted dust during the KOSI 2 experiment obtained from rotating sampler data.

near the center of the diagram). The dust concentration directly underneath the sample reflects material that has fallen down due to the 45°-tilt of the sample container. The analysis of the asymmetric dust emission pattern suggests the existence of a double peak (peaks near 0°-20° and 30° azimuth), which has to be reinvestigated, however, in the next experiments.

3.4.2. Size frequency distribution. The dust residuals collected in the rotating sampler allow detailed off-line studies. In Fig. 5 the number of particles per 25-μm-diameter interval are plotted for the first hour of insolation during the KOSI 2 experiment. More than 90% of the residuals (by number) are smaller than 100 μm in diameter. In the following hours of irradiation, the size distributions are steadily shifted toward smaller particle diameters (*Thiel et al.,* 1989b). These observations are confirmed by all experiments with samples of "snow"-like texture performed so far (cf. section 3.5 and Tables 1 and 2). The smallest and the largest particles ever observed in emitted dust during KOSI experiments are <1 μm and >1 mm in diameter. In cases of very-fine-grained samples showing "mud"-like appearance in liquid nitrogen (*Thiel et al.,* 1989a), cohering parts of the mantle were ejected due to the local release of high pressure gas, forming flakes of ~100-200-μm thickness and >1 cm (see section 3.5).

3.4.3. Mass frequency distribution. By assuming spherical shape and an average mass density of the dust residuals of 0.2 g/cm^3, the mass of the residuals in a given size range can be estimated from the diameter distribution. In Fig. 6 the percentage of the total residual dust mass is plotted for grain sizes exceeding a given diameter D (KOSI 2 experiment, rotating sampler data). The plot shows that considerable mass fractions are contained in the larger size fractions (e.g., particles >100 μm carry 94% of the total mass, and even for D> 300 μm the mass fraction is still 44%). This is in disagreement with observations of interplanetary dust and may reflect unsatisfactory simulation in the experiments.

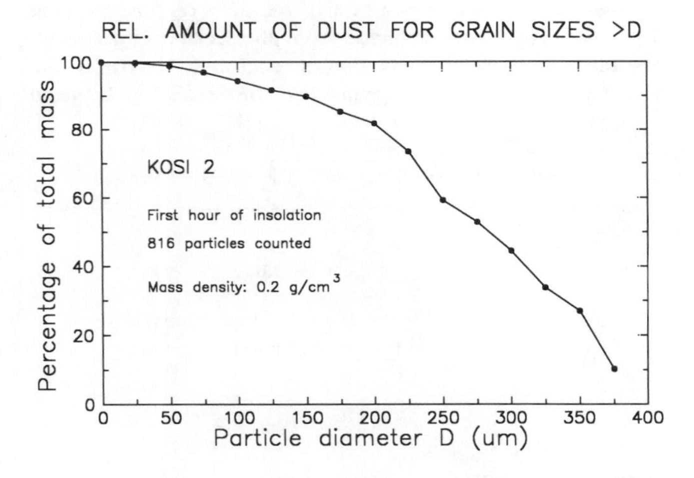

Fig. 6. Percentage mass fraction of dust residuals for different grain-size fractions containing particles with diameters exceeding a given value D. This mass frequency distribution is derived from the measured grain size distribution recorded by the rotating sampler during the first hour of insolation in the KOSI 2 experiment by assuming a mean mass density for single residual grains of 0.2 g/cm^3.

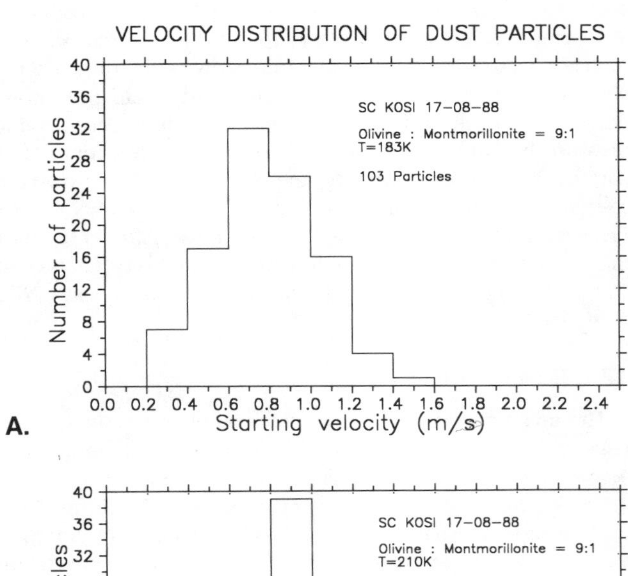

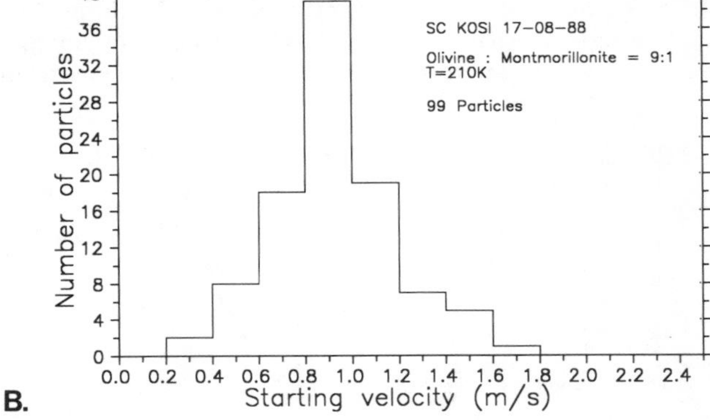

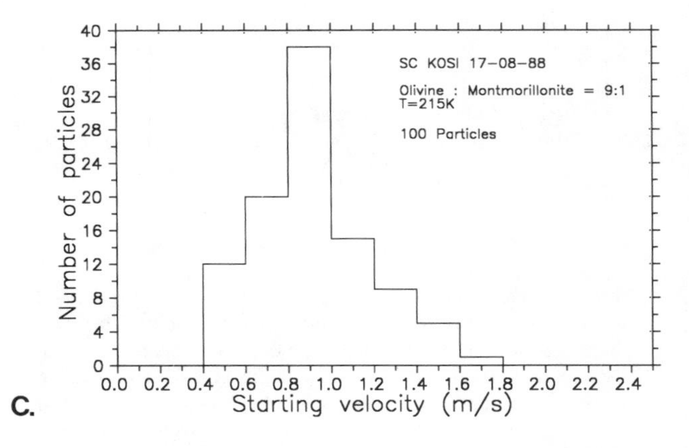

Fig. 7. Velocity distribution of emitted particles deduced from video records for three different values of the sample temperature ~1.5 cm below the surface of the target. **(a)** T = 183 K; **(b)** T = 210 K; **(c)** T = 215 K. All measurements were carried out in the small simulation chamber (experiments KOSI SC 16-08-88 and KOSI SC 17-08-88). For further experimental data compare Table 2.

3.4.4. Velocity frequency distribution. By means of video records of flying particles, the velocity can be determined as a function of the various experimental parameters. In Fig. 7 the frequency distribution of particle velocities is shown for three different time periods (~0 hr, ~1 hr, and ~2 hr after start of the insolation) during an experiment in the small simulation chamber. The changing parameter in the three plots is the surface temperature (ca. 1.5 cm below the surface). The distributions are very similar for the covered period of time, except for a slight shift of the maximum from ~70 cm/sec to ~90 cm/sec with increasing temperature of the sample. The lack of very slow particles (<20 cm/sec) is mainly due to the effect of 1 g, causing slow particles to fall back to the sample or to merely roll down the 45°-slope of the surface (cf. dust deposition in Fig. 4).

In the Space Simulator another technique was used to determine emission velocities of particles (*Kohl,* 1989). As pointed out in section 2.3, the impact signals of the Piezo detectors can also be used for an estimate of the particle velocity (Fig. 8). Particle cut-off at the low speed end of the diagram is caused by 1-*g*-interference and deteriorating electronic sensitivity for slow dust grains. The maximum particle velocities observed in the small simulation chamber and in the Space Simulator are ~2 m/sec and ~5.2 m/sec, respectively. The differing values may be partly due to the measuring technique (the upper not very well defined speed still measurable by means of video records in the small simulation chamber is ~4 m/sec).

3.4.5. Angular distribution of particle velocities. Particle emission from comet nucleus analogues preferably occurs in the direction of the surface normal. Parameters that may influence the emission characteristics of particles are, for example, the surface roughness of the sample, the mineral composition of the dust component, the adhesive forces acting on particles sticking on the surface, and the mechanism of breaking mechanical bonds between cohering particles. The effect of the parameters will be investigated in more detail in the forthcoming KOSI experiments. As an example, the influence of the mineral dust composition is demonstrated in Figs. 9a,b for two different mixtures of olivine and montmorillonite. Temperature and surface roughness of both targets was comparable for the time period of data acquisition. The sample was tilted by 45° toward the light source at an azimuth angle of 0° (i.e., the tilted sample was facing the Xe-lamps). In both cases, the speed values cluster around 1 m/sec and the particle trajectories exhibit an angular straggling of ±20° around 0° azimuth. The sample containing the higher phyllosilicate fraction (three parts montmorillonite, seven parts olivine; Fig. 9a) shows a slight tendency toward higher emission velocities.

3.4.6. Microseismic activity. To correlate the occurrence of high-frequency cracks (presumably indicative of mechanical alterations of the sample material) and particle emission activity of the surface, sensitive accelerometers were attached to the sample container and to a steel baffle plate hanging in front of the model comet. Figure 10 summarizes measurements performed in the small simulation chamber with a sample material containing 10 wt.% of montmorillonite as the only dust component. The upper graph shows time variations of microseismic events in the target, and the lower plotter record gives the corresponding fluctuations of the particle emission rate as measured via particle impacts on the external baffle plate. The measurements as well as visual observations suggest a positive correlation between enhanced "seismic" activity and the burst-like emission of groups of particles. Particle ejection turned out to occur with a time lag of ~10-30 sec after the "seismic" events. These phenomena could be observed only in small chamber experiments due to the high electronic noise background in the Space Simulator. The signal-to-noise ratio is presently being improved by employing space-proof electronic components.

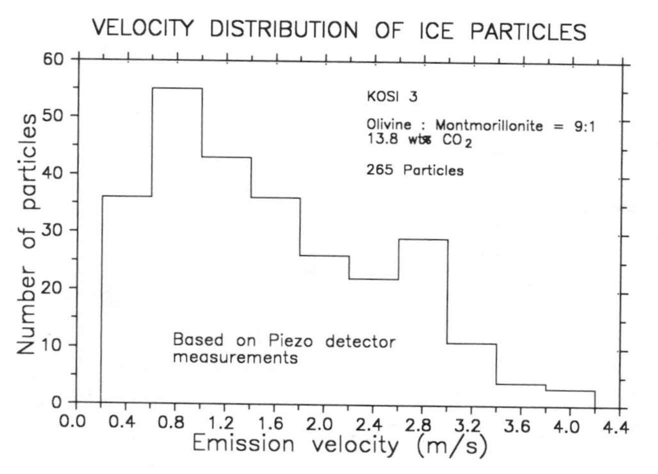

Fig. 8. Velocity distribution of emitted particles during the KOSI 3 experiment in the Space Simulator. The diagram is based on the analysis of Piezo detector signals. The cut-off at ~60 cm/sec is due to decreasing electronic sensitivity of the diagnostic device for low momentum particles.

3.5. Synopsis of Dynamic Parameters

The basic parameters controlling particle emission during insolation are interrelated in a way that allows us to draw some conclusions on the mechanism of particle ejection itself. Figure 11, which is derived from data of the experiments KOSI SC 16-08-88 and 17-08-88 (see Table 2), gives a summarizing plot of the particle diameter vs. emission velocity and emission angle with respect to the horizontal plane (azimuth). To use an average of both experiments seems to be justified since the differing amount of phyllosilicates (Table 2) mainly influences dust emission activity and texture of the dust residuals (see Fig. 3) but causes no significant changes of particle diameter and emission angle and velocity (cf., for example, Fig. 9). The diagrams of Fig. 11 show emission of large but low-velocity (<1.5 m/sec) particles preferably in the direction of the surface normal (which is tilted by 45° toward the light source, i.e., 0° azimuth). Angular straggling of the particle trajectories increases with decreasing particle size. The broad emission characteristics for small grains may be caused in part by the

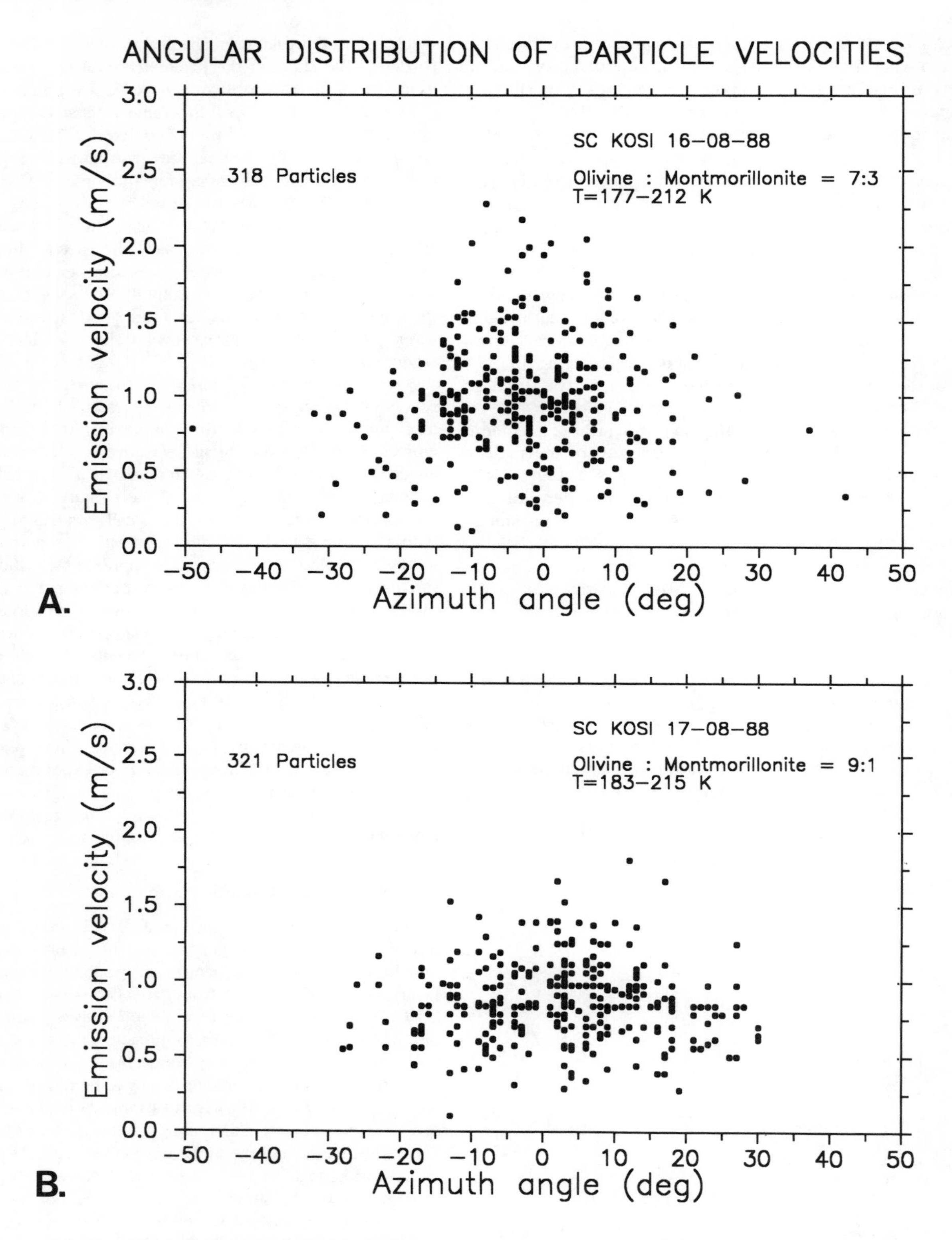

Fig. 9. Angular distribution of particle emission velocities for comet nucleus analogues with **(a)** high phyllosilicate, i.e., montmorillonite content (KOSI SC 16-08-88) and **(b)** low phyllosilicate content (KOSI SC 17-08-88). Velocity and angle measurements are based on high-shutter-speed video camera records (exposure time 1/2000 sec). For further data compare Table 2.

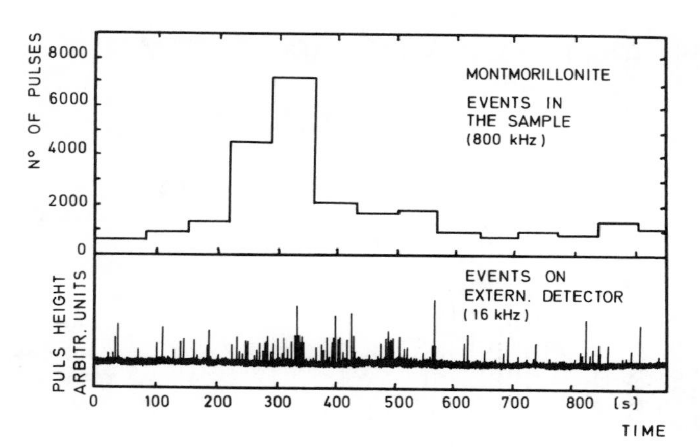

Fig. 10. Particle emission and microseismic events in the sample, measured with two Piezo accelerometers in the small simulation chamber. Sample material: 90% water ice, 10% montmorillonite dust. Preparation procedure: spraying of the aqueous mineral suspension into liquid nitrogen. The resonance frequencies of the accelerometers attached to the sample container and to the baffle plate in front of the target are 800 kHz and 16 kHz, respectively.

surface roughness of the sample (micro- and macro-relief). The highest emission velocities (>5 m/sec) are reached by the smallest grains emitted in the direction of the surface normal. Results obtained with the rotating dust sampler revealed a decrease in the mean grain size of emitted particles with increasing duration of insolation (cf. section 3.4.2). This means that the plot shown in Fig. 11 should also be time dependent.

Video records confirm that for high porosity samples of "snow"-like texture the superficial gas drag carries away a continuous flux of small particles ($\lesssim 100\ \mu m$), thus removing parts of the dust mantle. Larger dust grains due to gravity remain on the surface and lead to a net *increase* of the mantle thickness. The growth of the dust mantle reduces heat transport into the interior and quenches the gas flow through the dust cover. The decreasing gas drag on the surface leads to decreasing grain sizes in dust emission with increasing duration of insolation.

In cases of low-porosity samples of "mud"-like appearance, a less permeable dust mantle is formed that quenches the gas stream of the sublimating ice/dust layers underneath. The mechanically coherent dust cover prevents the lift-off even of small particles and causes a pressure buildup that may in turn lead to local blasting off of mechanically unstable centimeter-sized mantle areas. This is confirmed by the observation that the phenomenon can be triggered by emitted particles falling back to the surface or by touching the sample surface with a pin.

The interdependence of the three parameters in Fig. 11 indicates that particle emission from comet nucleus analogues cannot be dominated by a "rocket effect" caused by sublimating ice on the rear side of the particles, which would lead to roughly grain-size independent velocities. A mechanism that is in agreement with Fig. 11 and is supported by high-magnification video observation of the sample surface (*Kochan et al.,* 1990) is particle emission caused by the drag of a continuous gas flow emanating from the sample ("snow"-textured material) or by local gas pressure release ("mud"-textured material). A detailed study of the particle acceleration process within the first few centimeters above the surface is planned for the forthcoming KOSI experiments using specially designed laser particle counters.

4. CONCLUSIONS

Crust dynamics and dust emission of comet nucleus analogues under artificial insolation turned out to be unexpectedly complex. A wealth of parameters has to be measured and taken into account to allow a better unerstanding of phenomena like particle emission as a function of surface relief and surface temperature, or grain size of the emitted particles as a function of duration of insolation or thickness of the dust mantle. In particular, an extension of the experiment duration should help to clarify effects that are not very pronounced after 30-50 hours of irradiation and a loss of the sample volume of only 1/6 to 1/3 of the original volume. Better defined experimental parameters and improved diagnostics for the eight KOSI experiments to come should help to address most of the questions concerning terrestrial comet nucleus simulation under 1 g. In spite of this gravity interference, one can draw certain conclusions from the results obtained so far that in turn may bear some implications for real comets.

4.1. Mantle Formation

In all experiments the formation of a dry dust mantle was observed. Although the emanating gas stream is able to carry away large particles (e.g., >1-cm dust flakes under 1 g), cohesion forces may considerably prevail over gravity forces, as was shown in close-up video records, and thus retain a certain fraction of surface particles that gradually contribute to a net dust-mantle *growth.* A dry dust layer prevents heat penetration by gas diffusion and reduces the thermal conductivity and the dust emission activity of the surface. If such a layer is of low porosity and mechanically coherent but still sufficiently "energy transparent," it may induce the building up of considerable gas pressure near the ice-sublimating zone and allow the ejection of larger mechanically unstable mantle areas. The local removal of the heat-shielding mantle material exposes fresh ice to rapid sublimation, giving rise to dust "jets." The experiments support this scenario and shed new light on the "aging" of comet nuclei.

4.2. Dust Emission Regimes

The experiments reveal essentially two regimes of dust emission from cometary analogues: (1) gas drag of continuously emanating vapor from near-surface ice sublimation, and (2) temporary and local depressurizing of a steep pressure gradient within the superficial dust mantle and lift-off of larger particles. The first process predominantly occurs with fresh,

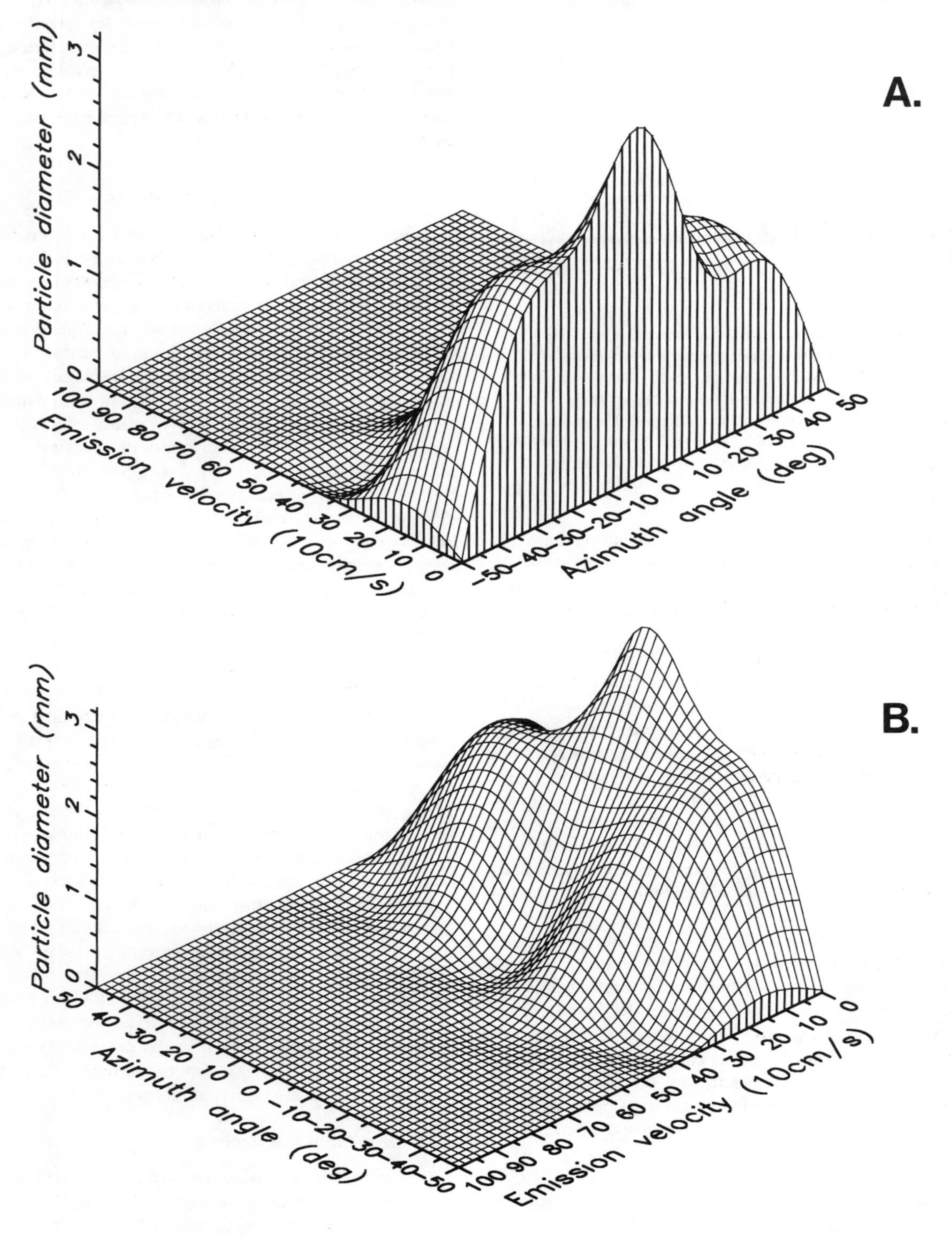

Fig. 11. Two different views of a 3D-plot of the grain size vs. emission speed (dm/s) and emission angle. The grid was calculated using the raw data of more than 300 measurements in the small simulation chamber (KOSI SC 16-08-88 and KOSI SC 17-08-88; cf. Table 2) and a smoothing interpolation of the z-data (particle diameter).

virgin ice/dust mixtures in the beginning of the experiments and may apply also to the first inner solar system encounters of new comets. The second process is typical of samples after dust-mantle formation and shows analogy to old comets. Taking into account a ratio of gravitational forces of the Earth and of typical cometary nuclei in the order of $\sim 10^4$, size distributions measured in terrestrial experiments have to be scaled up accordingly. This, together with the observed velocity distributions, is in agreement with the lift-off of decimeter-sized bodies from real comets.

Both emission mechanisms described above may be subdivided according to the retaining forces involved. Depending on whether a particle is loosely lying or sticking on the surface, one may distinguish between predominantly gravity-controlled and predominantly cohesion-controlled dust emission. Close-up video records show that both mechanisms occur and contribute to the total dust emission, which should also apply to real comets.

4.3. Dust Jets

The measured angular distributions (cf., for example, Figs. 4, 9, and 11) demonstrate that large particle emission mainly occurs in a very narrow solid angle of a few tens of degrees diameter compared to the gas emission, which shows a much broader angular pattern (D. Krankowsky, P. Lämmerzahl, and P. Hesselbarth, personal communication, 1989). This is in agreement with the observed shape of gas/dust jets of comets and is presently being investigated in more detail.

4.4. Support of CNSR Mission Sampler Design

The soil mechanical parameters of comet nucleus analogues as they are obtained in the KOSI experiments may narrow the parameter ranges presently being adopted for the planning of a comet nucleus sample return mission. In particular, hardness data of the dust/ice crust, which may reach considerable values, could define the boundary conditions for the design of more flexible sampler devices.

Acknowledgments. This project (KOSI) is financially supported within the Schwerpunktprogramm "Kleine Körper im Sonnensystem" by Deutsche Forschungsgemeinschaft, Bonn, under eight different contracts. The KOSI consortium wishes to thank all members of the technical staff of the Institute of Space Simulation at DLR/Köln for their qualified and effective assisance. The help of all team members working nightshift during the past chamber runs is gratefully acknowledged.

REFERENCES

Donn B. and Urey H. C. (1956) On the mechanism of comet outburst and the chemical compositon of comets. *Astrophys. J., 123,* 339-342.

Grün E., Kochan H., Roessler K., and Stöffler D. (1987) Simulation of cometary nuclei. *Symposium on the Diversity and Similarity of Comets,* April 6-9, 1987, Brussels, Belgium, ESA SP-278, pp. 501-508.

Grün E., Benkhoff J., Fechtig H., Hesselbarth P., Klinger J., Kochan H., Kohl H., Krankowsky D., Lämmerzahl P., Seboldt W., Spohn T., and Thiel K. (1988) Mechanisms of dust emission from the surface of a cometary nucleus. *Proceedings of the Twenty-seventh Plenary Meeting of the Committee on Space Research (COSPAR),* printed in *Advances in Space Research, No. 3* (1989), (3)133-(3)137, Pergamon, Oxford.

Grün E., Benkhoff J., Bischoff A., Düren H., Hellmann H., Hesselbarth P., Hsiung P., Keller H. U., Klinger J., Knölker J., Kochan H., Neukum G., Oehler A., Roessler K., Spohn T., Stöffler D., and Thiel K. (1989) Modifications of comet materials by the sublimation process: Results from simulation experiments (abstract). In *Papers Presented to the Workshop on Analysis of Returned Comet Nucleus Samples,* pp. 24-25. Lunar and Planetary Institute, Houston.

Kajmakov E. A. and Sharkov V. I. (1967a) Behaviour of water ice in vacuum at low temperatures. *Komety Meteory, 15,* 16-20.

Kajmakov E. A. and Sharkov V. I. (1967b) Sublimation of water ice containing dust particles. *Komety Meteory 15,* 21-24.

Klinger J., Benkhoff J., Espinasse S., Grün E., Ip W., Joó F., Keller H. U., Kochan H., Kohl H., Roessler K., Seboldt W., Spohn T., and Thiel K. (1989) How far do results of recent simulation experiments fit current models of cometary nuclei? *Proc. Lunar Planet. Sci. Conf. 19th,* pp. 493-497.

Kochan H., Benkhoff J., Bischoff A., Fechtig H. Feuerbacher B., Grün E., Joó F., Klinger J., Kohl H., Krankowsky D., Roessler K., Seboldt W., Thiel K., Schwehm G., and Weishaupt U. (1989a) Laboratory simulation of a cometary nucleus: Experimental setup and first results. *Proc. Lunar Planet Sci. Conf. 19th,* pp. 487-492.

Kochan H., Ratke L., Hellmann H., Thiel K., and Grün E. (1989b) Particle emission from artificial cometary surfaces: Material science aspects (abstract). In *Lunar and Planetary Science XX,* pp. 524-525. Lunar and Planetary Institute, Houston.

Kochan H., Ratke L., Hellmann H., Thiel K., and Grün E. (1990) Particle emission from artificial cometary surfaces: Material science aspects. *Proc. Lunar Planet. Sci. Conf. 20th,* this volume.

Kohl H. (1989) Messung und Charakterisierung von emittierten Partikeln sublimierender Eis-Staub-Gemische mit Hilfe piezokeramischer Impaktdetektoren. Diplomarbeit, Max-Planck-Institut für Kernphysik, Heidelberg. 59 pp.

Roessler K., Hsiung P., Kochan H., Hellmann H., Düren H., Thiel K., and Kölzer G. (1989) A model comet made from mineral dust and H_2O-CO_2 ice: Sample preparation development (abstract). In *Lunar and Planetary Science XX,* pp. 920-921. Lunar and Planetary Institute, Houston.

Roessler K., Hsiung P., Kochan H., Hellmann H., Düren H., Thiel K., and Kölzer G. (1990) A model comet made from mineral dust and H_2O-CO_2 ice: Sample preparation development. *Proc. Lunar Planet. Sci. Conf. 20th,* this volume.

Saunders R. S., Fanale F. P., Parker T. J., Stephens I. B., and Sutton S. (1986) Properties of filamentary sublimation residues from dispersions of clay in ice. *Icarus, 66,* 94-104.

Thiel K., Kochan H., Roessler K., Grün E., Schwehm G., Hellmann H., Hsiung P., and Kölzer G. (1989a) Mechanical and SEM analysis of articifical comet nucleus samples (abstract). In *Papers Presented to the Workshop on Analysis of Returned Comet Nucleus Samples,* pp. 75-76. Lunar and Planetary Institute, Houston.

Thiel K., Kölzer G., Kochan H. Grün E., and Kohl H. (1989b) Crustal evolution and dust emission of artifical cometary nuclei (abstract). In *Lunar and Planetary Science XX,* pp. 1113-1114. Lunar and Planetary Institute, Houston.

Proceedings of the 20th Lunar and Planetary Science Conference, pp. 401-411
Lunar and Planetary Institute, Houston, 1990

Particle Emission from Artificial Cometary Surfaces: Material Science Aspects

H. Kochan and L. Ratke

Institut für Raumsimulation, DLR, Postfach 906058, D-5000 Köln 90, Federal Republic of Germany

K. Thiel

Abteilung Nuklearchemie der Universität Köln, D-5000 Köln 1, Federal Republic of Germany

E. Grün

Max Planck Institüt für Kernphysik, Postfach 103980, D-6990 Heidelberg, Federal Republic of Germany

---

The dust emission mechanism from the surface of model comets is investigated using a high-resolution, high-speed shutter video technique. Video records taken from an insolated model comet show a highly structured and rough surface that is not motionless. Different types of particle emission from such surfaces have been observed for the first time. Many particles or porous aggregates of small particles vibrate within the gas jet originating from deeper ice layers. After a certain number of vibration cycles they lift off from the surface. Other particles were observed that, after a certain time, obviously lose some of their bonds to the surface beneath. When enough contact bridges are eroded, the partly free particles move around the remaining bonds until even these break. The particles then are dragged away by the outstreaming gas. The observations are interpreted with fracture mechanical models using very simplified assumptions concerning the structures of a model comet's surface. The objectives of this paper are to show that the apparently loose, open structered, rough surface of a comet (at least a model comet) is built up by bonded particles. Dust particle emission by the drag interaction with the outstreaming gas is simply not possible. Erosion of the bonds and fatigue fractures in the case of the vibrating particles have to first set the particles free. Then they can be lifted off by the gas jet. The erosion and the fatigue fracture mechanisms are described in models by simple assumptions. The model calculations yield numerical values in the order of magnitude of the observed data.

INTRODUCTION

Two years ago laboratory simulation experiments of a cometary nucleus started in a Space Simulator at DLR Cologne, Institute for Space Simulation (*Grün et al.,* 1987; *Kochan et al.,* 1989). Normally two experiments per year are performed in this Space Simulator and a lot of experiments are run in parallel in a small simulation facility. Their objectives are closely related to the investigations in the Space Simulator, e.g., as preliminary experiments investigating the action of various parameters. During preparatory studies for an experiment in the Space Simulator the surface of an artificial comet was observed with a high-resolution video technique. New features of processes at the fresh surface of such "comets" were seen that may be relevant for real comets as well. They are at least important for the dust emission processes from cometary analogs. The observations made and a possible interpretation of the results are presented here.

EXPERIMENTAL SETUP

Figure 1 shows the internal instrumentation of the small simulation chamber. All important systems are fixed to the cover to shorten the handling procedures during the experiments. The white box in the middle is the sample holder ($10 \times 7 \times 6$ cm^3), tilted 45° against the insolation direction and cooled from the back by LN_2. The sample holder is partly surrounded by LN_2-cooled cold walls. In front of the box is the rotating dust collector and above the opening of the sample container the hardness tester, which can be moved aside, is positioned. Through a window in the cover the artificial comet material within the sample holder can be observed visually, by photographic and video camera systems.

The sample preparation method, described in more detail by Roessler and coworkers (*Roessler et al.,* 1990), commonly starts with a suspension of different minerals in water (grain sizes of the mineral particles are 1-10 μm). This suspension is injected with a propellant gas (N_2, CO_2) into liquid nitrogen following the method of *Saunders et al.* (1986). In the sample preparation procedure three parameters can be varied: the propellant gas pressure, the flow rate of the suspension, and the distance between the injection nozzle and the LN_2 surface in the dewar.

The gas pressure emerged as the dominant parameter for the size of the frozen water-mineral grains. High pressure (>2 bar) led to very small grains and an overall "mud"-like sample material (Fig. 2). Fluffy material with much larger

Fig. 1. Internal instrumentation of the small simulation chamber with LN_2 cooling system and diagnostics devices (hardness tester, dust collector). The white box in the middle is the sample holder.

grains resulted from low-pressure (<2 bar) injection ("snow"-like material, Fig. 3). Suspensions with higher admixtures of phyllosilicates (sheet silicates) showed a more distinct tendency to yield muddy material.

The picture of Fig. 2 was taken 1.5 hours after switch-on of the solar irradiation with approximately 1 SC. The surface of the "mud"-like sample appeared very smooth. The crater visible in Fig. 2 was formed in the very beginning of the insolation as a result of an eruptive dust emission. Later on, this model comet was inactive. Compared to the fluffy "snow"-like samples, the "mud"-like material always showed a much lower dust emission activity. This is related to the microstructure. "Mud"-like material consists of very small ice-mineral grains and therefore has more contact points to neighboring grains than a larger, irregular-shaped grain in an environment of large irregular grains. Smaller and more densely packed grains are therefore more strongly bonded to their neighbors than larger and more loosely packed grains. As will be discussed later (cf. equation (1)), in addition to the higher degree of fixation, the small grains of the "mud"-like samples undergo smaller interaction with the drag force exerted by the outstreaming water vapour originating during insolation.

The sample photographed in Fig. 3 consists of larger and irregular-shaped ice-mineral grains ("snow"-like) as can be seen from the higher surface roughness. The picture was taken at the beginning of the insolation. The sample showed a high dust emission activity during the entire experiment time of three hours. This continuous activity is caused by the weaker fixation of the particles to their environment, resulting from the fewer contact points and the stronger interaction with the gas jet (the drag force is proportional to the square of the particle radius). All experimental results presented in the following sections were obtained from "snow"-like samples that had a composition of 9% dunite, 1% montmorillonite, and 0.083% carbon. The overall porosity was 0.54 and the mean density 400 kg/m^3.

EXPERIMÈNTAL EVIDENCE

Figures 4 and 5 show video pictures of the leading edge of the model comet's sample holder during an experiment with a dust particle shower. As can clearly be seen, these particles, which are somewhat smaller than one millimeter, are accelerated differentially in the outflowing H_2O gas jet (cf. the group of five particles near the edge of the sample holder in Fig. 4).

Model calculations (*Marconi and Keller,* 1988; *Klinger et al.,* 1989) of dust emission in a 1-*g* environment demonstrated that only particles of a maximum size between 10 μm and a few 100 μm (depending on the gas velocity, the particle density, and the drag coefficient) will be lifted off from the

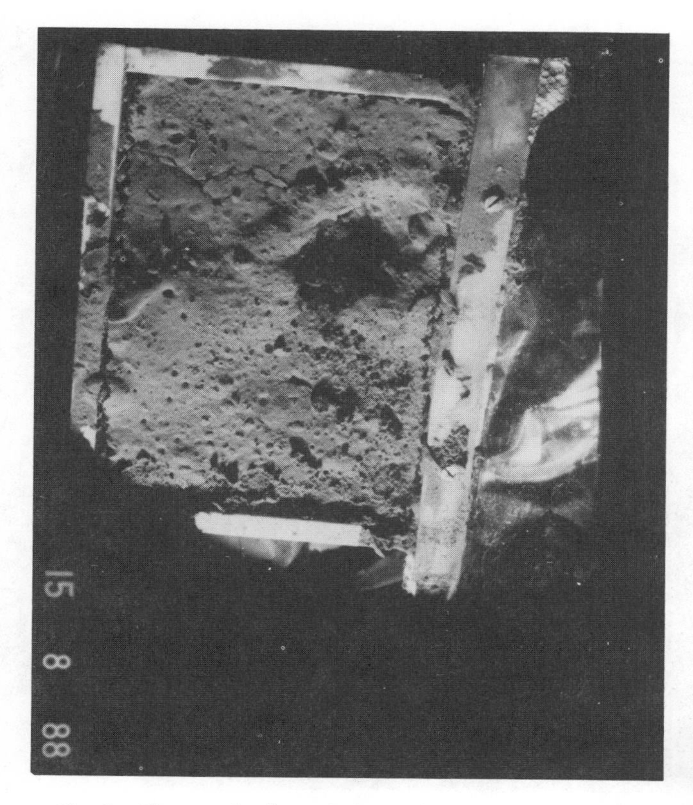

Fig. 2. Photograph of a typical model comet's surface with "mud"-like sample material.

surface by the gas flow (*Grün et al.,* 1989). Obviously, the video records show numerous particles emitted from the model comet's surface that are appreciably larger.

To overcome this discrepancy we searched for possible mechnisms of particle acceleration giving the particles initially a nonzero velocity. We therefore monitored the surface of the model comet with a magnification up to 25×. Magnified that much, fresh surfaces showed a number of strange phenomena.

1. First we shall make some comments as to the nature of the ice-mineral grains that form the surface of the model comet exposed to the insolation. Until now it has not been completely certain that the emitted "dust" particles are merely the dust residuals that the icy components have sublimated from or whether they are icy-mineral grains. During the last KOSI 4 experiment (May 1989) in the Space Simulator, K. Mauersberger, D. Krankowsky, and P. Lämmerzahl operated two newly developed dust cages that were heated and equipped with ionivacs. Following a communication at an experimenter's workshop they recorded some events with water vapor emission. This implies that particles with ice admixture are emitted from the model comet's surface. Our high-resolution video observations, especially of fresh surfaces during the early stages of insolation, strongly indicate that at least some particles with ice residuals are emitted. In a microscopic view these particles may be partly open-structured. They are definitely nonspherical. In contrast to the long-time sublimation experiments of *Saunders et al.* (1986) and *Storrs et al.* (1988), which were carried out without insolation, we do not observe the mineral residuals. Shortly after switch-on of the irradiation source, a rapid sublimation starts followed by an active dust emission.

2. The surface itself is not calm but partially in motion. Many small particles vibrate with various amplitudes and frequencies. In Figs. 6 and 7 the encircled areas show a vibrating particle. The amplitude is in the range of the particle diameter. After a certain time, differing from particle to particle, the particles lift off. This first observations show that the vibration frequencies can vary between 1 and 100 Hz and the time until particles leave the surface is between a few seconds and 10 minutes. Exact measurements are now performed to relate the vibration frequencies and amplitudes to the particle sizes and shapes and to obtain the number of cycles a particle can sustain until it is free to leave the surface.

3. Another type of dust emission process starts with particles obviously bonded to the surface. Suddenly some of these bonds break. The particle moves along the surface guided by the remaining bonds until even these break and the particle is free to leave the surface.

Fig. 3. Photograph of a typical model comet's surface with "snow"-like sample material.

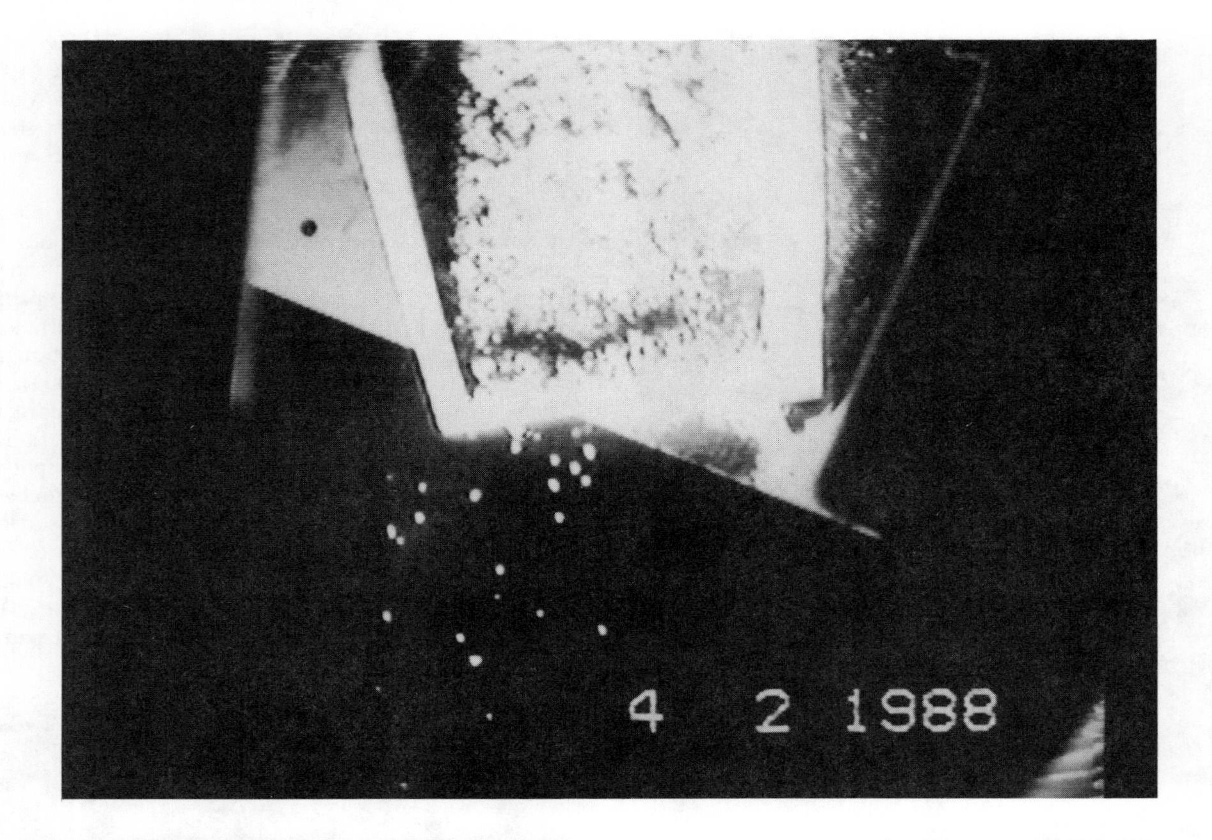

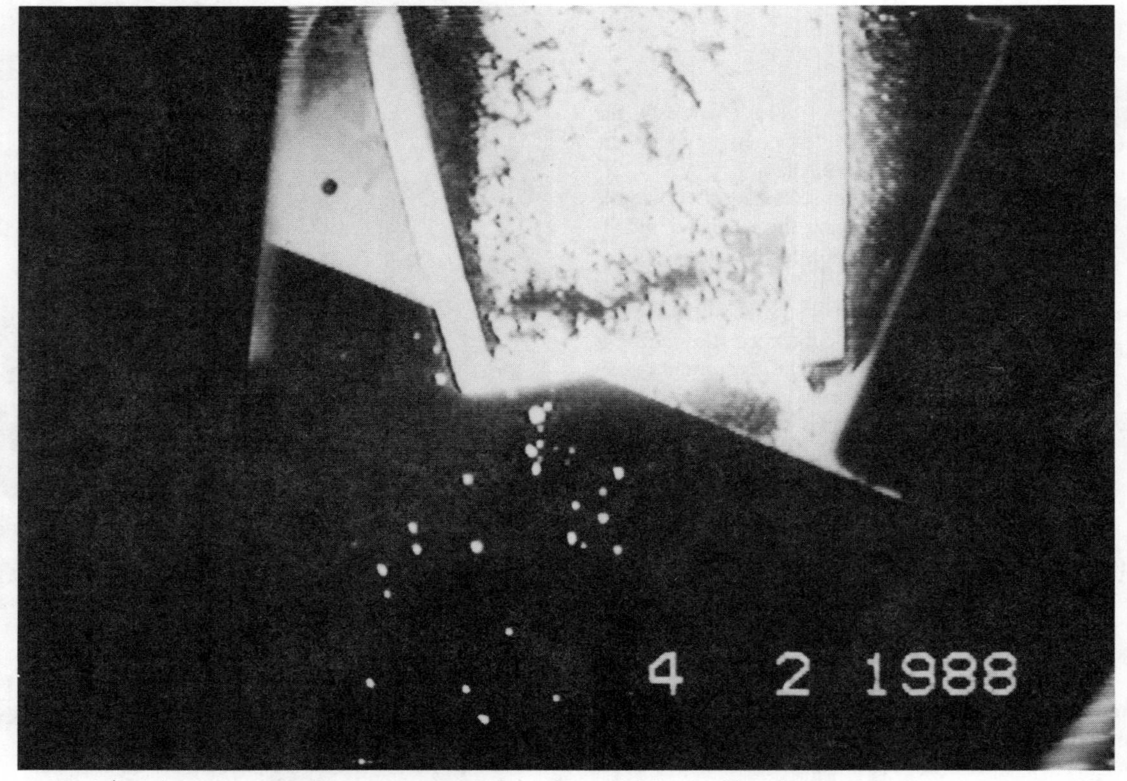

Figs. 4 and 5. Dust particle shower emitted from the model comet. Time between the two pictures is 1/50 sec.

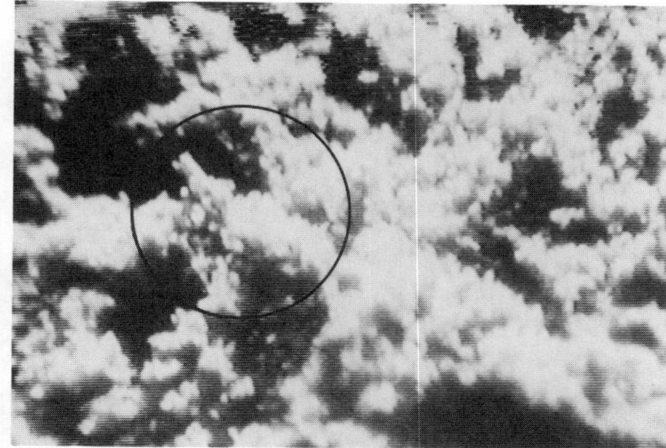

Figs. 6 and 7. Enlarged (25×) view of the model comet's surface with a vibrating particle in the encircled area. Length of the particle is approximately 1 mm.

4. Sometimes a protrusive particle is hit by a particle moving in a flat parabola that was emitted some moments before at another place on the surface because its velocity is not high enough for it to leave the sample's surface. The same mechanism may apply to the surface of a real comet although its gravity is essentially smaller (approximately 10^{-4} g_{Earth}).

5. Investigating the lift-off process with high resolution in space and time, first some cloud formation can be seen, comparable to a rocket launch, and then the particle is lifted off (see Figs. 8a-c).

6. During some small simulation chamber experiments when the model comet was very inactive, the insolation intensity was increased very rapidly by a factor of two. In an immediate reaction, the sample seemed to explode and parts of the crust flew away.

PHYSICAL INTERPRETATION

In the following discussion of the experimental observations we assume a structural model of the artificial comet's surface. The observations have shown that the surface of the "snow"-like material is very rough. The ice-mineral particles (which are partly aggregates of smaller grains) build up an interconnected network with pores and channels. They are connected to each other by contact bridges whose compositions are yet unknown but may vary between purely volatile and purely mineral components. Since we observe with high-resolution video techniques that contacts of grains to deeper surface layers are eroded, we conclude that the bridges are most likely a mixture of ice and minerals. In modeling the observations, we further simplify the sketch of a cross-section of a sample surface shown in Fig. 9 (Fig. 10). We use a spherical model of the grains [as also done by *Gombosi et al.* (1986) and *Marconi and Keller* (1988)], although such a simplified model may underestimate the coupling between the gas jet and the particles (see below).

The driving force for the particle vibration is assumed to be the gas flow originating from sublimation processes of deeper ice layers during insolation. The coupling between the ice-mineral grains and the gas jet surely depends on their shape, surface roughness, homogeneity (i.e., hard or open lattice structure), as well as on the texture of their environment, etc. This will be discussed in more detail later.

The particles bonded to the porous surface are exposed to the outstreaming volatiles. The bonds at the bottom side, especially, are eroded by the gas flow. The particles lift off in the moment when the drag force of the gas jet exceeds the force of gravity and the bonding forces. Figure 11 shows the model cross-section of the cometary analog during irradiation in the outstreaming gas jet and the enlarged view of two bonded particles. The mechanism of erosion of bonds by the gas flow is suggested by observations shown in Figs. 8a-c.

During insolation phases of remarkably higher intensity, it seems that the vapor pressure beneath the somewhat more compact dust mantle suddenly increases and breaks up this layer. Even the fracture of this porous mantle obeys the laws of fracture mechanics. The picture of a particle with broken bonds has to be enlarged to more extended aggregates.

Further and more detailed experiments to investigate the erosion and the lift-off process of a particle are planned.

FRACTURE MECHANICAL MODELING OF DUST EMISSION

The high-resolution video observations suggest that the particles are not loosely stacked together at the surface, but bonded to their neighbors. These exert mutual forces that must be overcome in order to loosen a particle so that it can be accelerated by the gas stream. Fracture mechnical models are used to describe such observed processes.

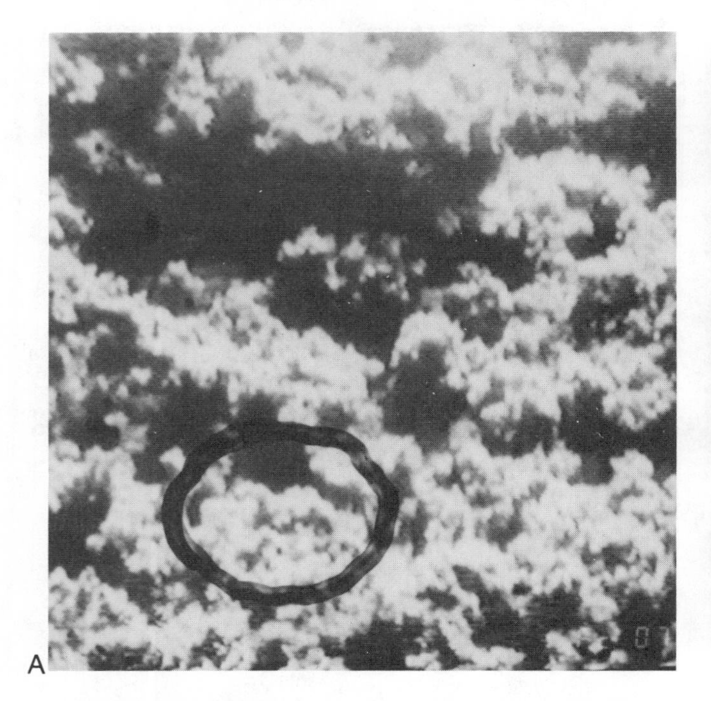

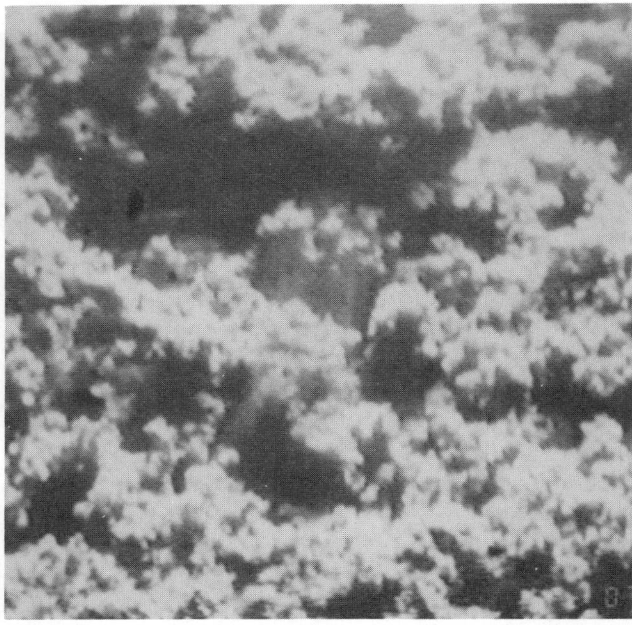

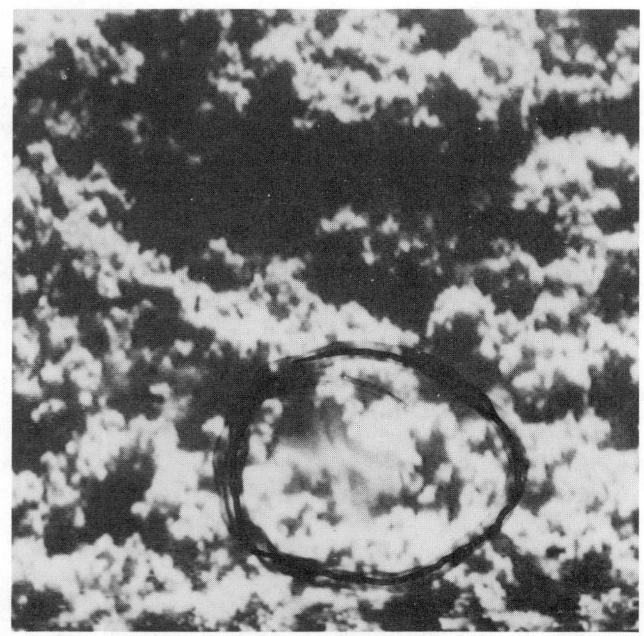

Fig. 8. **(a)** Particle at rest, **(b)** cloud plume, and **(c)** lift off.

Force Balance of Particles Bonded to the Surface

The simple fracture mechanical model described in the following starts with a description of the artificial comet's surface, where the particles are at least partially bonded to their neighbors via contact bridges. Figure 12 shows a schematic sketch of a contact bridge. Such "cohesive" contacts between particles, i.e., bonds that can sustain a mechanical load, may have various origins: The particles may be simply hooked together, since their surface is rough, or a part of the gas is recondensed at contact points of particles. They also may be electrically bonded, e.g., as a consequence of the admixture of phyllosilicates. The experimental observations presented above show that such "cohesive" bridges exist independent of their origin. These drastically change the force balance for a particle at a surface. This is schematically depicted in Fig. 13. The gas jet originating from the porous underground exerts a frictional force on a spherical particle F_D, which is equivalent to the drag force on a free particle at rest ($u_{particle} = 0$)

$$F_d = ½\, C_D \pi r^2 \rho u^2 \qquad (1)$$

where C_D is the drag coefficient (assumed here to be 2), r is the particle radius, ρ is the gas density, and u is the gas velocity. The assumption of spherical particles drastically

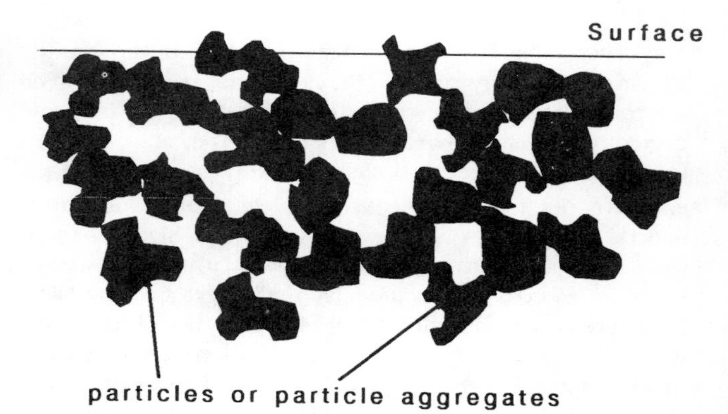

Fig. 9. Sketch of a realistic cross-section of the sample.

Fig. 10. Simplified model of Fig. 9.

influences the results of the calculations. Nonspherical particles with a high surface roughness and perhaps partially open lattice structure couple better to the gas jet, so the drag force increases considerably. We can include such an effect into the modeling by replacing the radius of a particle by the effective surface area exposed to the gas jet, which will

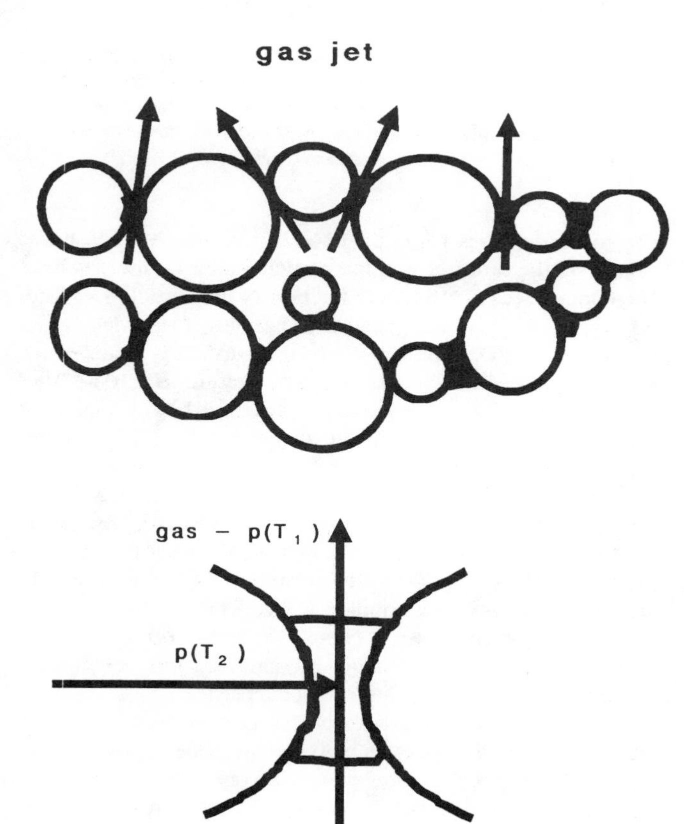

Fig. 11. Simplified model cross-section of the cometary analog during irradiation in the outstreaming gas (top) and the enlarged view of two bonded particles (bottom).

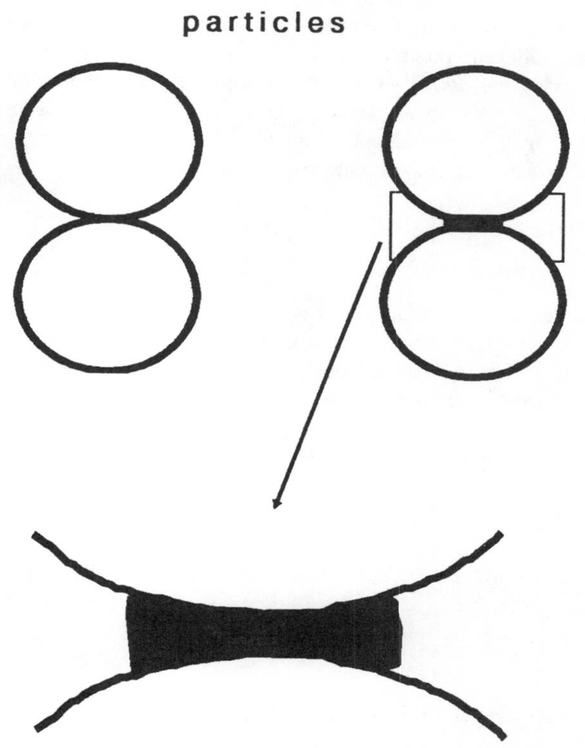

Fig. 12. Schematic picture of loosely arranged particles and a pair bonded by a contact bridge.

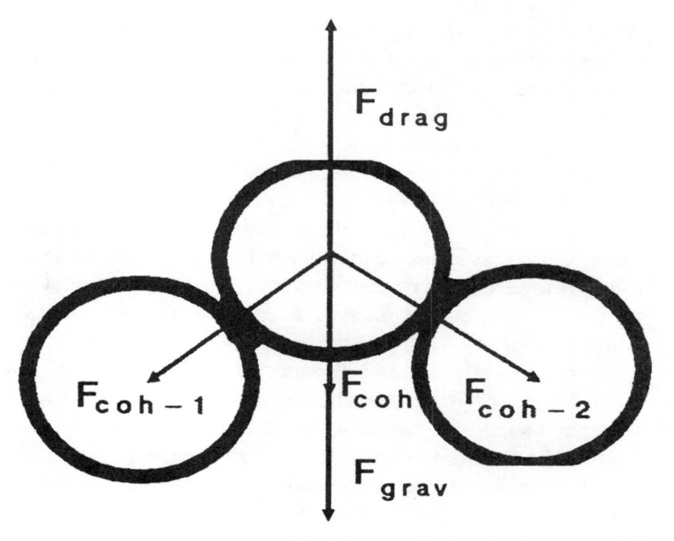

Fig. 13. Force balance at a triple of particles bonded by contact bridges. The bridges sustain a force exerted by the gas flow as a frictional force F_{Drag} minus the force of gravity F_{grav}.

generally be different from that of a sphere. This, however, would make sense only if we know the particles' microstructure well enough, i.e., the complete statistics of particle sizes, shapes, surface roughness, porosity, composition, etc. This is a very difficult task and seems even then unfruitful. The calculations done here should only deliver a rough estimate of the different parameters of the interaction between the outflowing gas and the bonded surface particles. For this first estimation, only mean values are used. The radius of the particle may be considered as an effective one. The frictional force (equation (1)) is counterbalanced by gravity F_g

$$F_g = \frac{4\pi}{3} r^3 \rho_p g \qquad (2)$$

where ρ_p is the particle density and g is the acceleration of gravity. The difference between F_D and F_g is balanced by stresses inside the contact bridges. If the bridge area is S_b the tensile stress σ_b induced in the bridge is

$$\sigma_b = \frac{(F_D - F_g)}{S_b} \qquad (3)$$

The bonds will break when the fracture stress (=ultimate tensile strength) σ_F of the material that builds up the contact bridge is reached

$$F_{bmax} = S_b \cdot \sigma_F \qquad (4)$$

Generally we get the inequality valid for bonded particles

$$F_{bmax} + F_g \geq F_D \qquad (5)$$

The particles lift off from the surface if the bonds break and $F_g < F_D$ is valid. In the calculations we assume furthermore that the contact bridges are water ice and have the shape of a cylinder with height h_b and contact areas S_b. The tensile strength of water ice depends on temperature, grain size, porosity and type, shape, and number density of inclusions such as minerals. A suitable dataset of the mechanical properties of water ice can be obtained from *Weeks and Assur* (1972), *Schulson* (1987), and *Nixon and Schulson* (1987). From these references we use for an estimate $\sigma_F = 1$ MPa. The fracture stress of a contact bridge between these particles at the model comet's surface may differ due to an irregular shape of the bridge, which lowers this stress. The size, on the other hand, will increase the fracture stress according to the Hall-Petch relation $\sigma_F = \sigma_0 + K/d^{1/2}$, where σ_0 is some reference fracture stress and d is the diameter of the contact bridge. If the bridge itself is porous, the fracture stress will be lower than for a solid one. According to *Ashby* (1983) and *Gibson* (1989) the tensile strength decreases proportional to the ratio of the densities of the actual and the solid bridge raised to the power 3/2. The fracture stress (or tensile strength) increases with decreasing temperature of the bridge.

Both sides of equation (5) are graphically presented in Fig. 14, where the forces are normalized by the particle volume.

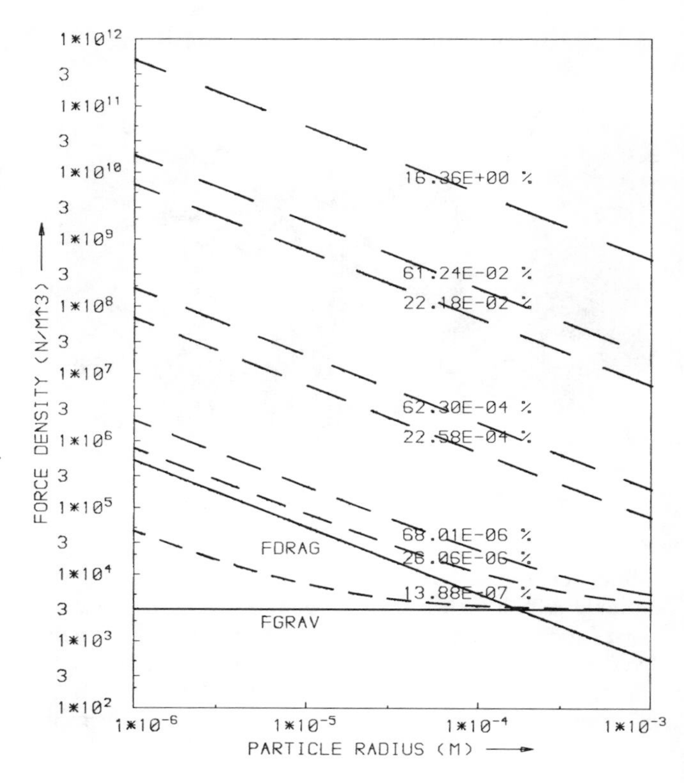

Fig. 14. Force densities on a particle. Solid horizontal line is due to gravity, inclined solid line due to the gas drag, and broken lines represent the sum of gravity and "cohesive" forces. The parameter given at the curves is the percentage of the particle surface in percent bonded to neighbors.

The force densities (F_D/V_p, F_g/V_p, $F_b = S_b\sigma_F/V_p + F_g/V_p$) are plotted on the ordinate, and the particle radius on the abscissa. The contact area is varied as a fraction of the particle's surface ($O_p = 4\pi r^2$). Then the maximum bonding force density is $f_{bmax} = 3\,\sigma_F \cdot S_p/O_p$. The density of the particle is assumed to be $\rho_p = 300$ kg/m^3. The surface temperature is 210 K. The partial pressure of water ice was calculated by the equation given by Washburn (see *Cowan and A'Hearn,* 1979) and the thermal gas velocity was corrected by a factor of 0.62 according to *Gombosi et al.* (1986). The horizontal line is the constant force density due to gravity (on Earth); the other solid line is due to the gas drag $f_D = F_D/V_p$. They intersect at a particle radius of approximately 150 μm. The other lines are the sum of gravity and bonding force densities. Even in the case of very small areas of contact to neighboring particles ($\geq 3 \cdot 10^{-5}$ of the particle surface is connected with neighbors) it is impossible for the gas jet to lift off a particle.

These model calculations show that dust emission is nearly impossible without processes reducing the bonding forces. The dust emission itself is, however, a unique feature of active comets and also cometary analogs. The observations in apparent contradiction to the calculations made above can be explained with the help of two models suggested from the high-resolution video observations: (1) contact bridges are really eroded in the gas stream and (2) the particles vibrate within the gas flow and the bonds break by fatigue fracture at considerably lower stresses than under a static tensile or compressive force.

Erosion of Contact Bridges

A surface element is considered containing particles as shown in Fig. 11. Some contact bridges are exposed to the gas stream. The partial pressure of water in the gas corresponds to a temperature T_1. The bridges are assumed to have a temperature T_2. The net gas sublimation rate per unit area from the bridges is given by (using simply kinetic gas theory)

$$z_{12} = \frac{p_1(T_1)}{\sqrt{2\pi\mu_{H_2O}RT_1}} - \frac{p_2(T_2)}{\sqrt{2\pi\mu_{H_2O}RT_2}} = z(T_1) - z(T_2) = z_{12} \quad (6)$$

where μ_{H_2O} is the molecular weight of water (kg/mol), and R is the universal gas constant. This sublimation rate leads to a reduction of the contact bridges until both sides of equation (5) are equal and the particle can be dragged away by the gas. In order to calculate this time we assume, as before, that the single contact bridge is cylindrical in shape. The bridge volume is then $V_b = h_b S_b$. Its surface is $M_b = 2\,h_b\sqrt{\pi S_b}$. The change in bridge volume per unit of time is given by

$$\frac{dV_b}{dt} = \Omega M_b z_{12} \quad (7)$$

Here Ω is the molecular volume of water. The contact area varies with time as

$$\frac{dS_b}{dt} = \sqrt{\pi}\,\Omega z_{12}\sqrt{S_b} \quad (8)$$

or integrated

$$S_b^{1/2} = S_o^{1/2} + 1/2\sqrt{\pi}\,z_{12}\Omega \cdot t \quad (9)$$

where S_o is the initial contact area. S_b increases if $z_{12} > 0$, and otherwise decreases. Here we assume that $z_{12} = -|z_{12}|$. The force balance of equation (5) can be written now (neglecting the effect of sublimation on the particles themselves, i.e., their change in radius)

$$S_b(t) \cdot \sigma_F + \frac{4\pi}{3} r^3 \rho_p g \geq 1/2\, C_D \pi r^2 \rho(T_1) \cdot u(T_1)^2 \quad (10)$$

Fracture of the bonds occurs when both sides of equation (10) are equal. We call the time t when this occurs the critical time t_{er}. From equation (10) we calculate

$$t_{er} = \left[S_o^{1/2} - \left(\frac{F_D - F_g}{\sigma_F}\right)^{1/2}\right] \frac{2}{\sqrt{\pi}} \frac{1}{|z_{12}|\Omega} \quad (11)$$

The result of such a calculation is shown in Fig. 15, where the time is plotted on the ordinate against particle diameter on the x-axis. Depending on the relative portion of the particle surface connected to neighbors by contact bridges, the time until fracture occurs varies between a few milliseconds to some ten seconds in this example. The actual value may differ depending on the parameters chosen (such as temperature of the gas jet and of the contact bridge). It is not the aim of this simple model to describe the physical process in total. It shows, however, that any process leading to a reduction of the contact areas between the particles, such as sublimation, will lead to a fracture of the bonds within a reasonable time and open the possibility for the particles to be lifted off by the gas jet. The calculations nevertheless yield times for the erosion of the contacts that are within the range observed by the video technique.

Fatigue Fracture

The observed vibrations of particles in the gas jet after a certain time lead to particle lift off. The time varies between a few seconds and ten minutes. The frequencies range from a few Hz to 100 Hz. The particles perform bending and

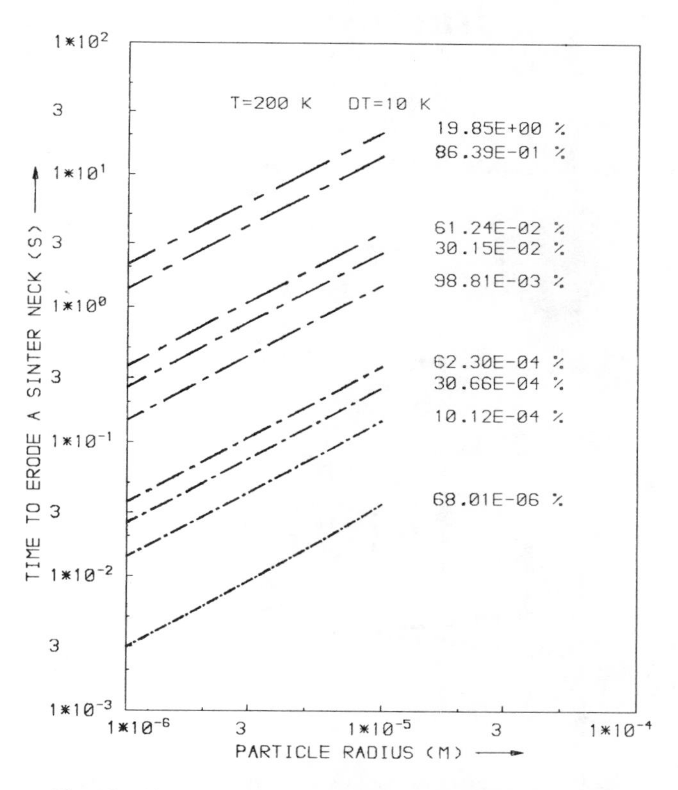

Fig. 15. Time needed to erode a contact bridge until the gas is able to accelerate the particle. The parameter at the curves gives the percentage of surface bonded to neighbors. The contact bridge is assumed to be at a temperature of 210 K and the gas jet flowing around it has a partial pressure of water corresponding to 200 K. The particle has a density of 1000 kg/m³.

torsional vibrations. These observed features are similar to the well-known fatigue fracture of metals and ceramics. For a first quantitative approach we use a simple picture of such a particle. Figure 16 shows a comparison of reality and model. We assume the particle to have a square cross-section of width h and height L. The vibration frequency is ν and the amplitude is assumed to be equal to the width h (independent of L). At the bottom the cross-section is only $[h-a(t)]h$, where $a(t)$ is the crack length. The vibrations induce a bending moment and therefore stresses in the bar are increased at the bottom due to the presence of a crack-like defect of length a ($a = a_0$ for $t = 0$).

Fatigue fracture occurs because the crack length a is not constant with time but changes due to the vibrations (whenever the material is not exclusively elastic, but viscous or plastic relaxations of stresses are possible). In the vicinity of the crack tip the external stresses are drastically enhanced (Fig. 16). This increase is controlled by the stress intensity factor K_I, which is, in the case of an infinitely extended material, proportional to the externally applied stress and the square root of the crack length. The growth under vibrating stresses is described by the Paris-law, which is used here in simplified version, following *Cherepanov* (1979)

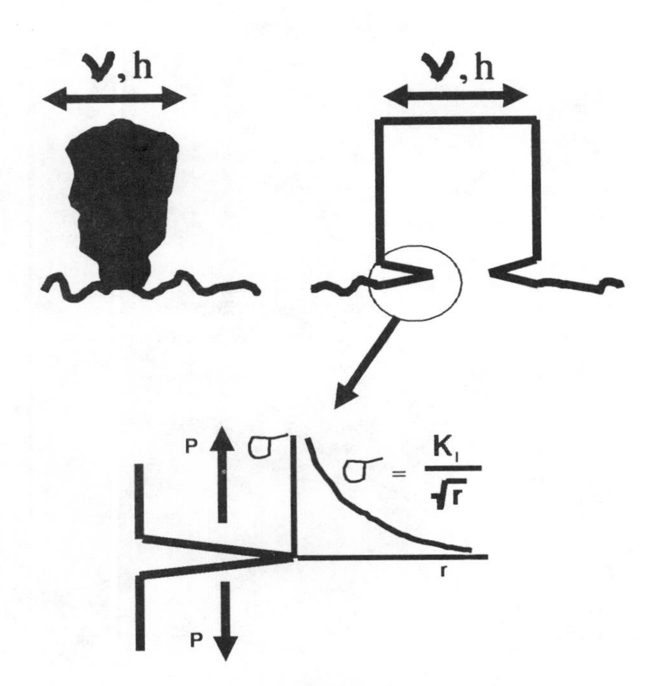

Fig. 16. Sketch of a real particle vibrating in the gas jet (left) with amplitude h (equal to width) and frequency ν, and model (right), representing a rectangular bar of cross-section h^2 with a crack-like defect at the side of contact to the surface. The stress distribution resulting from alternating bending of such an assembly is shown schematically below. P is the external load, r is the distance from the crack tip, and σ is the stress.

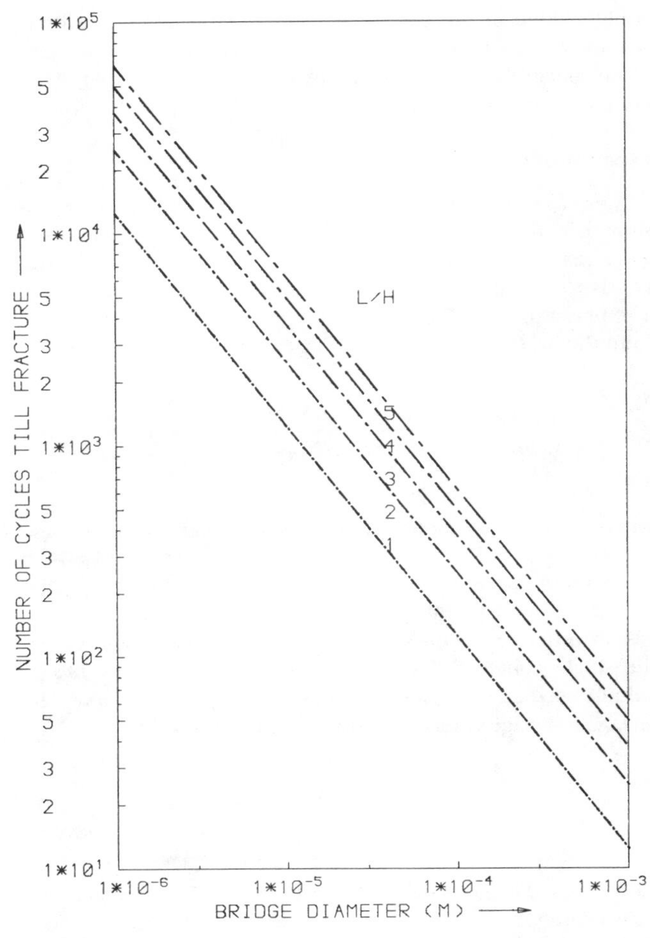

Fig. 17. Calculated number of vibration cycles until fatigue fracture occurs for a particle vibrating with constant amplitude, related to the initial bridge width (diameter).

$$\frac{da}{dN} = \beta \cdot K(h,L,a,M_b)^4_{max} \tag{12}$$

where K_{max} is the maximum stress intensity factor occurring during one cycle due to the bending vibrations, β is a constant explained below, and M is the bending moment. For the chosen geometry of the particle (=rectangular bar) and a vibration with constant amplitude, K_{max} can be calculated following *Cherepanov* (1979). Fracture of the contact bridge occurs under these conditions if the critical stress intensity factor K_{Ic} is reached at the crack tip. Then the crack length increases spontaneously under the gain of elastic energy and the particle is free to be accelerated by the gas jet. K_{Ic} is a material constant that can be measured, such as ultimate tensile strength. The number of cycles till fatigue fracture occurs can be obtained from the known expression for K_{max}

$$N_f \nu = t_f \cong \frac{1}{\beta} \left(\frac{K_{Ic}}{E} \right)^4 \left(\frac{L}{h} \right)^8 \frac{1}{h} \qquad (13)$$

The value of β depends on the model used to describe the nonelastic behavior of the material near a crack tip. Since this is a difficult task, the parameter is usually obtained experimentally. In order to be free from a fit parameter, here we use the model of a so-called Dugdale crack (for a definition of this see *Cherepanov*, 1979), where a region near the crack tip has a constant stress that depends on the flow stress of the material σ_s. Then β can be expressed analytically as as $\beta = \pi K_{Ic}^2 / 12 \sigma_s^2$. With the known material parameters for water ice we can calculate the number of vibration cycles until fatigue fracture occurs depending on the initial bridge width $h - a_o$. A result of such a calculation for different length to width ratios of particles is shown in Fig. 17. Depending on the bridge width, the number of cycles is in the range of 100 to 10,000. The increase length-to-width ratio is understandable since we assumed constant amplitude. The longer the bar, the smaller the stresses due to bending under this condition. The same holds for the increase in N, with decreasing bridge width.

From this model calculation we deduce a range of lifetimes for contact bridges t_r at a vibration frequency of the particle of 50 Hz from a few to a few hundred seconds. This is the observed range. The agreement only shows that fatigue fracture, i.e., the growth of crack-like defects in contact bridges (mineral inclusions or pores) or at their surface, may grow during cyclic loading and thus lead to the fracture of the bridges.

SUMMARY

High-resolution video investigations of dust emission mechanisms from artificial cometary surfaces were reported. For the first time the gas-jet-triggered emission of particles direct from the surface of a model icy body was recorded in an enlarged view (40×). It was observed that the ice-mineral particles are not dominated by gravity force at the surface. Every grain is sintered, bonded, or hooked via its rough surface with its neighbors. Before lift off, an erosive process has to set the particles free.

A discussion of the experiments in relevance with the current observations is at the moment impossible. 10^7 orders of magnitude separate the results of the space missions to comet Halley (point of closest approach 1000 km) from the camera observations reported here (distance to the surface 10 cm). If the icy body of the model comet is reasonably similar to real comets in structure, composition, and physical parameters, the physical processes should be similar. Future space missions to or near the surface of a real comet will deliver better comparisons with model comet experiments.

The observed dust emission phenomena are explained with simple geometrical and fracture mechanical assumptions. The model calculations fit within one order of magnitude.

Acknowledgments. These investigations are financially supported by the Deutsche Forschungsgemeinschaft.

REFERENCES

Ashby M. F. (1983) The mechanical properties of cellular solids. *Met. Trans., A14,* 1755-1769.

Cherepanov G. P. (1979) *Mechanics of Brittle Fracture,* pp. 342-397. MacGraw-Hill, New York.

Cowan J. J. and A'Hearn M. F. (1979) Dust release and mantle development in comets. *Moon and Planets, 21,* 155-171.

Gibson L. J. (1989) Modelling the mechanical behaviour of cellular materials. *Mat. Sci. Eng., A110,* 1-36.

Gombosi T. I., Nagy A. F., and Cravens T. E. (1986) Dust and neutral gas modelling of the inner atmospheres of comets. *Rev. Geophys., 24,* 667-700.

Grün E., Kochan H., Roessler K., and Stöffler D. (1987) Simulation of cometary nuclei. *Symposium on the Diversity and Similarity of Comets,* pp. 501-508. ESA SP-278, Brussels, Belgium.

Grün E., Benkhoff J., Bischoff A., Düren H., Hellmann H., Hesselbarth P., Hsiung P., Keller H. U., Klinger J., Knölker J., Kochan H., Neukum G., Oehler A., Roessler K., Spohn T., Stöffler D., and Thiel K. (1989) Modifications of comet materials by the sublimation process: Results from simulation experiments (abstract). In *Papers Presented to the Workshop on Analysis of Returned Comet Nucleus Samples,* pp. 24-25. Lunar and Planetary Institute, Houston.

Kochan H., Benkhoff J., Bischoff A., Fechtig H., Feuerbacher B., Grün E., Joó F., Klinger J., Kohl H., Krankowsky D., Roessler K., Seboldt W., Thiel K., Schwehm G., and Weishaupt U. (1989) Laboratory simulation of a cometary nucleus: Experimental setup and first results. *Proc. Lunar Planet. Sci. Conf.,* pp. 487-492. Lunar and Planetary Institute, Houston.

Klinger J., Benkhoff J., Espinasse S., Grün E., Ip W., Joó F., Kochan H., Kohl H., Roessler K., Seboldt W., Spohn T., and Thiel K. (1989) How far do recent simulation experiments fit current models of cometary nuclei? *Proc. Lunar Planet. Sci. Conf. 19th,* pp. 493-497. Lunar and Planetary Institute, Houston.

Marconi M. and Keller H.U. (1988) Outflow from a simulated cometary surface under laboratory conditions. *Report of the Max-Planck Institute for Aeronomy,* MPAE-W-100-88-31.

Nixon W. A. and Schulson E. M. (1987) A micromechanical view of the fracture toughness of ice. *J. de Physique, 48,* 313-319.

Roessler K., Hsiung P., Kochan H., Hellmann H., Düren H., Thiel K., and Kölzer G. (1990) A model comet made from mineral dust and H_2O-CO_2 ice: Sample preparation development. *Proc. Lunar Planet. Sci. Conf. 20th,* this volume.

Saunders R. S., Fanale F. P., Parker T. J., Stephens J. B., and Sutton S. (1986) Properties of filamentary sublimation residues from dispersions of clay in ice. *Icarus, 66,* 94-104.

Schulson E. M. (1987) The fracture of ice. *J. de Physique, 48,* 207-220.

Weeks W. F. and Assur A. (1972) Fracture of lake and sea ice. In *Fracture* (H. Liebowitz, ed.), pp. 880-979. Academic, New York.

Shock and Terrestrial Cratering

Proceedings of the 20th Lunar and Planetary Science Conference, pp. 415-420
Lunar and Planetary Institute, Houston, 1990

A Layered Moldavite Containing Baddeleyite

B. P. Glass

Geology Department, University of Delaware, Newark, DE 19716

J. T. Wasson

Institute of Geophysics and Planetary Science, University of California, Los Angeles, CA 90024

D. S. Futrell

6222 Haviland Avenue, Whittier, CA 90201

A layered moldavite weighing 11 g was found in Jakule, Czechoslovakia, which is just southeast of the Bohemian strewn field as previously defined. Although the specimen was found closer to the Bohemian strewn field, the major oxide composition indicates a closer affinity with the Moravian tektites. The specimen differs from typical splash-form moldavites in that it is compositionally and petrographically more heterogeneous and it contains baddeleyite produced by the breakdown of zircon. The shape, layered appearance, heterogeneity, and presence of baddeleyite suggest that the Jakule moldavite is a layered or Muong Nong-type tektite; however, instrumental neutron activation analysis (INAA) data indicate little if any enrichment of the volatile trace elements. If the Jakule specimen is a layered or Muong Nong-type tektite, its occurrence southeast of the Bohemian part of the moldavite strewn field is not that expected from simple distribution models if the Ries crater is the source of the moldavites.

INTRODUCTION

Tektites can be divided into three groups based on their morphologies: (1) splash forms, (2) ablated forms, and (3) layered forms. The splash forms are the most common; ablated forms are splash forms that have undergone a second period of melting to form a flange around the perimeter of the specimen. Layered forms from near Muong Nong, Laos, were first described by *Lacroix* (1935), who referred to them as unshaped tektites in contrast to the "shaped" splash and ablated forms. Layered or Muong Nong-type tektites occur as blocky chunks of glass up to 12.8 kg in size with subparallel layering (*Barnes,* 1961, 1971).

Muong Nong-type tektites can also be distinguished from splash forms petrographically and texturally. In the former, vesicles are more common and lechatelierite particles (SiO_2 glass) are generally frothy. Some Muong Nong-type tektites from Southeast Asia contain relict mineral grains (*Barnes,* 1963) showing evidence of shock metamorphism (*Glass and Barlow,* 1979) and coesite (*Walter,* 1965; *Glass et al.,* 1986). The assemblage of relict minerals and their size and shape suggest that the parent material was a fine-grained, well-sorted, sedimentary deposit (*Glass and Barlow,* 1979).

Muong Nong-type tektites also differ from splash-form tektites compositionally. Although they have similar concentrations of nonvolatile elements, the Muong Nong-type tektites are more heterogeneous and have slightly higher water and volatile element contents than the splash forms (*Koeberl,* 1986; *Koeberl and Beran,* 1988).

Most investigators agree that tektites were formed by impact melting (*King,* 1977; *Taylor,* 1973). The large size, unshaped forms, presence of frothy lechatelierite and relict mineral grains, heterogeneity, and volatile enrichment suggest that the Muong Nong-type tektites were deposited closer to the source crater than were the splash forms. In Southeast Asia layered tektites having masses >300 g are found across an area 1200 km long (*Wasson,* 1987). It has been inferred that the layered tektites formed as puddles melted in place by compression from a cometary coma (*Barnes and Pitakpaivan,* 1962; *Barnes,* 1989) or by local deposition of melt near a multitude of impact sites (*Wasson,* 1987).

In 1987, one of us (D. S. Futrell) obtained a layered moldavite (Czechoslovakian tektite) weighing 11 g from Milan Prchal of Ćeské Budějovice, Czechoslovakia. The specimen was collected from a gravel quarry in the vicinity of Jakule beyond the southeastern extreme of the Bohemia subfield as previously defined (Fig. 1). This site is 90 km west-southwest of the Moravian subfield and 60 km west-northwest of the recently described Austrian subfield.

The purpose of this study was to determine if the Jakule specimen shows the characteristics observed in layered or Muong Nong-type tektites from Southeast Asia, i.e., the presence of relict or shock metamorphosed mineral grains and an enrichment in volatile elements compared with splash-form moldavites.

SAMPLES AND ANALYTICAL PROCEDURES

The 11-g sample of Jakule was broken into three pieces. One 3.2-g piece was searched for mineral inclusions. A second, 2.9-g, piece was used to prepare a polished section and for neutron activation analysis.

The 3.2-g sample was crushed and sieved. The 149-250-μm-size fraction was put through a heavy liquid separation using mixtures of s-tetrabromoethane and methanol. Fragments from each specific gravity fraction were mounted in epoxy on glass slides, ground down to produce a flat surface, polished, and coated with carbon for major element analysis.

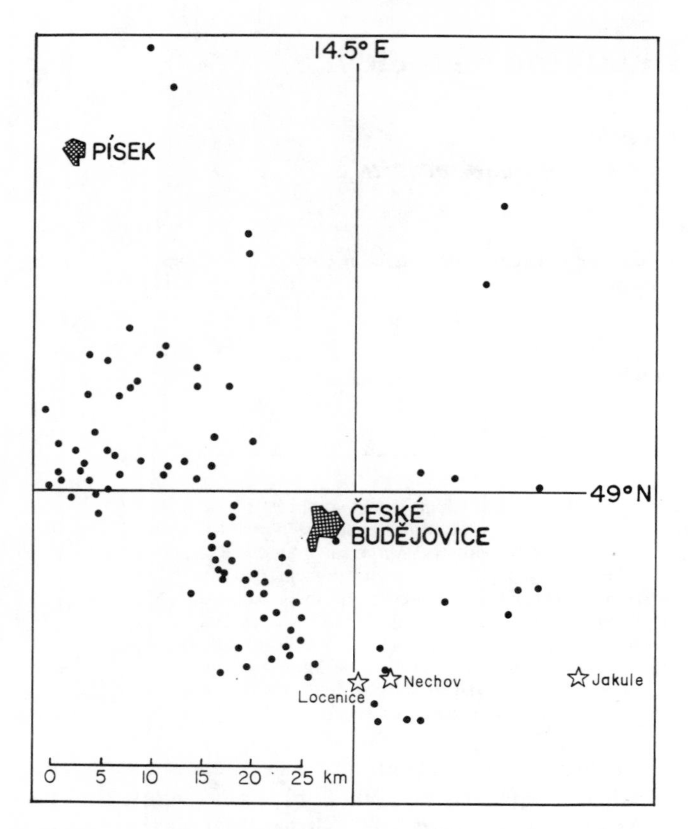

Fig. 1. Map showing moldavite sample locations in the Bohemian subfield (after *Bouška*, 1964). The locations of the three samples included in the present study are shown by stars. The Moravian subfield is about 90 km east-northeast, the Austrian subfield about 60 km east-southeast of Jakule.

Glass fragments from the denser specific gravity fractions were studied with a binocular microscope with up to 50× magnification. Fragments that appeared to contain mineral inclusions were X-rayed using a Gandolfi camera and then mounted for scanning electron microscope studies and energy dispersive X-ray analysis.

The 2.9-g piece was sawed into three slices. One slice was used to prepare a polished section for petrographic studies and major element analysis; the other two slices were used for neutron activation analysis. Splash-form tektites from nearby Bohemian localities, Něchov and Ločenice (Fig. 1), were also sawed into slices used for polished thin sections and for neutron activation analysis.

Instrumental neutron activation analyses (INAA) of the three tektites were carried out according to procedures similar to those described by *Kallemeyn et al.* (1989). All samples were slices 3.2 ± 0.3 mm thick having masses of 350-500 mg. Samples were counted 4 times, approximately 0.3, 1.5, 7, and 24 days following a 4-hour irradiation at the University of California, Irvine, reactor.

Standards consisted of solutions pipetted onto absorbent polyethylene. A sample of the U.S.G.S. standard rock powder Cody Shale Sco-1 was included in each radiation. Estimated 95% confidence uncertain limits on the means of duplicate determinations are ≤5% of the reported values with the exception of Ga (±8%); Ni, Zn, Rb, Ta, Th, Dy (single determination), and U (±10%); As and Sb (±20%); and Br (only upper limits could be determined).

Major element contents were determined using a Princeton Gamma Tech System 4 energy dispersive X-ray analyzer (EDS) in combination with a Cambridge S90B scanning electron microscope. Artificial glasses prepared by Corning Glass Company and analyzed by the U.S.G.S. were used as standards. The data were corrected for background, atomic number, absorption, and fluorescence effects using a computer program by Princeton Gamma Tech.

RESULTS AND DISCUSSION

Major and Minor Elements

Differences in mean composition between moldavites from Bohemian and Moravian subfields have been recognized by *Bouška et al.* (1973), *Delano and Lindsley* (1982), *Engelhardt et al.* (1987), and *Koeberl et al.* (1988). Moravian tektites have lower contents of MgO and CaO and generally higher contents of FeO, Na_2O, and TiO_2 than Bohemian tektites with similar SiO_2 contents (see Table 1).

The Jakule specimen appears to have higher FeO and Na_2O contents and lower MgO and CaO contents than Bohemian tektites with similar SiO_2 contents (Table 1). On the other hand, the major oxide contents are indistinguishable from those of the Moravian tektites. The Jakule specimen has a greater range in composition than the specimens from Ločenice and Něchov (Table 1).

Neutron Activation Results

The two Jakule INAA samples were from parallel slices but differed in that the first included three layers whereas the second only two. The good agreement between the analyses shows that compositional differences between layers are relatively minor.

With the exception of K, Ca, Mn, and Sr, concentrations of most elements are higher in Jakule relative to the two splash-

TABLE 1. Major oxide compositions (wt.%) of moldavites used in this study (Jakule, Ločenice, Něchov) plus other moldavites for comparison.

	Jakule*	Ločenice*	Něchov*	Bohemia†	Moravia†
SiO_2	79.8 (1.43)	76.9 (0.23)	78.8 (0.44)	79.5-80.12	79.4-80.2
Al_2O_3	10.4 (0.94)	9.37 (0.11)	10.1 (0.12)	9.62-10.73	10.03-11.43
FeO	1.85 (0.35)	1.49 (0.07)	1.56 (0.07)	1.62-1.80	1.77-2.26
MgO	1.36 (0.21)	2.19 (0.05)	2.01 (0.16)	1.57-1.91	1.13-1.54
CaO	1.54 (0.38)	5.17 (0.07)	2.77 (0.29)	2.16-2.78	1.46-2.20
Na_2O	0.71 (0.04)	0.41 (0.06)	0.65 (0.04)	0.37-0.51	0.49-0.75
K_2O	3.62 (0.16)	3.64 (0.05)	3.46 (0.11)	3.32-3.71	3.06-3.69
TiO_2	0.33 (0.07)	0.23 (0.04)	0.26 (0.04)	0.27-0.32	0.34-0.51

*Average of 30 analyses (determined by energy dispersive X-ray analysis) plus/minus one standard deviation in parentheses.
†Range of six analyses with SiO_2 contents within 0.5% of the average for the Jakule specimen (data from *Rost*, 1972).

TABLE 2. Duplicate instrumental neutron activation analyses of 33 elements in Jakule, and in two other moldavites from southern Bohemia and in USGS standard SCo-1.

	Na	K	Ca	Sc	Cr	Mn	Fe	Co	Ni	Zn	Ga	As	Se	Br	Rb	Sr	Zr
Jakule	3.96	28.5	11.3	4.96	29.3	230	14.3	5.61	21	23.3	8.5	0.55	1.2	<0.13	143	154	260
	3.71	28.9	11.5	4.93	30.2	260	13.9	5.58	19	18.4	7.4	0.24	1.0	<0.08	114	142	247
Ločenice	2.65	29.1	36.2	4.08	24.2	728	11.5	5.01	18	25.2	6.4	0.35	1.0	<0.11	131	176	172
	2.67	30.0	33.6	4.23	26.0	741	12.1	5.27	12	26.9	6.2	0.33	0.9	<0.06	138	164	191
Něchov	3.03	27.9	18.6	4.31	22.4	523	11.7	4.94	18	11.8	5.5	0.22	1.1	<0.08	121	163	267
	3.11	29.2	20.5	4.83	27.5	566	13.2	5.66	21	13.8	4.8	0.16	1.3	<0.08	140	161	267
SCo-1	5.72	21.8	17.9	10.3	64.4	368	31.4	9.9	36	99.0	16.9	11.8	2.0	0.58	102	168	336
	5.99	22.0	18.2	11.1	69.0	378	33.1	10.6	36	99.5	17.4	11.5	1.9	0.80	107	181	177

	Sb	Cs	Ba	La	Ce	Nd	Sm	Eu	Tb	Dy	Yb	Lu	Hf	Ta	Th	U
Jakule	105	15.1	813	31.1	58.1	31.1	4.93	1.18	0.61	—	1.84	0.31	5.97	0.75	11.7	2.38
	78	14.6	740	29.8	52.6	30.6	4.67	1.20	0.61	3.2	1.72	0.27	5.64	0.63	12.5	2.28
Ločenice	83	14.9	794	26.6	49.4	27.6	4.29	1.02	0.54	—	1.56	0.26	4.93	0.62	10.1	2.13
	86	15.0	780	26.1	48.0	26.6	4.24	1.04	0.54	2.6	1.47	0.24	4.79	0.55	10.8	2.17
Něchov	<70	14.1	786	29.2	55.4	29.7	4.65	1.11	0.58	—	1.74	0.28	6.22	0.59	11.3	2.23
	32	15.2	805	31.3	52.6	31.7	4.98	1.22	0.64	3.5	1.82	0.29	6.58	0.65	13.4	2.52
SCo-1	2230	7.5	574	28.9	50.6	26.2	4.44	1.18	0.65	—	2.14	0.34	4.30	0.78	8.5	2.70
	2270	7.8	565	30.0	52.5	25.8	4.65	1.16	0.75	4.8	2.23	0.35	4.31	0.75	9.2	2.83

Concentrations in μg/g except Na, K, Ca, Fe, mg/g; and Sb, ng/g.

form moldavites (Table 2). The Něchov specimen is more closely related to Jakule and for Co, Ni, and the incompatible elements these tektites are not resolvable. The concentrations of some volatile trace elements (Ga, As, Sb) in Jakule are somewhat variable (in agreement with *Meisel et al.,* 1989) and are moderately higher than those in the splash-form tektites, but the differences between the splash-forms Ločenice and Něchov are appreciably greater than those between Jakule and Ločenice. In tektites from Southeast Asia, Zn is about four times higher in layered tektites relative to splash-form tektites with only marginal overlap, but Zn in Ločenice is about 10% higher than our higher Jakule replicate. Thus, volatile element concentrations do not resolve Jakule from the local splash-form tektites.

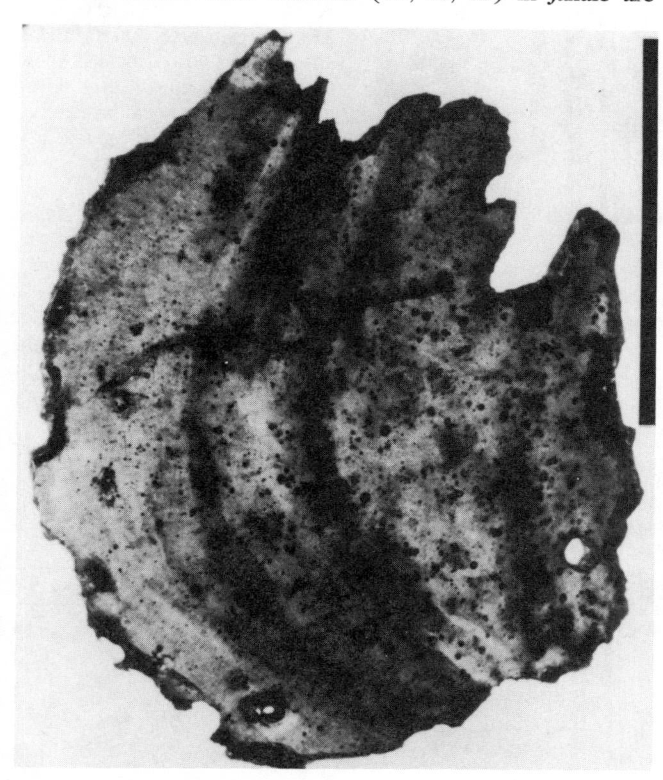

Fig. 2. Photomicrograph of a thin section of the Jakule specimen (taken in transmitted light) showing light and dark layers and numerous vesicles. Scale bar equals 1 cm.

In Southeast Asia the compositions of large layered or Muong Nong-type tektites can be used to establish lower limits on the mean loss fractions for certain volatiles from splash-form tektites. For example, Ga is ~2, Zn ~4, and Sb ~3 times lower in splash form relative to layered tektites. For these Bohemian samples, we do not yet have samples that are known to approximate the composition of the unfractionated precursor. We can only state that volatile losses in Jakule and in the splash forms were similar. The most plausible interpretation would seem to be that the degree of devolatilization of Jakule was more extensive than that of the Southeast Asian layered tektites, and thus that physical conditions (grain size, duration of heating, etc.) were more conducive to outgassing.

Petrography

The Jakule specimen exhibits layers of dark and light glass (Fig. 2). Bubble cavities are apparently elongated normal to the plane of Fig. 2. Furthermore, the Jakule specimen has more abundant, more pronounced, and more contorted schlieren than the two normal splash forms from the same area. In addition, the Jakule specimen is more vesicular and contains more lechatelierite particles than the two splash forms. One thousand point counts made on polished sections indicate that the Jakule specimen contains 2.2% and 1.4% bubbles and

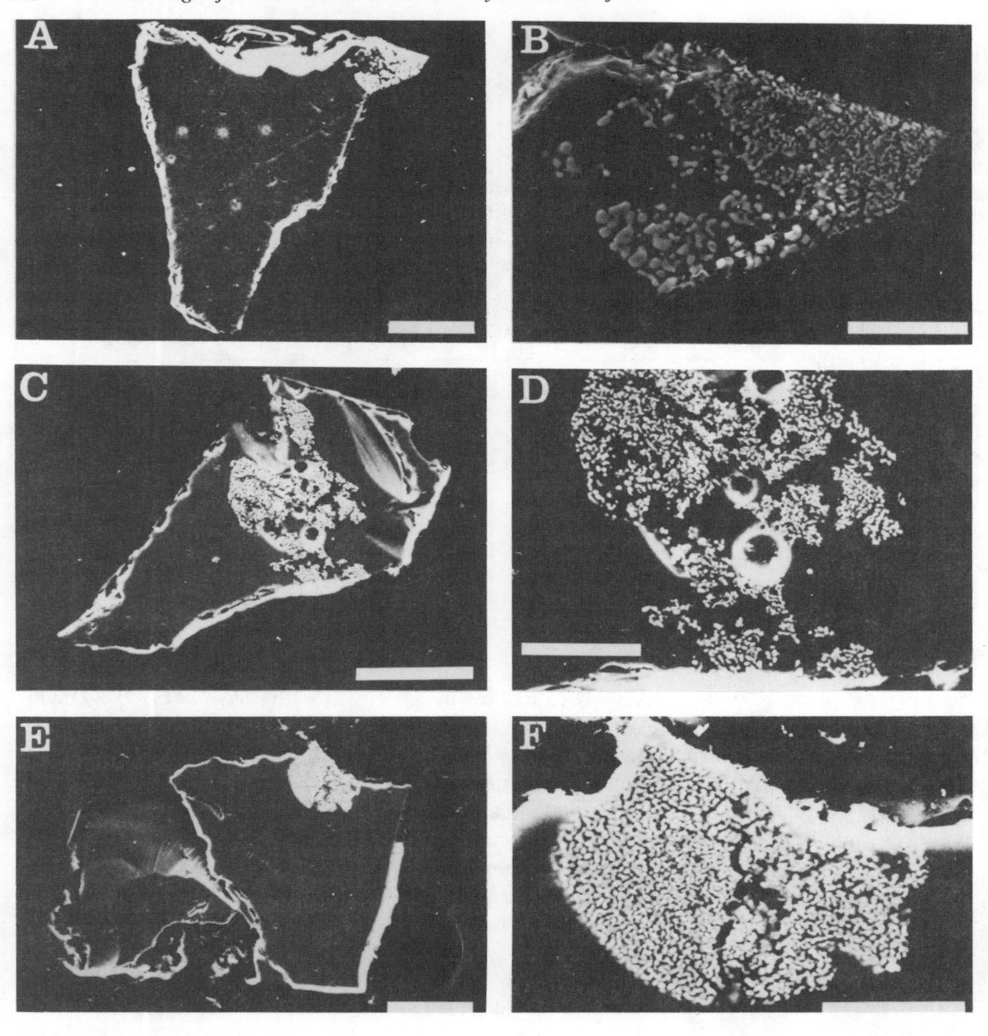

Fig. 3. Scanning electron microscope photomicrographs of polished sections of glass fragments from the Jakule moldavite showing baddeleyite inclusions. B, D, and F are enlargements of A, C, and E, respectively. For A, C, and E the black area is epoxy, gray is the glass, and white is the baddeleyite. For B, D, and F the light area is baddeleyite. The scale bars for A, C, and E equal 50 μm; for B equals 10 μm; for D and F equals 20 μm.

lechatelierite, respectively, whereas the Nechov and Locenice specimens both contain <0.1% bubbles and lechatelierite. The lechatelierite in the Jakule specimen is not frothy as it commonly is in the layered tektites from Southeast Asia, particularly those that contain relict mineral inclusions.

Baddeleyite. Four particles from the heavy specific gravity fractions each contained one white opaque inclusion. X-ray diffraction patterns show that the inclusions contain baddeleyite (ZrO_2). In addition to the baddeleyite lines, one pattern contains the strongest line for zircon. The inclusions range in major dimension from 35 to 85 μm. Two are partly dispersed, but with some straight sides suggestive of original crystal outlines; the third is rounded (Fig. 3). The baddeleyite occurs as rounded to elongate blebs (up to 2 μm in size) in a glass matrix (Fig. 3). Energy dispersive X-ray analysis indicates that the inclusions have a ZrO_2/SiO_2 ratio similar to zircon, but with much higher concentrations of Al_2O_3, FeO, and other oxides (Table 3).

The morphologies of the baddeleyite inclusions are similar to those reported by *El Goresy* (1965) for baddeleyite grains that formed by decomposition of zircon in impact glasses. The presence of the strongest zircon line in one of the X-ray diffraction patterns, as well as the bulk compositions of the inclusions, also suggest that the baddeleyite formed by the breakdown of zircon. The angular to rounded shapes of the inclusions and the small size suggest that the parent material was a fine-grained sediment.

TABLE 3. Major oxide composition (determined by EDS) of baddeleyite-bearing inclusion from layered moldavite from Jakule.

Analyses	(5)
SiO_2	33
ZrO_2	57
Al_2O_3	4.5
FeO*	2.3
MgO	0.9
CaO	0.4
Na_2O	0.8
K_2O	0.9
TiO_2	0.2

*All iron reported as FeO.

Baddeleyite has been reported in one Georgia tektite (*King*, 1966) and in the Martha's Vineyard tektite (*Clarke and Henderson*, 1961). Both these North American tektites are splash forms. *King* (1966) concluded that the baddeleyite inclusion that he studied was formed by decomposition of a zircon. Zircon is one of the most abundant relict mineral grains found in Muong Nong-type tektites from Southeast Asia. This suggests that the Southeast Asian Layered Tektites were not shock heated as intensely as the Jakule specimen.

IMPLICATIONS

The overall shape, layered structure, general petrography, chemical heterogeneity, and presence of mineral inclusions suggest that the Jakule specimen is a layered or Muong Nong-type moldavite. However, it is not appreciably enriched in volatiles, and the lechaterlierite is not frothy, observations that contrast with those made on layered tektites from Southeast Asia. Additional studies such as the determination of water content and the FeO/Fe_2O_3 ratio would add useful clues regarding the true nature of this specimen. The currently available evidence indicates that the properties of Jakule are intermediate between those of splash-form and the large layered tektites from Southeast Asia.

Cohen (1963) suggested that the Ries crater in Germany may be the source crater of the moldavites. *Konta* (1971) suggested that the variations between the Moravian and Bohemian moldavites could be explained by different ejection angles from the Ries crater that resulted in different atmospheric heating histories, with the lesser heated glass landing in Bohemia and the more heated glass landing in Moravia. On the other hand, based on petrographic studies of moldavites, *Barnes* (1964, 1969) suggested that moldavites are not from the Ries crater but from a source area centered in the vicinity of the Moravian area of moldavite occurrence. The discovery of layered moldavite containing baddeleyite in the southeasternmost part of the Bohemian strewn field appears to be more consistent with a source region between the Bohemian and Moravian strewn fields rather than to the west in the direction of the Ries crater. However, additional Muong Nong-type moldavites need to be found in order to test this conclusion.

CONCLUSIONS

The petrographic and compositional data reported here are consistent with the conclusion that a moldavite collected in Jakule, southeastern Bohemia, Czechoslovakia, is a layered or Muong Nong-type moldavite. The presence of baddeleyite after zircon in the Jakule specimen supports an impact origin for the moldavites. However, the relatively low volatile content of the Jakule specimen and the fact that it contains inclusions of baddeleyite that appear to have formed by decomposition of zircon indicate that it was subjected to more intense shock heating during formation than were those in the typical Southeast Asian layered tektites.

Acknowledgments. We thank M. Prchal for the Jakule specimen and X. Ouyang for carrying out the neutron activation determinations. One of us (D. S. Futrell) disagrees with those aspects of the above text that imply acceptance of a terrestrial impact origin for tektites. In his opinion the evidence for a lunar volcanic origin (see, e.g., *O'Keefe*, 1976; *Futrell*, 1986) is stronger than that for a terrestrial origin. This research was supported in part by NSF Grant EAR 88-16895.

REFERENCES

Barnes V. E. (1961) A world-wide investigation of tektites. *Geotimes, 6*, 8-12, 38.

Barnes V. E. (1963) Detrital mineral grains in tektites. *Science, 142*, 1651-1652.

Barnes V. E. (1964) Variations of petrographic and chemical characteristics of indochinite tektites within their strewn-field. *Geochim. Cosmochim. Acta, 28*, 893-913.

Barnes V. E. (1969) Petrology of moldavites. *Geochim. Cosmochim. Acta, 33*, 1121-1134.

Barnes V. E. (1971) Description and origin of large tektite from Thailand. *Chemie der Erde, 30*, 13-19.

Barnes V. E. (1989) Origin of tektites. *Texas J. Sci., 41*, 5-33.

Barnes V. E. and Pitakpaivan K. (1962) Origin of indochinite tektites. *Proc. Nat. Acad. Sci., 48*, 947-955.

Bouska V. (1964) Geology and stratigraphy of moldavite occurrences. *Geochim. Cosmochim. Acta, 28*, 921-930.

Bouska V., Benada J., Randa Z., and Kunciv J. (1973) Geochemical evidence for the origin of moldavites. *Geochim. Cosmochim. Acta, 37*, 121-131.

Clarke R. S. Jr. and Henderson E. P. (1961) Georgia tektites and related glasses. *Ga. Miner. Newsl., 14*, 90-114.

Cohen A. J. (1963) Asteroid- or comet-impact hypothesis of tektite origin: The moldavite strewn-fields. In *Tektites* (J. A. O'Keefe, ed.), pp. 189-211. Univ. of Chicago, Chicago.

Delano J. W. and Lindsley D. H. (1982) Chemical systematics among moldavite tektites. *Geochim. Cosmochim. Acta, 46*, 2447-2452.

El Goresy A. (1965) Baddeleyite and its significance in impact glasses. *J. Geophys. Res., 70*, 3453-3456.

Engelhardt W. V., Luft E., Arndt J., Schock H., and Weiskirchner W. (1987) Origin of moldavites. *Geochim. Cosmochim. Acta, 51*, 1425-1443.

Futrell D. S. (1986) Implication of welded breccia in Muong Nong-type tektites. *Nature, 219*, 663-665.

Glass B. P. and Barlow R. A. (1979) Mineral inclusions in Muong Nong-type indochinite: Implications concerning parent material and process of formation. *Meteoritics, 14*, 55-67.

Glass B. P., Muenow D. W., and Aggrey K. E. (1986) Further evidence for the impact origin of tektites (abstract). *Meteoritics, 21*, 369-370.

Kallemeyn G. W., Rubin A. E., Wang D., and Wasson J. T. (1989) Ordinary chondrite: Bulk compositions, classification, lithophile-element fractionations and composition-petrographic type relationships. *Geochim. Cosmochim. Acta*, in press.

King E. A. (1966) Baddeleyite inclusion in a Georgia tektite (abstract). *Eos Trans AGU, 47*, 145.

King E. A. (1977) The origin of tektites: A brief review. *Am. Sci., 64*, 212-218.

Koeberl C. (1986) Geochemistry of tektites and impact glasses. *Annu. Rev. Earth Planet. Sci., 14*, 323-350.

Koeberl C. and Beran A. (1988) Water content of tektites and impact glasses and related chemical studies. *Proc. Lunar Planet. Sci. Conf. 18th*, pp. 403-408.

Koeberl C., Brandstatter F., Niedermayr G., and Kurat G. (1988) Moldavites from Austria. *Meteoritics, 23,* 325-333.

Konta J. (1971) Shape analyses of moldavites and their impact origin. *Mineral. Mag., 38,* 408-417.

Lacroix A. (1935) Les tectites sans formes figurées de l'Indochine. *Acad. Sci. Paris, C. R., 200,* 2129-2132.

Meisel T., Koeberl C., and Jedlicka J. (1989) Geochemical studies of Muong Nong-type indochinites and possible Muong Nong-type moldavites (abstract). *Meteoritics,* in press.

O'Keefe J. A. (1976) *Tektites and Their Origin.* Elsevier, Amsterdam. 254 pp.

Rost R. (1972) *Vltaviny a Tektity* (Moldavites and Tektites). Czech. Acad. Sci. Publ., Academia Press, Prague. 241 pp.

Taylor S. R. (1973) Tektites: A post-Apollo view. *Earth Sci. Rev., 9,* 101-123.

Walter L. S. (1965) Coesite discovered in tektites. *Science, 147,* 1029-1032.

Wasson J. T. (1987) A multiple-impact origin of Southeast Asian tektites (abstract). In *Lunar and Planetary Science XVIII,* pp. 1062-1063. Lunar and Planetary Institute, Houston.

Proceedings of the 20th Lunar and Planetary Science Conference, pp. 421-431
Lunar and Planetary Institute, Houston, 1990

^{40}Argon-^{39}Argon Dating of Impact Craters

R. J. Bottomley[1] and D. York

Department of Physics, University of Toronto, Toronto, Ontario, Canada M5S 1A7

R. A. F. Grieve

Geophysics Division, Geological Survey of Canada, Ottawa, Ontario, Canada K1A 0Y3

The spectra from an ^{40}Ar-^{39}Ar study of 13 impact craters from Canada, Scandinavia, and Australia can be categorized broadly according to shape. Despite the fact that samples from eight of the craters show plateaus or near plateaus, the determination of ages for the cratering events is not straightforward. Melt rocks from Siljan (best age estimate from this study of the event, one sigma, 368 ± 1.1 Ma), Mien (121 ± 2.3 Ma), Pilot (445 ± 2 Ma), and Carswell (115 ± 10 Ma) all show near plateaus with varying degrees of structure. Only Clearwater West (445 ± 2 Ma) and Dellen (102 ± 1.6 Ma), however, give good to excellent plateaus over most of the gas released. Samples from Strangways, (<470Ma) Saaksjarvi (<330 Ma), and possibly Moinerie (400 ± 50 Ma) and Couture (430 ±25 Ma) all show spectra whose fraction ages rise with increasing temperature. Clearwater East (<460 Ma) and Nicholson (<400 Ma) give saddle-shaped spectra with greater ages at both the lowest and highest temperature steps. In addition, not all samples from the same crater site agree in age or spectral type. This study indicates that while ^{40}Ar-^{39}Ar dating is preferable to K-Ar for crater studies, multiple samples need to be analyzed from each site. Typical analytical errors quoted with age determinations on melt rocks are often misleading because the complex argon signatures do not lend themselves to simple interpretations and other information may be necessary to ascertain the correct age.

INTRODUCTION

Impact melt rocks typically appear as glassy to fine-grained, igneous textured rocks, often containing minerals and lithic clasts of the original target rock. In theory, these melt rocks should give a K-Ar age that closely approximates that of the impact for several reasons. The target rocks are melted instantaneously, allowing previous accumulations of radiogenic argon in the rock to escape. The melt usually occurs in surface or near-surface locations, often as melted pools in breccia lenses at simple craters or as annular sheets surrounding the central uplift at complex craters. As the melt rapidly cools, it once again begins to accumulate radiogenic argon. Since many of the currently recognized impact craters are located in stable Precambrian shield areas, the argon clock often has not been disturbed by other major geological events since the time of impact.

There are several factors that could potentially affect this simple description. Since the clasts in the melt rocks predate crater formation, they must be thoroughly outgassed to ensure that the melt rock is giving a geologically meaningful age. For example, the melt matrix could record the time of impact but the clasts could remember their own age of formation, with the result that the whole rock age will be a mixed age, greater than the true age of impact. However, the melt rocks are superheated at the time of impact and clasts interact with the matrix, eventually coming into thermal and often chemical equilibrium (*Grieve,* 1975). This gives the argon favorable opportunities to diffuse from the clasts and escape the melt-clast mixture.

In some cases, however, the argon degassed by these clasts may not fully escape and could be adsorbed elsewhere in the melt rock unit, where it will appear as inherited argon, possibly confusing the apparent age. If this component of argon achieves a locally uniform distribution in the impact melt, and the potassium is also uniformly distributed, a false ^{40}Ar-^{39}Ar plateau might be obtained. Since this would be a local phenomenon dependent on the type of clasts and the temperature in the immediate area, it may be possible to identify these false plateaus by dating several samples of spatially separated melt rocks from the same crater, and by dating separated mineral phases with different potassium concentrations.

If a melt rock comes into equilibrium with a sizable amount of inherited ^{40}Ar uniformly distributed throughout the rock, shapes other than false plateaus could be generated. Consider a melt rock that consists of a mixture of K-poor clasts (e.g., quartz) in a potassium rich matrix. The matrix will be the site of most of the ^{40}Ar production after impact. Upon analysis, the K-poor clasts will release only the initially inherited ^{40}Ar, which it shared with the matrix at the time of cooling. If the matrix and K-poor clasts release at different temperatures during heating in the laboratory, then the K-poor phase will give a false, elevated age because it has little offsetting K to match its relatively large inherited ^{40}Ar content. The argon from the K-rich phase could produce another false plateau segment, which would correspond to a mixture of the true radiogenic argon trapped since impact and the inherited argon. This false plateau would approach the true age of impact if the ratio of the radiogenic to inherited argon is very large. A similar effect could be produced if the melt has partially cooled by the time the inherited argon is baked out of the clast or the surrounding country rock by residual heating. Only that portion of the melt

---

[1]Now at Physics Department, Canadian Union College, College Heights, Alberta, Canada T0C 020

rock that was still above the blocking temperature for argon diffusion would show the effects of this inherited argon. This could result in a spectrum displaying two plateaus or some other indication of a mixture of two argon phases.

Another factor influencing spectral shape is the degree of devitrification. Since these rocks cool quite rapidly there can be a fair amount of glass in the matrix. Devitrification can be a serious problem as argon can be lost during this process (*Fleck et al.,* 1977), which may result in ages that are different from the ages of fresh glass.

EXPERIMENTAL TECHNIQUES

Melt rocks from 13 craters in Europe, North America, and Australia were selected for this project (Table 1). To minimize potential interpretation problems, samples with the freshest glass matrix and fewest clasts were chosen. Almost all of these clasts showed evidence of being reequilibrated with the matrix material. The ages of most of these craters were not well known.

This project utilized our established laboratory ^{40}Ar-^{39}Ar procedures using RF induction heating in a high vacuum followed by mass analysis on an MS10 mass spectrometer. Whole rock samples were weighed and wrapped in aluminum foil before being loaded into the irradiation container. A few samples, consisting of powdered mineral separates, were sealed in quartz vials and placed around the periphery of the can. The samples were irradiated in the McMaster University nuclear reactor, which typically ran at a power level of 1 MW during the irradiations. After irradiation the samples were allowed to cool for a period of several weeks to allow short-lived isotopes to decay. The samples were then unwrapped and placed in a degassed molybdenum crucible. The crucible was loaded into the fusion system and the entire system was then baked at 200°C for several days to reduce the atmospheric argon blank. After a typical fusion run of eight one hour steps, each fraction was analyzed for the five isotopes of argon in the mass range 36-40.

In the accompanying figures, the error for each individual fraction is indicated by the vertical width of the corresponding step on the plateau diagrams. This is the uncertainty in the argon measurement and associated corrections. The error in the integrated ages (Table 2) includes both the error in the fractions plus the experimental error assigned to the measurement of the standards. The standards used in this study included the hb3gr hornblende (1071 ± 7 Ma) and the Obedjiwan biotite (963 ± 9 Ma). All the laboratory errors are quoted at the one sigma level. These errors are not the geological uncertainty in the age of the crater; they represent the accuracy and reproducibility with which the argon ratio can be measured in the rock sample and standard. Detailed tables of all the relevant argon data are available in *Bottomley* (1982).

TABLE 1. A listing of the impact craters dated in this study, along with the sample code used for sample numbering on the figures.

Code	Crater Name	Location
CA	Carswell	Alberta, Canada
CWE	Clearwater East	Quebec, Canada
CWW	Clearwater West	Quebec, Canada
D	Dellen	Sweden
GW	Gow Lake	Saskatchewan, Canada
LC	Lac Couture	Quebec, Canada
LM	Lac La Moinerie	Quebec, Canada
M	Mien	Sweden
NC	Nicholson	Northwest Territories, Canada
P	Pilot	Northwest Territories, Canada
S	Siljan	Sweden
SJ	Saaksjarvi	Finland
ST	Strangways	Northern Territory, Australia

TABLE 2. The best interpretation of the age of impact as determined by this study.

Crater	Age (Ma)	Comment
Carswell	115 ± 10	error represents scatter
Clearwater East	460	maximum age estimate
Clearwater West	280 ± 2	plateau
Dellen	102 ± 1.6	plateau
Gow Lake	250	possible maximum age
Lac Couture	430 ± 25	error represents scatter
Lac La Moinerie	400 ± 50	error represents scatter
Mien	121 ± 2.3	structured plateau
Nicholson	400	maximum age estimate
Pilot	445 ±2	plateau
Siljan	368 ± 1.1	integrated age
Saaksjarvi	330	maximum age estimate
Strangways	470	maximum age estimate

RESULTS

In this study, certain spectra were found to have a straightforward interpretation, and the corresponding age of impact can be ascertained with some confidence. However, other samples display complex spectra such that only an estimate based on the best interpretation of the data can be put on the age of the impact event. While some researchers restrict the use of the term "plateau" to only those cases where the age of fractions are identical to each other, within the stated errors, in reality "plateau" is a subjective criterion that can vary depending on the type of samples being analyzed, heating schedule, and the number of contributing fractions. While the narrow definition is useful when dating single mineral grains, it is unnecessarily restrictive when dating whole rock samples as complex as melt rocks and ignores much of the chronological information available in the sample. We use the term "plateau" in this study to indicate a general agreement in age between adjacent fractions representing a significant portion of the argon released during the fusion. We also use the term "structured plateau" to describe samples where there is broad concordance of fraction ages in a general range that appears to have significance in terms of a thermal event such as impact but where there is also a definite pattern such as a general tilting or hump shape in the spectrum.

Lake Mien and the Siljan Ring Complex, Sweden

Interpretations of the data from these two craters have previously been published (*Bottomley et al.,* 1978), and will be summarized here with ages updated to the decay constants of *Steiger and Jager* (1977). Lake Mien (56°25′N, 14°52′E) is a rhombic-shaped lake with an average diameter of 5 km excavated in Precambrian granite and granitic gneiss of Gothian (1.7-1.3 Ga) and Dalslandian (1 Ga) age. Previously Whelin (1975) had found a spread in K-Ar ages from 103 to 114 Ma but suggested that the crater was formed between 112 and 122 Ma ago. *Storzer and Wagner* (1977) reported a fission track age of about 92 Ma.

Two samples of impact melt were dated and gave slightly different spectral shapes (Fig. 1). The basic petrographic character of the samples is described in *Bottomley et al.* (1978). Sample M15 presents a fairly well defined plateau over most of its temperature range. It also shows the slight drop with age, which gives a slightly tilted structural plateau that is sometimes characteristic of impact melt samples. Sample M17, on the other hand, shows a shallow hump shape with a high-temperature dip and recovery in the last few fractions. Both samples show concordant ages of 118 Ma in that portion of the spectra, which represents 50-90% of the gas released. The first fraction of each sample is distinctly lower in age than the plateau. Thus, an integrated ^{40}Ar-^{39}Ar age, which is equivalent to a conventional K-Ar date, would be 5% lower than the plateau age for the same rock. These low first fraction ages probably represent weakly bound argon from some of the matrix microlites and alteration products, which may have incompletely held their argon since the formation of the melt (*Bottomley et al.,* 1978).

Excluding the first fraction of each spectrum and the last three tiny fractions of M17, all ages are in the 115-124 Ma range. The average age of the plateau fractions M15 and M17 is 121 Ma. The integrated ages are slightly lower (116 and 114 Ma respectively) due to the low-temperature variable age gas fraction that mixes with the older argon to yield a mixed age in the first fraction. Although the argon spectra show some structure, there is general agreement between the two samples of melt rock dated and the best estimate of the age is 121 ± 2.3 Ma.

Siljan (61°02′N, 14°52′E) is a large (45 km) ring-shaped feature in central Sweden, which has a pronounced central uplift of shocked Precambrian (1.7 Ga) granite. An annular depression that surrounds the uplift is partially filled with lakes and contains Silurian to Ordovician sediments. Two impact-melt samples from a small dikelet were analyzed. In both spectra (Fig. 2), the low-temperature fractions make up the bulk of the gas released (600-730°C) and are generally concordant in age at ~365 Ma. The most distinctive feature of the Siljan samples is their high-temperature dropoff. ^{40}Ar-^{39}Ar dating of some fine-grained lunar samples gave monotonically decreasing spectra, even though their total gas-integrated ^{40}Ar-^{39}Ar was in agreement with Rb-Sr ages on the same rock. It was suggested that this was due to a closed system redistribution of argon caused by ^{39}Ar recoil during irradiation in the reactor (*Huneke and Smith,* 1976; *Turner and Cadogan,* 1974). Further experimental work (*Villa et al.,* 1983) has convinced most researchers that recoil is a real effect. Due to the short recoil range (~0.1 μm), the effect will be dominant in micron- to submicron-size grains and will decrease as the grain size increases. The matrices of the Siljan samples contain feldspars, which average 5 μm in diameter, and lesser amounts of pyroxene, which tend to be about 30 μm in diameter. The samples are described in more detail in *Bottomley et al.* (1978). The glassy K-rich matrix components are about the same size as the feldspars. It seems likely, therefore, that neutron-induced ^{39}Ar recoil out of the potassium-rich glass and into the pyroxene is responsible for the plateau shapes seen in Fig. 2. These K-poor phases tend to release their argon at higher temperatures. In both samples, the first three fractions make up 80-90% of the gas released (600°-730°C). In the absence of more spectral resolution in these early fractions, the best estimate of the age of impact would be either the age of the plateau fractions (365 Ma) or the total integrated age (368 Ma). As recoil is probably involved in the high-temperature age decline, the integrated age of 368 ± 1.1 Ma is the best estimate of the formation age.

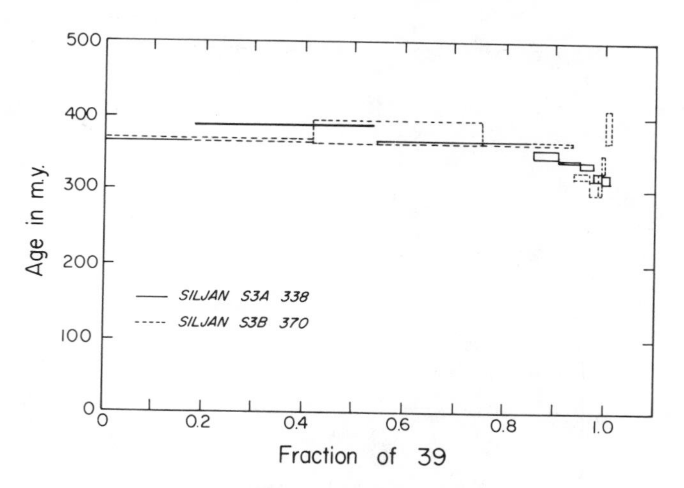

Fig. 2. Spectra from Siljan, Sweden.

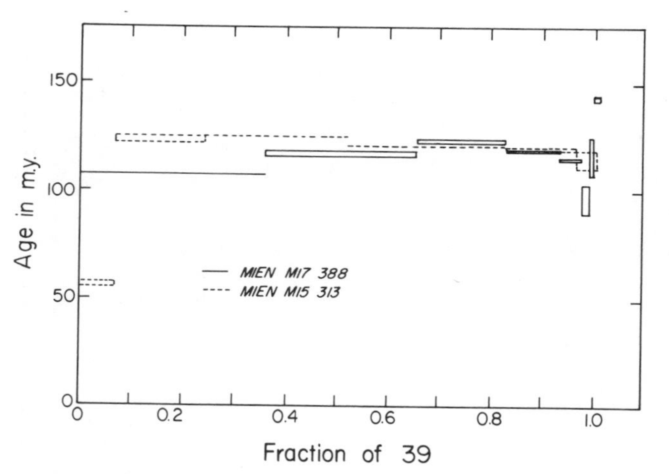

Fig. 1. Spectra from Lake Mien, Sweden.

Strangways, Australia

The Strangways crater (15°12′S, 138°35′E) is 20 km in diameter in Precambrian terrain of Proterozoic age. There is a 5-km core of granitic gneiss containing breccia and melt rock. There are no previous age determinations, but there are

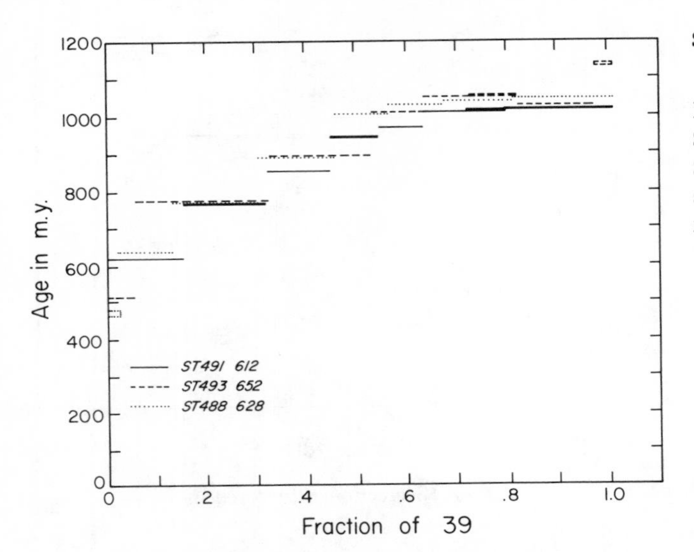

Fig. 3. Spectra from Strangways crater, Australia.

clean, mid-Cambrian limestones in the area, which are probably precrater sediments (M. R. Dence, personal communication, 1979). Three samples from sites some kilometers apart present very similar spectra (Fig. 3). All three samples are very fine-grained melt rocks. They have 10% clasts, which are relatively unrecrystallized. The samples are heavily stained with hematite. Their spectra show an apparent plateau (30-40% gas) at the highest temperature. These plateau segments indicate an age of 1000-1020 Ma. At the present time the age of the Precambrian country rock surrounding the crater is not known. It is possible that this high-temperature age records the age of the undisturbed country rock rather than the impact itself. The low-temperature portions of the Strangways spectra display a monotonic increase in age with temperature and are also quite similar in age except for the first fraction, which varies from 470-620 Ma between the samples.

Similar monotonically rising spectral shapes were modeled theoretically by *Turner* (1968). He calculated theoretical age spectra for a sample of spherical mineral grains that had undergone various degrees of argon loss. In situations like this, the low-temperature age can reflect a maximum estimate of the time of the reheating event. The "plateau" in the higher temperature fractions could be either a real plateau that reflects the undisturbed age of the rock before the reheating event or a false plateau representing a partial resetting of the most argon-retentive sites. These Strangways samples seem to be examples of thermally overprinted mineral systems that have not been totally reset by the impact. The lowest first fraction age (472 Ma) comes from the run on ST488, which had the smallest first fraction volume (2.5%), while the highest first fraction age (620 Ma) belongs to ST491, which has the largest first fraction volume (15%). Such a correlation between age and size of the first fraction is consistent with Turner's model, which suggests that the 472-Ma age is an upper limit for the age of the Strangways cratering event. This interpretation is also supported by local stratigraphic relationships, which have led Dence to believe that the crater is post-Cambrian (M. R. Dence, personal communication, 1982). Thus, the best estimate is a maximum age of 470 Ma based on the age of the lowest temperature fraction.

Saaksjarvi, Finland

Saaksjarvi (61°23′N, 22°25′E) is a 5-km-diameter crater in Precambrian gneissic and granodioritic terrain of Karelian-Svecofennian age (1.7-2.6 Ga) (*Papunen,* 1969). Four Saaksjarvi spectra show a similar pattern to the Strangways samples (Fig. 4). However, three of these are samples of one fine-grained melt rock in which clasts of quartz and feldspar make up 15% of the sample and are relatively unrecrystallized. (SJ125 354 and SJ125 439 are replicates and SJ125L 69L is a light density mineral separate from the same sample.) The two whole-rock spectra from SJ125 show good reproducibility and level off at higher temperatures, with 30-40% gas giving an age of 500 Ma. The age of the low-temperature fractions of both these samples are quite reproducible. However, unlike the Strangways samples, SJ125 439 has two small fractions reflecting the same age as a larger first fraction of SJ125 354, indicating that this may be a significant minimum age. The mineral separate SJ125L 691 has a larger first fraction and gives a slightly higher age (Fig. 4). The fourth stepwise rising sample, SJ105, is distinct in that its ages at every point are higher than the corresponding fractions of the SJ125 family. Its high-temperature plateau is marginally older than SJ125 at about 510 Ma, with a 530 Ma fraction at the end. The fact that it is higher in age at every temperature indicates that it was less completely reset than SJ125 and its low-temperature minimum age may not be as reliable as the low-temperature age of SJ125 for estimation of the time of impact. These four spectra can be interpreted in a fashion analogous to the Strangways spectra. This would indicate a maximum age of 330 Ma for impact and a high temperature approach to a plateau that, if real, would indicate a lower limit to the age of the country rock of about 510 Ma.

Sample SJ106 431 (Fig. 5) gives a totally different shape. Unlike most impact melt samples, this rock released most of its gas (70%) above 1000°C. The spectrum has a hump in the low-temperature fractions with an initial age of 430 Ma, which

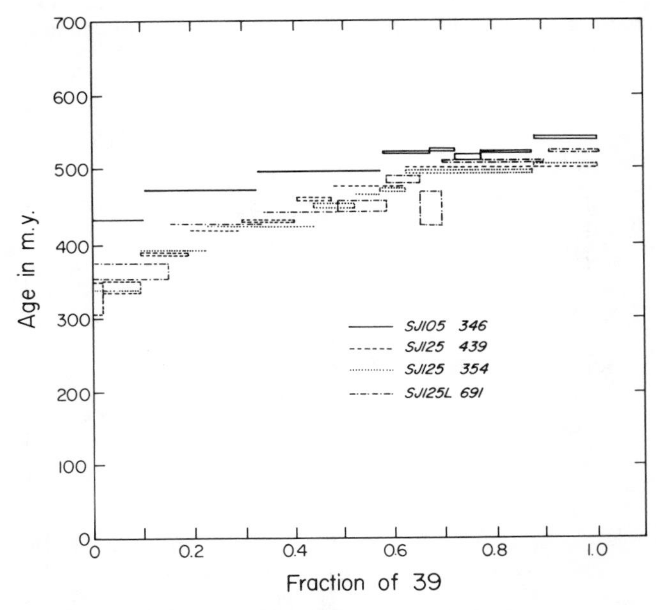

Fig. 4. Spectra from Saaksjarvi, Finland.

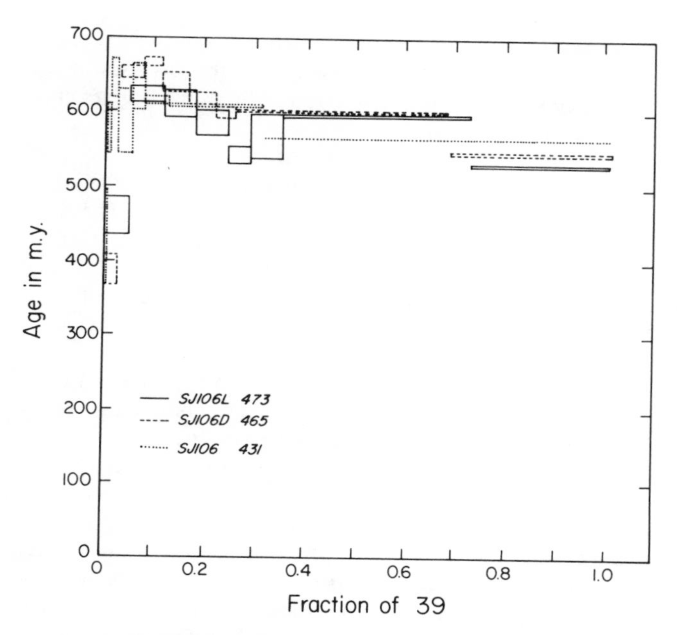

Fig. 5. Spectra from Saaksjarvi sample SJ106 and light and dark mineral separates.

quickly peaks at 640 Ma before falling to a small plateau at 600 Ma, then it releases 65% of its gas in one large high-temperature fraction (560 Ma). Two mineral separates SJ106L (light colored) and SJ106D 465 (dark colored) showed very similar behavior, with SJ106D showing the greatest exaggeration (Fig. 5). Its low-temperature age was 390 Ma and its peak 670 Ma, while SJ106L had 460 Ma and 625 Ma respectively. Both showed larger 600-Ma plateaus than the whole rock, and their 600-Ma fractions were larger in total volume than the corresponding high-temperature 550-Ma fraction. The integrated ages were all within 580 ± 10 Ma. This type of shape is common in melt rocks and when less exaggerated gives a structured plateau (e.g., Mien M15). The 550-600-Ma range of ages evident in the higher temperature fractions of SJ106 may be recording the same age or event as the high-temperature steps in the SJ125 spectra. It is not known whether the basement rocks of this area are in this age range.

The two large high-temperature fractions in the SJ106 family are associated with a high K/Ca phase, which typically has a K/Ca ratio that is 3-4 times higher than the preceding fraction. In spite of this, in all three samples the age of the last low K/Ca fraction and the next high K/Ca fraction are identical. The drop in age with temperature in the mid-temperature fractions may be correlated with a drop in the K/Ca ratio. The main mineralogical difference between the Saaksjarvi samples seems to be in the amount of quartz clasts. SJ105 and SJ125 contain approximately 20-25% inclusions. About 50-60% of these inclusions are reequilibrated quartz clasts. SJ106 has about 15% inclusions but only 5% are quartz. SJ105 and SJ106 have devitrified glass in their matrix, while SJ125 may not.

If we assume that the spectra from samples SJ105 and SJ125 are evidence of a partial thermal overprinting similar to the Strangways samples, a best estimate of the age of formation is 330 Ma or younger. It should be remembered, however, that the upper temperature fractions (even from the SJ106 samples, which do not show stepwise rising shapes) seem to be recording a 500-Ma age, considerably younger than the Precambrian country rock. Since the fractions making up the low-temperature hump and the high-temperature plateau are related to different K/Ca regimes, they probably represent release from mineral phases with different argon blocking temperatures. This suggests an alternate explanation for the SJ106 spectra, where they represent melt rock in which the lower temperature fractions cooled below their argon blocking temperatures in an environment of excess argon. The excess ^{40}Ar could have diffused from the nonmelted but severely heated target rock, which originally enclosed the melt rock unit. By the time the argon had diffused enough to produce a significant excess concentration, the higher temperature sites in the melt rock had already cooled below their blocking temperature. They were unaffected by the excess argon diffusing into the still open lower temperature sites. The very lowest temperature sites with lower apparent ages in SJ106 may just represent loss from later postimpact minor alteration. However, until more data are available from this crater, the preferred interpretation of these spectra is that the impact took place 330 Ma ago or less and that the country rock gives a K-Ar age of between 500 and 600 Ma.

Dellen, Sweden

Dellen (61°55′N, 16°32′E) consists of two arcuate-shaped lakes partially filling an annular depression. Separating the lakes is a long peninsula. The structure is a complex crater some 15 km across excavated in crystalline Precambrian gneiss and granodiorite of Karelian and Svecofennian age (1.7-2.6 Ga) (*Svensson,* 1968). Dellen spectra fall naturally into two families. Both show good plateau, but one is about 100 Ma and the other about 240 Ma. Samples Dl 423, D3 404, D3 396, and D5 362 (Fig. 6) all overlap at 100 Ma with about 90% of their gas giving a clear plateau. All these samples are petrographically similar, consisting of partially devitrified glass with feldspar phenocryst laths up to 1 mm long and rare pyroxene phenocrysts. Clasts are relatively rare. The spectra have small low-temperature fractions with ages between 50-70 Ma. The

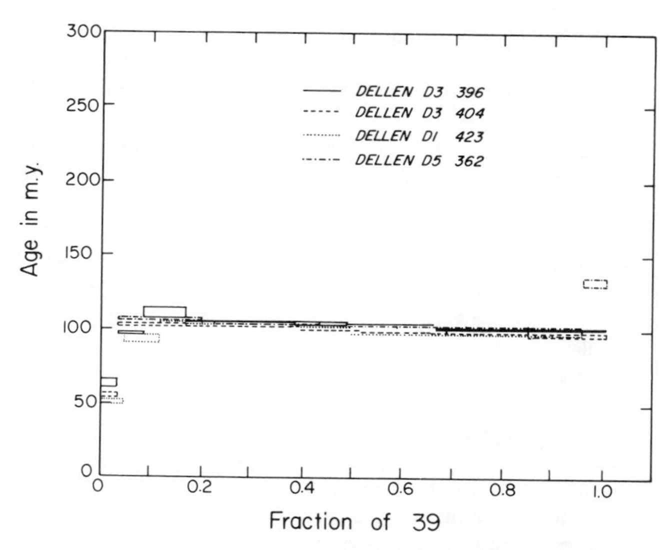

Fig. 6. Spectra from Dellen, Sweden, which define the primary plateau.

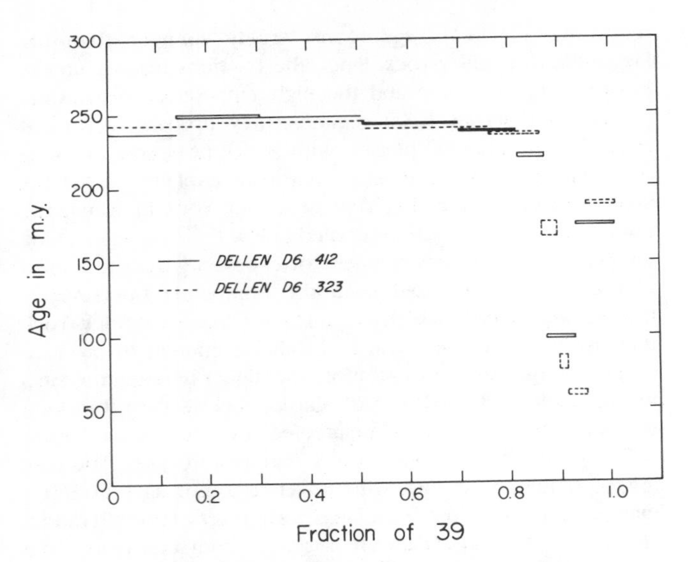

Fig. 7. Spectra from whole rock sample Dellen D6.

spectra rise quickly to plateaus, which are slightly tilted towards the right, showing an incipient monotonic decrease with higher temperature.

The concordance of the plateaus from samples Dl, D3, and D5 indicates an age of 100 Ma for the impact event. However, the spectra of D6 (D6 323 and D6 412) are totally different (Fig. 7). Petrographically D6 is similar to the other Dellen samples, with slightly more mineral clasts (up to 10%). The first 80% of the gas gives a structured plateau at 240 Ma. This is followed by a precipitous drop to lower ages (60-90 Ma) over the last 10-15% of the gas released. This shape is remarkably similar to some spectra observed by *Mak et al.* (1976) in Mistastin samples. Their high-temperature dip always fell to about 30 Ma at 85-97% of the gas released. The authors noted that this was accompanied by a drop in the 39/37 ratio (by a factor of 3 to 5), and attributed this dip to apparent recoil effects in pyroxene. In D6, the K/Ca dropoff is by a factor of about 2. In this case, however, the plateau portion gives a much higher age than the best estimate of the age of impact based on the other melt samples. In addition, the high-temperature dip drops to ages relatively close to the inferred age of impact.

In an attempt to choose the best interpretation, three density separates were prepared. These have been designated dark (D), intermediate (I), and light (L). The spectrum from each of these separates is distinctive (Fig. 8). All three have released 80% of their gas by the 800°C step. Sample D6D 505 most resembles the whole rock spectra, except that it displays a more distinct hump shape with its maximum at only 230 Ma. It is also apparent from the volume of radiogenic argon that it is this dark fraction that is releasing the bulk of argon in the whole rock sample. The intermediate sample (D6I 530) displays a very pronounced hump shape starting at 200 Ma and rising to 240 Ma before dropping to a minimum at 70 Ma. The last 10% of the gas then gives an age of 350 Ma. This phase contributes five times less gas than the dark phase to the whole rock age. Light sample D6L 456 gives an enormous gas pulse (50%, 180 Ma) at the first 500°C fraction, then rises marginally before plunging to 5 Ma. The last 25% of the gas rises in age, with three small volume steps followed by a final iarge fraction (18%) at an age of 1300 Ma.

The 240 Ma age seems associated with a single low-temperature mineral, which is totally degassed by 800°C. This phase has a characteristic 39/37 ratio of about five. The small amount of gas released in the temperature range 900-1600°C seems to be both lower in K/Ca and age. In general, the minerals in Dl, D3, and D5 show a much wider range of K/Ca and the 100-Ma age is found in both high and low K/Ca regimes.

Two interpretations of D6 spectra seem plausible: (1) The low-temperature mineral has not been completely reset, yielding a false plateau at 240 Ma. Alternatively, since it cooled through its argon blocking temperature after the high-temperature phases, it may have been exposed to an argon atmosphere that had much more excess ^{40}Ar. The high-temperature dip reflects a phase that has either been approximately reset to the impact age (~100 Ma) or has lost ^{40}Ar (through geological or reactor processes) or has gained ^{39}Ar (by recoil in the reactor). (2) The 240-Ma age is a correct plateau and the high temperature dip reflects perturbations either through geological or reactor processes.

While interpretation (2) would explain the spectra of D6, we would then have to assume that the three concordant samples with 100-Ma plateaus were in error. As they represent three separate samples and a variety of K/Ca regimes, it seems unlikely that D6 is a more reliable age.

Interpretation (1) seems to explain the anomalous age more satisfactorily, if we assume that during irradiation some ^{39}Ar recoiled out of the fine-grained potassium-rich matrix into phases that released argon at higher temperature. There appear to be two distinct argon regimes in sample D6. The first is a partially reset mineral that releases all its gas by 800°C. The second phase is a refractory mineral that is concentrated

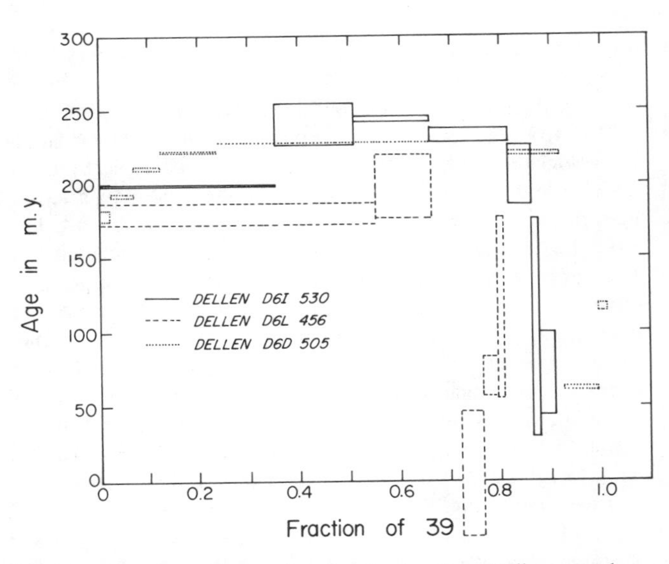

Fig. 8. Spectra from mineral separates from Dellen, sample D6.

mainly in the intermediate and light fractions, probably a partially reset country rock clast of about 1300-Ma age. As the first phase ends its degassing at about 800°C and its contribution gets smaller and smaller, more and more of the high-temperature phase's argon is released. The initial part of this release has a lower age, because of the excess ^{39}Ar it received via recoil. As more retentive sites release their argon, the age rises reflecting the older ages still retained by the clasts. However, only in the light fraction do we see it rise to its final age of 1300 Ma. In the whole rock and dark separate, slight amounts of residual low-temperature argon mix with the clast argon to produce an intermediate mixed age. In the separates of D6 the low-temperature fractions display, in varying degrees, the stepwise rising pattern found in the Strangways samples. This supports the suggestion that the low-temperature mineral was only partially reset by the impact.

Thus, the best age of Dellen would appear to be the average of the plateaus of the younger samples at about 102 ± 1.6 Ma. The results, when taken in total, show the remarkable complexity of the argon systematics in melt rocks and strongly emphasize the importance of examining a variety of melt rocks from a single crater.

Lac La Moinerie, Quebec, Canada

Moinerie (57°26′N, 66°36′W) is a roughly circular lake 8 km in diameter with a central uplift, represented by a circular ring of islands some 4 km in diameter. The crater is found in Precambrian gneissic terrain of Hudsonian age and is deeply eroded. Samples of melt rock along with impact breccias were found as glacial float on the northern side of the lake. Also recovered were apparently unshocked Silurian limestones, presumably scoured by the ice from the now submerged crater floor. If these represent postcrater sedimentation, then the crater may be pre-Silurian (>400 Ma) (*Robertson and Grieve,* 1975).

Three impact-melt samples were dated. All samples have glassy to cryptocrystalline matrices, with approximately 15% mineral clasts (mostly quartz and feldspar). The clasts show only minor thermal effects from the matrix. Sample LM15 209 gives a monotonically increasing age spectrum, with a low-temperature age of 230 Ma and a high-temperature age of 430 Ma (Fig. 9). There is no high-temperature plateau. There is, however, a very large third fraction with an age of 400 Ma (55% gas, 830°C) and a small last fraction of 750 Ma (2.5% gas, 1600°C). The remaining two samples show shallow hump-shaped spectra. Sample LM6 232 starts at 320 Ma and rises to a maximum age of 400 Ma before falling back to a final age of 370 Ma. It also has a high last fraction age of 660 Ma (1.5% gas). Sample LMll 271 has a similar shape but is higher in age at every step, starting at 350 Ma, and rising to 420 Ma before falling back to 400 Ma. It has a much larger last fraction (16% gas), which is 460 Ma in age.

The interpretation of these spectra is difficult. There is a general overlapping of most of the gas fractions in the range of 400 ± 20 m.y, which suggests a common event. All three samples have 80-90% of their gas giving ages in this range. However, LM15 clearly seems to be similar to a Strangways type of sample, and the two hump-shaped samples may be less severe examples of the Dellen D6 phenomena, which did not reflect the time of impact. If LM15 is a partially reset sample, then the crater should be no older than 230 Ma, the age of the lowest temperature fraction. Yet, if the general concordance of all three samples means anything, then it would appear that 400 Ma might be a better age. The major difference between these spectra and a true Strangways type is that only the first few fractions rise steadily and in LM6 and LM4 this rise is correlated with an increase in K/Ca. The shallow hump shape in the plateau portion of the spectra can probably be explained by minor amounts of ^{39}Ar recoil, especially in LM6 and LMll, as these are extremely fine-grained rocks. Although there is a step-like structure in the first fractions of these spectra, the best estimate of the age would be about 400 Ma based on the 80% of the argon in the more retentive sites. There is a fair degree of scatter in these structured plateaus and a fairly broad age of 400 ± 50 Ma is the preferred estimate of the age of impact. However, the possibility that the crater may be 230 Ma or younger should be kept in mind until better samples are dated.

Lac Couture, Quebec, Canada

The Couture crater (60°08′N, 75°18′W) is the island-free, deep-water portion of a slightly larger lake. It has a diameter of 10 km. The country rock consists of Precambrian gneissic terrain of Hudsonian age. There is no evidence of sedimentary crater fill and no published ages (*Beals et al.,* 1967). Three samples of Lac Couture melt rock were dated. The samples contain approximately 10% clasts, mostly quartz and feldspar, of which 30% are recrystallized. The melt matrix is generally cryptocrystalline, although in sample LC5 287 feldspar microlites are identifiable.

Samples LC12A 240, LC12B 280, and LC5 287 show similar ages and spectral shapes (Fig. 10). All present structured plateaus whose ages show a very shallow monotonic rise with temperature. Sample LC12A has the lowest first fraction age of 340 Ma. This is followed by a large fraction (65%) at 415 Ma and another (18%) at the slightly higher age of 440 Ma. The remaining gas executes a small high-temperature dip (min-

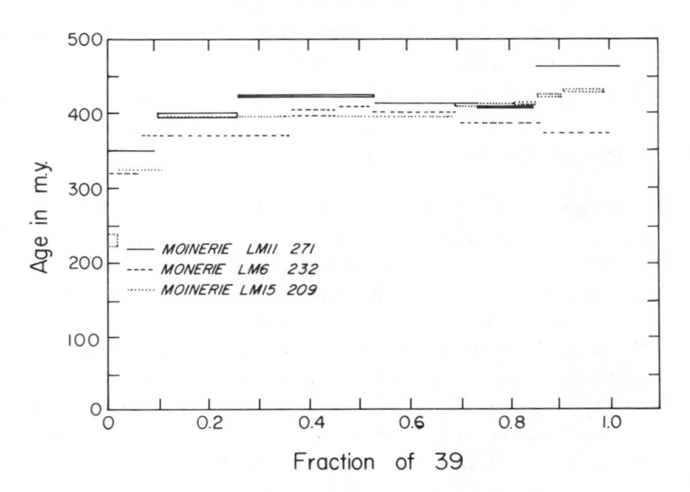

Fig. 9. Spectra from Lac la Moinerie, Canada.

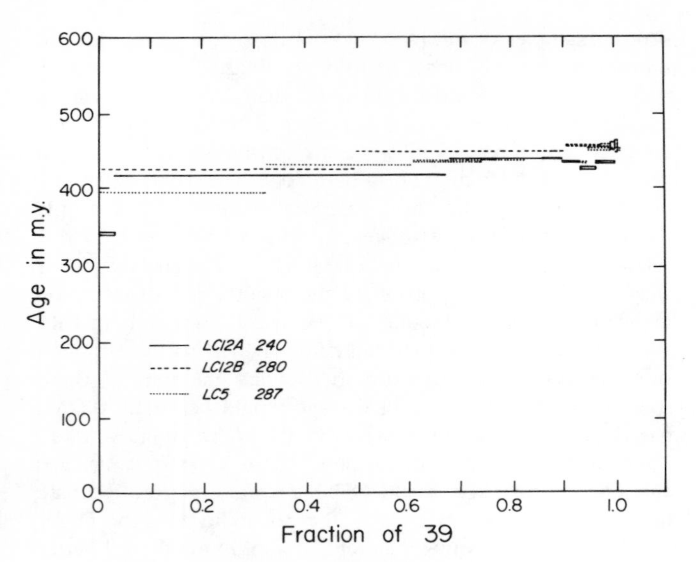

Fig. 10. Spectra from Lac Couture, Canada.

imum at 423 Ma), before shooting up to 700 Ma in the final fraction. LC12B has two large fractions (430 Ma, 50% and 450 Ma, 40%) that are followed by four small fractions (total 10%) with an age of 455 Ma. The last fraction contains less than 1% of the gas and yields an 800-Ma age. Sample LC5 287 starts with an age of 395 Ma (30%) followed by seven fractions at about 435 Ma before a slight rise to 450 Ma (6%).

These samples, like Moinerie, could have two interpretations. They show a very shallow stepwise rise from an age in the 400-Ma to 450-Ma range. One small first fraction (LC12A) gives an age of 350 Ma but the first fractions of the other two samples, as well as the second fraction of LC12A, are extremely large (40-70% of the total ^{39}Ar released). This very large volume in the argon sites of lowest retentivity sets these samples apart from the more steeply rising step-wise shapes of Strangways and Saaksjarvi. Although the rocks show no true plateaus, 95-100% of their gas gives an age between 400 and 450 Ma. The best age estimate for the cratering event is 425 ± 25 Ma based on this analysis. However, the data can permit the secondary interpretation that the crater has a maximum age of 350 Ma, based on these spectra being interpreted as Strangways types.

Clearwater, Quebec, Canada

Clearwater (56°10'N, 74°15'W) is a pair of circular lakes (West - 32 km, East - 22 km) located in Precambrian granodiorites, quartz monzonites, and granite gneisses of Archean age. A central uplift represented by a ring of islands is found in the west lake, where melt rock gives ages of 270-290 Ma by both K-Ar and Rb-Sr (*Wanless et al.,* 1966 and unpublished data). The central uplift at Clearwater East is submerged and not readily accessible. However, drilling has recovered samples from a melt rock zone at a depth of some 1000 ft. Two samples of melt rock from one core (separated by 20 ft) were dated along with one sample of melt rock from Clearwater West. A Rb-Sr date on this former material gives an isochron of 287 ± 26 Ma (*Reimold et al.,* 1981), supporting the widely held belief that both craters were formed simultaneously.

The Clearwater West sample, which is a fine-grained melt rock with 10% clasts set in a feldspar pyroxene matrix, gave a good plateau age of 280 Ma, in clear agreement with the previous determinations (Fig. 11). The last three fractions are indistinguishable in age and comprise some 85% of the released gas. The first fractions, although quite small, show the characteristic hump-shaped pattern of many melt rocks. However, the two samples of melt rock from Clearwater East (Fig. 11) show pronounced U-shape spectra. These samples are texturally similar to each other and the West Clearwater sample. Spectra with this U-shape have been identified as indicating excess argon (*Lanphere and Dalyrmple,* 1976; *Harrison and McDougall,* 1981) or contamination by a small amount of older material (*Lo Bello et al.,* 1987). This would appear to be the case with these samples. Sample CWE1090 starts with a small fraction at 1075 Ma (1%), then falls to a minimum at 465 Ma before increasing to 640 Ma in the final step. CWE1070 644 has a much broader U-shape with four fraction plateaus at the mid-temperature fractions (460 Ma, 80% gas) with a low-temperature age of 590 Ma and a high-temperature age of 1150 Ma. The general concordance of the ages at the bottom of the U-shaped trough at about 460 Ma indicates that the excess argon is rather uniformly distributed throughout the melt sheet, as sampled by this core.

It appears that an Ar isotopic age of Clearwater East cannot be recovered from these two samples and until melt rock from near the edge of the melt sheet or different cores can be sampled, the best age for both craters is the 280-Ma age given by the melt rock from Clearwater West and the Rb-Sr age from Clearwater East. The excess argon was apparently trapped in the melt rock as the hot mixture degassed after impact. Its quick burial and cover by fallback breccia apparently did not allow all the gas to escape before cooling through the argon blocking temperature.

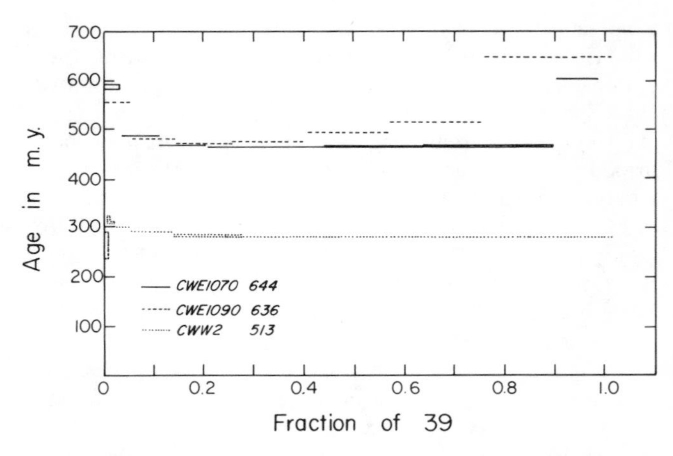

Fig. 11. Spectra from Clearwater Lakes, Canada.

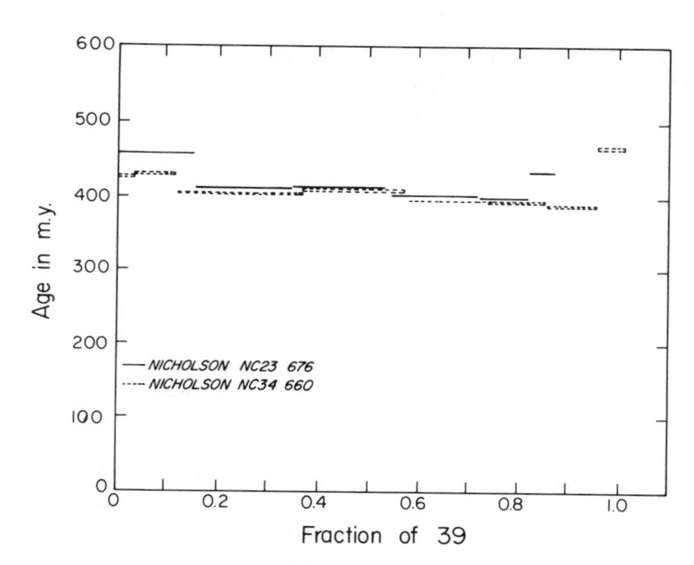

Fig. 12. Spectra from Nicholson Lake, Canada.

Nicholson, Northwest Territories, Canada

Nicholson Lake (62°40′N, 102°41′W) is an oval shaped lake 12.5 km in diameter. There is a large central island along with a sizable promontory jutting into the lake from the western side. The lake occupies a shallow depression in Precambrian gneissic and granitic terrain of Hudsonian age. Several small islands in the northern end of the lake preserve outcrops of Middle Ordovician dolomite. These are the only Palaeozoic rocks within a radius of 320 km, and were apparently protected from erosion by the crater structure. In addition, fragments of similar rock are found as glacial float around the lake (*Dence et al.,* 1968).

Two samples (Fig. 12) gave almost identical release spectra, despite the fact that they are petrographically distinct. Sample NC23 is more of a breccia than a true melt rock, with clasts of melt glass cemented by recrystallized matrix, which comprises 30% of the rock. NC34 is a cryptocrystalline melt rock with 35% clasts (mostly recrystallized quartz and feldspars). In both samples, 70-90% of the gas gave a shallow hump-shaped plateau in the mid-temperature range (400 Ma). However, this structured plateau itself appears to be the bottom of an even broader saddle-shaped curve that encompasses the low- and high-temperature fractions. NC34 660 starts out at 425 Ma before dropping to about 400 Ma in its mid-temperature fractions. The last 5% of the gas rises to 465 Ma. NC23 676 begins with an age of 450 Ma, dropping to its 400 Ma plateau before rising to a final age of 620 Ma. The third fraction of NC23 and the fourth fraction of NC34 rise in age very slightly, while the remaining plateau fractions drop slightly thereafter. This happens at about the same temperature in both samples (800-850°C). This would appear to be related to the release behavior of a specific but unidentified mineral phase. Although the saddle shape is much shallower than in the case of the Clearwater samples, it indicates excess argon, suggesting that the 400-Ma age of the structured plateau is a maximum estimate of the age of the cratering event.

Carswell, Saskatchewan, Canada

Carswell (58°27′N, 109°30′W) is a large (37-km-diameter) structure lying within the Athabasca Basin of northern Saskatchewan. It consists of an outer ring of Precambrian dolomites and siltstones (~1.5 Ga) that surround an inner ring of deformed Athabasca sandstone with ages of 1.9-2.3 Ga, and a central core of uplifted granitic basement (*Bell,* 1985).

Sample CA15 683 is a slightly vesicular impact melt rock. Clasts make up 5% of the rock and are generally recrystallized quartz and feldspar. The melt matrix contains feldspar microlites up to 0.1 mm long with minor partially chloritized pyroxene set in a very fine-grained quartz-feldspar matrix. The spectrum displays a structured plateau very similar in shape to Mien M17. If interpreted as such, it indicates an age of about 115 Ma. However, it is not clear that this is the age of the impact event. K-Ar ages reported for breccia dikes, which appear to cut all other structural aspects of the crater, have been reported as 485 ± 50 Ma (*Currie,* 1969). The disagreement between the CA15 683 age and that of the breccia might be explained by the breccia being only partially reset, as we have found to be the case with the breccias at the Slate Islands impact site (*Bottomley,* 1982). Carswell is of some commercial interest because of a large uranium find at Cluff Lake, and there is a large body of uranium-lead data on the minerals from various mines in the Athabasca Basin area. *Koeppel* (1968) found evidence of six episodes of uranium mobilization in deposits near Beaverlodge, Saskatchewan. These were 2350 Ma, 1920 Ma, 1780 Ma, 1125 Ma, 270 Ma, and 0-100 Ma.

Thus, it appears that there could have been some event in the Athabasca basin near 100 Ma that remobilized uranium. It is possible that the argon system in the Carswell sample is also recording this event. It is clear that more samples of both melt rock and breccia should be dated by the ^{40}Ar-^{39}Ar method in an effort to place the formation of this structure accurately within the framework of U-Pb dates for events in the Athabasca Basin.

Gow Lake, Saskatchewan, Canada

Gow Lake (56°27′N, 104°29′W) is a deeply eroded lake 5 km in diameter with a central uplift, represented by a 1-km-diameter island. The target rock is Precambrian granites and gneisses of Hudsonian age. *Thomas and Innes* (1977) have estimated its age as 100 Ma, based on depth of erosion and comparisons with the nearby Deep Bay crater. Sample GW12 668 (Fig. 13) starts with a large (30%) pulse of gas giving an age of 250 Ma. It displays a stepwise rising release pattern over the first 70% of its gas, with a short plateau segment of four fractions (20%) and a last fraction age of 540 Ma. In the absence of other samples, we interpret this as a Strangways type of sample and estimate the maximum age of formation as 250 Ma or less based on the first fraction. The sample has a cryptocrystalline melt matrix with approximately 15% lithic and mineral clasts. Most of the clasts are not recrystallized.

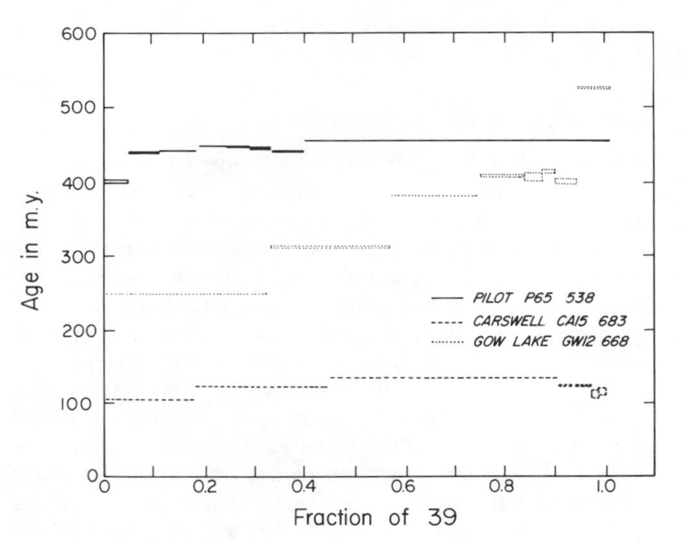

Fig. 13. Spectra from Pilot Lake, Carswell, and Gow Lake, Canada.

Thus, they could easily retain some memory of ages older than impact and, thereby, be responsible for the older high-temperature ages.

Pilot Lake, Northwest Territories, Canada

Pilot Lake (60°17′N, 111°W) is about 6 km in diameter in granitic and gneissic terrain of Hudsonian age. Melt rock was obtained from glacial float on the western side of the lake (*Dence et al.,* 1968). No sedimentary fill has been found at Pilot Lake. *Dence et al.* (1968) believed the crater was middle to late Palaeozoic on the basis of erosion level. Sample P65 538 is rather atypical of most melt rocks, which release the bulk of their argon below 900°C. In the first seven temperature steps, up to 1100°C, only 40% of the gas is released (Fig. 13). This portion of the spectrum appears to be a slightly structured plateau. The eighth and last fraction contains 60% of the gas with an age of 450 Ma, which is not that different from the average age of 440 Ma given by the previous six fractions. It appears that this sample contains more high-temperature phases than a typical melt rock. Indeed, when examined petrographically it contained 50% clasts, of which only some 30% had been recrystallized. The clasts are set in a devitrified and hematized glassy matrix and the sample is intermediate between a melt matrix breccia and a well defined impact melt rock. The simplest interpretation of the age spectrum is that the Pilot Lake crater was excavated about 445 ± 2 Ma ago. However, dates on single samples from craters should be viewed with caution.

DISCUSSION

Our best estimate of the age of formation of the 13 craters studied is given in Table 2. This study has demonstrated, however, that melt rocks are not the simple material for argon age determinations that they might appear to be in principle. There remain significant ambiguities, which need to be investigated further. The spectra of melt rocks often fall into distinct groups based on shape. The variety of shapes often makes it difficult to interpret the age of impact with any degree of certainty. However, in all cases studied, the rocks do show ages considerably younger than the country rock that surrounds them. The various spectral shapes identified in this melt rock study and their interpretation are

1. *Classical plateau spectra.* The age interpretation of this plateau pattern is straightforward and represents an ideal, undisturbed, loss-free K-Ar system that has not been disturbed since it last cooled through its blocking temperature (Clearwater West, Pilot).

2. *Structured plateaus.* The structured plateau, seen in samples such as Mien, results when most of the age fractions appear to be about the same, yet the individual ages of contiguous fractions do not overlap at the one sigma level. This is an understandable evolution of the classical plateau, when many different effects begin to modify the argon system in melt rocks. Spectra with a small amount of structure apparently can also give reliable ages. While, in some cases, a portion of this structure is undoubtedly due to ^{39}Ar recoil during irradiation, it is primarily due to the nature of the rock that has target rock clasts imbedded in a finer grained matrix (Mien, Siljan, Carswell, Couture, Moinerie).

3. *Dual plateau spectra.* At Dellen there appeared to be samples that gave reasonable, structured plateaus, but at two distinctly different ages. Although we call this "dual" plateau, there could presumably be any number of different plateaus from any given site (Dellen).

4. *Stepwise rising spectra.* The stepwise rising spectra appear to be the result of partial thermal overprinting of less retentive sites in the rock, wherein progressively higher temperature phases retain more of their initial pre-impact ^{40}Ar. In cases such as this, the low-temperature fractions are probably the best indications of the age of impact (Gow, Saaksjarvi, Strangways).

5. *Hump-shaped spectra.* The hump-shaped spectra seen in some melt rocks are the most difficult to explain. The most common type (e.g., SJ106) has a low age in the first fraction rapidly rising to a high age at moderate temperature before leveling off to an intermediate age in the high-temperature fractions. The lowest temperature fractions are probably affected by postformation argon loss from the lowest retentivity sites and this same effect is observable on the age spectra of almost any type of rock. The hump at low- to medium-temperature may be caused by several effects, including exposure of the low-temperature sites to an argon environment different from that seen by the high-temperature sites (which close earlier to argon diffusion), and ^{39}Ar recoil from the intermediate temperature minerals into the phases that release at higher temperature. The complex clast-matrix relationships that exist in any given sample could also contribute to this shape. These different factors are not independent of each other and make it difficult to assign an unambiguous cause in all cases (Saaksjarvi SJ106).

6. *Saddle-shaped spectra.* Saddle-shaped spectra have been shown to indicate excess argon and the lowest part of the saddle gives a maximum estimate of the age of the resulting event (Clearwater East, Nicholson).

If a sample suite does not yield good plateaus, the errors assigned to the crater ages, when published, should represent the geological scatter in the sample ages, as well as the analytical errors assigned to the mass analysis. In Table 2 we have indicated the type of assigned error in the comments column. Only in the case of reasonable plateaus have the usual analytical errors been used. The complexities in the dating of impact events should introduce a cautionary note to the interpretation and use of crater ages. This is particularly relevant to those workers who take them at face value when constructing hypotheses of events such as periodic cometary showers that may affect the biological evolution of the earth (*Alvarez and Muller,* 1984; *Grieve et al.,* 1985).

It is clear from the variety of spectra observed in terrestrial impact melt rocks that ^{40}Ar-^{39}Ar dating should be used in preference to conventional K-Ar technique. Conventional K-Ar ages on many of the rocks dated during this project would be quite different than the best age as determined by analysis of the argon spectra. As samples from the same impact can, at times, give different ages and different spectral signatures, it is important that as many different samples as possible from the same crater be dated. In some cases, as at Dellen, the multiplicity of samples may enable the aberrant samples to be identified. By the same token, as much other evidence on the crater age as is available should be used to try to identify which samples are the most reliable. This would include petrographic study, stratigraphy of sedimentary units involved in the crater structure, and other isotopic dating methods. It appears that only through fully integrated studies will reliable ages of impact be determined for many craters with any degree of confidence.

REFERENCES

Alvarez W. and Müller R. A. (1984) Evidence from crater ages for periodic impacts on Earth. *Nature, 308,* 718-720.

Beals C. S., Dence M. R. and Cohen A. J. (1967) Evidence for the impact origin of Lac Couture. *Publ. Dominion Observ. Ottawa, 31.*

Bell K. (1985) Geochronology of the Carswell Area, North Saskachewan. In *The Carswell Structure Uranium Deposits, Geol. Assoc. Canada Spec. Pap., 29,* 33-46.

Bottomley R. J. (1982) ^{40}Ar-^{39}Ar Dating of Melt Rock From Impact Craters. Ph.D. thesis, Univ. of Toronto, Toronto, Ontario, Canada.

Bottomley R. J., York D. and Grieve R. A. F. (1978) ^{40}Ar-^{39}Ar ages of Scandinavian impact structures: I Mien and Siljan. *Contrib. Mineral. Petrol., 68,* 79-84.

Currie K. L. (1969) Geological notes on the Carswell circular structure Saskatchewan (74K). *Geol. Surv. Canad. Pap.,* pp. 67-32.

Dence M. R., Innes M. J. S., and Robertson P. B. (1968) Recent Geological and Geophysical Studies of Canadian Craters. In *Shock Metamorphism of Natural Materials* (B. M. French and N. M. Short, eds.). Mono, Baltimore. 644 pp.

Fleck R. J., Sutter J. F., and Elliot D. H. (1977) Interpretation of discordant $^{40}Ar/^{39}Ar$ age-spectra of Mesozoic tholeiites from Antarctica. *Geochim. Cosmochim. Acta, 41,* 15-32.

Grieve R. A. F. (1975) Petrology and chemistry of the impact melt at Mistastin Lake crater, Labrador. *Bull. Geol. Soc. Am., 86,* 1617-1629.

Grieve R. A. F., Sharpton V. L., Goodacre A. K., and Garvin J. B. (1985) A perspective on the evidence for periodic cometary impacts on Earth. *Earth Planet. Sci. Lett., 76,* 1-9.

Harrison T. M. and McDougall I. (1981) Excess ^{40}Ar in metamorphic rocks from Broken Hill, New South Wales: Implications for $^{40}Ar/^{39}Ar$ age spectra and the thermal history of the region. *Earth Planet. Sci. Lett., 55,* 123-149.

Huneke J. C. and Smith S. P. (1976) The realities of recoil: ^{39}Ar recoil out of small grains and anomalous age patterns in ^{39}Ar-^{40}Ar dating. *Proc. Lunar Sci. Conf. 7th,* pp. 1987-2008.

Koeppel V. (1968) Age and history of the uranium mineralization of the Beaverlode area, Saskatchewan. *Geol. Surv. Canada Paper 67-31.*

Lanphere M. A. and Dalyrmple G. B. (1976) Identification of excess ^{40}Ar by the $^{40}Ar/^{39}Ar$ age spectrum technique. *Earth Planet. Sci. Lett., 32,* 141-148.

Lo Bello Ph., Fernaud G., Hall C. M., York D., Lavina P., and Bernat M. (1987) $^{40}Ar/^{39}Ar$ step-heating and laser fusion dating of a Quaternary pumice from Nechers, Massif Central, France: The defeat of zenocrystic contamination. *Chem. Geol. (Isot. Geosci. Sect.), 66,* 61-71.

Mak E. K., York D., Grieve R. A. F., and Dence, M. R. (1976) The age of the Mistastin Lake Crater, Labrador, Canada. *Earth Planet. Sci. Lett., 31,* 345-357.

Papunen H. (1969) Possible impact metamorphic textures in the erratics of the Lake Saaksjarvi area in southwestern Finland. *Bull. Geol. Soc. Finland, 41,* 151-155.

Reimold W. U., Grieve R. A. F., and Palme, H. (1981) Rb-Sr dating of the impact melt from East Clearwater, Quebec. *Contrib. Mineral. Petrol., 76,* 73-76.

Robertson P. B. and Grieve R. A. F. (1975) Impact structures in Canada: Their recognition and characteristics. *Roy. Astron. Soc. Can. J.,* 1-21.

Steiger R. H. and Jager E. (1977) Subcommission on geochronology: Convention on the use of decay constants in geo- and cosmochronology. *Earth Planet. Sci. Lett., 36,* 359.

Storzer R. and Wagner G. A. (1977) Fission track dating of meteorite impacts. *Meteoritics, 12,* 368-369.

Svensson N. B. (1968) The Dellen Lakes - a probably meteorite impact in central Sweden. *Geol. Foeren Stockholm Foerh, 90,* 309-316.

Thomas M. D. and Innes M. J. S. (1977) The Gow Lake impact structure, northern Saskatchewan. *Can. J. Earth Sci., 14,* 1788-1795.

Turner G. (1968) The distribution of potassium and argon in chrondrites. In *Origin and Distribution of the Elements* (L. H. Ahrens, ed.). Pergamon, New York. 1178 pp.

Turner G. and Cadogan P. H. (1974) Possible effects of ^{39}Ar recoil in ^{40}Ar-^{39}Ar dating. *Proc. Lunar Sci. Conf. 5th,* pp. 1601-1615.

Villa I. M., Huneke J. C., and Wasserburg G. J. (1983) ^{39}Ar recoil losses and presolar ages in Allende inclusions. *Earth Planet. Sci. Lett., 63,* 1-12.

Wanless R. K., Stevens R. D., Lachance G. R., and Rimsaite J. Y. H. (1966) Age determinations and geologic studies, K-Ar isotopic ages. *Rep. 6, Geol. Surv. Canada Paper, 65-71.*

Whelin E. (1975) K-Ar dating and Sr-isotope composition of rhyolitic rocks from Lake Mien in southern Sweden. *Geol. Foren. Stockh. Forhande, 97,* 307-311.

Proceedings of the 20th Lunar and Planetary Science Conference, pp. 433-450
Lunar and Planetary Institute, Houston, 1990

The "Bronzite"- Granophyre from the Vredefort Structure— A Detailed Analytical Study and Reflections on the Genesis of One of Vredefort's Enigmas

W. U. Reimold

Schonland Research Centre, University of the Witwatersrand, WITS 2050, Johannesburg, Republic of South Africa

H. Horsch

Bushveld Research Institute, University of Pretoria, Pretoria 0001, Republic of South Africa

R. J. Durrheim

Bernard Price Institute for Geophysical Research, University of the Witwatersrand, WITS 2050, Johannesburg, Republic of South Africa

---

Granophyric rocks occur in the Vredefort structure as vertical ring dykes along the contact between sedimentary collar and Archean Basement, thereby crosscutting overturned and/or faulted strata, and as vertical dykes extending northwest-southeast and northeast-southwest in the gneissic core. The Vredefort granophyre is unique in the following aspects: its unusual chemical composition (~67 wt.% SiO_2, ~7 wt.% total Fe as Fe_2O_3), ~3.5 wt.% MgO, ~3.7 wt.% CaO), its regional homogeneity, its high content of mainly recrystallized sedimentary inclusions, and its apparent lack of deformation phenomena (in contrast to all other lithologies in the Vredefort structure). Hypotheses for the granophyre origin are closely linked to the theories for the origin of the Vredefort Dome; those who favor an origin by impact view the granophyre as a manifestation of impact melt, whereas those who believe in an internal origin for the structure have suggested assimilation models. We present new observations of deformation phenomena (multiple fracturing and pseudotachylite occurrences) in granophyre, as well as new major- and trace-element data on the granophyre and on sedimentary inclusions therein. Harmonic least-squares mixing calculations achieved good matches for the granophyre composition for mixtures of ~25% shale — <3.5% quartzite — 75% granite and 9-14% shale — 9% quartzite — 60% granite — 20% lamprophyric magma, respectively. It is not possible to recognize a lamprophyric contribution of only 20% within the currently available trace-element database; thus, it is not possible at this stage to favor one of the present hypotheses for the origin of the granophyre. The pros and cons of the current genetic hypotheses for Vredefort granophyre are discussed and areas that warrant further work are indicated.

INTRODUCTION

The origin of the Vredefort structure, located in a near-central position to the Witwatersrand basin in its presently known margins, has been debated since the early 1960s. The formation of the so-called bronzite (or enstatite) granophyre in the central basement of the dome structure has been similarly contested. Those who favor an impact origin for the dome (e.g., *French and Nielsen*, 1988, 1990; *French et al.*, 1989) regard this rock type as an impact melt breccia, whereas advocates for a catastrophic internal origin (e.g., *Bisschoff*, 1972; *Nicolaysen*, 1987) regard the granophyre as an igneous rock formed at a late stage in the evolution of the dome by assimilation of granitic and supracrustal components by an intrusive mafic magma. *French et al.* (1989) recently presented a detailed account of this controversy. In the impact melt theory, the major component is envisioned to be of basaltic composition and in the intrusive magma theory the granophyre source is seen as a magma of probably dioritic lamprophyre composition (*Bisschoff*, 1972). A third hypothesis for granophyre formation suggests that it is a well-crystallized variety of pseudotachylite (*Hall and Molengraaff*, 1925).

Since the early Vredefort studies (e.g., *Nel*, 1927; *Hall and Molengraaff*, 1925), the bronzite granophyre has been regarded as unique because of its extremely homogeneous and unusual composition. Despite the fact that the SiO_2 content is that of a granodiorite (~67 wt.%), the granophyre is highly enriched in "mafic" elements, namely iron (~7 wt.%, i.e., total Fe calculated as Fe_2O_3) and MgO (~3.5 wt.%), and also in CaO (~3.7 wt.%). Macroscopically it contains numerous inclusions, most of which are apparently quartzitic in composition. Its field appearance (numerous clasts set into a dense, fine-grained matrix) closely resembles that of pseudotachylite. Granophyre was thought to be the only rock type in the Vredefort area that was not affected by deformation (e.g., it was not transected by pseudotachylite). What is more, *Bisschoff* (1972) described a granophyre dyke cutting across—and thus postdating—a major pseudotachylite occurrence. Macroscopic and petrographic observations were summarized by *French et al.* (1989).

This enigmatic rock type has been rather neglected by analysts, and only few analytical results are available to date, most of which were obtained early in this century (*Hall and

Molengraaff, 1925; *Nel,* 1927; *Willemse,* 1927; *Wilshire,* 1971). These authors also provide field and petrographic descriptions that are summarized and discussed by *Bisschoff* (1969, 1972) and *French et al.* (1989). The latter authors presented trace-element data and also called for further detailed analysis of this unique rock type. *French and Nielsen* (1988, 1990) contributed a numerical model showing that mixing of basalt, quartzite, shale, and granite can result in a mixture compositionally similar to Vredefort granophyre and concluded that this favored the impact melt hypothesis. In contrast, *Reimold et al.* (1987) provided new analytical results, including a compilation of accurate trace-element data indicative for generation of granophyre from felsic (granitic-dioritic) crust and assimilated supracrustal rocks. It should also be noted that the REE pattern and concentrations of the granophyre match the composition of average Proterozoic clastic sediment rather well (C. Hatton, personal communication, 1987, and *Hatton,* 1988). As shale inclusions are indeed frequently observed in Vredefort granophyre, the possible importance of West Rand Group shales as parent material for the granophyre needed to be further investigated. *French et al.* (1989) reported that the granophyre is significantly more enriched in Ir and other siderophile elements than is the granitic basement of the Vredefort Dome. However, these authors note that this enrichment probably is a consequence of incorporation of siderophile-rich West Rand shales into the melt mixture. They could not provide evidence for contamination with extraterrestrial material.

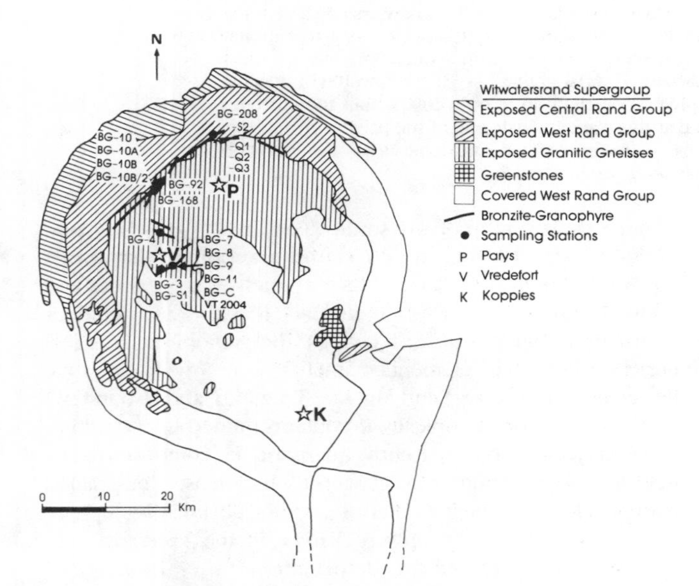

Fig. 1. Schematic representation of the geology of the Vredefort structure with occurrences of bronzite-granophyre and sampling sites for this study: BG-208, etc.—farm Rietpoort (~27°25′/20°50′); BG-92, BG-168, BG-10, etc.—farms Rensburgsdrift/Zuid Witbank (~27°21′/26°54′); BG-4—farm Lesutoskraal 72 (~27°24′/26°57′); BG-7, etc.—farm Holfontein (27°23′/27°01′); BG-3 and BG-S1—farm Lewensbron 125 (27°22′/27°01.5′).

In the following, we review mineralogical and chemical analysis of the granophyre, provide some new morphological, petrographical, and chemical results, and discuss the various models for the formation of Vredefort granophyre.

FIELD OBSERVATIONS

Figure 1 summarizes the geology of the Vredefort Dome and illustrates the known occurrences of granophyre dykes. Near vertical (*Hall and Molengraaff,* 1925; this work) dykes of up to 50 m in width and several kilometers in length occur in two modes: (1) as ring dykes more or less parallel to the periphery of the central granitic basement and thereby crosscutting the contact with the collar and, significantly, intruding up/overturned and, in part, faulted (*Nel,* 1927) horizons composed of quartzite, shale, and mafic intrusives (so-called epidiorite), as well as the Archean gneiss and (2) dykes running either along northeast-southwest or northwest-southeast directions through the gneissic basement. Even at localities several kilometers from the nearest sediment outcrop in the collar these dykes are laden with metasedimentary inclusions (*Manton,* 1964). The northwest-southeast, northeast-southwest structural trends are the most prominently displayed structural trends in the core of the Vredefort Dome (*Poldervaart,* 1962; *Reimold et al.,* 1985; *Colliston et al.,* 1987).

Macroscopically three varieties of granophyre can be distinguished: (1) a fine-grained, dense, and apparently clast-poor variety (Fig. 2a) that often displays a spherulitic (spherulites up to 2 cm) growth of matrix pyroxene (Fig. 2c) and that indeed contains numerous microclasts, (2) a slightly less fine-grained, darker gray, granular variety containing numerous large clasts (Fig. 2b), and (3) a vitrophyric variety (cf. *French et al.,* 1989, Fig. 2c). According to *Manton* (1964), 99% of the studied inclusions consist of quartzite, and only 1% consist of granite. In contrast, recent field and petrographical studies (e.g., *Reimold and Reid,* 1989) have shown that the actual proportion of granitic inclusions is much higher (estimated at 10-15%) in all dykes. There is, however, no doubt that sedimentary material (quartzite >> siltstone, shale, Fig. 2a) constitutes the major portion of the granophyre clast content. Coarse-grained inclusions consisting of at least two mineral phases of grayish and whitish colors (probably quartz and feldspar) are extremely abundant, and are here interpreted as partially recrystallized granite. An example is shown in Fig. 2c. *French et al.* (1989) state that Witwatersrand shale inclusions have not been observed in Vredefort granophyre, while we can report (admittedly rare) findings of shale and siltstone clasts.

A wide variety of different metasediment clasts can be recognized in granophyre; they range from highly siliceous to highly argillaceous, from extremely fine-grained to very coarse-grained, and are of homogeneous or strongly layered texture, with some showing cross-bedding, some graded bedding or folds. The nature and variety of sediment types recognized among the clasts of Vredefort granophyre suggest that a considerable proportion of Lower Witwatersrand sediment has been incorporated into the granophyre. Clasts are generally

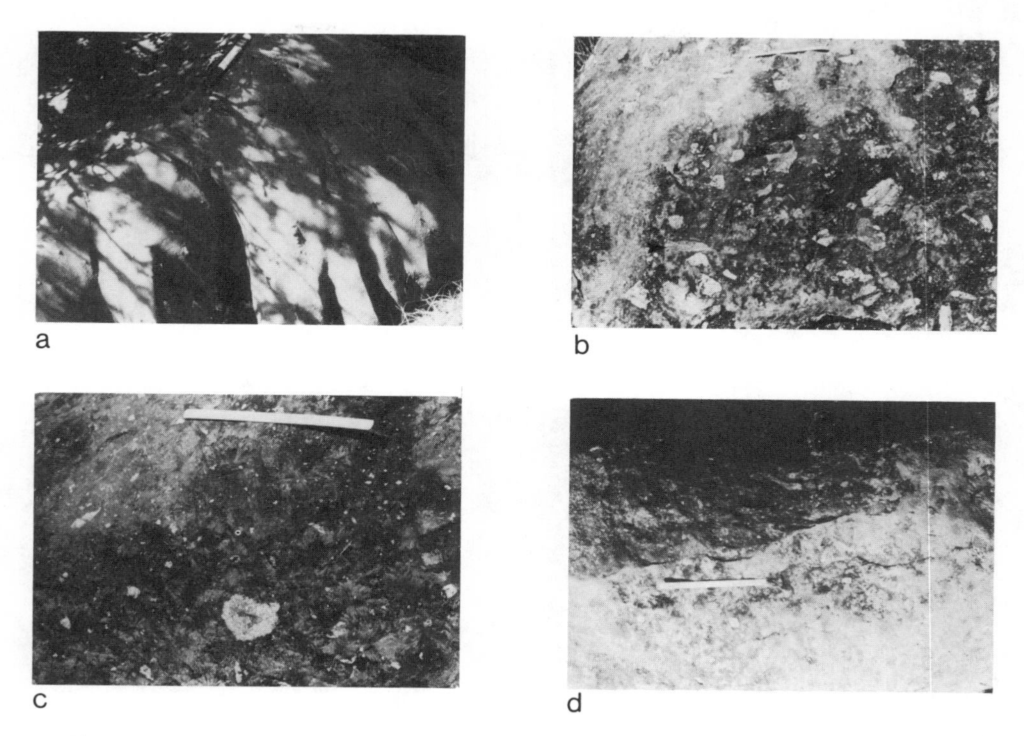

Fig. 2. Field aspects of granophyre: **(a)** Fine-grained, intersertal-spherulitic variety that apparently is clast-poor, but contains numerous microsopic inclusions (farm Holfontein). Pen length = 14 cm. **(b)** Granular variety with large clasts, up to 8 cm in diameter. Note several elongated clasts with evidence of sedimentary layering (arrows). Other inclusions are coarser grained and sometimes surficially pitted (preferential weathering of an individual mineral phase, probably feldspar). These clasts are most probably granite inclusions. Several subparallel fractures crosscut matrix and some of the large inclusions. Scale = 11 cm. **(c)** A coarse-grained, probably granite-derived inclusion in spherulitic granophyre. Note decrease in abundance of visible clasts (compare with **(a)**) and high ratio of small to large clasts. Scale = 11 cm. **(d)** Curviplanar multiple fractures crosscutting Lewensbron granophyre. These fractures are very reminiscent of the MSJS phenomenon discussed by *Nicolaysen and Reimold* (1987). Scale = 11 cm.

angular, but it is evident that the more argillaceous the inclusion, the more rounded is its outline. This could imply that argillaceous clasts are preferentially resorbed into the melt and thus are rarely observed.

There are no recorded observations of pseudotachylite cutting granophyre, and thus the granophyre was thought to be of postdeformational (i.e., postcatastrophic) origin. Based on a Rb-Sr whole-rock and biotite study by *Nicolaysen et al.* (1963) the granophyre age is approximately 1.93 ± 0.03 aeons (recalculated for $\lambda = 0.0142$ aeons-1) (*Walraven et al.,* 1990). These latter authors also discuss ^{40}Ar-^{39}Ar stepheating results by the late H. L. Allsop from granophyre whole rock and biotite samples, as well as ion microprobe analyses of zircon from granophyre (preferred age: 2.002 ± 0.52 aeons), which lend further support to the conclusion that the granophyre formed at ca. 2.0 aeons ago. The youngest age obtained for a deformed lithology in the Vredefort structure is 1.96 aeons, determined for pseudotachylite and shatter-cone-bearing alkali granite in the northwest collar. Thus *Nicolaysen et al.* (1963) concluded that the Vredefort catastrophe occurred between 1.93 and 1.6 aeons ago.

In this study, evidence for postformational deformation in granophyre was found as multiple, often curviplanar joints of subparallel orientations pervasively cross-cutting the granophyre dyke suite [Fig. 2d and *Reimold* (1987)]. This jointing deformation effect can be observed throughout the gneiss of the central basement as well, and has been associated with shatter cones from the collar of the dome by *Nicolaysen and Reimold* (1987). Microfracturing was observed in matrix feldspar of several granophyre samples (Fig. 3d). A specimen from the granophyre dyke on the farm Holfontein (Fig. 1) displays a sharp contact between granophyre and a very fine-grained light gray rock resembling pseudotachylite. As this sample was collected at the extreme margin of the dyke, it was originally assumed that it represented chilled granophyre from the contact with the host rock. Figure 4a schematically shows the original position of the specimen. The light gray lithology consists of brecciated (ultracataclastic) gneiss in which the original banded gneiss fabric can still be recognized. The contact to the granophyre is very sharp, with the exception of a small indentation (Fig. 4a, right) further discussed in the petrography chapter on deformation studies.

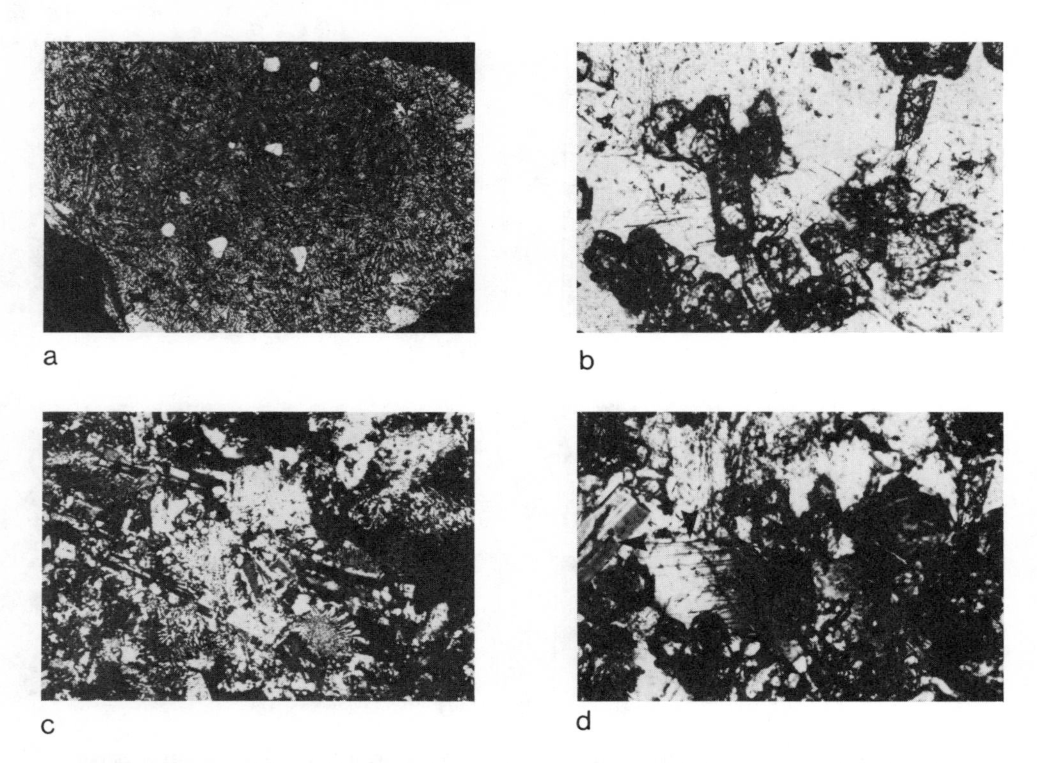

Fig. 3. Microscopic aspects of typical granophyre matrices: **(a)** Thin section of the spherulitic variety; note orthopyroxene needles and scarcity of relatively large clasts (all visible clasts are completely recrystallized SiO_2). Width: 3.5 cm, parallel nicols. **(b)** Granular variety; note prismatic orthopyroxene. Width: about 700 μm, parallel nicols. **(c)** Graphic and subradial growths of micropegmatite, often originating from marginally assimilated plagioclase laths. Granular granophyre variety (tiny roundish crystals of high relief are orthopyroxene). Width: ~2 mm, crossed nicols. **(d)** Two plagioclase crystals (arrows) between prismatic pyroxene; both crystals displaying fracturing. Width: 750 μm, crossed nicols.

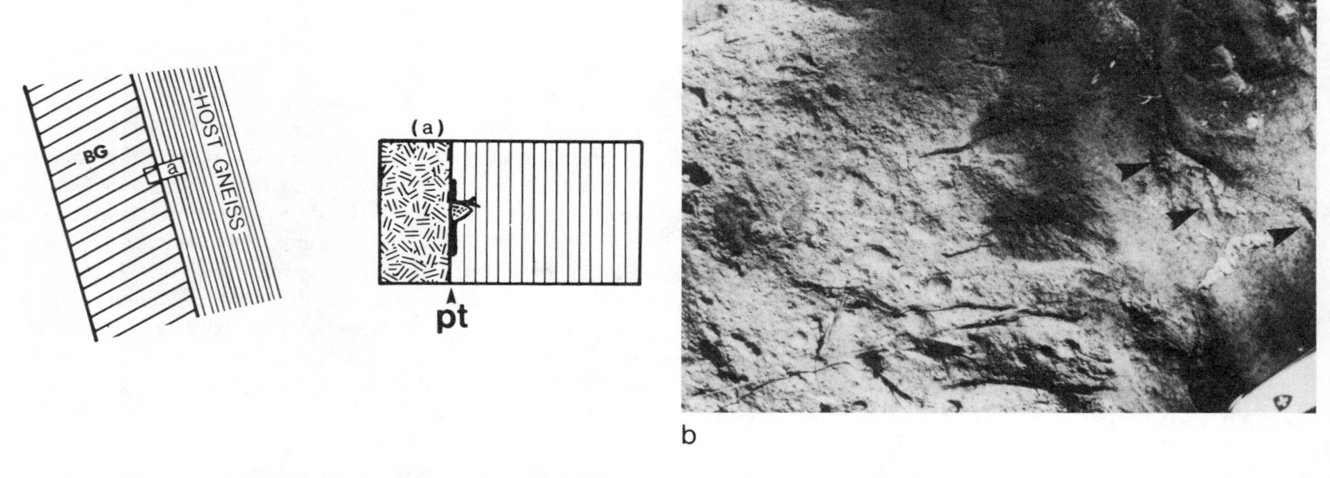

Fig. 4. **(a)** Left: Setting of pseudotachylite-containing specimen BG-C at contact between spherulitic granophyre and mylonitized host gneiss. Right: Schematic representation of thin section: pseudotachylite (pt) dark, granophyre on left, mylonitic breccia on right; stippled area represents local pocket of pseudotachylite formed as mixture from granophyric and gneissic components. **(b)** Pseudotachylite (arrows) crosscutting a granophyre dyke on farm Rensburgsdrift. Also note the multiple curviplanar fractures parallel to the strike of the dyke. Width: 25 cm.

PETROGRAPHY

Sampling and Petrographic Analysis

Samples from four different granophyre dykes (Fig. 1) were studied by polarized and reflected light microscopy, and selected samples were studied with the SEM and semiquantitatively analyzed by EDAX at the CSIR (Council for Scientific and Industrial Research), Pretoria. Some orthopyroxene and feldspar analyses were obtained by A. M. Reid on the CAMECA CAMEBAX electron microprobe in the Department of Geochemistry, University of Cape Town, at 15 KV accerating potential. Further wavelength-dispersive mineral analyses were carried out with the JEOL 733 SUPERPROBE at the University of Pretoria at 20 KV for pyroxene analysis and 15 KV for feldspar and pseudotachylite analysis. Beam diameters were 10 μm for feldspar, 20 μm (defocused beam, DFB) for pseudotachylite, and 1-2 μm for pyroxene. In Pretoria, feldspar standards were used for Ca, Na, and K calibrations, and pure oxide standards for other elements.

Petrographic Description

According to *Bisschoff* (1972) pyroxene (Fs_{26-28}) comprises 15-20% of the granophyre. In spherulitic granophyre pyroxene is generally lath- or needle-shaped (Fig. 3a), whereas in the granular variety short prismatic forms prevail (Fig. 3b). Pyroxene is commonly zoned. *Hall and Molengraaff* (1925) mention clinopyroxene overgrowths on orthopyroxene, and *Bisschoff* (1972) mentions augite (and hornblende) and biotite, magnetite, and ilmenite as common accessory minerals.

Plagioclase is equally abundant as pyroxene and according to *Bisschoff* (1972) apparently forms strongly zoned (An_{60-35}, *Bisschoff,* 1972) euhedral to subhedral lath-shaped crystals. Both *Hall and Molengraaff* (1925) and *Bisschoff* (1969, 1972) regard plagioclase crystals as nuclei for the micropegmatite-granophyric growths that compose about 30% to 45% of the matrix (Figs. 3c, 5a,d,6). This granophyric matrix consists of eutectic intergrowths of quartz, orthoclase, and some Na/K-feldspar. Also mentioned by Bisschoff is that

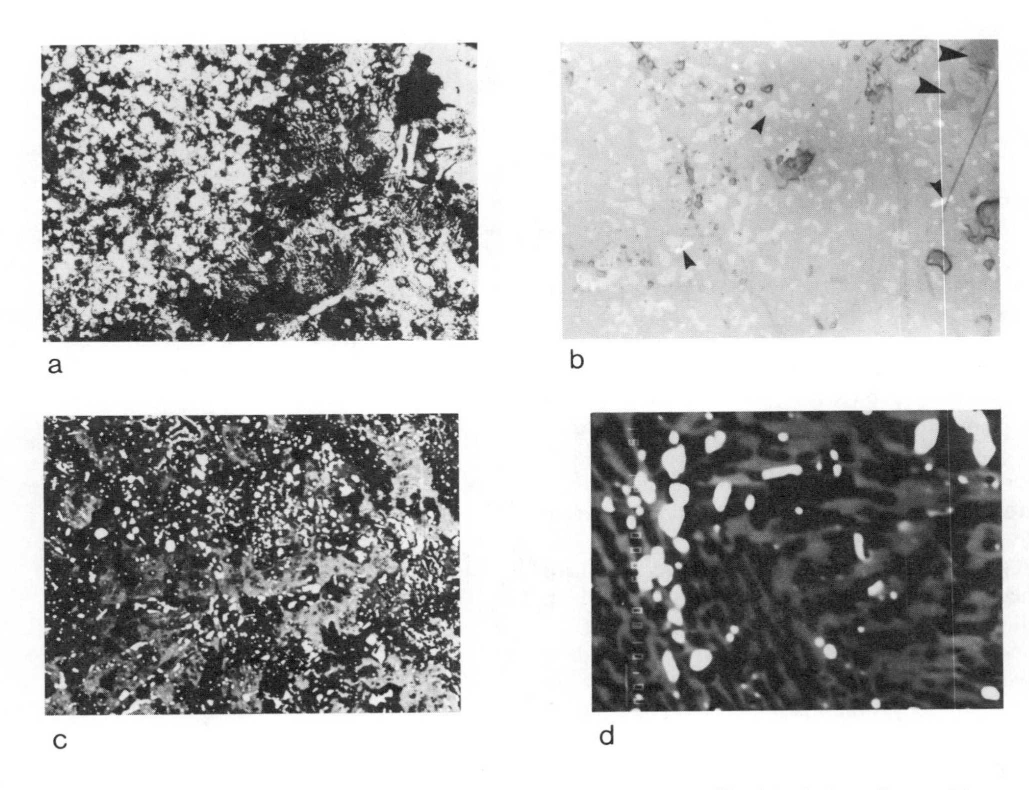

Fig. 5. (a) Contact between micropegmatite (right) and recrystallized inclusion of quartzitic composition. Width: 2 mm, crossed nicols. **(b)** Sample BG-C, Holfontein: High magnification photo of pseudotachy-lite matrix (cf. text). Note euhedral growth of ilmenite and magnetite (small arrows) from the somewhat heterogeneous matrix [note dark, SiO_2-Al_2O_3-rich areas (larger arrows)]. Width: 140 μm, reflected light. **(c)** Backscatter electron image (BEI) of pseudotachylite in sample BG-C (width: 400 μm). Note different gray-values of matrix areas (light: Mg, Al, Si, K, Ti, Fe; medium: Na, Al, Si, K; dark: Al, Si, Na (Al>>Si), always in order of decreasing importance. **(d)** BEI of granophyric matrix in BG-10A. Dark: quartz; gray: plagioclase; light: ilmenite. Scale bar = 10 μm. It is obvious that even with extremely focused electron beam, microprobe analyses of the feldspar phase are difficult to obtain.

Fig. 6. Mosaic of BE-images of matrix from granophyre BG-3 (cf. Fig. 1); scale = 100 μm; spots analyzed semiquantitatively (EDAX) are (1) orthopyroxene (Mg:Fe = 3:2); (2) orthopyroxene (Mg:Fe = 1.3:2); (3) SiO_2 plus little Ca; (4) plagioclase (Ca:Na = 2.7:1.2); (5) hornblende rim (Si:Al:Mg:Ca:Ti:Fe = 22:6:1.5:2.3:0.7:1.6); (6) orthopyroxene (Fe:Mg:Ca = 1.7:1.4:0.3); (7) orthopyroxene core (Fe:Mg:Ca = 2:3.2:0.4); (8) ilmenite; (9) ilmenite; (10) hornblende (similar to 5); (11) orthopyroxene (Fe:Mg:Ca = 3:1.8:0.3); and (12) clinopyroxene rim (Fe:Mg:Ca = 1.3:1.4:2).

growth of granophyric patches commonly seems to originate from clasts (Fig. 5a). *Hall and Molengraaff* (1925) describe partial incorporation of clast material into the micropegmatite. *Reimold and Reid* (1989) recently found micropegmatite development in partially melted granitic inclusions, that consist of rounded relic grains of completely annealed quartz in between micropegmatitic or frequently spherulitic or fine-grained granular melt.

Where the matrix of Vredefort granophyre is well crystallized (i.e., in between granophyric patches), it is either subophitic (Fig. 6), with varying degrees of prismatic pyroxene growth (mainly in the granular granophyre variety), or intersertal-spherulitic. Deformation of matrix minerals is restricted to rare irregular fracturing or cleavage in plagioclase (Fig. 3d). Two samples of granular to subophitic texture (BG-3, BG-92, Fig. 1) were studied by SEM-EDAX (Fig. 6): Major matrix constituents are zoned orthopyroxene, some of which displays thin clinopyroxene or hornblende overgrowths, and plagioclase. Some K- or K/Na-feldspar occurs as rounded or platy grains; their appearance seems to indicate that they are clasts. The granophyric groundmass consists of graphic intergrowths of quartz and K-feldspar with minor Na/K-feldspar. Abundant euhedral ilmenite grains are present.

Mineral Chemistry

Quantitative results of pyroxene and plagioclase analysis are presented in Figs. 7 and 8 respectively. Plagioclase in granular (BG-3 and -92) and spherulitic (BG-9A and -10A) granophyre is bimodally varied ($An_{23\text{-}65}$) in composition (Fig. 7). Plagioclase more sodic than Ab_{52} occurs in the finely intergrown matrix, whereas early crystallized larger laths display higher An content. Plagioclase in the granophyre appears unzoned, contrary to the information given by *Bisschoff* (1972). Only one K-feldspar analysis was obtained in the groundmass. Orthopyroxene (Fig. 8) analyzed was found to be generally of *hypersthene* composition. Both normal and reverse zoning (En_{66}-En_{49}, En_{60}-En_{72}) occur in orthopyroxene. Only two clinopyroxene overgrowths on orthopyroxene were encountered. Pyroxene compositions obtained in rapidly cooled spherulitic granophyre are less variable than those from granular granophyre, with only 2 out of 19 analyses indicating possible zoning (Fig. 8).

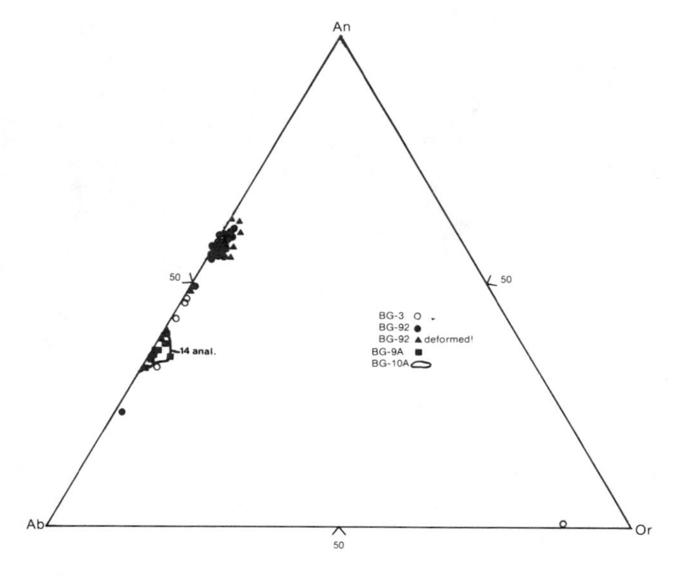

Fig. 7. Microprobe data of feldspar in matrix of granular (BG-3 and BG-92) and spherulitic (BG-9A and BG-10A) granophyre (mol.% Or-Ab-An).

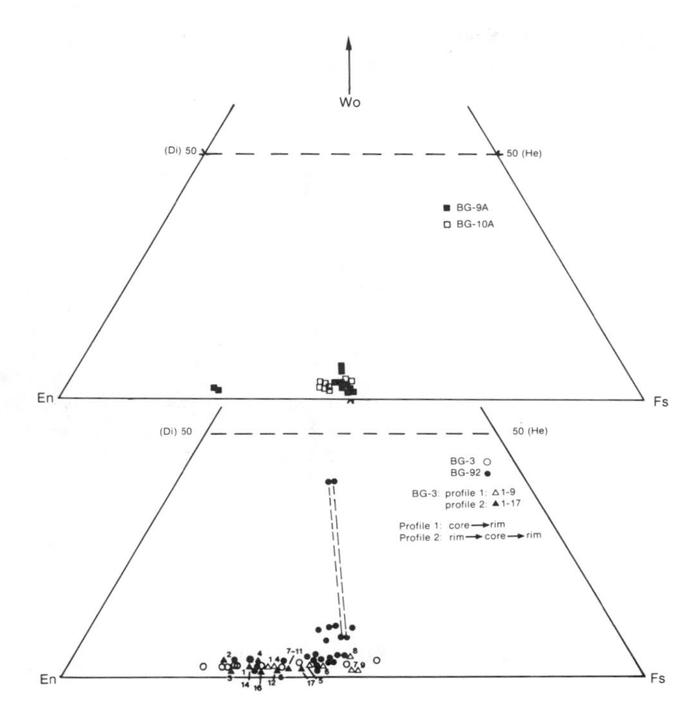

Fig. 8. Pyroxene compositions in granular and spherulitic granophyre. Two large pyroxene laths in BG-3 were studied for internal compositional zoning, which is responsible for the spread in orthopyroxene compositions.

Deformation Studies

Several deformed (microfractured) plagioclase laths were also analyzed with the microprobe and yielded the same compositions as other matrix plagioclase.

The small pocket of black material (Fig. 4a) found in sample BG-C from Holfontein was studied by SEM and also analyzed quantitatively. A low-magnification backscatter electron image (Fig. 5d) shows a material of highly varied composition (Table 1) containing numerous Fe- and Fe-Ti-oxide particles. Their partially euhedral growth (Fig. 5c, small arrows) is consistent with crystallization from a melt phase. The average composition calculated from 20 defocused beam microprobe analyses (Table 1) probably represents a nonrepresentative average of compositions produced by local melting of distinct mineral phases of host gneiss origin (e.g., feldspar, feldspar plus quartz, hornblende or biotite). This assumption is derived from comparison of individual analyses with ideal compositions of respective host rock (granite-gneiss) minerals. The compositions analyzed in the more homogeneous stippled area (only) of Fig. 4a may indicate mixing of molten granophyre and host gneiss. In the thin veinlet (Fig. 4a) three distinct phases were recognized: (1) a mafic type of high Ti and Fe contents with variable Mg, Al, Si, and K concentrations, (2) a Na/K-feldspathic phase with additional variable amounts of SiO_2, and (3) a highly Al-enriched phase with minor Si and Na. In the latter case (dark phase, Fig. 5d) small grains of ~1 μm diameter of >60 wt.% Al_2O_3 were identified (high-temperature Al-oxide?).

These observations are suggestive of generation of pseudotachylitic melt along the contact plane between granophyre and host rock. The chemical composition suggests that granophyre contributed as parent material to this later melt. In Fig. 4b a

TABLE 1. Pseudotachylite in granophyre BG-C (data in wt.%), DFB analysis, 20 analyses.

	Mean composition	Analysed range
SiO_2	50.25	33.04-72.37
TiO_2	0.61	0.05-1.98
Al_2O_3	18.61	9.56-27.06
FeO	6.67	0.15-19.09
MnO	0.07	0.00-0.18
MgO	1.25	0.00-3.09
CaO	0.52	0.00-1.58
Na_2O	2.12	0.19-6.82
K_2O	3.98	0.00-11.06
Cr_2O_3	0.11	0.19-0.47

TABLE 2a. Major element analyses of Vredefort "bronzite"-granophyre—literature data (wt.%).

	(A)	W1	W2	W3	W4	W5	I	II	III
SiO_2	66.4	67.0	62.9	67.1	68.8	67.0	67.4	63.7	67.45
TiO_2	0.5	0.4	0.5	0.4	0.8	0.5	0.4	0.5	0.4
Al_2O_3	14.6	12.4	12.6	12.1	12.4	12.6	12.5	12.75	12.2
Fe_2O_3*	6.8	7.8	9.3	7.7	6.1	7.1	7.9	9.4	7.7
MnO	0.14	0.15	0.1	0.15	0.14	0.1	0.15	0.1	0.15
MgO	3.5	3.9	4.15	4.1	3.1	3.35	3.9	4.2	4.1
CaO	3.6	4.5	5.7	4.4	3.5	3.9	4.5	5.8	4.4
Na_2O	2.3	1.6	2.4	1.7	2.4	2.8	1.85	1.8	2.0
K_2O	1.9	1.8	1.8	2.0	2.4	2.3	1.6	2.4	1.7
P_2O_5	0.1	0.3	0.2	0.1	0.25	0.2	0.3	0.2	0.1
CO_2	nd	0.05	0.05	0.05	bd	bd	0.05	0.05	0.05
LoI	0.3	0.1	0.3	0.3	0.3	0.25	0.7	0.3	0.3
Total	100.14	100.15	100.00	100.10	100.19	100.1	99.90	100.40	99.90

*Total Fe as Fe_2O_3.
(A) *Wilshire* (1971); (W1-W5) *Willemse* (1937); (I-III) *Hall and Molengraaff* (1925); nd = not determined, bd = below detection limit.

TABLE 2b. Major element analyses (XRF) of Vredefort "bronzite"-granophyre (BG-3—BG-10, USA-209 is epidiorite host rock to BG-208) and granite (BG-CM), siltstone (BG-S1, BG-S2), and quartzite (Q1-Q3) inclusions therein.

	BG-3	VT 2004	BG-4	BG-92	BG-168	BG-208 (1)	BG-208 (2)	BG-CB	G-7	BG-8	BG-9	BG-10	USA-209 (1)	USA-209 (2)	BG-CM	BG-S1	BG-S2	Q1	Q2	Q3
SiO_2	67.0	66.2	67.5	67.3	66.2	67.3	65.7	67.9	66.4	66.5	67.4	67.6	50.3	49.8	82.0	79.7	84.3	81.8	88.6	82.7
TiO_2	0.5	0.2	0.5	0.5	0.5	0.2	0.5	0.5	0.55	0.57	0.52	0.49	0.3	0.7	0.24	0.3	0.1	0.1	bd	0.1
Al_2O_3	13.0	12.5	12.7	12.4	12.8	12.5	12.5	12.6	12.8	12.7	12.6	12.6	11.6	12.2	8.9	11.5	7.9	8.6	7.0	10.2
Fe_2O_3*	7.0	7.4	7.2	7.1	7.4	7.7	7.0	7.2	7.1	7.2	6.9	6.8	12.1	12.0	2.8	1.5	2.6	3.7	1.5	3.8
MnO	0.14	0.25	0.14	0.14	0.15	0.09	0.14	0.13	0.13	0.14	0.13	0.13	0.1	0.2	0.05	0.04	0.1	0.1	bd	0.1
MgO	3.5	3.1	3.5	3.9	3.7	3.5	3.7	3.2	3.5	3.3	3.4	3.4	12.8	12.7	1.15	0.5	0.7	0.6	0.1	0.3
CaO	3.7	3.6	3.8	3.9	4.2	4.1	4.0	3.3	3.7	3.8	3.6	3.6	10.2	10.0	0.8	0.7	0.8	0.5	bd	0.1
Na_2O	2.7	3.4	2.5	2.6	2.6	2.9	2.9	2.5	3.3	3.3	3.1	3.0	1.7	1.9	1.8	2.3	0.6	1.0	0.4	0.5
K_2O	2.4	2.65	2.1	2.2	2.3	?1.3	2.2	2.4	2.3	2.3	2.3	2.3	bd	0.1	2.2	2.5	1.5	2.1	1.1	1.0
P_2O_5	0.1	0.3	0.1	0.1	0.1	0.2	0.1	0.13	0.11	0.13	0.1	0.1	0.15	0.1	0.04	0.1	0.3	0.2	0.4	0.3
CO_2	nd	0.1	nd	nd	nd	0.1	nd	nd	nd	nd	nd	nd	0.1	nd	nd	nd	0.3	0.2	0.5	0.3
LoI	0.05	0.2	0	0	0.1	0.25	1.3	0.09	0.12	0.4	0	0	0.5	0.3	0	0.8	0.75	0.9	0.5	0.7
Total	100.09	99.90	100.04	100.14	100.05	100.14	100.04	99.95	100.01	99.98	100.05	100.02	99.98	100.0	99.98	99.94	99.95	99.8	100.1	99.8

*Total Fe as Fe_2O_3. Data in weight percent.

second known occurrence of pseudotachylite in "bronzite"-granophyre is shown. There can be no doubt that this pseudotachylite veinlet in granophyre from farm Rensburgsdrift (northwest contact collar/basement) cross-cuts the host dyke. *Reimold et al.* (1990) present further evidence for the possible formation of pseudotachylite in the Vredefort structure at times later than 1.95 aeons, the assumed age of the Vredefort "event." Based on these findings we conclude that the possibility that major deformation events took place at Vredefort at postgranophyre formation times ought to be further investigated.

Numerous quartz, quartzite, and granite inclusions in granophyre were studied for possible occurrences of microdeformations and, in particular, of shock effects. In Vredefort granophyre more than 98% of all inclusions studied are completely recrystallized. The remaining ~2% of inclusions observed [including several completely unannealed granitic (Fig. 5b) clasts] do not exhibit deformation features at all, with a few exceptions that contain irregularly fractured quartz grains. No fluid inclusions (or fluid inclusion trails) were observed in granophyre clasts. No clasts with Vredefort-type (sub)planar fractures were observed in the granophyre either. In conclusion, the granophyre from the Vredefort Dome does not contain shock-metamorphosed clasts, as reported also by *French et al.* (1989). These authors describe a range of thermal alteration textures in granophyre inclusions, which, they conclude, are comparable to alteration textures from clasts in impact melts, but not diagnostic for the origin of these effects.

TABLE 2c. Composition of West Rand Group shales, after *Phillips* (1986).

	1	2	3
SiO_2	53.3 ± 2.0	48.0 ± 2.1	54.1 ± 4
TiO_2	0.7 ± 0.1	0.4 ± 0.07	1.1 ± 1.1
Al_2O_3	14 ± 1.4	9.7 ± 1.4	16 ± 10.8
Fe_2O_3	16.3 ± 3.4	30.1 ± 4.4	18.5 ± 15.5
MnO	0.1 ± 0.02	3.4 ± 1.0	0.1 ± 0.03
MgO	7.5 ± 1.0	4.3 ± 0.6	3.7 ± 1.6
CaO	1.3 ± 1.5	0.7 ± 0.3	0.1 ± 0.03
Na_2O	0.3 ± 0.2	0.3 ± 0.3	0.2 ± 0.2
K_2O	0.9 ± 0.7	1.6 ± 0.4	1.1 ± 0.8
P_2O_5	0.1 ± 0.03	0.1 ± 0.01	0.04 ± 0.04
LoI	7.2 ± 0.8	4.8 ± 0.4	7.5 ± 1.3

Listed are average compositons and standard deviations (1 σ); 1 = Parktown shale, 2 = Coronation shale, 3 = Jeppestown shale.

GEOCHEMISTRY

Major-element analysis for bulk granophyre samples (free of any visible inclusions) and selected inclusions were obtained by XRF at the Geological Survey laboratories in Pretoria and at the Department of Geology, University of the Witwatersrand. Trace-element analyses were carried out by instrumental neutron activation analysis (INAA) at the Schonland Research Centre, following the technique outlined by *Erasmus et al.* (1977). The results are presented in Tables 2b and 3. Table 2a summarizes chemical data on granophyre available from the literature, and Table 2c presents compositions of Lower Witwatersrand shales (West Rand Group; after *Phillips,* 1986) that were used for comparison with granophyre and its possible parent rocks.

Major-element Results

We report new major-element data for several granophyre samples and for the host rock (USA-209, epidiorite) of granophyre sample BG-208. Outer Granite Gneiss (OGG) and Inlandsee Leucogranofels (ILG) samples, as well as gneisses from farm Broodkop, southeast Vredefort Dome, as principal components of the Archean basement, were analyzed for comparison. In addition, quartzite, siltstone, and granite inclusions in granophyre were analyzed. The ternary diagrams of Figs. 9a,b represent the granophyre composition in terms

TABLE 3. INAA data for "bronzite"-granophyre and inclusions therein (sample nos. as in Tables 2a and b).

	BG-3	BG-4	BG-92	BG-168	BG-208	USA-209	BG-S1	BG-S2	Q1	Q2	Q3	(A)
Ni	93	100	105	134	111	439	44	45	77	21	48	89
Co	25	27	25	25	23	66	12	12	11	3	11	24
Cr	87	89	85	89	210*	261*	67	350*	188*	241*	493*	30
Au	0.2	0.2	2.8	1.8	nd	bd	6.7	nd	bd	bd	bd	nd
La	36	34	29	33	36	6.9	76	21	18	22	27	nd
Ce	46	44	41	42	56	nd	136	nd	2.4	38	43	nd
Nd	21	29	23	19	18.5	bd	37	nd	bd	bd	bd	nd
Sm	4.5	4.5	4.1	4.4	5.1	2.2	8.6	2.9	2.4	3.1	3.4	nd
Eu	0.9	1.0	0.8	0.9	0.6	0.4	0.95	0.2	0.4	0.4	0.4	nd
Tb	0.5	0.5	0.45	0.5	0.6	0.49	0.6	0.3	bd	0.3	0.3	nd
Yb	0.8	0.8	0.85	1.7	nd	bd	1.4	nd	bd	bd	bd	nd
Lu	0.13	0.13	0.12	0.26	0.2	bd	0.3	bd	bd	bd	bd	nd
Hf	4.6	3.8	3.7	3.7	2.7	0.9	4.5	2.3	1.6	1.2	1.8	nd
Ta	0.5	0.5	0.6	0.6	1.1	0.1	0.7	0.3	0.3	0.1	0.3	nd
Sc	13	14	13	14	12	24	5.2	bd	0.5	0.6	0.7	15
As	0.02	0.02	0.02	<0.01	1.8	0.1	2.5	4.3	3.8	2.4	3.3	nd
Sb	0.3	0.3	0.3	0.4	bd	bd	0.5	0.3	bd	bd	bd	nd
Rb	78	65	73	84	78	4	126	nd	nd	nd	nd	90
Zr	137	132	128	117	165	nd	nd	nd	nd	nd	nd	140
Cs	3.4	2.6	3.4	3.5	2.8	0.5	1.5	1.8	1.3	0.8	1.5	3
Ba	497	512	438	470	451	44	774	nd	nd	nd	nd	465
U	1.2	1.1	1.3	1.5	1.8	bd	3.3	1.4	1.6	0.8	0.7	nd
Th	6.7	5.8	6.1	6.8	6.0	0.7	13.2	nd	<0.05	bd	0.1	nd

*Possibly contaminated for Cr during sample crushing; Ir always <1 ppb = detection limit.
Data in ppm, only Au in ppb. bd = below detection limit, nd = not determined. (A) *Wilshire,* 1971.

of the most important major-element abundances and clearly show how the granophyre is homogeneous on a regional scale. Iron and Mg contents of granophyre are higher than those of basement gneiss. However, previously reported data seem to indicate some possibly significant sample-to-sample differences, e.g., in SiO_2 (W2 and W4, Table 2a) or Fe_2O_3 (W2, Table 2a) contents. Thus, it has been suggested that more or less mafic granophyre compositions were generated either by (impact) mixing of different rock components or by assimilation of different amounts of crustal lithologies (cf. *French and Nielsen,* 1987, 1988, 1990). However, sample BG-208, collected at a collar location where the sampled granophyre dyke cuts through epidiorite (that could have been assimilated), shows no obvious chemical difference to other newly analyzed specimens. All new analyses, from four different dykes and seven different sites, are extremely similar. An alternative

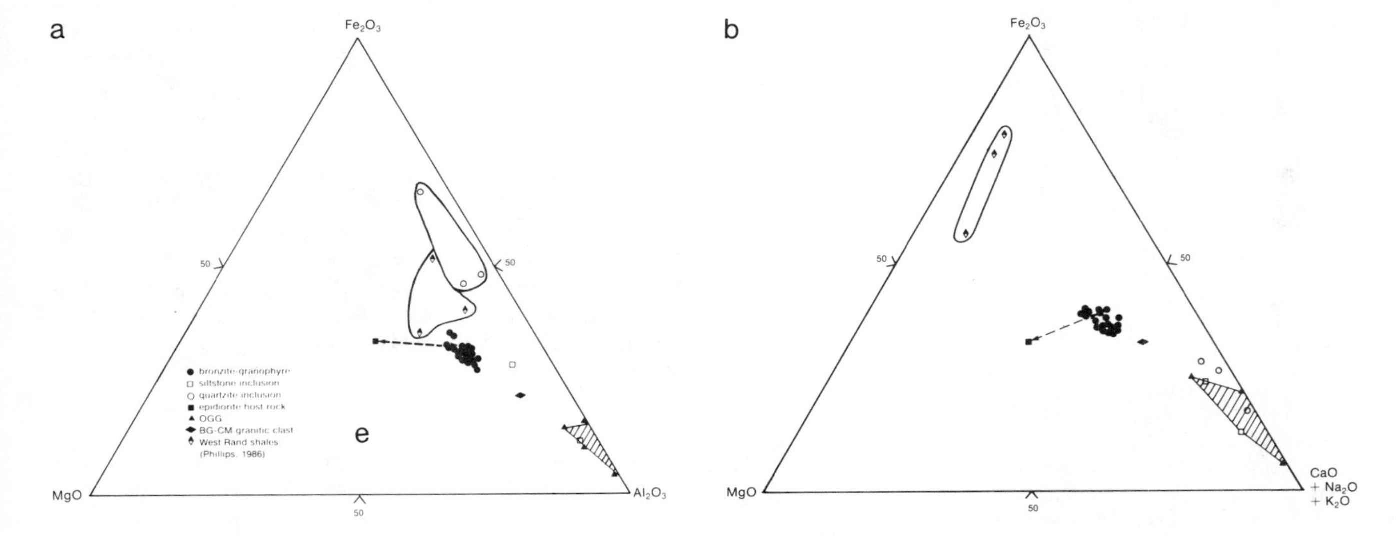

Fig. 9. Major-element compositions of granophyre, inclusions therein, epidioritic host rock to BG-208, and compositions of typical Outer Granite Gneiss (OGG) from the basement core of the Vredefort structure and of West Rand Group shales. **(a)** Fe_2O_3-MgO-Al_2O_3 representation; **(b)** Fe_2O_3-MgO-CaO+Na_2O+K_2O ternary diagram.

explanation of the "spread" in older data could therefore be that the wet chemical analyses may not have provided sufficiently accurate results. The granophyre from Vredefort is a very *homogeneous* rock type (see also section on trace-element chemistry, and discussion by *Reimold and Reid,* 1989).

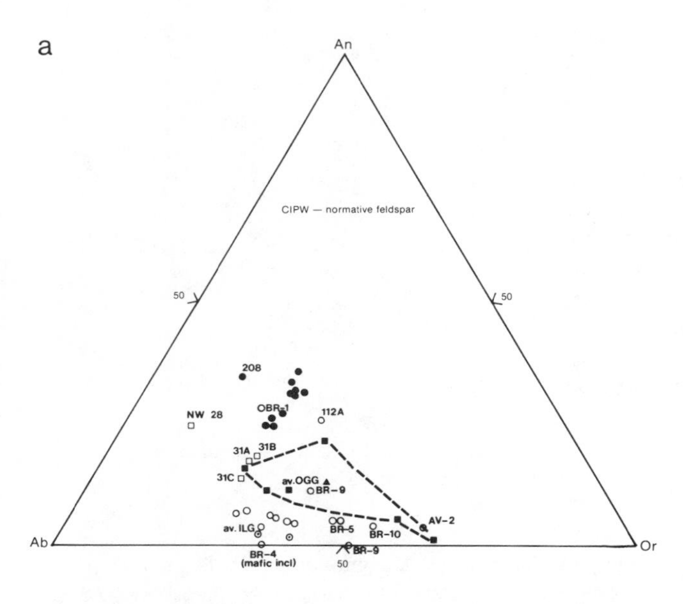

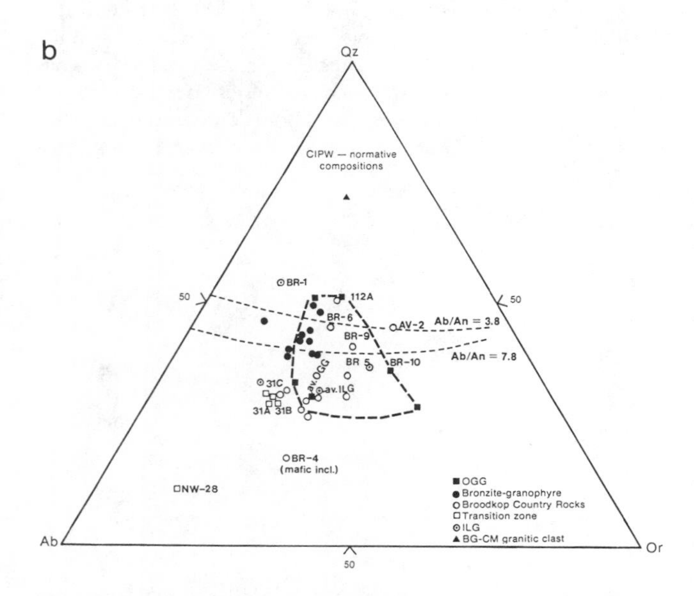

Fig. 10. Representations of CIPW normative rock and feldspar compositions for granophyre and selected basement gneiss samples [Outer Granite Gneiss (OGG), Inlandsee Leucogranofels (ILG), Broodkop gneiss from the southeastern basement]. **(a)** Anorthite-Albite-Orthoclase feldspar norms; **(b)** Quartz-Albite-Orthoclase normative granite system. Also shown are the cotectic lines at $pH_2O = 2$ kbar, for Ab/An = 3.8 and 7.8, respectively. Most data scatter around the minimum melting temperature regime.

Data on analyzed inclusions and Lower Witwatersrand shales are also given in Figs. 9a,b, suggesting that mixing of granite and shale, and possibly quartzite, could generate a mixture of granophyre composition (cf. below). Analyses of very fine-grained inclusions BG-S1 and BG-S2 (Table 2b) that were, on sampling, thought to represent shale clasts and for West Rand shales (Table 2c) demonstrate that the inclusions are siltstones.

Figure 10a is a CIPW normative representation of granophyre and selected basement rock compositions. Figure 10b shows normative feldspar compositions for these samples. Considering the higher normative Ab/An ratio of most basement granite samples, no implications can be drawn from Fig. 10a on the possible melting temperatures for granophyric melt. Both granophyre and basement granite data scatter around T_{min} for a granitic system at 2 kbar (*Winkler,* 1976). Lower normative orthoclase and higher normative anorthite contents of granophyre as compared with basement granite are obvious from the An-Ab-Or diagram of Fig. 10b. Therefore, no further information, e.g., on melting temperatures of granophyre samples, can be drawn from Fig. 10b.

Trace-element Results

The neutron activation results (Table 3) are presented in Figs. 11 and 12 in comparison with trace-elemental abundances in the granitic basement of Vredefort. The REE patterns (Fig. 11a) for granophyre samples display the typical shape of OGG or ILG from the central basement (data from W. U. Reimold, manuscript in preparation, and R. Hart, personal communication), with a distinct negative Eu anomaly. Such anomalies are also typical for the metasedimentary inclusions analyzed. Figure 11c represents the range and shape of lamprophyre REE patterns (after *Cullers and Graf,* 1984). It is obvious that only a minor contribution of such a component could be in agreement with the REE abundances determined for the Vredefort granophyre (cf. mixing calculations and discussion below). The narrow range of granophyre analyses shows that this rock type is also homogeneous with respect to nearly all the analyzed trace elements. Significant variabilities are observed for some siderophile elements (Au, Ni, possibly Cr), and for Zr and Ba, in agreement with some data reported by *French et al.* (1988, 1989), who found variable concentrations for Au, Co, and Cr. Whereas their Co and Ni data match our analyses rather well, their Cr (239-394 ppm) and Au (2.2-7.5 ppb) values significantly exceed the respective ranges given in Table 4 for our granophyre data.

French et al. (1988, 1989) also give Ir concentrations for granophyre and possible parent rock lithologies. Granophyre contains between 57 and 130 ppt Ir (below our detection limit of 1 ppb), which is comparable with Ir contents of Witwatersrand quartzites (30-270 ppt), much higher than in basement granite (<5 ppt) and much lower than in Witwatersrand shales (160-330 ppt). The concentrations of siderophile elements are different for each dyke, but also vary within a given dyke. As the less siderophile element Co is less variable, this may be due to heterogeneous distribution of Ir-, Au-, and Cr-containing trace minerals (ilmenite?).

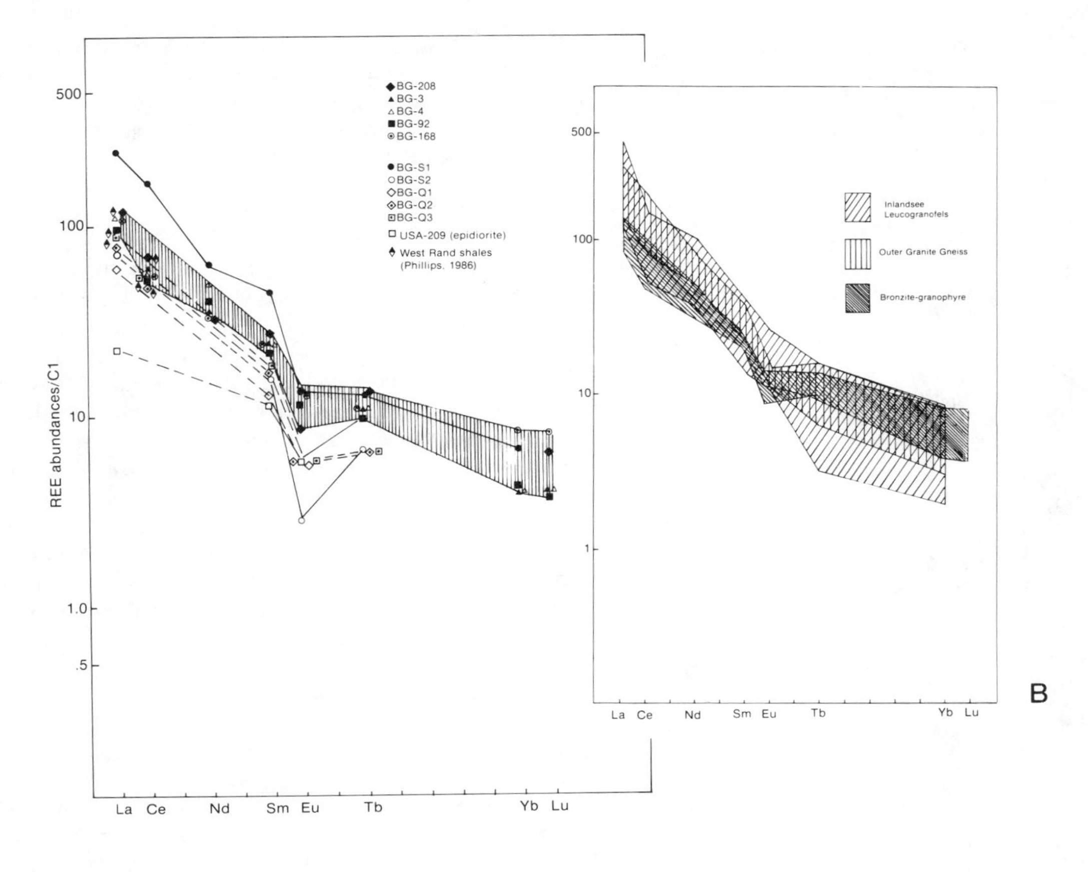

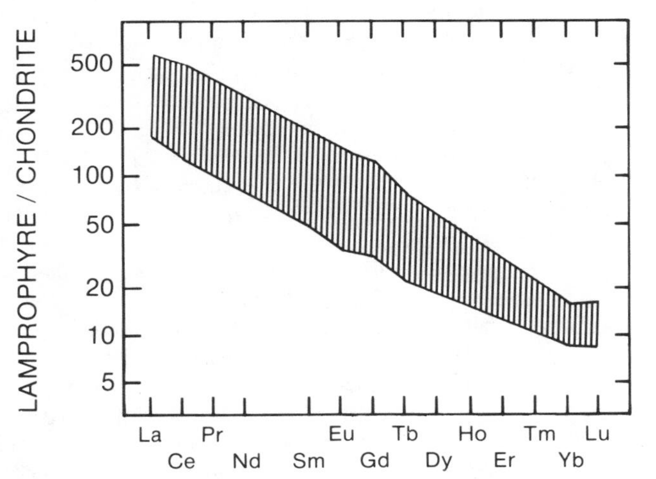

Fig. 11. **(a)** REE abundances (C1 normalized) in granophyre bulk samples (hatched area) and inclusion of sedimentary origin (BG-S1,2: siltstones; BG-Q1-3: quartzite), and epidiorite host rock (USA-209) to BG-208. Only La and Ce data are available for Lower Witwatersrand shales; **(b)** Comparison of REE distribution ranges (C1-normalized) for granophyre and basement gneiss (data after R. Hart, personal communication, and results from this study); and **(c)** lamprophyre (after *Culler and Graf,* 1984).

Discussing the mixing calculations by *French and Nielsen* (1988, 1990) and considering the lack of evidence for extraterrestrial Ir, *French et al.* (1989) conclude that Vredefort granophyre could nevertheless be understood as an impact melt derived by impact melting (and homogenization) of 25-80% Ventersdorp Lava (basalt), 10-30% Witwatersrand shale/hornfels, 5-25% granite, and 20% Witwatersrand quartzite.

Figure 12 compares selected lithophile and siderophile elements in granophyre, in OGG and ILG, and in Lower Witwatersrand shale. In general, there is a good match between granophyre and granitic rocks from Vredefort, except for Sc. Cesium could be enriched in granophyre as well. The very slight differences betweeen granophyre and granitic basement (e.g., slight depletions of granophyre in Ba and Rb contents, or increases in relative Co, Ni, and Cr contents) can be explained if shale were a major constituent of the granophyre parent material. A comparison of analytical ranges for lithophile and siderophile elements in granophyre,

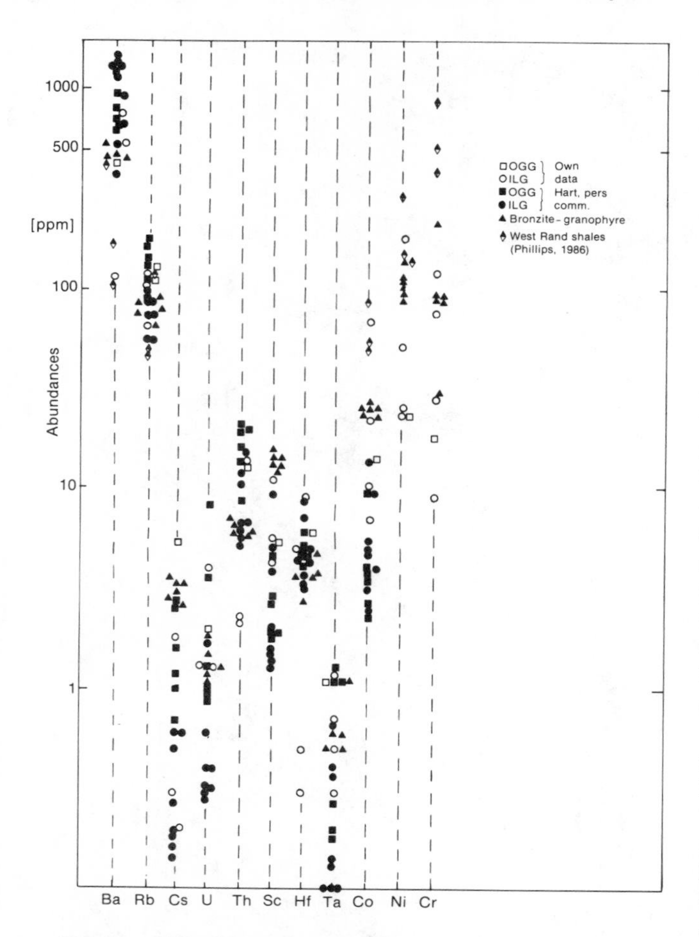

Fig. 12. Lithophile and siderophile element abundances in granophyre and OGG- and ILG-type basement gneiss.

basement gneiss, continental basalt, and West Rand shales is given in Table 4. The apparent Sc anomaly could thus also be explained, as Sc concentrations for shales in excess of 20 ppb have been reported (*Wronkiewicz and Condie,* 1987).

Mixing Calculations

In order to test the possibility that granite and supracrustals alone could be the parent components for a granophyric melt mixture, we performed HMX harmonic least-squares mixing calculations (*Stöckelmann and Reimold,* 1989; *Reimold,* 1982; *Stöffler et al.,* 1985). This method takes into account errors that obscure both the component and mixture parameters.

Parameters used to describe the possible components (shale, quartzite, granite) and granophyre mixture in HMX calculations are presented in Table 5a. Results are presented in Table 5b for calculations with (1) major-element parameters only, (2) major- and trace-element data, (3) different parameter sets for the shale component (from either *Phillips,* 1986 or *Wronkiewicz and Condie,* 1987), and (4) testing whether other components that were not considered would be required to bring the total of calculated component proportions to 100%. For each calculation a discrepancy factor was calculated (by integration of the individual differences between observed and calculated parameter values) as a relative indicator for goodness of fit. Perfectly modeled mixtures should yield a discrepancy factor of 1; values up to 10 are regarded as describing reasonable fit. All results of granophyre HMX calculations are listed in Table 5b.

Runs 1-3 and 6 show good discrepancy factors, whereas the results of other runs are less well defined. This is due to discrepancies between observed and calculated values for single mixture parameters (i.e., elements, as listed in the last column of Table 5b). However, when actual discrepancies [i.e., differences (observed concentration-calculated concentration)] for badly matched elements are compared, these individual elemental discrepancies are found to be mere fractions of the actual analytical inaccuracies for these elements. For example, the difference (observed-calculated) for Ni in run 4 is 7.5 ppm [which by a factor of 5 is the worst relative difference (observed-calculated) calculated for all parameters in this run], whereas the analytical uncertainty for every measured Ni concentration in granophyre or possible components is approximately 11 ppm, and standard errors for Ni concentrations of individual components range up to 76% of the respective component parameter.

The results presented in Table 5b stress that it is possible to achieve granophyre composition by mixing of ~25-30% Lower Witwatersrand shale, less than 3.5% quartzite, and ~65-75% OGG-type granitic crust without requiring a mafic source (assimilation theory) or parent rock (impact melt theory). The fact that only a minor quartzite contribution was computed is still consistent with the high proportion of xenoliths of such refractory material in the granophyre. Also, the observation that the more argillaceous clasts are rounded due to marginal absorption into the melt shows that easily more shale than quartzite would be molten.

Major-element discrepancies (diffs. X_{obs}-X_{calc}) are generally <0.06 wt.%-<0.01 wt.%, which in terms of analytical uncertainties is insignificant. It appears as if the CaO/K_2O ratios of granophyre (Tables 2a-c) cannot be modeled by simple mixing of granite and shale. However, as the HMX results have also shown, the wider variations for these elements (Table 5a) allow for reasonably good mixing results.

We also tested the possibility that a lamprophyric magma contributed (as parent magma?) to the Vredefort granophyre, as *Willemse* (1937) and *Bisschoff* (1969, 1972) suggested for their assimilation models. *Willemse* (1937) proposed that the granophyre originated from an alkaline lamprophyric magma (genetically associated with the alkali granitic intrusions into the northwest and north portions of the collar) contaminated with granite and quartzite. *Bisschoff* (1969, 1972), however, showed that a peralkaline magma is an unlikely parent for the granophyre for mineralogical reasons, but that a combination of dioritic lamprophyre, shale, and granite would be more reasonable.

The compositions of the lamprophyric components (peralkaline and dioritic, respectively) used in the HMX mixing calculations are given in Table 5c and the results of the calculations are presented in Table 5d. Statistically very good

TABLE 4. Comparison of lithophile and siderophile element abundances determined for Vredefort granophyre (BG), Outer Granite Gneiss (OGG) and Inlandsee Leucogranofels (ILG), and West Rand Group shale horizons.

Element	BG*	Otavi OGG*	Other OGG*	ILG*	avg. OGG†	avg. ILG†	Cont. Basalt	West Rand Group Shales‡			Shales§
								1	2	3	
Ni	93-134	4-11	23	24-181	—	—	120-170	302 ± 53	156 ± 24	151 ± 44	364.4 ± 109.7
Co	25-27	0.7-5	14	10-68	2.7-9.2	2.5-13.3	40-50	63 ± 11	94 ± 14	59 ± 46	42.0 ± 14.6
Cr	85-89 (210)	51-85	18	28-122	—	—	100-320	891 ± 125	463 ± 103	519 ± 133	762 ± 322
Au	0.2-2.8	1.4-4	8	1.7-3	0.8-1.8	0.8-1.8	0.4-10	—	—	—	—
Hf	2.7-4.6	2-7	6	0.3-9	4.2-6.0	3.3-8.6	20-80	—	—	—	3.8 ± 1.6
Ta	0.5-1.1	1-.7	1.1	0.3-1.2	0.2-1.1	0.1-4	0.5-6.0	—	—	—	0.7 ± 0.3
Sc	12-14	0.4-3	5.3	5.7-11.2	1.8-4.6	1.3-9.4	± 30	—	—	—	20.4 ± 6.8
As	0.02	bd	0.13	—	—	—	0.2-6.5	—	—	—	—
Sb	0.3-0.4	0.1	0.6	0.07-0.09	0.7-0.13	0.1-0.18	0.1-1.4	—	—	—	—
Rb	65-84	143-332	130	66-120	111-173	56-91	11-31	46 ± 28	124 ± 24	51 ± 20	76.2 ± 26.6
Zr	117-165	—	—	—	153-223	119-392	35->600	123 ± 9	67 ± 22	224 ± 219	54 ± 23.9
Cs	2.6-3.5	0.5-1.1	5.2	0.2-0.3	0.7-2.7	0.14-0.60	0.3-2.0 (4.0)	—	—	-	5.9 ± 3.0
Ba	438-512	251-1154	431	527-761	609-1159	377-1572	av. 246	179 ± 131	462 ± 83	112 ± 86	340 ± 123
U	1.1-1.8	0-2.4	2.0	1.3-4	0.9-3.6	0.3-1.7	<1	—	—	—	2.0 ± 1.4
Th	5.8-6.8	3.6-10.9	12.8	2.25	8.8-20.2	5.2-12.0	<2	—	—	—	5.5 ± 2.5

*Own data.
†R. Hart, personal communication.
‡*Phillips* (1986).
§*Wronkiewicz and Condie* (1987—West Rand and Central Rand Group shales; cf. also Table 5a).
Data in ppm, Au in ppb; bd = below detection limit. Basalt data after *Wedepohl* (1969) and *Rösler and Lange* (1972). 1 = Jeppestown shales, 2 = Coronation shales, 3 = Parktown shales; averages and standard deviations.

results (discrepancy factors of 5.4 and 7.5) were obtained for calculations in the system shale-quartzite-granite-lamprophyre. Near-perfect matches between observed and calculated mixture compositions were achieved with diffs. (X_{obs}-X_{calc}) for all parameters of the order of 0.01-0.02 or less. However, the results for the calculation with a dioritic lamprophyre composition are obscured by smaller errors and in this case the discrepancy factor (5.4) is also better (Table 5d). A calculation with lamprophyre, but without shale (Run III, Table 5d), resulted in a considerably worse discrepancy factor of 11.5. These results appear to favor the Bisschoff assimilation model.

In conclusion, the unique composition of the Vredefort granophyre can be modeled in two ways (in addition to the impact mixing model, *French and Nielsen,* 1990) by mixing of shale, granite, and (possibly) some quartzite, or by mixing of a lamprophyre component with granite and supracrustals. On the basis of mixing calculation results alone it cannot be determined which model is statistically more valid, or if any of them is valid at all.

Isotope Studies

According to *O'Neil et al.* (1987), $\delta^{18}O$ values for granophyre are 7.3‰ and 7.9‰, while Lower Witwatersrand metasediments range from 7.2-11‰ (10 data), and samples from the granitic basement (7 data) have $\delta^{18}O$ values from 7.6-8.3‰ (OGG) or 7.8-10.0‰ (ILG). If compared with various felsic or (ultra)mafic rock types (*Faure,* 1986), this limited database does not yield conclusive evidence whether the formation of granophyre involved mafic (i.e., basaltic) lithologies. The oxygen isotopic composition of granophyre is similar to the upper limit isotopic compositions of mafic rocks and near the lower limit of the range for granite lithologies. However, a large proportion of mafic parent rock (up to 80%, *French and Nielsen,* 1990) would most probably result in the lowering of $\delta^{18}O$ for granophyre below 7‰.

DISCUSSION

For a discussion of granophyre origin, the following results need to be considered:

1. Evidence is presented for formation of pseudotachylite melt from granophyre and for a fracturing event *postdating* granophyre intrusion.

2. Clasts found in the granophyre consist of a significant proportion of granite relics besides sedimentary lithologies (quartzite, siltstone, shale). No *mafic* inclusions are evident.

3. Characteristic shock metamorphic effects were not detected in inclusions. In fact, deformation recorded in granophyre clasts is minimal (rare irregular fracturing).

TABLE 5a. Compositions used for HMX mixing calculations (major elements in weight percent, trace elements in ppm).

	(a) Shales 1σ	(b)	Quartzite 1σ	OGG[†] 1σ	BG 1σ
SiO_2	51.8 ± 3.3	55.4 ± 7.3	83.4 ± 3.3	71.4 ± 7.6	66.9 ± 0.7
TiO_2	0.7 ± 0.4	0.6 ± 0.2	0.1 ± 0.2	0.3 ± 0.35	0.5 ± 0.25
Al_2O_3	13.2 ± 3.2	15.5 ± 5.8	9.0 ± 1.8	12.0 ± 3.5	12.6 ± 0.2
Fe_2O_3	21.6 ± 13*	15.8 ± 11.4	2.6 ± 1.1	2.5 ± 3.3	7.2 ± 0.3
MnO	1.2 ± 1.9	0.6 ± 1.0	0.1 ± 0.1	0.05 ± 0.05	0.14 ± 0.05
MgO	5.2 ± 2.0	4.6 ± 2.3	0.4 ± 0.25	0.8 ± 1.6	3.5 ± 0.2
CaO	0.7 ± 0.6	0.4 ± 0.3	0.4 ± 0.4	1.9 ± 2.2	3.7 ± 0.25
Na_2O	0.3 ± 0.3*	0.4 ± 0.3	1.0 ± 0.8	3.7 ± 1.1	2.9 ± 0.3
K_2O	1.2 ± 0.4	1.6 ± 0.5	1.6 ± 0.65	4.0 ± 1.7	2.2 ± 0.3
P_2O_5	0.1 ± 0.6*	0.1 ± 0.1	0.3 ± 0.1	0.1 ± 0.2	0.1 ± 0.06
Rb	73.3 ± 43.8	76.2 ± 26.6	63 ± 63	194.5 ± 93.4	75.6 ± 7.1
Ba	251.1 ± 186.2	339.8 ± 123.3	387 ± 387	720.8 ± 417.4	473.6 ± 30.9
La	32.6 ± 28.7*	27.4 ± 14.4	32.8 ± 24.4	47.7 ± 30.9	33.6 ± 2.9
Ce	46.4 ± 57.8*	55.0 ± 28.9	54.9 ± 57.0	64 ± 32.3	45.8 ± 6.0
Co	71.9 ± 46.3*	42.0 ± 14.6	9.8 ± 3.8	6.3 ± 5.3	25.0 ± 1.4
Ni	203.0 ± 86	364.4 ± 109.7	47.0 ± 19.9	12.7 ± 9.6	108.6 ± 15.7
Cr	624.0 ± 232.4	761.8 ± 321.6	267.8 ± 161.9	51.3 ± 33.5	87.5 ± 1.9
Cs	—	5.9 ± 3.0	1.4 ± 0.4	2.0 ± 2.0	3.1 ± 0.4
Th	—	5.5 ± 2.5	2.7 ± 5.9	11.3 ± 6.1	6.3 ± 0.4
U	—	2.0 ± 1.4	1.6 ± 1.0	1.8 ± 1.4	1.4 ± 0.3
Sc	—	20.4 ± 6.8	1.4 ± 2.1	3.0 ± 2.0	13.2 ± 0.8
Hf	—	3.8 ± 1.6	2.3 ± 1.3	5.0 ± 2.0	3.7 ± 0.7
Ta	—	0.7 ± 0.3	0.3 ± 0.2	0.8 ± 0.4	0.7 ± 0.25
Sm	—	4.7 ± 2.1	4.1 ± 2.6	4.4 ± 1.4	4.5 ± 0.4
Yb	—	2.4 ± 1.0	0.7 ± 0.7	1.6 ± 0.9	1.0 ± 0.4

*1σ standard deviations calculated obviously did not at all represent the full range of analytical data available, or was significantly smaller than analytical errors given by analysts, so maximum deviations were adopted from raw data.
†Typical analyses (XRF, Dept. of Geology, WITS) of Outer Granite Gneiss from farms Otavi and Broodkop; the average was calculated from 10 granite-gneiss analyses plus 1 amphibolite analysis (from Otavi) that was included to account for mafic contributions to OGG crust estimated at ca. 10% of the total volume.
(a) Data from *Phillips* (1986), (b) data from *Wronkiewicz and Condie* (1987), averages and 1σ standard deviatons calculated from means for various West Rand (a) or West and Central Rand (b) Group shale formations.

TABLE 5b. Results of HMX mixing calculations (in percent).

		Shale	Quartzite	OGG	Discr. Fact*
Run 1:	Major elements only, $\Sigma = 100\%$	26.3 ± 3.8	0 ± 3.4	73.7 ± 4.7	6.5
Run 2:	Major elements only, Σ may be $\neq$ 100%	26.3 ± 3.8	0 ± 3.4	73.7 ± 4.7	6.2
Run 3:	Major elements only, Σ may be $\neq$ 100% and quartzite is pivot comp.†	25.0 ± 3.7	0.04 ± 3.4	75.0 ± 4.6	6.2
Run 4: cond. as 3	Major elements, shale data from (a), also Rb, Ce, Ba, La, Co, Ni, Cr	33.0 ± 2.9	0 ± 0.02	67.0 ± 2.9	15.7 (Rb, Ce, Ba, Ni badly fit)
Run 5: cond. as 3	All major and trace element data of Table 5a, shale data from (b)	38.4 ± 0.01	0.01 ± 0	61.6 ± 0	32.9 (Rb, Ce, Ba, Ni, Sc, Hf)
Run 6: cond. as 3	Major element data only, but shale data from (b)	28.2 ± 4.5	0.01 ± 2.6	71.8 ± 4.5	11.5
Run 7: cond. as 3	as Run 4, but shale data from (b)	32.4 ± 2.8	2.6 ± 0.9	65.0 ± 2.9	28.4 (Rb, Ce, Ba, Ni)

*The discrepancy factor as relative indicator for goodness of fits is calculated as a measure relative to the ratio betwen calculated mixture parameters and observed parameters. A perfectly modeled mixture run should have a minimum discrepancy factor (close to 1); method described by *Stöckelmann and Reimold* (1989). A pivot component is an arbitrarily chosen component, assumed to be a definite contributor to the mixture.

TABLE 5c. Average composition (plus 1σ standard deviation) of lamprophyres.

	(a)	(b)
SiO_2	41.1 ± 2.9	47.0 ± 6.2
TiO_2	2.75 ± 0.23	2.45 ± 1.5
Al_2O_3	14.1 ± 1.2	14.4 ± 0.5
Fe_2O_3	13.8 ± 1.95	12.8 ± 8.1
Mno	0 ± 0 (n.o.)	0 ± 0 (n.o.)
MgO	7.8 ± 2.1	8.2 ± 2.6
CaO	10.5 ± 2.1	8.25 ± 2.6
Na_2O	3.1 ± 0.9	2.9 ± 0.7
K_2O	1.46 ± 0.9	1.6 ± 1.4
P_2O_5	1.45 ± 0.9	0.35 ± 0.18

(a) After *Rock* (1977, Table VIII); (b) dioritic lamprophyres, after *Correns* (1968); Fe_2O_3 = total Fe; n.o. = not obtainable.

4. The Vredefort granophyre is of extremely homogeneous composition, in terms of both major and trace elements. Strong chemical affinity to the granitic basement of the Vredefort Dome has been documented, and the chemical data suggest that shale was a major parent component.

5. Mixing calculation results show that mixing of 25-30% shale, less than 3.5% quartzite, and ~65-75% granite could produce a granophyre-like composition. The same could be achieved with ~9-14% shale, 9% quarzite, 60% granite, and 20% lamprophyric magma.

6. No enrichment of siderophile elements derived from an extraterrestrial source can be detected or is necessary to account for the concentrations analyzed in granophyre (*French et al.,* 1988, 1989; this work).

In the following, the current models for granophyre genesis will be discussed in the light of these findings.

The Impact Melt Model

This theory is mainly based on the fact that the Vredefort granophyre is an extremely homogeneous clast-laden melt rock. The model, as dicussed by *French and Nielsen* (1990), assumes that besides quartzite, shale, and granite, a basaltic component was part of the pre-impact stratigraphy as well (namely, the up to 5 km thick Ventersdorp Supergroup, cf. discussion by *French et al.,* 1989). However, considering the assumed age for an impact of ~2 Ga, Transvaal Sequence strata including large volumes of carbonate rocks (dolomite) would have been part of the target stratigraphy as well. It is rather difficult to understand in the context of this model that no relics of the mafic component (not to be confused with epidiorite inclusions that occasionally can be *locally* observed where granophyre cuts through an epidiorite horizon in the collar) have been observed to date in the granophyre. The chemical data presented in this study do not necessitate a basaltic component at all; supracrustals and granite alone can account for the trace-element data obtained. A mafic component should also be expected to be more resistant against resorption by the melt than felsic components melting at lower temperature. Furthermore, the impact melt model would be greatly strengthened if shock metamorphosed inclusions had been reported from Vredefort granophyre. This is not the case (cf. also *French et al.,* 1989), and independent evidence for an extraterrestrial component could not be found either (*French et al.,* 1989). This lack of shock metamorphosed inclusions is equally difficult to explain by the advocates of an internal gas explosion (e.g., *Nicolaysen,* 1972, 1986) who propose formation of granophyre melt from a gas-rich mafic magma by catastrophic gas explosion (*Nicolaysen,* 1987).

French et al. (1989) stress the fact that the well-known stratigraphy in the Witwatersrand Basin "presents a unique opportunity to use chemical information from the Granophyre to compare the geochemical modeling results against the well-established local geology; . . ." They then note "major differences between the relative amounts of lithologies in the Vredefort stratigraphic column and in the compositions calculated for the average Granophyre as an impact melt." However, in *French and Nielsen's* (1987, 1990) mixing model the various lithologies comprising the Transvaal Sequence (dolomites, shales, quartzites, and volcanics) overlying Witwatersrand and Ventersdorp strata have not been taken into account. This dilemma is aggravated by the fact that stratigraphic thicknesses over the area of the Vredefort Dome can only be approximated, as the opinion is widely held that the dome represented a basement-high already in Early Proterozoic times. Therefore, not too much emphasis should be given to currently available mixing models as fingerprints for the genesis of the granophyre.

Finally, the question has to be addressed whether the impact melt theory can explain the distinct structural control of granophyre dyke emplacement. This problem is difficult to evaluate because no other impact structure of this size is available for structural study and no computer modeling of the structural deformation caused in a 100-km impact structure has been attempted. Nevertheless, one should (?) expect that the main impact melt volume be generated at or around the central uplift and not further outside, and that major basement deformation should take place in rather central areas of the subcrater basement as well. It appears likely that melt dykes formed by gravitational settling of impact melt into fractures (*French and Nielsen,* 1990) should form with radial or irregular orientation rather close to the center of the structure. However, Vredefort granophyre is emplaced near the contact between collar and basement and extends in the directions determined by the regional stress field (since Archean times) only.

TABLE 5d. Results of HMX mixing calculations employing a lamprophyric component.

	Shale	Quartzite	OGG	Lamprophyre	Discr. Fact.
Run I, Lampr.(a)	13.5 ± 2.8	8.4 ± 3.1	60.3 ± 4.3	17.8 ± 2.4	7.5
Run II, Lampr.(b)	9.4 ± 1.9	9.3 ± 2.1	56.7 ± 3.2	24.5 ± 2.4	5.4
Run III w/o shale, Lampr.(a)	—	0 ± 1.5	84.4 ± 3.9	15.6 ± 3.0	11.5

The Assimilation Models

As pointed out by *Bisschoff* (1969, 1972), the assimilation models could explain the mineralogical characteristics of the major granophyre constituents (orthopyroxene and plagioclase) and the major-element composition of this rock type (cf. mixing models of this study). Alkali intrusives and associated lamprophyric dykes (also of dioritic affinity) are known from the environs of the Vredefort Dome.

A maximum contribution of about 25% lamprophyric component is allowed by our mixing calculations. However, it is difficult to see how mixing of such a minor component with ~75% of assimilated granite or supracrustals can produce a regionally homogeneous melt.

The trace-element data at hand can be used to test the validity of the mixing calculation results. The REE abundances for lamprophyre and granite do not differ sufficiently and cover ranges too wide to allow recognition of a ~20-25% lamprophyre contribution to the granophyre. Nevertheless, the distinct Eu anomaly of the granophyre is consistent with an origin from mainly crustal components. Also, the trace-element data summarized in Fig. 12 do not present a clear picture, as the ranges for concentrations in granite—that emerged as the major component in all calculations—are wider than the possible shift of a granophyre composition with a lamprophyre contribution. The concentrations of Ba, Rb, Cs, U, Th, Sc, Hf, and Co in the granophyre would be in agreement with a lamprophyre contribution, while Ta, Ni, and Ir abundances are far too low (according to a comparison with trace-element concentrations for kimberlitic rocks compiled by *Wedepohl and Muramatsu,* 1979).

The Pseudotachylite Hypothesis

First mentioned by *Hall and Molengraaff* (1925), this idea has not been followed up in more recent times. Pseudotachylite in the Vredefort structure is more or less homogeneous at a given locality, but heterogeneous from locality to locality. Microscopic textures, lack of shock effects, and high degree of recrystallization of clasts are very similar for Vredefort pseudotachylite and granophyre. Both lithologies have been emplaced according to the same structural control (compare, e.g., *Reimold et al.,* 1985). The regional homogeneity of the granophyre dykes seems to be a major shortcoming of the "pseudotachylite hypothesis." *Schwarzman et al.* (1983) suggested that Vredefort pseudotachylite was formed due to shock brecciation and subsequent thermal metamorphism. Other workers (e.g., *Killick and Reimold,* 1989) stress the similarities between pseudotachylite from Vredefort and other localities (Witwatersrand Basin and elsewhere, in tectonic settings) with regard to structural styles and mineralogy. At least some pseudotachylite occurrences in Vredefort can be reconciled with formation on faults; however, the tremendous volume of this breccia type at Vredefort cannot be easily explained with an origin by friction. *Killick and Reimold* (1989) noted the major shortcomings in the available database on Vredefort pseudotachylite and called for further multidisciplinary studies.

Large-scale fault rock formation (mylonites, ultracataclasites, pseudotachylites) is known from several crustal discontinuities in the Witwatersrand basin (e.g., Black Reef Décollement Zone, Master Bedding Fault—cf. *Fletcher and Reimold,* 1987, 1989), and pseudotachylite can comprise up to 95% of these breccias. It is proposed here to consider the possibility that the Vredefort granophyre could be a tectonite generated at a significant crustal discontinuity, possibly at the granite-supercrustal interface. This could explain why granophyre of the same composition and of the same clast content and population is found along the contact collar/basement *and* up to 10 km into the granitic core. It is to date not known whether the vast amounts of pseudotachylite (and other breccias) associated with Witwatersrand detachment zones are *in situ* formations or whether they have to be regarded as intrusions generated kilometers away from their present intersections. This is undoubtedly an interesting and necessary subject for future research with regard to pseudotachylite formation in general and to understanding of fault development in the basin.

In earlier single-stage shock event hypotheses for the origin of the Vredefort structure, the formation of granophyre was always considered the latest phase in the evolution of the dome (e.g., discussion in *Nicolaysen and Reimold,* 1985 and references therein; *Bisschoff,* 1972, 1982). We agree that one (or more?) major deformation event(s) occurred at Vredefort prior to the emplacement of granophyre dykes, but we propose that the tectonic development of the Vredefort Dome did not end with granophyre formation. Evidence presented here and elsewhere (*Reimold et al.,* 1990; *Colliston et al.,* 1987), suggests that pseudotachylite was formed at several times during the evolution of the Vredefort Dome and, in particular, also at times younger than ca. 1.9 aeons.

We agree with *Bisschoff* (1988) that "with the present state of our knowledge, any firm genetic description of this enigmatic structure (i.e., the Vredefort Dome) must be viewed as somewhat premature." Bisschoff asks, "Did the deformation take place during a nearly instantaneous episode (a cryptoexplosion caused by gases from within the Earth or by meteorite impact), one relatively short period of deformation (vertical and/or horizontal tectonic processes related to orogenesis and/or magmatism), or over an extended period (ca. 500-600 Ma), in which one or more phases of deformation were active (diapirism)?" Or were tectonic processes such as thrusting and gravity sliding, possibly in connection with diapirism (e.g., *Colliston,* 1990; *Du Toit,* 1954; *Fletcher and Reimold,* 1987, 1989), responsible?

In conclusion, the new data and observations presented here could not resolve the enigmatic origin of the Vredefort granophyre. It appears as if the deformation history recorded in Vredefort rocks did not terminate prior to granophyre emplacement. The new chemical data stress the regional homogeneity of the granophyre, a major dilemma for the hypotheses advocating an internal origin for the granophyre. Mixing calculations have shown that it is possible to model granophyre composition with granite and supracrustal rock types alone, which are the lithologies recognized in the clast content of the granophyre. In a sample collection covering the regional extent of Vredefort granophyre, no characteristic shock deformation was observed. Both the impact melt and the assimilation hypothesis are model-dependent and not founded on unequivocal natural evidence from the granophyre. While the impact melt model is favored by the regional homogeneity, the proof for a mafic precursor rock is lacking. Independent evidence, shock metamorphosed clasts or an extraterrestrial siderophile element contribution, has not been demonstrated. Assimilation models have to contend with the regional homogeneity of Vredefort granophyre, and the results of chemical analysis and modeling that require an impossibly high proportion of assimilated material (~75%). The third currently debated genetic model, the pseudotachylite hypothesis, does not (yet?) provide the answer, as the formation of Vredefort and Witwatersrand pseudotachylites itself is not yet sufficiently understood.

RECOMMENDATIONS

Detailed studies with regard to clast composition and deformation (TEM strain analysis) and to the mineralogy of feldspars are necessary, in particular with regard to the role granite played as a parent rock (micropegmatite formation in granite inclusions). Additional major- and trace-element analysis is needed for the various types of clasts, and the lamprophyric and alkaline granitic intrusions into the collar of the Vredefort structure. The degree of homogeneity of fault breccias associated with major detachments in the Witwatersrand basin has to be assessed. Further mixing tests, as well as petrogenetic modeling that considers, besides formation of total melts, the possibility that granophyric melt could be the product of partial melting or fractionization, and assimilation studies in the three-component quartzite-shale-granite and four-component lamprophyre-quartzite-shale-granite systems, are to be carried out. The observation that some minor pseudotachylite was formed after formation of granophyre needs to be further investigated by detailed field and thin-section studies. A project of friction experiments with various Vredefort lithologies, including granophyre, is already underway, in order to test whether the highly brittle granophyre could be a lithology less suitable for pseudotachylite formation than other rock types in the dome. Samarium-neodymium, in addition to further Rb-Sr, isotope data may be helpful to elucidate the origin of granophyre melts and conditions at its formation. However, the Vredefort problem has to be further attacked on a broad front: Detailed structural work in collar and basement core is warranted, together with further chemical and chronological analysis, the latter because it is vital to improve the knowledge of the sequence of deformation events that affected the central portion of the Witwatersrand Basin since ca. 2.3 Ga ago.

Acknowledgments. The authors are indebted to Prof. v. Gruenewald and Prof. McCarthy for permission to obtain data on the electron microprobe at the Department of Geology, Pretoria, and the X-ray fluorescence spectrometer of the Department of Geology, University of the Witwatersrand, respectively. With the support of Dr. Walraven, further XRF results were obtained at the Geological Survey in Pretoria. We are most grateful to A. M. Reid who granted permission to include his microprobe analyses of feldspar and pyroxene in this study. Ms. N. Day (Department of Geology) and Ms. L. de Matos of the SRC secured the XRF and INAA data. SEM work was performed at the CSIR (Council for Scientific and Industrial Research) in Pretoria while the senior author was a Senior Research Fellow with this institution. Central Graphics at the University of the Witwatersrand assisted with figure preparation and photographic work. The manuscript was expertly typed by Ms. D. D. Mthembu of the Schonland Research Center. W.U.R. is indebted to the Director of the Schonland Research Centre for permission to complete the research reported here. Comments by A. A. Bisschoff, R. G. Cawthorne, S. Grimmer, R. B. Hargraves, C. Hatton, R. Merkle, and L. O. Nicolaysen on earlier versions of this manuscript are much appreciated, and C. Koeberl and G. Ryder are thanked for their helpful reviews.

REFERENCES

Bisschoff A. A. (1969) The petrology of the igneous and metamorphic rocks in the Vredefort Dome and the adjoining parts of the Potchefstroom syncline. D. Sc. thesis (unpublished), Univ. Potchefstroom, R.S.A.

Bisschoff A. A. (1972) The dioritic rocks of the Vredefort Dome. *Trans. Geol. Soc. S. A., 75,* 31-45.

Bisschoff A. A. (1982) Thermal metamorphism in the Vredefort Dome. *Trans. Geol. Soc. S. A., 85,* 43-57.

Bisschoff A. A. (1988) The history and origin of the Vredefort Dome. *S. Afr. J. Sci., 84,* 413-417.

Colliston W. P. (1990) A model of compressional tectonics for the origin of the Vredefort Structure. *Proc. Intl. Workshop on Cryptoexpl. and Catastr. in the Geol. Rec.* Special issue of *Tectonophysics,* in press.

Colliston W. P., Reimold W. U., and Robertson A. S. (1987) A preliminary report on a detailed structural, geochemical, and isotopic study of the Broodkop Migmatite Complex, Southeast Vredefort Dome (working paper). *Contrib. to Intl. Workshop on Cryptoexpl. and Catastr. in the Geol. Rec., Parys, Section C3.* 22 pp.

Correns C. W. (1968) *Einführung in die Mineralogie (Kristallographie und Petrologie),* 2nd edition. Springer-Verlag, New York. 458 pp.

Cullers R. L. and Graf J. L. (1984) Rare earth elements in igneous rocks of the continental crust: predominantly basic and ultra-basic rocks. In *Rare Earth Element Geochemistry* (P. Henderson, ed.), pp. 237-274. Elsevier, Amsterdam.

Du Toit A. L. (1954) *The Geology of South Africa.* Oliver and Boyd, Edinburgh. 511 pp.

Erasmus C. S., Fesq H. W., Kable E. J. D., Rasmussen S. E., and Sellschop J. P. F. (1977) The NIMROC samples as reference materials for neutron activation analysis. *J. Radioanal. Chem., 39,* 323-334.

Faure G. (1986) *Principles of Isotope Geology,* 2nd edition. Wiley, New York.

Fletcher P. and Reimold W. U. (1987) The pseudotachylite problem and a few notes on the structural evolution of the central portion of the Witwatersrand Basin (working paper). *Contrib. to Intl. Workshop on Cryptoexpl. and Catastr. in the Geol. Rec., Parys, Section F4.* 4 pp.

Fletcher P. and Reimold W. U. (1989) Some notes and speculations on the pseudotachylites in the Witwatersrand Basin and the Vredefort Dome, South Africa. *S. Afr. J. Geol., 92,* in press.

French B. M. and Nielsen R. L. (1987) Vredefort Bronzite Granophyre: chemical evidence relating to its origin (working paper). *Contrib. to Intl. Workshop on Cryptoexpl. and Catastr. in the Geol. Rec., Parys, Section F5.* 29 pp.

French B. M. and Nielsen R. L. (1988) Vredefort bronzite granophyre: Chemical evidence for origin as an impact melt (abstract). In *Lunar and Planetary Science XIX*, pp. 354-355. Lunar and Planetary Institute, Houston.

French B. M. and Nielsen R. L. (1990) Vredefort bronzite granophyre: Chemical evidence for origin as an impact melt. *Proc. Intl. Workshop on Cryptoexpl. and Catastr. in the Geol. Rec.* Special issue of *Tectonophysics*, in press.

French B. M., Orth C. J., and Quintana L. R. (1988) Iridium in the Vredefort bronzite granophyre: Impact melting and limits on a possible extraterrestrial component (abstract). In *Lunar and Planetary Science XIX*, pp. 356-357. Lunar and Planetary Institute, Houston.

French B. M., Orth C. J., and Quintana L. R. (1989) Iridium in the Vredefort Bronzite Granophyre: Impact melting and limits on a possible extraterrestrial component. *Proc Lunar Planet. Sci. Conf. 19th*, pp. 733-744.

Hall A. L. and Molengraaff G. A. F. (1925) The Vredefort Mountain Land in the Southern Transvaal and the Northern Orange Free State. Shaler Memorial Series. *Verh. K. Akad. Wet. Amst., 2.* Sectie, Deel 24, No. 3. 183 pp.

Hatton C. J. (1988) Formation of the Bushveld Complex at a plate margin (extended abstract). *Geocongress '88, 22nd Bienn. Congr., Geol. Soc. S. Afr.*, 251-254.

Killick A. M. and Reimold W. U. (1989) Review of the pseudotachylites in and around the Vredefort "Dome," South Africa. *S. Afr. J. Geol., 92,* in press.

Manton W. I. (1964) The orientation and implication of shatter cones in the Vredefort ring structure. M.Sc. thesis (unpublished), Univ. Witwatersrand.

Nel L. T. (1927) The geology of the country around Vredefort. *Explan. Geol. Map, Geol. Surv. S.A., Dept. Min. Industr.* Govt. Printer, Pretoria. 134 pp.

Nicolaysen L. O. (1972) North American cryptoexplosion structures: Interpreted as diapirs which obtain release from strong lateral confinement. *Geol. Soc. Am. Mem., 132,* 605-620.

Nicolaysen L. O. (1986) Renewed ferment in the earth sciences—especially about power supplies for the core, for the mantle and for crises in the faunal record. *S. A. J. Sci., 81,* 210-132.

Nicolaysen L. O. (1987) Tektites: ejecta from massive cratering events, caused by periodic escape and detonation of deep mantle fluids (working paper). *Contrib. to Intl. Workshop on Cryptoexpl. and Catastr. in the Geol. Rec., Parys, Section N2.* 15 pp.

Nicolaysen L. O. and Reimold W. U. (1985) Shock deformation, shatter cones, and pseudotachylite at Vredefort: A review of major unsolved problems and current efforts to resolve them (abstract). In *Lunar and Planetary Science XVI*, pp. 618-619. Lunar and Planetary Institute, Houston.

Nicolaysen L. O. and Reimold W. U. (1987) Shatter cones revisited (working paper). *Contrib. to Intl. Workshop on Cryptoexpl. and Catastr. in the Geol. Rec., Parys, Section N2.* 8 pp.

Nicolaysen L. O., Burger A. J., and van Niekerk C. B. (1963) The origin of the Vredefort Structure in the light of new isotopic data (abstract). *Intl. U. Geodesy and Geophys., 13th Gen. Ass.,* Berkeley, California.

O'Neil J. R., Reimold W. U., and Nicolaysen L. O. (1987) Reconnaissance determinations of oxygen and hydrogen isotopic compositions of selected rocks from the Witwatersrand Basin and the Vredefort Structure, South Africa (working paper). *Contrib. to Intl. Workshop on Cryptoexpl. and Catastr. in the Geol. Rec., Parys, Section 01.* 4 pp.

Phillips G. N. (1986) Metamorphism of shales in the Witwatersrand Goldfields. *Inform. Circular No. 192,* Econ. Geol. Res. Unit, Univ. of the Witwatersrand, Johannesburg. 25 pp.

Poldervaart A. (1962) Notes on the Vredefort Dome. *Trans. Geol. Soc. S. Afr., 65,* 231-247.

Reimold W. U. (1982) The impact melt rocks of the Lappajarvi meteorite crater, Finland: petrography, Rb-Sr isotope, major and trace element geochemistry. *Geochim. Cosmochim. Acta, 46,* 1203-1225.

Reimold W. U. (1987) The bronzite-granophyre. *Field Guide to Intl. Workshop on Cryptoexp. and Catastr. in the Geol. Rec.,* 6-10 July, 1987, Parys, R.S.A. 17 pp.

Reimold W. U. and Reid A. M. (1989) Petrographic observations on granitic inclusions in granophyre from the Vredefort Structure (extended abstract). *Bienn. Symposium, Mineral. Assoc. S. Afr.,* Pretoria. 6 pp.

Reimold W. U., Andreoli M. A. G., and Hart R. (1985) Pseudotachylite from the Vredefort Dome (abstract). In *Lunar and Planetary Science XVI*, pp. 691-692. Lunar and Planetary Institute, Houston.

Reimold W. U., Horsch H., and Reid A. M. (1987) New facts on the bronzite granophyre from the Vredefort Structure and implications for the genesis of this enigmatic rock type (working paper). *Contrib. to Intl. Workshop on Cryptoexpl. and Catastr. in the Geol. Rec., Parys, Section R1.* 9 pp.

Reimold W. U., Jessberger E. K., and Stephan T. (1990) ^{40}Ar-^{39}Ar dating of pseudotachylite from the Vredefort Dome, South Africa—a progress report. *Proc. Intl. Workshop on Cryptoexpl. and Catastr. in the Geol. Rec.* Special issue of *Tectonophysics,* in press.

Rock N. M. S. (1977) The nature and origin of lamprophyre: Some definitions, distinctions, and derivations. *Earth Sci. Rev., 13,* 123-169.

Rösler H. J. and Lange H. (1972) *Geochemical Tables.* Elsevier, Amsterdam.

Schwarzman E. C., Meyer C. E., and Wilshire H. G. (1983) Pseudotachylite from the Vredefort Ring, South Africa, and the origins of some lunar breccias. *Bull. Geol. Soc. Am., 94,* 926-935.

Stöckelmann D. and Reimold W. U. (1989) The HMX mixing calculation program. *J. Math. Geol., 21.*

Stöffler D., Bischoff A., Borchardt R., Burghele A., Deutsch A., Jessberger E. K., Ostertag R., Palme H., Spettel B., Reimold W. U., Wacker K., and Wänke H. (1985) Composition and evolution of the lunar crust in the Descartes highlands, Apollo 16. *Proc. Lunar Planet. Sci. Conf. 15th,* in *J. Geophys. Res., 90,* C449-C506.

Walraven F., Armstrong R. A., and Kruger F. J. (1990) A chronostratigraphic framework for the north-central Kapvaal Craton, the Bushveld Complex and Vredefort Structure. *Proc. Intl. Workshop on Cryptoexpl. and Catastr. in the Geol. Rec.* Special issue of *Tectonophysics,* in press.

Wedepohl K. H., ed. (1969) *Handbook of Geochemistry.* Springer-Verlag, New York.

Wedepohl K. H. and Muramatsu Y. (1979) The chemical composition of kimberlites compared with the average composition of three basaltic magma types. In *Kimberlites, Diatremes, and Diamonds: Their Geology, Petrology, and Geochemistry* (F. R. Boyd and H. O. A. Meyers, eds.), pp. 330-312.

Willemse J. (1937) On the Old Granite of the Vredefort region and some of its associated rocks. *Trans. Geol. Soc. S. A., 40,* 43-119.

Wilshire H. G. (1971) Pseudotachylite from the Vredefort Ring, South Africa. *J. Geol., 79,* 195-206.

Winkler H. G. F. (1976) *Petrogenesis of Metamorphic Rocks.* Springer-Verlag, New York. 320 pp.

Wronkievicz D. J. and Condie K. C. (1987) Geochemistry of Archean shales from the Witwatersrand Supergroup, South Africa: Source-area weathering and provenance. *Geochim. Cosmochim. Acta, 51,* 2401-2416.

Proceedings of the 20th Lunar and Planetary Science Conference, pp. 451-457
Lunar and Planetary Institute, Houston, 1990

NMR Spectroscopy of Experimentally Shocked Quartz: Shock Wave Barometry

R. T. Cygan

Geochemistry Division, Sandia National Laboratories, Albuquerque, NM 87185

M. B. Boslough

Shock Wave and Structural Physics Division, Sandia National Laboratories, Albuquerque, NM 87185

R. J. Kirkpatrick

Department of Geology, University of Illinois, Urbana, IL 61801

---

^{29}Si MAS NMR spectra of synthetic quartz powders recovered from shock experiments at mean pressures of 7.5, 16.5, and 22 GPa indicate that the NMR technique is very sensitive to shock pressure. All of the shocked powders and the starting material have similar values for the refractive indices. The (101) X-ray diffraction peak of quartz broadens only slightly with shock pressure, at most by a factor of 1.7. By comparison, the ^{29}Si NMR spectra exhibit a nearly fivefold increase in relative peak width at the highest shock pressure. The broadening of the NMR peak is most likely the result of disordering and residual strain in the quartz lattice and the formation of an amorphous silica phase. The calibration of the NMR peak widths with shock pressure provides a very sensitive shock barometer for the determination of pressures associated with natural impacts, as well as materials subjected to shock by nuclear tests.

INTRODUCTION

The shock-loading of natural materials by impact can result in the formation of highly modified phases. Shocked minerals typically exhibit fracturing, planar deformation, comminution, disordering, lamellae, and, frequently, new phases. These new phases may be glass (fused or diaplectic) or other reaction products that include high-pressure polymorphs of the original minerals (see *Stöffler,* 1984). For a given material the extent to which these shock-induced features form depends upon the peak shock pressures and temperatures. These variables, in turn, are related to velocity, composition, density, and angle of impact associated with the projectile, and the original composition and density (porosity) of the target material. The modifications and changes in the original material induced by the shock are referred to as shock metamorphism.

The effects of shock metamorphism on minerals, in particular the presence of the high-pressure polymorphs of SiO_2 (coesite and stishovite), have been used to identify impact events (*Stöffler,* 1971). For the Meteor Crater, Arizona, terrestrial impact site, the occurrence and textural relationships of glass and the SiO_2 polymorphs have provided a basis for estimating the shock pressure (*Kieffer,* 1971; *Kieffer et al.,* 1976).

Recent interest in an extraterrestrial cause for the Cretaceous-Tertiary extinction (*Alvarez et al.,* 1980) has prompted an increase in research on shock features in quartz (for example, *Bohor et al.,* 1984; *Carter et al.,* 1986; *Alexopoulos et al.,* 1988). The *Alvarez et al.* (1980) theory proposed that a large asteroid collided with the Earth 65 m.y. ago, resulting in mass extinctions. Observed shock features in mineral grains and possible stishovite in clay layers from near the Cretaceous-Tertiary boundary support this idea (*Bohor et al.,* 1984; *McHone et al.,* 1989). However, several researchers interpret shock features in minerals to be the result of volcanic eruption processes (*Carter et al.,* 1986), but this interpretation remains controversial and is not generally accepted (*Izett and Bohor,* 1987; *Sharpton and Schuraytz,* 1989). Multiple sets of shock lamellae and reduced values of the refractive indices are widely accepted as diagnostic of a shock event in the minerals.

Studies of the effect of shock in minerals have traditionally relied on qualitative examination of microstructures by optical and electron microscopies. Refractivity, density, X-ray diffraction, infrared absorption, X-ray photoelectron spectroscopy, and electron paramagnetic resonance (EPR) have been used to quantify the physical state of shocked minerals and glasses (see *Stöffler,* 1984). Optical refractivity has proven to be one of the most useful diagnostic techniques of shock metamorphism, especially for quartz (*Chao,* 1968; *Hörz,* 1968). Its usefulness as a shock barometer was determined by recent spindle stage measurements, which show a decrease in birefringence and refractive index for material shocked to pressures above 22 GPa (*Grothues et al.,* 1989). Above shock pressures of 28 GPa, quartz becomes increasingly isotropic and completely transforms to diaplectic glass at 35 GPa, but at pressures below 22 GPa there is little change in the refractive indices (Fig. 1). Detailed X-ray line broadening studies of shocked powder and single crystal quartz have also been performed and indicate a decrease in domain size and an increase in microstrain effects with increasing shock pressure (*Hörz and Quaide,* 1973; *Schneider et al.,* 1984; *Ashworth and Schneider,* 1985). Because these microstructural properties are apparently a function of shock pressure, X-ray line broadening may also be used as a shock barometer.

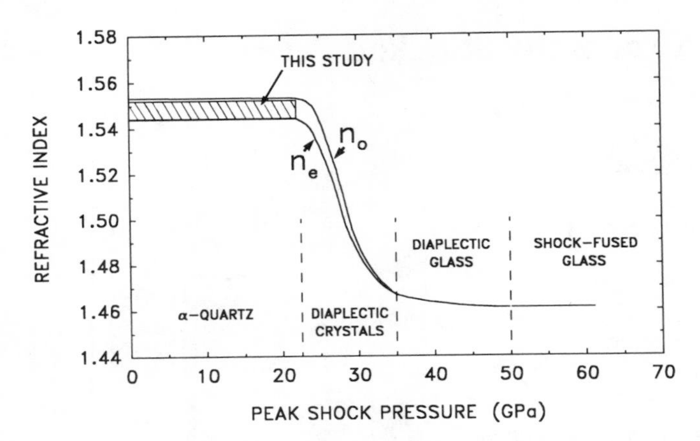

Fig. 1. Refractive index as a function of peak shock pressure, demonstrating the decrease in refractivity and birefringence with increasing shock pressure. The curves labeled n_o and n_e define the birefringence envelope for shocked single crystals of quartz (*Grothues et al.,* 1989). The ordinary index, n_o, and extraordinary index, n_e, begin to converge at shock pressures greater than 22 GPa and eventually coalesce at a shock pressure of about 35 GPa to indicate the formation of an isotropic phase (diaplectic glass). The refractivity measurements from the present study are for shocked powder samples.

In an attempt to better understand the structural changes in quartz with shock-loading and to develop a better measure of shock pressure, we use solid state ^{29}Si nuclear magnetic resonance (NMR) spectroscopy to examine quartz powders shocked to pressures of 7.5-22 GPa. This NMR technique involves monitoring radio frequency reemissions from ^{29}Si nuclei while the sample is in a very strong magnetic field (see *Kirkpatrick,* 1988). The resonance frequencies are sensitive to the local electronic (that is, structural) environment of silicon in the material, and thus provide information about the range of silicon environments resulting from, for instance, shock modification. One aim of this research is to provide a sensitive technique for monitoring shock pressure by calibrating a ^{29}Si NMR shock wave barometer.

NMR spectroscopy is useful in the analysis of shocked quartz because of the sensitivity of the NMR frequencies to the local (nearest and next-nearest neighbor) environment of the ^{29}Si. The resonance frequencies are reported as chemical shifts, which are ppm deviations of the resonance frequency from that of a standard (tetramethylsilane in this case). The ^{29}Si resonances of SiO_2 have been examined by *Smith and Blackwell* (1983), *Oestrike et al.* (1987), and *Murdoch et al.* (1985) among many others. These studies have determined the chemical shifts for the silicon sites in quartz, fused glass, coesite, and stishovite. *Yang et al.* (1986) examined shocked Coconino sandstone from Meteor Crater, Arizona, and identified the characteristic ^{29}Si resonances of quartz, coesite, and stishovite. Figure 2 provides several NMR spectra of these materials for reference. Stishovite has a very different chemical shift because its silicon is in sixfold coordination, and the electronic shielding is much different. The coesite structure has two tetrahedral silicon sites, one of which is has a chemical shift close to that of quartz.

Recently, *McHone et al.* (1989) used ^{29}Si MAS NMR to identify stishovite in a concentrated sample of a Cretaceous-Tertiary boundary clay from Raton, New Mexico. This result provides support for the hypothesis that the impact of an extraterrestrial object led to the Cretaceous-Tertiary extinction. *McHone and Nieman* (1988) also identified stishovite by ^{29}Si NMR in material from the Vredefort Dome of South Africa, which has been proposed to be an impact feature.

Application of the NMR technique to the shocked quartz powders is thus attractive due to the potential of NMR to distinguish among the amorphous, cryptocrystalline, and crystalline phases that may exist in bulk samples. Early studies of the shock-loading of quartz (*De Carli and Jamieson,* 1959; *De Carli and Milton,* 1965; *Deribas et al.,* 1965) recognized the presence of stable and metastable silica phases including fused and diaplectic glasses and the high-pressure polymorphs. NMR analysis provides a unique capability for identifying these phases based upon the local structural environment of silicon.

EXPERIMENTAL METHODS

The shock recovery experiments on quartz powder were performed using the Sandia "Bear" explosive loading fixtures to provide well-characterized shock states. The recovery fixtures allow samples to be shocked in a controlled and reproducible manner. Peak shock pressures and temperatures are determined by numerical simulations. *Graham and Webb* (1984, 1986) provide details of the experimental shock-loading experiment and the numerical simulations. Mean-bulk shock temperatures are determined in part by the initial packing density of the mineral powders. The mean peak shock pressures are 7.5, 16.5, and 22 GPa, and the mean bulk

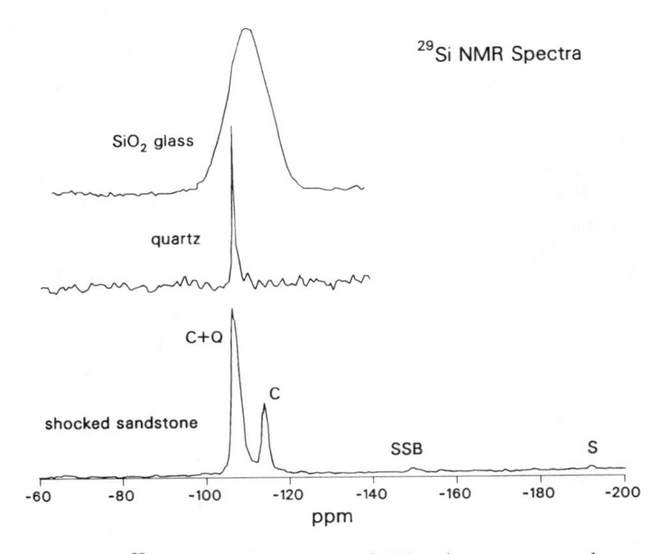

Fig. 2. ^{29}Si MAS NMR spectra of SiO_2 glass, quartz, and a naturally shocked sandstone from Meteor Crater, Arizona (modified from *Yang et al.,* 1986 and *Oestrike et al.,* 1987). The resonance frequencies are presented in ppm relative to the standard resonance of ^{29}Si in tetramethylsilane. Coesite (C), quartz (Q), and stishovite (S) are noted in the shocked sandstone spectrum. Spinning sidebands (SSB) are the result of the MAS technique.

TABLE 1. Schedule of shock recovery experiments.

Shot	Fixture	Explosive	Sample Compact Density (Mg/m^3)	Sample Compact Density (%)	Peak Pressure (GPa)	Estimated Mean Bulk Temperature (°C)
7B866	Momma Bear	Baratol	1.6	60	5-10	150-175
8B866	Momma Bear-A	Comp B	1.6	60	18-26	325-575
9B866	Momma Bear	Comp B	1.6	60	13-20	325-475
10B866	Momma Bear	Baratol	1.3	50	5-10	325-350
11B866	Momma Bear-A	Comp B	1.3	50	18-26	475-700
12B866	Momma Bear	Comp B	1.3	50	13-20	475-600

temperatures range from 150°C to 700°C. The range of pressures and temperatures presented in Table 1 are due to nonuniaxial loading; the variables have a radial dependence as determined by the numerical simulations. Samples to be analyzed were taken from a region with weak radial dependence (according to the simulations) and with pressures and temperatures close to the mean values. The use of two different packing densities for each shock pressure provided a low and a high temperature set of shocked samples.

The starting material for the shock-loading experiments was synthetic α-quartz powder (<325 mesh) obtained from Alfa Products (lot #88316). The material was briefly washed with deionized water and dried before insertion in the shock recovery sample holder; approximately 8 g of the quartz powder was required for each experiment. X-ray diffraction analysis of the starting material indicated pure, single phase α-quartz with no amorphous or glassy material present. Scanning electron microscopic (SEM) examination of the starting material determined a size distribution of approximately 30 to 90 μm. After removal from the recovery fixtures, the shocked material was lightly disaggregated with a mortar and pestle. Approximately 200-300 mg of powder sample are required for NMR analysis, and an additional 1-g sample was used for optical, SEM, and X-ray diffraction examinations.

Refractive index measurements were performed using a standard optical microscope with sodium light, standard immersion oils at 23°C, and the method of central illumination. The refractivity values are presented as a range of values because of variable orientation of the anisotropic grains. The restricted values of immersion oil standards limit the accuracy of the refractivity measurements to 0.004.

The X-ray diffraction patterns were obtained using Cu K_α X-ray radiation and a Phillips diffractometer with powders prepared on aluminum slides. Diffraction scans were performed from 10° to 70° 2θ at a scan rate of 2° 2θ/min and from 26° to 27° 2θ at a scan rate of 0.25° 2θ/min. The latter scan provides more detail of the shape of the quartz (101) peak.

The ^{29}Si NMR spectra were obtained at a frequency of 71.5 MHz (H_o = 8.45 T) under magic-angle spinning (MAS) conditions (see *Yang et al.,* 1986; *Kirkpatrick et al.,* 1986; *Oestrike et al.,* 1987; *Kirkpatrick,* 1988). MAS frequencies were typically 2.8 kHz. A pulse Fourier transform method was used to obtain the NMR spectra by first detecting signals in the time domain, then transforming the data to yield spectra in the frequency domain.

RESULTS

SEM examination of our shocked quartz powders indicates that substantial brittle disaggregation occurred due to shock-loading. Significant comminution, deformation, and fracturing is apparent. The mean grain size was reduced from 30-90 μm to 10-60 μm in all cases. Fine-grained (less than 1 μm) material is more abundant in the shocked quartz samples than in the unshocked sample. Small blebs (5-15 μm) of apparently amorphous material occur in the 22 GPa samples. Planar features (shock lamellae) were observed in only the low-temperature 22-GPa sample. One quartz grain in the sample contains three sets of parallel structural voids visible after hydrofluoric acid treatment. Presumably, the HF preferentially etches the diaplectic glass that fills the shock lamellae (*Bohor et al.,* 1984).

Within the precision of the measurement technique, there is no variation in the refractive indices of the shocked samples, and the measured refractive indices were all identical to the unshocked quartz sample (Fig. 1). Mean peak shock pressures for these samples are below that required to affect the refractive index.

The full X-ray diffraction scans for each of the shocked powders indicate the presence of only quartz. There is no evidence of a broad glass peak near 25° 2θ. The detailed scans of the most intense quartz peak (101) at 26.6° 2θ show a small but measurable broadening with increasing shock pressure for both the low- and high-temperature runs (Fig. 3). There is also a shift in peak position toward lower 2θ values (larger d-spacing) with increased shock pressure up to 16 GPa. There is no shift in the peak position for either of the 22 GPa samples.

The NMR spectra are characterized by a single peak with a maximum of about -108 ppm for four-coordinated ^{29}Si in quartz. However, this peak broadens substantially with increasing shock pressure (Fig. 4). The unshocked quartz has a full-width-at-half-height (FWHH) of 0.8 ppm, whereas that of the high-temperature 22-GPa sample is 3.8 ppm. The peaks of the high-temperature samples are somewhat more broadened than those of the low-temperature samples. There is no peak for stishovite, which would occur at about -192 ppm (*Smith and Blackwell,* 1983; *Yang et al.,* 1986), consistent with the relatively low pressures of our shock experiments. In addition, the NMR peak maxima tend to shift upfield (to lower frequencies or less negative chemical shifts) with increasing shock pressure. The only exception to this trend is the lower

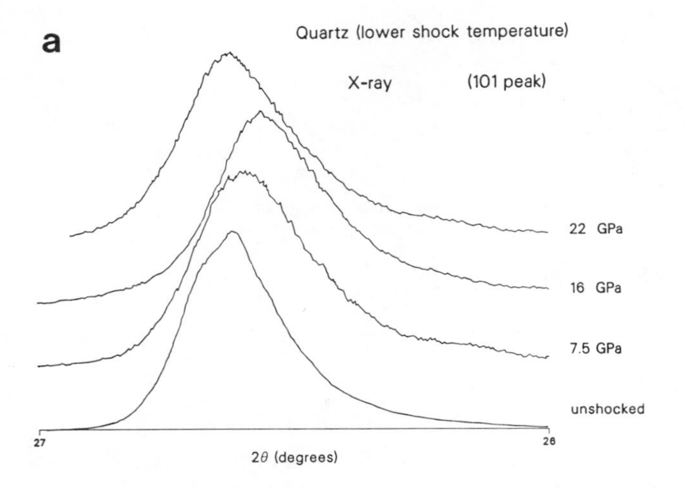

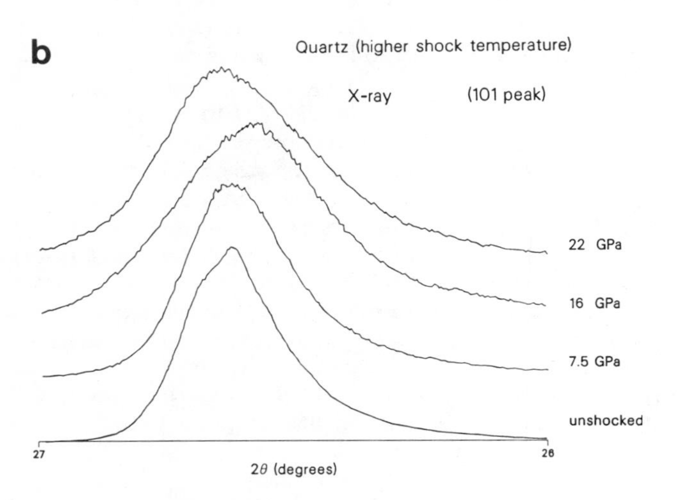

Fig. 3. Comparison of the X-ray diffraction patterns for the 26.6° 2θ peak (101) for the shocked quartz powders relative to that for the unshocked starting material: **(a)** low shock temperature samples and **(b)** high shock temperature samples. Diffraction patterns were obtained at a scan rate of 0.25° 2θ/min.

temperature 22-GPa sample, which is shifted upfield relative to the unshocked sample but not as far as the 16-GPa sample. This behavior is similar to the shifts in 2θ observed in the X-ray diffraction patterns for the same 22-GPa shocked samples.

DISCUSSION

X-ray Diffraction

Broadening of X-ray diffraction lines usually is associated with residual strain, disorder of the crystal, and decrease of crystal domain size (*Hörz and Quaide,* 1973; *Schneider et al.,* 1984). Shifts in diffraction peaks by lattice strain effects usually represent uniform strain whereas broadening of the peaks indicates a nonuniform strain (*Cullity,* 1978). We are uncertain how to interpret the X-ray diffraction data of the 22-GPa samples (see Fig. 3). It is possible that these samples experienced high enough temperatures (greater than 570°C) during (and after) shock-loading that an inversion to β-quartz occurred that later reverted to the α-quartz structure upon cooling.

The unshocked starting material possesses a fairly broad (101) X-ray peak indicating some inherent strain prior to shock-loading (F. Hörz, personal communication, 1989). This strain may have been generated during the initial synthesis and grinding of the material. However, all of the X-ray and NMR results are normalized relative to the starting material. A comparison of the sensitivity of the two techniques in quantifying the degree of disorder can therefore be made regardless of the initial state of the material. We hope to study this preshock effect in future research by examining a variety of synthetic, natural, and thermally annealed quartz samples.

Nuclear Magnetic Resonance

The broadening of the NMR peak of the shocked quartz may be due to several different causes, but must be due to an increase in the range of mean Si-O-Si bond angle per silicon tetrahedron in the samples. Many workers, including *Smith and Blackwell* (1983), *Dupree and Pettifer* (1984), and *Oestrike et al.* (1987), have described the excellent relationship between this and the ^{29}Si chemical shift for SiO_2 phases.

One possible cause for this broadening is extensive disorder and residual strain introduced into the quartz lattice by shock-loading. *Dupree and Pettifer* (1984) and *Devine et al.* (1987) have suggested that such NMR peak broadening ultimately results from the redistribution of Si-O-Si bond angles that forms in response to stress. Although static pressures were used in the previous studies of amorphous silica phases, a similar argument can be made for the dynamic stressing of our study.

A second possible cause for the observed NMR peak broadening is formation of an amorphous phase. Vitreous silica has a ^{29}Si resonance maximum at -111 to -112 ppm, and a FWHH of about 20 ppm (*Dupree and Pettifer,* 1984; *Oestrike et al.,* 1987) (Fig. 2). The extent of such a contribution to the NMR spectra in Fig. 4 is uncertain. The asymmetrical shape of the peaks and the intensity in the -109 to -118 ppm range in the spectra of the high-pressure shock samples is consistent with some contribution from a silica glass phase. The intensity of this broad component increases with increasing shock pressure. Although none of the X-ray diffraction scans provide evidence of glass, our SEM observations confirm the existence of a glass phase in the shocked samples. NMR spectroscopy probably cannot discern between fused and diaplectic glasses of similar compositions or silicon coordinations.

A third possible cause of the peak broadening could be the presence of coesite, one of the high-pressure polymorphs of silica (*Stöffler,* 1971, 1984). *Smith and Blackwell* (1983) and *Yang et al.* (1986) observed two ^{29}Si NMR peaks of -108.1 ppm and -113.9 ppm for this phase. Although no coesite is detected by X-ray diffraction, if coesite were formed it may have been in an amount below the detection level. It is possible that some coesite contributes to the peak broadening observed in Fig. 4.

An example of a possible deconvolution of the broadened NMR peak is presented in Fig. 5 where the relative contributions of the disordered quartz and silica glass may be graphically examined. An unidentified peak of low intensity

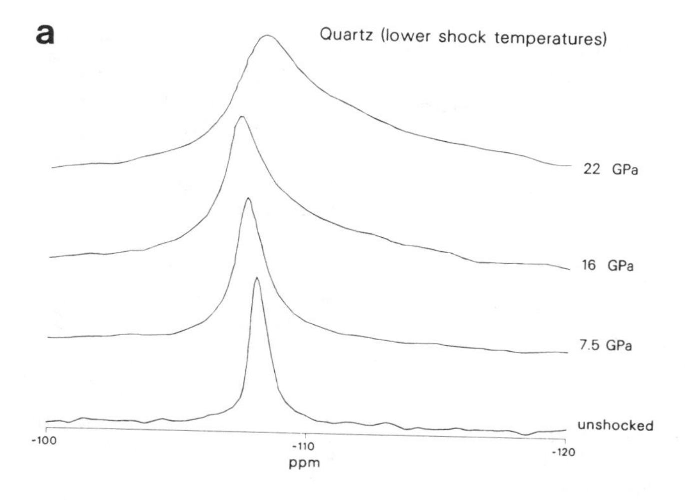

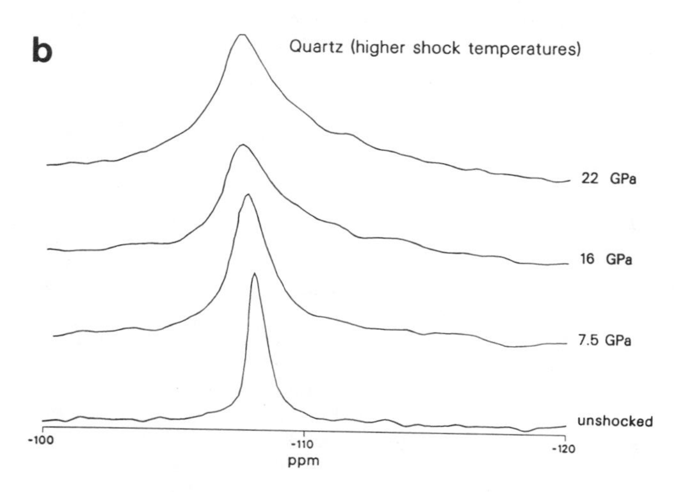

Fig. 4. ^{29}Si MAS NMR spectra for shocked quartz powders relative to that for the unshocked starting material: **(a)** low shock temperature samples and **(b)** high shock temperature samples.

occurs upfield as a result of the optimum fitting of Gaussian curves to the NMR spectrum. In this model deconvolution, the optimum fit attributes about 60% of the ^{29}Si NMR signal to the glass phase, whereas the disordered quartz accounts for 25% of the observed NMR peak. Because of the lack of singularities for constraining the model, the resulting curves do not necessarily represent a unique solution.

Relative Peak Changes and Shock Wave Barometry

Although the origin of the NMR peak broadening in not entirely clear, this broadening appears to be a useful empirical measure of peak shock pressures. Figure 6 shows that the NMR peak broadening increases much more rapidly with increasing shock pressure than X-ray diffraction peak broadening. The NMR peaks of the 22-GPa samples are almost five times broader than that of the unshocked quartz sample. The broadening of the NMR peaks shows no strong or consistent correlation with shock temperature.

The X-ray peak broadening increases at most by a factor of 1.7 relative to the unshocked samples and is much less than for the NMR peak widths (Fig. 6). This result suggests that NMR is a more sensitive measure of shock pressure. However, the synthetic quartz we used already had measurable line broadening in its unshocked state. For unstrained starting material, the X-ray line broadening may exhibit a stronger and more consistent trend than what we observed, but the NMR trend may also be enhanced.

The variation of the NMR peak widths with temperature is much smaller than that with pressure. Thus, the NMR peak width measurement can provide a shock barometer over a wide range of unknown shock temperatures. It may therefore be particularly useful in studies of naturally shocked quartz where the precise initial conditions and loading path (which dictate the shock temperature) are unknown.

One of the limitations in assessing the suitability of the NMR shock wave barometer is the uncertainty in the amount of glass and coesite generated during the experimental shock event. SEM examination of the shocked samples suggests that both fused and diaplectic glass are present. The NMR results indicate that these glasses contain only four-coordinated silicon and contribute to a broad peak with a mean chemical shift of approximately -110 ppm. Any higher coordinated silicon, such as a six-coordinated silicon in a stishovite-like diaplectic glass (*Chaabildas and Miller,* 1985; *Boslough,* 1988), would yield a resonance in the -190 to -200 ppm range and is not observed. Coesite or a disordered coesite-like material with mean Si-O-Si bond angles larger than quartz, and then more negative chemical shift, might also contribute to the observed peaks. Our NMR analysis of the low-temperature 22-GPa sample after treatment with hydrofluoric acid indicated an almost twofold decrease in FWHH relative to the untreated 22-GPa sample. The ^{29}Si NMR spectrum exhibits some spectral intensity in the -114 ppm region where coesite would have a resonance (*Smith and Blackwell,* 1983; *Yang et al.,* 1986). Future research will require improved NMR spectra with better signal-to-noise ratios and an analysis of standard mixtures of the quartz, coesite, and glass.

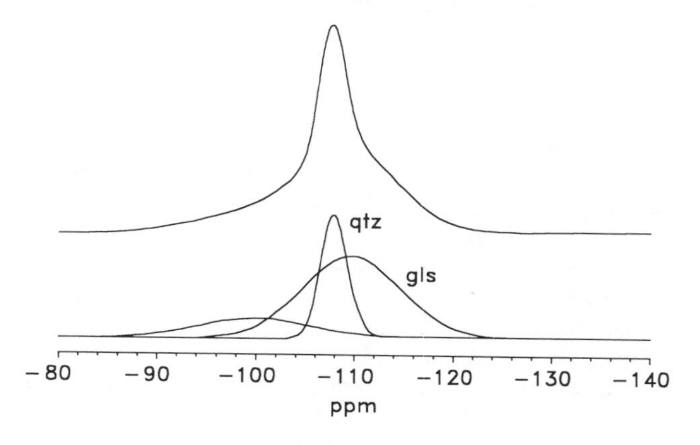

Fig. 5. Optimum curve fits for the ^{29}Si MAS NMR spectrum for the high-temperature 22-GPa shocked quartz sample. The upper curve is a smoothed peak derived from the raw data. The lower curves provide the results of a possible deconvolution assuming 100% Gaussian peak shape for quartz, glass, and a third unidentified peak.

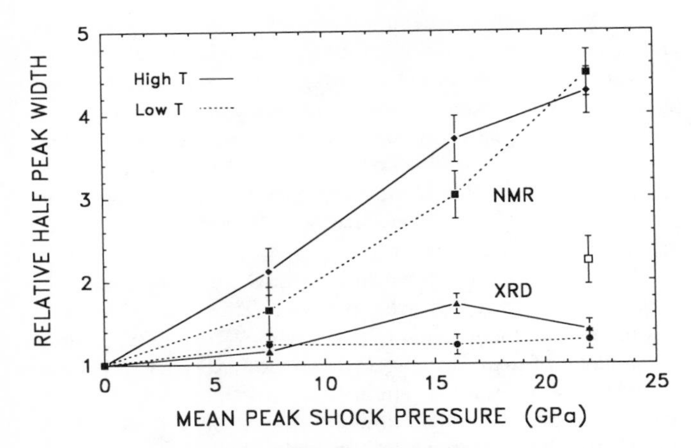

Fig. 6. Comparison of relative half-peak widths (full-width-at-half-height) obtained for X-ray diffraction (XRD) patterns and MAS NMR spectra of the shocked quartz powders as functions of the mean peak shock pressure. The open square indicates the NMR result for a low-temperature sample that was improperly selected.

Applications of the NMR Shock Wave Barometer

In applying an NMR shock barometer to terrestrial impact events, it is important to recognize the way the impact occurs and the amounts of material that are shocked by the impact. The amount of terrestrial material shocked to a given pressure by a meteoroid impact is a complex function of projectile mass, velocity, density, angle of impact, and other parameters. The highest shock pressures are in the small central region of the impact, and the material associated with the lower shock pressures is radially distant from the impact point, but represents a larger volume. The significant radial decay of shock pressure from the impact point therefore produces large volumes of material subjected to pressures less than 22 GPa (for example, *Stöffler,* 1984). An NMR shock barometer would be useful in determining the low shock pressures associated with this material. This is a significant advantage because refractive index-based shock barometers are limited to pressures greater than 22 GPa.

Another potential application of shock barometry using the NMR spectroscopic approach is the accurate scaling of stress decay associated with nuclear testing. The technique is amenable to postdetonation analysis of shock pressure exposure of rock samples. This method can be used in conjunction with real-time stress probes and sensors that collect data during the detonation, or for postdetonation analysis where real-time evaluation has failed or is lacking. Consequently, data from nuclear test sites can be used to calibrate natural shock events. This is similar to the approach of *Vizgirda et al.* (1980), who examined the hyperfine EPR splitting of Mn^{2+} in calcite from shocked coral samples from Cactus crater at the Pacific test site. These results, however, do not conclusively define a reproducible trend between hyperfine peak splitting and shock pressure. We believe an NMR shock barometer would be more useful, because of the larger observed spectral changes and the widespread occurrence of quartz and other silicate minerals.

Future research on shocked material using MAS NMR spectroscopy should emphasize the structural differences associated with ^{29}Si in diaplectic and fused glasses. Mineral samples should also be subjected to greater shock pressures in order to examine any coordination changes that may result. Although the present study has emphasized ^{29}Si NMR spectroscopy, there are other NMR-active nuclei (^{23}Na, ^{27}Al, and ^{39}K) that could provide detailed information on the disordered state of other shocked minerals that may be present on the Moon and terrestrial planets. The maskelynitization of feldspar minerals could be examined in this fashion.

CONCLUSIONS

This study demonstrates a significant positive correlation of ^{29}Si MAS NMR peak width with mean peak shock pressure for samples of synthetic quartz shocked up to 22 GPa and 700°C. The NMR technique is more sensitive to shock pressure than X-ray diffraction methods, with the (101) quartz X-ray peak broadening much less than the NMR peak for the synthetic quartz samples. The quartz refractive indices do not change in this low-pressure range, in agreement with the results of *Grothues et al.* (1989). The results of this study suggest the potential usefulness of NMR spectroscopy in providing a shock barometer that is sensitive and accurate for shock pressures below 22 GPa. No other observable parameter appears to change as much or as consistently in this pressure range.

Acknowledgments. We wish to acknowledge the comments and careful reviews provided by F. Hörz and V. L. Sharpton. We also thank B. Montez for his exacting method in obtaining the MAS NMR spectra and C. Daniel, K. Elsner, and M. Anderson for their technical support with the shock recovery experiments. Additional technical support and scientific comments were provided by W. Casey, M. Dvorack, and B. Morosin. This research was performed at Sandia National Laboratories and was supported by the U.S. Department of Energy under contract DE-AC04-76DP00789.

REFERENCES

Alexopoulos J. S., Grieve R. A. F., and Robertson P. B. (1988) Microscopic lamellar deformation features in quartz: Discriminative characteristics of shock-generated varieties. *Geology, 16,* 796-799.

Alvarez L. W., Alvarez W., Asaro F., and Michel H. V. (1980) Extraterrestrial cause for the Cretaceous-Tertiary extinction. *Science, 208,* 1095-1108.

Ashworth J. R. and Schneider H. (1985) Deformation and transformation in experimentally shock-loaded quartz. *Phys. Chem. Minerals, 11,* 241-249.

Bohor B. F., Foord E. E., Modreski P. J., and Triplehorn D. M. (1984) Mineralogic evidence for an impact event at the Cretaceous-Tertiary boundary. *Science, 224,* 867-869.

Boslough M. B. (1988) Postshock temperatures in silica. *J. Geophys. Res., 93,* 6477-6484.

Carter N. L., Officer C. B., Chesner C. A., and Rose W. I. (1986) Dynamic deformation of volcanic ejecta from the Toba caldera: Possible relevance to Cretaceous/Tertiary boundary phenomena. *Geology, 14,* 380-383.

Chaabildas L. C. and Miller J. M. (1985) Release-adiabat measurements in crystalline quartz. *Sandia Rep. SAND85-1092.* Sandia National Laboratories, New Mexico. 26 pp.

Chao E. C. T. (1968) Pressure and temperature histories of impact metamorphosed rocks—based on petrographic observations. In *Shock Metamorphism of Natural Materials* (B. M. French and N. M. Short, eds.), pp. 135-158. Mono, Baltimore.

Cullity B. D. (1978) *Elements of X-Ray Diffraction.* Addison-Wesley, Reading, Massachusetts. 555 pp.

De Carli P. S. and Jamieson J. C. (1959) Formation of an amorphous form of quartz under shock conditions. *J. Chem. Phys., 31,* 1675-1676.

De Carli P. S. and Milton D. J. (1965) Stishovite: Synthesis by shock wave. *Science, 147,* 144-145.

Deribas A. A., Dobretsov N. L., Kudinov V. M., and Zyuzin N. I. (1965) Shock compression of SiO_2 powders. *Dok. Akad. Nauk SSSR, 168,* 127-130.

Devine R. A. B., Dupree R., Farnan I., and Capponi J. J. (1987) Pressure-induced bond-angle variation in amorphous SiO_2. *Phys. Rev. B, 35,* 2560-2562.

Dupree E. and Pettifer R. F. (1984) Determination of Si-O-Si bond angle distribution in vitreous silica by magic angle spinning NMR. *Nature, 308,* 523-525.

Graham R. A. and Webb D. M. (1984) Fixtures for controlled explosive loading and preservation of powder samples. In *Shock Waves in Condensed Matter—1983* (J. R. Asay, R. A. Graham, and G. K. Straub, eds.), pp. 211-214. North Holland, New York.

Graham R. A. and Webb D. M. (1986) Shock-induced temperature distributions in powder compact recovery fixtures. In *Shock Waves in Condensed Matter—1985* (Y. M. Gupta, ed.), pp. 831-836. Plenum, New York.

Grothues J., Hornemann U., and Stöffler D. (1989) Mineralogical shock wave barometry: (I) Calibration of refractive index data of experimentally shocked α-quartz (abstract). In *Lunar and Planetary Science XX,* pp. 365-366. Lunar and Planetary Institute, Houston.

Hörz F. (1968) Statistical measurements of deformation structures and refractive indices in experimentally shock loaded quartz. In *Shock Metamorphism of Natural Materials* (B. M. French and N. M. Short, eds.), pp. 243-253. Mono, Baltimore.

Hörz F. and Quaide W. L. (1973) Debye-Scherrer investigations of experimentally shocked silicates. *The Moon, 6,* 45-82. Izett G. A. and Bohor B. F. (1987) Comment and reply on "Dynamic deformation of volcanic ejecta from the Toba caldera: Possible relevance to Cretaceous/Tertiary boundary phenomena." *Geology, 15,* 90-91.

Kieffer S. W. (1971) Shock metamorphism of the Coconino Sandstone at Meteor Crater, Arizona. *J. Geophys. Res., 76,* 5449-5473.

Kieffer S. W., Phakey P. P., and Christie J. M. (1976) Shock processes in porous quartzite: Transmission electron microscope observations and theory. *Contrib. Mineral. Petrol., 59,* 41-93.

Kirkpatrick R. J. (1988) MAS NMR spectroscopy of minerals and glasses. In *Spectroscopic Methods in Mineralogy and Geology, Reviews in Mineralogy, Vol. 18* (F. C. Hawthorne, ed.), pp. 341-403. Mineralogical Society of America, Washington, DC.

Kirkpatrick R. J., Dunn T., Schramm S., Smith K. A., Oestrike R., and Turner G. (1986) Magic-angle sample-spinning nuclear magnetic resonance spectroscopy of silicate glasses: A review. In *Structure and Bonding in Noncrystalline Solids* (G. E. Walrafen and A. G. Revesz, eds.), pp. 303-327. Plenum, New York.

McHone J. F. and Nieman R. A. (1988) Vredefort stishovite confirmed using solid-state silicon-29 nuclear magnetic resonance. *Meteoritics, 23,* 289.

McHone J. F., Nieman R. A., Lewis C. F., and Yates A. M. (1989) Stishovite at the Cretaceous-Tertiary boundary, Raton, New Mexico. *Science, 243,* 1182-1184.

Murdoch J. B., Stebbins J. F., and Carmichael I. S. E. (1985) High-resolution ^{29}Si NMR study of silicate and aluminosilicate glasses: The effect of network-modifying cations. *Am. Mineral., 70,* 332-343.

Oestrike R., Yang W. H., Kirkpatrick R. J., Hervig R. L., Navrotsky A., and Montez B. (1987) High-resolution ^{23}Na, ^{27}Al, and ^{29}Si NMR spectroscopy of framework aluminosilicate glasses. *Geochim. Cosmochim. Acta, 51,* 2199-2209.

Schneider H., Vasudevan R., and Hornemann U. (1984) Deformation of experimentally shock-loaded quartz powders: X-ray line broadening studies. *Phys. Chem. Minerals, 10,* 142-147.

Sharpton V. L. and Schuraytz B. C. (1989) On reported occurrences of shock-deformed clasts in the volcanic ejecta from Toba caldera, Sumatra (abstract). In *Lunar and Planetary Science XX,* pp. 992-993. Lunar and Planetary Institute, Houston.

Smith J. V. and Blackwell C. S. (1983) Nuclear magnetic resonance of silica polymorphs. *Nature, 303,* 223-225.

Stöffler D. (1971) Coesite and stishovite in shocked crystalline rocks. *J. Geophys. Res., 76,* 5474-5488.

Stöffler D. (1984) Glasses formed by hypervelocity impact. *J. Non-Cryst. Solids, 67,* 465-502.

Vizgirda J., Ahrens T. J., and Tsay F. D. (1980) Shock-induced effects in calcite from Cactus Crater. *Geochim. Cosmochim. Acta, 44,* 1059-1069.

Yang W. H., Kirkpatrick R. J., Vergo N., McHone J., Emilsson T. I., and Oldfield E. (1986) Detection of high-pressure silica polymorphs in whole-rock samples from a Meteor Crater, Arizona, impact sample using solid-state silicon-29 nuclear magnetic resonance spectroscopy. *Meteoritics, 21,* 117-124.

Geology of Mars

Proceedings of the 20th Lunar and Planetary Science Conference, pp. 461-471
Lunar and Planetary Institute, Houston, 1990

ISM Observations of Mars and Phobos: First Results

J.-P. Bibring[1], M. Combes[2], Y. Langevin[1], C. Cara[2], P. Drossart[2], T. Encrenaz[2], S. Erard[1], O. Forni[1], B. Gondet[1], L. Ksanfomaliti[3], E. Lellouch[2], P. Masson[4], V. Moroz[3], F. Rocard[1], J. Rosenqvist[2], C. Sotin[4], and A. Soufflot[1]

The infrared spectro-imager ISM, flown onboard the Soviet Phobos spacecraft, acquired ≈40,000 spectra of Mars and ≈600 spectra of Phobos, from February 5 to March 27, 1989. These infrared images are the first ever taken from space of a planet and a small solar system body. The pixel size varies from 5 to 30 km on Mars, and is ≈0.7 km on Phobos. For each resolved pixel, the infrared spectrum (0.76-3.16 μm) gives information on the martian atmosphere (CO, H_2O, CO_2), and the related measurements of altitudes and mineralogical characteristics of the surface of Mars and Phobos. A strong hydration feature is observed on Mars, and not on Phobos. Large variations in albedo and hydration are observed on a scale of a few tens of kilometers on Mars.

INTRODUCTION

Presented here is a first analysis of the data provided by the infrared imaging spectrometer ISM on the Soviet Phobos spacecraft. The major scientific objectives of ISM were to provide a mineralogical mapping of the low latitude zone of Mars in order to better understand the history and evolution of Mars. Significant data on the time/space variability of the martian atmosphere (pressure, dust, minor components) were also expected, although the low spectroscopic resolution of the instrument was not favorable for such an analysis. For Phobos, the ISM observing goals were related to the mineralogical composition of its surface, including observations at very high spatial resolution (~20 cm), with the aim of assessing the mean composition and the level of heterogeneity of this small body.

THE INSTRUMENT

ISM is an imaging spectrometer in the near infrared spectral range. Imaging spectrometry for remote sensing represents an important improvement with respect to broad spectral band mappers like SPOT or Landsat-TM by allowing efficient spectroscopic diagnostics unavailable to multicolor mappers. ISM is the first imaging spectrometer flown onboard a spacecraft for planetary observations (*Bibring et al.,* 1989). NIMS will be flown on Galileo and, hopefully, OMEGA/VIMS on Mars 94, CRAF, and Cassini. Similar instruments are expected to be implemented on Earth-orbiting platforms, following airborne instruments (AIS since 1983 and recently AVIRIS).

The spectroscopic spectral range (0.76 to 3.16 μm) is divided into 128 channels by means of a grating spectrometer used in the first (1.65-3.16 μm) and second (0.76-1.54 μm) orders (Fig. 1). The two spectral orders are separately focused onto two 64-pixel PbS arrays by a dichroic beamsplitter. The resulting spectroscopic resolution is 0.025 and 0.0125 μm per pixel in the first and second order respectively. A 256 Hz chopper into the entrance slit (focal plane) and an onboard calibration source allow a very efficient rejection of the instrumental background that is relatively low in this nonthermal spectral range. The imaging capability is provided through a small telescope 2.5 cm in diameter, by means of a scanning mirror. The instrument field of view (IFOV) is 12 arcmin, including geometrical aberrations of the optics. The scanning mirror can be moved in a ±20° range with respect to the optical axis, by 1 arcmin increments. The second direction of scanning for imaging is provided by the displacement of the spacecraft itself. The detectors are cooled down to a temperature close to 200 K, by means of a passive radiator on top of the instrument. The temperature is monitored with an accuracy better than 0.1 K.

The detection limits of ISM are actually very good due to the high performances of the PbS arrays manufactured by SAT (France) and also to the efficient overall design of the instrument. The random signal-to-noise ratio, measured during the pre-launch calibration sequences, is $\sim 10^3$ per spectrum for a 1 sec observing time, at 2.3 μm and for a martian albedo of 25%. This value is verified also for inflight data (onboard calibration source, empty sky, atmospheric data at limb). At this high sensitivity level, instrumental systematic effects show up easily above the noise level, and then become the actual limits of the instrumental performances. A detailed analysis of the data shows that the noise of the detector and of the electronics is significant for exposure times shorter than 1/4 sec per spectrum. The digitalization noise is dominant for larger exposure times. The small spatial scale variations of Mars reflectivity are usually larger than noise, even for "uniform" regions.

The detector arrays are formed of two staggered 32 linear arrays separated by 23 arcmin, while the IFOV is 12 arcmin. In this configuration, the "odd" and "even" pixels of each array are not imaging exactly the same area of Mars or Phobos at the same time. The odd and even components of the spectra give half the spectral resolution if treated separately. In

---

[1]Institut d'Astrophysique Spatiale, 91406 Orsay, France
[2]Departement de Recherches Spatiales, Observatoire de Paris Meudon, 92195,France
[3]Institute for Space Research, Moscow, USSR
[4]Laboratoire de Geodynamique Interne, 91406 Orsay, France

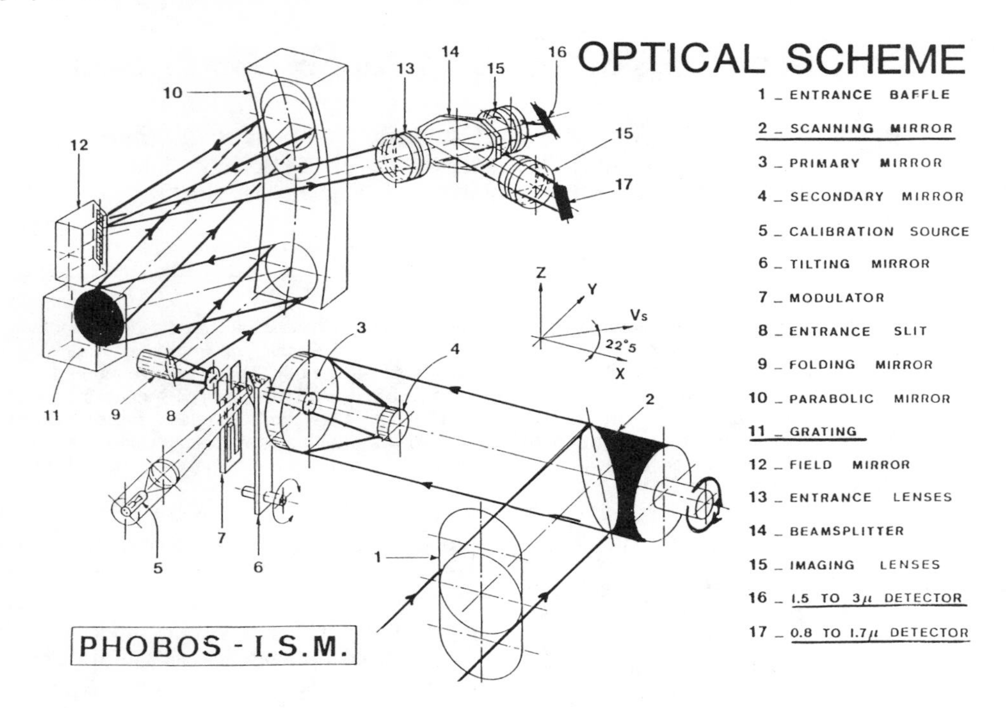

Fig. 1. Optical design of the ISM spectrometer.

principle, the complete spectrum from each surface element can be reconstructed by a numerical treatment. This has been performed for most atmospheric studies.

The Dataset for Mars and Phobos

The data consist of "cubes" (x,y,λ), constituted as follows: Each resolved area on Mars, or "spatial pixel," is defined by its position (x,y) on the martian surface and its size, which corresponds to the spatial resolution. For each spatial pixel, ISM gives an infrared spectrum in the form of the 128 data extracted from the PbS detectors. One cube corresponds to one observational sequence, containing thousands of spectra; the width of the region mapped depends on the range we chose for the scanning mirror (1 to 25 pixel large), while its length is a function of the duration of the sequence, typically a few tens of minutes.

ISM acquired spectra from February 5 to March 26, 1989. Up to February 12, the spacecraft was in an elliptical orbit, with an orbital period of three days and a low pericenter altitude (880 km). On February 8 and 11, ISM was turned on, when the spacecraft was three-axis stabilized and its altitude lower than 2000 km. This led to two high-resolution tracks of $20 \times 1650\,km^2$; each track is constituted of 8×300 pixels, with a pixel size of the order of $5 \times 5\,km^2$. The integration time is 0.125 sec. The first track is centered at 0.5° N in latitude, and extends from approximately 95° to 125° in longitude. It thus crosses the Pavonis Mons volcano, and extends within the Tharsis region. The second track is slightly further north (2.5°) and extends farther toward the west, from about 110° to 140° in longitude. It crosses two additional volcanoes, Ulysses and Biblis Paterae.

From February 21 to March 26, we obtained nine data cubes from a circular orbit at an altitude of 6300 km, the spacecraft being three-axis stabilized. Each cube maps a region of about $400 \times 3000\,km^2$, made of 25×125 pixels with a size of $20 \times 30\,km^2$ each. The integration time is 0.5 sec. Seven of these regions are located within ±20° in latitude, and between 20° and 180° in longitude, the region containing the large volcanoes of the Tharsis Montes, as well as most of the Valles Marineris formations (Plate 1). Altogether, ISM data cover more than 25% of this area. The additional regions mapped are located on the eastern hemisphere, within the old cratered terrains of Arabia Terra for one, Isidis Planitia, and Syrtis Major for the other (Plate 2). Finally, two additional sequences were obtained, one at high resolution and the other from the circular orbit, when the spacecraft had a spinning stabilization mode ("gyroscopic" mode), with no good precision on the position of the track onto the martian surface.

The total number of spectra obtained on Mars is close to 40,000 (6000 at high resolution and 30,000 at lower resolution, with a three-axis stabilization, 4000 during the spinning mode). Although the mission stopped on March 27, ISM has sampled most major geological formations on Mars, with the noticeable exception of the polar caps, which could not be observed from an equatorial orbit.

On March 25 the spacecraft was on an orbit close to that of Phobos, remaining at a distance of about 200 km from the small body. During two successive sequences, the spacecraft was slightly rotated in order to point toward Phobos, with the TV camera and ISM turned on. The resolution achieved by ISM was ~0.7 km per pixel. During the first observation of Phobos, we chose not to scan the mirror of ISM. We then obtained a track 200 pixels long, which crosses Phobos close to its equator, over about 20 km. Phobos was then visible against the background of the martian disk. During the second observation, on a dark background, we chose to map a region of Phobos 25 pixels in width and 23 pixels in length (20×20 km^2); the Phobos cube (x,y,λ) consists of ~400 spectra, representing about one third of its lighted hemisphere, with a spatial resolution better than 1 km. This observation constitutes the first high resolution infrared image of a small body of the solar system.

DATA REDUCTION

The infrared spectrum of Mars can be assumed to be due to reflected sunlight at wavelengths less than 3.2 μm. Groundbased observations between 1 and 5 μm show that at 3.2 μm the thermal component could add a small contribution to the total flux; but even for the highest possible surface temperatures, this contribution is expected to be less than 10% at 3.2 μm, and becomes completely negligible below 2.7 μm. All spectra were therefore divided by the solar flux at Mars, so as to infer the combined transmission coefficient of the atmosphere (for Mars) and reflectivity of the surface. The sun-Mars distance increased from 1.53 to 1.58 A.U. during observations, due to the eccentricity of the orbit of Mars.

The spectra must be corrected for the transfer function of the instrument, which combines its optical transmission, the detection limits of the PbS pixels, and the amplifier characteristics. The preflight absolute calibration of the instrument was performed using a 400 K black body, a GLOBAR source, and a Xenon lamp. The sensitivity of PbS detectors depends on their temperature. Average inflight detector temperature was slightly lower than expected and ground calibrated (by ~5 K). At the present stage, the absolute accuracy on the transfer function can be evaluated as better than 5% in the first order and better than 10% in the second order. Flat-fielding (the correction in efficiency between adjacent pixels) can be considered accurate to a few percent across the whole spectral range. The linearity of the response of the detectors is better than 1%. Nonlinear amplification and analog-to-digital conversion only become significant at very high digitization levels (above 3900 out of our 4095 range). As a consequence, we used absolute spectra mainly to identify major spectral features (larger than 10%). Most results presented in this paper are derived from ratioed spectra. In general, we treated separately the "even" and "odd" parts of the spectra so as to obtain the nominal spatial resolution.

To discriminate between atmospheric and surface signatures, we checked the correlation with the airmass factor. For analyzing atmospheric features, spectra obtained during the same observing session were divided by one of them (for example, in the Olympus area, the spectrum of the summit). In such a case, instrumental effects like pixel-to-pixel gain variations are eliminated. The remaining features are variations in band depths between the analyzed points. They are weaker than raw features but more safely identified. For studying the weakest spectral signatures, a three-point convolution was performed on the complete spectra. The signal-to-noise ratio can be increased by averaging spectra in a uniform area, at the cost of a loss in spatial resolution.

For surface features, we mapped weighted ratios of groups of spectral channels selected so as to be representative of the variations of major and minor spectral features with an accuracy limited by the signal-to-noise ratio.

Finally, the observations of Phobos make possible an inflight evaluation of the stray light: Martian atmospheric features are observed when Phobos is viewed against the martian disk. Taking into account the larger albedo of Mars, the contribution of stray light is smaller than 2% for a uniformly bright field. In this first analysis, we neglect this contribution for martian observations. However, it could play a role for small features observed near dark/light boundaries.

FIRST RESULTS ON THE ATMOSPHERE OF MARS

Between 0.8 and 3.2 μm, the spectrum of Mars exhibits absorption signatures of three gaseous atmospheric constituents: carbon dioxide, water vapor, and carbon monoxide. Calculations were made with a band model assuming a 1/s distribution of the lines, s being the intensity of an individual line (*Goody,* 1964). The spectroscopic data (line positions, intensities, and energy levels) were taken from the GEISA atmospheric data bank (*Husson,* 1986; *Rothman,* 1986) with CO_2-broadened collisional widths. Band model calculations were performed in the effective pressure approximation, where the atmosphere is equivalent to a single layer with an effective pressure $P_e = P_s/2$, P_s being the surface pressure, and an effective temperature $T_e = T(P_e)$, using the formulation of *Wallace et al.* (1974). The parameters of the model are (1) the surface pressure, (2) the CO and H_2O integrated column densities, (3) the atmospheric temperature near the surface T_s, and (4) the airmass factor. T_e was derived from T_s and from the thermal profile, defined by the following assumptions (thermal profile "A"): the atmospheric tempera-

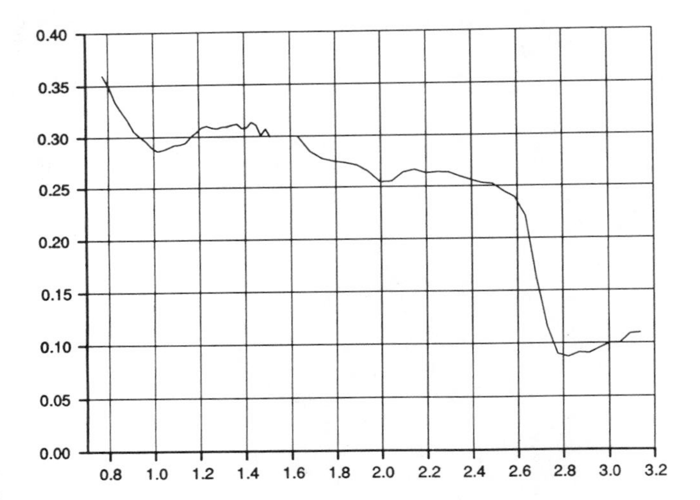

Fig. 2. Olympus Mons ISM spectrum; a three-point triangular convolution has been applied to the calibrated spectrum.

ture at a given altitude is spatially constant, and the lapse rate is -2.0 K/km (*Seiff,* 1982). In order to check the validity of these assumptions, the results discussed below have been compared with those obtained assuming a constant atmospheric temperature T_s near the surface, whatever its altitude (thermal profile "B"). The band model was checked against a line-by-line computation in two different cases, in the CO_2 2.0 μm and in the CO 2.3 μm bands. Agreement was found within 1% for CO_2 and 20% for CO.

Figure 2 shows a typical ISM spectrum of Mars, recorded near the summit of Olympus Mons. Two strong CO_2 bands are prominent at 2.0 and 2.7 μm respectively. Below 1.1 μm the spectrum of Mars is basically free from atmospheric signatures at the resolution of ISM. Above 1.1 μm, several weak atmospheric bands appear at other wavelengths, but they are more difficult to separate from possible mineralogic contribution. In order to discriminate between the atmospheric and mineralogic features, we have studied their variations as a function of the airmass factor during the observing sequence of February 5 (spinning mode). The bands showing the same strength dependence as the airmass variations (and, obviously, the 2.0 and 2.7 μm CO_2 bands) are most likely atmospheric signatures, whereas the bands showing no strong correlation with the airmass can be considered to be mineralogic signatures. A confirmation of the atmospheric origin of a band is the correlation with the local topography.

Minor constituents (CO and H_2O) were searched for in ratioed spectra obtained at different altitudes, with the highest possible difference in atmospheric path. A selection of 10 averaged spectra (between 10 and 30 individual spectra) was chosen on Olympus Mons (data from March 13), at various altitudes, and the ratios of pairs of spectra at different altitudes were studied.

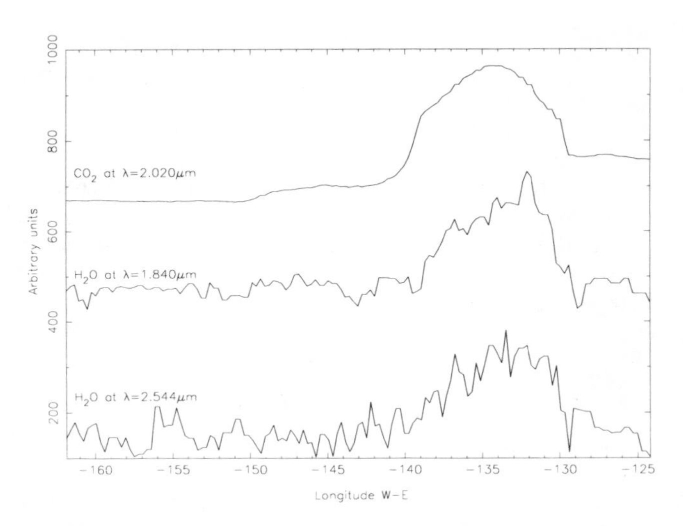

Fig. 4. Correlation between the H_2O 2.544 μm band (bottom), the H_2O 1.84 μm band (middle), and the CO_2 2.02 μm band (top) through the Olympus Mons region. Each curve represents the longitudinal profile of a ratio of the band intensity to the nearby continuum (without any atmospheric absorption), scaled arbitrarily, and vertically shifted.

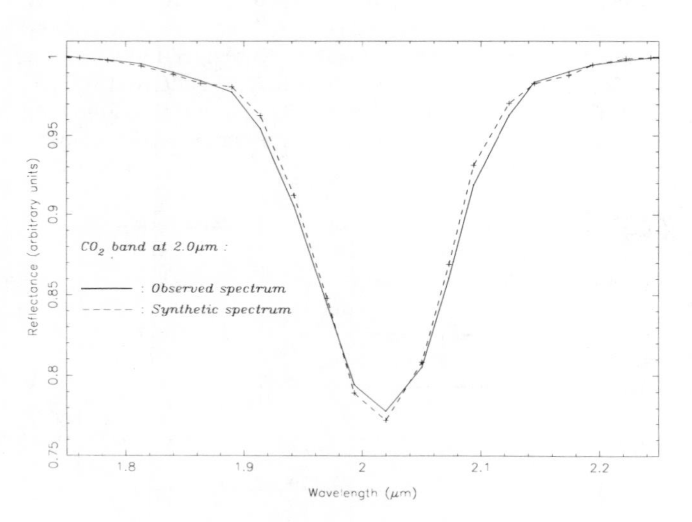

Fig. 3. CO_2 band at 2.0 μm: observed spectrum is a spectrum of the plateau of Olympus Mons divided by the spectrum observed at the summit; synthetic spectrum is calculated with T_e = 230 K, an airmass factor μm = 2.13 and a pressure of 6.1 mbar corresponding to the plateau of Olympus Mons, divided by another synthetic spectrum with T_e = 185 K, an airmass factor μm = 2.36, and a pressure of 0.72 mbar corresponding to the summit.

CO_2

The strongest CO_2 signature is the $(\nu_1 + \nu_3, 2\nu_2 + \nu_3)$ centered at 2.7-2.8 μm. According to synthetic models, a 7 mbar CO_2 pressure would cause a 97% absorption at the center of this band; however, in the observed spectra, absorption in this band never exceeds 85% (Fig. 2). We have checked that a contribution of thermal origin cannot be responsible for this effect; indeed, for the thermal component, the flux at the band center is expected to originate from the high atmospheric levels above 50 km, where the thermal profile is nearly isothermal (*Seiff,* 1982). Assuming a temperature lower than 170 K at this altitude, we derive that the flux at the band center should be less than 0.1% of the continuum flux. More likely, the residual flux in the band center is due to scattering effects by atmospheric particles. This scattering effect is also evident in a spectrum of the atmosphere alone, taken above the limb on February 5; it also explains why the depth of the 2.7 μm CO_2 band does not increase, even for large airmass factors, in the February 5 sequence. As a consequence, this band will be disregarded for a proper estimate of the CO_2 abundance. In contrast, it is expected that this band will provide valuable information on the distribution of atmospheric dust.

The 1.44 μm band appears to be too blended with mineralogic signatures to be used for a CO_2 determination; the 1.2 μm band could be usable, but is very weak and should lead to less precise CO_2 abundances. The 2.0 μm CO_2 band is strong and well fitted by our synthetic model (Fig. 3). It will be used for CO_2 abundances and related altimetry determinations. In order to safely eliminate contributions of mineralogical signatures and/or instrumental residual effects, ratioed spectra of different altitudes are used. As CO_2 is, by far, the dominant atmospheric component, the total CO_2 abundance is a direct determination of the ground pressure that can be used for determining the altimetry (see section on altimetry).

H_2O

Water vapor is detectable in four bands, at 2.55, 1.85, 1.38, and 1.14 μm. The 2.55 μm band is the strongest one and has been used for the H_2O abundance determination. Although the 1.85 μm band apparently is blended with a mineralogic feature that could be centered around 1.6-1.7 μm, the selection of spectra of similar mineralogic composition allows this band to be used in addition to the 2.55 μm band. The depth of these features (measured by the ratios of the flux in the band and in the continuum in individual spectra) is correlated to the atmospheric pressure (measured by the ratioed intensities at 2.02 and 2.22 μm; see Fig. 4). Therefore, the 1.85 and 2.55 μm are not of mineralogical origin. The 1.38 and 1.14 μm bands may be usable but are weak and may be contaminated by residual instrumental effects. However, the detection of these bands in the spectrum is clearly above the noise level.

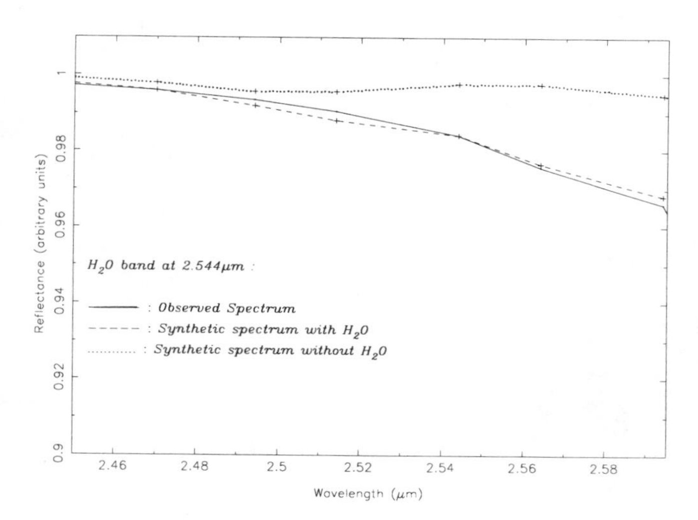

Fig. 5. H_2O band at 2.544 μm: Observed spectrum is a spectrum of the plateau of Olympus Mons divided by the spectrum observed at the summit. The continuum flux has been readjusted so that the observed band fits with the calculated band; synthetic spectrum with H_2O corresponds to a spectrum calculated with $T_e = 230$ K, μm = 2.13, a pressure of 6.1 mbar, and an integrated column density of H_2O of 0.40 cm-am, corresponding to the plateau near Olympus Mons, divided by another synthetic spectrum with $T_e = 185$, μm = 2.36, a pressure of 0.72 mbar, and a null integrated column density of H_2O, corresponding to the summit of Olympus Mons (1 cm-am = 8.032 μm of precipitable water). Synthetic spectrum without H_2O: Similar to above, but assuming no H_2O.

For each of the studied ratioed spectra, a constant H_2O mixing ratio was used within the corresponding layers, except that no H_2O was assumed to be present at the top of Olympus Mons, due to saturation ($T \leq 200$ K leading to $p_{H_2O} \leq 2 \times 10^{-6}$ bar; *Farmer et al.,* 1977). The measurements give $H_2O/CO_2 = 5 \times 10^{-5}$ with a factor of 2 uncertainty in the analyzed layers over Olympus Mons (2 to 26 km) and Valles Marineris (1.5 to 8 km). This value corresponds to 4 pr-μm (μm of precipitable water), and is compatible with the dry limit of the water measurement of *Davies* (1981) (2 to 60 pr-μm), and with the mearurements of *Farmer et al.* (1977) on Tharsis volcanoes. Further data reduction using the 1.38 μm band should improve the determination. Figure 5 shows an example of a fit on Olympus Mons with a synthetic spectrum. Further studies will be devoted to systematic search of horizontal and vertical variations of H_2O by selecting spectra at different places and altitudes.

CO

CO is detectable at 2.35 μm through its (2-0) vibrational band. The band is very weak and difficult to detect because of the low resolving power of the ISM instrument, but the high signal-to-noise allows a firm identification. It has been suggested (*Clark et al.,* 1988) from groundbased observations that the 2.35 μm band might be due to scapolite. However, the shape of the absorption band better fits the gaseous CO band (T. Encrenaz and E. Lellouch, unpublished data, 1989).

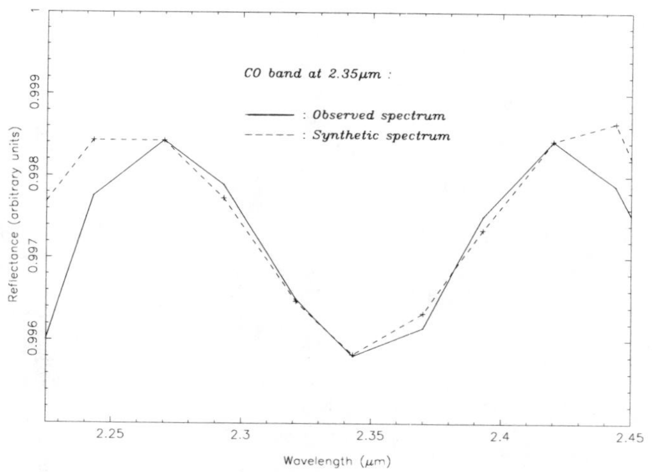

Fig. 6. CO band at 2.35 μm: Observed spectrum is a spectrum of the plateau of Olympus Mons divided by the spectrum observed at the summit. The continuum flux has been readjusted so that the observed band coincides with the calculated band. Synthetic spectrum corresponds to a spectrum calculated with $T_e = 230$ K, μm = 2.13, a pressure of 6.1 mbar, and an integrated column of CO of 2.0 cm-am (corresponding to the plateau near Olympus Mons), divided by an other synthetic spectrum with $T_e = 185$, μm = 2.36, a pressure of 0.72 mbar, and a CO integrated column density of 0.52 cm-am (corresponding to the summit of Olympus Mons).

Because of the weakness of the band, it is necessary to search for systematic errors. First, we checked that "even" and "odd" spectra showed the same absorption depth as the full, convoluted spectra. The signal-to-noise was improved by addition of individual spectra in homogeneous regions (between 10 and 60 spectra) and by three-point triangular convolution. The very good fit to the synthetic spectrum of CO, the positive correlation of the 2.35 μm feature with altitude, and the negative evidence for a similar feature on ratioed spectra at the same altitude confirm that the origin of the band is mostly atmospheric. We can also exclude any CO_2 contamination, as the synthetic spectra take into account all the weak bands of CO_2 and the wavelengths for the measurements of CO and H_2O are chosen to minimize CO_2 absorptions.

From the determination of the CO/CO_2 ratio in regions of different altitudes, it is possible to estimate the CO vertical distribution. The depth of the 2.35 μm feature is a few times 10^{-3}, and the signal-to-noise for an average of more than 10 individual spectra is higher than 2×10^3, allowing the measurement of the CO abundance. Figure 6 shows a fit of a synthetic spectrum to the ratio of the spectra of the plateau and the summit of Olympus Mons. The abundance of CO is found to lie between 2×10^{-4} and 10^{-3} in this region. This abundance measurement is compatible with previous infrared (*Kaplan et al.,* 1969) and millimeter (e.g., *Clancy et al.,* 1983) wavelength measurements. The large uncertainty comes from the unknown CO column density at the top of Olympus Mons, which affects the CO measurement in the spectrum of the lower levels in the ratioed spectra. The vertical variations of CO on Olympus Mons will be studied systematically.

ALTIMETRY OF PAVONIS MONS, OLYMPUS MONS, AND VALLES MARINERIS

The observing sequences of February 8, March 7, and March 13 covered the regions including Pavonis Mons, Valles Marineris, and Olympus Mons respectively. Individual spectra were calibrated, divided by a reference solar spectrum, and spatially interpolated in latitude and longitude. The spatial resolution is about 6 km in the Pavonis Mons region and 23 km for the other two. A straightforward retrieval of the absolute pressures and altitudes is not possible from the ratioed spectra used in this preliminary analysis. An independent knowledge

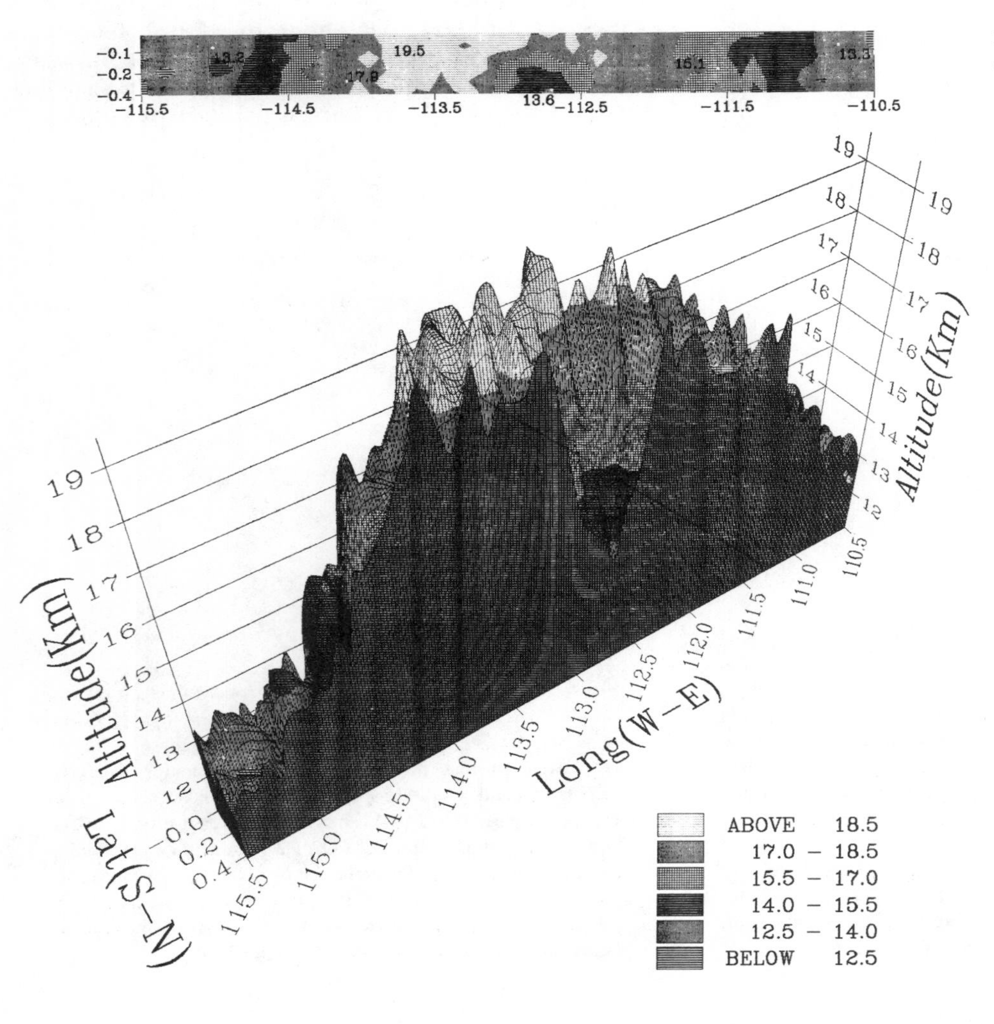

Fig. 7. Altimetry of Pavonis Mons (see text).

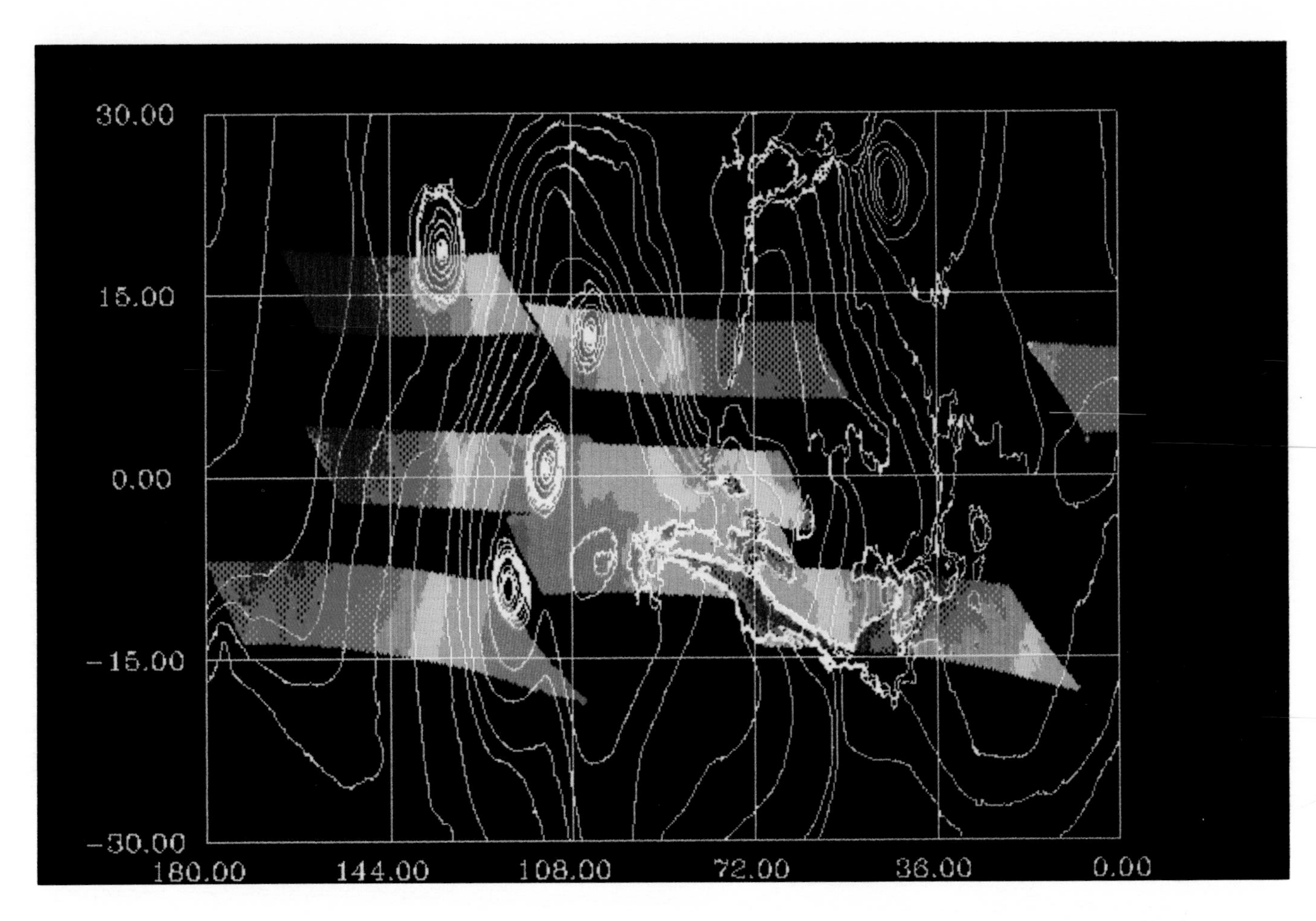

Plate 1. Observed regions in the western hemisphere during three-axis stabilized observations. Contour lines indicate altitudes as determined by Mariner 9; the altitudes determined by ISM are indicated on a blue to red scale from 0 to 25 km (see section on altimetry).

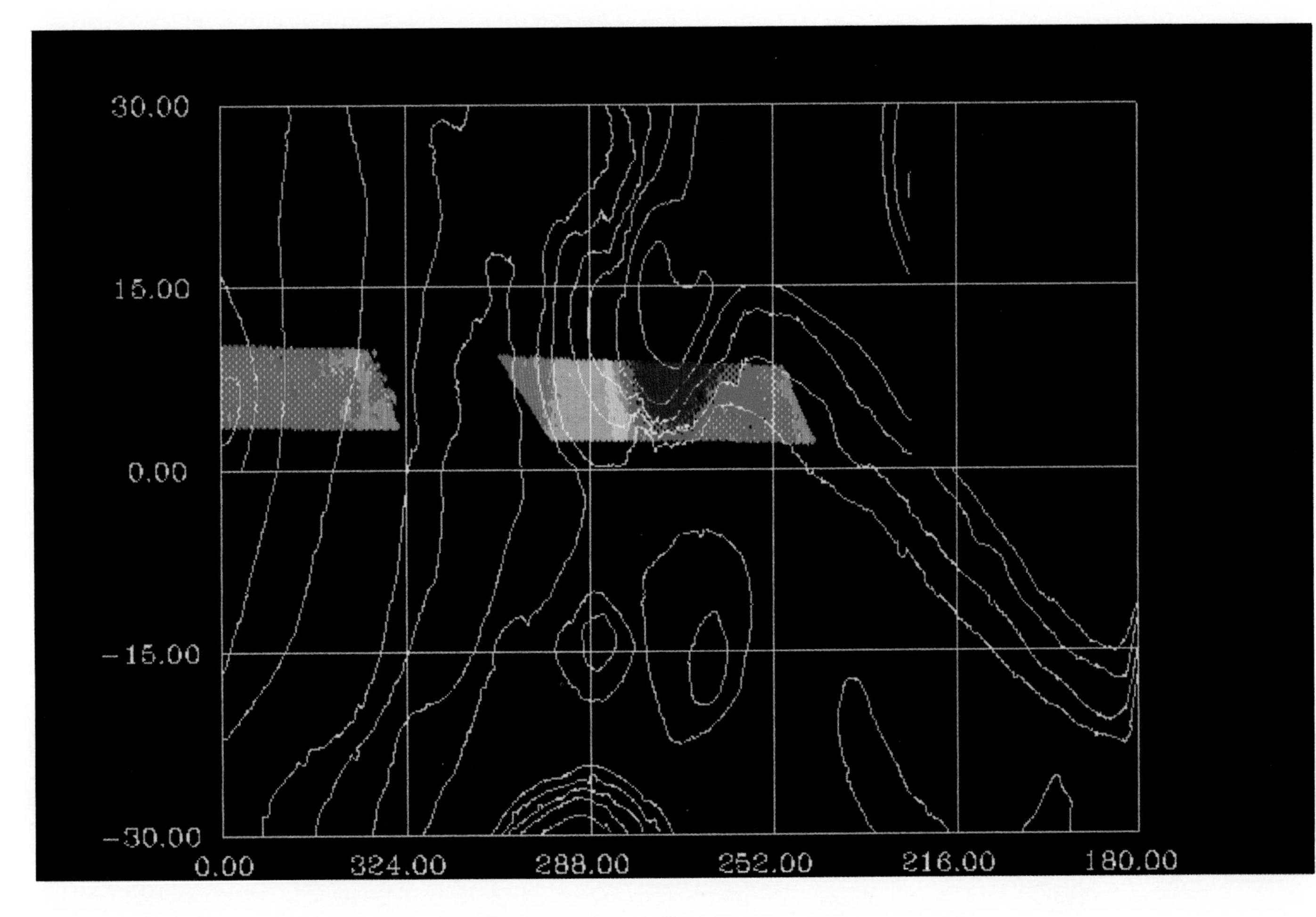

Plate 2. Observed regions in the eastern hemisphere of Mars; same as Plate 1.

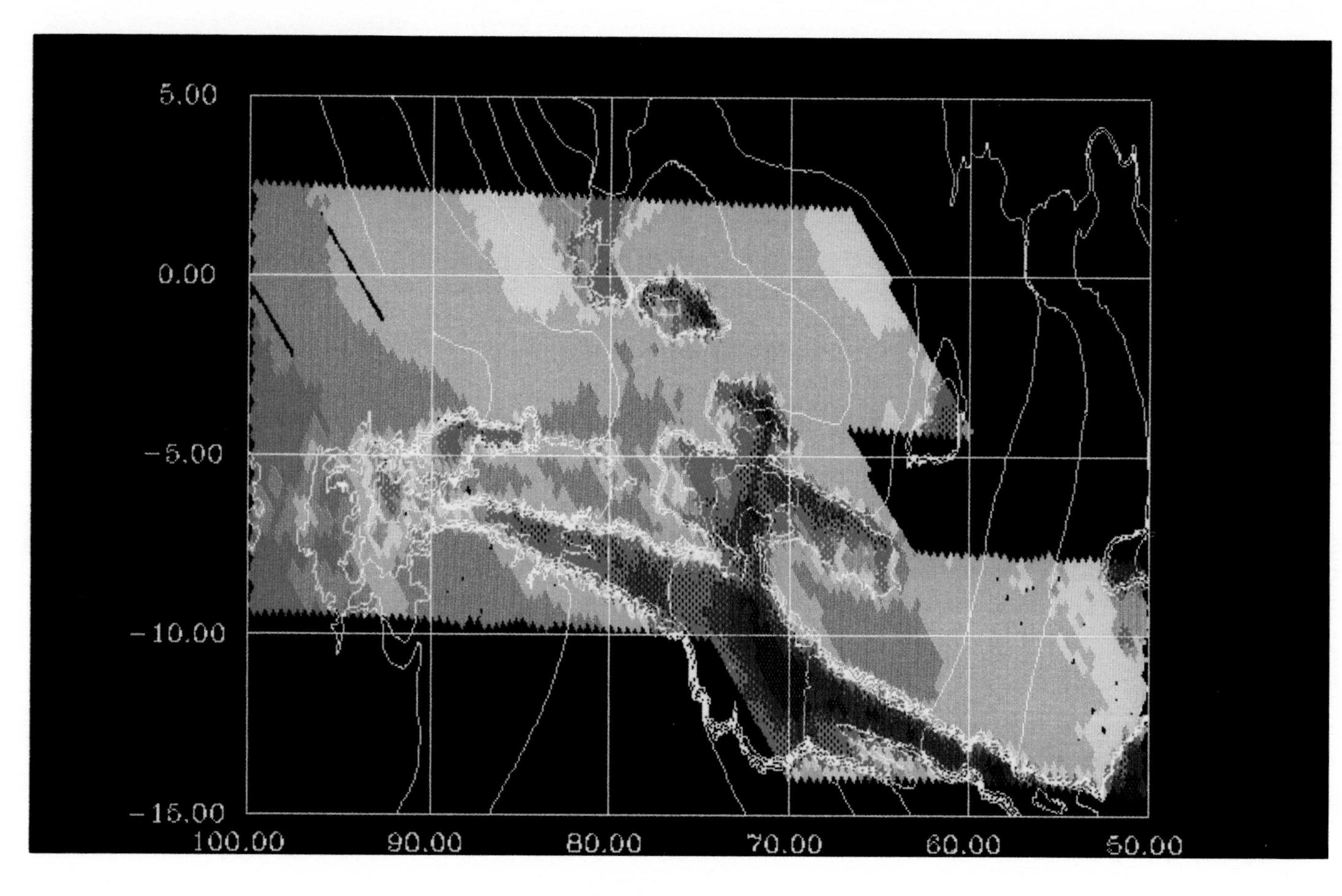

Plate 3. Altimetry of the Valles Marineris region (the blue to red scale corresponds to an altitude scale of 0 to 12 km).

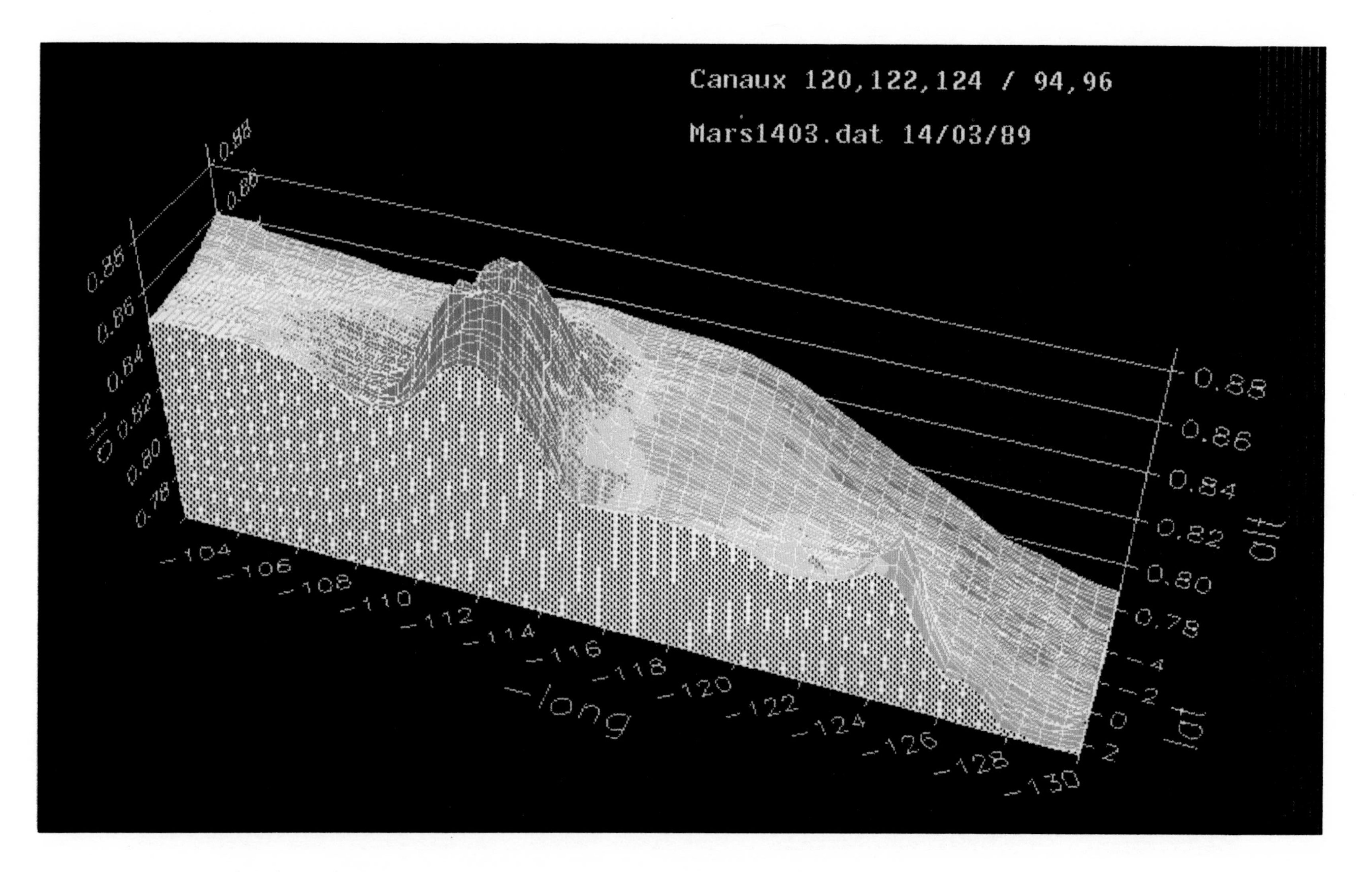

Plate 4. Block diagram of the Pavonis Mons region, observed on March 14. The z axis corresponds to the altitude as derived from the ground pressure. The colors indicate increasing strength of the hydration feature from red to blue.

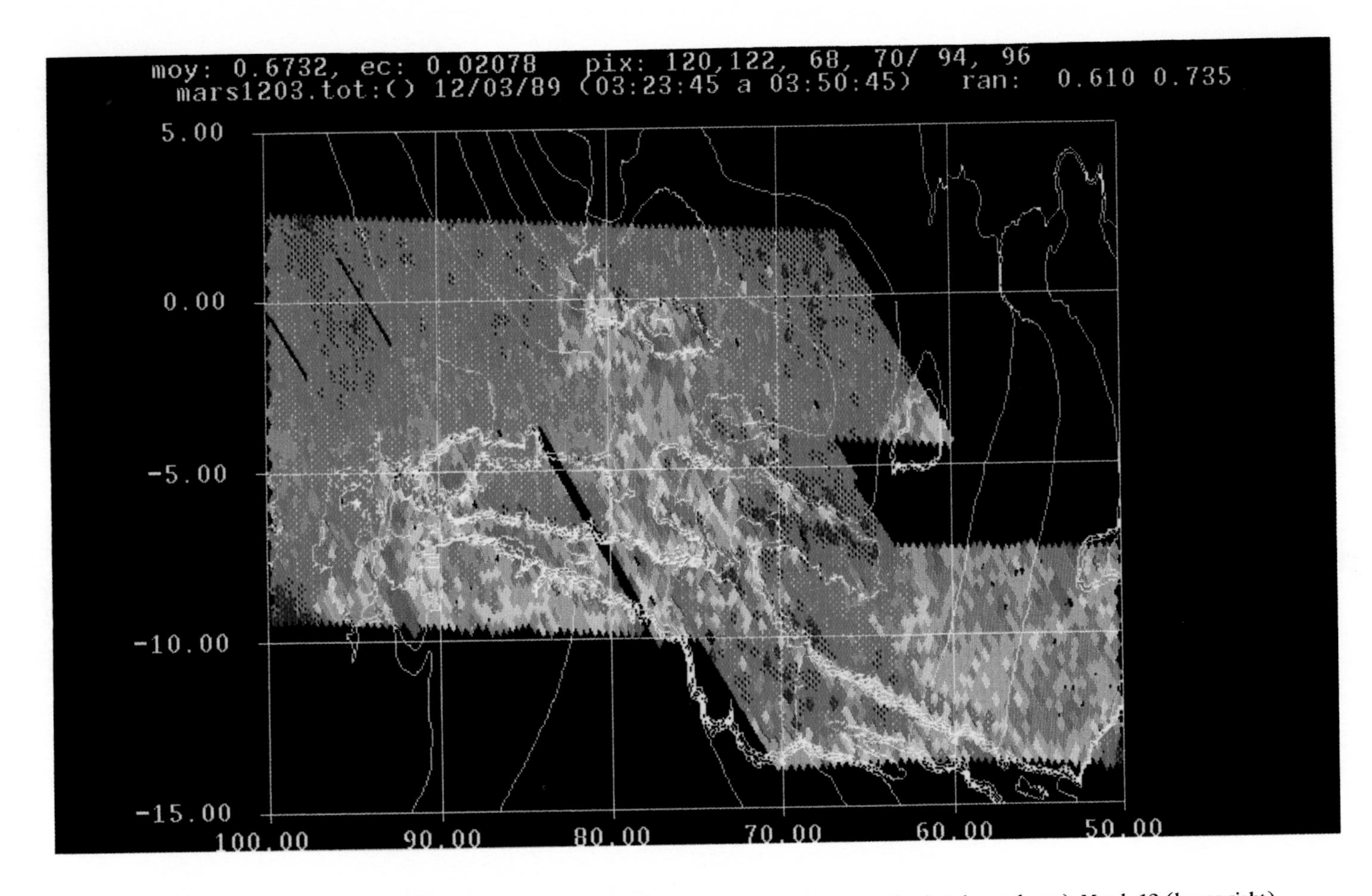

Plate 5. Variations of the hydration feature in the Valles Marineris region, obtained on March 7 (central part), March 12 (lower right), and March 26 (upper part). The red to blue scale corresponds to an absorption of 25% to 40% at 3 μm.

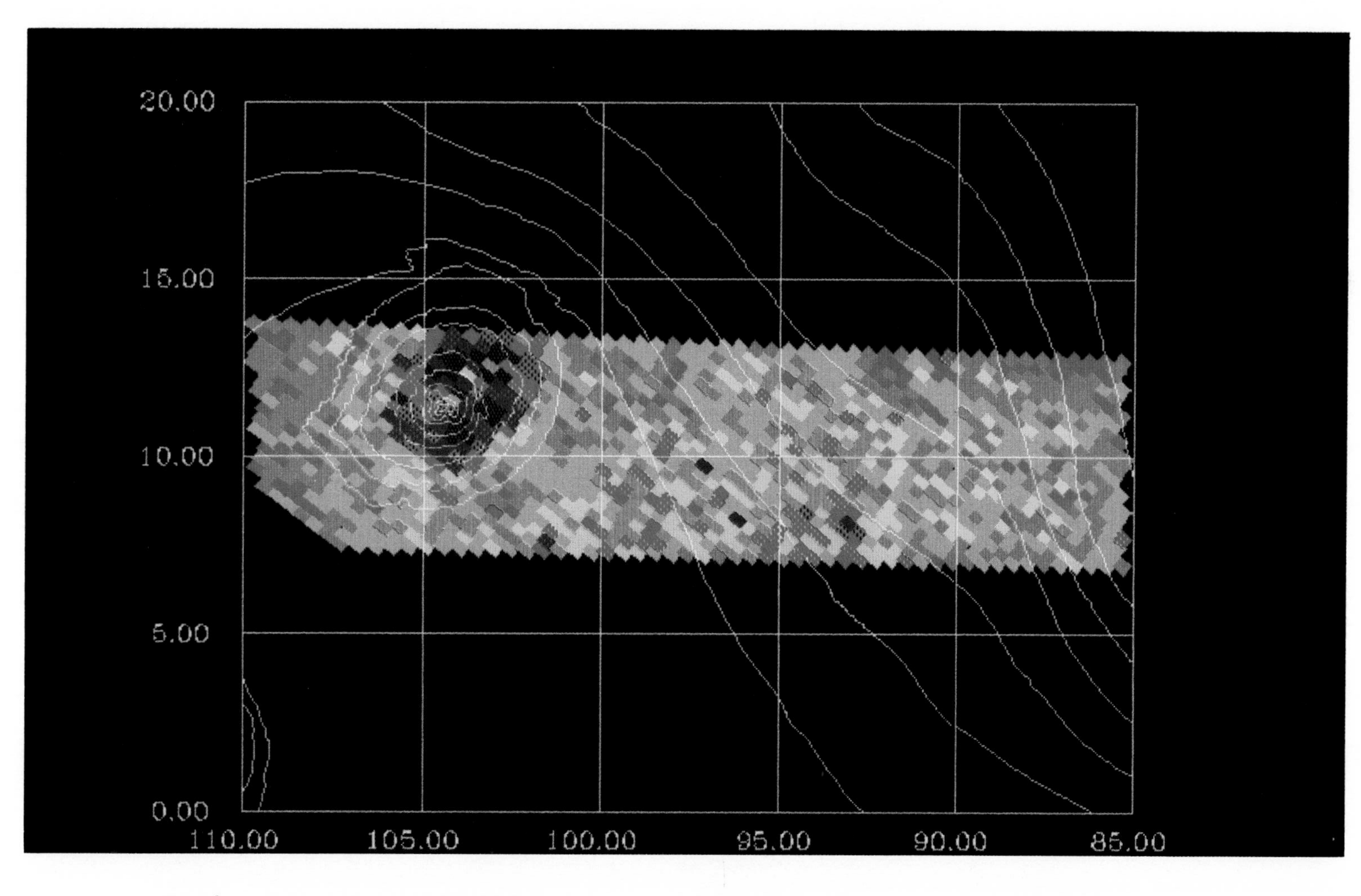

Plate 6. Variations of the depth of the absorption feature at 1 μm associated with silicates in the region of Ascraeus Mons. The red to blue color scale corresponds to increasing absorption from 7% to 8%. Contour lines indicate Mariner altimetry.

of the absolute altitudes of a few reference points and an estimate of the temperature profile was therefore needed. The following method was used: On Pavonis Mons, two spectra were selected, where absolute altitudes could be inferred from previous work with reasonable safety. The first reference point lies at the top of Pavonis Mons, and the second one is located on the equator at 109° longitude, on the plateau east of Pavonis Mons, which shows remarkable smoothness. Using preliminary topographic maps of Tharsis South-West Quadrangle of Mars (U.S.G.S. 1:2,000,000 scale, M 2M 7/124 T, 1981) derived from photogrammetric measurements based on Viking-Orbiter imaging (*Wu,* 1979; *Wu et al.,* 1982), the altitude of the reference points were found to be 19.5 km and 10.6 km respectively, above the standard 0 km level.

An estimate of the atmospheric thermal profile was derived from Viking IRTM and lander measurements (*Kieffer et al.,* 1977; *Martin,* 1981; *Seiff,* 1982) under the assumption of seasonal repeatibility. From the altitude of these reference points, the pressure at the top of Pavonis Mons is found to be 1.35 mbar and the near-surface temperature at an altitude of 0 km is found to be 250 K with the thermal profile "A" (see above); with the thermal profile "B", assuming T_s = 238 K, the pressure at the top is found to be 1.42 mbar. With the thermal profile "A", this corresponds to a pressure of 7.0 mbar at the standard 0 km level, which is entirely consistent with the Viking lander's measurements for this epoch of martian year (L_s ~0°; *Seiff,* 1982). Then, using the spectrum at the top of Pavonis Mons as a reference, 10 spectra that satisfactorily sampled the altitude range were selected and the corresponding pressures were measured from a fit of the 2.0 μm band. In this manner, the caldera of Pavonis Mons was measured at a depth of 5000 ± 300 m with respect to the mean altitude of the rim (the highest point of the rim was 500 m higher).

This time-consuming procedure was not used for all other (~3000) spectra of the Pavonis Mons region, and a fast, approximate method was developed. It basically consists of a pressure interpolation on the equivalent width of the CO_2 band in the ratioed spectra in the pressure range defined by the 10 previous spectra. An example of the synthetic spectrum compared to the ratioed spectrum is shown in Fig. 3 for Olympus Mons. This method led to the altitude map of Figs. 7 and 8 and Plate 3. The Western plateau of Pavonis Mons (longitude 117°-119°) is about 10 km high but shows large topographic variations.

Because of the need for reference points, the absolute accuracy in Fig. 7 is essentially limited by that of previous absolute determinations, about 1 km (i.e., 5%). Our altitude measurements depend linearly on the altitude of the reference

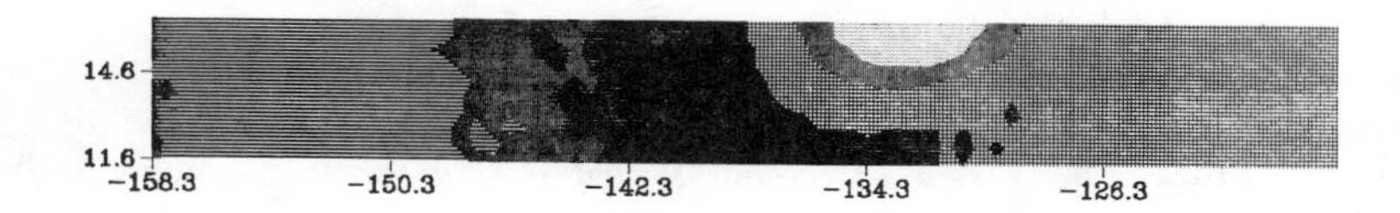

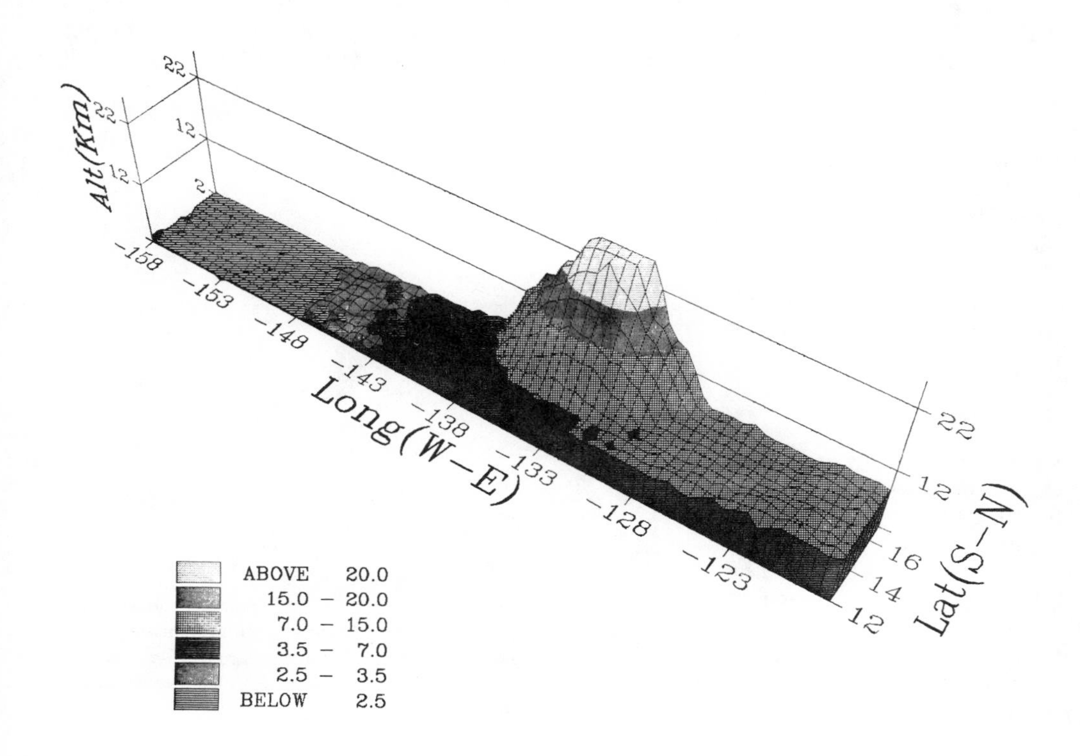

Fig. 8. Altimetry of Olympus Mons.

points, so the altitude scale in Figs. 7 and 8 and Plate 3 is defined within a multiplicative coefficient (1 ± 0.05). The difference in altitude derived using the two thermal profiles described above (thermal profile "A" and "B") are smaller than 400 m. Uncertainties on the band model (including the absolute line intensities) add a small uncertainty of ~100 m in the absolute altitudes. In relative terms, due to the high signal-to-noise ratio, local topographic features are in principle detectable down to 100 m (and measurable with a 5% precision). However, the interpolation method described above downgrades the vertical resolution to ~700 m. For any particular region, a detailed fit of the ratioed spectra should allow recovery of the S/N-limited resolution of 100 m in altitude, if mineralogical variations do not strongly affect the 2 μm region.

The same method was used for the Olympus Mons (Fig. 8) and Valles Marineris (Plate 3) sequences. For Olympus Mons, a reference point was selected near the top, at an altitude of 26.5 km (*Wu,* 1979). Its pressure was estimated from the pressure at the standard 0 km level (7.0 mbar) and with the same thermal profile as for Pavonis Mons. This assumes that, for the same local time, the pressure and temperature variations are negligible on a month timescale and an ~2200 km distance. In the case of Valles Marineris, a reference spectrum in the March 7 sequence was selected on the northern plateau (-4.9° lat., 112.6° long.). Its pressure and altitude were determined using spectra of the March 14 sequence, which covers (at lower spatial resolution than the February 8 sequence) both the Pavonis Mons and Valles Marineris regions. The Valles Marineris canyon is measured at an average depth of 6.5 km.

FIRST RESULTS ON THE INFRARED SPECTRAL REFLECTANCE OF THE SURFACE OF MARS AND PHOBOS

When analysing the spectral reflectance of the martian surface, the major difficulty is to take into account the atmospheric signatures, in particular at 2 and 2.7 μm. These large bands are saturated, which prevents a complete correction of the atmospheric contribution using atmospheric models. Smaller atmospheric signatures are also present (see above). When a feature could not be unambiguously identified from its spectral position, the correlation with altitude and/or airmass factor was used to discriminate between atmospheric and surface contributions.

The Continuum of the Martian Surface

The infrared characteristics of a surface in the continuum are its brightness and its redness, defined as the ratio of the reflected solar flux at a high and a low wavelength in the continuum. Telescopic observations of Mars reveal albedo variations; these variations increase from the red to the near infrared (*Arvidson,* 1989; J. F. Bell and T. B. McCord, unpublished data, 1988). Near 1 μm, the albedo variation is close to a factor of 2, but the actual range was expected to be larger since the limited spatial resolution of Earth observations (a few hundred kilometers) mixed regions of different albedo.

The ISM optical axis was pointed in the antisolar direction during three-axis stabilization. The scanning mirror allowed observations to be made up to 20° away from this direction. Consequently, the phase angle of our observations of the martian surface were always smaller than 20°. At such small phase angles, the opposition effect accounts for significant brightness variations (see, e.g., *Hapke,* 1986). At first order, this effect is similar at all wavelengths and can thus be identified, yielding information on surface scattering properties. A significant limb darkening was observed for regions observed at large incidence angle (referred to the surface normal).

We derived an average brightness from four wavelengths free of atmospheric signatures, at 1.1, 1.29, 1.76, and 2.45 μm. We observed bright and dark regions, with reflectances near opposition ranging from 12% to more than 40%. The increased dynamic range as compared to Earth-based observations (J. F. Bell and T. B. McCord, unpublished data, 1988) results from the improved spatial resolution (20 km); the bright/dark boundaries can be very sharp, and small dark regions are observed down to the resolution of the instrument, in particular in the regions of Valles Marineris and Isidis Planitia. After correction of limb-darkening and opposition effects, it will be interesting to compare medium resolution albedo maps obtained in the infrared-to-visible albedo maps obtained by Viking and the camera onboard the Phobos spacecraft.

The redness of the continuum in the first order was derived from the ratio of fluxes measured at 2.45 and 1.76 μm. We observed variations in this ratio by more than 10%, at scales down to our spatial resolution. As a general trend, bright regions are redder than dark regions. However, marked departures from this correlation are observed, in particular in the region of Valles Marineris. The redness of a spectrum can be associated with small grain size (e.g., *Arnold and Wagner,* 1988) or a high Fe^{3+} content (*Arvidson,* 1989; *McCord et al.,* 1982).

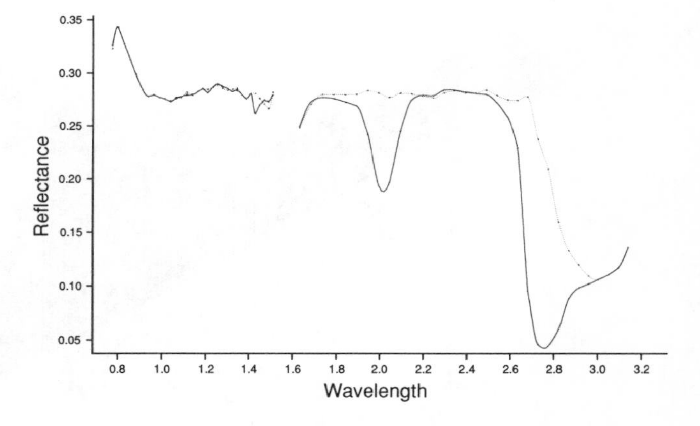

Fig. 9. Full line: ISM spectrum of Mars obtained on March 1 (Isidis Planitia) exhibiting a complex signature between 2.6 and 3.1 μm. Dashed line: The same spectrum after linear removal of the atmospheric contribution. The large remaining signature is attributed to surface hydration.

The Hydration Signature

The ISM absolute spectra (Fig. 9) exhibit a large absorption band from 2.7 to 3.15 μm. This feature is attributed to the hydration band of surface minerals. It is shifted with respect to the signature of water frost, centered at 3.1 μm. In laboratory spectra, hydrated minerals exhibit a strong decrease in spectral reflectance at 2.7 μm, followed by a slow increase up to 3.7 μm. In the ISM spectra of Mars, the very strong CO_2 band centered at 2.7 μm is superimposed onto the low wavelengh part of the hydration signature. The major problem in evaluating the strengh of this feature is the lack of data for the continuum above 3.7 μm. Consequently, the ratio of the flux at 3.0 μm (where the CO_2 band is negligible) to the flux at 2.45 μm (in the continuum) was used, corrected for the average slope (redness) of the continuum in this spectral region (see above). On the average, the hydration feature is strong (55% in absorption at 3 μm), as already observed from the ground.

Large variations (by up to 20%) are observed on a scale corresponding to the resolution of ISM. These variations do not exhibit a systematic correlation with altitude, which demonstrates that these features are free from atmospheric contamination. Plate 4 shows a mapping of the water band, superimposed on the altimetric mapping obtained through the 2 μm CO_2 band, on the region including Pavonis Mons. It clearly exhibits an increase in the degree of hydration along the slopes of the volcano, as compared to that of the surrounding plateau. Such a trend is also found for the south flank of Arsia Mons, observed on February 27. It is interesting to note that these two observations were performed at a relatively high incidence angle (50° to 60°). The increased hydration is less pronounced for Ascraeus Mons, observed closer to normal incidence.

If further analysis confirms this increased hydration of volcanic material, it may be interpreted as follows: Either hydration originates from the volcanic activity itself, which would suggest that the magma erupted through an ice-rich upper crust, similar to a permafrost, the water being incorporated to the silicates before flowing off the volcano, or hydration comes from the interaction with atmospheric water. In that case, the enrichment along the flanks of the volcanoe would require a finer grained material to be responsible for the higher degree of adsorption, through a larger surface/volume of the fines toward the top. This would be consistent with the fine grain size inferred for all of the Tharsis volcanoes from Viking imaging and IRTM measurements (*Zimbelman,* 1985). We should be able to discriminate between these two interpretations in two ways: First, adsorbed water and water blocked within the matrices might be distinguished spectrally after appropriate laboratory calibrations; second, we shall look for correlations of the degree of hydration with other spectral features (albedo, redness, mineral signatures) that could provide information on their origin.

The variation of the degree of hydration is not limited to the slopes of the volcanoes. Plate 5 shows similar maps obtained in the Valles Marineris area. It will be interesting to correlate surface hydration to morphological structures linked to water such as lobate craters or flows (e.g., *Lucchitta,* 1985; *Carr,* 1987). It is still too early to give a detailed interpretation of theses variations, from one slope to the other, and between the canyon and the surrounding plain. However, it is obvious that variations as low as 1% are straightforwardly identified with ISM, and that the hydration of martian surface minerals is highly nonuniform, on a scale of a few tens of kilometers.

Other Mineralogical Signatures

A wide, shallow spectral feature is observed in the region 0.9 to 1.1 μm. It most likely corresponds to the Fe^{2+} signature of pyroxene and olivine (*Hunt and Salisbury,* 1970). The depth of this band is evaluated by the ratio of the flux at 1 μm divided by the mean of the flux at 0.9 and 1.1 μm. The variations observed are small in general. As an example, Plate 6 exhibits variations in the depth of the 1 μm band over a region including Ascraeus Mons and Tharsis Tholus. The total scale corresponds to less than 1%. It shows that well-identified contours are present, probably related to compositional variations linked to mineralogical signatures, at a level of a few 10^{-3} in absorption. The absolute spectrum can in principle be used to infer the composition of mixtures of olivines, feldspars, and pyroxenes (*Singer,* 1981; *Mustard and Pieters,* 1987). Observations of silicate and hydration bands near the limb performed by ISM during nonstabilized orbits will provide information on atmospheric dust, which could be linked to surface deposits (see, e.g., *Huguenin,* 1987).

A very interesting problem is the detection of carbonates that, if present, would exhibit spectral features close to 2.35 μm (*Hunt and Salisbury,* 1971; *Gaffey,* 1987). Unfortunately, the CO band is centered at a similar value. Consequently, one has to carefully eliminate any possible contribution of atmospheric constituents before assessing the presence of carbonates. The observed signatures are very small, 2 to 5 times the noise level. However, in most spectra it is not

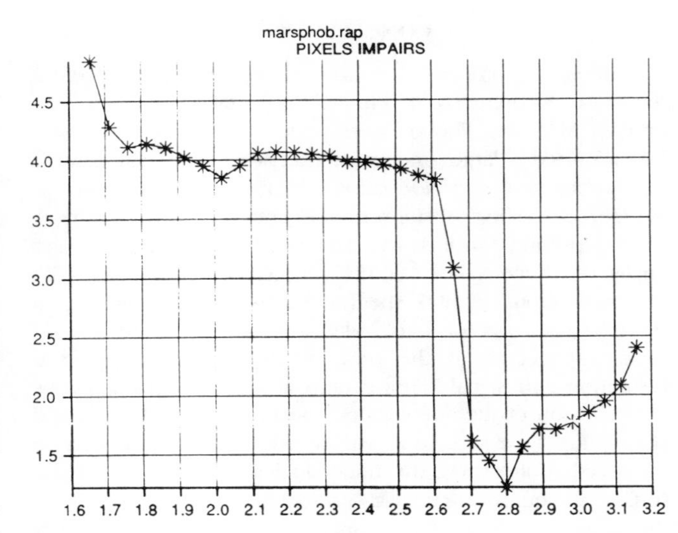

Fig. 10. Ratio of the spectra of Mars and Phobos from 0.8 to 3.2 μm.

possible to fit the observed signature with the single contribution of atmospheric CO. A detailed examination of this problem, involving filtering of altimetric correlations and noise reduction techniques, is underway.

Observations of Phobos

ISM acquired data of Phobos twice on March 25, during the last two maneuvers that tilted the spacecraft in the direction of the largest moon of Mars. The average distance from Phobos was of the order of 200 km, which led to a spatial resolution close to 0.7 km per pixel for ISM. At the time of the first encounter Phobos was observed in front of Mars, and on a dark background during the second observation. We chose not to scan the mirror this first session, so that we have a track ~700 m wide across the Phobos disk. The final part of the track continues on the martian disk, allowing direct comparisons between the spectral characteristics of the surface of these two bodies. The ratio of two characteristic spectra (Fig. 10) demonstrates that the albedo ratio is evaluated as ~4 between Phobos and the observed region of Mars, west of Pavonis Mons. On this ratioed spectrum, the atmospheric martian signatures at 2 and 2.75 μm can be readily observed, together with the hydration feature, which is clearly much stronger on Mars than on Phobos. The low hydration and low albedo of Phobos suggest a relationship with dark asteroids to relatively evolved carbonaceous chondrites (C4), as opposed to water-rich carbonaceous chondrites.

The second observation, obtained on a dark background, yields a map of ~30% of the disk of Phobos, between 180° and 250° longitude, -30° and +20° latitude. The albedo map can be associated with the visible images taken during the same session by the camera, allowing identification of the observed region. Although the signal-to-noise ratio is lower than for Mars, the hydration and silicate spectral features exhibit significant variations (by up to 10%) at a kilometer scale.

CONCLUSIONS

Although the Phobos 2 mission ended prematurely before the close encounter with Phobos could take place, observations of Mars and Phobos were performed before this event. ISM acquired a large amount of data from Mars that sample the various geological formations discovered by the previous Mariner and Viking missions: the old cratered plains, the more recent lowland terrains, the Tharsis Montes volcanoes, and finally the huge Valles Marineris structures. Altogether, ISM obtained about 40,000 spectra, each corresponding to a resolved pixel size of 4 to 25 km, depending on the altitude of the spacecraft at the time of exposure. The spectral resolution and signal-to-noise ratio allowed an unambiguous identification of the atmospheric signatures for CO_2, CO, and H_2O. Thus, their variations with location and altitude can be measured. Concerning the mineralogical composition of the surface, the quick-look analysis presented in this paper stressed the detection of hydrated minerals. These minerals, which appear all over the regions mapped by ISM, present large variations in the degree of their hydration, clearly linked to the geological formations to which they belong. Other mineralogical signatures, such as that of silicates near 1 μm, also vary to a lesser extent with the geological context. The observation of Phobos took place when the spacecraft was at a distance close to 200 km, leading to a spatial resolution of 0.7 km. About one third of the lighted hemisphere has been mapped. Phobos is a dark object, much drier than Mars, that exhibits large spectral variations on a kilometer scale.

REFERENCES

Arnold G. and Wagner C. (1988) Grain size influence on the mid-IR spectra of the minerals. *Earth, Moon and Planets, 41,* 163-172.

Arvidson R. E. (1989) Nature and distribution of surficial deposits in Chryse Planitia and vicinity, Mars. *J. Geophys. Res., 94,* 1573.

Bibring J-P., Combes M., Drossart P., Encrenaz Th., Erard S., Forni O., Gondet B., Ksanfomaliti L., Langevin Y., Lellouch E., Masson Ph., Moroz V., Rocard F., Rosenqvist J., Sotin C., and Soufflot A. (1989) First results of the ISM experiment. *Nature, 341,* 591-592.

Carr M. H. (1987) Water on Mars. *Nature, 326,* 30.

Clancy R. T., Muhleman D. O., and Jakosky B. M. (1983). Variability of carbon monoxide in the Mars atmosphere. *Icarus, 55,* 282-301.

Clark R. N., Swayze G. A., and Singer R. (1988) Mineralogy indicated by the Martian 2.36 μm band. *Bull. Am. Astron. Soc., 20,* 849.

Davies D. W. (1981) The Mars water cycle. *Icarus, 45,* 398-414.

Farmer C. D., Davies D. W., Holland A. L., Laporte D. D., and Doms P. E. (1977) Mars: water vapor observations from the Viking Orbiters. *J. Geophys. Res., 82,* 42-25.

Gaffey S. J. (1987) Spectral reflectance of carbonate minerals in the visible and near IR (0.35-2.55 μm): 2) Anhydrous carbonate minerals. *J. Geophys. Res., 92,* 1429-1437.

Goody R. M. (1964) *Atmospheric Radiation. I. Theoretical Basis.* Oxford Univ., London. 436 pp.

Hapke B. W. (1986) Bidirectional reflectance spectroscopy: IV. The opposition effect. *Icarus, 67,* 264-275.

Huguenin R. L. (1987) The silicate component of martian dust. *Icarus, 70,* 162.

Hunt G. R. and Salisbury J. W. (1970) Visible and near IR spectra of minerals and rocks: I. Silicate minerals. *Mod. Geol.,* 283-300.

Hunt G. R. and Salisbury J. W. (1971) Visible and near IR spectra of minerals and rocks: II carbonates. *Mod. Geol., 2,* 23-30.

Husson N. (1986) The GEISA spectroscopic lines parameter data bank in 1984. *Ann. Geophys. 4,* 185-190.

Kaplan L. D., Connes J., and Connes P. (1969) Carbon monoxide in the martian atmosphere. *Astrophys. J., 157,* L187-L192.

Kieffer H. H., Martin T. Z., Peterfreund A. R., Jakosky B. M., Miner E. D., and Palluconi F. D. (1977) Thermal and albedo mapping of Mars during the Viking primary mission. *J. Geophys. Res., 82,* 42-49.

Lucchitta B. K. (1985) Valles Marineris; wet debris flows and ground ice. *Icarus, 72,* 411-419.

Martin T. Z. (1981) Mean thermal and albedo behavior of the Mars surface and atmosphere over a martian year. *Icarus, 45,* 427-446.

McCord T. B., Clark R. N., and Singer R. B. (1982) Mars: near-IR spectral reflectance of surface regions and compositional implications. *J. Geophys. Res., 87,* 3021.

Mustard J. F. and Pieters C. M. (1987) Quantitative abundance estimates from bidirectional reflectance measurements. *Proc. Lunar Planet. Sci. Conf. 17th,* in *J. Geophys. Res., 92,* E617-E626.

Rothman L. S. (1986) Infrared energy levels and intensities of carbon dioxide. *Appl. Opt., 25,* 1795-1816.

Seiff A. (1982) Post-Viking models of the structure of the summer atmosphere of Mars. *Adv. Space Res., 2,* 3-17.

Singer R. B. (1981) Near IR spectral reflectance of mineral mixtures: systematic combinaisons of pyroxenes, olivine and iron oxides. *J. Geophys. Res., 86, B9,* 7967.

Wallace L., Prather M., and Belton M. J. S. (1974) The thermal structure of the atmosphere of Mars. *Astrophys. J., 193,* 481-493.

Wu S. S. C. (1979) Photogrammetric portrayal of Mars topography. *J. Geophys. Res., 84,* 7955-7959.

Wu S. S. C., El Assal A. A., Jordan R., and Shafer F. J. (1982) Photogrammetric applications of Viking orbital photography. *Planet. Space Sci., 30,* 45-55.

Zimbelman J. R. (1985) Surface properties of Ascraeus Mons: Dust deposits on a Tharsis Volcano (abstract). In *Lunar and Planetary Science XVI,* pp. 936-937. Lunar and Planetary Institute, Houston.

Proceedings of the 20th Lunar and Planetary Science Conference, pp. 473-478
Lunar and Planetary Institute, Houston, 1990

On the Possibility of Life on Early Mars

V. R. Oberbeck

NASA Ames Research Center, Moffett Field, CA 94035

G. Fogleman

SETI Institute, Moffett Field, CA 94035

Prebiotic reactants, liquid water, and temperatures low enough for organic compounds to be stable are requirements for the origination of life as we know it. Prebiotic reactants and sufficiently low temperatures were present on Mars before liquid water vanished. Early in this time period, however, large planetesimal impacts may have periodically sterilized Mars, pyrolyzed organic compounds, and interrupted chemical origination of life. However, the calculated time interval between such impacts on Mars was larger just before liquid water vanished 3.8 Gyr (billion years) ago than it was on Earth just before life originated. Therefore, there should have been sufficient time for life to originate on Mars. Ideal sites to search for microfossils are in the heavily cratered terrain of Upper Noachian age. Craters and channels in this terrain may have been the sites of ancient lakes and streams that could have provided habitats for the first microorganisms. These organisms might be analogs of the earliest microorganisms on Earth, the fossils of which have been destroyed by geologic activity. The possibility that fossils of the first microorganisms produced by chemical evolution might be discovered on Mars is an important justification for pursuit of exobiologic studies on Mars.

INTRODUCTION

There are certain requirements for the origin of life on any planet. First, prebiotic reactants would have been required. Second, because the origin of life and life as we know it depend upon water, liquid surface water was required. Liquid water at the surface would have provided a means of wetting and drying of prebiotic reactants and permitted redistribution of reaction products. This facilitated polymerization, a key step in chemical evolution (*Lahav et al.,* 1978). Third, global temperatures must have been lower than about 200°C in order that prebiotic synthesis could have occurred and organic compounds would have been stable (*Miller and Bada,* 1988).

Study of the primordial conditions on Earth and Mars as they affected these requirements can help determine the universal possibility of origination of life on any planet subject to the same physical and chemical laws. For example, the time between large sterilizing impacts that vaporized water and raised global temperatures to pyrolyze organic compounds must have exceeded the time required to originate life (*Maher and Stevenson,* 1988). In this model, 2×10^{34} erg impacts were assumed to be energetic enough to sterilize the planet, vaporize all water, and interrupt the process leading to the origin of life. *Oberbeck and Fogleman* (1989b) also investigated the effects these and larger bodies (2×10^{35} ergs) had on constraining the origin of life and concluded that impact could have constrained the origin of life on Earth. Therefore, we assume that early in the geologic history of Mars, large impacting objects interfered with the origination of life just as they did on Earth, and that the process of chemical evolution would have required the same length of time on Mars as on Earth. We further assume that the primordial cratering rates on Mars are those given by *Hartmann* (1977).

Supply of prebiotic reactants at the correct geologic time is another key process. Prebiotic reactants may have been components of the primordial atmosphere as has been assumed for Earth by *Miller* (1954) and *Pinto et al.* (1980). Alternatively, cometary impact may have supplied prebiotic reactants to Earth (*Oro,* 1961) or they may have arrived in carbonaceous chondrites (*Mukhin,* 1989). Fluvial processes may also have been important for chemical evolution on Mars. We assume that martian valley networks reflect the presence of surface liquid water that was beneficial to chemical evolution on Mars. Thus, we consider the requirements for origination of life on Mars within the framework of planetary surface processes. In the next three sections we discuss each of these topics.

PREBIOTIC REACTANTS ON MARS

Fanale (1971) considered the possibility of suitable prebiotic reactants and the oxidation state of the early martian atmosphere. He concluded that a reduced atmosphere would have been produced by accretion and that it was retained for about 10^8 years. Methane and ammonia would then have been replaced by CO_2. Thus, reactions between methane, ammonia, and water may have occurred for the first 10^8 years. It has also been pointed out (*Levine,* 1985) that an early terrestrial atmosphere of methane and ammonia would have also been replaced by CO_2.

CO_2, which may have reached partial pressures of 20 bars on the primordial Earth (*Holland,* 1984), could have been a useful prebiotic reactant. *Pinto et al.* (1980) suggested a photochemical mechanism for production of formaldehyde on primordial Earth from CO_2 and trace amounts of H_2 from volcanic eruptions. If only trace amounts of H_2 were supplied to the early martian atmosphere by volcanism after the first 100 m.y., formaldehyde may have formed on Mars. Mixtures of formaldehyde and nitrates exposed to UV radiation form amino acids (*Pavlovskaya and Pasynskii,* 1959). Aqueous reactions involving HNO could have produced nitrates in water on Mars (*Mancinelli,* 1988; *Mancinelli and McKay,* 1988). Thus, prebiotic chemistry involving atmospheric gases may have been possible from the time of accretion.

Other sources of prebiotic reactants are comets (*Oro,* 1961; *Clark,* 1988; *Oberbeck et al.,* 1989) and planetesimals containing carbonaceous material (*Mukhin,* 1989). *Fernandez and Ip* (1983) conclude that whether the comet reservoir is the outer region of the solar system or the Oort cloud, comets began to arrive in the vicinity of the terrestrial planets 4.5 b.y. ago. Thus, prebiotic reactants produced by either photochemical reactions or supplied or produced by cometary entry could have been present in the martian atmosphere from the time of accretion. This satisfies the first requirement in order that life might have originated on Mars.

LIQUID WATER ON MARS

Liquid water cannot now exist at the surface of Mars because the surface pressure is too low. However, based on a radiative convective climate model, *Pollack et al.* (1987) proposed that a dense atmosphere of CO_2 sustained on early Mars by volcanism permitted the existence of liquid water. According to this theory, intense volcanism may have recycled CO_2 by melting carbonate rocks. Greenhouse heating from CO_2 could have kept the surface temperature above the freezing point of water for the first billion years of martian history. Only after the first billion years, when the planetary internal heat declined enough to prevent volcanism from recycling CO_2, was liquid water no longer stable at the surface.

Spacecraft imagery has revealed features that could have been formed by running surface water. Valley networks resemble certain terrestrial channels formed by groundwater sapping (*Carr and Clow,* 1981). They formed until the end of the intense bombardment about 3.8 Gyr ago (*Carr and Clow,* 1981). These authors give crater counts of the youngest surfaces cut by valley networks that show about $(3.5 \pm 2) \times 10^{-5}$ craters greater than 20-km diameter per square km occur on these surfaces. The cratering rate was twice as high as on Earth after 3.8 Gyr ago (*Hartmann,* 1977). The terrestrial cratering rate for craters equal or larger than 20-km diameter was 0.35×10^{-14} km^{-2} yr^{-1} on the Earth after the heavy bombardment (*Grieve and Dence,* 1979). Thus, the calculated expected number of craters larger than 20-km diameter per km^2 on a 3.8-b.y.-old martian surface is 2.7×10^{-5}, in agreement with the observed number of craters on the youngest martian surfaces cut by runoff channels. Based on these results and the theoretical model results of *Pollack et al.* (1987), the second requirement for the origination of life on Mars, the presence of liquid surface water, appears to have been satisfied until 3.8 Gyr ago.

IMPACTS AND THE ORIGIN OF LIFE ON EARTH AND MARS

Miller (1984) noted that less than 1 Gyr was available for origination of life between the formation of Earth and the age of the oldest fossils, in part because time was required for the Earth to cool after accretion. This is probable because it has been reported that magma oceans formed until Earth reached 90% of its present radius (*Matsui and Abe,* 1986). *Hartmann and Davis* (1975) proposed that a giant impact formed the Moon from matter ejected during collision of a large planetesimal with the Earth. *Stevenson* (1987) concluded that this impact would have raised the global temperature by a few thousand degrees C and cooling time would have been as long as 10^3 to 10^4 years. After growth of 90% of the Earth, requiring about 35 m.y., hotspots may have formed only around impact sites, but regions away from the impacting objects would not have been molten (*Arrhenius,* 1987). *Kasting and Ackerman* (1986) point out that near-surface atmospheric temperatures would have been about 100°C after the first several hundred million years of Earth's geologic history. This is well below the 200°C threshold above which organic compounds would have been unstable (*Miller and Bada,* 1988).

Typical temperatures during most of terrestrial geologic history were probably low enough that organic compounds would have been stable and prebiotic chemical evolution could have proceeded. However, transient high temperature pulses from impacts could have constrained the origin of life for hundreds of millions of years. Life may not have originated until the time between sporadic impacts that raised global temperature enough that prebiotic chemistry would have been interrupted exceeded the time required to originate life (*Maher and Stevenson,* 1988). In this model, impacts with kinetic energy of about 2×10^{34} ergs would have been large enough to sterilize the planet. Time between these impact events was calculated as a function of geologic history. These time intervals were refined by *Oberbeck and Fogleman* (1989a,b) who used them to estimate the maximum time required to originate life on Earth. In the remainder of this paper we review the time required to originate life on Earth and compare it with the time interval between sterilizing impacts on Mars before liquid surface water vanished to test the possibility that life may have originated on Mars.

MAXIMUM TIME REQUIRED FOR ORIGIN OF LIFE ON EARTH

In order that life could have originated on Mars, the time between sterilizing impacts before liquid water vanished must have been greater than the maximum time between sterilizing impacts before the earliest appearance of life on Earth (maximum time required to originate life). Thus, it is necessary to estimate the time required to originate life on Earth before we examine the possibility that life originated on Mars. The time available between planetary sterilization impacts just before the appearance of life on Earth is also an approximation of the maximum time needed for origination of life; this is because life would originate when the time between sterilizing impacts exceeds the time required to originate life (*Maher and Stevenson,* 1988) and because the rate of impact cratering on the Moon and Earth is known to have declined with the passage of time (*Baldwin,* 1987; *Carr et al.,* 1984). *Oberbeck and Fogleman* (1989a) obtained the time interval, T_w, (years) between single impact events larger than a given size, D_{min}, capable of sterilizing Earth and stopping the process leading to the origin of life

$$T_w = C\, f(t)^{-1} \cdot [(1\ km/D_{min})^{1.8} - (1\ km/D_{max})^{1.8}]^{-1} \qquad (1)$$

where C is the reciprocal of the product of the surface area of the Earth and the terrestrial cratering rate constant, k, in the inverse power law ($N = k f(t)D^{-1.8}$) giving the cumulative number of craters (N) larger than D formed per km^2 per year on Earth, and C = 3735. The quantity $f(t) = 1 + \exp[(t-t_o)/T]$ gives the time dependence of the cratering rate (*Maher and Stevenson,* 1988). The parameter t is time measured backward from the present and t_o = 3.4 Ga, the approximate time at which the cratering rate becomes constant on the Moon (*Carr et al.,* 1984). When T, a decay constant, is 70 m.y., f(t) gives the observed lunar cumulative crater distributions observed on lunar surfaces of various ages (*Maher and Stevenson,* 1988). D_{min} is the minimum crater size capable of causing planetary sterilization and D_{max} is the maximum crater size expected on Earth.

Time between impacts is a strong function of T in f(t) and k. $k = 5 \times 10^{-13}$ appears to be well defined because it gives agreement with values determined by *Grieve and Dence* (1979) over the last few hundred million years. Maher and Stevenson assumed T = 70 m.y. However, there is some uncertainty in the time dependence of the impact flux before 3.8 Gyr ago and this could affect our estimate of the maximum time required to originate life. For example, *Baldwin* (1987) estimated the absolute ages of lunar basins using a viscosity model, counted the craters formed on the basin deposits, and obtained ages for different cumulative crater size distributions. *Oberbeck and Fogleman* (1989b) used Baldwin's data to estimate $k = 1.26 \times 10^{-12}$, T = 150 m.y., and t_o = 3.2 Gyr in f(t) for the impactor flux between 3.79 and 4.24 Gyr ago. These parameters are different from those used by Maher and Stevenson for time periods preceding 3.8 Gyr. Therefore, impactor decay laws with these impact parameters have also been used to study the sensitivity of estimates of origination time for life to impactor flux uncertainties.

The maximum time required for origination of life on Earth has been computed using the equation above for T_w for both time impactor flux laws and assuming life originated 3.5 or 3.8 Gyr ago (*Oberbeck and Fogleman,* 1989b). For example, that time, t, was determined using equation (1) such that the time interval between sterilizing impacts, T_w, capable of stopping the origin of life, when subtracted from t, would give 3.5 Gyr ago, the time at which there is the oldest evidence of life on Earth (*Schopf,* 1983). This value of T_w is the maximum time required for origination of life 3.5 Gyr ago because life must have originated during a time interval equal to the time between impact events capable of stopping it and, at times greater than t, these time intervals were less than T_w. However, there is no fossil evidence that life originated at the end of these earlier time intervals. At times less than t (more recent), the time intervals between adverse impacts were larger, but life had already originated. Note that we make no claim that first life survived later impacts or that all early life evolved from the first successful origination of life. Giant impacts occurring later may have killed off the first life. We only compute the maximum time needed to produce the first observed fossilized life and this is an estimate of the maximum time needed because earlier sterilizing impacts were more closely spaced and later ones were more widely spaced. First, using T = 70 m.y., we found T_w = 87 m.y. when t = 3.5 Gyr + 67 m.y., where D_{min} is 850 km and D_{max} is 2600 km (*Oberbeck and Fogleman,* 1989b). The calculation indicates that 3.567 Gyr ago the time interval between formation of 850-km-diameter and larger impact basins that may have been able to interfere with the origin of life was 67 m.y. If life required no more than 67 m.y. to originate, life could have originated 3.5 Gyr ago, 67 m.y. after a sterilizing event 3.567 Gyr ago. In this analysis, we computed basin size, D_{min}, from the kinetic energy of the impacting object (2×10^{34} ergs).

It is difficult to determine the exact level of impact energy and minimum basin diameter that was required to vaporize the oceans and sterilize the Earth. However, the estimated maximum time required to originate life can be constrained rather well even with this uncertainty. *Maher and Stevenson* (1988) assumed 2×10^{34} ergs was sufficient to vaporize the oceans (energy required to vaporize 1 g of water times the number of grams of water in the ocean). However, due to energy lost to space by radiation and other factors, it may have taken impacts with higher kinetic energy to interrupt the chemical evolution processes leading to the origin of life. Therefore, we calculated an additional value of maximum origination time by assuming that impacts with kinetic energy of 2×10^{35} ergs may have been required to stop the process leading to the origin of life. In this case, for T = 70 m.y., when D_{min} = 1611 km (2×10^{35} ergs) and t in f(t) = 3.5 Gyr + 133 m.y. is substituted into the equation, the calculated value of T_w = 133 m.y. (*Oberbeck and Fogleman,* 1989b). This, then, would be the maximum time required to originate life 3.5 Gyr ago.

It has been suggested that organisms may have existed as early as 3.8 Gyr ago. *Schidlowski* (1988) has reported that the ratio of ^{12}C to ^{13}C in 3.8×10^9-year-old sedimentary material is an indicator of photosynthesis. *Carlin* (1980) described a linear relationship between the natural logarithm of the number of adenylate units at the 3′ end of a mRNA from different species of organisms and the time of origin of each species. Extrapolation back to one adenylate unit implies the origin of mRNA (origin of life?) occurred 3.85 ± 0.2 Gyr ago. Therefore, it is appropriate to estimate the maximum time required to originate life if life first existed 3.8 Gyr ago. Still using f(t) with T = 70 m.y., but assuming that life first originated 3.8 Gyr ago and that impacts with kinetic energy of 2×10^{34} ergs were required to stop the processes leading to the origin of life, we found that 2.5 m.y. is the maximum time required to originate life (*Oberbeck and Fogleman,* 1989b). If impacts with kinetic energy of 2×10^{35} ergs were required to stop the processes leading to the origin of life and life originated 3.8 Gyr ago, we find that 11 m.y. is the maximum time required to originate life.

Next, assuming a more conservative impactor flux estimate, *Oberbeck and Fogleman* (1989b) used equation (1) with $k = 1.26 \times 10^{-12}$, T = 150 m.y., and t_o = 3.2 Gyr to obtain another estimate of the maximum origination time for life, if life first originated 3.8 Gyr ago. We found that if formation of 850-km-diameter and larger impact basins were able to stop prebiotic chemical evolution, then the maximum time between such impacts and the maximum time needed to originate life is 5.9 m.y. If impacts equal or larger than 1611-km diameter were required to stop prebiotic chemistry, the maximum time needed to originate life was 25 m.y. (*Oberbeck and Fogleman,* 1989b). Now that we have estimated the maximum time available for origination of life on Earth with a variety of assumptions we can use a similar type of analysis to determine the time available between sterilizing impacts on Mars in order to test the possibility that life might have had sufficient time to originate on Mars.

TIME AVAILABLE FOR ORIGIN OF LIFE ON MARS

The time interval between impact events capable of stopping the origination of life on Mars can be derived in the same way as it was for Earth by using the equation for T_w above and by using appropriate values of the constant, C, and D_{min} and D_{max} for Mars. C is the reciprocal of the product of the surface area of Mars and the cratering rate constant for Mars. The cratering rate on primordial Mars is given as half the cratering rate on the Moon (*Hartmann,* 1977). Thus, the appropriate value of k is 8.7×10^{-14} and $C = 7.9 \times 10^4$. The choice of D_{min} necessary to sterilize Mars is more difficult than for the Earth. While oceans may have been present on primordial Mars, they may have been shallower than terrestrial oceans (*Pollack et al.,* 1987) and the surface area of Mars is less than on Earth. We therefore attempt to allow for these differences and other parameters in the following analysis for an estimate of D_{min} and note, in advance, that the final conclusions allow for considerable uncertainty in the choice of D_{min}.

We note that different size impacting bodies may have been required to produce the same degree of heating of the two planets. The size of impact basin considered necessary to sterilize Earth's surface was related to the energy required to vaporize the oceans. This energy is proportional to the surface area of Earth. The surface area of Mars is 28% that of Earth so that the energy required to sterilize the surface of Mars should be at most 28% that required on Earth. Using the same crater scaling laws as in the earlier development, we now derive an estimate for the size of an impact crater having the same sterilization effect of a terrestrial basin of a given size.

If D_e is the basin diameter necessary to sterilize the Earth, we now derive D_m, the basin diameter necessary to sterilize Mars. Setting the sterilization energy equal to a fraction of that required on Earth and solving for projectile diameter D_m on Mars in terms of D_e on Earth

$$KE_m = C(KE_e),\ C = 0.28$$

and

$$d_m = C^{1/3}[v_e/v_m]^{2/3}d_e = 1.12\ d_e;\ v_e = 12.5\ \text{km/sec},\ v_m = 5.6\ \text{km/sec}$$

From *Schmidt and Holsapple* (1982) crater diameter scales with impact velocity and projectile size, d, according to

$$D_e = 3.31 v^{1/3}\,(d_e)^{5/6}$$

Modifying this to account for different gravity g_m on Mars, we obtain

$$D_m = 3.31\ [g_m/g_e]^{-1/6} \cdot v^{1/3}(d_m)^{5/6}$$

Upon substitution of d_m into the equation for D_m and rearrangement, we obtain

$$D_m = 0.986\ D_e$$

Using this relationship and the terrestrial basin diameters 850 and 1611 km corresponding to 2×10^{34} and 2×10^{35} ergs, we obtain for the analog martian sterilization basin sizes 838 km and 1588 km, respectively. Similar considerations suggest that if the largest size event expected on Earth, D_{max}, is 2600 km, then the largest size event expected on Mars is 2340 km.

First we estimate the time between planetary sterilization impacts on Mars using the impactor flux laws f(t) given by *Maher and Stevenson* (1988) where T = 70 m.y. and t_o = 3.4 Gyr. The values of C, D_{min} = 838 km, D_{max}, and f(t) computed using t = 3800 m.y. + 35 m.y., when substituted into the equation for T_w, give T_w = 35 m.y. Thus, the time between 838-km diameter and larger basins was 35 m.y., just before liquid water vanished on Mars 3.8 Gyr ago. Thus, if only 2.5 m.y. were required to originate life on Earth between comparable impacts on Earth (larger than 850 km basins), there should have been sufficient time for life to originate on Mars between sterlizing impacts while liquid water was present.

The conclusion regarding the possibility of origination of life on Mars is quite insensitive to the minimum size of the impact basin required to sterilize Mars. For example, it may be shown that the time interval between impacts with kinetic energy as small as 0.2% of the kinetic energy required to sterilize Earth would still be as large on Mars as the time between sterilizing impacts on Earth. In the event that impact events as large as 1588 km were required to sterilize Mars, then it may be shown that substitution of t = 3.8 Gyr + 81 m.y., D_{min} = 1588 km, and D_{max} = 2340 km gives 81 m.y. This exceeds the 11 m.y. between the analog 1611 km terrestrial impacts.

Let us assume that the terrestrial martian and lunar impact flux rates decay according to T = 150 m.y. and t_o = 3.2 Gyr, i.e., the parameters obtained from the crater size distribution-age relations given by Baldwin. While the time between terrestrial impacts just before 3.8 Gyr ago increases relative to the times obtained using T = 70 m.y., there is a corresponding increase in the time between analog impacts on Mars. Thus, the time available on Mars between giant impacts before 3.8 Gyr ago remains greater than the time between analog planetary sterilization giant impacts on Earth before 3.8 Gyr ago. Use of a more conservative flux does not change the conclusion that there could have been sufficient time to originate life on Mars before 3.8 Gyr ago if life originated on Earth at least 3.8 Gyr ago.

We realize that crater formation is a random process (*Hartmann,* 1984), the computed time between impacts is an average value, and, in reality, impact events are not evenly spaced in time. Uncertainties of the order of a factor of 2 can result from this (W. K. Hartmann, personal communication, 1989). After more extensive analysis, we have found that the maximum origination times could be up to three times the above values if T_w is much less than T. For the case where T_w is of the same order as T, the effect is not as great. However, there is high probability that the time available for origination of life on Mars was longer than the time available between sterilizing impacts on Earth before life originated. For example, for T = 70 m.y., the average maximum time between 850-km-diameter and larger impact basin formations just before the oldest evidence of life on Earth was 7%, as long as the time between analog 838-km-diameter and larger basin formation events on Mars just before liquid water vanished. The average time intervals between 1611-km events on Earth was about 14% of the average maximum time between analog 1588-km diameter basins formation events on Mars. Thus, even with the uncertainty introduced by randomness, if life originated on Earth by 3.8 Gyr ago, there would have been a better chance on Mars than on Earth that prebiotic processes escaped sterilization by impactors before liquid water vanished. The effect of allowing for the random nature of impact is at most to triple the time estimates required to originate life on Earth and the time available on Mars. This will be discussed in more detail in a future paper.

If life first originated on Earth 3.5 Gyr ago, the conclusion regarding origin of life on Mars is not as clear. T_w, the time between formation of impact craters equal or larger than 850 km in diameter 3567 m.y. ago on Earth, was 67 m.y., and the time between 1611-km basins 3633 m.y. ago was 133 m.y. These times, in that event, would represent the calculated maximum times required to originate life on Earth and they are about 1.9 and 1.6 times the lengths of the time intervals available between the analog sterilizing impact events on Mars before the liquid water vanished 3.8 Gyr ago. Thus, if life on Earth first originated 3.5 Gyr ago and required the maximum times to originate, there would not have been sufficient time for life to originate on Mars before liquid water vanished 3.8 Gyr ago. However, the time required to originate life on Earth may have been less than the maximum calculated times of 67 m.y. or 133 m.y. Thus, even under the assumption that life on Earth originated as late as 3.5 Gyr ago, there is a reasonable chance that life may have originated on Mars and this is sufficient reason to explore Mars for evidence of such microorganisms.

DISCUSSION

After considering the requirements for origination of life on Earth and Mars, we conclude that if life first existed on Earth 3.8 Gyr ago, conditions could have been favorable for the origin of life on Mars before 3.8 Gyr ago. We assumed that giant impacts were capable of frustrating the origination of life. We found that impacts associated with the heavy bombardment would not have been as adverse to the origination of life on Mars as on Earth because less debris impacted Mars before 3.8 Gyr ago than impacted the Earth. The impact analysis presented leads to the conclusion that there should have been time enough between sterilizing impacts for life to originate on Mars. However, even if certain assumptions in the model are not valid, this type of analysis leads to important conclusions. For example, if one assumes that giant impacts of any energy were incapable of stopping the process leading to the origin of life, then the time between accretion of Earth and the first microfossils is a gross estimate of the maximum time required to originate life. The time available for origination of life on Mars before liquid water vanished was approximately equal to this interval. This, alone, is sufficient justification for further exploration of Mars for evidence that life originated on that planet.

The discovery of fossil microorganisms or life on Mars would have profound influence upon our thinking about the possibility of the universal origination of life elsewhere. The process of solar-system development may be intimately connected to the origin of life on planetary surfaces by constraining the origin of life, by forming planets large enough so that the internal heat perpetuates an atmospheric pressure high enough to sustain liquid water at the surface, and by providing prebiotic reactants from planetary outgassing or cometary impact. Study of the possibility that life may have originated on Mars is a case study that may apply to planets in other solar systems. Comparisons of primordial conditions on Mars and Earth might also help us understand the origin of life on Earth where the record has been destroyed by subsequent geologic processes.

The evidence for life on Earth prior to 3.5 Gyr ago is indirect. Actual fossil evidence may never be found because Earth was geologically very active before 3.5 Gyr ago. For example, the entire cratering record of the heavy bombardment of Earth has been obliterated by geologic processes. Apparently, geologic processes were not so active on Mars after 3.8 Gyr ago because many of the craters formed during the heavy bombardment still remain. Thus, there is the exciting possibility that life that originated before 3.8 Gyr ago on Mars may have left fossil evidence that may still be preserved in the heavily cratered terrain on Mars in formations of the Upper Noachian period and perhaps earlier. Formations of the Upper Noachian period are excellent deposits to search for fossil evidence of first life. Sedimentary deposits may exist on the floors of craters and channels that may have been the sites of lakes and streams before 3.8 Gyr ago.

There has been considerable interest in the nature of the earliest terrestrial microorganisms. Based upon sequencing of DNA and RNA, *Woese* (1981) discovered a third kingdom of microorganisms called the archaebacteria. Whereas before this work there were considered to be only prokaryotic (without nucleus) and eurkaryotic (with nucleus) bacteria, Woese identified a new form of bacteria without a nucleus. All three forms are now believed to have evolved from a simpler lifeform that was previously considered to be a prokaryotic form. The new universal ancestor, believed to be present before 3.5 Gyr ago, is referred to as the progenote, which is also without a nucleus, but may have differed from prokaryots because it may have been in the process of developing the cell wall. Fossils of ancient lifeforms on Mars may thus provide glimpses of what missing terrestrial ancestral microorganisms may have been like. Perhaps one of the greatest values of Mars exploration is to try to learn about the nature of the most primitive lifeforms that were the first products of chemical evolution because such studies are not possible on Earth.

If life originated on Mars, it might have become extinct or been forced underground after 3.8 Gyr ago because liquid water vanished from the surface soon after the heavy bombardment ended. *Carr* (1989) has studied the role of large impacts and volcanism in recycling CO_2 to the martian atmosphere. He concluded that both processes may have been required to keep an atmosphere intermittently present until 3.8 Gyr ago. This would have kept liquid water intermittently present. *Pollack et al.* (1987) has estimated a weathering time constant of the order of several times 10^7 years for removal of CO_2 from the atmosphere by formation of carbonate rocks. During these periods, liquid water could have existed at the surface. There would have been sufficient time [$2.5-11 \times 10^6$ years, assuming 10^{34}-10^{35} erg impacts sterilized the planet, T = 70 m.y. in f(t) and life first existed on Earth 3.8 Gyr ago] for life to originate at the surface of Mars. However, after the heavy bombardment ended, impact would no longer have contributed to recycling CO_2 (*Carr,* 1989). CO_2 partial pressures would have decreased, and liquid water and microorganisms may have existed only beneath the surface under ice, and life at the surface may have become extinct.

Impacts associated with the heavy bombardment on Mars may have at first prevented the origin of life; then when impact rates declined there was time enough for life to originate between adverse impacts. During this period, impact rates were still high enough to sustain the CO_2 atmosphere so liquid water existed at the surface and life may have survived there. Impact rates then declined enough so that the CO_2 atmosphere became greatly diminished, liquid water vanished from the surface, and surface life would have become extinct. After this time, life may have existed under protective layers of ice.

REFERENCES

Arrhenius G. (1987) The first 800 million years: Environmental models for Earth. *Earth, Moon, and Planets, 37,* 187-199.

Baldwin R. B. (1987) On the relative and absoute ages of seven lunar front face basins, II from crater counts. *Icarus, 71,* 19-29.

Carlin R. K. (1980) Poly (A): A new evolutionary principle. *J. Theor. Biol., 82* 353.

Carr M. H. (1989) Recharge of the early atmosphere of Mars by impact-induced release of CO_2. *Icarus, 79,* 311.

Carr M. H. and Clow G. D. (1981) Martian channels and valleys: Their characteristics, distribution, and age. *Icarus, 48,* 91-117.

Carr M. H., Saunders R. S., Strom R. G., and Wilhelms D. E. (1984) The geology of the terrestrial planets. *NASA SP-469.* 317 pp.

Clark B. C. (1988) Primeval procreative comet pond. *Origins of Life and Evolution of the Biosphere, 18,* 209.

Fanale F. P. (1971) History of martian volatiles: Implications for organic synthesis. *Icarus, 15,* 279-303.

Fernandez J. A. and Ip W. H. (1983) On the time evolution of the cometary influx in the region of the terrestrial planets. *Icarus, 54,* 377-387.

Fox S. W. and Dose K. (1977) *Molecular Evolution and the Origin of Life.* Marcel Dekker, New York. 370 pp.

Grieve R. A. F. and Dence M. R. (1979) The terrestrial cratering record II. The crater production rate. *Icarus, 38,* 230-242.

Hartmann W.K. (1977) Relative crater production rates on planets. *Icarus, 31,* 260-276.

Hartmann W. K. (1984) Stochastic does not equal ad hoc (abstract). In *Papers Presented to the Conference on the Origin of the Moon,* p. 3. Lunar and Planetary Institute, Houston.

Hartmann W. K. and Davis D. R. (1975) Satellite-sized planetesimals and lunar origin. *Icarus, 24,* 504-515.

Holland H. D. (1984) *The Chemical Evolution of the Atmosphere and Oceans.* Princeton Univ., Princeton, New Jersey. 582 pp.

Kasting J. F. and Ackerman T. P. (1986) Climatic consequences of very high carbon dioxide levels in the Earth's early atmosphere. *Science, 234,* 1383.

Lahav N., White D., and Change S. (1978) Peptide formation in the prebiotic era: Therman condensation of glycine in fluctuating clay environments. *Science, 201,* 67.

Levine J. S. (1985) *The Photochemistry of Atmospheres, Earth, the Other Planets, and Comets.* Academic, New York. 518 pp.

Maher K. A. and Stevenson D. J. (1988) Impact frustration of life. *Nature, 331,* 612-614.

Mancinelli R. L. (1988) The nitrogen cycle on Mars. In *Exobiology and Future Mars Missions,* p. 42. NASA Conf. Publ. 10027.

Mancinelli R. L. and McKay C. P. (1988) The evolution of nitrogen cycling. *Origins of Life and Evolution of the Biosphere, 18,* 311.

Miller S. L. (1954) Production of some organic compounds under possible primitive Earth conditions. *J. Amer. Chem. Soc. 77,* 2351-2361.

Miller S. L. (1984) The prebiotic synthesis of organic molecules and polymers. In *Abstract of Chemical Evolution,* p. 85. Wiley, New York.

Miller S. L. and Bada J. L. (1988) Submarine hot springs and the origin of life. *Nature, 334,* 609-611.

Matsui T. and Abe Y. (1986) Evolution of an impact-induced atmosphere and magma ocean on the accreting Earth. *Nature, 319,* 303-305.

Mukhin L. M. (1989) Hypervelocity impacts of planetesimals as a source of organic molecules and of their precursors on the early Earth (abstract). In *Lunar and Planetary Science XX,* pp. 737-738. Lunar and Planetary Institute, Houston.

Oberbeck V. R. and Fogleman G. (1989a) Impacts and the origin of life. *Nature, 339,* 577.

Oberbeck V. R. and Fogleman G. (1989b) Estimates of the maximum time required to originate life. *Origins of Life and Evolution of the Biosphere,* in press.

Oberbeck V. R., McKay C. P., Scattergood T. W., Carle G. C., and Valentin J. R. (1989) The role of cometary particle coalescence in chemical evolution. *Origins of Life and Evolution of the Biosphere, 19,* 39.

Pavlovskaya T. E. and Pasynskii A. G. (1959) The original formation of amino acids under the action of ultraviolet rays and electric discharges. In *The Origin of Life on Earth* (A. I. Oparin, A. G. Pasynskii, and A. E. Brounshtein, eds.), p. 151. Pergamon, New York.

Pinto J. P., Gladstone G. R., and Yung Y. L. (1980) Photochemical production of formaldehyde in Earth's primitive atmosphere. *Science, 210,* 183-185.

Pollack J. B., Kasting J. F., and Richardson S. M. (1987) The case for a wet, warm climate on early Mars. *Icarus, 71,* 203-224.

Schidlowski M. (1988) A 3,800-million-year isotopic record of life from carbon in sedimentary rock. *Nature, 333,* 313.

Schmidt R. M. and Holsapple K. A. (1982) Estimates of crater size for large-body impact: Gravity scaling results. In *Geological Implications of Impacts of Large Asteroids and Comets on the Earth,* (L. T. Silver and P. H. Schultz, eds.), pp. 93-102. Geol. Soc. Amer. Spec. Pap. 190.

Schopf J. W. (1983) *Earth's Earliest Biosphere: Its Origin and Evolution,* Princeton Univ., Princeton, New Jersey. 543 pp.

Stevenson D. J. (1987) Origin of the Moon-The collision hypothesis. *Annu. Rev. Earth Planet. Sci., 15,* 271-315.

Woese C. R. (1981) Archaebacteria. *Sci. Am., 244,* 98-125.

Proceedings of the 20th Lunar and Planetary Science Conference, pp. 479-486
Lunar and Planetary Institute, Houston, 1990

Imaging Spectroscopy of Mars (0.4-1.1 μm) During the 1988 Opposition

J. F. Bell III, T. B. McCord, and P. G. Lucey

Planetary Geosciences Division, Hawaii Institute of Geophysics, University of Hawaii, Honolulu, HI 96822

Spectral reflectance data for Mars have previously been obtained at low to moderate spectral resolution ($R = \lambda/\Delta\lambda \approx 20$) in the visible to near-IR. Reported here are the initial results from a comprehensive series of high spectral resolution ($R \approx 350$) visible to near-IR imaging spectroscopic observations of Mars during the 1988 opposition. New 0.4-1.05-μm spectra show absorption features near 0.6-0.7 and 0.80-0.95 μm that have been attributed to Fe^{3+} crystal field transitions, confirming measurements made by J. F. Bell, T. B. McCord, and P. D. Owensby (unpublished manuscript, 1989) at lower spectral resolution. Attempts to reconcile the overall spectral shape with those of laboratory minerals or Mars analog soils spectra affirm the recent interpretations of Morris et al. (1989) that hematite, occurring in a wide span of particle sizes down to "crystals" only a few unit cells across, may account for nearly all of the ~18 wt.% Fe_2O_3 measured by the Viking Landers. The presence of other ferric oxides/oxyhydroxides, Fe-bearing smectite clays, or ferric sulfates cannot be ruled out from these data, however, without further laboratory study and data analysis.

INTRODUCTION

Imaging spectroscopy is a powerful tool for identifying mineralogy and for mapping its spatial distribution across a planetary surface. The Viking and Voyager missions showed the improved application of multispectral imaging to mapping material units determined by defining spectral units in only a few spectral bands. The result was color as well as morphology being used to define spatial units (e.g., *Soderblom et al.,* 1978; *Nelson et al.,* 1986). However, the spectral resolution and sampling did not allow characterization of various color units in terms of composition or mineralogy. Detailed compositional mapping requires high spatial *and* spectral resolution so that mineralogic differences that exist on a planetary surface can be directly correlated with observed morphologic change. Such information has recently been obtained for Mars by the ISM spectrometer on the Soviet Phobos mission, and hopefully similar data will be obtained for the Jovian satellites by the NIMS spectrometer on project Galileo (*Carlson,* 1981) and for terrestrial remote-sensing applications by the HIRIS instrument (*Goetz et al.,* 1985).

Imaging spectroscopy from groundbased telescopes is also currently possible because of the availability of two-dimensional detector array technology that allows simultaneous spatial and spectral measurements to be made across a two-dimensional grid of detector elements. Using conventional astronomical instruments such as spectrographs and imaging interferometers, high spectral resolution measurements ($R = \lambda/\Delta\lambda \approx 300$-$800$), similar to those proposed for the above planetary missions, can be made across a wide range of wavelengths. Groundbased observations, of course, suffer from the typically poor spatial resolution obtained on planetary targets. This can be alleviated somewhat by choosing (1) a large diameter telescope that affords good light gathering power despite the high magnification, (2) a high altitude or similarly stable observing site that minimizes atmospheric seeing difficulties, and (3) the time of observation to coincide with extremely favorable "close" passes between the Earth and the object being viewed. An observational program of imaging spectroscopy of Mars was carried out during the summer of 1988 that satisfied all of the above criteria.

This paper presents imaging spectroscopic data and some preliminary interpretations acquired at Mauna Kea Observatory (MKO) during the 1988 perihelic opposition of Mars at the highest spatial resolution achieved from Earth for spectrometric measurements (all contiguous spatial sampling for nearly 70% of the planet) and at a higher spectral resolution ($R \approx 350$) than most previous measurements in the 0.4-1.1-μm wavelength range. The goals of this paper are to present and discuss the highest spectral resolution data yet obtained in the visible to near-IR of the surface of Mars and to show the advantages of the imaging spectroscopic method in carrying out planetary observations.

OBSERVATIONS

A wide-field grism (grating-prism) spectrograph mounted on the University of Hawaii 2.24-m telescope with an f/35 secondary at MKO was used to obtain the data. The techniques used to obtain this imaging spectroscopic dataset are nearly identical to and are based on those used to obtain similar data of the lunar surface by *Lucey* (1988). Images and spectra were acquired in nearly 250 bandpasses from 0.4-1.05 μm by slowly scanning a 0.25 arcsec slit across the planet and imaging dispersed spectral information for each slit position onto an 800×800 Si-CCD (Fig. 1). The slit images were later assembled into full-disk images of Mars. Sampling was 5.2 pixels/arcsec along the slit and 1.6 pixels/arcsec in the cross-slit (scan) direction. The effective spatial resolution on Mars, as determined by scanning a standard star (theoretical point source), was ~150-250 km (limited by atmospheric seeing). The estimate of this effective spatial resolution is derived by observing the effective size (in pixels) of the standard star, then defining the size of that pixel "box" as one resolution element. A resolution element within our data was typically 2×6 pixels (FWHM), or ~1.0 arcsec square, which at f/35 corresponded to ~150 km for regions near the martian center-of-disk during the times of observation. Standard data reduction was performed using the star η Psc (G8III) as a solar analog.

These image cubes [three dimensional data arrays with dimensions (x,y,λ)—two spatial and one spectral; see Fig. 2] were obtained during four separate observing runs in May, August, and September 1988 (see Table 1). Weather conditions, both terrestrial and martian, were excellent throughout most of the summer.

BACKGROUND

Previous observations in the visible to near-IR have shown an apparent, near-total lack of crystalline structure in the martian spectra (see *McCord and Adams,* 1969, and *Singer et al.,* 1979, for reviews of early observations). Interpretations of groundbased reflectance spectra by *Evans and Adams* (1979) and *Singer* (1982) and Viking Lander geochemical data by *Clark et al.* (1982), *Soderblom and Wenner* (1978), and *Gooding and Keil* (1978) have suggested that the altered volcanic glass palagonite may be an abundant component of martian surface fines. Palagonites and palagonite-like materials have not been able to reproduce the Viking Lander LR and PR surface chemistry results, however (*Oyama and Berdahl,* 1977; *Banin and Margulies,* 1983). Banin and Margulies suggested that Fe-rich smectite clays provide a better match to the Lander surface chemistry data, and *Banin et al.* (1985) showed such materials to also be reasonable spectral analogs based on the available data. *Toulmin et al.* (1977) pointed out that palagonites and smectites may form simultaneously so that these interpretations are not mutually exclusive. Additionally, recent reinterpretation of Viking Lander photometry by *Guinness et al.* (1987) has noted that amorphous iron-bearing palagonite-like materials do not have the steeper slopes and greater spectral curvature necessary to explain the Viking data. *Singer et al.* (1979) reported evidence for a weak 0.87 μm Fe^{3+} band in the telescopic data but not of any other diagnostic bands or well-quantified slope inflections shortward of 0.7 μm. Recently, *Morris et al.* (1989) and Morris and Lauer (R. V. Morris and H. V. Lauer Jr., unpublished manuscript, 1989) have reinterpreted these data and suggested that the weak 0.87 μm band seen in the martian bright region spectrum is indicative of crystalline Fe^{3+} minerals, most likely hematite (αFe_2O_3). Telescopic measurements during the 1988 opposition at

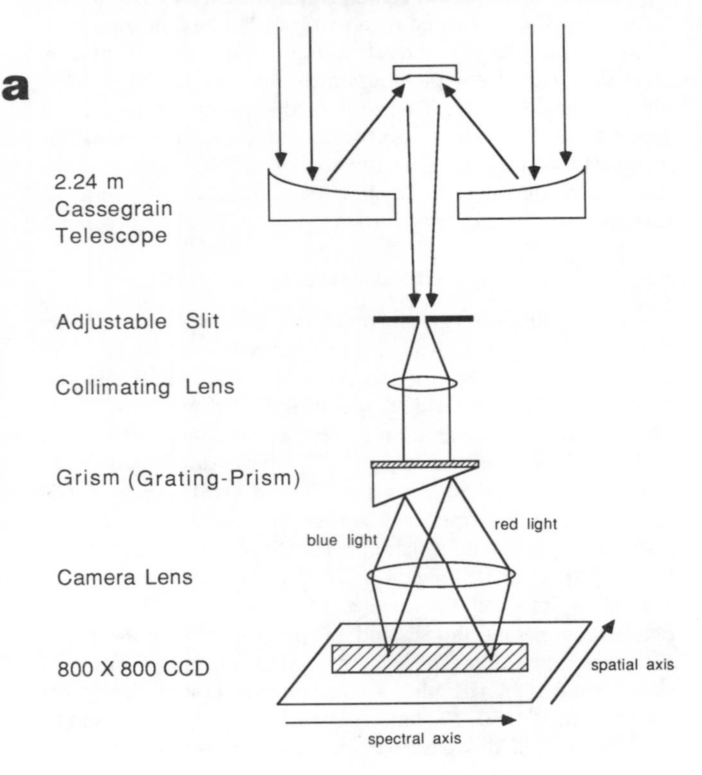

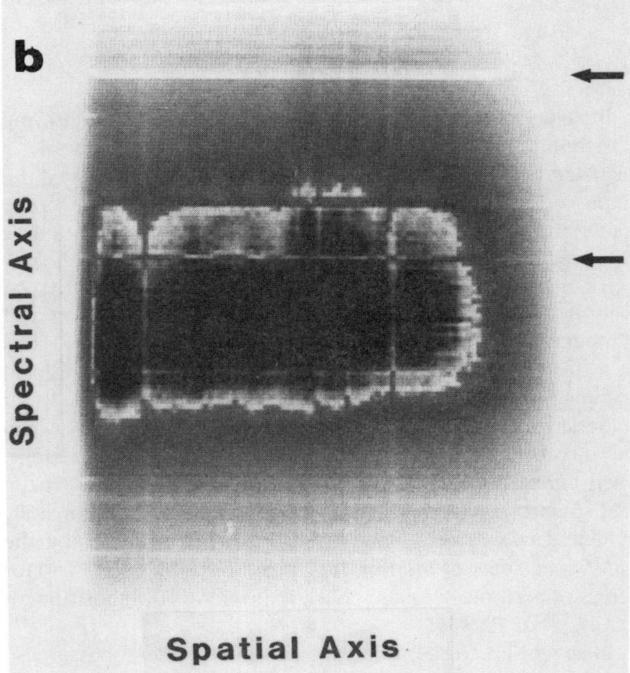

Fig. 1. (a) Schematic representation of the high-resolution grating-prism ("grism") instrument used to obtain imaging spectroscopic data of Mars. The 0.25 arcsec adjustable slit is slowly scanned across the planet ("pushbrooming"), producing ~50 two-dimensional "planes" of data representing individual point spectra for each pixel across the slit. **(b)** Example of raw imaging spectrometer data, existing as a spatial vs. spectral image plane. Even in the raw data, distinct spectral features (vertical axis) can be seen, such as the 0.76-μm and 0.69-μm terrestrial O_2 A and B absorptions (top and bottom arrows, respectively). Spatial variation (horizontal axis) across the slit is evident. Vertical lines running through the image are associated with bad CCD columns or slight imperfections in the slit.

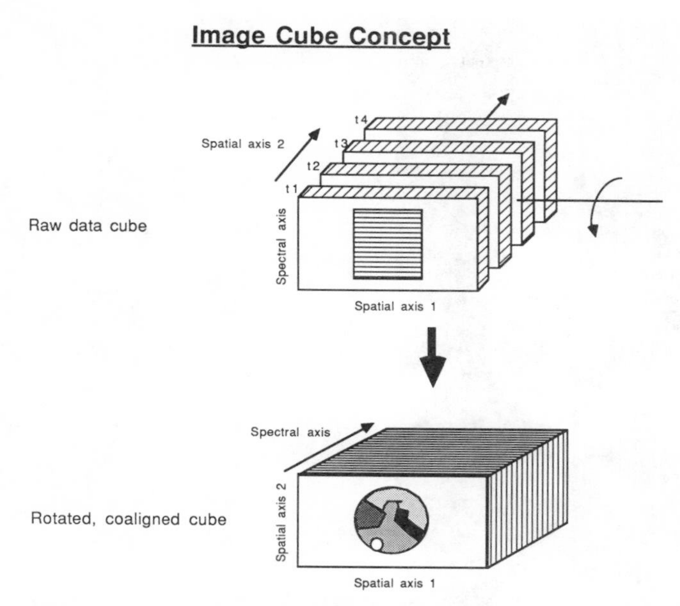

Fig. 2. The raw images produced as in Fig. 1 are stacked together through software into a three-dimensional "cube" with x,y,z axes corresponding to the spatial axis along the slit, the spectral axis, and the spatial axis of different slit positions on the planet (different pushbroom positions). This cube is then "rotated" through software so that the facing image planes represent the two spatial axes (thus an image can be built) and the spectra are produced by plotting data through the axis into the page. Examples of images produced can be found in Fig. 4.

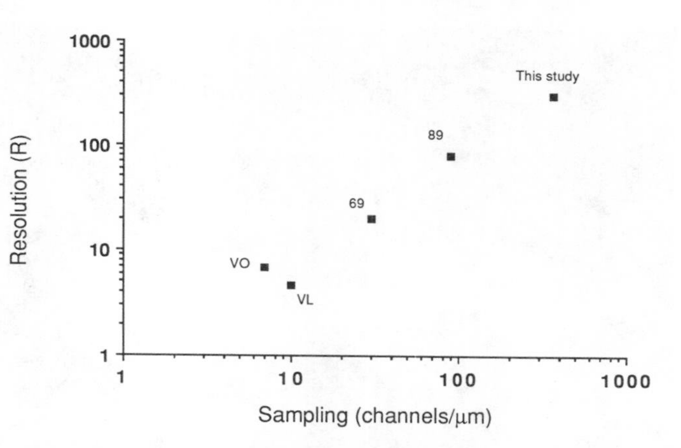

Fig. 3. Spectral resolution vs. spectral sampling for previously published Mars spectral data in the visible to near-IR (0.31-1.1 μm). Data points correspond to Viking Orbiter (VO) and Viking Lander (VL) multispectral imaging, 1969 (69) narrow band filter data of *McCord and Westphal* (1971), 1988 (89) Circular variable filter data of Bell et al. (J. F. Bell III, T. B. McCord, and P. D. Owensby, unpublished manuscript, 1989) and this study.

slightly higher spatial and spectral resolution and much finer spectral sampling by Bell, McCord, and Owensby (J. F. Bell III, T. B. MCord, and P. D. Owensby, unpublished manuscript, 1989) have confirmed this interpretation.

Much effort has been expended on the interpretation of the data of *McCord and Westphal* (1971), *McCord et al.* (1977, 1978), and *Singer et al.* (1979) to resolve the complex issues of mineralogy and degree of crystallinity of the martian surface soils. Though the capability to acquire much improved spectral data in the near-UV to near-IR has existed for some time, no new studies have been published prior to 1988 that present higher spectral resolution and sampling data with which to address the problem (see Fig. 3). It was clear that the major efforts in interpretation and identification would benefit greatly from further higher spectral resolution data. This has been the motivation for this project.

TABLE 1. Manua Kea 2.24-m telescope observations, 1988.

Date (UT)	L_s	Phase angle	λ Cover (μm)	No. of Images	No. of Spectra
5/1/88	188°	44°	0.55-1.1	~200	~300
8/25/88	259°	28°	0.55-1.1	400	1000
8/26/88	260°	27°	0.4-0.8	750	5000
9/24/88	278°	5°	0.55-1.1	200	800
9/25/88	278°	4°	0.4-1.1	550	1500
9/28/88	280°	3°	0.4-0.8	300	1000

INITIAL RESULTS

Some representative images and spectra from the new 1988 high resolution imaging spectroscopy dataset are presented in Figs. 4-6. The volume of data obtained by the imaging spectroscopic method is huge: 14 full image cubes from 6 different nights representing ~2500 full-disk images and ~10,000 individual point spectra in ~250 bandpasses from 0.36-1.05 μm (>500 Mbytes of data!). Several features of note in the spectra of Figs. 5 and 6 include: (1) spectral slopes, inflection points, and the near-UV absorption edge are well characterized; (2) a prominent wide band centered near 0.8-1.0 μm is evident and is most likely the combination of a strong 0.8-0.9 μm Fe^{3+} band [$^6A_1 \rightarrow {}^4T_1(^4G)$] and possibly the short wavelength wing of a 1.0-1.1 μm Fe^{2+} band that may be associated with exposed bedrock mafic mineralogies (*Adams*, 1968; Table 2); (3) a prominent 0.62-0.73 μm "cusp" also due to an Fe^{3+} electronic transition band [$^6A_1 \rightarrow {}^4T_2(^4G)$] that has not been evident in any lower resolution studies prior to 1988; (4) a relative reflectivity maximum near 0.75 μm; (5) apparent (though weak) variations between the spectra of classical bright and dark regions on the planet, specifically in the variation of spectral slopes in the visible and in the depths of the ~0.86-μm and ~0.65-μm Fe^{3+} bands; however, the lack of absolution flux calibration for these spectra precludes further quantification of bright/dark region similarities or differences without further data reduction; and (6) terrestrial atmospheric absorption features near 0.69 μm and 0.76 μm (Oxygen B and A lines) and 0.94 μm (water vapor; see Table 2).

Indications of this spectral variation are more fully evident in Fig. 7. Here, a distinct advantage of the imaging spectroscopic method is seen in its ability to produce contiguous spatial information across the planet while at the same time showing apparent variations in the 0.62-0.73-μm band. A great deal of information, equivalent to several hundred relative

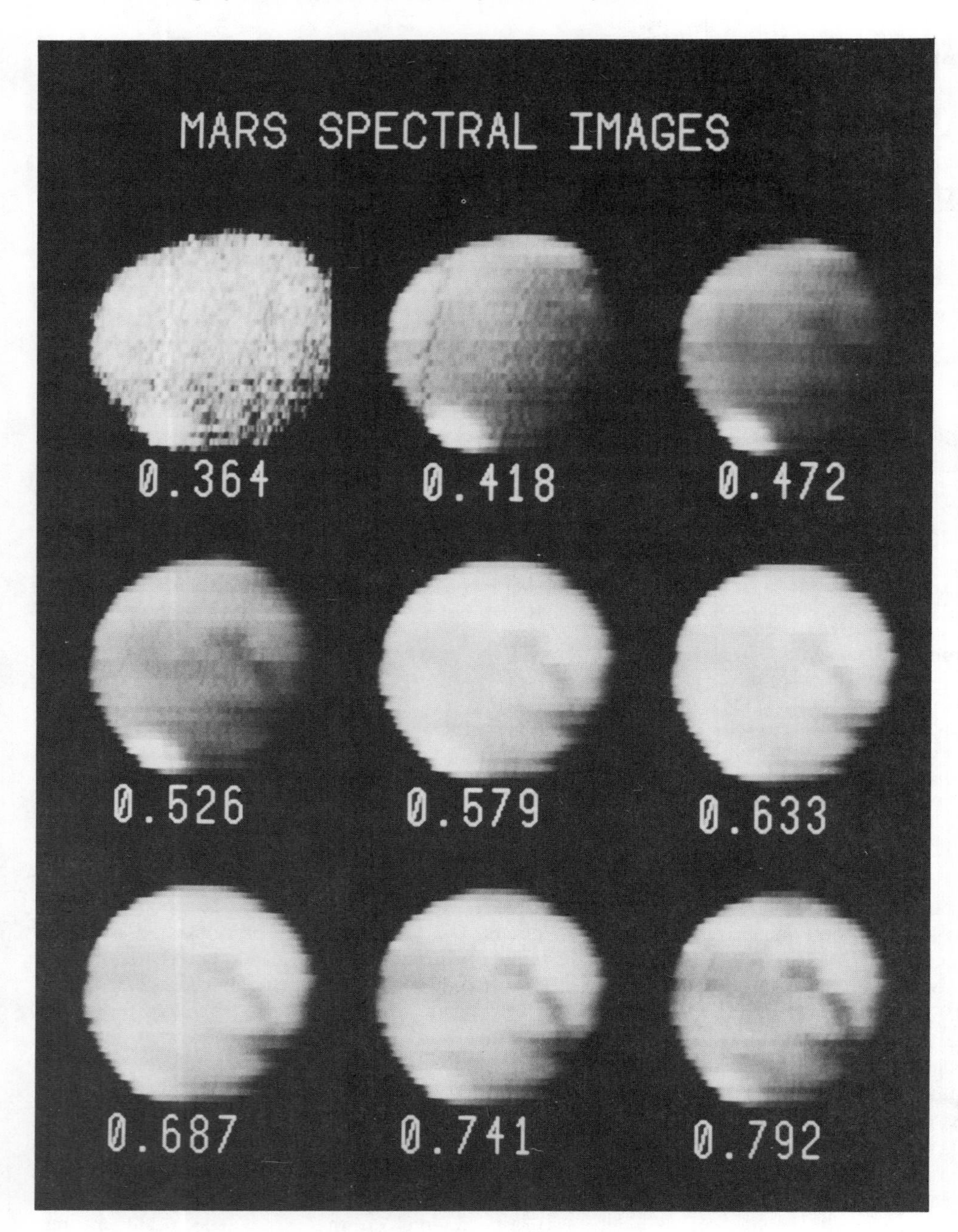

Fig. 4. Examples of Mars spectral images produced by the methods outlined in this paper. These photographs show Mars at nine of the 250 wavelengths sampled (in micrometers). The central meridian is approximately 0°. The prominent dark region visible in the redder images comprises Sinus Meridiani and Sinus Sabaeus. Contrast of surface features drops markedly into the near-UV, as does S/N. In the images shortward of 0.5 μm, the north polar hood can be clearly distinguished. For further examples of such images, see also *Bell et al.* (1989).

reflectance spectra, is presented in a straightforward, understandable format. Much of the future reduction of this dataset will rely heavily on such techniques.

The spectral data are presented as scaled relative "reflectance" (ratio of Mars flux to solar analog stellar flux), and when enough standard star observations were possible (for example, on the photometric night of 8/26/88), atmospheric extinction corrections were performed. Further photometric corrections are needed on much of the data to yield normal albedo or absolute reflectance. Detailed comparisons of spectra for widely separated areas on the planet or data taken on different nights must await these corrections. Nonetheless, the data in their current stage of reduction show unmistakable evidence, based on interpretations of laboratory and other telescopic data discussed below, of crystalline Fe^{3+} absorption, manifested as bands and inflections in the visible to near-IR spectra of the martian surface and airborne dust.

DISCUSSION

These data show more spectral structure attributable to the presence of crystalline material on the martian surface than do previous datasets [with the exception of CVF spectrophotometer data taken by Bell, McCord, and Owensby (J. F. Bell III, T. B.

McCord, and P. D. Owensby, unpublished manuscript, 1989) at MKO during the same opposition] because the spectral resolution, spectral sampling, and data precision are sufficient in these data to reveal this structure. No 1% or greater spectral resolution data for the martian surface have been published. Previous studies in the visible to near-IR in the last few decades have apparently been at a spectral sampling interval too coarse to unambiguously show the types of absorption features reported here (see Fig. 3). Additionally, the excellent 1988 perihelic opposition and extremely favorable terrestrial atmospheric seeing quality at MKO combined with the lack of major global dust storm activity on Mars may have effectively increased the spatial resolution to below the 150-km scale, reducing areal mixing of different mineralogic units (*Singer and McCord,* 1979; *Bell and McCord,* 1989).

Somewhat of a controversy currently exists over (1) what the iron-bearing phases are in the martian fines and (2) what is their degree of crystallinity. Obviously, these two issues are intimately connected, and the two most favored interpretations of the previously available data are that the fine materials either consist largely of Fe-rich smectite clays or that they are predominantly palagonitic. Clearly, these new data partially

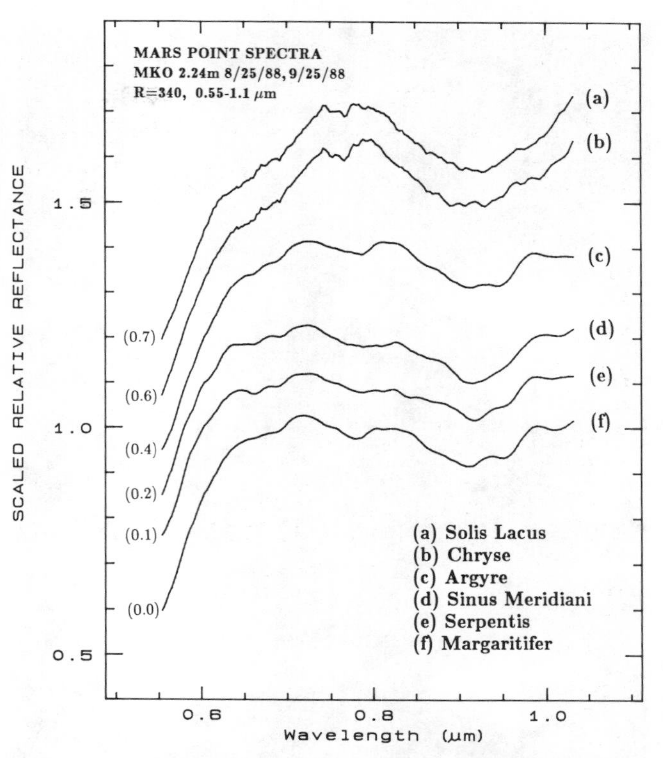

Fig. 6. Relative reflectivity spectra (Mars/ηPsc) from 0.55 to 1.03 μm of six representative regions (~200 km) on Mars. Scaling and smoothing as in Fig. 5. Most notable among these spectra are the 0.65-μm "cusp" (as in Fig. 5), the relative reflectivity maximum at 0.75 μm (cf. Fig. 8), and the prominent band at ~0.81-0.94 μm, which is most likely a combination of a ~0.86-μm Fe^{3+} band and possibly the short wavelength wing of a 1.0-1.1-μm Fe^{2+} band (see text). The features near 0.69 μm, 0.76 μm, and 0.94 μm are caused by telluric atmospheric oxygen and water vapor absorptions (Table 2) and can act as excellent wavelength calibration sources.

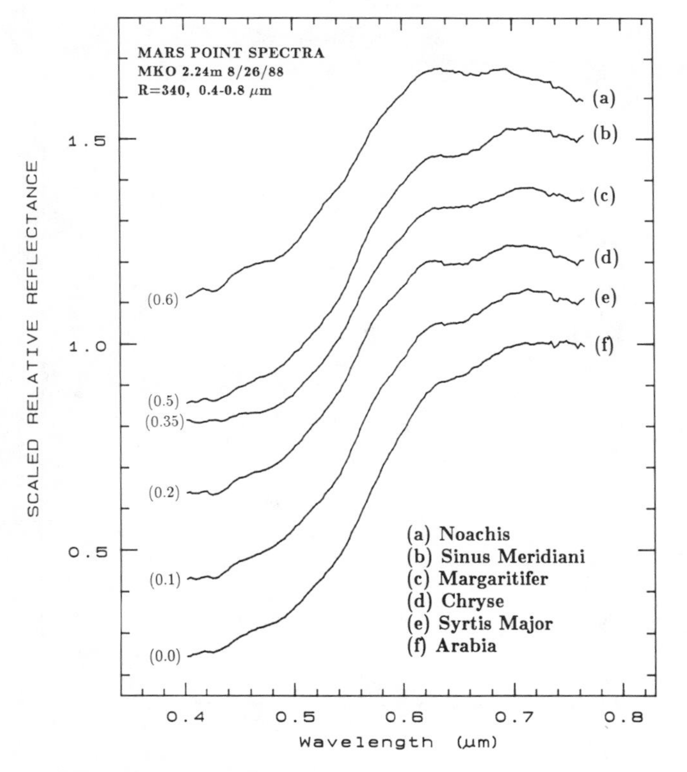

Fig. 5. Relative reflectivity spectra (Mars/ηPsc) from 0.4 to 0.8 μm of six representative regions (~200 km) on Mars. Data have been smoothed using a narrow Gaussian filter and are scaled to 1.0 at 0.75 μm. The spectra are stacked for clarity with the offsets given to the left of each spectrum. Note the strong dropoff in reflectivity into the near-UV and the "cusp" near 0.65 μm. Both these features are characteristic of the martian spectrum and of Fe^{3+} mineral laboratory spectra (cf. Fig. 8).

TABLE 2. Major absorption bands from 0.4-1.1 μm.

Absorber	Spectral Region(μm)	Band Centers (μm)
H_2O	0.64-0.66	0.650
	0.70-0.74	0.718
	0.79-0.84	0.810
	0.926-0.978	0.935
	1.095-1.165	1.130
O_2	0.75-0.77	0.76
	0.68-0.70	0.69
O_3	0.44-0.75	~0.60
Fe^{3+} (Hematite)	—	~0.405
	—	~0.44
	—	0.53-0.56
	~0.6-0.7	0.64-0.65
	~0.76-0.94	0.86-0.88
Fe^{2+} (ol,px)	0.9-1.25	~1.0

Gas phases from *Kondratyev* (1969), *Fleagle and Businger* (1980), and *Tull* (1966, Fig. 2). Fe bands from *Hunt et al.* (1971, 1973), *Sherman and Waite* (1985), and *Morris et al.* (1985).

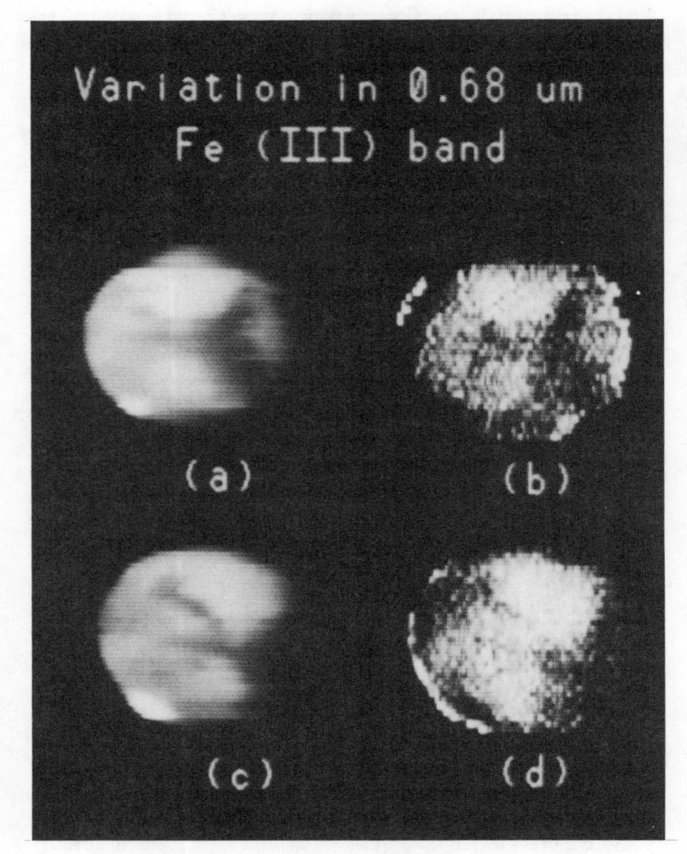

Fig. 7. Spectral ratio images, demonstrating one of the powerful advantages of imaging spectroscopy over point spectroscopy in its ability to display contiguous spatial information in a large number of bandpasses. Here, variations in the 0.65-μm Fe^{3+} band are examined. On the left are two images **(a,c)** at 0.62 μm showing the appearance on the surface at two different times on the night of 8/26/88 UT. On the right **(b,d)** are ratios between 0.68-μm images and an averaged image across the 0.61-0.72-μm band (continuum removal). In the ratio images bright regions have a deeper absorption (by about 5%) at 0.68 μm than dark regions. In this preliminary exercise Fe^{3+} band variations seem to correlate well with observed classical albedo boundaries, although the correlation is not complete (vis. the Sinus Sabaeus region).

resolve the controversy over the degree of crystallization of iron minerals in the Mars soil and airborne dust by confirming the presence of structure due to crystalline material. These new data also support the interpretations of *Morris et al.* (1989) and *Bell et al.* (1989) that suggest that at least hematite is present as one of the iron-bearing phases in the surface fines. However, the question of which other iron phases (if any) are possibly present awaits further data reduction. Additionally, other datasets in this wavelength region (e.g., *Singer et al.,* 1988) may also help resolve these issues.

Laboratory studies of ferric oxide/hydroxide minerals show that they exhibit a number of diagnostic spectral features in the 0.40-1.0-μm region (Fig. 8). Specifically, Fe^{3+} electronic transition bands at 0.6-0.7 μm and 0.80-0.95 μm and an intense near-UV absorption edge caused by overlapping Fe^{3+} electronic transfer and Fe^{3+}—Fe^{3+} pair transition bands, as well as some contribution from higher energy charge transfer absorptions, especially the intense $Fe^{3+}\rightarrow O^{2-}$ UV band, characterize these spectra and form the basis of the interpretations of the previous section (e.g., *Sherman et al.,* 1982; *Sherman and Waite,* 1985; *Morris et al.,* 1985).

Recent work on iron oxides and oxyhydroxides as potential components of the martian surface by *Morris et al.* (1989) and Morris and Lauer (R. V. Morris and H. V. Lauer Jr., unpublished manuscript, 1989) has focused on mixtures of bulk and superparamagnetic or nanophase (particle sizes $\leq$10 nm) hematite that have high Fe_2O_3 content (in accordance with the Viking XRF results of *Clark et al.,* 1982) and also match the martian spectral data. It has been shown that nanophase hematite is much more magnetic than the bulk (larger particle size) phase hematite (*Morris et al.,* 1989), thus potentially also explaining the results of the Viking magnetic properties experiment that concluded that there is 1-7% strongly magnetic phase (previously interpreted as maghemite, γ-Fe_2O_3) in the martian surface materials (*Hargraves et al.,*

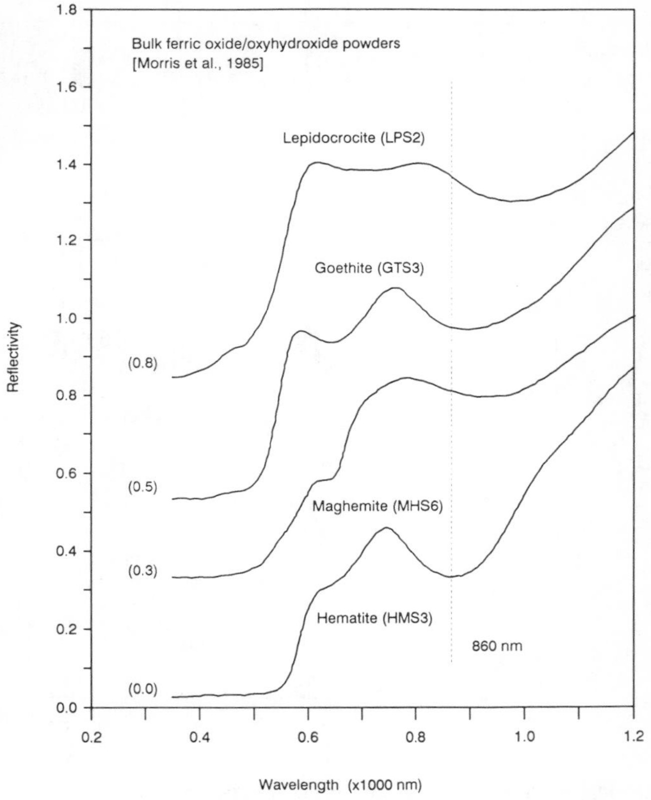

Fig. 8. Laboratory reflectance data for bulk ferric oxide/oxyhydroxide powders (particle sizes typically 20-1000 nm). Of interest are Fe^{3+} crystal field bands at ~0.86 μm and ~0.65 μm and the steep near-UV absorption edge. All of these features are now recognized to varying degrees in the martian spectral data. Data adapted from *Morris et al.* (1985).

1977). Also, as discussed above, Fe-doped smectite clays have been shown to accurately reproduce the Viking Lander LR and PR experimental results, while other potential soil analogs such as palagonite do not (*Oyama and Berdahl,* 1977; *Banin and Margulies,* 1983). Optical, X-ray, and Mössbauer studies of these smectite clay analogs have shown that the data best match the LR and PR results when the interlayer clay structure contains nanophase hematite (*Banin,* 1988). These laboratory analog studies suggest that a single iron oxide—hematite—that occurs in a wide range of particle sizes from conventional bulk forms down to "crystals" only a few unit cells across, may account for much if not all of the ~18 wt.% Fe_2O_3 content measured in the surface soils by Viking (*Toulmin et al.,* 1977; *Clark et al.,* 1982) as well as the spectral properties of the new telescopic data presented here. Since the spectral properties of mixtures of various particle sizes of the ferric mineral phase vary in perhaps systematic ways (*Sherman and Waite,* 1985; *Morris et al.,* 1985, 1989; R. V. Morris and H. V. Lauer Jr., unpublished manuscript, 1989), further analysis of the telescopic dataset to determine exact band depths and positions may provide a clue as to the partitioning of the iron among different phases and/or among different size ranges.

As mentioned above, the new data also do not rule out the possibility of some of the ferric phase occurring in other ferric oxides or oxyhydroxides. Data presented here along with previous telescopic data and Viking surface studies provide constraints on which phases should be present. Thus, hematite as the only ferric phases is consistent with the available data as is hematite with a small amount of some other highly magnetic phase such as maghemite. However, maghemite as the single ferric phase would yield a soil that is much more magnetic than observed. Also, single phases of goethite or lepidocrocite would show near-UV reflectance dropoffs at shorter wavelengths (Fig. 8), inconsistent with the observations. In short, the presence of other ferric oxides/oxyhydroxides cannot be positively ruled out, as their presence can be masked or explained by hematite.

The positive identification of finely dispersed hematite on Mars does not provide an unambiguous answer to the question of whether the martian surface fines are predominantly palagonitic or predominantly clay-like, since bulk and nanophase hematite may be dispersed throughout both palagonites and smectites. Further analysis of the data is being conducted to determine more quantitatively the abundances of bulk crystalline vs. nanophase hematite. Also, the presence of ferric sulfates such as jarosite (*Burns,* 1987), cannot be ruled out from these data. Ferric sulfate mineral content has important implications for the early evolution of the martian crust and mantle (*Burns and Fisher,* 1989), as well as subsequent volcanically induced evolution of the atmosphere and regolith. The telescopic spectral results must be combined with additional laboratory studies that emphasize the structural and mineralogic nature of other ferric oxide/hydroxide minerals or other spectral analog ferric mineral phases as discrete sediments or inclusions or rinds in weathered minerals (such as smectite clays or ferric sulfates) to yield a complete, synthesized picture of the state, distribution, and abundance of various iron-bearing minerals in the martian surface and airborne dust.

SUMMARY

1. New data in the visible and near-IR obtained during the 1988 opposition presented here (and in an unpublished manuscript by J. F. Bell III, T. B. McCord, and P. D. Owensby, 1989) have been shown to be of high enough spectral resolution and sampling to allow fairly detailed assessment of iron-oxide mineralogies and the degree of crystallinity of the martian suface soils.

2. A 0.62-0.72-μm band has been identified in the martian data presented here (and in the unpublished manuscript mentioned above). Both this new band and a 0.80-0.95-μm band also seen that has been reported in previous datasets demonstrate unmistakable presence of structure in the martian surface spectra due to crystalline Fe^{3+} materials.

3. These new data support and strengthen interpretations of *Morris et al.* (1989) that hematite is present in crystalline form as at least one of the ferric-iron phases on the martian surface. Laboratory and spectral data indicate that the bulk of the ≈18 wt.% Fe_2O_3 in the martian surface soils and airborne dust may exist as hematite spanning an extended range of particle sizes, from bulk crystals down to nanophase forms ("crystals" ≤10 nm).

4. Further data reduction and laboratory work must be carried out, in particular to address the possibility of the presence of other ferric oxides/oxyhydroxides, Fe-rich smectite clays, and ferric sulfates in the Mars soils.

Acknowledgments. The authors wish to extend their appreciation to the telescope operators, day crew, and support staff of the University of Hawaii 2.24-m telescope for their efforts in helping to make these observations possible. Also, we are grateful to R. Morris for providing useful data on iron-oxide laboratory spectra and J. Lo for much-needed help with the data reduction software. We feel that this paper was significantly improved by the in-depth reviews of R. Morris and R. Burns. This work was supported under NASA grants NSG-7312 and NSG-7323. This paper is Planetary Geosciences Division contribution number 565.

REFERENCES

Adams J. B. (1968) Lunar and martian surfaces: Petrologic significance of absorption bands in the near-infrared. *Science, 159,* 1453-1455.

Banin A. (1988) The mineral components of dust on Mars (abstract). In *MECA Workshop on Dust on Mars III* (S. Lee, ed.), p. 15. LPI Tech. Rpt. 89-01, Lunar and Planetary Institute, Houston.

Banin A. and Margulies L. (1983) Simulation of Viking biology experiments suggests smectites not palagonites, as martian soil analogs. *Nature, 305,* 523-525.

Banin A., Margulies L., and Chen Y. (1985) Iron-montmorillonite: A spectral analog of martian soil. *Proc. Lunar Planet. Sci. Conf. 15th,* in *J. Geophys. Res., 90,* C771-C774.

Bell J. F. III and McCord T. B. (1989) Mars: Near-infrared comparative spectroscopy during the 1986 opposition. *Icarus, 77,* 21-34.

Bell J. F. III, McCord T. B., and Lucey P. G. (1989) Mars during the 1988 opposition: High resolution imaging and spectroscopy. *Eos Trans. AGU, 70,* 50.

Burns R. G. (1987) Ferric sulfates on Mars. *Proc. Lunar Planet. Sci. Conf. 17th,* in *J. Geophys. Res., 92,* E570-E574.

Burns R. G. and Fisher D. S. (1989) Sulfide mineralization related to early crustal evolution of Mars. *J. Geophys. Res.,* in press.

Carlson R. W. (1981) Spectral mapping of Jupiter and the Galilean satellites in the near-infrared, Proceedings of the SPIE. *Imaging Spectroscopy, 268,* 29-34.

Clark B. C., Baird A. K., Weldon R. J., Tsusaki D. M., Schnabel L., and Candelaria M. P. (1982) Chemical composition of martian fines. *J. Geophys. Res., 87,* 10,059-10,067.

Evans D. L. and Adams J. B. (1979) Comparison of Viking Lander multispectral images and laboratory reflectance spectra of terrestrial samples. *Proc. Lunar Planet. Sci. Conf. 10th,* pp. 1829-1834.

Fleagle R. G. and Businger J. A. (1980) *An Introduction to Atmospheric Physics,* pp. 231-232. Academic, New York.

Goetz A. F. H., Vane G., Solomon J. E., and Rock B. N. (1985) Imaging spectrometry for Earth remote sensing. *Science, 228,* 1147-1153.

Gooding J. L. and Keil K. (1978) Alteration of glass as a possible source of clay minerals on Mars. *Geophys. Res. Lett., 5,* 727-730.

Guinness E. A., Arvidson R. E., Dale-Bannister M. A., Singer R. B., and Bruckenthal E. A. (1987) On the spectral reflectance properties of materials exposed at the Viking landing sites. *Proc. Lunar Planet. Sci. Conf. 17th,* in *J. Geophys. Res., 92,* E575-E587.

Hargraves R. B., Collinson D. W., Arvidson R. E., and Spitzer C. R. (1977) The Viking magnetic properties experiment: Primary mission results. *J. Geophys. Res., 82,* 4547-4558.

Hunt G. R., Salisbury J. W., and Lenhoff C. J. (1971) Visible and near-infrared spectra of minerals and rocks. III. Oxides and hydroxides. *Mod. Geol., 2* 195-205.

Hunt G. R., Salisbury J. W., and Lenhoff C. J. (1973) Visible and near-infrared spectra of minerals and rocks. VI. Additional silicates. *Mod. Geol., 4,* 85-106.

Kondratyev K. Y. (1969) *Radiation in the Atmosphere,* pp. 107-139. International Geophysics Series, vol. 12, Academic, New York.

Lucey P. G. (1988) Ground-based imaging spectroscopy of the Moon: On the threshold (abstract). In *Lunar and Planetary Science XIX,* pp. 703-704. Lunar and Planetary Institute, Houston.

McCord T. B. and Adams J. B. (1969) Spectral reflectivity of Mars. *Science, 163* 1058-1060.

McCord T. B. and Westphal J. A. (1971) Mars: Narrowband photometry, from 0.3 to 2.5 microns, of surface regions during the 1969 apparition. *Astrophys. J., 168,* 141-153.

McCord T. B., Huguenin R. L., Mink D., and Pieters C. (1977) Spectral reflectance of martian areas during the 1973 opposition: Photoelectric filter photometry, 0.33-1.10 μm. *Icarus, 31,* 25-39.

McCord T. B., Clark R. N., and Huguenin R. L. (1978) Mars: Near-infrared spectral reflectance and compositional implications. *J. Geophys. Res., 83,* 5433-5441.

Morris R. V., Lauer H. V. Jr., Lawson C. A., Gibson E. K. Jr., Nace G. A., and Stewart C. (1985) Spectral and other physicochemical properties of submicron powders of hematite ($\gamma-Fe_2O_3$), maghemite ($\gamma-Fe_2O_3$), magnetite (Fe_3O_4), goethite ($\alpha-FeOOH$), and lepidocrocite ($\gamma-FeOOH$). *J. Geophys. Res., 90,* 3126-3144.

Morris R. V., Agresti D. G., Lauer H. V. Jr., Newcomb J. A., Shelfer T. D., and Murali A. V. (1989) Evidence for pigmentary hematite on Mars based on optical, magnetic, and Mössbauer studies of superparamagnetic (nanocrystalline) hematite. *J. Geophys. Res., 94,* 2760-2778.

Nelson M. L., McCord T. B., Clark R. N., Johnson T. V., Matson D. L., Mosher J. A., and Soderblom L. A. (1986) Europa: Characterization and interpretation of global spectral surface units. *Icarus, 65,* 129-151.

Oyama V. I. and Berdahl B. J. (1977) The Viking gas exchange experiment results from Chryse and Utopia surface samples. *J. Geophys. Res., 82,* 4669-4676.

Sherman D. M. and Waite T. D. (1985) Electronic spectra of Fe^{3+} oxides and oxide hydroxides in the near IR to near UV. *Amer. Mineral., 70,* 1262-1269.

Sherman D. M., Burns R. G., and Burns V. M. (1982) Spectral characteristics of the iron oxides with application to the martian bright region mineralogy. *J. Geophys. Res., 87,* 10,169-10,180.

Singer R. B. (1982) Spectral evidence for the mineralogy of high albedo soils and dust on Mars. *J. Geophys. Res., 87,* 10,159-10,168.

Singer R. B. and McCord T. B. (1979) Mars: Large scale mixing of bright and dark surface materials and implications for analysis of spectral reflectance. *Proc. Lunar Planet. Sci. Conf. 10th,* pp. 1835-1848.

Singer R. B., McCord T. B., Clark R. N., Adams J. B., and Huguenin R. L. (1979) Mars surface composition from reflectance spectroscopy: A summary. *J. Geophys. Res., 84,* 8415-8426.

Singer R. B., Bus E. S., Wells K., and Swift C. (1988) Visible and near-IR spectral imaging of Mars during 1988 (abstract). *Bull. Am. Astron. Soc., 20,* 848.

Soderblom L. A. and Wenner D. B. (1978) Possible fossil H_2O liquid ice interfaces in the martian crust. *Icarus, 34,* 622-637.

Soderblom L. A., Edwards K., Eliason E. M., Sanchez E. M., and Charette M. P. (1978) Global color variations on the martian surface. *Icarus, 34,* 446-464.

Toulmin P. III, Baird A. K., Clark B. C., Keil K., Rose H. J. Jr., Christian R. P., Evans P. H., and Kelliher W. C. (1977) Geochemical and mineralogical interpretation of the Viking inorganic chemical results. *J. Geophys Res., 82,* 4625-4634.

Tull R. G. (1966) The reflectivity spectrum of Mars in the near-infrared. *Icarus, 5,* 505-514.

Proceedings of the 20th Lunar and Planetary Science Conference, pp. 487-501
Lunar and Planetary Institute, Houston, 1990

Chronology and Global Distribution of Fault and Ridge Systems on Mars

D. H. Scott and J. M. Dohm

U.S. Geological Survey, 2255 North Gemini Drive, Flagstaff, AZ 86001

A series of paleotectonic maps has been made showing fault and ridge systems on Mars as they existed during the geologic past. These structures have been classified chronologically by mapping their occurrence in rock units emplaced during successive geologic periods. The three martian systems—Noachian, Hesperian, and Amazonian—and their eight subdivisions (series) are the time-stratigraphic referents for assigning ages to the faults and ridges. System and series boundaries and the structures mapped within the chronostratigraphic units were extracted from recently completed global geologic maps of Mars. The paleotectonic maps for the western, eastern, and polar regions were compiled at 1:15,000,000 scale and photographically reduced for inclusion in this paper. Results of this study show that most major fault systems are associated with large volcanic centers in the western equatorial region of Mars and that both faulting and volcanism probably decreased with time but with a pulse of renewed activity during the Early Hesperian. Concentric faults and grabens are not as distinguishable around large martian impact basins (with the exception of Isidis Planitia) as they are around lunar basins of similar size. In places the highland-lowland boundary is bracketed by parallel systems of faults and ridges. The formation of ridges appears to have culminated during Hesperian time. Unlike faults, ridges are ubiquitous; most occur in extensive, plains-forming lava flows that erupted largely from fissure vents.

INTRODUCTION

The evolution of tectonic activity on Mars has been traced by mapping the areal distribution of fault and ridge systems as they formed through time during successive geologic periods: the Noachian, Hesperian, and Amazonian Periods, and their epoch subdivisions. Abundant evidence of these structures is preserved in the crustal rocks of the planet, whose detailed stratigraphy has been largely developed by *Scott and Carr* (1978) and *Tanaka* (1986) and by global geologic mapping based on Viking images (*Scott and Tanaka,* 1986; *Greeley and Guest,* 1987; *Tanaka and Scott,* 1987). Regional paleotectonic maps have been compiled at 1:15,000,000 scale for each of the three time-stratigraphic systems (Noachian, Hesperian, Amazonian) in which recognizable structures occur for the western, eastern, north polar, and south polar regions of Mars—a total of 10 maps covering the entire planet. (Maps of the Noachian System for the north polar region and the Amazonian System for the south polar region are omitted, because these systems contain few or no rock units or structures in these areas.)

METHODS USED AND MAP PORTRAYALS

The three martian stratigraphic systems (*Scott and Carr,* 1978) and their eight series subdivisions (*Tanaka,* 1986) are based on chronostratigraphic referents for age classifications; series names are adapted from system names and qualified by the prefix "Lower," "Middle," or "Upper." Their corresponding time equivalents (geochronologic units) are, respectively, periods and epochs with the prefix "Early," "Middle," or "Late." System and series boundaries and structures were extracted from geologic maps of the western (*Scott and Tanaka,* 1986), eastern (*Greeley and Guest,* 1987), and polar regions (*Tanaka and Scott,* 1987) of Mars. These data were compiled at 1:15,000,000 scale and reduced to about 1:60,000,000 for inclusion in this paper. Where the geologic base maps (1:15,000,000 scale) did not clearly reveal structural and stratigraphic relations, Viking photomosaics (1:2,000,000 scale) were examined to supplement the regional data.

The positioning of structures on the maps adhered to basic geologic principles: (1) faults and ridges of younger age (Amazonian) may extend across boundaries of older rock units, (2) structures of intermediate age (Hesperian) may extend into older units (Noachian) but not into Amazonian rocks, and (3) Noachian structures occur only within Noachian rocks. Older rock units, however, may contain younger faults and ridges whose ages with respect to their host rocks cannot be determined where these structures are not in contact with younger units. Where slip sense or separations are known, only normal faulting has been recognized on Mars, although hypotheses have been developed for the occurrence of both thrust (*Plescia and Golombek,* 1986; *Watters,* 1988;

TABLE 1. Symbols used in Figs. 1-10.

Symbol	Explanation
(hachured line)	System boundary; unhachured lines are contacts between series (l, Lower; M, Middle; U, Upper; Hesperian System divided into Lower and Upper Series only); lines dashed where approximately located
(line with dot)	Normal fault or graben
(line with diamond)	Ridge
(black shape)	Volcanic center
(stippled band)	Highland-lowland boundary

Golombek et al., 1989) and strike-slip dislocations (*Schultz,* 1989). The normal faults are commonly associated with grabens that formed from extensional stresses. Ridge formation also involves some type of faulting, either normal, reverse, or possibly strike slip.

The map units and structures of each series are shown in Figs. 1-10; map symbols are listed in Table 1.

WESTERN EQUATORIAL REGION

Geologic-Physiographic Setting

The western equatorial region (Figs. 1-3) has many of the major geologic and physiographic features on Mars. The Tharsis rise (topographic high) extends northeasterly across the central part of the map area and includes three large, young volcanoes: Arsia, Pavonis, and Ascraeus Montes. Adjacent to them lie the huge volcanic edifices of Olympus Mons and Alba Patera; the latter, though of relative low relief, has the most extensive shield of any volcano on Mars. The boundary between the southern cratered highlands and the relatively smooth and younger northern plains is referred to as the martian crustal dichotomy; this geologic and physiographic boundary is partly buried in the western region by young lava fields surrounding the Tharsis Montes. Where the boundary is not obscured by lava flows, it appears as a low-relief, irregular scarp that is bordered by clusters of low, knobby hills interspersed among mesas and larger remnants of the highland plateau.

Valles Marineris, the greatest canyon system on Mars, extends easterly for several thousand kilometers across the equatorial region. Associated with the system are large outflow channels that originated in aeons past from canyons and chaotic terrain in the highlands; these channels once discharged huge volumes of water into lowland areas (*Carr,* 1979). Many of these channels, like Valles Marineris, appear to have developed along fault systems (no longer apparent)

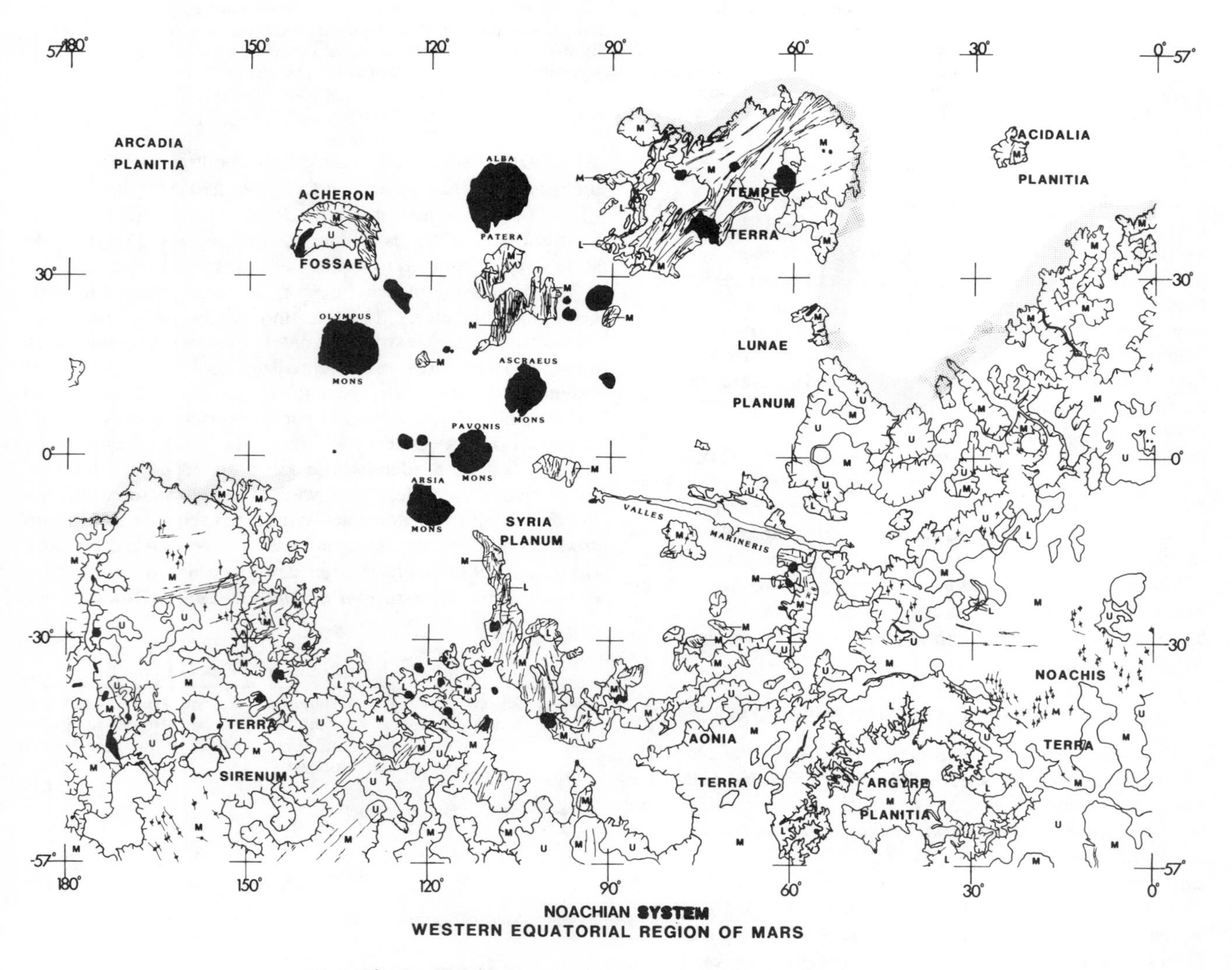

Fig. 1. Noachian system western equatorial region of Mars.

associated with the Tharsis topographic rise (*Frey,* 1979; *Plescia and Saunders,* 1982; *Tanaka and Golombek,* 1989).

Some of the most ancient Noachian rocks are exposed in the mountainous terrains south of Syria Planum and around Argyre Planitia, an impact basin more than 1000 km in diameter (*Scott and Tanaka,* 1986). Hesperian materials largely occupy the central part of the western region and encircle the Tharsis rise and the Olympus Mons and Alba Patera volcanoes. Amazonian deposits are mostly concentrated in Tharsis Montes and in the lowlands to the northwest.

Noachian System

Lower Noachian series. Materials in this series are exposed in highland areas mostly as scattered occurrences along the Tharsis Montes axial trend, south of Syria Planum, and around Argyre Planitia. Although no morphological evidence of their origin and composition is recognizable on Viking images, these old rocks probably consist of volcanic materials and impact breccias. Tectonic activity, expressed by faulting, was most intense in southern Syria Planum (*Tanaka and Davis,* 1988). Here the oldest appearing faults occur in a highly fractured basement complex characterized by highly degraded crater remnants and prominent relief. Evidence of faulting in the hilly and mountainous rim materials surrounding the large impact basin of Argyre Planitia is shown by linear, faceted scarps both radial and concentric to the basin center. These scarps outline fault blocks of large structured massifs that were formed during or shortly after the impact event and are not related to postcrater slumping. The faults and massifs are buried and embayed by Middle Noachian and Lower Hesperian materials. With the exception of a single occurrence in southeastern Argyre Planitia, no ridges have been recognized in Lower Noachian rocks.

Structural deformation was probably widespread and intense during the Early Noachian, particularly around Syria Planum and Argyre Planitia and possibly along the axial trend of the

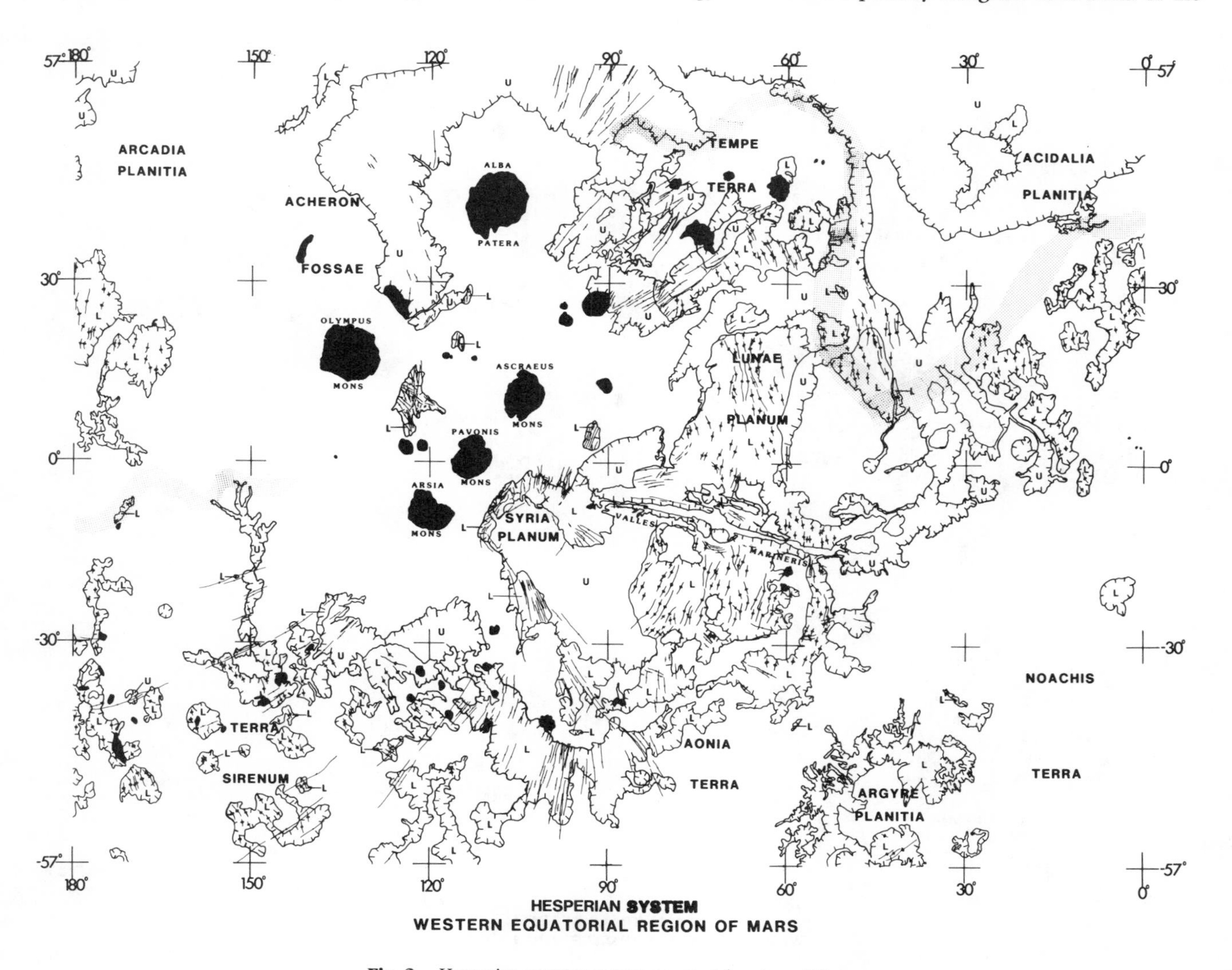

Fig. 2. Hesperian system western equatorial region of Mars.

Tharsis rise; large parts of Tharsis, however, have been covered by younger materials, and thus direct evidence of tectonism is lacking.

Middle Noachian series. A diverse assemblage of cratered, dissected, etched, and ridged materials characterizes the Middle Noachian. Their origins, though obscure, are believed to be volcanic but highly modified by cratering, aeolian, and possibly fluvial processes. They are the most extensive Noachian deposits in the western equatorial region. Many faults and grabens transect rocks of the series along the northeast and southwest extensions of the Tharsis axial trend, radial to Syria Planum and Alba Patera, and concentric to Acheron Fossae and Argyre Planitia. Southwest of Tharsis, faults diverge from their axial alignments with the Tharsis volcanoes to form a broad, fan-shaped pattern. Their traces approach parallelism with the highland-lowland boundary near the equator and a structural relation seems likely. As will be seen later, this relation between the dichotomy boundary and faults and ridges continues in the eastern equatorial region where these structures are expressed in rock series ranging from Middle Noachian to Upper Amazonian. Whether an endogenic (*Wise et al.,* 1979) or exogenic (*Wilhelms and Squyres,* 1984; *Frey and Schultz,* 1988) origin for the martian crustal dichotomy is invoked, an important constraint on the process is suggested by the long time interval over which the deformation appears to have continued. The oldest abundant parallel systems of ridges occur in the southeast and southwest quadrants of the map area; in the southwest they are concentric to the southwestern part of the Tharsis rise, probably indicating a causal relation to the structural growth of this topographic feature during the Middle Noachian. Generally, the ridges are larger and farther apart than the wrinkle-type ridges that characterize the lava plains of Early Hesperian age; many are asymmetric in profile and have a steep front suggesting normal faulting. That some wrinkle ridges did develop during the Middle Noachian is shown by their

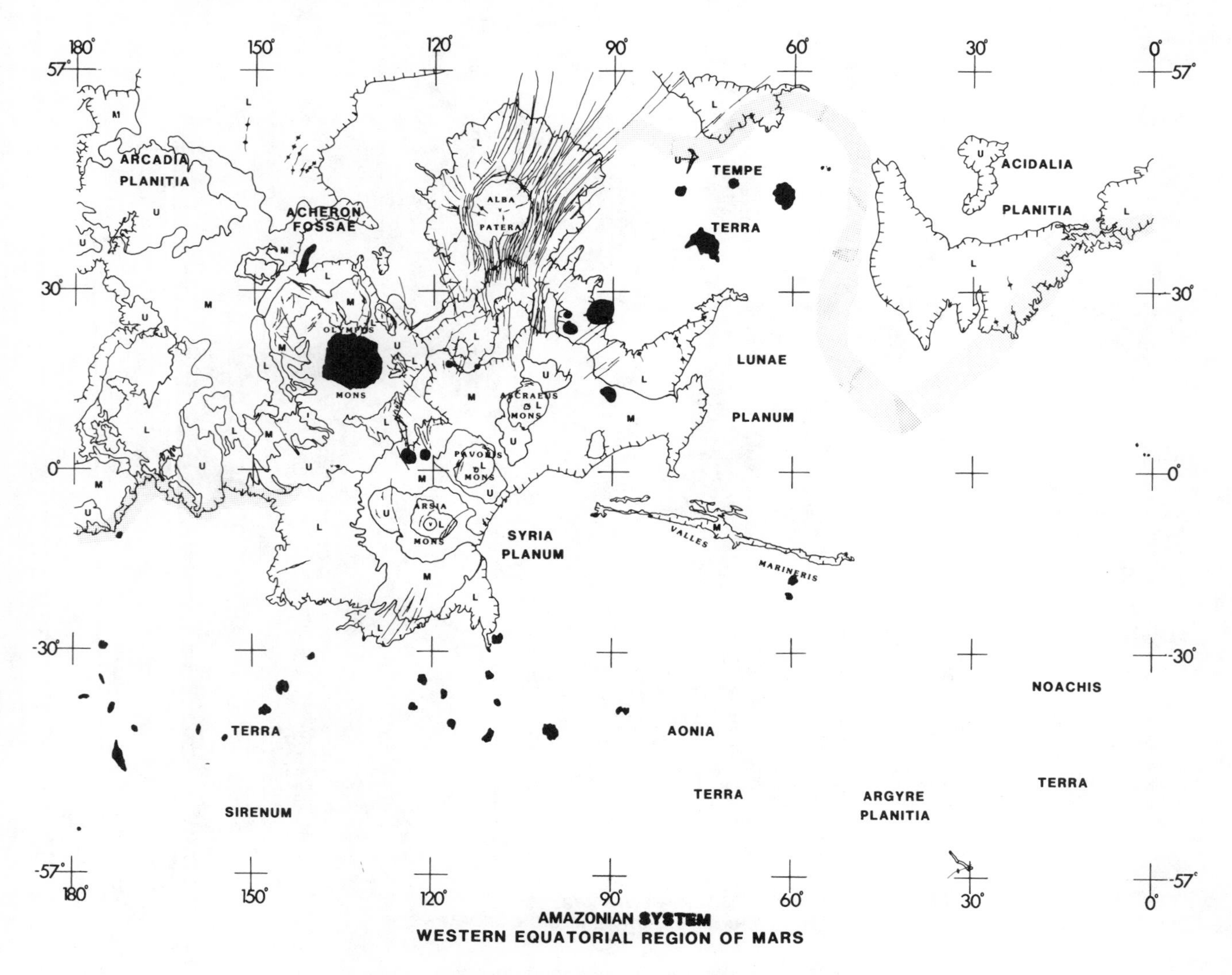

Fig. 3. Amazonian system western equatorial region of Mars.

occurrence, though sparse, in Tempe Terra (*Scott and Dohm,* 1990), where their relative age is established stratigraphically and by fault sets that have been relatively dated by other methods.

Upper Noachian series. Materials included in this series form subdued plains that are interpreted to consist of interbedded lava flows and eolian deposits (*Scott and Tanaka,* 1986). They overlie large areas of older, more rugged highland materials and crater rims, and they form relatively smooth intercrater plains. Fault density is lower than in Middle Noachian rocks, but this observation may be more apparent than real because Upper Noachian exposures are sparse around tectonically active centers. At Acheron Fossae, however, very few of the numerous faults in the complexly structured ring of Middle Noachian material extend into the relatively smooth Upper Noachian core. Most Late Noachian faulting in the Terra Sirenum region is parallel to faults in the Early and Middle Noachian; minor faulting south and southeast of Valles Marineris follows the trend of the canyon system. Mare-type wrinkle ridges are common southeast of Valles Marineris, but their time of origin is uncertain. Most occur only within Upper Noachian units and their stratigraphic position is not constrained; thus, they may have been formed during the Early Hesperian, an epoch of abundant ridge development, or later.

Hesperian System

Lower Hesperian series. The largest accumulations of Lower Hesperian materials are the ridged plains lava flows (*Scott and Carr,* 1978; *Greeley and Spudis,* 1981; *Tanaka et al.,* 1988), which cover most of the Lunae Planum plateau. Lobate flow fronts, though not common, are visible in places on Lunae Planum and where the ridged plains unit broadly encircles the southwestern part of the Tharsis rise. Wrinkle ridges are distinguishing features that characterize the lava plains planetwide. Ridges of a similar type are prevalent on the lunar maria and the plains of Mercury. Although they are recognized as structural features, interpretations of their origin are varied and controversial (*Colton et al.,* 1972; *Howard and Muehlberger,* 1973; *Scott,* 1973; *Lucchitta,* 1976; *Plescia and Golombek,* 1986; *Watters and Maxwell,* 1986). Recent evidence (*Scott,* 1989) suggests a causal relation between wrinkle ridges and extensional structures such as grabens and pit craters. Also, *Maxwell* (1989) has observed that not all ridges are of compressional origin and that the origin of many ridges is problematic.

Many faults transect Lower Hesperian rocks but are most common in materials other than the ridged plains unit. These materials constitute relatively featureless plains in the southern highlands and complexly fractured material around Syria Planum. The fault patterns and trends are very similar to those in Noachian rocks, as are the wrinkle ridges, a commonality suggesting that either (1) structural growth of the Syria Planum and Tharsis rises continued over a long period of geologic time, or (2) deformation in Noachian materials was largely initiated during the Early Hesperian. In the latter case, faults and ridges occurring solely within Noachian rocks and thus attributed to this period may actually have been formed during the Hesperian or later. A more plausible explanation, however, is that tectonic activity waxed and waned within different areas throughout Noachian and Hesperian times; this is illustrated at both Acheron Fossae and at Alba Patera as discussed in the following section.

Upper Hesperian series. The great decline in tectonic activity throughout the western equatorial region is strikingly illustrated at Syria Planum, Lunae Planum, and on the Tharsis rise, where Upper Hesperian materials have largely resurfaced and obliterated fault and ridge structures in the underlying Lower Hesperian rocks. Northeast of Alba Patera, however, faulting appears to have been initiated radial to this volcano during its development in Late Hesperian or possibly earlier times (*Rotto and Tanaka,* 1989; *Tanaka,* 1990). These observations support the conclusions of an earlier study of volcanic and tectonic events in the Tharsis region of Mars (*Scott and Tanaka,* 1980). This study suggested that although tectonic activity decreased with time in the Tharsis region, it was episodic and was concentrated in different localities at different times. The large outflow channels that terminate in Chryse Planitia (lat. 25°N, long. 40°) and those of Mangala Valles (lat. 30°S, long. 150°) were formed during the Late Hesperian (*Scott and Tanaka,* 1986; *Tanaka,* 1986). Flood-plain materials from the channels are mostly unaffected by faulting, but a few ridges appear to be nearly contemporaneous with or to postdate the flooding events (J. W. Rice, work in progress, Maja Valles region).

Amazonian System

Lower Amazonian series. Tectonism was largely restricted to the upper flanks of Alba Patera, where faults and large grabens are radial, tangential, and concentric to the central core of the large volcano. Minor faulting occurred on the central constructs of the major Tharsis Montes volcanoes and on some of the older aureole deposits on the east side of Olympus Mons. Elsewhere, materials emplaced during the entire Amazonian Period are restricted to the northern lowland plains and show little evidence of deformation. Only a few ridges occur in Lower Amazonian materials, mostly in Arcadia Planitia.

Middle Amazonian series. The decreasing areal extent of materials deposited and structures formed during Middle Amazonian time reflects the interdependency of volcanic and tectonic activity on Mars. Minor faulting continued radial to the Tharsis rise and was initiated in the younger aureole materials northwest of Olympus Mons. No ridges appear to have been formed during this time and the smooth plains of Arcadia Planitia are unfaulted.

Upper Amazonian series. Tectonism and volcanism had almost ceased in the western equatorial region by Late Amazonian time. Small lava flows cut by a few faults encircle the Tharsis Montes volcanoes and Olympus Mons. Some minor faulting occurred in central Valles Marineris.

EASTERN EQUATORIAL REGION

Geologic-Physiographic Setting

The eastern equatorial region (Figs. 4-6) is more clearly divided than the western region by the boundary scarp that separates the lowland and highland provinces. The scarp is several kilometers high in places (*Greeley and Guest,* 1987), and the highland side of this boundary contains grabens and fractures that may have been modified by channels of fluvial or sapping origin. Large outflow channels, however, have not been generally recognized. A possible exception is Ma'adim Vallis (lat. 20°S, long. 183°), which may have contributed to the filling of a huge paleolacustrine basin centered near lat. 0°, long. 200° in Elysium Planitia (*Scott et al.,* 1989).

A major volcanic province lies within the Elysium region of the northern lowlands, but its volcanoes and lava flows are smaller and generally older than those of the Tharsis volcanic province in the western hemisphere. Other dominant features in the eastern region are Hellas and Isidis Planitiae, basins believed to have been formed by giant impacts during an early period in Mars' history.

Noachian System

Lower Noachian series. The paucity of faults and ridges in Lower Noachian rocks indicates that little tectonic activity occurred, in contrast to the western equatorial region. Faults on the southeast margin of Isidis Planitia are concentric to this impact basin, and a few subdued scarps (unmapped), possibly associated with faulting, occur in older rocks around the Hellas impact basin (*Wichman and Schultz,* 1988). Although large areas of the eastern equatorial region were resurfaced in the Middle Noachian and older structures concealed, the relative absence of dense faulting in Lower Noachian rim materials of Hellas basin also suggests an epoch of low tectonic activity.

Middle and Upper Noachian series. Northwest-trending ridges, the dominant structures in these series, are concen-

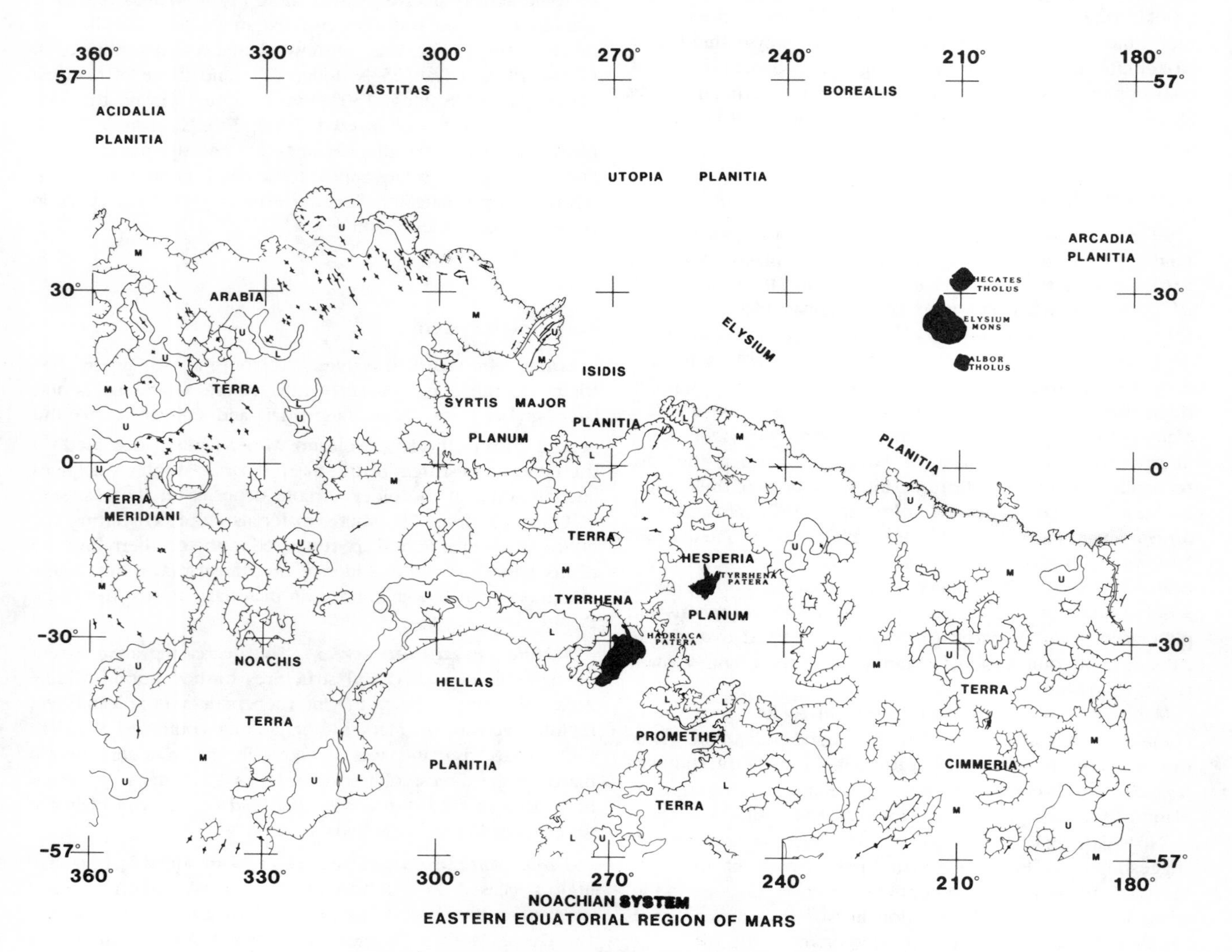

Fig. 4. Noachian system eastern equatorial region of Mars.

trated mostly in the Arabia Terra and Terra Meridiani areas. They generally follow the trend of the highland-lowland boundary to the border between the eastern and western regions where they trend more northerly. A few widely separated Middle Noachian ridges broadly encircle Hellas Planitia and extend into the south polar region (Fig. 9). Middle and Upper Noachian concentric faults cut the rim material of Isidis Planitia, showing that structural deformation continued around the basin over a protracted interval of time. Compared with the western equatorial region, however, tectonism was generally mild throughout the eastern part of Mars in the Noachian Period.

Hesperian System

Lower Hesperian series. Ridge formation continued during the Early Hesperian and followed the Noachian northwesterly trends in the Arabia Terra and Terra Meridiana areas, but faulting at Isidis ceased. Elsewhere, ridges, though widely distributed in Lower Hesperian lava flows, exhibit a variety of orientations that generally have consistent patterns within a particular locality. Exceptions occur in large areas covered by ridged plains lava flows at Elysium Planitia and Hesperia Planum. In these areas series of ridges trend northeasterly, unlike other local ridges, and follow the regional alignment extending more than 5000 km between the volcanoes Hadriaca Patera, Tyrrhena Patera, Elysium Mons, and Hecates Tholus; the alignment of volcanoes and ridges closely follows the trace of a great circle, somewhat similar to that of the Tharsis volcanoes recognized by *Wise et al.* (1979). Only a few widely scattered faults paralleling the highland-lowland boundary have been recognized.

Upper Hesperian series. Lava flows in Syrtis Major Planum, a large volcanic construct, contain many ridges having northwesterly trends similar to trends of the older ridge systems in Arabia Terra. In places these trends generally follow the direction of the highland boundary scarp. Elsewhere, other

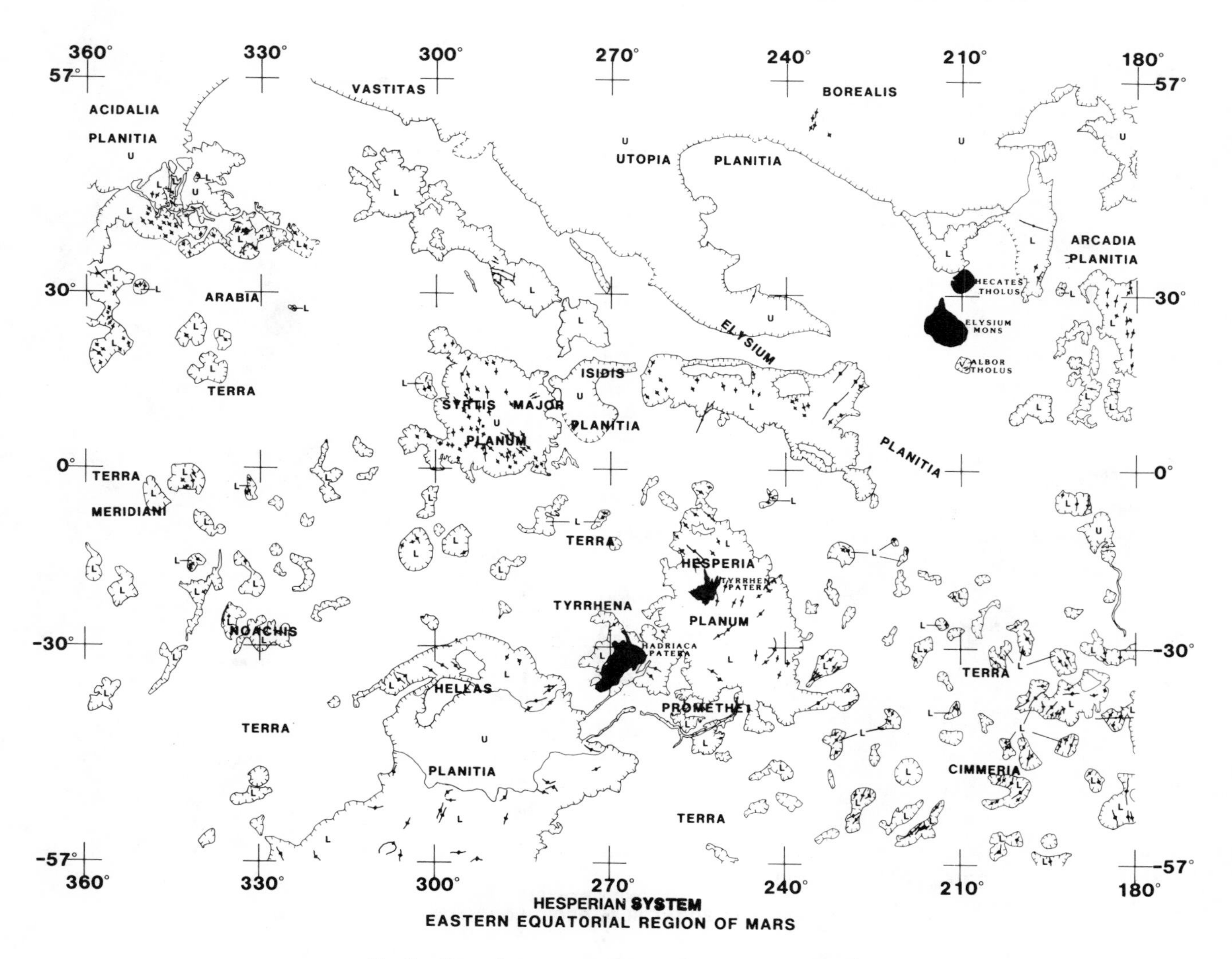

Fig. 5. Hesperian system eastern equatorial region of Mars.

large areas covered by this rock series are nearly devoid of structures and it is evident that in the absence of volcanic edifice building little tectonic activity occurred.

Amazonian System

Lower Amazonian series. Major faulting was concentrated in extensive Lower Amazonian lava flows that cover the plains around Elysium Mons and embay the older, smaller volcanoes of Albor and Hecates Tholi. Although some faults and grabens are concentric to Elysium Mons, most have northwesterly trends that closely parallel the highland-lowland boundary scarp from about long. 185° to 225°, strongly suggesting a causal relation between these features. North-trending ridges are common in the Arcadia Planitia-Elysium region, where they follow alignments of hills and knobby terrain that form the Phlegra Montes (lat. 40°N, long. 195°), an older (Hesperian-Noachian) range of mountains of possible tectonic origin (*Elston,* 1979).

Middle and Upper Amazonian series. Tectonism and volcanism declined during the Middle and Late Amazonian Epochs in this region as in the western region. A few ridges and faults occur in these rock series in Elysium Planitia, as well as sparse ridges in Isidis Planitia that are largely oriented normal to the boundary scarp. Perhaps the most significant volcanic event was the extrusion of possible ash-flow tuffs in southern Elysium Planitia (*Scott and Tanaka,* 1981, 1986). The close association between faulting and tectonic activity is also demonstrated around Elysium Mons, the largest volcanic complex in the eastern equatorial region during the Amazonian Period.

NORTH POLAR REGION

Geologic-Physiographic Setting

The north polar region (Figs. 7 and 8) lies wholly within the lowland plains. Most of the region is below the martian datum, but north of about lat. 80°N elevations rise toward the

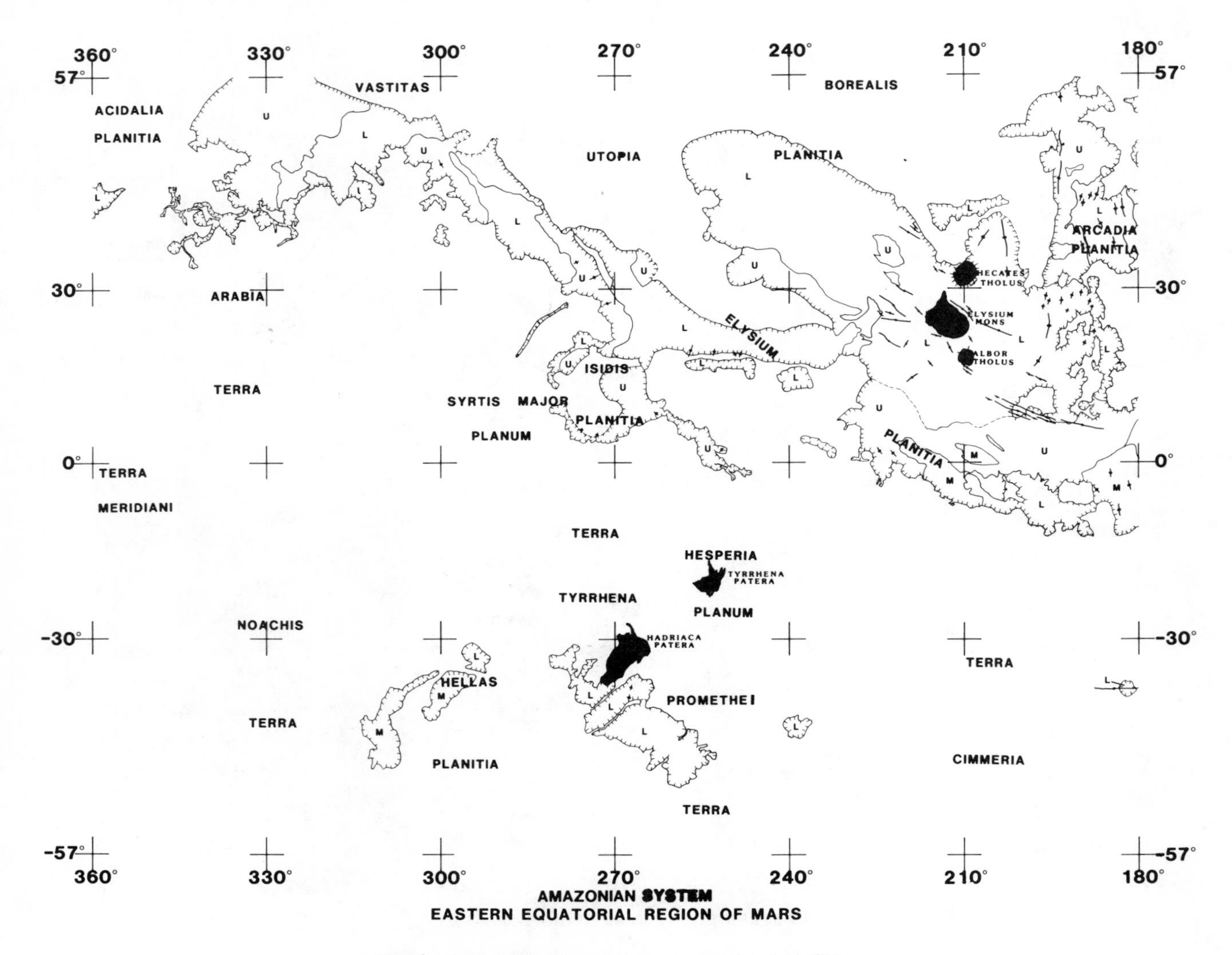

Fig. 6. Amazonian system eastern equatorial region of Mars.

polar ice cap (*U.S. Geological Survey,* 1989). The polar ice forms a residual cap of high albedo during northern late spring and summer (L_s = 92 to 154) that is probably composed of water ice and dust (*Kieffer et al.,* 1976). Other materials of Late Amazonian age surround and extend outward from the ice cap to about lat. 70°N. In approximate order from the pole, they are light and dark layered deposits, dune materials, and a smooth to hummocky mantle of eolian deposits that covers the northern part of the Upper Hesperian Vastitas Borealis Formation. The oldest rocks occur in the upper left quadrant of the map area and are assigned to the Lower Hesperian series, but they may be older remnants of ancient highland material projecting above the plains (*Tanaka and Scott,* 1987). Although lava flows cover a large part of the circumpolar plains below lat. 70°N, no large volcanic constructs have been recognized.

Hesperian System

Lower Hesperian series. No evidence of tectonic activity has been recognized in the rocks of this series, which cover

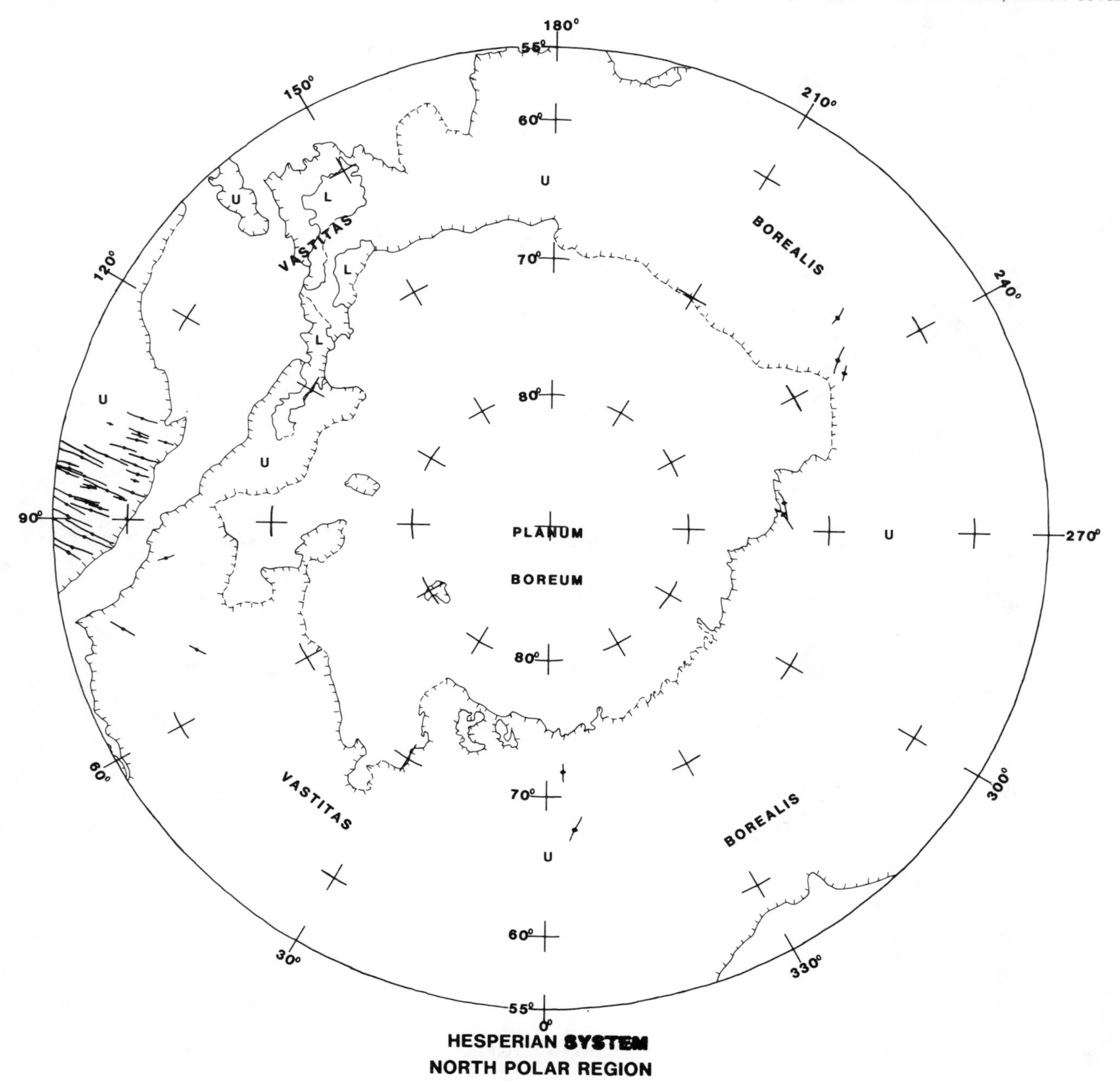

Fig. 7. Hesperian system north polar region.

only small areas in this region. The material forms hills and mesas that may be remnants of the highlands or old crater rims; they are highly altered by mass wasting and structural dislocations would be difficult to observe.

Upper Hesperian series. Many faults and grabens from the volcano Alba Patera (Fig. 2) extend radially northeast across lava flows and subpolar plains deposits, which consist of knobby and grooved members of the Vastitas Borealis Formation (*Tanaka and Scott,* 1987). Only a few minor ridges occur throughout the north polar region. Large parts of the region are either highly degraded by aeolian erosion or deeply mantled by the products of this erosion and small structures likely would be obscured. The absence of large volcanic constructs in the northern plains, however, suggests that tectonic activity was mild.

Amazonian System

Lower Amazonian series. Faults and grabens in the lower member of the Arcadia formation strike northeast and represent the continuation of faulting along the same trends

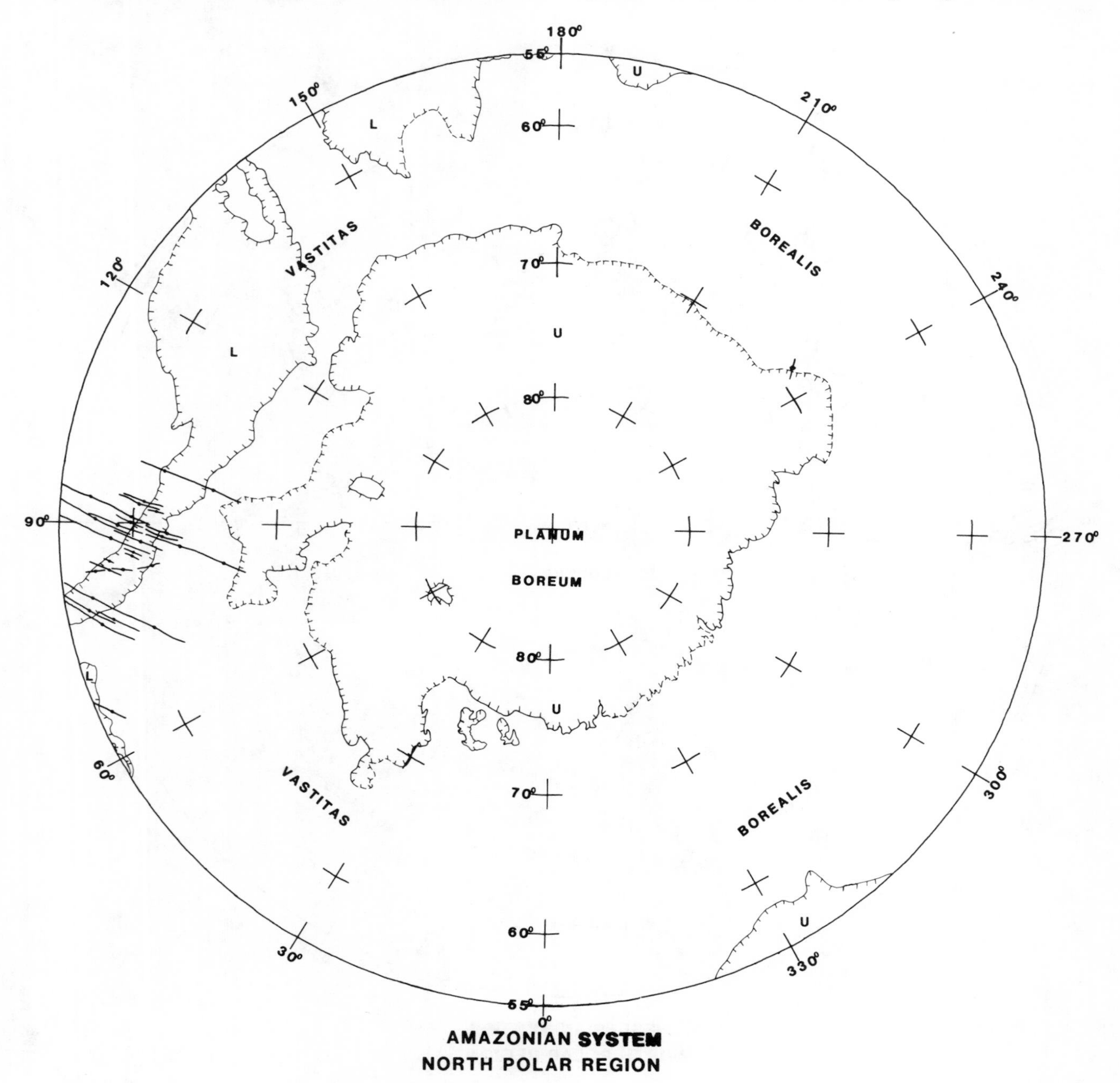

Fig. 8. Amazonian system north polar region.

as those of faults radial to Alba Patera in Upper Hesperian rocks (*Rotto and Tanaka,* 1989).

Middle and Upper Amazonian series. Materials specifically designated as Middle Amazonian in age are absent; some smooth plains material may contain minor deposits of Middle Amazonian age, but these deposits are mapped with those of Upper Amazonian age. The material forms a mantle that buries underlying members of the Vastitas Borealis Formation. Ridges have not been recognized and only a few of the faults from Alba Patera extend into the Upper Amazonian rocks. This may possibly be due to a decrease of fault activity in Late Amazonian time or to the greater distance of Upper Amazonian exposures from the volcanic center.

SOUTH POLAR REGION

Geologic-Physiographic Setting

The south polar region (Figs. 9 and 10) is unlike the boreal pole both physiographically and geologically, in that it consists almost entirely of older highland terrain that is several

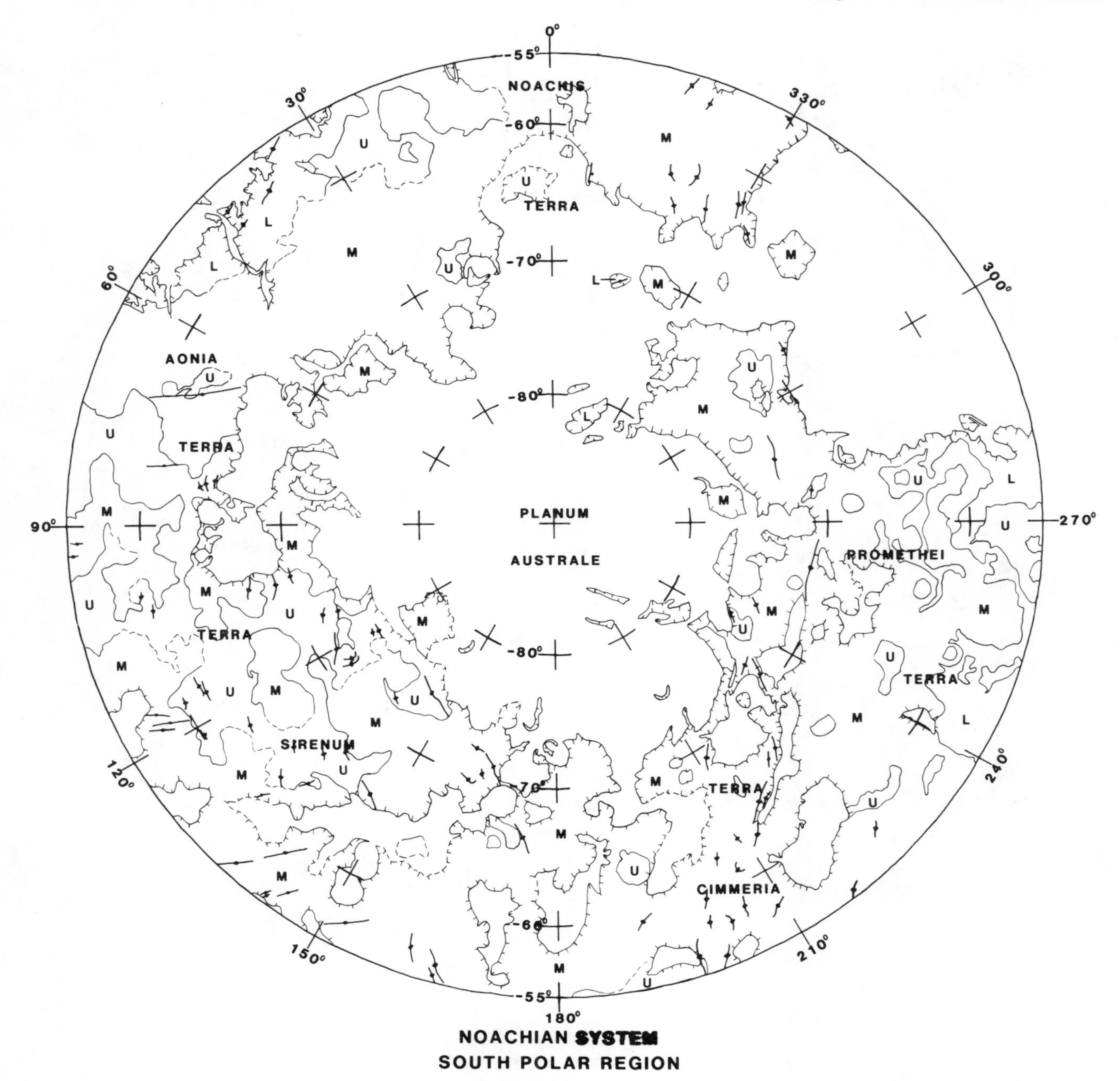

Fig. 9. Noachian system south polar region.

kilometers above the martian datum (*U.S. Geological Survey,* 1989). The south pole is capped by CO_2 ice that usually covers water ice, because temperatures are lower than at the north pole (*Kieffer,* 1979). The ice cap is surrounded by layered deposits of Late Amazonian age that form Planum Australe, a very large (1600×1200 km) plateau that is offset from the geographic pole along the prime meridian. No structures are evident in these layered deposits other than scarps and troughs carved by the wind. The layered deposits partly cover the rim of Promethei Rupes (lat. 75°S, long. 270°), a large (850-km-diameter) impact basin (*Wilhelms,* 1973). No other Amazonian materials are present in the region. Along the edge of the map area, rugged terrains are formed by the rims of Argyre and Hellas impact basins (centered at about long. 45° and 255°, respectively).

Noachian System

Lower Noachian series. Ridges in basin rim material around the southern part of Argyre Planitia occur between long. 30° and 60° along the margin of the map area. They

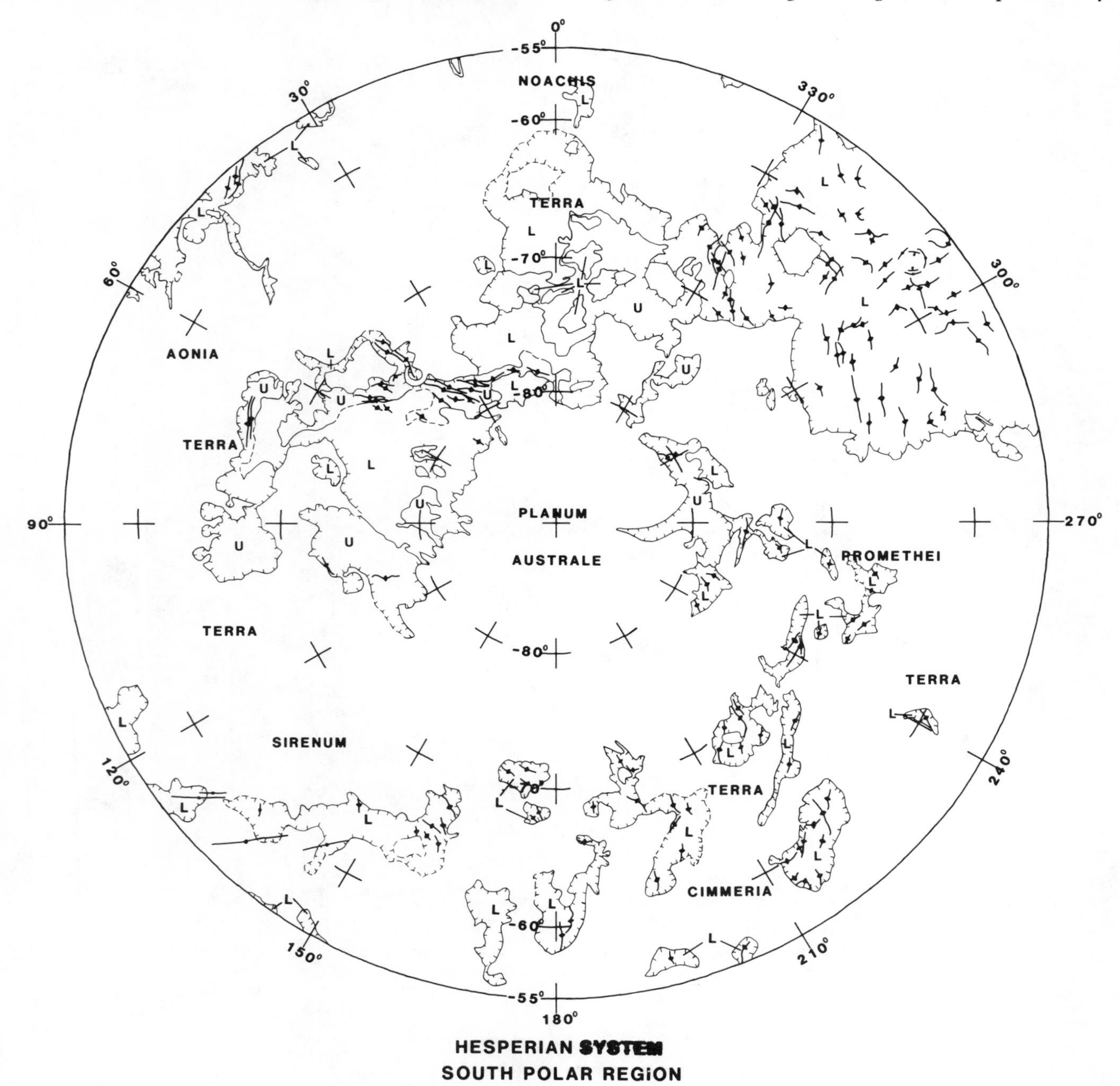

Fig. 10. Hesperian system south polar region.

were probably formed by block faulting and folding as a result of the impact that formed the basin and rim material. Structures of a similar type have not been recognized in the rim material of Hellas Planitia in the map area between long. 240° and 270°. A possible explanation for the disparity between these impact structures may be related to the older age of Hellas based on its more degraded appearance compared with that of Argyre.

Middle Noachian series. Rocks in this series cover much of the south polar region. Faulting is largely concentrated between lat. 55° and 60°S. and long. 120° and 150° and is the southern continuation of dense fault systems radial to Tharsis and Syria Planum. A series of ridges extends from the map margin near long. 340° in Noachis Terra southward in a broad arc across Promethei Terra to the vicinity of lat. 55°S, long. 210° in Terra Cimmeria. Together with the ridge systems in Noachian rocks of the eastern equatorial region (Fig. 4), they encircle Hellas Planitia.

Upper Noachian series. Scattered exposures of interbedded lava flows and aeolian deposits form intercrater plains throughout the region. Ridges are common in the Terra Sirenum area and are mostly oriented east-west; relatively few faults from the Tharsis-Syria Planum radial systems transect rocks of this series. This lends support to the reality of decreasing tectonic activity observed in the western equatorial region (Fig. 1).

Hesperian System

Lower Hesperian series. Clusters of closely spaced ridges, and lesser numbers of faults, occur in lava plains throughout the south polar region. These plains and their characteristic ridges are similar to those of Early Hesperian age that cover large areas in the western and eastern equatorial regions (Figs. 2 and 5). Faults and ridges have trends similar to those in Middle Noachian rocks. It is difficult to determine to what extent these structures represent a continuation of Noachian tectonic activity or Early Hesperian tectonism. In only a few places can definitive stratigraphic evidence be found where Hesperian plains materials directly overlap ridges and faults mapped in Noachian rocks.

Upper Hesperian series. Polar, plains-forming materials of the upper member of the Dorsa Argentea Formation (*Tanaka and Scott,* 1987) contain a dense array of northwest-trending ridges in the upper left quadrant of the map area. Their origin and structural significance in relation to other features in this region is unknown; no faulting has been recognized.

CONCLUSIONS: CONCEPTS DEVELOPED

Most major fault systems on Mars are associated with six large volcanic centers, two impact basins, and the highland-lowland boundary in the equatorial zone. The volcanic centers have been the primary contributors to faulting. In order of importance, they are (1) the Tharsis axial trend extending from southwest of Arsia Mons into Tempe Terra, (2) the Syria Planum rise, (3) Alba Patera, (4) Acheron Fossae, (5) Elysium Mons, and (6) Olympus Mons. All six volcanic centers are in the western region (Figs. 1-3) except Elysium, which is in the eastern region (Figs. 4-6). The Tharsis trend appears to have undergone the longest duration of faulting, extending possibly from the Early Noachian to the Late Amazonian. It seems probable that structural growth at Syria Planum also continued over a long period (*Tanaka and Davis,* 1988).

Of the three clearly recognized martian impact basins, only Isidis Planitia and, to a lesser extent, Argyre Planitia are partly encircled by grabens that are so common around many lunar basins of comparable size (e.g., Imbrium, Serenitatis, Orientale). Hellas, the largest well-preserved martian impact basin, is encircled by ridges, scarps, and troughs that were recognized in early stages of geologic mapping (*Binder and McCarthy,* 1972; *Scott et al.,* 1972; *Schultz and Ingerson,* 1973). However, these features have been largely subdued and obliterated by intense cratering (*Peterson,* 1977) and were not mapped at 1:15,000,000 scale (*Greeley and Guest,* 1987). The basins were created by giant impacts during an early stage in Mars' history, and their rims consist of Lower Noachian rocks. However, like the ancient lunar basins, tectonic adjustments continued for some time—into the Late Noachian and possibly Early Hesperian.

It is difficult to determine precisely where faults deviate from the southwest extension of the Tharsis trend and follow the highland-lowland boundary escarpment. This change probably occurs near long. 150°, where faults trend more westerly in the equatorial highland zone and continue to follow the boundary, mostly in the lowlands, to the west and northwest as far as long. 225°. The association of the boundary scarp with extensive parallel faulting in both the highlands and lowlands strongly supports a tectonic origin—possibly by a giant impact (*Wilhelms and Squyres,* 1984) or impacts (*Frey and Schultz,* 1988) in the subpolar lowland plains. Alternatively, subcrustal erosion and crustal lowering (*Wise et al.,* 1979) may have been responsible.

Both faulting and volcanism have progressively decreased with time from the Noachian to the Amazonian Periods. In this respect they agree with the declining impact flux through these periods. The future return of martian rock samples will allow these tectonic and volcanic processes to be directly correlated in time with cratering rates from radiometric age determinations.

Martian ridge systems are of two types: wrinkle ridges, similar to those on lunar maria, and larger ridges with steep flanks resembling fault scarps in places. Wrinkle ridges are everywhere associated with plains-forming materials interpreted to be lava flows. They range in age from Middle Noachian to Early Amazonian but had their greatest development and culmination during the Early Hesperian, not in the Late Noachian as indicated by *Chicarro* (1989). Larger ridges are mostly in Noachian materials. Both types of ridges encircle the Tharsis rise, supporting the concept of structural uplift that largely occurred during the Noachian and Hesperian Periods.

Acknowledgments. This paper represents a research effort supported by National Aeronautics and Space Administration contract

W-15, 814 from the Planetary Geology and Geophysics Program. We are indepted to K. L. Tanaka and B. K. Lucchitta for their excellent reviews of the manuscript and thoughtful suggestions.

REFERENCES

Binder A. B. and McCarthy D. W. (1972) Mars: The lineament systems. *Science, 176,* 279-281.

Carr M. H. (1979) Formation of Martian flood features by release of water from confined aquifers. *J. Geophys. Res., 84,* 2995-3007.

Chicarro A. F. (1989) Towards a chronology of compressive tectonics on Mars (abstract). In *MEVTV Workshop on Early Tectonic and Volcanic Evolution of Mars* (H. Frey, ed.), pp. 23-25. LPI Tech. Rpt. 89-04, Lunar and Planetary Institute, Houston.

Colton G. W., Howard K. A., and Moore H. J. (1972) Mare ridges and arches in southern Oceanus Procellarum. *Apollo 16 Preliminary Science Report,* pp. 90-93. NASA SP-315.

Elston W. E. (1979) Geologic map of the Cebrenia quadrangle of Mars. *U.S. Geol. Surv. Misc. Inv. Ser. Map I-1140,* scale 1:5,000,000.

Frey H. (1979) Martian canyons and African rifts—Structural comparisons and implications. *Icarus, 37,* 142-155.

Frey H. and Schultz R. (1988) Large impact basins and the mega-impact origin for the crustal dichotomy on Mars. *Geophys. Res. Lett., 15,* 229-232.

Frey H. and Schultz R. (1989) Overlapping large impacts and the origin of the northern lowlands of Mars (abstract). In *Lunar and Planetary Science XX,* pp. 315-316. Lunar and Planetary Institute, Houston.

Golombek M., Suppe J., Narr W., Plescia J., and Banerdt B. (1989) Involvement of the lithosphere in the formation of wrinkle ridges on Mars (abstract). In *MEVTV Workshop on Tectonic Features on Mars* (T. R. Watters and M. P. Golombek, eds.). LPI Tech. Rpt. 89-06, Lunar and Planetary Institute, Houston, in press.

Greeley R. and Guest J. E. (1987) Geologic map of the eastern equatorial region of Mars. *U.S. Geol. Surv. Misc. Inv. Ser. Map I-1082B,* scale 1:15,000,000.

Greeley R. and Spudis P. D. (1981) Volcanism on Mars. *Rev. Geophys. Space Phys., 19,* 13-41.

Howard K. A. and Muehlberger W. R. (1973) Lunar thrust faults in the Taurus-Littrow region. *Apollo 17 Preliminary Science Report,* pp. 22-25. NASA SP-330.

Kieffer H. H. (1979) Mars south polar spring and summer temperatures: A residual CO_2 frost. *J. Geophys. Res., 84,* 8263-8288.

Kieffer H. H., Chase S. C. Jr., Martin T. Z., Miner E. D., and Palluconi F. D. (1976) Martian north pole summer temperatures: Dirty water ice. *Science, 194,* 1341-1344.

Lucchitta B. K. (1976) Mare ridges and related highland scarps—Result of vertical tectonism? *Proc. Lunar Sci. Conf. 7th,* pp. 2761-2782.

Maxwell T. A. (1989) Origin of planetary wrinkle-ridges—An overview (abstract). In *MEVTV Workshop on Tectonic Features on Mars* (T. R. Watters and M. P. Golombek, eds.). LPI Tech. Rpt. 89-06, Lunar and Planetary Institute, Houston, in press.

Peterson J. E. (1977) Geologic map of the Noachis quadrangle of Mars. *U.S. Geol. Surv. Misc. Inv. Ser. Map I-910,* scale 1:5,000,000.

Plescia J. B. and Golombek M. P. (1986) Origin of planetary wrinkle ridges based on the study of terrestrial analogs. *Geol. Soc. Am. Bull., 97,* 1289-1299.

Plescia J. B. and Saunders R. S. (1982) Tectonic history of the Tharsis region of Mars. *J. Geophys. Res., 87,* 9775-9791.

Rotto S. L. and Tanaka K. L. (1989) Faulting history of the Alba Patera-Ceraunius Fossae region of Mars (abstract). In *Lunar and Planetary Science XX,* pp. 926-927. Lunar and Planetary Institute, Houston.

Schultz P. H. and Ingerson F. E. (1973) Martian lineaments from Mariner 6 and 7 images. *J. Geophys. Res., 78,* 8415-8427.

Schultz R. A. (1989) Strike-slip faulting in the ridged plains of Mars (abstract). In *MEVTV Workshop on Tectonic Features on Mars* (T. R. Watters and M. P. Golombek, eds.). LPI Tech. Rpt. 89-06, Lunar and Planetary Institute, Houston, in press.

Scott D. H. (1973) Small structures of the Taurus-Littrow region. *Apollo 17 Preliminary Science Report,* pp. 25-28. NASA SP-330.

Scott D. H. (1989) New evidence—old problems: Wrinkle ridge origin (abstract). In *MEVTV Workshop on Tectonic Features on Mars* (T. R. Watters and M. P. Golombek, eds.). LPI Tech. Rpt. 89-06, Lunar and Planetary Institute, Houston, in press.

Scott D. H. and Carr M. H. (1978) Geologic map of Mars. *U.S. Geol. Surv. Misc. Inv. Ser. Map I-1083,* scale 1:25,000,000.

Scott D. H. and Dohm J. M. (1990) Faults and ridges: Historical development in Tempe Terra and Ulysses Patera regions of Mars. *Proc. Lunar Planet. Sci. Conf. 20th,* this volume.

Scott D. H. and Tanaka K. L. (1980) Mars Tharsis region: Volcanotectonic events in the stratigraphic record. *Proc. Lunar Planet. Sci. Conf. 11th,* pp. 2403-2421.

Scott D. H. and Tanaka K. L. (1981) Ignimbrites of Amazonis Planitia region of Mars. *J. Geophys. Res., 87,* 1179-1190.

Scott D. H. and Tanaka K. L. (1986) Geologic map of the western equatorial region of Mars. *U.S. Geol. Surv. Misc. Inv. Ser. Map I-1802-A,* scale 1:15,000,000.

Scott D. H., Peterson J. E., and Wilhelms D. E. (1972) Prototype and preliminary geologic mapping of Mars from Mariner 6 and 7 data, part 2: Preliminary small-scale regional mapping (1:5,000,000). *U.S. Geol. Surv. Interagency Rpt., Astrogeology, 45,* 11-17.

Scott D. H., Tanaka K. L., and Chapman M. G. (1989) Water of the Elysium basin, Mars: Volumetric analysis and sources (abstract). In *Reports of Planetary Geology and Geophysics Program, NASA TM,* in press.

Tanaka K. L. (1986) The stratigraphy of Mars. *Proc. Lunar Planet. Sci. Conf. 17th,* in *J. Geophys. Res., 91,* E139-E158.

Tanaka K. L. (1990) Tectonic history of the Alba Patera-Ceraunius Fossae region of Mars. *Proc. Lunar Planet Sci. Conf. 20th,* this volume.

Tanaka K. L. and Davis P. A. (1988) Tectonic history of the Syria Planum province of Mars. *J. Geophys. Res., 93,* 14,893-14,917.

Tanaka K. L. and Golombek M. P. (1989) Martian tension fractures and the formation of grabens and collapse features at Valles Marineris. *Proc. Lunar Planet. Sci. Conf. 19th,* pp. 383-396.

Tanaka K. L. and Scott D. H. (1987) Geologic map of the polar regions of Mars. *U.S. Geol. Surv. Misc. Inv. Ser. Map I-1802C,* scale 1:15,000,000.

Tanaka K. L., Isbell N. K., and Scott D. H. (1988) The resurfacing history of Mars: A synthesis of digitized, Viking-based geology. *Proc. Lunar Planet. Sci. Conf. 18th,* pp. 665-678.

U.S. Geological Survey (1989) Topographic maps of the western equatorial, eastern equatorial, and polar regions of Mars. *U.S. Geol. Surv. Misc. Inv. Ser. Map I-2030,* 3 sheets, scale 1:15,000,000.

Watters T. R. (1988) Wrinkle ridge assemblages on the terrestrial planets. *J. Geophys. Res., 93,* 10236-10254.

Watters T. R. and Maxwell T. A. (1986) Orientation, relative age, and extent of the Tharsis plateau ridge system. *J. Geophys. Res., 91,* 113-125.

Wichman R. W. and Schultz P. H. (1988) An ancient Valles Marineris (abstract). In *MEVTV Workshop on Early Tectonic and Volcanic Evolution of Mars* (H. Frey, ed.), pp. 88-90. LPI Tech. Rpt. 89-04, Lunar and Planetary Institute, Houston.

Wilhelms D. E. (1973) Comparison of martian and lunar multiringed circular basins. *J. Geophys. Res., 78,* 4084-4095.

Wilhelms D. E. and Squyres S. W. (1984) The martian hemispheric dichotomy may be due to a giant impact. *Nature, 309,* 7934-7939.

Wise D. U., Golombek M. P., and McGill G. E. (1979) Tharsis province of Mars: Geologic sequence, geometry, and a deformation mechanism. *Icarus, 38,* 4456-472.

Proceedings of the 20th Lunar and Planetary Science Conference, pp. 503-513
Lunar and Planetary Institute, Houston, 1990

Faults and Ridges: Historical Development in Tempe Terra and Ulysses Patera Regions of Mars

D. H. Scott and J. M. Dohm

U.S. Geological Survey, 2255 North Gemini Drive, Flagstaff, AZ 86001

Tectonism has been active, though episodic and generally declining with time, throughout the geologic history of Mars; it has been concentrated mostly in the western equatorial region, especially along the regional topographic rise of the Tharsis Montes volcanic province. Tectonic activity as expressed by faults and ridges has been traced from the oldest to the youngest rocks. Multiple episodes of intense structural deformation, however, have occurred in some areas where stratigraphic information is insufficient to assess their chronology. Tempe Terra and Ulysses Patera, because of their structural complexity, have been selected for detailed studies to determine the distribution, orientation, and sequence of fault- and ridge-system development within relatively short intervals of time. This study is primarily a record of fault and ridge chronology that should be useful in discriminating between different theories proposed for the structural evolution of Tharsis. Results of the work show that from the Early Noachian through the Early Amazonian Epochs, at least eight episodes of faulting occurred at Tempe Terra and six at Ulysses Patera. Tectonic activity at Tempe Terra was expressed mainly by densely spaced faults along the northeast extension of the Tharsis rise; faulting culminated in the Middle and Late Noachian and was superseded by transverse fault systems from the Alba Patera region during the Hesperian. Ridge formation, however, was most active in the Early Hesperian. At Ulysses Patera, an early history of tectonism is recorded by complex arrays of faults in a relatively small area of Noachian rocks. Later deformation appears to have peaked in the Middle Hesperian: abundant faults are concentrated along a structural axis between Olympus and Pavonis Montes.

INTRODUCTION

The purpose of this work is to map the sequence of structural deformation that occurred within two intricately faulted areas of the Tharsis region of Mars. Although it is primarily an observational record of fault and ridge chronology, the map data should be considered when models and theories are proposed for the evolution of the Tharsis rise. The maps should be useful in discriminating between some presently advanced theories and possibly in promoting new ones.

Tectonic activity on Mars expressed by fault and wrinkle-ridge systems is recorded in rock units that were emplaced during all eight epochs of the three martian periods (Noachian, Hesperian, and Amazonian). The time of occurrence and global distribution of these tectonic features have been determined by mapping faults and ridges that either transect or are covered by rock units of known age (*Scott and Dohm,* 1990). Several areas in the western equatorial region of Mars, however, have undergone multiple episodes of intense structural deformation that are not decipherable from superposition relations with rock units alone. This is largely due to complicated, intricate patterns of structures manifest at a scale (1:2,000,000) much larger than that of the regional geologic maps (1:15,000,000) used for compilation of the global data. To show the various stages of faulting and ridge development in local areas, we have selected Tempe Terra and the area north of Ulysses Patera for detailed mapping because of their structural complexity and geologic setting relative to the Tharsis rise. This work has been accomplished by using Viking photomosaics (1:2,000,000 scale) to determine crosscutting relations of structures as well as their morphology and trend orientations. Similar types of analyses have been made of fault histories at Alba Patera-Ceraunius Fossae (*Rotto and Tanaka,* 1989; *Tanaka,* 1990) and Syria Planum (*Tanaka and Davis,* 1988).

METHODS USED

As on the smaller scale regional maps (*Scott and Dohm,* 1990) faults and ridges are assigned time-stratigraphic positions according to their occurrence in rocks of known relative age (*Scott and Tanaka,* 1986). However, by using Viking photomosaics at 1:2,000,000 scale, more detailed age discriminations could be made by considering (1) crosscutting relations among faults and between faults and ridges, (2) morphologic appearance, including degree of degradation, and (3) concurrence of structural trends that might be indicative of similar stress fields and closely contemporaneous

TABLE 1. Symbols used on figures.

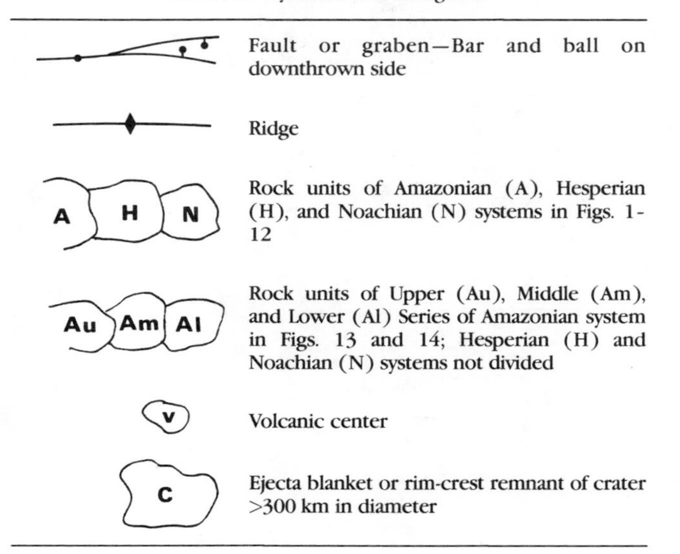

Symbol	Explanation
	Fault or graben—Bar and ball on downthrown side
	Ridge
A H N	Rock units of Amazonian (A), Hesperian (H), and Noachian (N) systems in Figs. 1-12
Au Am Al	Rock units of Upper (Au), Middle (Am), and Lower (Al) Series of Amazonian system in Figs. 13 and 14; Hesperian (H) and Noachian (N) systems not divided
v	Volcanic center
C	Ejecta blanket or rim-crest remnant of crater >300 km in diameter

periods of deformation. Although crosscutting relations are the most definitive age criteria, they are not clearly discernible in places, and some structures (mostly faults) cannot be relatively dated with assurance; ambiguities may also arise where older faults are rejuvenated. Lithostratigraphic units are shown for the martian time-stratigraphic systems (Table 1), but most are not subdivided into series because of the uncertainty of locating some of these boundaries at larger scales; wherever practicable, however, the occurrence of structures within a particular series is indicated in the discussion of the various fault sets and ridges. The sets are numbered from oldest to youngest (1 = oldest).

Although the maps have been photographically reduced in scale, details in the portrayal of the structures have been maintained by extracting fault sets and ridges of about the same age and showing them on individual figures.

TEMPE TERRA REGION

Tempe Terra is a large plateau of cratered highland terrain that projects into the northern lowland plains along the northeast extension of the Tharsis rise. Low, rugged mountain ranges and hills consisting of Lower Noachian basement rocks form the western margin of the plateau (*Scott and King,* 1984; *Scott and Tanaka,* 1986). Most of Tempe Terra, however, is covered by cratered and fractured rock units that have a wide range of ages—probably extending from the lower to the upper part of the Middle Noachian Epoch. The southeastern part of the plateau consists of Lower Hesperian ridged plains lava flows that are separated from the Lunae Planum region to the south by the broad channel of Kasei Valles, as much as 400 km wide in this area. Upper Hesperian lava flows of the Tempe Terra Formation (*Scott and Tanaka,* 1986) cover the southwestern part of the map area. Faults and grabens transect the plateau along the Tharsis axial trend in a zone 2000 km long and as much as 500 km wide. Small- to moderate-size shield volcanoes and volcanotectonic structures having concentric fault patterns occur in several places.

Boundaries between rock units are not clearly defined in many places, and thus the assignment of individual faults and, to a lesser extent, ridges to a particular rock series (lower, middle, upper) or epoch (early, middle, late) is not everywhere feasible. A more reliable relative age classification has been achieved by placing faults and ridges in sets, based on criteria discussed above, that better reflect their position in sequence than does their occurrence in a particular rock unit.

Noachian System

Fault set 1—Lower to Middle Noachian (Fig. 1a). Fault density is low and no ridges are recognized. Some north-trending faults cut Lower Noachian hills and mountains in the western part of the map area but do not extend into Middle Noachian rocks. However, the apparent paucity of Lower Noachian faults may be directly related to the relatively small areal extent of rock exposures of this age. Faults transecting Middle Noachian rocks are cut by some faults in sets 2 and 3 and appear more subdued and have different trends; they may belong in the lower part of the Middle Noachian Series.

Fault set 2—Middle Noachian (Fig. 1b). Fault density remains low and the style of faulting has now changed in places: Whereas fault set 1 contains long, linear traces, faults of set 2 are curvilinear and comparatively short in the southwestern and west-central map areas. These faults partly encircle and may be the earliest surface expression of probable

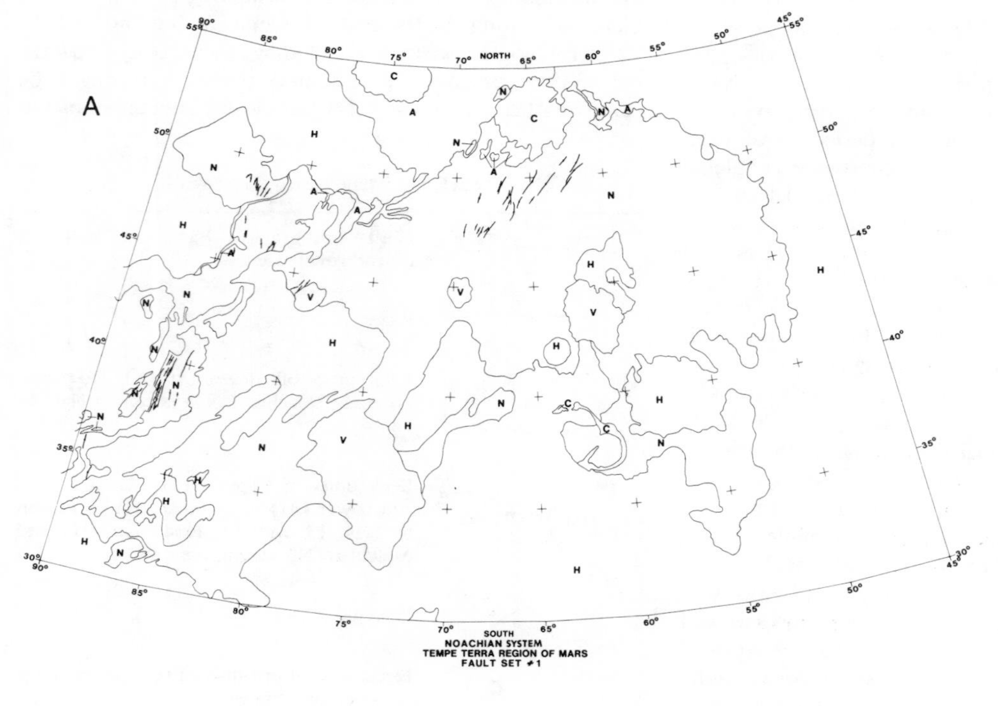

Fig. 1. Fault sets of the Noachian System, Tempe Terra region of Mars. Fault sets 1-5 are indicated by letters a-e. Map symbols shown in Table 1.

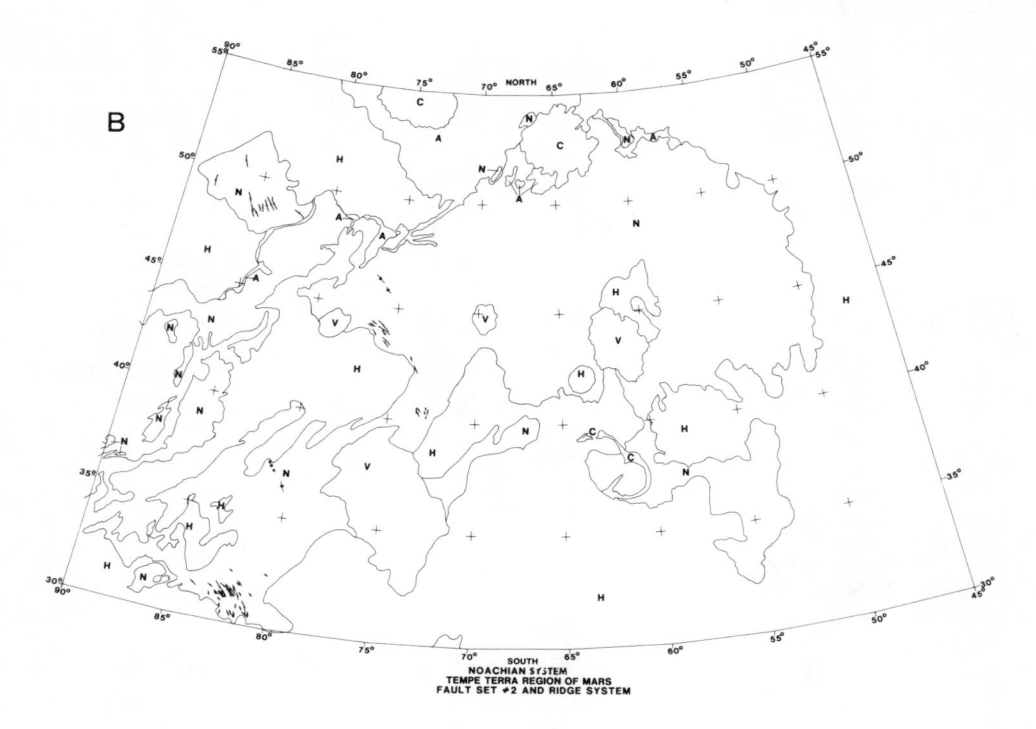

Fig. 1. (continued).

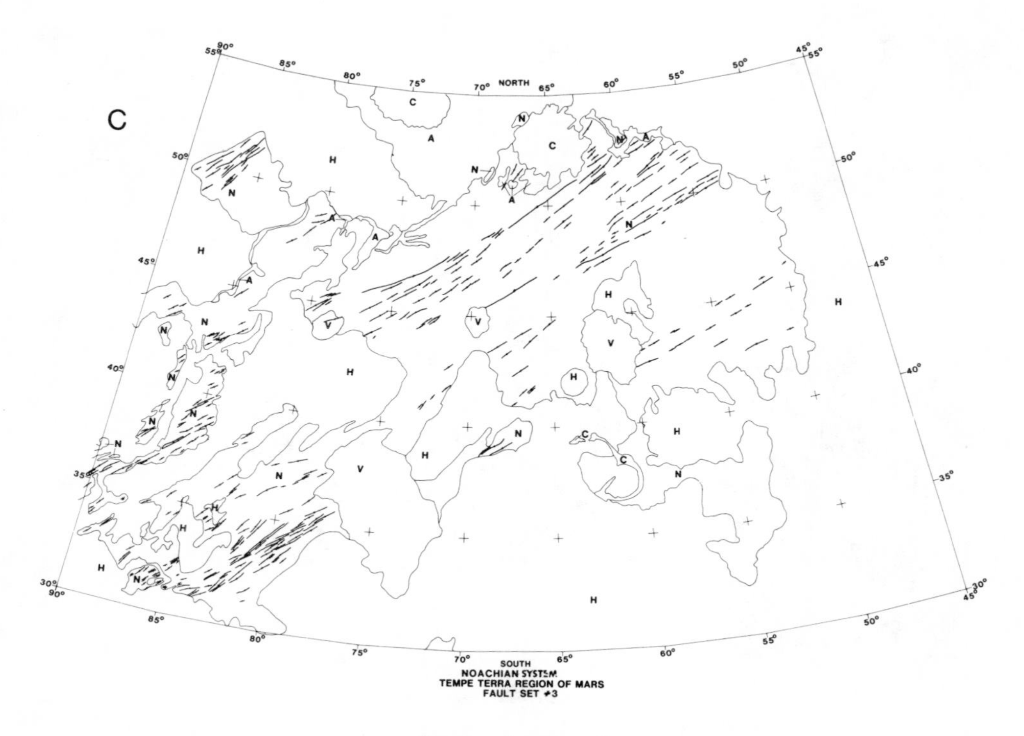

Fig. 1. (continued).

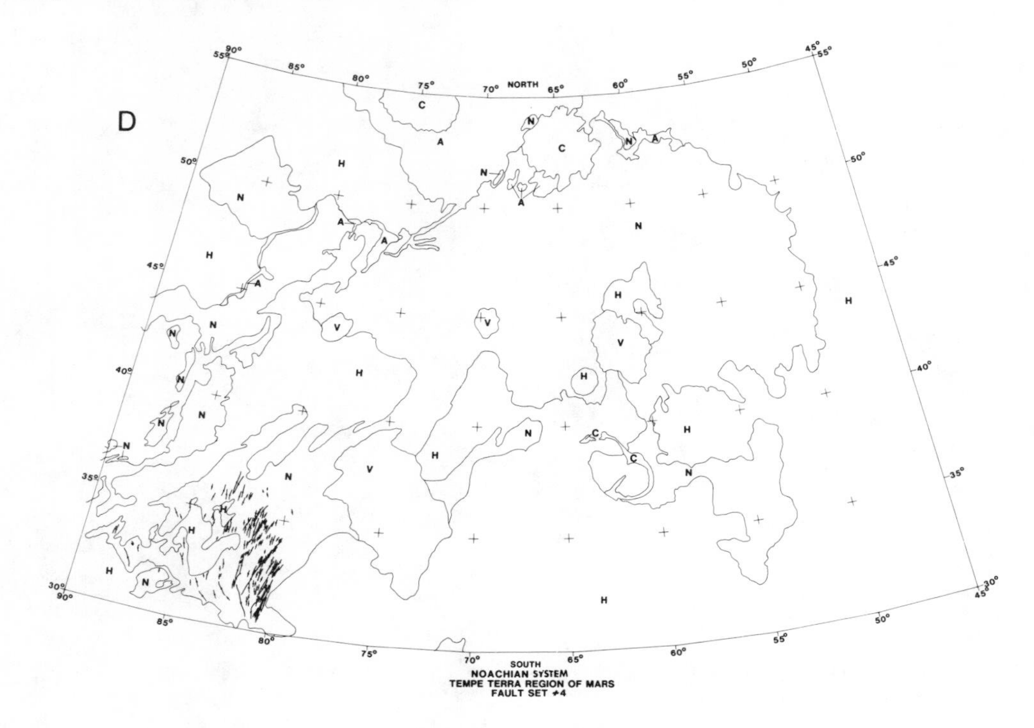

Fig. 1. (continued).

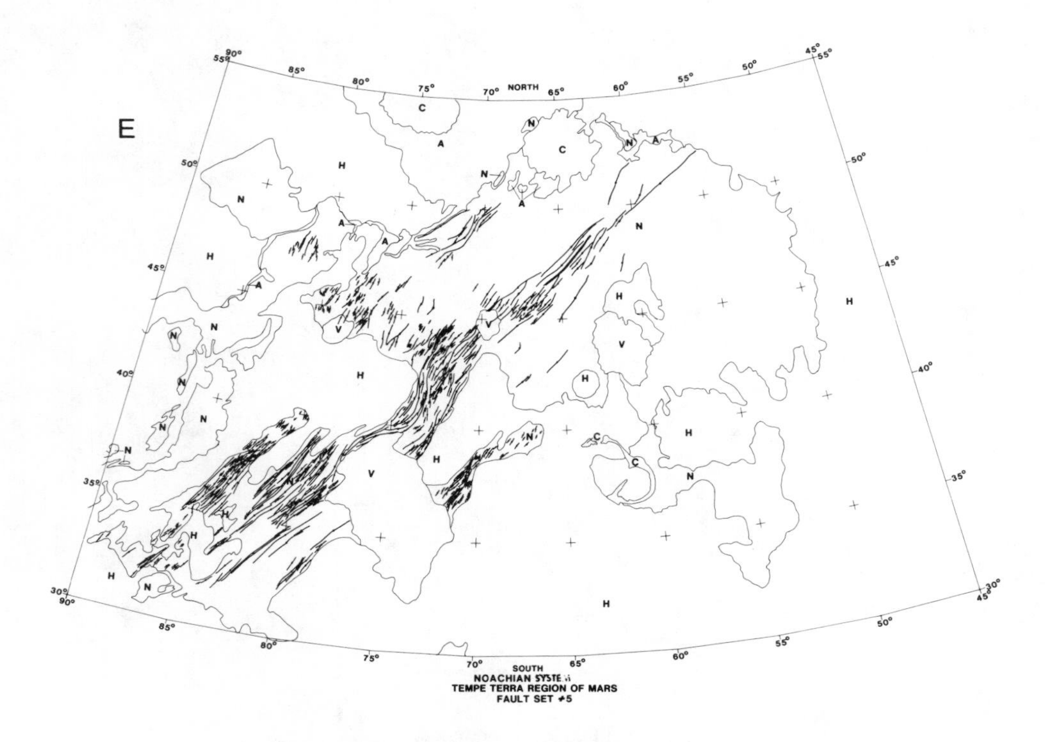

Fig. 1. (continued).

deep-seated volcanotectonic centers near lat. 42°N, long. 77° and lat. 30°N, long. 84°; both centers lie along the Tharsis trend. A few northwest-trending faults continue into the Alba Patera area adjacent to the west (*Rotto and Tanaka,* 1989; *Tanaka,* 1990). Sparse northwest-oriented ridges are cut by faults of set 3; these ridges are highly degraded and may be among the oldest in the Tempe Terra-Lunae Planum areas.

Fault set 3—Middle Noachian (Fig. 1c). A profusion of faults and grabens extends in a broad swath from the southwest to the northeast margins of Tempe Terra plateau. In the southwestern part of the map area they are cut by faults of sets 4 and 5, elsewhere by faults of set 5; they are overlapped throughout the region by lava flows of Early to Late Hesperian age. These faults represent the earliest major expression of the Tharsis extensional stress field in its continuation far to the northeast along the line of centers of the Tharsis Montes volcanoes. Although deviations from the Tharsis trend are minor, faults near the west border of the map area may be transitional with faulting and early tectonism at Alba Patera (*Rotto and Tanaka,* 1989; *Tanaka,* 1990).

Fault set 4—Middle Noachian (Fig. 1d). Clusters of curvilinear faults in the southwestern part of Tempe Terra may have developed, like those of set 2, in response to stresses associated with volcanotectonic intrusive centers. The distributions and curvatures of the faults suggest that centers may be located in the vicinity of lat. 33°N, long. 80° or possibly coincident with the volcano at lat. 37°N, long. 77°. Lava flows from this volcano are Hesperian in age, but the onset of magma intrusion into the core of the structure probably occurred during the Middle Noachian, as better indicated by faults of set 5.

Fault set 5—Middle to Upper Noachian (Fig. 1e). Dense arrays of northeast-trending faults, similar to those of set 3 but more closely spaced, are concentrated in a long, relatively narrow zone extending across the center of the plateau. This faulting may represent the culmination of intense structural deformation along the extension of the Tharsis axis, because Hesperian faulting is either greatly diminished along this trend (Fig. 2a) or overridden by swarms of radial and tangential faults (Fig. 2b) from the Alba Patera-Ceraunius Fossae region to the west (*Rotto and Tanaka,* 1989; *Tanaka,* 1990). Fault curvatures are strongly developed around the northwest side of the volcanic center near lat. 37°N, long. 77° and on the southeast side of the probable intrusive near lat. 42°N, long. 77° first recognized by fault curvatures in set 2 (Fig. 1b).

Hesperian System

Fault set 1 and ridges—Lower Hesperian (Fig. 2a). Lava flows extending north from Lunae Planum are the oldest Hesperian materials recognized on Tempe Terra. They probably originated from fissure vents and may be characterized by northwest-trending wrinkle ridges; in places the association of grabens, fissures, and wrinkle ridges has been demonstrated (*Scott,* 1989). The ridges are cut by Hesperian fault sets 2 and 3 (*Watters and Maxwell,* 1983); most ridges, however, do not extend into areas where Noachian rocks occur, and thus the most active epoch of ridge formation appears to have been the Early Hesperian. Although the major ridge-forming events were coeval with some Hesperian faulting, they did not precede major extensional faulting in the Tempe Terra area as suggested by *Watters and Maxwell* (1983). Faults are concentrated in the west-central part of the map area, where they cut Middle Hesperian lava flows of the Tempe Terra Formation (*Scott and Tanaka,* 1986). Most faults

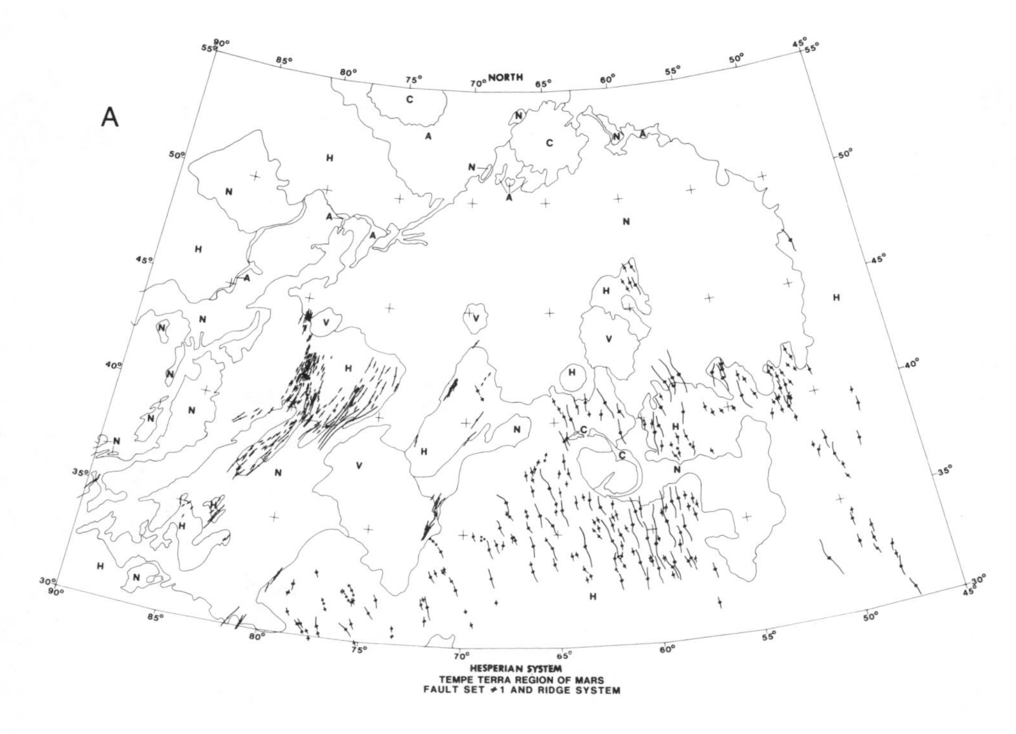

Fig. 2. Fault sets of the Hesperian and Amazonian Systems, Tempe Terra region of Mars. Fault sets 1-3 are indicated by letters a-c. Map symbols shown in Table 1.

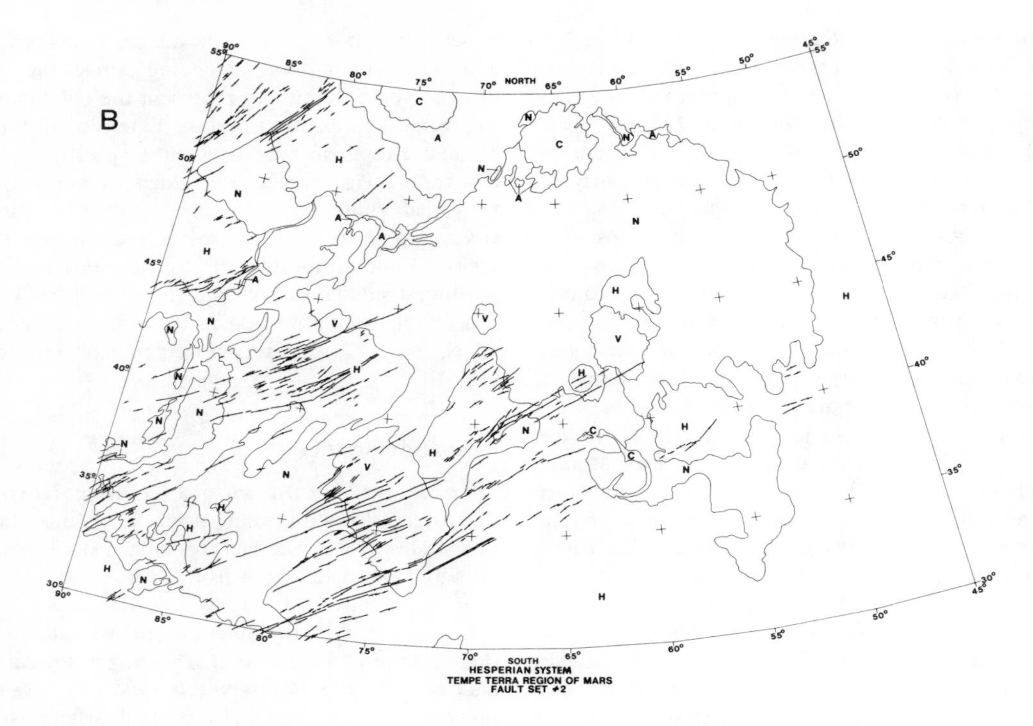

Fig. 2. (continued).

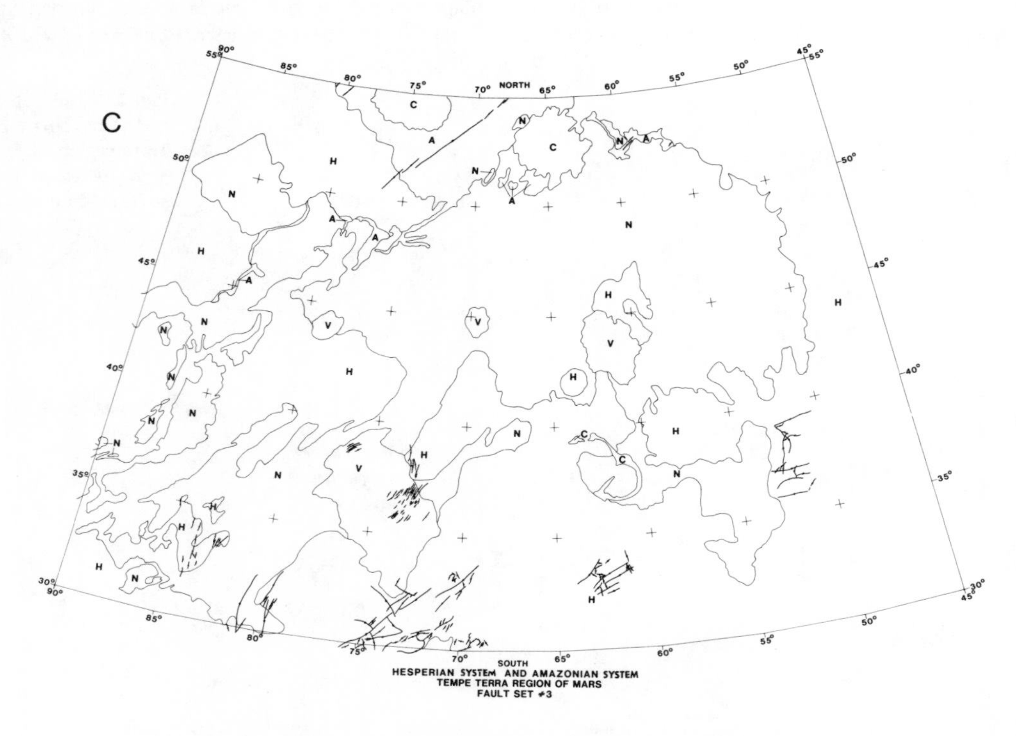

Fig. 2. (continued).

trend northeast, but some partly curve around the probable volcanotectonic center near lat. 42°N, long. 77°, similar to the Noachian faults of set 2 (Fig. 1b) and set 5 (Fig. 1e); also, a larger center may exist farther west, perhaps near lat. 42°N, long. 83°.

Fault set 2—Upper Hesperian (Fig. 2b). Swarms of faults extend northeasterly across the southwest half of Tempe Terra. They cut across Lower Hesperian fault set 1 (Fig. 2a) and Noachian fault sets 1-5 (Figs. 1a-e) that are mostly oriented in the direction of the Tharsis trend. Their distribution, trends transverse to Tharsis, and continuation to the west (*Rotto and Tanaka,* 1989; *Tanaka,* 1990) indicate origins mostly associated with the stress fields at Alba Patera-Ceraunius Fossae rather than the Tharsis rise.

Fault set 3—Upper Hesperian to Middle Amazonian (Fig. 2c). Faults that form fresh-appearing canyons occur in scattered clusters across the southern part of the plateau, which consists of Hesperian lava flows. In places the faults exhibit orthogonal patterns formed by northeast- and northwest-oriented intersecting systems. Faults and grabens concentric to the volcano at lat. 37°N, long. 77° suggest that tectonism associated with magma intrusions probably extended from the Middle Noachian (Fig. 1e) to the Late Hesperian. Exposures of Lower and Middle Amazonian rocks are virtually limited to a relatively small northern area in the lowland plains bordering the plateau; faults here follow northeast trends but are notably fewer than in the adjoining area of older Hesperian rocks (Fig. 2b).

Discussion

Tectonic activity at Tempe Terra throughout the Noachian Period is primarily expressed by dense arrays of faults along the northeast extension of the Tharsis rise. This Tharsis-related faulting culminated during Middle to Late Noachian time (Fig. 1e), and it was superseded by swarms of transverse faults tangential to and originating from the Alba Patera and Ceraunius Fossae regions (*Rotto and Tanaka,* 1989; *Tanaka,* 1990) during the latter part of the Hesperian (Fig. 2b). Faults and grabens concentric to known and postulated volcanotectonic centers are common in Middle Noachian and Lower Hesperian rocks (Figs. 1b,d,e and 2b). These fault-bounded

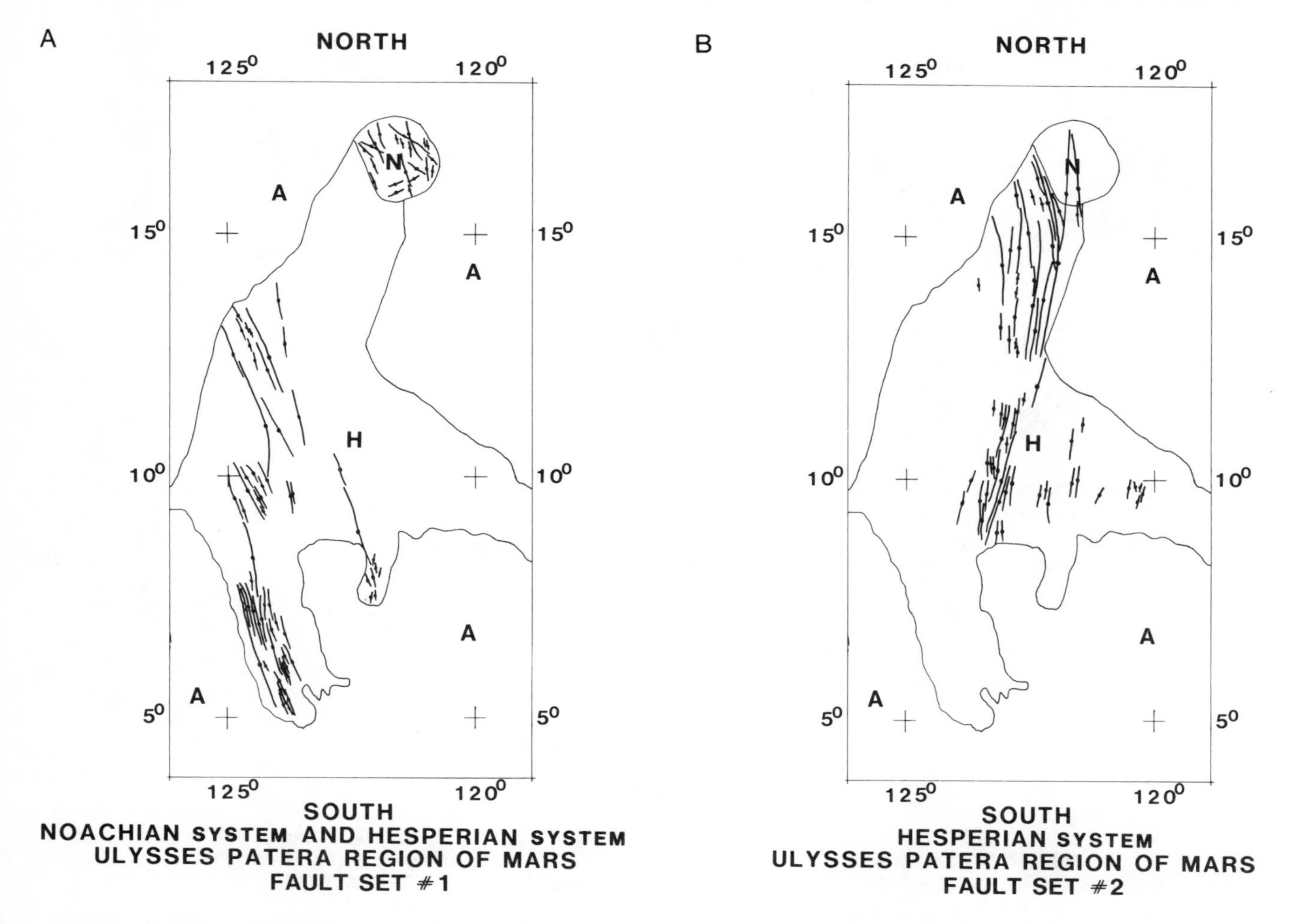

Fig. 3. Fault sets of the Noachian and Hesperian Systems, Ulysses Patera region of Mars. Fault sets 1-4 are indicated by letters a-d. Map symbols shown in Table 1.

centers probably indicate the presence of intrusive bodies at depth. A few ridges, trending northwest across the structural grain of Tharsis faulting, first occur in cratered terrain of Middle Noachian age (Fig. 1b). The most active epoch of ridge development, however, was the Early Hesperian (Fig. 2a); ridges characterize the plains-forming lava flows of Lunae Planum and their extension into Tempe Terra. Tectonism diminished from the Late Hesperian to the Early Amazonian; scattered occurrences of fresh-appearing faults transect Hesperian lava flows and relatively small exposures of Lower Amazonian rocks along the north margin of Tempe Terra plateau (Fig. 2c).

The history of faulting, ridge formation, and volcanism at Tempe Terra is partly in accord with the concept of an early development of the Tharsis rise accompanied by a vast radial fault system, and a long-lived volcanic stage (*Wise et al.,* 1979). The tectonic pattern also shows the result of several distinct episodes of faulting, originating from different sources, that have been superposed on one another; other fault centers have been recognized in several places within the Tharsis region (*Plescia and Saunders,* 1982). The youngest appearing faults transect lava flows in the southern part of Tempe Terra; they are probably contemporaneous and suggest that an isotropic distribution of stresses may have existed in this area during relatively recent times.

ULYSSES PATERA REGION

This second study area is centered on Noachian and Hesperian, highly deformed materials that are surrounded and embayed by lava flows and aureole deposits of Olympus Mons ranging in age from Early to Late Amazonian. The Noachian rocks are limited to a small exposure in the northern part of the map area; they are more highly faulted and fractured than adjacent Hesperian materials, and their boundary is readily distinguished by differences in fault density and complexity. Both of these highly deformed rock units may have a wide range of ages in their respective periods (*Scott and Tanaka,* 1986), and their further subdivision has not been attempted. To simplify the portrayal of chronologic relations between faults (no ridges are present) and rock stratigraphic units, subdivisions (series) are shown only on maps of the Amazonian System (Figs. 4a,b).

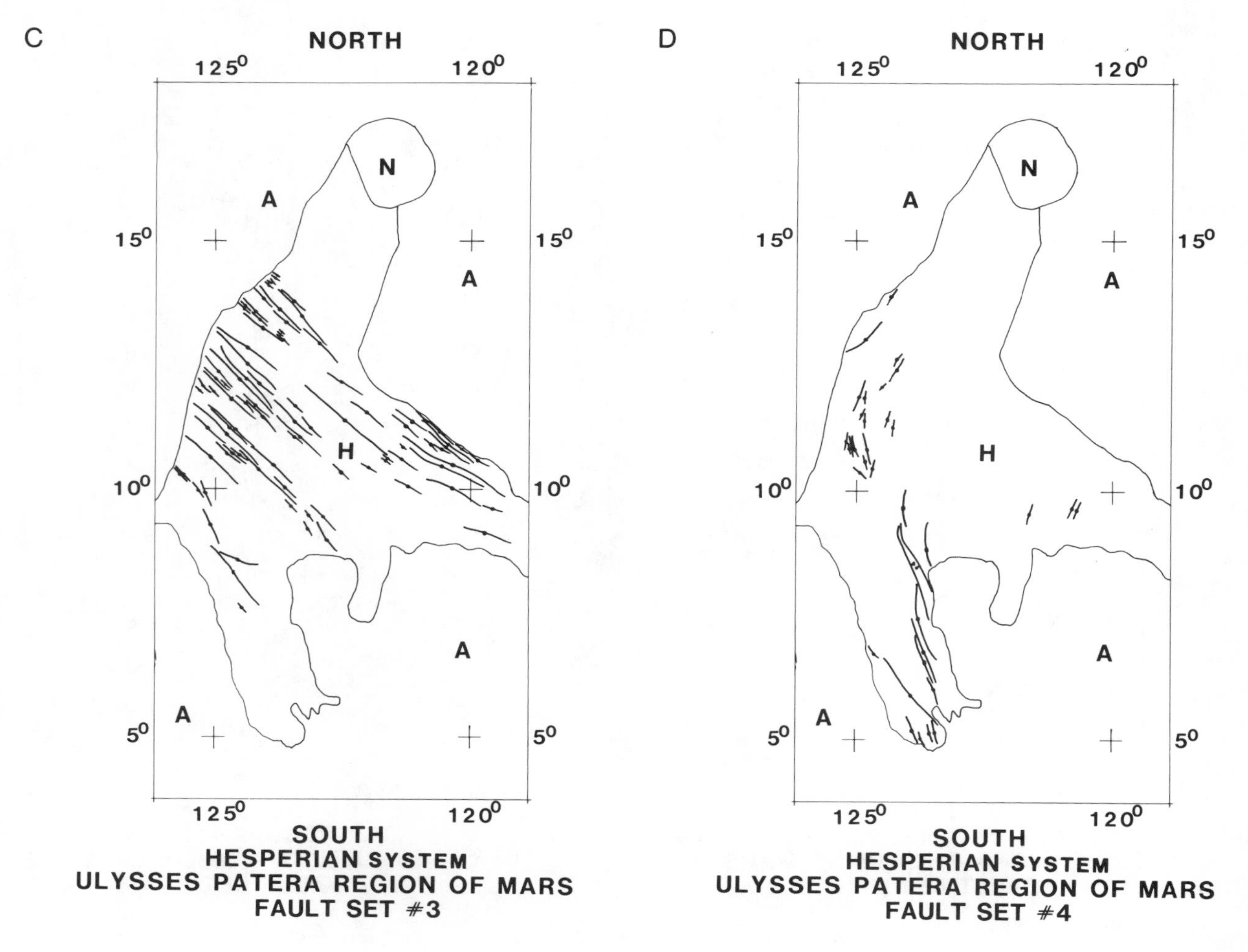

Fig. 3. (continued).

Noachian and Hesperian Systems

Fault set 1 (Fig. 3a). Highly subdued crosscutting traces and close spacing characterize Noachian age faults in a small area consisting of older fractured material (*Scott and Tanaka,* 1986) centered near lat. 15.5°N, long. 122°. They may be related to a local volcanotectonic center or to a confluence of fault systems originating from various centers in Tharsis Montes (*Plescia and Saunders,* 1982). Hesperian faults trend north-northeast radial to the volcano Ulysses Patera, immediately south of the map area; the flanks of this volcano are covered by lava flows from Pavonis Mons. Both the Noachian and Hesperian faults of this set are overlapped by Middle and Upper Amazonian lava flows and are crosscut in places by Hesperian faults of set 2 (Fig. 3b).

Hesperian System

Fault set 2 (Fig. 3b). Faults have subdued morphologies but are distinguished from Hesperian faults of set 1 by transection relations and more northerly trends; *en echelon* traces are common. Their orientation is similar to those of Lower Hesperian faults in Ceraunius Fossae (*Rotto and Tanaka,* 1989; *Tanaka,* 1990) and to faults in isolated patches of Hesperian and Noachian highly deformed terrain north of Noctis Labyrinthus and between Ceraunius Fossae and Ulysses Patera. Some may belong to regional fault systems associated with Tharsis that have been locally affected by Olympus Mons to the west as indicated by their curvature; the southernmost faults of this group, however, are radial to the volcano Biblis Patera immediately west of Ulysses Patera.

Fault set 3 (Fig. 3c). Great numbers of northwest-trending faults crosscut sets 1 and 2; they lie between and are radial to both Olympus Mons and Pavonis Mons, the central volcano of Tharsis Montes (each volcano is about 800 km from the map area). Their transection of fault sets 1 and 2 originating from Ulysses and Biblis Paterae suggests that Pavonis Mons may be the youngest volcano of this group.

Fault set 4 (Fig. 3d). Scattered groups of curvilinear faults and a large graben in places transect older fault sets. Their curvatures and patterns appear to be concentric to Pavonis Mons and they may represent reactivation of faulting around this volcanic center as a result of loading of the lithosphere by individual volcanic constructs (*Solomon et al.,* 1979; *Comer et al.,* 1980; *Solomon and Head,* 1982).

Amazonian System

Fault set 1 (Fig. 4a). Lower Amazonian aureole deposits (unit Al) are cut by many northwest-trending, small faults that, in turn, are buried by Upper Amazonian lava flows (unit Au). This fault set, however, is not clearly distinguishable from set 2 (Fig. 4b) on the basis of morphology, trends, or transections. For this reason, some of the faults of set 1, where not clearly terminated against Upper Amazonian rocks, may well have originated in Middle to Late Amazonian time.

Fault set 2 (Fig. 4b). All of these faults cut rock units of the Middle (unit Am) or Upper Amazonian (unit Au) Series. Like the faults of Amazonian set 1 and Hesperian set 3, they are radial to both Olympus Mons and Pavonis Mons and possibly, in the southern part of the map area, to Ulysses Patera. It seems likely that each of these volcanic centers contributed to tectonism in this region and that tectonic episodes partly overlapped in time (*Solomon and Head,* 1982).

Discussion

In the Ulysses Patera region, the earliest tectonism is recorded by complex sets of intersecting faults and grabens in a small, isolated area (lat. 16.5°N, long. 122°) of highly deformed rocks (Fig. 3a) that may range in age from Early to Late Noachian. Trends of these faults span 90° of arc, and the

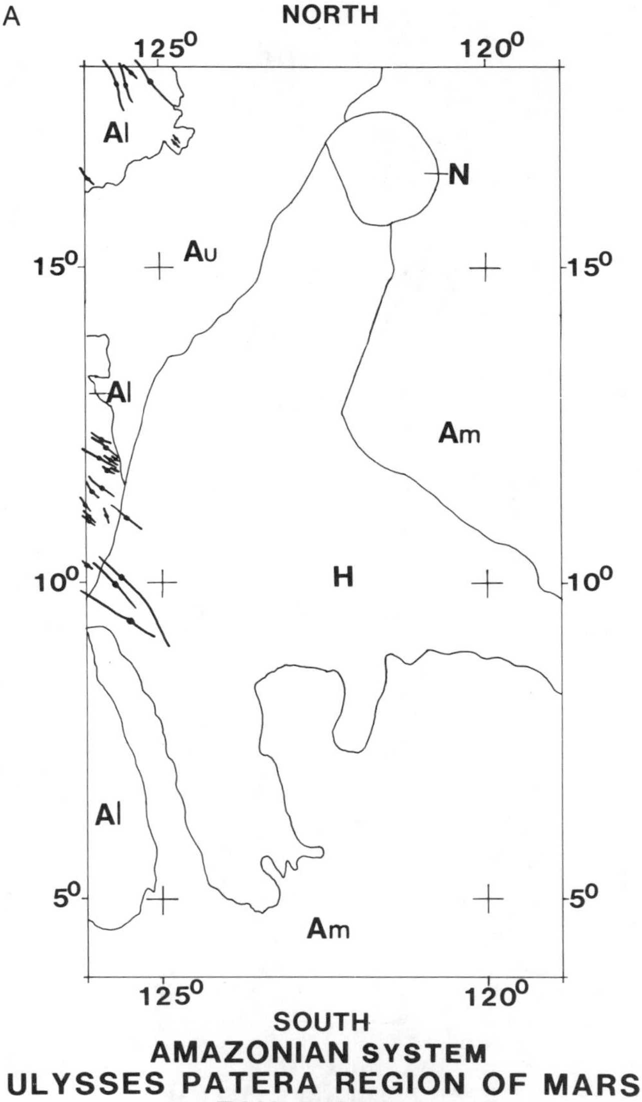

Fig. 4. Fault sets of the Amazonian System, Ulysses Patera region of Mars. Fault sets 1 and 2 are indicated by letters a and b. Map symbols shown in Table 1.

faults cannot clearly be identified with any single volcanotectonic center. The oldest recognized Hesperian faults are radial to Ulysses and Biblis paterae, whereas somewhat younger faults (Fig. 3b) follow more northerly trends similar to those of Lower Hesperian faults in Ceraunius Fossae (*Rotto and Tanaka,* 1989; *Tanaka,* 1990). Large numbers of northwest-trending Hesperian faults reflect structural control along an Olympus Mons-Pavonis Mons axis (Fig. 3c); their abundance suggests a peak in tectonic activity during this period and probably major growth of the Pavonis Mons volcanic edifice. Thereafter, faulting waned until the Amazonian Period, when northwest-trending faults were rejuvenated but on a smaller scale (Fig. 4a). Faulting in the Ulysses Patera region seems to follow, in part, the general order of tectonism at Tharsis described by *Banerdt et al.* (1982) where flexural loading and radial graben formation are followed by a last stage of loading with little regional deformation.

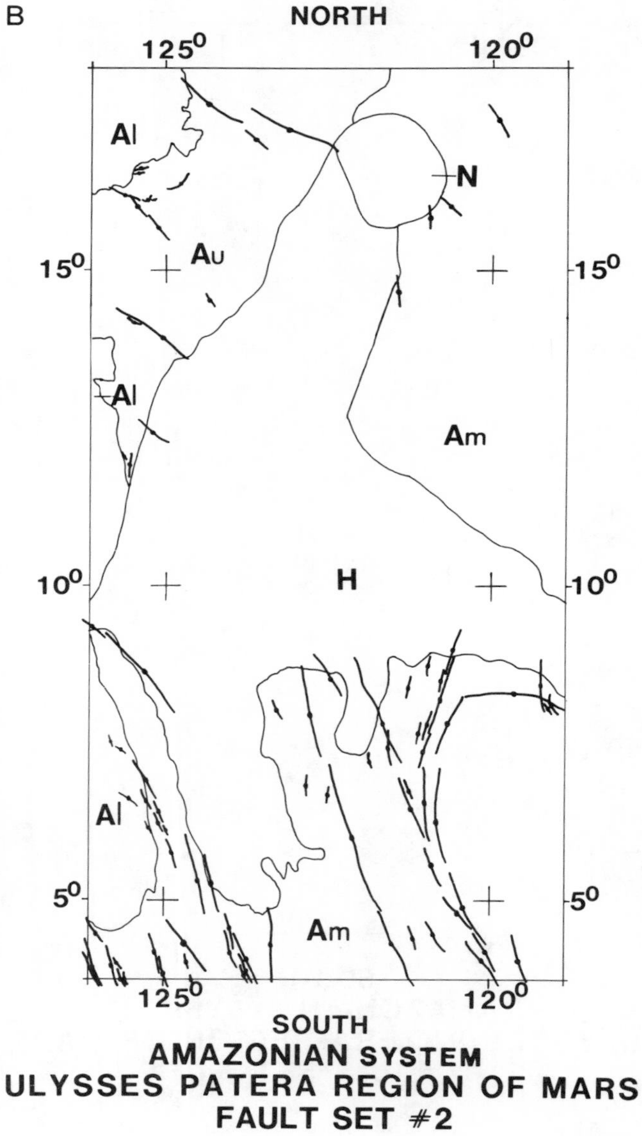

B NORTH

125° 120°

Al

N

15° Au 15°

Al

Am

10° H 10°

Al

5° 5°

Am

125° 120°

SOUTH

AMAZONIAN SYSTEM
ULYSSES PATERA REGION OF MARS
FAULT SET #2

Fig. 4. (continued).

CONCLUDING REMARKS

The evolution of the Tharsis rise, whether by uplift or the accumulation of volcanic piles, is only part of the complex history of this province. The volcanic and structural development of the entire region is recorded in the growth of its various parts including Syria Planum, Olympus Mons, Alba Patera, Acheron Fossae, and Tempe Terra as well as many smaller volcanotectonic centers. Age relations of these features are fairly well known from geologic mapping; however, the historical record is difficult to decipher in places because of complicating effects caused by the growth and development of structures, both large and small, that occurred along with the Tharsis rise. The extent of the problem may sometimes not be recognized, or may be distorted, due to overattention to Tharsis proper. Our work at Tempe Terra and Ulysses Patera, together with other studies in the Tharsis region, should help to reconstruct a long series of geological events leading to the present form of this large province.

Acknowledgments. This study was supported by National Aeronautics and Space Administration contract W-15,814 from the Planetary Geology and Geophysics Program. We thank K. L. Tanaka and M. G. Chapman for their reviews and comments.

REFERENCES

Banerdt W. B., Phillips R. J., Sleep N. H., and Saunders R. S. (1982) Thick shell tectonics on one-plate planets: Applications to Mars. *J. Geophys. Res., 87,* 9723-9733.

Comer R. P., Solomon S. C., and Head J. W. (1980) Thickness of the Martian lithosphere beneath volcanic loads: A consideration of the time dependent effects (abstract). In *Lunar and Planetary Science XI,* pp. 171-173. Lunar and Planetary Institute, Houston.

Plescia J. B. and Saunders R. S. (1982) Tectonic history of the Tharsis region, Mars. *J. Geophys. Res., 87,* 9775-9791.

Rotto S. L. and Tanaka K. L. (1989) Faulting history of the Alba Patera-Ceraunius Fossae region of Mars (abstract). In *Lunar and Planetary Science XX,* pp. 926-927. Lunar and Planetary Institute, Houston.

Scott D. H. (1989) New evidence-old problem: Wrinkle ridge origin (abstract). In *MEVTV Workshop on Tectonic Features on Mars* (T. R. Watter and M. P. Golombek, eds.). Lunar and Planetary Institute, Houston, in press.

Scott D. H. and King J. S. (1984) Ancient surfaces of Mars: The basement complex (abstract). In *Lunar and Planetary Science XV,* pp. 736-737. Lunar and Planetary Institute, Houston.

Scott D. H. and Tanaka K. L. (1986) Geologic map of the western equatorial region of Mars. *U.S. Geol. Surv. Misc. Inv. Ser. Map I-1802-A,* scale 1:15,000,000.

Scott D. H. and Dohm J. M. (1990) Chronology and global distribution of fault and ridge systems on Mars. *Proc. Lunar Planet. Sci. Conf. 20th,* this volume.

Solomon S. C. and Head J. W. (1982) Evolution of Tharsis province of Mars: The importance of heterogeneous lithospheric thickness and volcanic construction. *J. Geophys. Res., 87,* 9755-9774.

Solomon S. C., Head J. W., and Comer R. P. (1979) Thickness of the Martian lithosphere from tectonic features: Evidence for lithospheric thinning beneath volcanic provinces (abstract). In *NASA TM-80339,* pp. 60-62.

Tanaka K. L. (1990) Tectonic history of the Alba Patera-Ceraunius Fossae region of Mars. *Proc. Lunar Planet. Sci. Conf. 20th,* this volume.

Tanaka K. L. and Davis P. A. (1988) Tectonic history of the Syria Planum province of Mars. *J. Geophys. Res., 93,* 14,893-14,917.

Watters T. R. and Maxwell T. A. (1983) Crosscutting relations and relative age of ridges and faults in the Tharsis region of Mars. *Icarus, 56,* 278-298.

Wise D. U., Golombek M. P., and McGill G. E. (1979) Tharsis province of Mars: Geologic sequence, geometry, and a deformation mechanism. *Icarus, 38,* 456-472.

Proceedings of the 20th Lunar and Planetary Science Conference, pp. 515-523
Lunar and Planetary Institute, Houston, 1990

Tectonic History of the Alba Patera-Ceraunius Fossae Region of Mars

K. L. Tanaka

Branch of Astrogeology, U.S. Geological Survey, 2255 North Gemini Drive, Flagstaff, AZ 86001

Local volcanic and regional Tharsis tectonic activity produced long-lived, extensive, and varied types of faulting in the Alba Patera-Ceraunius Fossae region of Mars. Detailed mapping of fault patterns and determination of stratigraphic relations among faults and geologic units in the region indicate four main stages of faulting from Noachian to Late Amazonian time: (1) Noachian stage I faults dominantly trend northeast and may be related to a tectonic center near Ulysses Patera; (2) stage II faults, of Late Noachian and Hesperian age, were centered at Syria Planum, as indicated by north-trending faults of Ceraunius Fossae; (3) following extensive volcanism from Alba Patera, intense stage III faulting during the Early Amazonian produced north-trending faults of Ceraunius Fossae (reactivated from stage II), circumferential faults surrounding Alba Patera, calderas and wrinkle ridges on the patera, and north- to northeast-trending faults north of the patera (possibly reactivated buried structures). This intense faulting may have been caused by a combination of renewed Tharsis-centered activity and stresses induced by Alba Patera magmatism; (4) north-trending grabens of stage IV, which deflect to the northeast north of Ceraunius Fossae, may have been generated by tectonism in association with Middle to Late Amazonian volcanism at Ascraeus Mons. The broad pattern of volcanism and tectonism at Alba Patera is similar to those of Syria Planum and Tempe Terra; this pattern may reflect a style characteristic of intermediate-age volcano tectonic centers in heavily fractured areas of the Tharsis region. Lahars and channels associated with Ceraunius Fossae indicate episodic release of ground water during fracturing of stages II and IV. The extension of northeast-trending faults for over 1000 km from Alba Patera and Ascraeus Mons may have been governed by a crustal weakness along the highland/lowland boundary, consistent with proposed mega-impact and tectonic origins of the northern lowlands.

INTRODUCTION

Alba Patera and nearby plateaus and high plains fractured by Ceraunius Fossae form a volcanotectonic province within the Tharsis region of Mars. The planet's highland/lowland boundary apparently passes through or near this region, but it is buried by local volcanic rocks (*Wilhelms and Squyres,* 1984; *Scott and Tanaka,* 1986). Alba Patera's broad form, extensive and varied volcanic deposits, and abundant faults indicate a complex geologic history. The fault history of Alba Patera has broadly been examined by regional studies (*Scott and Tanaka,* 1980; *Plescia and Saunders,* 1982; *Tanaka,* 1986; *Scott and Dohm,* 1990), but it has not been thoroughly described or explained in detail. A particular observation not adequately explained by previous regional stress models and fault histories is the northeastward deflection of north-trending faults north of Ceraunius Fossae. Here, the structural history of Alba Patera and Ceraunius Fossae is presented in the context of regional and local tectonic controls. Although speculative, the following models are generally consistent with photogeologic evidence.

STRATIGRAPHY

As shown by recent geologic mapping and volcanologic studies of Alba Patera (e.g., *Cattermole,* 1986, 1987; *Scott and Tanaka,* 1986; *Mouginis-Mark et al.,* 1989; *Schneeberger and Pieri,* 1989), most materials in the study area (Fig. 1) have volcanic origins. Relative ages have also been established through overlap relations and crater densities. In this paper, stratigraphic assignments are made according to units defined by *Tanaka* (1986).

The generalized geologic map (Fig. 2) was compiled on parts of 1:5,000,000-scale shaded-relief maps of MC-3 and 9. This map and the correlation chart (Fig. 3) are modified from the 1:15,000,000-scale geologic map of the western equatorial region of Mars (*Scott and Tanaka,* 1986) for the purpose of accurately ascertaining the fault history of the region. Modifications include (1) differences in placement of contacts and amount of detail shown, (2) grouping of most units of the Alba Patera and Ceraunius Fossae Formations, and (3) reassignment of some stratigraphic positions. These changes are based on scale of mapping, crater densities, and relevance of unit ages to fault history.

Basement rocks (unit Nbv), although generally too rugged and limited in area for meaningful crater counts, are commonly embayed by Middle Noachian cratered terrain material (*Scott and Tanaka,* 1986), indicating that they are of Early Noachian age. Fractured materials (units HNf and Hf) also are so rugged that some crater obliteration is expected. Many large grabens in these units are filled by patches of younger Hesperian and Amazonian units (too small to be shown in Fig. 2). Therefore, crater counts for these units (*Plescia and Saunders,* 1982; *Schneeberger and Pieri,* 1989) are considered lower-limit values. Nevertheless, the fractured materials of Ceraunius Fossae have been assigned ages (Noachian and Hesperian, undivided, and Hesperian) that are younger than the strictly Noachian age of *Scott and Tanaka* (1986) because of the absence of large craters and the relative smoothness of their surfaces in most areas. Noachian faualted material at Claritas Fossae, for example, is markedly more cratered and degraded than Hesperian faulted material (*Tanaka and Davis,* 1988).

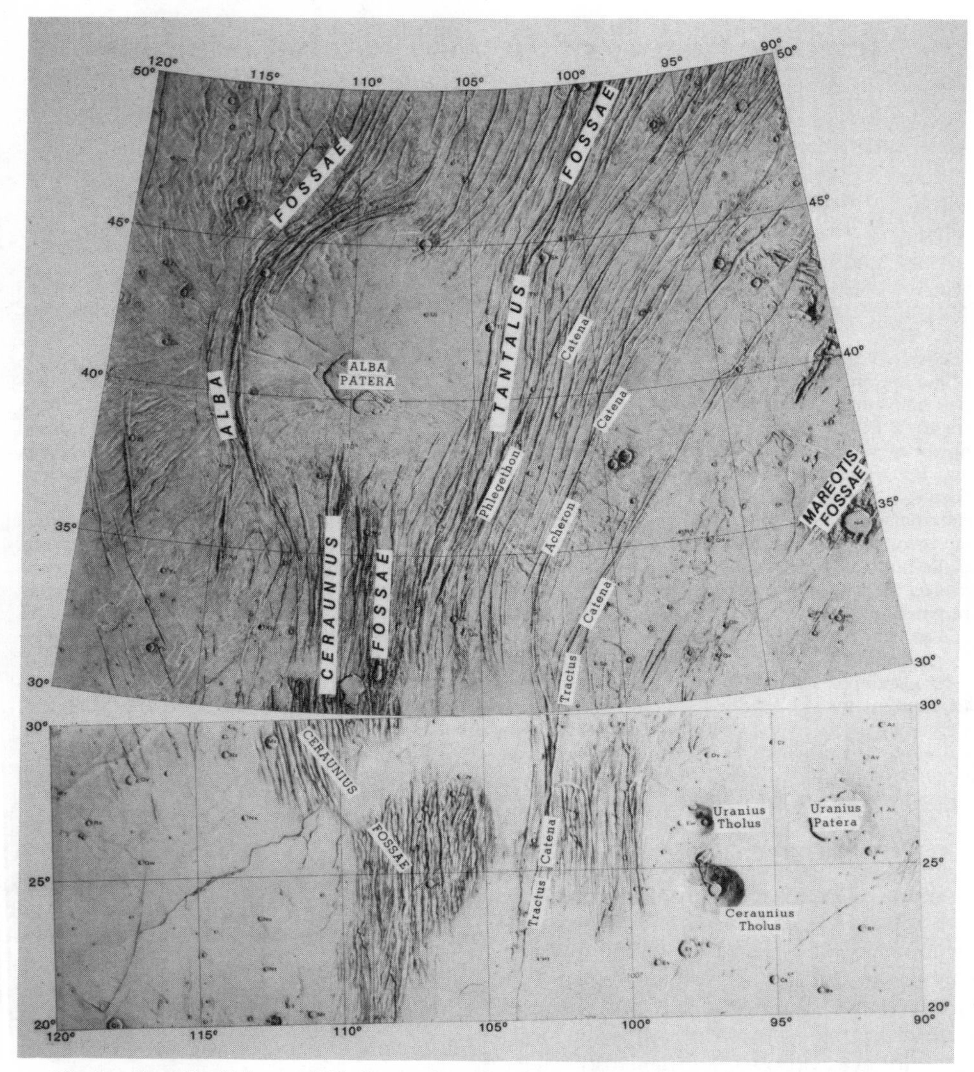

Fig. 1. Shaded-relief map of Alba Patera-Ceraunius Fossae region of Mars showing major features.

Uranius Patera (Fig. 1) is composed of Hesperian volcanic material (unit Hv); it apparently formed at the end of the Early Hesperian, according to the crater-density data of *Plescia and Saunders* (1979). Materials of the Alba Patera Formation on the lower flanks of Alba Patera and flows of the Ceraunius Formation east of Alba Patera were emplaced during the Late Hesperian, whereas flows on the upper parts of Alba Patera formed during earliest Amazonian time (*Scott and Tanaka,* 1986; *Schneeberger and Pieri,* 1989). Because very few structures appear to have formed in the region during these epochs, I grouped these units into undivided material of the Alba Patera and Ceraunius Fossae Formations (unit AHac). The earliest recognized Tharsis flows (unit Ht_2) in the area probably originated from Ascraeus Mons or associated fissures formed during Alba Patera volcanism; they appear to be onlapped by Alba Patera-Ceraunius Fossae flows (unit AHac).

According to crater counts (Fig. 3), local flows from Ascraeus Mons (unit At_4) formed at the beginning of Amazonian time, a slightly older age designation than that based on regional counts by *Scott and Tanaka* (1981, 1986). Younger flows of the Ceraunius Fossae Formation (unit Acf) are lightly cratered where they embay the fossae (in MC 9NE, eight craters larger than 2 km in diameter occur in 95,600 km^2, which yields $N(2) = 84 \pm 30$), but the flows become markedly more cratered to the west in MC 9NW ($N(2) = 200\text{-}350$). However, many of the craters in the west appear embayed by the flows, and so the crater density is uncertain (Fig. 3). Young flows from Ascraeus Mons (unit At_5) were erupted during Middle Amazonian time; these flows bury the older part of the Ceraunius Fossae Formation (unit AHac) south of the map area (*Scott and Tanaka,* 1986).

FAULT HISTORY

In this section, I describe the inferred fault history of the Alba Patera-Ceraunius Fossae region and show additional relations and details not reported in the preliminary results of *Rotto and Tanaka* (1989). Relative ages of structures were

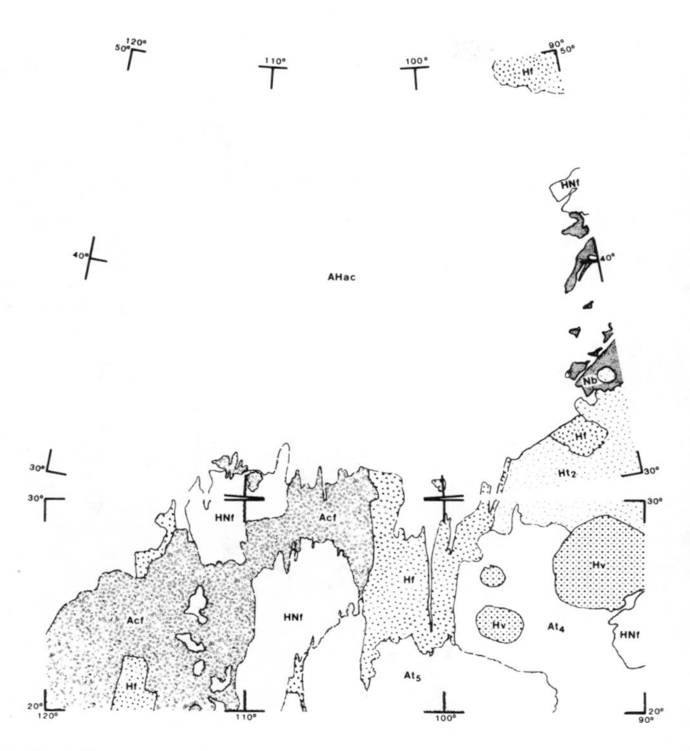

Fig. 2. Geologic map of the Alba Patera-Ceraunius Fossae region of Mars; modified from mapping of *Scott and Tanaka* (1986). Nb = basement rocks, HNf and Hf = fractured rocks, Hv = volcanic material, Ht_2 = member 2 of the Tharsis Montes Formation, AHac = Alba Patera-Ceraunius Fossae flows, At_4 = member 4 of the Tharsis Montes Formation, Acf = flows of the Ceraunius Fossae Formation, At_5 = member 5 of the Tharsis Montes Formation; see Fig. 3 for correlation of map units.

determined by crosscutting and overlap relations among geologic units and structures as described by *Tanaka and Davis* (1988).

Stage I—Noachian

Most of these faults (Fig. 4a) trend N20°-60°E, roughly parallel with the northeast trends of the Tharsis Montes to the south and Tempe Terra to the east. The faults generally are paired into grabens that range in length from tens of kilometers to more than 100 km and in width from about one to several kilometers. Fault scarps in basement material have particularly high relief and considerable width.

Stage I faults cut basement and fractured materials of Noachian and Hesperian age (units Nb, HNf, and Hf) and are overlain by Upper Hesperian and Amazonian materials. Faulting in Hesperian material, however, is probably due to the rejuvenation of Noachian faults. For example, a patch of northeast-trending faults in Ceraunius Fossae (lat. 27°N, long. 105°) is crosscut by north-trending stage II fractures (Fig. 4b) that partly formed during the Late Noachian (see below). Therefore, the initial faulting along northeast trends must be Noachian in age; however, the specific epochs of the Noachian are not precisely known (Fig. 3).

These northeast-trending stage I faults are generally coarrelative in age, orientation, and broad distribution with "Middle Noachian set 3" faults in Tempe Terra mapped by *Scott and Dohm* (1990). Although the stage I faults are possibly correlative in time with some early Syria Planum-centered radial faulting (*Plescia and Saunders,* 1982; *Tanaka and Davis,* 1988), their trends indicate a different stress center or structural control. A small group of faults that trend northwest along the east edge of the map area at lat. 43°N may correlate with similar faults in "Middle Noachian set 2" in Tempe Terra (*Scott and Dohm,* 1990).

Stage II—Late Noachian and Hesperian

These faults (Fig. 5a) cut Noachian and Hesperian materials, including large tracts of fractured material (units HNf and Hf, Fig. 2), Hesperian Alba Patera flows (unit AHac) in the northeastern part of the map area, and flows (unit Hv) making up the east flank of Uranius patera. The material of western Ceraunius Fossae (unit HNf) is more densely faulted and cratered than that of eastern Ceraunius Fossae (unit Hf), indicating that Early Hesperian resurfacing of eastern Ceraunius Fossae buried faults and craters produced during the Noachian period.

North-trending grabens of Ceranius Fossae, commonly hundreds of kilometers long and more than 10 km wide, dominate three broad plateaus; a few fractures of this group

Epoch	Fault stage	Geologic unit	Crater density N(2) N(5)	Age (b.y.)
Late Amazonian				
				0.25 0.70
Middle Amazonian	IV	Acf, At_5	84±30 (Acf) 92±11 (At_5) 18±5	
				0.70 2.50
Early Amazonian	III	At_4	380±92 (At_4) 70±39	
				1.80 3.55
Late Hesperian		AHac, Ht_2	(AHac) 79±21 (Ht_2) 113±36	
				3.10 3.70
Early Hesperian	II	Hv, Hf	~1000 (Hv) ~180 190±61	
				3.50 3.80
Late Noachian		HNf		
				3.85 4.20
Middle Noachian	I			
				3.92 4.30
Early Noachian		Nb		

Fig. 3. Correlation of martian epochs, fault stages, geologic units, and crater densities for the Alba Patera-Ceraunius Fossae region of Mars. Epochs from *Tanaka* (1986), crater counts from *Plescia and Saunders* (1979, 1982) and *Scott and Tanaka* (1981); N(2) and N(5) are numbers of craters larger than 2 and 5 km diameter per 10^6 km^2, respectively. Ages from crater-flux models described by *Tanaka* (1986) and *Tanaka et al.* (1988); left column Hartmann-Tanaka model, right column Neukum-Wise model.

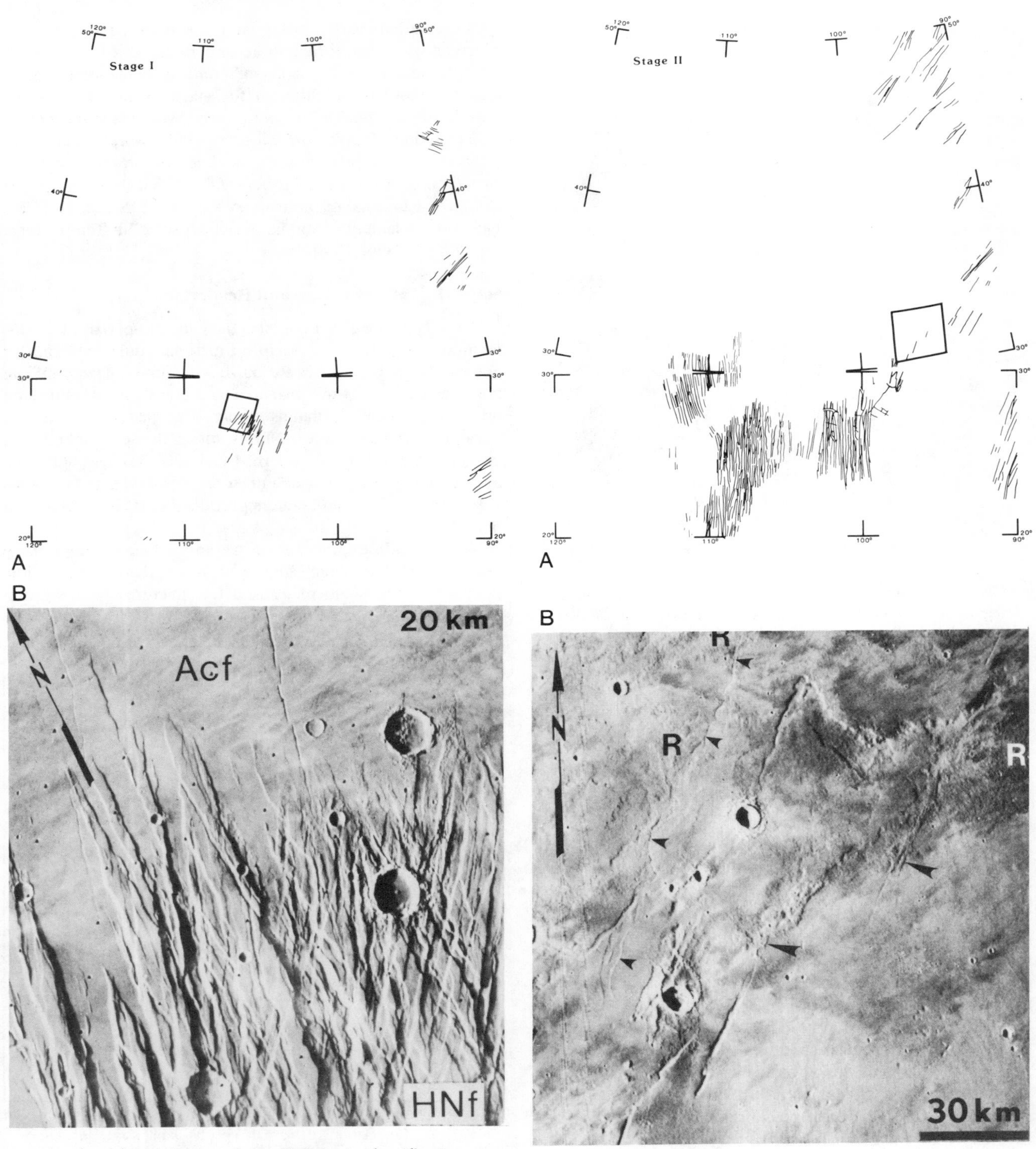

Fig. 4. **(a)** Distribution of stage I faults in the Alba Patera-Ceraunius Fossae region of Mars. Box shows location of **(b)**. **(b)** Part of northern Ceraunius Fossae showing north-trending faults of stage II crosscutting northeast-trending stage I faults in unit HNf. Stage I faults have regular traces, whereas scarps of stage II faults are irregular, because they partly follow stage I faults. Unit Acf forms smooth lava plains locally faulted along northerly trends. (Viking image 38B55.)

Fig. 5. **(a)** Distribution of stage II faults in the Alba Patera-Ceraunius Fossae region of Mars. Box shows location of **(b)**. **(b)** Area in northeastern Ceraunius Fossae showing flows emanating from fractures (large arrows). Locally the flows have a rugged, degraded surface (R) and are cut by a channel (small arrows). These observations suggest that the flows may be lahars (see text). (Viking image 516A02.)

are exposed in lower-lying fractured material west of Ceraunius Fossae. Most of these wide grabens appear to have nested within the narrower grabens that exhibit shallower offsets. In western Ceraunius Fossae (lat. 27°N, long. 110°), some of the fractures bend around lower areas buried by subsequent flows. Although the cause of these bends is not known, a possible explanation is that the fault trends are deflected by local volcanotectonic centers; this type of relation is observed at Tempe Terra (*Scott and Dohm,* 1990). In northeastern Ceraunius Fossae, some of the grabens are crosscut by narrow to broad linear troughs of various orientations. These troughs are similar in morphology to fretted troughs elsewhere on Mars (*Sharp,* 1973), and they and other nearby fractures may be the source of large, broad, degraded lobes or flows (Fig. 5b) similar to flows in the Elysium region interpreted to be lahars produced by volcano/ground-ice interactions (*Christiansen,* 1989).

Along the east edge of the map area, faults trending N30°-60°E formed in rocks that postdate stage I faulting (units Hf and AHac). It seems likely that the younger faults were produced by rejuvenation and upward propagation of buried stage I faults, which have similar occurrence and trend. East of Uranius Patera, stage II faults trend more northerly (about N20°E) than earlier ones. Some faults of similar age that trend N0°-20°E occur just north of the map area and east of about long. 105°; they extend to lat. 61°-63°N, where they are buried by Lower Amazonian plains material (*Tanaka and Scott,* 1987).

The north-trending faults of this stage belong to the episode of Syria Planum-centered radial faulting identified by *Plescia and Saunders* (1982). Northeast-trending faults along the east edge of the map area correspond in age and distribution to faults of "Middle to Upper Noachian set 5" and "Hesperian set 2" in Tempe Terra (*Scott and Dohm,* 1990).

Stage III—Early Amazonian

Following emplacment of extensive flows of the Alba Patera Formation and part of the Ceranius Fossae Formation, the majority of the Alba and Tantalus Fossae were formed (Figs. 1 and 6a). These grabens are long, wide, and linear to curved where aligned concentric with the broad oval crest of Alba Patera. They extend beyond the north edge of the map past lat. 60°N, where they splay along N5°-35°E trends and cut Lower Amazonian plains material (*Scott and Tanaka,* 1986; *Tanaka and Scott,* 1987); their bounding faults are perhaps rejuvenated buried faults of stage II. Alba Fossae, on the west side of Alba Patera, consist of fewer and more widely spaced grabens than Tantalus Fossae on the east side; however, some fractures of Alba Fossae occur twice as far away from the center of the volcano (nearly 1000 km) as those of Tantalus Fossae. Some of these more distant faults are overlain by what may be Lower Amazonian flows of Ceraunius Fossae (unit Acf), and thus a few stage III faults may be as old as Late Hesperian (Fig. 3). North-trending faults of Ceraunius Fossae north of about lat. 30°N cut flows of the Alba Patera Formation. No doubt, many of the faults of southern Ceraunius Fossae formed during stage II were reactivated; however, it is not possible to

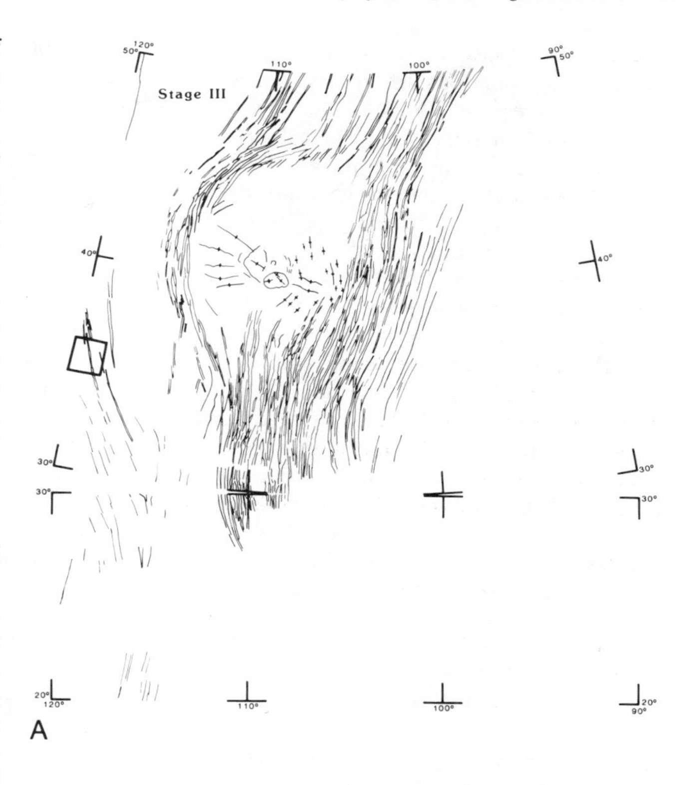

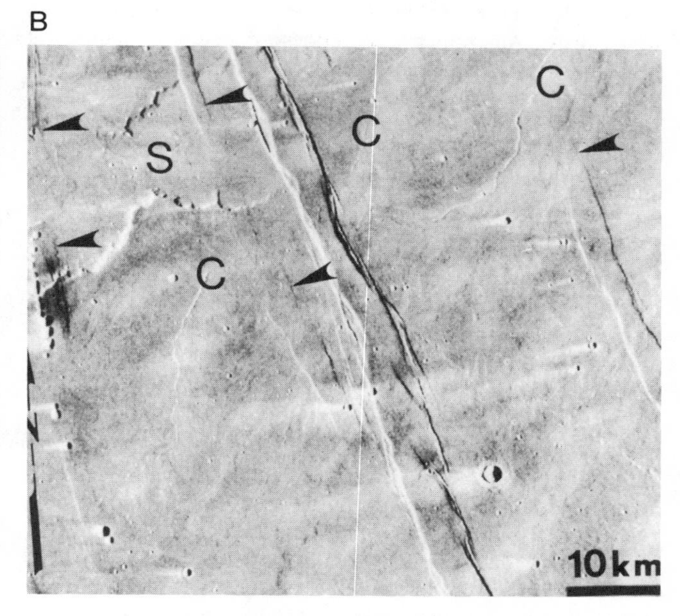

Fig. 6. (a) Distribution of stage III faults in the Alba Patera-Ceraunius Fossae region of Mars. Ridges shown by lines with diamond symbols; west- to northwest-trending ridges are wrinkle ridges and north- to northeast-trending ridges may be spatter ridges (see text); box shows location of **(b)**. **(b)** Area southwest of Alba Patera showing Alba Fossae grabens cutting Alba Patera formation lava flows. Grabens with minor offset and small pit chain (arrows) commonly terminate in prominent sheet (S) and crested (c) flows; major graben through center of picture appears unaffected. (Viking image 252S22.)

discern which ones. A few pit chains formed in grabens south of Alba Patera; the pits are generally about a kilometer wide and their centers are a few kilometers apart. No clear examples are seen of fault scarps or graben floors buried by flows of unit AHac, although in some places (Fig. 6b) thick sheet and crested flows appear either to have inhibited graben formation or to have been cut by limited, postflow faulting (implying that some preflow faulting occurred).

Within the area of Alba Patera encircled by large grabens, three other types of structures occur: (1) Several calderas lie in the center of the volcano. One caldera forms a complete, irregular oval having dimensions of 45 × 65 km. A prominent east-facing scarp may be a remnant of a partly buried 150-km-diameter caldera. On the basis of nested-caldera relations, at least four episodes of caldera formation can be identified (*Cattermole and Reid,* 1984; *Schneeberger and Pieri,* 1989). (2) East of the calderas, kilometer-wide linear ridges tens of kilometers to about 100 km long, trending north to northeast, are aligned concentrically with the broad shape of Alba Patera (Fig. 6a). The ridges appear to have a simpler form than wrinkle ridges, and they have been interpreted as spatter ridges (*Cattermole,* 1986). (3) Wrinkle ridges transect Alba Patera along N60°-90°E trends. The ridges are 100-200 km long on the outer flanks and somewhat shorter within the caldera complex.

Concentric faulting around Syria Planum occurred during the Late Hesperian (*Tanaka and Davis,* 1988). Similar concentric faulting at Alba Patera occurred at about the same time (Early Amazonian) as modest concentric faulting at Elysium Mons (Tanaka et al., unpublished map) and development of Noctis Labyrinthus (*Tanaka and Davis,* 1988). The Early Amazonian was a quiescent epoch for Tharsis-centered radial faulting, between the Pavonis I and II episodes (*Plescia and Saunders,* 1982).

Stage IV—Middle to Late Amazonian

Structures of this stage (Fig. 7) cut Middle Amazonian flows of the Ceraunius Fossae Formation (unit Acf) and Lower Amazonian flows from Ascraeus Mons (unit At_4). Some of the Middle Amazonian flows (unit At_5) from Ascraeus Mons (centered at lat. 11°N, long. 104°, south of the map area) are cut by stage IV faults, whereas some of the flows bury the faults. All of these flow units bury fractures of stage III.

Most of the fractures of stage IV trend north in Ceraunius Fossae but bend to the northeast south of Uranius Patera (N40°-50°E) and northeast of Ceraunius Fosse (N30°-40°E). Many can be distinguished from stage III faults on the basis of their linearity (in Tantalus Fossae) and their narrow, sharp appearance in Ceraunius Fossae. Many of the stage IV faults probably result from reactivation of buried faults of stages II and III. Some of the fractures that cut Ceraunius Fossae flows have irregular trends and connect with apparent channels that may originate from melt-water release (*Mouginis-Mark,* 1989).

Three main fracture systems—containing Phlegethon, Acheron, and Tractus Catenae (Fig. 1)—are parallel and evenly spaced about 130 km apart. They form grabens along which series of pit chains (the catenae) have formed. The pits are commonly a kilometer or two across, but in some graben segments they have coalesced to form linear depressions several kilometers across. Tractus Catena is part of the longest fracture system in the map area. This system originates in northern flank flows of Ascraeus Mons (lat. 20°N, long. 104°) and divides into two fractures crosscut by minor east-trending fractures. At lat. 30°N, the system has broadened into several faults, of which the western ones branch off to form Acheron Catena farther north. Tractus and Phlegethon extend northeast out of the map area into Vastitas Borealis.

Another, somewhat smaller, unnamed pit chain occurs about 150 km east of Tractus Catena (lat. 32°N, long. 98°), and many minor pit chains are associated with stage IV fractures in Ceraunius Fossae. The pit chains have been proposed to be collapse features due to drainage of material into deep tension fractures (*Tanaka and Golombek,* 1989; *Tanaka et al.,* 1989).

Stage IV faualts are part of the Pavonis II episode of Tharsis-centered tectonism (*Plescia and Saunders,* 1982). The northeast-trending fractures south of Uranius Patera extend into southern Tempe Terra, where they are assigned to "Hesperian and Amazonian set 3" of *Scott and Dohm* (1990).

TECTONIC CONTROLS ON FAULTING

Possible tectonic controls on faulting range from well documented to highly speculative scenarios that need further investigation. In order for plausible correlations to be made, fault trends and ages should correspond to the stress field produced by the proposed tectonic event; preexisting structurally weak trends may also influence fault trends.

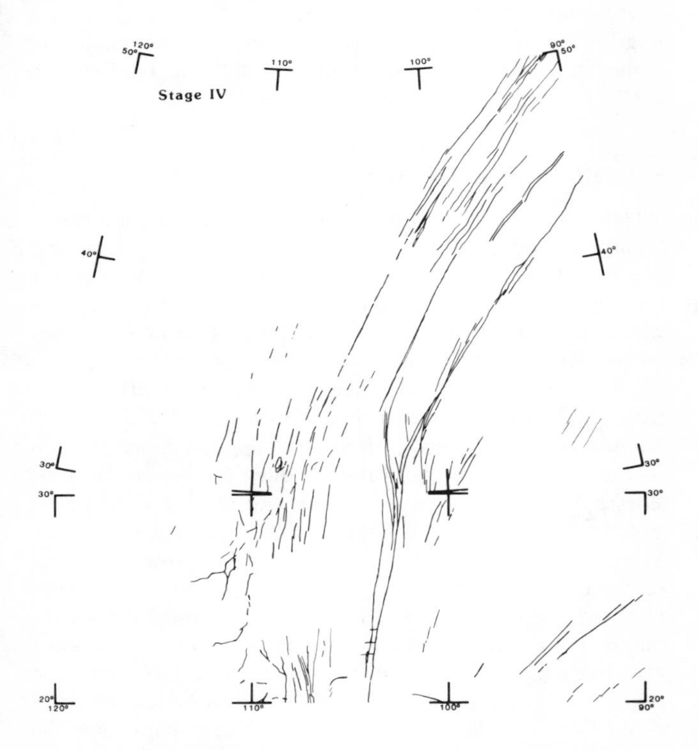

Fig. 7. Distribution of stage IV faults in the Alba Patera-Ceraunius Fossae region of Mars.

Alba Patera Magmatism

Several volcanotectonic processes may produce substantial stress, causing failure of crustal rocks. The nested, arcuate caldera structures suggest multiple stages of magma-chamber collapse (*Wise,* 1979; *Cattermole and Reid,* 1984). The broad, arcuate faults surrounding Alba Patera may be due to deflection of Tharsis-centered faulting surrounding a high-strength plutonic complex underlying Alba Patera (*Wise,* 1979). Alternatively, these faults may have been produced by the load of Alba Patera, which could have generated a radially oriented extensional stress pattern at a range of distances from the load center (*Comer et al.,* 1985). The absence of concentric faults on the south side of Alba Patera may be due to a dominant east-west extensional stress component on this side, which is closer to the Tharsis center. Loading may also have caused centralized compression, producing the east-trending wrinkle ridges (parallel with the least principal stress direction). On the Moon, loading by lava flows in impact basins and subsequent subsidence have produced circumferential grabens and interior wrinkle ridges (see *Wilhelms,* 1987, p. 112).

The style and timing of volcanism and tectonism of Alba Patera are broadly similar to that of Syria Planum (*Tanaka and Davis,* 1988) and Tempe Terra (*Scott and Dohm,* 1990). These areas were active during an intermediate stage of Tharsis development. Earlier Tharsis activity, such as in Thaumasia, did not produce well-defined fault and volcanic centers (e.g., *Plescia and Saunders,* 1982), whereas later activity resulted in huge shield volcanoes but only minor faulting (e.g., *Scott and Tanaka,* (1980). Therefore during the intermediate period of Tharsis evolution in which Alba Patera developed, the lithosphere was apparently thick enough to produce broadly centralized volcanic eruptions, yet weak enough to undergo extensive tectonic deformation. Also, eruptions probably involved fluid lavas at high discharge rates, resulting in longer flows and less edifice construction than at later, major Tharsis volcanoes. The key relation may be that Alba Patera-type centers result from volcanism in areas weakened by intense fracturing. If so, each of the Tharsis Montes may have originated similarly to Alba Patera, but as the lithosphere was strengthened by intrusive activity, less faulting and more centralized volcanism evolved.

Tharsis-centered Tectonism

North-trending faults in the Ceraunius Fossae region are consistent with stress patterns that may have been generated by isostatic loading centered near Syria Planum (*Banerdt et al.,* 1982). This type of faulting may have occurred in three distinct episodes, producing radial faults surrounding centers at Syria Planum and Pavonis Mons (*Plescia and Saunders,* 1982). The north-trending stage II faults are part of a group

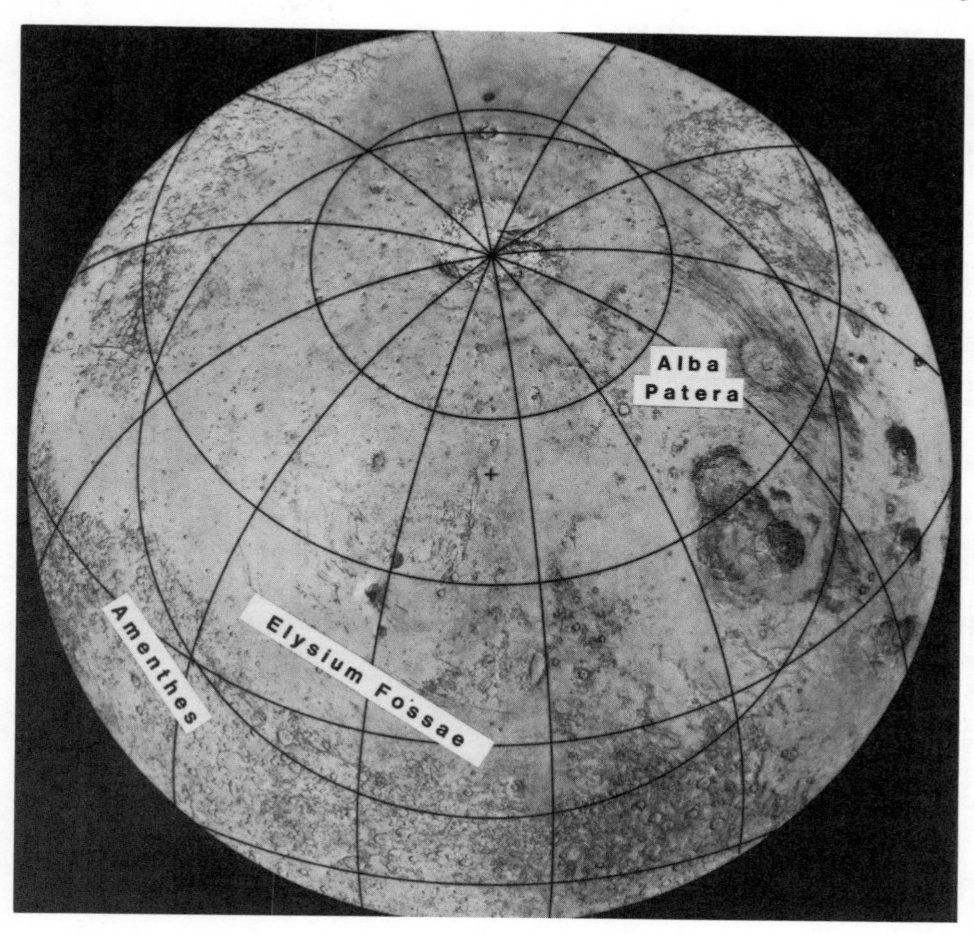

Fig. 8. Shaded-relief base (Lambert Equal Area projection; 30° grid) showing the hemisphere of Mars centered at lat. 50°N, long. 190°, in the northern plains. Circle indicates boundary of mega-impact basin proposed by *Wilhelms and Squyres* (1984); note parallelism of graben trends northeast of Alba Patera with boundary.

of regional faults centered at Syria Planum, and stage IV faults are part of Pavonis II tectonism. The latter faults may just as well be centered on Ascraeus Mons, the closest of the Tharsis Montes, which was active during Pavonis II tectonism.

Wise (1979) proposed an eastward migration of fault centers, and some lines of evidence (*Wise et al.,* 1979) seem to support this idea. This migration would require an early regional center west of Tharsis Montes. Tharsis fault-center studies (*Wise et al.,* 1979; *Plescia and Saunders,* 1982) have excluded the problematic Alba and Tantalus Fossae. However, *Wise et al.* (1979) noted that trends of Mareotis Fossae (faults parallel to Tantalus Fossae in northwestern Tempe Terra) have a center west of Pavonis Mons. Older faults of Ulysses Fossae (*Scott and Dohm,* 1990) also may be consistent with a possible early, western fault center near Ulysses Patera. However, if such a center existed, it did not produce faults in other areas of Tharsis.

Wise (1979) noted that Ceraunius Fossae-Alba Patera form a broad, north-trending arch along which volcanism and tectonism are centered; *Raitala* (1988) proposed that the arch is a peripheral rift zone of Tharsis. These explanations are certainly consistent with the regional topography (S. S. C. Wu, unpublished map, 1989), the strain pattern indicated by the grabens, and the peripheral location of Alba Patera relative to the crest of the Tharsis rise.

Dichotomy Boundary

An influence other than Tharsis appears to be necessary to explain eastward deflection of Alba and Tantalus Fossae east and north of Alba Patera for every stage of faulting. This deflection of fault trends from north to northeast roughly coincides with the boundary between the martian highlands and lowlands inferred in the Tharsis region by *Wilhelms and Squyres* (1984) and commonly referred to as the dichotomy boundary. In this region, the boundary is buried or embayed by the Alba Patera lava flows (Fig. 8). The boundary probably runs along the west edge of Tempe Terra, represented by isolated massifs along the east edge of the map area. Farther west, the boundary cuts across Ceraunius Fossae, perhaps partly along a scarp in the northwestern part of the fossae that lies near the 30° parallel. Although the origin of the boundary is controversial, an hypothesized mega-impact (*Wilhelms and Squyres,* 1984) or mantle overturn (*Wise et al.,* 1979) would likely produce circumferential faults. In this regard, the northeast-trending axis along which Tharsis Montes, Uranius Patera, and Tempe Fossae are aligned has been explained as a deformation zone controlled by the crustal boundary (*Wise et al.,* 1979). Possible circumferential features of Late Noachian age have been found in the Amenthes region of Mars (*Maxwell and McGill,* 1988); Elysium Fossae also are aligned with the boundary (Fig. 8). Grabens east and north of Alba Patera are generally circumferential to the first-order circle described by the northern plains (Fig. 8), and thus they may also have been influenced by the dichotomy boundary. Stage I faults may have followed dichotomy boundary trends farther south in Ceraunius Fossae, perhaps when an early Tharsis tectonic center west of Pavonis Mons was active. Thus tectonic fracturing within 20°-30° of the topographic highland/lowland boundary commonly is parallel to the boundary (Fig. 8), suggesting that the boundary itself is structurally controlled. This possibility is consistent with proposed mega-impact and tectonic origins of the northern lowlands.

CONCLUSIONS

The results of this study concur with conclusions by *Wise* (1979), *Plescia and Saunders* (1982), and *Tanaka and Davis* (1988) that both regional tectonic centers and major volcanic centers in the Tharsis region have imposed control on local faulting. The broad volcanic and structural character of Alba Patera typifies other Tharsis volcanotectonic centers at Syria Planum and Tempe Terera formed in intensely fractured terrains. Additionally, the dichotomy boundary, whose origin remains controversial, appears to have controlled northeast fault trends north and east of Alba Patera, which extends the original proposal by *Wise et al.* (1979) that the Tharsis axial trend was controlled by the boundary.

The following is the general sequence of fault patterns and associated volcanism, tectonism, and volcano/ground-ice interactions in the Alba Patera-Ceraunius Fossae region of Mars:

Stage I. A moderate intensity of northeast-trending Noachian faulting in the study region may have been induced by a tectonic center near Ulysses Patera, activating a preexisting complementary structural fabric parallel to the highland/lowland boundary.

Stage II. Initiation of north-trending faults in Ceraunius Fossae during the Late Noachian to Late Hesperian epochs generally coincided with Syria Planum-centered tectonism. Lahars issued from northeastern Ceraunius Fossae.

Stage III. Following Alba Patera volcanism in the Early Amazonian, north-trending faulting continued at Ceraunius Fossae; farther north, these faults terminated or deflected concentrically to Alba Patera. North of Alba Patera, most faults are deflected northeast. The concentric faults probably originated from loading stresses or increased crustal strength associated with Alba Patera plutonism. Calderas and east-trending wrinkle ridges developed on the patera as it subsided.

Stage IV. North-trending grabens and catenae deflected to a northeast trend north of Ceraunius Fossae. These faults appear to have been produced by stresses generated by the then-active Ascraeus Mons. Melt-water release along some fractures carved channels hundreds of kilometers long.

Acknowledgments. I extend my thanks to S. L. Rotto for preliminary detailed mapping of faults in the Alba Patera-Ceraunius Fossae region during an internship sponsored by USGS/NAGT. D. H. Scott and J. M. Dohm provided helpful reviews of an early version of the manuscript, and L. S. Crumpler, C. D. Condit, and M. P. Golombek added insightful comments that improved the paper.

REFERENCES

Banerdt W. B., Phillips R. J., Sleep N. H., and Saunders R. S. (1982) Thick shell tectonics on one-plate planets: Applications to Mars. *J. Geophys. Res., 87,* 9723-9733.

Cattermole P. (1986) Linear volcanic features at Alba Patera, Mars—Probable spatter ridges. *Proc. Lunar Planet. Sci. Conf. 17th,* in *J. Geophys. Res., 91,* E159-E165.

Catermole P. (1987) Sequence, rheological properties, and effusion rates of volcanic flows at Alba Patera, Mars. *Proc. Lunar Planet. Sci. Conf. 17th,* in *J. Geophys. Res., 92,* E553-E560.

Cattermole P. and Reid C. (1984) The summit calderas of Alba Patera, Mars (abstract). In *Lunar and Planetary Science XV,* pp. 142-143. Lunar and Planetary Institute, Houston.

Christiansen E. H. (1989) Lahars in the Elysium region of Mars. *Geology, 17,* 203-206.

Comer R. P., Solomon S. C., and Head J. W. (1985) Mars: Thickness of the lithosphere from the tectonic response to volcanic loads. *Rev. Geophys., 23,* 61-92.

Maxwell T. A. and McGill G. E. (1988) Ages of fracturing and resurfacing in the Amenthes region, Mars. *Proc. Lunar Planet. Sci. Conf. 18th,* pp. 701-711.

Mouginis-Mark P. J. (1989) Recent water release in the Tharsis region of Mars (abstract). In *Lunar and Planetary Science XX,* pp. 727-728. Lunar and Planetary Institute, Houston.

Mouginis-Mark P. J., Wilson L., and Zimbelman J. R. (1989) Polygenic eruptions on Alba Patera, Mars: Evidence of channel erosion on pyroclastic flows. *Bull. Volcanol.,* in press.

Plescia J. B. and Saunders R. S. (1979) The chronology of the martian volcanoes. *Proc. Lunar Planet. Sci. Conf. 10th,* pp. 2841-2859.

Plescia J. B. and Saunders R. S. (1982) Tectonic history of the Tharsis region, Mars. *J. Geophys. Res., 87,* 9775-9791.

Raitala J. (1988) Composite graben tectonics of Alba Patera on Mars. *Earth, Moon and Planets, 42,* 277-291.

Rotto S. L. and Tanaka K. L. (1989) Faulting history of the Alba Patera-Ceraunius Fossae region of Mars (abstract). In *Lunar and Planetary Science XX,* pp. 926-927. Lunar and Planetary Institute, Houston.

Schneeberger D. M. and Pieri D. C. (1989) Geomorphic-stratigraphic implications for the volcanic history of Alba Patera, Mars. *J. Geophys. Res.,* in press.

Scott D. H. and Dohm J. M. (1990) Faults and ridges: Historical development in Tempe Terra and Ulysses Patera regions of Mars. *Proc. Lunar Planet. Sci. Conf. 20th,* this volume.

Scott D. H. and Tanaka K. L. (1980) Mars Tharsis region: Volcanotectonic events in the stratigraphic record. *Proc. Lunar Planet. Sci. Conf. 11th,* pp. 2403-2421.

Scott D. H. and Tanaka K. L. (1981) Paleostratigraphic restoration of buried surfaces in Tharsis Montes. *Icarus, 45,* 304-319.

Scott D. H. and Tanaka K. L. (1986) Geologic map of the western equatorial region of Mars. *U.S. Geol. Surv. Misc. Inv. Ser. Map I-1802-A,* scale 1:15,000,000.

Sharp R. P. (1973) Mars: Fretted and chaotic terrains. *J. Geophys. Res., 78,* 4073-4083.

Tanaka K. L. (1986) The stratigraphy of Mars. *Proc. Lunar Planet. Sci. Conf. 17th,* in *J. Geophys. Res., 91,* E139-E158.

Tanaka K. L. and Davis P. A. (1988) Tectonic history of the Syria Planum province of Mars. *J. Geophys. Res., 93,* 14,893-14,917.

Tanaka K. L. and Golombek M. P. (1989) Martian tension fractures and the formation of grabens and collapse features at Valles Marineris. *Proc. Lunar Planet. Sci. Conf. 19th,* pp. 383-396.

Tanaka K. L. and Scott D. H. (1987) Geologic maps of the polar regions of Mars. *U.S. Geol. Surv. Misc. Inv. Ser. Map I-1802-C,* scale 1:15,000,000.

Tanaka K. L., Isbell N. K., Scott D. H., Greeley R., and Guest J. E. (1988) The resurfacing history of Mars: A synthesis of digitized, Viking-based geology. *Proc. Lunar Planet Sci. Conf. 18th,* pp. 665-678.

Tanaka K. L., Davis P. A., and Golombek M. P. (1989) Development of grabens, tension cracks, and pits southeast of Alba Patera, Mars. In *MEVTV Workshop on Tectonic Features on Mars* (T. R. Watters and M. P. Golombek, eds.). LPI Tech. Rpt. 89-06, Lunar and Planetary Institute, Houston, in press.

Wilhelms D. E. (1987) *The Geologic History of the Moon.* U.S. Geol. Surv. Prof. Paper 1348. 302 pp.

Wilhelms D. E. and Squyres S. W. (1984) The martian hemispheric dichotomy may be due to a giant impact. *Nature, 309,* 138-140.

Wise D. U. (1979) Geologic map of the Arcadia quadrangle of Mars. *U.S. Geol. Surv. Misc. Inv. Serv. Map I-1154,* scale 1:5,000,000.

Wise D. U., Golombek M. P., and McGill G. E. (1979) Tharsis province of Mars: Geologic sequence, geometry, and a deformation mechanism. *Icarus, 38,* 456-472.

Proceedings of the 20th Lunar and Planetary Science Conference, pp. 525-530
Lunar and Planetary Institute, Houston, 1990

Outliers of Dust Along the Southern Margin of the Tharsis Region, Mars

J. R. Zimbelman

Center for Earth and Planetary Studies, National Air and Space Museum, Smithsonian Institution, Washington, DC 20560

This work documents the distinctive characteristics of two locations near the southeastern margin of the Tharsis low thermal inertia region on Mars that support the inferred south-to-north migration of dust deposits under present climatic conditions (Christensen, 1986, 1988). Low thermal inertia and high albedo indicate the presence of fine-grained surface materials on the highly fractured terrain of northern Claritas Fossae (-16°, 110°), with a narrow zone of fine material connecting this location to the southern margin of the Tharsis low thermal inertia region. A similar deposit occurs in Sinai Planum (-16°, 80°) on an isolated exposure of fractured terrain. In comparison to their surroundings, both locations have higher visual albedo, lower thermal inertia, and higher topographic relief coupled with lower reflectivity and higher roughness at radar wavelengths. The Sinai Planum location has less contrast with its surroundings at all wavelengths than the Claritas Fossae location. The remote-sensing data are interpreted to indicate erosional stripping of dust deposits that progressed in a northerly direction, with dust preferentially preserved on rough surfaces due to reduced efficiency of aeolian erosion.

INTRODUCTION

Global mapping of the thermal properties of Mars identified three continent-scale regions of low thermal inertia, interpreted to consist of very fine (silt-sized) particles at the surface (*Kieffer et al.,* 1977; *Zimbelman and Kieffer,* 1979; *Palluconi and Kieffer,* 1981). Recent synthesis of global properties at visual, thermal, and radar wavelengths led to the hypothesis that the low thermal inertia regions consist of dust deposits, on the order of meters in thickness, that are redistributed around the planet as changing insolation conditions alter the global wind regime (*Christensen,* 1986). Variations in the pattern of atmospheric clearing following global dust storms indicate a net south-to-north transport of dust under present climatic conditions (*Christensen,* 1988). However, the lack of large, identifiable dust deposits in the southern hemisphere at present makes it difficult to relate the dust transportation to the evolution of the regional dust deposits. In this paper the distinctive characteristics of two locations near the southeastern margin of the Tharsis low thermal inertia region are documented and are shown to support the inferred south-to-north migration of dust deposits under present climatic conditions.

BACKGROUND

The Tharsis and Elysium volcanic provinces and cratered highlands in the Arabia region all have distinctive remote-sensing characteristics. Comparison of the physical properties in these three regions provide consistent results of low thermal inertia, high visual albedo, and considerable roughness at radar wavelengths, interpreted to result from the penetration of the radar signal through a comparatively thin dust mantle with subsequent scattering by roughness elements on the underlying volcanic and cratered terrains (*Christensen,* 1986). Imaging observations of wind streaks and variable features indicate that significant sediment transport within the Tharsis region occurred during the Viking mission, with downslope winds carrying materials away from the Tharsis sh[illegible] volcanoes (*Lee et al.,* 1982). However, changes in the overall distribution of the large regional dust deposits are likely driven by changes in climate and global circulation (*Christensen,* 1986, 1988).

It seems unlikely that the migration of the dust deposits can be observed directly, due to the long timescales (ranging from tens of thousands to millions of years) thought to be associated

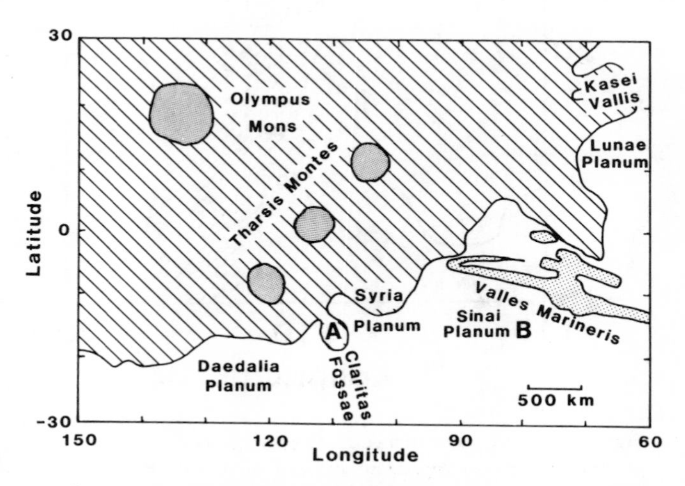

Fig. 1. Location map for the Claritas Fossae (A) and Sinai Planum (B) features described in the text. The large shield volcanoes of Tharsis (dark pattern) and Valles Marineris (stippled pattern) are shown for reference, as well as the names of several prominent features. The lined pattern indicates the extent of the Tharsis low thermal inertia region within the map area. The boundary of the Tharsis region corresponds to the -10 K contour (equivalent to a thermal inertia of about 4×10^{-3} cal cm^{-2} sec$^{-1/2}$ K^{-1}) of the predawn residual temperature maps of *Kieffer et al.* (1977) and *Zimbelman and Kieffer* (1979).

with climatic changes on Mars (e.g., *Ward,* 1974). This conclusion is consistent with thermal measurements made throughout the Viking mission that failed to detect evidence of a change in the position of the margins of the three low thermal inertia regions (*Zimbelman and Kieffer,* 1979), implying that the present margins of the dust deposits are stable on a scale of tens of kilometers. Even if the migration of the dust deposits cannot be observed directly, there may be remnants of a former dust mantle that could indicate the general direction in which the dust deposits have moved under recent climatic conditions. The approach adopted here is to look for evidence of a previous dust mantle that can be related to the present dust deposits.

The largest of the three regional dust deposits surrounds the Tharsis Montes extending over 130° of longitude and 60° of latitude and encompassing an area of 22 million km^2, only slightly smaller than the area of the North American continent (*Zimbelman and Kieffer,* 1979). The southern margin of the Tharsis region skirts the northern edge of the Valles Marineris canyon system and follows an irregular course across Sinai, Syria, and Daedalia Planae (Fig. 1). The eastern margin includes a pronounced embayment into the low thermal inertia material along Kasei Vallis, the site of considerable aeolian activity and abundant sand-sized material confined within the channel (*Christensen and Kieffer,* 1979). The south-to-north transport of dust observed under present climatic conditions (*Christensen,* 1988) implies that evidence for previous dust deposits should exist along the southern margin of the deposit.

Claritas Fossae is a complex assemblage of grabens and fractured terrain associated with a projection of low thermal inertia material into the higher thermal inertia materials of the southern hemisphere (A in Fig. 1). The distinctive thermal properties of Claritas Fossae were first noted early in the Viking mission (*Kieffer et al.,* 1976; see Fig. 2), but this area received less attention than did the unusual thermal characteristics on and around Arsia Mons. During an examination of the published physical properties of Claritas Fossae, a second location of fractured terrain in Sinai Planum (B in Fig. 1) was observed to have many characteristics similar to Claritas Fossae but generally with less contrast in comparison to its surroundings (*Zimbelman,* 1989). Both locations have bedrock interpreted to be highly deformed, fractured materials of Noachian age, the oldest of the martian stratigraphic systems (*Scott and Tanaka,* 1986). Many additional outcrops of this old, fractured terrain are present in the cratered highlands south of the Valles Marineris (*Scott and Tanaka,* 1986), but only the two locations described above have remote-sensing characteristics comparable to those of the regional dust deposits. The variations in the remotely observed properties of the ancient, fractured materials provide strong clues to the direction in which a migrating dust mantle could have moved across Mars.

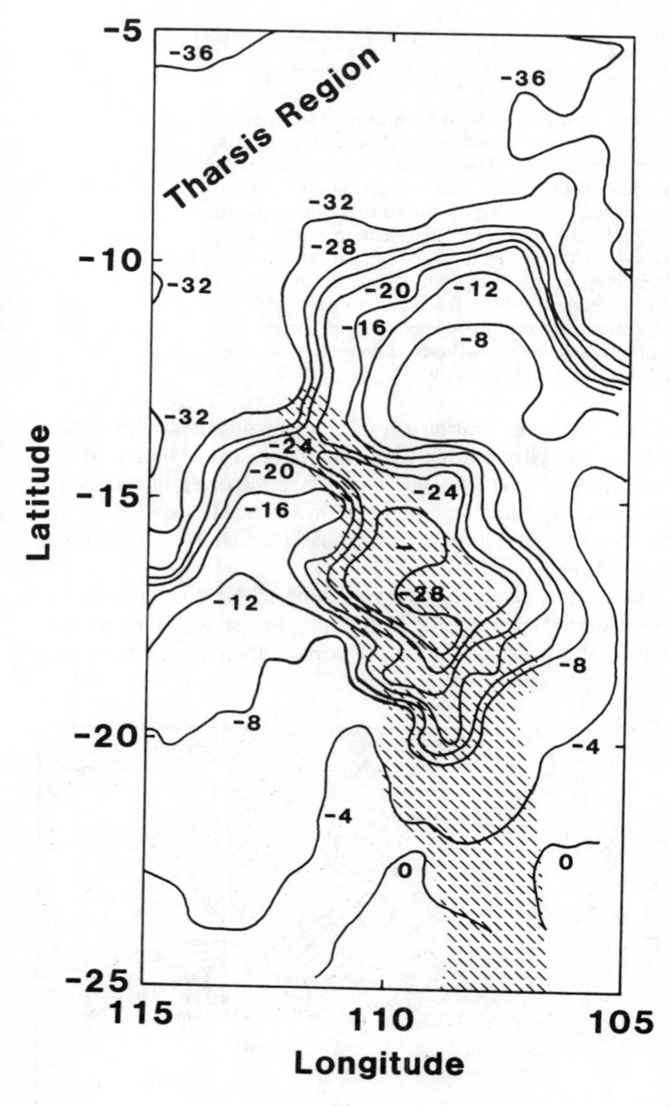

Fig. 2. Late afternoon temperature residuals (the difference between observed temperature and that predicted for a Mars average thermal inertia of 6.5×10^{-3} cal cm^{-2} sec$^{-1/2}$ K^{-1} and albedo of 0.25) for Claritas Fossae (modified from Fig. 4B of *Kieffer et al.,* 1976). Dashed pattern indicates the extent of ancient, fractured terrain (from *Scott and Tanaka,* 1986). Temperature residuals of -28 K, -8 K, and 0 K are equivalent to thermal inertias of 2.5, 4.5, and 6.5×10^{-3} cal cm^{-2} sec$^{-1/2}$ K^{-1}, respectively. The low temperature residuals (low thermal inertia) on Claritas Fossae are concentrated north of -20° lat., and they are comparable to the temperature residuals within the nearby Tharsis region.

DATA

The physical properties of both the Claritas Fossae and Sinai Planum locations of ancient, fractured terrain are summarized below. All of the data have been published previously, but here the results for Claritas Fossae and Sinai Planum are examined within the context of the proposed global transport of dust on Mars.

Thermal Inertia

The northern portion of Claritas Fossae (Fig. 2) has temperature residuals indicative of a thermal inertia much lower ($I < 3$) than its immediate surroundings ($I > 5$). A zone

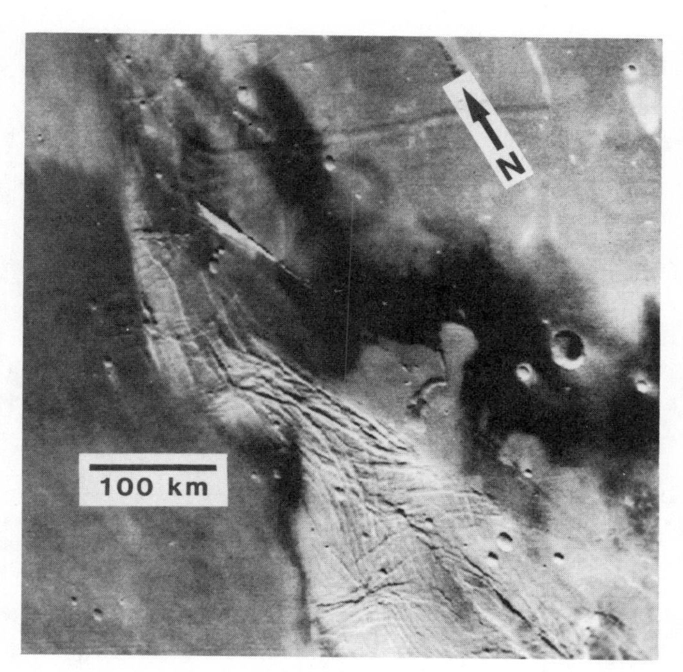

Fig. 3. Northern portion of Claritas Fossae. The area of low thermal inertia corresponds to the high albedo material visible here (compare with Fig. 2). The lowest thermal inertias occur on the highly fractured terrain (-16° lat., 110° long.) below the center of the scene shown here. Portion of Viking Orbiter frame 334S12 (centered on -14° lat., 110° long.), red filter, 790 m/pixel, shading corrected rectilinear version.

of low thermal inertia material connects the Tharsis region with the low thermal inertia material on the northern portion of Claritas Fossae, centered approximately on -16° lat., 110° long. The thermal properties of northern Claritas Fossae are comparable to the properties of the large regions of low thermal inertia, interpreted to be a surface mantled by silt-size material (*Kieffer et al.,* 1977). It is significant that the fractured terrain of Claritas Fossae south of -20° lat. has thermal properties very similar to the moderate thermal inertias typical of the southern midlatitudes of Mars, in great contrast to the distinctive thermal properties of northern Claritas Fossae.

Fractured terrain in Sinai Planum (centered approximately on -16° lat., 80° long.) also has temperature residuals indicative of a thermal inertia lower than its surroundings, although the thermal contrast is considerably reduced compared to that of northern Claritas Fossae. Global mapping of predawn temperature residuals (Fig. 9b of *Kieffer et al.,* 1977) shows the Sinai Planum location to be between 0 and 2 K (I = 6.5), while the surrounding plains have temperature residuals between 4 and 6 K (I = 7.5). Such a small thermal contrast might seem insignificant if it did not also correlate with distinct properties at both visual and radar wavelengths.

Visual Albedo

Subtle albedo variations in the vicinity of Valles Marineris are portrayed dramatically in a recent color mosaic (*McEwen,* 1987). Northern Claritas Fossae has a higher albedo than its surroundings, particularly at red wavelengths (Fig. 3). It is not a coincidence that the planimetric shape of the high albedo region on northern Claritas Fossae corresponds to the general shape of the low thermal inertia material at this same location (compare Figs. 2 and 3). Global mapping has revealed a planetwide correlation between low thermal inertia and high visual albedo (*Kieffer et al.,* 1977; *Arvidson et al.,* 1982), a relationship that holds down to a scale of a few kilometers, the highest resolution of the Viking Infrared Thermal Mapper data (*Zimbelman and Leshin,* 1987). Results for northern Claritas Fossae are very consistent with this well-documented global trend. South of -20° lat. Claritas Fossae lacks a prominent albedo contrast with its surroundings, consistent with the reduced contrast of this area at thermal wavelengths.

The Sinai Planum location has a slightly higher albedo on the southern portion of the fractured terrain (Fig. 4), coincident with the area that shows the subtle thermal contrast. As was the case at thermal wavelengths, the high albedo portion of the Sinai Planum location has considerably less contrast with its surroundings at visual wavelengths than does northern Claritas Fossae. The reduced contrast does not keep this location from being visible in the color mosaic (*McEwen,* 1987).

Radar Topography and Surface Roughness

Both the northern Claritas Fossae and the Sinai Planum dust deposits are located on terrain possessing considerable

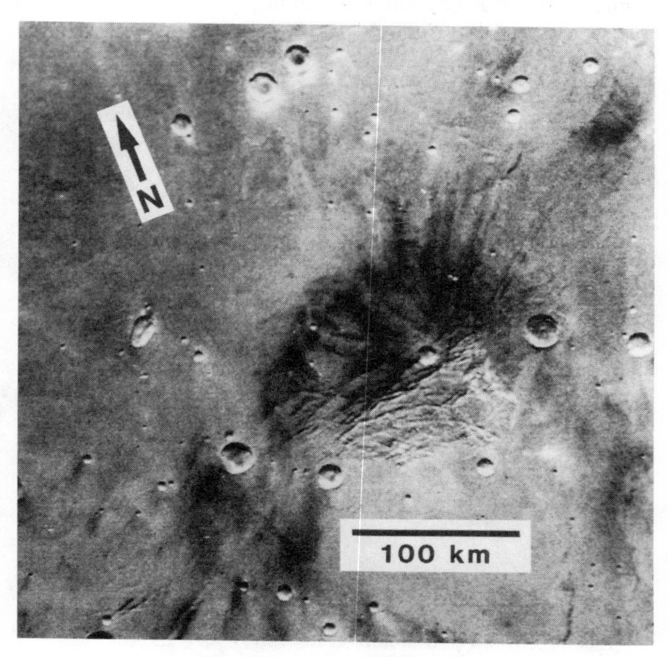

Fig. 4. Fractured terrain in Sinai Planum. The high albedo material on the southern portion of the fractured terrain has physical properties similar to the northern portion of Claritas Fossae (see Fig. 3), but the Sinai Planum location generally has less contrast with its surroundings than does the Claritas Fossae location. Portion of Viking Orbiter frame 334S46 (centered on -15° lat., 80° long.), red filter, 790 m/pixel, shading corrected rectilinear version.

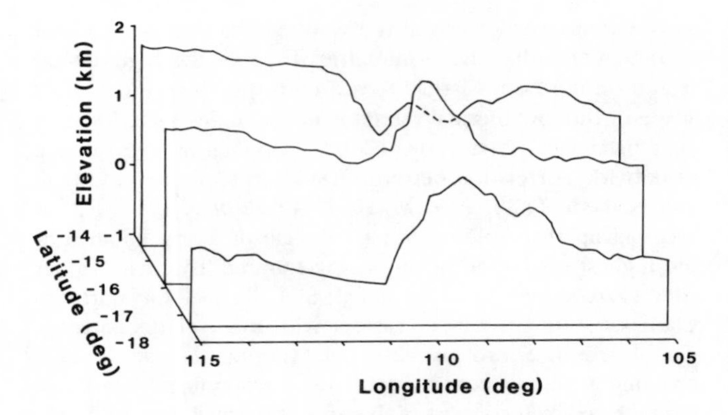

Fig. 5. Topographic profiles across northern Claritas Fossae obtained from Earth-based radar measurements (modified from Fig. 7 of *Roth et al.,* 1980). The dust deposit is associated with the prominent ridge possessing relief on the order of one kilometer, in contrast to the more subdued plains on either side. Dashed line shows the portion of a profile hidden by an intervening profile in this particular projection. The base of each profile is plotted at the latitude of the corresponding groundtrack (from north to south these latitudes are -14.50°, -15.91°, and -17.46°). Profiles are shown at 100× vertical exaggeration.

topographic relief. Earth-based radar measurements indicate that northern Claritas Fossae consists of a broad ridge that stands more than one kilometer above the uniform plains on either side of the ridge (Fig. 5). The topographic ridge corresponds closely with the location of the dust deposit inferred from the visual and thermal measurements. Similarly, the fractured terrain in Sinai Planum also possesses relief greater than one kilometer, again in contrast to the relatively flat plains surrounding the topographic high (Fig. 6). The area of greatest relief within the fractured terrain of Sinai Planum occurs at a latitude of -16°, precisely where the dust deposit is inferred from the other remote sensing data.

Radar measurements also provide information on the scattering characteristics of the martian surface on a scale comparable to or larger than the radar wavelength (12 cm for the data described below). Northern Claritas Fossae and the fractured terrain in Sinai Planum both stand out from their surroundings in several parameters obtained from the characteristics of the reflected radar signal (Fig. 7). Both locations have low reflectivity and high roughness and RMS slopes, relative to the surrounding terrain, but the contrast is

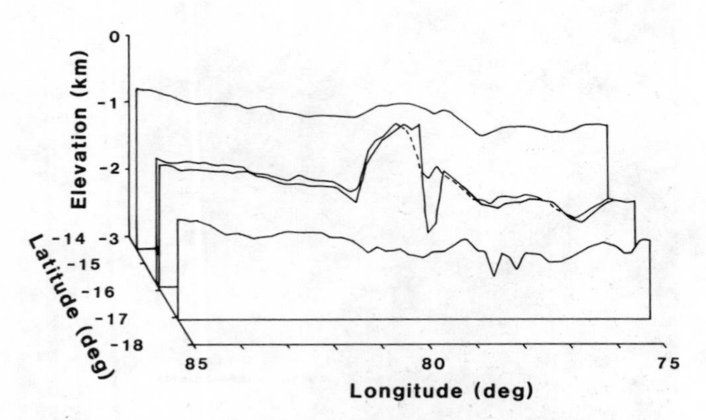

Fig. 6. Topographic profiles across the Sinai Planum fractured terrain obtained from Earth-based radar measurements (modified from Fig. 8 of *Roth et al.,* 1980). As is the case with Claritas Fossae (see Fig. 5), the dust deposit is associated with relief on the order of one kilometer, in contrast to the more subdued plains on either side. Dashed line shows the portion of a profile hidden by an intervening profile in this particular projection. The base of each profile is plotted at the latitude of the corresponding groundtrack (from north to south these latitudes are -14.50°, -15.80°, -15.91°, and -17.08°). Profiles are shown at 100× vertical exaggeration.

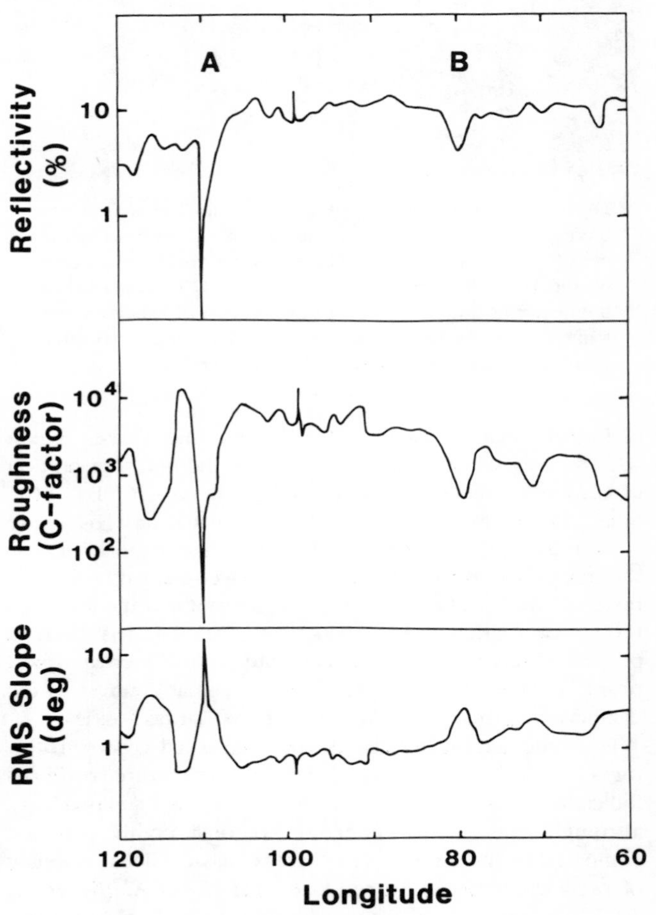

Fig. 7. Surface properties obtained from Earth-based radar measurements (modified from Fig. 7 of *Downs et al.,* 1975). The data are for a latitude of -16°, where A and B indicate the locations of fractured terrain in northern Claritas Fossae and Sinai Planum, respectively. Note that the C-factor (center diagram) is smaller for rough surfaces than for smooth surfaces. Both locations have decreased reflectivity and increased roughness and RMS slope, relative to their surroundings, but the magnitude of the contrast is much greater at northern Claritas Fossae than at Sinai Planum.

much more pronounced at Claritas Fossae than at Sinai Planum. The RMS slope values are particularly instructive; Claritas Fossae has a maximum RMS slope greater than 10° at long. 110°, while Sinai Planum has a fairly uniform RMS slope of 3°, and both locations are surrounded by plains with slopes of 1° or less (Fig. 7). These distinctive radar properties are unique within the longitude range of 0° to 120°, at a latitude of -16°, except for Valles Marineris and Arsia Mons.

Enhanced roughness and/or decreased reflectivity in the Tharsis region correlate well with lava flows that are observed to have a relatively thin aeolian mantle when examined in Viking images with a resolution of 15 m/pixel or better (*Schaber,* 1980). Although images of this resolution do not exist for the areas of Claritas Fossae and Sinai Planum under discussion here, it seems reasonable to conclude that the radar data are sensing a rough, irregular surface (on a scale of 1 to 100 m) that do not contribute to the remote-sensing characteristics at visual and thermal wavelengths.

DISCUSSION

All of the remote-sensing data for northern Claritas Fossae and the fractured terrain in Sinai Planum are consistent with a surface that is rough and irregular at the meter scale but that is mantled by fine dust that dominates the surface properties at the centimeter (and smaller) scale. This is precisely the surface condition inferred by *Christensen* (1986) to characterize the large, low thermal inertia dust deposits of the northern midlatitudes. Of equal importance is the fact that several other nearby locations of ancient, fractured materials do not have anomalous remote-sensing properties, but rather these features have properties quite similar to their surroundings. The fractured terrains having characteristics of dust deposits are at a latitude comparable to the current southern margin of the nearby Tharsis dust deposit (see Fig. 1), while the fractured terrain locations lacking evidence of a dust covering are south of -20° lat., strengthening the likelihood of a causal relationship between the large and small dust deposits.

The requirement that the surface properties be dominated by dust at thermal wavelengths implies an average dust thickness of at least several centimeters, while the penetration of the dust by the radar signals, to the underlying rough surface, implies an average dust thickness that is probably not much greater than a few meters (*Christensen,* 1986). A similar analysis procedure indicates that northern Claritas Fossae has a dust deposit comparable to the large regional dust deposits, while the fractured terrain at Sinai Planum probably has a dust cover on the order of centimeters thick, still sufficient to affect visual albedo and the diurnal variation of surface temperatures but thin enough to cause reduced contrast between the dust deposit and the surrounding plains.

How did this nonuniform distribution of dust on isolated areas of rough terrain come about? It is possible that the dominance of roughness on the fractured terrain may be so strong that subsequent aeolian erosion may be unimportant. If surface roughness is the primary agent responsible for producing dust deposits, then one should expect to find dust trapped on all of the numerous outcrops of fractured terrain mapped south of -20° lat. between 70° and 110° long. (*Scott and Tanaka,* 1986). The lack of anomalous remote sensing properties for Nectaris Fossae (fractured terrain between -18° and -25° lat. along long. 58°) indicates that the lack of preserved dust deposits on fractured terrains is not solely due to latitude but also is related to increased distance from the present margins of the Tharsis dust deposit. These factors make it highly unlikely that roughness alone can explain the isolated dust deposits.

Is there evidence that the isolated dust deposits are the result of aeolian activity? The available image resolution is not sufficient to provide diagnostic evidence for the process of deposit formation, but temporal variations provide an important clue. Images from both Mariner 9 and Viking show considerable variation in the occurrence and orientation of bright streaks in Syria Planum just north of the dust deposit on northern Claritas Fossae (*Thomas and Veverka,* 1979). These same images show virtually no change in the high albedo material on northern Claritas Fossae. Also, streak orientations indicate winds capable of moving surface materials are blowing toward the north on the western side of Claritas Fossae, while winds blow to the south on the eastern side of Claritas Fossae (*Thomas and Veverka,* 1979). Thus, there is evidence of considerable aeolian activity in the vicinity of Claritas Fossae and the nonuniform wind direction may contribute to the relative stability of surface materials on Claritas Fossae itself. It seems reasonable to infer an aeolian origin for the dust deposit on northern Claritas Fossae and, by analogy, for the thinner dust deposit in Sinai Planum.

Why are the dust deposits located on rough terrain? Field studies, laboratory simulations, and theoretical considerations indicate that increased roughness at the surface increases the difficulty of moving material by the wind by decreasing the wind shear at the particulate surface (*Bagnold,* 1941, pp. 53-55). It is important to realize that surface roughness can vary both on small (less than one meter) and large (greater than many meters) scales where small-scale roughness can protect erodible elements, such as dust, while large-scale roughness tends to enhance wind shear and the possibility of erosion (*Lee et al.,* 1982). It is the small-scale roughness that affects the wind shear at the surface and creates an aerodynamic regime in which mobile particles (dust or sand) are protected from the lifting forces of the wind. There is no reason to expect dust to be preferentially deposited on the rough terrains on Mars, but once on the surface it is more difficult to remove the dust from rough terrains without reducing the local roughness.

Are there isolated dust deposits near the southern margins of the other regional dust deposits on Mars? There do not appear to be isolated deposits elsewhere along the southern margin of the Tharsis region nor along the southern margins of the Elysium or Arabia regions. This result may be partly due to the lack of isolated regions with properties as distinctive as those observed at northern Claritas Fossae, but it may also be related to the lack of exposures of ancient, fractured terrain near the southern margins of the regional dust deposits except at the locations discussed here. Isolated deposits smaller than about 25 km probably are present, but these deposits likely will be detectable only with the highest resolution data.

The interpretations described here do not provide conclusive verification of the migration of regional dust deposits on Mars, but the spatial relationship and the local contrast of the remote-sensing properties at northern Claritas Fossae and Sinai Planum provide some compelling circumstantial evidence for the regional migration. Erosional removal of a previously widely distributed dust deposit seems the simplest explanation for the occurrence of isolated dust deposits on rough terrain near the margin of a regional dust deposit, while other exposures of similar rough terrain do not display evidence of the dust deposit. Alternative mechanisms can be proposed for isolated dust accumulations, but it seems unlikely that a nonaeolian process would be as consistent with the spatial distribution and the observed remote-sensing characteristics. Since Sinai Planum is both further removed from the present margin of the Tharsis dust deposit (see Fig. 1) and has less dust preserved than at northern Claritas Fossae, the inferred removal of the hypothesized dust mantle should have progressed from south to north, as inferred from the hypothesis of *Christensen* (1986).

CONCLUSIONS

Remote-sensing characteristics at visual, thermal, and radar wavelengths are used to infer the presence of isolated dust deposits on areas of fractured terrain in northern Claritas Fossae and Sinai Planum. The dust deposit at Sinai Planum is probably less thick than the deposit at northern Claritas Fossae, which is comparable to the dust present within the nearby Tharsis low thermal inertia region. Numerous exposures of fractured terrain are present in this area south of -20° lat. but these locations lack distinctive contrast with their surroundings in the remote-sensing data. Aeolian removal of a previously widely distributed dust deposit is interpreted to be the most likely explanation for the isolated dust deposits that have been preserved because they exist on rough terrain that decreases the local efficiency of aeolian erosion. The different distances of the Claritas Fossae and Sinai Planum dust deposits from the southern margin of the Tharsis low thermal inertia region, in combination with a thinner dust deposit at Sinai Planum, indicate that the removal of a widely distributed dust deposit progressed from the south to the north, in agreement with the global movement of dust proposed by *Christensen* (1986).

Acknowledgments. The comments of S. Lee, S. Williams, and an anonymous reviewer were very helpful during the revision of the manuscript. This work was supported by NASA grant NAGW-1390.

REFERENCES

Arvidson R. E., Guinness E. A., and Zent A. P. (1982) Classification of surface units in the equatorial region of Mars based on Viking Orbiter color, albedo, and thermal data. *J. Geophys. Res., 87,* 10149-10157.

Bagnold R. A. (1941) *The Physics of Blown Sand and Desert Dunes.* Chapman and Hall, London. 265 pp.

Christensen P. R. (1986) Regional dust deposits on Mars: Physical properties, age, and history. *J. Geophys. Res., 91,* 3533-3545.

Christensen P. R. (1988) Global albedo variations on Mars: Implications for active aeolian transport, deposition, and erosion. *J. Geophys. Res., 93,* 7611-7624.

Christensen P. R. and Kieffer H. H. (1979) Moderate resolution thermal mapping of Mars: The channel terrain around the Chryse basin. *J. Geophys. Res., 84,* 8233-8238.

Downs G. S., Reichley P. E., and Green R. R. (1975) Radar measurements of martian topography and surface properties: The 1971 and 1973 oppositions. *Icarus, 26,* 273-312.

Kieffer H. H., Christensen P. R., Martin T. Z., Miner E. D., and Palluconi F. D. (1976) Temperatures of the martian surface and atmosphere: Viking observation of diurnal and geometric variations. *Science, 194,* 1346-1351.

Kieffer H. H., Martin T. Z., Peterfreund A. R., Jakosky B. M., Miner E. D., and Palluconi F. D. (1977) Thermal and albedo mapping of Mars during the Viking primary mission. *J. Geophys. Res., 82,* 4249-4292.

Lee S. W., Thomas P. C., and Veverka J. (1982) Wind streaks in Tharsis and Elysium: Implications for sediment transport by slope winds. *J. Geophys. Res., 87,* 10,025-10,041.

McEwen A. S. (1987) Mars as a planet (abstract). In *Lunar and Planetary Science XVIII,* pp. 612-613. Lunar and Planetary Institute, Houston.

Palluconi F. D. and Kieffer H. H. (1981) Thermal inertia mapping of Mars from 60°S to 60°N. *Icarus, 45,* 415-426.

Roth L. E., Downs G. S., Saunders R. S., and Schubert G. (1980) Radar altimetry of south Tharsis, Mars. *Icarus, 42,* 287-316.

Schaber G. G. (1980) Radar, visual and thermal characteristics of Mars: Rough planar surfaces. *Icarus, 42,* 159-184.

Scott D. H. and Tanaka K. L. (1986) Geologic map of the western equatorial region of Mars. *U.S. Geol. Surv. Map I-1802-A.*

Thomas P. and Veverka J. (1979) Seasonal and secular variation of wind streaks on Mars: An analysis of Mariner 9 and Viking data. *J. Geophys. Res., 84,* 8131-8146.

Ward W. R. (1974) Climatic variations on Mars. 1. Astronomical theory of insolation. *J. Geophys. Res., 79,* 3375-3386.

Zimbelman J. R. (1989) Erosional outliers of dust along the southern margin of the Tharsis region, Mars (abstract). In *Lunar and Planetary Science XX,* pp. 1237-1238. Lunar and Planetary Institute, Houston.

Zimbelman J. R. and Kieffer H. H. (1979) Thermal mapping of the northern equatorial and temperate latitudes of Mars. *J. Geophys. Res., 84,* 8239-8251.

Zimbelman J. R. and Leshin L. A. (1987) A geologic evaluation of thermal properties for the Elysium and Aeolis quadrangles of Mars. *Proc. Lunar Planet. Sci. Conf. 17th,* in *J. Geophys. Res., 92,* E588-E586.

Proceedings of the 20th Lunar and Planetary Science Conference, pp. 531-539
Lunar and Planetary Institute, Houston, 1990

Small Valleys and Hydrologic History of the Lower Mangala Valles Region, Mars

M. G. Chapman and K. L. Tanaka

Branch of Astrogeology, U.S. Geological Survey, 2255 North Gemini Drive, Flagstaff, AZ 86001

Three diverse types of small valleys (degraded, theater-headed, and ribbon) in the lower Mangala Valles region of Mars indicate a complex hydrologic history involving two main episodes of valley and channel formation. Discontinuous, shallow (<100 m deep) pits and degraded valleys of Late Noachian age, associated with knobby plateau material in lower plateau areas west of Mangala Valles, are interpreted to be thermokarst alases. They may have originated when climate change or local volcanic heating melted relatively shallow, extensive ground ice. Much later, Late Hesperian catastrophic flooding that formed Mangala Valles may have percolated into the ground and in the lower Mangala region charged aquifers, leading to groundwater flow along north-trending structures east of Mangala Valles. The water apparently sapped out along the highland/lowland scarp, controlling formation of theater-headed valleys 50 to 500 m deep. Fans cut by distributary channels at some of the mouths of theater-headed valleys indicate erosion by runoff, whereas concentrically ridged fans at other mouths indicate a process such as mud flow or glacial flow. Some of the lobate plains material that was later deposited (during Late Hesperian/Early Amazonian time) in channels of Mangala Valles may have been mud flows or sediments laid down by floods that formed runoff streams. These streams produced small outflow channels that narrow into shallow (10 to 25 m deep), single or anastamosing ribbon valleys that cut across older plateau materials and into theater-headed valleys, further eroding them. Ages of Mangala Valles and of subsequent valley formation are bracketed by lava flows from Tharsis and the northern plains.

INTRODUCTION

Two principal types of martian channels were recognized in Mariner 9 and Viking images: large outflow channels and small runoff channels that are commonly referred to as valley networks (*McCauley et al.,* 1972; *Sharp and Malin,* 1975; *Mars Channel Working Group,* 1983). Most authors have suggested that the smaller runoff valleys (largely dendritic) are due to sapping or local precipitation and that most formed early in martian history (*Sharp and Malin,* 1975; *Masursky et al.,* 1977; *Pieri,* 1980; *Carr and Clow,* 1981; *Baker and Partridge,* 1986).

Recent, detailed geologic mapping of two areas at 1:500,000 scale (*Chapman et al.,* 1989, 1990) reveals that three types of small nondendritic valleys cut into highland terrain on either side of lower Mangala Valles (Figs. 1-3). These two map areas include the scarp that marks the highland/lowland boundary (*Scott and Tanaka,* 1986), as well as areas of the northern lowlands into which Mangala Valles and many of the small valleys debouch. The three types of valleys—degraded, theater-headed, and ribbon—have distinct morphologies. Crater counts and stratigraphic relations (Fig. 4) indicate that development of the theater-headed and ribbon valleys followed shortly the formation of the Mangala Valles outflow channels in Late Hesperian time, whereas the degraded valleys date back to the Noachian Period.

Mars is now an arid planet incapable of sustaining liquid water on or near its surface. However, even under these conditions, streams as shallow as 1 to 2 m may flow as much as a few hundred kilometers before freezing (*Carr,* 1983). The past recurrence of valley formation near Mangala Valles suggests intermittent hydrologic activity over a long time span (except during a considerable hiatus between the Noachian and the late Hesperian), and the different types of valley morphology suggest diverse processes of formation. Here we attempt to establish (1) precise ages for the small valleys relative to regional climatic, volcanic, and outflow-channel activity; (2) origins of the small valleys; and (3) the hydrologic history of the lower Mangala Valles region.

STRATIGRAPHY

Our map units (from *Chapman et al.,* 1989, 1990) correspond or are partly equivalent to some units on smaller scale maps (*Mutch and Morris,* 1979; *Scott and Schaber,* 1981; *Scott and Tanaka,* 1986). However, in many places interpre-

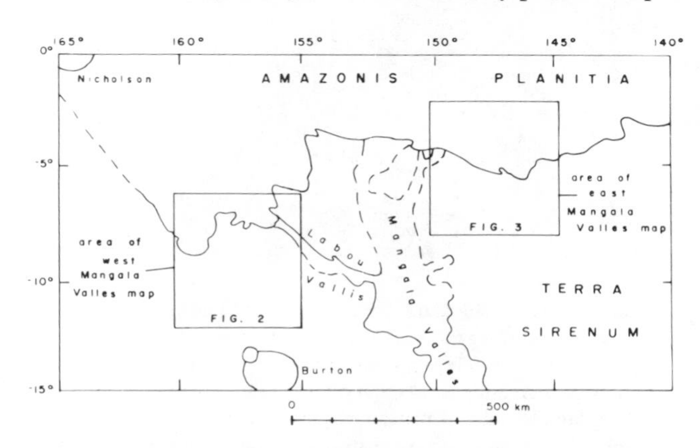

Fig. 1. Index map showing location of map areas in relation to Mangala Valles, Amazonis Planitia, Terra Sirenum, and large impact craters. Dashed lines indicate approximate boundaries of features.

tations and contacts have been revised to reflect information visible on high-resolution Viking Orbiter images. Formal geologic and stratigraphic terms follow the schemes by *Scott and Tanaka* (1986) and *Tanaka* (1986), which are based on global-scale geologic mapping.

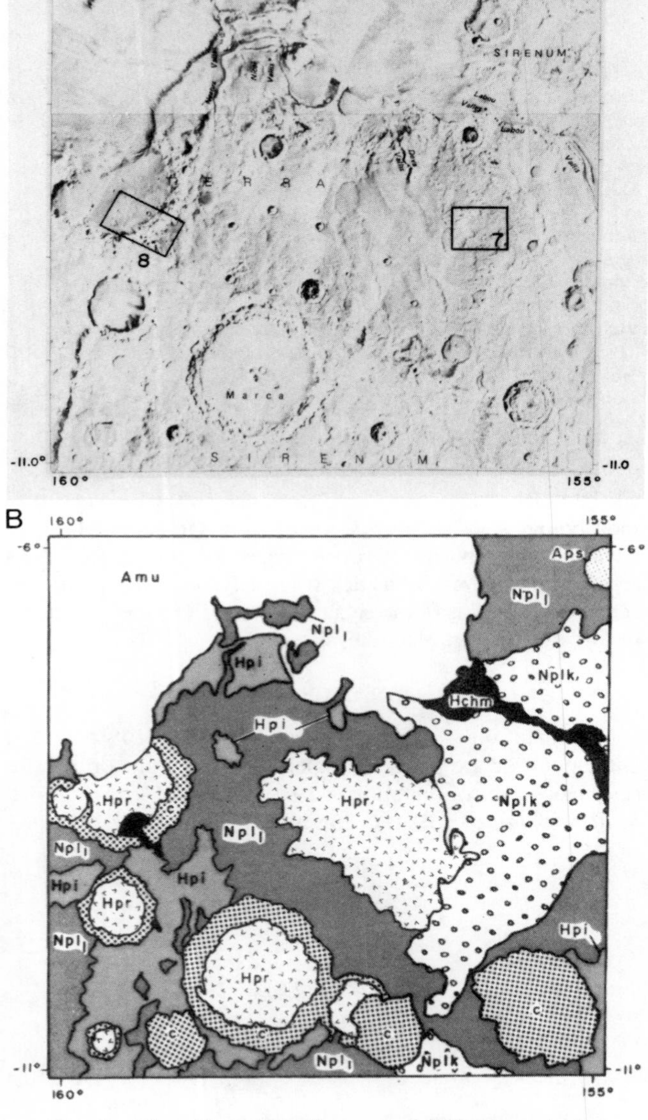

Fig. 2. West Mangala Valles area. **(a)** Photomosaic base showing locations of Figs. 7 and 8 (boxes). **(b)** Generalized geologic map. Symbol c, crater material. Other units listed in order of decreasing age. Npl_1, cratered unit of plateau sequence; Nplk, knobby unit of plateau sequence; Hpi, intercrater plains unit; Hpr, ridged plains material; Hchm, floor material of Mangala Valles branches; Aps, smooth plains material; Amu, upper member of Medusae Fossae Formation. Theater-headed valleys shown in black (degraded valleys too small to show).

Plateau and Plains Material

Three units of Noachian age occur in the Terra Sirenum highlands of the lower Mangala Valles region (Figs. 2b and 3b). West of Mangala Valles, the cratered unit (unit Npl_1) and

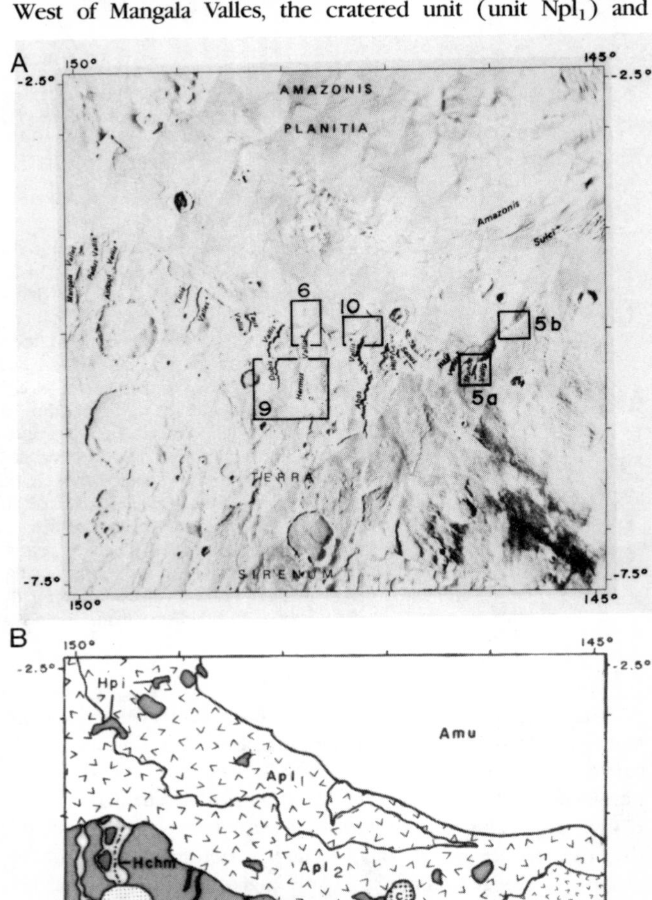

Fig. 3. East Mangala Valles area. **(a)** Photomosaic base showing locations of Figs. 5, 6, 9, and 10 (boxes). **(b)** Generalized geologic map. Symbol c, crater material. Other units listed in order of decreasing age. Nplr, ridged unit of plateau sequence; Hpi, intercrater plains unit; Hchm, floor material of Mangala Valles branches; AHpl, old lobate plains material; AHt_3, member 3 of Tharsis Montes Formation; Apl_1, intermediate lobate plains material; Apl_2, young lobate plains material; Aps, smooth plains material; Amu, upper member of Medusae Fossae Formation, Theater-headed valleys shown in black; ribbon valleys outside theater-headed valleys shown by dash-dot pattern.

knobby material (unit Nplk, as defined by *Chapman et al.,* 1990) of the plateau sequence make up rugged, heavily cratered highland terrain. South of Labou Vallis, the cratered unit surrounds local areas of the knobby unit containing irregular hummocks and hollows. These heavily modified materials lack geomorphic indications of original lithology. The hollows appear similar to terrestrial thermokarst alases—circular or oval depressions formed by thawing of permafrost (*Czudek and Demek,* 1970). The hummocks and hollows of the knobby member may have resulted from local endogenic processes that melted ground ice within brecciated regolith or erosion of a localized mantle superposed on low areas of the cratered unit. This mantle may have consisted of lavas or sediments that degraded when ground ice sublimated producing thermokarst landforms. East of Mangala Valles (Fig. 3b) the northern part of a large exposure of the ridged unit of the plateau sequence (unit Nplr) forms rugged highlands containing large, irregular, north-trending ridges.

Also covering parts of Terra Sirenum in both map areas (Figs. 2b and 3b) are two units whose crater counts indicate Late Hesperian ages (Fig. 4). The intercrater plains unit (unit Hpi) extends over broad areas at an apparent elevation intermediate between that of Mangala Valles and of higher Noachian materials; the unit contains subdued craters and low

SERIES	GEOLOGIC UNITS: VALLEY-FLOOR MATERIAL	GEOLOGIC UNITS: PLATEAU & PLAINS MATERIAL	CRATER DENSITY	AGE (b.y.) 1	AGE (b.y.) 2
Upper Amazonian		Amu			
— 160 —					
Middle Amazonian		Aps Apl2	340 ± 90 440 ± 90		
— 600 —				0.70	2.50
Lower Amazonian	Achr	Apl1	1,300 ± 350		
Amazonian-Hesperian 1,600 Undivided	AHcht	AHt3 AHpl	1,800 ± 700 1,900 ± 900	1.80	3.55
Upper Hesperian	Hchm	Hpr Hpi	2,400 ± 300 2,900 ± 500		
— 3,000 —				3.10	3.70
Lower Hesperian					
— 4,800 —				3.50	3.80
Upper Noachian	Nchd	Nplk			
				3.85	4.20
Middle Noachian		Npl1 Nplr			

Fig. 4. Correlation chart for geologic units in lower Mangala Valles region. From left to right, columns indicate series [with boundaries for the number of craters larger than 1-km diameter per 10^6 km^2 according to *Tanaka* (1986)]; geologic units (area of bars roughly proportional to areal exposure of unit; symbols explained in Figs. 2b and 3b); crater density (number of craters larger than 1-km diameter per 10^6 km^2) of plateau and plains units (data from *Chapman et al.,* 1989, 1990; placement of Middle Noachian units based on *Scott and Tanaka,* 1986); and absolute ages according to Hartmann-Tanaka (age 1) and Neukum-Wise (age 2) crater-flux chronologies (*Tanaka,* 1986; *Tanaka et al.,* 1988).

A

B

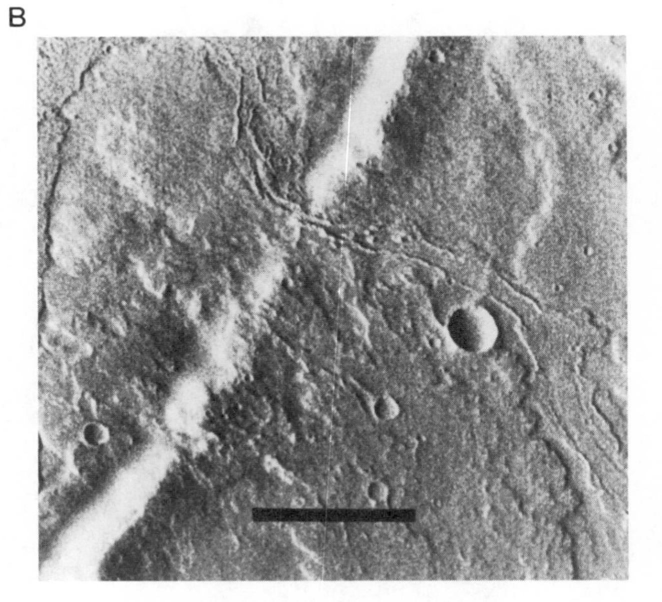

Fig. 5. Images showing stratigraphic relations of member 3 of Tharsis Montes Formation. North at top; scale bars = 5 km. **(a)** Flows of member 3 entering theater-headed valley (Viking image 454S12). **(b)** Flows of member 3 overlying highland/lowland boundary scarp; note leveed lava channel in center of image. Flows of young lobate plains material overlie the member at left (Viking image 454S14).

scarps and is located on either side of the broad channel. It is interpreted to be reworked Noachian volcaniclastic material, possibly related to early Mangala flooding. West of Mangala Valles (Fig. 2b) a low area and several large Noachian craters are filled by ridged plains material [unit Hpr, a ridged unit that has a younger crater age than the ridged plains material (unit Hr) of *Scott and Tanaka,* 1986] that may be made up of lavas based on numerous lobate flow fronts and wrinkle ridges. East of Mangala Valles (Fig. 3b) the intercrater plains unit is overlain by lobate plains material (unit AHpl), which is composed of possible lava flows within and surrounding Mangala Valles (as far south as Memnonia Fossae). The lobate plains unit most likely overlies terraced areas resulting from early Mangala flooding. Sparse lobate scarps and its subdued and channeled appearance in some areas suggests that in addition to lava flows it consists of mud flows or even fluvial deposits. However, in many areas in the Mangala channel it forms a highly resistant unit (lat. 7.8°S, long. 151.5°; image 452S21) and contains a conspicuous leveed channel (lat. 10.6°S, long. 150.4°; image 450S31); hence, a large part of the unit may be volcanic. This unit may be slightly younger than or similar in age to overlapping lava tongues of member 3 of the Tharsis Montes Formation (unit AHt_3) that originated east of the study areas, as the stratigraphic relations are not obvious. The flows of member 3 enter a theater-headed valley at lat. 5.5°S, long. 146.2° (Fig. 5a) and overlie the highland/lowland boundary scarp (Fig. 5b).

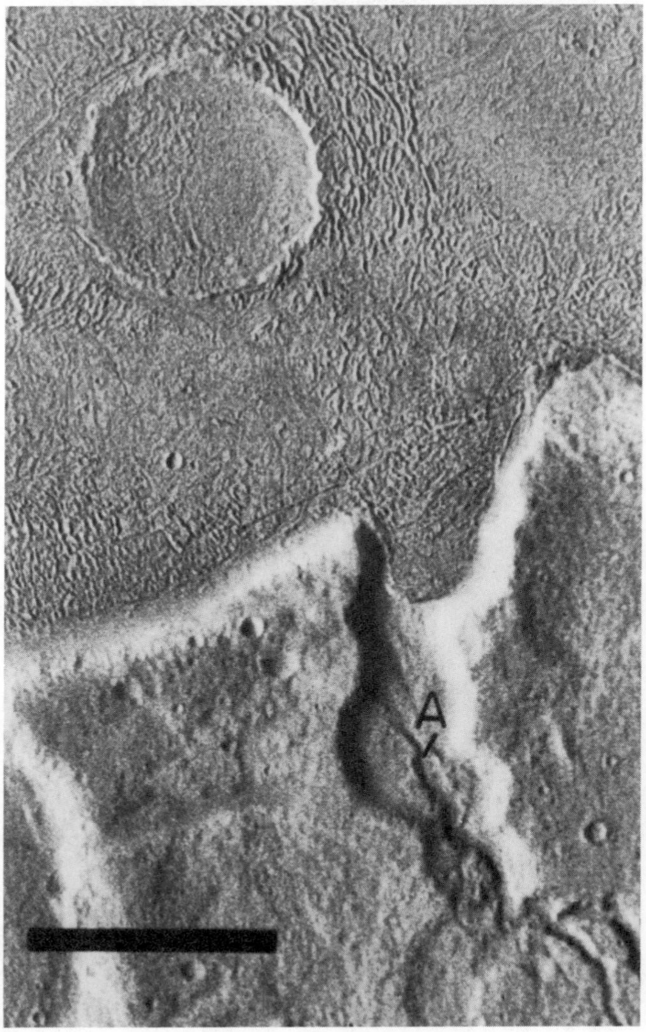

Fig. 6. Flow tongue of young lobate plains material that embays mouth of theater-headed valley. Note narrow ribbon valley (arrow at A) incised in floor of valley. Circular feature near top is buried impact-crater rim. North at top; scale bar = 5 km (Viking image 456S14).

Intermediate (unit Apl_1) and young (unit Apl_2) lobate plains materials are Amazonian lava deposits that fill the lowlands in the eastern map area (Fig. 3). Tongues of the young lobate plains material extend into several theater-headed valleys that cut the boundary scarp (Fig. 6). Amazonian smooth plains material (unit Aps) occurs locally on Terra Sirenum, partly covering the floors of most large impact craters. On the basis of its age, occurrence, and smooth, featureless surface, the unit is interpreted as a young, aeolian deposit.

The youngest unit in both map areas is the upper member (unit Amu) of the Medusae Fossae Formation. The unit has no visible bedding, and its surface is marked by streamedlined ridges that were interpreted by *Ward* (1979) to be yardangs eroded into friable material. The unit is as much as 2 to 3 km thick in areas of southern Amazonis Planitia (*Scott and Tanaka,* 1982) and about 1 km thick where it partly buries a crater rim at lat. 6.5°S, long. 156.1° (Fig. 2b). The origin of this material is uncertain; suggested possibilities include ignimbrites (*Scott and Tanaka,* 1982), paleopolar deposits

Fig. 7. Degraded valleys (marked by arrows) at lat. 9°S, long. 156°. Note hummocks and hollows on surrounding surface of knobby unit of the plateau squence. North at top; scale bar = 3 km (Viking image 450S23).

(*Schultz and Lutz,* 1988), aeolian deposits (*Greeley and Guest,* 1987), and exhumed highland materials (*Forsythe and Zimbelman,* 1988).

Valley-Floor Materials

The two geologic maps of part of the Mangala region (Figs. 2b and 3b) and relations described below indicate that materials of small valley floors were emplaced during two separate episodes: Material of degraded valley networks (unit Nchd) was deposited in the Late Noachian Epoch, and materials of theater-headed valleys (unit AHcht) and ribbon valleys (unit Achr) were emplaced at the end of the Hesperian and beginning of the Amazonian. The relative ages of valley- and channel-floor materials, and thus the episodes of valley erosion, were derived by stratigraphic relations with plateau and plains materials. The small extents of valley-floor materials prohibited derivation of meaningful crater counts. For some critical areas, the Viking images were enhanced by spatial filtering techniques (*Condit and Chavez,* 1979) that amplify details relevant to geologic interpretation. A summary of the morphology and stratigraphic relations of the small valleys follows.

Discontinuous linear depressions west of lower Mangala Valles form valley networks that include Deva, Nestus, and Nicer Valles (Figs. 2a and 7). These networks extend for tens of kilometers and incise the Noachian cratered unit (unit Npl_1) and knobby material (unit Nplk) of the plateau sequence. The main valleys commonly connect with first- and second-order tributaries. These networks were mapped as degraded channels (and material within them shown as unit Nchd) by *Chapman et al.* (1990). The networks are overlain in places by the intercrater plains unit (unit Hpi) and by ridged plains material (unit Hpr). Some networks are cut by Labou Vallis. Their degraded appearance and possible genetic assocaition with the

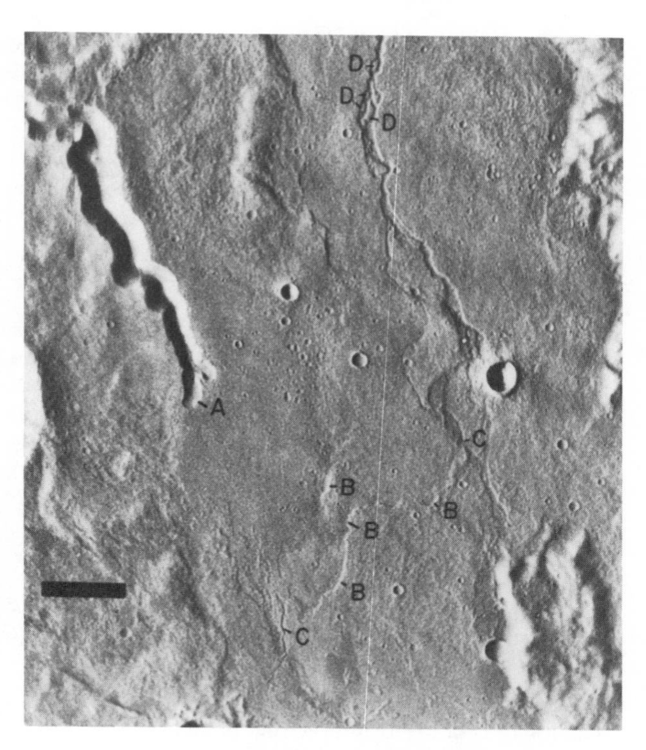

Fig. 9. Image of an area east of lower Mangala Valles showing Dubis Vallis (A), a theater-headed valley; lobate scarps (B) on older lobate plains material; ribbon valleys (C) cutting older lobate plains material. Ribbon valley anastamoses, forming intervalley polygonal blocks (D). Rugged material in ridge and scarp on right edge is ridged unit of plateau sequence. North at top; scale bar = 5 km (Viking image 455S09).

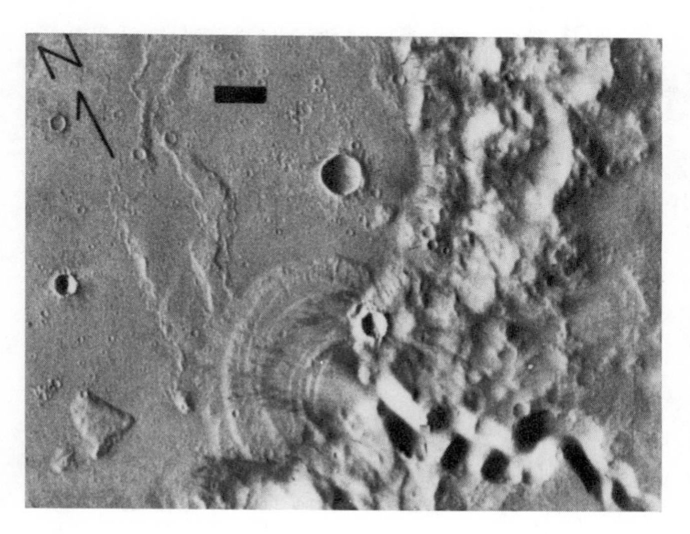

Fig. 8. Fan associated with theater-headed valley overlies Upper Hesperian ridged plains material at lat. 8.7°S, long. 159.4°. Note concentric rings on fan. Scale bar = 3 km (Viking image 450S04).

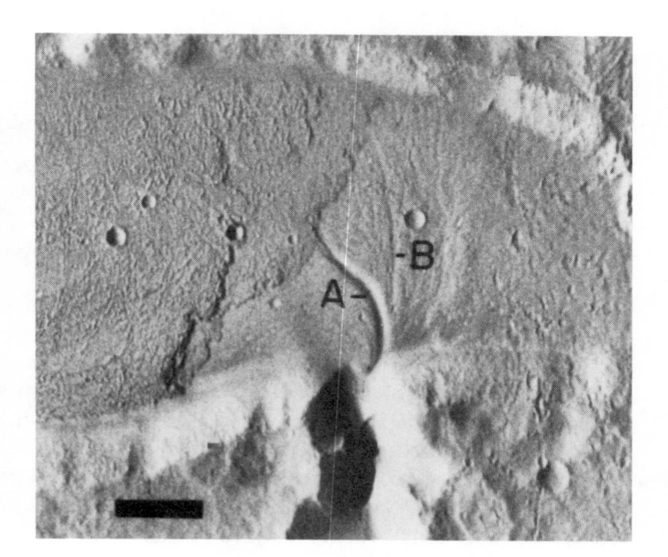

Fig. 10. Mouth of Abus Vallis, a theater-headed valley. Ribbon valley (A) incised into fan at mouth of Abus. Faint, abandoned ribbon valley traces can be seen in fan (e.g., at B). Note lava flows of young lobate plains unit overlying fan. North at top; scale bar = 3 km (Viking image 455S11).

knobby material are consistent with Upper Noachian surfaces elsewhere on Mars (*Tanaka,* 1986), and thus the networks are assigned a Late Noachian age.

Material of Mangala Valles (unit Hchm) covers the floors of three outflow channels—Asopus, Padus, and Labou—that cut highland materials and debouched into Amazonis Planitia. Their Late Hesperian age is constrained by the intercrater plains unit (unit Hpi) of the plateau sequence that they cut and by the oldest lobate plains material (unit AHpl) that partly buries Padus Vallis.

Material of theater-headed valleys (unit AHcht) fills 12 deep valleys incised in plateau materials. Most of the valleys end in the lowland plains; they have nearly constant widths and short lengths (5 to 60 km) and lack tributaries. A few of these valleys end at the base of scarps within plateau areas. The valley floors are at about the same level as the lowland plains with which they connect. Fan deposits occur at the mouths of some of these valleys (Fig. 8). The valleys cut the intercrater plains unit and oldest lobate plains material and are partly buried by young lobate plains material. Some Tharsis flows (unit AHt_3) are cut by the valleys, whereas others have flowed into them (Fig. 5a), indicating that theater-headed valley formation was coeval with local lava flooding.

The narrow ribbon valleys (whose floor material is denoted as unit Achr) cut materials of theater-headed valleys (unit Acht; Fig. 6), Mangala Valles channels (unit Hchm), Tharsis flows (unit AHt_3), and older lobate plains material (unit AHpl; Fig. 9). These valleys are sinuous and short (4 to 50 km long), have fairly constant widths, and locally separate into anastamosing branches, producing polygonal interchannel blocks (Fig. 9). The ribbon valleys at the mouths of Abus (Fig. 10), Asopus, and Munda Valles are buried by young lobate plains flows (unit Apl_2), and the head of the ribbon valley in Asopus Vallis is buried by smooth plains material (unit Aps). Many of the ribbon valleys are better developed on older lobate plains materials, particularly to the south, and may have formed at the same time (see section on "Hydrologic History"). If so, these valleys formed over a considerable period of active valley formation and volcanism (Fig. 4).

VALLEY MORPHOMETRY

Photoclinometric methods of *Davis and Soderblom* (1984) were used to derive the relative depths of valleys and channels of the three youngest episodes, the thickness of geologic units, and the displacements along faults. By these methods integral slope information is calculated along linear profiles from single calibrated Viking Orbiter images, using a phase-dependent photomegtric function and atmospheric-scattering information (*Tanaka and Davis,* 1988, Appendix).

Plots (Fig. 11) of measured widths vs. depths of some of the small valleys (Table 1) indicate that each valley type has a fairly distinct cross-sectional size. Ribbon and theater-headed valleys have similar cross-sectional relations and show a positive relation between depth and width, but the theater-headed valleys are distinctly larger and more irregular in width-to-depth ratio. The degraded valleys overlap in width with theater-headed valleys but are shallower for a given width and

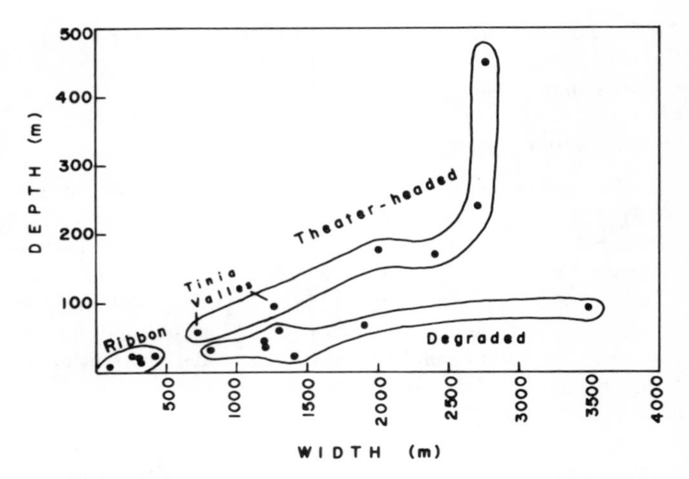

Fig. 11. Plot of depth vs. width for small valleys in lower Mangala Valles region. Each valley type forms distinct field in plot. Data from Table 1.

therefore have a more uniform depth regardless of width. This relation may be a result of erosion and deposition, but perhaps the degraded valleys, unlike the young valleys, reached a base level of development. Width-to-depth ratio may be a function of fluid velocity vs. time, the process of formation, the thickness of confining lithologic layers, or some combination of these.

HYDROLOGIC HISTORY

The three types of small valleys in the lower Mangala Valles region, as well as the outflow channels, developed during different times ranging from the Late Noachian to the Early Amazonian (Fig. 4). A hiatus in regional fluvial activity occurred during most of the Hesperian Period, and then

TABLE 1. Widths and depths (measured by photoclinometry) of small valleys in the lower Mangala Valles region of Mars.

Valley Type	Location	Width (m)	Depth (m)
Ribbon	Tinia Valles (upstream)	100	10
	Tinia Valles (downstream)	302	19
	Asopus Vallis	264	23
	Hermus Vallis (head)	300	20
	Hermus Vallis (mouth)	400	25
Theater-headed	Abus Vallis	2700	240
	West Mangala Valles	2750	450
	East Tinia Valles	2400	167
	West Tinia Valles (head)	706	56
	West Tinia Valles (mouth)	1258	97
	Hermus Vallis	2000	175
Degraded (all south of Labou Vallis)	lat. 9°S, long. 156° (Fig. 7)	800	30
	same	1400	25
	same	1200	45
	same	1900	70
	lat. 11.6°S, long. 156.6°	1200	40
	same	3500	90
	same	1300	60

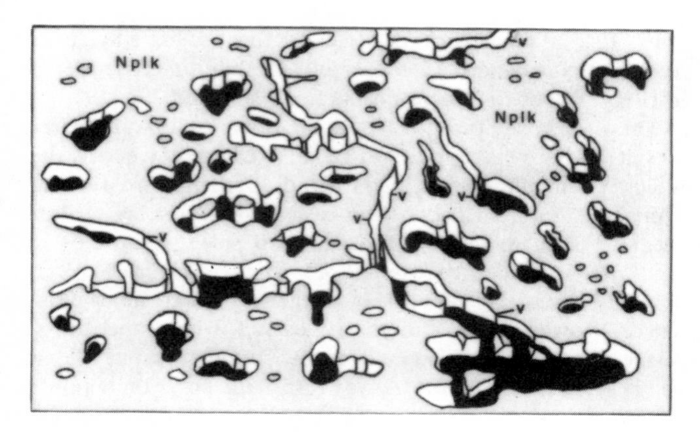

Fig. 12. Schematic interpretation of formation of degraded valleys and knobby unit of the plateau sequence (unit Nplk). Valleys (v) result from coalescence of thermokarst alas pits (valley and pit bottoms shown in black).

activity was intermittent from near the end of the Late Hesperian Epoch on into the Early Amazonian. The three types of valleys and channels are also geomorphically distinct, indicating different origins.

Degraded valleys (Fig. 7) have unconnected sections and tributaries. Although the valleys are locally mantled, postvalley burial does not appear to account for their discontinuities. This observation excludes a runoff origin. Instead, the proximity of the valleys in many places to hummocks and hollows indicates that they formed by a karst-like process in which intermittent collapse occurred above subsurface drainages. Such drainages may have been fed by melted or sublimated ground ice, perhaps generated by local impact or volcanic heating (*Brakenridge et al.,* 1985; *Squyres et al.,* 1987; *Wilhelms and Baldwin,* 1989) or by near-surface ice instability (*Fanale et al.,* 1986) perhaps accelerated by climatic change, as occurred on Earth (*Czudek and Demek,* 1970). The widening and narrowing of the channels could be due to coalescence of alases related to the formation of thermokarst topogrphy (Fig. 12). Terrestrial alas valleys (and these martian channels) differ from river valleys in that they display unexpected turns, blind spurs, irregular tributaries, and, in places, trends against the general inclination of the relief almost independent of slope (*Czudek and Demek,* 1970).

If a thermokarst origin is correct, Noachian materials in the region contained considerable ground ice. Other valley networks of Late Noachian to Early Hesperian age are widespread elsewhere on Mars (*Carr and Clow,* 1981; *Baker and Partridge,* 1986; *Tanaka,* 1986; *Wilhelms and Baldwin,* 1989). Degraded valleys occur only in relatively low-lying highland materials. Perhaps other highland Noachian materials in the region were not similarly degraded because the ground ice was at greater depth in these more elevated areas (*Fanale et al.,* 1986). Also, the degraded valleys and the knobby unit, if due to volcanic heating, may constitute a type of hybrid terrain produced by shallow intrusion of sills (*Wilhelms and Baldwin,* 1989). In the lowlands north of both map areas, ancient knobby terrain may represent lower lying materials that were more severely and pervasively degraded by ground-ice removal (*Scott and Tanaka,* 1986).

During the Late Hesperian Epoch, large, low-lying areas were covered by plains materials that include lava flows (unit Hpr) and materials possibly related to deposition or erosion by Mangala floods (unit Hpi). These areas are not visibly degraded, indicating that little if any interstitial water was initially deposited in the materials, or perhaps it was removed by evaporation (sublimation) rather than sapping. Other rocks, however, may have become water saturated during Mangala Valles flooding at the end of the Hesperian period. The floods originated from a major graben of Memnonia Fossae more than 500 km south of the study area, which, through a system of connected fractures, released water from aquifers possibly disturbed by Tharsis tectonism (*Tanaka and MacKinnon,* 1989). This activity may have saturated highland rocks in the upper and middle Mangala Valles regions and groundwater percolation from floods and temporary lakes (*DeHon,* 1988) in those areas may have charged aquifers in the lower Mangala region.

Shortly thereafter, when Tharsis volcanic activity strongly affected the region, theater-headed valleys formed, apparently by groundwater sapping of the highland aquifers along the base of scarps, particularly the highland/lowland boundary. In their steep walls and lack of tributaries, they are similar to terrestrial canyons whose morphology is controlled by ground-water sapping (*Laity,* 1983; *Laity and Malin,* 1985; *Kochel and Piper,* 1986; *Schumm and Phillips,* 1986). On Earth, removal of the canyon material is generally accomplished by runoff. Some of the martian theater-headed valleys contain ribbon valleys, indicating that subsequent surface runoff (sapping generated

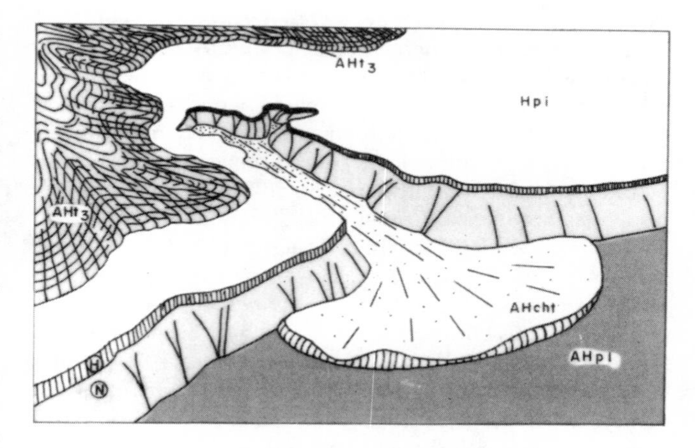

Fig. 13. Schematic diagram showing stratigraphic relations of theater-headed valleys. Intercrater plains material (unit Hpl_3) is cut by theater-headed valley, which contains floor material (unit AHcht) that overlies old lobate plains material (unit AHpl). Formation of valleys roughly coeval with emplacement of Tharsis flows (unit AHt_3). Noachian plateau material (N) exposed along highland/lowland boundary scarp and valley walls apparently formed a temporary aquifer from which sapping flows emanated to form theater-headed falleys; caprock is Hesperian material (H).

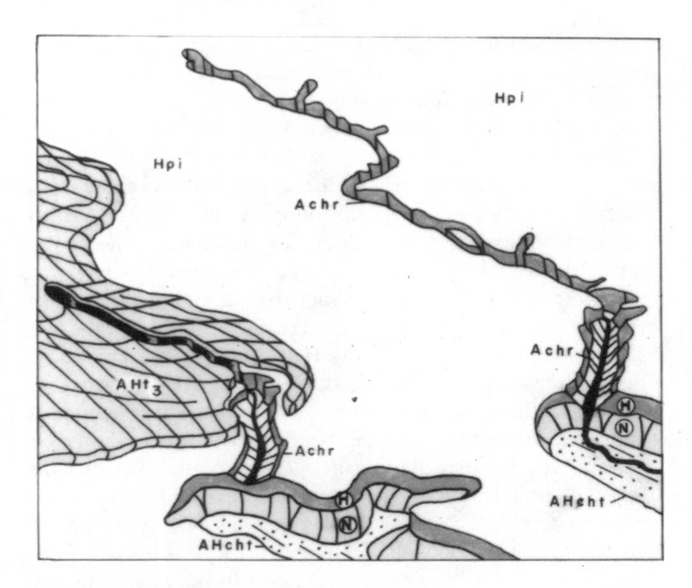

Fig. 14. Schematic diagram showing stratigraphic relations of ribbon valleys and their floor deposits (unit Achr). Ribbon valleys cut plateau and plains materials (such as unit Hpi) and enter theater-headed valleys where they erode older valley deposits (unit AHcht) and valley-wall material (N, Noachian and H, Hesperian).

overland flow) also may have been an important erosional process. Additionally, some of the valleys have fan deposits at their mouths, indicating removal of material by debris flow or glaciation. One fan deposit has concentric ridges (Fig. 8) that may be solifluction lobes or recessional moraines, similar to those on terrestrial glacial outwash fans such as the Northern Iliamna glacier in Alaska (*Press and Siever,* 1986, p. 244).

The dominant northerly trends of theater-headed valleys probably reflect structural control. The longest valleys, Abus and Dubis Valles, extend southward along the northern projections of major north-trending scarps and ridges in the ridged unit (unit Nplr) of the plateau sequence (Fig. 3). This relation suggests that the ridges and scarps are surface manifestations of faults along which groundwater flow was enhanced. The outflows from Mangala may have charged aquifers in which northward flow, toward the highland/lowland boundary, took place. Theater-headed valleys formed chiefly along this boundary where it may have been intersected by north-trending faults (Fig. 13; *Zimbelman,* 1989). Valley depths range from about 50 to 500 m (Table 1), indicating that sapping horizons were at depths similar to or somewhat shallower than those of valley floors.

Ribbon valleys generally appear to have a fluvial origin, although some may be volcanic; they appear to postdate plateau lava flows or to be unassociated with them. The valleys are continuous for tens of kilometers and lack levees, which is unusual for lava channels or collapsed lava tubes. Additionally, several of the ribbon valleys cut floor and fan deposits of theater-headed valleys and may have contributed to valley erosion. In fact, some of the ribbon valleys resemble interdistributary channels on a fan or delta. One ribbon valley cuts Tharsis plains material, enters Abus Vallis, and at its terminus cuts material at the mouth of Abus, where traces of earlier, abandoned, narrow channels are observed (Fig. 10).

On older lobate plains material, the ribbon valleys are better developed on what appears to be inclined areas, whereas the valleys commonly disappear in flatlands. In their upper reaches along the southern part of the eastern map area, as well as farther south, the valleys broaden into small, anastomosing outflow channels. If the source of water was external to the lobate flows, the valleys likely would be carved along flow edges. However, the valleys are carved within the flows, suggesting an internal source of water. This implies that at least in these areas the lobate flows were mud flows or deposits of floods heavily laden with sediments. Water may have seeped through flat areas of the flows but sapped out to form streams in steeper areas. Also, as streams flowed north, some may have lost water due to percolation and formed narrower valleys. The ribbon valleys cut across the north edges of older lobate plains material into older plateau material and traveled down theater-headed channels, further eroding them (Fig. 14). The easternmost ribbon valley cuts through Tharsis flows (Fig. 3b), and it may originate from upland materials poorly resolved by available images. The stream that cut this valley may have resulted from sapping of the upland material. Farther west, the southern extent of the ribbon valley (and the older lobate plains material) is also not well defined because of poor image resolution. The valleys possibly connect with the edge of southern Mangala Valles and grabens of Memnonia Fossae.

SUMMARY

Geologic maps of the lower Mangala Valles region of Mars indicate a complex hydrologic history. West of Mangala Valles, local areas of Noachian cratered and knobby plateau material contain degraded, shallow, discontinuous valleys. Construction of degraded channels may be due to coalescence of alases related to the formation of thermokarst, which would imply that Noachian materials in low-lying plateau areas may have once had a considerable ice content.

On either side of Mangala, theater-headed valleys formed along the highland/lowland scarp and other highland scarps, following the main period of outflow-flooding of Mangala Valles. Their morphology is consistent with a groundwater sapping origin in which debris was removed by runoff, mud flow, or glacial flow. These valleys are similar in age to member 3 (unit AHt_3) of the Tharsis Montes formation. The sapping water may have been derived from an aquifer (50 to 500 m deep) charged by percolation during Mangala Valles flooding; sapping was enhanced along north-trending fault structures.

Ribbon valleys, which mostly appear to have sources in old lobate plains material (unit AHpl), may have resulted from sapping and channeling of wet debris flows or sediment-laden floods, as well as from sappping of some upland material. This indicates that a relatively minor outflow event occurred at Mangala Valles following formation of the main channels.

Acknowledgments. The authors would like to thank P. Davis, D. H. Scott, R. Craddock, and J. Rice for timely and invaluable review comments.

REFERENCES

Baker V. R. and Partridge J. B. (1986) Small Martian valleys: Pristine and degraded morphology. *J. Geophys. Res., 91,* 3561-3572.

Brakenridge G. R., Newsom H. E., and Baker V. R. (1985) Ancient hot springs on Mars: Origins and paleoenvironmental significance of small Martian valleys. *Geology, 13,* 859-862.

Carr M. H. (1983) Stability of streams and lakes on Mars. *Icarus, 56,* 476-495.

Carr M. H. and Clow G. D. (1981) Martian channels and valleys: Their characteristics, distribution, and age. *Icarus, 48,* 91-117.

Chapman M. G., Masursky H., and Dial A. L. Jr. (1989) Geologic maps of science study area 1A, east Mangala Valles, Mars. *U.S. Geol. Surv. Misc. Inv. Ser. Map I-1962,* scale 1:500,000.

Chapman M. G., Masursky H., and Dial A. L. Jr. (1990) Geologic map of science study area 1B, west Mangala Valles, Mars. *U.S. Geol. Surv. Misc. Inv. Ser. Map I-2087,* scale 1:500,000, in press.

Condit C. D. and Chavez P. S. Jr. (1979) Basic concepts of computerized digital image processing for geologists. *U.S. Geol. Surv. Bull. 1462.* 16 pp.

Czudek T. and Demek J. (1970) Thermokarst in Siberia and its influence on the development of lowland relief. *Quat. Res., 1,* 103-120.

Davis P. A. and Soderblom L. A. (1984) Modeling crater topography and albedo from monoscopic Viking orbiter images, 1. Methodology. *J. Geophys. Res., 89,* 9449-9457.

DeHon R. A. (1988) Ephemeral Martian lakes: Temporary ponding and local sedimentation (abstract). In *Lunar and Planetary Science XIX,* p. 261. Lunar and Planetary Institute, Houston.

Fanale F. P., Salvail J. R., Zent A. P., and Postawko S. E. (1986) Global distribution and migration of subsurface ice on Mars. *Icarus, 67,* 1-18.

Forsythe R. D. and Zimbelman J. R. (1988) Is the Gordii Dorsum escarpment on Mars an exhumed transcurrent fault? *Nature, 336,* 143-146.

Greeley R. and Guest J. E. (1987) Geologic map of the eastern equatorial region of Mars. *U.S. Geol. Surv. Misc. Inv. Map I-1802-B,* scale 1:15,000,000.

Kochel R. C. and Piper J. F. (1986) Morphology of large valleys on Hawaii: Evidence for groundwater sapping and comparison with Martian valleys. *Proc. Lunar Planet. Sci. Conf. 17th,* in *J. Geophys. Res., 91,* E175-E192.

Laity J. E. (1983) Control on groundwater sapping and valley formation, Colorado Plateau. *Phys. Geograph., 4,* 103-125.

Laity J. E. and Malin M. C. (1985) Sapping processes and the development of theater-headed valley networks on the Colorado Plateau. *Geol. Soc. Am. Bull., 96,* 203-217.

Mars Channel Working Group (1983) Channels and valleys on Mars. *Geol. Soc. Am. Bull., 94,* 1035-1054.

Masurksky H., Boyce J. M., Dial A. L. Jr., Schaber G. G., and Strobell M E. (1977) Classification and time of formation of Martian channels based on Viking data. *J. Geophys. Res., 82,* 4016-4038.

McCauley J. F., Carr M. H., Cutts J. A., Hartmann W. K., Masursky H., Milton D. J., Sharp R. P., and Wilhelms D. E. (1972) Preliminary Mariner 9 report on the geology of Mars. *Icarus, 17,* 289-327.

Mutch T. A. and Morris E. C. 91979) Geologic map of the Memnonia quadrangle of Mars. *U.S. Geol. Surv. Misc. Inv. Ser. Map I-1137,* scale 1:5,000,000.

Pieri D. C. (1980) Geomorphology of Martian valleys. In *Advances in Planetary Geology* (A. Woronow, ed.), pp. 1-160. NASA TM 81979, Washington D.C.

Press F. and Siever R. (1986) *The Earth* (4th edition). W. H. Freeman, New York. 656 pp.

Schultz P. H. and Lutz A. B. (1988) Polar wandering of Mars. *Icarus, 73,* 91-141.

Schumm S. A. and Phillips L. (1986) Composite channels of the Canterbury Plain, New Zealand: A Martian anolog? *Geology, 14,* 326-329.

Scott D. H. and Schaber G. G. (1981) Map showing lava flows in the northeast part of the Memnonia quadrangle of Mars. *U.S. Geol. Surv. Misc. Inv. Ser. Map I-1270,* scale 1:2,000,000.

Scott D. H. and Tanaka K. L. (1982) Ignimbrites of the Amazonis Planitia region of Mars. *J. Geophys. Res., 87,* 1179-1190.

Scott D. H. and Tanaka K. L. (1986) Geologic map of the western equatorial region of Mars. *U.S. Geol. Surv. Misc. Inv. Ser. Map I-1802-A,* scale 1:15,000,000.

Sharp R. P. and Malin M. C. (1975) Channels on Mars. *Geol. Soc. Am. Bull., 86,* 593-609.

Squyres S. W., Wilhelms D. E., and Moosman A. C. (1987) Large-scale volcano-ground ice interactions on Mars. *Icarus, 70,* 385-408.

Tanaka K. L. (1986) The stratigraphy of Mars. *Proc. Lunar Planet. Sci. Conf. 17th,* in *J. Geophys. Res., 91,* E139-E158.

Tanaka K. L. and Davis P. A. (1988) Tectonic history of the Syria Planum province of Mars. *J. Geophys. Res., 93,* 14,893-14,917.

Tanaka K. L. and MacKinnon D. J. (1989) Release of Martian catastrophic floods by fracture discharge from volcanotectonic regions (abstract). In *Fourth International Conference on Mars,* pp. 200-201. Tucson, Arizona.

Tanaka K. L., Isbell N. K., Scott D. H., Greeley R., and Guest J. E. (1988) The resurfacing history of Mars: A synthesis of digitized, Viking-based geology. *Proc. Lunar Planet. Sci. Conf. 18th,* pp. 665-678.

Ward A. W. (1979) Yardangs on Mars: Evidence of recent wind erosion. *J. Geophys. Res., 84,* 8147-8166.

Wilhelms D. E. and Baldwin R. J. (1989) The role of igneous sills in shaping the martian uplands. *Proc. Lunar Planet. Sci. Conf. 19th,* pp. 355-365.

Zimbelman J. R. (1989) Geologic mapping of southern Mangala Valles, Mars (abstract). In *Lunar and Planetary Science XX,* pp. 1239-1240. Lunar and Planetary Institute, Houston.

Proceedings of the 20th Lunar and Planetary Science Conference, pp. 541-554
Lunar and Planetary Institute, Houston, 1990

Geologic Setting of Diverse Volcanic Materials in Northern Elysium Planitia, Mars

P. J. Mouginis-Mark

Planetary Geosciences Division, Hawaii Institute of Geophysics, University of Hawaii, Honolulu, HI 96822

Geologic mapping of high-resolution (30-50 m/pixel) Viking Orbiter images of northern Elysium Planitia has identified seven sites where current problems in martian volcanology, chronology, and stratigraphy can be resolved. These sites, which are discussed in the context of a potential Mars rover/sample return mission, would permit the following investigations: (1) the dating of Lower Amazonian lava flows from Elysium Mons (thereby providing absolute calibration for global crater size/frequency relative chronologies, (2) the petrologic investigation of long run-out lava flows, (3) the geologic interpretation of materials that may either be lava flows or lahar deposits, (4) the analysis of materials believed to be ash deposits produced by explosive eruptions of Hecates Tholus, and (5) the investigation of the stratigraphy of fractured terrain along the boundary between northern Elysium Planitia and southern Utopia Planitia.

INTRODUCTION

For more than a decade, analyses of the Viking Orbiter image data set have revealed the diversity of volcanic materials on Mars from scales ranging from <1 km to >1000s km. Global mapping at 1:25M (*Scott and Carr,* 1978) and 1:15M (*Scott and Tanaka,* 1986; *Greeley and Guest,* 1987), as well as numerous regional geological studies at 1:2M, have motivated detailed mapping at 1:500K under NASA's Mars Geologic Mappers Program in order to resolve local geologic relationships. One of several areas identified as important to the interpretation of the volcanic history of Mars is northern Elysium Planitia. Materials resulting from both lava-producing and explosive volcanic eruptions are believed to exist here (*Mouginis-Mark et al.,* 1982, 1984; K. L. Tanaka, M. G. Chapman, and D. H. Scott, unpublished data, 1989), as well as the products of complex interactions between volcanic materials and subsurface volatiles (*Mouginis-Mark,* 1985). To the west and north of the volcano Hecates Tholus, numerous lava flows from Elysium Mons, collapse features, plains units, and possible melt water deposits have been identified by these mapping efforts. In addition, the morphology of Hecates Tholus indicates that explosive volcanism (rather than the eruption of lava flows) most likely dominated the eruptive history of the volcano (*Mouginis-Mark et al.,* 1982); many valley networks and the absence of well-preserved lava flows make this volcano significantly different from those found within the Tharsis region (cf. *Carr et al.,* 1977a).

By virtue of the variety of volcanic materials, northern Elysium should be considered as a candidate landing site for a future Mars rover mission. Only by the *in situ* investigation of surface phenomena, and by the return of samples to Earth for dating and petrologic study, can details of the emplacement processes and chronology be resolved (*Gooding et al.,* 1989). This study therefore presents the geologic objectives that such a rover misssion to northern Elysium Planitia would address. The goals of this mission would be: (1) to provide an age for plains materials that are sufficiently extensive that crater size/frequency curves derived from orbital data can be absolutely calibrated, (2) to permit petrologic studies of Elysium Mons lava flows, (3) the geologic interpretation of materials that may either be lava flows or lahar deposits, (4) to study the physical properties (the degree of cementation and particle size distribution of the surface, stratigraphic relationships, and microtopography) of the volcano Hecates Tholus in order to better define its eruptive history, and (5) to study stratigraphic sections of the boundary scarp that separates Elysium Mons lava flows from the northern plains.

ASSUMED ROVER CAPABILITIES

For the purpose of this investigation, it is assumed that the rover will land at the same locality as the sample return vehicle and that selected samples can be returned to Earth in addition to the rover performing *in situ* investigations. Alternative configurations have also been discussed (*Blanchard et al.,* 1985; *Mars Study Team,* 1987; *Gooding et al.,* 1989), but these configurations would not significantly affect the science objectives as they are described here, provided that the rover landed within ~10 km of the sample return vehicle. In this analysis the rover is assumed to have the following capabilities:

1. The spacecraft will have the ability to land a roving vehicle within a target ellipse approximately 10×6 km in size. As with the site selection procedure for the Viking Landers (*Masursky and Craybill,* 1976a,b), possible hazards within the landing site ellipse are considered here to be either craters and their ejecta blankets, steep slopes, or areas of differential erosion (stripping, mantling, and texturing). From the analysis of 40 m/pixel Viking Orbiter images, it is assumed that areas that are free of resolvable impact craters and other hazardous features represent "safe" landing sites. However, as has been shown by the analysis of Viking Lander images (cf. *Mutch et al.,* 1977; *Garvin et al.,* 1981) and the interpretation of very high resolution orbital images (*Zimbelman,* 1987), the morphology of Mars appears to be very different at the 50 m/pixel, 10 m/pixel, and centimeter-scales. It is thus likely that

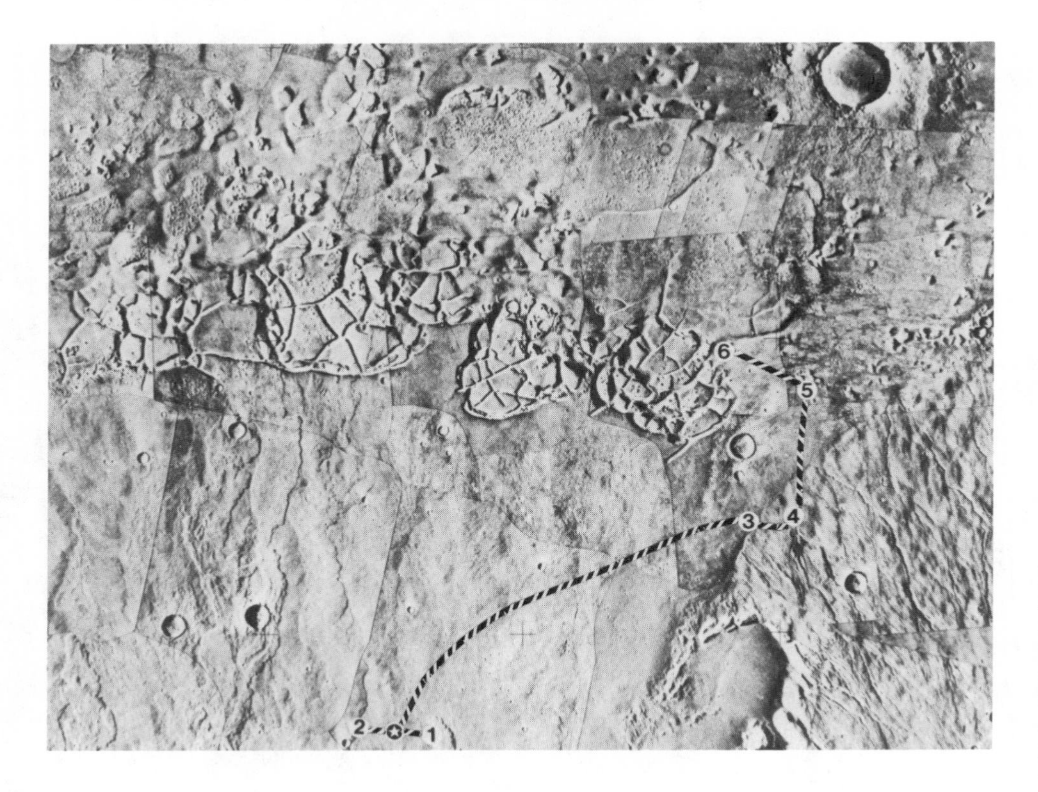

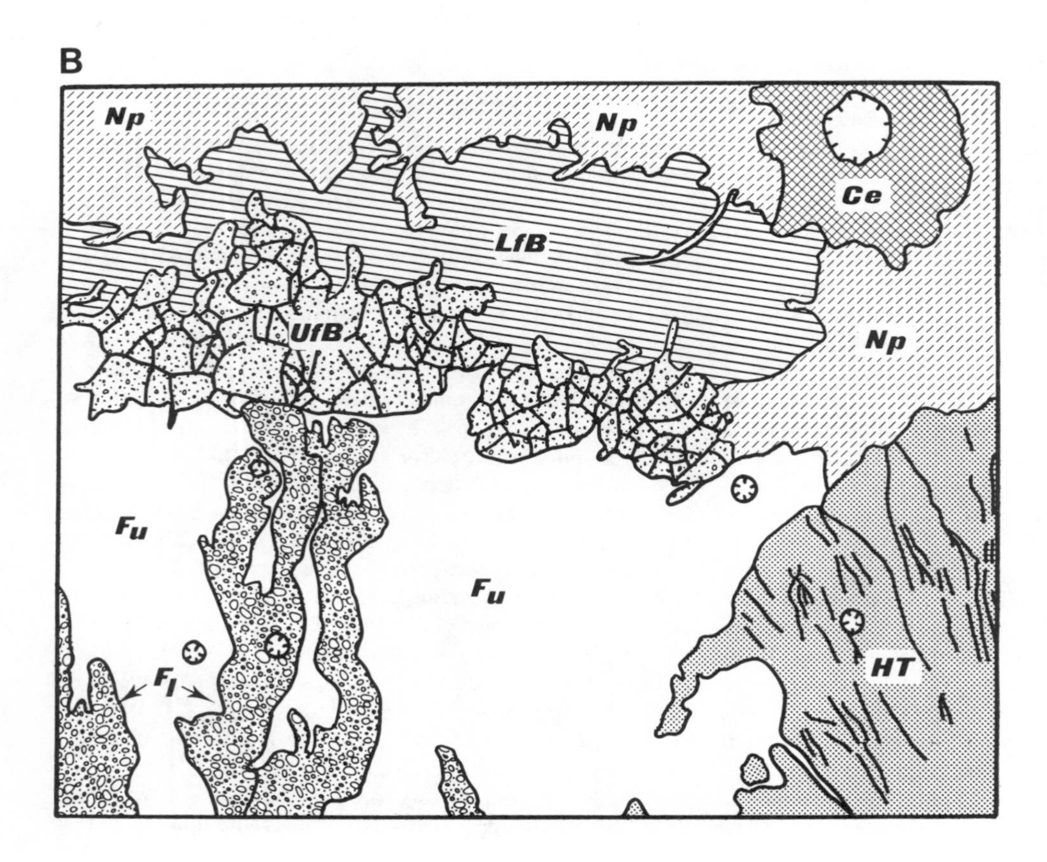

Fig. 1. **(a)** Overview of the proposed rover traverse site in northern Elysium Planitia. Proposed landing site is marked by a star, and the six science stations are numbered 1 through 6. Portion of U.S. Geological Survey photomosaic MTM-35212. Width of image equivalent to 190 km, north is to the top. **(b)** Geologic map of the proposed rover traverse site in northern Elysium Planitia (area shown is the same as that in **a**). Unit abbreviations are as follows: "HT" — Hecates Tholus, "F_u" — undifferentiad flows, "F_l" — lobate flows, "N_p" — Northern Plains materials, "UfB" — upper fractured boundary materials, "LfB" — lower fractured boundary materials, and "Ce" — crater ejecta. Rim crests of meteorite craters are marked by barbed circles.

candidate landing sites will have to come under greater scrutiny for more detailed site selection during the Mars Observer (MO) Mission, when such sites could be imaged at a resolution of ~1.5 m with the MO high-resolution camera.

2. A key aspect of the mission would be the return to Earth of at least three selected samples that are obtained from documented *in situ* localities. The rover must therefore be capable of visiting suitable localities identified either from orbit or from terminal-descent imaging. It is assumed here that each of these sample localities may be up to 10 km from the return of the sapcecraft. An extended surface traverse will begin once the collected samples are placed within the sample return vehicle.

3. The rover vehicle must have a range capability of ~200 km over surfaces expected to locally possess a few meters of vertical relief (i.e., comparable to the topography on terrestrial aa and pahoehoe flows; B. Campbell, S. H. Zisk, and P. Mouginis-Mark, unpublished data, 1989) and have boulder populations comparable to those seen at the Viking Lander sites (*Garvin et al.,* 1981). The rover will also have the ability to climb (or descend) slopes of ~2°-3° for traverses perhaps 10-15 km in horizontal extent and to communicate directly with Earth (i.e., there will be no need for the rover to communicate with Earth via the landing craft). An alternative way to achieve this goal is for the rover to communicate with Earth via a Mars-orbiting relay station (*Mars Study Team,* 1987).

4. The instrument complement of the rover will depend upon the specific mission science requirements, but, as discussed by *Singer* (1988), it should carry suitable instruments that will enable it both to navigate across the martian surface and to perform sample identification.

For the proposed northern Elysium traverse the minimum capabilities and payload of the rover would comprise: (1) an onboard computer that has sufficient artificial intelligence to permit terrain navigation without frequent delays due to the round-trip time involved with communications with Earth (*Solar System Exploration Committee,* 1986, pp. 226-227); (2) a video camera system (probably stereographic) for terrain navigation; (3) an imaging spectrometer that has moderate- to high-spectral resolution for mineral characterization and sample selection with a spatial resolution sufficient to identify individual rock samples and, preferably, mineral grains at close proximity; (4) a drill capable of obtaining unweathered samples from rock outcrops; (5) a surface sample that will be able to conduct trenching experiments comparable to those by the Viking Landers (*Moore et al.,* 1982) to enable the investigation of materials at depths of 10-20 cm beneath the surface; (6) a gravimeter for measuring local gravity anomalies; and (7) an active seismic experiment for the investigation of shallow crustal structure (within ~5 km of the surface) along the rover traverse.

The life expectancy of the rover is not defined as part of this analysis, but it is presumed that the vehicle will have the ability to operate for a sufficient duration on the martian surface to complete the traverse. The traverse described here is ~165 km in length, and science investigations at each station would involve drilling, sampling, and imaging, so that a one Earth year life expectancy should be planned for the rover. Assuming 10 hours of traverse time per martian day, and two days spent at each science station and the landing site (including three visits to return samples), the rover would have to travel at a mean speed of ~50 m/hour when traversing the martian surface.

THE LANDING SITE

The landing site chosen for this study is located at 33.0°N, 212.4°W and is devoid of meteorite craters larger than ~300 m in diameter (Fig. 1). Although no good estimates of absolute elevation exist for this site, the elevation is believed to be about 5 km (*U.S. Geological Survey Map I-1120,* 1978). However, stereogrammetrically derived data for the area surrounding Elysium Mons (*Blasius and Cutts,* 1981) show that the elevation of the proposed landing site may be less than 3 km above the 6.1 mb mean Mars datum [Earth-based radar topography data of *Downs et al.* (1982) can be used to provide an absolute reference datum for the stereographically derived elevation].

In addition to representing a safe landing site, the primary criterion for selecting this landing site is the collection and return to Earth of a sample of the plains materials that underlie the lobate lava flows from northern Elysium Mons. The plains materials are interpreted to be undifferentiated lava flows of Lower Amazonian age that form part of the Elysium formation (Unit Ael_1 of *Greeley and Guest,* 1987) and have a crater frequency of ~250 craters larger than 2-km diameter per 10^6 km^2. Because these Unit Ael_1 flows cover a large portion of Elysium Planitia, obtaining a sample would enable an absolute age-calibration point to be added to the crater-frequency curves that are currently used to interpret relative martian ages (*Neukum and Hiller,* 1981; *Albee,* 1988). *Neukum* (1988) has argued that it is particularly important to obtain an absolute age for Lower Amazonian to Lower Hesperian materials, because these materials evidently were erupted during a transition from a decaying crater flux to a steady-state crater flux. Determining the radiometric age for Lower Amazonian samples may thus aid in the determination of the absolute age of the inflection point in the martian cratering rate, thereby defining much of the absolute chronology for the planet.

GEOLOGIC OBJECTIVES

In order to define the range of volcanic materials in northern Elysium, six science stations have been investigated via analysis of the 30-50 m/pixel Viking Orbiter images of this area. The geologic objectives at each station are as follows.

Station 1

Seven kilometers to the east of the landing site is a prominent lobate lava flow that appears to have been erupted from the northern flanks of Elysium Mons (Fig. 2). This flow can be traced for ~150 km toward the summit of Elysium Mons before Viking Orbiter image resolution is insufficient to resolve individual flow lobes. This lava flow is one of many

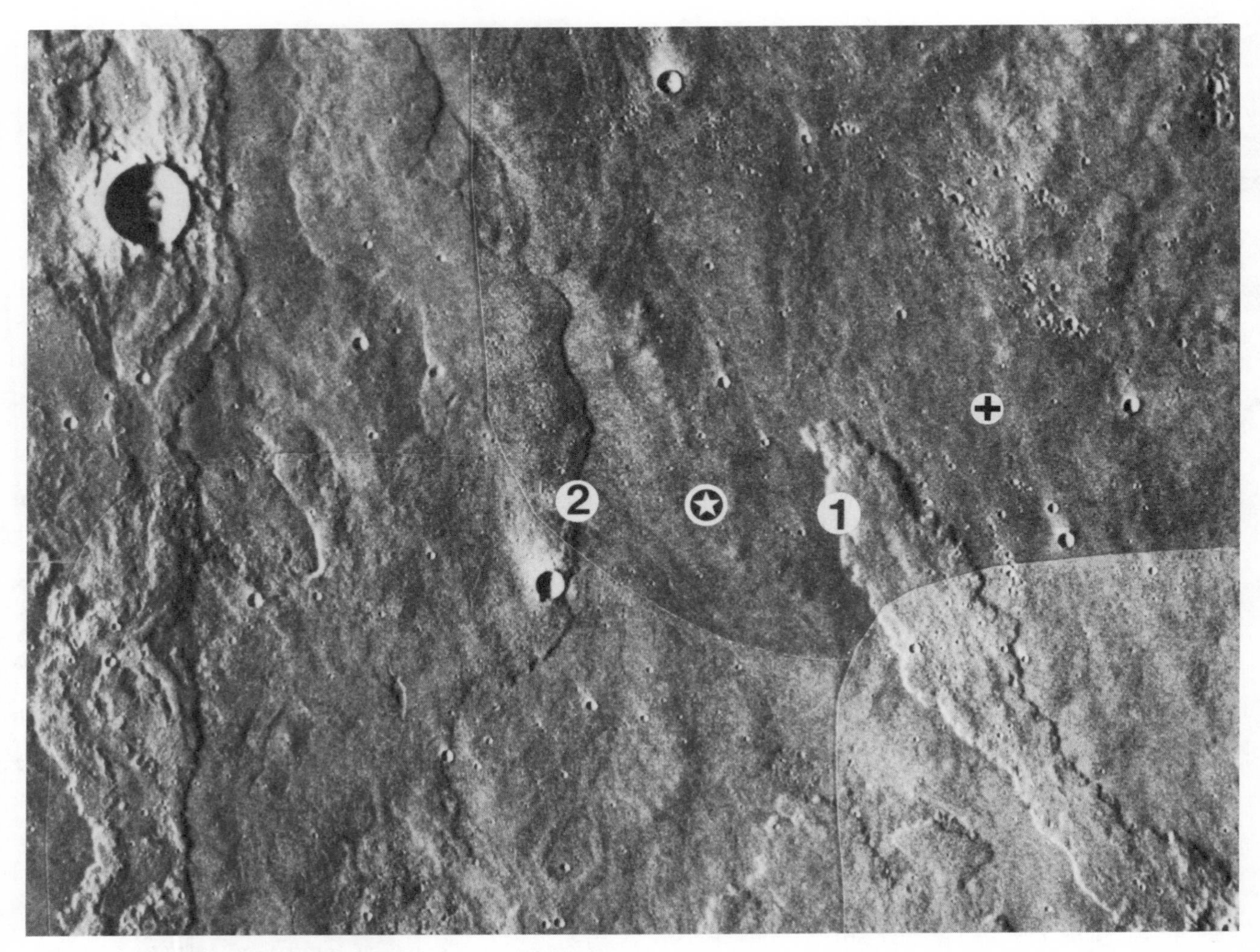

Fig. 2. View of the proposed landing site (star) and Science Stations 1 and 2. The alternate landing site, which would reduce the total traverse distance by ~35 km (but does not include Station 2) is marked by "+". Station 1 (32.70°N, 211.35°W) is located at the edge of a lobate lava flow that was erupted from the northern flank of Elysium Mons. Station 2 (32.72°N, 211.64°W) is at the edge of an enigmatic flow that could be the product of volcano/ground-ice interactions (*Mouginis-Mark,* 1985). Mosaic of Viking Orbiter images 651A10-15, which have a resolution of 52 m/pixel. Image width is equivalent to 65 km and north is to the top.

such flows associated with Elysium Mons (and other martian volcanoes) that are very long compared to terrestrial examples (e.g., *Wood,* 1984; *Cattermole,* 1987). The reason these flows are so long is poorly known. By analogy with lava flows on Earth, martian flows may attain such a size due to high effusion rate (*Walker,* 1973), mafic chemistry and its effects on flow viscosity (*Cattermole,* 1987), or other flow properties such as their propensity to form tube-fed flows (*Greeley,* 1987a). Additionally, super-elevated eruption temperatures similar to those inferred for terrestrial komatiites (*Basaltic Volcanism Study Project,* 1981) may lead to high effusion rates. A returned sample from this lava flow thus not only would provide an age date for the late-stage eruptions of Elysium Mons, but also would permit the geochemistry of such long-runout lava flows to be investigated. It is expected that the collected sample from Station 1 would be returned to the ascent vehicle prior to visiting Station 2, in order to minimize mission loss should the rover fail during the visit to Station 2.

Station 2

This station is located 8 km to the west of the landing site where there is an unusual low-relief flow that does not appear to be a lava flow based on its subdued morphology (Fig. 2). Such flows not only lack lobate edges and the rugged surface texture of lava flows on Mars (*Theilig and Greeley,* 1986) but are also evidently thinner than the lava flows in this area. *Mouginis-Mark* (1985) proposed that these unusual flows are similar to Icelandic jokulhalups, produced by the release of sediment-laden melt water by the interaction of magma and layers of ground ice.

There is increasing morphologic evidence that interactions between magma and subsurface volatiles took place over an extended period of martian geologic history (*Allen,* 1979; *Mouginis-Mark,* 1985; *Squyres et al.,* 1987). *Mouginis-Mark* (1985) hypothesized that within northern Elysium Planitia this release of melt water may have been related to late-stage

Fig. 3. Station 3 (located at 33.37°N, 211.12°W) is at the boundary between Hecates Tholus (lower right) and the undifferentiated lava flows from Elysium Mons. This station is key to identifying the relative age of the channel systems that are found on Hecates Tholus; channel deposits on the plains would indicate a relatively recent period of erosion long after the formation of the volcano and would considerably extend the time over which the volcano may have been active. Station 4 (located at 33.30°N, 210.93°W) is located on the flanks of Hecates Tholus, at a location where the lack of major channels would permit the rover to ascend and descend the flanks. Either drill core or trenching experiments could be performed at this location in order to investigate the possible explosively derived origin of these volcanic materials. Viking Orbiter frame 86A38, resolution 38 m/pixel. Image width is equivalent to 40 km and north is to the top.

intrusive events rather than the emplacement of lava flows at the surface (i.e., comparable to events hypothesized to occur in other areas of Mars by *Squyres et al.,* 1987), and so the age of the jokulhalup deposits may be considerably younger than the adjacent lava flows. The goals for collecting and returning to Earth a sample from Site 2 are therefore: (1) to identify the process by which the flow was emplaced, specifically distinguishing (via visual observations made from the rover's video camera) between a volcanic origin and a chaotic debris flow produced by volcano/ground-ice interactions; and (2) to obtain a relative age determination for this material (via an analysis of the degree of development of weathering rinds on materials within the jokulhalup deposit and the adjacent lava flows) to see how recently the hypothesized subsurface volatiles may have been released to the surface. Should the material at Station 2 indeed prove to be related to melt-water release, obtaining an age estimate for the clays and/or other fine particles expected to be within the jokulhalup deposit would help constrain the poleward migration of subsurface volatiles and the probable depth of ice as a function of geologic time (as modeled by *Fanale et al.,* 1986). The sample from Station 2 would also be returned to the ascent vehicle prior to the rover starting the traverse to Station 3. It is assumed that after the collection of samples from the landing site, Station 1, and Station 2 the return stage of the lander would begin its return to Earth while the rover initiates a more far-ranging exploration of the region.

Station 3

Two scientific sites are located on the northwest flank of the volcano Hecates Tholus (Fig. 3), which lies ~80 km northeast of the landing site. Of particularly scientific interest at Station 3 would be the analysis of materials that are expected to lie at the distal ends of numerous digitate channels that are very numerous on all flanks of the volcano. Although the lower slopes of the Hecates Tholus edifice are embayed by Elysium Mons lava flows, these channels are of uncertain age and origin and thus could have formed after emplacement of the lavas. Channels of this type are quite common features on the older highland patera of Mars, such as Hadriaca, Tyrrhena, and Apollinaris Paterae (*Reimers and Komar,* 1979), and the two most likely modes of formation are that they were carved by volcanic debris flows (*Reimers and Komar,* 1979) or that they were water-carved by volatiles released from the volcano (*Mouginis-Mark et al.,* 1982). The implications of either mechanism for martian volcanology are of basic importance not only for identifying the range of eruption styles (*Wilson et al.,* 1982), but also for investigating the role that volcanoes may have played in providing inputs to the early martian atmosphere (*Postawko and Kuhn,* 1986; *Greeley,* 1987b; *Wilson and Mouginis-Mark,* 1987).

It is expected that *in situ* morphologic investigations of the distal ends of these channels will resolve not only whether they were carved by flowing water, water-laden debris avalanches (lahars), or volcanic debris avalanches, but also whether they pre- or postdate the emplacement of the lava flows from Elysium Mons. Recognizing the morphologic differences between these types of deposits will depend upon analyses of the grain size distribution, thickness, and detailed stratigraphy within the deposits. Comparable facies studies have been conducted on Monte Vulture Volcano, Italy, to discriminate between the ignimbrites and lahar deposits (*Guest et al.,* 1988); cross-bedded surge deposits were found to be associated with variable thicknesses of ignimbrites (analogs to the volcanic debris avalanches described by *Reimers and Komar* (1979), while lahars (water-laden flows) tended to fill preexisting depressions and have relatively uniform thickness in longitudinal section. In addition, terrestrial lahars (and probably martian examples, too) may be distinguished from other volcaniclastic and fluvial deposits by a greater abundance of clay-sized particles and the presence of extremely large boulders (*Fisher and Schmincke,* 1984, pp. 307-311). Water-carved channels are expected to have fine-grained lacustrine deposits at the margins of the flows. In the case of the Elysium examples, morphologic evidence for late (post-Elysium Mons lava flows) channel formation on the flanks of the volcano would be particularly important to search for with the rover. If such evidence were found, this would imply that volatile release took place from Hecates Tholus in the relatively recent history of Mars (i.e., postdating the age of the lava flows sampled at the landing site).

Station 4

Hecates Tholus has a basal diameter of ~190 km, and is believed to rise ~6 km above the surrounding plain (*Hord et al.,* 1974). The science objective at Station 4 is primarily to investigate the origin of the materials on the flanks of the highland paterae. In order to access Station 4, the rover must be able to ascend and descent slopes of 3-4 degrees for a distance of ~5 km (Fig. 3). Although the physical strength of the flanks is unknown, it is likely that the volcano's surface materials are less consolidated than the adjacent Elysium Mons lava flows, as the flanks appear to be relatively easy to erode and have permittted the formation of numerous digitate channels. Thus, although vehicle traction and surface gullying on these slopes may reduce the ability of the rover to reach Station 4, it is likely that the flanks of Hecates Tholus are probably sufficiently consolidated to support a rover as no evidence of wind erosion of friable materials (e.g., yardangs) can be found on the flanks.

Hecates Tholus is comparable to Hadriaca, Tyrrhena, and Apollinaris Paterae in that each of these volcanoes have channelized flanks but lack any lobate lava flows (*Reimers and Komar,* 1979). It has been proposed by *Mouginis-Mark et al.* (1982) that Hecates Tholus experienced numerous explosive (plinian) eruptions that produced flanks comprised of ash deposits rather than lava flows. Similar models for the early histories of the volcanoes Tyrrhena Patera (*Greeley and Spudis,* 1981) and Alba Patera (*Mouginis-Mark et al.,* 1988) have also been proposed. Rover investigations of the physical properties of the flank deposits of Hecates Tholus, particularly studies that involve the use of a trenching tool to dig a few tens of centimeters beneath the surface of a (possible) duricrust layer, would help resolve a crucial issue pertaining to the diversity

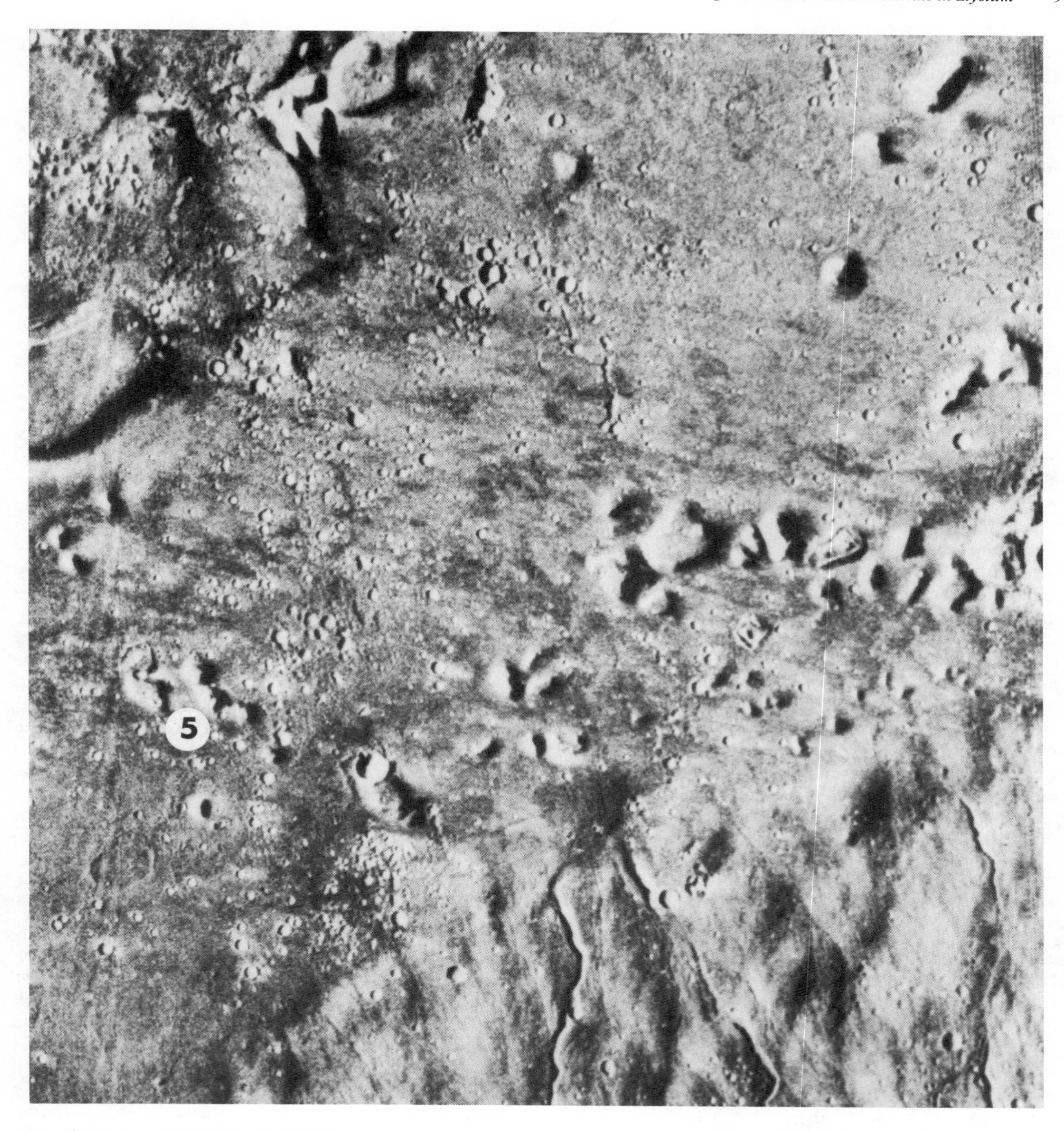

Fig. 4. Station 5 (33.84°N, 210.88°W) is selected to provide sampling opportunities for the analysis of the isolated massifs that characterize the northern plains in Utopia Planitia. Currently unresolved is the issue of whether these massifs are remnants of a once more regionally extensive cap-rock of Elysium Mons lava flows or, alternatively, basement materials that have now been almost completely buried by sediments within the northern plains. Viking Orbiter image 86A39, resolution 38 m/pixel. Image width is equivalent to 40 km and north is to the top.

of styles of volcanic eruptions on Mars. If the flanks of Hecates Tholus were indeed produced by plinian eruptions as proposed by *Mouginis-Mark et al.* (1982), then layers of ash deposits should be found, thereby indicating that the early volcanic history of Mars was significantly different from the activity that characterized the late-stage (Amazonian) lava-producing eruptions on the Tharsis shield volcanoes (*Carr et al.,* 1977a; *Scott and Tanaka,* 1986). Alternative interpretations for the Hecates Tholus flank deposits include pyroclastic flows [as hypothesized for Alba Patera (*Mouginis-Mark et al.,* 1988), which would also imply explosive volcanic eruptions] or, possibly, hydrothermally altered lava flows.

A separate issue relating to the possible ash deposits on the flanks of Hecates Tholus is the determination of the magma chemistry. *Francis and Wood* (1982) have argued against silicic explosive volcanism on Mars, but it is theoretically possible for Mars to have experienced basaltic plinian eruptions (*Wilson et al.,* 1982). Identifying silicic materials on Mars would thus have significant implications for crustal differentiation and would also be relevant to the current debate about the mode of formation of the enigmatic deposits within the Amazonis region of Mars, which *Scott and Tanaka* (1982) hypothesize are ignimbrite deposits. Although surficial deposits may be interpretable from orbit, by using the imaging spectrometer onboard the rover it would be possible to investigate the geochemical diversity and thickness of individual strata within the drainage valleys upon the flanks of Hecates Tholus without returning samples of these materials to Earth. Particularly if this spectrometer were to operate in the 5-12-μm wavelength region (where a number of characteristic silicate emission features exist), it should be possible to determine whether the inferred explosive eruptions were due to variations in magma chemistry or the inclusion of nonjuvenile volatiles within the magma (*Mouginis-Mark et al.,* 1982), and the approximate magnitude of the eruptions (*Wilson et al.,* 1982).

Station 5

Upon leaving the lower flanks of Hecates Tholus, the rover will traverse the boundary between the lava flows of Elysium Mons and the plains units that compose part of Utopia Planitia in the northern plains of Mars (Fig. 1.) The boundary between these two surface units cannot be located to better than ~10 km on the available Viking Orbiter images due to insufficient image resolution, but it is clear that during a traverse of ~40 km the morphology of the surface changes from subdued lava flows in the southwest to smoother plains materials in the northeast. These materials in the northeast appear to overlie a unit that is now exposed only as a series of isolated massifs (Fig. 4), as well as to possess numerous enigmatic sinuous ridges. These ridges have been hypothesized by *Lucchitta et al.* (1986) to be compressional features formed in sediments, while *Parker et al.* (1987) have documented morphologically similar features elsewhere on Mars that they compare to tombolos on Earth (bars created by wave action on open bodies of water that deposit sediments at the shoreline).

The primary science objective at Station 5 is to investigate the morphology of the surficial deposits at this site and, via the imaging spectrometer, attempt a spectroscopic determination of the composition of the isolated massifs that project through the surficial cover of the plains materials. On stratigraphic grounds these massifs may be some of the oldest materials exposed in this region of Mars, or, alternatively, they may be remnants of the Elysium Mons lava flows that have been down-dropped in the rest of this area and almost completely buried by the younger, presumably sedimentary, deposits.

Station 6

Exploration of an area of chaotic terrain commences at Station 6, where the rover can gain easy entrance to a series of interconnected canyons (Fig. 5). These canyons measure about 10-15 km in length, have widths of ~500 m-1 km, and are estimated from shadow length measurements to be about 400-600 m deep. This area was described by *Carr and Schaber* (1977) as a likely area for the collapse of surface materials following the release of subsurface volatiles, and further mapping of the boundary scarp by *Mouginis-Mark* (1985) supports the idea of melt-water release from the scarp following the emplacement of lava flows from Elysium Mons. At Station 6, depending upon the degree of talus covering the slope face, the stratigraphy of the scarp may be interpretable using the rover's imaging spectrometer insofar as a section that cuts through the surface layers of Elysium Mons lava flows and the underlying megaregolith may be visible. In addition to this compositional mapping, which may identify volatile-rich sediments that are important for the analysis of paleoclimates (*Markun,* 1988), it is also possible that geomorphic evidence (e.g., stream channels, moraines, or freeze/thaw phenomena) might be found that would help resolve the mode of formation of the canyon system.

After completion of the geological objectives at Station 6, it is possible that the rover might be able to explore further the canyon system along the base of the boundary scarp and investigate the local stratigraphy of the wall rock. Such a task may be more hazardous than the earlier parts of the traverse, so that this exploration is reserved until the end of the proposed mission in case the rover is lost. Within the canyon system it is possible that in addition to Elysium Mons lava flows and the topmost layers of the megaregolith, distal layers of ash from Hecates Tholus also may be found as the summit of the volcano is only ~120 km from the scarp. If this is the case, such ash deposits may be spectroscopically identified, permitting further assessments of the eruptive history of the volcano to be made [specifically, placing constaints on the dispersal of ash from the volcano, which pertains to the height of the eruption column in plinian eruptions; *Wilson et al.* (1982), *Mouginis-Mark et al.* (1988)].

SURFACE PROPERTIES

En route Topography

While little is known about the meter-scale topography to be expected at this locality on Mars, it is likely that a combination of information from the two Viking Landers

Fig. 5. The distal end of the canyon system associated with the breakup of the cap-rock along the boundary of Elsyium Planitia and Utopia Planitia marks the location of Station 6 (located at 33.97°N, 211.19°W). It is expected that at this location the stratigraphy of the escarpment could be observable to the rover's imaging device, and morphological evidence may be found to help in the interpretation of volcano/ground-ice interactions such as the release of basal melt water following the emplacement of the lava flows from Elysium Mons. Viking Orbiter image 86A37, image resolution 38 m/pixel. Image width is equivalent to 40 km and north is to the top.

(*Mutch et al.,* 1977; *Garvin et al.* 1981), Viking Orbiter thermal inertia measurements (*Christensen,* 1986, 1988), and common experience with terrestrial volcanic plains (*Aubele and Crumpler,* 1988; *Garvin and Zuber,* 1988) can be used to predict the topography that a rover would have to traverse during the collection of samples for return to Earth (i.e., at the landing site and Stations 1 and 2) and during the traverse from the landing site to Station 3 and beyond.

From Viking images that have a resolution of 50 m/pixel, the morphology of the Elysium Mons lava flows in this area appears to be similar to that of flows found in the Tharsis region, where small-scale (~50 m horizontal spacing) pressure ridges and flow channels have been found (*Theilig and Greeley,* 1986). Analysis of the Earth-based radar data for some of the lava flows in Tharsis (*Schaber,* 1980) shows that these Tharsis flows have very high surface roughness values (rms slopes in excess of 3°-4°), so it is possible that the Elysium lava flows may be quite rough. However, because the Elysium flows are apparently older than the Tharsis lava flows (*Scott and Tanaka,* 1986; *Greeley and Guest,* 1987), it is also possible that the surfaces of the flows along the proposed rover traverse may not have such a high surface roughness provided that the weathering rates in Tharsis and Elysium are comparable.

The travel distances between each science station are presented in Table 1. From inspection of Fig. 1, it is evident that in order to find a "safe" landing site and sample the possible jokulhalup deposit at Station 2, a total rover range of the order of at least 164 km is required. In addition, this range estimate ignores possible diversions that may be needed to navigate across the (possibly) rugged surface of the lava flows in this area. From the perspective of providing an absolute age date of Lower Amazonian mateials, determining the petrology of the lava flows and gaining an understanding the cause(s) of very long run-out lava flows on Mars, returned samples from the landing site and the lava flow at Station 1 would be the highest priorities. The need for sampling the jokulhalup deposit is less pressing. If confirmed as a mudflow, this deposit would provide evidence of relatively recent near-surface volatiles. Moving the landing site 18 km to the east of the proposed location and deleting Station 2 would reduce the total length of the traverse to Station 6 to 135 km without significantly affecting the hazard level of the landing site.

En route Science

Key geological observations can also be made during the drive between several of the primary science stations. *Aubele and Crumpler* (1988) have proposed a model for the *in situ* weathering of martian lava flows, based on observations of terrestrial flows. If this model is correct, it should be possible to obtain, in a relatively short period of time, imaging spectrometer data of chip fragments of the flows that could provide information on the heterogeneity of the lava flows during the traverse from Station 2 to Station 3. Depending upon the science instrumentation onboard the rover, important observations could also be made at the boundaries between the Elysium Mons lava flows, Hecates Tholus, and the northern plains. For example, from 20 km west of Station 3, continuing up the flank of Hecates Tholus to Station 4, and then on to Stations 5 and 6, both gravity measurements and active seismic experiments could help resolve the subsurface structure of the area and provide information on the original structure of Hecates Tholus volcano.

TABLE 1. Travel distances between science stations.

Location	Distance
Landing Site to Station 1 (round trip)	15 km
Landing Site to Station 2 (round trip)	13 km
Landing Site to Station 3	81 km
Station 3 to Station 4 (upslope)	11 km
Station 4 to Station 5 (downslope)	26 km
Station 5 to Station 6	18 km
Beyond Station 6 (canyon exploration)	?
Total Distance	164 km

Based on numerical modeling of martian pyroclastic flows (*Wilson et al.,* 1982; *Mouginis-Mark et al.,* 1988), it is expected that such volcanic flows may have traveled about 350-400 km from the vent. Although it seems likely that Hecates Tholus has experienced explosive eruptions (*Mouginis-Mark et al.,* 1982), geomorphic evidence only exists to support the hypothesis that plinian eruptions (producing ash falls rather than pyroclastic flows) took place on this volcano. Such ash fall materials are likely to be widely distributed at distances in excess of ~100 km from the vent, and their distribution should place an upper limit on the basal diameter of the volcano. Thus, if a buried lower flank of Hecates Tholus were revealed by seismic measurements from the rover, it would indicate that another form of volcanism, either subplinian pyroclastic flows or effusive activity, took place.

Observational data and theoretical models indicate that the shallow martian crust is likely to contain several discontinuities (*Davis and Golombek,* 1989). From the analysis of the available Viking Orbiter images, it is possible to construct an idealized cross-section from the lower exposed flanks of Hecates Tholus to the isolated mesas north of the boundary scarp (Fig. 6). The seismic and gravity data from the rover should provide measurements of the basal diameter of Hecates Tholus and the thickness of the Elysium Mons lavas in this area. In addition, spectroscopic measurements made from the rover on leaving Station 6 may also help resolve whether outliers of Elysium Mons lavas are preserved as cap-rock on the mesas north of the boundary scarp, improving our knowledge of the maximum radial extent of effusive volcanism from Elysium Mons.

A further science target during the drive from Station 4 to Station 5 is the ejecta blanket of the 5-km-diameter impact crater located at 33.6°N, 211.1°W. Like most fresh martian meteorite craters of this size, this crater has fluidized, lobate ejecta deposits (cf. *Carr et al.,* 1977b). Several secondary craters can also be found around this crater in addition to the ejecta lobes, which is somewhat unusual for martian craters of this diameter (*Schultz and Singer,* 1980), and suggests that the primary crater ejected both coherent blocks and more

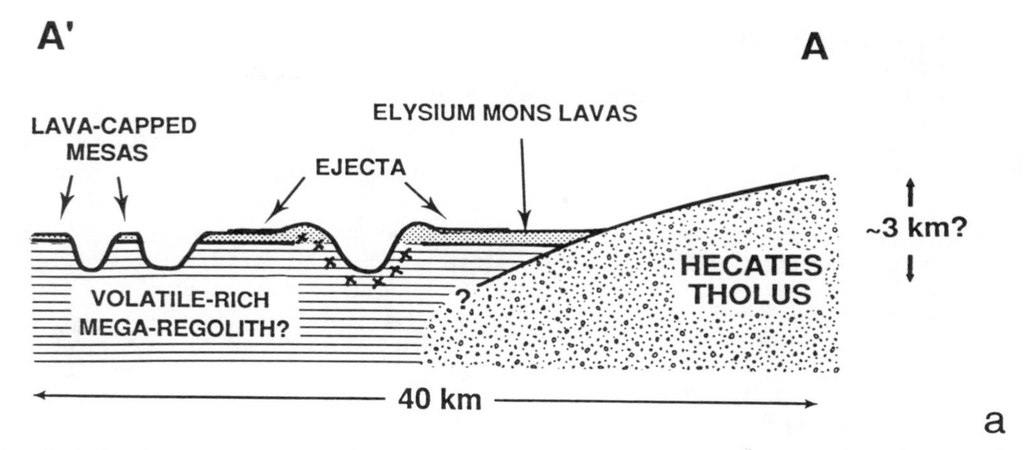

Fig. 6. **(a)** Hypothetical cross-section extending from Station 4 to the boundary scarp beyond Station 6 (see **b** for location). Of particular interest is the determination (via active seismic or gravimeter experiments) of the maximum horizontal extent of the flanks of Hecates Tholus volcano, and the thickness of the lava flows from Elysium Mons. In this sketch the depth of excavation of the 5-km-diameter crater is schematic, but the determination of the depth of the breccia lens could be of value in the analysis of martian impact craters with fluidized ejecta deposits (cf. *Mouginis-Mark,* 1987). **(b)** Location of the cross-section (shown by dashed line) illustrated in **(a).** The positions of Stations 3-6 are also shown. Mosaic of Viking Orbiter images 86A35-40, resolution 38 m/pixel. Image width is equivalent to 68 km and north is to the top.

fluidized materials. The opportunity to study the morphology of the deposits around this crater may help resolve some of the issues relating to the role of target volatiles (*Mouginis-Mark,* 1987) or atmospheric effects (*Schultz and Gault,* 1979) in the formation and emplacement of martian ejecta blankets.

CONCLUSIONS AND RECOMMENDATIONS FOR A FUTURE MARS ROVER

This study was motivated by the need to understand further the local diversity of volcanic materials on the martian surface in order to resolve differences in the styles of volcanism and the role of volcano/ground ice interactions. In many instances the geologic objectives within Elysium Planitia would require a surface rover to adequately interpret the emplacement process and stratigraphy of units identified. The landing site described here does not, however, permit samples of very old (or very young) martian rocks to be collected, based upon our current knowledge of martian stratigraphy and chronology (*Scott and Tanaka,* 1986; *Greeley and Guest,* 1987). Nevertheless, returning samples that will permit an absolute age for the Unit Ael_1 lava flows surrounding Elysium Mons will be of major value in defining the martian crater chronology curve. Thus, the northern Elysium site described here should be considered as an example of an area where fundamental geologic and volcanologic research could be conducted by a rover with a 200-km range and smaple return capability, and it is emphasized that this site should not be the only one considered during planning for a Mars rover/sample return mission.

What is clear from this study, however, is the need for the rover to reach specific exposures or surface materials, even though such sites may require traverses over relatively rugged lava flows and an ascent of several hundred meters. As the trafficability potential of the rover increases, a greater diversity of terrains could be studied on Mars that would permit several different geologic goals to be reached during the same mission. One hundred kilometers is often cited as the limit for a long-duration Mars rover (cf. *Masursky et al.,* 1987, 1988; *Scott,* 1988), but for this Elysium site this maximum range capability would permit only the return-sample collection or the exploration of Sites 3-6 to be conducted. Not surprisingly, a traverse of ~200 km is likely to be significantly more rewarding than a traverse of ~100 km, due to the greater diversification of terrains accessible to the rover.

Rover life expectancy will become important during the course of the mission, due to the travel distance required and the rover's ability to autonomously negotiate potential obstacles. For the traverse outlined here, a mean speed of 50 m/hr would be required to visit all six science stations and return the three collected samples to the Earth-return ascent stage. However, it is evident from this preliminary planning of the rover traverse that very little is known either about the topography and other obstacles that exist on the martian surface at a scale of a roving vehicle, or about the physical properties of the surface materials. Local topography and variable surface lithologies may thus prevent the rover from crossing the martian surface at this desired speed.

Some information on physical properties of the martian surface was obtained by the Viking Landers (e.g., *Shorthill et al.,* 1976; *Moore et al.,* 1982), but remote-sensing data indicate that the Viking Lander sites are distinctive from most of the rest of Mars (*Jakosky and Christensen,* 1986). It is inferred here that it is likely that the undifferentiated lava flows that make up the northern Elysium Planitia landing site will be able to support the lander and permit the rover to reach the other two return sample sites (Stations 1 and 2). What is not so obvious is the bearing strength of the flanks of Hecates Tholus (Station 4), which on morphologic grounds could comprise relatively unconsolidated ash deposits or nonwelded pyroclastic flows (*Mouginis-Mark et al.,* 1982, 1988). In order to answer these and other pertinent engineering and geologic issues for a variety of candidate landing sites on Mars, it therefore appears appropriate that new efforts be made to provide such information (via photogeologic and remote-sensing means) at this early stage of planning for a rover mission.

Acknowledgments. This research was supported by a grant from the Mars Geologic Mappers Program, NASA Grant No. NAGW-1345. I thank J. F. Bell for his comments on an earlier version of this manuscript, K. Tanaka and R. De Hon for formal reviews, and J. Zimbelman for editorial assistance. This is Hawaii Institute of Geophysics Contribution number 2188.

REFERENCES

Albee A. L. (1988) The sample site on Mars (abstract). In *Workshop on Mars Sample Return Science* (M. J. Drake et al., eds.), pp. 28-29. LPI Tech. Rpt. 88-07, Lunar and Planetary Institute, Houston.

Allen C. C. (1979) Volcano-ice interactions on Mars. *J. Geophys. Res., 84,* 8048-8059.

Aubele J. C. and Crumpler L. S. (1988) Constraints on Mars sampling based on models of basaltic flow surfaces and interiors (abstract). In *Workshop on Mars Sample Return Science* (M. J. Drake et al., eds.), pp. 33-34. LPI Tech. Rpt. 88-07, Lunar and Planetary Institute, Houston.

Basaltic Volcanism Study Project (1981) *Basaltic Volcanism on the Terrestrial Planets,* pp. 24-28. Pergamon, New York.

Blanchard D. P., Gooding J. L., and Clanton U. S. (1985) Scientific objectives for a 1996 Mars Sample Return Mission. In *The Case for Mars II, 62,* Science and Technology (C. P. McKay, ed.), pp. 99-119. American Astronomical Society.

Blausius K. R. and Cutts J. A. (1981) Topography of martian central volcanoes. *Icarus, 45,* 87-112.

Carr M. H. and Schaber G. G. (1977) Martian permafrost features. *J. Geophys. Res., 82,* 4039-4054.

Carr M. H., Greeley R., Blasius K. R., Guest J. E., and Murray J. B. (1977a) Some martian volcanic features as viewed from the Viking Orbiters. *J. Geophys. Res., 82,* 3985-4015.

Carr M. H., Crumpler L. S., Cutts J. A., Greeley R., Guest J. E., and Masursky H. (1977b) Martian impact craters and emplacement of ejecta by surface flow. *J. Geophys. Res., 82,* 4055-4065.

Cattermole P. (1987) Sequence, rheological properties, and effusion rates of volcanic flows at Alba Patera, Mars. *Proc. Lunar Planet. Sci. Conf. 17th,* in *J. Geophys Res., 92,* E553-E560.

Christensen P. R. (1986) The spatial distribution of rocks on Mars. *Icarus, 68,* 217-238.

Christensen P. R. (1988) High-resolution thermal imaging of Mars (abstract). In *Lunar and Planetary Science XIX,* pp. 180-181. Lunar and Planetary Institute, Houston.

Davis P. A. and Golombek M. P. (1989) Discontinuities in the shallow martian crust (abstract). In *Lunar and Planetary Science XX,* pp. 224-225. Lunar and Planetary Institute, Houston.

Downs G. S., Mouginis-Mark P. J., Zisk S. H., and Thompson T. W. (1982) New radar-derived topography for the Northern Hemisphere of Mars. *J. Geophys. Res., 87,* 9747-9754.

Fanale E. P., Salvail J. R., Zent A. P., and Postawko S. E. (1986) Global distribution and migration of subsurface ice on Mars. *Icarus, 67,* 1-18.

Fisher R. V. and Schmincke H.-U. (1984) *Pyroclastic Rocks.* Springer-Verlag, New York. 472 pp.

Francis P. and Wood C. A. (1982) Absence of silicic volcanism on Mars: Implications for crustal composition and volatile abundance. *J. Geophys. Res., 87,* 9881-9889.

Garvin J. B. and Zuber M. T. (1988) A Mars orbital laser altimeter for rover trafficability: Instrument concept and science potential (abstract). In *Workshop on Mars Sample Return Science* (M. J. Drake et al., eds.), pp. 81-82. LPI Tech. Rpt. 88-07, Lunar and Planetary Institute, Houston.

Garvin J. B., Mouginis-Mark P. J., and Head J. W. (1981) Characterization of rock populations on planetary surfaces: Techniques and preliminary analysis of Venus and Mars. *Moon and Planets, 24,* 355-387.

Gooding J. L., Carr M. H., and McKay C. P. (1989) The case for planetary sample return missions: 2. History of Mars. *Eos Trans AGU, 70,* 745 and 754-755.

Greeley R. (1987a) The role of lava tubes in Hawaiian volcanoes. *U. S. Geol. Surv. Prof. Pap. 1350,* 1589-1602.

Greeley R. (1987b) Release of juvenile water on Mars: Estimated amounts and timing associated with volcanism. *Science, 236,* 1653-1654.

Greeley R. and Guest J. E. (1987) Geologic Map of the Eastern Equatorial Region of Mars. *U.S. Geol. Surv. Misc. Map I-1802-B.*

Greeley R. and Spudis P. (1981) Volcanism on Mars. *Rev. Geophys. Space Phys., 19,* 13-41.

Guest J. E., Duncan A. M., and Chester D. K. (1988) Monte Vulture Volcano (Basilicata, Italy): An analysis of morphology and volcaniclastic facies. *Bull Volcanol., 50,* 244-257.

Hord C. W., Simmons K. E., and McLaughlin L. K. (1974) *Pressure-Altitude Measurements on Mars — An Atlas of Mars Local Topography.* Laboratory for Atmospheric and Space Physics, Univ. of Colorado, Boulder.

Jakosky B. M. and Christensen P. R. (1986) Are the Viking Lander sites representative of the surface of Mars? *Icarus, 66,* 125-133.

Lucchitta B. K., Ferguson H. M., and Summers C. (1986) Sedimentary deposits in the northern lowland plains, Mars. *Proc. Lunar Planet. Sci. Conf. 17th,* in *J. Geophys. Res., 91,* E166-E174.

Markun C. D. (1988) Martian sediments and sedimentary rocks (abstract). In *Workshop on Mars Sample Return Science* (M. J. Drake et al., eds.), pp. 117-118. LPI Tech. Rpt. 88-07, Lunar and Planetary Institute, Houston.

Mars Study Team (1987) *A Preliminary Study of Mars Rover Sample Return Missions.* Report presented to NASA Headquarters, January 1987. 687 pp.

Masursky H. and Crabill N. L. (1976a) Search for the Viking Lander 1 landing site. *Science, 194,* 62-68.

Masursky H. and Crabill N. L. (1976b) Search for the Viking Lander 2 landing site. *Science, 194,* 809-812.

Masursky H., Chapman M. G., Davis P. A., Dial A. L., and Strobel M. E. (1987) Mars Lander/rover/returned sample sites (abstract). In *Lunar and Planetary Science XVIII,* pp. 600-601. Lunar and Planetary Institute, Houston.

Masursky H., Dial A. L., Strobell M. E., and Applebee D. J. (1988) Geology of six possible martian landing sites. *Rpts. Planet. Geol. Geophys. Program, 1987,* NASA TM-4041, 531-533.

Moore H. J., Clow G. D., and Hutton R. E. (1982) A summary of Viking sample-trench analyses for angles of internal friction and cohesion. *J. Geophys. Res., 87,* 10,043-10,050.

Mouginis-Mark P. J. (1985) Volcano/ground ice interactions in Elysium Planitia, Mars. *Icarus, 64,* 265-284.

Mouginis-Mark P. J. (1987) Water or ice in the martian regolith? Clues from rampart craters seen at very high resolution. *Icarus, 71,* 268-286.

Mouginis-Mark P. J., Wilson L., and Head J. W. (1982) Explosive volcanism on Hecates Tholus, Mars: Investigation of eruption conditions. *J. Geophys. Res., 87,* 9890-9904.

Mouginis-Mark P. J., Wilson L., Head J. W., Brown S. H., Hall J. L., and Sullivan K. D. (1984) Elysium Planitia, Mars: Regional geology, volcanology, and evidence for volcano-ground ice interactions. *Earth, Moon and Planets, 30,* 149-173.

Mouginis-Mark P. J., Wilson L., and Zimbelman J. R. (1988) Polygenic eruptions on Alba Patera, Mars. *Bull. Volcanol., 50,* 361-379.

Mutch T. A., Arvidson R. E., Binder A. B., Guinness E. A., and Morris E. C. (1977) The geology of the Viking Lander 2 site. *J. Geophys. Res., 82,* 4452-4467.

Neukum G. (1988) Absolute ages from crater statistics: Using radiometric ages of martian samples for determining the martian cratering chronology (abstract). In *Workshop on Mars Sample Return Science* (M. J. Drake et al., eds.), pp. 128-129. LPI Tech. Rpt. 88-07, Lunar and Planetary Institute, Houston.

Neukum G. and Hiller K. (1981) Martian ages. *J. Geophys. Res., 86,* 3097-3121.

Parker T. J., Schneeberger D. M., Pieri D. C., and Saunders R. S. (1987) Geomorphic evidence for ancient seas on Mars (abstract). In *MECA Symposium on Mars: Evolution of Its Climate and Atmosphere* (V. Baker et al., eds.), pp. 96-98. LPI Tech. Rpt. 87-01, Lunar and Planetary Institute, Houston.

Postawko S. E. and Kuhn W. R. (1986) Effect of the greenhouse gases (CO_2, H_2O, SO_2) on martian paleoclimate. *Proc. Lunar Planet. Sci. Conf. 16th,* in *J. Geophys. Res., 91,* D431-D438.

Reimers C. E. and Komar P. D. (1979) Evidence for explosive volcanic density currents on certain martian volcanoes. *Icarus, 39,* 88-110.

Schaber G. G. (1980) Radar, visual and thermal characteristics of Mars: Rough planar surfaces. *Icarus, 42,* 159-184.

Schultz P. H. and Gault D. E. (1979) Atmospheric effects on martian ejecta emplacement. *J. Geophys. Res., 84,* 7669-7687.

Schultz P. H. and Singer J. (1980) A comparison of secondary craters on the Moon, Mercury and Mars. *Proc. Lunar Planet Sci. Conf. 11th,* pp. 2243-2259.

Scott D. H. (1988) Mars sample return: recommended sites. *Rpts. Planet. Geol. Geophys. Program, 1987,* NASA TM-4041, 534-536.

Scott D. H. and Carr M. H. (1978) Geological Map of Mars. *U.S. Geol. Surv. Misc. Map I-1083.*

Scott D. H. and Tanaka K. L. (1982) Ignimbrites of Amazonis Planitia region of Mars. *J. Geophys. Res., 87,* 1179-1190.

Scott D. H. and Tanaka K. L. (1986) Geologic Map of the Western Equatorial Region of Mars. *U.S. Geol. Surv. Misc. Map I-1802-A.*

Shorthill R. W., Moore H. J., Hutton R. E., Scott R. F., and Spitzer C. R. (1976) The environs of Viking Lander 2. *Science, 194,* 1309-1318.

Singer R. B. (1988) Sampling strategies on Mars: Remote and not-so-remote observations from a surface rover (abstract). In *Workshop on Mars Sample Return Science* (M. J. Drake et al., eds.), pp. 156-157. LPI Tech. Rpt. 88-07, Lunar and Planetary Institute, Houston.

Solar System Exploration Committee (1986) *Planetary Exploration Through the Year 2000: An Augmented Program.* U.S. Government Printing Office. 239 pp.

Squyres S. W., Wilhelms D E., and Mooseman A. C. (1987) Large-scale volcano-ground ice interactions on Mars. *Icarus, 70,* 385-408.

Theilig E. and Greeley R. (1986) Lava flows on Mars: Analysis of small surface features and comparisons with terrestrial analogs. *Proc. Lunar Planet. Sci. Conf. 17th,* in *J. Geophys. Res., 91,* E193-E206.

Walker G. P. L. (1973) Lengths of lava flows. *Philos. Trans. R. Soc. London A, 274,* 107-118.

Wilson L. and Mouginis-Mark P. J. (1987) Volcanic input to the atmosphere from Alba Patera on Mars. *Nature, 330,* 354-357.

Wilson L., Head J. W., and Mouginis-Mark P. (1982) Theoretical analysis of martian volcanic eruption mechanisms. In *Proceedings of the Workshop on The Planet Mars,* pp. 107-113. ESA SP-185.

Wood C. A. (1984) Why martian lava flows are so long (abstract). In *Lunar and Planetary Science XV,* pp. 929-930. Lunar and Planetary Institute, Houston.

Zimbelman J. R. (1987) Spatial resolution and the geologic interpretation of martian morphology. *Icarus, 71,* 257-267.

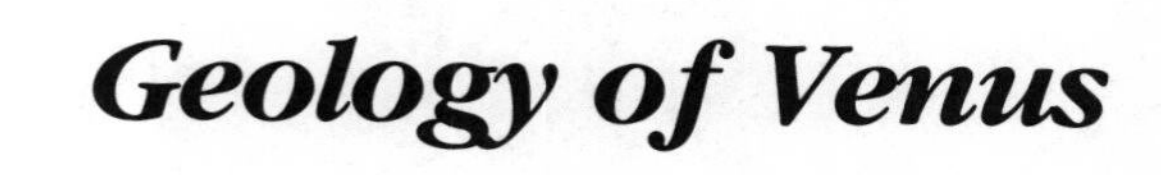
Geology of Venus

Proceedings of the 20th Lunar and Planetary Science Conference, pp. 557-572
Lunar and Planetary Institute, Houston, 1990

Geology of Southern Guinevere Planitia, Venus, Based on Analyses of Goldstone Radar Data

R. E. Arvidson and J. J. Plaut

McDonnell Center for the Space Sciences, Department of Earth and Planetary Sciences, Washington University, St. Louis, MO 63130

R. F. Jurgens, R. S. Saunders, and M. A. Slade

Jet Propulsion Laboratory, 4800 Oak Grove Drive, Pasadena, CA 91103

The 41 backscatter images of Venus acquired by the Goldstone 12.5-cm wavelength radar system cover approximately 32 million square kilometers and include the equatorial portion of Guinevere Planitia, parts of Devana Chasma and Phoebe Regio, Navka Planitia, Heng-O Chasma, a portion of Eisila Regio, and Tinatin Planitia. The images and associated altimetry data combine relatively high spatial resolution with small incidence angles for regions not covered by either Venera Orbiter or Arecibo radar data. Our analyses of the Goldstone data suggest that: (1) Volcanic plains dominate, including groups of volcanic constructs, lava flows, and circular depressions that are probably coronae and arachnoids. The areal distribution and appearance of features is similar to that found in Venera radar images in northern Guinevere Planitia and in Sedna Planitia. (2) A rift valley is located on the western flank of northern Phoebe Regio and extends northward as Devana Chasma. The floor has lower backscatter cross-sections than surrounding plains in the Goldstone data and higher values than plains in Pioneer-Venus scanning altimetry radar (SAR) data. These trends imply that the floor is rougher than plains at both length-scales (>radar wavelength) characteristic of quasispecular scattering and length-scales (~radar wavelength) characteristic of diffuse scattering. Groups of hills adjacent to the rift system show the same scattering patterns. (3) A discontinuous, circular ridge system with a diameter of approximately 1000 km may be the largest corona yet identified on Venus. The northern section corresponds to Heng-O Chasma. (4) A ridge belt can be discerned that straddles the equator and trends E-W from longitudes 0° to at least 30°E. Other ridge and scarp-like features observed in the Goldstone data mainly trend northwest-southeast, as do numerous topographic elements observed in Venera 15 and 16 data to the north. These are directions similar to those discerned in Pioneer-Venus topography throughout the equatorial region, suggesting regional-scale deformation. The northwest-southeast structures seem to predate rifting associated with Beta Regio, Devana Chasma, and Phoebe Regio. (5) Terrain imaged by Goldstone probably has typical impact crater retention ages, measured in hundreds of millions of years. We estimate that the resurfacing rate is no greater than 2 to 4 km/Ga and that resurfacing has clearly involved volcanic processes. (6) Calibrated versions of Goldstone images show regions with relatively high radar cross-sections that are not associated with obvious topography and that retain high cross-sections over a range of incidence angles. These areas expose materials with high dielectric constants, implying that processes that resurface Venus produce high dielectric materials in low plains in addition to highlands.

INTRODUCTION

The geology of Venus is receiving considerable attention, both because of the availability of Earth-based and orbital radar backscatter and altimetry data that allow inferences to be made about the cloud-covered surface and the interior (e.g., *Pettengill et al.,* 1980, 1988; *Jurgens et al.,* 1980; *Campbell et al.,* 1984; *Barsukov et al.,* 1986) and because of the importance of Venus in comparative planetology. Beginning in 1972 and continuing until the present, the Goldstone radar system has been used during inferior conjunctions to acquire S-band (12.5 cm) backscatter images and altimetry data for the equatorial region located between approximately 260° and 30°E longitudes (e.g., *Rumsey et al.,* 1974; *Goldstein et al.,* 1978; *Jurgens et al.,* 1980, 1988a,b,c; Table 1). Goldstone data offer a unique view of the planet, since the regions observed are located to the south of Venera Orbiter radar coverage and occupy a latitudinal belt that is not accessible by the Arecibo radar system. Goldstone observations cover approximately 32 million square kilometers or the equivalent of about 30% of the total area covered by Venera observations. For reference, Fig. 1 shows both the coverage of Goldstone data and the southern limit of Venera data. Figure 2 shows a regional-scale SAR image and elevation trends from Pioneer-Venus data, and Fig. 3 shows an ensemble of Pioneer-Venus, Goldstone, and Venera radar images.

Some of the early Goldstone data were analyzed by *Saunders and Malin* (1976, 1977) and *Malin and Saunders* (1977). A description of possible landforms observed in the 1972 to 1978 data has been given by *Clark et al.* (1986) and *Stooke* (1988). *Jurgens et al.* (1988a,b) described evidence for highly reflecting surfaces, based on the calibrated Goldstone observations acquired in 1980, 1982, and 1986. *Jurgens et al.* (1988c), and *Arvidson et al.* (1989) described a mosaic of Goldstone data and made some preliminary photogeological interpretations. The intent of this paper is to systematically analyze the entire Goldstone data ensemble, including

TABLE 1. Goldstone Venus radar images.

Date	Ctr. Lat.	Ctr. E. Long.	Radius, deg.	Pixel Size, km	Date	Ctr. Lat.	Ctr. E. Long.	Radius, deg.	Pixel Size, km
72Jun20	1.8	325.4	7.2	5.3	77Mar30	-8.9	324.7	21.7	10.5
73Dec23	0.1	286.4	7.2	5.3	77Apr08	-8.1	332.5	11.9	7.0
73Dec28	-1.0	295.6	7.2	5.3	77Apr12	-7.4	336.1	8.2	5.3
74Jan13	-4.8	317.3	7.2	5.3	78Oct28	7.1	309.8	8.2	5.3
74Jan20	-6.3	323.9	7.2	5.3	78Nov03	6.4	315.7	8.2	5.3
74Feb01	-7.8	334.7	7.2	5.3	78Nov10	4.9	321.8	8.2	5.3
74Feb18	-6.9	357.8	7.2	5.3	78Nov17	3.2	328.4	8.2	5.3
74Feb27	-5.9	14.4	7.2	5.3	80Jun08	-0.7	319.9	7.0	3.5
75Jul21	2.5	266.6	7.2	5.3	80Jun14	0.7	325.2	7.0	3.5
75Jul26	3.4	276.1	7.2	5.3	80Jun22	2.3	332.3	7.0	3.5
75Aug03	5.1	289.7	27.7	10.5	80Jun27	3.2	337.5	7.0	3.5
75Aug08	6.2	297.2	7.2	5.3	80Jul02	3.8	343.6	7.0	3.5
75Aug17	8.0	308.0	7.2	5.3	80Jul11	4.4	358.0	7.0	3.5
75Sep04	9.0	324.2	7.2	5.3	82Jan16	-5.7	325.0	7.0	3.5
75Oct01	4.8	4.3	7.2	5.3	82Jan23	-6.9	331.1	7.0	3.5
75Oct07	3.8	16.3	7.2	5.3	82Jan31	-7.6	227.7	7.0	3.5
75Oct11	3.2	25.0	7.2	5.3	82Feb04	-7.6	343.4	7.0	3.5
77Mar13	-7.8	303.7	7.2	5.3	86Oct21	7.5	307.2	4.5	1.0
77Mar18	-8.4	311.1	7.2	5.3	86Oct31	6.8	317.7	4.5	1.0
77Mar24	-8.8	318.7	7.2	5.3	86Nov16	3.2	332.6	4.5	1.0
					86Nov27	0.6	347.2	6.0	2.1

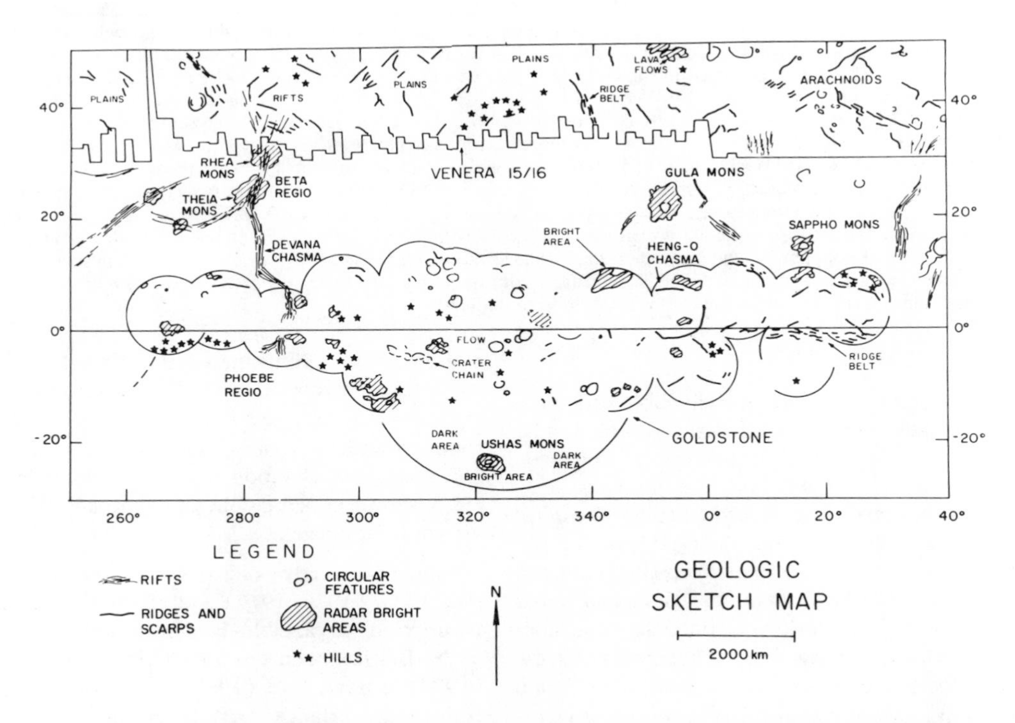

Fig. 1. Sketch map showing major features in areas covered by the ensemble of Goldstone backscatter images, together with features seen in Venera 15 and 16 images to the north. Pioneer-Venus radar image and altimetry data were used to map features in areas not covered by Goldstone or Venera data. Guinevere Planitia extends roughly from 300° to 355°E and from 45°N to the equator. Navka Planitia is centered at approximately 8°S, 320°E and has a radius of about 1000 km. Eisila Regio is a set of northwest-southeast-trending hills that includes Sappho and Gula Mons. Finally, Tinatin Planitia extends southeast from Heng-O Chasma.

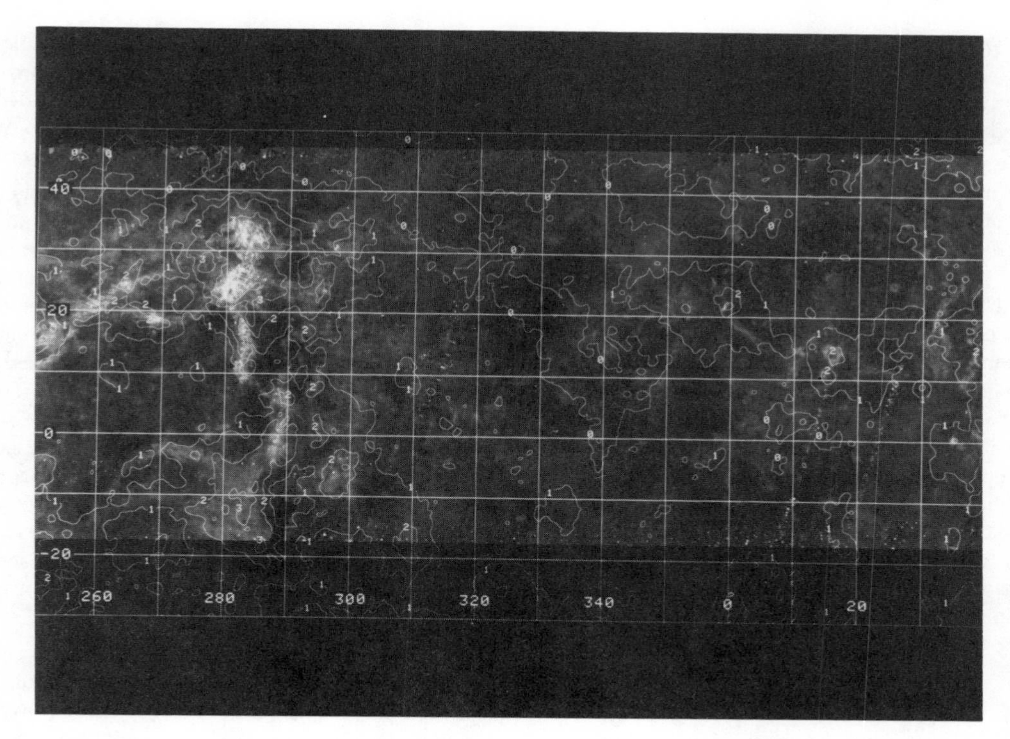

Fig. 2. Pioneer-Venus SAR image data overlain with elevation contours. Data have been normalized to cross-sections that would occur at a 45° incidence angle. Beta Regio corresponds to the north-south highlands with high cross-sections, located between about 20° to 35°N, 280° to 285°E. Phoebe Regio is the highland located south of the equator between 275° and 285°E. The two features are connected by Devana Chasma, which appears as north-south and northwest-southeast bright features. Sappho Mons is located about 5°N, 15°E. Simple cylindrical projection in IAU 1985 coordinates. Contours in kilometers above planetary mean radius (6051 km). Data from *Pettengill et al.* (1980, 1988).

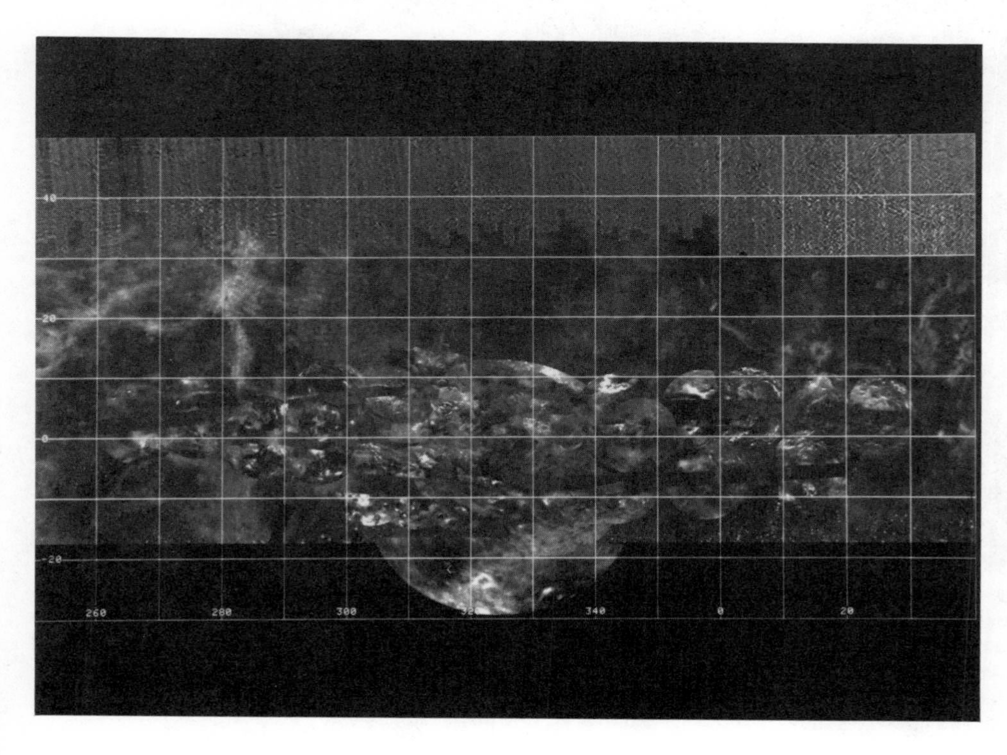

Fig. 3. Mosaic of Goldstone radar backscatter images and Venera 15 and 16 backscatter data overlain on Pioneer-Venus SAR image. The highest resolution Goldstone data are shown in areas where frames overlap.

inventorying the geological features evident in the data, analyzing cross-section variations observed for selected regions, and using the information to understand the processes that have shaped the Venusian surface.

As noted by *Arvidson et al.* (1988), the ability to use radar images to identify endogenic and exogenic processes (e.g., volcanism, aeolian processes, mass wasting, etc.) based on morphological criteria (e.g., identifying lava flows, dunes, landslides, etc.) critically depends on spatial resolution. Even the best Goldstone data have a spatial resolution (approximately 1-km pixels) that allows characterization of larger volcanic, tectonic, and impact features, but probably not the fine morphological features needed to delineate the nature and importance of such exogenic processes as chemical corrosion by reactions with the atmosphere or the extent to which wind erosion and deposition has shaped the surface. Thus, this paper will focus on the volcanic and tectonic evolution of the regions covered by Goldstone data. Magellan images, expected in 1990, should allow detailed characterization of both exogenic and endogenic processes involved in shaping the Venusian surface, since the spatial resolution will be an order of magnitude better than the resolution of existing data.

METHODOLOGY

In this section the methodology used in analyses of the Goldstone data is described in two parts. First, system and target parameters that describe the observation geometry and controls on radar cross-section are summarized to familiarize the reader with the unique characteristics of the Goldstone data ensemble. Second, the methods used to extract geological information from the Goldstone data are described.

Figure 4 shows a Goldstone radar image and associated altimetry data acquired on March 18, 1977, over part of Navka Planitia. Radar image brightness in these and subsequent figures is shown in proportion to backscatter cross-section, normalized to the average cross section for each range gate. The data were acquired at a 12.5-cm wavelength using circular polarization. The transmitter and receiver were set for opposite sense circular polarizations, so echoes are controlled by the quasispecular component scattered from the surface. The dark "runways" in the middle of the image and altimetry data are related to the use of time-delay, Doppler mapping procedures to isolate returns and are a consequence of the poor spatial resolution near the radar equator (*Rumsey et al.,* 1974). The best spatial resolution occurs in the upper and lower parts of the frame, regions that correspond to the best time-delay and Doppler discrimination.

The backscatter and altimetry data shown in Fig. 4 illustrate both the geometry of Goldstone observations and the strong control that slopes have on the returned power. The sub-Earth point, corresponding to a 0° incidence angle for a smooth planet, is located in the center of each circular image. This

Fig. 4. Goldstone backscatter image (left) and associated altimetry data (right) acquired on 3/18/77 between Phoebe Regio and Ushas Mons. Frames are approximately 1500 km in diameter. About 2.3 km of relief are shown in the altimetry data, with brighter pixels corresponding to higher areas. Backscatter cross-section is controlled by local slopes, as shown by backscatter patterns for coalesced craters and conical hills (white arrows). Variations due to surface properties are also evident for the bright flow-like features near the top of the backscatter image (black arrow).

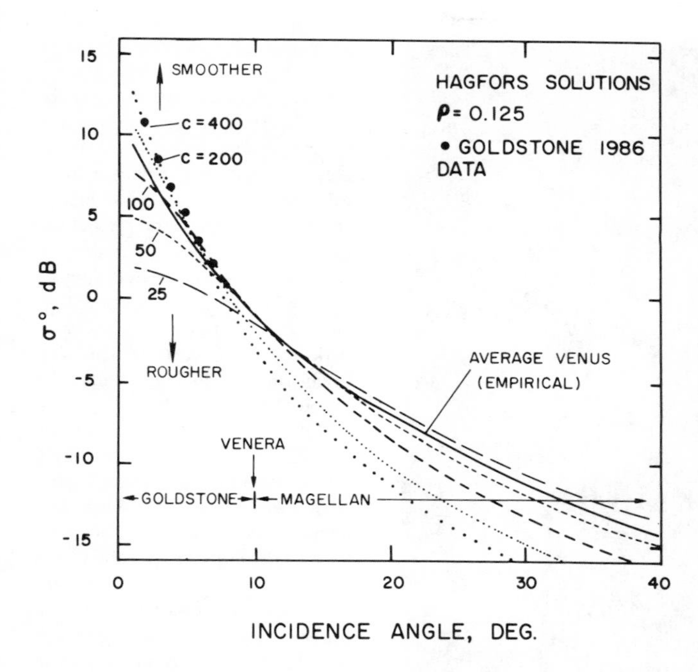

Fig. 5. Backscatter cross-section (σ°) vs. incidence angle predicted from *Hagfors* (1964) model for surfaces with reflection coefficient (ρ) of 0.125 and varying roughness parameter, C. Also shown is average observed scattering behavior for Venus derived from Pioneer-Venus data (*Pettengill et al.,* 1988) and observations from 1986 Goldstone data (*Jurgens et al.,* 1988a,b). Note that at Goldstone incidence angles rough surfaces have low cross-sections as compared to smooth surfaces.

local incidence angle increases with radial distance from the sub-Earth point to a value of approximately 7° for this 1977 observation. Large-scale topographic slopes that have strike azimuths that are not aligned with radial directions from the sub-Earth point alter local incidence angles and thus modulate backscatter strength. The incidence angle control on backscatter is illustrated by Fig. 5, which shows average cross section data for Venus based on 1986 Goldstone observations (*Jurgens et al.,* 1988a,b) and analyses of Pioneer-Venus data (*Pettengill et al.,* 1988). As shown in Fig. 5, because of the large change in cross-section with incidence angle for angles less than about 10°, surfaces with tilts toward or away from the sub-Earth point will cause large changes in backscattered power. For example, the altimetry data shown in Fig. 4 exhibit a series of coalescing circular depressions with raised rims in the northwest section of the frame. The backscatter image clearly depicts this topographic arrangement as enhanced returns when the slopes face toward the sub-Earth point and lower returns when the slopes face away from this point. The same situation holds for the hills at the bottom of frames shown in Fig. 4.

A second factor to consider is that for the small incidence angles typical for Goldstone data, radar cross-sections are largely controlled by quasispecular scattering, i.e., by Fresnel reflections from surface facets that have radii of curvature much greater than the radar wavelength. Shown in Fig. 5 are theoretical scattering curves derived from the *Hagfors* (1964) quasispecular scattering model for a constant Fresnel reflection coefficient (0.125) and varying roughness parameter, C; C is equal to the inverse of the square of rms slope in radians, and thus high C values correspond to smooth surfaces. At small incidence angles, backscatter behavior is in the "reversed" regime in which smoother surfaces of a given Fresnel reflectivity produce brighter returns as compared to rougher surfaces (e.g., *Jurgens et al.,* 1988a,b). A radar-bright signature in a Goldstone image may therefore be caused by (1) quasispecular reflection from a smooth surface, (2) surfaces tilted toward the sub-Earth point, (3) surfaces with high Fresnel reflection coefficients, or (4) some combination of these effects. As an example, the bright region in the upper right portion of the backscatter image shown in Fig. 4 has no obvious topographic control, implying that the surface is either smooth or has a high Fresnel reflection coefficient. Data shown in Fig. 4 have not yet been processed to be in proportion to actual cross-section values, so it is not possible to tell if roughness or Fresnel reflectivities control the high returns.

In summary, both modulation of radar backscatter by topographic slopes dipping toward or away from the sub-Earth point and control by local scattering and electrical properties must be taken into account when interpreting Goldstone data. Our procedure has been to utilize the Goldstone altimetry data to separate topographically modulated returns from those due to variations in surface properties. For each Goldstone observation, ridges, valleys, craters, and other features were delineated and plotted in appropriate locations. Pioneer-Venus SAR and altimetry data (Figs. 2 and 3) were also used to understand how the regions covered by the Goldstone system fit within a global context of elevation and surface properties. The Pioneer-Venus radar system operated at 17 cm and the SAR data were acquired using HH linear polarization (*Pettengill et al.,* 1988). Venera Orbiter 15 and 16 data were used to compare geological features found in the plains to the north with those observed in Goldstone data. The Venera system operated in an HH mode at a wavelength of 8 cm (*Barsukov et al.,* 1986). Comparison of Goldstone and Venera data is facilitated by the low incidence angles (10°) and the relatively high spatial resolution (1 to 2 km) for the Venera coverage. Finally, for the 1980, 1982, and 1986 Goldstone data, which are calibrated to be in proportion to cross-section values, we also considered the geological ramifications of the regions with high cross-sections noted by *Jurgens et al.* (1988a,b). For reference, Figs. 6 to 13 are mosaics of Goldstone data and Fig. 1 includes a geological interpretation map generated from the ensemble of Goldstone observations.

REGIONAL CONTEXT

According to *Barsukov et al.* (1986), Venera Orbiter 15 and 16 radar observations show that approximately 70% of the region to the north of approximately 30° in latitude consists of volcanic plains. For example, regions to the north of Goldstone coverage, including Sedna and northern Guinevere Planitiae, are interpreted to be volcanic plains with circular

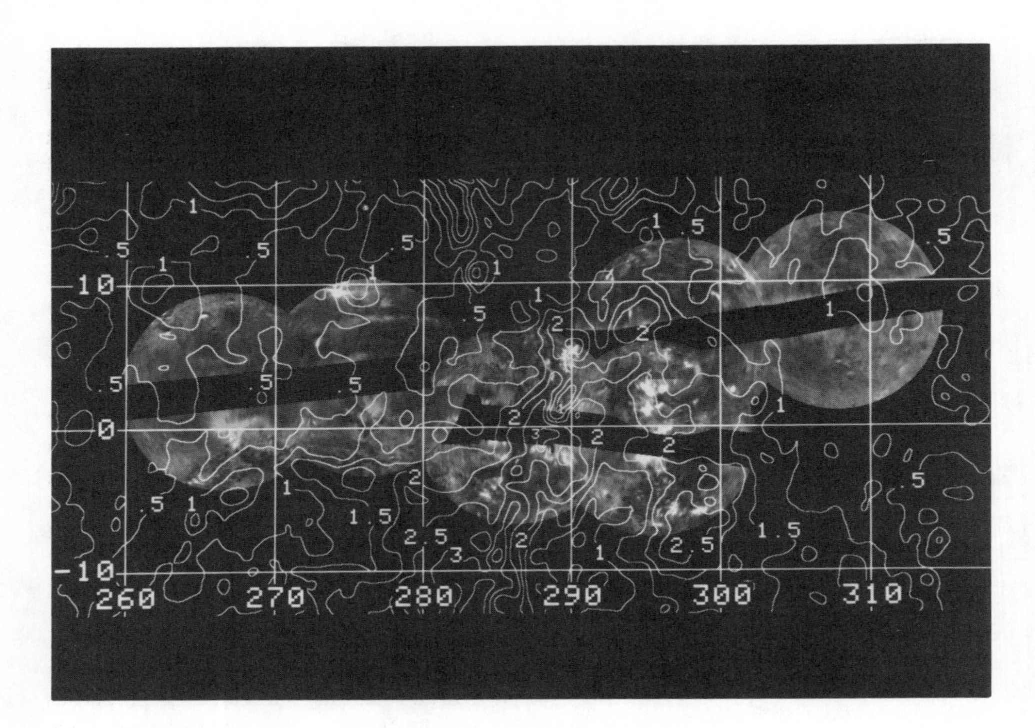

Fig. 6. Mosaic of western portion of 1972-1975 Goldstone backscatter image data with Pioneer-Venus elevation contours. Contour intervals are 500 m. Goldstone images are shown in Mercator projections using IAU 1985 coordinates. Devana Chasma is located approximately 7°N to 7°S, 282° to 288°E. Note that (1) Devana Chasma is dark in the Goldstone data but bright in the Pioneer-Venus SAR image (Fig. 2); (2) the dark groups of hills located just south of the equator between 260° to 280°E correspond to an area of high return in the Pioneer Venus SAR image, as does the region between 0° to 10°S and 295° to 302°E; and (3) the bright features located to the east of Devana Chasma do not show distinctly different signatures relative to surroundings in the Pioneer-Venus SAR image.

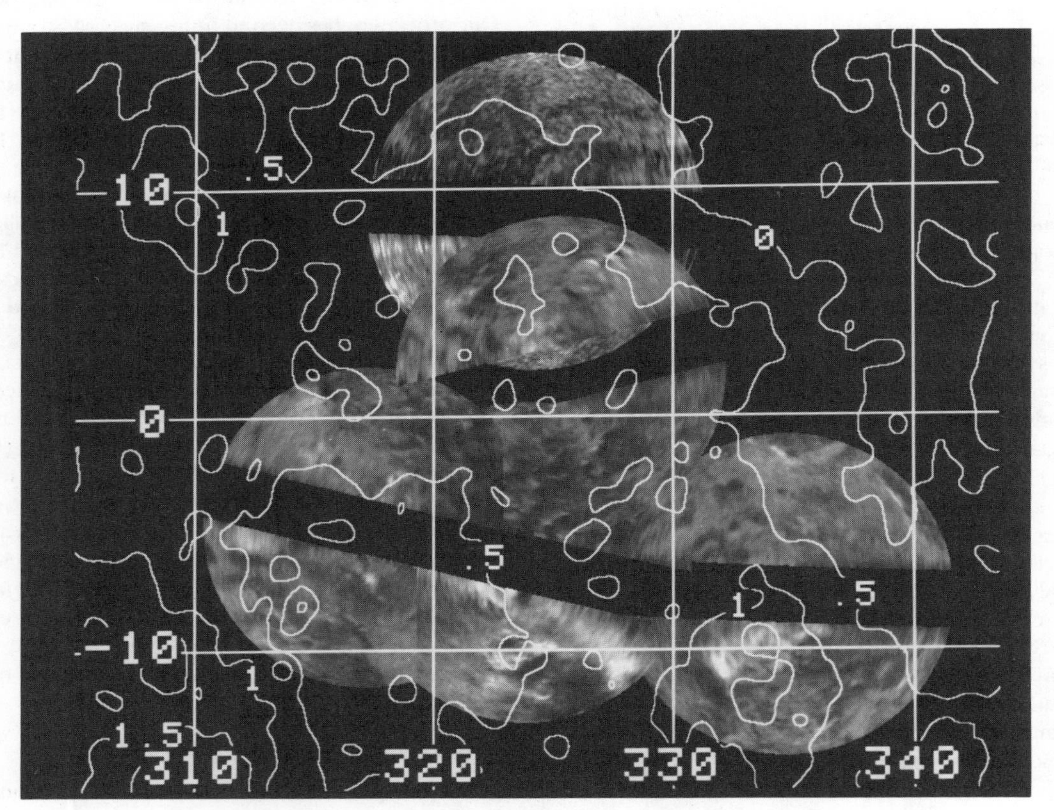

Fig. 7. Mosaic of middle part of 1972-1975 Goldstone backscatter image data with Pioneer-Venus elevation contours. Note circular depressions and groups of hills located just below the equator between 320° and 330°E. The circular feature located at 1°S, 328°E overlaps with a smaller circular feature centered to the southwest of the larger structure. Note the flanking hills on the smaller structure.

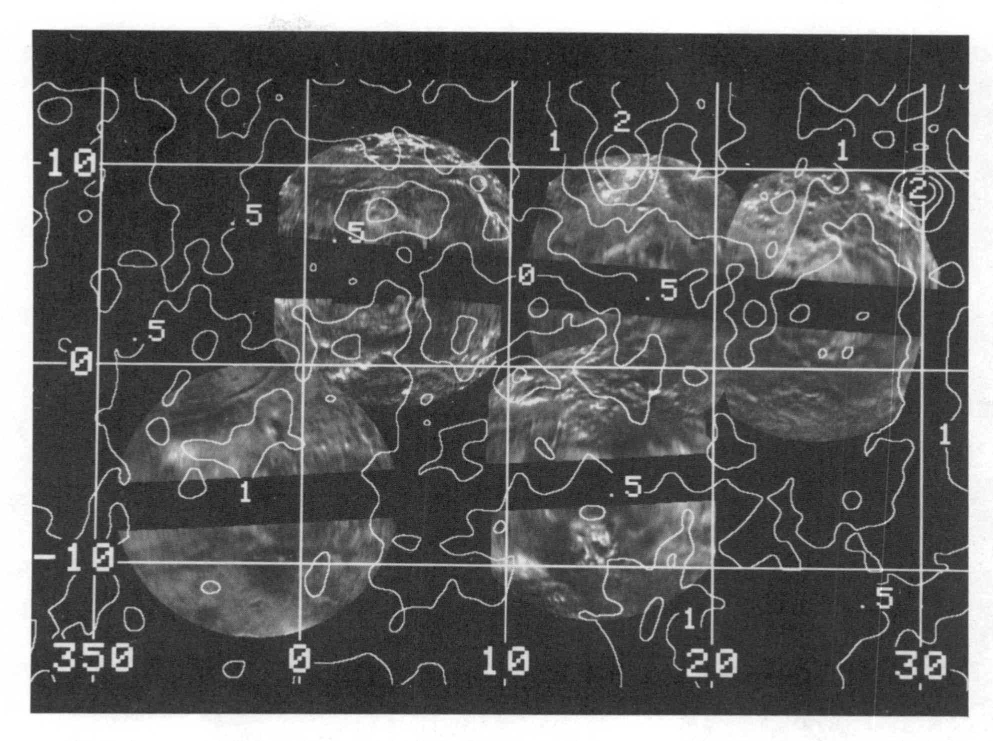

Fig. 8. Mosaic of easternmost 1972-1975 Goldstone backscatter image data with Pioneer-Venus elevation contours. The curved ridge located just below the equator between 350° to 0°E is part of the 1000-km circular structure that includes Heng-O Chasma. The east-west-trending hills straddling the equator between 10° and 30°E may be equivalent to ridge belts seen in Venera 15 and 16 data. Hills in the northeast are part of Eisila Regio.

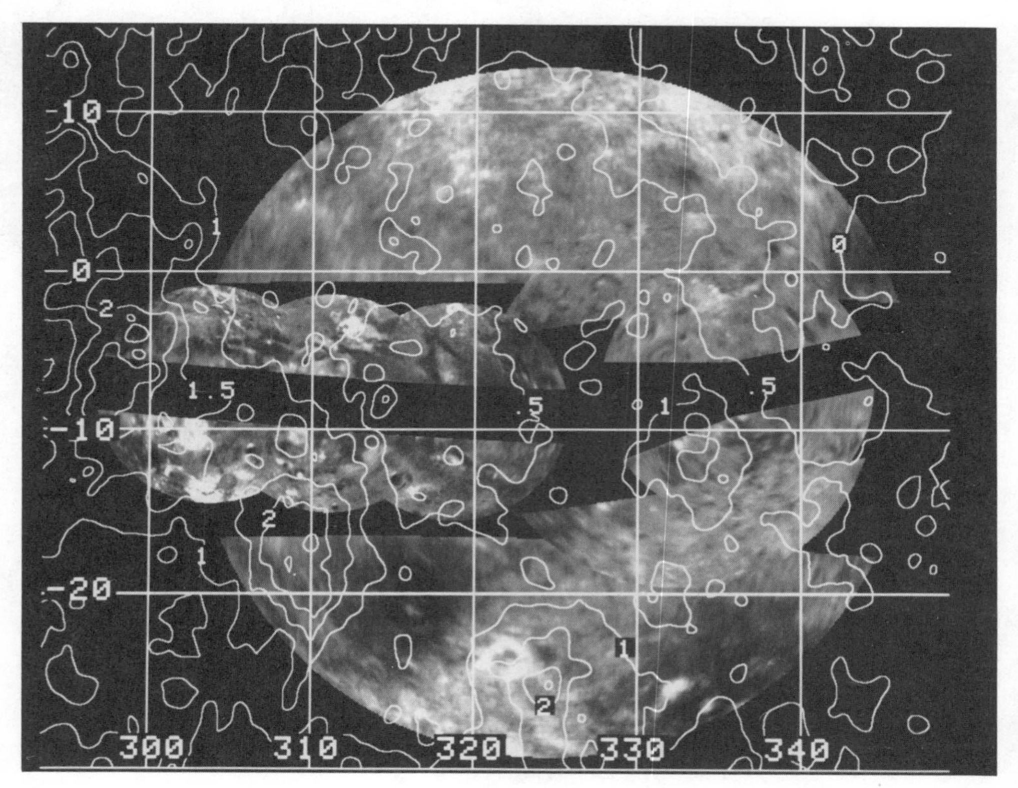

Fig. 9. Mosaic of 1977 Goldstone backscatter images with Pioneer-Venus elevation contours. The frame at left-center is shown in Fig. 4. Ushas Mons is the bright circular feature at about 24°S, 322°E acquired with a 14° incidence angle.

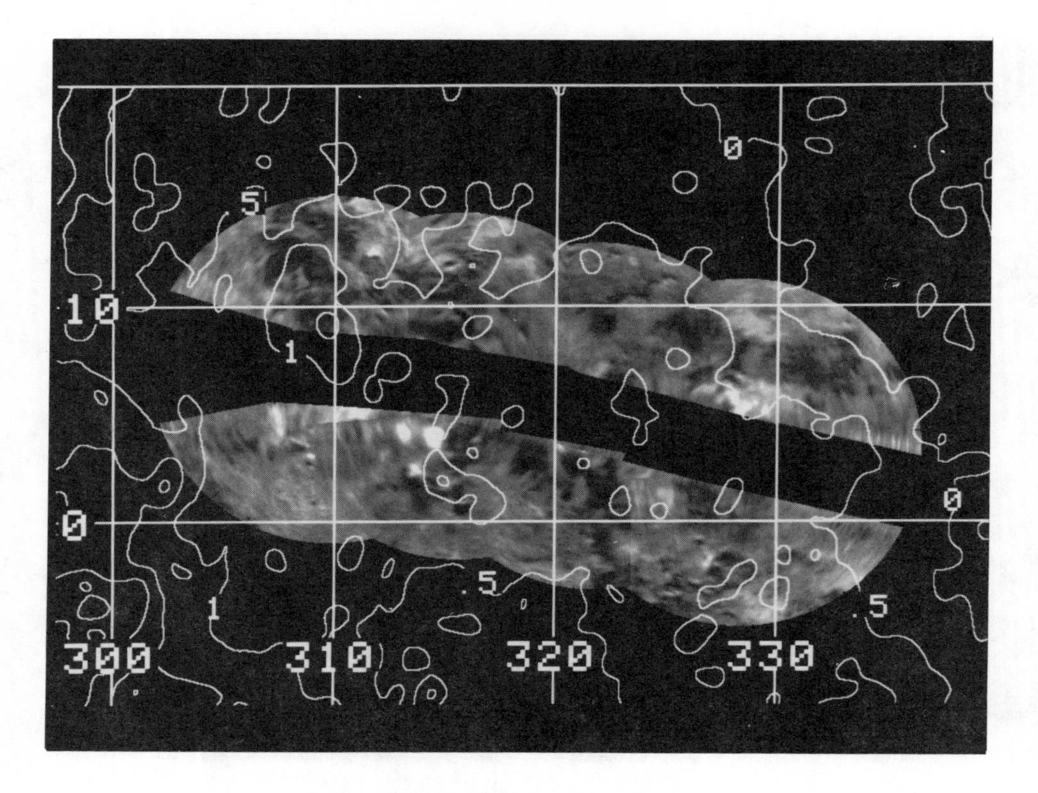

Fig. 10. Mosaic of 1978 Goldstone backscatter images with Pioneer-Venus elevation contours.

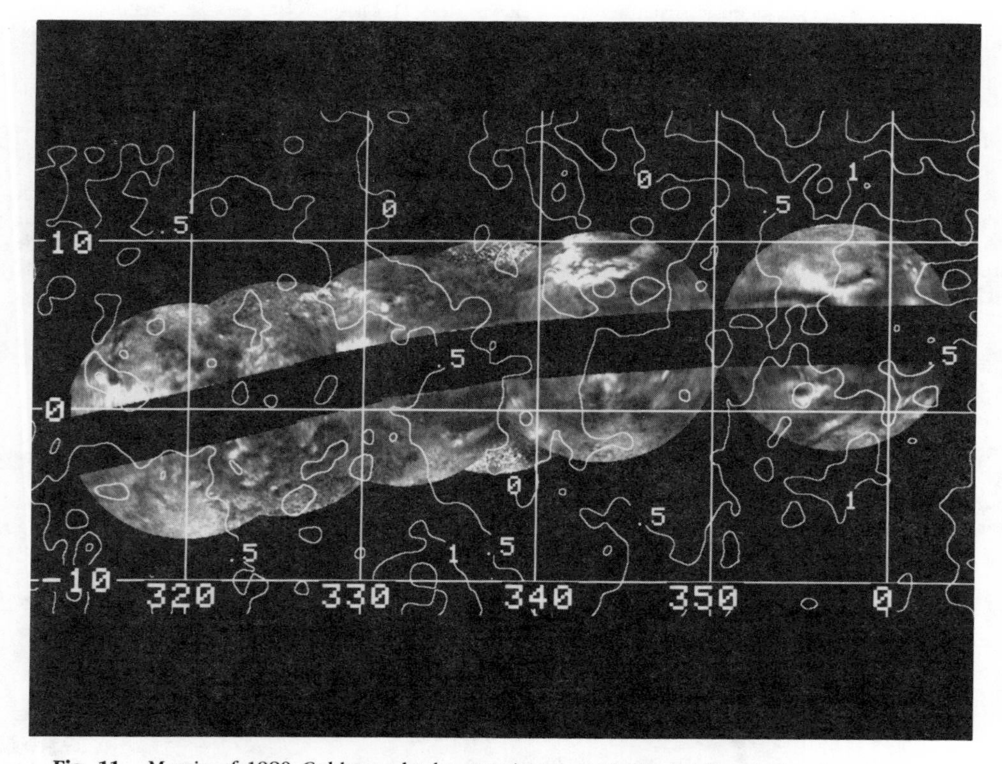

Fig. 11. Mosaic of 1980 Goldstone backscatter images with Pioneer-Venus elevation contours. Note the group of hills south of the equator between 320° to 330°E, the extended bright area just south of 10°N between 340° to 350°E, and the circular ridge system in the easternmost frame. Incidence angle range for extended bright feature is 4° to 7°. Heng-O Chasma is located at about 6°N, 352° to 356°E.

volcano-tectonic depressions (coronae, arachnoids), lava flows, and groups of volcanic constructs. Our analyses of features seen in the Goldstone data suggest that Navka Planitia, southern Guinevere Planitia, and Tinatin Planitia are also volcanic plains, replete with circular features that appear to be coronae and arachnoids, lava flows, and hills that we interpret to be volcanoes (Fig. 1). Based on visual comparisons of Venera and Goldstone data we also find that the areal distribution of features is similar in the two data sets. For example, Fig. 14 is a Goldstone/Venera comparison of groups of hills that are interpreted to be volcanic constructs, and Fig. 15 is a comparison of sizes and geometries of what are interpreted to be lava flows.

The conclusion that volcanic plains dominate the region covered by Goldstone data is reinforced by consideration of Pioneer-Venus SAR data [Fig. 2 and work by *Senske and Head* (1989)]. These data and estimates of Fresnel reflection coefficients and Hagfors C parameters derived by *Pettengill et al.,* (1988) from Pioneer-Venus altimetry data show that the plains covered by Goldstone observations have lower cross-sections, lower reflection coefficients, and are smoother than uplifted areas such as Aphrodite Terra, Beta Regio, Devana Chasma, and Phoebe Regio. Specifically, Navka, Guinevere, and Tinatin Planitiae have reflection coefficients values (corrected for diffuse scattering) of 0.1 to 0.2 and rms slopes of 1° to 5° as derived from Pioneer-Venus data, values that are typical for regions interpreted from Venera data to be volcanic plains (*Bindschadler and Head,* 1988).

Finally, our analysis of the Pioneer-Venus SAR image (Fig. 2) shows that northwest- and northeast-trending structures coincide with the northwest-southeast grid-like structure evident in the altimetry maps generated from the Pioneer-Venus data. This pattern implies control by a regional-scale tectonic framework. We now consider in more detail each area covered by Goldstone observations, beginning with Devana Chasma and Phoebe Regio regions. We focus on evidence for volcanic and tectonic processes that have operated in regions covered by Goldstone and return later to a discussion of the origin of the northwest-northeast regional structural pattern evident in the Pioneer-Venus data.

DEVANA CHASMA AND PHOEBE REGIO REGIONS

Saunders and Malin (1976) and *Malin and Saunders* (1977) utilize 1972 to 1975 Goldstone data to describe a 150-km-wide, 2-km-deep, 1400-km-long rift valley that bifurcates to the south and becomes a series of coalescing craters to the north. The data are shown in Fig. 6. Comparison of Figs. 1-3 shows that the rift valley is broadly aligned with topography associated with Phoebe Regio and extends to the north to become Devana Chasma. Further, the Pioneer-Venus SAR and elevation data show that Devana Chasma is offset by a northwest-southeast structure that is located between approximately 10°N, 283°E and 5°N, 290°E (Fig. 2). This northwest-southeast feature also appears in high resolution 1988 Arecibo data (D. Campbell, personal communication, 1989) that clearly show that the faults within the rift also extend in a north-south direction across the nothwest-southeast trending structure. This implies that at least some rifting events occurred after formation of the NW-SE structure. In fact, our detailed mapping, as well as work of *Stofan et al.* (1989), show that rifts associated with the Beta Regio uplift to the north also cross-cut many of the local northwest-southeast-trending features (see Fig. 1).

Figure 6 shows the low returns associated with Goldstone observations of the rift valley and of a suite of hills that are found to the west and east of the rift valley. The low returns from the valley floor and hills indicate that the surfaces are rough at the length-scales (>radar wavelength) characteristic of quasispecular scattering for the low incidence angle Goldstone data (e.g., Fig. 5). This conclusion is also consistent with relatively high roughness at quasispecular length-scales for these areas derived from analyses of scattering of the Pioneer-Venus altimetry data (*Pettengill et al.,* 1988). Note that the contrast between the rift valley and hills as opposed to surrounding plains is reversed in the Pioneer-Venus SAR image (Fig. 2). That is, the valley floor and hills are bright as compared to surrounding plains. The SAR image is presented with data normalized to a 45° incidence angle. At these angles, diffuse scattering dominates the radar return. Diffuse scattering is controlled by radar wavelength scale roughness elements. Thus, the rift valley and hills are interpreted to be rough at both long and short length-scales. The higher degree of roughness on the rift valley floor may be due to relatively recent fracturing and volcanism. Likewise, the enhanced roughness associated with the hills implies that such exogenic processes as mass wasting, chemical corrosion, and wind action have not yet had a chance to remove the distinctive signatures associated with volcanism.

Figure 6 also shows a number of regions to the east of the rift valley that have relatively high cross-sections as compared to surrounding plains. For example, the bright feature located at 5°N, 290°E is associated with a circular region that rises ~1 km above surrounding areas. On the other hand, these bright features are not much different than the regional cross-sections seen in the Pioneer-Venus SAR image. The high cross-section values at low incidence angles (i.e., Goldstone), combined with the lack of a distinct signature (relative to surrounding) at the higher angle Pioneer-Venus SAR observations, suggest that material covering these areas is relatively smooth at quasispecular scattering length-scales. The association with an equidimensional hill suggests that the materials may have a volcanic origin, i.e., the topographic feature may be a volcano.

GUINEVERE AND NAVKA PLANITIAE REGIONS

The southern part of Guinevere Planitia, together with Navka Planitia, occupy the middle section of the broad longitudinal swath covered by the ensemble of Goldstone data (Figs. 1 and 3). In both Planitiae, regions of relatively low cross-section variation are interspersed with hills, groups of hills, brighter regions, circular features, and circular depressions. In addition, a chain of circular depressions trending west-northwest can be observed in Figs. 4 and 9.

There are several flow-like structures with relatively high cross-sections that are associated with circular features on the

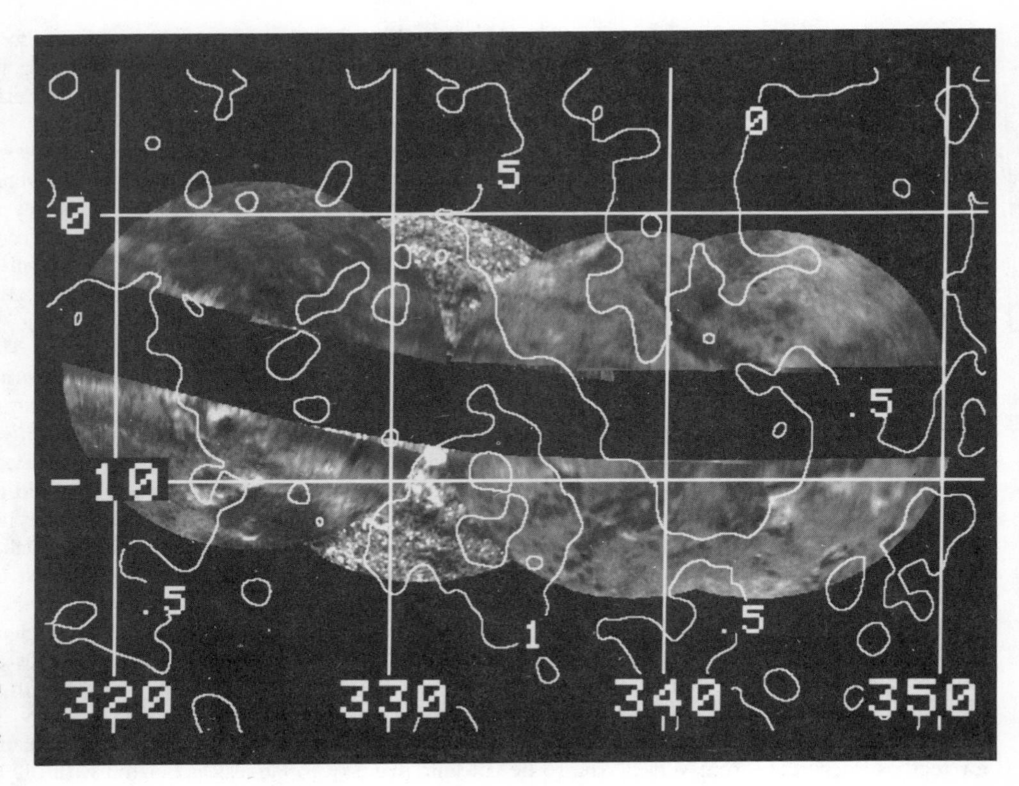

Fig. 12. Mosaic of 1982 Goldstone backscatter images with Pioneer-Venus elevation contours. Note the northwest-trending ridge at about 4°S, 340°E.

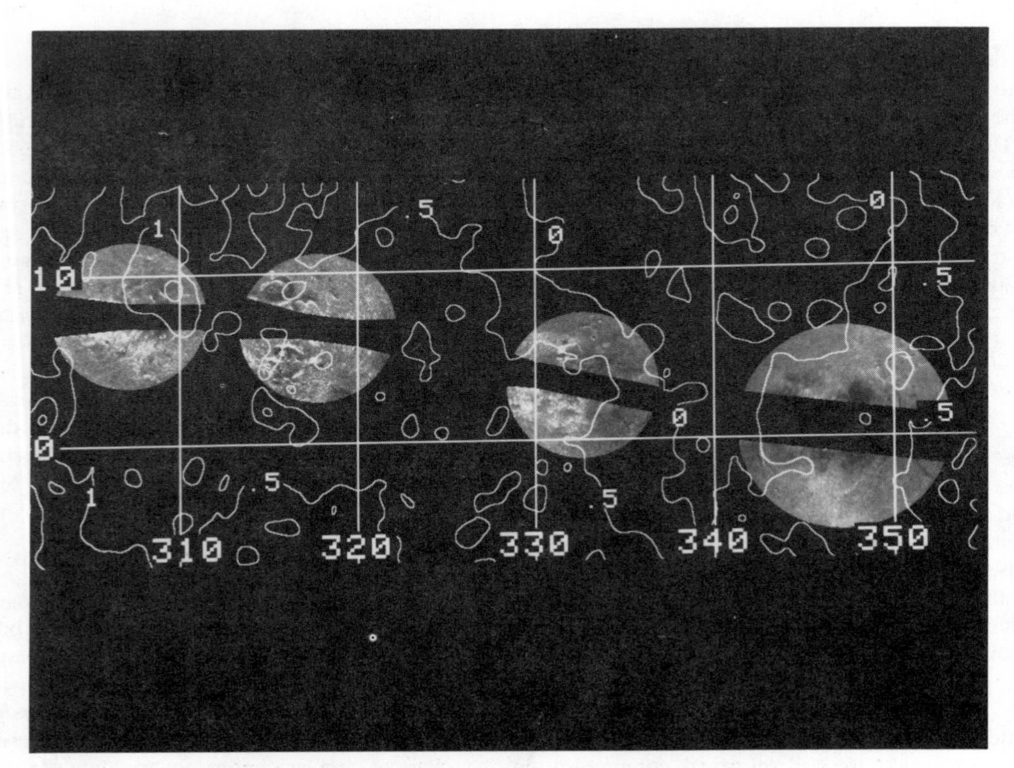

Fig. 13. Mosaic of 1986 Goldstone backscatter images with Pioneer-Venus elevation contours. Two circular features can be seen in the second from left frame. Some of the areas with higher cross-sections suggest to *Jurgens et al.* (1988a,b) that materials with high dielectric constants are exposed.

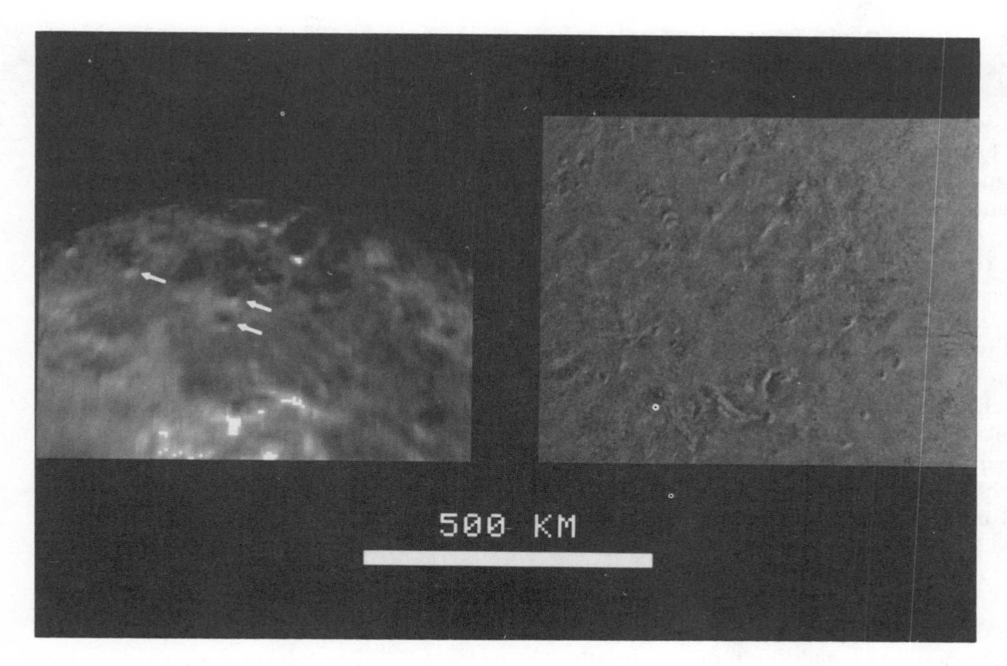

Fig. 14. Portion of Goldstone backscatter data acquired in 1980 (left) and portion of Venera 15 and 16 polar stereographic mosaic (right). North is toward the top. Arrows on Goldstone image point to features interpreted to be equidimensional hills on the basis of bright returns toward the radar sub-Earth point and low returns away from the sub-Earth point. For the Venera image, radar illumination is from the east. A number of conical hills of similar size and spacing to those in the Goldstone data are seen. The features are interpreted to be volcanic constructs. Goldstone data centered at 6°N, 332°E; Venera data centered at 40°N, 325°E.

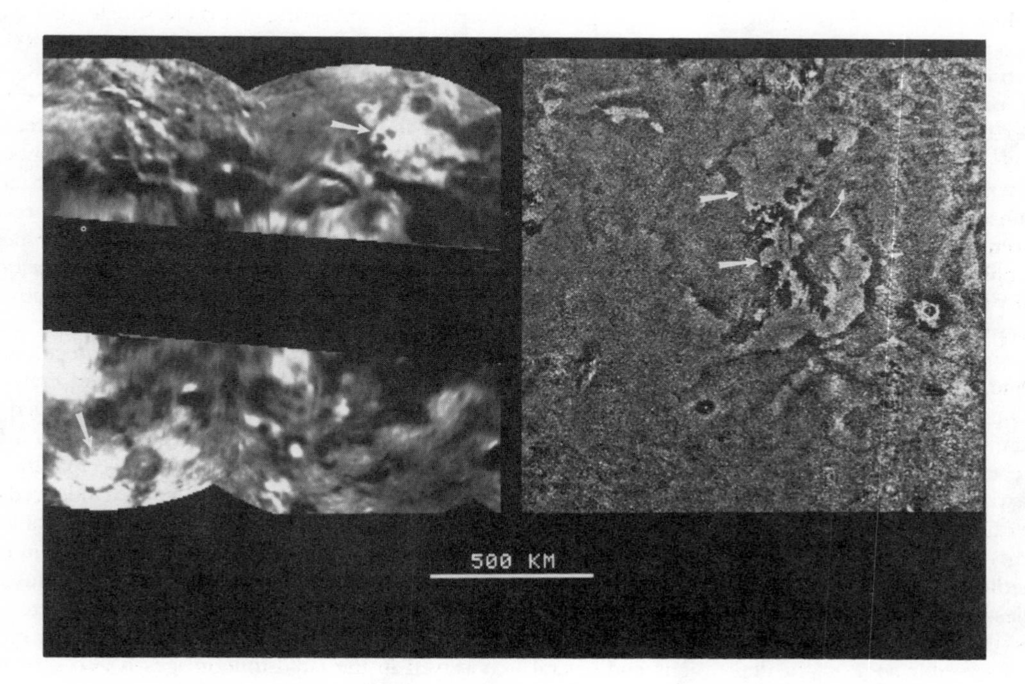

Fig. 15. Portion of 1978 Goldstone backscatter data (left) and portion of Venera 15 and 16 polar stereographic mosaic (right) in Sedna Planitia. Arrows on Goldstone image point to flow-like features of variable brightness associated with circular features. Arrows on Venera image outline region with flow structures of similar scales that are also discernable of Arecibo images of the same area. Goldstone data centered at 8°S, 308°E; Venera data centered at 52°N, 352°E.

plains to the southwest of Navka Planitia (Figs. 3, 4, and 9). Based on analysis of Pioneer-Venus altimetry (Fig. 2 and U.S. Geological Survey I-series charts), the flow-like features occur on a northwest-southeast-trending arm of the Phoebe Regio highlands. The bright circular feature associated with Ushas Mons is also aligned with this arm (Figs. 3 and 9). These areas are not discernable as distinct signatures in the Pioneer-Venus SAR data or in roughness or reflectivity data derived from Pioneer-Venus altimetry data. Arecibo data also show a set of northwest-southeast features with relatively high cross-sections in these areas (*Campbell and Burns,* 1980), although they do not correspond with the locations of bright, flow-like features seen in Goldstone data. Thus, the flow-like features are probably relatively smooth at the long length-scales associated with quasispecular scattering. We interpret these features to be lava flows. Support for this interpretation is also found in comparison of the plan-form shape and size of the flows with those identified in Venera and Arecibo observations in Sedna Planitia (Fig. 14).

Other regions that are bright in Goldstone data and that do not have discernably different signatures than surroundings in Pioneer-Venus SAR data are located throughout the midregion covered by Goldstone (Figs. 1 and 3). Prominent regions are located at 12°N, 333°E; 8°N, 343°E. The latter area has been analyzed using calibrated Goldstone data by *Jurgens et al.* (1988a,b), who suggest that the high cross-sections are due to high Fresnel reflection coefficients. Thus, some of these areas are not only relatively smooth, they also expose materials with high dielectric constants. The observation that materials of high dielectric constant are found in plains of moderate elevation implies that more is going on than a global weathering pattern whereby chemical reactions and subsequent wind erosion keep high dielectric materials preferentially exposed on the highlands (e.g., *Nozette and Lewis,* 1982). The Goldstone data suggest that materials with high reflectivity may be generated wherever there is ongoing tectonic and volcanic activity or relatively recent impact excavation. Subsequent weathering would then alter the materials and remove the high reflectivity signatures. Uplifted regions may simply have a greater areal concentration of these high reflectivity materials as compared to other regions because tectonism, volcanism, and/or appropriate weathering processes occur more rapidly in uplifted areas.

The Goldstone data also reveal 28 circular or quasicircular depressions, ranging from 25 to 1039 km in diameter, largely in the Guinevere and Navka Planitiae regions (Fig. 1; Table 2). The negative relief of the features was discerned either directly from the Goldstone altimetry data or indirectly from variations in backscatter caused by local slopes. Raised rims, evidenced by enhancements on the edge of features facing the sub-Earth point, are sometimes observed, although most circular features appear as rimless depressions. Interiors typically have lower cross-sections than surrounding areas. Knobby interior structures are observed in some of the depressions, and several display a single central peak. Evidence for deposits or structures surrounding the depressions is lacking in the Goldstone data. In fact, we were unable to unequivocally identify any of the circular features as impact structures. The

TABLE 2. Circular depressions observed in Goldstone backscatter images of Venus.

Center Latitude (deg.)	Center Longitude (deg. E)	Diameter (km)
2.3	354.4	1039
-3.3	323.1	249
-23.7	321.9	236
11.4	315.0	232
4.7	316.0	207
-1.4	327.9	182
-10.2	323.7	166
6.0	326.4	165
-11.2	339.2	152
8.9	315.3	150
11.1	312.3	127
4.8	310.7	114
-2.1	327.0	106
-11.7	324.9	93
5.2	315.7	93
11.9	311.5	89
-2.4	325.2	88
10.6	5.9	87
-11.2	338.0	80
7.6	265.2	76
6.0	273.0	72
8.0	262.8	66
7.0	266.5	46
12.6	296.5	46
-2.0	340.9	34
-1.5	14.1	34
6.0	333.3	25

lack of unequivocal impact craters in Goldstone data is not surprising, considering that there should be, on average, about one impact crater larger than 100 km (about 1° in latitude) in diameter, if the impact crater size frequency distribution is similar to that observed for plains covered by Venera 15 and 16 data. In our opinion, the majority of the circular depressions observed in the Goldstone data are of volcanotectonic origin and are probably analogous to the coronae and arachnoids seen in Venera data. For example, the feature at 1°S 328°E (Fig. 7) consists of a linked pair of circular depressions and a pair of flanking circular hills. This association suggests a complex volcanotectonic depression with subsidiary volcanic constructs. The crater chain shown in Fig. 9 at approximately 5°S, 307°E is also suggestive of a volcanotectonic origin.

Figure 16 shows the postulated impact crater size distribution for northern hemisphere plains as derived by analysis of Venera 15 and 16 radar data (*Basilevsky et al,* 1987). The abundance is relatively low and suggests that the average crater retention age is measured in several hundred million years (*Schaber et al.,* 1987). This value is comparable to the crater retention age of Earth's cratons and implies, in contrast to the Moon, Mercury, or even Mars, that the Venusian plains have been actively resurfaced even in relatively recent geological time. The size frequency distribution of the circular features observed in the Goldstone images is also shown in Fig. 16 to be intermediate between that of unequivocal impacts and all circular features seen by Venera 15 and 16 in the northern plains. The lower abundance of circular features in the Goldstone data may be indicative of genuine differences in

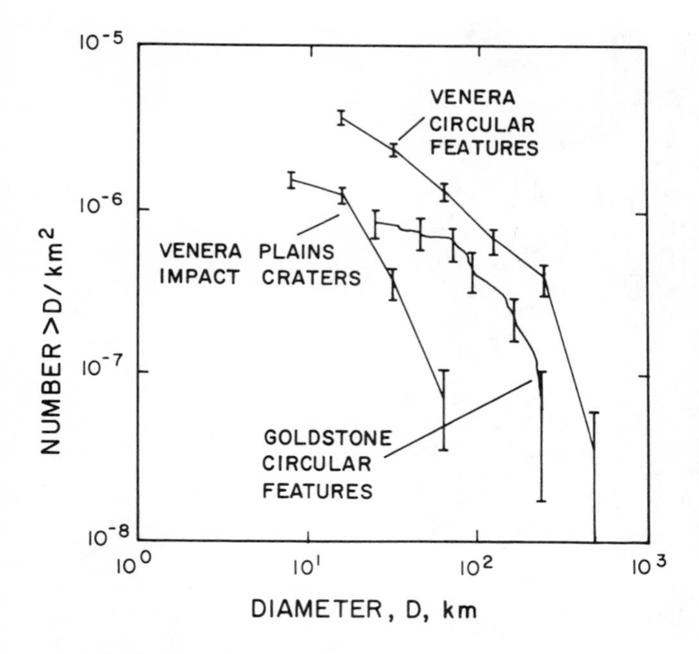

Fig. 16. Cumulative size frequency distribution of circular features inventoried from Goldstone data (Table 2), compared to population of impact craters found on plains and all circular features inventoried from Venera 15 and 16 data (*Basilevsky et al.,* 1987).

character between the equatorial and northern plains, or it may be an artifact of the lower resolution and peculiar observing geometry of the Goldstone observations.

HENG-O CHASMA, EISILA REGIO, AND TINATIN PLANITIA REGIONS

Heng-O Chasma is an arcuate feature located at approximately 7°N, 355°E. It appears as paired bright and dark bands that suggest the presence of a ridge in the Goldstone data shown in Fig. 11. Figure 17 is a mosaic of Goldstone backscatter images covering the Heng-O Chasma area and immediate surroundings. A discontinuous, ridge-like circular structure of approximately 1000-km diameter can be discerned, with Heng-O Chasma occupying the northern perimeter. We term the overall circular structure that includes Heng-O Chasma the Heng-O structure. Pioneer-Venus altimetry along-track (i.e., not gridded) data show an overall north-facing slope to the structure, with relief of ~1 km, but are not detailed enough to portray topography directly associated with the overall structure. The Heng-O structure is similar in some respects to coronae observed by Venera 15 and 16 (see *Barsukov et al.,* 1986). For example, patches of materials with relatively low and high cross-sections occur in the interior and along the northeast rim of the structure; these patches are similar in appearance to what are interpreted to be volcanic deposits and structures commonly associated with coronae observed in Venera 15 and 16 data. However, the singular, relatively narrow form of the circular ridge system is unusual among coronae and the 1000-km diameter is larger than coronae observed in Venera 15 and 16 data. Given the resolution limitations of the Goldstone data, a variety of possibilities must be entertained for the origin of the Heng-O structure, including viscous relaxation, erosion, and burial of a multiringed impact basin.

The easternmost Goldstone coverage includes Tinatin Planitia and parts of a group of northwest-southeast-trending hills that extend to the northwest to include Eisila Regio, Gula Mons, and Sappho Patera. The Goldstone data that cover the southwestern portion of Eisila Regio region show clusters of hills and regions of variable backscatter (Fig. 8). A 2000-km-long, 300-km-wide set of elongate hills straddles the equator and trends east-west (Figs. 3 and 8). Goldstone altimetry data suggest that relief approaches 2 km for the larger hills. These features can be discerned quite readily because the strike direction is perpendicular to the radar look direction. We interpret this system of hills to be the equivalent of the ridge belt systems seen in Venera 15 and 16 data and they are mapped as such in Fig. 1.

GEOLOGICAL EVOLUTION

In this section we discuss selected aspects of the geological evolution of the area covered by Goldstone data. If we assume that the impact crater population is comparable to that found in the northern plains, then crater retention ages would be measured in hundreds of millions of years and a resurfacing rate of less than 2 to 4 km/Ga can be invoked (*Grimm and Solomon,* 1987; *Arvidson and Plaut,* 1988). The evidence for volcanic activity presented in previous sections implies that plains-style volcanism has been responsible for at least part of the resurfacing, although the relative importance of viscous relaxation and exogenic processes (weathering, aeolian erosion, and deposition) in removing craters remains unknown. For example, *Fegley and Prinn* (1989) estimate that the rate of anhydrite formation on Venus is 1 km/Ga, which would account for a significant fraction of the resurfacing rate quoted above. We suggest that volcanism occurs both in association with local volcanic constructs, volcanotectonic depressions, and from linear vents. Deciphering stratigraphic relations is difficult, although some of the younger flows may correspond to the radar-bright regions discussed in a previous section.

The ensemble of available data also imply that a regional structure, symmetrical about the equator, developed on Venus before at least some of the rifting associated with Beta Regio, Devana Chasma, and Phoebe Regio. We note that *Schaber* (1982) also described a global set of linears in the Pioneer-Venus data that trend northwest-southeast and northeast-southwest and that are symmetrical about the equator. A possible scenario for this regional structure is tidal despinning. *Melosh* (1977) showed that bodies with thin lithospheres, upon slowing down from a faster spin state, would develop strike-slip faults that are symmetrical about the equator and that extend to about 48° in latitude. A gradual despinning can generate stresses in the Venusian lithosphere sufficient to cause recurrent failure in the upper few kilometers (R. Grimm, personal communication, 1989). However, high-latitude east-

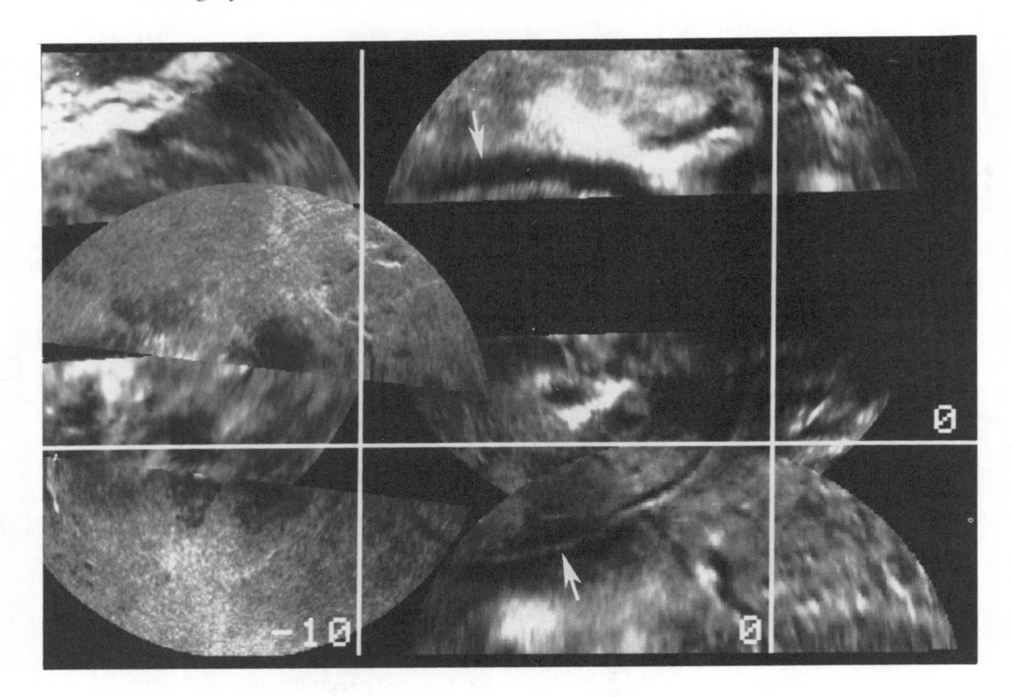

Fig. 17. Goldstone mosaic of Heng-O Chasma circular ridge system. Arrows denote ridges.

west-trending extensional structures are also predicted in such a model and have not been observed in either Venera 15 and 16 or Arecibo data. Also, this model does not explain why northwest-southeast structures in the equatorial region dominate over northeast-southwest-trending features. We note that the gridded structure and topography in Aphrodite Terra have been interpreted by *Head and Crumpler* (1987) and *Crumpler and Head* (1988) as due to the presence of a divergent plate boundary. Divergent plate boundaries must end along strike in a triple junction. Yet, the Goldstone and Pioneer-Venus data show that the same northwest-southeast, northeast-southwest grid structure exists in both Aphrodite Terra and the plains to the west. No evidence can be found for a triple junction in the Goldstone data. In summary, we feel that the origin of this regional scale northwest-southeast, northeast-southwest structure remains an open issue. Whatever caused the uplift in Aphrodite may simply have reactivated that structure in a manner similar to the way features may have formed along the northwest-southeast arm of Phoebe Regio (Fig. 2) and in Eislila Regio.

CONCLUSIONS

The ensemble of 41 backscatter images of Venus acquired by the S-band (12.5 cm) Goldstone radar system covers approximately 35 million square kilometers and includes the equatorial portion of Guinevere Planitia, parts of Devana Chasma and Phoebe Regio, Navka Planitia, Heng-O Chasma, a portion of Eisila Regio, and Tinatin Planitia. The images and associated altimetry data combine relatively high spatial resolution (best data have ~1 km pixels) with small incidence angles (less than 10°) for regions not covered by either Venera Orbiter or Arecibo radar data. Systematic photogeological analyses of the Goldstone data were done using Pioneer-Venus data (17-cm wavelength) to provide a regional context, and Venera data (8-cm wavelength) to provide information on the geology to the north of regions covered by Goldstone data. Our analyses show that:

1. Volcanic plains dominate the area covered by Goldstone data, including groups of hills interpreted to be volcanic constructs, lava flows between Devana Chasma and Ushas Mons, and circular depressions that are analogous to the coronae and arachnoids seen in Venera 15 and 16 data. The areal distribution and appearance of features is similar to that found in Venera radar images in northern Guinevere Planitia and in Sedna Planitia.

2. A 150-km-wide, 2-km-deep, 1400-km-long rift valley is located on the western flank of Phoebe Regio and extends northward as Devana Chasma. The floor has lower backscatter cross-sections than surrounding plains in the Goldstone data and higher values than plains in Pioneer-Venus SAR data. These trends imply that the floor is rougher than plains at both length-scales (>radar wavelength) characteristic of quasispecular scattering and length-scales (~radar wavelength) characteristic of the diffuse scattering. Groups of hills adjacent to the rift system show the same trends. Relatively recent rifting, volcanism, and perhaps mass wasting are probably responsible for the enhanced roughness.

3. A discontinuous, circular ridge system with a diameter of approximately 1000 km may be the largest corona yet identified on Venus. The northern section corresponds to Heng-O Chasma.

4. The Goldstone observing geometry makes identification of long, linear structures difficult. However, a ridge belt can be discerned that straddles the equator and trends east-west from longitudes 0° to at least 30°E. Other ridge and scarp-like features observed in the Goldstone data mainly trend northwest-southeast or northeast-southwest, as do numerous topographic elements observed in Venera 15 and 16 data to the

north. These are directions similar to those discerned in Pioneer-Venus topography throughout the equatorial region, suggesting regional-scale deformation. The structures seem to predate rifting associated with Beta Regio, Devana Chasma, and Phoebe Regio.

5. The cumulative size frequency distribution of circular features lies between the impact crater distribution defined from Venera and Arecibo data for plains and the population of all circular features observed in Venera data. Most likely, terrain imaged by Goldstone has typical impact crater retention ages, measured in hundreds of millions of years. The resurfacing rate is probably no greater than 4 km/Ga. Resurfacing is probably dominated by volcanic processes, although data are not of sufficiently high resolution to evaluate the efficacy of weathering or aeolian erosion and deposition.

6. Calibrated versions of Goldstone images show regions with relatively high radar cross-sections that are not associated with obvious topography and that retain high cross-sections over a range of incidence angles. These areas probably expose materials with high dielectric constants, implying that processes that resurface Venus produce high dielectric materials in low plains in addition to highlands.

Acknowledgments. One of the authors (R.E.A.) was supported by NASA Planetary Geology and Geophysics Program Grant NSG-7087 to Washington University; J.J.P. was supported by the NASA Graduate Fellowship Program.

REFERENCES

Arvidson R. E. and Plaut J. J. (1988) On the rate of resurfacing on Venus. *Eos Trans. AGU, 69,* 1294.

Arvidson R. E., Plaut J. J., Jurgens R. F., Saunders R. S., and Slade M. A. (1989) Geology of Southern Guinevere Planitia, Venus, based on analyses of Goldstone radar data (abstract). In *Lunar and Planetary Science XX,* p. 25. Lunar and Planetary Institute, Houston.

Arvidson R. E., Schulte M., Kwok R., Curlander J., Elachi C., Ford J. P., and Saunders R. S. (1988) Construction and analysis of simulated Venera and Magellan images of Venus. *Icarus, 75,* 163-181.

Barsukov V. L., Basilevsky A. T., Burba G. A., Bobina N. N., Kryuchkov V. P., Kuzmin R. O., Nikolaeva O. V., Pronin A. A, Ronca L. B, Chernaya I. M., Sashkina V. P., Garanin A. V., Kushky E. R., Markov M. S., Sukhanov A. L., Kotelnikov V. A., Rzhiga O. N., Petrov G. M., Alexandrov Yu. N., Sidorenko A. I., Bogomolov A. F., Skrypnik G. I., Bergman M. Yu., Kudrin L. V., Bokshtein I. M., Kronrod M. A., Chochia P. A., Tyuflin Yu. S., Kadnichansky S. A., and Akim E. L. (1986) The geology and geomorphology of the Venus surface as revealed by the radar images obtained by Veneras 15 and 16. *Proc. Lunar Planet. Sci. Conf. 17th,* in *J. Geophys. Res., 91,* D378-D398.

Basilevsky A. T., Ivanov B. A., Burga G. A., Chernaya I. M., Kryuchkov V. P., Nikolaeva O. V., Campbell D. B., and Ronca L. B. (1987) Impact craters of Venus: A continuation of the analysis of data from the Venera 15 and 16 spacecraft. *J. Geophys. Res., 92,* 12,869-12,901.

Bindschadler D. L. and Head J. W. (1988) Characterization of Venera 15/16 geologic units from Pioneer-Venus reflectivity and roughness data. *Icarus, 77,* 3-20.

Campbell D. B. and Burns B. A. (1980) Earth-based radar imagery of Venus. *J. Geophys. Res., 85,* 8271-8281.

Campbell D. B., Head J. W., Harmon J. H., and Hine A. A. (1984) Venus: Volcanism and rift formation in Beta Regio. *Science, 226,* 167-170.

Clark P. E., Jurgens R. F., and Kobrick M. (1986) Venus: Images of rolling hills terrane from tristatic S-Band radar (abstract). In *Lunar and Planetary Science XVII,* pp. 133-134. Lunar and Planetary Institute, Houston.

Crumpler L. S. and Head J. W. (1988) Bilateral topographic symmetry patterns across Aphrodite Terra, Venus. *J. Geophys. Res., 93,* 301-312.

Fegley B. and Prinn R.G. (1989) Estimates of the rate of volcanism on Venus from reaction rate measurements. *Nature, 337,* 55-57.

Goldstein R. M., Green R. R., and Rumsey H. C. (1978) Venus radar brightness and altimetry images. *Icarus, 36,* 334-352.

Grimm R. E. and Solomon S. C. (1987) Limits on modes of lithospheric heat transport of Venus from impact crater density. *Geophys. Res. Lett., 14,* 538-541.

Hagfors T. (1964) Backscattering from an undulating surface with applications to radar returns from the Moon. *J. Geophys. Res., 69,* 3779-3784.

Head J. W. and Crumpler L. S. (1987) Evidence for divergent plate boundary characteristics and crustal spreading: Aphrodite Terra. *Science, 238,* 1380-1385.

Jurgens R. F., Goldstein R. M., Rumsey H. C., and Green R. R. (1980) Images of Venus by three-station radar interferometry—1977 results. *J. Geophys. Res., 85,* 8282-8294.

Jurgens R. F., Slade M. A., and Saunders R. S. (1988a) Evidence for highly reflecting materials on the surface and subsurface of Venus. *Science, 240,* 1021-1023.

Jurgens R. F., Slade M. A., Robinett L., Brokl S., Downs G. S., Frank C., Morris G. A., Farazian K. H., and Chan F. P. (1988b) High resolution images of Venus from ground-based radar. *Geophys. Res. Lett., 15,* 577-580.

Jurgens R. F., Arvidson R. E., Plaut J. J., Saunders R.S., and Slade M. A. (1988c) New ground-based radar images of Venus (abstract). In *Lunar and Planetary Science XIX,* pp. 575-576. Lunar and Planetary Institute, Houston.

Malin M. C. and Saunders R. S. (1977) Surface of Venus: Evidence of diverse landforms from radar observations. *Science, 196,* 987-990.

Melosh H. J. (1977) Global tectonics of a despun planet. *Icarus, 31,* 221-243.

Nozette S. and Lewis J. S. (1982) Venus: Chemical weathering of igneous rocks and buffering of atmospheric composition. *Science, 216,* 181-183.

Pettengill G. H., Eliason E., Ford P. G., Loriot G. B., Masursky H., and McGill G. E. (1980) Pioneer Venus radar results: Altimetry and surface properties. *J. Geophys. Res., 85,* 8261-8270.

Pettengill G. H., Ford P. G., and Chapman B. D. (1988) Venus: Surface electromagnetic properties. *J. Geophys. Res., 92,* 14,881-14,892.

Rumsey H. C., Morris G. A., Green R. R., and Goldstein R. M. (1974) A radar brightness and altitude image of a portion of Venus. *Icarus, 23,* 1-7.

Saunders R. S. and Malin M. (1976) Venus: Geologic analysis of radar images. *Geol. Rom., 15,* 507-515.

Saunders R. S. and Malin M. C. (1977) Geologic interpretation of new observations of the surface of Venus. *Geophys. Res. Lett., 4,* 547-549.

Schaber G. G. (1982) Venus: Limited extension and volcanism along zones of lithospheric weakness. *Geophys. Res. Lett., 9,* 499-502.

Schaber G. G., Shoemaker E. M., and Kozak R. C. (1987) The surface age of Venus: Use of the terrestrial cratering record. *Solar Sys. Res., 21,* 89-94.

Senske D. A. and Head J. W. (1989) Synthesis of Venus equatorial geology: Variations in styles of tectonism and volcanism and comparison with northern high latitudes (abstract). In *Lunar and Planetary Science XX,* pp. 984-985. Lunar and Planetary Institute, Houston.

Stofan E. R., Head J. W., Campbell D. B., Zisk S. H., Bogomolov A. F., Rzhiga O. N., Basilevsky A. T., and Armand N. (1989) Geology of a rift zone on Venus: Beta Regio and Devana Chasma. *Geol. Soc. Am. Bull., 101,* 143-156.

Stooke P. J. (1988) Venus: Geologic mapping from Goldstone images (abstract). In *Lunar and Planetary Science XIX,* pp. 1135-1136. Lunar and Planetary Institute, Houston.

Proceedings of the 20th Lunar and Planetary Science Conference, pp. 573-584
Lunar and Planetary Institute, Houston, 1990

Incidence Angle and Resolution: Potential Effects on Interpreting Venusian Impact Craters in Magellan Radar Images

J. P. Ford

Jet Propulsion Laboratory, California Institute of Technology, Pasadena, CA 91109

The effects of incidence angle and spatial resolution were analyzed to evaluate the types of information that Magellan imaging radar may provide about impact craters on Venus. Seasat SAR images of small terrestrial craters at high resolution (~25 m) and Soviet Venera images of comparatively large venusian craters at low resolution (~1000 to 2000 m) and shorter wavelength show comparable radar responses to crater morphology. At low incidence angles ($\lesssim$15°), it is difficult to locate the precise position of crater rims on images. Abrupt contrasts in radar response to changing slope (hence incidence angle) across a crater produce sharp tonal boundaries normal to the illumination. Radially symmetrical crater morphology appears on images to be bilaterally symmetrical parallel to the illumination vector. Craters are compressed in the distal sector and drawn out in the proximal sector. Only the medial sector may yield an accurate measure of the diameter. At higher incidence angles ($\gtrsim$35°), SIR-A images show less distortion of crater morphology. Radar-bright halos that surround some craters imaged by SIR-A and Venera 15 and 16 denote the small-scale surface roughness of the ejecta blankets. Halos that are bilaterally symmetrical show reduced brightness from the foreslopes to the backslopes that probably denotes contrasting incidence angle effects. Similarities in the radar responses of small terrestrial impact craters and volcanic craters of comparable dimensions (~2-km diameter) emphasize the need to distinguish the geologic associations of craters in venusian images so as to discriminate impact from volcanic origins. Magellan will obtain images through a range of incidence angles that varies from about 45° at low latitudes to 16° at polar latitudes. Radially symmetrical landforms such as impact craters are expected to show outlines that vary with the incidence angle and to appear bilaterally symmetrical and increasingly compressed at higher latitudes. The breakpoints for small-scale surface roughness will range from small-pebble size (definitely smooth) to very-coarse-pebble size (definitely rough). The smooth and rough breakpoints will vary 20% to 30% from the low to the high latitudes. The comparatively high spatial resolution of Magellan radar should enable the discrimination of small craters with diameters below the limits of reliable Venera resolution (~8 km) if they exist.

INTRODUCTION

Investigators have speculated on the presence of impact features on Venus (*Saunders and Malin,* 1976). Earth-based radar observations show features on Venus that are morphologically akin to impact craters (*Burns and Campbell,* 1985). Radar images obtained by Soviet Venera missions 15 and 16 cover about 25% of the planetary surface and display a large number of craters, many of which resemble impact craters on Mercury, Moon, Mars, and other planetary bodies. However, the resolution of Venera images limits perception to craters with a diameter greater than about 8 km.

The Magellan orbiter is scheduled to arrive at Venus in August 1990. It includes an imaging radar that will cover 70% to 90% of the venusian surface at a resolution an order of magnitude higher than that of the Venera missions (*Dallas and Nickle,* 1987). Thus, Magellan will provide extensive regional coverage and a more detailed view of the surface. Closer study of the crater origins will be possible and smaller craters may be observed. However, the interpretation of craters and structures of impact origin in synthetic-aperture radar (SAR) images depends not only on resolution but also on contrasts in surface reflectivity and such imaging characteristics as the illumination geometry and wavelength. To understand the types of information about impact craters on Venus that Magellan radar is expected to provide, it is instructive to review existing spaceborne radar image observations of impact features on Earth where there is good groundtruth data and compare them with impact features on the Venera images.

RADAR IMAGING CHARACTERISTICS

The importance of large-scale slope, small-scale surface roughness, and incidence angle in interpreting crater morphology and structure from radar images is examined below by reference to terrestrial examples obtained by Seasat SAR and SIR-A. The observations are compared with images obtained by Venera 15 and 16 and discussed in the context of the Magellan imaging radar. The essential radar parameters of these imaging systems are shown in Table 1. Seasat SAR and SIR-A were operated at a comparatively long radar wavelength (23.5 cm) and high spatial resolution (25 to 40 m), with antenna geometries that produced different nominal incidence

TABLE 1. Spaceborne imaging radar system parameters.

Radar System	Nominal Incidence Angle (deg.)	Wavelength (cm)	Spatial Resolution (m)
Seasat SAR	23 ± 3	23.5	25
SIR-A	50 ± 3	23.5	40
Venera 15,16	11-13 ± 1.5	8.0	1000-2000
Magellan	16-45	12.6	120-300

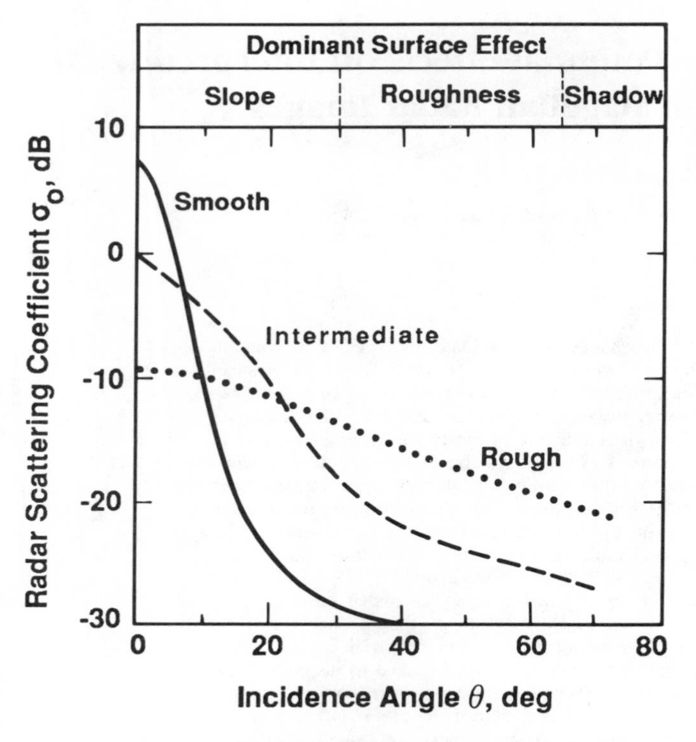

Fig. 1. Variation of radar backscatter with incidence angle for surfaces from rough, through intermediate, to smooth, showing the ranges of dominant surface effects.

angles. The Venera system obtained images at a shorter wavelength (8 cm) and lower resolution (1000 to 2000 m), with a very low nominal incidence angle. Magellan will operate at an intermediate wavelength (12.6 cm) and spatial resolution (120 to 300 m). However, the nominal incidence angle of Magellan will vary significantly with latitude on Venus (*Johnson,* 1989).

For each system, the specified incidence angle (Table 1) is nominal because it is valid only for a level horizontal surface. On sloping surfaces, the incidence angle decreases at foreslopes or increases at backslopes by the amount of the slope in the direction of the illumination vector. Thus, foreslopes are negative and backslopes are positive relative to a nominal incidence angle.

At low incidence angles, up to about 30°, backscatter is dominated by large-scale surface slope, on the order of meters. Small changes of slope, and hence incidence angle, produce relatively large changes of backscatter. This yields strongly contrasting gray levels on images. At incidence angles from about 30° to 60° backscatter is dominated mostly by the small-scale roughness of a surface, on the order of the radar wavelength. At this scale smooth surfaces produce specular reflection and are dark on images; rough surfaces produce diffuse scattering and are bright on images. Intermediate surfaces produce both types of scattering and are gray on images. Backscatter curves for rough, intermediate, and smooth surfaces (Fig. 1) show that in the range of incidence angles from 30° to 60° changes in surface roughness have a greater effect on backscatter intensity than changes in incidence angle. At incidence angles greater than about 60° backscatter is weak and surfaces are dark to very dark on images.

A breakpoint between smooth and rough surfaces (and the corresponding dominant radar scattering mechanisms) is loosely implied by the Rayleigh criterion. According to this criterion a surface is considered smooth if the root-mean-square (rms) height of the microrelief is less than ⅛ of the radar wavelength divided by the cosine of the incidence angle. This is given by

$$h < \lambda/8 \cos\theta \qquad (1)$$

where h is the rms height, λ is the wavelength, and θ is the incidence angle. Because this criterion does not consider an intermediate category of surfaces between definitely smooth and definitely rough, it was modified by *Peake and Oliver* (1971) to include factors that define the upper and lower values of rms surface smoothness or roughness. The modified Rayleigh criterion considers a surface smooth where

$$h < \lambda/25 \cos\theta \qquad (2)$$

and rough where

$$h > \lambda/4.4 \cos\theta \qquad (3)$$

Field measurements of different types of surfaces have led researchers to experiment with a variety of different modifiers [e.g., *Schaber et al.,* 1976 (geologic surfaces); *Ulaby et al.,* 1982 (agricultural surfaces)]. This illustrates one of the difficulties associated with modeling the radar behavior of natural surfaces. Regardless of the modifying factor that is used, it follows from the equations that at any incidence angle a given surface is rougher as the wavelength decreases and, independent of wavelength, a given surface is smoother as the incidence angle increases.

Using the modified Rayleigh criterion for example, Table 2 shows that there is about 29% increase in rms surface smoothness and 25% decrease in rms surface roughness through the range of Magellan nominal incidence angles from about 45° to 16°. This has implications for the interpretation of Magellan radar images that are discussed further under the section entitled Radar Halos.

TABLE 2. Breakpoints in rms surface smoothness and roughness for spaceborne SAR systems through the range of Magellan nominal incidence angles (from modified Rayleigh criterion).

Incidence Angle (deg.)		Wavelength (cm) 23.5 (Seasat, SIR)	12.6 (Magellan)	8.0 (Venera)
16	Smooth	<1.0	<0.5	<0.3
	Rough	>5.6	>3.0	>1.9
45	Smooth	<1.4	<0.7	<0.5
	Rough	>7.6	>4.0	>2.6

IMPACT CRATERS AND STRUCTURES

Characteristically, an impact crater has a circular rim and a small depth relative to its diameter. Because impact craters are radially symmetrical, the landforms are ideal for relating image responses to radar illumination geometry. Terrestrial craters range in form from simple bowl-shaped features to complex structures with central peaks and/or rings (*Grieve,* 1987). They also exhibit a wide range of degradation states from relatively fresh structures, some with portions of preserved ejecta, to erosional scars known as astroblemes (*Dietz,* 1961). Simple bowl-shaped craters are confined to diameters up to about 2 km in sedimentary rocks and about 4 km in crystalline rocks. At greater diameters, up to about 150 km, terrestrial impact structures have a complex form that commonly includes a central uplift. Astroblemes expose a root level of crater substructure from which the overlying craterform has been completely removed by erosion.

Measurements indicate that the circularity of terrestrial impact craters is generally higher than that of craters of other origins (*Pike,* 1974). At low incidence angles pronounced relief displacement creates a major difficulty in determining the precise position of crater rims on images. It is convenient here to consider the radar responses from crater surfaces in an orthogonal frame of reference. Abrupt contrasts in radar response to changing slope conditions (hence incidence angle) produce sharp tonal boundaries normal to the illumination vector; these contrasts divide a crater into distal, medial, and proximal sectors. This image characteristic provides a basis for analysis and interpretation in the following examples.

SIMPLE TERRESTRIAL STRUCTURES

The best currently available spaceborne radar coverage of simple terrestrial structures includes Barringer (Meteor) crater, Arizona, U.S.A. and Talemzane (Daiet el Maadna) crater, Algeria (Fig. 2). The craters display similar morphology and an approximate radial symmetry.

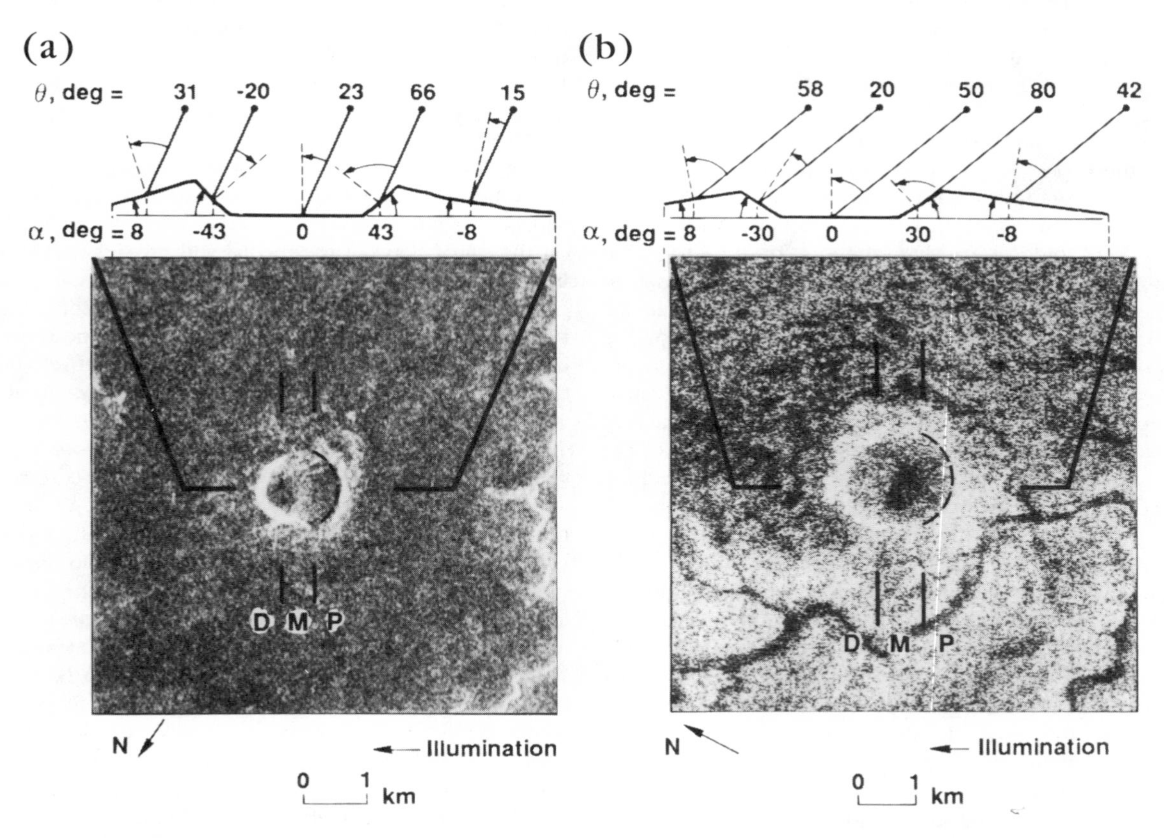

Fig. 2. Simple terrestrial impact structures: **(a)** Seasat SAR image of Barringer crater, Arizona, U.S.A.; **(b)** SIR-A image of Talemzane crater, Algeria. The craters are divided by strips into sectors that are distal (D), medial (M), or proximal (P) relative to the illumination vector. The proximal sectors of the crater rims are indicated by dashed lines. The schematics show the average surface slope angle (α) and corresponding radar incidence angle (θ) at five locations across the diameter of each crater in the direction of radar illumination. Negative slope angles represent foreslopes. The negative incidence angle in **(a)** indicates radar foldover of the crater wall and rim in the distal sector. Note that the contrasting intensities of responses around the halo in **(b)** are bilaterally symmetrical in the direction of illumination. There is no vertical exaggeration.

Barringer

Barringer crater is located on a plateau underlain by sandstone and dolomitic limestone formations. The crater is essentially bowl shaped, about 1.2 km in diameter (D) and 180 m deep (d). In plan view it appears to be square sided with rounded corners (low roundness but high sphericity), which may represent some structural control by the regional joint sets. The depth-to-diameter ratio (d/D) is 0.15. The crater has an encompassing rim that rises 30 to 60 m above the surrounding plateau. The rim is locally overturned and ejecta blocks up to 30 m are scattered about the surface close to the rim (*Shoemaker and Kieffer,* 1974). The crater has a comparatively level floor and a surrounding interior wall composed of talus slopes up to about 43°. Bedrock outcrops toward the top are locally vertical or overturned at the rim. The outer surface slopes away from the rim at 8° to 10°. The rim is outlined by tonal contrasts on the Seasat SAR image (Fig. 2a). The contrasts are pronounced in the distal and proximal sectors but subdued in the medial sector. The crater floor and the undisturbed plateau surface show medium to dark image tones because they are mostly level and comparatively smooth to the 23.5-cm wavelength (L-band) radar. There is no evidence of a radar halo on the image.

The illumination geometry shown in Fig. 2a falls mostly in the range where backscatter is dominated by large-scale slope effects, which result in notable relief displacement. For instance, the negative incidence angle on the crater wall in the distal sector indicates that the upper part of the slope was imaged by the radar before the base. This compresses the slope to the point of foldover, causing saturation on the image and obscuring part of the crater floor. Further contributions to saturation are provided by corner reflections from within the crater. Forward scatter from relatively smooth surfaces on the crater floor and from the proximal sector of the wall are doubly reflected as backscatter from appropriately oriented facets on the wall and rim in the distal sector. Consequently, in this sector the wall and the interior facing portion of the rim appear very bright, rim structure is imperceptible, and the position of the rim on the image is indeterminate.

In the proximal sector, the outline of the crater rim is more clearly marked. The rim structure on the outer slope is bright from the inclined surfaces of overturned strata that locally produce specular returns and from randomly distributed meter- and decameter-size ejecta blocks that produce multiple forward returns. The crater wall slopes away from the radar illumination in this sector and is dark because of low backscatter at the large incidence angle (~66°). The rim shows a significantly greater radius of curvature than that in the opposite distal sector.

In the medial sector, the surfaces of the crater wall and the outer slope on each side are aligned essentially parallel to the illumination vector. In this orientation, slope changes produce lower backscatter contrasts. Tonal contrasts are subdued and the edge of the rim is not sharply defined on the image. The flanks of the distal sector appear compressed relative to the flanks of the proximal sector. This lowers the apparent circularity of the crater outline in plan view and imparts a bilateral symmetry in the direction of the radar illumination.

Talemzane

Talemzane crater is located on a plateau underlain by limestone strata. It forms an approximately circular depression about 1.75 km in diameter and 67 m in depth (d/D ratio = 0.04). It is encompassed by a rim uplifted some 27 m above the surrounding plateau. The crater floor contains an asymmetrical area of fine-grained alluvial sediments to the south. The crater wall slopes all around at angles between 22° and 35°. Limestone beds are more steeply inclined or vertical in the upper part of the wall and in places they are overturned (*Lambert et al.,* 1980). However, the wall is not as steep and the edge of the rim is not as sharply defined as the wall and rim of Barringer crater. The outer surface slopes away from the rim at 8° to 10°. It shows a radar halo on the SIR-A image (Fig. 2b) that varies systematically in brightness around the rim (*McHone and Greeley,* 1987). Rocks exposed on this surface are mainly mixed breccias that closely resemble those in the crater. They form a nearly continuous annular blanket that is concentric with the crater and that extends about 400 m from the rim. The size of the fragments ranges up to 10 m and decreases with distance from the rim. The outline of the rim (Fig. 2b) is marked by less pronounced tonal contrasts than those of Barringer crater observed on the Seasat SAR image (Fig. 2a). Dark image tones on the crater floor and most of the plateau surface north and east of the crater represent low backscatter from level and relatively smooth sandy surfaces.

The illumination geometry shown in Fig. 2b lies mostly in the range where backscatter is dominated by small-scale surface roughness. Moderate to relatively high incidence angles on the outer slope in the medial and proximal sectors result in bright image tones dominated by the small-scale roughness of the coarse-grained ejecta fragments. This produces the radar-bright halo seen on the image out to about one third of the crater diameter (Fig. 2b). The halo is comparatively bright in the proximal sector but appears darker in the distal sector where there is less radar signal return as the incidence angle approaches 60°. Note that equivalently rough, blocky, weathered surfaces of undisturbed limestone bedrock on the level surface of the plateau south and west of the crater produce image tones comparable to those in the radar-bright halo. However, the low incidence angle on the distal sector of the wall and on adjacent segments of the uplifted rim results in bright image tones dominated here by large-scale slope. Additional radar brightness may be provided from energy scattered forward off the crater floor and then backscattered from appropriately oriented facets on the rim.

In the medial sector where the walls are essentially in line with the illumination vector and slope changes provide little contribution to backscatter, the rim and the outer slope appear equally bright on the image because of diffuse backscatter from the rough surface. This roughness-related brightness merges imperceptibly with slope-related brightness on the image toward the distal sector of the crater wall. The transition between dominance of backscatter by surface-roughness and dominance by slope is not readily apparent from the image. However, the curvature of the rim appears to be comparable in both proximal and distal sectors and the outline of the crater retains a radially symmetrical form on the image.

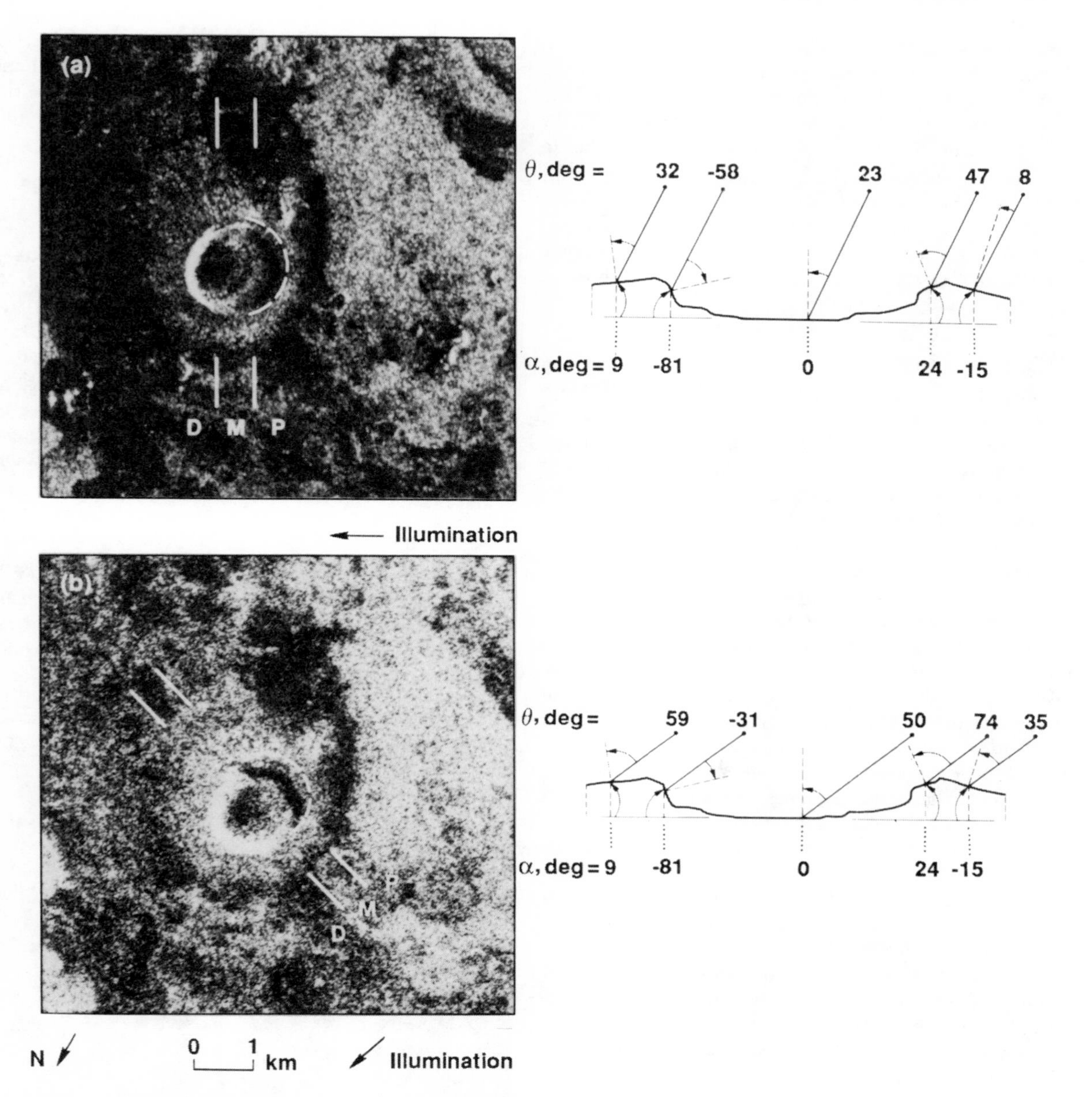

Fig. 3. Elegante crater, Pinacate volcanic field, Sonora, Mexico: **(a)** Seasat SAR image; **(b)** SIR-A image. The crater is divided by strips into sectors that are distal (D), medial (M), or proximal (P) relative to the illumination vector. The proximal sector of the crater rim is indicated by a dashed line on each image. The schematic sections are in the direction of the Seasat SAR illumination; the average slope angles (α) and incidence angles (θ) correspond to five locations across the crater diameter. Negative slope angles represent foreslopes. On both images, the radar brightness that extends to the south from the gentle outer slope denotes the steeper erosional slopes of a small breached cinder cone. There is no vertical exaggeration.

VOLCANIC MAAR CRATERS

General

Though the morphology of the small craters described above is typical of fresh simple impact structures on Earth, it is not unique to an origin by impact. Among the various crater types that do not originate by impact, volcanic maar craters have morphologic characteristics that most closely resemble those of small impact craters. Maar craters are features that originate from volcanic eruption and collapse in association with shallow groundwater. They have an uplifted rim structure and a circular form in plan view that approaches the circularity of impact craters. In the absence of regional geologic information, the two types of craters can be confused on radar images (*Greeley et al.,* 1987).

Maar craters were imaged by both Seasat SAR and SIR-A in the Pinacate volcanic field, Sonora, Mexico. This is a Quaternary basaltic shield that consists of numerous lava flows and cinder cones, as well as eight maar craters and a partially collapsed tuff cone. The craters, which range up to 1.6 km in diameter, expose steeply sloping to vertical interior walls with stratified volcanic rocks (*Jahns,* 1959; *Gutmann and Sheridan,* 1978). Though the maars in the Pinacate field display equal or greater depth relative to diameter (d/D ratio) and steeper interior slopes, they show radar responses similar to those of the impact craters discussed above.

Elegante

Elegante is the largest maar crater in the Pinacate field and the most nearly circular in outline; it also has the simplest rim with a symmetrical outer slope. It is a flat floored essentially cylindrical depression approximately 1.6 km in diameter and 244 m deep (d/D ratio = 0.15). The crater walls form a continuous ring of cliffs up to 130 m high with an average slope of about 80°. At the base of the cliffs an apron of talus with a more gentle slope (about 24°) extends to the base of the crater. The rim rises approximately 50 m above the surrounding plain and slopes outward from its crest at 9° to 15°. It exposes a steeper inner slope, with tuff breccia up to 70 m thick. The tuff breccia thins as it encompasses the crater out to about one half the diameter. The circular pattern of the rim crest is interrupted to the southeast by a depression known as the Scallop (*Gutmann,* 1976).

The Seasat SAR and corresponding SIR-A images of Elegante crater (Fig. 3) provide good examples for contrasting the dual-incidence-angle viewing effects and for comparing these effects to the observations by Seasat SAR at Barringer crater and SIR-A at Talemzane crater. Though the Seasat SAR and the SIR-A images of Elegante crater were obtained from azimuthal directions that are 40° apart the essentially radial symmetry of the crater renders it virtually isotropic with respect to the illumination azimuth. Thus, the contrasts in radar signal intensity around the crater relate mostly to different Seasat SAR and SIR-A incidence-angle effects.

The slope of the crater wall produces negative incidence in both Seasat SAR and SIR-A; the negative incidence results in foldover and saturation around the distal sector of the wall in both images. The saturation obliterates signal from the inner slope of the upturned rim and the position of the rim crest in this sector is indeterminate on both images. In the proximal sector, the steep crater wall is in shadow on both images. Here the rim crest is outlined by the subdued tonal contrast between its comparatively steep inner slope and its more gentle outer slope. Foldover in the distal sector and shadowing in the proximal sector combine to distort the radial symmetry of the rim crest on both Seasat SAR and SIR-A images and to produce an apparent bilateral symmetry parallel to the illumination vector.

In the medial sector, the Seasat SAR image (Fig. 3a) shows the outline of the rim crest, which is normal to the craterward limit of numerous small radial gullies that extend across the outer slope. The gullies are highlighted in this sector, where they are normal or highly oblique to the illumination vector. However, they are obscure in the proximal and distal sectors, where they are parallel or gently inclined to the illumination vector. At the lower resolution of SIR-A (Table 1) the rim crest is unclear and individual gullies are imperceptible on the SIR-A image (Fig. 3b).

COMPLEX TERRESTRIAL STRUCTURES

A small but varied assemblage of complex terrestrial structures including astroblemes was imaged by Seasat SAR and SIR-A. The structures range widely in size and level of preservation, but mostly the craterforms have been removed by postimpact erosion. The best spaceborne radar image coverage of known structures includes Elgygytkhyn, Chukotka, U.S.S.R, and about 60% of Manicouagan, Quebec, Canada. Elgygytkhyn is the largest terrestrial impact structure that has the erosional remains of the original crater rim. Manicouagan presents topography that has been profoundly modified by glacial erosion.

Elgygytkhyn

Elgygytkhyn crater (also referred to as Elgygytgyn) is located on the upland of the Anadyr' Plateau in eastern Siberia. A possible impact origin was suggested by *Zotkin and Tsvetkov* (1970). An origin by impact is supported on morphological grounds by the remarkable circularity of the ring of eroded

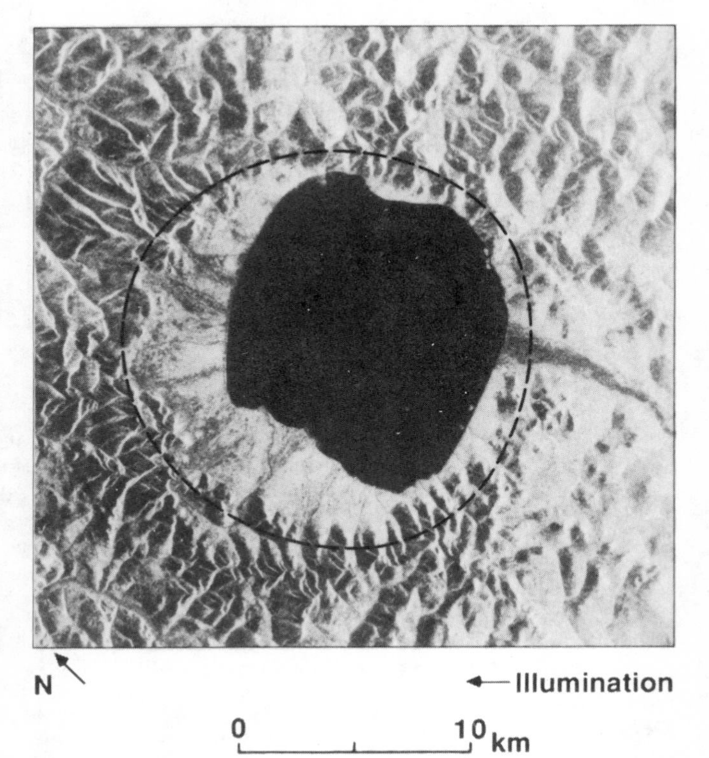

Fig. 4. Seasat SAR image of Elgygytkhyn crater, Chukotka, U.S.S.R. The dashed line indicates the crater rim. The form of the mountains and valleys on the image gives no evidence of significant glacial erosion in the area outside the crater.

mountain peaks that surround the crater, and the lack of collapse scalloping around the crater margin. These crater features have been shown to be in sharp contrast with caldera morphology as observed on Landsat images. Elgygytkhyn does not display the geomorphic aspect of a caldera (*Dietz and McHone,* 1976).

The Seasat SAR image of Elgygytkhyn clearly shows the ring of mountains that surrounds the crater. The ring has a diameter of about 18 km (Fig. 4). Nevertheless, the rim crest is so deeply eroded that significant contrasts of radar response in different sectors of the crater are not apparent. If a central uplift is associated with the structure, it is obscured entirely by the lake.

Manicouagan

Manicouagan is one of the more intensively studied complex impact structures on Earth. It has concentrically arranged morphological elements that were defined by *Floran and Dence* (1976), and have more recently been elaborated by other authors. The morphology consists of an inner central region, an inner plateau, and an annular trough, and a group of outer elements that include an inner fractured zone, outer disturbed zone, and outer circumferential depression. These morphological elements represent structural levels much below the original form of the crater (*Grieve and Head,* 1983). The Seasat SAR image (Fig. 5) covers all but the western sector of the Manicouagan structure. However, it displays varying proportions of each of the morphological elements shown on the sketch map.

The three outer morphological elements of the structure are best defined to the west, in the portion not covered by the Seasat SAR image. The distal boundary of the inner fracture zone is perceptible on the image (Fig. 5a) to the north where it coincides with a sharp bend in most streams that flow inward to the annular trough. East of the annular trough the outer elements have not been defined. An area of distinctive coarse-grained texture on the image east of the trough corresponds to massive crystalline rocks (gabbro). Conjugate sets of linear topographic features denote fracturing in this area. The higher density and closer spacing of these features for about 20 km outward from the trough probably indicate the inner fractured zone here.

Concentrically aligned valley segments are prominent from 10 to 50 km east and southeast of the annular trough. The

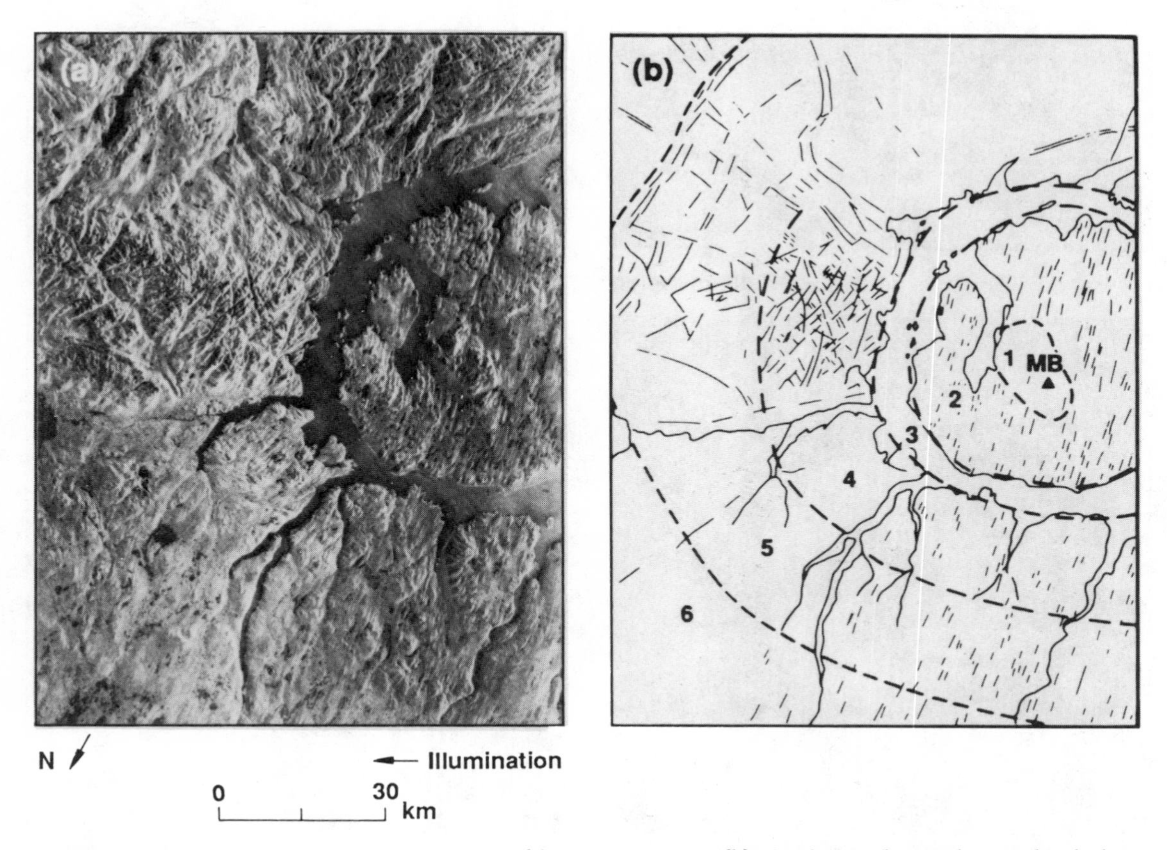

Fig. 5. Manicouagan structure, Quebec, Canada: **(a)** Seasat SAR image; **(b)** sketchmap. The annular trough, which contains the Manicouagan Reservoir, is about 15 km wide and 65 km in diameter. The sketchmap shows (1) Mont de Babel (MB) in the central peaks; (2) the inner plateau; (3) the annular trough; (4) the inner fractured zone; (5) the outer disturbed zone; and (6) the outer circumferential depression. Note the north-south orientation of glacial topography at the center, the high density of conjugate linear features east of the annular trough, and the extensive radial alignments of valley segments outward from the trough.

orientation suggests that they are structurally related to the Manicouagan feature. Locally, the concentric valleys are truncated against prominent topographic alignments that radiate outward from the annular trough for tens of kilometers. The radially aligned features denote the extensive fracture halo that is reported to have a diameter of about 150 km (*Grieve et al.,* 1988).

VENUSIAN IMPACT CRATERS

Venera radar images reveal about 150 impact craters with diameters from 8 to 144 km that are distributed over less than 20% of the surface (*Ivanov et al.,* 1986; *Basilevsky et al.,* 1987). Smaller craters have been reported, but diagnostic features are not resolved. Calculations by *Ivanov et al.* (1986) have shown that under the present venusian atmosphere the smallest craters that can be created by low-strength meteoroids are in the range of 6 to 13 km diameter. The paucity of small craters on Venera 15/16 images has been assumed to result from atmospheric shielding effects; it may also be due to marginal resolution in the Venera images. The size-frequency distribution of the craters shows a maximum in the 16-to-32-km-diameter interval. The craters have been classified into morphological types that change progressively with increasing diameter. Most morphological types show contrasts in levels

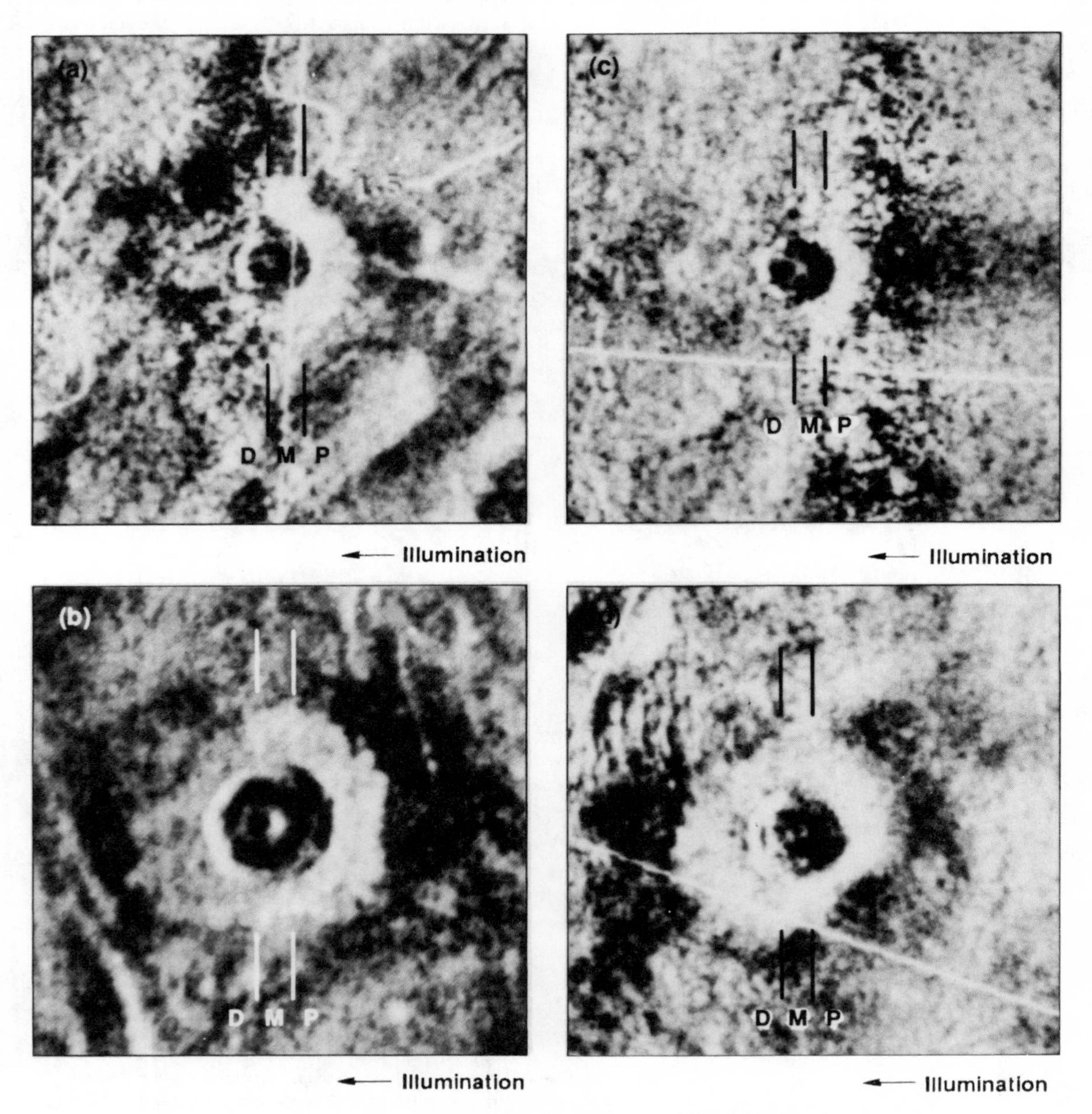

Fig. 6. Soviet Venera 15 and 16 radar images of central-peak-type impact craters with sharp rims, radar-bright halos **(a)**, **(b)**, **(d)**, and missing a halo **(c)**. The craters are divided by strips into sectors that are distal (D), medial (M), or proximal (P) relative to the illumination vector. Identities and diameters (km) are: **(a)** Kemble, 24; **(b)** Rudneva, 34; **(c)** Zvereva, 24; and **(d)** Golubkina, 28.

of preservation from fresh through mature to degraded. Apparently this represents a degradational sequence that is independent of crater size (*Basilevsky et al.,* 1987).

Venusian impact craters range from bowl-shape types, which are the smallest observed on Venera images, through knobby-bottom to central-peak and multiple-ring types, which are the largest. Each of the first three types is represented in the maximum size-frequency interval, though the central-peak type predominates (*Basilevsky et al.,* 1987). Fresh craters have sharp rims and outlines that suggest many of them are radially symmetrical. The diameters of craters in the maximum size-frequency interval range from 10 to 20 times larger than those of Barringer, Talemzane, and Elegante observed by Seasat SAR and/or SIR-A, but because the spatial resolution of Venera images ranges from 70 to 20 times lower than that of Seasat SAR or SIR-A the scales of perception of the small terrestrial craters and the larger venusian craters are comparable.

Examples of prominent-central-peak type craters were selected from the Venera 15/16 coverage for this analysis. The craters range in diameter from 24 to 34 km (Fig. 6) and exhibit fresh to mature morphology with sharp, presumably circular, rims (*Basilevsky et al.,* 1987). Though it is not possible to determine surface slope values from the images, the slope-dominated outlines of the crater walls are analogous in each case. In the distal sectors, the walls appear saturated and compressed due to foldover. Evidently the slope of the walls equals or exceeds the nominal incidence angle here. In the proximal sectors, the crater walls are dark and the sharp rim crests are drawn toward the illumination vector. In the medial sectors, the distortion of the crater rims is minimal but the outlines are weak. These responses impart an appearance of bilateral symmetry to the craters that is further enhanced by the outlines of the central peaks, notably on Rudneva and Zvereva (Figs. 6b,c). Comparable image characteristics were observed on the Seasat SAR images of Barringer and Elegante craters (Figs. 2a and 3a).

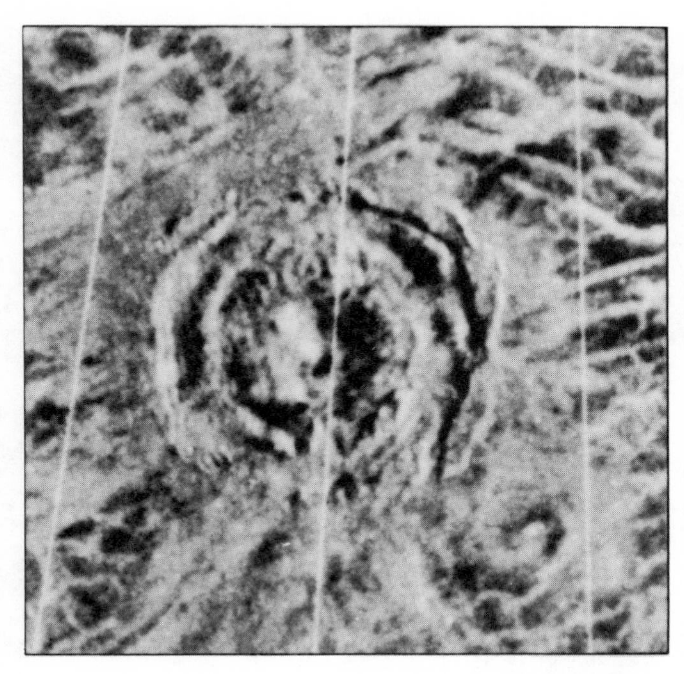

Fig. 7. Klenova shown in this Soviet Venera 16 radar image is the largest and morphologically most prominent multiple-ring type of impact crater observed on Venus. Outer rings with diameters of 144 and 105 km consist of discontinuous ridges 10 to 20 km wide. The surface inside the inner ring is knobby at the perimeter and smooth toward the 15-km diameter central hill (*Ivanov et al.,* 1986).

Fresh craters are reported to have radar-bright halos that commonly show bilateral symmetry on the images. The halos are thought to denote surface roughness that is related to ejecta blankets. Mature craters show no apparent halo. This condition is thought to represent terrain modification, perhaps by eolian erosion and deposition (*Ivanov et al.,* 1986). On images of Kemble, Rudneva, and Golubkina (Figs. 6a,b,d) the radar halos are brightest in the proximal sectors but comparatively subdued in the distal sectors. A comparable effect was noted on the SIR-A image of Talemzane crater (Fig. 2b), where angular fragments of mixed breccia are reported to form a nearly continuous annular blanket that extends up to 400 m from the crater rim (*Lambert et al.,* 1980).

In the case of Zvereva (Fig. 6c), the proximal sector appears bright, but there is no evidence of a halo in the medial and distal sectors. As this brightness contrast occurs along a mosaic boundary on the image, it is probably a processing artifact.

In addition to obvious impact craters many circular features have been noted on the Venera 15 and 16 images (*Nikolaeva et al.,* 1986; *Basilevsky et al.,* 1987). While some of these features have morphology suggestive of a volcanic or tectonic origin, a number of circular features are believed to be highly degraded impact craters. Nevertheless, a clear venusian analog to the degraded terrestrial Elgygytkhyn crater has not been reported.

The largest and morphologically most prominent impact crater reported on Venus is a double-ring type named Klenova. It has outer and inner rings with diameters of 144 and 105 km (Fig. 7). The crater is surrounded by a ring-like zone of smooth terrain 40 to 60 km wide. This is probably the ejecta blanket, which overlies the ridge-and-band pattern of the target terrain (*Ivanov et al.,* 1986; *Basilevsky et al.,* 1987). The overall dimensions of Klenova are comparable to the terrestrial Manicouagan structure (Fig. 5). However, the two features expose different structural levels. At the Venera resolution, there is insufficient morphological detail in the image (Fig. 7) to make further comparison.

DISCUSSION

At low incidence angles, Seasat SAR images of terrestrial impact craters and Venera 15 and 16 images of venusian craters (obtained at a shorter wavelength) show similarities in the radar responses to crater morphology. SIR-A images show notable differences in the responses at high incidence angles. These observations have important implications for the interpretation of Magellan images, which will be obtained through a wide range of incidence angles at a wavelength that is intermediate between that of Seasat SAR/SIR-A at one extreme and Venera 15/16 at the other.

Due to orbital considerations, the nominal incidence angle of Magellan radar at the center of the swath will vary from about 45° at 10°N lat. to about 16° at the north pole and at 70°S lat. (*Johnson,* 1989). Figure 8 shows that at latitudes from 20°N to 10°S the viewing geometry will approach the SIR-A configuration. North of 60°N and south of 40°S the viewing geometry will be analogous to Seasat SAR. At the latitudinal extremities of imaging in both hemispheres, the viewing geometry will approach the Venera 15 and 16 configuration. Thus, radially symmetrical landforms such as impact craters, central peaks, cones, and other circular features can be expected to show varied outlines on Magellan images at different latitudes.

Impact craters commonly occur as circular features. Degradation or multiple impact may impose a secondary asymmetry on the landforms. It is important on radar images to distinguish between departures from crater circularity that are real and image distortion that relates to the scene illumination.

Crater Morphology

At any radar incidence angle, the distortion of a crater outline is minimal across the medial sector, in a direction normal to the illumination. Crater diameters can be measured accurately only in this direction, provided the outline of the rim is distinguishable. This yields the radius of the rim and the center of a circle for comparison to the crater outline on each image. In cases where the slope of the crater wall equals or exceeds the nominal incidence angle, the imaging beam produces foldover, as at Barringer and Elegante on the terrestrial images (Figs. 2a and 3) and Kemble, Rudneva, Zvereva, and Golubkina on the venusian images (Figs. 6a-d). In such cases, the outline of the crater corresponds to the circle in the median sector, but is comparatively prolate in the proximal sectors and oblate in the distal sectors. The former are drawn out and the latter are compressed, so the craters appear to show bilateral symmetry parallel to the illumination vector. Where the incidence of the beam does not produce foldover, as at Talemzane (Fig. 2b), the outline of the crater rim corresponds to the inscribed circle with comparatively little distortion. The variable viewing geometry of Magellan indicates that impact crater distortion will be greater on images obtained at higher latitudes. Essentially similar morphologic features that appear circular at low latitudes can be expected to show a distinctly different outline at high latitudes. If the outline of a crater at high latitudes appears bilaterally symmetrical parallel to the illumination vector, it is likely to be an artifact of the viewing geometry.

Radar Halos

The radar-brightness of crater halos varies with the rms surface roughness, which is dependent on both the wavelength and the incidence angle (e.g., Table 2). Radar brightness may also vary with the reflectivity properties (dielectric) of surface and near-surface materials. The radar-bright halo on the SIR-A image (23.5-cm wavelength) of Talemzane (Fig. 2b) represents a surface with rms roughness about three times greater than the radar-bright halos on the Venera images (8-cm wavelength) of Kemble, Rudneva, and Golubkina (Figs. 6a,b,d). Small-scale surface roughness or high Fresnel reflectivity provide equally plausible alternatives to account for the radar-bright halos. However, the halo brightness varies systematically around the craters and shows bilateral symmetry relative to the illumination vector in each case. If the radar-bright halos denote the rms roughness of the ejecta blankets it is likely that either the ejecta are not uniformly rough around the craters, or the slopes of the ejecta blankets provide a sufficient range of incidence angles to produce a notable difference in backscatter intensity from the foreslopes to the backslopes.

At Talemzane crater, the ejecta are reported to form a nearly continuous annular blanket of mixed breccia that extends out 400 m from the crater rim (*Lambert et al.,* 1980). In this instance, the symmetrical variation of the halo brightness around the crater probably represents contrasted incidence angle effects on the image (Fig. 2b).

Assuming the slopes around Kemble, Rudneva, and Golubkina craters are radially symmetrical, the relations between

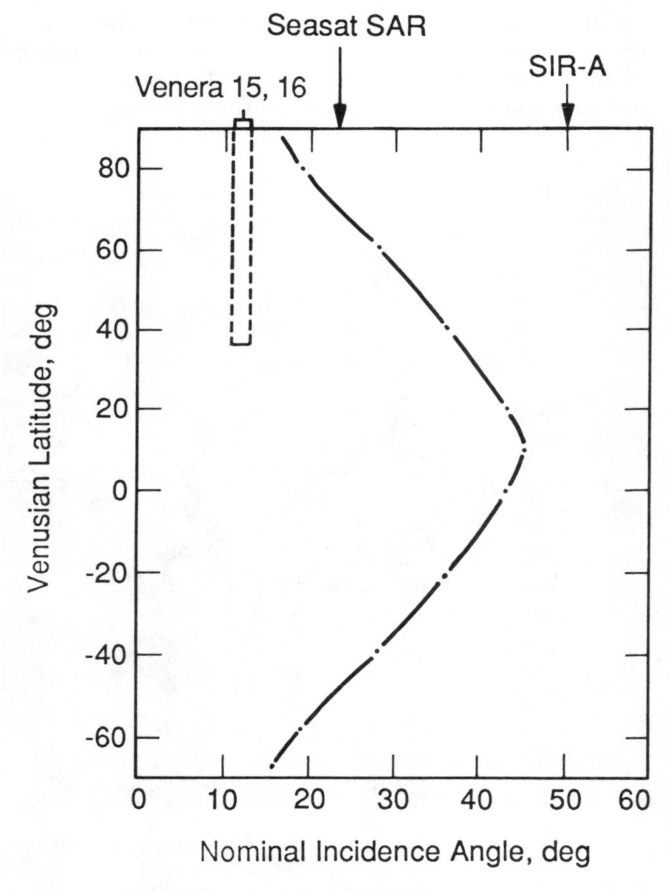

Fig. 8. Variation of Magellan nominal incidence angle with Venus latitude, showing corresponding nominal values at the beam centers of Seasat SAR, SIR-A, and Venera 15 and 16. Short dashed lines indicate the latitudinal range of the Venera coverage. Negative latitudes denote the southern hemisphere.

symmetry, brightness, and the illumination vector suggest in each case that there may be no significant difference in roughness around the ejecta blanket. The brightest responses from the proximal sector of the halos suggest foreslopes where the incidence angle approaches zero and the backscatter is dominantly specular; consequently, the radar return has the greatest intensity here. Less intense responses from the backslopes probably indicate a contrasting incidence angle effect.

It has been proposed that bright halos may result from fresh ejecta material with a comparatively high Fresnel reflectivity at the low Venera incidence angles (*Jurgens et al.,* 1988). Iron-sulfide inclusions in the rocks have been suggested as a mechanism that would increase the conductivity of the material sufficiently to account for the increased radar reflectivity. While this is possible, it would be reasonable in such circumstances to find that the halos were uniformly bright around the craters and that the crater floors (the source areas of the high-reflectivity ejecta) showed an approximately equivalent radar brighness. On the contrary, the venusian craters discussed above not only show symmetrical reduction in brightness around the halos from the foreslopes to the backslopes in the direction of illumination but also the radar responses from the crater floors in each case are comparatively dark.

From Table 2 and Fig. 8, it is evident that the breakpoint of rms smoothness for high incidence angles in the low to middle latitudes will be rougher by about 28% at lower incidence angles in the polar latitudes. However, both breakpoints fall within the range of fine-pebble-size particles (0.4 to 0.8 cm). Similarly, the breakpoint of rms roughness for polar latitudes will be smoother by about 25% at the low to middle latitudes. These breakpoints both fall within the range of very-coarse-pebble-size particles (3.2 to 6.4 cm). Comparisons of the properties of ejecta blankets at widely separated latitudes must therefore allow for the contrasting effects of the different incidence angles on the smoothness/roughness breakpoints.

The criteria used to evaluate rms surface roughness breakpoints assume that reflection occurs at a free-space boundary without modification by fine-grained unconsolidated deposits at the surface. The conditions that permit radar penetration of such fine-grained deposits and the perception of subsurface roughness cannot be discriminated from images obtained at a single radar wavelength. Corresponding pairs of images obtained at widely different wavelengths are needed to verify radar penetration (*Blom,* 1988). Thus, in addition, the distinction between surface reflections and possible subsurface returns will not be feasible in the Magellan images.

Resemblance of Volcanic Maars

The similarities in the radar responses of small terrestrial impact craters and volcanic maar craters of comparable dimensions indicate difficulty in distinguishing the contrasted origins of these craters from their morphology in images. On Earth, the regional setting of the craters in volcanic or sedimentary terrains can provide helpful discriminatory information. On Venus, the absence of groundwater inhibits the formation of maar craters and under present atmospheric conditions it is thought that features formed by explosive volcanism are unlikely to occur (*Garvin et al.,* 1982; *Head and Wilson,* 1986). However, small cone and dome structures are known to exist on Venus that require study with higher-resolution images than are currently available. If small circular craters are found in Magellan images their geologic setting, including possible tectonic alignments and the presence or absence of associated lava flows, as well as their shape, size, and distribution should be evaluated to distinguish between an origin by impact or by volcanism.

In all the terrestrial craters mentioned above, the radar responses have been related to field observations of slope, surface roughness, and structure. In venusian examples, where direct field observations cannot be made, slope characteristics can be approximated from the image responses by analogy with terrestrial examples, and roughness characteristics can be related to the observing wavelength and the incidence angle (Table 2).

Spatial Resolution

Large terrestrial impact structures with diameters up to 150 km retain little or none of their original crater form. At the high resolution of Seasat SAR, they show details of the structurally related morphologic elements. Correspondingly large venusian impact features at the Venera resolution show only the broadest morphologic outlines. The spatial resolution of the Magellan radar system is coarser than Seasat SAR or SIR-A, but finer than Venera by about an order of magnitude (Table 1). Magellan images will provide the morphologic detail necessary to support an impact origin or to identify a plausible alternative for large and degraded venusian craters comparable to those observed on Venera images.

Magellan images should enable the discrimination of craters with diameters below the limits of reliable Venera resolution (about 8 km) and possibly as small as 2 km, if they exist. This in turn would provide a more definitive indication of the ability of comparatively small bolides to penetrate the venusian atmosphere. However, the difficulties in discriminating an impact origin from a volcanic origin for small terrestrial craters on higher resolution Seasat SAR and SIR-A images have been described above. Similar difficulties will probably apply in discriminating the origin of small venusian craters. In any event, the extended coverage of Magellan radar, up to 90% of the venusian surface, will yield a more comprehensive understanding of the size-frequency distribution of impact craters on Venus.

Acknowledgments. This study was performed at the Jet Propulsion Laboratory, California Institute of Technology, under contract to the National Aeronautics and Space Administration. The author acknowledges helpful technical discussions with R. Blom, L. Roth, and R. S. Saunders. Editorial assistance was provided by D. Fulton. Thanks are expressed to J. Curlander, S. Pang, and J. Weirick for Seasat SAR digital data correlations.

REFERENCES

Basilevsky A. T., Ivanov B. A., Burba G. A., Chernaya I. M., Kryuchkov V. P., Nikolaeva O. V., Campbell D. B., and Ronca L. B. (1987) Impact craters of Venus: A continuation of the analysis of data from the Venera 15 and 16 spacecraft. *J. Geophys. Res., 92,* 12,869-12,901.

Blom R. G. (1988) Effects of variation in look angle and wavelength in radar images of volcanic and aeolian terrains, or Now you see it, now you don't. *Int. J. Remote Sensing, 9,* 945-965.

Burns B. A. and Campbell D. B. (1985) Radar evidence for cratering on Venus. *J. Geophys. Res., 90,* 3037-3047.

Dallas S. S. and Nickle N. L. (1987) The Magellan Mission to Venus, Advances in Astronautical Sciences; Part 1: Aerospace Century 21. *AAS 86-331,* 64.

Dietz R. S. (1961) Astroblemes. *Sci. Am., 205,* 50-58.

Dietz R. S. and McHone J. F. (1976) Elgygytgyn: Probably world's largest meteorite crater. *Geology, 4,* 391-392.

Floran R. J. and Dence M. R. (1976) Morphology of the Manicouagan ring-structure, Quebec, and some comparisons with lunar basins and craters. *Proc. Lunar Sci. Conf. 7th,* pp. 2845-2865.

Garvin J. R., Head J. W., and Wilson L. (1982) Magma vesiculation and pyroclastic volcanism on Venus. *Icarus, 52,* 365-372.

Greeley R., Christensen P. R., and McHone J. F. (1987) Radar characteristics of small craters: implications for Venus. *Earth, Moon and Planets, 37,* 89-111.

Grieve R. A. F. (1987) Terrestrial impact structures. *Annu. Rev. Earth Planet. Sci., 15,* 245-270.

Grieve R. A. F. and Head J. W. (1983) The Manicouagan impact structure: An analysis of its original dimensions and form. *Proc. Lunar Planet. Sci. Conf. 13th,* in *J. Geophys. Res., 88,* A807-A818.

Grieve R. A. F., Wood C. A., Garvin J. B., McLaughlin G., and McHone J. F., eds. (1988) *Astronauts' Guide to Terrestrial Impact Craters.* LPI Tech. Rpt. 88-03, Lunar and Planetary Institute, Houston. 89 pp.

Gutmann J. T. (1976) Geology of Crater Elegante, Sonora, Mexico. *Geol. Soc. Am. Bull., 87,* 1718-1729.

Gutmann J. T. and Sheridan M. F. (1978) Geology of the Pinacate volcanic field. In *Guidebook to the Geology of Central Arizona,* (D. M.Burt and T. L. Pewe, eds.), pp. 47-59. Arizona Bur. Geol. and Min. Technology Spec. Pap. 2.

Head J. W. and Wilson L. (1986) Volcanic processes and landforms on Venus: Theory, predictions and observations. *J. Geophys. Res., 91,* 9407-9446.

Ivanov B. A., Basilevsky A. T., Kryuchkov V. P., and Chernaya I. M. (1986) Impact craters on Venus: Analysis of Venera 15 and 16 data. *Proc. Lunar Planet. Sci. Conf. 16th,* in *J. Geophys. Res., 91,* D413-D430.

Jahns R. H. (1959) Collapse depressions of the Pinacate volcanic field, Sonora, Mexico. In *Southern Arizona Guidebook, 2,* pp. 165-184. Arizona Geol. Society.

Johnson W. T. K. (1989) Radar system design for the Magellan radar mission to Venus. *ISPRS, Proc.,* Kyoto, Japan, in press.

Jurgens R. F., Slade M. A., and Saunders R. S. (1988) Evidence for highly reflecting materials on the surface and subsurface of Venus. *Science, 240,* 1021-1023.

Lambert Ph., McHone J. F. Jr., Dietz R. S., and Houfani M. (1980) Impact and impact-like structures in Algeria, Part I. Four bowl-shaped depressions. *Meteoritics, 15,* 157-178.

McHone J. F. and Greeley R. (1987) Talemzane: Algerian impact crater detected on SIR-A orbital imaging radar. *Meteoritics, 22,* 253-264.

Nikolaeva O. V., Ronca L. B., and Basilevsky A. T. (1986) Circular features on the plains of Venus as an indication of their geologic history (in Russian). *Geochimia,* 579-589.

Peake W. H. and Oliver T. L. (1971) *The Response of Terrestrial Surfaces at Microwave Frequencies.* U.S. Air Force Avionics Lab. Report AFAL-TR-70-301. 255 pp.

Pike R. J. (1974) Craters on Earth, Moon and Mars. *Earth Planet. Sci. Lett., 22,* 245-255.

Saunders R. S. and Malin M. C. (1976) Venus: Geologic analysis of radar images. *Geol. Rom., 15,* 507-515.

Schaber G. G., Berlin G. L., and Brown W. E. (1976) Variations in surface roughness within Death Valley, California: Geologic evaluation of 25-cm-wavelength radar images. *Geol. Soc. Am. Bull, 87,* 29-41.

Shoemaker E. M. and Kieffer S. W. (1974) *Guidebook to the Geology of Meteor Crater, Arizona.* 37th Annual Meeting of the Meteoritical Society, 66 pp. incl. geol. map.

Ulaby F. T., Moore R. K., and Fung A. K. (1982) *Microwave Remote Sensing: Active and Passive, vol. 2, Radar Remote Sensing and Surface Scattering and Emission Theory.* Addison-Wesley, Reading, Massachusetts.

Zotkin I. T. and Tsvetkov V. I. (1970) Searches for meteorite craters on Earth. *Astron. Vestnik., 4,* 55-65.

Indexes

Author Index

Subject Index

Sample Index

Meteorite Index